Topley and Wilson's

Principles of bacteriology, virology and immunity

Seventh edition in four volumes

Volume 2

Edward Arnold

General Editors

Sir Graham Wilson
MD, LLD, FRCP, FRCPath, DPH, FRS

Formerly Professor of Bacteriology as Applied to Hygiene, University of London, and Director of Public Health Laboratory Service, England and Wales.

Sir Ashley Miles CBE
MD, FRCP, FRCPath, FRS

Deputy Director, Department of Medical Microbiology, London Hospital Medical College, London.
Emeritus Professor of Experimental Pathology, University of London, and formerly Director of the Lister Institute of Preventive Medicine, London.

M. T. Parker
MD, FRCPath, Dip Bact

Formerly Director, Cross-Infection Reference Laboratory, Central Public Health Laboratory, Colindale, London.

Principles of bacteriology, virology and immunity

W. W. C. Topley, 1886–1944

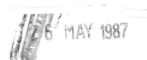

Volume 2

Systematic bacteriology

Edited by

M. T. Parker

First published 1929
by Edward Arnold (Publishers) Ltd
41 Bedford Square, London WC1B 3DQ.

Reprinted 1931, 1932, 1934
Second edition 1936
Reprinted 1937, 1938, 1941 (twice), 1943, 1944
Third edition 1946
Reprinted 1948, 1949
Fourth edition 1955
Reprinted 1957, 1961
Fifth edition 1964
Reprinted 1966
Sixth edition 1975
Seventh edition in four volumes 1983 and 1984

Volume 1 ISBN 0 7131 4424 6
Volume 2 ISBN 0 7131 4425 4
Volume 3 ISBN 0 7131 4426 2
Volume 4 ISBN 0 7131 4427 0

British Library Cataloguing in Publication Data

Topley, William Whiteman Carlton
Topley and Wilson's principles of bacteriology, virology and immunity. - 7th ed.
Vol. 2: Systematic bacteriology
1. Medical microbiology
I. Title II. Wilson, *Sir* Graham
III. Miles, *Sir* Ashley IV. Parker, M. T.
616'.01 QR46

ISBN 0-7131-4425-4

To EM, BRP and the memory of JW

Filmset in 9/10 Times New Roman
and printed and bound in Great Britain by
Butler & Tanner Ltd
Frome and London

Volume Editor

M. T. Parker MD, FRCPath, Dip Bact
Formerly Director, Cross-Infection Reference Laboratory, Central Public Health Laboratory, Colindale, London.

Contributors

Joyce D. Coghlan BSc, PhD
Formerly Director, Leptospira Reference Laboratory of the Public Health Laboratory Service, Colindale, London.

L. H. Collier MD, DSc, FRCP, FRCPath
Professor of Virology at The London Hospital Medical College, Honorary Consultant in Virology in the Tower Hamlets District Health Authority.

Brian I. Duerden BSc, MD, MRCPath
Professor of Medical Microbiology, University of Sheffield Medical School and Honorary Consultant Microbiologist, Children's Hospital, Sheffield.

John M. Grange MD, MSc
Reader in Microbiology, Cardiothoracic Institute, University of London.

Roger J. Gross MA, MSc, MIBiol
Principal Microbiologist, Division of Enteric Pathogens, Central Public Health Laboratory, Colindale Avenue, London.

L. R. Hill BSc, DSc, FIBiol
Curator, National Collection of Type Cultures, Central Public Health Laboratory, Colindale Avenue, London.

Barry Holmes BSc, MSc
Senior Microbiologist, Computer Identification Laboratory, National Collection of Type Cultures, Central Public Health Laboratory, Colindale Avenue, London.

B. P. Marmion DSc (Lond), MD (Lond), FRCP, FRCPath (UK), FRSE
Clinical Professor (Virology) University of Adelaide, South Australia; Late Professor of Bacteriology, Edinburgh University, Scotland.

M. T. Parker MD, FRCPath, Dip Bact
Formerly Director, Cross-Infection Reference Laboratory, Central Public Health Laboratory, Colindale, London.

Bernard Rowe MA, MB, FRCPath, DTM&H
Director, Division of Enteric Pathogens, Central Public Health Laboratory, Colindale Avenue, London.

G. R. Smith PhD, MRCVS, DVSM, Dip Bact
Head, Department of Infectious Diseases, Nuffield Laboratories of Comparative Medicine, Institute of Zoology, the Zoological Society of London, Regent's Park, London.

J. W. G. Smith MD, FRCPath, FFCM, Dip Bact
Director, National Institute for Biological Standards and Control, London.

A. E. Wilkinson MB, BS, FRCPath
Formerly Director, Venereal Diseases Reference Laboratory (PHLS), The London Hospital, London.

A. Trevor Willis MD, DSc, PhD, FRCPA, FRCPath
Director, Luton Public Health Laboratory and PHLS Anaerobe Reference Unit. Consultant Microbiologist and Medical Administrator, Luton and Dunstable Hospital, Luton.

Sir Graham Wilson MD, LLD, FRCP, FRCPath, DPH, FRS
Formerly Professor of Bacteriology as Applied to Hygiene, University of London, and Director of Public Health Laboratory Service, England and Wales.

General Editors' Preface to 7th edition

After the publication of the 6th edition in 1975 we had to decide whether it would be desirable to embark on a further edition and, if so, what form it should take. Except for the single-volume edition of 1936, the book had always appeared in two volumes. We hesitated to alter this arrangement but reflection made us realize that a change would be necessary.

If due attention was to be paid to the increase in knowledge that had occurred during the previous ten years two volumes would no longer be sufficient. Not only had the whole subject of microbiology expanded greatly, but some portions of it had assumed a disciplinary status of their own. Remembering always that our primary concern was with the causation and prevention of microbial disease, we had to select that part of the newer knowledge that was of sufficient relevance to be incorporated in the next edition without substantial enlargement of the book as a whole.

One of the subjects that demanded consideration was virology, which would have to be dealt with more fully than in the 6th edition. Another was immunology. Important as this subject is, much of it is not directly concerned with immunity to infectious disease. Moreover, numerous books, reviews and reports were readily available for the student to consult. What was required by the microbiologist and allied workers was a knowledge of serology, and by the medical and veterinary student a knowledge of the mechanisms by which the body defends itself against attack by bacteria and viruses. We resolved, therefore to provide a plain straightforward account of these two aspects of immunity similar to but less detailed than that in the 6th edition.

The book we now present consists of four volumes. The first serves as a general introduction to bacteriology including an account of the morphology, physiology, and variability of bacteria, disinfection, antibiotic agents, bacterial genetics and bacteriophages, together with immunity to infections, ecology, the bacteriology of air, water, and milk, and the normal flora of the body. Volume 2 deals entirely with systematic bacteriology, volume 3 with bacterial disease, and volume 4 with virology.

To the last volume we would draw special attention. It contains 27 chapters describing the viruses in detail and the diseases in man and animals to which they give rise, and is a compendium of information suitable alike for the general reader and the specialist virologist.

The first two editions of this book were written by Topley and Wilson, and the third and fourth by Wilson and Miles. For the next two editions a few outside contributors were brought in to bridge the gap that neither of us could fill. For the present edition we enlisted a total of over fifty contributors. With their help every chapter in the book has been either rewritten or extensively revised. This has led to certain innovations. The author's name is given at the head of each chapter, and each chapter is prefaced by a detailed contents list so as to afford the reader a conspectus of the subject matter. This, in turn, has led to a shortening of the index, which is now used principally to show where subjects not obviously related to any particular chapter may be found. A separate but consequently shorter index is provided for volumes 1, 2 and 3, and a cumulative index for all four volumes at the end of volume 4. Each volume will be on sale separately. As a result of these changes we shall no longer be able to ensure the uniformity of style and presentation for which we have always striven, or to take responsibility for the truth of every factual statement.

We are fortunate in having Dr Parker, who has been associated with the 5th and 6th editions of the book, as the third general editor of all four parts of this edition and as editor of volume 2. Dr Geoffrey Smith with his extensive knowledge of animal disease has greatly assisted us both as a contributor and as editor of volume 3. Dr Fred Brown, of the Animal Virus Research Institute, has organized the production of volume 4, and Professor Heather Dick the immunity section of volume 1.

Two small technical matters may be mentioned. Firstly, in volume 2 we have retained many of the original photomicrographs and added others at similar magnifications because they portray what the student sees when he looks down an ordinary light microscope in the course of identifying bacteria. Electronmicrographs have been used mainly to illustrate general statements about the structure of the organisms under consideration. Secondly, all temperatures are given in degrees Celsius unless otherwise stated.

Apart from those to whom we have just expressed our thanks, and the authors and revisers of individual chapters, we are grateful to the numerous workers who have generously supplied us with illustrations; to Dr N. S. Galbraith and Mrs Hepner at Colindale for furnishing us with recent epidemiological information; to Dr Dorothy Jones at Leicester for advice on the *Corynebacterium* chapter and Dr Elizabeth Sharpe at Reading for information about *Lactobacilius*, to Dr R. Redfern at Tolworth for his opinion on the value of different rodent baits; to Mr C. J. Webb of the Visual Aids Department of the London School of Hygiene and Tropical Medicine for the reproduction of various photographs and diagrams, and finally to the Library staff at the London School and Miss Betty Whyte, until recently chief librarian of the Central Public Health Laboratory at Colindale, for the continuous and unstinted help they have given us in putting their bibliographical experience at our disposal.

GSW
AAM

Volume Editor's Preface

In this Volume we present descriptions of bacteria, and of other micro-organisms bearing a close resemblance to them, such as the spirochaetes, chlamydiae, rickettsiae and Mycoplasmatales, that may be encountered by medical and veterinary microbiologists. Our treatment of systematic bacteriology resembles fairly closely that of earlier editions of this book. The authors of individual chapters have endeavoured to strike a reasonable balance between giving suitable recognition to important earlier work and summarizing adequately the very large amount of information that has accumulated in recent years; with each succeeding edition this becomes a more difficult task. In their descriptions of individual bacteria, authors have given prominence to characters useful in classification and identification that can be detected by workers in moderately well equipped non-specialized laboratories, and to those that may be of interest to microbiologists concerned with disease in man or animals.

When presenting facts that may throw light on the ability of organisms to cause disease it has not always been easy to make consistent decisions about what should appear in this volume and what should be included in the accounts of the individual diseases in Volume 3. The general principle we have followed on the subject of pathogenicity is to describe the effect on small animals in this volume, and in Volume 3 the diseases to which the organisms give rise in man and the larger animals. Exceptions, of course, have had to be made for such organisms as the diphtheria bacillus, the tetanus bacillus, and the cholera vibrio, where toxin production and pathogenicity are so closely related as to render some repetition inevitable. In such cases cross-references are given so as to ensure that the reader is made fully aware of all the information we provide on the subject.

In order for the reader to trace any particular organism we have included in the index a list of bacterial genera and the chapter number in which each will be found, together with a list of bacterial species with a note of the genus to which each belongs.

In separating our description of the organisms themselves from the diseases to which they give rise we have adhered to the practice decided upon in the first edition of the book in 1929. At that time medical and veterinary workers were interested in bacteria almost entirely on account of their ability to cause disease. Little attention was paid to the nature and properties of the bacteria themselves. Topley, however, influenced to some extent by the botanists, concluded that bacteriology should be regarded as a branch of biological science and that bacteria should be studied in their own right. This decision, novel as it then appeared to the writers of medical textbooks, attracted some criticism, but it was soon justified by the rapid growth of bacterial chemistry and the many contributions this made to our knowledge of the growth, survival and death of bacteria, of their pathogenic activities, and of ways of identifying them. Opinion has now moved so much further in the same direction that it is important to remember that bacteria are not simply collections of enzymes and other macromolecules, or carriers of interesting pieces of nucleic acid, but are fellow members of the living world. In this volume we have endeavoured to keep this delicate balance.

1983 MTP

Contents of volume 2

Systematic bacteriology

Contents of volumes 1, 3 and 4

Contents of volume 1
General microbiology and immunity

Contents of volume 3
Bacterial diseases

Contents of volume 4
Virology

20

Isolation, description, and identification of bacteria

M. T. Parker

Obtaining pure cultures of bacteria

The first essential in the study of bacteria is to obtain them in pure culture. During the sixties and seventies of the last century, micro-organisms were perforce cultivated in liquid media, and the preparation of pure cultures under these conditions was difficult. The only method available was to dilute the cultures with a sterile fluid until there was only about one organism in each two drops and to seed a number of tubes of liquid medium each with one drop (Lister 1878); but this afforded no certainty that the cultures so obtained were pure. The introduction of solid media by Robert Koch in 1881 rendered possible the easy separation of different organisms, because most of them formed discrete colonies, and serial subculture could be made from single colonies. Originally the medium was spread in the melted state on glass slides and allowed to set. Later, the petri dish was introduced and agar replaced gelatin as the solidifying agent in the medium; the resultant technique has since remained virtually unchanged.

Collecting suitable samples

To isolate the desired organisms, we must ensure that a suitable sample of the material to be examined is provided. This will often not be collected by a laboratory worker; we should therefore be prepared to spend time educating clinical and environmentalist colleagues in the correct methods of obtaining samples and transmitting them to the laboratory. These methods depend to some extent on the material sampled, the bacteria being sought and the practicability of rapid transport to the laboratory.

In many instances we are seeking to isolate pathogenic bacteria, which may not be present in large numbers. The site sampled must be one at which the pathogen is known to be harboured, and the sample should be of adequate size; these are matters upon which the laboratory worker should advise his clinical colleagues. In general, a quantity of pus or a piece of tissue, when obtainable, is preferable to a swab. When swabbing a dry surface, for example, the anterior nares or the skin, it is advisable first to moisten the swab (with peptone water, not water or saline). The swab should be rubbed vigorously on the site to be sampled, but care should be taken to avoid contact with adjacent areas likely to yield large numbers of irrelevant organisms; for example, when swabbing the throat, contact with the mouth and tongue should be avoided.

Blood for culture should be collected from a vein by means of a dry-sterilized needle and syringe or similar closed device immediately after disinfection of the skin (Chapter 4), and never through an existing intravenous catheter. It may be transferred to a specimen tube containing an anticoagulant, preferably liquid (sodium poylanethol sulphonate), because this neutralizes the bactericidal power of fresh blood (von Haebler and Miles 1938). Alternatively, bottles of medium may be taken to the bedside; several bottles should be inoculated with a total of 10 ml or more of blood, and the volume of medium should always be at least 10 times the volume of blood added. In most instances, one of these bottles should contain a medium suitable for the growth of fastidious anaerobes. (For blood-culture media, see Chapter 57). Specimen tubes and bottles of medium should be provided with a diaphragm covered by a removable plastic cap, so that the blood can be added without opening them.

At necropsy, solid organs should be sampled *in situ* after searing the surface with a hot iron; portions of tissue should be removed with sterile instruments. When this is impracticable, reasonably good results can be obtained by removing a large portion of tissue (at least a 30 mm cube), taking this immediately to the laboratory, plunging it into boiling water for 30 sec and sampling it from the centre. A method for sampling vegetations on heart valves removed at necropsy is described by Tyrrell and his colleagues (1979).

The difficulty of obtaining samples uncontaminated by normal flora has led to the introduction of various 'invasive' techniques of specimen collection. For example, lung puncture and transtracheal aspiration are used to sample the lower respiratory tract, and suprapubic puncture of the bladder is used to obtain uncontaminated samples of urine. These are valuable methods, but are not entirely without risk, so they should be used only when the information to be obtained is likely to benefit the patient, and by skilled operators.

For the collection of urine samples, see Chapter 57, and for further information about the collection of samples from patients, see Tyrrell *et al.* (1979) and Stokes and Ridgway (1980).

Samples for quantitative culture

For solid or liquid materials, the absolute and relative numbers of bacteria may be measured either by plating measured dilutions on to solid media and counting the colonies or by diluting to extinction by various modifications of Lister's (1878) method. Quantitative sampling of surfaces by means of swabs does not give results in absolute numbers, because only a proportion of the organisms is picked up and many of these remain attached to the fibres of the swab on plating or elution; nevertheless, valuable comparative results can be obtained if the method is carefully standardized. Alternatively, surfaces can be sampled by applying a measured volume of fluid, detaching the organisms by a standard procedure, and performing colony counts. A quite different method of sampling surfaces, by making impression-preparations, is widely used; agar medium may be applied directly to the surface (ten Cate 1959), or a moistened carrier, such as velvet, may be used to transfer the inoculum to an agar surface (Holt 1966). The resultant colony counts bear no relation to those obtained by swabbing or elution; they give some indication of the distribution of organisms on the surface, but none about the numbers present at the point at which a colony appears.

None of these cultural methods measures the number of viable bacterial cells, because bacteria tend to form clumps or chains. What we are measuring is the significant viable unit, which for colony-counting methods is the *colony-forming unit*. Attempts should therefore be made to standardize the degree of bacterial aggregation by using a constant method of shaking and a dispersing agent such as Triton X100.

(For a critical assessment of quantitative methods for the sampling of surfaces, see Favero *et al.* 1968. Methods for sampling air, water, milk, foodstuffs and environmental surfaces were discussed in Chapter 9).

Transport to the laboratory

Every effort should be made to ensure that the desired organism is viable when it reaches the laboratory. Organisms may die out in specimens for a variety of reasons: bacterial multiplication may cause an adverse change in pH; a number of organisms are very sensitive to oxygen and some to drying. Thus, no one set of conditions is likely to be optimal for all pathogens. Rapid transport of samples to the laboratory is always desirable, and in a few instances there is no alternative to it except the inoculation of media directly from the sampling site. Swabs, whether made of cotton or synthetic fibres, tend to contain inhibitory material (see Chapter 4 and Dadd *et al.* 1970); these may be removed or neutralized by boiling in phosphate buffer, by impregnating with horse serum (Rubbo and Benjamin 1951) or bovine albumen when viruses are also being sought (Bartlett and Hughes 1969), or by coating the swab with powdered charcoal.

Several general-purpose transport media have been designed in which most pathogens will survive without being overgrown by more robust organisms. The most successful are semisolid media containing thioglycollate but little nutrient (Stuart 1959, Amies 1967, Gästrin *et al.* 1968). For the survival of the more fastidious anaerobes, swabs should be submerged deeply in a long column of semisolid thioglycollate medium (see also Hill 1978). Material from transport media is seldom suitable for direct microscopic examination. Some pathogens, such as streptococci, survive better when completely dry than at intermediate humidities. Strips of filter paper may be seeded from a swab and air-dried before despatch (Hollinger *et al.* 1960) or the swab may be placed in contact with a dehydrating agent (Redys *et al.* 1968). Virus-transport media often contain antibiotics and are thus unsuitable for the isolation of bacteria, rickettsiae and chlamydiae. There is little advantage in refrigerating specimens during transit to the laboratory, except when quantitative culture is to be performed after significant delay. Specimens for the culture of *Chlamydia* should be transported in liquid nitrogen if it is impracticable to inoculate media directly from the patient (Tyrrell *et al.* 1979).

Plating on solid medium

This is usually the first step in obtaining a pure culture. The media chosen will depend upon what bacteria are sought; guidance on this is given in the appropriate chapters of this book. After the inoculum has been seeded on to a plate, careful streaking with a loop or bent glass rod to spread it in an even gradient of decreasing concentration is essential if well spaced colonies are to be obtained.

Attention must be given to providing the optimal temperature and gaseous atmosphere for growth. Most pathogens will grow at 37°, but some do rather better at a slightly lower temperature; 36° is therefore preferable for routine diagnostic bacteriology. Some organisms that grow in the presence of oxygen do so only when CO_2 is added; now that CO_2 incubators are generally available, an increasing proportion of primary cultures are made in air + 5–10 per cent of CO_2. Growing interest in the non-sporing anaerobes, and the recognition that some of them die rapidly when exposed to oxygen, have led to the design of very elaborate systems for anaerobic culture, in which all operations are performed in a special cabinet that has been repeatedly flushed with oxygen-free gases and in which media are pre-reduced and sterilized anaerobically (Moore 1966). Some workers (e.g., Drasar 1967, Moore *et al.* 1969, Peach and Hayek 1974) claim to have obtained much better results in this way than by conventional anaerobic-jar techniques. Others report equally good results with anaerobic jars when sufficient attention is given to the maintenance and regular testing of jars (Watt *et al.* 1976), when freshly poured medium is used and CO_2 is added to the oxygen-free atmosphere (Watt *et al.* 1974), and when the jar is not opened until it has been incubated for 48 hr (Wren 1977).

After the primary plate has been incubated, it is inspected under magnification and a representative of each colonial type is subcultured with a needle. Careful picking usually results in pure cultures, but this cannot be guaranteed. It is therefore advisable to replate supposedly pure cultures at least once and to inspect the resultant colonies for homogeneity. The plates should be fairly dry; if there is a layer of moisture on the medium, organisms are apt to form a film of confluent growth over the whole surface. This often leads to difficulties in obtaining pure cultures of organisms, such as many of the clostridia, that tend to spread unless the surface of the medium is very dry. The pour-plate method is, in general, of less value. Deep colonies are not usually as characteristic as surface colonies, and subculture from them carries greater risks of contamination.

Single-cell methods Growth of a culture from a single cell provides certainty that it is pure. Briefly, single cells are identified microscopically in tiny liquid droplets or on an agar surface, and are transferred to a liquid medium or removed to a sterile area of the agar remote from all other bacteria by means of a micro-manipulator (see Johnstone 1973).

Isolating organisms from mixtures

Numerous methods have been devised to isolate organisms that form a minority of a bacterial population. These methods seldom yield pure cultures in the first instance; their use must be followed by subculture from a single well spaced colony.

Firstly, the material itself may be treated in such a way that the unwanted organisms are destroyed or are separated physically from the organisms to be cultured. *Heating* a mixture to 80° for 10 min will destroy all the vegetative bacteria but not the spores. This method is used in the purification of clostridia. *Chemicals* with a germicidal action are useful in destroying susceptible organisms while leaving the more resistant unaffected. For example, tubercle bacilli are more resistant than are most other vegetative organisms to chemical disinfectants. A common method of isolating them is to treat sputum with acid or alkali strong enough to kill the accompanying bacteria. *Filtration* may be used to separate viruses from bacteria. L-forms and small bacteria such a *Campylobacter* can grow slowly through a membrane filter and so be separated from larger organisms. The membrane is placed on the surface of a suitable medium and a drop of fluid containing the organisms to be isolated is placed on it; after incubation, the membrane is removed and the plate re-incubated.

Various methods have been devised for separating motile from non-motile or less motile organisms. The device most often used is that described by Craigie (1931). It consists of a test-tube of semisolid agar medium containing a piece of glass tubing which projects above the surface of the agar. The medium in the inner tube is seeded with the culture under test. Motile organisms pass down the inner tube and up the outer tube, from the top of which they may be subcultured after incubation (see also Tulloch 1939). Flagellar agglutination may be used to inhibit the motility of one organism in the presence of another, as in separating the phases of diphasic salmonellae (see Chapter 37). Although strict aerobes will not migrate through semisolid agar, these methods can be applied to pseudomonads if nitrate is added to the medium (see Chapter 31).

Secondly, the conditions of growth may be so arranged that the desired organism is favoured. It is sometimes possible to make use of the *optimum temperature for growth* of an organism. Thermophilic bacteria may be separated from others by growth at 60°. Again, certain bacteria will not grow at 22°, whereas others will. In this way *Branhamella catarrhalis* may be separated from the meningococcus. *Aerobic and anaerobic* culture may be used to separate some groups of bacteria. Incubated aerobically the strict anaerobes will not grow, but many of them, e.g. most clostridia, will survive; incubated anaerobically most of the strict aerobes will not grow, but some of them (see Chapter 31) can obtain oxygen by denitrification and will therefore grow unless the nitrate content of the medium is very low. Many facultative anaerobes do not grow as well under anaerobic as under aerobic conditions.

The addition of *chemical substances* to a medium may facilitate the isolation of an organism by stimulating its growth or by inhibiting the growth of other organisms that may be present in larger numbers. If either type of substance is added to a liquid medium, the desired organism increases in numbers both absolutely and relatively to other organisms. Such a liquid is known as an *enrichment medium*. If, on the other hand, a substance that inhibits the growth of some organisms is added to a solid medium, the colonies that form will be mainly of other organisms that are resistant to the substance. Such a solid medium is known as a *selective medium*. It is common to use the two types of medium in conjunction, inoculating the material first into enrichment medium and later subculturing from this on to a selective medium. The colonies on a selective medium are, however, not necessarily pure. At the base of the colony other organisms may be present which, though unable to develop on the selective medium, will grow when transferred to a non-selective medium. It is therefore wise to replate organisms from a selective medium on to a plain medium before beginning to study them.

Enrichment media containing substances that are a source of energy only for some organisms have been widely used by soil microbiologists. Another example of the same principle is the use of tetrathionate broth for the enrichment of salmonellae, which obtain energy by means of a tetrathionate reductase (Pollock *et al.* 1942); but some other enterobacteria, including *Proteus* and *Citrobacter*, also form this enzyme (Le Minor and Pichinoty 1963), so it is advisable also to include selective chemicals in tetrathionate broth.

A novel selective procedure is to add to the medium a substance that is not harmful to bacteria but may be metabolized by some strains with the production intracellularly of a toxic substance. Thus *Esch. coli* and other organisms that form β-galactosidase will hydrolyse phenylethyl β-D-galactopyranoside with the release of phenylethyl alcohol; organisms, such as salmonellae of sub-genus I, which lack this enzyme, are unaffected (Johnston and Thatcher 1967, Johnston and Pivnick 1970).

Selective substances used either in enrichment broths or in solid selective media include aniline dyes, metallic salts and bile salts. In recent years, the use of antibiotics in selective media has greatly extended their range and effectiveness. Many examples will be found in the chapters devoted to the description of individual organisms.

Thirdly, an *indicator* may be added to the medium, which changes colour when a certain organism or group of organisms develops. Thus, the diphtheria bacillus reduces sodium tellurite, whereas many of the organisms likely to be associated with it in a throat swab do not. When this substance is added to the medium, the colonies of the diphtheria and of the diphtheroid bacilli are coloured black; those of the streptococci and numerous other organisms are colourless. Indicators are frequently used to detect the production of acid from a carbohydrate incorporated in the medium. Blood is a very useful indicator. Some organisms produce no alteration in it, others form from it a green pigment, and others lyse it completely.

Selective agents and indicators are frequently included in the same medium. Thus in MacConkey's medium, bile salts are added to inhibit the growth of organisms other than those capable of multiplying freely in the intestine, and lactose and neutral red are added to distinguish the lactose-fermenting coliform organisms from the non-lactose-fermenting group.

Finally, use may be made of the fact that certain *pathogenic organisms* are invasive and when injected into a susceptible laboratory animal can be isolated from an organ distant from the site of injection. The introduction of this method of separating organisms from one another we owe to Koch (1880). As examples, we may quote the tubercle bacillus in pus, the pneumococcus in sputum, and Whitmore's bacillus in surface waters. The contaminating organisms are rapidly killed in the animal body, whereas the pathogenic organism multiplies and can be recovered in pure culture from the tissues. This method is not without danger. If the animal happens to be suffering from a latent infection with some organism, this organism may be isolated in culture and thought mistakenly to have been derived from the material injected into the animal (see Wilson 1959).

Maintenance of pure cultures

Having obtained an organism in pure culture, the next task is to keep it in that state while it is being studied. If a reasonably good technical standard is maintained this should seldom prove difficult. But the possibility of contamination, or the inadvertent substitution of one culture by another, should always be kept in mind. When an unexpected result is obtained, this should be an indication for replating the culture and comparing the appearances of the colonies with that recorded earlier.

Protracted biochemical and genetic studies are particularly liable to be marred by contamination. Before they are begun, therefore, freeze-dried stocks of the cultures to be used should be prepared, samples of them should be checked for purity, and a full cultural and biochemical examination of the organisms should be made. When possible, they should be typed serologically or by means of phage. These tests should be repeated at intervals during the experiments. Particular care should be taken to avoid errors due to contamination or substitution when organisms are re-isolated after injection into animals and when they are being repeatedly selected on plates containing inhibitory agents. Suspicion should be aroused when 'variants' are isolated that appear to differ from the parent organism in several phenotypic characters.

Identification of bacteria

This is performed within the framework of *bacterial classification*, in which species and genera have been carefully defined and comprehensively described, and are each represented by a type culture; order is preserved by the use of an agreed system of *bacterial nomenclature* (Chapter 21). Much help can be obtained by carrying out a direct comparison of the unknown organism with type cultures of the organisms it appears to resemble, but we must always remember that no two bacterial strains are identical in all respects.

Ideally, the characters of the unknown organism should be compared with those of all recognized groups of bacteria, but limitations of personal experience and time make this impracticable. Most of us work with material from a limited environment; the medical bacteriologist is familiar with organisms isolated from man and mammals, and from their immediate surroundings, but the plant pathologist and the marine bacteriologist, for example, work with quite a different range of organisms. Therefore, a number of almost independent systems of bacterial identification are used in practice. This is satisfactory most of the time, but occasionally an 'intrusion' from an unfamiliar environment may not be recognized, such as the soil and water organisms that are now an important cause of infection in hospital patients.

In the classification of organisms (Chapter 21) we take into consideration all their ascertainable characters, but in bacterial identification we are more selective. We learn from experience, or we are taught, which tests are most useful in distinguishing one organism from another, and we deploy these tests in such a way as to arrive at an identification in the most economical manner. We are usually identifying an organism for a purpose, and the information is often needed quickly. Sometimes we need to know only whether the organism in question does or does not belong to a particular group or species. In these cases our system of 'identification' may be very simple indeed. If we are dealing with a culture from a familiar source we may pick out a characteristic colony from a primary plate and apply to it a small number of tests—or even a single test.

When confronted with a quite unfamiliar organism we have to engage in a more comprehensive process. We begin by examining a few of the basic characters of the organism. In medical bacteriology these would probably include: a Gram and an acid-fast stain; a motility test; a spore stain and a test for resistance to 80° for 30 min if the organism was a gram-positive rod; a comparison of growth in the presence and absence of air; an oxidase and a catalase test; and a test for the ability to oxidize or ferment glucose. With this information we should be able to allot the organism tentatively to one of the main groups of organisms of medical importance and then proceed to a series of secondary tests appropriate for the group in question (see Cowan 1974).

We shall now discuss in turn the various sorts of tests that may be used for the identification of bacteria. It is no longer practicable to include technical details of the tests; reference should be made to one of the books devoted to this subject, for example, those of Skerman (1969) and Cowan (1974). Only widely applied tests are mentioned; others used for the identification of particular organisms are described in the appropriate chapters in this volume.

Morphological appearances

Under this heading we include the shape and size of the organism, its arrangement and its motility, the number, distribution and shape of flagella, the presence of fimbriae, the shape and situation of spores, capsule formation, and staining behaviour. It is impossible to study all these properties on a single medium; motility, for example, should be looked for in a young, rapidly growing broth culture, and in some species is evident only at temperatures at the lower end of the growth range; flagella are sought for on a young agar culture; fimbriae after several subcultures in liquid medium; spores in old cultures, and so on. The shape and size of organisms vary. With a few exceptions, such as the corynebacteria, organisms are larger in young than in old cultures (Henrici 1926, Wilson 1926); and some organisms, for example, *Acinetobacter*, are bacillary in young cultures but may be mistaken for diplococci in stationary cultures. Even in one preparation the individual cells may vary greatly in size and shape, and this may be so pronounced as to justify the term 'pleomorphic'. Moreover, the appearance of organisms is often influenced by the type of medium on which they are grown. Chain formation, for example, is more evident in liquid media than on solid. The nature of the medium often influences the production of spores and capsules; *Bac. anthracis* forms spores readily in culture but never in the animal body; capsules, on the other hand, are frequently formed in the animal body but less often in culture. The arrangement of the organisms should be studied; if they are cocci, they may be arranged singly, in pairs, tetrads, packets, clusters or chains; if bacillary they may be arranged singly, in pairs end-to-end, in bundles, chains, clusters, or in Chinese-letter forms in which the individual bacilli lie more or less at right angles to each other; if vibrios, they may be arranged singly, in S-forms, semi-circles or wavy chains composed of S-forms.

The size and shape of organisms may be studied in preparations that have been stained lightly with a weak dye, such as methylene blue or safranin. Heavy staining (e.g. with crystal violet) tends to increase the apparent size of the organism and may obscure fine detail. Phase-contrast microscopy shows more of the internal structure of bacteria than does conventional light microscopy; and electron microscopy reveals many more differences in morphology between members of the groups of bacteria. It is not used regularly in bacterial identification, but may be necessary to decide whether flagella are attached at the pole or at the side of the organism. Fimbriae can be demonstrated only by electron microscopy, but their presence can often be inferred by other means (Chapter 33). Gram's stain serves to divide bacteria into two classes, the gram-positive and the gram-negative, and Ziehl-Neelsen's stain into the acid-fast and the non-acid-fast. Numerous other stains are used for special purposes, such as the demonstration of flagella, capsules, spores, nuclei, and metachromatic granules. Fluorescent dyes, such as acridine orange, enable ribonucleic and deoxyribonucleic acids within the cell to be distinguished from each other (Armstrong and Niven 1957); tubercle bacilli fluoresce when stained with auramine (Chapter 24). Among the organisms that take up the ordinary aniline dyes irregularly are some that contain granules of poly-β-hydroxybutyrate, which stain with Sudan black (Forsyth *et al.* 1958). For technical details of the main staining processes, see Cowan (1974).

A scheme for the systematic recording of the morphological appearances of bacteria is given at the end of this Chapter (see p. 15; for more information about bacterial morphology, see Chapter 2).

Cultural appearances

Under this heading we record the appearances of colonies formed on solid media and the type of growth in liquid media. Nutrient agar and nutrient broth are the usual media for this purpose if they support the growth of the organism; if not, others must be chosen. Observations are usually made after 24 hours' incubation under optimal conditions for growth and again at intervals. The colonial forms of many groups of bacteria are fairly distinctive, and within a group there may be important differences between members. Bacteriologists pay a great deal of attention to colonial morphology because this is often the only means available to them of picking out clones for further investigation, but the appearances of colonies may be influenced by quite small differences in the composition of media. Greater attention should therefore be attached to the observation of an experienced worker in one laboratory than to a careful description—even a picture—of colonies seen in another laboratory. Nevertheless, such descriptions are important and must be made systematically (pp. 15–18).

The appearances of colonies on special media—including indicator media that give evidence of biochemical characters—are of great assistance in arriving at a preliminary identification. Changes produced in blood agar help in the recognition of many organisms, but the animal species of origin of the blood, the composition of the basal medium and the presence or absence of oxygen all affect the results. No one type of blood agar medium is suitable for all purposes. Observations on the type of growth obtained by streak culture on an agar slope, by 'stab' or 'shake' culture in agar, and on Loeffler's serum, coagulated egg, potato, and litmus milk are now seldom recorded, and the use of the gelatin 'stab' has been generally superseded by plate tests for gelatinase.

Pigmentation The production of coloured substances is often of value in the recognition of individual bacteria, but non-pigmented strains occur in most pigmented species. Bacterial pigments are of many different colours, and may be retained in the cell or diffuse into the surrounding medium. Optimal conditions for their production vary widely, many being formed only under certain conditions of temperature or atmosphere, or in media of a particular composition. Many examples will be found in subsequent chapters.

Conditions for growth

Under this heading we study the partial pressure of oxygen and CO_2 required for growth, the limits of temperature and the optimum temperature for growth, the nutritional requirements. It is usual to divide bacteria into three classes according to their oxygen requirements: (1) *strict aerobes* grow only in the presence of oxygen; (2) *strict anaerobes* grow only in the absence of oxygen; and (3) *facultative anaerobes* grow under aerobic and anaerobic conditions, though usually better in the presence of oxygen. To these may be added a fourth class, the *microaerophiles*, which grow best under a pressure of oxygen less than that of the atmosphere. The distinction between microaerophiles and strict anaerobes is, however, one of degree. Among the strict anaerobes there are wide differences in the maximum O–R potential at which growth is initiated. Because some strict aerobes can obtain oxygen from nitrate, tests for the inability to grow anaerobically should be made in media that contain very little nitrate. Both facultative and strict anaerobes may require CO_2. According to their temperature requirements, bacteria may be divided into (1) *psychrotrophic* organisms that can grow at low temperatures (e.g. 2–5°), and usually have a temperature optimum in the range 20–30°; (2) *mesophilic* organisms, that grow within the range 10–45° and have a temperature optimum between 30 and 40°; and (3) *thermophilic* organisms, that grow poorly or not at all at 37° and have a temperature optimum between 50 and 60°. The term *psychrophilic* should not be used, because cold-growing organisms almost invariably grow faster above 10° than below this temperature.

A detailed study of the nutrients required for growth seldom forms part of the identification process, but individual characters are often used. These include the ability to grow on ordinary nutrient medium, and the effect of adding blood, serum or glucose. But the composition of nutrient media varies greatly between laboratories, and allowance must be made for this. Requirement for some nutrients, such as the X and V factors, can be tested in a complex medium known not to contain it (Chapter 39), but with other more widely distributed substances a synthetic or otherwise defined basal medium must be used. Tests for the ability to grow in mineral-salts medium with ammonium ions and a range of single organic compounds play a considerable part in the identification of gram-negative bacilli with simple requirements for nutrients. Koser's citrate-utilization test (1923, 1924) is of a similar character and is used in the identification of enterobacteria.

Resistance

Gross resistance to *heat* occurs only in spore-bearing bacteria; most vegetative bacteria are destroyed by moist heat at 60° in 1 hr. However, spores vary greatly in their susceptibility to heat; spores of some strains of *Cl. perfringens*, for example, are destroyed by boiling for a few minutes or even by lesser degrees of heat, whereas the spores of *Cl. botulinum* may withstand boiling for hours. Some clostridia rarely spore on ordinary media, and so may appear to be heat sensitive. Small degrees of heat resistance may be useful in the identification of individual species, for example, the ability of enterococci to survive 60° for 30 min. Resistance to other extreme conditions, e.g. of pH and salt concentration, are also useful characters. A general resistance to disinfectants characterizes the mycobacteria. Sensitivity to metronidazole is found only in strict anaerobes, and is a valuable means of distinguishing them from microaerophilic and CO_2-dependent organisms. Resistance to KCN (Braun and Guggenheim 1932, Report 1958) is useful for distinguishing between enterobacteria, and to the pteridine compound O/129 for separating the genera of polar-flagellated gram-negative bacilli (Chapter 27). Sensitivity or resistance to numerous other chemicals forms the basis of numerous differential tests, e.g. ethylhydrocupreine (optochin) sensitivity in pneumococci, tellurite resistance in corynebacteria and group-D streptococci, and dye sensitivities in the brucella group and among the gram-negative non-sporing anaerobes.

Antibiotic-susceptibility tests may be of value in identification, despite the fact that resistance to many antibiotics in common use is a character of individual strains rather than of species, and may be mediated by genetic determinants that are easily lost or gained. However, sensitivity or resistance of certain groups of bacteria to individual antibiotics has been shown empirically to be very stable, e.g. the sensitivity of moraxellae to penicillin, and of group-A streptococci to bacitracin, and the resistance of certain of the coagulase-negative staphylococci to novobiocin. The general pattern of suceptibility of anaerobic non-sporing gram-negative bacilli to antibiotics is a valuable aid to their identification (Chapter 26).

It is seldom justifiable to make an unqualified statement that a particular organism is sensitive or resistant to an antibiotic. The extent of its susceptibility should be specified in terms of the minimum inhibitory concentration of the agent or the size of the zone of inhibition in a disk-diffusion test. Zone sizes are often given as a 'diameter'; this may include the width of the disk, which may not be constant, or this may have been subtracted. The use of the term 'width', with an indication that this has been measured in one direction only, is preferable. Sensitivity tests for bacterial identification must be performed with as great care for technical detail as those used for more immediate clinical purposes

(Chapter 5). Some of the more valuable distinctions are based upon quite small differences in susceptibility, as in the bacitracin-sensitivity test.

Resistotyping methods (Elek and Higney 1970), in which bacteria are tested for susceptibility to a battery of chemicals and antibiotics in a gradient plate (Bryson and Szybalski 1952), have been advocated for defining 'types' within certain species. They are of some value in the subdivision of *Esch. coli* but less successful with *Staph. aureus*, in which resistance to some of the agents proved to be determined by plasmids and so was unstable (Elek and Moryson 1974).

Biochemical properties

The biochemical tests are a hotchpotch of empirical methods originally devised for the classification of individual groups of bacteria and later applied—often in many different modifications—for more general purposes. Bacteria are often grown with a substrate in an undefined medium, and the results of the test may therefore be complicated by secondary chemical reactions. A positive result may be the final outcome of a series of linked enzymic reactions, and if any one of these enzymes is missing the result may be negative. In other cases, several different tests may be influenced by a single enzymic reaction. It should be noted that some of the reactions traditionally recorded as 'biochemical characters' are in reality evidence of nutritional requirements, e.g. Koser's citrate test, or of resistance to single chemicals, e.g. the KCN test.

Respiratory function

The catalase test for the ability to release oxygen from H_2O_2 gives evidence of the presence of a functioning cytochrome system. The results are usually easy to interpret, but some lactobacilli and pediococci give weakly positive results that are not due to a true catalase (Whittenbury 1964). In cases of doubt, Deibel and Evans (1960) test for iron porphyrins by flooding an agar-plate culture first with benzidine dihydrochloride and then with H_2O_2, when a blue colour appears; a negative result in this test confirms the absence of catalase. Most anaerobic bacteria, and the Lactobacillaceae, are catalase negative. Some catalase-negative organisms form H_2O_2; tests for the production of this (Whittenbury 1964) are of value in the identification of certain species of streptococci; for a non-carcinogenic test reagent, see Marshall (1979).

The oxidase test (Gordon and McLeod 1928) is believed to detect a cytochrome oxidase that catalyses the oxidation of reduced cytochrome by molecular oxygen; most genera and some groups of related genera behave uniformly in this test. It is usually performed by the simplified method of Kovács (1956).

Nitrate reduction is usually tested for by a modification of the Griess-Ilosvay method (see Cowan 1974). The test determines (1) whether nitrite is formed from nitrate, and (2) whether all the nitrate is reduced beyond nitrite. A Durham's tube may be included to detect the formation of gaseous nitrogen.

The term *denitrification* is applied by soil bacteriologists to the reduction of nitrate to molecular nitrogen or nitrous oxide. In this process nitrate is used as an electron acceptor alternative to oxygen, so that strictly aerobic denitrifiers can grow anaerobically in the presence of nitrate. Denitrification defined as the ability to grow to the bottom of a lightly seeded tube of nitrate agar (Stanier *et al.* 1966) is of significance in the identification of certain pseudomonads. Strict aerobes that grow anaerobically in the presence of nitrate usually but not invariably form gaseous nitrogen from it.

The reduction of various dyes, tetrazolium salts, tellurites and selenites forms the basis of various empirical tests used in bacterial identification.

Action on carbohydrates and related compounds

Tests on the 'sugars' are of two sorts: those which reveal the general method of attack on these substances (usually exemplified by glucose); and those which determine the range of sugars from which the organism will produce acid. Some of the tests in the first category are useful means of making broad distinctions between groups of bacteria; others, and tests in the second category, are used mainly to characterize species and biotypes.

Method of attack on sugars The oxidation-fermentation (OF) test (Hugh and Leifson 1953) is used to distinguish between fermentation—for which oxygen is not necessary—and oxidation—for which it is essential. A semisolid agar medium, containing only 0.2 per cent of peptone, together with a sugar and an indicator, is dispensed in narrow tubes. Two of these are seeded with the organism, and the surface of the medium in one tube is covered with a thick layer of soft paraffin. In fermentation, acid is formed in both tubes; in oxidation, only in the open tube. Oxidation-fermentation tests are usually performed with glucose, but a few organisms oxidise maltose but not glucose. Oxidation-fermentation tests in Hugh and Leifson's medium usually give clear cut results with gram-negative aerobes, but they are less successful with some other groups of organisms, such as the staphylococci, for which an empirical modification of it has been proposed (see Chapter 30).

Fermentation of sugars may proceed along one of two main pathways: *homofermentative organisms*, such as streptococci, convert glucose almost entirely into lactic acid; *heterofermentative organisms*, such as the enterobacteria and the staphylococci, form other organic acids and alcohols, together with varying amounts of CO_2 and H_2. Large amounts of gas can be detected by means of a Durham's tube, but some heterofermentative organisms do not form sufficient gas for a bubble to appear in peptone-water sugars. Special tests (Langston and Bouma 1960) are therefore necessary to detect gas formation by some of the heterofermentative lactobacilli and by certain streptococcus-like cocci.

More detailed analysis of the products of fermentation by gas-chromatological methods, first used by Moore and his colleagues (1966) to characterize the species of clostridia by their production of volatile fatty acids and alcohols, has since been used for the classification of other groups of organisms, mainly but not exclusively anaerobes (Moore 1970, Holdeman and Moore 1972, Larsson and Mårdh 1977). The place of these methods in routine identification is not yet clear, but it is unlikely that they will replace traditional identification methods. (See, for example, in relation to the identification of gram-negative non-sporing anaerobes, Duerden *et al.* 1980.)

The methyl-red (MR) and Voges-Proskauer (VP) tests Some coliform organisms produce considerable amounts of acid from glucose, but the pH change is partially reversed on further incubation. This may occur if the pyruvic acid mainly responsible for the initial acidity is further degraded to neutral substances.

The MR test (Clark and Lubs 1915) distinguishes between strains that produce and maintain a high concentration of hydrogen ions and those in which reversion of pH occurs. The former give a red colour when the indicator is added to a culture in glucose phosphate peptone medium after incubation for 5 days at 30° and are referred to as MR positive; the latter give a yellow colour and are referred to as MR negative.

The VP test (Voges and Proskauer 1898) detects acetylmethylcarbinol, which is formed as an intermediate in the conversion of pyruvic acid to 2,3 butylene glycol. This reaction takes place only at an acid pH, and results in the formation of one molecule of a neutral substance from two molecules of acid. If a glucose phosphate peptone culture that has been incubated for 2 days at 30° is treated with strong alkali and exposed to air, the acetylmethylcarbinol is oxidized to diacetyl, which reacts with a constituent of the peptone to form a pink fluorescent substance. (For technical details of the MR and VP tests, see Cowan 1974.)

Among the enterobacteria, the expected negative correlation between the results of the two tests is fairly consistent, but in other bacterial groups many strains give a negative or a positive result in both tests. A test for the dehydrogenation of 2,3 butylene glycol has been used for the classification of pseudomonads, vibrios and enterobacteria (Schubert 1964, Schubert and Kexel 1964).

The acidification of sugars Conventional tests for the production of acid from sugars are made in tubes of a nutrient medium containing a sugar, usually at a 1 per cent concentration, and an indicator of pH—a method devised early in the century for the examination of enterobacteria. The nutrient may be either a peptone or a meat extract, and various indicators are employed. With bacteria of many groups—fortunately not including the enterobacteria—the results are greatly influenced by the composition of the medium, especially when the amount of acid formed is small and the organism also forms alkali from nitrogenous constituents of the medium. If the organism has simple nutrient requirements, more consistent results are obtained in appropriate synthetic media, but care must be taken to distinguish between failure to attack the sugar and failure to grow.

For tests with some groups of bacteria it may be necessary to add extra nutrients to peptone water or broth, for example, X and V factors for haemophilic bacilli or lipid for some anaerobic cocci; when natural products such as yeast extract, serum or ascitic fluid are added, these must be heated to inactivate enzymes.

Acidification of sugars may be the end result of the action of several enzymes, and a negative or delayed reaction may be due to the absence of any one of these. A single enzyme may be concerned with action on several sugars. Enzymes responsible for the acidification of sugars may be genetically determined by plasmids. For all of these reasons, individual sugar reactions, and even more, accounts of these performed in another laboratory, should not be accorded great significance in bacterial identification.

β-Galactosidase To ferment lactose, an organism must possess not only a β-galactosidase to split it into monosaccharides but also a permease to enable it to enter the cell. In the absence of the permease, acidification of lactose peptone water may be delayed or absent. The galactosidase can be detected by means of *o*-nitrophenyl-β-D-galactopyranoside (ONPG); *o*-nitrophenol is released from this compound by the enzyme, and a yellow colour appears (Lederberg 1950). The test can be done in one of two ways: the organism is incubated overnight in peptone water containing 0.15 per cent of ONPG (Lowe 1962); or a suspension of organisms in buffered saline is treated with toluene to release the enzyme and the ONPG is added (Le Minor and Ben Hamida 1962).

Oxidation of gluconate Haynes (1951) studied the ability of certain organisms to oxidize potassium gluconate to a reducing substance presumed to be potassium 2-keto-gluconate. The reaction is useful for the differentiation of organisms in the pseudomonas and coliform groups. The reducing substance is detected by heating with Benedict's solution.

Fermentation of organic acids The fermentation of organic acids must not be confused with their simple utilization as the sole source of carbon. In fermentation, the salt of the organic acid is broken down with the release of free sodium ions and the consequent production of an alkaline pH, which can be detected by means of a suitable dye. The salts usually employed are D-, L- and I-tartrate, citrate, mucate (Kauffmann and Petersen 1956) and malonate (Leifson 1933).

Action on nitrogenous compounds

Decarboxylation of amino acids The simplified methods introduced by Møller (1954, 1955) in which the presence of lysine, arginine, ornithine and glutamic acid decarboxylases is indicated by a pH change, and of arginine dihydrolase (or desiminase) by the addition of Nessler's reagent, have proved valuable in the identification of enterobacteria and many other groups of gram-negative bacteria. With the possible exception of the glutamic acid test, they present little practical difficulty. An alternative method for detecting the production of ammonia from arginine by pseudomonads is that of Thornley (1960); with streptococci, the special medium of Niven and his colleagues (1942) should be used.

The ability to produce ammonia from peptone water is still tested occasionally; Nessler's reagent is added to the culture after incubation.

Deamination of amino acids This is conveniently tested for by the conversion of phenylalanine to phenylpyruvic acid (Henriksen 1950, Ewing *et al.* 1957; see also Chapter 35).

Indole production The ability to form 'indole', in reality indole and a number of related substances (see Isenberg and Sundheim 1958), from a tryptophan-rich peptone may be detected by the appearance of a red colour in an extract of the culture, with ether, petroleum or preferably amyl alcohol, after treatment with Ehrlich's reagent (*p*-dimethylaminobenzaldehyde in acid-ethanol). This test is best performed after incubation of the culture for only one day. Alternatively, a paper impregnated with oxalic acid may be suspended over a peptone-water culture during incubation and observed for the appearance of a pink colour. (For technical methods, see Cowan 1974.)

Hydrogen sulphide Several methods are available for detecting the production of H_2S. Some are selected not because of their sensitivity but because they provide the basis for a useful taxonomic subdivision. The most delicate methods are those in which a rich source of a suitable sulphur compound is provided, and the H_2S is detected by suspending a strip of lead acetate paper over a culture in a test-tube. The amount of browning or blackening of the medium is measured in millimetres, and a fresh strip may be inserted daily. Liver agar slopes were much used formerly, but Clarke's (1953) standard broth supplemented with cysteine is to be recommended.

A metallic salt, such as lead acetate, ferric ammonium citrate, or ferrous acetate may be included in the medium (Levine *et al.* 1934, Zobell and Feltham 1934). Rather similar results are given by Kligler's iron sugar agar or ferrous chloride gelatin (Kauffmann 1954), both of which have been much used for the classification of enterobacteria. In fact, all members of this group produce H_2S in cysteine broth, and the organisms that give a positive result in these special media are not simply strong producers of H_2S, but also have the ability to form it in the presence of heavy metals and in a medium that is poor in cysteine (Brisou and Morand 1952).

Urea hydrolysis The hydrolysis of urea is detected either by the production of an alkaline reaction or by a chemical test for ammonia. A simple buffered medium containing urea as the sole source of nitrogen and an indicator of pH are suitable for use with members of the *Proteus* group (Ferguson and Hook 1943, Stuart *et al.* 1945). But other organisms hydrolyse urea only when provided with an additional source of nitrogen. Christensen (1946) therefore devised an agar medium containing peptone, besides urea and a small amount of glucose, with phenol red as indicator; this gives reliable results with many groups of bacteria. Some organisms can alkalinize this medium without attacking urea, however, and when there is doubt about the production of urease a method should be used in which ammonia is detected by means of Nessler's reagent, such as that of Elek (1948).

Hydrolysis of hippurate The ability to hydrolyse hippurate to benzoate is of importance in the classification of streptococci (Ayers and Rupp 1922). The organism is grown in hippurate broth, and a predetermined amount of ferric chloride is then added. Both hippurate and benzoate are precipitated by ferric chloride, but the hippurate is soluble in excess. The correct amount of ferric chloride to add is just sufficient to redissolve the precipitate in uninoculated broth. In the test, a precipitate indicates hippurate hydrolysis (see Hare and Colebrook 1934). A rather different test for hippurate hydrolysis has been used for the classification of enterobacteria (Hajna and Damon 1934). Here a mineral salts medium is employed, so that the test is one not only of the organism's ability to hydrolyse hippurate, but to use it as the sole source of carbon. (For rapid methods applicable to the identification of streptococci, see Hwang and Ederer 1975, and Edberg and Samuels 1976.)

Hydrolysis of complex biological substances

Proteinases Tests for the liquefaction of gelatin and coagulated serum or egg, and for the coagulation and peptonization of milk, are now seldom used. Gelatinase is more conveniently detected in gelatin-agar plates; after growth the plate is flooded with a protein precipitant, and a clear zone around the inoculated area indicates hydrolysis (Frazier 1926). A more sensitive method (Kohn 1953) makes use of disks of formalin-denatured gelatin containing powdered charcoal. These may be incubated in growing cultures or with washed suspensions of bacteria; hydrolysis of gelatin releases particles of charcoal. Numerous plate tests for the hydrolysis of other proteins are used; a positive result may be indicated, according to the nature of the protein attacked, either by clearing of an opaque plate, by a zone of opacity in a clear plate, or clearing revealed by the application of a protein precipitant.

Deoxyribonucleases Organisms may be incubated on plates of DNA agar, which is then flooded with acid to precipitate unchanged DNA. A more convenient and sensitive test is made in methyl green DNA agar (Smith *et al.* 1969) on which DNAase activity is indicated by a colourless zone around the bacterial growth.

Lipases Natural fats may be used as substrate; hydrolysis may be detected by the disappearance of fat globulcs, by pH change, or by the use of specific fat stains, but the results are often indefinite. Tests on the related Tween compounds, the hydrolysis of which gives rise to products that are precipitated in the presence of calcium ions (Sierra 1957) are much more easy to interpret. The production of opacity in plates or tubes of serum or egg-yolk media are used as empirical tests; these changes may be due to the action of lecithinases, lipoproteinases or lipases.

Diastases The presence of enzymes that hydrolyse starch by organisms that ferment glucose can be inferred from the production of acid in a conventional 'sugar' tube. A more sensitive and widely applicable test is to grow the organism on a plate of starch agar and flood this with a solution of iodine.

Many of the tests for hydrolytic enzymes give rather non-specific results. Their value can be enhanced in several ways, of which the following are examples: (1) the use of chemically defined substrates, such as the Tweens; (2) the detection of specific products of hydrolysis, e.g. of the lecithinases; (3) demonstrating specific neutralization of the enzyme by antiserum, as in the tests for clostridial lecithinases (Chapter 42); and (4) separating the enzymes electrophoretically, e.g. the staphylococcal proteinases (Chapter 30).

Newer methods of biochemical testing

Dissatisfaction with the imprecision of conventional biochemical tests, the length of time taken to obtain results, and the laborious processes of preparing and standardizing a variety of special media and setting up and reading the tests, have stimulated the development of rapid and automated procedures. Many reactions can be speeded up by adding very large inocula to small volumes of substrate. Early attempts at mechanization made use of *batch-testing* procedures, in which a multiple-point inoculator transferred a large number of different organisms simultaneously to clusters of small tubes, to wells in a plastic plate or to segments of a Petri dish that are separated by internal walls. These methods were labour-saving for testing many strains by a standard set of tests. They were of little help to the clinical microbiologist, who requires *set-testing* procedures, by means of which smaller numbers of strains can, after preliminary examination, be subjected to one of a series of alternative batteries of tests. Numerous attempts to meet this need have been made by commercial manufacturers (see Hedén and Illéni 1975), who have devised sets of papers, disks or tablets impregnated with substrate or medium, or of cupules containing these substances and moulded to a strip of plastic. Some of these devices seem to be useful in practice, but all of them need extensive comparison with conventional test-methods. One potentially valuable development is the introduction of tests for the rapid detection of a range of single enzymes (Buissière *et al.* 1967, Humble *et al.* 1977). Bascomb and Spencer (1980) describe an automated, continuous-flow system for the identification of gram-negative bacilli from urine by means of a pattern of enzymic activities; identification was completed within 6 hr of subculture from the primary plate. (For further information about rapid and automated methods see Johnson and Newsome 1976.)

Antigenic structure

The frequency of cross-reactions between unrelated organisms makes it unwise to place too much weight on the sharing of an antigen by otherwise dissimilar organisms. The place of serological tests, therefore, is for identification within bacterial groups that have already been defined by cultural and biochemical means and in which the accepted classification is based on antigenic relationships, for example, the salmonellae or the group A streptococci.

We also recognize the disadvantages of tests in which the whole bacterial cell is used as 'antigen,' e.g. agglutination tests, especially when whole cells are used to prepare the antiserum. Greater precison is obtained by a preliminary separation of classes of antigen: by physical or chemical treatment of the cells, and by various purification procedures. The use of precipitation tests in gel, with or without electrophoresis, now makes it possible to identify two or more antigen–antibody reactions in the same mixture, and to obtain a 'reaction of identity' between an antigen in the crude extract of an unknown organism and a purified antigen from a reference strain.

As well as testing separately for different classes of cellular antigens, e.g. for O, H, and K antigens in enterobacteria, or M and T antigens in group A streptococci, various extracellular products, including haemolysins, other toxins, and enzymes, may be similarly demonstrated, either by specific neutralization of the biological activity of the product, or by gel-precipitation tests with purified antigen. More precise identification of biologically active materials is possible if, after electrophoretic separation of the material in gel, neutralization of the activity by antiserum can be demonstrated at a point corresponding to the line of precipitation.

The use of immunofluorescence to demonstrate antigen-antibody reactions has greatly extended our ability to 'identify' organisms rapidly, even when they are present as only a minority of the bacterial population. Even more ingenious techniques have since been introduced: radioimmunoassay; enzyme-linked immunosorbent assay; and co-agglutination techniques, in which a protein-A-forming strain of *Staph. aureus* (Chapter 30) with immunoglobulin non-specifically adsorbed to its surface is mixed with the organism under examination. It must be emphasized, however, that the results are no more specific than the antiserum used, and that final identification can be made only by isolating the organism and performing conventional tests on it.

Chemical composition

Analysis of the cellular constituents is of great importance in bacterial classification. It has proved particularly useful in bringing order into groups of organisms that give indeterminate results in conventional tests. However, many of the analytical methods at present available require chemical expertise and apparatus not available in most bacteriological laboratories, and few of them are appropriate for the examination of a wide range of bacterial genera. Thus, their incorporation into existing general identification systems will take a long time.

The amino-acid composition of the cell walls of a number of bacterial groups is characteristic; even finer distinctions between gram-positive cocci can be made according to the structure of the interpeptide bridges in the cell-wall mucopeptide (see Chapter 30); chemical analysis of the menaquinones is helpful in the classification of staphylococci and enterococci; many similar examples will be found in later chapters of this volume. More complex molecules that cannot be recognized chemically may be characterized by their electrophoretic mobility; these include various proteins and enzymes (J. R. Norris 1964, Kersters and De Ley 1975, Gross *et al.* 1978, Feltham and Sneath 1979).

Even more sensitive methods applicable to whole organisms include infrared spectrometry (see K. P. Norris 1959), and gas-liquid chromatography and mass spectrometry of the products of pyrolysis (Reiner 1965, Larsson and Mårdh 1977, Gutteridge and J. R. Norris 1979). With such 'fingerprinting' methods it is possible to make very fine discriminations over a wide range of bacterial genera, and to obtain rapid results. They give promise, with further development, of providing an alternative to existing systems of bacterial identification.

Use of phages and bacteriocines for identification

The action of phages and bacteriocines on bacteria, and the carriage of phages and the production of bacteriocines, may be used in appropriate cases for the characterization of groups of related strains within a species ('types') and occasionally even of whole species.

Phages vary greatly in their range of lytic activity. Some lyse all members of a particular species, such as *Br. abortus* or *Bac. anthracis*. Others are specific in their action on biotypes within species, such as certain of the *Vibrio cholerae* phages (Chapter 27). Phages used in the infraspecific typing systems, such as those for *Ps. aeruginosa* and *Staph. aureus*, have narrow lytic spectra. These usually give patterns of lysis that are strain or type specific, and seldom attack organisms outside the species; thus, typability may be good evidence that an organism belongs to the species, but untypability does not exclude this possibility.

The range of activity of *bacteriocines* also varies widely. Many of those formed by gram-positive bacteria have a wide range of activity on members of other genera of gram-positive organisms, but some of those formed, for example, by haemolytic streptococci have serotype or serogroup specificity and may prove to be of value in identification (Chapter 29). Bacteriocines of gram-negative bacilli generally show narrow specificity within individual species, genera or groups of genera. Bacteriocine-typing systems are of two sorts: (1) 'active', in which a strain is characterized by the range of activity of its bacteriocines against a set of indicator strains; and (2) 'passive', in which it is characterized by the pattern of its susceptibility to the bacteriocines of a set of indicator strains. In both cases the indicator strains are usually but not invariably members of the same species as the strains being typed. Combined active-passive typing systems are often used. Bacteriocines are considerably less useful for identification at or above the level of species. Pyocines, though active on *Ps. aeruginosa* but not usually on other pseudomonads or enterobacteria, also attack gonococci and meningococci (Chapter 31). Colicines attack a wide range of enterobacteria.

Pathogenicity

Tests for pathogenicity are used in the final identification of certain organisms, and to distinguish between virulent and avirulent members of a genus or a species. They are of most value when the organism with which the unknown is to be identified produces a clear cut and characteristic disease process. Acute death following the injection intravenously or intraperitoneally of massive doses of organisms is seldom of much differential value. For the complete identification of certain pathogenic organisms it may be necessary to determine whether or not they produce an exotoxin; that is to say, whether the injection of a sterile filtrate of a culture grown under the appropriate conditions will cause death with characteristic lesions. In such cases, neutralization of this effect by a specific antitoxin may play an important part in identification.

Composition of the DNA

The results of the tests described in the preceding pages provide us with a picture of the *phenetic* characters

(Chapter 21) of the organism under investigation. Chemical studies of the DNA give information about the *genetic* determinants for these characters, and are now of great importance in bacterial classification (Chapter 21). It is seldom practicable at present to include analysis of the DNA among the methods used for the routine identification of bacteria, but they are now essential for the characterization of any previously undescribed organism.

The simplest and most widely used test is to determine the total base composition of the DNA, which is expressed as the moles per cent of guanine plus cytosine (G + C) in the DNA. (For information about the DNA base composition of individual bacterial species, see Hill 1966, Rosypal and Rosypalová 1966, and Holländer and Pohl 1980, as well as the relevant chapters of this book.) A wide gap between the DNA base composition of two organisms indicates that they are not closely related, but the converse is not true (Chapter 21).

Study of the sequence of bases in the DNA of organisms with a similar total base composition might be expected to define accurately the 'true' relations between them. We have so far very little detailed information about these sequences, but DNA–DNA pairing (Chapter 21) now provides quantitative estimates of the total similarity of the chromosomal DNA of bacteria ('genetic homology'). Results obtained by these pairing methods are very discriminating, and are revealing numerous instances of genetic dissimilarity between organisms with insignificant differences in phenetic characters when tested by current methods. As will be seen in subsequent chapters, the practice of naming new species defined almost entirely on evidence of genetic similarity is causing temporary difficulties in bacterial identification, but these should be overcome when their phenetic characters have been more intensively studied and the relation of genetic dissimilarity to taxonomic rank has been more clearly defined.

It is not clear to what extent the transfer of chromosomal material from one organism to another *in vitro* can be taken as evidence of relatedness, because it provides information only about homology in the part of the chromosome that is transferred. Nevertheless, the frequency of transfer of streptomycin resistance by means of transformation appears to be a useful means of assessing relations among streptococci, and between *Neisseria* and *Moraxella*. On the other hand, the transfer of characters determined by plasmids occurs over such a wide range of bacterial species that it is of limited value in bacterial identification.

Use of phenetic characters in bacterial identification

There is no standard process for the identification of an unknown bacterium. In some cases a few well chosen tests will suffice, but in others a lengthy investigation may be necessary and a mass of information will be accumulated. The problem then is how to compare this with the reported characters of the established taxa. Few organisms conform to the typical pattern in every character, and quantitative information about such deviations is often incomplete. The reproducibility of most tests is seldom total even in one laboratory, and is usually less so when the tests are made in different laboratories. Some commonly used tests are of low reproducibility, and others give fairly constant results with some bacterial groups but not with others. The potential value of a test also depends on its power to discriminate between pairs of organisms, and this may also vary widely in different bacterial groups.

The *sequential method* of identification by means of a key (e.g. Skerman 1967) has the disadvantage that a single discrepancy at an early stage may lead to serious misdirection. It may therefore be preferable to use a *simultaneous method*, in which all the characters of the unknown organism and of the taxa it resembles are set out in a table. In practice, a combination of the two methods is often used (e.g. Cowan 1974); for example, it may first be established that an organism is a gram-negative rod that grows on nutrient agar in air; its other characters are then compared in a table with those of all groups of similar organisms. This may be difficult when the table is large and complex, but the labour can be reduced by various mechanical aids (Cowan and Steel 1960, Olds 1966). Visual comparisons of this sort are often rather crude, because it is difficult to take into account the variability and the discriminating power of individual tests with particular groups of organisms. This can be done, however, by means of a probability model (see Lapage *et al.* 1970), but the process is complex and can be handled only on a computer. First, however, it is necessary to construct a probability matrix for all the relevant characters of the established taxa, and much of the necessary information cannot be obtained from the literature. (See also Gyllenberg 1965; Sneath 1969, Lapage *et al.* 1973, Bascomb *et al.* 1973, Willcox *et al.* 1973, Gyllenberg and Niemelä 1975, Lapage 1976.)

When an organism differs significantly from any described species, and when a description of it is to be published, a full record of all its characters should be made. Such a record should include careful comparisons with those species or types that most nearly resemble the newly isolated organism. It should be deposited in one of the national collections of type cultures.

Non-cultural methods of determining the cause of bacterial infections

Brief mention must be made of non-cultural methods that may be used to infer the presence of bacteria or their products in body tissues, fluids or exudates when a very rapid diagnosis is imperative, and sometimes even after living bacteria have been eliminated by antimicrobial treatment.

We may detect bacterial antigens by tube-precipitation tests (Ascoli 1911), and in much smaller amount by counter-current immunoelectrophoresis (Dorff *et al.* 1971) or co-agglutination (Christensen *et al.* 1973). We may also detect endotoxin in persons suffering from septicaemia due to gram-negative aerobic bacilli by means of the *Limulus* test (Chapter 57).

Characteristic changes in the body fluids of patients with certain bacterial infections can be recognized by means of gas-liquid chromatography (Mitruka 1975), and the products of fermentation by gram-negative non-sporing anaerobes may be similarly detected in pus (Phillips *et al.* 1976) and in blood cultures (Sondag *et al.* 1980). A raised lactic acid content in the cerebrospinal fluid in cases of meningitis, and in joint fluid in acute arthritis, is said to be a good indication of bacterial infection even when cultures are sterile (Controni *et al.* 1977, Brook and Controni 1978, but see Gästrin *et al.* 1979, Gold *et al.* 1980). Methods for identifying presumptively the causative organism in purulent meningitis by gas-liquid chromatography of the cerebrospinal fluid are being developed (Brice *et al.* 1979). Bacteria in the urine can be enumerated by a bioluminescence method of estimating adenosine triphosphate (Johnson *et al.* 1976), or by means of a particle counter (Alexander *et al.* 1981).

Systematic description of morphological and cultural characters of bacteria

Table 20.1 Check-list of the morphological characters of bacteria

Shape Cocci, spherical, oval or lanceolate; short rods, long rods, filaments, commas, or spirals.
Axis. Straight or curved.
Size Length and breadth.
Sides Parallel, bulging, concave, or irregular.
Ends Rounded, truncate, concave, or pointed.
Arrangement Singly, in pairs, in chains, in fours, in groups, in grape-like clusters, in cubical packets, in bundles, or in Chinese letters.
Irregular Forms Variations in shape and size; club, filamentous, branched, navicular, citron, fusiform, giant swollen forms and shadow forms.
Motility Motile or non-motile.
Flagella Polar (monotrichate, amphitrichate, lophotrichate) or peritrichate, or both (Fig. 20.1). (In electron micrographs, length, breadth, wave-length and amplitude.)
Fimbriae In electron micrographs, approximate number and size, polar or peritrichate.
Spores Spherical, oval, or ellipsoidal; equatorial, subterminal, or terminal; single or multiple; causing bulging of bacillus or not (Fig. 20.2).
Capsules Present or absent, indefinite mucoid sheath or envelope.
Staining Even, irregular, unipolar, bipolar, beaded, barred; and variations in depth between different organisms. Presence of metachromatic granules; reaction to Gram and to Ziehl-Neelsen stains.

Table 20.2 Check-list for description of the appearances of growth on solid or in liquid medium*

Surface colonies on solid media

Shape Circular, irregular, radiate, rhizoid.
Size In millimetres.
Elevation Effuse, raised, low convex, convex or dome-shaped, umbonate, umbilicate; with or without bevelled margin.
Structure Amorphous; fine, medium, or coarsely granular; filamentous, curled.
Surface Smooth; contoured; beaten-copper; rough; fine, medium, or coarsely granular; ringed; striated; papillate; dull or glistening.
Edge Entire, undulate, lobate, crenated, erose, fimbriate, curled, effuse, spreading.
Colour Colour by reflected and transmitted light; fluorescent, iridescent, opalescent, self-luminous.
Opacity Transparent, translucent, or opaque.
Consistency Butyrous, viscid, friable, cohesive, membranous, 'corroding'; growth down into medium.
Emulsifiability Easy or difficult; forms homogeneous or granular suspension, or remains membranous when rubbed up in a drop of water with a loop.
Differentiation Differentiated into a central and a peripheral portion (Fig. 20.3).

Growth in fluid medium

Degree None, scanty, moderate, abundant, or profuse.
Turbidity Present or absent; if present, slight, moderate, or dense; uniform, granular, or flocculent.
Deposit Present or absent; if present, slight, moderate, or abundant; powdery, granular, flocculent, membranous, or viscid; disintegrating completely or incompletely on shaking.
Surface Growth Present or absent; if present, ring growth around wall of tube; or surface pellicle, which is thin or thick, with a smooth, granular, or rough surface, and which disintegrates completely or incompletely on shaking.
Odour Absent, decided, resembling ——.

*On nutrient agar and in nutrient broth (if the organism will grow) under stated conditions of time, temperature and atmosphere. Record appearances also on special media that reveal characters of the group of organisms under investigations and growth, e.g. in ring form, below the surface in a semisolid medium.

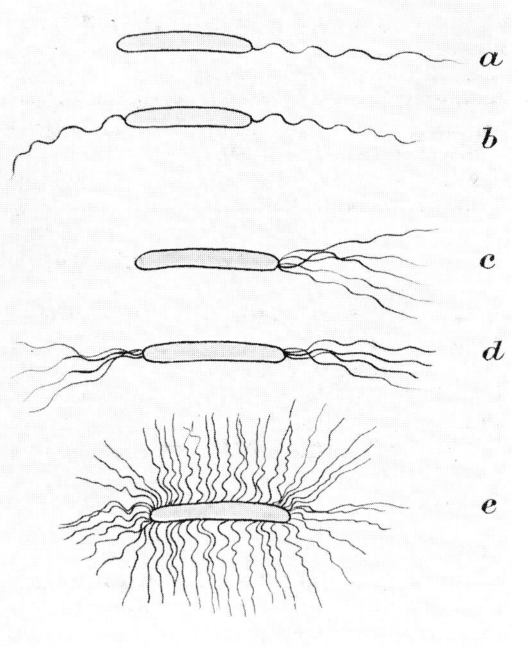

Fig. 20.1 *a–d.* Polar distribution of flagella. *a.* Monotrichate. *b.* Amphitrichate. *c* and *d.* Lophotrichate. *e.* Peritrichate.

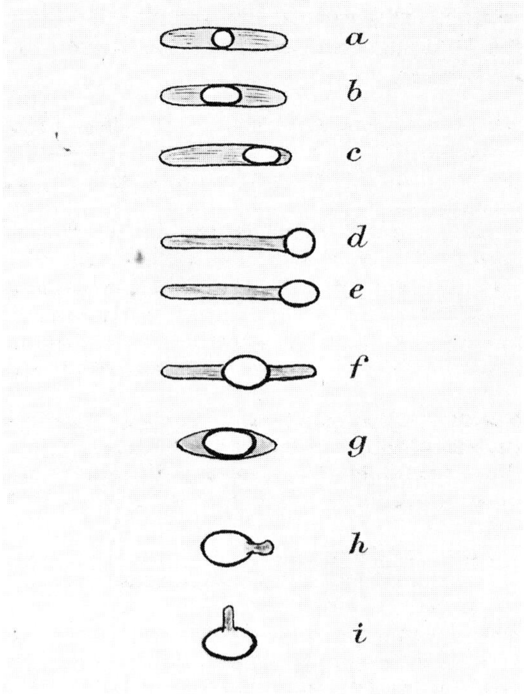

Fig. 20.2 Spores. *a–c.* Without distortion of bacterial cell. *a.* Spherical equatorial. *b.* Oval equatorial. *c.* Oval subterminal. *d–g.* With distortion of bacteria cell. *d.* Spherical terminal. *e.* Oval terminal. *f.* Oval equatorial. *g.* Oval equatorial. *h* and *i.* Germination of spores. *h.* Polar germination. *i.* Equatorial germination.

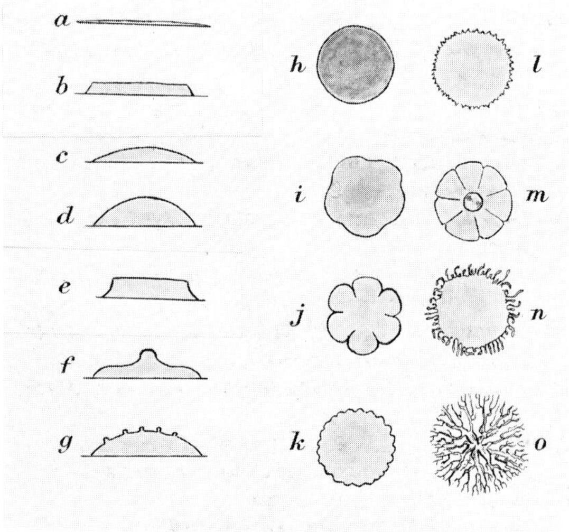

Fig. 20.3 Description of colonies. *a–g.* Elevation of colonies. *a.* Flat or effuse. *b.* Raised. *c.* Low convex. *d.* Convex or dome-shaped. *e.* Raised with concave bevelled edge. *f.* Umbonate. *g.* Convex with papillate surface. *h–o.* Edge of colonies. *h.* Entire. *i.* Undulate. *j.* Lobate. *k.* Crenated. *l.* Erose or dentate. *m.* Radially striated periphery with lobate edge. *n.* Fimbriate. *o.* Rhizoid or arborescent.

Glossary of terms used to describe the morphological and cultural appearances of bacteria

Amorphous (colonies): without visible differentiation in structure.
Amphitrichate: having one (or two) flagella at each pole.
Beaded (stained bacteria): deeply staining granules arranged at regular intervals along the course of the rod. (In stab or stroke culture): disjointed or semi-confluent colonies along the line of inoculation.
Beaten-copper: multiple small crateriform depressions on the surface of a growth, resembling beaten copper.
Bipolar: at both ends or poles of the bacterial cell.
Butyrous: growth of butter-like consistency.
Chains: four or more organisms attached end-to-end.
Citron: shaped like a lemon, having a small knob at each end.
Clavate: club-shaped.
Contoured: an irregular, smoothly undulating surface, or like a relief map.
Convex: the segment of a sphere of short radius; *low convex*, the segment of a sphere of long radius.
'*Corroding*': causes depressions in the medium under colonies.
Crenated: small, shallow indentations of the edge, which has a scalloped appearance.
Cuneate: wedge-shaped.
Curled: composed of parallel chains in wavy strands, as in anthrax colonies.
Effuse: growth thin, hardly raised at all from the medium.
Endospores: thick-walled spores formed within the bacterial cell.
Entire: with an even margin.
Equatorial: situated about equidistant from each end.
Erose: border showing fine, pointed, tooth-like projections.
Filaments: applied to morphology of bacteria, refers to thread-like forms, generally unsegmented; if segmented, to be distinguished from chains (*q.v.*) by the absence of constrictions between the segments.
Filamentous: growth composed of long, often interwoven threads.
Filiform: in stroke or stab cultures, a uniform growth confined to the line of inoculation.
Fimbriate: fine, sometimes recurved, processes projecting from the edge of the colony or growth. Also used to denote the presence of fimbriae.
Flocculent: containing small adherent masses of bacteria of various shapes floating in the culture fluid, or deposited at the bottom.
Fluorescent: having one colour by transmitted light and another by reflected light.
Friable: growth dry and brittle, when touched with a needle.
Granular: composed of granules; fine, medium or coarse.
Haemolysis: on blood agar plate; α-haemolysis: colonies surrounded by a greenish ring; β-haemolysis: colonies surrounded by an area of clearing, which is transparent (see Chapter 29).
Heaped-up: irregular, coarse processes projecting considerably above the level of the rest of the growth.
Iridescent: exhibiting changing rainbow colours in reflected light.
Lenticular: surface colony, which is convex and translucent, and which acts like a plano-convex lens, giving an inverted image of an object viewed through it. Deep colony, which is shaped like a lentil.
Lobate: having the margin deeply undulate, producing lobes (see *Undulate*).
Lophotrichate: having a tuft of flagella at one or both poles.
Luminous: glowing in the dark, phosphorescent.
Membranous: growth thin, coherent, like a membrane.
Mirror-like: having a smooth glistening surface, in which reflections of surrounding objects, e.g. window bars, can be seen.
Monotrichate: having a single flagellum at one pole.
Navicular: shaped like a boat.
Opalescent: coarsely iridescent, like an opal.
Opaque: objects, e.g. window bars, cannot be seen through growth.
Papillate: growth beset with small nipple-like processes.
Pellicle: bacterial growth forming either a continuous or an interrupted sheet on the surface of the culture fluid.
Peritrichate: having flagella disposed around the organism.
Punctiform: very small, but visible to naked eye; under 0.5 mm in diameter.
Radiate: showing fissures or ridges arranged in a radial manner.
Raised: growth thick, with a comparatively flat surface, and with abrupt or terraced edges.
Rhizoid: growth of an irregular branched or root-like character, as in *Bac. mycoides*.
Ring: growth at the upper margin of a liquid culture, adhering to the glass.
Ringed: having one or more circular depressions or elevations on the surface, sometimes giving a draughtsman-like appearance.
Rough: general term for an irregular surface, the irregularity being of a coarsely granulated type, or resembling morocco-leather.
Spreading: growth extending much beyond the line of inoculation, i.e. several millimetres or more; sometimes over an entire slope or plate.
Subterminal: situated towards the end.
Terminal: situated at the extreme end.

Translucent: objects, e.g. window bars, are visible through growth, but growth is not water-clear.

Transparent: growth is water-clear.

Truncate: ends abrupt, square.

Turbid: cloudy; may be a uniform, flocculent, or granular turbidity.

Umbonate: having a button-like, raised centre.

Undulate: border wavy, with shallow sinuses.

Unipolar: at one end only of the bacterial cell.

Viscid: sticky, semi-fluid; on withdrawal of the needle, the growth follows it in the form of a thread; sediment on shaking rises as a coherent swirl.

References

Alexander, M. K., Khan, M. S. and Dow, C. S. (1981) *J. clin. Path.* **34,** 194.

Amies, C. R. (1967) *Canad. J. publ. Hlth.* **58,** 296.

Armstrong, J. A. and Niven, J. S. F. (1957) *Nature, Lond.* **180,** 1335.

Ascoli, A. (1911) *Zbl. Bakt.* **58,** 63.

Ayers, S. H. and Rupp, P. (1922) *J. infect. Dis.* **30,** 388.

Bartlett, D. I. and Hughes, M. H. (1969) *Brit. med. J.* **iii,** 450.

Bascomb, S., Lapage, S. P., Curtis, M. A. and Willcox, W. R. (1973) *J. gen. Microbiol.* **77,** 291.

Bascomb, S. and Spencer, R. C. (1980) *J. clin. Path.* **33,** 36.

Braun, H. and Guggenheim, K. (1932) *Zbl. Bakt.* **127,** 97.

Brisou, J. and Morand, P. (1952) *Ann. Inst. Pasteur* **82,** 643.

Brook, I. and Controni, G. (1978) *J. clin. Microbiol.* **8,** 676.

Bryson, V. and Szybalski, W. (1952) *Science, N.Y.* **116,** 45.

Buisserière, J. (1972) *C.R. Acad. Sci., Paris* **274,** 1426.

Cate, L. ten (1959) *Fleischwirtschaft* **12,** 1011.

Christensen, P., Kahlmeter, G., Jonsson, S. and Kronvall, G. (1973) *Infect. Immun.* **7,** 881.

Christensen, W. B. (1946) *J. Bact.* **52,** 461.

Clark, W. M. and Lubs, H. A. (1915) *J. infect. Dis.* **17,** 160.

Clarke, P. H. (1953) *J. gen. Microbiol.* **8,** 397.

Controni, G. *et al.* (1977) *J. Pediat.* **91,** 379.

Cowan, S. T. (1974) *Cowan and Steel's Manual for the Identification of Medical Bacteria*, 2nd edn. Cambridge University Press, London.

Cowan, S. T. and Steel, K. J. (1960) *Lancet* **i,** 1172.

Craigie, J. (1931) *J. Immunol.* **21,** 417.

Dadd, A. H., Dagnall, V. P., Everall, P. H. and Jones, A. C. (1970) *J. med. Microbiol.* **3,** 561.

Deibel, R. H. and Evans, J. B. (1960) *J. Bact.* **79,** 356.

Dorff, G. J., Coonrod, J. D. and Rytel, M. W. (1971) *Lancet* **i,** 578.

Drasar, B. S. (1967) *J. Path. Bact.* **94,** 417.

Duerden, B. I., Collee, J. C., Brown, R., Deacon, A. G. and Holbrook, W. P. (1980) *J. med. Microbiol.* **13,** 231.

Edberg, S. C. and Samuels, S. (1976) *J. clin. Microbiol.* **3,** 49.

Elek, S. D. (1948) *J. Path. Bact.* **60,** 183.

Elek, S. D. and Higney, L. (1970) *J. med. Microbiol.* **3,** 103.

Elek, S. D. and Moryson, C. (1974) *J. med. Microbiol.* **7,** 237.

Ewing, W. H., Davis, B. R. and Reavis, R. W. (1957) *Publ. Hlth. Lab.* **15,** 153.

Favero, M. S., McDade, J. J., Robertson, J. A., Hoffman, R. K. and Edwards, R. W. (1968) *J. appl. Bact.* **31,** 336.

Feltham, R. K. A. and Sneath, P. H. A. (1979) *Comput. biomed. Res.* **12,** 247.

Ferguson, W. W. and Hook, A. E. (1943) *J. Lab. clin. Med.* **28,** 1715.

Forsyth, W. G. C., Hayward, A. C. and Roberts, J. B. (1958) *Nature, Lond.* **182,** 800.

Frazier, W. C. (1926) *J. infect. Dis.* **39,** 302.

Gästrin, B., Briem, H. and Rombo, L. (1979) *J. infect. Dis.* **139,** 529.

Gästrin, B., Kallings, L. O. and Marcetic, A. (1968) *Acta path. microbiol. scand.* **74,** 371.

Gordon, J. and McLeod, J. W. (1928) *J. Path. Bact.* **31,** 185.

Gould, I. M., Irwin, W. J. and Wadhwani, R. R. (1980) *Scand. J. infect. Dis.* **12,** 185.

Gross, C. S., Ferguson, D. A. Jr and Cummins, C. S. (1978) *Appl. environm. Microbiol.* **35,** 1102.

Gutteridge, C. S. and Norris, J. R. (1979) *J. appl. Bact.* **47,** 5.

Gyllenberg, H. G. (1965) *J. gen. Microbiol.* **39,** 401.

Gyllenberg, H. G. and Niemelä, T. K. (1975) In: *New Approaches to the Identification of Microorganisms*, p. 201. Ed. by C.-G. Hedén and T. Illéni, John Wiley and Sons, London.

Haebler, T. von and Miles, A. A. (1938) *J. Path. Bact.* **46,** 245.

Hajna, A. A. and Damon, S. R. (1934) *Amer. J. Hyg.* **19,** 545.

Hare, R. and Colebrook, L. (1934) *J. Path. Bact.* **39,** 429.

Haynes, W. C. (1951) *J. gen. Microbiol.* **5,** 939.

Hedén, C.-G. and Illéni, T. (1975) *New Approaches to the Identification of Microorganisms.* John Wiley and Sons, London.

Henrici, A. T. (1926) *J. infect. Dis.* **38,** 54.

Henriksen, S. D. (1950) *J. Bact.* **60,** 225.

Hill, G. B. (1978) *J. clin. Microbiol.* **8,** 680.

Hill, L. R. (1966) *J. gen. Microbiol.* **44,** 419.

Holdeman, L. V. and Moore, W. E. C. (1972) *Anaerobe Laboratory Manual.* Virginia Polytechnic Institute and State University Anaerobe Laboratory, Blacksburg, Va.

Holländer, R. and Pohl, S. (1980) *Zbl. Bakt.* **A246,** 236.

Hollinger, N. F. *et al.* (1960) *Publ. Hlth Rep., Wash.* **75,** 251.

Holt, R. J. (1966) *J. appl. Bact.* **29,** 625.

Hugh, R. and Leifson, E. (1953) *J. Bact.* **66,** 24.

Humble, M. W., King, A. and Phillips, I. (1977) *J. clin. Path.* **30,** 275.

Hwang, M. and Ederer, G. M. (1975) *J. clin. Microbiol.* **1,** 114.

Isenberg, H. D. and Sundheim, L. H. (1958) *J. Bact.* **75,** 682.

Johnston, H. H., Mitchell, C. J. and Curtis, G. D. W. (1976) *Lancet* **ii,** 400.

Johnson, H. H. and Newsom, S. W. B. (Eds) (1976) *Proc. 2nd Int. Symp. Rapid Methods and Automation in Microbiology.* Learned Information (Europe), Oxford.

Johnston, M. A. and Pivnick, H. (1970) *Canad. J. Microbiol.* **16,** 83.

Johnston, M. A. and Thatcher, F. S. (1967) *Appl. Microbiol.* **15,** 1223.

Johnstone, R. I. (1973) *Micromanipulation of Bacteria.* Churchill Livingstone, Edinburgh.

Kauffmann, F. (1954) *Enterobacteriaceae*, 2nd edn. Einar Munksgaard, Copenhagen.

Kauffmann, F. and Petersen, A. (1956) *Acta path. microbiol. scand.* **38,** 481.

Kersters, K. and De Ley, J. (1975) In: *New Approaches to the Identification of Microorganisms*, Ed. by C.-G. Hedén and T. Illéni, John Wiley and Sons, London.

Koch, R. (1880) *Investigations into the Etiology of the Traumatic Infective Diseases.* New Sydenham Soc., Lond.; (1881) *Mitt. ReichsgesundhAmt.* **1,** 1.

Kohn, J. (1953) *J. clin. Path.* **6,** 249.

Koser, S. A. (1923) *J. Bact.* **8,** 493; (1924) *Ibid.* **9,** 59.
Kovács, N. (1956) *Nature, Lond.* **178,** 703.
La Force, F. M., Brice, J. L. and Tornabene, T. G. (1979) *J. infect. Dis.* **140,** 453.
Langston, C. W. and Bouma, C. (1960) *Appl. Microbiol.* **8,** 212.
Lapage, S. P. (1976) In: *Proc. 2nd Int. Symp. Rapid Methods and Automation in Microbiology*. Ed. by H. H. Johnson and S. W. B. Newsom, Learned Information (Europe), Oxford.
Lapage, S. P., Bascomb, S., Willcox, W. R. and Curtis, M. A. (1970) In: *Automation, Mechanization and Data Handling in Microbiology*, p. 1. Ed. by A. Baille and R. J. Gilbert, Academic Press, London; (1973) *J. gen. Microbiol.* **77,** 273.
Larsson, L. and Mårdh, P.-A., (1977) *Acta path. microbiol. scand.* **B259** (Suppl), 5.
Lederberg, J. (1950) *J. Bact.* **60,** 381.
Leifson, E. (1933) *J. Bact.* **26,** 329.
Le Minor, L. and Ben Hamida, F. (1962) *Ann. Inst. Pasteur* **102,** 267.
Le Minor, L. and Pichinoty, F. (1963) *Ann. Inst. Pasteur* **104,** 384.
Levine, M., Epstein, S. S. and Vaughan, R. H. (1934) *Amer. J. publ. Hlth* **24,** 505.
Lister, J. (1878) *Quart. J. micr. Sci.* **18,** 177.
Lowe, G. H. (1962) *J. med. Lab. Tech.* **19,** 21.
Marshall, V. M. (1979) *J. appl. Bact.* **47,** 327.
Mitruka, B. M. (1975) In: *New Approaches to the Identification of Microorganisms*, p. 123. Ed. by C.-G. Hedén and T. Illéni. John Wiley and Sons, London.
Møller, V. (1954) *Acta path. microbiol. scand.* **34,** 102; (1955) *Ibid.* **36,** 158.
Moore, W. E. C. (1966) *Int. J. syst. Bact.* **16,** 173; (1970) *Ibid.* **20,** 535.
Moore, W. E. C., Cato, E. P. and Holdeman, L. V. (1966) *Int. J. syst. Bact.* **16,** 383; (1969) *J. infect. Dis.* **119,** 641.
Niven, C. F., Smiley, K. L. and Sherman, J. M. (1942) *J. Bact.* **43,** 651.
Norris, J. R. (1964) *J. appl. Bact.* **27,** 439.
Norris, K. P. (1959) *J. Hyg., Camb.* **57,** 326.
Olds, R. J. (1966) In: *Identification Methods for Microbiologists*, p. 131. Ed. by B. M. Gibbs and F. A. Skinner, Academic Press, London.
Peach, S. and Hayek, L. (1974) *J. clin. Path.* **27,** 578.
Phillips, K. D., Tearle, P. V. and Willis, A. T. (1976) *J. clin. Path.* **29,** 428.
Pollock, M. R., Knox, R. and Gell, P. G. H. (1942) *Nature, Lond.* **150,** 94.
Redys, J. J., Hibbard, E. W. and Borman, E. K. (1968) *Publ. Hlth. Rep., Wash.* **83,** 143.
Reiner, E. (1965) *Nature, Lond.* **206,** 1273.
Report (1958) *Int. Bull. bact. Nomencl.* **8,** 25.
Rosypal, S. and Rosypalová, A. (1966) *Folia biol., Brno* **7,** No. 3.
Rubbo, S. D. and Benjamin, M. (1951) *Brit. med. J.* **i,** 983.
Schubert, R. H. W. (1964) *Zbl. Bakt.* **195,** 61.
Schubert, R. H. W. and Kexel, G. (1964) *Zbl. Bakt.* **194,** 130.
Sierra, G. (1957) *Leeuwenhoek med. Tijdschr.* **23,** 15.
Skerman, V. B. D. (1967) *A Guide to the Identification of the Genera of Bacteria.* Williams and Wilkins, Baltimore; (1969) *Abstracts of Microbiological Methods.* Wiley, New York.
Smith, P. B., Hancock, G. A. and Rhoden, D. L. (1969) *Appl. Microbiol.* **18,** 991.
Sneath, P. H. A. (1969) *J. clin. Path.* **22** (Suppl. 3), 87.
Sondag, J. E., Ali, M. and Murray, P. R. (1980) *J. clin. Microbiol.* **11,** 274.
Stanier, R. Y., Palleroni, N. J. and Doudoroff, M. (1966) *J. gen. Microbiol.* **43,** 159.
Stokes, E. J. and Ridgway, G. L. (1980) *Clinical Bacteriology*, 5th edn. Edward Arnold, London.
Stuart, C. A., Stratum, E. van and Rustigian, R. (1945) *J. Bact.* **49,** 437.
Stuart, R. D. (1959) *Publ. Hlth Rep., Wash.* **74,** 431
Thornley, M. J. (1960) *J. appl. Bact.* **23,** 37.
Tulloch, W. J. (1939) *J. Hyg., Camb.* **39,** 324.
Tyrrell, D. A. J., Phillips, I., Goodwin, C. S. and Blowers, R. (1979) *Microbial Disease: the use of the laboratory in diagnosis, therapy and control.* Edward Arnold, London.
Voges, O. and Proskauer, B. (1898) *Z. Hyg. InfektKr.* **28,** 20.
Watt, B., Collee, J. G. and Brown, R. (1974) *J. med. Microbiol.* **7,** 315; (1976) *J. clin. Path.* **29,** 534.
Whittenbury, R. (1964) *J. gen. Microbiol.* **35,** 13.
Willcox, W. R., Lapage, S. P., Bascombe, S. and Curtis, M. A. (1973) *J. gen. Microbiol.* **77,** 317.
Wilson, G. S. (1926) *J. Hyg., Camb.* **25,** 150; (1959) *J. gen. Microbiol.* **21,** 1.
Wren, M. W. D. (1977) *J. med. Microbiol.* **10,** 195.
Zobell, C. E. and Feltham, C. B. (1934) *J. Bact.* **28,** 169.

21

Classification and nomenclature of bacteria

L. R. Hill

Introductory

By the end of the 19th century, many different kinds of bacteria had been described and named, and some general classifications of them had appeared. The controversy as to whether they were really distinct or were simply variants of one and the same organism (*Coccobacteria septicum*, Billroth 1874) had been resolved; the association of particular kinds of bacteria with individual diseases played a large part in bringing this about. Thus, for bacteria, as for plants and animals, the concept of the species came to be accepted, and from an early stage bacteriologists adopted the hierarchical system of classification established by Linnaeus for botany and zoology. In the hierarchical system, species are grouped into genera, genera into families, and so on, and species are designated by Latin, or Latinized, binomials. As evidenced by the great number of species listed by, for example, Migula (1900), the late-nineteenth-century bacterial taxonomists held in the main to a doctrine of fixity of characters; each new isolate that differed in however small a way from any previously described bacterium tended to acquire a separate name and be given the status of a species. Early twentieth century taxonomists reversed this trend and descriptions of species often included features stated to be 'usually positive', 'usually negative', or even 'variable'.

There is no need here to discuss the philosophical problems raised by the concept of the biological species, except to say that with bacteria it is even more apparent than with plants or animals that the definition of a species is simply a matter of practical convenience. The small size of bacteria and their asexual reproduction, and our scanty knowledge of the genetics of bacterial populations, necessitate a pragmatic approach. A bacterial species is an artibrarily assembled set of strains that has been found to be convenient for practical purposes. Whether a set of strains that possess a great number of common features but differ in a few respects should be considered a single species, or two or more closely related species of the same genus, or a single species subdivided into subspecies, is and always will be a matter for subjective judgement. In microbiology the rank of species is very unevenly defined over the whole range of organisms; in well studied groups, such as the enterobacteria, it tends to be more narrowly defined than in poorly studied groups. It is evident that different kinds of bacteria show varying degrees of similarity to each other; from this arose the need to construct hierarchical classifications, extending upwards from the species to successively broader groups.

Biological classifications are information storage systems, and economy in storage is achieved through the hierarchical arrangement. The variety in features shown by the different kinds of bacteria, and the numbers of bacterial species generally found convenient to recognize, combine to make it difficult if not impossible for the human mind to store all this information as separate items. By classifying similar strains into the same species, similar species into the same genus, similar genera into families, and so on, and by devising a nomenclature to reflect this classification, detailed information becomes compressed. General resemblances and differences between species and between genera can then be assimilated. The classification enables the allocation of fresh isolates to known species (the identification process; Chapter 20), or, if they cannot be so allocated, leads to the recognition of new species. By means of nomenclature, classification also permits ease of communication. Classification, nomenclature and identification thus comprise the trinity that Cowan (1965*a*) referred to as '*taxonomy*'.

Early bacterial classifications were based mainly on the opinion of individual taxonomists (see Buchanan 1925). As the science of bacteriology grew so did specialization within it, and individual taxonomists rapidly became unable to study the entire range of bacteria. The only general classification of bacteria that has persisted through the years is that contained in successive editions of *Bergey's Manual of Determinative Bacteriology* (1923, 1926, 1930, 1934, 1939, 1948, 1957, 1974). The earlier editions were written by as few as five authors; in later editions, a growing number of persons contributed sections, and the general planning and arrangement was undertaken by a small committee; the 8th Edition had 129 contributors. *Bergey's Manual* thus represents a consensus of opinion, and even at the specific level the individual authors try to reflect generally held views.

'Consensus opinion', but with a greater emphasis on uniform nomenclature than on uniform classification, is also encouraged through an international organization for bacterial taxonomy: The International Committee for Systematic Bacteriology (ICSB). This is a committee of the International Union of Microbiological Societies, itself a union within the International Council of Scientific Unions. The International Committee is served by 28 Taxonomic Subcommittees, each concerned with a major group of bacteria, and a Judicial Commission. Bacterial taxonomy has its own specialist journal, the *International Journal of Systematic Bacteriology*.

Principles of classification

Bacterial classification is an empirical process for which it is impossible to lay down general rules. Nevertheless, some classifications are 'better' than others, depending on their success as information-storage systems. A 'good' classification is one into which newly acquired information can be easily assimilated; when attempts to incorporate new information lead to disruption of the classification, this is evidence that the classification is 'bad'. Natural classification, and the assumption that this, if obtainable, would be the best possible classification, has engendered much debate (see Heywood and McNeill, 1964). Taken in the phylogenetic context of showing the evolution of organisms, classification based on the hypothetical sequence of events ('*cladistics*') leading to the present-day species reflects only that aspect of evolution and may not necessarily match a classification based on genetic similarities between organisms, because of the complex effects of divergent, parallel, and convergent evolution. Classification based upon empirical consideration of the general characters ('*phenetics*') is also an expression of evolution by reflecting its current end-product. Thus, natural classification can mean several different things.

In practice, phenetic classification is of paramount importance; nomenclature is derived from it (p. 27) and identification is performed in relation to it (Chapter 20). Several general principles can be recommended to arrive at a 'good' classification. (1) Classification should be based upon as wide a variety of characters as is practicable. When only certain types of characters are used, e.g. only morphological, or only biochemical, the resulting classification can only be a special one and of limited use. (2) Taxa should be defined so that the within-taxon variation in characters is less than the between-taxon variation. This principle should be maintained at each taxonomic rank (species, genus, family and so on). (3) Taxa of the lowest rank in the hierarchical system, the species, should be defined on the basis of several, not single, strains, because a single strain cannot give a measure of within-taxon variation. (4) Taxa should be defined by sets of characters, and no single character need necessarily be possessed by all members of the taxon. When, for example, production of catalase is one of the set of characters defining a particular species, a catalase-negative mutant should not be excluded from that taxon. Taxa so defined are sometimes called *polythetic*. (5) Classifications are only working hypotheses. As new knowledge is gained, this should be added to the classification and, if necessary, the classification should be changed. This may, in turn, require a change in nomenclature; non-taxonomists may find this irritating, but few would wish to maintain today the nomenclature used in, for example, the first edition of *Bergey's Manual*. (6) Continuity between successive classifications—and 'continuity between generations' of taxonomists, a phrase used by Gordon (see Cowan 1978)—should be established through type and other reference strains. Ironically, the nomenclatural type strain of a species (or indeed the type species of a genus) need not necessarily be typical of its taxon (see

p. 28), but no revision of classification is undertaken without reference to previously published work, which will indicate whether a type strain or species is typical or not.

Numerical taxonomy

The essential principles of numerical taxonomy are: (1) that similarities between strains should be assessed over a wide range of characters; (2) that each character should have equal weight (in practice, a value of 1 allocated to possession of a given character, 0 if not possessed); (3) that similarities should be expressed as percentages of shared characters among all the characters considered (in practice, percentage similarity coefficients are calculated between all pairs of strains); and (4) that the classification should be based on correlated characters (in practice, ensured by bringing together strains with high similarity coefficients with each other, and separating them from other groups of strains which each have lower similarity coefficients with strains of the first group and high coefficients among themselves).

Numerical taxonomy has had a particularly wide impact on bacterial classification. It is applicable to most, though not all, of the many different types of data used by bacterial taxonomists. When first proposed by Sneath (1957*a*, *b*) it was criticized, though more by botanists and zoologists than by microbiologists, on the grounds (1) it was non-evolutionary or, at best, would not yield phylogenetic classifications, and (2) it was based on the equal weighting of characters and so contradicted the idea implicit in evolutionary theory, that some characters are more important than others. There is no reason *a priori* why bacterial classification should reflect the *process* of evolution; nevertheless, the equal weighting of characters in bacterial taxonomy must be justified.

Equal weighting of characters

This seems at first sight to conflict with the general principles on which higher plants and animals are classified; for example, possession of hair among animals is taxonomically more important than colour of hair. In microbiology, too, certain characters have come to be regarded as more important than others, for example, the Gram reaction or catalase production. Part of the justification for equal weighting is to ask that if a particular character is indeed more important then how much more weight, quantitatively, should be given to it. The main justifications, however, are that by giving equal weight, mathematical-statistical methods can be applied readily, and that the mechanics of numerical taxonomy result in the classification being determined by the highly correlated characters; poorly correlated characters have little effect on the result. Though not immediately apparent in the original method of numerical taxonomy, the technique is in fact a way of determining how much weight different characters should have. This becomes more evident in some other methods of numerical taxonomy, such as Principal Component Analysis, which start by the calculation of variances and co-variances (or correlation coefficients) between characters, rather than of similarity coefficients between strains (see Hill *et al.* 1965).

The results of according equal importance *a priori* to all characters give an insight into the purposes underlying classification. The highly correlated characters determine the broad classification of the whole range of organisms included in the study and the less correlated characters make smaller contributions to this. Within taxa defined at a particular rank in that broad classification, the less highly correlated characters may assume greater importance; they may become highly correlated characters within a given subset of the whole range of organisms. Possession of hair, in the example given above from animals, is associated with a taxon, Mammalia, of high taxonomic rank (a Class); colour of hair, on the other hand, may sometimes be important in distinguishing divisions below the rank of species. The 'importance' of a character is meaningless except in relation to taxonomic rank and is only a reflection of the 'importance' attached to particular taxonomic ranks. The importance of particular ranks is itself, however, subject to the work in hand. For some tasks, the major divisions are important, but for others, for example, in epidemiological studies, very fine divisions within species are relevant. The importance of the ranks, in turn determining the importance of characters, is thus relative to the requirements of users, and no rank is intrinsically more, or less, important than another; and no character is intrinsically more important than another.

In practice, this argument is modified by two considerations. Firstly, the user-requirement at any taxonomic rank includes an essential need to be certain that the organisms being studied are indeed members of the taxon at the next higher rank. For example, before determining the phage-typing pattern of a strain of *Staph. aureus* it is essential to know that the strain indeed belongs to this species. Secondly, a part of the variation of characters can be shown not to be associated with taxonomic rank but is simply a result of lack of reproducibility in performing tests and reading the results. In studies of numerical taxonomy, some strains should be duplicated by making two subcultures and numbering them randomly; the calculated similarities between duplicated strains will give one measure of the reproducibility of tests within the laboratory. Reproducibility of tests performed in different laboratories can be controlled only by very precise and detailed descriptions of how tests are carried out, and by using control strains.

Handling of data

In studies of numerical taxonomy, a large number of tests (100–200) are performed on a considerable number of cultures (from 100 to several hundred). Recourse is made to computers to analyse the data, even when the particular method does not make use of any complex mathematics. The basic concepts of modern numerical taxonomy are to a large extent derived from the work of the eighteenth-century naturalist Adanson (1763) but their application became possible only with the development of computers. However, taxonomists must not be beguiled by the mathematical-statistical aspects of numerical taxonomy or by the use of computers: the quality of the results obtained cannot be greater than the quality of the data analysed.

The similarity coefficient most frequently used in numerical taxonomy can be expressed as

$$M = a/(a+b)$$

where, between a pair of strains, a is the number of characters for which both strains are scored the same (both 1s, or both 0s) and b the number of dissimilar characters. This coefficient takes account of negative matches as well as positive matches as similarities. The allocation of a positive ('present') or negative ('absent') sign to a test result is usually an indication of change with respect to an uninoculated control, but sometimes it is a matter of historical precedent or convention; for example, susceptibility to antimicrobial agents is habitually recorded as a positive character. With some types of information, however, exclusion of negative matches is advisable; for instance, when tests are made for susceptibility to a large set of bacteriophages and each strain is lysed by only a few of the phages, very many negative similarities will be recorded. The original similarity coefficient of Sneath might then be more suitable:

$$S = a/(a+b+c)$$

where a is the number of positive matches, b the number of characters scored positive for the first strain only, and c the number scored positive for the second strain only. Negative matches are thus excluded. Both coefficients can be multiplied by 100 to express the similarity as a percentage.

There are some difficulties associated with definition of unit characters, recording results, coding the information, or transforming quantitative data. Certain types of information cannot easily be incorporated with the usual range of characters, e.g., cell-wall structure, as opposed to composition; and the results of serological tests. Nevertheless, most characters can be accommodated, and in many cases quantitative characters can be reduced to sets of semi-quantitative characters to each of which present/absent scores can be applied or, better, more complex similarity coefficients can be used.

Similarity coefficients are calculated between all pairs of strains. Pairs of strains with the highest coefficients are selected as the nuclei of taxonomic groups. Pairs with the next highest coefficients are then added to existing nuclei if one of the pair had already been selected; if neither strain had been selected, the pair forms a new nucleus. This process ('*single linkage sorting*') is continued progressively from high coefficients to low ones until every strain has been linked in. During this progression, the numbers of strains in groups will increase, groups will merge together at various levels of similarity and the outline of a hierarchical taxonomic tree will be discerned. Various methods have been devised to help the taxonomist to decide the similarity level at which the most informative groupings occur and to draw up trees (or 'dendrograms') that indicate the different levels of similarity between groups. The essential feature of dendrograms is that they provide a hierarchical arrangement relative to a similarity scale (see, for example, Fig. 21.1).

Sneath and Sokal (1974) discuss a number of refinements of analytical procedure for use in the numerical taxonomy; these include (1) different sorting procedures, (2) calculation of the properties of 'average' organisms to represent groups and identification of real organisms closest to the average, and (3) alternatives to dendrograms as means of representing the relationships between groups (see also Dunn and Everitt, 1982).

By and large, classification by means of numerical taxonomy gives results that broadly coincide with non-numerical (or 'intuitive') classifications that are based on a wide range of characters. In addition, it has proved very useful for the study of groups previously thought to be difficult to classify, such as the 'Rhodochrous' group (Goodfellow and Alderson 1977). It increases greatly the quantity of data that can be handled, in terms of numbers of tests and of strains, and thus serves to increase progressively the information content of the taxonomic scheme. An important sequel to a study of numerical taxonomy is determining, from the original table of test results and from the groupings given by the analysis, the characters that best serve to distinguish between taxa and so should find a place in identification tables.

The wide application of numerical taxonomy has had several other beneficial effects. It has encouraged taxonomists to seek better original data, to express these quantitatively whenever possible, and to make use of many new categories of information. It has directed attention to the problems of test reproducibility and how this can be measured; this in turn has stimulated improvements in the methods of testing. Finally, it has aided progress towards consensus by making possible co-operative studies in which a number of taxonomists examine a given set of strains and all the information is pooled. This has proved very helpful, for example, with the mycobacteria (Wayne *et al.* 1971).

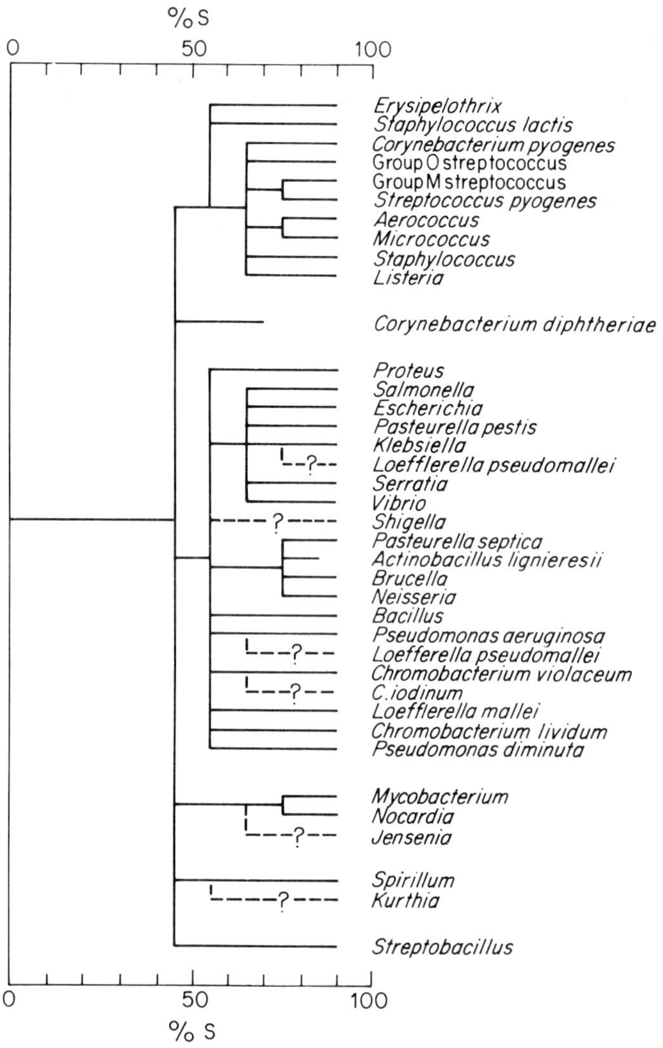

Fig. 21.1 Showing the outline of a classification of bacteria based on a numerical taxonomic analysis (from Sneath and Cowan 1958).

Composition of the DNA

Study of bacterial DNA for taxonomic purposes began at about the same time as the introduction of numerical taxonomy. The sequence of the four bases (guanine, cytosine, adenine and thymine) in the DNA, when read in triplets, forms a code that determines the phenetic characters of the organism. Those who advocated the direct study of the genome for taxonomic purposes pointed out that the data obtained by determining even as many as 200–300 phenetic characters represented only some 5–10 per cent of the information coded in the DNA (De Ley 1968). To make use of all the genetic information it would be necessary to read the sequences of the bases in the genome in its entirety—at present a task beyond our capabilities—and also to determine the precise beginning and ending of sentences and the control mechanisms. There would be a further requirement to understand the relationships between genome-encoded information and that contained in plasmids. Nevertheless, similarity between two organisms expressed in terms of DNA similarity, even if only partially understood, would necessarily contain more information than similarity expressed in phenetic terms. On the other hand, the techniques for measuring DNA similarity are complex and time-consuming, and are not yet suitable for the routine identification of fresh isolates, which is in practice the most important requirement of a system of classification of bacteria. Despite these practical limitations, however, studies of the DNA have already made important contributions to our understanding of the relationships between microorganisms.

It is generally not difficult to extract high-molecular weight DNA from bacteria and to purify it (Marmur 1961; Owen and Hill 1979). The DNA from different bacteria may show similarities or dissimilarities in several respects: (1) the amount per cell, or genome size; (2) the degree of molecular heterogeneity; (3) the base composition; and (4) the base sequence. For taxonomic purposes, the first three of these can be used only to express dissimilarity; it is important to remember that the absence of dissimilarity in the DNA from two bacteria in these three respects does not necessarily indicate similarity between the organisms. The only similarity that can be positively expressed is that of base sequence. Thus, if two DNAs are of different base composition, they are indeed dissimilar—and, by inference, the organisms from which the DNAs were obtained are dissimilar—but two DNAs of the same base composition may still be totally dissimilar in base sequence. Of course, when a group of organisms has a similar DNA base composition and shares very many phenetic characters, it is very likely that their base sequences are also similar.

Differences in genome size and in molecular heterogeneity have been noted between different bacteria, but have not yet had much impact on taxonomy. On the other hand, determination of base composition has become almost mandatory in any modern taxonomic work.

DNA base composition

Chargaff (1950) noted, from chemical analyses, that whatever the source of the DNA the relative amount of guanine (G) always equalled that of cytosine (C), and the amount of adenine (A) equalled that of thymine (T). This observation was one ingredient in the Watson-Crick double-helix model of DNA in which a G moiety on one strand is partnered by C on the other, and A is partnered by T. However the relative amounts of G to A (or T) and of C to T (or A) can vary from a ratio of 1 : 3 to the reverse, 3 : 1, according to the natural source of the DNA. It is convenient, and now a convention, to express the variable ratios of bases as the percentage of G + C pairs among total pairs. Thus 50 per cent G + C means equal proportions of all four bases; 25 per cent G + C means a composition of 12.5 per cent each of G and of C, and 37.5 per cent of A and of T; in other words 25 per cent of the base pairs are of the G + C type and 75 per cent are of the A + T type.

The range of DNA base compositions found in bacteria is wide, from a minimum of about 25 per cent G + C, found with certain mycoplasmas and clostridia, to a maximum of 75 per cent G + C with micrococci and *Streptomyces*. It presumably becomes impossible to code genetic information with two of the four-letter "alphabet" depleted beyond these limits.

It is technically quite simple to estimate the base composition of DNA from either (1) the so-called 'melting temperature' (T*m*), the temperature at which double-stranded DNA changes to single-stranded DNA with concomitant increase in ultraviolet absorbance, or (2) the buoyant density, the precise density of the DNA measured from the position the molecules take up in a cesium chloride density-gradient formed under ultracentrifugal conditions (see Marmur and Doty 1962, Schildkraut *et al.* 1962).

DNA homology

At present it is possible to determine the sequence of bases only in small nucleic acids, but similarities in the sequences in the genomic DNA of two organisms can be determined indirectly and without knowledge of the actual sequence in either. The physical conditions of ionic strength and temperature that permit denaturation of DNA in aqueous solutions (the 'melting' of DNA used to estimate base composition) can be reversed to allow single-strand DNA to re-form double-strand DNA ('renaturation'), provided that the base sequences are complementary. If the DNAs of two different organisms, AA and BB, are each denatured and then mixed together, and the conditions altered to permit renaturation, double-strand DNA of both original types will form (AA and BB); in addition, so-called 'molecular hybrids' will form (AB and BA) if the two DNAs have the same base sequence. When the sequences are not the same, only AA and BB molecules will form. Thus, if AB and BA molecules do form, the experiments will have demonstrated that the base sequences of the two DNAs are the same, without having established what the sequences actually are.

The first demonstration that this was possible was made by ultracentrifugation (Schildkraut *et al.* 1961). The buoyant density of one of the two original DNAs (say BB) was increased artificially by substitution of all the nitrogen atoms with a heavy-nitrogen isotope. When ultracentrifuged together in a density gradient, the two DNAs would form separate bands. After denaturation, followed by renaturation and again ultracentrifugation, three bands would form: the band of the least buoyant density was of AA molecules; that of the greatest buoyant density of BB molecules; and an intermediate band was of AB and BA molecules, since with these one strand was 'light', the other 'heavy'. The two DNAs Schildkraut and his colleagues (1961) first used were from subcultures of the same organism; one had been grown in minimal medium to substitute the nitrogen atoms. Therefore, at the outset it was known that the base sequences were the same and the experiment showed that the proportions of AA to AB + BA to BB molecules were in the expected ratio 1 : 2 : 1. Similar experiments were then performed with DNA from different organisms; the proportions of 'hybrid' molecules varied from the maximum as found in the first (or 'homologous') experiment to lesser amounts or none at all. Proportions of AB and BA molecules less than the maximum indicated that the base sequences of the two DNAs were only in part the same. It was possible therefore to say of pairs of DNAs that they were 100 per cent homologous in base sequence, or 0 per cent homologous, or had intermediate percentages of homology; a quantitative measure of total genetic relatedness between organisms was thus obtained.

The use of the term 'hybridization' for this type of experiment, even if qualified as 'molecular hybridization', is not to be recommended because it has other biological connotations; various other terms have been used (renaturation, annealing, reassociation, 'DNA binding') but it is more accurate to use a term that reflects directly what takes place experimentally: DNA-DNA pairing (Owen and Hill 1979). The ultracentrifugation method has been superseded by others that are more easily and quickly performed, and also make possible more accurate measurements of intermediate percentages of homology. Several of these methods are radioactive-isotope tracer-methods: the agar-gel method (Bolton and McCarthy 1962), the membrane-filter method (Gillespie and Spiegelman 1965), the hydroxypatite batch method (Brenner *et al.* 1969), the S1 endonuclease-assay method (Crosa *et al.* 1973); or are measures of rates of renaturation (the less homology in base sequence, the slower the rate of renaturation; De Ley *et al.* 1970).

The precise physical and chemical conditions for renaturation must be specified in DNA-DNA pairing experiments. When single strands of DNAs from different organisms have complementary base sequences in only some sections of the DNA, but not in other sections along the entire length of the molecules, an imperfect double-helix will form, in which the complementary sections will have renatured, but the non-complementary sections will not; such sections remain as 'looped-out' sections, or loose ends, of the hybrid molecule. Partly mismatched hybrid molecules will be thermally less stable and have a lower melting temperature than perfectly matched hybrid molecules. Some of the more recent DNA-DNA pairing methods measure the melting termperature of the hybrid molecules and, from these, estimates are made of how much, if any, mismatching is present.

The use of total DNA from the organisms gives a measure of general genetic relatedness, but does not allow one to state what percentage homology would be indicative of belonging to the same species, or the same genus, and so on, because taxonomic rank is a matter of convenience. Owen and Hill (1979) tabulate some data: levels of 85–100 per cent homology are quoted for some species of Enterobacteriaceae, e.g. *Esch. coli*, *Edwardsiella tarda*, *Klebsiella pneumoniae*; other well studied species also have high values, e.g. *Ps. aeruginosa* with 90–99 per cent homology, but some species have much wider ranges, e.g. *List. monocytogenes*, 40–100 per cent; *Cl. butyricum*, 15–100 per cent. These much wider ranges indicate that the species definition in different areas is at present uneven. Similarly, percentage homology between strains of different species within the same genus can range from very high values, e.g. *Bruc. abortus* and *Bruc. melitensis*, about 100 per cent, or *Salm. typhimurium* and *Salm. choleraesuis*, 91–94 per cent, to much lower levels, e.g. *Cl. botulinum* and *Cl. perfringens*, 18–24 per cent, *Haemophilus influenzae* and *H. parainfluenzae*, 44 per cent, *Ps. aeruginosa* and *Ps. stutzeri*, only 5–10 per cent.

It is also possible to study base sequence homologies of only parts of the DNA. Of special taxonomic interest are homologies among plasmids, which might indicate their histories (see, for example, Harwood 1980), and homologies in only those parts of the genomic DNA that code for ribosomal RNA (rRNA), since these sections of DNA are probably highly conserved, or in other words less changed during the course of evolution than the rest of the DNA. Levels of similarity found in the rRNA cistrons that are higher than levels found in the total DNA-DNA pairing are indicative of more distant genetic relatedness; De Smedt and his colleagues (1977) have proposed 'rRNA superfamilies' to describe these.

Other recent developments in taxonomy

In Chapter 20 we listed the methods traditionally used for determining the phenetic characters of bacteria for purposes of identification. A number of new categories of information of potential taxonomic value have become available in recent years (see also Goodfellow and Board 1980). These include finer morphological details revealed by electronmicroscopy, notably of the cell wall and the cell surface; the chemical composition and structure of the cell wall and the bacterial lipids; the electrophoretic patterns of cellular proteins, or of particular enzymes; the end-products of metabolism by gas-liquid chromatography; and the products of pyrolysis of whole organisms.

Some of the methods of 'chemotaxonomy' (taxonomy based on chemical data) now available make possible very fine distinctions between organisms and reveal dissimilarities not detected by examination at a more superficial level. For example, instead of determining the mere presence or absence of a particular enzymic activity, the physico-chemical properties, the amino-acid compositions, or even amino-acid sequences, in functionally similar enzymes present in different bacteria can be determined (Hartley 1974). Many of these methods yield quantitative results amenable to mathematical-statistical analysis. Analytical methods similar to those used in numerical taxonomy are sometimes used (for example, see Kersters and De Ley 1980).

Certain chemotaxonomic methods provide such precise 'fingerprints' of unknown organisms that they may in future form the basis of independent systems of bacterial identification (Chapter 20), by comparison, perhaps by computer, with a library of fingerprints of reference organisms belonging to known taxa. This is particularly true of electrophoretic patterns of cellular proteins and of the examination of the products of pyrolysis by gas-liquid chromatography (Chapter 20).

Nomenclature and the Bacteriological Code

In whatever way the methods of classification may change, there will remain a constant need for nomenclature to provide the means of communication and to reflect advances in classification as new knowledge is gained. Nomenclature is often considered the least important aspect of taxonomy. Most microbiologists think of it as a subject to be left to specialist taxonomists and even many of these regard it as something of a necessary chore. Cowan (1965*b*) even proposed a 'numericlature' (replacing names by a numerical code); such systems may well arise for practical purposes as more taxonomy, especially identification, is carried out aided by computers. Even in this case, however, the final means of communication will still be through names, as these convey concepts whereas numbers cannot, except on the smallest scale.

The correct naming of bacteria has been controlled since 1948 by the Bacteriological Code (current edition: Lapage *et al.* 1975, confusingly called the '1976 Revision'). The Code contains the rules of nomenclature and is periodically revised by the Judicial Commission of the ICSB; the Judicial Commission also monitors the application of the Code and pronounces Opinions (effectively, judgements) when disputes arise as to, for example, which of two or more synonyms for a particular taxon is correct. An Ad-hoc Committee acting on behalf of the Judicial Commission and ICSB has published Approved Lists of Bacterial Names (Skerman *et al.* 1980), as was provided for by the 1976 Revision of the Code. According to Rule 24a of the 1976 Revision of the Code, only names that appear in the Approved Lists are correct names; names not included in the Approved Lists have no standing in bacterial nomenclature. The 1976 Revision contains several appendices, one of which (Appendix 2) is simply a title ('Approved Lists of Bacterial Names') and a brief note; publication in 1980 of the Approved Lists themselves is a completion of the 1976 Revision. The 1976 Revision changed the starting date for bacterial nomenclature from 1st May 1753 (the date of Linnaeus's *Species Plantarum*, inherited by microbiologists from the Botanical Code) to 1st January 1980. By this mechanism, the correct naming of bacteria has been greatly simplified and possibly as many as eight thousand specific names have sunk into oblivion; the Approved Lists contain only about 2200 names of genera and species and 123 names of higher taxa.

The background to this recent development in bacterial nomenclature is as follows. Bacteria were being given latinized specific and generic names towards the end of the last century and even names for higher taxa were proposed. However, not surprisingly, these names were given not in accordance with any rules, since none existed specifically for bacteria; moreover, systems of classification were very much more fluid than they are now. Inevitably many confusing names gained usage, often applied to species that had already been named. Descriptions of species were, by modern standards, very brief, so that it is now often impossible to recognize a particular species from its original description; such names have thus become meaningless and the purpose of nomenclature, to convey concepts, is defeated. Botanists (or specifically Cohn, see Cowan 1978) took charge of bacterial nomenclature, and when eventually a separate Bacterial Code was created it was modelled on the Botanical Code. Two linked aspects of this development were (1) the inheritance of the rule that if a given taxon has more than one name, the correct name is the oldest one, provided that it conforms with other rules, such as having been validly published, and the other more recent names are synonyms, and (2) the starting date for considering such priorities was 1st May 1753. Although this 'priority rule' is a sensible way to settle questions of synonymy, the eighteenth century date gave rise to many difficulties. For example, the names of certain bacterial species had priority over more recent names even though no-one could be quite sure what the older names meant; indeed the later name might have been proposed deliberately in an attempt to avoid confusion. In the early days of naming bacteria, there were cases of the opposite effect: a perfectly 'good' species that is still recognizable today was given a name that did not conform to one or other of the rules of the present Bacteriological Code, and this name was widely and continuously used, yet a later 'synonym', not at all widely known but conforming to all the rules, would become the 'correct' name. In effect, the operation of the Bacteriological Code with the early starting date meant that any controversial question of nomenclature could be solved only by a meticulous search of the old literature and a subjective interpretation of what past microbiologists actually meant. The modern author was always uncertain whether a small unimportant publication might have been overlooked.

It was equally difficult to set a more recent starting date for the priority rule; for example, 1900 is a convenient date to remember, and was the year of publication of the second volume of Migula's compendium *System der Bakterien*, but most anaerobes were not then even in pure culture. In 1966, Buchanan and his colleagues published the *Index Bergeyana*, a listing of all the bacterial names used in successive editions of *Bergey's Manual* and of a sizeable proportion of other names that had appeared in the bacteriological literature, together with comments on the nomenclatural status of each name vis-à-vis the then current Bacteriological Code. However, even this Index cannot be guaranteed to be complete, and some Addenda have subsequently appeared in the *International Journal of Systematic Bacteriology*.

The 1976 Revision removed many of the anomalies of the previous editions of the Code in a novel manner by selecting a starting date in the then future, January 1980. The intervening period 1975–1980 was used to carry out a radical overhaul of all bacterial names in which only names thought to have a current meaning were retained for publication in the Approved Lists and the rest were discarded. For this overhaul, an Ad-hoc Committee, acting for ICSB, made use of the opinions of the ICSB Taxonomic Subcommittees and many other individual experts. No doubt some perfectly good specific names were omitted, but the 1976 Revision sets out the procedure for reviving them; the onus is on the interested worker to do this. It must be emphasized that the Approved Lists are of purely nomenclatural significance and do not give a seal of approval to any particular scheme of classification.

A further provision of the 1976 Revision was that all future proposals about nomenclature, if they are to have official status, must be made in the official journal of the ICSB (currently the *International Journal for Systematic Bacteriology*). New species, for example, are best described in that journal; if they are first published elsewhere, a note must be sent to this journal. Official recognition of the name will date from its appearance there, not in the original journal. Such announcements must give a reference to the first paper and specify the type strain of the species.

Two benefits should arise from these developments. (1) Users who are uncertain which of two or more alternative names is now correct can find out quickly by referring to the Approved Lists and subsequent issues of just one journal. If none of the alternatives appears there, none of them is correct; the user must then decide whether the issue is important enough for him to take action, such as to revive one of the names officially; alternatively, he may use one of the invalid names with the assurance that the 'error' is not of his own making. (2) Anyone wishing to name a new species need no longer search the old literature; all that is necessary is to ensure that the intended name does not already appear in the Approved Lists and in subsequent issues of the *International Journal of Systematic Bacteriology*.

Type strains

The Bacteriological Code differs from the Botanical and the Zoological Code in one important respect. To ensure the fixity of names, all three Codes use the type concept. In botany and zoology, the type for a specific name is a designated specimen in a known location (an herbarium or a museum); the name is attached to that specimen, and when other specimens are called by the same name, an actual or implied comparison is made with that specimen; the specimen is the name-bearer. Under the Bacteriological Code, the type is a designated viable strain; such strains must be deposited in one or more of the recognized culture collections. Although the ICSB does not specify which culture collections are, for these purposes, 'recognized', the larger supply-service culture collections are intended, such as the American Type Culture Collection (Washington), the National Collection of Type Cultures (London), and others of like standing. For the Approved Lists, each name considered was adjudicated, not only on the grounds of whether that name still had current meaning, but also on whether a type strain for that name was in existence in a culture collection.

The type concept is simply a name-bearing device; it is continued with higher taxa: for each genus there is a type species, for each family a type genus, and so on. Though desirable, there is no requirement that the type, at any taxonomic rank, has to be typical. This may appear surprising, but any other solution would be very restrictive. An author may designate a particular strain as the type, but later study may show that it is not typical of what becomes regarded as that species; in such an event, nomination of a new type, more typical than the original, would introduce an element of instability. Other workers may not agree with the new nomination and an undesirable precedent would have been set for continual changes of type strains.

Two kinds of type cultures are important in bacteriology: *holotypes* and *neotypes*. When an author originally describes a species, names it, and designates a type strain, that strain is called a holotype. If the original description was based on only one strain, this is automatically the holotype whether or not the author so designated it. However, many bacterial species were named before the existence of a bacteriological code and the requirement to designate a type strain. Also, in some cases, original holotype strains may have since been found to be contaminated, or were simply lost. In such cases, a later author may propose a type strain, a *proposed neotype*, which, if unchallenged, duly becomes the *neotype*.

A note by the Volume Editor

Classification and nomenclature in this book

The species and genera of bacteria that are encountered in the human or animal body form only a small part of those that have been described. It is therefore not profitable to consider the general classification of bacteria in the context of medical and veterinary bacteriology. Those who are interested in this subject should consult the most recent edition of *Bergey's Manual of Determinative Bacteriology* (1974).

In our accounts of bacteria of medical and veterinary importance (Chapters 22–43) we describe genera and species, and occasionally group them into families, for example, the Bacteroidaceae and the Enterobacteriaceae, but seldom refer to taxa of higher rank than this. Some chapters are devoted entirely or mainly to single well defined genera, and in others we group together two or more genera. The allocation of genera to chapters is, however, in some cases merely a matter of convenience and may not indicate that the genera described in the same chapter are closely related. At the end of many of the chapters we make brief mention of other genera that bear some resemblance to the main subject of the chapter but are not listed in its title. These may be located from the Index or from the table of contents that appears at the head of each chapter.

The reasons for accepting the existence of the genera we name are given in the appropriate chapters. We include in Chapter 43 a number of genera that, although almost certainly of bacteria, show little resemblance to other named bacterial genera; perhaps we should have increased this number considerably. Finally, we give accounts of the spirochaetes (Chapter 44), the chlamydiae (Chapter 45), the rickettsiae (Chapter 46), and the mycoplasmas (Chapter 47)—micro-organisms that, though not strictly bacteria, bear some resemblance to them.

Orthography of bacterial names

Linnaean binomials are printed in Italic, the generic name having an initial capital letter. The names of higher taxa also have an initial capital letter but are printed in Roman type.

Colloquial names Other well recognized designations for various bacteria are frequently used; these are printed in ordinary type without an initial capital. Thus, we refer to *Neisseria gonorrhoeae* as the gonococcus, *Streptococcus pneumoniae* as the pneumococcus, *Mycobacterium tuberculosis* as the tubercle bacillus and *Corynebacterium diphtheriae* as the diphtheria bacillus, and so on. It should be noted that when we use the word bacillus in Roman lower-case type we mean simply a rod-shaped bacterium and not necessarily a member of the genus *Bacillus* (see Chapter 41).

It is common practice to refer to a member or members of, for example, the genus *Streptococcus* as 'a streptococcus' or 'the streptococci', and the genus *Staphylococcus* as 'a staphylococcus' or 'the staphylococci', putting the words in lower-case type without an initial capital letter. These terms bear the same strict taxonomic connotations as their Linnaean counterparts. There is no reason why such colloquial names should not be used in the singular and the plural form, but in constructing plurals it may be necessary to sacrifice grammatical consistency to euphony. Thus we say 'rickettsiae', but 'mycoplasmas' rather than 'mycoplasmata'. Adjectival forms are also permissible, e.g. 'a salmonella antigen', but when a generic name is being used in a strictly hierarchical sense, as when referring to species in a genus, it is better to use the Italic form, such as 'some species of *Salmonella*', but this distinction is sometimes not easy to make.

Abbreviation of generic names Some system for the abbreviation of generic names is highly desirable. It is almost impossible to design a set of abbreviations that significantly shortens the words and is applicable to all the bacteria we need to mention. Our solution to this problem is as follows. In the general chapters of this book we have made use of a common set of abbreviations for frequently mentioned genera and spelled out the names of the rest. In certain of the chapters and sections of chapters concerned mainly with single genera, further abbreviation of the name of this genus, even to a single letter, is permitted when this does not cause ambiguity. Our list of common abbreviations is:

Achrom.	*Achromobacter*	*Lepto.*	*Leptospira*
Aerom.	*Aeromonas*	*List.*	*Listeria*
Alk.	*Alkaligenes*	*Micro.*	*Micrococcus*
Bac.	*Bacillus*	*Mor.*	*Moraxella*
Bart.	*Bartonella*	*Myco.*	*Mycobacterium*
Bord.	*Bordetella*	*Mycopl.*	*Mycoplasma*
Borr.	*Borrelia*	*Noc.*	*Nocardia*
Bruc.	*Brucella*	*Past.*	*Pasteurella*
Chlam.	*Chlamydia*	*Pr.*	*Proteus*
Chromo.	*Chromobacterium*	*Ps.*	*Pseudomonas*
Citro.	*Citrobacter*	*Salm.*	*Salmonella*
Cl.	*Clostridium*	*Ser.*	*Serratia*
Ery.	*Erysipelothrix*	*Sh.*	*Shigella*
Esch.	*Escherichia*	*Sp.*	*Spirochaeta*
Flavo.	*Flavobacterium*	*Staph.*	*Staphylococcus*
Fr.	*Francisella*	*Str.*	*Streptococcus*
Lacto.	*Lactobacillus*	*Tr.*	*Treponema*

Specific names In one respect our practice is at variance with the Approved Lists of Bacterial Names (Skerman *et al.* 1980). Since the 6th Edition of this book we have adopted the single 'i' termination of the masculine genitive of all personal names when they are used as specific epithets. The only exception is in respect of species named for a worker whose latinized name ends in and 'i'. Thus we say *'burneti'*, *'boydi'* and *'morgani'*. Our practice is forming specific epithets from the names of places is not consistent. For the species of *Salmonella* we follow traditional practice in omitting the genitive suffix, e.g. *Salm. dublin*, but we cannot persuade our colleagues to accept *Myco. kansas*.

Specific names Subspecies (subsp.) is the currently accepted name for a taxonomic subdivision below the species; one subspecies generally differs from others by several unrelated characters, for example, in biochemical reactions, antigenic constitution and susceptibility to lysis by phages. The term variety (var.) is a synonym for subspecies, but under the Bacteriological Code its use is illegitimate for new names given after 1975. We use the two terms interchangeably for organisms named at an earlier date. Subspecific divisions made by means of characters of a single class are termed 'infrasubspecific'. Practical bacteriologists have traditionally referred to such subdivisions as 'types', usually with a qualification to indicate the basis for their recognition, e.g. 'biotype', 'serotype' or 'phage type'. These terms are disliked by taxonomists, who would substitute respectively 'biovar', 'serovar' and 'phagovar'. We find these unacceptable because it is impossible to form a pronounceable verb from them. To avoid the use of such terms as 'a phagovaring system' we propose to accept the current use of the word 'type'.

References

Adanson, M. (1763) *Familles des Plantes.* Vincent, Paris.

Bergey's Manual of Determinitive Bacteriology (1923–1974); 1st edn 1923; 2nd edn 1926; 3rd edn 1930; 4th edn 1934; 5th edn 1939; 6th edn 1948; 7th edn 1957; 8th edn 1974. Williams and Wilkins, Baltimore.

Bolton, E. T. and McCarthy, B. J. (1962) *Proc. nat. Acad. Sci., N.Y.* **48,** 1390.

Brenner, D. J., Fanning, G. R., Rake, A. V. and Johnson, K. E. (1969) *Analyt. Biochem.* **28,** 447.

Buchanan, R. E. (1925) *General Systematic Bacteriology.* Williams and Wilkins, Baltimore.

Buchanan, R. E., Holt, J. C. and Lessel, E. F., Jr (1966) *Index Bergeyana.* Williams and Wilkins, Baltimore.

Chargaff, E. (1950) *Experientia* **6,** 201.

Cowan, S. T. (1965*a*) *J. gen. Microbiol.* **39,** 147; (1965*b*) *Advanc. appl. Microbiol.* **7,** 139; (1978) *A Dictionary of Microbial Taxonomy.* pp. 1, 11. Ed. by L. R. Hill. Cambridge University Press, Cambridge.

Crosa, J. M., Brenner, D. J., Ewing, W. H. and Falkow, S. (1973) *J. Bact.* **115,** 307.

De Ley, J. (1968) *Evol. Biol.* **2,** 103.

De Ley, J., Cattois, H. and Reynaerts, A. (1970) *Europ. J. Biochem.* **12,** 133.

De Smedt, J., Bauwens, M., Tytgat, R. and De Ley, J. (1980) *Int. J. syst. Bact.* **30,** 106.

Dunn, G. and Everitt, B. S. (1982) *An Introduction to Mathematical Taxonomy.* Cambridge University Press, Cambridge.

Gillespie, D. and Spiegelman, S. (1965) *J. molec. Biol.* **12,** 829.

Goodfellow, M. and Alderson, G. (1977) *J. gen Microbiol.* **100,** 99.

Goodfellow, M. and Board, R. G. (Eds.) (1980) *Microbiological Classification and Identification.* Academic Press, London.

Hartley, B. S. (1974) *Symp. Soc. gen. Microbiol.* **24,** 151.

Harwood, C. R. (1980) In: *Microbiological Classification and Identification,* p. 27. Ed. by M. Goodfellow and R. G. Board. Academic Press, London.

Heywood, V. H. and McNeill, J. (Eds.) (1964) *Phenetic and Phylogenetic Classification.* Systematics Assoc., London.

Hill, L. R., Silvestri, L. G., Ihm, P., Farchi, G. and Lanciani, P. (1965) *J. Bact.* **89,** 1393.

Kersters, K. and De Ley, J. (1980) In: *Microbiological Classification and Identification,* p. 273. Ed. by M. Goodfellow and R. G. Board. Academic Press, London.

Lapage, S. P., Sneath, P. H. A., Lessel, E. F., Skerman, V. B. D., Seeliger, H. P. R. and Clark, W. A. (1975) *International Code of Nomenclature of Bacteria* (*Bacteriological Code, 1976 Revision*). Amer. Soc. Microbiol., Washington, D.C.

Marmur, J. (1961) *J. molec. Biol.* **3,** 208.

Marmur, J. and Doty, P. (1962) *J. molec. Biol.* **5,** 109.

Migula, W. (1900) *System der Bakterien. 2. Specielle Systematik der Bakterien.* Gustav Fischer, Jena.

Owen, R. J. and Hill, L. R. (1979) In: *Identification Methods for Microbiologists,* 2nd edn, p. 277. Ed. by F. A. Skinner and D. W. Lovelock. Academic Press, London.

Schildkraut, C. L., Marmur, J. and Doty, P. (1961) *J. molec. Biol.* **3,** 595; (1962) *Ibid.* **4,** 430.

Skerman, V. B. D., McGowan, V. and Sneath, P. H. A. (Eds.) (1980) *Int. J. syst. Bact.* **30,** 225.

Sneath, P. H. A. (1957*a*) *J. gen. Microbiol.* **17,** 184; (1957*b*) *Ibid.* **17,** 201.

Sneath, P. H. A. and Cowan, S. T. (1958) *J. gen. Microbiol.* **19,** 551.

Sneath, P. H. A. and Sokal, R. R. (1973) *Numerical Taxonomy: the Principles and Practice of Numerical Classification.* Freeman, San Francisco.

Wayne, L. G. *et al.* (1971) *J. gen. Microbiol.* **66,** 255.

22

Actinomyces, *Nocardia* and *Actinobacillus*

Graham Wilson

Actinomyces

Definition

Gram-positive, non-motile, non-sporing, non-acid-fast organisms often growing in the form of a primary mycelium, which breaks up into coccoid bodies and rods of unequal length. No aerial hyphae produced. Most grow preferentially under anaerobic conditions in the presence of CO_2. Have little proteolytic activity, but produce acid from a variety of carbohydrates. Cell wall contains alanine, glutamic acid, lysine, ornithine, and in some species aspartic acid. Catalase negative (except *A. viscosus*). Parasitic, forming part of the oral flora, and potentially pathogenic. G + C content of DNA 60–63 moles per cent.

Type species: *Actinomyces bovis.*

Introductory

The name *Actinomyces bovis* was originally given by Harz (1877–78) to a mould-like organism found by Bollinger (1877) in the lesions of cattle suffering from a peculiar disease of the tongue and jaw, now known as actinomycosis. A similar organism, now known as *A. israeli*, was cultivated by Wolff and Israel in 1891 under anaerobic conditions. Since then a number of similar organisms have been isolated from a variety of diseases in man and animals. All of these organisms appear morphologically as jointed or unjointed filaments, which frequently show branching. In culture, rod forms are not uncommon. In the animal body, many of the pathogenic species are characterized by the formation of granules of varying size, which are found to consist of a filamentous mycelium surrounded by radiating clubs—a picture which is responsible for the term 'ray-fungus'. (Botanically the term 'ray' refers to the marginal portion of a composite flower, consisting of ligulate florets arranged radially.)

Nocardia is another genus of branched filaments that includes some species of medical importance. It shares many properties with *Actinomyces*, and we shall include it in our general description of this genus. Later in the chapter, however, we shall give a more particular account of *Nocardia* and its more important species, and shall refer briefly to some other genera of gram-positive filamentous organisms, including *Streptomyces*.

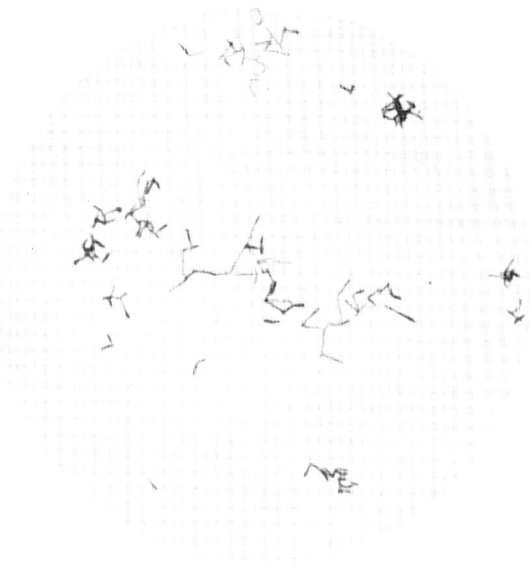

Fig. 22.1 *Actinomyces bovis* × 1000. From an agar slope culture, 14 days, 37° anaerobically.

Habitat

Members of the genus *Actinomyces* appear to be strict parasites: *A. bovis* is responsible for lesions in cattle, horses and dogs; *A. israeli* is found in the human mouth, and gives rise to actinomycosis, as does also occasionally *A. eriksoni*; *A. odontolyticus* occurs in human saliva and in carious teeth; *A. naeslundi*, also an inhabitant of the human mouth, is thought to be non-pathogenic; the closely related *A. viscosus* has been isolated from the oral cavity of man and hamsters, in which it produces gingival plaque and periodontal disease. The primary residence of the nocardiae is the soil. *N. asteroides* gives rise to abscesses and pulmonary disease in man; *N. farcinica* to farcy in cattle, *N. caprae* to pulmonary disease in goats, and *N. brasiliensis* to mycetoma in man.

Morphology

On culture media the morphology is variable. In ordinary film preparations *A. israeli* occurs chiefly as rods, 3–4 μm long by 0.6 μm broad, which from their arrangement, their clubbed ends, and their irregular staining bear a resemblance to certain members of the corynebacteria; careful search, however, will generally reveal a few definite filaments, some of which may show false branching (Fig. 22.1). Electron-microscopic studies show that true branching occurs by budding of one cell in a chain, with the formation of a lateral outgrowth at right angles to the axis of the chain (Duda and Slack 1972). The nocardiae in young cultures occur chiefly as long unsegmented straight or wavy filaments, which show simple or dichotomous branching (Fig. 22.2), and which not infrequently grow upwards from the surface as aerial hyphae. Later the mycelium undergoes segmentation leading to the formation of short rod forms and cocci. The aerial hyphae, which are rather wider than the filaments, break up into chains of bead-like spores (see Gordon and Mihm 1958), which on fresh media develop into mycelia. Reproduction is by fission and by budding.

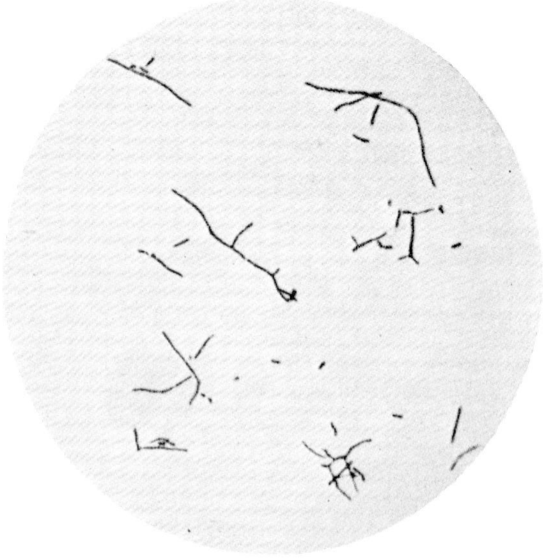

Fig. 22.2 *Nocardia asteroides* × 1000. From a broth culture, 24 hr, 37° aerobically.

In some strains segmentation is visible within 24 hr; in others it may not occur for 3 weeks or more. On soil media the filaments are arranged in loose groups or in a tangled mycelium, but in broth definite colonies occur consisting of a densely matted central core of filaments and a peripheral zone in which the filaments are more loosely disposed (Fig. 22.6). These colonies in liquid media are common to both the aerobic and the anaerobic types. The rods and filaments may stain evenly, but as a rule granular staining is evident. After growth for some time in liquid media both the aerobic and the anaerobic types may show involution forms, consisting mainly of spherical or club-shaped swellings on the ends of the filaments. Actinomyces and nocardiae are generally considered to be non-motile, though, according to Higgins and co-workers (1967), some actinomyces have one subpolar flagellum. Many strains of *Actinomyces* are shown by electron microscopy to be surrounded by fimbriae (Ellen *et al.* 1978).

Electron microscopy of *N. asteroides* shows the cell wall to consist of two dense layers with an electron-transparent layer in between. In the moderately dense cytoplasm are scattered denser polyribosomes, mesosomes and some electron-transparent lipid bodies which stain with Sudan Black B (Farshtchi and McClung 1967). The cell-wall constituents of *Actinomyces* do not include diaminopimelic acid; in *Nocardia* the meso-form and in *Streptomyces* the LL-form of this compound is regularly present (see p. 37). Cyclopropan fatty acids are found only in *Actinomyces*; oleic acid is characteristic of *Nocardia.*

With a few possible exceptions all species are gram positive. Actinomyces are uniformly non-acid-fast. The nocardiae may be divided into: (1) acid-fast; these resist decolorization with 1 per cent sulphuric acid for 5 min, but are usually decolorized by the application of 25 per cent H_2SO_4 for a similar length of time; there is, however, a considerable variation in the acid-fastness of different species; and (2) non-acid-fast. In nocardiae the aerial hyphae may be distinguished from the vegetative mycelium by their property of staining with Sudan iv in ethanol; this is due to their lipid content (Erikson 1947).

In the animal body the morphology often differs from that on culture media. *A. israeli* and *A. bovis* form colonies that appear in the pus as granules or *Drusen*. When crushed and examined microscopically these granules are seen to consist of a central filamentous gram-positive mycelium surrounded by a peripheral zone of large gram-negative clubs. In old colonies the mycelium is replaced by short rods and coccoid bodies.

The *clubs* vary in size, and may be as long as 10 μm and as broad as 5 μm. Their mode of origin has given rise to much discussion. Ørskov (1923) thought that there were two types of club: one representing the swollen end of the mycelial filament—the culture club; the other originating from lipid material deposited around the end of the filament by the tissues of the host—tissue club. More recently Pine and Overman (1963) have brought evidence to suggest that the club itself is formed entirely by the organism. Under the influence of acid phosphatase, secreted partly by the organism and partly by the host, calcium phosphate derived from the host is laid down in the mycelial mass. Analysis reveals that, except for the calcium phosphate, the granule has essentially the same composition as that of the organism grown *in vitro*. The granule may be regarded as a mass of mycelium cemented together by a polysaccharide-protein complex impregnated with about 50 per cent calcium phosphate. Tissue clubs are observed occasionally in lesions caused by other organisms, such as tubercle bacilli and staphylococci.

In sections of tissues the filaments may be distinguished from the clubs by a modified Ziehl-Neelsen stain. When a section is stained with carbol-fuchsin, decolorized for 20 to 30 sec with 1 per cent H_2SO_4, and counterstained with methylene blue, the clubs appear red and the filaments blue.

The nocardiae, when growing in the animal body, generally form a tangled mycelium without evidence of ray or of club formation.

The most striking feature of the actinomyces and nocardiae is their pleomorphism. All forms may be seen—filaments, rods, cocci and even spirilla. In actinomyces rod forms predominate in culture, in nocardiae filaments. But in most of the nocardiae all forms are seen, coexisting in a single culture. For a detailed description of their morphology, the reader is referred to a monograph by Lieske (1921) and for their fine structure to Williams *et al.* (1973).

Cultural reactions

In general, growth on artificial media is readily obtained. The usual media suffice, but the addition of glucose or glycerol is beneficial. The nocardiae, with a few exceptions, multiply rapidly, so that in 24 hr growth is visible on agar. The actinomyces, on the other hand, grow more slowly, taking 3 or 4 days to form macroscopic colonies. Great diversity of cultural appearance is noticeable, particularly in the nocardiae. The descriptions that follow refer only to some of the commoner types.

On an agar plate the nocardiae form round, low convex, opaque, finely granular filamentous colonies, which later undergo differentiation into a raised, knob-like, sometimes radially striated centre and an effuse, ground-glass-like periphery. The surface is finely granular and often has a 'chalk-powder' covering due to the formation of aerial spores: the edge is rhizoid, indented, or feathery, never entire (Fig. 22.3). Most strains form pigment—yellowish, pink or orange in colour—which becomes apparent after a few days' incubation, and which may show progressive alterations in tint. This is especially noticeable in cultures that have been incubated at 37°, and subsequently left in the dark at room temperature. After a variable time

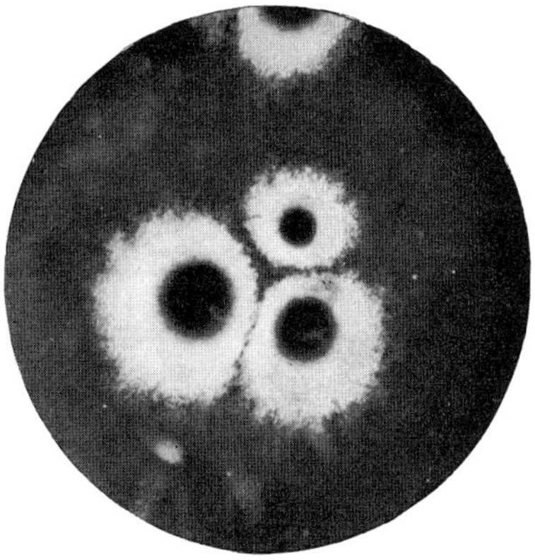

Fig. 22.3 *Nocardia asteroides.* Colonies on agar plate, 7 days, 37° aerobically: × 8.

Fig. 22.5 *Nocardia asteroides.* Culture, 14 days, 37° glycerine agar slope, aerobically.

under suitable conditions, aerial hyphae may develop, giving rise to a characteristic bloom on the surface of the colony—the chalk-powder appearance just described.

Actinomyces form smaller colonies, not apparent for 3 or 4 days; they are more compact, greyish or porcelain white in colour, and have a nodular surface (Fig. 22.4). (See also Slack *et al.* 1969).

On glycerol or glucose agar the norcardiae give a luxuriant, confluent, heaped-up, worm-cast, pigmented growth, adherent to the medium, of tough consistency, and difficult to emulsify (Fig. 22.5). The actinomyces grow in the form of discrete colonies, which are only slightly adherent to the medium and are much easier to emulsify.

In a glucose agar shake culture the nocardiae give a thick pigmented growth confined entirely or almost entirely to the surface. The actinomyces give a characteristic band-like growth situated about 0.5 to 1 cm below the surface, with a few larger discrete colonies scattered throughout the medium below. No growth at all occurs in the upper few millimetres.

In broth the nocardiae often form a thick, dry, dull, scaly or nodular, pigmented surface pellicle, which may extend for some distance up the sides of the tube. A ropy or membraneous, sometimes pigmented sediment forms, augmented by frequent deposits from the surface membrane. The broth remains clear, or at most shows a finely granular turbidity. Aerial hyphae may sprout from the surface pellicle. Sometimes growth commences at the bottom, and characteristic fluff-balls, resembling the head of a seeding dandelion,

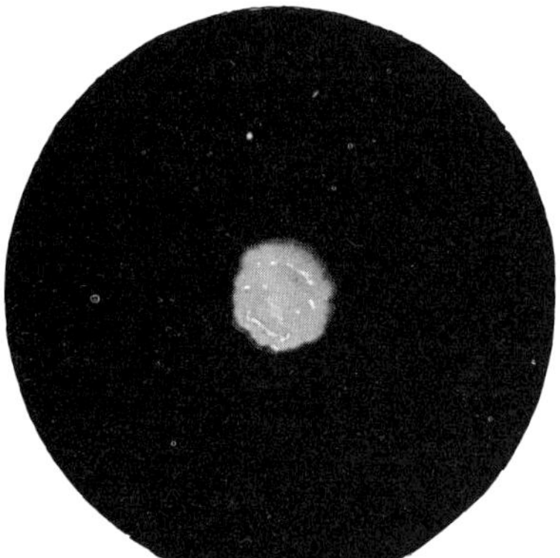

Fig. 22.4 *Actinomyces israeli.* Colony on agar plate, 14 days, 37° anerobically: × 8.

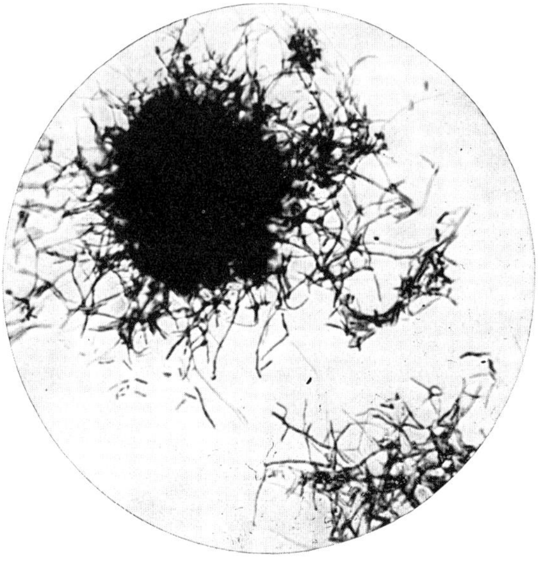

Fig. 22.6 *Nocardia asteroides* × 1000. A small granule composed of radiating filaments; from a broth culture, 24 hr, 37° aerobically.

develop. The actinomyces grow at the bottom of the tube in the form of compact whitish granules with a nodular contour; there is no turbidity and no surface growth. Nitrate broth is more favourable for growth than ordinary broth.

Resistance

The members of this group show no special resistance to heat or disinfectants. Most are killed in 15 min when exposed to moist heat at 60°. Owing presumably to their lipid coat, the so-called spores of the nocardiæ are rather more resistant to desiccation and to heat than the ordinary rods and filaments. Erikson (1955), for example, reported that *Nocardia sebivorans*—which is probably the same as *N. asteroides*—could withstand an exposure to 90° for 10 min when dispersed in a phosphate solution. Vincent (1894), working with *N. madurae*, found that the non-sporing forms were killed at 60° in 3–5 min, whereas the spores required a temperature of 85° for their destruction. Ørskov (1923) likewise found the spores to be more resistant than the plain mycelium; in his experience they sometimes survived exposure to 65° for 3 hr. *A. israeli* and *A. bovis* are sensitive to most antibiotics. Nocardiae, on the other hand, are resistant to practically all antibiotics, though sensitive to the sulphonamides (Lentze 1967). Reports on sensitivity to lysozyme are discrepant. According to Mordarska and her colleagues (1978), whose findings differ from those of Gordon and Barnett (1977), actinomyces are more or less sensitive and nocardiae mostly resistant. When kept at room temperature cultures of nocardiae remain viable for months: actinomyces usually die out in about 6–8 weeks at room temperature, but more rapidly at 37°.

Growth requirements

Actinomyces are facultative anaerobes; most of them grow best under anaerobic or microaerobic conditions, and many will not grow at all under normal atmospheric conditions, at least on first isolation. Growth is generally improved by the addition of *ca* 10 per cent of CO_2. Nocardiae, on the other hand, are strict aerobes. The optimum temperature for the growth of actinomyces is 35–37°; that for nocardiae is considerably lower (25–35°). Unlike the actinomyces, nocardiae grow over a wide range of temperature, many growing at 10–20° and some at 50° or more. The growth of actinomyces is improved by glucose and glycerol, and sometimes by blood or serum; nocardiae grow well on unenriched media. *A. israeli* may be grown on a partly defined medium; five growth factors are required—biotin, inositol, nicotinic acid, pyridoxal and riboflavin (Christie and Porteous 1962). The

Table 22.1 Some characters of the species of *Actinomyces*

	Cell wall content*	Acid in differential sugars	Nitrate reduction	Oxygen requirements	Pathogenicity
A. israeli	Ornithine	Mannose, raffinose, xylose, salicin, and often arabinose and mannitol	+	Anaerobic	Actinomycosis in man. Pathogenic to mice and hamsters on ip. injection
A. bovis	Aspartic acid	Sometimes in mannose and xylose, *not* in raffinose, salicin, arabinose or mannitol	usually −	Anaerobic to microaerophilic	Actinomycosis in cattle. Pathogenic to mice and hamsters on ip. injection, though less so than *A. israeli*
A. eriksoni	Neither in appreciable amount	Mannose, raffinose, xylose, salicin, arabinose and mannitol	−	Anaerobic; growth *not* improved by CO_2	Abscess formation in human chest or abdomen without granule formation. Less pathogenic for mice than *A. israeli*
A. naeslundi	Ornithine	Mannose, raffinose, and often salicin	+	Facultative aerobe	Non-pathogenic inhabitant of human mouth
A. odontolyticus	Ornithine	Often salicin, and sometimes raffinose, arabinose and mannitol	+	Facultative aerobe	Found in carious teeth
A. viscosus†	Ornithine	Mannose, raffinose and salicin	+	Facultative aerobe, growth improved by CO_2	Periodontitis in hamsters. Found in mouth of rats and man

* For other constituents, see p. 37.
† Catalase positive.

actinomyces cannot grow in the presence of bile salts; some of the nocardiae can do so.

Biochemical characters

Actinomyces ferment glucose, maltose, lactose and sucrose with the production of acid but not gas. Reactions with certain other sugars are of some diagnostic significance (see Table 22.1; for other sugars, see Slack 1974). Fermentation is slow and results should not be read for 14 days. *A. israeli* ferments mannose, raffinose, xylose and salicin, and often arabinose and mannitol. Of these sugars, *A. bovis* often ferments mannose and sometimes xylose, but it seldom attacks the others. Only *A. bovis* gives a wide zone of hydrolysis on starch-iodine plates, though many other strains form acid from starch. Nearly all strains hydrolyse aesculin and form H_2S; only *A. viscosus* forms catalase. Methyl-red reactions are positive and Voges-Proskauer reactions negative; urease and indole are not formed. Nitrate reduction is variable (see Table 22.1) Proteolytic activity is feeble; gelatin is not liquefied in tube tests, but some strains of *A. israeli* give a positive reaction in plate tests.

The nocardiae are on the whole less active on sugars. Some ferment only glucose and glycerol. Others ferment one or more of the following sugars: mannose, maltose, rhamnose, arabinose, lactose, sucrose, trehalose, xylose and mannitol. They can use acetate, malate, propionate, pyruvate and succinate, but vary in their ability to decompose xanthine, hypoxanthine, casein, tyrosine and urea. (For these and other biochemical reactions, see Table 22.2.) Nocardiae are catalase positive. Many of the nocardiae turn litmus milk slightly alkaline and peptonize it slowly; the actinomyces turn it acid (Negroni and Bonfiglioli 1937–8, Slack 1942).

Pigment formation is common among the nocardiae, the usual colour being some shade of pink or brown. Actinomyces are generally non-pigmented, but *A. odontolyticus* may form a dark-red pigment when grown anaerobically on horse-blood agar and subsequently exposed to air at room temperature.

Antigenic structure

The antigenic structure of the actinomyces and the nocardiae is still incompletely worked out. Aoki (1936*a*, *b*), using agglutination and complement-fixation tests, found that six strains of actinomyces fell into a single group and 19 strains of nocardiae into eight different groups. The antigenic homogeneity of *A. israeli* was confirmed by Cummins (1962). Erikson (1940) distinguished between actinomyces of human and of bovine origin. By means of the fluorescent antibody and the agar-gel precipitation techniques Slack and Gerencser (1966, 1970) distinguished six serogroups, as follows:

Serogroup	*Serotypes*	*Species*
A	1	*A. naeslundi*
B	2	*A. bovis*
C	1	*A. eriksoni*
D	2	*A. israeli*
E	2	*A. odontolyticus*
F	2	*A. viscosus*

By the use of similar techniques Lambert, Brown and Georg (1967) likewise found evidence of two serotypes in *A. israeli* and distinguished between this organism and *A. naeslundi*. These differences were confirmed by Brock and Georg (1969) and Slack and his colleagues (1969). *A. viscosus* is closely related to *A. naeslundi* (Cisar *et al.* 1978).

Among the nocardiae Cummins (1962) found a common antigenic component, shared with those strains of mycobacteria and of corynebacteria that contained arabinose and galactose as their principal cell-wall sugars. Kwapinski (1970), who studied so-called endoplasm preparations by an immunodiffusion technique, obtained evidence of the existence of multiple antigens, many of them shared, enabling him to distribute 75 strains of *Nocardia* into 15 serogroups. The observations of Castelnuovo and his colleagues (1964) by immunoelectrophoresis and the double diffusion technique in agar agree with those of Aoki (1936 *a*, *b*) and of Kwapinski (1970) in suggesting the existence among the nocardiae of several antigenic groups. In general, it would appear that the nocardiae are more antigenically heterogeneous than the actinomyces.

Both these groups of organisms are readily acted upon by *bacteriophages*, which are widespread in nature (Waksman 1967).

Pathogenicity

In man actinomycosis is produced by *A. israeli*, and less often by *A. eriksoni*; and in cattle, pigs and dogs by *A. bovis*. *N. pelletieri*, *N. brasiliensis*, *N. griseus*, and *N. madurae* are responsible for mycetoma in man. *A. viscosus* causes periodontal disease and bone loss in hamsters. This organism, like *A. naeslundi* and *A. odontolyticus*, occurs in the human mouth, and like them appears to cause lesions of the gums and teeth. As already mentioned, some species of *Nocardia* cause chronic granulomatous lesions in man and animals. *N. asteroides* gives rise to abscesses in man and animals; and in man to a severe disease of the lungs, often referred to as nocardiosis. (See also Chapter 49.)

For laboratory animals both actinomyces and nocardia have a low pathogenicity. Inoculated in a large dose subcutaneously into rabbits they give rise to a circumscribed abscess, which may persist for weeks or months; in the pus *Drusen* may often be found (Hassegawa *et al.* 1938).

Intraperitoneal injection of *A. israeli* into guinea-pigs and rabbits may give rise to small nodules containing typical granular pus, but the lesions are neither extensive nor fatal. By the repeated inoculation of massive doses intravenously Slack (1942) was able to kill rabbits in 6–10 weeks, and to demonstrate *post mortem* the presence of macrosopic or microscopic abscesses and focal necroses in the lungs and liver, sometimes containing granules with hyalinized clubs. According to Hazen, Little and Resnick (1952) the male hamster is the most susceptible laboratory animal. The intraperitoneal injection of a 1-week-old culture of *A. bovis* leads in the course of 4–6 weeks to abscess formation throughout the abdominal cavity; granules are found in the pus, and microscopical examination reveals the presence of clubs and gram-positive pleomorphic branching organisms. Meyer and Verges (1950) set up a progressive disease by the intraperitoneal injection into mice of *A. bovis* suspended in 5 per cent hog mucin; and Gale and Waldron (1955) reported similar findings with *A. israeli*. The lesions were greyish-white, hard, compact nodules resembling tubercles, 2–10 mm in diameter, and were distributed throughout the abdomen.

The acid-fast nocardiæ, such as *N. asteroides*, are more pathogenic. Injection of this organism by the subcutaneous, intravenous or intraperitoneal route leads to a progressive fatal infection of guinea-pigs and rabbits in 5 days to 4 weeks. *Post mortem*, small tubercles are found scattered throughout the organs, being especially numerous in the lungs, liver and spleen (Eppinger 1891, MacCallum 1902). Microscopically these nodules contain tangled filaments, sometimes arranged in the typical ray form.

Classification

In their septate nature, the diameter of their mycelium (ca. 1 μm), the composition of their cell wall, the absence of a nuclear membrane, the high G + C content of their DNA, their susceptibility to phage action and to antibacterial but not antifungal antibiotics, *Actinomyces* and *Nocardia* resemble bacteria rather than fungi (Lechevalier and Lechevalier 1967, Gottlieb 1973). Further reasons in support of this conclusion are listed by Waksman (1967; but see Memmesheimer 1967).

The classification of these organisms is fraught with difficulty. Ørskov (1923) and Erikson (1935) used a combination of morphological and physiological characters. Naeslund (1925) and Waksman and Henrici (1943), on the other hand, paid greater attention to their oxygen requirements. Waksman and Henrici distinguished between organisms having a mycelium that does not fragment—*Streptomyces*—and those in which the mycelium breaks up into bacillary or coccoid elements—*Actinomyces* and *Nocardia*. Too much attention, however, should not be paid to morphological differences. Gordon (1966*b*) points out, for instance, that fragmentation of the substrate hyphæ, abundance of the aerial hyphae and segmentation of the aerial hyphæ into chains of spores are variable properties in both *Nocardia* and *Streptomyces*. A firmer basis is afforded by the difference in cell-wall composition (Waksman 1967). According to Ballio and Barcellona (1968) *Streptomyces* is rich in branched fatty acids but practically devoid of unsaturated fatty acids, whereas *Nocardia* has far more straight-chained fatty acids and always some unsaturated fatty acids. Streptomyces tend to be more saccharolytic and proteolytic than nocardiae (Gordon and Smith, 1955).

According to Buchanan and Gibbons (1974, p. 657), *Actinomyces*, *Nocardia* and *Streptomyces* belong to the Order Actinomycetales, which contains gram-positive bacteria that form branching filaments, though by no means all of them form a true mycelium and some are normally bacillary in form. With a few exceptions – notably the actinomyces—members of this order live saprophytically in soil. Only four of the families in this order need concern us: (1) the Actinomycetaceae, which includes the genera *Actinomyces*, *Bifidobacterium* (Chapter 43) and *Rothia* (p. 42); (2) the Mycobacteriaceae, with a single genus *Mycobacterium* (Chapter 24); (3) the Nocardiaceae, which include *Nocardia*; and (4) the Streptomycetaceae.

Apart from the microaerophilic or anaerobic requirements of *Actinomyces* and the aerobic character of *Nocardia*, these two genera are distinguished by the nature of the constituents—sugars, amino sugars and amino acids—in their cell wall. The cell wall of *Actinomyces* contains alanine, glutamic acid and lysine, and either aspartic acid or ornithine, but never diaminopimelic acid; several sugars are present, but these do not include arabinose. In *Nocardia* the cell wall contains meso-diaminopimelic acid and arabinose or galactose, or both; in *Streptomyces* there is LL-diaminopimelic acid and glycine (Romano and Sohler 1956, Cummins and Harris 1958, Cummins 1962, Lechevalier and Lechevalier 1965, 1967, 1970*a*, Murray and Proctor 1965, DeWeese *et al.* 1968, Lechevalier *et al.* 1971*a*). The major cell-wall components of nocardiae and mycobacteria are identical (Chapter 24). Like mycobacteria and corynebacteria, nocardiae form mycolic acids (Chapter 24), but there are genus-specific differences in the chemical composition of these. Related to the mycolic acids is a lipid possessed by *Nocardia* but absent from *Actinomyces, Streptomyces, Mycobacterium and Corynebacterium* (see Mordarska *et al.* 1972). The relation between *Nocardia*, *Mycobacterium* and *Corynebacterium* is borne out by the demonstration of common antigens between them (Cummins 1962, Castelnuovo *et al.* 1964, Kwapinsky 1966, Slack and Gerencser 1966; see also Chapters 24 and 25).

The distinction between individual species in *Actinomyces* and *Nocardia* is based mainly on biochemical characters and cell-wall composition (Tables 22.1 and 22.2). As Gordon and Mihm (1962*a*), Gordon (1966*b*) and several other workers have found, the identity of many of these organisms is hard to establish. There is, for example, still some doubt whether *A. israeli* and *A. bovis* are specifically distinct, whether *N. farcinica* is the same organism as *N. asteroides*, and, if so, which of them should be regarded as the type species. Duda and Slack (1972), who examined five species of *Actinomyces* by the electronmicroscope, found no differences in the structure of the cytoplasm that could be used for taxonomic purposes. They did note, however, that the cell wall of *A. israeli* was twice as thick as that of *A. bovis*—65 nm as against 31 nm.

For further information about classification, and for descriptions of species of *Actinomyces* and *Nocardia*, see Lieske (1921), Naeslund (1925), Setti (1929), Rosebury (1944), Gordon and Mihm (1957), Waksman (1959, 1967, 1971), Cummins (1962), Juhlin (1967), Lechevalier and Lechevalier (1967), Tsukamura (1969), Goodfellow *et al.* (1972), Schofield and Schaal (1981), and the symposium volume edited by Sykes and Skinner (1973), notably the papers of Bowden and Hardie, Cross and Goodfellow, Gottlieb, and Williams and his colleagues. Kilian (1978) describes enzymic methods for the rapid identification of Actinomycetaceae and related organisms.

A detailed description of some of the more important species of *Actinomyces* follows.

Actinomyces israeli

Habitat Strict parasite found in lesions of actinomycosis in man. Frequent in human mouth and in salivary calculi.

Morphology *Glycerol agar, 7 days at 37°.* Long and short rods predominate; long continuous or segmented threads with a straight or curved axis, showing false or rarely true branching; S-shaped or spiral organisms; coccoid forms. The rods resemble, and are arranged like, certain members of the corynebacteria; sides parallel or irregular; ends rounded, clubbed or tapered; axis straight or curved; great variation in appearance; irregular staining is usual; granular and beaded forms are not uncommon. Non-motile. Non-sporing. Gram positive. Non-acid-fast.

Agar plate *7 days at 37° anaerobically.* Poor growth of round, 0.5–1.0 mm in diameter, convex, opaque, amorphous colonies with smooth dull surface and entire edge; greyish-white by transmitted, porcelain white by reflected light; butyrous or friable consistency; emulsifiability not difficult as a rule. 21 *days*, rather larger, 1–1.5 mm in diameter, umbonate, with slightly irregular nodular surface and lobate edge; differentiated into a glistening raised centre and a dull shelving periphery resembling a rosette. Colonies may grow into medium.

Agar slope *7 days at 37° anaerobically.* Moderate growth of discrete colonies similar to those described. Numerous greyish-white floccular masses of coarsely granular structure in water of condensation; they are irregular in shape, have an irregular edge, and are opaque. In the condensation water there is also a finely granular turbidity.

Gelatin stab No growth at 23°. After 12 days at 37°, the culture shows, when cooled, a band of growth 4 mm deep with its upper margin 1 mm below the surface. Growth consists of very fine greyish-white interlacing filaments, looking like cotton-wool. No liquefaction.

Broth *5 days at 37° anaerobically.* Poor to moderate growth; deposit of compact, white mulberry-like granules with nodular surface, often adherent to each other; not disintegrated on shaking. No turbidity; no surface growth; no odour.

Glucose agar shake *5 days at 37°.* No growth for 1 cm below surface. Then comes a turbid band, about 0.8 mm deep, consisting of large numbers of tiny colonies. Throughout the rest of the medium are scattered discrete, irregularly round, opaque, greyish-white colonies, about 0.1–1.0 mm in diameter, with smooth or slightly knobby surface.

Resistance At 37° cultures live for about 1 to 4 weeks, sometimes longer. Dried on glass and kept in the dark, organisms may live for 7 weeks or more. Killed by moist heat at 60° in 15 min. Susceptible to most antibiotics.

Metabolism Anaerobe of the microaerophilic type. Will not grow on surface culture exposed to the air. Optimum temperature for growth 37°; growth below 30° is either very slight or absent. Optimum pH 7.3–7.6. No haemolysis of horse red cells. No pigment formation. Growth is improved by nitrate, glycerol, blood, an increased partial pressure of CO_2, and sometimes by glucose.

Biochemical Acid, no gas, in glucose, maltose, lactose, sucrose, mannose, raffinose, xylose, salicin, and often arabinose, cellobiose, mannitol and inositol within 14 days under anaerobic conditions. LM acid. Indole−. MR+14 days. VP−. Catalase −. Oxidase −. Urease −. H_2S −. NH_3 +. MB reduction −. Gelatin liquefaction −. Nitrate reduction +. No growth in presence of bile salts.

Antigenic structure Two serotypes defined by fluorescent antibody technique. Type 1 overlaps with *A. naeslundi.*

Pathogenicity Responsible for actinomycosis in man, not cattle. Low pathogenicity for laboratory animals. Intraperitoneal inoculation of large doses into rabbits or guinea-pigs may be followed by the appearance of small nodules, chiefly in the great omentum, but the animals survive indefinitely. The male hamster is more susceptible; after intraperitoneal injection extensive abscesses develop throughout the abdominal cavity.

Acitinomyces bovis

The cause of actinomycosis in cattle and probably some other animals. Differential features from *A. israeli* are said to be: more rapid fragmentation of mycelium with predominance of rod forms; less polymorphic; scantier growth; smoother softer colonies with an entire edge and not adherent to the medium; sometimes slight turbidity in liquid media and a wispy or light flocculent deposit; thinner cell wall, and aspartic acid instead of ornithine in the cell wall; wide zones of hydrolysis on starch plates; usually failure to ferment raffinose, salicin, arabinose or mannitol, or to reduce nitrate; difference in antigenic structure; rather lower pathogenicity for mice and hamsters.

Actinomyces eriksoni

Isolated from pleural fluid and from abscesses in chest and abdomen, and named *A. eriksoni* by Georg and his colleagues (1965). Pleomorphic organism forming branched cells usually 2–4 × 1 μm, but sometimes 12–14 μm long. Clubbed or bifurcated ends common. Forms in 7–11 days creamy white, convex to conical colonies with smooth or pebbly surface and entire or scalloped edge. For cell-wall composition and sugar reactions, see Table 22.1. Does not reduce nitrate. Obligate anaerobe; growth not stimulated by 5 per cent CO_2. Optimum temperature 37°. Antigenically distinct. Does not form granules in human tissues. Low pathogenicity for mice. G+C content of DNA 62 moles per cent.

Actinomyces naeslundi

Described by Naeslund (1925) and named *A. naeslundi* by Thompson and Lovestedt (1951). Often found in the human mouth by Howell and his colleagues (1959). Differs from *A. israeli* in being isolated more often from blood agar plates incubated aerobically than anaerobically; in giving a rapid and diffuse or flaky growth in thioglycollate broth; in forming rough or smooth colonies or both; in its sugar reactions (see Table 22.1); in sometimes growing in media containing 10–20 per cent bile; in its antigenic structure; and in its non-pathogenicity to man and laboratory animals, though this is denied by Coleman and Georg (1969). Though a facultative aerobe, it is said to grow best anaerobically in 5 per cent CO_2. (For isolation from dental plaque, see Kornman and Loesche 1978.)

Actinomyces odontolyticus

This organism was described by Batty (1958). It is said to be always present in deep carious dentine, often present in the saliva, but seldom in superficial caries. It can be isolated by culture in Robertson's meat medium, at the bottom of which it forms a thin veil. Subculture on to blood agar yields very few colonies; these look at first like α-haemolytic streptococci, but after 3 or 4 days they take on a distinctive dark red colour. The base of the colony is adherent to the medium. Early subculture is necessary if the organisms are not to die out. Produces in yeastrel peptone water a glutinous ropy sediment which may be dispersed to give a uniform turbidity. Poor fermentative ability; a few strains form acid from sucrose, arabinose or mannitol (see Table 22.1). Grows equally well under aerobic and anaerobic conditions.

Actinomyces viscosus

This organism was first isolated from hamsters with spontaneous periodontal disease and subgingival plaque (Howell 1963; see also Howell and Jordan 1963) and originally named *Odontolyticus viscosus* (Howell *et al.* 1965). Has also been isolated from the oral cavity of rats and man. Experimentally causes periodontal disease in hamsters.

After 7 days' incubation, colonies are smooth, entire, convex or raised, whitish, glistening and opaque, and of a soft or mucoid consistency. Grows diffusely in liquid media, often with a viscous deposit. A facultative aerobe the growth of which is improved by CO_2. Unlike other actinomyces it forms catalase. For sugar fermentations and cell-wall composition, see Table 22.1. (See also Gerencser and Slack 1969; for isolation from dental plaque, see Kornman and Loesche 1978).

Actinomyces baudeti This organism was described by Brion (1939), and Brion and his colleagues (1952), as a cause of actinomycosis-like lesions in cats and dogs. However, their account of it was not sufficiently detailed to merit recognition as a new species of *Actinomyces*, and authentic cultures are not available for study.

Actinomyces propionicus Isolated by Pine and Hardin (1959) from a human case of lachrymal canaliculitis. Recognized by Buchanan and Pine (1962) to be a new species. Forms a branched mycelium in culture and is a facultative aerobe. Inoculated intraperitoneally into mice it gives rise to lesions similar to those caused by *A. israeli*. It should probably not be looked upon as an actinomyces because its cell wall contains diaminopimelic acid and it forms propionic acid from glucose (Yu-Ying and Georg 1968). Differs from *Propionibacterium* in the mycelial structure of its colonies, in having aspartic acid in its cell wall, in its pathogenicity, and in its failure to ferment lactic acid or glycerol (Pine and Georg 1969).

Nocardia

Definition

Aerobic, non-motile, gram-positive organisms growing as a primary mycelium that fragments into rod-shaped and coccoid elements. Some are weakly acid-fast. Usually form aerial hyphae. Contain mycolic acids. Have a cell wall containing meso-diaminopimelic acid, arabinose and galactose. Generally saprophytic, but may cause disease in man and animals. G+C content of DNA 64–68 moles per cent.

Type species hitherto given as *N. farcinica*, but this may be replaced by *N. asteroides*.

Introductory

The family Nocardiaceae contains a number of genera, such as *Nocardia*, *Rhodococcus*, *Oerskovia*, *Rothia*, and in the opinion of some workers *Actinomadura*, but we shall confine our attention here mainly to the genus *Nocardia*. The precise identity of even this genus is still in dispute. For example, strains of *N. farcinica* have been shown by Ridel (1975) to fall into two well separated clusters, one of which apparently belongs to the genus *Mycobacterium*. Likewise, Tsukamura (1977) found two varieties of *N. asteroides*, which he referred to as A and B (see also Mordarski *et al.* 1977). Since the original strain of *N. farcinica* has been lost, there is no means of establishing its identity; and should *N. asteroides* replace it as the type species (Lechevalier 1976) a decision will have to be taken as to whether the A or the B variety is chosen.

There is little to add to the description of *Nocardia* that has already been given. Growth occurs best between 25° and 37°. On plain tap-water agar, nocardiae form filamentous colonies. Most species form aerial hyphae, but inconstantly, and some species conidia. Acid-fastness is a variable and often unreliable criterion of species. Apart from the presence of meso-diaminopimelic acid in the cell wall of *Nocardia*, and of LL-diaminopimelic acid in that of *Streptomyces*, there is no reliable way of distinguishing between these two genera (Mishra *et al.* 1980). The genus *Actinomadura* was separated by Lechevalier and Lechevalier (1970b) from *Nocardia* because of the absence of arabinose and galactose in the cell wall, but Goodfellow and Minnikin (1977) and Mishra, Gordon and Barnett (1980) favour its retention in the genus *Nocardia*.

The species of *Nocardia* and *Streptomyces* we describe below are ones that are most frequently found in lesions of man and animals. For their identification Mishra, Gordon and Barnett (1980) devised a series of 14 tests—ten based on the acidification of carbohydrates, three on the decomposition of adenine, casein and hypoxanthine, and one on the reduction of nitrate to nitrite. They provided also a useful differential key for the benefit of the clinical pathologist. (For general information about the classification of nocardiae, see Lechevalier and Lechevalier 1967, Goodfellow 1971, Goodfellow *et al.* 1976, Goodfellow and Minnikin 1977, Minnikin *et al.* 1977, Orchard and Goodfellow 1980.) (See also Table 22.2.).

Nocardia farcinica

This organism is not identifiable as a single species and will not be described (but see the 6th edition of this book, Vol. I, p. 543).

Nocardia asteroides

Isolated by Eppinger in 1891 from a brain abscess in a glass grinder. Mycelium consists of threads showing true and false branching. In the body it forms long granular interlacing filaments with no ray or club formation. Gram positive. Moderately acid-fast. Aerobic. Destroyed by heat at 70° in 5 min. On glucose agar forms a yellow, ochre or orange, flat and spreading or heaped-up wrinkled growth. Colonies tend to be star-shaped; hence the name *asteroides*. Aerial hyphæ vary in abundance. In serum broth a delicate waxy pellicle with small round fluffy masses of mycelium at the bottom of the tube; no turbidity. Acid in glucose and glycerol. Forms urease. Reduces nitrate to nitrite. Uses paraffin and rubber as sources of carbon (Pier *et al.* 1961). Said to be antigenically heterogeneous (Castelnuovo *et al.* 1964, Pier and Fichtner 1971). After subcutaneous, intraperitoneal or intravenous injection rabbits and guinea-pigs die in 1 to 4 weeks. *Post mortem* the viscera, especially the lungs, liver and spleen, are studded with small white nodules. In the muscles, kidneys and other organs abscesses may develop that contain branching test-tube-brush forms with laterally radiating clubs. Dogs, cats, horses and asses are resistant to intraperitoneal or intravenous injection. Subcutaneous injection causes a slowly progressive abscess that ulcerates and heals. Pathogenic to mice when injected intravenously (Smith and Hayward 1971), and when injected intraperitoneally with mucin (Mason and Hathaway 1969). This organism is responsible in man for most nocardial diseases of the lung, brain and subcutaneous tissues, but not for mycetoma (González-Ochoa 1962). In cattle it may cause severe and persistent infection of the udder (Pier *et al.* 1961).

A similar or identical organism, sometimes known as *N. caprae*, may give rise to a tuberculosis-like disease in the lung of goats (Silberschmidt 1899, Galli-Valerio 1912, Erikson 1935).

Nocardia brasiliensis

Said to be responsible for 94 per cent of cases of mycetoma in Mexico (González-Ochoa 1962), and has been isolated both in Mexico and Australia from soil and dust (González-Ochoa and Sandoval 1960, Rodda 1964). Differs from *N. asteroides* in producing acid from mannitol and inositol, in decomposing casein, hypoxanthine and tyrosine, in liquefying gelatin, and in clotting milk. According to Bojalil and Cerbon (1959) the two organisms can be readily distinguished by growth for 20 days (temperature not stated) in a 0.4 per cent solution of gelatin in distilled water. *N. brasiliensis* forms rounded colonies that adhere to the wall, form a pellicle or sink to the bottom of the tube; it turns the medium alkaline and breaks down the gelatin with the production of amino acids detectable by the ninhydrin test. *N. asteroides*, on the other hand, gives a poor flaky growth, turns the medium slightly acid or leaves it neutral, and does not break down the gelatin.

Nocardia caviae

Described by Gordon and Mihm (1962*b*; see also Juhlin 1967). Uncommon organism. Differs from *N. asteroides*

Table 22.2 Some characters of the species of *Nocardia*, *Rhodococcus* and *Streptomyces*

Species	Acid-fast	Acid in sugars*	Cell-wall content†	Hydrolysis of: xanthine	hypoxanthine	casein	tyrosine	urea	Hydrolysis of aesculin	Nitrate reduction	Pathogenicity	DNA: G+C content (%)	Remarks
N. asteroides	±	Gly	meso-DAP, Ara, Gal	–	–	–	–	+	+	+	Abscess in lungs and brain	66.5	Pink, orange or red pigment
N. brasiliensis	±	Gly, Ino, Mnt, Tre	meso-DAP, Ara, Gal	–	+	+	+	+	+	+	One form of mycetoma in man	65	Liquefies gelatin; clots milk
N. caviae	±	Gly, Ino, Mnt, but Tre ±	meso-DAP, Ara, Gal	+	+	–	–	+	+	+	Abscess in man	66	Forms no pigment
N. pelletieri	–	Tre	meso-DAP, Ara, Gal.	–	+	+	+	–	–	+	Reddish mycetoma in man	65	Pink, orange or red pigment
N. madurae	–	Ado, Ara, Cel, Gly, Mno, Mnt, Tre, Xyl	meso-DAP; Gal‡; little or no Ara	–	+	+	+	–	+	+	Pale variety of Madura foot in man	68	Often brownish-red pigment
N. dassonvillei	–	Cel, Gly, Mal, Mno, Mnt	meso-DAP; no Gal	+	+	+	+	–	–	+	Lesions in man and animals	65	
Rhodococcus rhodochrous	–	Gly, Mno, Mnt, Sbl, Tre	meso-DAP, Ara, Gal	–	–	–	+	+	+	+	Non-pathogenic	63–69	Pink, orange or red pigment. Generic status questionable
Str. somaliensis	–	Mal	LL-DAP, glycine	–	+	+	+	–	–	–	Mycetoma in man	73	Yellow, brown or black pigment
Str. griseus	–	Cel, Gly, Mal, Mno, Mnt, Tre, Xyl	LL-DAP, Xyl	+	+	+	+	+	+	V	Mycetoma in man	71.5	Yellow pigment and streptomycin

Sugars: Ado = adonitol; Ara = arabinose; Cel = cellobiose; Gal = galactose; Gly = glycerol; In = inositol; Lac = lactose; Mal = maltose; Mno = mannose; Mnt = mannitol; Sbl = sorbitol; Tre = trehalose; Xyl = xylose.
DAP = diaminopimelic acid; V = variable.
*All species produce acid in glucose.
†All species contain glucosamine, alanine, muramic acid and glutamic acid.
‡And madurose.

mainly in its ability to decompose xanthine and hypoxanthine, and to form acid from mannitol and inositol; and from *N. brasiliensis* in its ability to decompose xanthine, its failure to decompose casein or tyrosine and its greater resistance to heat. Injected intravenously into mice it gives rise, like *N. asteroides*, to abscess and granuloma formation in several major organs, proving fatal within 9 days; or causes a chronic illness characterized by *spinning disease* due to lesions of the inner ear (Smith and Hayward 1971). According to these authors it lyses ox and rabbit blood.

The following three organisms are grouped by Lechevalier and Lechevalier (1970*b*) in a separate genus, *Actinomadura*, but Mishra, Gordon and Barnett (1980) prefer to regard them as nocardiae.

Nocardia pelletieri

This organism was first observed by Laveran (1906) in a fistulous tumour of the knee in Senegal. It appeared to be a micrococcus arranged in zoologloeal masses and short chains, and was called *Micrococcus pelletieri*. Later, it was isolated by Thiroux and Pelletier (1912) from a large tumour of the thorax which was ulcerating and discharging pus containing red granules like grains of sand. Similar granules were present in the purulent sputum. Culture on Sabouraud's glucose agar at 25–28° led to the appearance in 10 days of small convoluted red colonies looking like frog spawn. Later, they became drier and covered with a white efflorescence. The organism differed from *N. madurae* by restriction of its growth to Sabouraud's medium, the early red coloration of the colonies, the easy detachability of the colonies from the medium, and their mucilaginous consistency, as opposed to the dry scaly appearance of *N. madurae*. The name *Nocardia Pelletieri* was suggested for it by Pinoy (1912). According to Gordon (1966*b*) the organism is not acid-fast. Aerial hyphae are not formed in old laboratory strains. Its biochemical characters are listed in Table 22.2. Pigment produced is said to be prodiginine (Gerber 1969).

Nocardia madurae

Isolated from pale variety of Madura foot by Vincent in 1894. Long non-segmented filaments, 0.4–0.6 μm thick, showing true and false branching. Not acid-fast. Aerial hyphæ grow up from the mycelium on some media. Cell wall contains galactose and madurose. Forms greyish-yellow, beige or reddish-brown heaped-up colonies resembling worm casts and reminiscent of a rosette. Poor growth in broth without turbidity. Grows between 20 and 40°. Acid in glucose, rhamnose, arabinose, trehalose, xylose and mannitol. Hydrolyses aesculin and starch. Decomposes casein, hypoxanthine and sometimes tyrosine. Reduces nitrate to nitrite. Susceptible to lysozyme. Injected subcutaneously into rabbits, guinea-pigs and mice, it causes a local nodule that increases in size for a month and then regresses. Responsible for pale or ochroid variety of Madura foot in man (see Gordon 1966*b*). Pigment is said to be nonyl prodigiosin (Gerber 1969).

For its selective isolation, the addition of rifampicin 5 mg per l to the medium is recommended (Athalye *et al.* 1981).

Nocardia dassonvillei

This organism was isolated by Liégard and Landrieu (1911) from a case of conjunctival mycosis. Differs from *N. madurae* in the absence of galactose and madurose from the cell wall; in the more frequent formation of aerial hyphae; in decomposing adenine and xanthine; in the failure of most strains to hydrolyse aesculin; and in the type of lesions to which it gives rise.

Rhodococcus

This genus was suggested by Goodfellow and Alderson (1977) to accommodate bacteria variously classified as *Gordona*, *Jensenia*, *Mycobacterium rhodochrous*, and the 'rhodochrous complex'. The type species is *R. rhodochrous*. This organism was designated tentatively as a separate species by Gordon and Mihm (1959), but is rather a cluster of somewhat similar organisms (Goodfellow *et al.* 1972). Has often been regarded as a corynebacterium or a mycobacterium. Though able to use glucose oxidatively it is unable, as members of the *Corynebacterium* group are, to use it fermentatively. From *Mycobacterium* it differs in being only weakly acid-fast, in its lipid composition (Lanéelle *et al.* 1965), and in its mycolic-acid esters (Lechevalier *et al.* 1971*b*). Its lipid closely resembles that which is specific for *Nocardia* (Mordarska *et al.* 1972). Its characters are described by Gordon (1966*a*), and under the name of *N. corallina* by Juhlin (1967; see also Becker *et al.* 1965). Primary mycelium breaks up into rod and coccoid forms. Grows readily on ordinary media, giving rise to dense, red, sometimes ringed or filamentous colonies, usually with an irregular margin. Aerial hyphae are sparse and short. Grows freely at 10°, as a rule at 37°, often at 45°, but not at 52°. Produces acid from glucose, mannose, mannitol, sorbitol and trehalose, but not from rhamnose, arabinose, xylose, galactose, raffinose, dulcitol or erythritol. Does not grow on MacConkey's agar. Reduces nitrate and forms urease. Usually destroyed by heat at 60° within 4 hr. Is sensitive to penicillin and lysozyme. Non-pathogenic to man. Strains of *N. rhodochrous*, *N. corallina*, *N. rubra* and *N. pellegrino* are said to be antigenically alike and to be lysed by the same bacteriophage (Castelnuovo *et al.* 1964). (See Table 22.2).

Rothia

This is a monotypic genus erected by Georg and Brown (1967) to accommodate *Actinomyces dentocariosus*, *Nocardia dentocariosus*, and *Nocardia salivæ*. These organisms form a cluster clearly distinct from phena containing aerobic and anaerobic bacteria (Holmberg and Hallander 1973).

Streptomyces

Streptomyces somaliensis

Described by Brumpt in 1906 (see Brumpt 1927). Was first isolated by Bouffard in French Somaliland from patients affected with mycetoma. Consists of long branching filaments with truncate or sometimes tapering ends. Gram positive. Grows on agar, but better on blood agar. On this medium colonies are at first small, circular, convex and translucent, but after a few days they become irregularly heaped up, nodular, worm-cast or crateriform; they are opaque, vary in colour from white, through yellowish-orange, to brown or black, often show radial segmentation which gives them a stellate appearance, are extremely tough in consistency and adherent to the medium, and have a peculiar odour. In broth no turbidity or surface growth, but a deposit of little greyish-white puff balls. On potato a white folded layer of growth, which in 5 to 6 days becomes yellow. Aerial hyphæ formed, some with long chains of spores. Peptonizes milk. Some strains form acid in glucose and maltose. Decomposes casein and tyrosine. Differs from *N. pelletieri* in failing to reduce nitrate, to ferment trehalose or to decompose hypoxanthine, and in having a cell wall containing LL-diaminopimelic acid and glycine instead of meso-DAP. Gives rise in man to mycetoma of the hand or foot. Lesions contain hard smooth yellowish-red granules, 1 mm in diameter, not dissociated by caustic potash.

Streptomyces griseus

Described by Krainsky (1914). Cell wall contains LL-diaminopimelic acid and xylose, and sometimes glucose and mannose. Non-acid-fast. Forms aerial hyphae and conidia. Produces acid in glucose, cellobiose, glycerol, maltose, lactose, mannitol and xylose. Decomposes adenine, casein, hypoxanthine, tyrosine, and usually xanthine and urea. Hydrolyses starch. Reduction of nitrate to nitrite variable (see Mishra *et al.* 1980). Gives rise to mycetoma in man and animals.

Actinobacillus

Introductory

Lignières and Spitz (1902) isolated a non-motile, non-branching, gram-negative bacillus from the lesions of cattle suffering from a disease which in many respects resembled actinomycosis. They called the organism the actinobacillus, and the disease to which it gave rise actinobacillosis. Three other organisms have since been described, having some points of similarity with this bacillus, and it is therefore convenient to consider them as forming a group to which the generic name *Actinobacillus* may be applied. The systematic position of this group is doubtful. The genus is clearly distinct from *Actinomyces*, and our excuse for keeping it in this chapter is partly that the lesions to which it gives rise in the animal body resemble those caused by *Actinomyces*, and partly that there is no other genus with which it seems to be associated.

Gumbrell and Smith (1974) drew attention to the similarity of the three genera *Actinobacillus*, *Pasteurella*, and *Haemophilus*. Mannheim, Pohl, and Holländer (1980) agreed that these three genera formed a group that needed reclassifying (see also Kilian *et al.* 1981; and for a historical account see Zinnemann 1981).

The members are small non-motile, non-sporing, non-acid-fast coccoid or rod-shaped organisms, growing on blood agar under aerobic, microaerophilic, or anaerobic conditions, fermenting glucose with the production of acid, having a G+C content in their DNA of 38 to 47 moles per cent, and leading a parasitic existence. They are separated from the enterobacteria by their smaller genome, obligatory parasitism, and absence of motility. Though sensitive to the vibriostatic compound 0129 (Chatelain *et al.* 1979), they differ from the vibrios in their smaller genome, absence of flagella, and in various other phenetic properties. They are said to be common commensal organisms in animals but not in man.

The species we shall describe are:

Actinobacillus lignieresi Brumpt, the type species
" *actinomycetemcomitans*
" *actinoides*
" *equuli*
" *piliformis*
" *salpingitidis*
and for convenience
Streptobacillus moniliformis

Of these, according to Buchanan and Gibbons (1974, pp. 373 and 378), only *A. lignieresi* and *A. equuli* belong to the genus *Actinobacillus*; *A. actinomycetemcomitans* and *A. actinoides* are *species incertae sedis*. *A. piliformis* has not been grown in artificial culture, so that its position is likewise uncertain. *A. salpingitidis* is of questionable species; and *Streptobacillus monili-*

formis, which we referred to in previous editions as *Actinobacillus moniliformis*, is now generally recognized not to belong to this genus.

A new species, *Actinobacillus hominis*, giving rise to severe respiratory infections in Denmark was described by Friis-Møller in 1981 (see Kilian *et al.* 1981).

Actinobacillus lignieresi

Synonym Probably the same as *Bact. purifaciens* (see Tunnicliff 1941).

Habitat Considered by Phillips (1961) to play a commensal rôle in the rumen of cattle. By means of a selective medium consisting of 5 per cent blood agar with the addition of 1 μg oleandomycin and 200 units nystatin per ml he isolated strains of this organism from 10 per cent of specimens of ruminal contents. Also isolated by Phillips (1964) from 37 per cent of tongues of normal cattle. Found in abscesses in rats.

Morphology In young cultures it is a small rod-shaped organism; in older cultures it is coccobacillary, and various involution forms appear. In serum broth long streptobacillary forms are common. In glucose agar shake cultures long, tangled, unbranched filaments may be formed, accompanied by smaller bacilli and coccoid bodies (Griffith 1916). In all cultures small granules are found scattered among the bacilli

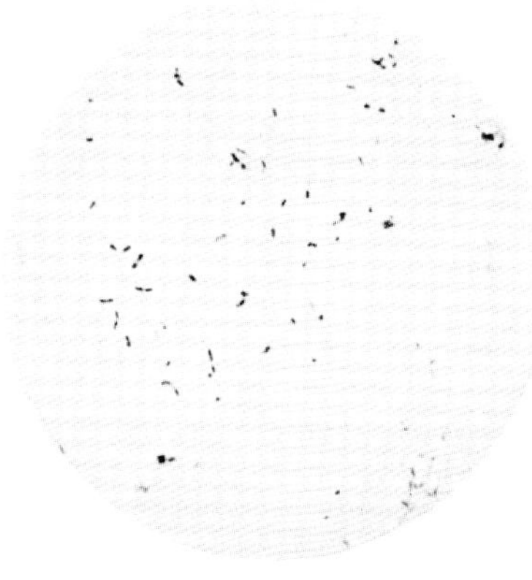

Fig. 22.7 *Actinobacillus lignieresi.* From a liver agar slope, 2 days, 37° aerobically (× 1000).

(Phillips 1960). Dimensions of the bacilli are given by Lignières and Spitz (1902) as 1.15–1.25 μm by 0.4 μm. Non-motile; non-sporing; non-acid-fast. Stains readily, especially with carbol-fuchsin, and is gram negative; frequently shows bipolar staining.

In lesions in the animal body small granules are found, which consist of tufts of radially disposed clubs similar to those in actinomycosis. An important point of difference is that the centre of the granule is occupied not by gram-positive filamentous mycelium, such as is formed by *Actinomyces bovis*, but by minute gram-negative bacilli, which may quite easily be overlooked. Though both the bacilli and the clubs formed by *Actinobacillus lignieresi* are gram negative, it is possible to distinguish between them by a modified Ziehl-Neelsen stain, as was pointed out by Bosworth (1923). If a section of affected tissue is stained with carbol-fuchsin, decolorized for 20 to 30 sec with 1 per cent H_2SO_4, and counter-stained with methylene blue, the clubs appear red and the bacilli blue. For pus, one of the best stains is glycerine picro-carmine, which stains the clubs yellow and the pus cells pink. The G + C content of the DNA is 41–44 moles per cent.

Cultivation Cultures are best obtained by grinding up infective pus in a mortar, and seeding on to agar. Growth occurs readily under aerobic conditions, and less readily under anaerobic conditions. The optimum temperature for growth is 37°; very slight growth occurs at 20°.

On agar in 24 hr at 37° small, circular, bluish-grey translucent colonies with a smooth surface and an entire edge, up to 1.5 mm in diameter, are formed; further incubation results in a considerable increase in size—up to 4 mm—due to peripheral extension of the colony. On an agar slope the growth of freshly isolated strains is poor, consisting of small, discrete, translucent bluish colonies, or of a thin, dry, confluent layer of growth adherent to the medium. After cultivation for some time in the laboratory, the organism grows more readily, giving a confluent, viscous growth with a thickened edge.

In gelatin stab growth is very poor and is not visible for some days. A small opaque spot appears at the surface; no growth occurs down the stab, and there is no liquefaction. Coagulated serum is not a very favourable medium; only a thin whitish growth is formed. Grows on MacConkey's agar. Diffuse and often flocculent turbidity in peptone broth; later a ring growth or surface pellicle, and a small sediment at the bottom of the tube. Growth improved by serum. On 5 per cent sheep blood agar partial lysis may be present in 3 days.

Resistance The organism is killed by heating to 62° in 10 min. It rapidly succumbs to drying. Cultures do not live long, and should be transplanted every few days. Infected pus preserved in sealed tubes may remain virulent for a month or two. Organisms are susceptible to chloramphenicol, and of variable susceptibility to streptomycin, chlortetracycline, and oxytetracycline; only weakly susceptible to sulphathiazole; resistant to penicillin.

Biochemical reactions Fermentation of carbohydrates is weak, ill defined and often delayed. Tests should not be prolonged beyond 14 days; otherwise non-specific positive reactions may result from the breaking down of peptone with the liberation of acid-reacting amino acids (Mráz 1969). Most strains produce acid without gas in glucose, maltose, mannitol, sucrose, xylose and often trehalose and dextrin. Lactose fermentation is variable. Acid is produced in litmus milk, but not clot. Indole −. Nitrate reduced to nitrite or gaseous nitrogen. Catalase +; oxidase reaction variable; H_2S+; NH_3−; urease positive in fluid Christensen's medium. MR −; VP +; MB reduction ±.

Antigenic structure Strains seem to share a common heat-stable polysaccharide antigen. In addition there are numerous simple or complex agglutinogens. Among 218 strains from cattle Phillips (1967) defined 6 types, 1–6, and 2 subtypes 1a and 4c, on basis of heat-stable somatic antigens. Most cattle strains belonged to type 1 and most sheep strains to types 2, 3 and 4. Also found heat-labile antigens common to different antigenic types.

Pathogenicity No exotoxin is formed. The organism is responsible for actinobacillosis in cattle and sheep (Chapter 49). The virulence of different strains seems to vary considerably and, though cattle inoculation experiments are successful with some strains, they are completely negative with others (Magnusson 1928). Subcutaneous inoculation of pure cultures into cattle produces an abscess identical with those occurring spontaneously; in the pus granules are found consisting of bacilli surrounded by clubs. Most workers, including ourselves, have been unable to produce any specific lesions in laboratory animals, and the statement of Lignières and Spitz (1902) that the organism produces a fatal infection when given intraperitoneally to guinea-pigs has not been confirmed. According to Mráz (1969) mice and chick embryos are more susceptible than guinea-pigs or rabbits. Mice inoculated intraperitoneally and chick embryos inoculated via the allantoic cavity usually die.

For an annotated bibliography of this organism see Mráz (1968).

Actinobacillus actinomycetemcomitans

Described by Klinger in 1912 under the name *Bact. actinomycetem comitans*. Found in lesions caused by *Actinomyces israeli*, as densely packed gram-negative coccobacilli (Colebrook 1920). In culture the rod forms are 1.0–1.5 μm long; the coccoid forms are 0.6–0.8 μm in diameter. Intermediate forms are frequent. The organism is non-motile. In broth or liquid gelatin at 37° it forms isolated, translucent granules, 0.5–1.0 mm in diameter, along the sides of the tube, most numerous near the surface. After some days they fuse into a greyish-white mass, forming a ring round the tube and a pellicle over the surface. The granules can be picked off the wall of the test-tube with a loop, but are very difficult to break up. Later they may become opaque and greyish-white. On agar it gives rise to small tough colonies, not unlike those of streptococci, adherent to the medium. Growth is improved by blood or ascitic fluid. Translucent colonies viewed under the microscope (× 100) have a characteristic crossed-cigar or starfish appearance (Heinrich and Pulverer 1959). The organism is microaerophilic, forming in shake agar tube cultures a band of growth 5 mm below the surface. Growth is favoured by CO_2. No growth below 25°. Fermentative reactions are recorded differently by different workers. Acid is usually formed in glucose and maltose, sometimes in mannitol and xylose (Goldsworthy 1938, King and Tatum 1962, Pulverer and Ko 1970). Nitrate is reduced. Weak H_2S production. No formation of indole or urease, and no liquefaction of gelatin. King and Tatum (1962) found three precipitating antigens distributed singly among different strains. Pulverer and Ko (1972) defined six heat-stable agglutinating antigens, and by the use of six absorbed agglutinating antisera established 24 agglutinating patterns among 100 strains. It is toxic on injection into rabbits, but does not set up a true infection. According to Thjøtta and Sydnes (1951) it differs from *Actinobacillus lignieresi* in growing as well anaerobically as aerobically, in not growing in milk, in failing to produce indole, and in being non-pathogenic to guinea-pigs. According to King and Tatum (1962) it somewhat resembles *Haemophilus aphrophilus*. Electronmicroscopic pictures show that both these organisms have numerous vesicular structures on their surface which may become detached and are found free in the external environment (Holt *et al.* 1980). The G+C content of the DNA is 47 moles per cent.

Actinobacillus actinoides

This organism was isolated by Smith in 1918 from the lungs of calves suffering from epizoötic pneumonia, and called by him *Bac. actinoides*. Also isolated from vesiculæ seminales of bulls (Jones *et al.* 1964). In the animal body it appears as a minute gram-negative bacillus arranged in groups. In the condensation water of coagulated serum it forms minute whitish flocculi, which consist of a central mass of radiating non-branching filaments ending peripherally in clubs. In tissue-agar cultures the organism grows as aggregations of rounded, ring-like bodies, 2 μm in diameter, having a minute refringent speck on the periphery or near the centre. There are thus three distinct forms in which this bacillus occurs. Most strains are capsulated (Smith 1921*b*), but the capsule does not stain with the usual dyes. Growth occurs only under a raised content of CO_2. On coagulated serum whitish flocculi appear in the condensation water in 3 days at 37° and after several weeks very tiny, elevated, pointed-like colonies may appear on the slant. Growth may be obtained on agar to which a piece of guinea-pig's spleen has been added and on chocolate agar. No growth on ordinary media or on MacConkey's medium. Forms acid in glucose and glycerol. Forms no catalase, oxidase, urease, indole, or H_2S, and does not reduce nitrate (Jones *et al.* 1964). Non-pathogenic for laboratory animals on experimental injection. Subcutaneous injection into calves causes a large necrotic swelling with caseous contents; ulceration occurs in 4 weeks. Intratracheal injection into calves causes small necrotic foci in the lungs, identical with those observed in the natural disease (Smith 1921*a*; see also Jones 1922). There is some doubt whether this organism is responsible for calf pneumonia (Levi and Cotchin 1950). The fact that it has been isolated from the lungs of white rats suffering from pneumonia (Jones 1922) suggests that it may be a natural parasite of these animals. For further references to this organism see Smith (1921*a*, *b*). Mannheim, Pohl, and Holländer (1980) would exclude it from the genus *Actinobacillus*. A similar organism, differing only in minor particulars, was isolated from the middle ear of white rats, in which it was causing suppuration (Nelson 1930, 1931).

Actinobacillus equuli

This organism is associated with joint-ill and sleepy disease of foals. It is referred to by a variety of names, such as *Bac. nephritidis equi*, *Bac. equirulis*, *Bacterium viscosum equi*, *Bac. pyosepticus equi* and *Shigella equirulis*. Morphologically it is a pleomorphic gram-negative bacillus occurring singly, in streptococcus-like chains and as filaments. Some authors describe it as capsulated, others as non-capsulated. It grows readily on ordinary media, but dies out rapidly unless it is subcultured frequently. On nutrient agar the colonies are usually circular with undulate edges, raised and opaque, and very tenacious. A smooth variant may appear in artificial culture, and has occasionally been isolated directly from foals. It attacks glucose, maltose, mannitol, lactose, and sucrose fermentatively with the production of acid but not gas, and gives a strong urease reaction. The MR, VP, Koser's citrate, indole, H_2S, KCN, gluconate and malonate tests are negative. It does not liquefy gelatin, or form decarboxylases for lysine, ornithine or arginine (see Cowan and Steel 1961). It is catalase positive but gives a variable oxidase reaction. It shares major antigens with *Actinobacillus lignieresi*. The taxonomic position of this organism is doubtful. Wetmore and her colleagues (1963) regard it as virtually identical with *A. lignieresi* and recommend that the specific name *equuli* be dropped. On the other hand Vallée, Thibault and Second

(1963) would separate the two organisms on the failure of *A. lignieresi* to ferment lactose and of *A. equuli* to produce H_2S—properties that are not constant with all strains; and Mráz (1967) on the different G + C content of their DNA (40.6 and 42 per cent), on the animals attacked and on the nature of the disease produced. (See also Edwards 1931, 1932, Report 1949, Maguire 1958.)

Actinobacillus piliformis

We suggest this name quite tentatively for the organism, *Bacillus piliformis*, that was described by Tyzzer (1917) in the United States as responsible for a disease of Japanese waltzing mice, by Gard (1944) in Sweden for summer diarrhoea in albino mice, by Allen and his colleagues (1965) in New Zealand for enteritis in rabbits, and by Sparrow and Naylor (1978) in England for necrotic hepatitis in guinea-pigs. The organism appears to be but mildly pathogenic to mice and to give rise to disease only when the diet or some other factor is disturbed. In our experience the liver disease was seen not more than once in about every 5000 mice. The organism may, however, after lying latent for a variable time, resume activity and cause havoc among mouse stocks. Indeed **Tyzzer's disease** is said to have interfered more with cancer research than any other disease in mice (Report 1960). In Gard's albino stock the brunt of the disease fell on mice 3–7 weeks old. Each summer, diarrhoea started and in individual animals became profuse; death often occurred within 2–3 days. The outbreak ceased within a fortnght when the moist stock diet was changed to a dry diet with water dispensed from bottles. In Tyzzer's series, diarrhoea was less striking. If the disease lasts long enough, the mesenteric lymph nodes are enlarged and hyperaemic, and the liver contains necrotic foci, circular, greyish, opalescent with sometimes a brown central spot, flush with the surface or sunken slightly below it, and ranging from a fraction of a millimetre to 2½ mm in diameter. Microscopical examination reveals the presence of long thin gram-negative rods, 10–40 μm × 0.5 μm, slightly curved, sometimes banded, and having tapered ends and often subterminal swellings. Natural spread of infection is by the gastro-intestinal route. The organism is an obligatory intracellular parasite. Inside the infected cell it forms spores that resist heating at 56° for one hour but are destroyed at 65° (Craigie 1966). It depends on its spores for extracellular survival; the vegetative form dies rapidly *in vitro*. Said to be motile by peritrichate flagella. The organism can be grown in the yolk sac of the developing chick embryo, where it multiplies in the endoderm (Craigie 1966); and also in primary monolayer cultures of adult mouse hepatocytes (Kawamura *et al.* 1983), in which peritrichate flagellation of the organism can be demonstrated. Mice can be infected experimentally by the intracerebral route (Fujiwara *et al.* 1964), and by the intraperitoneal route, particularly when they are treated with cortisone (Craigie 1966). Rats and hamsters are also susceptible. After intracerebral injection mice die in 4–6 days with necrotic lesions in the brain. After intraperitoneal injection they die with in a week showing extensive lesions in the liver. In mice tetracycline is the only drug that appears to be beneficial. (For review see Ganaway *et al.* 1971.)

Actinobacillus salpingitidis

This organism was first described by Kohlert (1968). It was isolated by Mráz and his colleagues (1976) in Czechoslovakia from fowls suffering from salpingitis or with other lesions in the internal organs.

Streptobacillus moniliformis

There seems little doubt that the organism *Streptobacillus moniliformis*, studied by Levaditi, Nicolau and Poincloux (1925), Parker and Hudson (1926), Levaditi, Selbie and Schoen (1932), Strangeways (1933), and Mackie, van Rooyen and Gilroy (1933), is the same as the organism isolated by Schottmüller (1914), Blake (1916), and Tileston (1916) from one type of rat-bite fever in man and called *Streptothrix muris ratti*. The resemblance of this organism to *Actinomyces bovis* led us in previous editions of this book to classify it with the actinomyces; but it now seems clear that it is more closely related to the actinobacilli. We propose to call it by its original name, *Streptobacillus moniliformis*.

Synonyms *Streptothrix muris ratti* Schottmüller; *Streptobacillus moniliformis* Levaditi, *Actinomyces muris* Lieska; *Haverhillia moniliformis* Levaditi *et al.*

Habitat Natural parasite inhabiting the nasopharynx of rats (Strangeways 1933).

Morphology Loeffler's serum at 37°: slender branching filaments, 0.4–0.6 μm wide, growing in interwoven masses. After 18–24 hr fragmentation of the filaments sets in, and many of the filaments are replaced by chains of bacillary or coccoid bodies. Very striking pleomorphism. Occasional filaments show spherical, oval, fusiform or club-shaped swell-

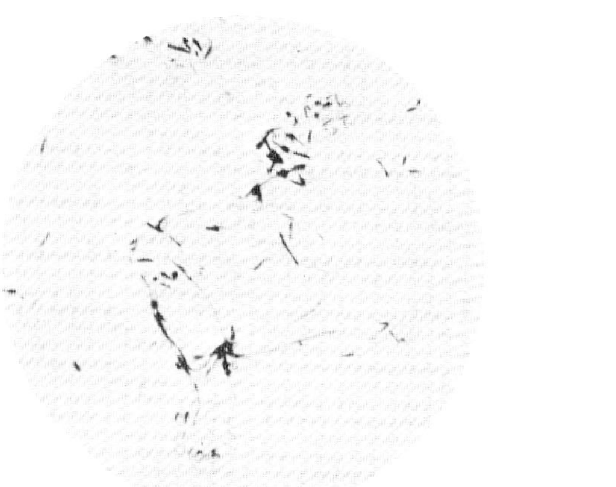

Fig. 22.8 *Streptobacillus moniliformis*. From a Loeffler's slope, 2 days, 37° aerobically (× 1000).

ings occurring terminally, sub-terminally, or in some other situation—hence the term '*moniliformis.*' These swellings may be 2–5 times the diameter of the filament, and may project from one side only. In the animal body the morphology is more regular and bacillary. Great irregularity in depth of staining. Non-motile. Usually described as gram negative, but may be gram positive in young cultures. Non-acid-fast. (See Fig. 22.8). Cultures often contain many small pleomorphic bodies that somewhat resemble mycoplasmas; at one time these were thought to be symbionts but are now recognized to be L-forms.

Cultural appearances No growth on nutrient agar, glucose agar or nutrient gelatin, or in nutrient broth.

Serum agar plate *2 days at* 37°. Circular, greyish-yellow, low convex, almost water-clear, amorphous colonies, 0.2–0.3 mm in diameter, with smooth glistening surface and entire edge; butyrous in consistency and easily emulsifiable. No differentiation. Little or no increase in size on further incubation.

Serum broth *2 days at* 37°. No turbidity. Abundant, greyish-white, coarsely granular sediment, looking like fluffy bread crumbs, miniature cotton balls, or tiny snowflakes, not disintegrating completely on shaking. No surface growth. No odour.

Loeffler's serum *2 days at* 37°. Discrete, circular, low convex colonies, similar to those on serum agar, but rather larger—0.5–0.7 mm in diameter. *7 days*; some colonies may show a differentiation into a slightly raised umbonate centre with a flatter periphery having an irregular or crenated edge; surface appears finely granular and rather dull. Growth may be confluent from the start, and appear slightly raised, colourless, with a glistening beaten-copper surface and a more or less entire edge. No liquefaction, even after 3 weeks.

Dorset egg *2 days at* 37°. Similar to colonies on Loeffler's serum, but perhaps slightly smaller—0.3–0.6 mm in diameter. No liquefaction, even after 3 weeks.

Horse blood agar *2 days at* 37°. Colonies resemble those on serum agar. No haemolysis.

Growth in developing egg Inoculated on to the chorio-allantoic membrane of the developing chick embryo, it invades the embryo and becomes localized almost exclusively in the synovial lining of the joints, where it appears to grow mainly as an intracellular parasite; the embryo dies within 4 days (Buddingh 1944).

Resistance Destroyed in serum broth by heating to 55° for 30 min. Dies out in culture very readily. Serum broth cultures may remain viable at 37° for a week.

Metabolism Grows aerobically, but grows equally well or better under anaerobic conditions. Growth improved by 10 per cent CO_2 especially on first isolation. Optimum temperature 37°; little or no growth at 22°. No haemolysin for horse red cells. No pigment formation. No proteolytic activity.

Biochemical Not thoroughly studied. In serum sugar media acid is produced within 3 days in glucose and salicin, sometimes in maltose and lactose. Litmus milk unchanged. Indole –; MR –; VP –; nitrate reduction –; H_2S –; catalase –; oxidase –; MB reduction –.

Antigenic structure Different strains appear to be antigenically homogeneous.

Pathogenicity Responsible in man for one type of rat-bite fever—sometimes described as infectious erythema or Haverhill fever. May give rise to an epizoötic disease in mice characterized by oedematous swelling of the feet and legs, arthritis, conjunctivitis, and lymphadentis. Intraperitoneal inoculation of 0.5 ml of a serum broth culture into mice is usually fatal in 1–2 days; no characteristic post-mortem appearances. Subcutaneous inoculation into one of the hind feet often leads to a more or less perfect reproduction of the natural disease. May also cause arrested pregnancy and abortions in white mice (Sawicki *et al.* 1962), and chronic abscesses in the cervical lymph nodes of guinea-pigs (Gledhill 1967). Comparatively avirulent for rats, guinea-pigs, and rabbits, though intravenous inoculation of culture into rabbits and rats may sometimes lead to arthritis.

Dick and Tunnicliff (1918) isolated a similar organism—*Actinobacillus putori*—from a boy bitten by a weasel.

References

Allen, A. M., Ganaway, J. R., Moore, T. D. and Kinard, R. F. (1965) *Amer. J. Path.* **46,** 859.

Aoki, M. (1936*a*) *Z. ImmunForsch.*, **87,** 196; (1936*b*) *Ibid.* **87,** 200.

Athalye, M., Lacey, J. and Goodfellow, M. (1981) *J. appl. Bact.* **51,** 289.

Ballio, A. and Barcellona, S. (1968) *Ann. Inst. Pasteur* **114,** 121.

Batty, I. (1958) *J. Path. Bact.* **75,** 455.

Becker, B., Lechevalier, M. P. and Lechevalier, H. A. (1965) *Appl. Microbiol.* **13,** 236.

Blake, F. G. (1916) *J. exp. Med.* **23,** 39.

Bojalil, L. F. and Cerbon, J. (1959) *J. Bact.* **78,** 852.

Bollinger, O. (1877) *Zbl. med. Wiss.* **15,** 481.

Bosworth, T. J. (1923) *J. comp. Path.* **36,** 1.

Bowden, G. H. and Hardie, J. M. (1973) See Sykes and Skinner, p. 277.

Brion, A. (1939) *Revue Méd vét.* **91,** 121.

Brion, A., Goret, P. and Joubert, L. (1952) *vi Congr. int. Patol. comp., Madrid* **i,** 47.

Brock, D. W. and Georg, L. K. (1969) *J. Bact.* **97,** 581, 589.

Brumpt, E. (1927) *Précis de Parasitologie*, p. 1201, Masson, Paris.

Buchanan, B. B. and Pine, L. (1962) *J. gen. Microbiol.* **28,** 305.

Buchanan, R. E. and Gibbons, N. E. (1974) In: *Bergey's Manual of Determinative Bacteriology*, 8th edn. Williams and Wilkins, Baltimore.

Buddingh, G. J. (1944) *J. exp. Med.* **80,** 49.

Castelnuovo, G., Bellezza, G., Duncan, M. E. and Asselineau, J. (1964) *Ann. Inst. Pasteur* **107,** 828.

Chatelain, R., Bercovier, H., Guiyoule, A., Richard, C. and Mollaret, H. H. (1979) *Ann. Microbiol. (Inst. Pasteur)* **A130,** 449.

Christie, A. O. and Porteous, J. W. (1962) *J. gen. Microbiol.* **28,** 443, 455.

Cisar, J. O., Valter, H. E. and McIntyre, F. C. (1978) *Infect. Immun.* **19,** 312.

Colebrook, L. (1920) *Brit. J. exp. Path.*, **1,** 197; (1921) *Lancet* **i,** 893.

Coleman, R. M. and Georg, L. K. (1969) *Appl. Microbiol.* **18,** 427.

Cowan, S. T. and Steel, K. J. (1961) *J. Hyg., Camb.* **59,** 357.

Craigie, J. (1966) *Proc. R. Soc. B.* **165,** 35, 61.

Cross, T. and Goodfellow, M. (1973) *See* Sykes and Skinner, p. 11.
Cummins, C. S. (1962) *J. gen. Microbiol.* **28,** 35.
Cummins, C. S. and Harris, H. (1958) *J. gen. Microbiol.* **18,** 173.
DeWeese, M. S., Gerencser, M. A. and Slack, J. M. (1968) *Appl. Microbiol.* **16,** 1713.
Dick, G. F. and Tunnicliff, R. (1918) *J. infect. Dis.* **23,** 183.
Duda, J. J. and Slack, J. M. (1972) *J. gen. Microbiol.* **71,** 63.
Edwards, P. R. (1931) *Kentucky agric. Exp. Sta. Bull.*, No. 320; (1932) *J. infect. Dis.* **51,** 268.
Ellen, R. P., Walker, D. L. and Chan, K. H. (1978) *J. Bact.* **134,** 1171.
Eppinger, H. (1891) *Beitr. path. Anat.* **9,** 287.
Erikson, D. (1935) *Spec. Rep. Ser. med. Res. Coun., Lond.* No. 203; (1940) *Ibid.* No. 240; (1947) *J. gen. Microbiol.* **1,** 39; (1954) *Ibid.* **11,** 198; (1955) *Ibid.* **13,** 127.
Farshtchi, D. and McClung, N. M. (1967) *J. Bact.* **94,** 255.
Fujiwara, K., Maéjima, K., Takagaki, Y., Naiki, M., Tajina, Y. and Takahashi, R. (1964) *C. R. Soc. Biol., Paris* **158,** 407.
Gale, D. and Waldron, C. A. (1955) *J. infect. Dis.* **97,** 251.
Galli-Valerio, B. (1912) *Zbl. Bakt.* **63,** 555.
Ganaway, J. R., Allen, A. M. and Moore, T. D. (1971) *Amer. J. Path.* **64,** 717.
Gard, S. (1944) *Acta path. microbiol. scand.* Suppl. No. 54, p. 123.
Georg, L. K. and Brown, J. M. (1967) *Inst. J. syst. Bact.* **17,** 79.
Georg. L. K., Robertstad, G. W., Brinkman, S. A. and Hicklin, M. D. (1965) *J. infect. Dis.* **115,** 88.
Gerber, N. N. (1969) *Appl. Microbiol.* **18,** 1.
Gerencser, M. A. and Slack, J. M. (1969) *Appl. Microbiol.* **18,** 80.
Gledhill, A. W. (1967) *Lab. Anim.* **1,** 73.
Goldsworthy, N. E. (1938) *J. Path. Bact.* **46,** 207.
González-Ochoa, A. (1962) *Lab. Invest.* **11,** 1118.
González-Ochoa, A. and Sandoval, M. de los Angeles (1960) *Revta Inst. Salubr. Enferm. trop., México* **20,** 147.
Goodfellow, M. (1971) *J. gen. Microbiol.* **69,** 33.
Goodfellow, M. and Alderson, G. (1977) *J. gen. Microbiol.* **100,** 99.
Goodfellow, M., Brownell, G. H. and Serrano, J. A. (Eds) (1976). *The Biology of the Nocardiae.* Academic Press, London.
Goodfellow, M., Fleming, A. and Sackin, M. J. (1972) *Int. J. syst. Bact.* **22,** 81.
Goodfellow, M. and Minnikin, D. E. (1977) *Annu. Rev. Microbiol.* **31,** 159.
Gordon, R. E. (1966*a*) *J. gen. Microbiol.* **43,** 329; (1966*b*) *Ibid.* **45,** 355.
Gordon, R. E. and Barnett, D. A. (1977) *Int. J. Syst. Bact.* **27,** 176.
Gordon, R. E. and Mihm, J. M. (1957) *J. Bact.* **73,** 15; (1958) *Ibid.* **75,** 239; (1959) *J. gen. Microbiol.* **20,** 129; (1962*a*) *Ibid.* **27,** 1; (1962*b*) *Ann. N.Y. Acad. Sci.* **98,** 628.
Gordon, R. E. and Smith, M. M. (1955) *J. Bact.* **69,** 147.
Gottlieb, D. (1973) See Sykes and Skinner, p. 1.
Griffith, F. (1916) *J. Hyg., Camb.* **15,** 195.
Gumbrell, R. C. and Smith, J. M. B. (1974) *J. gen. Microbiol.* **84,** 399.
Harz (1877–78) *Dtsch. Z. Thiermed.* **5,** 125.
Hassegawa, S., Nakamoto, T., Miyasaki, Y., Arimiti, T. and Akiyosi, M. (1938) *Jap. J. med. Sci.* V, **3,** 27.
Hazen, E. L., Little, G. N. and Resnick, H. (1952) *J. Lab. clin. Med.* **40,** 914.
Heinrich, S. and Pulverer, G. (1959) *Zbl. Bakt.* **174,** 123.
Higgins, M. L., Lechevalier, M. P. and Lechevalier, H. A. (1967) *J. Bact.* **93,** 1446.
Holmberg, K. and Hallander, H. O. (1973) *J. gen. Microbiol.* **76,** 43.
Holt, S. C., Tanner, A. C. R. and Socransky, S. S. (1980) *Infect. Immun.* **30,** 588.
Howell, A. (1963) *Sabouraudia* **3,** 81.
Howell, A. and Jordan, H. V. (1963) *Sabouraudia* **3,** 93.
Howell, A., Jordan, H. V. Georg, L. K. and Pine, L. (1965) *Sabouraudia* **4,** Pt 2, p. 65.
Howell, A., Murphy, W. C., Paul, F. and Stephan, R. M. (1959) *J. Bact.* **78,** 82.
Jones, F. S. (1922) *J. exp. Med.* **35,** 361.
Jones, T. H., Barrett, K. J., Greenham, L. W., Osborne, A. D. and Ashdown, R. R. (1964) *Vet. Rec.* **76,** 24.
Juhlin, I. (1967) *Acta path. microbiol. scand.* **70,** Suppl. No. 189.
Kawamura, S. *et al.* (1983) *J. gen. Microbiol.* **129,** 277.
Kilian, M. (1978) *J. clin. Microbiol.* **8,** 127
Kilian, M., Frederiksen, W. and Biberstein, E. L. (1981) *Haemophilus, Pasteurella and Actinobacillus.* Academic Press, London.
King, E. O. and Tatum, H. W. (1962) *J. infect. Dis.* **111,** 85.
Klinger, R. (1912) *Zbl. Bakt.*, **62,** 191.
Kohlert, R. (1968) *Mn. vet. Med.* **23,** 392.
Kornman, K. S. and Loesche, W. J. (1978) *J. clin. Microbiol.* **7,** 514.
Krainsky, A. (1914) *Zbl. Bact. IIte Abt.* **41,** 649.
Kwapinski, J. B. (1966) *Zbl. Bakt.*, **200,** 80; (1970) *Proc. int. Conf. Cult. Collect. Tokyo* 1968, p. 439.
Lambert, F. W., Brown, J. M. and Georg, L. K. (1967) *J. Bact.* **94,** 1287.
Lanéelle, M. A., Asselineau, J. and Castelnuovo, G. (1965) *Ann. Inst. Pasteur*, **108,** 169.
Laveran (1906) *C. R. Soc. Biol., Paris* **61,** 340.
Lechevalier, H. A. and Lechevalier, M. P. (1965) *Ann. Inst. Pasteur* **108,** 662; (1967) *Annu. Rev. Microbiol,* **21,** 71; (1970*b*) In *The Actinomycetales*, p. 393. Ed. by H. Prauser. Gustav Fischer, Jena.
Lechevalier, M. P. (1976) See Goodfellow, Brownell and Serrano, p. 1.
Lechevalier, M. P., Horan, A. C. and Lechevalier, H. (1971*b*) *J. Bact.* **105,** 313.
Lechevalier, M. P. and Lechevalier, H. A. (1970*a*) *Int. J. Syst. Bact.* **20,** 435.
Lentze, F. (1967) In: *Krankeiten durch Aktinomyzeten und verwandte Erreger*, p. 1. Ed. by H. J. Heite. Springer-Verlag, Berlin.
Levaditi, C., Nicolau, S. and Poincloux, P. (1925) *C. R. Acad. Sci.* **180,** 1188.
Levaditi, C., Selbie, R.-F. and Schoen, R. (1932) *Ann. Inst. Pasteur* **48,** 308.
Levi, M. L. and Cotchin, E. (1950) *J. comp. Path.* **60,** 17.
Liégard, H. and Landrieu, M. (1911) *Ann. Ocul.* **46,** 418.
Lieske, R. (1921) *Morphologie und Biologie der Strahlenpilze (Actinomyceten).* Gebrüder Borntraeger, Leipzig.
Lignières, J. and Spitz, G. (1902) *Bull. Soc. Méd. vét. Centre* **20,** 487, 546.
MacCallum, W. G. (1902) *Zbl. Bakt.* **31,** 529.

Mackie, T. J., Rooyen, C. E. van and Gilroy, E. (1933) *Brit., J. exp. Path.* **14,** 132.
Magnusson, H. (1928) *Acta. path. microbiol. scand.* **5,** 170.
Maguire, L. C. (1958) *Vet. Rec.* **70,** 989.
Mannheim, W., Pohl, S., and Holländer, R., (1980) *Zbl. Bakt.*, I Abt., Orig., **A246,** 512.
Mason, K. N. and Hathaway, B. M. (1969) *Arch. Path.* **87,** 389.
Memmesheimer, A. R. (1967) In: *Krankheiten durch Aktinomyzeten und verwandte Erreger*, p. 50. Ed. by H. J. Heite. Springer-Verlag, Berlin.
Meyer, E. and Verges, P. (1950) *J. Lab. clin. Med.* **36,** 667.
Minnikin, D. E., Patel, P. V., Alshamaony, L. and Goodfellow, M. (1977) *Int. J. Syst. Bact.* **27,** 104.
Mishra, S. K., Gordon, R. E. and Barnett, D. A. (1980). *J. clin. Microbiol.* **11,** 728.
Mordarska, H., Cebrat, S., Black, B. and Goodfellow, M. (1978) *J. gen. Microbiol.* **27,** 176.
Mordarska, H., Mordarski, M. and Goodfellow, M. (1972) *J. gen. Microbiol.* **71,** 77.
Mordarski, M. *et al.* (1977) *Int. J. syst. Bact.* **27,** 66.
Mráz, O. (1967) *Folia microbiol. Praha*, **12,** 403; (1968) *Zbl. Bakt., Ref.* **214,** 149; (1969) *Zbl. Bakt.* **209,** 212, 336, 349.
Mráz, O., Vladík, P., and Boháček, J. (1976) *Zbl. Bakt.* **236,** 294.
Murray, I. G. and Proctor, A. G. J. (1965) *J. gen. Microbiol.* **41,** 163.
Naeslund, C. (1925) *Acta path. microbiol. scand.* **2,** 110.
Negroni, P. and Bonfiglioli, H. (1937–38) *Folia biol.* No. 82, p. 351.
Nelson, J. B. (1930) *J. infect. Dis.*, **46,** 64; (1931) *J. Bact.* **21,** 183.
Orchard, V. A. and Goodfellow, M. (1980) *J. gen. Microbiol.* **118,** 295.
Ørskov, J. (1923) *Investigations into the Morphology of the Ray Fungi.* Levin and Munksgaard, Copenhagen.
Parker, F. and Hudson, N. P. (1926) *Amer. J. Path.* **2,** 357.
Phillips, J. E. (1960) *J. Path. Bact.* **79,** 331; (1961) *Ibid.* **82,** 205; (1964) *Ibid.* **87,** 442; (1967) *Ibid.* **93,** 463.
Pier, A. C. and Fichtner, R. E. (1971) *Amer. Rev. resp. Dis.* **103,** 698.
Pier, A. C., Mejia, M. J. and Willers, E. H. (1961) *Amer. J. vet. Res.* **22,** 502.
Pine, L. and Georg, L. K. (1969) *Int. J. syst. Bact.* **19,** 267.
Pine, L. and Hardin, H. (1959) *J. Bact.* **78,** 164.
Pine, L. and Overman, J. R. (1963) *J. gen. Microbiol.* **32,** 209.
Pinoy, E. (1912) *Bull. Soc. Path. exot.* **5,** 589.
Pulverer, G. and Ko, H. L. (1970) *Appl. Microbiol.* **20,** 693; (1972) *Ibid.* **23,** 207.
Report. (1949) *Annu. Rep. Animal Health Trust*, p. 27; (1960) 58th *Annu. Rep. Imper. Cancer Res. Fund. Lond.* p. 19.
Ridel, M. (1975) *Int. J. syst. Bact.* **25,** 124.
Rodda, G. M. J. (1964) *Med. J. Aust.* ii, **13.**
Romano, A. H. and Sohler, A. (1956) *J. Bact.* **72,** 865.
Rosebury, T. (1944) *Bact. Rev.* **8,** 189.
Sawicki, L., Bruce, H. M. and Andrews, C. H. (1962) *Brit. J. exp. Path.* **43,** 194.
Schofield, G. M. and Schaal, K. P. (1981) *J. gen. Microbiol.* **127,** 237.
Schottmüller, H. (1914) *Derm. Wschr.* **58,** Supp., p. 77.
Setti, C. (1929) *G. Batt. Immun.* **4,** 585.
Silberschmidt. (1899) *Ann. Inst. Pasteur* **8,** 841.
Slack, J. (1942) *J. Bact.* **43,** 193.
Slack, J. M. (1974) In: *Bergey's Manual of Determinative Bacteriology*, 8th edn, p. 660. Ed. by R. E. Buchanan and N. E. Gibbons. Williams and Wilkins, Baltimore.
Slack, J. M. and Gerencser, M. A. (1966) *J. Bact.* **91,** 2107; (1970) *Ibid.* **103,** 265.
Slack, J. M., Landfried, S. and Gerencser, M. A. (1969) *J. Bact.* **97,** 873.
Smith, I. M. and Hayward, A. H. S. (1971) *J. comp. Path.* **81,** 79.
Smith, T. (1918) *J. exp. Med.* **28,** 333; (1921*a*) *Ibid.* **33, 441;** (1921*b*) *Ibid.* **34,** 593.
Sparrow, S. and Naylor, P. (1978) *Vet. Rec,* **102,** 288.
Strangeways, W. I. (1933) *J. Path.* Bact. **37,** 45.
Sykes, G. and Skinner, F. A. (Eds) (1973) *Actinomycetales: characteristics and practical importance. Soc. appl. Bact. Sympos. Ser., No. 2.* Academic Press, London.
Thiroux, A. and Pelletier, J. (1912) *Bull, Soc. Path. exot.* **5,** 585.
Thjotta, T. and Sydnes, S. (1951) *Acta path. microbiol. scand.* **28,** 27.
Thompson, L. and Lovestedt, S. A. (1951) *Proc. Mayo Clin.* **26,** 169.
Tileston, W. (1916) *J. Amer. med. Ass.* **66,** 995.
Tsukamura, M. (1969) *J. gen. Microbiol.* **56,** 265; (1977) *Int. J. syst. Bact.* **27,** 311.
Tunnicliff, E. A. (1941) *J. infect. Dis.* **69,** 52.
Tyzzer, E. E. (1917) *J. med. Res.* **37,** 307.
Vallée, A., Thibault, P. and Second, L. (1963) *Ann. Inst. Pasteur* **104,** 108.
Vincent, H. (1894) *Ann. Inst. Pasteur* **8,** 129.
Waksman, S. A. (1959) *The Actinomycetes.* Williams and Wilkins, Baltimore; (1967) *The Actinomycetes. A Summary of Present Knowledge.* Ronald Press Co., New York; (1971) In: *Bronner and Bronner's Actinomycosis*, 2nd ed., pp. 124, 134. John Wright, Bristol.
Waksman, S. A. and Henrici, A. T. (1943) *J. Bact.* **46,** 337.
Wetmore, P. W., Thiel, J. F., Herman, Y. F. and Harr, J. R. (1963) *J. infect. Dis.* **113,** 186.
Williams, S. T., Sharples, G. P. and Bradshaw, R. M. (1973) *See* Sykes and Skinner, p. 113.
Wolff, M. and Israel, J. (1891) *Virchows Arch.* **126,** 11.
Yu-Ying, F. Li and Georg, L. K. (1968) *Canad. J. Microbiol.* **14,** 749.
Zinnemann, K. (1981) In Kilian *et al.* (1981), p. 1.

23

Erysipelothrix and *Listeria*

Geoffrey Smith

Erysipelothrix

Definition Thin rod-shaped organisms with a tendency to the formation of long filaments. Non-motile. Non-sporing. Gram positive. Aerobic, but growth improved by 5–10 per cent CO_2. Grow between 15° and 44°. Very slight fermentative activities. Catalase negative. Serotypes number at least 22. Naturally pathogenic to a fairly wide range of mammals including man, and birds. Produce septicaemia, arthritis, endocarditis, and skin lesions. G + C content of DNA: 38–40 moles per cent.

The only species is *Erysipelothrix rhusiopathiae* (*insidiosa*).

The first member of this group to be described was the bacillus of mouse septicaemia, *Erysipelothrix muriseptica*; it was found by Koch (1880) in the blood of mice that had been inoculated subcutaneously with putrefying blood. In 1882 Loeffler (1886) observed a similar bacillus in the blood vessels of the skin of a pig that had died of swine erysipelas. [It is possible that the bacillus observed four months previously by Thuillier (Pasteur and Thuillier 1883) in pigs dying of *rouget* was the same organism as that described by Loeffler, but this is not absolutely clear.] Another organism closely allied to *E. rhusiopathiae* was found by Rosenbach in cases of human erysipeloid. Subsequent workers recorded the presence of *Erysipelothrix* in outbreaks of polyarthritis in sheep and joint-ill in lambs, and in occasional infections of cattle, horses, turkeys, peacocks, and man (see Beaudette and Hudson 1936, Paterson and Heatley 1938, Greener 1939). Three species were distinguished by Rosenbach (1909)—*muriseptica*, *porci*, and *erysipeloides*. All strains are now generally considered to belong to a single species.

Habitat

Enough has already been said to indicate the wide range of animals infected by this organism. Though it is sometimes found in the slime surrounding the body of various fish (Klauder 1932, Schoop 1936), there is no evidence that fish are naturally infected during life (Schoop and Stoll 1966, Shewan 1971). The slime, however, is liable to contamination later from trawler holds, fish boxes and rats and mice in the docks (Smylie 1962). The organisms may also be found in the sewage effluent from abattoirs (Hettche 1937). They often occur in the tonsils and faeces of apparently normal pigs and other animals; a saprophytic existence in soil seems unlikely (Doyle 1960, Wood 1973, Lamont 1979).

Morphology

The work of Spryszak and Szymanowski (1929), Meyn (1931), Redlich (1932), and Barber (1939) has made it clear that *Erysipelothrix* occurs in a smooth and a rough form, each characterized by closely associated morphological and colonial appearances. In the *smooth* form the organisms appear as small, straight or slightly curved, gram-positive rods with rounded ends, about 0.8–2.5 μm long and 0.3–0.6 μm broad, arranged singly, in small packets or groups, or in short chains. In the *rough* form long filaments, up to 60 μm or more, predominate; some are seen breaking down to form chains of bacilli. The morphology, however, varies to some extent with the medium on which the organisms are growing. At pH 7.6–8.2 smooth forms are said to be in the ascendant, and at pH 5.2–7.0 rough forms (Žák *et al.* 1965). In ageing cultures core-like structures have been observed by electron microscopy (Stuart 1972). *E. rhusiopathiae* is non-motile, non-capsulated and non-sporing.

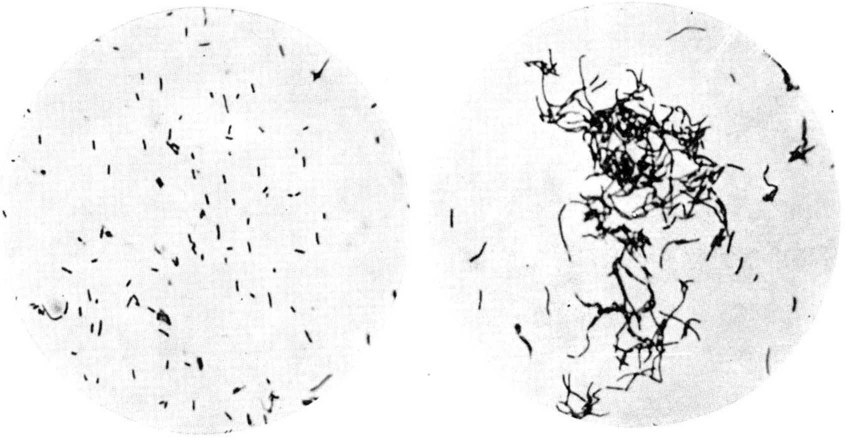

Fig. 23.1 *Erysipelothrix rhusiopathiae*. Left: smooth form. Right: rough form. From a surface agar culture, 3 days, 37° (× 1000).

Cultural characters

In the *smooth* form the colonies after 24 hours' incubation at 37° are very small, circular, convex, amorphous and water-clear, with a smooth glistening surface and entire edge. On further incubation they show little or no increase in size. In the *rough* form the colonies are rather larger and flatter; their matt surface, curled structure and fimbriate edge render them not unlike miniature anthrax colonies. Both forms produce colonies exhibiting a light blue colour and sometimes a shimmering green colour when viewed by oblique illumination. The distinction between smooth and rough colonies is not sharp; intermediate forms are common. Generally speaking, the smooth form tends to breed true, the rough form to give off occasional smooth colonies. The rough form is said to be changed into the smooth form by mouse passage (Meyn 1931). In gelatin stab culture the smooth form gives rise to a growth confined to the line of inoculation; with the rough form lateral branches, more numerous near the surface, grow out horizontally into the medium producing the so-called test-tube brush appearance. There is no liquefaction of the gelatin. In broth the smooth form produces a slight uniform turbidity and a small powdery deposit; with the rough form flocculi of varying size or tangled hair-like masses appear, settle on the sides or bottom of the tube and are difficult to disintegrate by shaking. No visible growth occurs on potato or MacConkey's medium. Some haemolysis may be seen around deep colonies in blood agar, but there is none in blood broth.

Resistance

Most strains of *Erysipelothrix* are killed by exposure to moist heat for 15 minutes at 55°. The organism is resistant to salting, pickling and smoking, and may remain alive in putrefying carcasses for months. Hettche (1937) found that it survived for 4 to 5 days in drinking water, and for 12 to 14 days in sewage and aquarium water. It grows in the presence of 0.05 per cent potassium tellurite and of 0.001 per cent crystal violet. It is said to be one of the most tolerant of all bacteria to sodium azide, growing in a concentration of even 0.1 per cent (Packer 1943). It is resistant to sulphonamides and neomycin (Füzi 1963), but susceptible *in vitro* to penicillin and streptomycin (Woodbine 1950).

Growth requirements and metabolism

Growth is scanty on ordinary media. It is improved by glucose, and to a less extent by blood and serum. It is said to be best in 0.1 per cent glucose broth and 0.5 per cent glucose agar; larger quantities of sugar are inhibitory (Colella 1936). Vawter (1937) recommends a liver digest medium containing 2–4 per cent of sterile horse serum. The organism is a facultative anaerobe. On first isolation it grows in the form of a band just below the surface of a shake agar culture; whether this is due to a preference for CO_2 or for a lowered partial pressure of oxygen is not clear. The growth requirements have not been fully worked out. All strains are said to need riboflavin and oleic acid; and in peptone water serum or thioglycollate must be added (Hutner 1942). Growth occurs between 15° and 44°, but is best at 30° to 37°. The optimum pH for growth is 7.2–7.6; the limits are pH 6.8–8.2 (Karlson and Merchant 1941). Blood cells are not lysed.

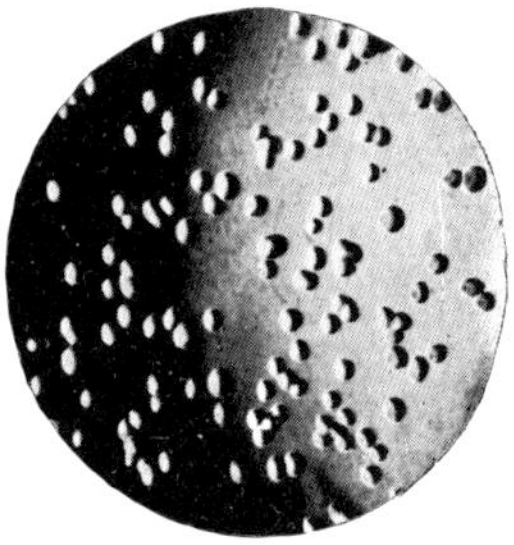

Fig. 23.2 *Erysipelothrix rhusiopathiae*. Smooth form. Surface colonies on agar, 24 hr 37° (×8).

Biochemical characters

Erysipelothrix has only weak fermentative activity. It produces acid without gas in glucose, lactose, fructose and galactose (Deem and Williams 1936, Karlson and Merchant 1941). There is little or no change in litmus milk, no indole formation and no decolorization of methylene blue. Both methyl-red and Voges-Proskauer reactions are negative. The production of catalase and the reduction of nitrate are weak or absent. Most strains are said to produce hyaluronidase (Ewald 1957), H_2S on suitable media (Nørrung 1970), and neuraminidase (Müller 1971), but not oxidase. Neuraminidase production is believed to be related to virulence (Krasemann and Müller 1975).

Antigenic structure

Watts (1940) and Atkinson and Collins (1940) showed that by agglutination and absorption tests strains of *Erysipelothrix* could be divided into two main types. Watts found that each type possessed a heat-stable specific antigen, and apparently two heat-labile antigens which were present in different proportions in the two types and were responsible for cross-agglutination. Dedié (1949) was able to extract the type-specific antigens by Lancefield's acid method and to show by precipitation tests that the strains fell into two types, A and B. This was confirmed by Roots (1953), who recognized in addition a group or species-specific antigen, G, common to all strains. Further confirmation of these findings was obtained by Truszcynski (1961) by means of the agar-gel diffusion test. All three antigens were found to be present in soluble form in liquid cultures, the G antigen being in excess. This explained why in plain filtrates strains of A and B types were precipitated by both A and B antisera. Erler (1968) obtained evidence of two subtypes in type B strains, and Nørrung (1970) of four subtypes. According to Kucsera (1971), who used type-specific acid-soluble antigens, or later autoclaved bacilli, in the double agar-gel precipitation test, 15 types labelled A, B, C, etc. were recognizable among strains from different sources in different parts of the world. Kucsera (1973) proposed that serotypes should be designated by Arabic numerals instead of capital letters, small letters being placed after numerals to differentiate subtypes; the letter N was retained for the designation of strains that lacked a type-specific antigen. Serotypes 1 and 2 correspond to the types A and B of Dedié (1949). There are at least 22 serotypes (Fábián *et al.* 1973, Wood *et al.* 1978, Nørrung 1979). Strains of human and animal origin are antigenically alike (Sneath *et al.* 1951). (For information on the complexity of the heat-labile and heat-stable antigens, see Gledhill 1945, 1947.) Brill and Polityńska (1961) found that type B strains were latently infected with a **bacteriophage** that lysed strains of type A.

G + C content of DNA 38–40 moles per cent (Flossmann and Erler 1972).

Pathogenicity

E. rhusiopathiae has a wide range of pathogenicity for animals under natural conditions, and may occasionally infect man. The diseases to which it gives rise will be dealt with in Chapter 50; here we shall concern ourselves mainly with its effect on laboratory animals. Three of the striking characters possessed by both *Erysipelothrix* and *Listeria*, namely the production of monocytosis in rabbits, of pin-point focal necroses in the liver of mice and of conjunctivitis in rabbits and mice, will be described more conveniently when the relations between these two organisms are discussed. The smooth form of *Erysipelothrix* has generally been found to be more virulent than the rough (Meyn 1931, Redlich 1932, Schoening, Gochenour and Grey 1938, Žák *et al.* 1965).

Inoculation into animals

Swine Loeffler (1886), who first isolated the swine erysipelas bacillus, failed to reproduce the disease in swine with pure cultures, but Schütz (1886) later succeeded in doing so. Broth cultures injected subcutaneously proved fatal to two pigs, one animal dying in 3, the other in 4 days; there were typical findings at necropsy, and the bacilli were recovered in pure culture from the blood and spleen, and from the pleural and peritoneal exudates. Artificial cultures rapidly lose their virulence for swine. Collins and Goldie (1940) produced polyarthritis by repeated intravenous inoculation of cultures. In addition there was a focal inflammatory polyarteritis, focal necrosis of the liver and myocardium, lymphadenopathy, a monocytosis and endocarditis, but skin lesions were never found.

The bacillus is pathogenic for mice, pigeons, and rabbits, but not for guinea-pigs.

Mice A dose of 0.001–0.1 ml of a 24-hr broth culture injected subcutaneously or intraperitoneally is usually fatal in 2 to 3 days. During life the mice develop conjunctivitis and their lids become glued together with a muco-purulent secretion; arching of the back is very common, and constipation is usual. *Post mortem*, the vessels of the skin and subcutaneous tissue are congested, the spleen is enlarged, and the lungs are bright red and oedematous. Bacilli are usually abundant in the blood and viscera; they are found particularly within the phagocytic cells, in which they appear to multiply (Tenbroeck 1920).

Rats Subcutaneous inoculation of a type B strain is said to give rise in white rats to a generalized infection accompanied by an acute serofibrinous synovitis of the limb joints causing lameness (Ajmal 1970).

Pigeons A dose of 0.001–0.1 ml of a 24-hr broth culture injected intramuscularly proves fatal in 3 or 4 days as a rule. Death is often preceded by paralysis of the legs, dyspnoea and convulsions. *Post mortem*, there is a black haemorrhagic mass in the muscle at the site of inoculation; the spleen is enlarged; there are often punctiform haemorrhages in the mucosae and viscera; and there is almost constantly a clear lemon-yellow exudate in the pericardium (Crimi 1914). The bacilli are fairly numerous in the blood and organs.

Rabbits A dose of 0.5 ml of a 24-hr broth culture injected intravenously sometimes proves fatal in 2 to 3 days. An oedematous swelling or erysipelatous rash develops in the injected ear, and there is a rise in temperature and a loss in weight. *Post mortem*, besides the rosy skin lesion, there is congestion of the viscera, and often a clear lemon-yellow pericardial exudate; there may be large haemorrhages into the lungs. The bacilli are scarce. If the disease is not acutely fatal, a monocytosis occurs, reaching its maximum in 3 to 7 days. In animals dying about this time occasional tiny focal necroses may be found in the liver, and areas of mononuclear cell reaction may be seen in sections of the spleen. Inoculation of the conjunctiva gives rise to conjunctivitis, which often proves fatal. After subcutaneous inoculation death seldom occurs.

Hamsters The subcutaneous injection of 1 ml of a 24-hr broth culture into the Syrian hamster (*Cricetus auratus*) sometimes proves fatal in about 10 days. At necropsy the organisms can be recovered from the heart blood, liver and spleen (Shuman and Lee 1950).

Fish Hettche (1937) and Brunner (1938) showed that both freshwater and sea fish can be readily infected by feeding or by intraperitoneal inoculation with the bacilli. The organisms are widely distributed in the tissues and may be recovered after several weeks. They are particularly abundant in the kidneys and are excreted in the urine. The infection appears to be of the covert type, the fish showing no evidence of illness.

Listeria

Definition Coccoid to rod-shaped organisms with a tendency to the formation of long filaments. Motile by peritrichate flagella. Non-sporing. Gram positive. Aerobic and microaerophilic. Grow between 4° and 42°. Usually β-haemolytic. Weak fermentative activities. A number of serotypes are distinguishable by flagellar and somatic antigens. Pathogenic to a wide range of animals including man. Produce monocytosis in rodents, keratoconjunctivitis in rabbits and septic lesions in various organs in warm-blooded animals. G + C content of DNA: 38 moles per cent.

Type species is *Listeria monocytogenes*.

In 1926 Murray, Webb and Swann described a disease of rabbits characterized by a large-mononuclear leucocytosis, and caused by a small gram-positive non-sporing bacillus which they termed *Bact. monocytogenes*. The same organism has since been isolated by a number of workers from various diseases in animals and man characterized most often by a generalized infection tending to localize in the liver, myocardium or central nervous system (see Chapter 50). For this organism Pirie (1927) suggested the generic name of *Listerella* presumably in honour of Lord Lister. Since this name, however, had been used by Jahn in 1906 for a mycetozoan (see Pirie 1940), it was later replaced by *Listeria*. Until recently the genus *Listeria* was usually said to contain only one species—*L. monocytogenes*. Seeliger and Welshimer (1974) listed four species. Wilkinson and Jones (1977) recognized three subgroups, namely *L. monocytogenes*, *L. grayi* (Errebo Larsen and Seeliger 1966), and non-haemolytic strains; *L. murrayi* (Welshimer and Meredith 1971) did not differ sufficiently from *L. grayi* to warrant the status of a separate species. The name *L. innocua* denotes strains that are non-haemolytic, possess antigenic factors xi and xv, and lack pathogenicity for experimental animals (Seeliger 1981). *L. monocytogenes* contains at least six serotypes and a number of subtypes.

Habitat

L. monocytogenes has been isolated from more than 50 species of wild and domestic animals, including birds (Murray 1955, Seeliger and Finger 1976); its distribution is worldwide, regardless of climate. Weis and Seeliger (1975) in south-west Germany examined 194 strains isolated from natural sources. The greatest number came from uncultivated fields, large numbers from plants, and some from mouldy fodder, the feeding grounds of wildlife, and birds. They concluded that *Listeria* was a saprophytic organism living in a plant-soil environment, rendering it possible for man and animals to be infected from many different sources by many different routes. In addition to silage (Gray 1960), plants and soil, it has often been isolated from the faeces of man (Ortel 1977), animals, and birds.

Morphology

In cultures of the *smooth* form short rods are seen, 1–3 μm long and 0.5 μm broad, often with slightly pointed ends. The organisms are arranged singly, in pairs end to end, side by side, in V-formation, in small packets or in short chains. Coccoid forms may predominate in young cultures; in older cultures slightly curved longer rods may be seen. In the *rough* form bacilli 6–20 μm long are found together with filaments stretching across the microscopic field. The organisms are motile by up to three flagella (Paterson 1939, Seeliger 1958). Motility is best seen at 20–25°; at 37° it may be so poor as to be imperceptible (Seastone 1935). In hanging-drop preparations, tumbling and rotatory movements may alternate with periods of comparative rest. Motility may also be demonstrated by the progressive clouding of the medium in soft agar. L-forms are readily induced by growth in penicillin agar (Edman *et al.* 1968). Electron microscopy reveals a cell wall having the usual three layers, beneath which is a complex cytoplasmic membrane the constitution of which is described differently by different observers (Edwards and Stevens 1963, Kawata 1963, Gray and Killinger 1966). Within the cytoplasm are dense polymorphic granular bodies.

Cultural characters

Growth on the surface of media is never abundant, even when blood, serum or a fermentable carbohydrate is added. On nutrient agar the *smooth* form gives rise to very small, translucent, slightly iridescent, dewdrop-like colonies, 0.2–0.4 mm in diameter, butyrous and easily emulsifiable. Viewed by oblique transmitted light they have a characteristic bluish-green colour (Gray 1957). Older colonies are larger, have a brownish, slightly granular centre and a transparent effuse periphery with a finely crenated margin; when magnified, they have a poached-egg appearance. The *rough* form gives rise to rather larger, flatter colonies, having an irregularly indented edge; they are umbonate with a central crater, friable, and impossible to emulsify evenly. On blood agar haemolysis is weaker than with the smooth form. Intermediate colonial forms occur (see Gray, Stafseth and Thorp 1957).

Growth in ordinary broth is very poor, but is greatly improved by the addition of 0.5–1 per cent of glucose. In this medium the *smooth* form gives rise within 24 hr to a fairly dense, even turbidity; after some days the organisms settle to the bottom in floccules. The *rough* form gives rise to a floccular turbidity, a slight surface pellicle, and a granular deposit. The organisms multiply rapidly in milk. A soluble haemolysin is formed in fluid media. In plain gelatin stab culture there is a filiform growth without lateral branches; no liquefaction is seen. Growth occurs on MacConkey's agar, but the colonies seldom reach more than 0.5 mm in diameter after three days.

Resistance

The organisms were said by Potel (1955) to be unusually resistant to heat, possibly even surviving pasteurization in milk. Barber (1939), however, found that they were destroyed at 55° in 60 min; and Dedié (1958) 65° in 30–40 seconds, and at 75° in 10 seconds. They grow freely in broth containing 0.1 per cent potassium tellurite, but not as a rule in 0.03 per cent sodium azide or 0.001 per cent crystal violet. They grow in the presence of 10 per cent salt and at pH 9.6. They remain viable in 16 per cent salt for a year, and, like staphylococci, can tolerate concentrations up to about 20 per cent. They withstand 40 per cent of bile. In culture media at 4° they survive for 3–4 years, and in hay, straw, sand, earth and milk for weeks or months (Warnecke 1963). Of the antibiotics, ampicillin is said to be the most effective both *in vitro* and *in vivo*, but penicillin, erythromycin, tetracycline and sulphadiazine all inhibit growth of the organisms (Lochman *et al.* 1968). The sensitivity to penicillin is less than that of most staphylococci and streptococci; and organisms rapidly become resistant to streptomycin (Foley *et al.* 1944, Zink *et al.* 1951, Linzenmeier and Seeliger 1954). The organism is resistant to polymyxin B and the aminoglycosides.

Growth requirements and metabolism

Growth is improved by adding to the medium a fermentable carbohydrate. Tryptose agar is the medium of choice. Growth is possible in a synthetic medium containing 19 amino acids, glucose, mineral salts, riboflavin, biotin, thiamine and thioctic acid (see Gray and Killinger 1966). The organisms grow aerobically and anaerobically, but best in the presence of CO_2 and a lowered partial pressure of oxygen. The temperature range is 3° to 45° with an optimum at 30° to 37°.

Listeria is one of the few pathogenic organisms that can form colonies at 4° (Gray *et al.* 1948). The organisms are more susceptible to acid than to alkali; they can grow at pH 9.6, but not below pH 6.0. In glucose broth fermentation lowers the pH to about 4.5 and leads to death of the organisms (Hartwigk 1958). Colonies on blood agar plates are surrounded by a narrow zone of β-lysis; the haemolytic zone produced by serotype 5 strains is wide (Cooper *et al.* 1973). A filtrable haemolysin is produced, attacking most mammalian red blood corpuscles. Faecal and environmental sources, but not pathological sources, sometimes yield non-haemolytic strains; such strains are non-pathogenic for experimental animals and often belong to serotypes not known to be associated with disease. Lipolytic activity and haemolytic activity are apparently not due to the same protein (Jenkins and Watson 1971, Siddique *et al.* 1974).

Biochemical characters

Listeria produces acid without gas in a fairly wide range of sugars. There is some variability between strains. As a rule glucose, trehalose and salicin are fermented in 24 hours at 37°. Arabinose, lactose, maltose, sucrose, rhamnose, melezitose, dextrin, aesculin, sorbitol and glycerol are fermented in 3 to 10 days or not at all. Raffinose, inositol, dulcitol, adonitol and inulin are not fermented; nor is mannitol, except by the particular variety referred to as *L. grayi* (Seeliger 1954, 1958, Potel 1955, Young 1956, Hartwigk 1958, Gray and Killinger 1966). Fewer carbohydrates are fermented by serotype 5 strains than by other strains (Ivanov 1962, Cooper *et al.* 1973, Cooper and Dennis 1978). Litmus milk is acidified and slowly decolorized. The methyl-red reaction is positive, and the Voges-Proskauer also when Barritt's method is used for testing. There is no production of indole or urease, no reduction of nitrate, no growth in citrate and no liquefaction of gelatin. Catalase is produced, though often weakly; ammonia is produced from arginine, and H_2S when lead acetate strips are used over a liver agar medium. Oxidase negative. Phosphatase is produced (Mrsvic *et al.* 1975). Methylene blue is decolorized. Formazan is said to be produced by the reduction of 2, 3, 5-triphenyltetrazolium chloride giving rise to an intense red colour in serum broth cultures (Dias and da Silva 1958).

Antigenic structure

Seastone (1935), Webb and Barber (1937), Schultz, Terry, Brice and Gebhardt (1938), Paterson (1939, 1940*a*), and Julianelle and Pons (1939) brought evidence to show the existence of some antigenic diversity among *Listeria* strains. Paterson recognized four types, the division being made primarily on the H and secondarily on the O antigens. Robbins and Griffin (1945) adopted Paterson's classification of O antigens, and found that antigenic factor iii, which is common to groups 1 and 2, was more susceptible than the rest to both heat and alcohol. Seeliger (1958) divided type 4 into two serotypes, 4a and 4b, and Donker-Voet (1957, 1959) recognized three further subtypes, 4c, 4d and 4e, on the basis of somatic antigens. She also divided types 1 and 3 into two by virtue of slight differences in flagellar antigens. Table 23.1 shows the antigenic structure of *L. monocytogenes* in the light of more recent studies; it should not be regarded as final. According to Ullmann and Cameron (1969) the cell walls of types 1 and 2 contain predominantly glucosamine and rhamnose; of type 3 galactose, rhamnose and glucosamine; and of types 4a and b glucose and galactose.

Table 23.1 Antigenic structure of *Listeria monocytogenes* (After Seeliger 1975)

Serotype	Somatic antigens	Flagellar antigens
1/2a	i, ii, (iii)	A, B
1/2b	i, ii, (iii)	A, B, C
1/2c	i, ii, (iii)	B, D
3a	ii, (iii), iv	A, B
3b	ii, (iii), iv	A, B, C
3c	ii, (iii), iv	B, D
4a	(iii), (v), vii, ix	A, B, C
4ab	(iii), v, vi, vii, ix	A, B, C
4b	(iii), v, vi	
4c	(iii), v, vii	A, B, C
4d	(iii), (v), vi, viii	A, B, C
4e	(iii), v, vi, (viii), (ix)	A, B, C
4f	(iii), v, xv	A, B, C
4g	(iii), v, vi, vii, x, xi	A, B, C
5	(iii), (v), vi, viii, x	A, B, C
6	(iii), vii, viii, xi	A, B, C
7	(iii), (xii), xiii	A, B, C
*L. grayi**	(iii), xii, xiv	E

* The organisms referred to as *L. murrayi* (Welshimer and Meredith 1971) are antigenically similar to *L. grayi*; they differ from *L. grayi* mainly in their ability to reduce nitrate to nitrite.

Listeria and *Erysipelothrix* are antigenically distinct, but *Listeria* shares some antigens with several other organisms, such as staphylococci, enterococci, *Corynebacterium pyogenes* and *Escherichia coli.* Distinction of specific from non-specific antibodies in human and animal sera may be aided by the complement-fixation-absorption test (Seeliger 1962), and the agar-gel-diffusion-precipitin test (Muraschi and Tompkins 1963). According to Schierz and Burger (1966), the antibodies that act on staphylococci and enterococci are polysaccharides, whereas the specific antibody in listeriae is a protein. This can be adsorbed on to the tanned red cells of the sheep and demonstrated by a haemagglutination test. From serotype 4b Delvallez and his colleagues (1979) isolated by freeze-pressing, centrifugation, and gel filtration three major antigens. One of these—antigen 2—was purified, and found to be shared by all serotypes of *L. monocytogenes* and by *L. grayi*. A serum against antigen 2 was used to prepare a serologically homogeneous antigen by immunoabsorption. This proved to be genus-specific, showing no cross-reactions with other organisms. Confirmation of this finding should add greatly to the ease of diagnosing listeria infections. The protein in the cell walls of living virulent bacilli is said to give rise to a type-specific immunity in mice (Klasky and Pickett 1968). According to Seeliger and Finger (1976) serotypes 1/2a and 4b account for 97 per cent of the strains from human and animal disease; serotypes other than 1/2a, 4a and 4b are usually obtained from faecal or environmental sources.

No haemagglutination is produced by living cells.

Phage-typing Sword and Pickett (1961) released **bacteriophage** particles from lysogenic strains of *L. monocytogenes* submitted to ultraviolet irradiation. By means of four separate phages they were able to classify 127 out of 149 strains into eight groups. The phages proved specific for *Listeria*, and the grouping they provided agreed closely with that obtained serologically. Phage-typing now permits the recognition of 88 per cent of type 4 and 57 per cent of type 1 strains (Taylor 1980).

Pathogenicity

Listeria monocytogenes is naturally pathogenic to a wide range of animals (see Chapter 50). It is a facultative intracellular parasite. Experimentally strains appear to differ in virulence, though when freshly isolated from pathological lesions they are always pathogenic (Seeliger 1958). The smooth form is more virulent than the rough. Like *Erysipelothrix*, it gives rise to monocytosis in rabbits, to pin-point focal necroses in the liver of mice, and to conjunctivitis in rabbits and mice (see below). The factor responsible for monocytosis can be extracted from smooth cultures by chloroform and appears to be a glyceride (Stanley 1949, Holder and Sword 1969). Durst (1975) noticed an increase in the virulence of strains subcultured repeatedly on media incubated at 4°.

The production of a heat-labile extracellular toxin producing haemorrhagic necrosis when injected intradermally in rabbits and death when injected intradermally in mice was described by Liu and Bates (1961); and a toxin that was lethal to chinchillas by McIlwain and Barnes (1968). DiCapua and his colleagues (1968), however, who implanted diffusion chambers in the peritoneal cavity of mice, could obtain no evidence of toxin formation. Experimentally monocytosis can be produced in rabbits, though strains vary in this respect, but not in mice or guinea-pigs.

Inoculation of *Listeria* into animals

Mice Subcutaneous or intraperitoneal inoculation of 10^8 living organisms of a highly virulent culture causes death in about 1 to 4 days. During life conjunctivitis sometimes occurs. At post-mortem examination multiple tiny focal necroses are found scattered throughout the liver. The organisms can be readily recovered from the spleen and heart blood, and sometimes from the faeces. Khan *et al.* (1973) examined 28 strains from man, cattle, sheep, and silage; the LD50 values for mice inoculated intraperitoneally varied from 3.9 to 4.5 million. Ralovich *et al.* (1972) found LD50 values of 10^6–10^7 for β-haemolytic strains, and LD50 values of *ca* 1.5×10^9 for non-haemolytic strains; these two groups of strains also differed in their ability to cause purulent keratoconjunctivitis in guinea-pigs, and in their virulence for chick embryos and day-old chicks. Further differences were associated with persistence in the internal organs of mice inoculated intravenously, and in the faeces of mice infected orally (Emödy and Ralovich 1975). Patočka and co-workers (1979) found that intracerebral inoculation of sucking mice with a strain of *L. innocua* produced encephalitis.

Rabbits Intravenous inoculation proves fatal in 24 hours or not for several days according to the dose and virulence of the strain. Animals surviving for some time develop a monocytosis, which reaches its maximum in about 3 to 7 days. Post-mortem examination of animals dying after a few days reveals the presence of multiple focal necroses in the liver, and rarely in the spleen and myocardium. Necrosis of the suprarenals is common. Occasionally abscesses in the myocardium and inflammation of the meninges may be met with (see Burn 1935). After intraperitoneal inoculation, much the same lesions are found, but in addition there is a sero-fibrinous peritonitis, with abscesses containing thick white pus in the rolled-up omentum. The organisms can rarely be recovered from the blood stream. In pregnant animals metritis may occur. Instillation of a pure culture into the conjunctiva, or swabbing of the everted lid, gives rise to a severe conjunctivitis within 24 hours followed by keratitis; the animal itself rarely dies.

Guinea-pigs These animals die after inoculation with large doses. The lesions at necropsy are similar to those in mice. The organisms may be recovered from the spleen and sometimes from the heart blood.

Rats These animals are fairly resistant, but may sometimes be infected with large doses.

Pigeons are refractory, but **turkey** poults are susceptible.

Chick embryos Paterson (1940*b*) showed that *Listeria* gave rise to focal lesions on the chorio-allantoic membrane of chick embryos.

Comparison of *Erysipelothrix* and *Listeria*

E. rhusiopathiae and *L. monocytogenes* resemble each other closely in several respects (Barber 1939, Julianelle 1941). Thus, both organisms are small, gram-positive rods, alike in the morphological and cultural appearances of their smooth and rough forms. Both have much the same growth requirements, degree of resistance, and G+C content of their DNA. Both have a wide range of pathogenicity for animals, and both occasionally give rise to disease in man. In rabbits, experimental inoculation of either organism

Table 23.2 Differences between *Erysipelothrix rhusiopathiae* and *Listeria monocytogenes*

Characteristic	*E. rhusiopathiae*	*L. monocytogenes*
Morphology	Slender and non-motile	Thicker and motile
Cell-wall content	Glycine and serine but not DL-diaminopimelic acid	DL-diaminopimelic acid but not glycine or serine
Gelatin stab	Test-tube brush growth	Filiform growth
Haemolysis	α -Haemolysis	β -Haemolysis
Soluble haemolysin	Not produced	Produced
Haemagglutination	Positive for human and some animal red cells	Negative
Growth at 4°	Negative	Positive
Biochemical activity	Ferments fewer sugars and has less active reducing powers	Ferments more sugars and has more active reducing powers
Fermentation of aesculin	Negative	Positive
Catalase production	Negative	Positive
MR and VP reactions	Both negative	Usually both positive
Antigenic structure	At least 22 serotypes; not related to *Listeria*	At least 6 serotypes; not related to *Erysipelothrix*
Pathogenicity	Kills pigeons but not guinea-pigs	Kills guinea-pigs but not pigeons

results in the development of a generalized infection accompanied by the appearance of conjunctivitis and, if the disease is not acutely fatal, of a considerable monocytosis; *post mortem*, focal necroses can be found in the liver. Against these similarities, it may be said that *Listeria* is thicker and is motile, attacks rather more sugars, is less resistant to azide, is said to reduce Prontosil (Kujumgiev 1959), is antigenically distinct, and is fatal experimentally to guinea-pigs but not to pigeons, in contrast to *Erysipelothrix* which is fatal to pigeons but not to guinea-pigs. The differences are listed in Table 23.2.

It might seem to the practical bacteriologist who has studied several strains of these organisms side by side that the differences between them are less than those between species of many other organisms that are placed in the same genus, for example, in *Corynebacterium*, *Pasteurella*, *Bacillus* or *Clostridium*. Nevertheless, the general view of taxonomists is that they should be classified in separate genera; to this conclusion we must submit. The dissimilarity of the two organisms is supported by the numerical taxonomic survey carried out by Wilkinson and Jones (1977), which showed that the two organisms belonged to quite separate genera. *Erysipelothrix* was related to *Streptococcus* and *Gemella*; *Listeria* to *Lactobacillus*. The opinion, however, was expressed that both organisms, together with *Streptococcus*, *Lactobacillus*, and *Gemella*, should be included in the family Lactobacillaceae. In a later taxonomic survey Piot and his colleagues (1980) found a sharp separation between *Erysipelothrix* and *Listeria*.

(For reviews of *Erysipelothrix rhusiopathiae* and *Listeria monocytogenes* see Woodbine 1950, Potel 1955, Roots and Strauch 1958, Seeliger 1958, Gledhill 1959, Füzi and Pillis 1962, Nestoresco *et al.* 1964, Vanini and Moro 1964, Report 1966, 1975, Stuart and Pease 1972, Seeliger and Finger 1976.)

References

Ajmal, M. (1970) *Res. vet. Sci.* **11,** 279.

Atkinson, N. and Collins, F. V. (1940) *Aust. vet. J.* **16,** 193.

Barber, M. (1939) *J. Path. Bact.* **48,** 11.

Beaudette, F. R. and Hudson, C. B. (1936) *J. Amer. vet. med. Ass.* **88,** 475.

Brill, J. and Polityńska, E. (1961) *Zbl. Bakt.* **181,** 473.

Brunner, G. (1938) *Zbl. Bakt.*, IIte Abt. **97,** 457.

Burn, C. G. (1935) *J. Bact.* **30,** 573.

Colella, C. (1936) *Nuova Vet.* **14,** 6.

Collins, D. H. and Goldie, W. (1940) *J. Path. Bact.* **50,** 323.

Cooper, R. F. and Dennis, S. M. (1978) *Canad. J. Microbiol.* **24,** 598.

Cooper, R. F., Dennis, S. M. and McMahon, K. J. (1973) *Amer. J. vet. Res.* **34,** 975.

Crimi, P. (1914) *Ann. Staz. Mal. Best. Napoli* **2,** 107.

Dedié, K. (1949) *Monatsh. Vet.-Med.* **4,** 7; (1958) *Listeriosen. Zbl. VetMed.*, Beiheft 1, p. 99. Paul Parey, Berlin.

Deem, A. W. and Williams, C. L. (1936) *J. Bact.* **32,** 303.

Delvallez, M., Carlier, Y., Bout, D., Capron, A. and Martin, G. R. (1979) *Infect. Immun.* **25,** 971.

Dias, V. M. and Silva, N. P. M. da (1958) *Zbl. Bakt.* **171,** 317.

DiCapua, R. A., Osebold, J. W. and Stone, K. R. (1968) *Amer. J. vet. Res.* **29,** 2023.

Donker-Voet, J. (1957) *Tijdschr. Diergeneesk.* **82,** 341; (1959) *Amer. J. vet. Res.* **20,** 176.

Doyle, T. M. (1960) *Vet. Rev. Annot.* **6,** 95.

Durst, J. (1975) *Zbl. Bakt.* I. Abt. Orig. **A233,** 72.

Edman, D. C., Pollock, M. B. and Hall, E. R. (1968) *J. Bact.* **96,** 352.

Edwards, M. R. and Stevens, R. W. (1963) *J. Bact.* **86,** 414.

Emödy, L. and Ralovich, B. (1975) See *Report* (1975), p. 131.

Erler, W. (1968) *Zbl. Bakt.* **207,** 340.

Errebo Larsen, H. and Seeliger, H. P. R. (1966) See Report (1966), p. 35.

Ewald, F. W. (1957) *Mh. Tierheilk.* **9,** 333.

Fábián, L., Kemenes, F., Kucsera, G. and Vetési, F. (1973) *Magy. Allatorv. Lap.* **28,** 515.

Flossmann, K. D. and Erler, W. (1972) *Arch. exp. vet. Med.* **26,** 817.

Foley, E. J., Epstein, J. A. and Lee, S. W. (1944) *J. Bact.* **47,** 110.

Füzi, M. (1963) *J. Path. Bact.* **85,** 524.

Füzi, M. and Pillis, I. (1962) *Zbl. Bakt.* **186,** 556.

Gledhill, A. W. (1945) *J. Path. Bact.* **57,** 179; (1947) *J. gen. Microbiol.* **1,** 211; (1959) In: *Infectious Diseases of Animals. Diseases due to Bacteria,* vol 2, p. 651. Ed. by A. W. Stableforth and I. A. Galloway. Butterworths, London.

Gray, M. L. (1957) *Zbl. Bakt.* **169,** 373; (1960) *Science* **132,** 1767.

Gray, M. L. and Killinger, A. H. (1966) *Bact. Rev.* **30,** 309.

Gray, M. L., Stafseth, H. J. and Thorp, F. (1957) *Zbl. Bakt.* **169,** 378.

Gray, M. L., Stafseth, H. J., Thorp, F., Sholl, L. B. and Riley, W. F. (1948) *J. Bact.* **55,** 471.

Greener, A. W. (1939) *Brit. J. Derm. Syph.* **51,** 372.

Hartwigk, H. (1958) *Listeriosen. Zbl. VetMed.,* Beiheft 1, p. 5. Paul Parey, Berlin.

Hettche, H. O. (1937) *Arch. Hyg.* **119,** 178.

Holder, I. A. and Sword, C. P. (1969) *J. Bact.* **97,** 603.

Hutner, S. H. (1942) *J. Bact.* **43,** 629.

Ivanov, I. (1962) *Mh. VetMed.* **17,** 729.

Jenkins, E. M. and Watson, B. B. (1971) *Infect. Immun.* **3,** 589.

Julianelle, L. A. (1941) *J. Bact.* **42,** 367, 385.

Julianelle, L. A. and Pons, C. A. (1939) *Proc. Soc. exp. Biol., N.Y.* **40,** 364.

Karlson, A. G. and Merchant, I. A. (1941) *Amer. J. vet. Res.* **2,** 5.

Kawata, T. (1963) *J. gen. appl. Microbiol.* **9,** 1.

Khan, M. A., Seaman, A. and Woodbine, M. (1973) *Zbl. Bakt.* I. Abt. Orig., **A224,** 355.

Klasky, S. and Pickett, M. J. (1968) *J. infect. Dis.* **118,** 65.

Klauder, J. V. (1932) *J. industr. Hyg.* **14,** 222.

Koch, R. (1880) *Investigations into the Etiology of Traumatic Infective Diseases.* New Sydenham Society, London.

Krasemann, C. and Müller, H. E. (1975) *Zbl. Bakt.* I. Abt. Orig. **A231,** 206.

Kucsera, G. (1971) *Acta vet. hung.* **21,** 211; (1973) *Int. J. syst. Bact.* **23,** 184.

Kujumgiev, I. (1959) *Zbl. Bakt.* **174,** 282.

Lamont, M. H. (1979) *Vet. Bull.* **49,** 479.

Linzenmeier, G. and Seeliger, H. (1954) *Zbl. Bakt.* **160,** 543.

Liu, P. V. and Bates, J. L. (1961) *Canad. J. Microbiol.* **7,** 107.

Lochman, O., Výmola, F. and Heizlar, M. (1968) *Ćsl. Epidem.* **17,** 179.

Loeffler (1886) *Arb. ReichsgesundhAmt.* **1,** 46.

McIlwain, P. K. and Barnes, R. W. (1968) *Amer. J. vet. Res.* **29,** 483.

Meyn, A. (1931) *Zbl. Bakt.* **122,** 507.

Mrsvic, S., Ulahovic, M. and Nedeljkovic, M. (1975) See Report (1975), p. 30.

Müller, H. E. (1971) *Path. et Microbiol., Basel* **37,** 241.

Muraschi, T. F. and Tompkins, V. N. (1963) *J. infect. Dis.* **113,** 151.

Murray, E. G. D. (1955) *Canad. med. Ass. J.* **72,** 99.

Murray, E. G. D., Webb, R. A. and Swann, M. B. R. (1926) *J. Path. Bact.* **29,** 407.

Nestoresco, N., Popovici, M., Bădulesco, E. and Rosca, V. (1964) *Archs roum. Path. exp. Microbiol.* **23,** 221.

Nørrung, V. (1970) *Acta vet. scand.* **11,** 577, 586; (1979) *Nord. VetMed.* **31,** 462.

Ortel, S. (1977) *Zbl. Bakt.* I. Abt. Orig. **A239,** 342.

Packer, R. A. (1943) *J. Bact.* **46,** 343.

Pasteur and Thuillier. (1883) *C. R. Acad. Sci.* **97,** 1163.

Paterson, J. S. (1939) *J. Path. Bact.* **48,** 25; (1940*a*) *Ibid.* **51,** 427; (1940*b*) *Ibid.* **51,** 437.

Paterson, J. S. and Heatley, T. G. (1938) *Vet. J.* **94,** 33.

Patŏcka, F., Menčíková, E., Seeliger, H. P. R. and Jirásek, A. (1979) *Zbl. Bakt.* I. Abt. Orig. **A243,** 490.

Piot, P., Dyck, E. van, Goodfellow, M. and Falkow, S. J. (1980) *J. gen. Microbiol.* **119,** 373.

Pirie, J. H. H. (1927) *Publ. S. Afr. Inst. med. Res.* **3,** 163; (1940) *Nature, Lond.* **145,** 264.

Potel, J. (1955) *Arch. Hyg.* **139,** 245.

Ralovich, B., Emödy, L. and Mérö, E. (1972) *Acta microbiol. hung.* **19,** 323.

Redlich, E. (1932) *Z. InfektKr. Haustiere* **42,** 300.

Report (1966) *Proc. 3rd int. Sympos. Listeriosis, Bilthoven*; (1975) *Problems of Listeriosis,* Proc. 6th int. Sympos. Listeriosis, Univ. Nottingham, Sept. 1974. Leicester Univ. Press.

Robbins, M. L. and Griffin, A. M. (1945) *J. Immunol.* **50,** 237.

Roots, E. (1953) *Proc. 15th int. vet. Congr., Stockholm,* p. 44.

Roots, E. and Strauch, D. (1958) *Listeriosen, Zbl. VetMed,* Beiheft 1. Paul Parey, Berlin.

Rosenbach, F. J. (1909) *Z. Hyg. InfektKr.* **58,** 343.

Schierz, G. and Burger, A. (1966) *Z. med. Mikrobiol. Immun.* **152,** 300.

Schoening, H. W., Gochenour, W. S. and Grey, C. G. (1938) *J. Amer. vet. med. Ass.* **92,** 61.

Schoop, G. (1936) *Dtsch. tierärztl. Wschr.* **44,** 371.

Schoop, G. and Stoll, L. (1966) *Z. med. Mikrobiol. Immun.* **152,** 188.

Schultz, E. W., Terry, M. C., Brice, A. T. and Gebhardt, L. P. (1938) *Proc. Soc. exp. Biol., N.Y.* **38,** 605.

Schütz. (1886) *Arb. ReichsgesundhAmt.* **1,** 56.

Seastone, C. V. (1935) *J. exp. Med.* **62,** 203.

Seeliger, H. P. R. (1954) *Z. Hyg. InfektKr.* **139,** 389; (1958) *Listeriose, Beit. Hyg. Epidem.* Heft 8, 2te Aufl.; (1962) *Derm. trop.* **1,** No. 1; (1975) *Acta microbiol. hung.* **22,** 179; (1981) *Zbl. Bakt.* I Abt. Orig. **A249,** 487.

Seeliger, H. P. R. and Finger, H. (1976) In: *Infectious Diseases of the Fetus and Newborn Infant,* p. 333. Ed. by J. S. Remington and J. O. Klein. Saunders, Philadelphia, London, Toronto.

Seeliger, H. P. R. and Welshimer H. J. (1974) In: *Bergey's Manual of Determinative Bacteriology,* 8th ed., p. 593. Ed. by R. E. Buchanan and N. E. Gibbons. Williams and Wilkins, Baltimore.

Shewan, J. M. (1971) *J. appl. Bact.* **34,** 299.

Shuman, R. D. and Lee, A. M. (1950) *J. Bact.* **60,** 677.

Siddique, I. H., Lin, I. F. and Chung, R. A. (1974) *Amer. J. vet. Res.* **35,** 289.

Smylie, H. G. (1962) Personal communication.

Sneath, P. H. A., Abbott, J. D. and Cunliffe, A. C. (1951) *Brit. med. J.* **ii,** 1063.

Spryszak, A. and Szymanowski, Z. (1929) *C. R. Soc. Biol.* **100,** 1151.

Stanley, N. F. (1949) *Aust. J. exp. Biol. med. Sci.* **27,** Pt. 2, 123.

Stuart, M. R. (1972) *J. gen. Microbiol.* **73,** 571.

Stuart, M. R. and Pease, P. E. (1972) *J. gen. Microbiol.* **73,** 551.

Sword, C. P. and Pickett, M. J. (1961) *J. gen. Microbiol.* **25,** 241.

Taylor, A. G. (1980) *Lancet* **i,** 1136.
Tenbroeck, C. (1920) *J. exp. Med.* **32,** 331.
Truszcynski, M. (1961) *Amer. J. vet. Res.* **22,** 846.
Ullmann, W. W. and Cameron, J. A. (1969) *J. Bact.* **98,** 486.
Vanini, G. C. and Moro, S. (1964) *G. Batt. Immun.* **57,** 87, 116, 129, 288, 298.
Vawter, L. R. (1937) *J. Amer. vet. med. Ass.* **90,** 635.
Warnecke, B. (1963) *Zbl. Bakt.* **189,** 162.
Watts, P. S. (1940) *J. Path. Bact.* **50,** 355.
Webb, R. A. and Barber, M. (1937) *J. Path. Bact.* **45,** 523.
Weis, J. and Seeliger, H. P. R. (1975) *Appl. Microbiol.* **30,** 29.
Welshimer, H. J. and Meredith, A. L. (1971) *Int. J. syst. Bact.* **21,** 3.
Wilkinson, B. J. and Jones, D. (1977) *J. gen. Microbiol.* **98,** 399.
Wood, R. L. (1973) *Cornell Vet.* **63,** 390.
Wood, R. L., Haubrich, D. R. and Harrington, R. (1978) *Amer. J. vet. Res.* **39,** 1958.
Woodbine, M. (1950) *Bact. Rev.* **14,** 161.
Young, S. (1956) *Vet. Rec.* **68,** 459.
Žák, O., Grígelová, K. and Cerník, K. (1965) *Folia microbiol., Praha* **10,** 211.
Zink, A., Mello, G. C. de and Burkhart, R. L. (1951) *Amer. J. vet. Res.* **12,** 194.

24

The mycobacteria

John M. Grange

Introductory

Definition The only genus in the family Mycobacteriaceae. Straight or slightly curved rods, but coccobacillary, filamentous and branched forms also occur. Cells are gram positive, acid-fast, non-motile and non-sporing. Some strains are encapsulated by peptidoglycolipids (mycosides). Some strains produce a yellow pigment either in the dark or after exposure to light. Aerobic or microaerophilic. Acid is produced from sugars oxidatively. Nutritional requirements and temperature range of growth vary considerably. Two major subdivisions recognized: rapid growers and slow growers. Cells contain large quantities of lipid. The genus is distinguished by characteristic antigenic patterns and mycolic acid structures. G + C content of DNA:66–72 moles per cent.

Type species is *Mycobacterium tuberculosis*.

The generic name *Mycobacterium* (Lehmann and Neumann 1896) was given to a group of bacteria which grew as mould-like pellicles on liquid media. The genus contains over 30 species, most of which are well defined, including the causative agents of tuberculosis (Chapter 51), leprosy (Chapter 52) and Johne's disease (Chapter 52). Members of other species, though occasionally the cause of disease in man and animals, usually lead a saprophytic existence in the natural environment. These species have been termed 'atypical', 'anonymous', 'paratubercle', 'tuberculoid' and 'MOTT' (mycobacteria other than typical tubercle) bacilli (see Francis and Abrahams 1982).

An important character of the mycobacteria is their ability to resist decolorization by acid after being stained by an arylmethane dye—*acid fastness* (p. 64)—but the genus may be more accurately defined by the chemical structure of its mycolic acids (Etémadi 1967, see p. 72) and its antigenic structure (Stanford 1973*a*, see p. 74). The cultivable members of the genus are divisible into two major groups, the slow growers and rapid growers (p. 79), which differ in biochemical properties (Tsukamura 1967), antigenic structures (Stanford and Grange 1974) and in DNA relatedness (Gross and Wayne 1970, Baess and Bentzon 1978). With very few exceptions, mycobacteria synthesize lipid-soluble iron-binding compounds, termed mycobactins, which differ in their chemical structure from functionally similar compounds in other genera (Patel and Ratledge 1973).

Definition of mycobacterial species

The mycobacteria were, for many years, in a state of taxonomic chaos but the situation has been considerably improved by recent investigations, including some organized by the International Working Group on Mycobacterial Taxonomy (Wayne *et al.* 1971, 1981, Kubica *et al.* 1972, Meissner *et al.* 1974, Saito *et al.* 1977). *Mycobacterium* is now probably one of the best classified of the bacterial genera (Stanford and Grange 1974).

Many studies on mycobacteria have been based on numerical or Adansonian taxonomy but three highly discriminative techniques have also been used, namely antigenic analysis (p. 73), DNA relatedness (Gross and Wayne 1970, Bradley 1975, Baess 1979; see also Chapter 21) and the chemical composition of whole cells as determined by pyrolysis mass spectrometry (Wieten *et al.* 1981).

The evidence suggests that the rapid and slow growers separated early in the evolution of the mycobacterial species (Stanford 1973*a*). Intraspecific variants show much smaller genetic differences which are often due to point mutations, deletional mutations or infection by bacteriophage (Grange 1973). The range of natural or experimentally induced mutational variation differs enormously from species to species. No variant of *M. smegmatis* differed in its properties from its parent strain by more than 5 per cent after exposure to mutagens, whereas some variants of a similarly treated strain of *M. vaccae* bore only a 40 per cent similarity to their parent strain (Tarnok and Tarnok 1978).

A list of 41 mycobacterial species has been included in the *Approved Lists of Bacterial Names* (Skerman *et al.* 1980; see Table 24.1). This list is not universally accepted; thus, many authorities regard *M. avium* and *M. intracellulare* as variants of a single species (Meissner *et al.* 1974); *M. kansasi* and *M. gastri* are serologically very closely related (Norlin *et al.* 1969) and

Table 24.1 Alphabetical list of approved mycobacterial names (from Skerman *et al.* 1980). (In parentheses: the common names of some pathogenic species.)

M. *africanum**	*M. malmoense*
M. asiaticum	*M. marinum*
M. aurum	*M. microti**
M. avium (the avian tubercle bacillus)	*M. nonchromogenicum*
*M. bovis**	*M. neoaurum*
M. chelonei	*M. parafortuitum*
M. chitae	*M. paratuberculosis* (Johne's bacillus)§
M. duvalii†	*M. phlei*
M. farcinogenes	*M. scrofulaceum*
M. flavescens	*M. senegalense*
M. fortuitum	*M. simiae*
M. gadium	*M. smegmatis*
M. gastri‡	*M. szulgai*
M. gilvum	*M. terrae*
M. gordonae	*M. thermoresistibile*
M. haemophilum	*M. triviale*
M. intracellulare§	*M. tuberculosis* (the tubercle bacillus)
M. kansasii†	*M. ulcerans*
M. komossense	*M. vaccae*
M. leprae (the leprosy bacillus)	*M. xenopi*
M. lepraemurium (the rat-leprosy bacillus)§	

* Variants of *M. tuberculosis*.
† Referred to respectively as *M. duvali* and *M. kansasi* in this book.
‡ Probably a variant of *M. kansasi*.
§ Variants of *M. avium*.

the separate specific status of *M. bovis*, *M. africanum* and *M. microti* has been challenged (Grange 1979, 1982). The non-cultivable species *M. leprae* is included in the genus *Mycobacterium* on account of its acid-fastness and mycolic acid structure (Etémadi and Convit 1974).

Habitat

The existence of non-pathogenic mycobacteria was demonstrated soon after the successful cultivation of *M. tuberculosis*. Moeller (1898) described the timothy grass bacillus (*M. phlei*) and isolated strains of *M. smegmatis* from cows' milk and compost. During the ensuing years the 'tuberculocentric' approach to the genus *Mycobacterium* obscured the fact that most of its constituent species are saprophytes that only rarely cause disease (Collins, 1965).

In recent years interest in the mycobacteria that live freely in the inanimate environment has increased considerably for several reasons: (1) as the incidence of tuberculosis has decreased in the developed countries the relative incidence of infections due to other species has increased; (2) there is increasing evidence that contact with non-pathogenic mycobacteria may alter considerably the response of the host when it is subsequently infected with a pathogenic mycobacterium (Abrahams 1970, Stanford 1976) and (3) present-day identification systems enable informative ecological studies to be undertaken. The distribution of mycobacteria may be studied directly by culturing them from the environment, or indirectly by observing the incidence of infections in man and animals in different regions or the frequency of reactions to a range of specific skin-test reagents in various human populations (see Chapter 51).

Mycobacteria occur in soil and water and on vegetation. In 1909 Brem isolated mycobacteria from cold-water taps and many other workers have subsequently isolated strains, usually of *M. kansasi* and *M. xenopi*, from water supplies. *Mycobacterium marinum*, as the name suggests, is particularly associated with aquatic environments. Beerwerth (1971) isolated mycobacteria from various natural sources in Germany, and subsequently (Beerwerth 1973) studied the ability of samples of soil and water to support their growth. Most samples, which varied in their acidity from pH 6.5 to 3.8, supported the growth of a wide range of mycobacteria but not *M. tuberculosis* even when incubated at 37°. Kazda (1978) reported that the deep and partially decomposed grey layer in *Sphagnum* vegetation contained many mycobacteria and that fluid from this layer supported the growth of most species, but not *M. tuberculosis*. In Norway the highest mycobacterial count was found in *Sphagnum rubellum* which, on account of its red pigmentation, can absorb solar energy and raise its temperature (Kazda *et al.* 1979). There is some evidence for the determining rôle of pH on the distribution of mycobacteria. Stanford and Paul (1973) noted that human infections with *M. ulcerans* in Uganda occurred in swampy areas with a surface pH of 6.1–6.9 but not in regions with a higher pH. Mycobacteria have been found on vegetation but it is not clear whether they multiply in this site or are merely contaminants (see Chapter 51).

Non-virulent mycobacteria may also live saprophytically in or on the human or animal body, for example in smegma (Pellegrino 1906), faeces (Bönicke and Juhasz 1964), tonsils (Beck 1905) and nasal secretions (Marchoux and Halphen 1912).

Morphology

The mycobacteria are usually straight or slightly curved rod-shaped organisms, with more or less parallel sides and rounded ends, usually 1–4 μm × 0.3–0.6 μm in size, which are often arranged in small groups. The morphology varies from species to species: cells of *M. xenopi* are often filamentous with occasional branching and aerial hyphae, those of *M. kansasi* (Fig. 24.2) are elongated with a beaded appearance while those of *M. avium* are very short, almost coccoid (Runyon 1974).

Fig. 24.1 *Mycobacterium tuberculosis.* Glycerine agar culture, 4 weeks, at 37°, showing some short straight forms, and some longer curved forms. Ziehl-Neelsen stain. (× 1000).

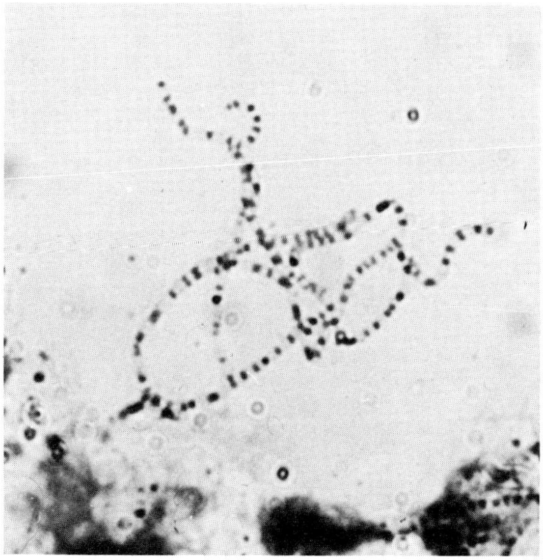

Fig. 24.2 *Mycobacterium kansasi* in sputum, showing the elongated beaded form of the organism. Ziehl-Neelsen stain (× 2500).

Electron microscopy shows that mycobacteria possess a relatively thick cell wall separated from the cell membrane by a thin electron-transparent zone (Imaeda *et al.* 1968*a*). The cytoplasm contains a fairly well defined nuclear body, and granular and electron-transparent bodies which are probably polyphosphate and lipid storage bodies respectively; the cell membrane frequently shows infoldings or mesosomes (Asano *et al.* 1973). The mesosomes are occasionally quite large and show a lamellar structure. Membranous structures resembling mesosomes are abundant in lysogenic mycobacteria (Robbins *et al.* 1970). The electronmicroscopic appearance of mycobacteria is shown in Figs. 24.3, 24.4 (see also Barksdale and Kim 1977).

Electronmicrographs show clearly that mycobacteria may divide by binary fission (Fig. 24.4). Some authors, nevertheless, have considered that they may display a more complicated life cycle in which bacilli fragment into a number of coccal forms which then elongate into rods (Chang and Andersen 1969, McCarthy 1971). Other workers have postulated even more elaborate life cycles, with the formation of cell-wall-defective organisms (Mattman 1970) that release mycoplasma-like viable forms. In 1907 Much reported that miliary tubercles in cattle often contained few acid-fast organisms but many granules, stainable by aniline gentian violet, which he considered to be forms of the tubercle bacillus. Some workers accepted that these **Much's granules** represented a stage in the life cycle of *M. tuberculosis* (Sweany 1928), but others considered them to be products of degeneration (Oerskov 1932), artefacts due to damage to the organisms (Yegian and Porter 1944), or mitochondria (Mudd *et al.* 1951).

Cell-wall-defective forms of mycobacteria (spheroplasts or protoplasts) may be induced experimentally by the use of lysozyme and other enzymes (Willett and Thacore 1966, 1967), mycobacteriophages (Millman 1958), glycine (Adámek *et al.* 1969) and cycloserine (Imaeda *et al.* 1968*b*). Although partly damaged cell walls may be reparable, there is no evidence that experimentally induced cell-wall-free mycobacteria are viable or capable of either reversion or multiplication in a cell-wall-free form.

Non-acid-fast or filtrable forms of mycobacteria may be involved in the aetiology of sarcoid and Crohn's disease. Transmissible agents present in the lesions in both these diseases will pass through 0.22-μm membrane filters (Mitchell and Rees, 1969, 1970, 1976, 1979, Mitchell *et al.* 1976). A mycobacterium with the cultural characteristics of *M. tuberculosis* was isolated from some of the mice inoculated with filtrates from sarcoid material though no mycobacteria had been cultured from the original homogenates (Mitchell *et al.* 1976). Burnham and his colleagues (1978) isolated *M. kansasi* from mesenteric lymph nodes of one patient with Crohn's disease and a small mycoplasma-like organism from 22 of 27

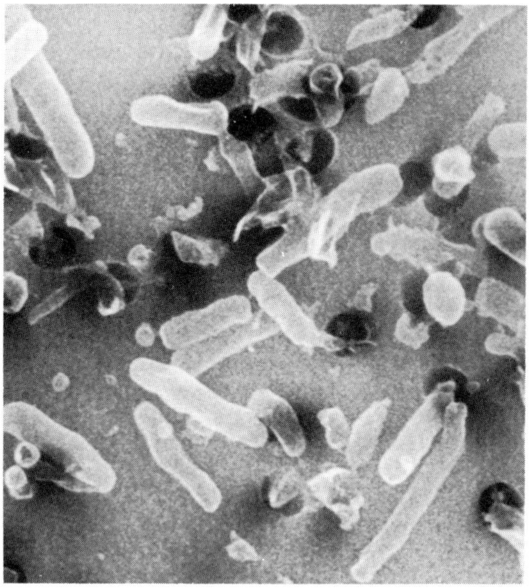

Fig. 24.3 Scanning electronmicrograph of *Mycobacterium kansasi* on a membrane filter (× 10 000).

other patients. This organism was acid-fast and, like *M. kansasi*, photochromogenic, but attempts to isolate a normal mycobacterium from these cultures failed.

Staining reactions

Koch (1882) stained the tubercle bacillus by immersing it in an alkaline solution of methylene blue for 24 hr. Shortly afterwards Ehrlich (1882) discovered the now well known acid-fast property; staining the organism with fuchsin in the presence of aniline oil as a mordant and destaining with dilute mineral acid. Ziehl (1882) changed the mordant to carbolic acid, and Neelsen (1883) increased the strength of the carbolic acid and combined it with the dye to form carbol fuchsin. Thus although the staining technique now bears the epithet 'Ziehl-Neelsen' it is really a modification of Ehrlich's method.

Other staining techniques have been described but most of these differ from the original Ziehl-Neelsen method only in minor details. In most methods, mineral acids are used for decolorization but some use alcohol in addition (Kinyoun 1915). The following method is recommended by the International Union Against Tuberculosis (1977).

Ziehl's carbol fuchsin is prepared by adding 10 ml of a saturated solution of basic fuchsin in 95 per cent ethanol to 90 ml of distilled water in which 5 g of phenol crystals have been dissolved. This mixture is filtered and poured on to heat-fixed smears on microscope slides. The slides are heated until steam rises, but boiling must be avoided as this causes the specimen to become detached from the glass. The slides are left for 5 min, rinsed in water and covered with 25 per cent sulphuric acid for 3 min. After thorough washing in water the smears are counterstained in 0.3 per cent methylene blue for 1 min. Variations of this method include destaining with 3 per cent hydrochloric acid in 95 per cent ethanol (Kinyoun 1915, Vestal 1975) and counterstaining with malachite green or picric acid, the latter of which is recommended for the detection of acid-fast bacilli in pus (Gardner 1926). Modifications of the Ziehl–Neelsen stain for the de-

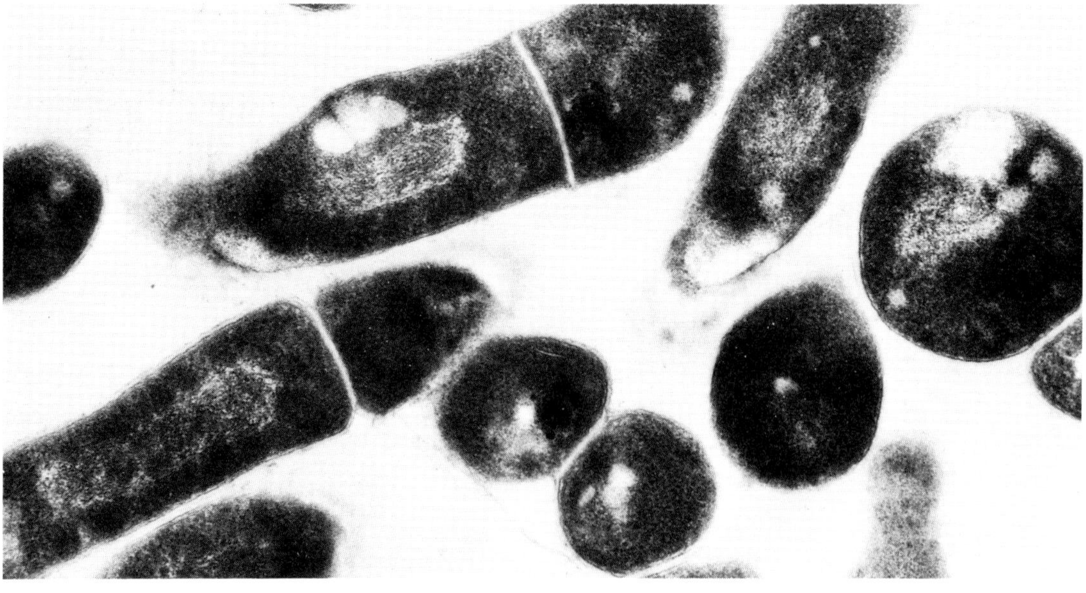

Fig. 24.4 Transmission electronmicrograph of a thin section of BCG. The section shows cell walls, septa, nuclear bodies and inclusion granules (× 50 000).

monstration of acid-fast bacilli in tissue sections have been described (Fite *et al.* 1947, Wade 1957, Ridley 1977). Methods in which the slides are not heated appear to be less efficient than the standard method (Moučka and Kaňkanovská 1967, Šlosárek and Mezensky 1974).

Fluorescent arylmethane dyes have been used to detect acid-fast bacilli. The dyes used are auramine O (Hagemann 1938) or a combination of auramine O and rhodamine B (Truant *et al.* 1962). The principle of this staining technique is identical with that of the Ziehl–Neelsen method: smears are treated with a phenolic solution of the dye or dyes, but without heating, and are then decolorized with acid or acid with ethanol. The counterstain may be either potassium permanganate, which suppresses non-specific fluorescence, or a fluorescent dye of a different colour than that used to stain the bacilli. Technical details of the methods are given by Vestal (1975). Although fluorescence microscopy requires more expensive equipment than standard light microscopy, it is much less demanding on the laboratory staff as bacilli are clearly visible at a relatively low magnification.

Acid-fastness appears to be due to the formation of complexes between the dye and mycolic acid, so trapping the dye within the cell (see Goren 1972, Barksdale and Kim 1977). It is not specific for mycobacteria; fungal and bacterial spores and, to some extent, the genera *Nocardia* and *Gordona* are also acid-fast.

Cultural appearances

Slowly growing strains of mycobacteria take 1–3 weeks or even longer to form colonies on a suitable medium, but rapidly growing strains do this in 2–3 days. Rapid growers are usually defined as organisms that give abundant growth on subculture on Löwenstein–Jensen medium within 7 days (Runyon 1959). Some of the saprophytic mycobacteria grow freely on ordinary nutrient agar or in broth; the tubercle bacilli and members of a number of other species require enriched media.

Colonies on solid media

Several different colonial forms of mycobacteria occur and attempts have been made to use colonial variation to identify species (Fregnan and Smith 1962, Kubica and Jones 1965, Jones and Kubica 1965, Runyon 1970). Many of the species of saprophytic mycobacteria have smooth and rough variants. Several investigators have observed more than one colonial type in strains of *M. tuberculosis*. Smithburn (1936) described a non-spreading (R) colony and a flat, spreading (S) colony in cultures of this species and found that the relative proportions of these forms in the culture depended on the pH of the medium. More recently Osborn (1976) has shown that BCG strains of bovine tubercle bacilli dissociate into spreading and non-spreading colonial types and that the relative proportion of these in a strain alters rapidly when the cultural conditions are changed The relevance of this variability to virulence and immunogenicity remains to be established. Colonial morphology may be further modified by the presence of bacteriophage (p. 77).

Pigmentation The pigments of mycobacteria are carotenoids. In some strains pigment synthesis is constitutive (*scotochromogens*) while in others it is induced by light (*photochromogens*). Many different carotene pigments have been isolated from mycobacteria by means of thin-layer chromatography. Tarnok and Tarnok (1970) isolated up to six different pigments from each strain and found that their distribution varied from species to species; the major pigments and carotenes were synthesized from colourless phytoenes via phytofluene, neurosporine and lycopene (Tarnok and Tarnok 1971). (For further information on pigment biosynthesis see Barksdale and Kim 1977).

Treatment with mutagens causes variation in pigment production by rapidly growing scotochromogens (Tarnok and Tarnok 1978, Levy–Frebault and David 1979), notably the appearance of extra pigments, many of which appear to be precursors of those present in the parent strain. In the absence of mutational change, pigmentation is a stable character and is not much influenced by the nature of the culture medium (Tsukamura 1970).

Growth in liquid media

Unless a detergent such as polyoxyethylene sorbitan mono-oleate (Tween 80) is included in the medium, the hydrophobic nature of the lipid-rich bacilli causes

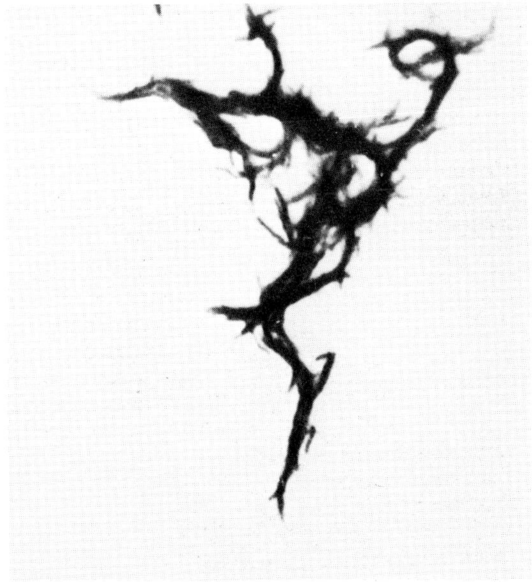

Fig. 24.5 Cord formation by *Mycobacterium tuberculosis* in slide-culture (× 3000).

them to clump together and float on the surface to form a fungus-like pellicle. Middlebrook, Dubos and Pierce (1947) pointed out that, when grown in suitable liquid media, virulent human and bovine strains of *M. tuberculosis* formed characteristic long 'serpentine cords' in which the bacilli were aligned in parallel (see Fig 24.5); attenuated strains, on the other hand, were arranged more or less indiscriminately. Bloch (1950) isolated a lipid from virulent strains and named it *cord factor* (see p. 73) although its relationship to cord formation is now considered to be fortuitous (Goren 1972).

Culture media

Koch's original medium for the growth of tubercle bacilli consisted of heat-coagulated bovine or sheep serum. Subsequently two main classes of media were developed: those containing egg and a group of synthetic or semi-synthetic media. The egg-containing media are solidified by heating to coagulate the egg proteins; other media are either used in the fluid form or are solidified with agar.

The original egg-based medium (Dorset 1902) has been considerably modified, especially by Löwenstein (1930) and Jensen (1955), whose Löwenstein-Jensen medium is in general use. This medium contains whole eggs, asparagine, glycerol and mineral salts. Stonebrink's (1957) medium contains sodium pyruvate instead of glycerol for the isolation of bovine strains of *M. tuberculosis*. The inclusion of a dye, usually malachite green, in egg-based media (Corper and Cohn 1946) enables colonies of mycobacteria to be more readily seen against a contrasting background and also tends to inhibit the growth of certain contaminants in primary cultures. Some workers have used egg yolk instead of whole eggs for the cultivation of the nutritionally more exacting species. Herrold's (1931) medium contains glycerol, peptone and egg yolk and is solidified with agar. This medium, supplemented with mycobactin, is advocated by the American Veterinary Services Laboratories for the isolation of *M. paratuberculosis* (Report 1974). A simple medium containing only eggs, malachite green and coconut water is a useful and cheap alternative to Löwenstein-Jensen medium in regions where coconuts are plentiful (Basaca-Sevilla *et al.* 1976).

Many non-egg-containing media have been described but none has achieved the popularity of Löwenstein-Jensen medium for routine use. The first synthetic media were described by von Schweinitz (1893) and Proskauer and Beck (1894). The latter medium, which is still in use, contains asparagine, glycerol and mineral salts. Minor modifications of this medium were made by Sauton (1912), Long (1926), Kirschner (1932) and Youmans (1944). The last two media were enriched by the addition of horse serum and plasma respectively. Dubos (1945) added Tween 80 to obtain a dispersed growth of bacteria. Dubos's oleic agar is a similar medium enriched with an oleic acid-bovine serum albumin complex and solidified with agar (Dubos and Middlebrook 1947). Middlebrook's 7H9 broth and 7H10 agar (Report 1955) closely resemble the above media but contain sodium glutamate as a nitrogen source. The use of serum albumin adds to the cost of the media but simpler media are unsuitable for primary isolation; they are, however, useful for preparing cultures for serological investigations because they are non-antigenic (Stanford and Beck 1968). Smith's (1953) medium is Dubos's broth solidified with agar and enriched with an ethanolic extract of *M. phlei* for the isolation of *M. paratuberculosis*.

The rapidly growing species grow on unenriched media such as nutrient agar.

Preparation of specimens for primary culture It is first necessary to destroy the other bacteria and fungi that are present. Use is made of the ability of mycobacteria to resist destruction by acid, alkali or a disinfectant (Chapters 4 and 20; for methods, see Chapter 51).

The conditions under which mycobacteria are cultured in the laboratory differ considerably from those in either the infected host or the environment. Dubos, in 1954, pointed out that mycobacteria were cultured by 'primitive bacteriological techniques worked out decades ago'; there has been virtually no progress since this statement was made. The metabolic activity of cells of *M. tuberculosis* grown *in vitro* differ considerably from that of cells obtained from the tissues of infected hosts (David 1978). Transfer of cells from one environment to a dissimilar one is fatal to some of them (David 1972).

Conditions for growth

Mycobacteria are *obligate aerobes* but some species and strains are microaerophilic and grow as a narrow band under the surface of a semi-solid medium (Fig. 24.6). Variation in oxygen requirements has been used as an aid to identification (Marks 1976); its existence explains the conflicting statements in the literature about the effects of varying the partial pressure of oxygen.

Carbon dioxide enhances the growth of *M. tuberculosis*: Rockwell and Highberger (1926) and Whitcomb and his colleagues (1962) observed a growth-stimulating effect with 5 or 10 per cent of CO_2; Knox and his co-workers (1961) obtained no growth in its complete absence. Rotation of a flask containing *M. tuberculosis* in a liquid medium increases the yield of growth 3 to 10 times over that in a stationary flask; when 6 per cent of CO_2 is added, the yield is increased 18-fold (Nayebi 1970). Some saprophytic mycobacteria appear to be able to use atmospheric CO_2 as a carbon source (Bruner 1934).

The *optimum temperature for growth* and the range of temperatures at which mycobacteria grow vary both between and within species; this variation is of value as a means of identifying strains (Käppler 1973, Marks 1976). By studying their ability to grow at 25, 37 and 45°, species may be placed into three groups: *psychrophiles*, *mesophiles* and *thermophiles*.

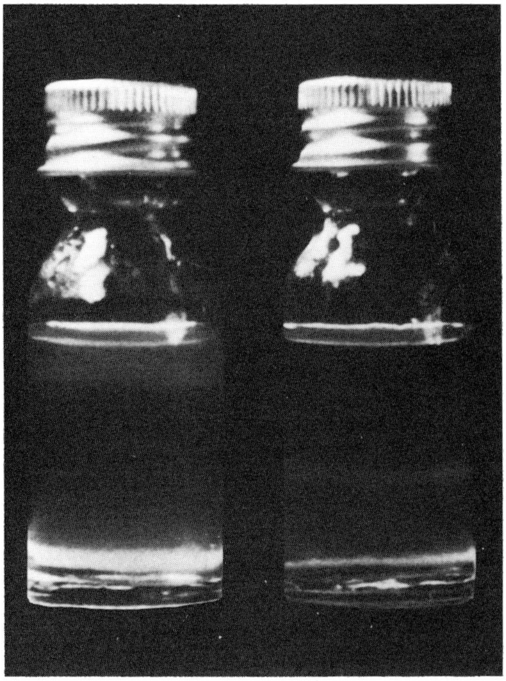

Fig. 24.6 Oxygen preference. The left hand bottle shows a band of growth at the surface (aerobic) and the right hand bottle shows a band of growth in the depths of the medium (microaerophilic).

There is little information concerning the effect of hydrogen-ion concentration on mycobacterial growth (see Drea and Andrejew 1953). The optimum pH for growth of *M. tuberculosis* in glycerol broth was reported to be pH 6.0–6.5 by Dernby and Näslund (1922).

Nutritional requirements and metabolic activities

The mycobacteria vary markedly in their metabolic activities, nutritional requirements and rate of growth. At one extreme *M. smegmatis* yields a profuse growth within a few days on very simple media; at the other, *M. ulcerans* requires several months' incubation on enriched medium to yield the same amount of growth. Much use has been made of tests for metabolic activity in the classification and identification of mycobacteria; the biochemical basis of these tests will be discussed in this section.

The basic nutritional requirements of mycobacteria are carbon, nitrogen, oxygen, phosphorus and sulphur together with metal ions, especially iron and magnesium.

Carbon sources

Carbohydrates, organic acids, *n*-alkanes and acyl esters are degradable by mycobacterial enzymes. (For the possible use of CO_2 see p. 66). Glycerol is used by most mycobacteria, but it inhibits the growth of the bovine, vole and some African strains of *M. tuberculosis*. Glucose and pyruvic acid are also metabolized (Dixon and Cuthbert 1967), the latter compound being included in media for the growth of strains of *M. tuberculosis* that are inhibited by glycerol. Action on other sugars varies widely between and sometimes within species, particularly among the rapidly growing mycobacteria. These differences have been used as the basis of identification tests. In the method of Gordon and Smith (1953), carbohydrates are incorporated at a concentration of 1 per cent in an ammonium phosphate-based agar containing bromocresol purple, which serves to detect acid formation. A more sensitive and rapid test is based on the inability of washed mycobacteria to reduce nitrite to ammonia in the absence of a usable carbon source (Bönicke and Kazda 1970).

All rapidly growing mycobacteria use glucose, mannose and trehalose; use of other sugars including those with alcoholic groups e.g. inositol, dulcitol, mannitol and sorbitol is variable. Mycobacteria hydrolyse glycosides; hydrolysis of β-galactoside (Tacquet *et al.* 1966) and β-glucoside (David and Jahan 1977) have been used for taxonomic purposes. Glycosidase activity of slowly growing species may be studied by the use of the sugars linked to the highly fluorescent compound 4-methylumbelliferone (4-methyl-7-hydroxycoumarin); very rapid hydrolysis of α-L-fucoside was found in *M. marinum* and a lesser activity in *M. szulgai* (Grange and McIntyre 1979).

Organic acids are used as carbon sources in some culture media. The use of pyruvic acid has been referred to above. Oleic acid is a useful carbon source (Hedgecock 1968) but is toxic when present in high concentration; this can be avoided by complexing the acid with albumin (Dubos and Middlebrook 1947) from which it is slowly released. Hydrolysis of Tween 80 releases oleic acid; this enzymic activity has been used to identify mycobacterial species. Tween 80 combines with neutral red to form an amber coloured complex; on hydrolysis of the Tween 80 the dye is released and regains its red colour (Wayne *et al.* 1964). Differences in the use of simple organic acids have been studied for taxonomic purposes (Gordon and Mihm, 1957); in particular, action on citric acid is used to distinguish two major types of *M. chelonei* (Stanford *et al.* 1972). All cultivable mycobacterial species hydrolyse esters of fatty acids, esters of butyric and heptanoic acids being most rapidly hydrolysed (Grange 1978). The acyl esterase activity of *M. fortuitum* is considerably more heat resistant than that of *M. chelonei*, thus providing a rapid means of distinguishing these two species (Grange 1977). Heat-stable acyl esterases have been detected by the use of the chromogenic substrate phenolphthalein dibutyrate in some strains of *M. kansasi*, particularly those associated with human disease (Nakayama and Takeya 1963).

Mycobacteria degrade *n*-alkanes (*n*-paraffins); those with nine or more carbon atoms are regularly degraded by all strains but there is variable degradation of shorter chains (van der Linden and Thijsse 1965, Grange 1974).

Nitrogen sources

Nitrogen is essential for mycobacterial growth and is obtainable from inorganic sources, including ammonium, nitrate and nitrite ions or organic nitrogenous compounds, including a wide range of amino acids, amides, amines, purines and pyrimidines. Ammonium ions afford one of the principal nitrogen sources (Tsukamura 1966*b*). Probably all species reduce nitrite to ammonia (Bönicke and Kazda 1970), but strains vary in their ability to reduce nitrate, a useful property for identification (Virtanen 1960, Bönicke 1962).

Considerable variation in the ability of mycobacteria to use organic nitrogen sources has been observed. The role of amino acids as nitrogen sources has been amply demonstrated by the use of asparagine or glutamic acid as the nitrogen source in Löwenstein–Jensen medium and many other media. The use of asparagine has been studied by several investigators (see Andrejew *et al.* 1974); that of other amino acids is variable and depends to some extent on the presence of carbon sources. In the absence of these, all species hydrolyse glycine, alanine and asparagine but not the aromatic amino acids tyrosine, tryptophan and phenylalanine or the sulphur containing compounds ethionine and methionine (Grange 1976). Other amino acids are variably hydrolysed, and some species differences are encountered; thus histidine was hydrolysed only by *M. smegmatis*, and *M. ulcerans* is unique in hydrolysing hydroxyproline.

The variable hydrolysis of amides has been widely applied in both taxonomic studies and routine identification. Bönicke (1962) described an 'amide-row' containing ten different amides namely acetamide, benzamide, carbamide (urea), isonicotinamide, nicotinamide, pyrazinamide, salicylamide, allantoin, succinamide and malonamide. Washed mycobacteria are added to phosphate buffer containing the amide, and ammonia liberated by hydrolysis is detected after overnight incubation by the addition of Russell's reagent. Nicotinamide and pyrazinamide are probably hydrolysed by the same enzyme (Tarnok *et al.* 1979). The relevance of pyrazinamidase to resistance to the chemotherapeutic agent pyrazinamide is discussed on p. 70.

Röhrscheidt and her colleagues (1970) have shown that some mycobacteria and most nocardiae are able to hydrolyse certain purines and pyrimidines; variation between species was demonstrated.

Iron

This metal is essential for the growth of mycobacteria and requires two classes of compounds synthesized by the bacteria for its uptake, namely *mycobactins* and *exochelins*.

Twort and Ingram (1912, 1913) found that a factor present in dead mycobacteria was essential for the growth of *M. paratuberculosis*. In 1953, Francis and his colleagues extracted this factor from mycobacteria and named it mycobactin. Subsequently its structure was elucidated (Snow and White 1969, Snow 1970) and minor variations in this were found to occur in different species of mycobacteria. Synthesis of mycobactins is greatly increased in iron-deficient media (Ratledge and Marshall 1972). Mycobactins transport ferric ions across the cell membrane, but only after they have been chelated from the medium by other substances (Ratledge and Marshall 1972). In certain circumstances salicylic acid can perform this function (Ratledge and Winder 1962). Macham and Ratledge (1975; see also Macham *et al.* 1977) discovered a class of chelating agents, termed exochelins, that are secreted by mycobacteria and increase in concentration in conditions of iron deficiency. After chelation by the water-soluble exochelins, the iron is transferred across the cell wall and cell membrane by the water-insoluble mycobactins.

Serum has a bacteristatic effect on mycobacteria because it contains transferrin which binds free iron thus rendering it unavailable to the bacteria (Kochan 1973). The presence of exochelin reverses this effect but under conditions of severe iron-limitation the induction of exochelin synthesis is itself repressed (Macham *et al.* 1977).

Almost all mycobacteria synthesise mycobactins. Important exceptions are *M. paratuberculosis* and some strains of *M. avium*, especially those isolated from the wood pigeon (McDiarmid 1962, Wheeler and Hanks 1965). The growth of many strains of *M. avium* is greatly stimulated on primary isolation by the inclusion of exogenous mycobactin in the medium (Matthews *et al.* 1977).

Other requirements for growth

The need for other trace elements including zinc (Winder and Dennery 1959) and manganese (Ratledge and Hall 1971) has been reported. Zinc limitation inhibits the growth of *M. smegmatis* (Winder and O'Hara 1962) and interferes with the activity of certain enzymes, notably glycerol dehydrogenase (Winder and O'Hara 1964). Magnesium is essential for growth (Sauton 1912) and plays an important role in the maintenance of ribosomal function (Trnka and Mison 1971).

Metabolic pathways in mycobacteria

The metabolic activities of mycobacteria may be divided into those unique to the genus—mostly concerned with the synthesis of lipids—and those shared with other forms of life. Studies on enzyme systems in mycobacteria have been reviewed by Goldman (1961),

Ramakrishnan and his colleagues (1972) and by Ratledge (1977).

The biochemical basis for the difference in the rate of bacterial multiplication between the rapidly and slowly growing mycobacteria has not been elucidated. Differences in the metabolism of glucose between *M. phlei* (a rapid grower) and *M. tuberculosis* (a slow grower) have been demonstrated (Le Cam-Sagniez and Sagniez 1973) but it is not known whether these affect their growth rates.

Most antimycobacterial agents interfere with metabolic processes that also occur in other bacterial genera. An exception is isoniazid, which inhibits mycolic acid synthesis (p. 70) and is active only against some mycobacterial species.

The observations of Norman and Williams (1962) that coenzyme Q stimulates the growth of *M. tuberculosis* and other slowly growing mycobacteria led Gangadharam and his colleagues (1978) to study the inhibitory action of various analogues of coenzyme Q. One of these, 6-cyclo-octylamino-5,8-quinolinequinone (CQQ) inhibited the growth of *M. tuberculosis* and *M. avium intracellulare* at concentrations of 1 μg and 8 μg per ml respectively but was inactive against rapidly growing mycobacteria or strains of other genera. This agent may prove to be of therapeutic value.

Other metabolic activities useful in identification tests

Other tests referred to elsewhere in this chapter include niacin production and arylsulphatase, phosphatase, catalase and reductase activity.

The production of *niacin* was originally thought to differentiate human strains of *M. tuberculosis* from all other mycobacteria (Konno 1956), but it also occurs in *M. simiae* and a few strains of *M. chelonei*. Niacin is detected by the formation of a yellow colour with cyanogen bromide and aniline (see Collins and Lyne 1979), by means of a bioassay with a strain of *Lactobacillus arabinosus* (Marks 1965) or by commercially available test strips.

Arylsulphatase (Whitehead *et al.* 1953) and *phosphatase* (Saito *et al.* 1968) activities are detected by the liberation of phenolphthalein from the sulphate and phosphate salts respectively. *Catalase activity*: *M. tuberculosis* is differentiated from most other mycobacteria by its loss of catalase activity after heating at 68° for 20 min at pH 7 (Kubica *et al.* 1966). *Reductase activity* is detected by the reduction of potassium tellurite to black metallic tellurium (Kilburn *et al.* 1969). The test is used to distinguish *M. avium* and its *intracellulare* variant from other slowly growing mycobacteria. (For full technical details of these tests see Vestal 1975, and Collins and Lyne 1979).

Resistance

Mycobacteria possess the same degree of susceptibility to heat as other non-sporing bacteria (Chapter 4); this is exploited for the destruction of tubercle bacilli in milk by pasteurization (see Wilson 1942 and Chapter 9). They have a higher degree of resistance to acids, alkalies and most chemical disinfectants, properties made use of in the decontamination of sputum and other clinical specimens before culture for mycobacteria (see Chapter 51).

Material contaminated with infected sputum is sterilized within 4 hr by 2 per cent phenol or 3 per cent formalin (Hailer 1938). Ethanol is also an effective disinfectant, an 80 per cent solution sterilizes pieces of cloth contaminated by mycobacteria in 10 min or less (Hailer and Heicken 1939, Smith 1947) and has been recommended as a disinfectant for skin, rubber gloves and clinical thermometers. Mycobacteria are also rapidly killed by acetone (Weiszfeiler *et al.* 1968) and tincture of iodine (Hailer and Heicken 1939). Formaldehyde and ethylene oxide are moderately effective, though their activity is diminished when the bacilli are embedded in sputum because of the poor penetrating power of these gases (Report 1958). Bergan and Lystad (1971), who compared different classes of disinfectants, insisted that only the phenolic compounds could be relied upon to disinfect sputum.

Mycobacteria are resistant to drying and survive for weeks or months on inanimate objects if protected from sunlight. *M. tuberculosis* does not appear to replicate in the environment outside the animal body, but it may survive for several months in soil or cow dung. Although long-term survivors on inanimate surfaces appear to play little part in the transmission of tuberculosis to man (American Thoracic Society 1969), there is now evidence that cattle and other animals, including badgers (Muirhead *et al.* 1974), may be infected from soil or dung.

Mycobacteria are rapidly destroyed by sunlight or ultraviolet light (S. Tsukamura 1964) and are from four to ten times more sensitive to ultraviolet light than *Esch. coli* (David 1973). Sensitivity to ultraviolet light is related to pigment content: scotochromogenic species are more resistant than non-pigmented strains, while uninduced photochromogenic strains are the most sensitive of all (S. Tsukamura 1964). (Exceptions occur; the scotochromogenic species *M. flavescens* is four times more sensitive to ultraviolet light than *M. tuberculosis*.) David (1973) showed that sensitivity depended on genome size and capacity for DNA repair as well as on pigment content. The pigments do not absorb ultraviolet light, but appear to neutralize photosensitized or photoexcited substances such as superoxide (see Barksdale and Kim 1977). Mycobacterial mutants that are very sensitive to, or very resistant to, ultraviolet irradiation owe their properties to modifications of the DNA repair mechanism (Mizuguchi 1974, Norgard and Imaeda 1978).

Susceptibility to antibiotics and other antimicrobial agents

Drugs that have been used to treat tuberculosis include antibiotics: aminoglycosides (streptomycin, kanamycin), rifampicin, viomycin, capreomycin and cycloserine; and synthetic chemicals such as isoniazid (INAH), ethambutol, ethionamide, pyrazinamide, thiacetazone and *p*-aminosalicylic acid (PAS). Among the most commonly used are INAH, rifampicin and ethambutol; PAS, although once widely used, is virtually obsolete. The occurrence of resistance to these drugs and the use of sensitivity tests are described in Chapter 51. Many of the other mycobacteria are resistant to most of these drugs *in vitro*; nevertheless combinations of these drugs appear to be effective *in vivo* (Hunter *et al.* 1981).

Antibiotics

The structure and mode of action of antibiotics on mycobacteria have been discussed in Chapter 5 and is reviewed in detail by Doub (1979).

The target structure for aminoglycoside antibiotics is the ribosome; mutations to resistance modify either the 30S or the 50S subunits, or both. Genetic analysis of *M. smegmatis* by means of conjugation (p. 76) has elucidated the basis of resistance and cross resistance (Mitzuguchi *et al.* 1978, Yamada *et al.* 1978). At least four genes are concerned. Those responsible for resistance to streptomycin and kanamycin-neomycin both affect the 30S subunit while two genes determining viomycin-capreomycin resistance affect the 50S and 30S subunits respectively. A third gene for viomycin-capreomycin resistance (*vicC*) does not modify the ribosome but alters the permeability of the cell to these drugs. Although viomycin and capreomycin are cyclic peptides completely unrelated in their structure to the aminoglycosides, the two groups of antibiotics are similar in their mode of action on the ribosomes (see Doub 1979).

Rifampicin is bactericidal and acts synergistically with pyrazinamide in sterilizing tuberculous lesions (Mitchison and Dickinson 1978). It inhibits the synthesis of messenger RNA by blocking the activity of DNA-dependent RNA polymerase; rifampicin resistance may arise by mutational alterations in this enzyme (Konno *et al.* 1973). These authors found no difference in the rate of uptake of ^{14}C-labelled rifampicin in resistant and sensitive strains of *M. phlei*. (For further information about the action of streptomycin, rifampicin and cycloserine, see Chapter 5.)

Isoniazid Isonicotinyl hydrazide (isoniazid, INAH) is a potent antituberculous agent (Report 1952, Vivien *et al.* 1972). Its therapeutic activity is limited to the mycobacteria, in particular *M. tuberculosis*, which may be killed by concentrations as low as 0.015 μg per ml (Winder 1964). Mycobacteria with heat-labile catalases are particularly susceptible to it; these include human and bovine strains of *M. tuberculosis* and some strains of *M. kansasi* (Wayne *et al.* 1968).

It is now virtually certain that isoniazid inhibits the synthesis of mycolic acid (Winder and Collins 1970, Wang and Takayama 1972). Isoniazid causes a rapid decrease in the cellular content of mycolic acids and their precursors, and sensitive mycobacteria cease to be acid-fast when exposed to this drug (Koch-Weser *et al.* 1953). Inactivation of mycolate synthetase (MS) by isoniazid has been demonstrated (Takayama *et al.* 1974). For isoniazid to be effective, the organisms must be actively metabolizing and multiplying (Holland and Ratledge 1971). Isoniazid is the most effective drug in reducing the viable mycobacterial population in a patient with tuberculosis but is much less effective than rifampicin or pyrazinamide in eliminating dormant or near-dormant bacilli (Mitchison and Dickinson 1978).

Mutations to isoniazid resistance do not modify the MS enzyme but cause an inhibition of the transport of the drug into the cell. Isoniazid-resistant strains of *M. tuberculosis* and other mycobacteria usually lack catalase (Middlebrook 1954, Winder 1960), which is thought to transport isoniazid into the bacterial cell (Wimpenny 1967). Catalase and peroxidase appear to reside on the same protein (Diaz and Wayne 1974) and a third H_2O_2-splitting function termed the Y (Youatt's) enzyme is also present on this molecule (Gayathri Devi *et al.* 1974). The catalase-peroxidase protein binds isoniazid at the site of activity of the Y enzyme and transports the drug to the interior of the cell (Gayathri Devi *et al.* 1975). Mutations affecting this site are responsible for isoniazid resistance; some mutations may, however, leave the catalase site intact, thus explaining the occasional occurrence of catalase-positive isoniazid-resistant strains (Winder 1964). Sriprakash and Ramakrishnan (1970) found that a second enzyme, NAD glycohydrolase, could also bind isoniazid; this ability was lost in some but not all isoniazid-resistant strains. The relation of this enzyme change to isoniazid resistance is, however, unknown. Isoniazid-resistant strains of *M. tuberculosis* become sensitive to the drug in the presence of dimethyl sulphoxide, an agent known to modify cell-wall permeability (Glasser 1978).

Pyrazinamide This is the most active against *M. tuberculosis* of a number of nocinic acid derivatives (Kushner *et al.* 1952, Schwartz and Moyer 1953). Unlike all other antituberculous agents it is virtually inactive at a neutral pH but very effective at an acid pH (around 5.5) and against bacilli within macrophages (Mackaness 1956, Mitchison and Dickinson 1978). At this low pH, pyrazinamide is hydrolysed to pyrazinoic acid by a bacterial enzyme, pyrazinamide aminohydrolase (pyrazinamidase) (Konno *et al.* 1967). Pyrazinoic acid is the actively bactericidal agent, hence strains lacking the enzyme are resistant to this agent. Most bovine strains of *M. tuberculosis* differ from human strains in being resistant to pyrazinamide and also in failing to hydrolyse this substance (Konno *et al.* 1959). A pyrazinamide-sensitive variant of the bovine type (the Afro-Asian bovine type) has been described (Collins *et al.* 1981). The mode of action of pyrazinamide is unknown; its similarity to nicotinamide suggests that it affects nicotinic acid metabolism.

Other synthetic substances Ethionamide (2-ethylthioisonicotinamide) is cross-resistant with thiosemicarbazones but not with isoniazid although it probably acts in a similar way to the latter (Winder *et al.* 1971). Ethambutol (2,2′-ethylenedimino-di-1-butanol) is a chelating agent; it has been said to interfere with the rôle of metal cations in the synthesis of RNA (see Pratt 1977). The mode of action of thiacetazone (*p*-acetaminobenzaldehyde thiosemicarbazone) is unknown (see Protivinsky 1971, Doub 1979). (More information about the action of PAS, ethambutol and thiacetazone will be found in Chapter 5.)

Susceptibility to certain other chemicals that are not used therapeutically is of considerable importance for purposes of identification. Examples include sensitivity to *p*-nitrobenzoic acid (500 mg per l), which distinguished *M. tuberculosis* from all other mycobacteria except for a few strains of *M. kansasi* and *M. marinum* (Tsukamura and Tsukamura 1964, see p. 79). Sensitivity or resistance to thiophen-2-carboxylic acid hydrazide (TCH) at a concentration of 1 mg per l, distinguishes between human and bovine strains of *M. tuberculosis* (Bönicke 1958), and at 5 mg per l it divides the human strains into two types (see p. 79).

Cell walls and lipids

Actinomycetales are divisible into four groups according to their major cell wall constituents (see Gottlieb 1974); mycobacterial cell walls, in common with those of *Nocardia*, *Thermomonospora* and *Micropolyspora*, are of type IV and contain arabinose, galactose and meso-diaminopimelic acid but not glycine. An exception is *M. leprae*, the cell wall of which contains glycine (Draper 1976).

The mycobacterial cell wall is complex and has a particularly high lipid content (Acharya and Goldman 1970). It is about 20 nm thick and appears to be separated from the cell membrane by a periplasmic space, 3–10 nm across, but this may be an artefact (Imaeda *et al.* 1968*a*). The cell wall consists of three fairly distinct layers: (1) an inner layer of murein (peptidoglycan) which gives the cell wall its shape and rigidity; (2) an intermediate layer containing rope-like structures, up to 10 nm in diameter, composed of complex lipopolysaccharides set in a homogeneous matrix (Imaeda *et al.* 1968*a*, Barksdale and Kim 1977); and (3) an outer layer of ribbon-like fibrils which extend into the matrix of the intermediate layer and consist of peptidoglycolipids termed mycosides (Draper and Rees 1973).

Murein

This resembles in its general structure the peptidoglycans of other gram-positive bacteria. It consists of chains of *N*-acetylglucosamine and *N*-glycollylmuramic acid linked through the latter molecules by tetrapeptides (L-alanine, D-glutamate, diaminopimelate and D-alanine) which are cross-linked by peptide

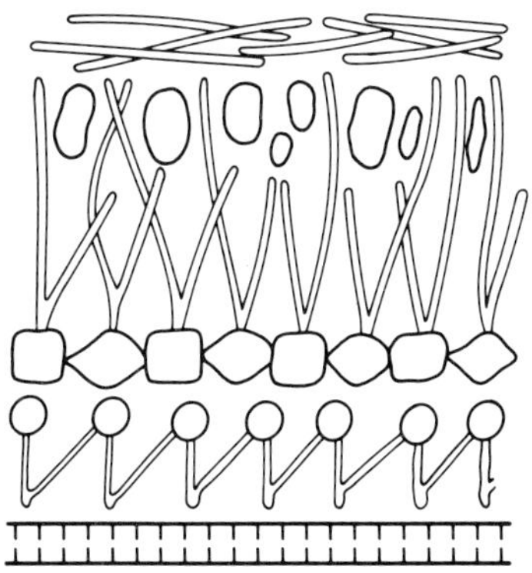

Fig. 24.7 Diagrammatic representation of the mycobacterial cell wall.

bonds between the diaminopimelate and D-alanine. (For further details on the structure of murein see Azuma *et al.* 1970, Lederer 1971, Vilkas *et al.* 1971.)

To this basic murein network is linked, through phosphodiester bonds attached to about one in ten of the *N*-glycollylmuramic acid molecules, a polysaccharide composed by arabinose and galactose (Kanetsuna and San Blas 1970, Lederer 1971). This arabinogalactan is a branching structure with mycolic acids (see below) attached by ester bonds to the terminal arabinose units (Kanetsuna *et al.* 1969).

Wax D is a complex molecule extractable by chloroform from old cultures of mycobacteria. It consists of a mycolic acid linked through a small arabinogalactan to a short *N*-acetylglucosamine-*N*-glycollylmuramic acid chain (about four units) to which are attached two tetrapeptide chains, one of which lacks a terminal alanine (Migliore and Jolles 1968, 1969, Markovitz *et al.* 1971). Units of wax D may arise by autolysis or may be building blocks from which the cell wall polymer is constructed (Goren 1972).

The peptidoglycans of mycobacteria contribute to the adjuvant properties of the cell wall (Azuma *et al.* 1971, Adam *et al.* 1972). Water-soluble compounds with adjuvant properties are extracted from cell walls by lysozyme treatment (Adam *et al.* 1972). These, and other water-soluble adjuvants, have been analysed, and the simplest one has been synthesized (Ellouz *et al.* 1974, Adam *et al.* 1975). This substance *N*-acetyl-muramyl-L-alanyl-D-isoglutamine (muramyl dipeptide, MDP) is the minimal active portion of peptidoglycan. It has potent adjuvant properties but neither sensitizes the animal to tuberculin nor induces an autoaggressive reaction (Audibert *et al.* 1976). It also activates macrophages and renders them able to destroy tumour cells (Juy and Chedid 1975).

The remainder of the cell wall consists of lipids, lipopolysaccharides and peptidoglycolipids surrounded by a matrix that contains polysaccharides including glucan (Stacey 1955, Draper 1971), polypeptides (Vilkas and Markovitz 1972) and proteins (Augier *et al.* 1971).

Lipids

These make up at least 30 per cent of the dry weight of the cell wall (Acharya and Goldman 1970) and are responsible for many of the biological properties of the mycobacteria. Their composition and metabolism have been reviewed by Goren (1972), Lederer (1971), Barksdale and Kim (1977) and Ratledge (1977).

The lipid content of mycobacteria varies from species to species and even in a single strain under different conditions of growth. With nitrogen limitation, lipids accumulate within the cells and probably serve as energy sources (Antonie and Tepper 1969). The total lipid content of the cell is not significantly affected by the temperature of growth, but the amounts of individual lipids vary considerably (Taneja *et al.* 1979). In *M. smegmatis* the total phospholipid content is 45 per cent higher in cells grown at 27° than at 37°, suggesting that this variation maintains a constant cell membrane fluidity (see below). The fatty-acid composition varies with the age of the culture, although the rate of synthesis of mycolic acids parallels that of mycobacterial growth. The synthesis of other fatty acids declines as the culture ages (Bennet and Asselineau 1970).

There are many different lipids in the mycobacteria, the principal ones being mycolic acids, glycolipids, mycosides and phospholipids (Goren 1972).

Mycolic acids These are β-hydroxy acids substituted at the α position with a moderately long aliphatic chain (see Goren 1972, Barksdale and Kim 1977). The acids are formed either by step-by-step elongation of the carbon chains or by the 'head-to-tail' fusion of shorter preformed chains (Claisen condensations).

The mycolic acids vary in structure from genus to genus particularly in respect of the lengths of the main and side chains; these differences are of taxonomic significance (Yano *et al.* 1978). The total number of carbon atoms in mycobacterial, nocardial and corynebacterial mycolic acids are respectively 70–90, 46–58 and 30–36 (Alshamaony *et al.* 1976*a,b*, Lechevalier *et al.* 1973). The side chains in the mycobacterial mycolic acids contain from 22 to 26 carbon atoms (see also Etémadi 1967 and Goren 1972). The mycobacterial mycolic acids also differ from their analogues in the other genera in containing additional oxygen atoms (Minnikin *et al.* 1975) and in being insoluble in ethanol-diethyl ether 1:1 (Kanetsuna and Bartoli 1972). The mycolic acids of *M. leprae* are identical in structure to those of other mycobacterial species (Etémadi and Convit 1974), an important indication that this species is a member of the genus *Mycobacterium*.

Mycosides These are superficially situated peptidoglycolipids consisting of fatty acids linked to an oligopeptide to which various sugar groups are attached; they are of importance in relation to colonial morphology, agglutination serotype and phage type, and possibly to virulence (Lanéelle and Asselineau 1968, Goren 1972, Brennan and Goren 1979). The mycosides of a single mycobacterial species were originally considered to be uniform in structure (Smith *et al.* 1960) but many small intraspecific differences are now known to occur.

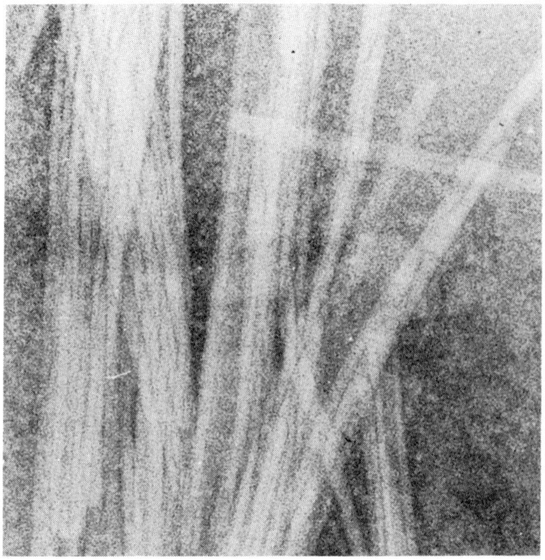

Fig. 24.8 Electron micrograph of mycoside fibres from a strain of *Mycobacterium fortuitum* (× 120 000).

The mycosides appear on electronmicroscopy as long ribbon-like fibrils wrapped around the bacilli (Kim *et al.* 1976, Barksdale and Kim 1977, Draper 1974). Such fibrillar structures composed of mycosides also surround cells of *M. lepraemurium* within vacuoles in macrophages (Draper and Rees 1973). As this organism is able to survive within the phagolysosomes of macrophages (Hart *et al.* 1972), a protective role for the mycoside integument seems probable. (See Fig. 24.8)

Phospholipids These occur in the cell wall and the cell membrane and include cardiolipin, phosphatidyl ethanolamine (PE) and phosphatidylinositol mannosides (PIM). Cardiolipin occurs principally in the cell membrane and has a high turnover rate (Dhariwal *et al.* 1978). As mentioned above, the phospholipid content in *M. smegmatis* varies according to the temperature of incubation: a low temperature increases the content of the total phospholipid and PE, but the content of PIM decreases while that of cardiolipin is unaffected (Dhariwal *et al.* 1977, Taneja *et al.* 1979). The structures of these lipids have been reviewed by Lederer (1967), Pangborn (1968), and Goren (1972).

Phospholipids have been implicated in granuloma formation but, as Goren (1972) points out, such tissue reactions occur only when large amounts are injected into sensitized animals. Purified phospholipids are haptens. Mycobacterial cardiolipin reacts similarly to beef heart cardiolipin in the Wassermann reaction (Akamatsu and Nojima 1965) and PIM has been used in some agglutination tests for the diagnosis of tuberculosis (Takahashi 1962, Tanaka *et al.* 1967).

Phospholipids are strongly anionic and may play a role in the inactivation of lysosomal hydrolases by interacting with cationic sites in these proteins (Goren *et al.* 1974*a*).

Sulpholipids and cord factor These glycolipids are somewhat similar in structure and consist of fatty acids linked to trehalose. Both have been considered to be associated with virulence in *M. tuberculosis*. Middlebrook and his colleagues (1959) observed that virulent strains bound neutral red to their surfaces and isolated a sulphated glycolipid responsible for this. Several different sulpholipids are synthesized by *M. tuberculosis*, the most abundant being sulpholipid I (2,3,6,6′-tetraacyl-α,α′-D-trehalose-2-sulphate) (Goren 1970, Goren *et al.* 1971). In contrast to cord factor, the fatty acids in this compound are not mycolic acids.

Cord factor (p. 66) was so named on account of the assumption, now known to be false, that this substance was responsible for the ability of *M. tuberculosis* to grow in culture as 'serpentine cords'. It is 6,6′-dimycoloyl-α-D-trehalose (Noll *et al.* 1956) and thus consists of two mycolic acids linked to a molecule of trehalose. Cord factor is not *per se* responsible for virulence in *M. tuberculosis*; indeed, it is present in some non-pathogenic species of mycobacteria (Azuma *et al.* 1962).

In a study of strains of *M. tuberculosis* of Indian origin, Goren and his colleagues (1974*a,b*) found a significant but incomplete correlation between virulence in the guinea-pig and a high content of sulpholipids. Strains of low virulence were characterized by the presence of a unique lipid termed the 'attenuation indicator lipid'. Subsequently it was found that the strains of low virulence differed from the virulent strains in many other respects, including phage type, susceptibility to H_2O_2 and resistance to thiophene-2-carbonic acid hydrazide (Grange *et al.* 1978). These two types of strain may therefore represent divergent variants of *M. tuberculosis* and the differences in lipid content may be coincidental. A few virulent strains of phage type B differed from the equally virulent type A strains in possessing a low content of sulpholipids.

It therefore seems unlikely that either sulpholipids or cord factor are major determinants of virulence in the guinea-pig, although they may be of greater relevance in man, particularly as a cause of toxicity in advanced disease. Cord factor is highly toxic for mice; it also causes swelling and loss of function in isolated mitochondria (Kato 1969). Sulpholipid I acts similarly on mitochondria but is non-toxic for mice because it is neutralized by proteins; however it enhances lethality when administered with cord factor (Kato and Goren 1974).

Antigenic structure

Mycobacterial antigens may be broadly classified as (1) soluble or particulate, (2) according to their chemical structure (polysaccharide, lipid or protein), (3) according to the nature of the immune response to them or (4) by their distribution within the genus.

Soluble antigens

The earliest preparations of mycobacterial antigens used experimentally were Koch's *tuberculins* which were filtrates of old cultures of tubercle bacilli concentrated by evaporation. Tuberculins have been widely used in studies on antigens but in recent years many workers have liberated antigens from whole cells by freeze-pressing or by ultrasonic disintegration. The first serious attempts to purify mycobacterial antigens were made by Florence Seibert, who described the well-known and widely used Purified Protein Derivative (PPD) (Seibert and Glenn 1941) for use in the tuberculin test (Chapter 51). This was prepared from old culture filtrates that had been concentrated and sterilized by heating to 100° and contained materials denatured by autolysis and heat. Precipitation of protein by 50 per cent ammonium sulphate, repeated several times, probably caused further denaturation.

In later studies, Seibert (1949) attempted to improve the specificity of the skin testing reagent by avoiding the denaturating effect of heat and using other means of fractionating the antigens. Sequential precipitation by acetic acid and ethanol at various concentrations yielded three protein and two polysaccharide fractions. Two of the protein fractions, A, and B, elicited delayed-type skin reactions (Vandiviere *et al.* 1961) but offered no advantage over PPD.

Immunoelectrophoretic analysis showed that these protein fractions contained a multiplicity of antigens, most of which were present in both preparations (Daniel and Affronti 1973). These studies illustrate the difficulty that many workers have encountered in the preparation of pure antigens, a difficulty probably due to the presence of many different antigenic determinants (epitopes), some species-specific and some shared, on the same moleucle (Chaparas 1979). Conversely one epitope may occur on a number of molecules with different physicochemical properties. Nassau and Nelstrop (1976), for example, found that an antigen specific for *M. tuberculosis* was present on four proteins of different electrophoretic mobility.

Attempts to use more advanced fractionation methods began with the studies of Kniker and LaBorde in 1964, in which antigens were separated by ion-exchange chromatography on diethylaminoethyl cellulose. The same separation system was used in many other studies (see Daniel and Janicki 1978) but in none were pure antigens obtainable. Likewise the use of carboxymethyl cellulose ion-exchange chromatography (Lind 1961*b*) proved unsuccessful. Gel filtration also enabled many fractions to be isolated but on serological analysis each fraction contained many antigens (Baer and Chaparas 1966). More effective separation was obtained by acrylamide gel electrophoresis (Wright *et al.* 1972) and small amounts of an antigen specific for *M. tuberculosis* were isolated (Nassau and Nelstrop 1976).

Many investigators have combined the various methods (Daniel and Janicki 1978) and have separated some individual antigens. In particular, Yoneda and Fukui (1965) isolated two antigens α and β; the β antigen was specific for *M. tuberculosis* and the α antigen occurred in slowly growing but not in rapidly growing species. Single antigens have also been isolated by isoelectric focusing (Moulton *et al.* 1971) and by affinity chromatography with either concanavalin A (Daniel 1974) or specific antisera (Daniel and Anderson 1977*a,b*) as adsorbents.

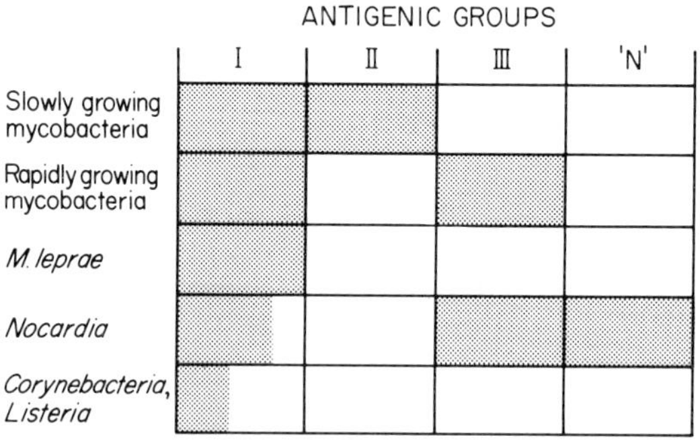

Fig. 24.9 Antigenic sharing between mycobacteria and related genera (species-specific, or group IV, antigens are not shown.)

Distribution: antigenic analysis

Antigenic analysis is now, with the possible exception of DNA hybridization, the most discriminating technique available to mycobacterial taxonomists. For this purpose diffusion-in-gel techniques have been widely used to analyse concentrated culture filtrates or disintegrated cells (Parlett and Youmans 1956, Lind 1959, 1960*a,b,c*, 1961*a*, Norlin 1965, Norlin and Navalkar 1966, Norlin *et al.* 1969). Stanford and his colleagues (reviewed by Stanford 1973*a*, Stanford and Grange 1974) demonstrated four groups of soluble antigens in ultrasonically disintegrated preparations of mycobacteria. Group i antigens, five or six in number, are common to all mycobacteria; group ii antigens, three in number, are found in slowly growing species; and group iii antigens, three or four in number, occur in rapidly growing species. Each species contains from two to eight antigens of its own—the group iv antigens. These findings suggest that mycobacteria evolved from a common ancestral form and that there was an early division of the genus into the slow and rapid growers. Three species, *M. leprae*, *M. vaccae* and *M. nonchromogenicum*, possess neither the group ii nor the group iii antigens.

Members of many species show some differences in their specific antigens that usually correspond with biochemical and cultural differences. Thus in *M. avium*, the *avium*, *intracellulare* and *lepraemurium* variants, and in *M. chelonei*, the *abscessus* and *chelonei* variants, show differences in their species-specific antigens. Seven immunodiffusion serotypes of *M. fortuitum* have been described, all of which show other phenetic differences (Grange and Stanford 1974). On the other hand, strains termed *M. gastri* and *M. kansasi* are identical by immunodiffusion (Norlin *et al.* 1969) yet differ enough phenetically for many investigators to regard them as separate species.

Immunodiffusion analysis of ultrasonic preparations has also demonstrated antigenic relations between *Mycobacterium*, *Corynebacterium*, *Nocardia* and *Rhodochrous*. Stanford and Wong (1974) showed that nocardiae shared the group iii antigens and most of the group i antigens with mycobacteria but possessed, in addition to species-specific antigens, a group of three 'N' antigens. At least three antigens common to the genera *Mycobacterium*, *Nocardia* and *Rhodochrous* (Ridell 1974, 1975) are shared to a variable extent with corynebacteria (Ridell 1977). Two of these intergeneric antigens resided in core particles from the 30S subunits of ribosomes; and antibodies to these particles were detected in sera from leprosy patients (Ridell *et al.* 1976, Ridell 1977). Wilhelm and Sellier (1976) reported that a ribonucleoprotein from *M. tuberculosis* reacted with sera from patients with lepromatous leprosy.

The immunodiffusion studies discussed above and the immunoelectrophoretic analyses of Janicki (see below) reveal from 11 to 15 antigens in each mycobacterium. The more discriminative technique of crossed or two-dimensional immunoelectrophoresis has revealed 60 precipitation lines for *M. tuberculosis* (Roberts *et al.* 1972) and over 40 antigens for other mycobacteria (Closs *et al.* 1975, Thorel 1976, Chaparas 1979). They do not, however, appear to have contributed any more to the classification of mycobacteria than the simpler immunodiffusion techniques.

Janicki and his colleagues (1971) allotted Arabic numerals to 11 antigens in culture filtrates of *M. tuberculosis* strain H37Rv according to their relative positions on electrophoretic analysis, but on this basis it has proved very difficult to ascertain the distribution of these antigens in other species (Chaparas 1975). Antigens 1,2,6,7 and 8 were found to be widely distributed among mycobacteria (Daniel and Janicki 1978) and thus almost certainly correspond to Stanford's group i antigens. Three of the antigens are identified as well-characterized polysaccharides. Antigen 1 is arabinomannan, antigen 2 is arabinogalactan (Daniel and Misaki 1976) and antigen 3 is a glucan (Janicki *et al.* 1971). Daniel and Affronti (1973) showed that Seibert's (1949) polysaccharide fraction I contained antigens 1 and 2 and polysaccharide fraction II contained antigen 3. Antigen 5, which is specific for *M. tuberculosis*, is a protein with a molecular weight of between 28 500 and 35 000 (Daniel and Anderson 1977*b*).

Insoluble (agglutination) antigens

Smooth suspensions are necessary for agglutination tests. Strains of *M. tuberculosis* can be studied by agglutination only with difficulty; nevertheless Wilson (1925) and Griffith (1925) who used the technique, found that human and bovine strains of *M. tubercul-*

osis were serologically indistinguishable but could be differentiated from *M. avium*.

The agglutination of *M. avium* was extensively investigated by Schaefer (1965, 1967). Originally many of Schaefer's serotypes were referred to by the names of patients from whom strains had been isolated but subsequently a numbering scheme was introduced (Wolinsky and Schaefer 1973). In this scheme *M. avium* var. *avium* is of serotype 1,2 or 3, *M. avium* var. *intracellulare* serotypes are numbered from 4 onwards and *M. scrofulaceum* serotypes are numbered from 40 onwards. In all, 31 serotypes are recognized in the so-called MAIS (*M. avium-M. intracellulare-M. scrofulaceum*) complex (Matthews *et al.* 1979). Non-agglutinable smooth strains in this group are also encountered (Reznikov and Dawson 1973). Juhlin and Winbland (1973) performed serotyping by co-agglutination tests with suspensions of protein-A-bearing *Staph. aureus* coated with appropriate antibodies.

The Schaefer serotypes agree closely with the thin-layer chromatographic patterns of glycolipids extracted from strains of *M. avium* (Marks *et al.* 1971, Sehrt *et al.* 1976), *M. scrofulaceum* (Jenkins *et al.* 1972) and *M. fortuitum* (Pattyn *et al.* 1974). Schaefer's antigens and the lipids demonstrable by thin-layer chromatography are the same compounds, namely mycosides (Brennan *et al.* 1978, Brennan and Goren 1979). Antigenic differences lie in the carbohydrates of the mycosides.

Agglutination serotypes have also been described in *M. flavescens*, *M. gordonae*, *M. marinum*, *M. xenopi* (Jenkins *et al.* 1972, Goslee *et al.* 1976), *M. kansasi* (Schröder and Magnusson 1968), *M. szulgai* (Schaefer *et al.* 1973) and rapidly growing species (Pattyn, 1970). The genetic basis of variation in the mycoside structure within species is unknown. A phage-induced antigenic modification was observed in *M. smegmatis* by Jones and Beam (1969) but its molecular basis was not determined.

Genetic characters

The genome

Mycobacterial chromosomes were found to have molecular weights of 2.5×10^9-4.9×10^9 by Bradley (1972) and 3.01×10^9-5.55×10^9 by Baess and Mansa (1978). The smallest genomes were found in *M. tuberculosis*. The G + C content of the DNA is high: 66.1-71.4 moles per cent (Baess and Mansa 1978). The modified bases 6-methylaminopurine and 5-methylcytosine have been detected (Johnson and Coghill 1925, Dunn and Smith 1958).

Little is known about the dynamics of mycobacterial DNA replication. Repair by a DNA polymerase similar to polymerase I of *Esch. coli* (MacNaughton and Winder 1977), other DNA polymerases (Campbell, Carty and Winder 1979) and ATP-dependent deoxyribonuclease (Winder and Levin 1971) have been identified. Mechanisms for DNA repair and genetic recombination are present in mycobacteria. Thus, in *M. smegmatis* recombination-deficient strains that are very sensitive to UV light (Mizuguchi 1974) and, conversely, mutants that are highly resistant to UV light and competent in DNA-mediated transformation (Norgard and Imaeda 1978) have been described. In the latter, treatment with low doses of UV light increased competence for transformation, indicating the presence of an inducible DNA repair system analogous to the 'SOS' repair system in *Esch. coli*.

The evidence for the existence of plasmids in mycobacteria is scanty. Jones and David (1972) suggested that the gene responsible for streptomycin resistance in a strain of *M. smegmatis* was located on a plasmid but no further evidence of plasmid-borne resistance determinants has been forthcoming. Extrachromosomal closed circular units of DNA that may be plasmids have been detected in lysates of some strains of *M. avium* (Crawford and Bates 1979).

Mutation

Mutations occur naturally at a low frequency and are inducible at a much higher frequency by ultraviolet light and chemical mutagens (Koníčková-Radochová *et al.* 1970) and by γ-radiation (Kolman and Ehrenberg 1978). Aminoglycoside antibiotics may be mutagenic when present in subinhibitory concentrations (Tsukamura 1979). The variation inducible in mycobacteria by the use of mutagens differs considerably from species to species, being much greater for example in *M. vaccae* than in *M. smegmatis* (Tarnok and Tarnok 1978).

Koníčková-Radochová and her colleagues (1970) found that nitrosoguanidine was one of the most effective agents for the induction of mutations. It acts principally on the DNA at the site of its replication and was used to determine the temporal sequence of replication of genes on the chromosome of *M. phlei* (Koníček and Koníčková-Radochová 1978).

The isolation of mycobacterial mutants generally presents no difficulties but if auxotrophic mutants are required bacterial clumping must be avoided or the mutants will be obscured and overgrown by prototrophic cells. Auxotrophic mutants may be concentrated by transferring the bacteria, after treatment with a mutagen, to a minimal medium containing an antibiotic that selectively kills the prototrophic cells which alone are able to grow in the medium (Holland and Ratledge 1971). Auxotrophic mutants of photochromogenic mycobacteria have also been detected by their failure to produce pigment on exposure to light after transfer on a cellophane membrane to a minimal medium (Vajda 1980).

Induced mutants have been used in genetic studies and also to investigate the biosynthesis of pigments (Levy-Frebault and David 1979). From the clinical point of view the most important mutational changes are those affecting virulence and susceptibility to antibacterial agents.

Genetic transfer

Genes have been transferred between mycobacteria only with difficulty, and only in a few rapidly growing strains.

Transduction Hubáček (1960) transferred streptomycin resistance from resistant to sensitive strains of *M. phlei* by phage-mediated transduction. Gelbart and Juhasz (1970) transferred the ability to use xylose (xyl^+) by means of a phage propagated on a xyl^+ strain. However, the addition of DNAase to the reaction mixture showed that about 20 per cent of the genetic transfer was by transformation (Gelbart and Juhasz 1973). Attempts to transduce other properties failed, suggesting that the phage mediated only restricted transduction. A generalized transducing phage for *M. smegmatis* was isolated from soil by Sundar Raj and Ramakrishnan (1970), and has been used to transduce a number of genes determining amino acid synthesis and also genes responsible for resistance to streptomycin and both resistance and sensitivity to isoniazid (Saroja and Gopinathan 1973).

Transformation Several early attempts to transfer genes by transformation failed (see Grange 1975*a*, Šlosárek 1978). In 1954, Katsunuma and Nakasato succeeded in transferring a streptomycin-resistance gene between strains of rapidly growing mycobacteria by this means, and this was later repeated with other strains (Tsukamura *et al.* 1960, Rytíř *et al* 1966). As we have seen, the ability to use xylose is transferred by transformation as well as by transduction in *M. phlei*.

Competence for transformation requires the ability both to take up 'foreign' DNA and to integrate it into the host genome. Uptake of free DNA is increased by pre-incubation in a medium containing serine or threonine (Tokunaga and Sellers 1964), and a low pH (Nakamura 1970). Calcium ions increase the uptake of free phage DNA (Tokunaga and Sellers 1964) and the binding of isotope-labelled DNA to the cells (Tarnok and Bönicke 1970).

Conjugation Redmond (1970*a*) reported that a smooth, non-chromogenic auxotrophic unidentified mycobacterium and a rough, prototrophic pigmented strain yielded smooth pigmented strains when plated together on minimal medium. This suggested that genes determining smooth colony morphology were transferred from the auxotrophic to the prototrophic strain. Mizuguchi and Tokunaga (1971) obtained prototrophic recombinants by mixing auxotrophic mutants of two different strains of *M. smegmatis*. Only certain strains of *M. smegmatis* were competent for conjugation. It was necessary to plate the organisms together on solid medium (Tokunaga *et al.* 1973); apparently the bacterial cells require prolonged contact before conjugation occurs. Mizuguchi (1974) isolated recombination-deficient mutants of these strains and used them to demonstrate the unidirectional nature of the genetic exchange. Mating compatibility was found to be controlled by a fairly complex genetic system situated on the chromosome. Nineteen strains of *M. smegmatis* could be placed into five compatibility groups (Mizuguchi *et al.* 1976). Attempts to map the chromosome of *M. smegmatis* have not been entirely successful owing to ambiguities in the observed sequence of genes; this may be attributable to limited genetic homologies between the recombining genomes (Suga and Mizuguchi 1974).

A search for recombining pairs of strains in *M. phlei* and *M. tuberculosis* (Koníček and Koníčková-Radochová 1975), and attempts to induce competence for conjugation by transferring F factors from *Esch. coli* to *M. phlei* (Šlosárek *et al.* 1978), were unsuccessful. The structural and biochemical basis of conjugation is unknown but a relation between mating type, phage type and colonial form has been noted (Mizuguchi *et al.* 1976), suggesting that a surface component, possibly a mycoside, plays a role.

Bacteriophages

In 1947 Gardner and Weiser isolated a bacteriophage capable of lysing a strain of *M. smegmatis*. During the following 15 years many phages were isolated from natural sources, such as soil and water, in the unfulfilled hope that these would assist in the classification of mycobacteria (Redmond 1963). The usual method employed was to mix the material with strains of mycobacteria, usually *M. smegmatis*, in order to allow any phages to which the strains were sensitive to replicate (Froman *et al.* 1954, Grant 1971). Phages have also been isolated from lysogenic strains by the cross-culture method (Juhasz and Bönicke 1965, Buraczewska *et al.* 1971, Grange and Bird 1975). Other lysogenic mycobacteria have been detected by the occurrence of properties induced in the host bacteria by the phage: production of deoxyribonuclease (Mankiewicz and Tamari 1972) and the formation of mucoid colonies (Grange and Bird 1978*a*).

Many mycobacteria infected with temperate phages are not truly lysogenic. Instead of being integrated into the host chromosome, the phage genome appears to exist independently within the bacterium, like a plasmid. As a consequence, the phage is not transmitted to all daughter cells, the culture remains susceptible to the phage and plaques often appear spontaneously on bacterial lawns. Such 'pseudolysogeny' has been described in *M. fortuitum* (Baess 1971, Grange and Bird 1975), *M. chelonei* (Jones and Greenberg 1977) and *M. diernhoferi* (Grange 1975*c*).

Lysogeny has been detected in many species of mycobacteria, including *M. fortuitum*, *M. chelonei*, *M. smegmatis*, *M. phlei* and *M. diernhoferi* among the rapid growers and *M. kansasi*, *M. marinum* and *M. scrofulaceum* among the slow growers (see Grange and Redmond 1978). Lysogeny occurs in 5 to 10 per cent of isolates of *M. fortuitum*, *M. chelonei* and *M. kansasi* (Grange and Bird 1978*b*); its frequency in other species is unknown. Some mycobacteria contain DNA, possibly derived from phages, that is able to recombine with infecting phages to give rise to phages with modified properties (Mankiewicz and Redmond 1968, Juhasz and Bönicke

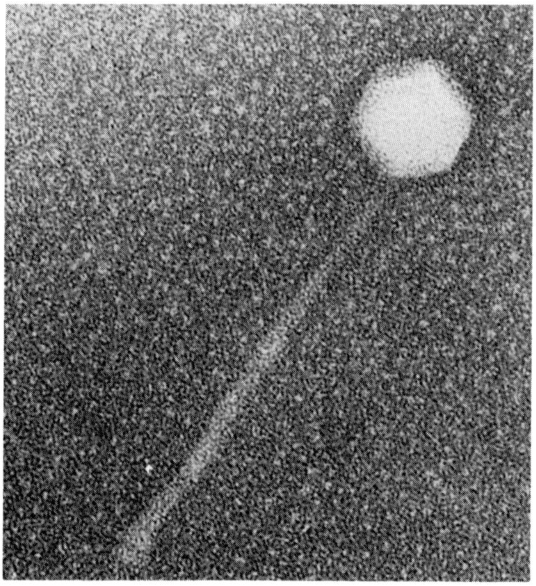

Fig. 24.10 Electron micrograph of a mycobacteriophage from a lysogenic strain of *Mycobacterium fortuitum* (× 180 000).

1970, Grange and Bird 1975). Lysogeny was first established experimentally in the mycobacteria by Russell, Jann and Froman (1960).

Bacteriophages of the mycobacteria, usually termed mycobacteriophages, are usually of type B in the classification of Bradley (1967). They have long non-contractile tails, and the heads are hexagonal, or less frequently cylindrical in shape. Only one mycobacteriophage with a contractile tail has been found (Kozloff *et al.* 1972). Illustrations of mycobacteriophages are given by Kolbel (1967), Buraczewska and her colleagues (1972) and Barksdale and Kim (1977). Mycosides serve as receptor sites for mycobacteriohages (Goren *et al.* 1972, Furuchi and Tokunaga 1972). (For a review of the biological properties of mycobacteriophages see Grange and Redmond 1978).

Phage conversion The establishment of lysogeny in mycobacteria modifies the activity of many enzymes (Juhasz *et al.* 1969, Bönicke and Saito 1970), that of some being increased and others decreased, suggesting an interference with the mechanisms for the regulation of enzymic activity. Lysogeny may modify the antigenic structure of the cell wall and so alter the agglutination serotype; the nature of this change varies from phage to phage (Jones and Beam 1969).

Other phage-induced properties are probably related to the mechanisms of phage replication and release; thus many lysogenic strains liberate lipases (Jones and David 1970) and DNAase (Mankiewicz and Tamari 1972). Many lysogenic strains have smoother colonies than non-lysogenic strains, and they are of a mucoid consistency (Grange and Bird 1978*a*). This appears to be due, at least in part, to DNA liberated from lysing cells (Grange and Bird 1978*b*). There is no definite evidence that lysogeny enhances virulence.

Phage typing

Many phages lyse, with equal ease, strains of several different species; they are therefore unsuitable for taxonomic studies but permit the subdivision of recognized species. Such a subdivision has been described in *M. tuberculosis* (see below), *M. avium* (Froman and Scammon 1964), *M. kansasi* (Engel *et al.* 1980*a*), *M. xenopi* (Gunnals and Bates 1972), *M. ulcerans* (Grant 1973) and *M. fortuitum* (Nordström and Grange 1974). The typing systems for species other than *M. tuberculosis* have found very limited application. Geographical variations in the distribution of *M. kansasi* strains in the natural environment have been demonstrated by phage typing but no type was particularly related to disease in man (Engel *et al.* 1980*a*).

Phage typing of tubercle bacilli The use of phage typing in epidemiological studies of tuberculosis was first attempted by Baess (1966), who divided strains of *M. tuberculosis* into two groups by means of phage BK1. The subsequent success of the method was largely due to the strict standardization of phage suspensions as advocated by Redmond and Ward (1966). (For technical details see Redmond and Ward 1966, Rado *et al.* 1975, Grange *et al.* 1976, 1977). No lysogenic strains of *M. tuberculosis* have been found, thus all phages used for the typing of this species were isolated from other mycobacteria or from the natural environment. Some are propagated on *M. tuberculosis* itself while others are propagated on other mycobacteria, principally *M. smegmatis* no. ATCC607.

In 1967 Bates and Fitzhugh described three phage types of *M. tuberculosis* designated A, B and C; two years later a further type, intermediate in its phage susceptibility between types A and B and designated type I (intermediate), was added (Bates and Mitchison 1969). With good techniques types A, I and B may be determined with great accuracy. A further subdivision of types A and I is obtained by the use of a group of phages isolated from sarcoid tissue by Mankiewicz (1972) but with a lesser degree of accuracy (Rado *et al.* 1975, Grange *et al.* 1977).

In early studies, phage typing was used to confirm the clinically suspected transmission of disease within small groups of patients (Baess 1966, 1969). Variations in the distribution of the phage types within countries (Ionesco 1972, Mankiewicz 1972), between countries (Bates and Mitchison 1969), and between indigenous and immigrant populations in the same region (Mankiewicz 1972, Grange *et al.* 1976, 1977), have been found. Phage type A is the commonest type and has a worldwide distribution; type B occurs in Europe and North America and type I in India and nearby countries. Strains of type C are rarely encountered (Grange and Redmond 1978).

Phage typing has also been used to demonstrate that infection by more than one strain of *M. tuberculosis* is common in regions where tuberculosis is rife but not in areas where its incidence is low (Mankiewicz and Liivak 1975). Thus 33 of 233 Eskimo patients were found to be infected by more than one phage type but multiple infection was not detected in 150 non-Eskimo patients from a region with a considerably lower incidence of disease. Raleigh and Wichelhausen (1973) and Raleigh and his co-workers (1975) used typing to distinguish between reactivation and reinfection in patients who relapsed after chemotherapy. In a series of 26 patients with recurrence of disease, nine were found to have strains differing in phage type from the original isolate (Raleigh *et al.* 1975).

Phage 33D isolated from a lysogenic environmental mycobacterium by Buraczewska and her colleagues (1970) separates BCG strains from other strains of *M. tuberculosis*. Strains of BCG are resistant to this phage whereas virtually all other variants of *M. tuberculosis* are sensitive (Yates *et al.* 1978).

Bacteriocines

Redmond (1970*b*) observed that *M. tuberculosis* and *M. kansasi* were mutually antagonistic when plated together on solid medium. Takeya and Tokiwa (1972) studied rapidly growing mycobacteria by a method similar to that used for the pyocine typing of *Ps. aeruginosa* and detected species-specific patterns of 'mycobacteriocine' activity. The agents responsible inhibited the growth of a wide range of bacterial species other than those from which they were derived. Takeya and Tokiwa (1974) reported that *M. tuberculosis* was divisible into 11 types by means of bacteriocines from rapidly growing mycobacteria. Growth-inhibitory substances from mycobacteria have also been reported by Adámek and his colleagues (1968) and by Imaeda and Rieber (1968).

Mycobacteriocines have been isolated from strains of *M. smegmatis* by gel-filtration and ion-exchange chromatography. The purified substances were found to be proteins with molecular weights of 75 000–85 000 and were active on a wide range of rapidly growing mycobacterial species but not on bacteria from other genera, including *Nocardia* and *Corynebacterium* (Takeya *et al.* 1978, Saito *et al.* 1979).

Experimental infection of animals

Mycobacterium tuberculosis

Differences in pathogenicity for various experimental animals were at one time widely used to distinguish between the various types of tubercle bacilli, and the injection of milk and various sorts of pathological material into guinea-pigs was a routine procedure for the detection of human and bovine bacilli. These tests are now seldom used; alternative cultural methods are available for both purposes (Chapter 51), and animal experiments expose the laboratory worker to a considerable hazard. Detailed information about the pathogenicity for animals when given experimentally by various routes will be found in the 6th Edition of this book; here we shall give only a brief summary of this.

Human and *bovine tubercle bacilli* cause progressive and ultimately fatal disease, with the formation of tubercles or areas of caseation in the viscera, after subcutaneous or intramuscular injection in the guinea-pig, the golden hamster, a few species of birds (notably parrots and cockatoos), and in monkeys. Bovine bacilli, but not human bacilli, cause a similar progressive disease also in rabbits, voles, cattle, goats, pigs and cats; dogs show little susceptibility to either. Rats and mice show an intermediate degree of susceptibility to both types of bacillus; after intravenous injection they may develop a more chronic form of the disease—the *Yersin type*—in which small necrotic foci packed with acid-fast bacilli appear in the spleen and liver, but macroscopic tubercles or areas of caseation are absent. The *vole bacillus* causes typical progressive disease only in voles, but it may give rise to the Yersin-type disease in rats and mice.

The *guinea-pig* is highly susceptible to fully virulent strains of human and bovine tubercle bacilli. As few as one bacillus may give rise to a lesion, but this is seldom progressive. With 10–100 bacilli given subcutaneously or intramuscularly, infection occurs regularly; a local caseous lesion develops, followed by spread to local and regional lymph glands, which also become caseous, and then by generalized infection and death in 6–15 weeks. The virulence of *M. tuberculosis* in the guinea-pig may be expressed as follows (Mitchison *et al.* 1960). The animals are inoculated intramuscularly with 1 mg moist weight of organism and killed after 6 or 12 weeks. The extent of disease is then scored according to its extent: liver, 0–40; spleen, 0–30; lungs, 0–20; inoculation site and draining lymph nodes, 0–10; giving a possible total of 100. This value, divided by the survival time in days (i.e. 42 or 84 days), gives the root index of virulence (RIV).

The virulence of strains of human and bovine tubercle bacilli may become attenuated. As we shall see (p. 80), the Asian type, which closely resembles the human type, is regularly less virulent for the guinea-pig than other tubercle bacilli that infect man. The RIV of the former is almost always less than 1; that of the latter is greater than 1 (see Grange *et al.* 1978). Isoniazid-resistant strains (p. 70) are also relatively avirulent. Under prolonged culture in the laboratory on certain media, strains may gradually lose their virulence, as for instance the BCG strain (Chapter 51) which was originally a virulent bovine strain.

Mycobacterium avium

Subcutaneous injection of the avian bacillus into domestic fowls, pigeons and other birds leads to progressive tuberculous disease in which death occurs after a variable period of emaciation. Numerous tubercles and areas of caseation appear in liver, spleen and kidneys. Mammalian species are in general less susceptible than birds. Cattle, sheep, pigs, rabbits and guinea-pigs infected subcutaneously develop a local lesion; the organism is disseminated and persists in the tissues for a long time, but death is unusual. However, young pigs fed avian bacilli may develop a generalized disease and eventually die. The intracellulare variant of *M. avium* is pathogenic for all these mammals but not for birds.

Mycobacterium paratuberculosis

Johne's disease can be reproduced in cattle and sheep by injecting the organism by various routes and by feeding, but large doses of bacilli are necessary, the disease takes 1–3 years to develop, not all of the animals become ill, and only young animals are affected. Young mice given the organism intravenously may develop lesions resembling Johne's disease after a latent period of many months. A mycobactin-dependent variant of *M. avium*, isolated from wood-pigeons also causes hypertrophic enteritis in calves (p. 82).

Mycobacterium leprae

The leprosy bacillus multiplies slowly in the footpads of mice (Shepard 1960); in irradiated mice infected by this route it produces a leproma-like lesion (Rees *et al.* 1967). In the nine-banded armadillo (*Dasypus novemcintus*) a disease resembling human lepromatous leprosy devleops. After several months lepromas containing up to 2×10^{10} bacilli per g of tissue are present, and bacilli are detectable in the blood (Kirchheimer and Storrs 1971, 1972).

Mycobacterium lepraemurium See Chapter 52.

Other mycobacteria *M. kansasi* may give rise to visceral lesions in mice when given intravenously. *M. ulcerans* and *M. marinum* produce ulcerative lesions in the footpads of mice. Among the rapidly growing mycobacteria, *M. fortuitum* and *M. chelonei* are pathogenic for amphibia and reptiles and may also cause kidney abscesses when given intravenously to mice.

The species of mycobacteria

The slowly growing mycobacteria

Mycobacterium tuberculosis

This species, the causative agent of tuberculosis, is an obligate pathogen; to the best of our knowledge none of its several variants exists saprophytically outside the animal body.

The original variants of *M. tuberculosis* were termed the human and bovine types, these names reflecting the usual source of the strains. In 1970 the latter type was given the separate specific epithet *M. bovis* (Karlson and Lessel 1970). Other variants that have been given separate specific status are the vole tubercle bacillus of Wells (1937), which has been named *M. microti* (Reed 1957), and some strains isolated in Africa termed *M. africanum* (Castets *et al.* 1969). Almost identical antigenic structures, as shown by immunodiffusion tests (Stanford and Grange 1974), and DNA hybridization (Baess 1979), give strong indications that the human and bovine types belong to the same species. The other variants are intermediate in their phenetic properties between the human and bovine tubercle bacilli so they do not qualify for separate specific status. The variants of ths species should therefore be termed *M. tuberculosis* human type, bovine type, vole type and African type. A fifth major variant, the Asian type, is described below.

The *human type* of *M. tuberculosis* grows well on Löwenstein–Jensen medium (*eugonic growth*). On primary culture from clinical material colonies are usually visible after 2–5 weeks' incubation; only rarely is their appearance delayed beyond 5 weeks. The colonies are of an off-white or cream colour and are often of a heaped-up or 'breadcrumb' appearance. Too much reliance should not be placed on this colonial appearance; on very moist media the colonies may assume a remarkably smooth consistency. Like other variants of *M. tuberculosis*, the human type is a strict mesophile, showing little or no growth below 30° and above 39°. This temperature range of growth, together with the failure of *M. tuberculosis* to grow on media containing 500 mg of *p*-nitrobenzoic acid per l, enables this species to be distinguished from other slowly-growing non-chromogens.

Human strains reduce nitrate to nitrite, produce large amounts of niacin and are strongly aerobic. Unless resistant to isoniazid, they form large amounts of catalase which, unlike its analogues in most other species, is inactivated by heating to 68° for 30 min. They are usually sensitive to pyrazinamide and resistant to TCH (Table 24.2).

In contrast, tubercle bacilli of the *bovine type* show a poor or *dysgonic* growth on Löwenstein–Jensen medium, but their growth is improved if the medium contains pyruvic acid in place of glycerol, as in Stonebrink's medium. Bovine strains do not reduce nitrate to nitrite, do not produce niacin, are microaerophilic and are resistant to pyrazinamide but sensitive to TCH (Yates and Collins 1979). Strains of BCG, although originally derived from a bovine strain, differ from the latter in growing well in the presence of glycerol and in being very aerobic (Yates *et al.* 1978). (For their other characters, see Table 24.2). Marks (1976) recognized two types of bovine strain, the European and the Afro-Asian. The former was resistant to pyrazinamide and the latter was not. Yates and Collins (1979) regarded '*M. africanum*' and 'Afro-Asian bovine strains' as synonymous, but later the term 'African' was introduced to describe strains which differed from classical bovine strains in being sensitive to pyrazinamide (Collins *et al.* 1981, 1982).

The vole strain was isolated by Wells (1937) from trapped wild voles (*Microtus agrestis*), some of which had a tuberculosis-like disease caused by an acid-fast bacillus similar to the tubercle bacillus. Unlike other types of tubercle bacilli the vole bacillus did not cause progressive disease in guinea-pigs and a range of other animals. It was later found to be non-virulent in man and has, in fact, been used for the preparation of a live vaccine (see Chapter 51). It has been isolated from other species of voles, wood mice and shrews; a similar

Table 24.2 Properties of the major variants of *Mycobacterium tuberculosis*

Variant	Nitrate reductase	Niacin production	Oxygen preference	Pyrazinamide 60 mg/l	TCH* 5 mg/l	Phage type
Classical human	+	+	Aer	S	R	A, B, C
Asian human	+	+	Aer	S	S	I
Classical bovine	–	–	Mic	R	S	A
African†	V	–	Mic	S	S	A
Vole	V	+	Mic	S	S	?
BCG	–	–	Aer	R	S	A

* Thiophen-2-carboxylic acid hydrazide.
† These strains have been divided into African I (nitratase negative) and African II (nitratase positive) (Collins *et al.* 1982).
+ = Positive; – = negative; V = Strain-to-strain variation; S = sensitive; R = resistant; Aer = aerobic; Mic = microaerophilic.

organism has been isolated from the dassie or Cape hyrax (*Procavia capensis*) (Smith 1960). The vole bacillus was originally termed 'the murine type of tubercle bacillus' or *M. tuberculosis* var. *muris* and was renamed *M. muris* in 1948 and *M. microti* by Reed in 1957. It is antigenically almost identical with the human and bovine types and, in its cultural and biochemical properties, is intermediate between them (see Table 24.2), so separate specific status is not justifiable.

Castets and his colleagues (1969) gave the name *M. africanum* to a heterogeneous group of tubercle bacilli isolated in Africa. Like the vole type, these bacilli possess properties intermediate between the human and bovine types. Strains isolated in West Africa have several properties in common with the human type; those from Rwanda and Burundi are phenetically more closely related to the bovine type (David *et al.* 1978*c*). Strains recognizable as *M. africanum* are not confined to Africa; they have occasionally been isolated in Great Britain (McLeod 1977, Yates and Collins 1979).

The *Asian type* is essentially a variant of the human type; it is of low virulence in the guinea-pig and was originally described in South India and nearby areas (Dhayagude and Shah 1948, Frimodt-Möller *et al.* 1956, Bhatia *et al.* 1961), in Thailand (Bhatia *et al.* 1963) and among Asian expatriates in East Africa (Mitchison 1970). These strains are isoniazid sensitive and are highly susceptible to hydrogen peroxide (Nair *et al.* 1964). They contain small amounts of sulpholipids (p. 73), and a characteristic 'attenuation indicator lipid' (Goren *et al.* 1974*a,b*). They are usually sensitive to TCH (Grange *et al.* 1977) and belong to a distinct phage type (type I, Grange *et al.* 1978). (For further details of the variants of *M. tuberculosis* see Grange 1982.)

Mycobacterium kansasi*

This is one of the two photochromogenic species included in Runyon's group I; the other is *M. marinum*. Originally isolated from an infected human lung by Buhler and Pollack (1953), it was named *M. kansasi* by Hauduroy in 1955. It is one of the faster growing members of the slowly growing mycobacteria, usually yielding a good growth on Löwenstein-Jensen medium within 2 weeks on subculture. The bacterial cells are usually elongated and show a barred or beaded appearance (Nassau and Hamilton 1957). This species differs from *M. marinum* in growing well at 37°, usually reducing nitrate to nitrite, failing to hydrolyse pyrazinamide (Schröder and Magnusson 1968) and lacking α-L-fucosidase activity (Grange and McIntyre 1979). Bacilli are agglutinated by a specific antiserum (Schaefer 1966) and contain a specific phenol-soluble antigen detectable by immunodiffusion (Wayne 1971).

Strains of this species are usually photochromogenic, but scoto- or non-chromogenic strains are occasionally encountered (Schröder and Magnusson 1968). By means of nine phages, 14 phage types are distinguishable (Engel *et al.* 1980*a*). The species is divisible into two groups by their quantitative catalase activity; those with high catalase activity are more virulent in both guinea-pigs and man (Wayne 1962).

The organism has been isolated from environmental sources, notably water systems (McSwiggan and Collins 1974, Engel *et al.* 1980*b*). Strains have also been associated with human disease; the lung is the usual site but many other organs have also been affected (Watanakunakorn and Trott 1973, Wolinsky 1979).

Mycobacterium gastri In 1966 Wayne isolated an apparently new species from gastric washings. Similar strains were obtained from sputum (Kestle *et al.* 1967) and from soil (Wolinsky and Rynearson 1968) but never from cases of disease. Strains of *M. gastri* closely resemble those of *M. kansasi* in their colonial and microscopic appearance, in their

* The 'approved name' of this species is *M. kansasii* (see Table 24.1).

Table 24.3 Identification of slowly growing mycobacteria other than the tubercle bacillus.

	Growth at				Pigment in light	Pigment in dark	Arylsulphatase 3 days	Tellurite reduction 3 days	Nitrate reductase 24 hr	Urease 24 hr	α-L-fucosidase 24 hr
	25°	33°	37°	45°							
*M. kansasi**	+	+	+	–	+	–	–	–	+	+	–
M. marinum	+	+	(+)	–	+	–	–	–	–	+	+
M. scrofulaceum	+	+	+	–	+	+	–	–	–	+	–
M. gordonae	+	+	+	–	+	+	–	–	–	–	–
M. szulgai	+	+	+	–	+	+	–	–	+	+	+
M. avium (*avium*)	+	+	+	+	–	–	–	+	–	–	–
M. avium (*intracellulare*)	+	+	+	–	–	–	–	+	–	–	–
M. xenopi	–	+	+	+	–	–	+	–	–	–	–
M. ulcerans	–	+	–	–	–	–	–	–	–	–	–

* Or *kansasii* (see Table 24.1).

rate of growth, and in many biochemical properties, but are non-pigmented, nitratase negative, lack heat-stable catalase activity and hydrolyse propionamide (Schröder and Magnusson 1968, Wayne *et al.* 1978). They also differ from *M. kansasi* in being sensitive to rifamicin SV and in using either *n*- or *iso*-butanol, or both (Tsukamura 1973). Two photochromogenic strains with the biochemical and antigenic properties of *M. gastri* have been isolated (Anz and Schröder 1970). The taxonomic status of *M. gastri* is subject to controversy. Although *M. gastri* is almost identical with *M. kansasi* on immunodiffusion analysis (Norlin *et al.* 1969, Stanford 1973*a*), the two taxa differ in their agglutination serotype (Wayne 1966) and possess different phenol-soluble antigens (Wayne 1971). Further evidence for antigenic differences has been obtained by a comparison of the specificity of delayed-type skin reactions in sensitized guinea-pigs (Magnusson 1971). These antigenic differences do not necessarily justify the allocation of the strains to separate species. On the basis of numerical analysis the strains of *M. gastri* formed a distinct cluster separate from *M. kansasi* (Wayne *et al.* 1978) although only seven strains of *M. gastri* were studied.

Mycobacterium marinum Isolated originally from diseased fish by Aronson in 1926, this species was found to be identical to *M. balnei* (Linell and Norden 1954), the causative agent of swimming-pool granuloma (see Bojalil 1959). It resembles *M. kansasi* in its colonial and microscopic appearance, its rate of growth and its light-induced pigmentation. On chromatographic analysis (Tarnok and Tarnok 1970) the pigments of both species are identical and differ from those of all other mycobacterial species. As with *M. kansasi*, non-chromogenic variants are occasionally found.

Biochemically *M. marinum* is distinguishable from *M. kansasi* in growing poorly at 37°, being nitratase negative and hydrolysing pyrazinamide (Wayne *et al.* 1978). Strains of *M. marinum* have strong α-L-fucosidase activity, a property that enables it to be positively identified within 2 hr (Grange and McIntyre 1979; see p. 67).

Antigenically *M. marinum* is of a distinct agglutination serotype and possesses a species-specific phenol-soluble antigen (Wayne 1971) and at least three specific antigens are demonstrable by immunodiffusion analysis (Stanford and Grange 1974).

This species appears to be favoured by an aquatic environment and has been isolated from salt-water fish (Aronson 1926), swimming-pools and aquaria. In man the commonest lesions are skin granulomas resembling those of sporotrichosis on the hands of aquarium keepers and the knees and elbows of swimmers. More generalized infections are extremely rare.

Mycobacterium xenopi This species was isolated by Schwabacher (1959) from a skin lesion in a South African toad (*Xenopus laevis*). The cells are long and slender; filamentous forms, pseudomycelia and small aerial hyphae may be observed (Runyon 1968). It grows more slowly than *M. tuberculosis*. On primary isolation colonies are not usually visible until after 6–8 weeks' incubation. A light yellow pigment is produced, especially in older colonies, although the species is usually classified as a non-chromogen. Strains grow well at 42° and possess strong arylsulphatase activity but are otherwise biochemically rather unreactive (Tsukamura 1967). At least four species-specific antigens are detectable by immunodiffusion analysis (Stanford 1973*a*, Stanford and Grange 1974).

This species is somewhat limited in its distribution, most isolations reported in Great Britain are from south-east England (Marks and Schwabacher 1965). Strains have been isolated from taps, mostly hot-water taps, in two hospitals whose water tanks were contaminated by bird droppings but not in a third hospital with a protected water tank (Bullin *et al.* 1970). A similar contamination of the plumbing of a hospital has been reported from America (Gross *et al.* 1976). Strains have also been isolated from main water supplies (McSwiggan and Collins 1974). *M. xenopi* is occasionally responsible for pulmonary lesions, usually in elderly males, but many clinical isolates prove to be non-significant (Marks and Schwabacher 1965, Bullin *et al.* 1970). Rare cases of renal disease have been reported (Engbaek *et al.* 1967).

Mycobacterium ulcerans Originally isolated from ulcerating lesions of the human skin in Australia (MacCallum *et al.* 1948). Strains isolated from similar lesions in Uganda and named *M. buruli* (Clancey 1964, Dodge 1964) show only minor phenetic differences from *M. ulcerans* and give identical results in serological tests by immunodiffusion (Stanford 1973*b*). *M. ulcerans* grows on Löwenstein–Jensen medium but is one of the slowest growing of the cultivable mycobacteria. It has a very limited temperature range of growth, 31–34°. Biochemically it is very unreactive but is unique among the mycobacteria in degrading hydroxyproline (Grange 1976).

In man it causes the disease known as *Mycobacterium ulcerans* infection, Buruli ulcer, Songololo ulcer or Bairnsdale disease (Chapter 51).

Mycobacterium avium and related organisms

In this group are organisms that have been named *M. avium*, *M. intracellulare*, *M. lepraemurium* and *M. paratuberculosis*.

The name *M. avium* was applied by Chester (1901) to strains termed '*Tuberculose des oiseaux*' by Strauss and Gamaléia in 1891, which were subsequently referred to as avian tubercle bacilli. A phenetically similar organism, *Nocardia intracellularis*, described by Cuttino and McCabe in 1949, was found to be a mycobacterium by Runyon (1965) and renamed *M. intracellulare*. It is identical with the so-called Battey bacillus, which was named after the sanatorium from which it was first obtained. This organism differs from the *M. avium* strains in a few biochemical properties and in its usual inability to cause progressive disease in birds. There has been much debate as to whether *M. avium* and *M. intracellulare* are separate species or variants of a single species. In an international study (Meissner *et al.* 1974) no clear-cut distinction could be made between the two organisms and a majority of the 22 participants recommended that *M. intracellulare* should be reduced to a synonym of *M. avium*. The six dissenting participants recommended the use of the term '*M. avium-intracellulare* complex' until further studies resolved the question. On immunodiffusion analysis this 'complex' appears to be a single species but with minor characteristic antigenic differences between the two members (Stanford and Muser 1969). In contrast, DNA hybridization (Baess 1979) reveals two distinct clusters, one comprising *M. avium* and one of the serotypes of *M. intracellulare* and the other comprising the remaining serotypes. Some authors include the species *M. scrofulaceum* with these organisms in the MAIS (*Mycobacterium avium-intracellulare-scrofulaceum*) complex (Reznikov and Dawson 1973, Wolinsky and Schaefer 1973). Although there is some phenetic similarity between

M. scrofulaceum and the other two members, immunodiffusion analysis (Stanford and Grange 1974) and studies on DNA relatedness (Baess 1979) show that it is a quite distinct species. *Mycobacterium lepraemurium*, the causative agent of rat leprosy, has been shown by immunodiffusion analysis to be closely related to *M. avium* (Stanford 1973*c*) and *M. paratuberculosis*, the causative agent of Johne's disease or chronic hypertrophic enteritis of cattle, is similarly related.

***Mycobacterium avium* and the intracellulare variant** The bacterial cells are usually coccoid or coccobacillary, but under some growth conditions filamentous forms appear. The organisms grow on Löwenstein–Jensen medium and the other media that are used for the cultivation of tubercle bacilli. Colonies are usually smooth and domed, though rough colonies are occasionally encountered. Although usually unpigmented, a few strains are scotochromogenic. Little or no catalase is formed. Most strains deaminate nicotinamide and pyrazinamide but not urea. Tween 80 is not hydrolysed in 5 days. Both variants are negative in the nitrate reductase test at 24 hr and the arylsulphatase test after 3 days; the *intracellulare* variant is usually positive in the latter test after 10 days. No biochemical test reliably distinguishes between *avium* and *intracellulare* strains (Meissner *et al.* 1974). Both variants grow over a wide range of temperature (25–37°), but only the *avium* strains grow well at 43°.

The agglutination serotypes of the *avium* and intracellulare strains have been extensively studied (Schaefer 1965, 1967, Wolinsky and Schaefer 1973, Matthews *et al.* 1979). They are referred to by Arabic numerals (Wolinsky and Schaefer 1973). Strains of the *avium* variant are of serotypes 1, 2 and 3, those of the *intracellulare* type from 4 to 24. Untypable strains are also encountered. Serotype 8 (originally termed the Davis serotype) more closely resembles the avian strains than do other *intracellulare* strains in DNA relatedness (Baess 1979), skin-test reactions (Anz *et al.* 1970), fluorescent-antibody reactions (Bennedsen 1968) and virulence (Engbaek *et al.* 1968). This serotype includes strains known as *M. brunense* (Kubin *et al.* 1969).

Of the three serotypes of the *avium* variant, Nos. 1 and 2 are widely distributed throughout the world (Wolinsky and Rynearson 1968) whereas No. 3 is virtually restricted to Europe (Marks *et al.* 1969). All three serotypes cause disease in wild and domesticated birds and in animals, particularly cattle and pigs. Serotype 2 is the most virulent and the commonest cause of disease in birds (Schaefer 1968). The serotypes of the *intracellulare* variant are much less virulent in birds than those of the *avium* variant.

Mycobacterium paratuberculosis (Johne's bacillus) This organism was found to be associated with chronic hypertrophic enteritis of cattle by Johne and Frothingham in 1895 and was considered to be a variant of the 'avian tubercle bacillus' (*M. avium*). Serological evidence (Lind and Norlin 1963, Tuboly 1965) tends to support this. It was isolated in pure culture on a medium containing dead tubercle bacilli (Twort 1910) or *M. phlei* (Twort and Ingram 1912). Dependence for growth on the presence of other mycobacteria is due to the inability of *M. paratuberculosis* to synthesize mycobactin (see p. 68).

M. paratuberculosis is a short, thick rod, 1–2 μm long, that usually stains evenly. On mycobactin-containing medium growth is slow, and on primary culture colonies may not be visible for 2 or 3 months. The colonies are initially colourless but later they assume a light yellow colour. Growth occurs between 28° and 43°. According to Taylor (1951), three different varieties of the organism occur in sheep: (1) the usual bovine variety: (2) a similar variety found in Iceland that requires for growth a medium containing at least 50 per cent of egg yolk; and (3) a variety found principally in Scotland that is also dependent on egg yolk and produces a bright orange pigment.

Mycobactin-dependent organisms similar to *M. paratuberculosis* have been isolated from wood pigeons (McDiarmid 1962). Experimentally they, like *M. avium*, cause severe and widespread disease in chickens but they also cause chronic hypertrophic enteritis in calves (Matthews and McDiarmid 1979). These strains, like those of *M. paratuberculosis*, were not agglutinated by any of the available mycobacterial typing antisera. They may represent an evolutionary link between the classical *M. avium* and *M. paratuberculosis* strains. Strains similar to those from wood pigeons have also been isolated from free-living deer in Great Britain (Rankin and McDiarmid 1969) and from a brown hare (Matthews and Sargent 1977) although the latter strain did not cause Johne's disease experimentally in calves. Mycobactin dependence has also been observed in some strains of *M. avium* serotypes 1, 2 and 3 particularly on primary isolation (Matthews *et al.* 1977). It was recommended that mycobactin be added to media used for the isolation of *M. avium* from clinical material.

Mycobacterium lepraemurium (the rat leprosy bacillus) This organism causes an indurating and ulcerating skin disease in rats, mice and occasionally cats. It was first described by Stefansky in 1903 who termed it *Bacillus der Rattenlepra* and its present epithet was published by Marchoux and Sorel in 1912. It is 3–5 μm long, often slightly curved, and may exhibit granular staining.

It is not easily cultivated *in vitro* but growth has been obtained on Ogawa egg medium (Ogawa and Hiraki 1972, Mori 1975) and in a modified Dubos or Kirchner medium containing cytochrome c and α-ketoglutaric acid (Nakamura 1972, 1975). Little is yet known of its biochemical properties. Six strains propagated in mice have been studied immunologically; they were antigenically homogeneous on immunodiffusion analysis (Stanford 1973*c*) and differed from *M. avium* by only two precipitation lines. *M. lepraemurium* and *M. leprae* are quite distinct species.

Slowly growing scotochromogens

Mycobacterium scrofulaceum The organism given this name was first isolated from diseased cervical lymph nodes (Prissick and Masson 1965) but has since been found in other clinical specimens and in soil (Kestle *et al.* 1967). Similar organisms were earlier called *M. marianum* by Suzanne and Penso (1953; see also Wayne and Lessel 1969), but this name has since been abandoned on account of its frequent confusion with *M. marinum* (p. 81).

Strains of *M. scrofulaceum* are phenetically similar to scotochromogenic strains of *M. avium* but differ from them in hydrolysing urea and in producing large amounts of catalase; they have been shown to form a distinct species by antigenic analysis (Castelnuovo and Morellini 1962, Stanford and Grange 1974) and by the specificity of delayed hypersensitivity reactions (Magnusson 1962).

Mycobacterium gordonae Named after Dr Ruth Gordon, a pioneer of mycobacterial taxonomy, this species was described by Bojalil and his colleagues in 1962 and further characterized by Wayne and Doubek (1968). Many strains termed *M. aquae* belong to this species but others have been identified as *M. scrofulaceum* and *M. flavescens* (Stanford and Grange 1974). This species is characterized by specific antigens detected in immunodiffusion tests. It differs from both *M. scrofulaceum* and the pigmented variants of *M. avium* by hydrolysing Tween 80 within 5 days and by failing to hydrolyse nicotinamide and pyrazinamide; and from *M. scrofulaceum* in the absence of urease activity. Strains of this species are frequently isolated from the natural environment but are rarely implicated as pathogens.

Mycobacterium szulgai was named after Dr T. Szulga who helped to develop techniques for the analysis of mycobacterial lipids that led to its discovery (Marks *et al.* 1972). Strains of this species superficially resemble those of *M. gordonae* but differ in immunodiffusion serotype, lipid structure and biochemical properties. In particular *M. szulgai* differs from *M. gordonae* and *M. scrofulaceum* in reducing nitrate to nitrite (Marks *et al.* 1972) and in possessing α-L-fucosidase activity (Grange and McIntyre 1979). Strains of these species have been isolated from sputum and from cases of olecranon bursitis. Although scotochromogenic when incubated at 37°, it is photochromogenic at 25° (Sommers 1977).

Slowly growing mycobacteria of uncertain taxonomic status

The following 'species' are of uncertain taxonomic status: *M. malmoense*, *M. simiae*, *M. asiaticum* and *M. habana*.

Mycobacterium malmoense This name was given to a group of seven strains isolated from patients from Malmö in Sweden (Schröder and Juhlin 1977). They were identical with a group of five strains described by Birn and his colleagues in 1967 that differed from *M. avium* in their lipid pattern on thin-layer chromatography, agglutination serotype and the temperature range for growth. *M. malmoense* has been isolated from several coalminers with pneumoconiosis in Great Britain (Jenkins and Tsukamura 1979). Infections may prove fatal (Schaefer *et al.* 1973).

Mycobacterium simiae* and *M. asiaticum In 1964, 50 strains of mycobacteria were isolated from two species of monkey, *Macacus rhesus* and *Cercopithecus ethiops* (Karasseva *et al.* 1965). They were heterogeneous in their properties (Report 1970), and among them were two groups of strains thought to constitute new species: *M. asiaticum*, which was photochromogenic, and did not form niacin or amidases and *M. simiae* which was weakly photochromogenic and formed niacin (Weiszfeiler *et al.* 1971).

Mycobacterium habana This name was given to a group of niacin-forming, slowly growing strains isolated in Havana, Cuba (Valdivia-Alvarez *et al.* 1971). They were originally not considered to be of clinical significance but some strains have since been isolated from patients with advanced cavitating pulmonary disease. The strains are photochromogenic (although the pigment develops slowly), nitratase negative and catalase positive. They cross-react with a serotype of *M. avium* (*intracellulare*) (Pattyn and Dommisse 1973).

Other slowly growing species A few other species have been named but are not considered to be pathogens. These include *M. nonchromogenicum*, *M. terrae* and *M. triviale*. The former two species have several properties in common; they are separable into two clusters by numerical analysis, but cannot be distinguished serologically by immunodiffusion tests (Meissner *et al.* 1974). Although often regarded as a slow grower, *M. nonchromogenicum* resembles *M. vaccae* and *M. leprae* in lacking the antigens common to either rapidly growing or slowly growing mycobacteria (Stanford *et al.* 1975). *Mycobacterium triviale* was isolated from sputum by Kubica and his colleagues (1970) but, as the name indicates, it is of little importance. Its lack of pigmentation and its colonial appearance may, however, cause it to be confused with *M. tuberculosis*.

The rapidly growing mycobacteria

Group 1: non-chromogenic strains with strong arylsulphatase activity.

The two species in this group, *M. fortuitum* and *M. chelonei*, are the principal pathogenic species among the rapid growers; infections due to other species are rare. All variants of *M. chelonei* may be encountered as pathogens, but within *M. fortuitum* virulence is mainly restricted to one serologically and biochem-

Table 24.4 Properties of *Mycobacterium fortuitum* and *M. chelonei*

Property	*M. fortuitum* biotype A	B	C	*M. chelonei* subtype *abscessus*	*chelonei*
Reduction of nitrate	+	+	+	–	–
Growth at 42°	+	–	–	–	–
Acid production from mannitol	–	+	+	–	–
inositol	–	–	+	–	–
Heat-stable esterase (100°/15 min)	+	+	+	–	–
Heat-stable acid phosphatase (70°/30 min)	+	+	+	+	–
β-D-galactosidase	–	–	–	+	+
Citrate utilization	–	–	–	–	+
Growth on media containing 5% NaCl	–	–	–	+	–

ically defined subspecies. Both species are non-chromogenic and strongly arylsulphatase positive but differ in many other respects.

Mycobacterium fortuitum In 1904 Rupprecht described a tuberculosis-like infection in a frog; shortly afterwards Küster (1905) cultivated mycobacteria from similar cases. This organism was termed the 'frog tubercle bacillus' and subsequently named *M. ranae* (Bergey *et al.* 1923). The epithet *M. fortuitum* was given by Cruz in 1938 to a supposedly new species but its identity with *M. ranae* was later established (Gordon and Mihm 1959, Stanford and Gunthorpe 1969). Although *ranae* clearly has chronological priority over *fortuitum* the Judicial Commission of the International Committee on Bacteriology accepted the recommendation of Runyon (1972) that the former epithet should be rejected in favour of the latter. Other organisms now known to be strains of *M. fortuitum* include *M. giae* (Darzins 1950), *M. minetti* (Penso *et al.* 1952) and *M. peregrinum* (Bojalil *et al.* 1962). The strain of *M. fortuitum* subspecies *runyoni* (Tsukamura 1967) has been shown by Stanford and his colleagues (1972) to belong to the species *M. chelonei*.

Strains of *M. fortuitum* are rapidly growing, non-chromogenic organisms that hydrolyse Tween 80, reduce nitrate to nitrite and possess marked arylsulphatase activity. Urea, allantoin and usually acetamide are hydrolysed. Isonicotinamide and pyrazinamide are weakly and irregularly hydrolysed (see Grange and Stanford 1974). Bönicke (1966) divided the species into biotypes A, B and C on the basis of their ability to produce acid from sugars (see Table 24.4).

Most strains of *M. fortuitum* isolated from cases of infection are of biotype A; the other biotypes are usually encountered as saprophytes in the natural environment (Grange and Stanford 1974). Biotype A strains are more virulent in mice (Kubica *et al.* 1972). The strains originally identified as *M. fortuitum*, *M. ranae* and *M. minetti* belong to biotype A whereas *M. peregrinum* and *M. giae* belong to biotype B.

Mycobacterium chelonei This species includes the original turtle tubercle bacillus (Friedmann 1903). Other organisms now known to belong to this species include *M. abscessus*, *M. runyoni*, *M. borstelense* and *M. borstelense* var. *niacinogenes* (Kubica *et al.* 1972). The *abscessus* variant (Table 24.4) grows in the presence of 1 per cent deoxycholate or 5 per cent NaCl and uses nitrite as a nitrogen source; var. *chelonei* does not grow under these conditions but can use citrate as a carbon source. The former is principally isolated in the USA and Africa, the latter in Europe (Stanford *et al.* 1972). Both variants have been isolated from a wide range of lesions in man and animals (Chapter 51).

Strains of *M. chelonei* differ from those of *M. fortuitum* in failing to reduce nitrate and to hydrolyse allantoin. The pattern of sugar use by *M. chelonei* is the same as that of biotype A strains of *M. fortuitum* but the latter differ in their ability to grow at 45° (Kubica *et al.* 1972). Differences between *M. fortuitum* and *M. chelonei* are also demonstrable on lipid chromatography (Jenkins *et al.* 1971) and by agglutination (Pattyn 1970) and immunodiffusion tests (Stanford 1973*a*). Additional differences between the two species include *β*-galactosidase activity in *M. chelonei* (Tsukamura 1975, Grange 1978) and heat-stable acyl esterase activity in *M. fortuitum* (Grange 1977).

Group 2: thermophilic strains

Mycobacterium smegmatis This species is widely used by biochemists and geneticists but is seldom encountered in either clinical specimens or the natural environment. The widely held view that it is a frequent contaminant of urine is not based on fact. The 'Smegma bacillus' was described by Alvarez and Tavel in 1885. Strains termed *M. stercoria* (Moeller's Mist bacillus), *M. butyricum* (the butter bacillus), *M. lacticola* and *M. jucho* also belong to this species. Strains grow at 45° and are either photochromogenic or non-chromogenic. All strains are very active biochemically and hydrolyse a wide range of amides, organic acids and sugars (Gordon and Smith 1953, Gordon and Mihm 1959, Gordon 1966, Bönicke 1962, Kubica *et al.* 1972). This species differs from *M. phlei* in its failure to grow at 52°, to produce acid phosphatase and to hydrolyse hippurate, in its ability to produce acid from inositol and rhamnose and to deaminate benzamide (Kubica *et al.* 1972) and in its lack of *α*- and *β*-esterase activity (Käppler 1965).

Mycobacterium phlei Isolated by Moeller (1898) from timothy grass (*Phleum pratense*) and named *M. phlei* by Lehmann and Neumann in 1899, this is a very uncommonly encountered species that has never been known to cause infections in man or animals. On Löwenstein–Jensen medium a salmon-pink pigment is produced in the dark; on non-egg-based media the strains produce a yellow pigment on exposure to light. Like *M. smegmatis*, it possesses a wide range of biochemical activities (Gordon and Smith 1953, Kubica *et al.* 1972). It grows at 52° and, unlike other mycobacteria, remains viable after heating at 60° for 4 hr.

Mycobacterium thermoresistibile This uncommon species was described by Tsukamura (1966*a*) and, like *M. phlei*, grows at 52°. It is usually non-chromogenic although brown coloration may develop. The strains are not so metabolically active as those of the other two species in this group and do not produce acid from arabinose, galactose, mannitol, trehalose and xylose (Tsukamura 1975). There are no reports of infection due to this species.

Table 24.5 Properties of *Mycobacterium flavescens*, *M. gilvum* and *M. duvali**

Property	*M. flavescens* biotype 1	2	3	*M. gilvum*	*M. duvali*
Acid production from inositol	–	–	–	+	–
Acid production from mannitol	–	–	+	+	+
Acid production from sorbitol	–	–	+	+	+
Urease	–	+	+	+	+
Arylsulphatase	+	+	+	+	–
Utilization of citrate	–	–	–	+	–

* Or *duvalii* (see Table 24.1).

Group 3: rapidly growing scotochromogenic mycobacteria with limited saccharolytic activity.

The species in this group include *M. flavescens*, *M. gilvum* and *M. duvali**. They produce acid from some or all of these sugars: glucose, mannose, trehalose, inositol, mannitol and sorbitol; their other properties are shown in Table 24.5. *M. flavescens* is one of several species delineated by Bojalil and his colleagues (1962) on the basis of numerical analysis. Another species described in the same paper is *M. acapulcense*, but this is now known to be synonymous with *M. flavescens* (Pattyn *et al.* 1968, Stanford and Gunthorpe 1971). *M. gilvum* and *M. duvali* are rarely encountered species described by Stanford and Gunthorpe (1971). The former differs from the other two species in this group in its ability to use citrate and inositol. Most of the strains of *M. gilvum* were isolated from pleural fluid or sputum, but were not considered to be pathogens. Two strains of *M. duvali* were isolated by Duval and Wellman (1912) and were erroneously thought to be *M. leprae*.

Group 4: other rapidly growing mycobacteria

This is an area of mycobacterial taxonomy that is still in considerable chaos. A large number of strains, mostly scotochromogenic, have been isolated from the natural environment and from clinical specimens, but they virtually never cause disease. Many specific epithets have been given to strains within this group. These 'species' include *M. aurum* (Tsukamura 1966*a*), *M. chitae* (Tsukamura 1966*b*), *M. diernhoferi* (Bönicke and Juhasz 1964), *M. neoaurum* (Tsukamura 1972), *M. parafortuitum* (Tsukamura *et al.* 1965) and *M. vaccae* (Bönicke and Juhasz 1964). Other 'species' referred to by Tsukamura (1975) include *M. aichiense*, *M. chubuense*, *M. agri*, *M. obuense*, *M. rhodesiae* and *M. tokaiense*. Most of these co-called species were delineated by numerical analysis and have been described in detail by Tsukamura (1975) and by Saito and his co-workers (1977).

Two of these merit special attention: *M. diernhoferi* and *M. vaccae*, both of which were characterized by Bönicke and Juhasz (1964). *M. diernhoferi* was isolated from the environment of animals and was named after Karl Diernhofer, a veterinary surgeon from Vienna. It was regarded as a distinct species by Kubica and his co-workers (1972) but later its specific status was questioned (Saito *et al.* 1977); it was not included in the list of approved mycobacterial species (Skerman *et al.* 1980, see p. 61 and Table 24.1). It is, however, one of the few species in this group that has been shown to possess species-specific antigens on immunodiffusion analysis (Stanford and Grange 1974, Stanford and Wong 1974). It is non-chromogenic, although colonies often become grey on prolonged incubation.

Mycobacterium vaccae was isolated, as the name suggests, from the environment of cows (Bönicke and Juhasz 1964). It is of particular interest because of its possible antigenic relationship to *M. leprae*, which is described below. Members of this species vary considerably in their properties.

Non-cultivable mycobacteria

Mycobacterium leprae

This is the only species that has never convincingly been cultivated *in vitro*. It was, however, the first mycobacterium to be described, having been observed microscopically by Hansen in 1868 (see Hansen 1874). It is included in the genus *Mycobacterium* on account of its acid-fastness and the structure of its mycolic acids (Etémadi and Convit 1974).

* The 'approved name' of this species is *M. duvalii* (see Table 24.1).

Leprosy bacilli, as seen in infected tissues, vary considerably in size and shape. They may be straight or slightly curved, 1–8 μm in length. Their arrangement and staining properties depend on the nature of the host's immune response to the pathogen (Chapter 52). Acid-fastness is removed by previous treatment with pyridine (Fisher and Barksdale, 1973), a property shared with some but not all strains of *M. vaccae* (Stanford *et al.* 1975) and some strains of other mycobacterial species (Šlosárek *et al.* 1978).

The use of the armadillo, in which lepromas containing large numbers of organisms develop after inoculation (Kirchheimer and Storrs 1971) has enabled enough organisms to be obtained for limited biochemical studies. Unlike all other mycobacterial cell walls, those of *M. leprae* contain substantial amounts of glycine (Draper 1976). Earlier, Prabhakaram and Kirchheimer (1966) detected *o*-diphenoloxidase in *M. leprae* and recommended this as a test for the identification of this species. Subsequently, however, the specificity and even the existence of this biochemical character have been doubted (Kato and Ishaque 1977). Likewise controversies exist as to the occurrence of other enzymes in *M. leprae* (Report 1979).

Stanford and his colleagues (1975) detected 12 antigens in organisms from human and armadillo tissues by immunodiffusion tests. As in *M. vaccae*, precipitating antigens confined to the rapidly growing species, and those confined to the slowly growing species, were not detected in *M. leprae* (Stanford 1976). Delayed-hypersensitivity reactions elicited by a range of tuberculins in leprosy patients, their contacts and healthy persons suggested a similarity between *M. leprae*, *M. vaccae* and *M. nonchromogenicum* (Paul *et al.* 1975) but there is no evidence that these organisms are taxonomically linked.

The adsorption of phage D29 to *M. leprae* cells causes patchy damage to the cell walls and the appearance of hexagonal crystalline structures within the cell (David *et al.* 1978*a*, *b*). The relationship of these structures to the phage is unknown; treatment with mitomycin C caused the appearance of similar structures.

The reason for the non-cultivability of *M. leprae* remains totally elusive. The inability to grow *M. leprae in vitro* does not necessarily imply that this species is unable to replicate outside the body. Kazda and his colleagues (1980) isolated non-cultivable acid-fast bacilli from sphagnum vegetation in Scandinavia by mouse-footpad inoculation.

The G + C content of the DNA of *M. leprae* is given as 55.8 moles per cent, less than that of other mycobacteria (Imaeda *et al.* 1982).

Identification of mycobacteria

Taxonomists have used many different techniques of varying degrees of complexity for subdividing the genus *Mycobacterium*. These include the biochemical techniques referred to earlier in this chapter; immunodiffusion analysis (Stanford and Grange 1974, Lind 1978); lipid chromatography (Jenkins *et al.* 1971); pyrolysis gas chromatography (Reiner *et al.* 1969); pyrolysis mass spectroscopy (Wieten *et al.* 1981); gel electrophoresis (Haas *et al.* 1972); DNA homology (Baess 1979) and specificity of skin-test reactions in guinea-pigs (Magnusson 1967, Takeya *et al.* 1970, Kazda 1977).

In contrast to the needs of the taxonomist, the clinical microbiologist requires a set of easy, reliable and reproducible tests to identify the mycobacteria he encounters, of which *M. tuberculosis* is usually the most important. Thus the first step is to determine whether an isolate is of this species. Strains of other species require a suitable battery of tests for their identification, and in many instances these tests will be performed by a regional or reference centre.

Mycobacterium tuberculosis is characterized by its lack of yellow pigmentation, its slow rate of growth, its failure to grow at 25° and its failure to grow on media containing 500 mg per l of *p*-nitrobenzoic acid (Yates and Collins 1979).

The methods for the identification of mycobacteria other than *M. tuberculosis* vary considerably from laboratory to laboratory (Report 1973). Some investigators carry the process to the level of species or subspecies; while others consider it sufficient to allocate strains to one of several clinically significant groups, complexes or clusters, depending on the workload, experience and ability of the technical staff, and financial resources. Käppler (1973) advocates the use of 25 tests to identify each strain. Marks (1976), by contrast, places mycobacterial strains into 14 groups by means of four simple tests, namely growth at 25, 37 and 45°, oxygen preference (aerobic or microaeraphilic), pigment production in light or dark, and Tween 80 hydrolysis. Confirmation of the identity is, in some cases, made by the observation of characteristic drug sensitivity patterns. The identification strategies employed in a number of European reference centres have been presented (Report 1973). The American Department of Health, Education and Welfare advocates the use of a work-flow chart with the use of different tests for rapid growers and the photochromogenic, scotochromogenic and non-chromogenic slow growers (see Vestal 1975). Kubica (1973) lists 12 simple characters that enable the great majority of species encountered in clinical practice to be reliably identified. The slowly growing and rapidly growing species may be considered separately from the point of view of identification.

Slow growers These are roughly divisible into three groups according to Runyon (1959) on the basis of pigmentation. Exceptions occur, however, such as scotochromogenic and non-chromogenic variants in the predominantly photochromogenic species and pigmented variants of *M. avium*. The species listed in Table 24.3 include the great majority of clinical isolates. A few organisms of uncertain specific status are omitted. These include *M. simiae*, *M. habana*, *M. malmoense* and *M. gastri*, which are discussed in more detail on p. 83.

The separation of these strains on the basis of the simple tests shown in Table 24.3 usually presents no difficulty. The species *M. avium* is complex and is here divided into the *avium* and *intracellulare* subspecies on the basis of growth at 45°. Although *M. xenopi* and *M. ulcerans* are both included as non-chromogens they often produce a pale yellow pigment especially on prolonged incubation. All the tests used in this scheme are well established, except for the hydrolysis of α-L-fucosidase (Grange and McIntyre 1979) which rapidly distinguishes *M. marinum* from *M. kansasi*, and *M. szulgai* from the other scotochromogens. The eight species in this table are thus distinguishable on the basis of temperature of growth, pigment production and five biochemical tests readable either after 24 hr or 3 days.

Rapid growers These are first subdivided into the four groups described on pp. 83–85 and subsequently identified by the properties shown in Tables 24.4 and 24.5.

The species *M. fortuitum* and *M. chelonei* are the most commonly encountered rapidly growing mycobacteria in clinical specimens and are the only species in this group that merit serious consideration as pathogens. Both are non-chromogenic and strongly arylsulphatase positive and are thereby differentiated from other rapid growers.

The strains in the fourth group are mostly scotochromogenic but a few are photo- or non-chromogenic. Until the taxonomy of this group of organisms has been considerably improved there is little point in attempting to identify them in a clinical laboratory.

The rapidly growing mycobacteria are, with the exception of the fourth group, identifiable at the specific level and, in the case of the two pathogenic species, at the subspecific level.

(A great deal of useful information will be found in the Proceedings of the International Conference on Atypical Mycobacteria held in 1979 at Denver, Colorado; Report 1982.)

References

Abrahams, E. W. (1970) *Tubercle* **51,** 316.

Acharya, P. V. N. and Goldman, D. S. (1970) *J. Bact.* **102,** 733.

Adam, A., Ciorbaru, R., Petit, J.-F. and Lederer, E. (1972) *Proc. nat. Acad. Sci., Wash.* **69,** 851.

Adam, A., Ellouz, F., Ciobaru, R., Petit, J.-F. and Lederer, E. (1975) *Z. ImmunForsch.* **149,** 341.

Adámek, L., Mišoň, P., Mohelská, H. and Trnka, L. (1969) *Arch. Mikrobiol.* **69,** 227.

Adámek, L., Trnka, L., Mišoń, P. and Gutova, M. (1968) *Beitr. klin. Tuberk.* **138,** 51.

Akamatsu, Y. and Nojima, S. (1965) *J. Biochem., Tokyo* **57,** 430.

Alshamaony, L., Goodfellow, M. and Minnikin, D. E. (1976*b*) *J. gen. Microbiol.* **92,** 188.

Alshamaony, L., Goodfellow, M., Minnikin, D. E. and Mordarska, H. (1976*a*) *J. gen. Microbiol.* **92,** 183.

Alvarez, E. and Tavel, E. (1885) *Arch. Phys. norm. Path.* **6,** 303.

American Thoracic Society (1969) *Amer. Rev. resp. Dis.* **99,** 631.

Andrejew, A., Orfanelli, M-T. and Desbordes, J. (1974) *Ann. Microbiol., Inst. Pasteur* **125,** 323.

Antonie, A. D. and Tepper, B. S. (1969) *Arch. Biochem. Biophys.* **134,** 207.

Anz, W., Lauterbach, D., Meissner, G. and Willers, I. (1970) *Zbl. Bakt.* **215,** 536.

Anz, W. and Schröder, K. H. (1970) *Zbl. Bakt.* **214,** 553.

Aronson, J. D. (1926) *J. infect. Dis.* **39,** 315.

Asano, A., Cohen, N. S., Baker, R. F. and Brodie, A. F. (1973) *J. biol. Chem.* **248,** 3386.

Audibert, F., Chedid, L., Lefrancier, P. and Choay, J. (1976) *J. cell. Immunol.* **21,** 243.

Augier, J., Augier-Gibory, S. and Lepault, F. (1971) *Ann. Inst. Pasteur.* **121,** 657.

Azuma, I., Kishimoto, S., Yamamura, Y. and Petit, J.-F. (1971) *Jap. J. Microbiol.* **15,** 193.

Azuma, I., Nagasuga, T. and Yamamura, Y. (1962) *J. Biochem., Tokyo* **52,** 92.

Azuma, I. *et al.* (1970) *Biochim. biophys. acta* **208,** 444.

Baer, H. and Chaparas, S. D. (1966) *Science, N.Y.* **146,** 245.

Baess, I. (1966) *Amer. Rev. resp. Dis.*, **93,** 622; (1969) *Acta path. microbiol. scand.* **76,** 464; (1971) *Ibid.* **79,** 428; (1979) *Ibid.* **B87,** 221.

Baess, I. and Bentzon, M. W. (1978) *Acta. path. microbiol. scand.* **B86,** 71.

Baess, I. and Mansa, B. (1978) *Acta. path. microbiol. scand.* **B86,** 309.

Barksdale, L. and Kim, K. S. (1977) *Bact. Rev.* **41,** 217.

Basaca-Sevilla, V., Sevilla, J. S., Faraon, P. C., Fernando, C. L. and Uvero, J. A. (1976) *J. Phil. med. Ass.* **52,** 251.

Bates, J. H. and Fitzhugh, J. K. (1967) *Amer. Rev. resp. Dis.* **96,** 7.

Bates, J. H. and Mitchison, D. A. (1969) *Amer. Rev. resp. Dis.* **100,** 189.

Beck, M. (1905) *Tuberk. Arzt.* **3.** 145.

Beerwerth, W. (1971) *Prax. Pneumonol.* **25,** 661; (1973) *Proc. 3rd Int. Coll. Mycobacteria*, p. 151. International Colloquia Series, Antwerp.

Bennedsen, J. (1968) *Acta. path. microbiol. scand.* **72,** 330.

Bennet, P. and Asselineau, J. (1970) *Ann. Inst. Pasteur* **118,** 324.

Bergan, T. and Lystad, A. (1971) *J. appl. Bact.* **34,** 751.

Bergey, D. H., Harrison, F. C., Breed, R. S., Hammer, B. W. and Huntoon, F. M. (1923) *Manual of Determinative Bacteriology*, 1st ed. Williams and Wilkins, Baltimore.

Bhatia, A. L., Csillag, A., Mitchison, D. A., Selkon, J. B., Somasundaram, P. R. and Subbaiah, T. V. (1961) *Bull. World Hlth Org.* **25,** 313.

Bhatia, A. L., Jacob, C. V., Hitze, K. L., Ramachandran, K. and Selkon, J. B. (1963) *Bull. World Hlth Org.* **29,** 483.

Birn, K. J., Schaefer, W. B., Jenkins, P. A., Szulga, T. and Marks, J. (1967) *J. Hyg., Camb.* **65,** 575.

Bloch, H. (1950) *J. exp. Med.* **91,** 197.

Bojalil, L. F. (1959) *Rev. Lat. Am. Microbiol.* **2,** 169.

Bojalil, L. F., Cerbon, J. and Trujillo, A. (1962) *J. gen. Microbiol.* **28,** 333.

Bönicke, R. (1958) *Z. Hyg. InfektKr.* **145,** 263; (1962) *Bull. Un. int. Tuberc.* **32,** 13; (1966) *Ibid.* **37,** 361.

Bönicke, R. and Juhasz, S. E. (1964) *Zbl. Bakt.* **192,** 133.

Bönicke, R. and Kazda, J. (1970) *Zbl. Bakt.* **213,** 68.

Bönicke, R. and Saito, H. (1970) *Bull. Un. int. Tuberc.* **43,** 217.

Bradley, D. E. (1967) *Bact. Rev.* **31,** 230.

Bradley, S. G. (1972) *Amer. Rev. resp. Dis.*, **106,** 122; (1975) *Adv. appl. Microbiol.* **19,** 59.

Brem, W. V. (1909) *J. Amer. med. Ass.* **53,** 909.

Brennan, P. J. and Goren, M. B. (1979) *J. biol. Chem.* **254,** 4205.

Brennan, P. J., Souhrada, M., Ullom, B., McClatchy, J. K. and Goren, M. B. (1978) *J. clin. Microbiol.* **8,** 374.

Bruner, D. W. (1934) *J. infect. Dis.* **55,** 26.

Buhler, V. B. and Pollack, A. (1953) *Amer. J. clin. Path.* **23,** 363.

Bullin, C. H., Tanner, E. I. and Collins, C. H. (1970) *J. Hyg., Camb.* **68,** 97.

Buraczewska, M., Kwiatkowski, B., Manowska, W. and Rdultowska, H. (1972) *Amer. Rev. resp. Dis.* **105,** 22.

Buraczewska, M., Manowska, W. and Rdultowska, H. (1970) *Amer. Rev. resp. Dis.* **103,** 116; (1971) *Ibid.* **104,** 760.

Burnham, W. R., Lennard-Jones, J. E., Stanford, J. L. and Bird, R. G. (1978) *Lancet* **ii,** 693.

Campbell, G. R., Carty, P. and Winder, F. G. (1979) *Biochem. Soc. Trans.* **7,** 23.

Castelnuovo, G. and Morellini, M. (1962) *Ann. Ist. Forlanini* **22,** 1.

Castets, M., Rist, N. and Boisvert, H. (1969) *Méd. Afr. noire* **4,** 321.

Chang, Y. T. and Andersen, R. N. (1969) *J. Bact.* **99,** 867.

Chaparas, S. D. (1975) *Amer. Rev. resp. Dis.* **112,** 135; (1979) *Bull. Un. int. Tuberc.* **54,** 156.

Chester, F. D. (1901) *A Manual of Determinative Bacteriology*. Macmillan, New York.

Clancey, J. K. (1964) *J. Path. Bact.* **88,** 175.

Closs, O., Harboe, M. and Wassum, A. M. (1975) *Scand. J. Immunol.* **4,** Suppl. 2, 173.

Collins, C. H. (1965) M. I. Biol. Thesis. Institute of Biology, London.

Collins, C. H. and Lyne, P. M. (1979) *Microbiological Methods*, 4th edn. (revised). Butterworth, London.

Collins, C. H., Yates, M. D. and Grange, J. M. (1981) *Tubercle* **62,** 113; (1982) *J. Hyg., Camb.* **89,** 235.

Corper, H. J. and Cohn, M. L. (1946) *Amer. J. Clin. Path.* **16,** 621.

Crawford, J. T. and Bates, J. H. (1979) *Infect. Immun.* **24,** 979.

Cruz, J. da C. (1938) *Acta. med. Rio de Janeiro* **1,** 297.

Cuttino, J. T. and McCabe, A. M. (1949) *Amer. J. clin. Path.* **25,** 1.

Daniel, T. M. (1974) *Amer. Rev. resp. Dis.* **110,** 634.

Daniel, T. M. and Affronti, L. F. (1973) *Amer. Rev. resp. Dis.* **108,** 1244.

Daniel T. M. and Anderson, P. A. (1977*a*) *J. Lab. clin. Med.* **90,** 354; (1977*b*) *Amer. Rev. resp. Dis.* **115,** Suppl., 258.

Daniel, T. M. and Janicki, B. W. (1978) *Microbiol. Rev.* **42,** 84.

Daniel, T. M. and Misaki, A. (1976) *Amer. Rev. resp. Dis.* **113,** 705.

Darzins, E. (1950) *Arch. Inst. bras. Invest. Tuberc.* **9,** 29.

David, H. L. (1972) *Amer. Rev. resp. Dis.* **105,** 944; (1973) *Ibid.* **108,** 1175; (1978) *Ann Microbiol., Inst. Pasteur* **A129,** 561.

David, H. L., Clavel, S., Clement, F., Meyer, L., Draper, P. and Burdett, I. D. J. (1978*a*) *Ann. Microbiol., Inst. Pasteur* **B129,** 561.

David, H. L., Clement, F. and Meyer, L. (1978*b*) *Ann. Microbiol., Inst. Pasteur* **B129,** 563.

David, H. L. and Jahan, M.-T. (1977) *J. clin. Microbiol.* **5,** 383.

David, H. L., Jahan, M.-T., Jumin, A., Grandry, J. and Lehman, E. H. (1978*c*) *Int. J. syst. Bact.* **28,** 467.

Dernby, K. G. and Näslund, C. (1922) *Biochem. Z.* **132,** 393.

Dhariwal, K. R., Chander, A. and Venkita-Subramanian, T. A. (1977) *Canad. J. Microbiol.* **23,** 7; (1978) *Arch. Mikrobiol.* **116,** 69.

Dhayagude, R. G. and Shah, B. R. (1948) *Ind. J. med. Res.* **36,** 79.

Diaz, G. A. and Wayne, L. G. (1974) *Amer. Rev. resp. Dis.* **110,** 312.

Dixon, J. M. S. and Cuthbert, E. H. (1967) *Amer. Rev. resp. Dis.* **96,** 119.

Dodge, O. C. (1964) *J. Path. Bact.* **88,** 167.

Dorset, H. (1902) *Amer. Med.* **3,** 555.

Doub, L. (1979) In: *Tuberculosis*, p. 435. Ed. by G. Youmans. Saunders, Philadelphia.

Draper, P. (1971) *J. gen. Microbiol.* **69,** 313; (1974) *Ibid.* **83,** 431; (1976) *Int. J. Leprosy* **44,** 95.

Draper, P. and Rees, R. J. W. (1973) *J. gen. Microbiol.* **77,** 79.

Drea, W. F. and Andrejew, A. (1953) *The Metabolism of the Tubercle Bacillus.* Charles C. Thomas, Springfield, Ill.

Dubos, R. J. (1945) *Proc. Soc. exp. Biol., N.Y.* **58,** 361; (1954) *Amer. Rev. Tuberc.* **70,** 391.

Dubos, R. J. and Middlebrook, G. (1947) *Amer. Rev. Tuberc.* **56,** 334.

Dunn, D. B. and Smith, J. F. (1958) *Biochem. J.* **68,** 627.

Duval, C. W. and Wellman, C. (1912) *J. infect. Dis.* **2,** 116.

Ehrlich, P. (1882) *Dtsch. med. Wschr.* **8,** 269.

Ellouz, F., Adam, A., Ciorbaru, R., and Lederer, E. (1974) *Biochem. biophys. Res. Commun.* **59,** 1317.

Engbaek, H. C., Vergmann, B., Baess, I. and Bentzon, M. W. (1968) *Acta path. microbiol. scand.* **72,** 295.

Engbaek, H. C., Vergmann, B., Baess, I. and Will, D. W. (1967) *Acta path. microbiol. scand.* **69,** 576.

Engel, H. W. B., Berwald, L. G., Grange, J. M. and Kubin, M. (1980*a*) *Tubercle* **61,** 11.

Engel, H. W. B., Berwald, L. G., and Havelaar, A. H. (1980*b*) *Tubercle* **61,** 21.

Etémadi, A. H. (1967) *J. gas. Chromat.* **5,** 447.

Etémadi, A. H. and Convit, J. (1974) *Infect. Immun.* **10,** 236.

Fisher, C. A. and Barksdale, L. (1973) *J. Bact.* **113,** 1389.

Fite, G. L., Cambre, P. J. and Tuner, M. H. (1947) *Arch. Path.* **43,** 624.

Francis, J. and Abrahams, E. W. (1982) *Tubercle* **62,** 309.

Francis, J., MacTurk, H. M., Madinaveitia, J. and Snow, G. A. (1953) *Biochem. J.* **55,** 596.

Fregnan, G. B. and Smith, D. W. (1962) *J. Bact.* **83,** 819.

Friedmann, F. F. (1903) *Zbl. Bakt.* **34,** 647, 793.

Frimodt-Möller, J., Matthew, K. T. and Barton, R. M. (1956) Proc. 13th Tuberc. Workers Conf., p. 157. Tuberculosis Association of India, New Delhi.

Froman, S. and Scammon, L. (1964) *Amer. Rev. resp. Dis.* **89,** 236.

Froman, S., Will, D. W. and Bogen, E. (1954) *Amer. J. publ. Hlth.* **44,** 1326.

Furuchi, A. and Tokunaga, T. (1972) *J. Bact.* **111,** 404.

Gangadharam, P. R., Pratt, P. F., Damle, P. B., Davidson, P. T., Porter, T. H. and Folkers, K. (1978) *Amer. Rev. resp. Dis.* **118,** 467.

Gardner, A. D. (1926) *Lancet* **i,** 1090.

Gardner, G. M. and Weiser, R. S. (1947) *Proc. Soc. exp. Biol. Med.* **66,** 205.

Gayathri Devi, B., Ramakrishnan, T. and Gopinathan, K.P. (1974) *Proc. Indian Acad. Sci.* **B80,** 240.

Gayathri Devi, B., Shaila, M. S., Ramakrishnan, T. and Gopinathan, K. P. (1975) *Biochem. J.* **149,** 187.

Gelbart, S. M. and Juhasz, S. E. (1970) *J. gen. Microbiol.* **64,** 253; (1973) *Leeuwenhoek ned. Tijdschr.* **39,** 1.

Glasser, D. (1978) *Amer. Rev. resp. Dis.* **118,** 969.

Goldman, D. (1961) *Adv. Tuberc. Res.* **11,** 1.

Gordon, R. E. (1966) *J. gen. Microbiol.* **43,** 329.

Gordon, R. E. and Mihm, J. M. (1957) *J. Bact.* **73,** 15; (1959) *J. gen. Microbiol.* **21,** 736.

Gordon, R. E. and Smith M. M. (1953) *J. Bact.* **66,** 41.

Goren, M. B. (1970) *Biochim. biophys. acta* **210,** 127; (1972) *Bact. Rev.* **36,** 33.

Goren, M. B., Brokl, O., Das, B. C. and Lederer, E. (1971) *Biochemistry* **10,** 72.

Goren, M. B., Brokl, O. and Schaefer, W. B. (1974*a*) *Infect. Immun.* **9,** 142; (1974*b*) *Ibid.* **9,** 150.

Goren, M. B., McClatchy, J. K., Martens, B. and Brokl, O. (1972) *J. Virol.* **9,** 999.

Goslee, S., Rynearson, K. and Wolinsky, E. (1976) *Int. J. syst. Bact.* **26,** 136.

Gottlieb, D. (1974) In: *Bergey's Manual of Determinative Bacteriology*, 8th edn, p. 657. Ed. by R. E. Buchanan, and N. E. Gibbons. Williams and Wilkins, Baltimore.

Grange, J. M. (1973) *Ann. Soc. Belge. Méd. trop.*, **53,** 339; (1974) *J. appl. Bact.* **37,** 465; (1975*a*) *Tubercle* **56,** 227; (1975*b*) *Ann. Sclavo* **17,** 594; (1975*c*) *J. gen. Microbiol.* **89,** 387; (1976) *J. appl. Bact.* **41,** 425; (1977) *Tubercle* **58,** 147; (1978) *J. clin. Path.* **31,** 378; (1979) *Brit. J. Hosp. Med.* **22,** 540; (1982) *Zbl. Bakt.* **A251,** 297.

Grange, J. M., Aber, V. R., Allen, B. W., Mitchison, D. A. and Goren, M. B. (1978) *J. gen. Microbiol.* **108,** 1.

Grange, J. M. and Bird, R. G. (1975) *J. med. Microbiol.* **8,** 215; (1978*a*) *Ibid.* **11,** 1; (1978*b*) In: *Genetics of the Actinomycetales*, p. 243. Ed. by E. Freerksen, I. Tarnok and J. H. Thumin. Gustav Fischer, Stuttgart.

Grange, J. M., Collins, C. H. and McSwiggan, D. (1976) *Tubercle* **57,** 59.

Grange, J. M. and McIntyre, G. (1979) *J. appl. Bact.* **47,** 285.

Grange, J. M. and Redmond, W. B. (1978) *Tubercle* **59,** 203.

Grange, J. M. and Stanford, J. L. (1974) *Int. J. syst. Bact.* **24,** 320.

Grange, J. M. *et al.*, (1977) *Tubercle* **58,** 207.

Grant, J. (1971) *Appl. Microbiol.* **21,** 1091; (1973) *Int. Res. Commun. Syst.*, **25,** 1.

Griffith, A. S. (1925) *Tubercle* **6,** 417.

Gross, W. M., Hawkins, J. E. and Murphy, D. B. (1976) *Bull. Un. int. Tuberc.* **51,** 267.

Gross, W. M. and Wayne, L. G. (1970) *J. Bact.* **104,** 630.

Gunnals, J. J. and Bates, J. H. (1972) *Amer. Rev. resp. Dis.* **105,** 288.

Haas, H., Davidson, Y. and Sacks, T. (1972) *J. med. Microbiol.* **5,** 31.

Hagemann, P. (1938) *Münch. med. Wschr.* **85,** 1066.

Hailer, E. (1938) *Beitr. Klin. Tuberk.* **92,** 81, 371.

Hailer, E. and Heicken, M. (1939) *Beitr. Klin. Tuberk.* **93,** 1.

Hansen, G. H. A. (1874) *Norsk. Mag. Laegevidensk.* **4,** Suppl. 9, 1.

Hart, P. D'A., Armstrong, J. A., Brown, C. A. and Draper, P. (1972) *Infect. Immun.* **5,** 803.

Hauduroy, P. (1955) *Derniers aspects du monde des mycobacteries.* Masson, Paris.

Hedgecock, L. W. (1968) *J. Bact.* **96,** 306.

Herrold, R. D. (1931) *J. infect. Dis.* **48,** 236.

Holland, K. T. and Ratledge, C. (1971) *J. gen. Microbiol.* **66,** 115.

Hubáček, J. (1960) *Fol. Microbiol., Praha* **5,** 171.

Hunter, A. M., Campbell, I. A., Jenkins, P. A. and Smith, A. P. (1981) *Thorax* **36,** 326.

Imaeda, T., Kanetsuna, F. and Galindo, B. (1968*a*) *J. ultrastruct. Res.* **25,** 46.

Imaeda, T., Kanetsuna, F. and Rieber, M. (1968*b*) *Tubercle* **49,** 385.

Imaeda, T., Kirchheimer, W. F. and Barksdale, L. (1982) *J. Bact.* **150,** 414.

Imaeda, T. and Rieber, M. (1968) *J. Bact.* **96,** 557.

International Union Against Tuberculosis (1977) *Technical guide for collection, storage and transport of sputum specimens and for examination for tuberculosis by direct microscopy.* IUAT, Paris.

Ionesco, H. (1972) *Rev. Tuberc. Pneumol.* **36,** 917.

Janicki, B. W., Chaparas, S. D., Daniel, T. M., Kubica, G. P., Wright, G. L. and Yee, G. S. (1971) *Amer. Rev. resp. Dis.* **104,** 602.

Jenkins, P. A., Marks J. and Schaefer, W. B. (1971) *Amer. Rev. resp. Dis.* **103,** 179; (1972) *Tubercle* **53,**118.

Jenkins, P. A. and Tsukamura, M. (1979) *Tubercle* **60,** 71.

Jensen, K. A. (1955) *Bull. Un. int. Tuberc.* **25,** 89.

Johne, H. A. and Frothingham, L. (1895) *Dtsch. Z. Tiermed.* **21,** 438.

Johnson, T. B. and Coghill, R. D. (1925) *J. Amer. chem. Soc.* **47,** 2838.

Jones, W. D. and Beam, R. E. (1969) *Canad. J. Microbiol.* **15,** 1112.

Jones, W. D. and David, H. L. (1970) *Amer. Rev. resp. Dis.* **102,** 818; (1972) *Tubercle* **53,** 35.

Jones, W. D. and Greenberg, J. (1977) *J. gen. Microbiol.* **99,** 389.

Jones, W. D. and Kubica, G. P. (1965) *Zbl. Bakt.* **196,** 68.

Juhasz, S. E. and Bönicke, R. (1965) *Canad. J. Microbiol.* **11,** 235; (1970) *Pneumonology* **142,** 181.

Juhasz, S. E., Gelbart, S. and Harize, M. (1969) *J. gen. Microbiol.* **56,** 251.

Juhlin, I. and Winblad, S. (1973) *Acta path. microbiol. scand.* **B81,** 179.

Juy, D. and Chedid, L. (1975) *Proc. nat. Acad. Sci., Wash.* **72,** 4105.

Kanetsuna, F. and Bartoli, A. (1972) *J. gen. Microbiol.* **70,** 209.

Kanetsuna, F., Imaeda, T. and Cunto, G. (1969) *Biochim. biophys. acta* **173,** 341.

Kanetsuna, F. and San Blas, G. (1970) *Biochim. biophys. acta* **208,** 434.

Käppler, W. (1965) *Bull. Un. int. Tuberc.* **38,** 46; (1973) *Proc. 3rd Int. Coll. Mycobacteria,* p. 11. International Colloquia Series, Antwerp.

Karasseva, V., Weiszfeiler, J. G. and Krasznay, E. (1965) *Acta microbiol. hung.* **12,** 275.

Karlson, A. G. and Lessel, E. F. (1970) *Int. J. syst. Bact.* **20,** 273.

Kato, L. and Ishaque, M. (1977) *Int. J. Leprosy* **45,** 38.

Kato, M. (1969) *Amer. Rev. resp. Dis.* **100,** 47.

Kato, M. and Goren, M. B. (1974) *Infect. Immun.* **10,** 733.

Katsunuma, N. and Nakasato, H. (1954) *Kekkaku* **29,** 19.

Kazda, J. (1977) *Zbl. Bakt.* **237,** 90; (1978) *Ibid.* **B166,** 463.

Kazda, J., Irgens, L. M. and Muller, K. (1980) *Int. J. Leprosy* **48,** 1.

Kazda, J., Muller, K. and Irgens, L. M. (1979) *Acta. path. microbiol. scand.* **B87,** 97.

Kestle, D. G., Abbot, V. D. and Kubica, G. P. (1967) *Amer. Rev. resp. Dis.* **95,** 1041.

Kilburn, J. D., Silcox, V. A. and Kubica, G. P. (1969) *Amer. Rev. resp. Dis.* **99,** 94.

Kim, K. S., Salton, R. J. and Barksdale, L. (1976) *J. Bact.* **125,** 739.

Kinyoun, J. J. (1915) *Amer. J. publ. Hlth.* **5,** 867.

Kirchheimer, W. F. and Storrs, E. H. (1971) *Int. J. Leprosy* **39,** 693; (1972) *Ibid.* **40,** 212.

Kirschner, O. (1932) *Zbl. Bakt.* **124,** 403.

Kniker, W. T. and LaBorde, J. B. (1964) *Amer. Rev. resp. Dis.* **89,** 20.

Knox, R., Thomas, C. G. A., Lister, A. J. and Saxby, C. (1961) *Guy's Hosp. Rep.* **110,** 174.

Koch, R. (1882) *Berl. klin. Wschr.* **19,** 221. (Translated by M. Pinner, 1932, *Amer. Rev. Tuberc.* **25,** 285).

Kochan, I. (1973) *Curr. Top. Microbiol. Immunol.* **60,** 1.

Koch-Weser, D., Ebert, R. H., Barclay, W. R. and Lee, V. S. (1953) *J. Lab. clin. Med.* **42,** 828.

Kolbel, H. (1967) *Beitr. Klin. Tuberc.* **136,** 107.

Kolman, A. and Ehrenberg, L. (1978) *Mutation Res.* **49,** 297.

Koníček, J. and Koníčková-Radochová, M. (1975) *Fol. Microbiol., Praha* **20,** 382; (1978) *Ibid.* **23,** 261.

Koníčková-Radochová, M., Koníček, J. and Málek, I. (1970) *Fol. Microbiol., Praha* **15,** 88.

Konno, K. (1956) *Science, N.Y.* **124,** 985.

Konno, K., Feldmann, F. M. and McDermott, W. (1967) *Amer. Rev. resp. Dis.* **95,** 461.

Konno, K., Nagayama, H. and Oka, S. (1959) *Nature, Lond.* **184,** 1743.

Konno, K., Oizumi, K. and Oka, S. (1973) *Amer. Rev. resp. Dis.* **107,** 1006.

Kozloff, L. M., Sundar Raj, C. V., Nagaraja Rao, R., Chapman, V. A. and DeLong, S. (1972) *J. Virol.* **9,** 309.

Kubica, G. P. (1973) *Amer. Rev. resp. Dis.* **107,** 9.

Kubica, G. P. and Jones, W. D. (1965) *Zbl. Bakt.* **196,** 53.

Kubica, G. P., Jones, W. D., Abbot, V. D., Beam, R. E., Kilburn, J. O. and Cater, J. C. (1966) *Amer. Rev. resp. Dis.* **94,** 400.

Kubica, G. P. *et al.*, (1970) *Int. J. syst. Bact.* **20,** 161.

Kubica, G. P. *et al.*, (1972) *J. gen. Microbiol.* **73,** 55.

Kubin, M., Matuskova, E. and Kazda, J. (1969) *Zbl. Bakt.* **210,** 207.

Kushner, S., Dalatian, H., Sanjwzjio, J. L., Bach, F. L., Safir, S. R., Smith, V. K. and Williams, J. H. (1952) *J. Amer. chem. Soc.* **74,** 3617.

Küster, E. (1905) *Münch. med. Wschr.* **52,** 57.

Lanéele, G. and Asselineau, J. (1968) *Europ. J. Biochem.* **5,** 487.

Le Cam-Sagniez, M. and Sagniez, G. (1973) *Ann. Soc. Belge Méd. trop.* **53,** 347.

Lechevalier, M. P., Lechevalier, H. A. and Horan, A. C. (1973) *Canad. J. Microbiol.* **19,** 965.

Lederer, E. (1967) *Chem. Phys. Lip.* **1,** 294; (1971) *Pure appl. Chem.* **25,** 135.

Lehmann, K. B. and Neumann, R. (1896) *Atlas und Grundriss der Bakteriologie und Lehrbuch der speciellen bakteriologischen Diagnostik*, 1st Edn. J. F. Lehmann, Munchen; (1899) *Ibid.* 2nd edn.

Levy-Frebault, V. and David, H. L. (1979) *J. gen. Microbiol.* **115,** 317.

Lind, A. (1959) *Int. Arch. Allergy* **14,** 264; (1960*a*) *Ibid.* **16,** 336; (1960*b*) *Ibid.* **17,** 1; (1960*c*) *Ibid.* **17,** 300; (1961*a*) *Ibid.* **18,** 305; (1961*b*) *Ibid.* **19,** 1; (1978) *Ann. Microbiol., Inst. Pasteur* **A129,** 99.

Lind, A. and Norlin, M. (1963) *Scand. J. clin. Lab. Invest.* **15,** 152.

Linden, A. C. van der and Thijsse, G. J. E. (1965) *Advanc. Enzymol.* **27,** 469.

Linell, F. and Norden, A. (1954) *Acta tuberc. scand.* **33,** Suppl., 1.

Long, E. R. (1926) *Amer. Rev. Tuberc.* **13,** 393.

Löwenstein, E. (1930) *Dtsch. med. Wschr.* **56,** 1010.

MacCallum, P., Tolhurst, J. C., Buckle, G. and Sissons, H. A. (1948) *J. Path. Bact.* **60,** 93.

McCarthy, C. (1971) *Appl. Microbiol.* **22,** 546.

McDiarmid, A. (1962) *Tuberculosis in Free-Living Wild Animals*, p. 10. FAO Agriculture Studies no. 57, Rome.

Macham, L. P. and Ratledge, C. (1975) *J. gen. Microbiol.* **89,** 379.

Macham, K. P., Stephenson, M. C. and Ratledge, C. (1977) *J. gen. Microbiol.* **101,** 41.

Mackaness, G. B. (1956) *Amer. Rev. Tuberc.* **74,** 718.

McLeod, I. M. (1977) *Tubercle* **58,** 39.

MacNaughton, A. W. and Winder, F. G. (1977) *Molec. gen. Genet.* **150,** 301.

McSwiggan, D. A. and Collins, C. H. (1974) *Tubercle* **55,** 291.

Magnusson, M. (1962) *Amer. Rev. resp. Dis.* **86,** 395; (1967) *Z. Tuberk.* **127,** 55; (1971) *Amer. Rev. resp. Dis.* **104,** 377.

Mankiewicz, E. (1972) *Canad. J. publ. Hlth.* **63,** 342.

Mankiewicz, E. and Liivak, M. (1975) *Amer. Rev. resp. Dis.* **111,** 307.

Mankiewicz, E. and Redmond, W. B. (1968) *Amer. Rev. resp. Dis.* **98,** 41.

Mankiewicz, E. and Tamari, M. G. (1972) *Amer. Rev. resp. Dis.* **106,** 609.

Marchoux, E. and Halphen, E. (1912) *C.R. Soc. Biol., Paris,* **73,** 249.

Marchoux, E. and Sorel, F. (1912) *Ann. Inst. Pasteur* **26,** 675.

Markovitz, J., Vilkas, E. and Lederer, E. (1971) *Eur. J. Biochem.* **18,** 287.

Marks, J. (1965) *Tubercle*, **46,** 65; (1976) *Ibid.*, **57,** 207.

Marks, J., Jenkins, P. A. and Schaefer, W. C. (1969) *Tubercle* **50,** 394; (1971) *Ibid.* **52,** 219.

Marks, J., Jenkins, P. A. and Tsukamura, M. (1972) *Tubercle* **53,** 210.

Marks, J. and Schwabacher, H. (1965) *Brit. med.* J. **i,** 32.

Matthews, P. R. J., Brown, A. and Collins, P. (1979) *J. appl. Bact.* **46,** 425.

Matthews, P. R. J. and McDiarmid, A. (1979) *Vet. Rec.* **104,** 286.

Matthews, P. R. J., McDiarmid, A., Collins, P. and Brown, A. (1977) *J. med. Microbiol.* **11,** 53.

Matthews, P. R. J. and Sargent, A. (1977) *Brit. vet. J.* **133,** 399.

Mattman, L. H. (1970) *Ann. N.Y. Acad. Sci.* **174,** 852.

Meissner, G. *et al.* (1974) *J. gen. Microbiol.* **83,** 207.

Middlebrook, G. (1954) *Amer. Rev. Tuberc.* **69,** 472.

Middlebrook, G., Coleman, C. and Schaefer, W. B. (1959) *Proc. nat. Acad. Sci., Wash.* **45,** 1801.

Middlebrook, G., Dubos, R. J. and Pierce, C. H. (1947) *J. exp. Med.* **86,** 175.

Migliore, D. and Jolles, P. (1968) *FEBS Lett.* **2,** 7; (1969) *C.R. Acad. Sci., Paris,* **D269,** 2268.

Millman, I. (1958) *Proc. Soc. exp. Biol., N.Y.* **99,** 216.

Minnikin, D. E., Alshamaony, L. and Goodfellow, M. (1975) *J. gen. Microbiol.* **88,** 200.

Mitchell, D. N. and Rees, R. J. W. (1969) *Lancet* **ii,** 81; (1970) *Ibid.* **ii,** 168; (1976) *Ann. N.Y. Acad. Sci.* **278,** 233; (1979) *Proc. 8th Int. Conf. Sarcoidosis* Cardiff.

Mitchell, D. N., Rees, R. J. W. and Goswami, K. K. A. (1976) *Lancet* **ii,** 761.

Mitchison, D. A. (1970) *Pneumonology* **142,** 131.

Mitchison, D. A. and Dickinson, J. M. (1978) *Bull. Un. int. Tuberc.* **53,** 254.

Mitchison, D. A., Wallace, J. G., Bhatia, A. L., Selkon, J. B., Subbaiah, T. V. and Lancaster, M. C. (1960) *Tubercle* **41,** 1.

Mizuguchi, Y. (1974) *J. Bact.* **117,** 914.

Mizuguchi, Y., Suga, K. and Tokunaga, T. (1976) *Jap. J. Microbiol.* **20,** 435.

Mizuguchi, Y., Suga, K., Yamada, T. and Kawaguchi, K. (1978) In: *Genetics of the Actinomycetales*, p. 73. Ed. by E. Freerksen, I. Tarnok and J. H. Thumin. Gustav Fischer, Stuttgart.

Mizuguchi, Y. and Tokunaga, T. (1971) *Jap. J. Microbiol.* **15,** 359.

Moeller, A. (1898) *Dtsch. med. Wschr.* **24,** 376.

Mori, T. (1975) *La Lepro* **44,** 49.

Moučka, C. and Kaňkanovská, F. (1967) *Rozhl. Tuberk.* **27,** 16.

Moulton, R. G., Dietz, T. M. and Marcus, S. (1971) *Infect. Immun.* **3,** 378.

Much, H. (1907) *Beitr. Klin. Tuberk.* **8,** 85.

Mudd, S., Winterscheid, L. C. DeLamater, E. D. and Henderson, H. J. (1951) *J. Bact.* **62,** 459.

Muirhead, R. H., Gallacher, J. and Birn, K. J. (1974) *Vet. Rec.* **95,** 552.

Nair, N. C. *et al.* (1964) *Tubercle* **45,** 345.

Nakamura, M. (1972) *J. gen. Microbiol.* **73,** 193; (1975) *Kurume med. J.* **22,** 67.

Nakamura, R. M. (1970) In: *Host-virus relationships in Mycobacterium, Nocardia and Actinomyces*, p. 166. Ed. by S. E. Juhasz and G. Plummer. Charles C Thomas, Springfield, Ill.

Nakayama, Y. and Takeya, K. (1963) *Nature, Lond.* **198,** 1113.

Nassau, E. and Hamilton, G. (1957) *Tubercle* **38,** 387.

Nassau, E. and Nelstrop, A. E. (1976) *Tubercle* **57,** 197.
Nayebi, M. (1970) *J. med. Lab. Technol.* **27,** 218.
Neelsen, F. (1883) *Zbl. med. Wiss., Berlin* **21,** 497.
Noll, H., Bloch, H. Asselineau, J. and Lederer, E. (1956) *Biochim. biophys. acta* **20,** 299.
Nordström, G. and Grange, J. M. (1974) *Acta path. microbiol. scand.* **B82,** 87.
Norgard, M. and Imaeda, T. (1978) *J. Bact.* **133,** 1254.
Norlin, M. (1965) *Bull. Un. int. Tuberc.* **36,** 25.
Norlin, M., Lind, A. and Ouchterlony, Ö. (1969) *Z. Immun.-Forsch.* **137,** 241.
Norlin, M., and Navalkar, R. (1966) *Bull. Un. int. Tuberc.* **38,** 52.
Norman, J. O. and Williams, R. P. (1962) *Nature, Lond.* **193,** 702.
Oerskov, J. (1932) *Zbl. Bakt.* **123,** 271.
Ogawa, T. and Hiraki, M. (1972) *La Lepro* **41,** 113.
Osborn, T. W. (1976) *Tubercle* **57,** 181.
Pangborn, M. C. (1968) *Ann. N.Y. Acad. Sci.* **154,** 133.
Parlett, R. and Youmans, G. P. (1956) *Amer. Rev. Tuberc.* **73,** 637.
Patel, P. V. and Ratledge, C. (1973) *Trans. biochem. Soc.* **1,** 886.
Pattyn, S. R. (1970) *Zbl. Bakt.* **215,** 99.
Pattyn, S. R. and Dommisse, R. (1973) *Proc. 3rd Int. Coll. Mycobacteria*, p. 63. International Colloquia Series, Antwerp.
Pattyn, S. R. Hermans-Boveroulle, M. T. and Van Ermengem, K. (1968) *Zbl. Bakt.* **207,** 509.
Pattyn, S. R., Magnusson, M., Stanford, J. L. and Grange, J. M. (1974) *J. med. Microbiol.* **7,** 67.
Paul, R. C., Stanford, J. L. and Carswell, J. W. (1975) *J. Hyg., Camb.* **75,** 57.
Pellegrino, P. L. (1906) *Ann. Igiene* **16,** 163.
Penso, G., Castelnuovo, G., Gaudiano, A., Princivalle, M., Vella, L. and Zampieri, A. (1952) *R.C. Ist sup. Sanitá.* **15,** 491.
Prabhakaram, K. and Kirchheimer, W. F. (1966) *J. Bact.* **92,** 1267.
Pratt, W. B. (1977) *The chemotherapy of infection.* Oxford University Press.
Prissick, F. H. and Masson, A. M. (1956) *Canad. med. Ass. J.* **75,** 798.
Proskauer, B. and Beck, M. (1894) *Z. Hyg. InfektKr.* **18,** 128.
Protivinsky, R. (1971) *Antibiot. Agents Chemother.* **17,** 101.
Rado, T. A. *et al.* (1975) *Amer. Rev. resp. Dis.* **111,** 459.
Raleigh, J. W. and Wichelhausen, R. (1973) *Amer. Rev. resp. Dis.* **108,** 639.
Raleigh, J. W., Wichelhausen, R. W., Rado, T. A. and Bates, J. H. (1975) *Amer. Rev. resp. Dis.* **112,** 497.
Ramakrishnan, T., Suryanarayana Murthy, P. and Gopinathan, K. P. (1972) *Bact. Rev.* **36,** 65.
Rankin, J. D. and McDiarmid, A. (1969) *Symp. Zool. Soc., Lond.* **24,** 119.
Ratledge, C. (1977) *The Mycobacteria.* Meadowfield Press, England.
Ratledge, C. and Hall, M. J. (1971) *J. Bact.* **108,** 314.
Ratledge, C. and Marshall, B. J. (1972) *Biochim. biophys. acta* **279,** 58.
Ratledge, C. and Winder, F. G. (1962) *Biochem. J.* **84,** 501.
Redmond, W. B. (1963) *Advance. Tuberc. Res.* **12,** 191; (1970*a*) *Bull. Un. int. Tuberc.* **43,** 214; (1970*b*) *Pneumonology* **142,** 191.
Redmond, W. B. and Ward, D. M. (1966) *Bull. World Hlth. Org.* **35,** 563.
Reed, G. B. (1957) In: *Bergey's Manual of Determinative Bacteriology*, 7th edn, p. 695. Ed. by R. S. Breed, R. G. E. Murray and L. D. Smith. Williams and Wilkins, Baltimore.
Rees, R. J. W., Waters, N. F. R., Wedell, A. G. M. and Palmer, E. (1967) *Nature, Lond.* **215,** 559.
Reiner, E., Beam, R. E. and Kubica, G. P. (1969) *Amer. Rev. resp. Dis.* **99,** 750.
Report (1952) *Brit. med. J.* **ii,** 735; (1955) *Fitzsimmons Army Hosp. Rep.* no. 1; (1958) *J. Hyg., Camb.* **56,** 488; (1970) *Proc. Microbiol. Res. Group, Acad. Sci. Hung.* **3,** 41; (1973) *Proc. 3rd Coll. Mycobacterium.* International Colloquia Series, Antwerp; (1974) *Laboratory Methods in Veterinary Mycobacteriology*, U.S. Dept. Agric., Veterinary Services Lab., Ames, Iowa; (1979) *Int. J. Leprosy* **47,** Suppl., 291. (1982) *Rev. infect. Dis* **3,** 813.
Reznikov, M. and Dawson, D. J. (1973) *Appl. Microbiol.* **26,** 470.
Ridell, M., (1974) *Int. J. syst. Bact.* **24,** 64; (1975) *Ibid.* **25,** 124; (1977) *Serological Relationships in Nocardia, Mycobacteria, Corynebacteria and the Rhodochrous Taxon with Special Reference to Taxonomy*, University of Göteborg, Sweden.
Ridell, M., Baker, R., Lind, A., Norlin, M. and Ouchterlony, Ö. (1976) *Int. Arch. Allergy, Basel* **52,** 297.
Ridley, D. S. (1977) *Skin Biopsy in Leprosy.* Documenta Geigy, Ciba-Geigy, Basle.
Robbins, M. L., Amako, K., Muraoka, S. and Takeya, K. (1970) In: *Host-virus relationships in Mycobacterium, Nocardia and Actinomyces*, p. 103. Ed. by S. E. Juhasz and G. Plummer. Charles C Thomas, Springfield, Ill.
Roberts, D. B., Wright, G. L., Affronti, L. F. and Reich, M. (1972) *Infect. Immun.* **6,** 564.
Rockwell, G. E. and Highberger, J. H. (1926) *J. infect. Dis.* **38,** 92.
Röhrscheidt, E., Tarnok, Z. and Tarnok, I. (1970) *Zbl. Bakt.* **215,** 550.
Runyon, E. H. (1959) *Med. Clin. N. Amer.* **43,** 273; (1965) *Advanc. Tuberc. Res.* **14,** 235; (1968) *J. Bact.* **95,** 734; (1970) *Amer. J. clin. Path.* **54,** 578; (1972) *Int. J. syst. Bact.* **22,** 50; (1974) *Proc. 1st Int. Conf. Nocardia.* McCowan, Augusta, Ga.
Rupprecht, J. (1904) *Über saurefeste Bazillen nebst. Beschreibung eines Falles von Spontaner Froschtuberculose.* Ernst Heinrich Moritz, Stuttgart.
Russell, R. L., Jann, C. J. and Froman, S. (1960) *Amer. Rev. resp. Dis.* **82,** 384.
Rytíř, V., Hubáček, J. and Málek, I. (1966) *Fol. microbiol., Praka* **11,** 257.
Saito, H., Hosakawa, H. and Tasaka, H. (1968) *Amer. Rev. resp. Dis.* **97,** 474.
Saito, H., Watanabe, T. and Tomioka, H. (1979) *Antimicrob. Agents Chemother.* **15,** 504.
Saito, H. *et al.* (1977) *Int. J. syst. Bact.* **27,** 75.
Saroja, D. and Gopinathan, K. P. (1973) *Antimicrob. Agents Chemother.* **6,** 643.
Sauton, B. (1912) *C.R. Acad. Sci., Paris* **155,** 860.
Schaefer, W. B. (1965) *Amer. Rev. resp. Dis.* **92,** Suppl., 85; (1966) *Ibid.* **94,** 478; (1967) *Ibid.* **96,** 115; (1968) *Ibid.* **97,** 18.
Schaefer, W. B., Wolinsky, E., Jenkins, P. A. and Marks, J. (1973) *Amer. Rev. resp. Dis.* **108,** 1320.

Schröder, K. H. and Juhlin, I. (1977) *Int. J. syst. Bact.* **27,** 241.
Schröder, K. H. and Magnusson, M. (1968) *Zbl. Bakt.* **207,** 498.
Schwabacher, H. (1959) *J. Hyg., Camb.* **57,** 57.
Schwartz, W. S. and Moyer, R. E. (1953) *Proc. 12th Conf. Chemother. Tuberc.* p. 296. Veterans Adm., Washington.
Schweinitz, E. von (1893) *Zbl. Bakt.* **14,** 330.
Sehrt, I., Käppler, W. and Lange, A. (1976) *Z. Erkr. Atmungsorgane* **144,** 146.
Seibert, F. B. (1949) *Amer. Rev. Tuberc.* **59,** 86.
Seibert, F. B. and Glenn, J. T. (1941) *Amer. Rev. Tuberc.* **44,** 9.
Shepard, C. C. (1960) *J. exp. Med.* **112,** 454.
Skerman, V. B. D., McGowan, V. and Sneath, P. H. A. (1980) *Int. J. syst. Bact.* **30,** 225.
Šlosárek, M. (1978) *Fol. Microbiol., Praha* **23,** 140.
Šlosárek, M. and Mezensky, L. (1974) *J. Hyg. Epidem., Praha* **18,** 22.
Šlosárek, M., Sula, L., Theophilus, S. and Hruby, L. (1978) *Int. J. Leprosy* **46,** 154.
Smith, C. R. (1947) *Publ. Hlth Rep., Wash.* **62,** 1285.
Smith, D. W., Randall, H. M., MacLennan, A. P. and Lederer, E. (1960) *Nature, Lond.* **186,** 887.
Smith, H. W. (1953) *J. Path. Bact.* **66,** 375.
Smith, N. (1960) *Tubercle* **41,** 203.
Smithburn, K. C. (1936) *J. exp. Med.* **63,** 95.
Snow, G. A. (1970) *Bact. Rev.* **34,** 99.
Snow, G. A. and White, A. J. (1969) *Biochem. J.* **115,** 1031.
Sommers, H. M. (1977) The identification of mycobacteria. In: *Technical Improvement Service*, no. 28, p. 1. Ed. by D. M. Baer. American Society of Clinical Pathologists.
Sriprakash, K. S. and Ramakrishnan, T. (1970) *J. gen. Microbiol.* **60,** 125.
Stacey, M. (1955) *Advanc. Tuberc.* Res. **6,** 7.
Stanford, J. L. (1973*a*) *Ann. Soc. Belge. Méd. trop.* **53,** 339; (1973*b*) *J. med. Microbiol.* **6,** 405; (1973*c*) *Ibid.* **6,** 435; (1976) *Leprosy Rev.* **47,** 87.
Stanford, J. L. and Beck, A. (1968) *J. Path. Bact.* **95,** 131.
Stanford, J. L. and Grange, J. M. (1974) *Tubercle* **55,** 143.
Stanford, J. L. and Gunthorpe, W. J. (1969) *J. Bact.* **98,** 375; (1971) *Brit. J. exp. Path.* **52,** 627.
Stanford, J. L. and Muser, R. (1969) *Tubercle* **50,** Suppl. 80.
Stanford, J. L., Pattyn, S. R., Portaels, F. and Gunthorpe, W. J. (1972) *J. med. Microbiol.* **5,** 177.
Stanford, J. L. and Paul, R. C. (1973) *Ann. Soc. Belge. Méd. trop.* **53,** 389.
Stanford, J. L. and Wong, J. K. C. (1974) *Brit. J. exp. Path.* **55,** 291.
Stanford, J. L. *et al.* (1975) *Brit. J. exp. Path.* **56,** 579.
Stefansky, W. K. (1903) *Zbl. Bakt.* **33,** 481.
Stonebrink, B. (1957) *Proc. Tuberc. Res. Coun.* **44,** 67.
Strauss, I. and Gamaléia, N. (1891) *Arch. Med. exp.* **3,** 457.
Suga, K. and Mizuguchi, Y. (1974) *Jap. J. Microbiol.* **18,** 139.
Sundar Raj, C. V. and Ramakrishnan, T. (1970) *Nature, Lond.* **228,** 280.
Suzanne, M. and Penso, G. (1953) *R.C. 6th Int. Congr. Microbiol., Roma* **2,** 382.
Sweany, H. C. (1928) *Amer. Rev. Tuberc.* **17,** 53.
Tacquet, A., Tison, F., Polspoel, B., Roos, P. and Devulder, B. (1966) *Ann. Inst. Pasteur* **111,** 86.
Takahashi, Y. (1962) *Amer. Rev. resp. Dis.* **85,** 708.
Takayama, K., Armstrong, E. L. and David, H. L. (1974) *Amer. Rev. resp. Dis.* **110,** 43.
Takeya, K., Nakayama, Y. and Muraoka, S. (1970) *Amer. Rev. resp. Dis.* **102,** 982.
Takeya, K., Shimamoto, M. and Mizuguchi, Y. (1978) *J. gen. Microbiol.* **109,** 215.
Takeya, K. and Tokiwa, H. (1972) *Int. J. syst. Bact.*, **22,** 178; (1974) *Amer. Rev. resp. Dis.* **109,** 304.
Tanaka, A., Hirota, N. and Sugiyama, K. (1967) *Int. Arch. Allergy, Basel* **32,** 349.
Taneja, R., Malik, U. and Khuller, G. K. (1979) *J. gen. Microbiol.* **113,** 413.
Tarnok, I. and Bönicke, R. (1970) *Bull. Un. int. Tuberc.* **43,** 210.
Tarnok, I., Pechmann, H., Krallmann-Wenzel, U., Röhrscheidt, E. and Tarnok, Z. (1979) *Zbl. Bakt.* **244,** 302.
Tarnok, I. and Tarnok, Z. (1970) *Tubercle* **51,** 305; (1971) *Ibid.* **52,** 127; (1978) In: *Genetics of the Actinomycetales*, p. 243. Ed. by E. Freerksen, I. Tarnok and J. H. Thumin. Gustav Fischer, Stuttgart.
Taylor, A. W. (1951) *J. Path. Bact.* **63,** 333.
Thorel, M. F. (1976) *Ann. Microbiol., Inst. Pasteur* **B127,** 41.
Tokunaga, T., Mizuguchi, Y. and Suga, K. (1973) *J. Bact.* **113,** 1104.
Tokunaga, T. and Sellers, M. I. (1964) *J. exp. Med.* **119,** 139.
Trnka, L. and Mison, P. (1971) *Fol. Microbiol., Praha* **16,** 97.
Truant, J. P., Brett, W. A. and Thomas, W. (1962) *Bull. Henry Ford Hosp.* **10,** 287.
Tsukamura, M. (1966*a*) *J. gen. Microbiol.* **45,** 253; (1966*b*) *Med. Biol., Tokyo* **73,** 203; (1967) *Tubercle* **48,** 311; (1970) *Pneumonology* **142,** 93; (1972) *Med. Biol., Tokyo* **85,** 229; (1973) *J. gen. Microbiol.* **74,** 193; (1975) *Identification of mycobacteria.* National Sanatorium, Chubu Chest Hosp., Obu, Aichi-ken, Japan; (1979) *Actinomyces* **14,** 57.
Tsukamura, M., Hashimoto, H. and Noda, Y. (1960) *Amer. Rev. resp. Dis.* **81,** 403.
Tsukamura, M., Toyama, H. and Mizuno, S. (1965) *Med. Biol., Tokyo* **70,** 232.
Tsukamura, M. and Tsukamura, S. (1964) *Tubercle* **45,** 64.
Tsukamura, S. (1964) *Jap. J. Tuberc.* **12,** 1, 7.
Tuboly, S. (1965) *Acta microbiol. hung.* **12,** 233.
Twort, F. W. (1910) *Proc. roy. Soc. B.* **83,** 156.
Twort, F. W. and Ingram, G. L. Y. (1912) *Proc. roy. Soc. B.* **84,** 517; (1913) *A monograph on Johne's disease.* London.
Vajda, B. P. (1980) *J. gen. Microbiol.* **116,** 253.
Valdivia-Alvarez, J., Suarez-Mendez, R. and Echemendia-Font. M. (1971) *Bol. Hig. Epidem.* **9,** 65.
Vandiviere, H. M., Vandiviere, M. R. and Seibert, F. B. (1961) *J. infect Dis.* **108,** 45.
Vestal, A. (1975) *Procedures for the isolation and identification of mycobacteria*, US Dept. of Health, Education and Welfare, Publ. no. (CDC) 79-8230.
Vilkas, E. and Markovitz, J. (1972) *C.R. Acad. Sci., Paris* **C275,** 913.
Vilkas, E., Markovitz, J., Amar-Nacasch, C. and Lederer, E. (1971) *C.R. Acad. Sci., Paris* **C273,** 845.
Virtanen, S. (1960) *Acta tuberc. scand.* **48,** Suppl. 1.
Vivien, J. N., Thibier, R. and Lepeuple, A. (1972) *Advanc. Tuberc. Res.* **18,** 148.
Wade, H. W. (1957) *Stain Technol.* **32,** 287.
Wang, L. and Takayama, K. (1972) *Antimicrob. Agents. Chemother.* **2,** 438.
Watanakunakorn, C. and Trott, A. (1973) *Amer. Rev. resp. Dis.* **107,** 846.
Wayne, L. G. (1962) *Amer. Rev. resp. Dis.* **86,** 651; (1966) *Ibid.* **93,** 919; (1971) *Infect. Immun.* **3,** 36.

Wayne, L. G., Diaz, G. A. and Doubek, J. R. (1968) *Amer. Rev. resp. Dis.* **97,** 909.

Wayne, L. G. and Doubek, J. R. (1968) *Appl. Microbiol.* **16,** 925.

Wayne, L. G., Doubek, J. R. and Russell, R. L. (1964) *Amer. Rev. resp. Dis.* **90,** 588.

Wayne, L. G. and Lessel, E. F. (1969) *Int. J. syst. Bact.* **19,** 257.

Wayne, L. G. *et al.* (1971) *J. gen. Microbiol.* **66,** 255; (1978) *Ibid.* **109,** 319; (1981) *Int. J. syst. Bact.* **31,** 1.

Weiszfeiler, J. G., Czanik, P. and Baktai, L. (1968) *Zbl. Bakt.* **207,** 517.

Weiszfeiler, J. G., Karasseva, V. and Karczag, E. (1971) *Acta microbiol. hung.* **18,** 247.

Wells, A. Q. (1937) *Lancet.* **i.** 1221.

Wheeler, W. C. and Hanks, J. H. (1965) *J. Bact.* **89,** 889.

Whitcomb, F. C., Foster, M. C. and Dukes, C. D. (1962) *Texas Rep. Biol. Med.* **20,** 508.

Whitehead, J. E. M., Wildy, P. and Engbaek, H. C. (1953) *J. Path. Bact.* **28,** 69.

Wieten, G., Haverkamp, J., Meuzelaar, H. L. C., Engel, H. W. B. and Berwald, L. G. (1981) *J. gen. Microbiol.* **122,** 109.

Wilhelm, G. and Sellier, J. L. (1976) *Zbl. Bakt.* **A234,** 69.

Willett, H. P. and Thacore, H. (1966) *Canad. J. Microbiol.* **12,** 11; (1967) *Ibid.* **13,** 481.

Wilson, G. S. (1925) *J. Path. Bact.* **28,** 69; (1942) *The pasteurization of milk.* Edward Arnold, London.

Wimpenny, J. W. T. (1967) *J. gen. Microbiol.* **47,** 389.

Winder, F. G. (1960) *Amer. Rev. resp. Dis.* **81,** 68; (1964) In: *Chemotherapy of Tuberculosis*, p. 111. Ed. by V. C. Barney. Butterworth, Lond.

Winder, F. G. and Collins, P. B. (1970) *J. gen. Microbiol.* **63,** 41.

Winder, F. G., Collins, P. B. and Whelan, D. (1971) *J. gen. Microbiol.* **66,** 379.

Winder, F. G. and Dennery, J. (1959) *Nature, Lond.* **184,** 742.

Winder, F. G. and O'Hara, C. (1962) *Biochem. J.* **82,** 98; (1964) *Ibid.* **90,** 122.

Winder, F. G. and Levin, M. F. (1971) *Biochim. biophys. acta.* **247,** 542.

Wolinsky, E. (1979) *Amer. Rev. resp. Dis.* **119,** 107.

Wolinsky, E. and Rynearson, T. K. (1968) *Amer. Rev. resp. Dis.* **97,** 1032.

Wolinsky, E. and Schaefer, W. B. (1973) *Int. J. syst. Bact.* **23,** 182.

Wright, G. L., Affronti, L. F. and Reich, M. (1972) *Infect. Immun.* **5,** 432.

Yamada, T., Masuda, K., Kawaguchi, K., Mizuguchi, Y. and Suga, K. (1978) In: *Genetics of the Actinomycetales*, p. 81. Ed. by E. Freerksen, I. Tarnok, and J. H. Thumin, Gustav Fischer, Stuttgart.

Yano, I. *et al.* (1978) *Biomed. Mass. Spectrometry* **5,** 14.

Yates, M. D. and Collins, C. H. (1979) *Ann. Microbiol., Inst. Pasteur*, **B130,** 13.

Yates, M. D., Collins, C. H. and Grange, J. M. (1978) *Tubercle* **59,** 143.

Yegian, D. and Porter, K. R. (1944) *J. Bact.* **48,** 83.

Yoneda, M. and Fukui, Y. (1965) *Amer. Rev. resp. Dis.* **92,** Suppl., 9.

Youmans, G. P. (1944) *Proc. Soc. exp. Biol., N.Y.* **57,** 119.

Ziehl, F. (1882) *Dtsch. med. Wschr.* **8,** 451.

25

Corynebacterium and other coryneform organisms

Graham Wilson

Introductory

The generic name *Corynebacterium* was allotted by Lehmann and Neumann in 1896 to the group of bacteria containing the diphtheria bacillus and other species resembling it in morphology. By its derivation the name emphasizes the tendency to the formation of club-like forms that is characteristic of the type species and of several other species within the genus.

The diphtheria bacillus was observed in the diphtheritic false membrane by Klebs in 1883, and was cultivated the following year by Loeffler, whose classical paper in 1884 provided a description of the causative organism that afforded a standard of reference for all subsequent studies on this bacterial group. Though an organism that was probably *C. xerosis* was isolated from the conjunctival sac by Reymond and his colleagues in 1881, the diphtheria bacillus has been generally accepted as the type species, partly because of its importance as a human pathogen and partly because the exact identity of Reymond's organism is doubtful. Numerous so-called diphtheroid bacilli have been described from time to time. Many of these have been given specific names, probably without adequate justification (Graham-Smith 1908). Here as elsewhere, we propose to adopt a conservative view, and to list as species only those organisms which have been adequately described and appear from this description to be reasonably well differentiated.

The diphtheria bacillus is generally regarded as a single species, but it may be mentioned here that it is divisible into three types—*gravis*, *intermedius* and *mitis*—to which reference will frequently be made throughout this chapter; they are described in detail in a later section.

The term 'coryneform' was used originally to describe the club-shaped or wedge-shaped cells of *C. diphtheriae* and related animal parasites (Jones 1975). It is now generally applied to gram-positive, non-mycelial, non-sporing bacteria that exhibit a pleomorphic morphology. The group of coryneform bacteria is a large one, including the genera *Corynebacterium*, *Arthrobacter*, *Brevibacterium*, *Microbacterium*, *Cellulomonas*, *Listeria*, *Erysipelothrix*, *Mycobacterium*, and some species of *Nocardia* (Bousfield 1972). Here we are concerned only with the genus *Corynebacterium* as defined below. The numerical taxonomic studies that have been made by Bousfield (1972) and Jones (1975) lead to the conclusion that this genus should be confined to *C. diphtheriae*, *C. ulcerans* and certain animal parasites, together with *Microbacterium flavum*. Excluded from it are several well known organisms, notably '*C.*' *pyogenes*, '*C.*' *haemolyticum*, '*C.*' *equi*, '*C.*' *kutscheri* and *Microbacterium lacticum* (see also Barksdale 1970). Since, however, the taxonomic position of these organisms has not yet been determined, it will be convenient to describe them in a separate section of this chapter, indicating by inverted commas that they do not belong strictly to the genus. The term 'diphtheroid' we shall use for descriptive purposes to include organisms other than *C. diphtheriae* that fall into the genus *Corynebacterium* (see p. 103). (For a review of coryneform bacteria as a whole, see Sebald *et al.* 1965, Jensen 1866, Bousfield 1972, Bousfield and Callely 1978.)

Corynebacterium

Definition Gram-positive rods arranged usually in Chinese letter or palisade forms with little tendency to branching. Clubbed ends frequent. Staining usually segmental or granular. Non-sporing, non-acid-fast. Generally non-motile. Predominantly parasitic. Facultative anaerobes. Tend to be exacting in their nutritional requirements. May or may not form acid in carbohydrate media, but never gas. Contain corynemycolic acids; and in the cell wall *meso*-diaminopimelic acid, arabinose and galactose. G + C content of DNA: 55–60 moles per cent. Some species pathogenic and form exotoxins.

Type species: *Corynebacterium diphtheriae*.

Morphology

The club-form, from which the name is derived, is only one of many shapes which may be assumed by the individual cells of the type species, *C. diphtheriae*. This organism is indeed characteristically pleomorphic. One of the most typical forms (see Fig. 25.1) in films prepared from a 24-hr culture on Loeffler's serum is that of a long, rather slender bacillus, often slightly curved, with rounded, somewhat swollen ends and sometimes with localized swellings elsewhere, and staining unevenly with such dyes as methylene blue: but in the same, or in other, cultures there will also be found much shorter forms, cells which stain solidly and evenly, cells in which the irregular staining takes the form of a series of transverse bars, and cells in which the combination of uneven staining and localized swellings gives to a single bacillus the appearance of a short chain of streptococci. As Goldsworthy and Wilson (1942) showed, the morphology of diphtheria bacilli varies greatly on different batches of Loeffler's serum, depending on its mode of preparation. A strain which, on one batch, develops into typical long curved granular forms may appear on another batch as short rods devoid of granules, simulating a diphtheroid bacillus.

Fig. 25.1 *C. diphtheriae*. From 24-hr culture on Loeffler's serum (× 1000).

Another feature that characterizes *C. diphtheriae* as a species, and serves to distinguish it from some, but by no means all, of the related diphtheroids, is the presence of the metachromatic granules described by Babes (1886) and by Ernst (1888, 1889). These granules are coloured a reddish purple when a film preparation is stained with a suitable sample of methylene blue. They may be demonstrated more clearly by the differential stain devised by Neisser, or by one of its many modifications. A single cell may contain one or more of these granules, seldom more than half a dozen, usually two or three. When only one or two are present they show a tendency to be situated at one or both poles. The granules consist of an electron-dense network filled with a substance of little electron density that is soluble in ribonuclease, leaving intracytoplasmic membrane systems—mesosomes (Hentrich *et al.* 1977).

The arrangement of the bacilli in film preparations is at least as characteristic as the form of the individual cells. Adjacent cells tend to lie at any angle to one another, forming a V or an L according to the degree of angular displacement; and groups of such pairs form characteristic clusters, resembling Chinese letters, or cuneiform writing. It would appear, from the observations of Hill (1898, 1899, 1902*a*, *b*), that this particular arrangement results from incomplete separation at the moment of division, the daughter cells remaining attached at one point, and bending on this attachment as on a hinge as growth proceeds (see also Bisset 1950).

Finally, it may be noted that *C. diphtheriae* provided the first instance in which true branching was demonstrated in a bacillary species. The observations of Hill showed that this appearance was not an artefact, but could be observed to take place during the growth of the living cell.

The morphological appearance of the diphtheroid bacilli varies considerably, and is influenced by the nature of the environmental conditions under which they are grown (see Müller 1957). Some, such as *C. ovis* and '*C.*' *kutscheri*, bear a close resemblance to *C. diphtheriae*; others, such as *C. hofmanni*, have their own characteristic morphology. On the whole, the diphtheroid bacilli are far more regular in their appearance than is the diphtheria bacillus; and though a given species of diphtheroid may be composed of long curved or of barred forms similar to those seen in diphtheria cultures, pleomorphism within the species is slight or absent.

All members of the *Corynebacterium* group are gram positive, but there is a good deal of variation in the ease with which they are decolorized. The diphtheria bacillus—apart from the metachromatic granules, which retain the stain tenaciously—is only weakly gram positive, so that if it is overdecolorized and counterstained with a red or brown dye, it may present a picture similar to that resulting from the use of Neisser's stain. Hofmann's bacillus, on the other hand, is very resistant to decolorization. The Gram stain is often of value in distinguishing between the weakly positive diphtheria bacillus and the strongly positive diphtheroid bacilli (Tomlinson 1966).

Most corynebacteria are non-capsulated and, apart from a few species, non-motile and non-flagellated. The occurrence of fat-like droplets in some strains of *C. diphtheriae*, seen in agar-block cultures treated with chloroform, was described by Ørskov (1948). Pili occur on *C. renale* (Yanagawa *et al.* 1968) and on numerous other species including *C. diphtheriae* (Yanagawa and Honda (1976). Their number varies greatly from one species to another. The cell wall contains *meso*-diaminopimelic acid (DAP), arabinose and galactose (Cummins and Harris 1956, Cummins 1971); and within the cell are mycolic acids of the *Corynebacterium* type as distinct from those of the *rhodochrous* type (Keddie and Cure 1977).

Cultural characters

The members of this group grow on ordinary nutrient agar, though to a variable degree. Many of the diphtheroids, including *C. hofmanni*, develop freely, but the growth of all members is improved by the presence of natural animal protein. For several decades Loeffler's serum constituted the medium of choice. In stroke cultures on this medium there is a fairly abundant growth within 24 hr, having a moist, slightly creamy and sometimes faintly pigmented appearance. The degree of development, however, varies, and all transitions are seen betweeen a slightly raised colourless film and a profuse succulent pigmented growth.

McLeod and his colleagues at Leeds (Anderson *et al.* 1931) recorded the differentiation of two types of the diphtheria bacillus on a blood tellurite agar medium. One, which was prevalent in severe cases, they called *gravis*; the other, which was isolated from milder cases, they called *mitis*. In a subsequent report (Anderson *et al.* 1933) they extended these observations, and concluded that certain strains which they had found previously to correspond to neither the *gravis* nor the *mitis* type constituted a third type—now generally known as *intermedius*. Robinson and Marshall (1934) at Manchester, and numerous subsequent workers, confirmed the accuracy of these observations.

A general description of the colonial appearance of the *gravis*, *intermedius* and *mitis* types is rendered peculiarly difficult because of the absence of a standard medium. McLeod and his colleagues originally used a heated rabbit-blood tellurite agar, on which the differentiation of the three variants was excellent. Later, however, it was found that some *mitis* and occasional *gravis* strains failed to develop on this medium. Numerous modifications have therefore been introduced. Rabbit and guinea-pig blood in 5–10 per cent concentration give the best type differentiation, but are difficult to obtain in quantity. Sheep, horse and ox blood are less satisfactory, but they are improved either by heating the medium in which they are contained or by lysis of the red cells (Neill 1937, Hoyle 1941). Against the first method is the objection, noted above, that some strains, particularly of the *mitis* type, are inhibited by heated blood (Glass 1937, 1939*a*); against lysis there is the objection that the ability of the different variants to cause haemolysis on blood agar plates is obscured. Some workers therefore favour a medium made with unheated and unlysed blood containing sufficient potassium tellurite to prevent or restrain the growth of other organisms that are likely to be present in the nose or throat. A 10 per cent sheep blood 0·5 per cent glucose agar medium containing a final concentration of 0.03–0.04 per cent potassium tellurite is satisfactory in practice, though the type differentiation is not so good as on McLeod's medium. For general diagnostic purposes Hoyle's medium has proved excellent in practice.

The colonial appearances of the three types will be summarized more conveniently, along with their morphological, biochemical and other characters, in a later section of this chapter (p. 101). Suffice it to say here that diphtheroid bacilli also develop on a tellurite medium, forming colonies which are usually low convex, undifferentiated, about 1–2 mm in diameter, varying in colour from pearl grey to jet black, with an entire edge, and a smooth, finely granular, or liquorice type of surface. Most of them can be easily recognized, but some strains form colonies so closely resembling the *gravis*, *intermedius* or *mitis* types of the diphtheria bacillus that they can be distinguished from them only after careful study of their other characters. The grey, brown or black coloration of colonies on tellurite media results from the reduction of the metallic salt. Besides tellurite, the diphtheria bacillus also reduces tetrazolium salts. If a drop of nitro blue tetrazolium is placed on a culture, the colonies rapidly turn grey, then dark blue, and within two minutes jet black (Monis and Reback 1962).

Growth occurs readily in broth, but is seldom abundant. The degree of turbidity, pellicle formation, and the amount and nature of the deposit vary with different members of the group; with some species, as for instance the three types of diphtheria bacillus, they are of value in identification.

Growth in *gelatin* at room temperature is generally poor to moderate. A few members of the group, such as *C. ovis*, '*C.*' *pyogenes*, and *C. ulcerans*, liquefy the medium, but most members, including the diphtheria bacillus, do not.

Resistance

C. diphtheriae is readily killed by heat, suspensions of the bacilli failing to survive 10 minutes' heating at 58°. It is also easily destroyed by most of the usual disinfectants. It would appear to be relatively resistant to drying, though the evidence on this point is somewhat conflicting. So far as our information goes most diphtheroid bacilli resemble the type species. On the whole the diphtheria bacillus is said to be more resistant than the dipheroids to the action of sodium fluoride, and Rantasalo (1948) recommended the inclusion of this substance in 0.1 per cent concentration in a tellurite serum medium for routine diagnosis. Most species of corynebacteria, including *C. diphtheriae*, are highly sensitive to penicillin, erythromycin and broad-spectrum antibiotics such as the tetracyclines and chloramphenicol (see also Chapter 53). Plasmid-determined resistance to erythromycin has been described in *C. diphtheriae* (Schiller *et al.* 1980). (For staphylococcal bacteriocines active on corynebacteria, see Chapter 30.)

Metabolism and growth requirements

Though all species are apparently able to grow in the absence of gaseous oxygen, some species, including *C. diphtheriae* itself, develop far more freely under aerobic conditions, and display this preference by growing as a film or veil over the surface of a liquid medium. The temperature range over which most members grow in artificial media extends from about 15° to 40°, with an optimum at about 37°. Growth of all species is improved by the addition to the medium of a natural animal protein, such as blood or serum. Some members, particularly the *mitis* type of *C. diphtheriae*, are inhibited by heated blood, as well as by blood treated with acid or alkali. According to Glass (1939*a*, *b*) this effect is apparent only in aerobic culture.

The nutritional requirements of the diphtheria bacillus have been studied by several workers and have been reviewed by Knight (1936). In general, it may be said (1) that the organism cannot use ammonia as the sole source of nitrogen; (2) that growth and toxin production often occur when amino acids are added to a synthetic medium and that, among these, cystine, aspartic acid, and tryptophan are apparently indispensable, though cystine is said to be replaceable by sodium sulphide (Braun 1938); (3) that it is only certain 'non-exacting' strains that are capable of growing in a synthetic medium, and even with these strains toxin production is usually much less abundant than with a more adequate food supply. Considerable progress has been made in attempts to identify the additional substances required for good growth and toxin production by fractionating peptone, meat extract, and other forms of complex protein extracts (see Mueller 1935*a*, *b*, *c*, Gordon and Zinnemann 1949).

Biochemical reactions

The carbohydrates commonly used as test substrates for distinguishing between the species of this genus are glucose, maltose and sucrose. Of these, *C. diphtheriae* ferments the first two; sucrose is attacked only by occasional strains. *C. xerosis* ferments all three sugars, and *C. hofmanni* none. No species produces gas. Within the diphtheria bacillus species the *gravis* type ferments starch, glycogen and dextrin; the *mitis* and *intermedius* types have no action on starch or glycogen and give irregular results with dextrin. In practice great care is necessary in the preparation of the medium if reliable results are to be obtained. A medium containing serum is almost indispensable, but it has certain drawbacks. The serum, for example, may contain sufficient fermentable carboydrate to give a false reaction; this can be overcome by suitable buffering. Unheated horse serum may hydrolyse starch owing to the presence of a natural diastase (Hendry 1938); this may be destroyed by heating. Not all samples of soluble starch are satisfactory; each sample should be tested before use.

A suitable medium may be prepared by dissolving 0·5 per cent peptone and 0·1 per cent Na_2HPO_4 in 1400 ml of distilled water, steaming for 15 min, filtering, adjusting to pH 7·4, adding 250 ml of horse serum, steaming for 20 min, adding 11 ml of Andrade's indicator, adjusting to pH 7·6–7·8, tubing in 3 ml quantities, autoclaving at 10 lb for 10 min, and adding to each tube separately sufficient of a sterile solution of the sugar in distilled water to give a final concentration of 0·4 per cent starch or 1 per cent of other sugars (Robinson 1940).

A more detailed and extended table of the fermentation reactions of eleven named species within this genus is appended to this chapter (Table 25.3), but experience has revealed the existence of several more fermentative types (see Andrewes *et al.* 1923).

Though the fementation of sucrose has generally been held to exclude the diphtheria bacillus, Christovão (1957) brought conclusive evidence to show that some morphologically typical fully virulent organisms isolated from cases of clinical diphtheria are able to attack this substance. In a series of 193 virulent strains isolated at São Paulo, no fewer than 54 (28 per cent) possessed this property. Their distinction from *C. xerosis*, which also attacks sucrose, was aided by their fermentation of dextrin and glycerol.

The corynebacteria attack glucose by the fermentative method (Hugh and Leifson 1953), producing acetic, lactic, and propionic acids as end products. All produce catalase, none but *C. bovis* oxidase. Of the true corynebacteria, only *C. ulcerans* liquefies gelatin; most of them, including *C. diphtheriae* but not *C. ulcerans*, reduce nitrate to nitrite. Most strains of the *mitis* type of *C. diphtheriae* are *haemolytic* on blood agar; *intermedius* strains are not; *gravis* strains vary in their activity. The haemolytic strains lyse the red cells of various animals—sheep, rabbit, guinea-pig, horse and man, when these are added to a broth culture. Though Goldie (1933) brought evidence in favour of the formation of a soluble thermostable haemolysin, most workers (Schwoner 1904, Costa *et al.* 1918) have failed to find any; and Hewitt (1974*a*) maintains that haemolysis occurs only in the presence of living organisms. Several diphtheroids produce haemolysis on blood agar plates.

Püschel (1936) drew attention to the ability of most diphtheroid bacilli to hydrolyse *urea* with the production of ammonia. The diphtheria bacillus, on the other hand, is almost uniformly negative in this respect. Among the animal diphtheroids, *C. renale* is a potent urease producer. *C. ovis* also hydrolyses urea.

Some diphtheroid strains form phosphatase; the true diphtheria bacillus never does. The production of this enzyme may be tested for on a medium containing phenolphthalein phosphate (see Bray 1944).

Antigenic structure

The antigenic structure of the corynebacteria is still a little uncertain. Numerous observations have been made by different workers but have not yet been sufficiently co-ordinated to present us with a complete picture.

Specific alkali-labile mainly heat-labile protein K antigens situated on the surface of the bacillus are detectable by precipitin tests in strains of *C. diphtheriae* (Wong and T'ung 1940). Huang (1942), using agglutination, classified 246 out of 286 strains of living virulent diphtheria bacilli into 8 serotypes labelled D1 to D8. At a deeper level in the cell, as in other gram-positive organisms, are heat-stable group-specific O antigens shared by all strains of *C. diphtheriae*, *mitis* strains having an additional antigen of their own (Lautrop 1950). In living and formolized suspensions O agglutination is inhibited by the surface K antigens. A similar finding of surface and deep antigens was reported by Cummins (1954) for a *mitis* strain. Barber and co-workers (1965; see also Lazar 1968) studied the endo-antigens of corynebacteria obtained by treatment of acetone-extracted organisms with a 2 per cent solution of sodium deoxycholate. By the double-diffusion agar-gel technique they found evidence of (1) nucleoprotein antigens common to diphtheria and diphtheroid bacilli; (2) protein antigens that were specific for the *gravis* type of *C. diphtheriae*; and (3) polysaccharide antigens that were species-specific for diphtheria bacilli, and others that were specific for certain diphtheroid bacilli

Previously Ewing (1933), by agglutination of 0·2 per cent formolized suspensions heated to 60° for 1 hr, had distinguished 4 serotypes, A–D, among *gravis* strains; Robinson and Peeney (1936), using bacilli suspended in N/250 NaOH solution and heated at 56° for 4 hr, had distinguished 5 antigenic types i–v, the first four of which corresponded to Ewing's serotypes; and Hewitt (1947*b*), using unheated bacilli suspended in 10 per cent of 0·02N NaOH, had added 8 more serotypes, vi–xiii. Hewitt recognized 4 serotypes among *intermedius* strains, and 40 among the antigenically heterogeneous *mitis* strains. Presumably the antigens concerned in the results obtained by these three groups of workers were of the alkali-soluble K type. In Australia, Gibson (1975) noted an association between the antigenic structure and the colonial appearance on tellurite blood agar.

The presence of deep-seated antigens common not only to different species of corynebacteria but to *Myco. tuberculosis* as well was demonstrated by Krah and Witebsky (1930) using alcoholic extracts, and confirmed by Cummins (1962), who found a common antigen in all strains of corynebacteria, mycobacteria and nocardiae that had arabinose and galactose as the principal sugars in their cell wall (see also Oeding 1950*b*).

Toxin production

The type species, *C. diphtheriae*, is an important human pathogen, giving rise to a characteristic and often fatal disease, the lesions of which are, in the main, produced by the action of a powerful exotoxin. This diffuses throughout the body from the primary focus of infection, which is most frequently situated in the tonsillar region. The pathogenesis of diphtheria in man, and its diagnosis, prevention and treatment so far as these depend on bacteriological methods, are dealt with in Chapter 53.

Though the conditions necessary for optimal production of toxin appear to differ from those for optimal growth, there is, according to Nishida (1954), a parallelism between the concentration of toxin in the medium and the total number of organisms alive and dead. The toxin is formed by the young rapidly growing cells and liberated into the medium. As the cells die and are replaced by fresh ones, so the amount of toxin increases.

The early work on the mode of preparation of diphtheria toxin in the laboratory is described on pages 619 and 620 of the 6th edition, to which the reader may be referred.*

The present method of production is to use a medium prepared from an acid hydrolysate of casein or a tryptic digest of muscle, together with a balanced mixture of salts, cystine, glutamic acid, maltose and iron (see Barr *et al.* 1941, Norlin 1943, Holt 1950). The toxin is labile, and is readily inactivated or destroyed by heat and strong chemical reagents. It may be concentrated by precipitation with weak acids, ammonium sulphate or acetone, or by dialysis, by fractional filtration through graded collodion membranes, by adsorption on to aluminium hydroxide followed by elution, or by a combination of these methods (see Glenny and Walpole 1915, Eaton and Bayne-Jones 1934, Relyfeld and Raynaud 1964, Barksdale 1970). Its molecular weight is 62 000–63 000.

The evidence so far obtained suggests that diphtheria toxin is a heat-coagulable protein. Some of the purest preparations contain about 16 per cent nitrogen, 0.75 per cent sulphur, 9 per cent tyrosine, and 1.4 per cent tryptophan. The isoelectric point is pH 4.1. The toxin is extremely sensitive to denaturation by solutions more acid than pH 6 and by moderate heat. The amount of nitrogen per flocculation (Lf) unit is 0.00045 mg, and the MLD for guinea-pigs is about 0.0001 mg (Eaton 1936, Pappenheimer 1937, Pappenheimer and Robinson 1937). An even purer protein was isolated in crystalline form by Pope and Stevens (1953). It gave a single diffusion line and, judged by the Lf/mg nitrogen, was about 50 per cent purer than Pappenheimer and Robinson's product. Even so the purest crystalline diphtheria toxin-protein so far obtained contains three or four distinct antigens demonstrable by gel-diffusion methods with antitoxin prepared against it (Pope and Stevens 1958, Pope *et al.* 1966). The toxin is highly lethal for certain animals (see below), and is cytopathic in tissue culture for monkey kidney, human amnion, and Hela cells (Sousa and Evans 1957). It can be converted into toxoid by suitable treatment with formalin; in this state it will still cause a precipitate when mixed with specific antiserum and give rise to antibody production in animals, but is no longer toxic. It enters the cell by endocytosis, and is broken down by enzymes. It inhibits protein synthesis and affects other cellular processes, causing gross pathological changes including necrosis (Bonventre and Imhoff 1966, Collier 1975). Goor and Pappenheimer (1967) suggest that in the presence of nicotinamide adenine dinucleotide it inactivates one of the soluble enzymes required for the transfer of amino acids from aminoacyl-sRNA to the growing polypeptide chain. (See also Pappenheimer and Brown 1968 and Collier 1975.)

*Also for an account of the effect of iron concentration in the medium on the production of toxin.

It may be added that, in addition to toxin, the diphtheria bacillus forms neuraminidase—of a type similar to that produced by the clostridia but different from that of the influenza virus and *V. cholerae* (Warren and Spearing 1963).

Pathogenicity of the diphtheria bacillus for laboratory animals

The classical paper in which Loeffler (1884) first described the isolation and characters of the diphtheria bacillus, and the report by Roux and Yersin (1888) of the separation of the filtrable toxin, contain descriptions of the lesions produced by the living organism, or by its separated toxin, in a variety of laboratory animals. These original observations have since been extended by a host of experimental studies. Among laboratory animals, the guinea-pig and the rabbit are the most susceptible; rats and mice are extremely resistant. Dogs, cats, pigeons, and other birds appear to occupy an intermediate position. (See Loeffler 1884, 1890, Roux and Yersin 1888, Wernicke 1893, Goodman 1907, Coca *et al.* 1921, Glenny and Allen 1922, Andrewes *et al.* 1923.) The bacillus has little power of tissue invasion; whether the inoculum consists of a living culture, or of a toxic filtrate, death occurs as the result of a toxaemia in the strict sense.

When a guinea-pig is inoculated subcutaneously into the flank with a suitable dose of a virulent culture or of a toxic filtrate, a soft oedematous swelling usually appears at the site of inoculation within 12 to 18 hr, and gradually extends. The animal becomes obviously ill, developing a staring coat and sitting crouched in its cage. Death usually occurs between 18 and 96 hr, according to the size of the dose of culture or filtrate inoculated. With very large doses the time to death may be even shorter, but is never less than 10 to 14 hr (see Glenny 1925*b*). Animals that survive beyond the 4th day may suffer from cachexia and paralysis, and die at some later period; but the pathogenesis of this form seems to be essentially different from that of the acutely fatal toxaemia, and it is with the latter that we are here concerned. When a guinea-pig that has died within 4 days after a subcutaneous inoculation is examined *post mortem*, the typical findings are as follows.

At the site of inoculation is found an extensive area of gelatinous haemorrhagic oedema, extending to the skin superficially, and deeply to the muscles or to the parietal peritoneal membrane. If the animal has survived for several days, the tissues in the more central parts of the oedematous area may be obviously necrotic. The regional lymph glands are usually swollen and congested. The peritoneum may contain a varying amount of fluid, which may be clear, cloudy or blood-stained. The abdominal viscera as a whole are congested; but the most striking lesion is the swelling and congestion of the adrenal glands. On macroscopical section there are seen to be scattered haemorrhages, situated in the medulla, in the cortex, or in both. Sometimes all naked-eye distinction between cortex and medulla is lost. On opening the thorax a serous exudate will often be found in the pleural cavities, usually clear, sometimes cloudy or blood-stained. A pericardial effusion may or may not be present. Films prepared from such effusions reveal a preponderance of mononuclear cells.

Mollard and Regaud (1895) recorded the occurrence of degenerative changes in the myocardium in experimental diphtheria, and Flexner (1897) noted that fatty degeneration of the cardiac muscle was almost constantly present in animals which died within a short time after inoculation. Dudgeon (1906) brought evidence to support the suggestion of Bolton (1905), that the direct action of diphtheria toxin on the cardiac muscle is the most important cause of acute cardiac failure in human diphtheria. Examining a large series of guinea-pigs, killed or dying in various stages of acute diphtheritic toxaemia, Dudgeon demonstrated the occurrence of fatty degeneration of the diaphragmatic muscle within 4 hr after inoculation, and of the cardiac muscle within 16 hr. Similar results were later recorded by Jaffé (1920).

In his original experiments Loeffler (1884) showed that a false membrane could sometimes be produced experimentally by the intratracheal injection into rabbits and pigeons of living organisms. More recently Ørskov (1948) described how fatal cutaneous diphtheria could be reproduced in very young guinea-pigs by rubbing a small quantity of a virulent broth culture into the shaven skin of the abdominal wall; haemorrhagic local oedema occurs and death follows in 3–4 days.

When living bacilli or the toxin are injected into the skin itself, a localized erythematous lesion results followed by necrosis (Römer 1909). The reaction reaches its height after 48 hr in guinea-pigs and 72 hr in rabbits. The intracutaneous method is very useful in practice for determining the virulence of diphtheria bacilli. Several strains can be tested on the same animal. Either (*a*) two animals can be injected simultaneously, one of which has been given a protective dose of antitoxin beforehand to prevent the development of the skin lesions; or (*b*) the same animal may be given a series of injections down one side of the body followed 5 hr later by an intravenous dose of antitoxin and a repetition of the injections of the same cultures on the opposite side of the body (see Report 1950); in this animal the antitoxin prevents the development of skin lesions resulting from the second but not from the first series of injections.

Guinea-pigs injected intracutaneously with living virulent bacilli and protected by antitoxin acquire delayed sensitivity of the tuberculin type to diphtheritic toxin and associated proteins (Uhr *et al.* 1957).

According to Ts'un T'ung (1945) both virulent and avirulent diphtheria bacilli set up a generalized infection in the chick embryo after inoculation into the chorio-allantoic membrane; the bacilli can be recovered from the blood and viscera.

Although the production of the filtrable toxin, with its characteristic action in the guinea-pig and its property of being specifically neutralized by the homologous antitoxin, is one of the most important characters by which *C. diphtheriae* is identified, there exist strains of bacilli which, though conforming in other respects with the diphtheria bacillus, fail to form toxin. These strains are commonly classed as non-toxigenic, or avirulent, diphtheria bacilli. Whether they should all be assigned to this species is perhaps open to question; but there can be no reasonable doubt that many of them are actually non-toxigenic variants of *C. diphtheriae*. We have mentioned above that certain avirulent strains can be shown to be antigenically related to typical toxigenic strains, and the actual emergence of an avirulent variant from a virulent organism, under laboratory conditions, has been recorded by several observers.

Relation between lysogenicity and virulence Both Crowell (1926) and Cowan (1927) were able by ordinary colonial selection to separate off avirulent variants from virulent strains of diphtheria bacilli. More recent work has demonstrated a relation between lysogenicity and virulence. It is possible by submitting an avirulent strain to the action of a suitable bacteriophage to convert it into a virulent strain (Freeman 1951, Freeman and Morse 1952, Hewitt 1952, Groman 1953, 1955, Groman and Memmer 1958). The reverse procedure, that of converting a lysogenic virulent into a non-virulent strain, is more difficult, but can be accomplished by growing a virulent strain for several passages in broth containing an anti-phage serum (Anderson and Cowles 1958). All virulent or toxigenic strains of diphtheria bacilli appear to be lysogenic, but the opposite does not hold; Christensen (1957) found that quite a number of non-toxigenic strains harboured bacteriophage. A kinetic study revealed a close parallelism between the formation of toxin and the synthesis of new phage particles. Addition of iron to the medium inhibited toxin production without affecting the synthesis of bacteriophage (Matsuda and Barksdale 1966). The relation of lysogenicity and toxin formation is discussed by Collier (1975).

The bacteriophages of diphtheria bacilli fall into at least nine types. Toshach and his colleagues (1977) in Canada found that the phage types of *C. diphtheriae* correspond well with the biotypes. Phage- and bacteriocine-typing systems for *C. diphtheriae*, *C. ulcerans* and *C. ovis* are described by Saragea and coworkers (1979).

Table 25.1 Characters of the types of diphtheria bacillus

	Gravis	*Intermedius*	*Mitis*
Morphology on tellurite blood agar	Usually short rods, resembling irregular forms of *C. hofmanni*. Staining fairly uniform, few or no granules, and often a narrow unstained bar dividing the rod unequally. Some degree of pleomorphism with irregularly barred, snow-shoe and tear-drop forms. May be coccoid on first isolation. Occasional strains resemble *intermedius* or *mitis* type.	Usually long, irregularly barred rods, often with terminal clubbing. Granulation generally poor. Pleomorphism always present. Distinguished from diphtheroids, which may closely resemble them, by irregularity of size, shape, barring and arrangement. Some strains indistinguishable from *mitis*.	Usually long, curved, pleomorphic rods with prominent metachromatic granules. Except for some shadow areas, protoplasm stains evenly. Some strains show barring, with or without granules. Occasional strains are coccoid and others yeast-like.
Colonial appearance on telurite blood agar	An 18-hr colony is 1–2 mm *q*, circular, low convex, pearly grey or with greyish-black centre and paler semitranslucent periphery, with a smooth matt or rarely liquorice type of surface, and a commencing crenation of the edge. The colony is coherent, tending, when touched, to move as a whole on the surface of the medium; has a consistency of cold margarine, and breaks up radially into small masses that are not easily emulsified. Slight haemolysis around colonies of some strains. In 2–3 days colony reaches 3–5 mm in diameter, is flattened with a slightly raised centre, is slaty-grey or greyish-black in colour, often darker at the centre than at the periphery, has a frosted surface and a crenated edge, and shows radial striation, especially towards the margin. When striation and differentiation are well developed, the term 'daisy-head' colony is applicable.	Colonies are uniform, small, discrete, delicate, almost misty in appearance, and undergo little increase in size between 24 and 48 hr. At 18 hr the colony is less than 1 mm *q*, is slightly raised, with or without umbonation, or is of sugar-loaf appearance; centre is greyish black and generally darker than periphery; surface smooth or very finely granular, and edge entire or slightly spiky. Consistency is intermediate between the brittle *gravis* and the butyrous *mitis* type. Haemolysis never seen. At 48 hr colony is not much larger, has a dull granular centre and a smoother more glistening periphery, and is dark in colour except for a lighter ring near the edge—frog's-egg appearance. Edge may be entire or finely crenated. On further incubation, colony may enlarge and come to resemble a daisy-head colony of the *gravis* type.	Very variable in size, usually ranging between *intermedius* and *gravis*. At 18 hr colonies may be less than 1 mm *q* up to 1·5 mm *q*. They are circular, convex, usually of a mushroom-grey colour, darker than that of *gravis*, though varying considerably with a smooth, glistening surface and entire edge. Consistency is of soft butter, and emulsifiability is easy. Small ring of haemolysis is usual. At 48 hr colony is 2–4 mm *q*, undifferentiated, and dark greyish-black with sometimes a narrow paler margin. On further incubation colony may become flatter with a central elevation—poached-egg appearance—and the surface may become granular and contoured, or develop concentric rings or papular excrescences.
Growth in broth	Appearance variable. Usually surface pelicle and coarse granular deposit with little or no turbidity.	Appearance very constant. In 24 hr there is slight turbidity with little or no deposit. In 48 hr broth has cleared, and there is a very finely granular sediment, which can be easily dispersed on shaking.	Appearance variable. Usually diffuse even turbidity, denser than that of *intermedius*, and moderate non-granular deposit. Soft pellicle may form on further incubation.
Fermentation of starch and glycogen.	+	–	–
Antigenic structure	At least 13 types recognizable by direct agglutination. Types I and II form the classical daisy-head colonies; types III, IV and V approach nearer to the *mitis* type.	Little known, but probably at least 4 types with specific antigens.	Little known, but at least 40 distinct types.
Animal pathogenicity	Almost invariably virulent to guinea-pigs.	Almost invariably virulent to guinea-pigs.	Usually virulent to guinea-pigs, but strains from carriers are often avirulent.

Note. *q* = diameter.

The *gravis*, *mitis* and *intermedius* types of *Corynebacterium diphtheriae*

Attention has already been drawn on p. 96 to the recognition by McLeod and his colleagues of three stable varieties of the diphtheria bacillus – the *gravis*, *intermedius* and *mitis* types. The characters of these three types, as recorded by Robinson (1934, 1940), and Cooper, Happold, McLeod and Woodcock (1936), are summarized in Table 25.1.

This summary can do no more than serve as a guide to the recognition of the three types. On the whole the characters of the *intermedius* type are the most constant. In the identification of *gravis* strains starch fermentation is of particular value since, though it may be delayed beyond 24 hr, it is seldom absent. It is interesting to note that in 1898 Wesbrook and his colleagues recorded the isolation of atypical virulent bacilli from cases of diphtheria in a school in Minneapolis. Their description corresponds in many respects to that of the *gravis* type of McLeod.

Electronmicrographs show that the cell wall of the *gravis* and *mitis* types has three layers. The cell wall of the *intermedius* type has only two layers, but is 2–3 times as thick as that of the other types. The *mitis* and *intermedius* types are rich in mesosomes. The space between the cytoplasmic membrane and the cell wall contains granules that may be associated with toxin formation. The barring of the bacilli, when it occurs, is due to partition walls, resulting in the presence of daughter cells within what appears to be a single bacillus (Lickfeld 1967).

Sometimes it is difficult to distinguish not only between the three types, but between diphtheria bacilli and diphtheroids. Each of the types has one or more species of diphtheroid bacilli that closely resemble it. In general, the diphtheroid bacilli tend morphologically to be more regular in shape, size, depth of staining, distribution of bars or granules, and arrangement; diphtheroids lie often in palisades, whereas true diphtheria bacilli show the Chinese-letter type of distribution, or, when they are in bundles, they have seldom the same regularity of arrangement as that of diphtheroids. The colonial differences are often very slight and, as they vary from one medium to another, they can be learned only by close observation and long experience.

Fermentation reactions are often of help, since many diphtheroids either ferment sucrose or have no action at all on sugars, though it must be remembered that occasional strains of true diphtheria bacilli ferment sucrose. Distinction is sometimes impossible except by virulence tests, and even this method leaves the true nature of an avirulent *mitis*-like organism doubtful. Whether in identifying the individual types of diphtheria bacilli, or in distinguishing between diphtheria bacilli and diphtheroid bacilli that closely resemble them, too much attention should never be paid to any one character. Often it is only by observing all the characters of a particular strain, and assigning to each character a weight which experience alone can provide, that a sound conclusion on its identity can be reached. Valuable help will be obtained by a careful study of McLeod's (1943) review.

A so-called *minimus* type was described by Frobisher, Adams and Kuhns (1945) and studied further by Frobisher (1946). There seems little justification, however, for the use of this term, since the *minimus* strains are for all practical

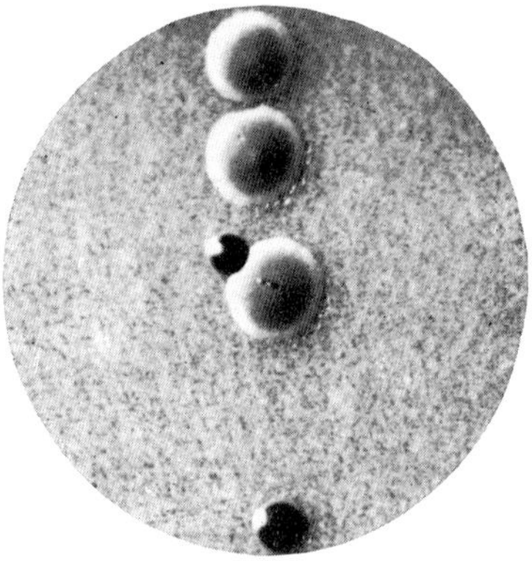

Fig. 25.2 *C. diphtheriae*. Three colonies of *gravis* type and two of *mitis* type, on blood-tellurite-agar (× 8).

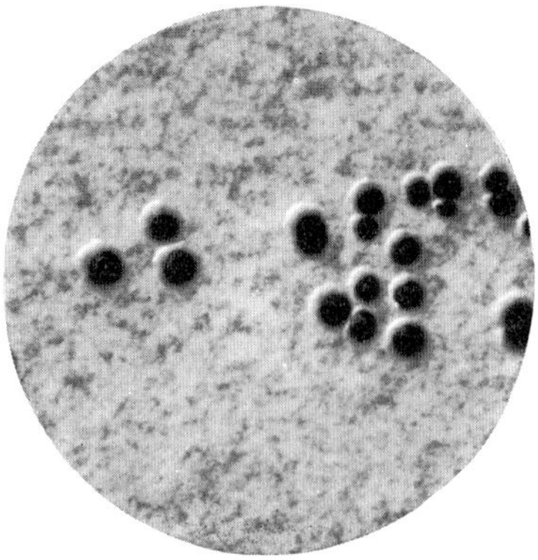

Fig. 25.3 *C. diphtheriae*. Colonies of *intermedius* type on blood-tellurite-agar (× 8).

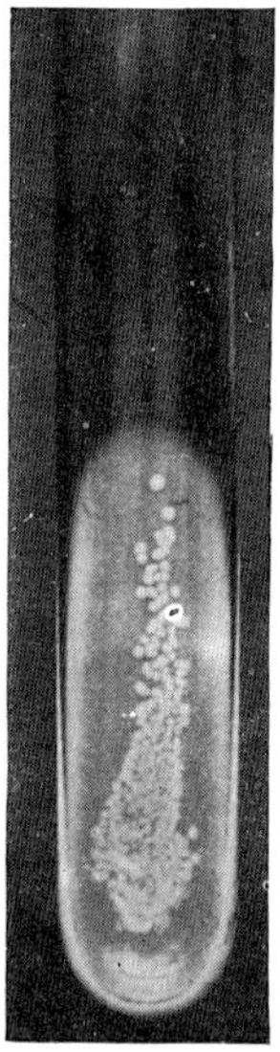

Fig. 25.4 *C. diphtheriae*. 24 hr culture on Loeffler's serum.

purposes identical with the *intermedius* (Johnstone and McLeod 1949, Freeman and Minzel 1950). The '*minimus*' strains met with by Galbraith, Fraser and Bramhall (1948) during an outbreak of diphtheria in Utah were apparently strains of diphtheroid and not of true diphtheria bacilli.

In the body the three main types of *C. diphtheriae* appear to be fairly stable, though minor modifications are seen in strains from different localities (Siemens 1938). In the test-tube, however, some degree of transformation may perhaps occur. Oeding (1950*a*), for example, by in-vitro cultivation on poor media, by serial transfers in penicillin broth or in 0.1 per cent phenol broth, and by other means, claimed to have effected several type transformations—*gravis* into *mitis* and *mitis* into *gravis*, though not *gravis* or *mitis* into *intermedius*.

Summarized description of corynebacterial species

Corynebacterium diphtheriae

The morphological, cultural and biochemical characters of *C. diphtheriae* are summarized in Tables 25.1 and 25.2.

The diphtheroid bacilli

Corynebacterium ulcerans

This organism, which was isolated by Mair (1928) from a bovine udder and described by Barratt (1933) and Jebb (1948), is intermediate between the diphtheria bacillus and the animal diphtheroids. In man it gives rise to a disease indistinguishable from diphtheria. Morphologically, it is a pleomorphic bacillus forming a heavy creamy growth on Loeffler's serum in 24 hr. On unheated blood tellurite agar the colonies after 24 hr are low convex with a matt black surface and a 'cold margarine' consistency; they are usually surrounded by a narrow zone of haemolysis, and when touched move en masse. *C. ulcerans* differs from the *gravis* type of diphtheria bacillus not only in minor colonial appearances, but in liquefying gelatin, in fermenting trehalose (though not for some days), in failing to reduce nitrate, in its inagglutinability with gravis antisera, and in the production of two toxins—one immunologically identical with diphtheria toxin, and one related to the toxin of *C. ovis* (Petrie and McClean 1934) or identical with it (Carne and Onon 1982). It differs from *C. ovis* in its colonial appearances, its fermentation of starch and trehalose, and its ability to

Table 25.2 Differential reactions of the corynebacteria and diphtheroid-like bacilli

	Production of acid from														
	glucose	maltose	sucrose	starch	dextrin	lactose	mannitol	Gelatin liquefaction	Haemolysis on blood agar	Filtrable haemolysin	Nitrate reduction	Hydrolysis of urea	Toxin production	Pathogenicity for man(M) or animals(A)
C. diphtheriae	+	+	−	±*	±*	−	−	−	±	−	+	−	+	+M
C. ulcerans†	+	+	−	+	+	−	−	+	+	−	−	+	+	+M
C. hofmanni	−	−	−	−	−	−	−	−	−	−	+	+	−	−
C. xerosis	+	±	+	−	−	−	−	−	−	−	+	−	−	−
C. ovis	+	+	∓	∓	±	∓	∓	−	+	−	∓	+	+	+A
C. renale	+	±	−	−	−	−	−	−	−	±	+	+	−	+A
C. flavum	+	±	−	−	−	∓	−	−	?	?	−	?	−	−
'*C.*' *pyogenes*	+	±	∓	±	±	±	−	+	+	+	−	−	+	+A
'*C.*' *haemolyticum*	+	+	+	±	+	+	−	+	+	−	−	−	−	±M
'*C.*' *equi*	−	−	−	−	−	−	−	−	−	−	+	±	−	+A
'*C.*' *kutscheri*†	+	+	+	+	−	−	−	−	−	−	+	±	+	+A
P. acnes†‡	+	±	−	−	±	−	±	+	−	−	±	−	−	±M

* For distinction between *gravis*, *intermedius* and *mitis* strains, see Table 25.1.
† Ferments trehalose.
‡ Microaerophilic.

kill guinea-pigs when injected subcutaneously, and with the production of lesions similar to those of the diphtheria bacillus. Intradermal injection causes a purulent necrotic ulcerating lesion against which diphtheria antitoxin affords no protection. In man it causes tonsillitis and skin lesions with little evidence of toxaemia. In the 4th edition (1955) of this book we suggested for it the name of *C. ulcerans* – the name originally applied by Gilbert and Stewart (1926–7) to a similar, if not identical, organism isolated from cases of diphtheria. It is now generally regarded as a variant of the diphtheria bacillus.

Corynebacterium hofmanni

This organism was isolated in 1888 by von Hofmann from the throat of healthy persons. Morphologically it has a far more regular appearance than *C. diphtheriae*, forming short straight rods with parallel sides and a single unstained septum in the middle. The bacilli are arranged in parallel rows, or irregular groups with the usual angular displacement of adjacent cells. Like many other diphtheroids they are more tenacious of the Gram stain than the diphtheria bacillus. Colonies on Loeffler's serum are larger and whiter than those of *C. diphtheriae*. It grows readily on agar, ferments no carbohydrates, does not liquefy gelatin, form indole, or cause haemolysis. Nitrate is reduced, and urea is hydrolysed. Antigenic structure differs from that of other corynebacteria. It is said to contain two antigens, of which the major is composed of polysaccharide and nucleic acid. It produces no toxin and is non-pathogenic. G + C content of DNA is 55–57 moles per cent.

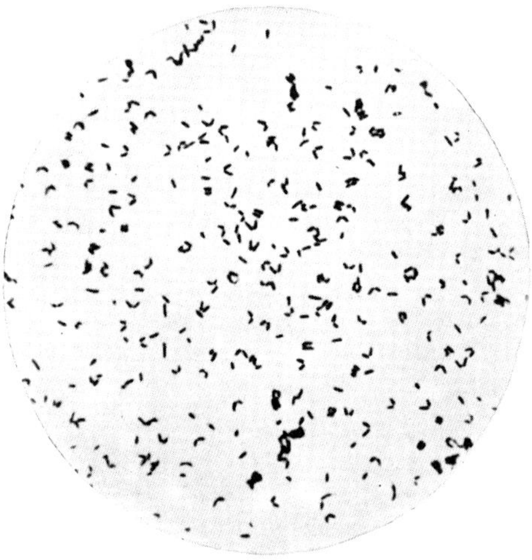

Fig. 25.5 *C. hofmanni*. From 24 hr culture on Loeffler's serum (× 1000).

Corynebacterium xerosis

This organism was isolated from the conjunctiva by Reymond and his colleagues in 1881, and was described more fully by Kuschbert and Neisser in 1883. Morphologically it resembles the diphtheria bacillus, but differs from it in its greater regularity, its infrequency of club formation and granularity, and in the predominance of barred and segmented forms. It ferments glucose, maltose and sucrose, but not mannitol or dextrin. Nitrate is reduced. It forms no indole and does not liquefy gelatin. It hydrolyses tributyrin but not urea, and fails to cause haemolysis. Antigenically it differs from other corynebacteria. It appears to be non-pathogenic. G + C content of DNA is 55–57 moles per cent.

Corynebacterium ovis

This organism, often referred to as *C. pseudo-tuberculosis* or the Preisz-Nocard bacillus (see Nocard 1889, Preisz 1894), causes caseous lymphadenitis in sheep and ulcerative lymphangitis in horses. It differs sharply from *C. diphtheriae* in that it is a pyogenic organism and invades the tissues, but resembles it in producing a filtrable toxin.

On Loeffler's serum medium colonies are umbonate, opaque, friable, and slightly yellowish. As they enlarge, they often display a series of concentric rings around the raised centre. Growth on agar and in broth is poor. On blood agar colonies are surrounded by a zone of haemolysis. Forms acid from glucose, maltose, glycerol, and weakly in dextrin, not usually in mannitol, lactose or sucrose, though Carne (1939) in Australia disputes this. Urea and arginine are hydrolysed; litmus milk is unchanged; gelatin is not liquefied. Two biotypes are recognized on the basis of nitrate reduction (Miers and Ley 1980).

The exotoxin of *C. ovis* differs from that of *C. diphtheriae*. Nicolle and his colleagues (1912) showed that an antitoxin prepared against *C. ovis* protected against *C. ovis* toxin, but that diphtheria antitoxin did not. Dassonville (1907), Hall and Stone (1916), Minett (1922*a*, *b*), and Barratt (1933) recorded some degree of protection by diphtheria antitoxin against the toxin of *C. ovis*; but the detailed study of Petrie and McClean (1934) leaves little doubt that these effects were due to the fact that the sera of normal horses may contain varying amounts of *C. ovis* antitoxin—*C. ovis* being a natural pathogen of the horse—and that the two toxins are immunologically distinct.

Besides its lethal effect, the toxin is haemolytic and dermonecrotic. It potentiates the action of staphylococcal δ-lysin and inhibits the effects of the β-lysin of *Staph. aureus*. Haemolytic activity is apparent below pH6, and haemagglutinating activity at or slightly above pH7 (Burrell 1979, 1980). The toxin is inactivated by heat at 75° for 15 min (Lovell and Zaki 1966). It is a phospholipid D that is thought to attack the sphingomyelin of the vascular endothelium (Carne and Onon 1978). Its molecular weight is 14500 ± 1000 (Onon 1979). Among the corynebacteria, a toxic phospholipase is found only in *C. ovis* and *C. ulcerans* (Barksdale *et al.* 1981).

Nicolle, Loiseau and Forgeot (1912) carefully recorded the lesions met with in guinea-pigs that died as the result of inoculating either living cultures of *C. ovis* or bacteria-free filtrates subcutaneously. When living organisms are given in such a dose as to kill the animal at about the 25th day, subcutaneous abscesses develop in various situations during life. At necropsy, in addition to these superficial lesions, small granulomatous masses are found in the liver, spleen and lungs, and beneath the parietal peritoneum. In the male guinea-pig similar lesions are found in the tunica of the testis and epididymis. Some of these lesions may have developed into large caseous or caseo-purulent masses.

When a *guinea-pig* is injected subcutaneously with a fatal dose of a toxic broth filtrate, death occurs within a few days, often in less than 24 hr, from an acute toxaemia. The necropsy findings in such cases are entirely different from those described above. There is a local, subcutaneous, inflammatory, gelatinous oedema at the site of inoculation, often haemorrhagic. The abdominal viscera are congested and often show small haemorrhages, particularly in the stomach, large intestine, and in the kidneys, which may be almost black in colour. There is, however, no congestion of the adrenals, and no exudation into the pleura. Hall and Stone (1916) record a very similar picture.

The effect of injecting *C. ovis* toxin into the skin of a guinea-pig likewise differs from that produced by the injection of diphtheria toxin (Petrie and McClean 1934). A papular lesion results, which may become pustular if the dose of toxin is large enough. According to Carne (1940), who defines the optimal conditions for toxin production, guinea-pigs, rabbits, sheep, goats, pigs, horses, oxen, dogs and cats are all sensitive to the toxin. The pyogenic substance is largely contained in the bacterial cells and is fairly thermostable (Bull and Dickinson 1935). Suspensions of *C. ovis*, killed by heat at 60° for 1 hr, no longer produce toxic death in susceptible animals, but they give rise to sterile abscesses when injected in adequate dosage. *Mice* are less susceptible to the toxin than guinea-pigs, but they die, usually in 8 to 24 hr depending on the dose; histological examination reveals degenerative changes in the liver cells and in the epithelial lining of the convoluted tubules of the kidneys.

C. ovis is a natural pathogen of horses, sheep and perhaps cattle, and is pathogenic for rabbits and guinea-pigs, but not for pigeons or fowls. A few infections of man, causing mainly lymphadenitis, are on record (see McCoy *et al.* 1978).

Corynebacterium renale

This organism, which was first recognized by Enderlen (1890–1), named by Ernst (1905, 1906), and studied by Jones and Little (1925, 1930) and several subsequent workers (see Lovell 1946, 1951, 1956), gives rise to cystitis and pyelitis in cattle. It is said to be common in the vagina of healthy cows (Weitz 1947). It is a large diphtheroid organism showing metachromatic granules. Among the ordinary sugars it ferments only glucose. It produces strong alkali with saponification and a zone of clearing on milk agar and slow digestion. It gives positive methyl-red and Voges-Proskauer reactions, and is a potent producer of urease, which is thought by some to play a part in the pathology of the disease (see Smith 1966). Experimentally it is mildly pathogenic to mice and rabbits, in which a fatal pyelonephritis may be set up if a sufficient number of organisms are injected intravenously (Lovell and Cotchin 1946, Feenstra *et al.* 1949, Kuzdas *et al.* 1951). Guinea-pigs appear to be more resistant. Three serological types are recognized (Yanagawa *et al.* 1967, Yanagawa and Honda 1978). Pili, found on all three types, appear to assist in attaching the bacteria to mucous membranes (Yanagawa *et al.* 1968, Honda and Yanagawa 1978). The G + C content of the DNA is 53 moles per cent. The diphtheroid organisms associated with ovine posthitis in Australia are thought to be identical with *C. renale* (Rojas and Biberstein 1974).

Corynebacterium bovis

This organism, which was studied particularly by Jayne-Williams and Skerman (1966), is found in milk. The distinctive features are its requirement for Tween 80 as a necessary supplement to most culture media, and its ability to hydrolyse butter-fat and grow in nutrient broth containing 9 per cent of NaCl. It is non-haemolytic, non-proteolytic, has no action on gelatin or starch, ferments glucose and maltose but not mannitol or sucrose, and is oxidase, urease, catalase and VP positive. Though apparently a commensal organism of the cow's udder, it is suspected of giving rise to clinical mastitis (Cobb and Walley 1962). The G + C content of DNA is 58 moles per cent.

Corynebacterium flavum

This organism, which was isolated from milk and described by Orla-Jensen (1919), is 1–2 μm × 0.5 μm in size. It has an irregular outline, displaying prominent angular division and often bipolar staining. In broth it produces a granular turbidity with a filmy pellicle and a diffuse flocculent precipitate. Surface colonies on Yeastrel milk agar are cream-coloured and of waxy consistency. It grows best at 30° and very poorly, if at all, over 35°. It is said by Orla-Jensen (1919) and by McKenzie, Morrison and Lambert (1946) to survive heating to 75°, but by Robinson (1966) to be destroyed at 65° within 10 min. It forms no toxin, and is non-pathogenic to animals. Its DNA base content of 56 moles per cent brings it into line with other genuine corynebacteria (see Jones 1975). The descriptions given by different workers are not always in agreement (see e.g. Abd-el-Malek and Gibson 1952).

Other corynebacteria

Besides these, there are a few other organisms that appear to be diphtheroid bacilli but have not yet been studied sufficiently to classify them with certainty in the *Corynebacterium* genus.*

Corynebacterium belfanti This organism was isolated by Belfanti and della Vedova in 1896 from the nasal secretion of a patient suffering from ozaena. Resembles a *mitis* strain of *C. diphtheriae* but does not reduce nitrate. It is a weak toxin producer, but is said to be responsible for a small proportion of cases of diphtheria (Oehring 1963). It should probably be regarded as a diphtheria bacillus, not as a separate species.

Corynebacterium erythrasmae Isolated by Sarkany, Taplin and Blank (1961) from cases of erythrasma. On a special culture medium containing fetal bovine serum the colonies after 24 hours' incubation at 34° are circular, slightly convex, 1–2 mm in diameter, shiny, moist, greyish-white and translucent, and show a coral-red to orange fluorescence under long ultraviolet light. Forms acid from glucose, maltose and sucrose; produces catalase, but not H_2S or urease. No formation of indole or reduction of nitrate. Growth is inhibited by erythromycin.

*For highly antibiotic-resistant skin diphtheroids ('JK organisms') that occasionally cause systemic disease in man, see Chapter 58.

Cornebacterium mycetoides Isolated by Castellani (1948, 1951) from patients suffering from tropical ulcer in North Africa, and studied by Ortali and Capocaccia (1957). In the body it appears as a coccus or coccobacillus, generally gram negative, but in culture gram-positive bacillary forms are present in addition containing prominent dark granules. Growth is poor at first, but in subculture on rich media becomes more abundant and is characterized by a yellow or yellowish-green non-diffusible pigment. Glucose is slowly fermented. Non-pathogenic for laboratory animals, but causes ulceration when implanted on the scarified leg of man. Ferments no other sugars. Catalase +; urease −; nitrate not reduced.

Corynebacteria resembling *C. ulcerans* Ilukewitsch (1948) isolated an organism from horses dying of nephritis in Germany. It differs from *C. ulcerans* in not liquefying gelatin. It is very pathogenic for horses, guinea-pigs, rats, mice, cats and pigeons, but only slightly so for rabbits and dogs. It haemolyses human and rabbit blood, forms a soluble toxin not completely neutralized by diphtheria antitoxin, and ferments only glucose and maltose among the common sugars.

Rountree and Carne (1967) isolated an organism from an abscess in the thigh of a man suffering from a progressive pyogenic infection of the lymph nodes in the leg. This organism is toxic to guinea-pigs in large doses and pyogenic in small. The toxin is neutralized by *C. ovis* antitoxin but not by diphtheria antitoxin. The organism differs from *C. ulcerans* in reducing nitrate and in failing to ferment starch or trehalose or to liquefy gelatin. It differs from *C. ovis* in reducing nitrate, in being non-pathogenic to sheep and in producing a dermal lesion after intracutaneous injection not neutralizable by *C. ovis* antitoxin.

Diphtheroid bacilli not falling into the genus *Corynebacterium*

'Corynebacterium' pyogenes

This organism, first described by Lucet (1893), is a small pleomorphic bacillus, frequently almost coccal. Grows scantily on unenriched media. On Loeffler's serum it forms minute colonies in 24 hr at 37°; which gradually enlarge to a diameter of 2–3 mm; the centre becomes granular and the medium is slowly liquefied, the liquefaction beginning as a small pit beneath each colony. On blood agar, colonies are small and surrounded by a zone of β-haemolysis. Ferments several sugars including lactose, liquefies Loeffler's serum slopes, forms an acid clot with peptonization in litmus milk, hydrolyses gelatin, dissolves egg yolk, forms no catalase or urease, and is VP negative. Nitrate is not reduced. Strains isolated in Nigeria differed from the typical organisms in fermenting trehalose but not maltose, and in usually failing to grow in a 0.1 per cent tellurite medium (Mohan and Uzoukwu 1980).

'C' pyogenes is rapidly killed at 56°, and is very sensitive to disinfectants (Brown and Orcutt 1920). It is antigenically homogeneous (Lovell 1937). Unlike true corynebacteria, the cell wall contains lysine, rhamnose and glucose (see also Roberts 1968*b*). The G + C content of the DNA is 48.5 moles per cent.

The organism forms a filtrable haemolysin and an exotoxin that are apparently identical (Roberts 1968*a*). It is adsorbed to and is haemolytic for rabbit erythrocytes, is lethal to rabbits and mice injected intravenously, is dermonecrotic for guinea-pigs and rabbits, dissolves egg yolk, and is inactivated by heat at 56° in 1 hr (Souček *et al.* 1965, Smith 1966).

The injection of living cultures of *'C' pyogenes* into the rabbit is followed by the development of localized abscesses if the injection is given subcutaneously. When given intravenously, or if generalization occurs after a subcutaneous injection, abscesses develop in the bones and joints, less frequently in other organs.

The organism is a widespread parasite of domestic animals, affecting particularly cattle, sheep, goats, and pigs. Under natural conditions it gives rise to suppurative pneumonia, suppurative arthritis, and other pyogenic infections including mastitis (see Merchant 1935, Magnusson 1938). Among laboratory animals the rabbit appears to be the most susceptible; guinea-pigs are less so, and mice are relatively resistant (see Brown and Orcutt 1920). It has occasionally been isolated from the throat of human patients suffering from a disease associated with a scarlatiniform rash (see Gärtner and Knothe 1960).

'Corynebacterium' haemolyticum

This organism was isolated from the nasopharynx and skin of American soldiers and natives on islands in the South and West Pacific Ocean, and described by Maclean, Liebow and Rosenberg (1946). It produces a wide zone of haemolysis on *human* blood agar plates. On horse blood agar plates after 48 hr, colonies are 1–1.5 mm in diameter, surrounded by a narrow zone of haemolysis. A characteristic feature is the presence of a central opaque dot; when the colony is lightly scraped aside, the central dot is left behind and the colony is seen to have etched the medium (Fell *et al.* 1977). It ferments glucose, maltose, lactose, sucrose and dextrin, liquefies gelatin slowly, fails to reduce nitrate or to form a soluble toxin. It produces a plateau-like abscess in guinea-pigs and rabbits when injected intradermally against which diphtheria antitoxin affords no protection. In man, it may give rise to sore throat, often accompanied by a maculopapular rash (Fell *et al.* 1977). It is sensitive to penicillin, erythromycin, and tetracycline. According to Barksdale and his colleagues (1957) this organism is a variant of *'C. pyogenes'*, differing from it in the larger size of its colonies on blood agar, the absence of glucose from its cell wall, and its failure to ferment xylose or produce a soluble haemolysin. According to Keddie and Cure (1978), however, the cell wall, like that of *'C. pyogenes'*, does contain glucose. Its difference from the type species, *C. diphtheriae*, is so great that Collins, Jones and Schofield (1982) would remove it to a new genus, *Arcanobacterium*, call it *A. haemolyticum*, and place it tentatively within the 'coryneform' group of bacteria.

'Corynebacterium' equi

An organism that has sometimes been confused with *'C'. pyogenes* is *'C'. equi*, which was isolated by Magnusson (1923) in Sweden from foals affected with suppurative pneumonia. It has since been isolated from tuberculous-like lesions in the submaxillary nodes of pigs (see Woodroofe 1950). Occasionally it infects cattle and goats. It appears to have little pathogenicity for rabbits or guinea-pigs, but the natural disease of foals can be reproduced experimentally by respiratory infection. *'C'. equi* is frequently present in the faeces of horses and other animals (Woolcock *et al.* 1980).

It is a fairly large pleomorphic bacillus showing metachromatic granules; in pus and in surface colonies it may appear coccoid. Grows freely on ordinary media forming irregular, large, succulent, pinkish colonies, owing to the presence of a polysaccharide capsule having by electronmicrography a laminated appearance (Smith 1966, Woolcock and Mutimer 1978). Slate to black colonies on tellurite media. Grows well between 18° and 27°. Said to be unusually resistant to oxalic acid used for the destruction of non-acid-fast bacilli in pathological material (Karlson *et al.* 1940). Ferments no sugars. Nitrate reduced to nitrite. Slight H_2S production. Hydrolysis of urea variable. Catalase formed. No liquefaction of gelatin. Contains a species-specific antigen demonstrable by complement fixation with acid-extracted organisms. Five type-specific antigens recognizable by agglutination and precipitation tests (see Woodroofe 1950). Seven capsular serotypes have been described (Prescott 1981).

'C'. equi differs from *'C'. pyogenes* in its possession of a capsule, its abundant growth on ordinary media, its highly mucoid colonies, its pigment formation, its resistance to oxalic acid, its failure to liquefy coagulated serum or gelatin, to lyse blood, ferment carbohydrates or produce an exotoxin, and in its pathogenicity for horses. (See also the review by Barton and Hughes 1980.)

'Corynebacterium' lacticum

Found in pasteurized milk and on dairy equipment (Orla-Jensen 1919). Forms short irregular rods, sometimes granulated, about 0.5 μm thick (Abd-el-Malek and Gibson 1952). Grows best at 24–30°, and hardly at all over 35°. Colonies may be faintly yellow. May endure heating to 80–85°; and may sometimes be isolated in pure culture from pasteurized milk. Acidifies glucose, maltose, lactose, dextrin, and sometimes sucrose, starch and mannitol. No liquefaction of gelatin or formation of urease. Some strains reduce nitrate. Should probably be classified as a *Microbacterium*.

'Corynebacterium' kutscheri

This organism, often known as *C. murium*, was isolated from a mouse by Kutscher in 1894 and by Bongert in 1901, and since then by other observers (Andrewes *et al.* 1923).

Morphologically and culturally there is nothing distinctive about the bacillus. Biochemically it produces acid from glucose, maltose and sucrose, reduces nitrate, forms no indole, does not liquefy gelatin, and gives a variable urease reaction. So far, its antigenic structure and its toxin do not seem to have been investigated. Its G + C content of DNA is 58.5 moles per cent.

The natural disease was described by Kutscher and by Bongert, and has been observed on many occasions by the present authors during necropsies on mice, though it is certainly not very common. The most characteristic lesion in the naturally occurring disease is the presence of large, firm, caseous areas in the lung. In sections or films from these lesions the bacilli are usually abundant. Caseous nodules

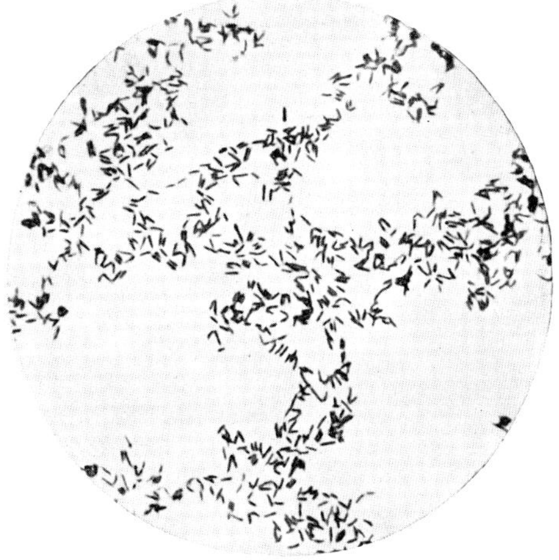

Fig. 25.6 *'C.' kutscheri.* From 24 hr culture on Loeffler's serum (× 1000).

may be found in the liver, though they are less frequent; when present they project from the surface, in contradistinction to the necrotic areas seen in mouse typhoid. The lymphatic nodes of the axilla, neck, mediastinum and mesentery may be enlarged and caseous; but the pulmonary lesions are often the only obvious sign of disease. Occasionally the bacillus may be isolated from a single caseous node, found at necropsy without any other detectable lesion. Latent infection with this organism is said to occur in many strains of mice. It can be wakened into activity by a single injection of 10 mg of cortisone (Fauve *et al.* 1964).

The disease may readily be reproduced by inoculating mice with pure cultures of *'C.' kutscheri*, or by administration *per os*. The findings at necropsy depend largely on the route of administration. After feeding, lesions develop in the mesenteric nodes and in the liver. After intraperitoneal inoculation, which usually leads to death within a week, the peritoneum is found to be studded with minute tubercles, and there is a spreading granulomatosis, of varying extent, affecting the regional lymphatic nodes, the liver, and less often the spleen. In our experience pulmonary lesions are much less frequent in the experimental than in the natural disease, though they occasionally occur. Intravenous injection is fatal to mice of some genetic types but not to others (Pierce-Chase *et al.* 1964). In animals dying after subcutaneous inoculation the only detectable lesion may be a small caseous abscess at the site of inoculation. It would appear that this organism produces an exotoxin which is fatal for mice. According to Bongert it is relatively heat-stable, since it withstands heating for 2 h at 55°, or for a few minutes at 74°; but the particulars given are not sufficiently precise to allow of any definite conclusion on the time or temperature required for inactivation.

The toxin of this organism has not, so far as we are aware, been compared with that of *C. diphtheriae* or of *C. ovis*, but it seems exceedingly unlikely that there is any relationship, since the mouse is resistant to diphtheria toxin, and the guinea-pig and rabbit, which are very susceptible to *C. ovis*, are resistant to *'C'. kutscheri.*

Other Coryneform bacteria

Besides the animal pathogens already described, there is a large miscellaneous collection of mainly club-shaped bacteria whose taxonomic position is still uncertain. Some of these are found in the intestine, and are occasionally isolated from such diseases as endocarditis, septicaemia, suppurating adenitis or arthritis, necrotic lesions of the mouth and teeth, the skin, and bone marrow (Prévot 1960, Saino *et al.* 1976). Others, many of which are plant pathogens, are found chiefly in the soil. We shall refer to these organisms as the 'coryneform group'.

Propionibacterium

These organisms, which are found in milk and cheese, are non-motile gram-positive coccobacilli, often resembling diphtheroids. They ferment carbohydrates and polyalcohols with the formation of propionic and acetic acids and CO_2. They form catalase but do not usually reduce nitrate to nitrite. They prefer anaerobic conditions and grow best at about 30°. Surface growth may be pigmented. Their nutritive requirements vary, but they generally grow in a synthetic medium containing ammonium sulphate as the source of nitrogen (Wood *et al.* 1938). Their vitamin B requirements are fairly simple; most strains need only pantothenic acid and biotin; but the most exacting need thiamine and *p*-aminobenzoic acid in addition (Thompson 1943, Delwiche 1949). (For reference, see van Niel 1928.) Their G + C content is 66–68 moles per cent.

The *acne bacillus*, which is believed to be pathogenic for man, is thought to belong to this group, though its G+C content of 57–60 moles per cent differs widely from that of other members. Moreover, it forms delicate regular rods staining unevenly, whereas *Propionibacterium* forms coarse pleomorphic rods staining uniformly (Zierdt *et al.* 1968).

Propionibacterium acnes Found mainly on the human skin. Forms delicate rods staining unevenly. Non-motile. Grows best on a slightly acid medium under anaerobic conditions. After 5 days at 35°, colonies are circular, 1–2 mm in diameter, convex, and cream or yellow, but there is some variation according to whether they belong to group I or II (Marples and McGinley 1974). Growth is favoured by glucose, glycerol, blood, Fildes's extract of blood, serum and, as Fleming (1909) showed, oleic acid (see also Pollock *et al.* 1949). Cultures have a sour smell. Forms acid in glucose, dextrin, glycerol, and trehalose, but not in lactose or sucrose. Propionic acid is formed during the fermentation of glucose. Liquefies gelatin, but not Loeffler's serum. Gives α-lysis of human and rabbit blood. Oxidase positive, urease negative. Cell wall contains L-DAP, not *meso*-DAP, and mannose in place of arabinose and galactose (Barksdale 1970). Two groups are distinguishable, defined by biochemical properties, agglutination (Craddock 1942), and susceptibility to bacteriophages (Marples and McGinley 1974). Group II is subdivided into types A and B. G+C content of DNA variously given, but probably about 57–60 moles per cent.

Described originally by Unna (1896) in the lesions of cutaneous acne; isolated by Sabouraud (1897); and studied by numerous workers of whom the more recent are Moss and his colleagues (1967), Puhvel (1968), Smith and Bodily (1968), Johnson and Cummins (1972), and Marples and McGinley (1974). Generally considered to be responsible, in part at least, for acne. The organism known as *'C. parvum'* is apparently identical with *P. acnes* (Cummins and Johnson 1974). It has obtained some notoriety as a means of preventing the development of fibrosarcoma in mice (Woodruff and Speedy 1978).

Propionibacterium avidum An anaerobic or aerotolerant bacillus described by Moore and Holdemann (1969; see also Prévot 1957). Cell wall contains L-DAP, galactose, glucose, and mannose. G+C content of DNA 62–63 moles per cent. Isolated occasionally from blood, pus, and faeces, but found mainly on the skin.

Propionibacterium granulosum An anaerobic or aerotolerant bacillus described by Moore and Holdeman (1970). Cell wall contains L-DAP and galactose. Isolated from the skin, intestine and occasionally abscesses. G+C content 61–63 moles per cent.

(For a description of some other anaerobic coryneform bacteria, see Beerens and Demont 1949, Tanner 1949.)

'C.' *suis* Under the term *Corynebacterium suis* Soltys (1961) described an anaerobic diphtheroid bacillus that he isolated from the kidneys of ten sows suffering from pyelonephritis and cystitis and from the semen of nine healthy boars. It was innocuous to laboratory animals but, in combination with saponin it damaged the renalcells, it gave rise to pyelonephritis when injected into the kidneys of young pigs. The taxonomic position of this organism is in doubt. Morphologically it is a corynebacterium; and though it is said to be anaerobic it does multiply slowly under aerobic conditions. Chromotographically it differs from *Propionibacterium* (Wegienek and Reddy 1977). According to Wegienek and Reddy (1982) it is a *Eubacterium.* On blood agar anaerobically colonies appear in 3 to 4 days; they are non-haemolytic. It produces urease, ferments maltose and xylose, and hydrolyses starch, but gives negative reactions in the catalase, MR, VP, indole, and nitrate-reduction tests. *C. suis* is present in the prepuce of a large proportion of male pigs, but the cystitis and pyelonephritis it causes occur mainly in adult females. The organism is sensitive to several antibiotics, including penicillin, the tetracyclines and chloramphenicol. For a fuller account of the organism itself and of its pathogenic properties the reader should consult the various papers by J. E. T. Jones (1981), who gives a full series of references to the subject.

'C.' *typhi*, which was isolated by Plotz (1914), from the blood of typhus fever patients, is now generally admitted to be a harmless parasitic diphtheroid organism (Olitsky 1921). (For numerical classification of coryneform bacteria, see Seiler 1983).

Kurthia

The organism, first described by Kurth (1883) as *Bacterium zopfii*, and by Hauser (1885) as *Proteus zenkeri*, belongs to a group of motile, non-sporing, aerobic, gram-positive bacilli found in decomposing materials. These organisms are long rods with somewhat rounded ends, 2–12 μm × 0.8 μm (Fig. 25.7(*a*)). In liquid media some form filaments which break up into evenly curved chains of coccoid bodies (Fig. 25.7(*b*)). They form no capsules and are not acid-fast. The flagella are peritrichate.

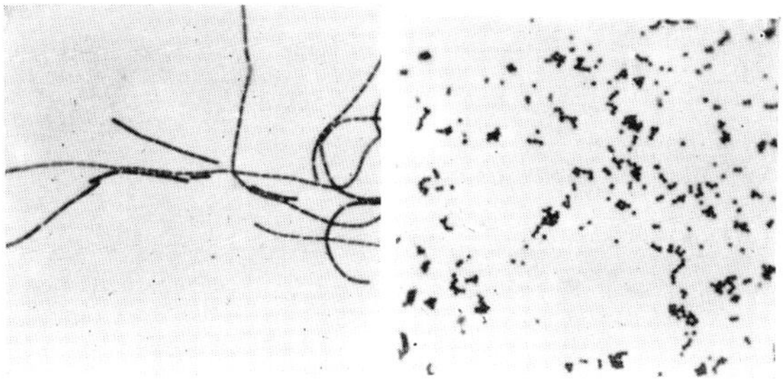

(*a*) (*b*)

Fig. 25.7 (a) *Kurthia* spp. Gram-stained preparation after 24 hr on TCM agar. (b) The same after 7 days. (Photomicrographs kindly supplied by Mr G. A. Gardner.)

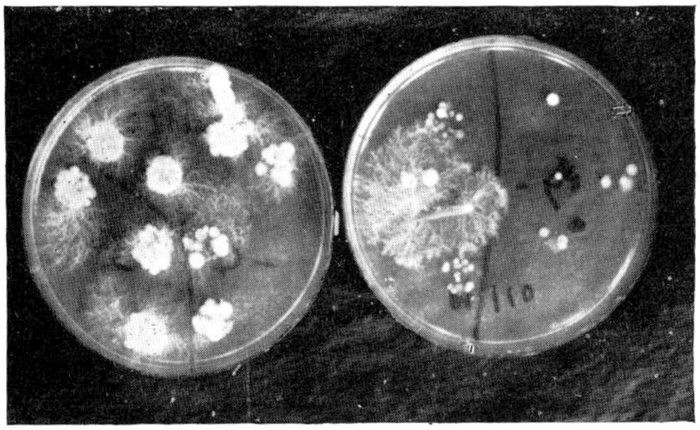

Fig. 25.8 *Kurthia* spp. Typical outgrowth on an agar plate. (Photograph kindly supplied by Mr G. A. Gardner.)

They form catalase but not oxidase, ferment no sugars, and fail to produce indole or H_2S or to reduce nitrate. Growth occurs between 5° and 37°, best at 25°. Some strains, however, can grow at 44°.

Kurth's original organism, now known as *Kurthia zopfi*, forms small, indistinct grey rhizoid or spider-web colonies on agar with a Medusa-head appearance (Fig. 25.8). In gelatin stab it forms interlacing arborescent branches; the gelatin is not liquefied. On a gelatin slope it produces a feather-like growth (Fig. 25.9). It gives off a putrid ammoniacal odour, but is not proteolytic and does not form H_2S. Gardner (1969), who found it in about 80 per cent of meat products, distinguished two sub-species on the basis of growth at 44°, acid production from ethanol or glycerol, and yellow or cream pigmentation.

Two other species are recognized (see Breed *et al.* 1957). *K. variabilis* differs in producing H_2S and in not giving a putrid odour. On agar it forms smooth, circular, grey colonies, 1–2 mm in diameter. There are no lateral outgrowths in gelatin stab cultures. Like *K. zopfi*, it does not liquefy gelatin. *K. bessoni* liquefies gelatin rapidly and causes alkaline digestion of litmus milk. It forms flat, spreading colonies with a lacerate edge and fimbriate outgrowths in gelatin but the filaments do not break up into chains in liquid media. Reports of its H_2S production are discrepant. Elston (1961) isolated it from human sputum, and from a cyst.

Brevibacterium

Breed (1953) established this genus. According to Hester and Weeks (1969), it is a conglomeration of non-motile, gram-positive, aerobic, non-sporing, usually pigmented rods. Analysis revealed three major and three minor clusters among 42 strains studied in 104 tests. These authors conclude that the retention of this genus is questionable. Little can be said about its taxonomic position (Jones 1975).

Microbacterium

This genus was established by Orla-Jensen (1919). Its members are small, non-motile, non-sporing, gram-positive bacilli which are unusually heat-resistant, withstanding a temperature of 70° for 5–10 min, and form D-lactic acid from glucose.

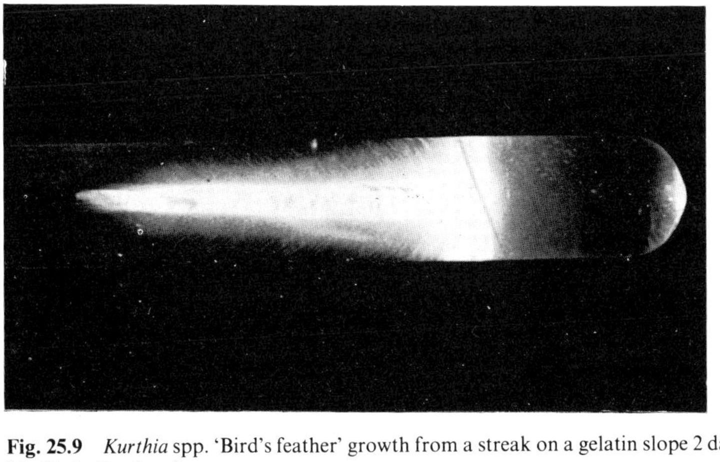

Fig. 25.9 *Kurthia* spp. 'Bird's feather' growth from a streak on a gelatin slope 2 days at 22°. (Photograph kindly supplied by Mr G. A. Gardner.)

They cannot use citrate as the sole source of carbon, or inorganic compounds as the sole sources of nitrogen (Robinson 1966). They ferment glucose, mannose and lactose, but not the pentoses raffinose or inulin. Some strains form acid from dextrin, salicin, starch and mannitol. In milk they produce 0.5 per cent lactic acid. They grow between 10 and 39°, best between 30 and 37°. Some strains liquefy gelatin and digest casein; some form a yellow surface growth on milk agar; and some fail to grow anaerobically. They are found in milk products, faeces and soil, and are non-pathogenic. They differ from the lactobacilli in being aerobic, in forming catalase, in generally reducing nitrate to nitrite, and in using an iron-containing enzyme instead of lactoflavin for their respiration (Orla-Jensen 1943). The type species is *Microbacterium lacticum*. Of the few other species described, the two best known are *M. flavum* and *M. liquefaciens*. According to Jones (1975), however, *M. flavum* is a *Corynebacterium* and must be classed as such (see p. 105): and *M. thermosphactum*, now known as *Brochothrix thermosphacta*, does not belong to the family of Corynebacteriaceae at all.

Brochothrix thermosphacta, described by McLean and Sulzbacher (1953) and Davidson, Mobbs ad Stubbs (1968), is of interest to food bacteriologists. It is said to be the commonest organism in British sausages (Dowdell and Board 1968). It differs from *Kurthia* spp. in being non-motile, facultatively anaerobic, and acidifying glucose.

Arthrobacter is a genus of strict aerobes, sometimes motile, that undergo complete morphological changes from the bacillary to the coccal form, and are among the dominant organisms met with in cultures of soil on ordinary bacteriological media. Like *Brochothrix thermosphacta* it should be excluded from the family of Corynebacteriaceae (Jones 1975).

References

Abd-el-Malek, Y. and Gibson, T. (1952) *J. Dairy Res.* **19,** 153.
Amies, C. R. and Jones, S. A. (1957) *Canad. J. Microbiol.* **3,** 579.
Anderson, J. S., Cooper, K. E., Happold, F. C. and McLeod, J. W. (1933) *J. Path. Bact.* **36,** 169.
Anderson, J. S., Happold, F. C., McLeod, J. W. and Thomson, J. G. (1931) *J. Path. Bact.* **34,** 667.
Anderson, P. S. and Cowles, P. B. (1958) *J. Bact.* **76,** 272.
Andrewes, F. W. *et al.* (1923) Med. Res. Coun., *Monograph on Diphtheria.* London.
Babes, V. (1886) *Bull. Soc. Anat., Paris* **61,** 72.
Bailey, G. H. (1925) *J. Immunol.* **10,** 791.
Banach, T. M. and Hawirko, R. Z. (1966) *J. Bact.* **92,** 1304.
Barber, C., Meitert, E. and Saragea, A. (1965) *Path. et Microbiol., Basel* **28,** 274.
Barksdale, L. (1970) *Bact. Rev.* **34,** 378.
Barksdale, L., Lanéelle, M. A., Pollice, M. C., Asselineau, J., Welby, M. and Norgard, M. V. (1979) *Int. J. Syst. Bact.* **29,** 222.
Barksdale, L., Linder, R., Sulea, I. T. and Pollice, M. (1981) *J. clin. Microbiol.* **13,** 335
Barksdale, W. L., Li, K., Cummins, C. S. and Harris, H. (1957) *J. gen. Microbiol.* **16,** 749.
Barr, M., Glenny, A. T., Pope, C. G. and Linggood, F. V. (1941) *Lancet* **ii,** 301.
Barratt, M. M. (1923) see Andrewes *et al.*, p. 174; (1933) *J. Path. Bact.* **36,** 369.
Bartholomew, L. E. and Nelson, F. R. (1972) *Ann. rheum. Dis.* **31,** 22, 28.
Barton, M. D. and Hughes, K. L. (1980) *Vet. Bull.* **50,** 65.
Beerens, H. and Demont, F. (1949) *C. R. Soc. Biol., Paris.* **143,** 1200.
Belfanti, S. and della Vedova. (1896) *G. Accad. Med. Torino* **3,** 149.
Bisset, K. A. (1950) *The Cytology and Life-history of Bacteria.* E. & S. Livingstone, Edinburgh.
Bolton, C. (1905) *Lancet* **i,** 278.
Bongert. (1901) *Z. Hyg. InfektKr.* **37,** 449.
Bonventre, P. F. and Imhoff, J. G. (1966) *J. exp. Med.* **124,** 1107.
Bousfield, I. J. (1972) *J. gen. Microbiol.* **71,** 441.
Bousfield, I. J. and Callely, A. G. (1978) *Coryneform Bacteria.* Academic Press, London.
Braun, H. (1938) *Schweiz. Z. allg. Path.* **1,** 113.
Bray, J. (1944) *J. Path, Bact.* **56,** 497.
Breed, R. S. (1953) See *Bergey's Manual of Determinative Bacteriology*, 7th edn. 1957, p. 490. Baillière, Tindall & Cox, London.
Breed, R. S., Murray, E. G. D. and Smith, N. R. (1957) *Bergey's Manual of Determinative Bacteriology*, 7th edn. Baillière, Tindall & Cox, London.
Brown, J. H. and Orcutt, M. L. (1920) *J. exp. Med.* **32,** 219.
Bull, L. B. and Dickinson, C. G. (1935) *Aust. vet. J.* **11,** 126.
Burrell, D. H. (1979) *Res. vet. Sci.* **26,** 333; (1980) *Ibid.* **28,** 51.
Carne, H. R. (1939) *J. Path. Bact.* **49,** 313; (1940) *Ibid.* **51,** 199.
Carne, H. R. and Onon, E. O. (1978) *Nature, Lond.* **271,** 246; (1982) *J. Hyg., Camb.* **88,** 173.
Carpenter, C. M., Howard, D. H. and Lehman, E. L. (1956) *J. Lab. clin. Med.* **47,** 194.
Castellani, A. (1948) *J. trop. Med. Hyg.* **51,** 245; (1951) *Ann. Inst. Pasteur* **80,** 83.
Christensen, P. E. (1957) *Acta path. microbiol. scand.* **41,** 67.
Christovão, D. de A. (1957) *Arch. Faculd. Hig. Saúd. públ. Univ. São Paulo* **11,** 97, 115.
Cobb, R. W. and Walley, J. K. (1962) *Vet. Rec.* **74,** 101.
Coca, A. E., Russell, E. F. and Baughman, W. H. (1921) *J. Immunol.* **6,** 887.
Collier, R. J. (1975) *Bact. Rev.* **39,** 54.
Collins, M. D., Jones, D. and Schofield, G. M. (1982) *J. gen. Microbiol.* **128,** 1279.
Collins-Thompson, D. L., Sørhaug, T., Witter, L. D. and Ordal, Z. J. (1972) *Int. J. syst. Bact.* **22,** 65.
Cooper, K. E., Happold, F. C., McLeod, J. W. and Woodcock, H. E. de C. (1936) *Proc. R. Soc. Med.* **29,** 1029.
Costa, S., Troisier, J. and Dauvergne, J. (1918) *C. R. Soc. Biol., Paris* **81,** 89.
Cowan, M. L. (1927) *Brit. J. exp. Path.* **8,** 6.
Craddock, S. (1942) *Lancet* **i,** 558.
Crowell, M. J. (1926) *J. Bact.* **11,** 65.
Cummins, C. S. (1954) *Brit. J. exp. Path.* **35,** 166; (1962) *J. gen. Microbiol.* **28,** 35; (1971) *J. Bact.* **105,** 1227.
Cummins, C. S. and Harris, H. (1956) *J. gen. Microbiol.* **14,** 583.
Cummins, C. S. and Johnson, J. L. (1974) *J. gen. Microbiol.* **80,** 433.
Dassonville. (1907) *Bull. Soc. Méd. vét d. Centre* **61,** 576.
Davidson, C. M., Mobbs, P. and Stubbs, J. M. (1968) *J. appl. Bact.* **31,** 551.
Delwiche, E. A. (1949) *J. Bact.* **58,** 395.

Douglas, H. C. and Gunter, S. E. (1946) *J. Bact.* **52,** 15.
Dowdell, M. J. and Board, R. G. (1968) *J. appl. Bact.* **31,** 378.
Dudgeon, L. S. (1906) *Brain* **29,** 227.
Dudley, S. F., May, P. M. and O'Flynn, J. A. (1934) *Spec. Rep. Ser. med. Res. Coun., Lond.* No. 195.
Eaton, M. D. (1936) *J. Bact.* **31,** 367.
Eaton, M. D. and Bayne-Jones, S. (1934) *J. Bact.* **29,** 56.
Elston, H. R. (1961) *J. Path. Bact.* **81,** 245.
Enderlen, E. (1890–91) *Dtsch. Z. Thierheilk.* **17,** 325.
Ernst, W. (1888) *Z. Hyg. InfektKr.* **4,** 25; (1889) *Ibid.* **5,** 428; (1905) *Zbl. Bakt.* **39,** 549, 660; (1906) *Ibid.* **40,** 79.
Evans, A. C. (1916) *J. infect. Dis.* **18,** 437.
Ewing, J. O. (1933) *J. Path. Bact.* **37,** 345.
Fauve, R. M., Pierce-Chase, C. H. and Dubos, R. (1964) *J. exp. Med.* **120,** 283.
Feenstra, E. S., Thorp, F. and Clark, C. F. (1945) *J. Bact.* **50,** 497.
Feenstra, E. S., Thorp, F. and Gray, M. L. (1949) *Amer. J. vet. Res.* **10,** 12.
Fell, H. W. K., Nagington, J. and Naylor, G. R. E. (1977) *J. Hyg., Camb.* **79,** 269.
Fischl, V., Koech, M. and Kussat, E. (1931) *Z. Hyg. InfektKr.* **112,** 421.
Fleming, A. (1909) *Lancet* **i,** 1035.
Flexner, S. (1897) *Rep. Johns Hopk. Hosp.* **6,** 259.
Freeman, V. J. (1951) *J. Bact.* **61,** 675.
Freeman, V. J. and Minzel, G. H. (1950) *Amer. J. Hyg.* **51,** 305.
Freeman, V. J. and Morse, I. U. (1952) *J. Bact.* **63,** 407.
Frieber, W. (1921) *Zbl. Bakt.* **87,** 254.
Frobisher, M. (1946) *Proc. Soc. exp. Biol., N.Y.* **62,** 304.
Frobisher, M., Adams, M. L. and Kuhns, W. J. (1945) *Proc. Soc. exp. Biol., N.Y.* **58,** 330.
Galbraith, T. W., Fraser, R. S. and Bramhall, E. H. (1948) *Publ. Hlth. Rep., Wash.* **63,** 577.
Gardner, G. A. (1969) *J. appl. Bact.* **32,** 371.
Gärtner, H. and Knothe, H. (1960) *Arch. Hyg., Berl.* **144,** 308.
Gibson, L. F. (1975) *J. Hyg., Camb.* **75,** 413.
Gilbert, R. and Stewart, F. C. (1926–7) *J. Lab. clin. Med.* **12,** 756.
Glass, V. (1937) *J. Path. Bact.* **44,** 235; (1939*a*) *Ibid.* **48,** 507; (1939*b*) *Ibid.* **49,** 549.
Glenny, A. T. (1925*a*) *J. Hyg., Camb.* **24,** 301; (1925*b*) *J. Path. Bact.* **28,** 251.
Glenny, A. T. and Allen, K. (1922) *J. Path. Bact.* **24,** 61.
Glenny, A. T. and Walpole, G. S. (1915) *Biochem. J.* **9,** 298.
Goldie, H. (1933) *C. R. Soc. Biol.,* **112,** 1210; (1934) *Ibid.* **116,** 17.
Goldsworthy, N. E. and Wilson, H. (1942) *J. Path. Bact.* **54,** 183.
Goodman, H. M. (1907) *J. infect. Dis.* **4,** 509.
Goor, R. S. and Pappenheimer, A. M. (1967) *J. exp. Med.* **126,** 899, 913.
Gordon, M. and Zinnemann, K. (1949) *J. clin. Path.* **2,** 209.
Goudswaard, J. and Budhai, S. (1975) *Zbl. VetMed.* **B22,** 473.
Graham-Smith, G. S. (1908) *Diphtheria.* Nuttall and Graham-Smith, Cambridge.
Griffith, A. S. (1901). *Rep. Thomps. Yates Lab. Univ. L'pool.* **4,** 99.
Greenwood, J. R. and Pickett, M. J. (1980) *Int. J. syst. Bact.* **30,** 170.
Groman, N. B. (1953) *J. Bact.,* **66,** 184; (1955) *Ibid.* **69,** 9.
Groman, N. B. and Memmer, R. (1958) *J. gen. Microbiol.* **19,** 634.
Hall, I. C. and Stone, R. V. (1916) *J. infect. Dis.* **18,** 195.
Hartley, P. (1922) *J. Path. Bact.* **25,** 479.
Hauser, G. (1885) *Ueber Fäulnisbakterien.* Leipzig.
Hendry, C. B. (1938) *J. Path. Bact.* **46,** 383.
Hentrich, F., Menge, B. and Lickfeld, K. G. (1977) *Zbl. Bakt.,* I Abt. Orig. **A239,** 24a.
Hester, D. J. and Weeks, O. B. (1969) *Bact. Proc.* p. 19.
Hewitt, L. F. (1947*a*) *J. Path. Bact.* **59,** 145; (1947*b*) *Brit. J. exp. Path.* **28,** 338; (1952) *Lancet* **ii,** 272.
Hill, H. W. (1898) *J. Boston. Soc. med. Sci.* **3,** 86; (1899) *Ibid.* **4,** 78; (1902*a*) *J. med. Res.* **7,** 115; (1902*b*) *Ibid.* **7,** 202.
Hill, L. R. (1966) *J. gen. Microbiol.* **44,** 419.
Hofmann, G. von. (1888) *Wien. med. Wschr.* **38,** 65, 108.
Holt, L. B. (1950) *Developments in Diphtheria Prophylaxis.* Heinemann. London.
Holth, H. (1908) *Z. InfektKr. Haustiere* **3,** 155.
Honda, E. and Yanagawa, R. (1978) *Amer. J. vet. Res.* **39,** 155.
Hoyle, L. (1941) *Lancet,* **i.** 175; (1942) *J. Hyg., Camb.* **42,** 416.
Huang, C. H. (1942) *Amer. J. Hyg.* **35,** 325.
Hugh, R. and Leifson, E. (1953) *J. Bact.* **66,** 24.
Ilukewitsch, A. (1948) *Tierarztl. Umsch.* **3,** 135.
Jaffé, R. (1920) *Arb. Inst. exp. Ther. Georg Speyer Hause, Frank. am. M.* **11,** 5.
Jayne-Williams, D. J. and Skerman, T. M. (1966) *J. appl. Bact.* **29,** 72.
Jebb, W. H. H. (1948) *J. Path. Bact.* **60,** 403.
Jensen, H. L. (1966) *J. appl. Bact.* **29,** 13.
Johnson, J. L. (1970) *Int. J. syst. Bact.* **20,** 421.
Johnson, J. L. and Cummins, C. S. (1972) *J. Bact.* **109,** 1047.
Johnstone, K. I. and McLeod, J. W. (1949) *Publ. Hlth Rep., Wash.* **64,** 1181.
Jones, D. (1975) *J. gen. Microbiol.* **87,** 52
Jones, F. S. and Little, R. B. (1925) *J. exp. Med.* **42,** 593; (1930) *Ibid.* **51,** 909.
Jones, J. E. T. (1981) In: *Diseases of Swine,* 5th edn., pp. 149, 530. Ed. by A. D. Leman, R. D. Gloek, W. L. Mengeling, R. H. C. Penny, E. Scholl, and B. Straw. Iowa State University Press, Ames, Iowa.
Karlson, A. G., Moses, H. E. and Feldman, W. H. (1940) *J. infect. Dis.* **67,** 243.
Keddie, R. M. and Cure, G. L. (1977) *J. appl. Bact.,* **42,** 229; (1978) *Coryneform bacteria,* p. 47. Ed. by I. J. Bousfield and A. G. Callely. Academic Press, London.
Keogh, E. V., Simmons, R. T. and Anderson, G. (1938) *J. Path. Bact.* **46,** 565.
Klebs, E. (1883) *Verh. Cong. inn. Med., Wiesbaden* 139.
Knight, B. C. J. G. (1936) *Spec. Rep. Ser. med. Res. Coun., Lond.* No. 210.
Krah, E. and Witebsky, E. (1930) *Z. ImmunForsch.* **66,** 59.
Kurth, H. (1883) *Bot. Ztg.* **41,** 369, 393, 409, 425.
Kuschbert and Neisser. (1883) *Jber. schles. Ges. vaterl. Kult.* **60,** 50.
Kutscher. (1894) *Z. Hyg. InfektKr.* **18,** 327.
Kuzdas, C. D., Morse, E. V. and Ellis, R. Hl (1951) *J. Bact.* **62,** 763.
Lautrop, H. (1950) *Acta path. microbiol. scand.* **27,** 443.
Lazar, I. (1968) *J. gen. Microbiol.* **52,** 77.
Lehmann, K. B. and Neumann, R. O. (1896) *Atlas u. Grund-*

riss. d. Bakt. u. Lehrb. d. spez. bakt. Diagnostik. 6th edn. Munich.
Leopold, S. (1953) *U.S. Forces med. J.* **4,** 263.
Lickfeld, K. G. (1967) *Z. med. Mikrobiol. Immunol.* **153,** 326.
Linzenmeier, G. (1957) *Zbl. Bakt.* **170,** 85.
Loeffler, F. (1884) *Mitt. ReichsgesundhAmt.* **2,** 421; (1890) *Zbl. Bakt.* **7,** 528.
Lovell, R. (1937) *J. Path. Bact.* **45,** 339; (1941) *Ibid.* **52,** 295; (1944) *Ibid.* **56,** 525; (1946) *J. comp. Path.* **56,** 196; (1951) *Vet. Rec.* **63,** 645; (1956) *J. comp. Path.* **66,** 332.
Lovell, R. and Cotchin, E. (1946) *J. comp. Path.* **56,** 205.
Lovell, R. and Harvey, D. G. (1950) *J. gen. Microbiol.* **4,** 493.
Lovell, R. and Zaki, M. M. (1966) *Res. vet. Sci.* **7,** 302.
Lucet, A. (1893) *Ann. Inst. Pasteur* **7,** 325.
Maclean, P. D., Liebow, A. A., and Rosenberg, A. A. (1946) *J. infect. Dis.* **79,** 69.
McCoy, E. L., McCusker, J. J., Keslin, M. Lutch, J. S. and Biberstein, E. (1978) *Abstracts Annu. Meetg. Amer. Soc. Microbiol.* p. 292.
McKenzie, D. A., Morrison, M. and Lambert, J. (1946) *Proc. Soc. appl. Bact.* p. 37.
McLean, R. A. and Sulzbacher, W. L. (1953) *J. Bact.* **68,** 428.
McLeod, J. W. (1943) *Bact. Rev.* **7,** 1.
Magnusson, H. (1923) *Arch. wiss. prakt. Tierheilk.* **50,** 22; (1938) *Vet. Rec.* **50,** 1459.
Mair, W. (1928) *J. Path. Bact.* **31,** 136.
Marples, R. R. and McGinley, K. J. (1974) *J. med. Microbiol.* **7,** 349.
Matsuda, M. and Barksdale, L. (1966) *Nature, Lond.* **210,** 911.
Maximescu, P. (1968) *J. gen. Microbiol.* **53,** 125.
Merchant, I. H. (1935) *J. Bact.* **30,** 95.
Minett, F. C. (1922*a*) *J. comp. Path.* **35,** 71; (1922*b*) *Ibid.* **35,** 291.
Miers, K. C. and Ley, W. B. (1980) *J. Amer. Vet. Med. Ass.* **177,** 250.
Mollard, J. and Regaud, C. (1895) *C. R. Soc. Biol., Paris* **2,** 828.
Mohan, K. and Uzoukwu, M. (1980) *Vet. Rec.* **107,** 252.
Monis, B. and Reback, J. F. (1962) *Proc. Soc. exp. Biol. Med.* **111,** 81.
Moore, W. E. C. and Cato, E. P. (1963) *J. Bact.* **85,** 870.
Moore, W. E. C. and Holdeman, L. V. (1969) In: *Outline of clinical methods in anaerobic bacteriology*, Ed. by E. V. Cato *et al.* Blacksburg, Va; (1970) *Ibid.* 2nd rev.
Moss, C. W. and Cherry, W. B. (1968) *J. Bact.* **95,** 241.
Moss, C. W., Dowell, V. R., Lewis, V. J. and Schekter, M. A. (1967) *J. Bact.* **94,** 1300.
Mueller, J. H. (1935*a*) *Science* **81,** 50; (1935*b*) *J. Bact.* **29,** 383; (1935*c*) *Ibid.* **29,** 515; (1939) *J. Immunol.* **37,** 103.
Müller, J. (1957) *Arch. Microbiol.* **27,** 105.
Neill, G. A. W. (1937) *J. Hyg., Camb.* **37,** 552.
Niel, C. B. van (1928) *The Propionic Acid Bacteria.* J. W. Boissevain & Co., Haarlem.
Nicolle, M., Loiseau, G. and Forgeot, P. (1912) *Ann. Inst. Pasteur* **26,** 83.
Nishida, S. (1954) *Jap. J. med. Sci. Biol.* **7,** 453, 495, 505.
Nocard, E. (1889) *C. R. Soc. Biol. Paris,* **1,** 608.
Norlin, G. (1943) *Acta path. microbiol. scand.* Suppl. No. 50.
Oeding, P. (1950*a*) *Acta path. microbiol. scand.* **27,** 16; (1950*b*) *Ibid.* **27,** 427, 597.
Oehring, H. (1963) *Arch. Hyg. Bakt.* **147,** 432.
Olitsky, P. K. (1921) *J. exp. Med.* **34,** 525.
Onon, E. O. (1979) *Biochem. J.* **177,** 181.
Orla-Jensen, S. (1919) *The Loctic Acid Bacteria.* A. F. Høst, Copenhagen; (1943) *Die echten Milchsäurebakterien.* Munksgaard, Copenhagen.
Ørskov, J. (1948) *Acta path. microbiol. scand.* **25,** 829.
Ortali, V. and Capocaccia, L. (1957) *Ann. Inst. Pasteur* **93,** 786.
Pappenheimer, A. M. (1937) *J. biol. Chem.* **120,** 543; (1947) *Ibid.* **167,** 251; (1955) *Symp. Soc. gen. Microbiol.* **5,** 40.
Pappenheimer, A. M. and Brown, R. (1968) *J. exp. Med.* **127,** 1073.
Pappenheimer, A. M. and Robinson, E. S. (1937) *J. Immunol.* **32,** 291.
Partridge, B. M. and Jackson, F. L. (1962) *Lancet* **i,** 591.
Petrie, G. F. and McClean, D. (1934) *J. Path. Bact.* **39,** 635.
Pierce-Chase, C. H., Fauve, R. M. and Dubos, R. (1964) *J. exp. Med.* **120,** 267.
Piot, P., Dyck, E. V., Goodfellow, M. and Falkow, S. (1980) *J. gen. Microbiol.* **119,** 373.
Plotz, H. (1914) *J. Amer. med. Ass.* **62,** 1556.
Pollock, M. R., Howard, G. A. and Boughton, B. W. (1949) *Biochem. J.* **45,** 417.
Pope, C. G. and Stevens, M. F. (1953) *Lancet,* ii. 1190; (1958) *Brit. J. exp. Path.* **39,** 139, 150.
Pope, C. G., Stevens, M. F., Caspary, E. A. and Fenton, E. L. (1951) *Brit. J. exp. Path.* **32,** 246.
Pope, C. G., Stevens, M. F. and Thomas, D. (1966) *Brit. J. exp. Path.* **47,** 45.
Preisz, H. (1894) *Ann. Inst. Pasteur* **8,** 231.
Prescott, J. F. (1981) *Canad. J. comp. Med.* **45,** 130.
Prévot, A. R. (1957) *Manuel de classification et détermination des bactéries anaérobies*, 3rd. edn. Masson, Paris; (1960) *Ergebn. Hyg.* **33,** 1.
Prévot, A. R. and Courdurier, J. (1949) *Ann. Inst. Pasteur* **76,** 232.
Puhvel, S. M. (1968) *J. gen. Microbiol.* **50,** 313.
Püschel, J. (1936) *Klin. Wschr.* **15,** 375.
Rantasalo, I. (1948) *Über die Einwirkung von Natriumfluorid bei Züchtung von Diphtheriebazillen.* Helsinki.
Ray, L. F. and Kellum, R. E. (1970) *Arch. Derm., N.Y.,* **101,** 36.
Reid, J. D. and Joya, M. A. (1969) *Int. J. syst. Bact.* **19,** 273.
Relyveld, E. H. and Raynaud, M. (1964) *Ann Inst. Pasteur* **107,** 618.
Report. (1950) *Diagnostic Procedures and Reagents*, 3rd edn. p. 168. Amer. publ. Hlth Ass., New York.
Reymond, Colomiatti and Perroncito. (1881) *Congr. period. int. ophthal. C. R.* 1880, Milano, Annexes 48; (1883) *G. Accad. Med. Torino* **31,** 519.
Roberts, R. J. (1968*a*) *Res. vet. Sci.* **9,** 350; (1968*b*) *J. Path. Bact.* 95, 127.
Robinson, D. T. (1934) *J. Path. Bact.* **38,** 551; (1940) *Pers. Comm.*
Robinson, D. T. and Marshall, F. N. (1934) *J. Path. Bact.* **38,** 73.
Robinson, D. T. and Peeney, A. L. P. (1936) *J. Path. Bact.* **43,** 403.
Robinson, K. (1966) *J. appl. Bact.* **29,** 607.
Rojas, J. A. B., and Biberstein, E. L. (1974) *J. comp. Path.* **84,** 301.
Römer, P. H. (1909) *Z. ImmunForsch.* **3,** 208.
Rountree, P. M. and Carne, H. R. (1967) *J. Path. Bact.* **94,** 19.

Roux, E. and Yersin, A. (1888) *Ann. Inst. Pasteur* **2,** 629.

Sabouraud, R. (1897) *Ann. Inst. Pasteur* **11,** 134.

Saino, Y. *et al.* (1976) *Jap. J. Microbiol.* **20,** 17.

Saragea, A. and Maximescu, P. (1966) *Bull. Wld Hlth Org.* **35,** 681.

Saragea, A., Maximescu, P. and Meitert, E. (1979) In: *Methods in microbiology*, vol 13, p. 61. Ed. by T. Bergan and J. R. Norris. Academic Press, London.

Schiller, J., Groman, N. and Coyle, M. (1980) *Antimicrob. Agents Chemother.* **18,** 814.

Sarkany, I., Taplin, D. and Blank, H. (1961) *J. invest. Derm.* **37,** 283.

Schwoner, J. (1904) *Zbl. Bakt.* **35,** 608.

Scott, W. M. (1923) *Rep. publ. Hlth med. Subj. Lond.*, No. 22.

Sebald, M., Gasser, F. and Werner, H. (1965) *Ann Inst. Pasteur* **109,** 251.

Seiler, H. (1983) *J. gen. Microbiol*, **129,** 1433.

Siemens, B. W. L. (1938) *Bijdrage tot de Kennis der Typen van het Corynebacterium Diphtheriae.* Van Corcum and Co., Assen.

Smith, J. E. (1966) *J. appl. Bact.* **29,** 119.

Smith, R. F. and Bodily, H. L. (1968) *Hlth Lab. Sci.* **5,** 95.

Soltys, M. A. (1961) *J. Path. Bact.* **81,** 441.

Somerville, D. A. (1972) *Brit. J. Derm.* **86,** Suppl. No. 8, p. 16.

Souček, A., Mára, M. and Součková, A. (1965) *Folia microbiol., Praha* **10,** 210.

Sousa, C. P. and Evans, D. G. (1957) *Brit. J. exp. Path.* **38,** 644.

Tanner, E. (1949) *Ann. Inst. Pasteur* **76,** 541.

Tasman, A. and Waasbergen, J. P. van. (1932) *Z. Immun.-Forsch.* **75,** 164.

Thompson, R. C. (1943) *J. Bact.* **46,** 99.

Tomlinson, A. J. H. (1966) *J. appl. Bact.* **29,** 131.

Toshach, S., Valentine A. and Sigurdson, S. (1977) *J. infect. Dis.* **136,** 655.

Ts'un T'ung. (1945) *Amer. J. Hyg.* **41,** 57.

Uhr, J. W., Pappenheimer, A. M. and Yoneda, M. (1957) *J. exp. Med.* **105,** 1.

Unna, P. G. (1896) *The Histopathology of Diseases of the Skin.* Eng. Transl. by N. Walker, Edin.

Voss, J. G. (1970) *J. Bact.* **101,** 392.

Ward, A. R. (1917) *J. Bact.* **2,** 619.

Warren, L. and Spearing, C. W. (1963) *J. Bact.* **86,** 950.

Wegienek, J. G., and Reddy, C. A. (1977) *Abstr. annu. Mtg Amer. Soc. Microbiol.* **77,** 178; (1982) *Int. J. syst. Bact.* **32,** 218.

Weitz, B. (1947) *J. comp. Path.* **57,** 191.

Werner, H. (1966) *J. appl. Bact.* **29,** 138; (1967) *Zbl. Bakt.* **205,** 210.

Wernicke. (1893) *Arch. Hyg.* **18,** 192.

Wesbrook, F. F., Wilson, L. B., McDaniel, O. and Adair, J. H. (1898) *Brit. med. J.* **i,** 1008.

Wong, S. C. and T'ung, T. (1940) *Proc. Soc. exp. Biol., N.Y.* **43,** 749.

Wood, H. G., Andersen, A. A. and Werkman, C. H. (1938) *J. Bact.* **36,** 201.

Woodroofe, G. M. (1950) *Aust. J. exp. Biol. med. Sci.* **28,** 399.

Woodruff, M. and Speedy, G. (1978) *Proc. roy. Soc., Lond., B* **201,** 209.

Woolcock, J. B. and Mutimer, M. D. (1978) *J. gen. Microbiol.* **109,** 127.

Woolcock, J. B., Mutimer, M.D. and Farmer, A. M. T. (1980) *Res. vet. Sci.* **28,** 87.

Wurch, T. and Lutz, A. (1955) *Rev. franç. Gynéc.* **50,** 289.

Yanagawa, R., Bari, H. and Otsuki, K. (1967) *Jap. J. vet. Res.* **15,** 111.

Yanagawa, R. and Honda, E. (1976) *Infect. Immun.* **13,** 1293; (1978) *Int. J. syst. Bact.* **28,** 217.

Yanagawa, R., Otsuki, K. and Tokui, T. (1968) *Jap. J. vet. Res.* **16,** 31.

Zamiri, I. and McEntegart, M. G. (1973) *J. med. Microbiol.* 128th Mtg Path. Soc. GBI, Synopsis, p. 5.

Zierdt, C. H., Webster, C. and Rude, W. S. (1968) *Int. J. syst. Bact.* **18,** 33.

26

The Bacteroidaceae: *Bacteroides*, *Fusobacterium*, *and Leptotrichia*

Brian I. Duerden

Introductory

Gram-negative, non-sporing, anaerobic bacilli with rounded or pointed ends, sometimes fusiform, sometimes filamentous and often pleomorphic, are classified in the family Bacteroidaceae. They have been recognized as important causes of human and animal infection since the latter part of the nineteenth century, when several members of the group were isolated from a variety of necrotic lesions (see Chapter 62). They are also a major component of the normal bacterial flora of man and animals. They colonize the mucous membranes of the mouth (Gibbons *et al.* 1963, Socransky and Manganiello 1971); lower gastro-intestinal tract (Eggerth and Gagnon 1933; Drasar *et al.* 1969; Holdeman *et al.* 1976); and vagina (Gorbach *et al.* 1973; Sanders et al. 1975) in man. In animals they are found in the mouth and on the teeth of many species and at various sites in the alimentary tract. Certain members of the group are concerned in the digestion of cellulose and other polysaccharides in the rumen of some herbivores; some of the most oxygen-sensitive species have been found in the caecum of chickens and other poultry.

Classification

The isolation and classification of these organisms has presented particular difficulties. These were partly technical, because reliable anaerobic conditions were difficult to achieve and many workers were unable to obtain pure cultures, but they also reflected the diverse properties of members of the group. There was confusion in nomenclature and disagreement between different observers over the description of the same organisms.

The classification of gram-negative anaerobic bacilli has undergone many changes since Veillon and Zuber (1898) named their isolates *Bacillus fragilis*, *Bacillus fusiformis*, etc. Castellani and Chalmers (1919) proposed a genus *Bacteroides* to contain species of obligately anaerobic non-sporing bacilli. Weiss and Rettger (1937) redefined the genus to exclude gram-positive organisms. Knorr (1922) introduced the genus *Fusobacterium* for the spindle-shaped gram-negative anaerobic bacilli. The generic term *Fusiformis* used by Veillon and Zuber (1898) is illegitimate because it was originally given to an entirely different organism, *F. termitidis*, which belongs to the Myxobacteriales (Hoelling 1910). The generic names *Sphaerophorus* (Prévot 1938) and *Necrobacterium* (Lahelle and Thjotta 1945, Bergey's Manual 1957), and *Bacteroides* as used by Beerens and Tahon-Castel (1965), are illegitimate names for species in the genus *Fusobacterium* Knorr.

Gram-negative anaerobic bacilli that produced black-pigmented colonies on media containing blood were described by Oliver and Wherry (1921) and called *Bacterium* (later *Bacteroides*) *melaninogenicum*. Eggerth and Gagnon (1933) produced a scheme based upon the classification of Castellani and Chalmers (1919) for the identification of strains isolated from the gastro-intestinal tract; they defined 18 species on the basis of morphology and carbohydrate-fermentation tests, but all of their strains belonged to the *B. fragilis* group (Spiers 1971, Holdeman and Moore 1974*a*).

Prévot (1938) produced a detailed classification but employed generic names that differed from those in general use and are taxonomically invalid. He divided gram-negative anaerobic bacilli into two families, Ristellaceae with five genera (*Ristella*, *Pasteurella*, *Dialister*, *Capsularis*, *Zuberella*) and Sphaerophoraceae with two genera (*Sphaerophorus* and *Sphaerocillus*). The present genus *Bacteroides* includes most of Prévot's Ristellaceae, and *Fusobacterium* is the legitimate name for the organisms he classified as *Sphaerophorus*. In the sixth edition of Bergey's Manual (1948) the Bacteroideae, comprising the two genera *Bacteroides* and *Fusobacterium*, was a tribe of the family Parvobacteriaceae. Most of the 23 species of *Bacteroides* were members of the *B. fragilis* group except for *B. melaninogenicus*, which was described as saccharolytic, and one asaccharolytic species, *B. caviae*. The genus *Fusobacterium* comprised four species: *F. plauti-vincenti*, *F. biacutum*, *F. nucleatum* and *F. polymorphum*. In the seventh edition of Bergey's Manual (1957), gram-negative, anaerobic bacilli were re-classified as the family Bacteroidaceae with five genera. The genus *Bacteroides* contained 30 species distinguished by gas production, gelatin liquefaction, cellular morphology and the production of acid from carbohydrates; two species were motile and two—*B. coagulans* and *B. putredinis*—were non-fermentative. There were six species of *Fusobacterium*: *F. fusiforme* (incorrectly regarded as Vincent's organism), *F. polymorphum*, *F. praeacutum*, *F. nucleatum*, *F. vescum* and *F. biacutum*—and two of *Dialister*. The second largest genus was *Sphaerophorus* with 14 species, and aerobic but facultatively anaerobic organisms of the genus *Streptobacillus* were also included in the Bacteroidaceae. The classification was further confused when Beerens and Tahon-Castel (1965) used the term *Bacteroides* as synonymous with *Necrobacterium* for organisms that are now classified as *Fusobacterium* and introduced the term *Eggerthella* for organisms that are generally regarded as *Bacteroides*. In the sixth edition of this book (1974), we described only three species of *Fusobacterium* and five of *Bacteroides*, and we included a third genus, *Dialister*, to contain the minute gram-negative bacillus *D. pneumosintes*. Our distinction between *Fusobacterium* and *Leptotrichia* was confused, and *Leptotrichia* was regarded as gram positive.

The first general agreement on taxonomy and classification was reached at the meeting of the International Commission for Systematic Bacteriology Sub-committee for gram-negative anaerobic rods at Lille in 1967, when new principles for classification were defined (Beerens 1970). These were embodied in the eighth edition of Bergey's Manual (Holdeman and Moore 1974*a*). The Bacteroidaceae were divided into three genera: *Bacteroides*, *Fusobacterium* and *Leptotrichia*. The genus *Bacteroides* contained 22 species in five groups.

(i) *B. fragilis* grew well in 20 per cent bile, was saccharolytic and produced succinic acid as a major metabolic product. This species included most of those described by Eggerth and Gagnon (1933) and previous workers, and was divided into five subspecies (subsp.): *fragilis*, *vulgatus*, *distasonis*, *ovatus* and *thetaiotaomicron*. (ii) Similar strains that were inhibited by bile included *B. ruminicola* (Bryant *et al.* 1958), *B. oralis* (Loesche *et al.* 1964), *B. ochraceus* and *B. amylophilus* (Hamlin and Hungate 1956). (iii) A group of six species that did not produce succinic acid but were otherwise unrelated: *B. hypermegas*, *B. serpens*, *B. termitidis*, *B. biacutus*, *B. clostridiiforme* and *B. constellatus*. (iv) Non-saccharolytic non-pigmented strains were divided into nine species: *B. putredinis*, *B. coagulans*, *B. praeacutus*, *B. corrodens* (Eiken 1958), *B. nodosus*, *B. furcosus*, *B. capillosus*, *B. succinogenes*, and *B. pneumosintes*. (v) *B. melaninogenicus* produced black-pigmented colonies on laked-blood agar and was divided into two saccharolytic subspecies: *melaninogenicus* and *intermedius*, and one non-saccharolytic subspecies, *asaccharolyticus*.

The genus *Fusobacterium* contained 16 species that formed *n*-butyric acid as a major metabolic product; it included most of the species that had previously been designated *Fusobacterium* or *Sphaerophorus* spp. Vincent's organism (*F. plauti-vincenti*) was now assigned to the third genus, *Leptrotrichia* (Gilmour *et al.* 1961), under the name *L. buccalis*, because it produced lactic but not *n*-butyric acid as a major product.

Criteria for classification

Early workers relied almost exclusively upon observations of microscopic and colonial morphology, but microscopic morphology in particular is an unreliable criterion that varies with the conditions of incubation. Eggerth and Gagnon (1933) established the first biochemical key for the group, but their tests were performed under varied and often unsuitable conditions. Prévot (1938) also used an extensive series of biochemical tests. Conventional bacteriological tests remain an important part of identification schemes for gram-negative anaerobic bacilli (Dowell 1972; Holdeman and Moore 1974*a*; Duerden *et al.* 1976, 1980), but they are now selected to discriminate between species or groups that have been defined on the basis of more complex tests that are believed to have a greater taxonomic significance.

The short-chain fatty acids formed as end-products of protein or carbohydrate metabolism are particularly important in the current system of classification. They were first detected by distillation and paper chromatography (Guillaumie *et al.* 1956), but this was superseded by gas-liquid chromatography (GLC; Werner 1969*a*, Cato *et al.* 1970, Moore 1970, Carlsson 1973). The family Bacteroidaceae has been assigned to two principal genera, *Bacteroides* and *Fusobacterium*, on the grounds that *Fusobacterium* spp. produce *n*-butyric acid as a major product whereas most *Bacteroides* spp. do not. The third genus, *Leptotrichia*, comprises oral fusiform bacteria that produce lactic acid as the only major product.

Several groups of workers have used comparisons of the G + C content of the DNA to determine similarities and differences between members of the groups of Bacteroidaceae (Sébald 1962, Williams *et al.* 1974, 1975); a more precise comparison is obtained by studies of DNA homology (see, for example, Johnson 1973). Cato and Johnson (1976) established specific status for the subgroups within the fragilis group of *Bacteroides* by demonstrating poor homology between the DNA preparations from representative strains.

Barnes and Goldberg (1968*a, b*) used the methods of numerical taxonomy and concluded that the most useful tests were cellular morphology, terminal pH in glucose broth, production of formic, acetic, propionic and butyric acids, deamination of threonine, stimulation of growth by 20 per cent bile, and the effect of various inhibitors and antibiotics. More recent work has shown that the basic classification should be based upon studies of the DNA and the analysis of the fatty acid end-products of metabolism. Other techniques for the analysis of the cellular components have also proved useful. Proteins in whole-cell extracts of *Bacteroides* cultures analysed by discontinuous gradient polyacrylamide-gel electrophoresis (PAGE) gave specific, subspecific, and strain-specific patterns (Strom *et al.* 1976); and cell-surface proteins released from members of the fragilis group by EDTA treatment and mild sonic disintegration and treated by sodium dodecyl sulphate PAGE (SDS-PAGE) had species-specific patterns (Poxton and Brown 1979).

The complex methods used in research laboratories are unsuitable for routine use, and several simpler approaches to the identification of the Bacteroidaceae have been devised. Beerens and Tahon-Castel (1965) showed that *Fusobacterium* spp. produced propionic acid from threonine but *Bacteroides* spp. did not. Suzuki and his colleagues (1966) distinguished between *Bacteroides*, *Fusobacterium* and *Sphaerophorus* by differential inhibition tests with ethyl violet, Victoria blue 4R, brilliant green and gentian violet, together with tests for the decarboxylation of glutamic acid and the deamination of threonine. Susceptibility to kanamycin, neomycin, penicillin, rifampicin, colistin and erythromycin in disk tests and the ability to grow in the presence of 20 per cent bile or the bile salts sodium deoxycholate and sodium taurocholate have proved useful in the identification of several groups of *Bacteroides* and *Fusobacterium* (Finegold *et al.* 1967, Shimada *et al.* 1970, Sutter and Finegold 1971). These simple procedures are useful in assigning strains to the appropriate major group or genus; specific identification requires the performance of fermentation and other conventional bacteriological tests.

The family Bacteroidaceae

Definition

Gram-negative rods with pointed or rounded ends; often pleomorphic; some are fusiform, some filamentous and some produce bizarre cells that degenerate into L-forms and sphaeroplasts. Do not form spores; most species are non-motile. Strict anaerobes; growth often improved by 5–10 per cent CO_2. Growth often poor or absent without the addition of blood, serum, yeast extract and other growth factors in a rich, complex medium.

The family comprises three genera: *Bacteroides*, *Fusobacterium* and *Leptotrichia*.

Bacteroides

Definition

Gram-negative bacilli and coccobacilli that are often pleomorphic and occur singly, in pairs or in short chains; some predominantly coccobacillary, but in some strains a minority of cells form short filaments. May form a polysaccharide capsule. The genus includes fermentative and asaccharolytic species. In the fermentative species, acetic and succinic acids are major metabolic products; in general, *n*-butyric acid is not produced; in the asaccharolytic species, *n*-butyric acid is the main product. Members of three species may form black-pigmented colonies on media that contain blood. Threonine not deaminated to propionic acid. G+C content of the DNA: 40–55 moles per cent.

Type species is *B. fragilis*.

Fusobacterium

Definition

Gram-negative rods that are often spindle-shaped with pointed ends ('fusiform'); many strains are filamentous. *n*-Butyric acid is a major metabolic product. Most species are asaccharolytic or ferment glucose and a few other carbohydrates only weakly and after prolonged incubation. Threonine deaminated to propionic acid. G+C content of DNA: 26–34 moles per cent.

Type species is *F. nucleatum*.

Leptotrichia

Definition

Oral fusiform gram-negative bacteria; the cells are usually long filaments with pointed ends. Lactic acid is the only major fatty acid produced. G+C content of the DNA: 34 moles per cent.

The type species is *L. buccalis*.

We shall now describe the morphological, cultural, metabolic and biochemical characters of the various species of *Bacteroides*, *Fusobacterium* and *Leptotrichia*, and give some indication of their importance as pathogens. (For further information about this, see Chapter 62.) Then we summarize, in respect of all three genera, information about (1) cell-wall composition and antigenic constitution, (2) susceptibility to antimicrobial agents at concentrations attainable in the body, (3) genetic mechanisms, (4) bacteriocines, and (5) phages. Finally, we give an account of the distribution of Bacteroidaceae in the flora of man and animals.

Bacteroides

The genus *Bacteroides* is a heterogeneous group of gram-negative anaerobic bacilli. The common species of human origin, and some from animals, can be divided into three main groups—the fragilis group, the melaninogenicus-oralis group and the asaccharolytic group (Table 26.1)—but the relation of some of the less common animal strains to these groups is uncertain.

Table 26.1 Species in the genus *Bacteroides*

Fragilis group	
B. fragilis	*B. eggerthi*
B. vulgatus	*B. variabilis*
B. distasonis	*B. uniformis*
B. ovatus	*B. splanchnicus*
B. thetaiotaomicron	
Melaninogenicus-oralis group	
B. melaninogenicus	*B. oralis*
subsp. *melaninogenicus*	*B. bivius*
subsp. *intermedius*	*B. disiens*
subsp. *levi*	*B. ruminicola*
Asaccharolytic group	
B. asaccharolyticus	*B. pneumosintes*
B. gingivalis	*B. coagulans*
B. ureolyticus	*B. nodosus*
B. putredinis	*B. furcosus*
B. praeacutus	*B. capillosus*
Others	
B. succinogenes	*B. biacutus*
B. amylophilus	*B. serpens*
B. hypermegas	*B. constellatus*
B. multiacidus	*B. termitidis*

The fragilis group

This group contains a variety of closely related species that share many common properties and a common ecology; they are tolerant of bile and resistant to penicillin and are commensals in the lower gastro-intestinal tract in man. They form a major part of the normal human faecal flora and some species are important pathogens, causing infections particularly after accidental or surgical injury to the gastro-intestinal tract or in association with pathological lesions of it. Most of the species described by Veillon and Zuber (1898), and by Eggerth and Gagnon (1933) were members of this group, but its classification was confused.

In the eighth edition of Bergey's Manual, Holdeman and Moore (1974*a*) gathered all the members of the group into a single species—*B. fragilis*—with five subspecies: *fragilis*, *vulgatus*, *distasonis*, *ovatus* and *thetaiotaomicron*. They believed that the species represented a continuum of variants with clusters of strains that were designated subspecies and smaller numbers of intermediate strains. Subsequently other clusters of strains that fell within the group were given specific status; these included *B. splanchnicus*, *B. eggerthi*, *B. variabilis* and *B. uniformis*. However, Cato and Johnson (1976) found poor DNA homology between the reference strains of the five original subspecies of *B. fragilis*. They proposed that they should be reinstated to specific rank and this was accepted by the International Committee on Systematic Bacteriology. The relationship between the species remains the subject of considerable debate. They share many properties, and ecological and epidemiological studies support the concept of a closely related group. The results of a wide range of phenetic tests on members of the fragilis group form a continuous spectrum with clusters of strains that represent the named species, but there remain some intermediate organisms that clearly belong to the fragilis group but cannot be allocated to a recognized species.

The fragilis group comprises small non-motile gram-negative bacilli and coccobacilli; they are moderately pleomorphic but long filaments, bizarre shapes, L-forms and sphaeroplasts are rare (Fig. 26.1).They grow well in 24–48 hr on horse blood agar plates incubated anaerobically at 37° to form circular, low-convex colonies, 1–3 mm *q* with an entire edge; the colonies are usually smooth, shiny, translucent or semi-opaque and grey; colonies of fresh isolates are often moist and some are mucoid. Most strains do not produce haemolysis but a few strains are slightly haemolytic and a very small proportion (< 1 per cent) are frankly β-haemolytic. The nutritional requirements of the fragilis group are simple. Most strains will grow on a medium that contains glucose, haemin, vitamin B_{12}, minerals, ammonium chloride and a sulphide and provided with a CO_2/CO_3^- buffer system; they do not use organic nitrogen compounds. They have generally been regarded as catalase negative but have been shown to produce small amounts of catalase and also a second enzyme that was thought to be absent in anaerobes, superoxide dismutase (Gregory *et al.* 1977). The G + C content of the DNA (40–44 moles per cent) is similar for all species in the group, but there is poor DNA homology between the species. All members of the fragilis group ferment a range of carbohydrates, including glucose, with the production of acid and gas. The major volatile fatty-acid products of metabolism are acetic and succinic acids; *n*-butyric acid is not produced except by *B. splanchnicus*. Most strains are stimulated by 20 per cent bile and are tolerant of the bile salt sodium taurocholate but inhibited by sodium deoxycholate. They will grow in the presence of Victoria blue 4R (1 in 80 000) but most are inhibited by gentian violet (1 in 100 000) and ethyl violet (1 in 100 000). All strains are resistant to penicillin and to high concentrations (1000 μg disks) of neomycin and kanamycin. They decarboxylate glutamic acid but threonine is not dehydrogenated, nitrate is not reduced and urease is not produced. Most strains hydrolyse aesculin rapidly and produce acid from xylose. The species within the fragilis group are identified by a small number of variable characteristics that include indole production, aesculin hydrolysis and the fermentation of lactose, sucrose, rhamnose, trehalose, mannitol and arabinose (Table 26.2).

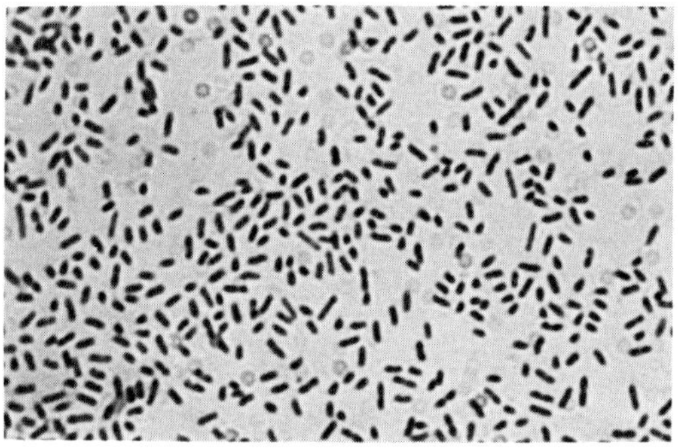

Fig. 26.1 *B. fragilis*: gram-stained film of 48-hr culture on blood agar (× 1000).

B. fragilis Members of this species cannot be distinguished from other members of the fragilis group by their colonial form, or by their microscopic morphology, except that freshly isolated strains usually form a capsule (see below). *B. fragilis* hydrolyses aesculin and produces acid from glucose, lactose, sucrose, maltose and usually xylose, but not from rhamnose, trehalose, mannitol or arabinose. Indole is not formed and charcoal-gelatin disks are digested slowly or not at all.

B. fragilis forms less than 10 per cent of the fragilis-group organisms in normal human faeces, yet it is by far the most common species of *Bacteroides* to be isolated from infections related to the large intestine (Duerden 1980*d*; see also Chapter 62). It thus appears to be particularly pathogenic for man. This may be attributable to its possession of a polysaccharide capsule (Kasper 1976*a*, *b*) or to the action of one or more of the extracellular or membrane-associated enzymes that it forms: proteinases, including collagenase, fibri-

Table 26.2 The fragilis group of *Bacteroides*

Test	Results* obtained with the stated species
Antibiotic-disk-resistance tests	
neomycin (1000 μg)	R
kanamycin (1000 μg)	R
penicillin (2 units)	R
rifampicin (15 μg)	S
Tolerance tests	
taurocholate	+
Victoria blue 4R	+
gentian violet	−

Test	*B. fragilis*	*B. vulgatus*	*B. distasonis*	*B. ovatus*	*B. thetaiotaomicron*	*B. eggerthi*	*B. variabilis*	*B. uniformis*	*B. splanchnicus*
Indole production	−	−	−	+	+	+	+	+	+
Aesculin hydrolysis	+	+/−	+	+	+	+	+	+	+
Fermentation of:									
glucose	+	+	+	+	+	+	+	+	+
lactose	+	+	+	+	+	+	+	+	+
sucrose	+	+	+	+	+	−	+	+	−
rhamnose	−	+	+/−	+	+	+	+	−	−
trehalose	−	−	+	+	+	−	−	+	−
mannitol	−	−	−	+	−	−	−	−	−
xylose	+	+	+	+	+	+	+	+	+

* R = resistant; S = sensitive; + = positive result (growth in tolerance tests); − = negative result.

nolysin, haemolysin, neuraminidase, phosphatase, DNAase, hyaluronidase, chondroitin sulphatase, and heparinase (Gesner and Jenkin 1961, Müller and Werner 1970, Rudek and Haque 1976). These matters are discussed more fully in Chapter 62.

The cell wall of *B. fragilis* contains a lipopolysaccharide with weak endotoxic activity, the polysaccharide portion of which confers type-specific O-antigenic specificity (see p. 129). A species-specific protein component of the outer membrane has been identified; its presence may be used for the identification of *B. fragilis*. The pattern of proteins in the outer membrane shown by SDS-PAGE is also species specific. External to the outer membrane is a thick polysaccharide capsule 1.5–2 times the thickness of the cell wall; it is composed of a large-molecular-weight polysaccharide (mol. wt $> 7.5 \times 10^6$). It has anti-phagocytic properties and also protects the cell from complement-mediated lysis by antibodies against cell-wall antigens. The capsular antigen is species specific. This capsule is found only in *B. fragilis*. All clinical isolates of *B. fragilis* are capsulate, but the capsule is often lost on repeated subculture; it can be demonstrated in India ink preparations and identified by the *Quellung* reaction and in an indirect immunofluorescent assay.

We now give brief notes on other species in the fragilis group (see also Table 26.2).

B. vulgatus First isolated from human faeces by Eggerth and Gagnon (1933). The most common *Bacteroides* species in normal human faeces but only occasionally implicated in infections. Results of tolerance tests resemble those given by other members of the fragilis group, but *ca* 50 per cent of strains are tolerant of ethyl violet. Indole is not produced but charcoal-gelatin disks are digested within a few days. Unlike other members of the fragilis group, *ca* 50 per cent of *B. vulgatus* strains fail to hydrolyse aesculin and others do so only weakly.

B. distasonis Described by Eggerth and Gagnon (1933) in their studies of the faecal flora and named after the Romanian bacteriologist A. Distaso. Like *B. vulgatus* it is a common member of the normal human faecal flora but appears seldom to cause clinical infections. Most strains ferment rhamnose but a significant minority do not. Indole is not produced and charcoal-gelatin disks are digested slowly or not at all.

B. ovatus One of the less commonly encountered species in the fragilis group. It is not a major component of the normal faecal flora and is isolated only occasionally from clinical specimens; but when present in an infection it is usually in large numbers and appears to be a significant pathogen. Produces acid from a wider range of carbohydrates than other members of the fragilis group. The ability to produce acid from salicin and mannitol is used to identify *B. ovatus*. Indole positive; digests charcoal-gelatin disks within a few days.

B. thetaiotaomicron Named by Distaso (1912) as a combination of the Greek letters theta, iota and omicron. A common commensal in normal human faeces and the second commonest species in the fragilis group isolated from clinical infections, where it appears to have a significant pathogenic role. Acid not formed from mannitol. Gives a variable reaction with the charcoal-gelatin disk; some strains digest it readily but others do so only weakly or not at all. In some studies all indole-positive members of the fragilis group that were not *B. ovatus* have been classified as *B. thetaiotaomicron*. However, other indole-positive sub-groups or species can be recognized.

B. eggerthi Named after the American bacteriologist A. H. Eggerth and described by Holdeman and Moore (1974*b*) as a result of their studies on the human faecal flora. Acid formation from rhamnose but not sucrose distinguishes *B. eggerthi* from other indole-positive species in the fragilis group. Digests charcoal-gelatin disks.

B. variabilis Similar to *B. eggerthi* and to other indole-positive members of the fragilis group; described by Distaso (1912) but included in the species *B. thetaiotaomicron* in the eighth edition of Bergey's Manual (Holdeman and Moore 1974*a*). A common commensal in normal human faeces and isolated from a small proportion of infections where *Bacteroides* of the fragilis group are implicated. Some strains are inhibited by sodium taurocholate but grow in 20 per cent bile broth. No acid from mannitol or trehalose. Charcoal-gelatin disks are usually digested within a few days.

B. uniformis Found in normal human faeces but rarely implicated in clinical infections. Described by Eggerth and Gagnon (1933). In many studies strains of *B. uniformis* have been included with other indole-positive strains as *B. thetaiotaomicron*. Like *B. variabilis*, some strains that grow well in 20 per cent bile broth are inhibited by sodium taurocholate. Charcoal-gelatin disks are usually digested within a few days.

B. splanchnicus Isolated from normal human faeces and from several infections related to the lower gastro-intestinal tract (Werner *et al* 1975); in particular, several strains were isolated in significant numbers from infectious complications of appendicitis. It shares many characters with other members of the fragilis group, but differs from them in one important respect; it is the only fermentative species of *Bacteroides* to form significant amounts of *n*-butyric acid. This it does in addition to forming a variety of other acids including *iso*-valeric, *iso*-butyric and propionic acids as well as acetic and succinic acids that are produced by all members of the fragilis group. A minority of strains that grow well in 20 per cent bile broth are inhibited by sodium taurocholate. Some strains digest charcoal-gelatin disks but others do not.

The melaninogenicus-oralis group

Oliver and Wherry (1921) described gram-negative anaerobic bacilli that produced black pigmented colonies when grown on blood agar; they called their strains *Bacterium melaninogenicum*; subsequently all pigmented strains were assigned to the species *Bacteroides melaninogenicus*. This characteristic appearance was regarded as highly specific and was the sole criterion for differentiation from other *Bacteroides* spp. Oliver and Wherry (1921) thought that the pigment was extracellular melanin, but Schwabacher and her colleagues (1947) found that the pigment was a derivative of haemoglobin, spectroscopically identical with haematin and located intracellularly. Duerden (1975) and Shah and his colleagues (1976) concluded that the pigment was an intracellular or cell-associated derivative of haemoglobin assimilated from the medium. Confusion surrounded studies on *B. melaninogenicus* because of diversity of characters between pigmented strains and because of the specific requirements of some strains for growth factors such as vitamin K (Lev 1959), haemin and sodium succinate. Some early workers had difficulty in maintaining *B. melaninogenicus* strains in pure culture. Some *B. melaninogenicus* strains isolated from the normal flora, particularly the gingival crevice, are nutritionally demanding and require a variety of growth factors provided by co-cultivation with other bacteria but as yet unidentified.

The pigmented strains appeared to share several characters in addition to pigmentation that distinguished them from the non-pigmented *Bacteroides* spp., but there were major differences in metabolic and biochemical activity between groups of pigmented strains, and so *B. melaninogenicus* was divided into three subspecies: *melaninogenicus*, *intermedius* and *asaccharolyticus* (Sawyer *et al.* 1962, Moore and Holdeman 1973.) Studies of cell-wall composition and DNA base ratios supported this division (Williams *et al.* 1975). Lambe (1974) and Lambe and Jerris (1976) distinguished between the same groups by fluorescent-antibody staining and also subdivided subsp. *intermedius* into two serogroups. Werner and his colleagues (1971) found that pigmented strains from human faeces and from infections related to the lower gastro-intestinal tract formed a homogeneous group of asaccharolytic strains and recognized only these as *B. melaninogenicus*. The differences between the subspecies cast doubt upon the validity of assigning all pigmented strains to a single species. The asaccharolytic strains were distinguished from the saccharolytic strains by their production of *n*-butyric acid and also by differences in G+C content of the DNA and in cell-wall composition. The ICSB taxonomic sub-committee on gram-negative anaerobic rods (Finegold and Barnes 1977) therefore proposed that the asaccharolytic strains should be classified as a separate species, *B. asaccharolyticus*, now included in the asaccharolytic group of *Bacteroides*.

B. melaninogenicus subsp. *melaninogenicus* and subsp. *intermedius* remain as two subspecies of the fermentative pigmented species, and a third weakly fermentative subspecies, subsp. *levi*, has been added. It is now clear that the production of pigmented colonies on media that contain blood has less taxonomic significance than was previously thought. The melaninogenicus-oralis group includes the three fermentative subspecies of *B. melaninogenicus* and several non-pigmented species that share many characters with them. The species *B. oralis* was proposed for non-pigmented bile- and penicillin-sensitive strains of *Bacteroides* isolated from the human mouth (Loesche *et al.* 1964), but the type strain of *B. oralis* proved to be pigmented and was re-classified as *B. melaninogenicus* subsp. *melaninogenicus* (Holbrook and Duerden 1974). However, other non-pigmented strains have been described that correspond with the original description of *B. oralis*. Some strains isolated from the mouth, vagina and clinical specimens closely resemble strains of *B. melaninogenicus* subsp. *melaninogenicus* and are distinguished from them only by their failure to produce pigment. The distinction is even less clear with strains of *B. melaninogenicus* subsp. *melaninogenicus* that produce brown pigmentation slowly, and strains of *B. oralis* that produce buff-coloured colonies after prolonged incubation. The relation between *B. melaninogenicus* subsp. *melaninogenicus* and *B. oralis* remains unresolved.

Two other non-pigmented members of the melaninogenicus-oralis group, *B. disiens* and *B. bivius*, were isolated from human clinical infections by Holdeman and Johnson (1977) and have since been found in the normal mouth and vagina. The fifth member of the group, *B. ruminicola*, was isolated from the reticulo-rumen of cattle, sheep and elk (Bryant *et al.* 1958).

The melaninogenicus-oralis group of species appear to form a closely related group. It remains to be decided whether similar strains should be allocated to separate species on the basis of pigment production alone. The description of non-pigmented variants of *B. melaninogenicus* subsp. *melaningenicus* (Harding *et al.* 1976) suggests that this should not be the invariable rule. In general, however, differences in DNA constitution, cell-wall composition and enzyme mobilities (Shah *et al.* 1976) support the recognition of distinct pigmented and non-pigmented species within the group.

The species cannot be distinguished by cellular morphology. Most are short gram-negative bacilli with rounded ends or coccobacilli and many are pleomorphic (Fig. 26.2). Some strains of *B. melaninogenicus* and *B. oralis* are predominantly coccobacilli that may be arranged in short chains. The cells of *B. ruminicola* subsp. *ruminicola* and some strains of *B. oralis* are longer and more regular.

B. melaninogenicus colonies are distinguished by pigment production; they assimilate haemoglobin from blood in the medium and haemolysis is an essential accompaniment to pigmentation on whole-blood agar. Pigmentation develops more rapidly on lysed-blood agar and some non-haemolytic strains produce pigment only on lysed-blood agar (Fig. 26.3). Colonies of the non-pigmented species are indistinguishable. In 48-hr cultures they are 1–2 mm *q*, circular with an entire edge, convex, shiny, semi-opaque and light buff or grey. Colonies of some strains of *B. oralis* become larger after incubation for 5–7 days; they develop a spreading edge with an irregular outline and the colonies tend to coalesce.

All species produce acid from glucose; their energy metabolism is fermentative, and succinic and acetic acids are major metabolic products; they do not produce *n*-butyric acid except for the weakly fermentative *B. melaninogenicus* subsp. *levi*. In tolerance tests with bile salts and dyes, all are inhibited by sodium taurocholate and sodium deoxycholate; nor will they grow in 20 per cent bile broth. They are inhibited by most dyes but many strains are tolerant of Victoria blue 4R (1 in 80 000). In antibiotic-resistance tests, all strains are sensitive to neomycin 1000 μg and rifampicin 15 μg disks and resistant to kanamycin 1000 μg disks; many strains are sensitive to benzylpenicillin, but some strains in each species are resistant as a result of plasmid-coded β-lactamase production. Threonine is not dehydrogenated, nitrate is not reduced and urease is not produced. The species within the group are identified by pigment production and a small number of characters that include indole production, aesculin

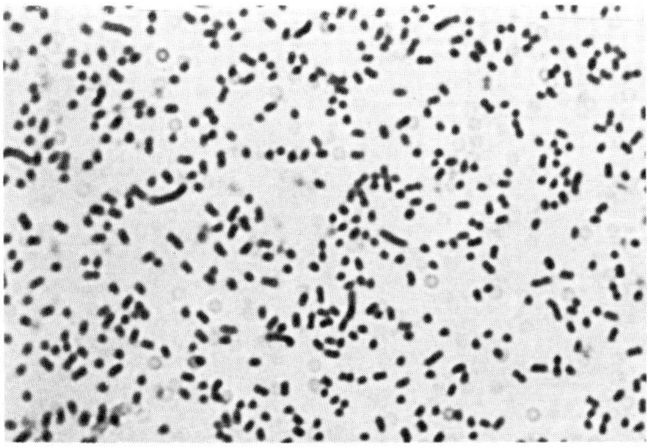

Fig. 26.2 *B. melaninogenicus* subsp. *intermedius*: gram-stained film of 72-hr culture on lysed-blood agar (× 1000).

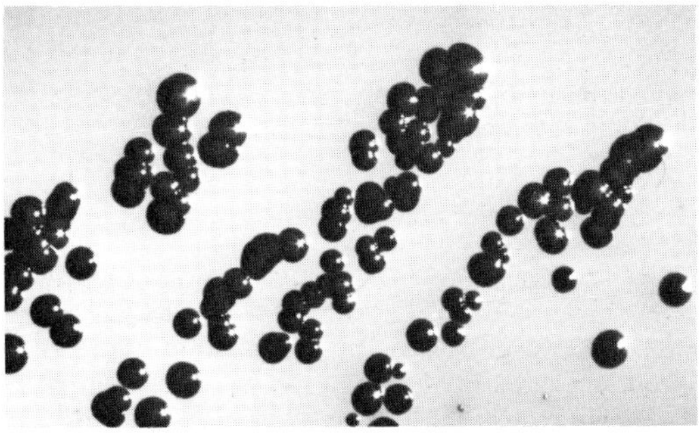

Fig. 26.3 *B. melaninogenicus* subsp. *intermedius*: black-pigmented colonies after incubation for 72 hr on lysed-blood agar (× 4).

Table 26.3 The melaninogenicus-oralis group of *Bacteroides*

Test	*B. melaninogenicus* Subsp. *melaninogenicus*	Subsp. *intermedius*	Subsp. *levi*	*B. oralis*	*B. bivius*	*B. disiens*	*B. ruminicola*
	Results* obtained with the given species						
Antibiotic-disk-resistance tests							
neomycin (1000 μg)				S			
kanamycin (1000 μg)				R			
penicillin (2 units)				S/R			
rifampicin (15 μg)				S			
Tolerance tests							
taurocholate				−			
Victoria blue 4R				−/+			
gentian violet				−			
Pigment production	+	+	+	−	−	−	−
Indole production	−	+	−	−	−	−	−
Aesculin production	−/+	−	−	+	−/+	−	+
Fermentation of:							
glucose	+	+	+	+	+	+	+
lactose	+	−	+	+	+	−	+
sucrose	+	+	−	+	−	−	+
rhamnose	−	−	−	−	−	−	+
trehalose	−	−	−	−	−	−	−
mannitol	−	−	−	−	−	−	−
xylose	−	−	−	−	−	−	+

* R = resistant; S = sensitive; + = positive result (growth in tolerance tests); − = negative result.

hydrolysis and the fermentation of lactose, sucrose, rhamnose, xylose and arabinose (Table 26.3).

B. melaninogenicus All strains of anaerobic gram-negative bacilli that produce black or brown pigmented colonies on media containing blood and produce acid from glucose are included in the species *B. melaninogenicus*, which is divided into three subspecies: *melaninogenicus*, *intermedius* and *levi*.

B.melaninogenicus subsp. ***melaninogenicus*** is a common commensal in the gingival crevice and in the vagina in women and is found in infections related to these sites. Most strains are short gram-negative bacilli or coccobacilli and many are moderately pleomorphic. The colonies of subsp. *melaninogenicus* are 1–2 mm q, round, convex and opaque. After incubation for 48 hr they are typically light grey, becoming brown after further incubation. Pigmentation develops more rapidly on lysed-blood agar and varies between strains from light brown to almost black; it begins in the centre of the colony and the colonies of many strains have a light-brown or pale annulus around a dark-brown centre. Biochemical characters are as shown in Table 26.3; only a few strains hydrolyse aesculin. Charcoal-gelatin disks are digested within a few days. The G+C content of the DNA is 40–42 moles per cent by the buoyant-density centrifugation method (van Steenbergen *et al.* 1979) and the electrophoretic mobility of malate dehydrogenase is 5.1–5.4 cm (Shah *et al.* 1976). Strains of subsp. *melaninogenicus* can be divided into two groups according to their cell wall composition—the mucopeptide of one group contains diaminopimelic acid (DAP) and that of the other group lysine—but SDS-PAGE of polypeptides extracted from subsp. *melaninogenicus* strains gives no evidence of differences between the two groups.

B. melaninogenicus subsp. ***intermedius*** is also a common commensal in the gingival crevice and is often isolated from cases of gingivitis and other purulent lesions related to the mouth. Most strains are short gram-negative bacilli and coccobacilli and many are pleomorphic; filamentous or bizarre forms are rare. The colonies are 1–2 mm *q*, round, convex and opaque. After incubation for 48 hr, isolated colonies are typically grey but confluent growth is usually turning black; the individual colonies become uniformly black after further incubation. Some strains are shiny and glistening, but others are dull with a rough, dry surface and these may be adherent to the underlying medium. Pigmentation develops more rapidly on lysed-blood agar. Most strains produce clear zones of complete haemolysis in blood agar. Most strains are sensitive to penicillin and inhibited by Victoria blue 4R. This is the only organism in the melaninogenicus-oralis group that produces indole; charcoal-gelatin disks are digested readily. Most strains fail to ferment lactose. The G + C content of the DNA is 40–42 moles per cent by the buoyant-density centrifugation method (van Steenbergen *et al.* 1979) and the electrophoretic mobility of malate dehydrogenase is 4.3–4.9 cm (Shah *et al.* 1976). The peptidoglycan contains DAP. SDS-PAGE of extracted polypeptides gives a uniform pattern, but Lambe and Jerris (1976) describe two distinct serogroups.

B. melaninogenicus subsp. ***levi*** is a recent addition to the species. The type strain produces dark-brown colonies slowly on blood agar and more promptly on lysed-blood agar. It is weakly fermentative and appears to be asaccharolytic after incubation of fermentation tests for 48 hr, but if the tests are continued for 4 days it ferments glucose and lactose but not other carbohydrates. Charcoal-gelatin disks are digested within a few days. Unlike other members of the melaninogenicus-oralis group, subsp. *levi* forms *n*-butyric acid as a major product of metabolism; significant quantities of acetic, propionic, *iso*-butyric and *iso*-valeric acids, but not of lactic or succinic acids are produced.

B. oralis Strains that correspond with the original description (Loesche *et al.* 1964) are an important component of the normal flora of the gingival crevice, and similar strains have been isolated from the normal vagina. *B. oralis* strains have also been implicated in infections derived from these sites (see Chapter 62). All these infections are associated with tissue necrosis and the production of foul-smelling pus. Colonies of *B. oralis* are 1–2 mm *q*, semi-opaque and light-buff or grey; after prolonged incubation some become pale brown and some develop a spreading edge with an irregular outline. Most strains are predominantly coccobacillary but a few produce longer, more regular bacilli. Most strains hydrolyse aesculin and digest charcoal-gelatin disks within a few days; some strains ferment rhamnose. Differences between strains in fatty acid and isoprenoid quinone composition (Shah and Collins 1980), and biochemical characteristics (Shah *et al.* 1980, Williams and Shah 1980), and DNA homology studies (van Steenbergen *et al.* 1980), suggest that *B. oralis* is a heterogeneous group of organisms; further studies are needed to establish its taxonomic status.

Brief descriptions of other members of the melaninogenicus-oralis group follow.

B. bivius Isolated from human infections by Holdeman and Johnson (1977). The commonest species of *Bacteroides* in the normal human vaginal flora (Duerden 1980*b*); most clinical isolates are from infections related to the female genital tract. Cannot be recognized by cellullar morphology or colonial appearance. Does not form pigment, but colonies often pale brown after prolonged incubation on lysed-blood agar. Few strains hydrolyse aesculin, but charcoal-gelatin disks are digested within a few days.

B. disiens Described by Holdeman and Johnson (1977) at the same time as *B. bivius* and isolated from similar infections. However, it has been recognized less commonly in the normal human flora. Differs from *B. bivius* only in not fermenting lactose.

B. ruminicola. This is a prominent member of the normal flora of the reticulo-rumen of most ruminants in which it plays an important part in the digestion of vegetable food. It has also been isolated from clinical infections and from normal faeces in man. The species is divided into two subspecies—*ruminicola* and *brevis*—and the subspecies are further subdivided into eight biotypes of subsp. *ruminicola* and three biotypes of subsp. *brevis* (Bryant *et al.* 1958). Rumen strains of *B. ruminicola* are among the most demanding of anaerobes and are very sensitive to oxygen; they have been handled successfully only in roll-tubes (Hungate 1950) or in an anaerobic cabinet. The G + C content of the DNA is 49–50 moles per cent. Under conditions of low Eh they ferment pectin, starch, many naturally occurring pentoses and a variety of other carbohydrates (Reddy and Bryant 1977). Haemin is an essential growth factor for many strains and the growth of others is stimulated by it. Human strains of *B. ruminicola* are generally of subsp. *brevis* and are less demanding than the rumen strains.

Two other saccharolytic *Bacteroides* species are common members of the normal flora of the rumen and are important for food digestion in ruminants; *B. succinogenes* and *B. amylophilus*. Their relation to each other and to the melaninogenicus-oralis group is not clear (Reddy and Bryant 1977).

B. succinogenes Described by Hungate (1950). It is one of the more numerous bacterial species in the rumen of most ruminants. A demanding anaerobe that requires rumen fluid for growth in laboratory media. Will grow on cellulose agar, forming colonies that are too small to be visible macroscopically but are surrounded by a clear zone of cellulose digestion. Acid produced by the fermentation of cellulose and its hydrolytic products and succinic acid one of the major metabolic products. G + C content of DNA: 47–49 moles per cent. Various volatile fatty acids, ammonium ions and bicarbonate ions required for growth but organic nitrogen sources and haemin not required.

B. amylophilus A less constant member of the rumen flora. A demanding anaerobe that may not grow on the surface of agar plates in an anaerobic jar; it requires rumen fluid and carbon dioxide but not haemin for growth. G + C content of DNA: 40–42 moles per cent. Under suitably reduced conditions ferments starch, dextrins and maltose; acetic and succinic acids are major metabolic products (see Hamlin and Hungate 1956).

The asaccharolytic group

The gram-negative anaerobic bacilli that fail to produce acid from glucose and other carbohydrates but are not classified as fusobacteria form a heterogeneous group in which taxonomic relations are uncertain. Among the pigmented members of this group the more commonly recognized species are *B. asaccharolyticus* and *B. gingivalis*, formerly known collectively as *B. melaninogenicus* subsp. *asaccharolyticus*. Non-pigmented members include *B. ureolyticus* (formerly *B. corrodens*), which produces 'pitting' or 'corroding' of the agar surface. Classification and identification of the other non-pigmented asaccharolytic strains present difficulties. Some are found in the normal body flora, but they are demanding organisms that produce only small, inconspicuous colonies that are easily overlooked. The relations of these organisms to the recognized asaccharolytic species of *Bacteroides* and to other members of the Bacteroidaceae is unclear; some can be distinguished from *B. asaccharolyticus* only by their failure to produce pigment. This casts further doubt upon the validity of pigment production as a major character in the classification of *Bacteroides*.

B. asaccharolyticus These organisms are mainly coccobacilli with only a few slightly longer bacilli. Very small colonies may be visible on blood agar after 24 hr but in some strains these do not appear for 48 hr; they are smooth, shiny and grey. Pigmentation begins to develop after 3–4 days on whole-blood agar but may be visible in the inoculum after 36–48 hr on lysed-blood agar. After 4–5 days colonies are 0.5–1 mm *q*, black or very dark brown. They usually have a clear halo of complete haemolysis, although a few strains that produce pigment on lysed-blood agar do not produce a haemolysin and are non-pigmented on whole-blood agar in the absence of a haemolytic contaminant. The colonies are smooth and shiny; they are often very moist and can be smeared over the agar surface, although the organisms do not produce large amounts of extracellular mucus. Most strains have a characteristically strong, putrid smell. They do not produce acid from any carbohydrate but may use glucose without fermentation. *n*-Butyric acid is the major metabolic product. The results of antibiotic-disk resistance tests and tolerance tests are similar to those of the melaninogenicus-oralis group; *B. asaccharolyticus* is inhibited by bile and bile salts and by dyes, resistant to kanamycin and sensitive to penicillin and rifampicin; most strains are sensitive to neomycin but some are resistant. *B. asaccharolyticus* is vigorously proteolytic and digests charcoal-gelatin disks within 24–48 hr; over 95 per cent of strains give strongly positive results in tests for indole production.

The cell-wall peptidoglycan contains lysine. The G + C content of the DNA is 53–54 moles per cent by buoyant density centrifugation and the electrophoretic mobility of malate dehydrogenase is 4.8–4.9 cm (Shah *et al.* 1976, van Steenbergen *et al.* 1979). *B. asaccharolyticus* does not agglutinate sheep or horse erythrocytes (Slots and Genco 1979).

B. asaccharolyticus is found in normal human faeces but less consistently and in smaller numbers than are members of the fragilis group; it also occurs in the normal vagina. It is an important pathogen in mixed infections but is rarely isolated alone; it is found in infections related to the lower gastro-intestinal tract and in destructive ulcers and gangrene in diabetics and others with peripheral vascular disease. Its pathogenicity may be related to the production of several extracellular enzymes; in particular, it is strongly proteolytic and has both clotting and fibrinolytic activity. It should be noted that early work on the experimental pathogenicity of '*B. melaninogenicus*' (Werner *et al.* 1971) was probably done with *B. asaccharolyticus*.

B. gingivalis Oral strains of asaccharolytic, pigmented *Bacteroides* are now classified as a distinct species—*B. gingivalis* (Coykendall *et al.* 1980). This species cannot be distinguished from *B. asaccharolyticus* by antibiotic resistance, tolerance or simple biochemical tests or by GLC determination of the metabolic products; *n*-butyric acid is the major product. However, the G + C content of the DNA from *B. gingivalis* strains is lower, at 45–47 moles per cent by thermal denaturation (48–50 moles per cent by buoyant density centrifugation) and the malate dehydrogenase mobility is slower at 3.7–3.9 cm (Shah *et al.* 1976, 1980, van Steenbergen *et al.* 1979, 1980). These differences are supported by distinct patterns of polypeptides extracted from the two species and separated by SDS-PAGE. The most useful tests to identify *B. gingivalis* are the production of phenylacetic acid (Mayrand 1979, Kaczmarek and Coykendall 1980) and the agglutination of horse and sheep erythrocytes (Slots and Genco 1979).

B. gingivalis is found irregularly and in only very small numbers in the normal gingival flora but is isolated commonly and in large numbers from periodontal pockets in patients with advanced destructive periodontitis.

B. ureolyticus (corrodens) This is a non-fermentative, non-pigmented organism the relation of which to other gram-negative anaerobic bacilli is uncertain. Eiken (1958) isolated 21 strains that he called *B. corrodens* from abscesses, principally from the buccal region. Khairat (1967) cultivated a similar organism from blood cultures after tooth extraction and also three strains of it from the gingival crevice. Reinhold (1966) found it in abscesses and Schröter and Stawru (1970) in diseased tonsils. It was re-named *B. ureolyticus* to clarify the distinction between it and *Eikenella corrodens* (see Chapter 43), an aerobic and facultatively anaerobic, carbon dioxide-dependent organism that also forms 'corroding' colonies; the two were often confused in early reports (Jackson and Goodman 1978).

B. ureolyticus forms small colonies, 0.5 mm *q*, depressed below the surface of an agar medium. After incubation for 2–3 days, it gives rise to a characteristic pitting or corroding of the agar surface around the colonies (Fig. 26.4). This pitted area has a rough surface and spreads radially; with prolonged incubation it is seen to represent not only erosion of the agar surface but also the spreading edge of the colony. The centre of the colony is translucent and colourless. Typically *B. ureolyticus* is more regular in microscopic morphology than most other species of *Bacteroides*. The cells are straight, slender rods of moderate length (1–3 μm) with rounded ends; they do not stain strongly (Fig. 26.5). It is sensitive to most antibiotics and very sensitive to the neomycin, kanamycin and penicillin

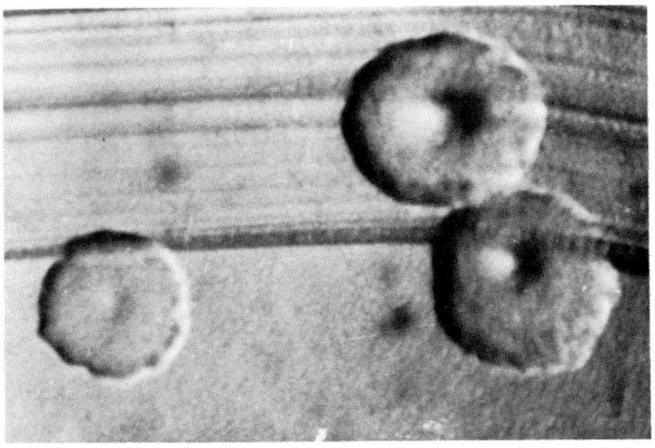

Fig. 26.4 *B. ureolyticus*: 'corroding' colonies after incubation for 3 days on blood agar; viewed by a plate microscope with a combination of transmitted and incident light (× 6).

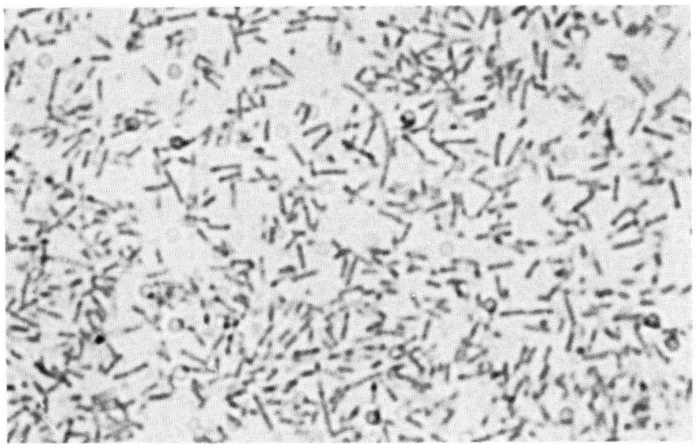

Fig. 26.5 *B. ureolyticus*: gram-stained film of 72-hr culture on blood agar (× 1000).

disks in the antibiotic resistance tests, but many strains show reduced sensitivity to rifampicin and some are frankly resistant. It is inhibited by bile and taurocholate and by most dyes but is tolerant of Victoria blue 4R. It does not produce indole and does not acidify any sugars, but it produces H_2S and shows moderate gelatinase activity. The oxidase test is positive, but catalase is not formed; decarboxylases for lysine and ornithine but not for glutamic acid are produced. Acetic and succinic acids are major metabolic products. It is unusual amongst *Bacteroides* spp. in reducing nitrate to nitrite and hydrolysing urea. The differences between *B. ureolyticus* and other species of *Bacteroides* are emphasized by the low G + C content of its DNA (28–30 moles per cent).

Other asaccharolytic species Definitions of putative species are tentative. *B. putredinis* was isolated from an inflamed appendix by Heyde (1911) and described by Weinberg and his colleagues (1937). It is a putrefactive organism that shares many characters with *B. asaccharolyticus* but does not form pigment. *B. putredinis* is a small non-motile gram-negative bacillus or coccobacillus that grows as very small colonies and gives few positive reactions in conventional tests except that most strains produce indole, H_2S and glutamic acid decarboxylase and are proteolytic; gelatin, milk and serum are digested with the production of a foul smell. Their metabolic products are similar to those of *B. asaccharolyticus* and include *n*-butyric acid. Most strains are sensitive to penicillin, rifampicin and high concentration of neomycin but resistant to kanamycin; they are inhibited by sodium taurocholate and by most dyes. *B. putredinis* is found in normal faeces and is occasionally implicated in abdominal and perianal abscesses and in infections related to appendicitis. In animals it has been isolated from foot rot in cattle and sheep.

B. praeacutus has also been isolated from human faeces and from clinical specimens; it differs from other species of *Bacteroides* in being motile by means of peritrichous flagella. It gives few positive results in conventional tests but most strains reduce nitrate completely.

B. pneumosintes is a very small bacillus (0.13–0.3 μm in length) that is filtrable through Berkfeld V and N filters and is unreactive in most conventional bacteriological tests. It was described by Olitsky and Gates (1921, 1922) in nasopharyngeal washings from patients in the 1918 influenza pandemic and called *Dialister pneumosintes*; these authors thought that it was a significant pathogen. It has since been isolated from normal persons, although it may be concerned in secondary infections of the respiratory tract, including lung abscesses, and in metastatic brain abscesses.

Three other asaccharolytic species of *Bacteroides* from the faeces of man and animals have been described, and have occasionally been isolated from abdominal infections: *B. furcosus*, *B. capillosus* and *B. coagulans*. The last of these is so named because it clots milk. In general they are unreactive and analysis of their metabolic products has not been particularly helpful. Of the remaining species in the asaccharolytic group, *B. nodosus* is a significant pathogen, causing foot rot in sheep (Egerton and Parsonson, 1966) and possibly also in goats and cattle (Wilkinson *et al.* 1970); the hoof seems to be its natural habitat. It is longer than most other *Bacteroides*, and the cells are irregular with terminal enlargements, especially in smears prepared directly from lesions. It is strongly proteolytic and will digest powdered hoof *in vitro*.

Other species of *Bacteroides*

There remain several species that cannot be allocated to one of the above groups; most are strains discovered in studies of the gastro-intestinal flora of a variety of animals from insects to mammals. *B. hypermegas* and *B. multiacidus* were originally thought to be closely related species; both are very large gram-negative bacilli that produce a variety of short-chain fatty acids, but further studies have shown many significant differences. *B. hypermegas* is found in the intestinal tract of poultry, particularly turkeys and chickens, where there may be 10^7–18^8 per g of caecal contents (Harrison and Hansen 1950, 1963, Barnes and Goldberg 1965, 1968*a*, *b*). The G + C content of DNA is 56–58 moles per cent. Propionic acid is a major metabolic product with only small amounts of acetic acid. Acid is produced from a variety of carbohydrates but nitrate is not reduced; *B. hypermegas* is sensitive to neomycin and resistant to penicillin. *B. multiacidus* has been isolated from the faeces of pigs and man in Japan. The G + C content of the DNA is 32–35 moles per cent. It produces acetic acid as a major metabolic product but not propionic acid. Acid is formed from many carbohydrates and nitrate is reduced to nitrite; *B. multiacidus* is sensitive to penicillin and resistant to neomycin (Mitsuoka *et al.* 1974).

B. biacutus is a strongly fermentative species that reduces nirtate to nitrite; acetic and formic acids are major metabolic products. It was originally isolated from an infected appendicectomy wound. Some strains have been confused with *Clostridium clostridiiforme* (formerly *B. clostridiiforme*). *B. serpens* and *B. constellatus* are two unusual motile gram-negative anaerobic bacilli for which little established information is available. *B. termitidis* is a very fine gram-negative bacillus that has been found only in the posterior intestinal contents of termites where it is a major part of the normal flora. It is strongly fermentative but non-proteolytic, and is thought to be concerned in the digestion of cellulose.

The organism described as *B. ochraceus* has been removed from the family Bacteroidaceae. It is not an anaerobe and has been assigned to the genus *Capnocytophaga* as *C. ochraceus*; it is a carboxyphilic organism that will grow on the surface of agar media in air plus 10 per cent CO_2 and it is resistant to metronidazole (Newman *et al.* 1979, Williams *et al.* 1979).

Fusobacterium

The term *Fusobacterium* was first applied by Knorr (1922) to long, spindle-shaped gram-negative bacilli with pointed ends which resembled Vincent's description of his fusiform bacillus (see p. 128). There has been confusion over the relation between fusobacteria, Vincent's organism and the fusiform anaerobic organisms called *Leptotrichia buccalis* (Thjötta *et al.* 1939).

In 1915, Kligler had applied a somewhat similar name to that used by Thjötta and his colleagues—*Leptothrix buccalis*—to an oral fusiform bacillus with very different properties. Organisms closely resembling those of Kligler were later described as *Cladothrix matruchoti* (Mendel 1919) and *Leptotrichia dentium* (Bibby and Berry 1939). *Leptothrix buccalis*, *Cladothrix matruchoti* and *Leptotrichia dentium* are now considered to be gram-positive branching filaments and not to be strict anaerobes; they should be included in the genus *Bacterionema* (Family Actinomycetaceae) under the name *B. matruchoti* (Gilmour *et al.* 1961*a*, *b*).

Knorr (1922) suggested the name *Fusobacterium plauti-vincenti* for organisms resembling Vincent's bacillus, and this name was used by Bøe (1941) and Berger (1956). It is now thought that *Leptotrichia buccalis* is the correct designation for Vincent's organism.

Fusobacteria are gram-negative anaerobic rods of varied size and morphology and have a typical gram-negative cell-wall structure. They range from short to very long and filamentous; some strains, such as the type strain of *F. varium*, are coccobacilli. Width is variable, and fusobacteria may be thin and delicate or thick and coarse; they are often wider near the middle than at the ends (Fig. 26.6). Many have tapered or pointed ends but others have rounded ends and some are extremely pleomorphic with long filaments and short coccobacilli in the same culture. The cells may be single, in pairs end-to-end, or may form long coiled filaments. Staining is often irregular, and L-forms and sphaeroplasts are so common in some species that earlier workers assigned them to a separate genus *Sphaerophorus*. Fusobacteria are usually non-motile. They are nutritionally demanding and many will not survive successive subculture when exposed to air. Growth is often improved by 5–10 per cent CO_2. The metabolism of fusobacteria varies, but carbohydrates are fermented only feebly or not at all. They are destroyed by moist heat at 53° within an hour, and they are generally sensitive to most antibiotics, although this varies between species and strains. *n*-Butyric and propionic acids are major metabolic products and threonine is deaminated to propionic acid. The G + C content of the DNA is 32–33 moles per cent.

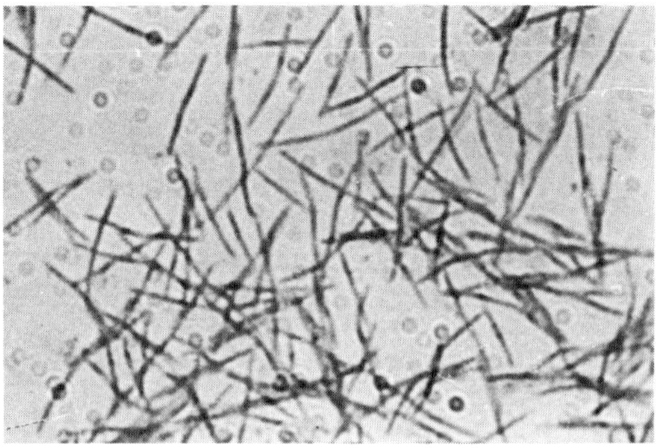

Fig. 26.6 *F. nucleatum*: gram-stained film of 72-hr culture on blood agar × 1000).

Classification within the genus *Fusobacterium* is uncertain. Sixteen species are included in the eighth edition of Bergey's Manual (Holdeman and Moore 1974*a*; Table 26.4) but the basis of the classification is often unsatisfactory. Many fusobacteria from the normal flora of man and animals cannot be identified as members of the named species. A few species are well recognized: *F. necrophorum* is an important pathogen; *F. nucleatum* (*polymorphum*) is a common commensal in the human mouth and is found in some infections related to it; and *F. varium* is found in animal and poultry faeces. The status and role of the others must await more detailed investigation. They are found in the mouth of man and animals and in various necrotic infections.

Table 26.4 Species of fusiform bacteria

Fusobacterium	
F. necrophorum	*F. gonidiaformis*
F. nucleatum (*polymorphum*)	*F. naviforme*
F. varium	*F. bullosum*
F. necrogenes	*F. aquatile*
F. symbiosum	*F. plauti*
F. mortiferum	*F. perfoetus*
F. russi	*F. glutinosum*
F. prausnitzi	*F. stabile*
Leptotrichia	
L. buccalis	

F. necrophorum This organism was first recognized by Loeffler in 1884 as the agent of calf diphtheria and was subsequently implicated in a wide variety of infections in man and animals under several synonyms that included *Bacillus funduliformis* (Henthorne *et al.* 1936), *Bacteroides funduliformis*, *Necrobacterium funduliforme* and *Sphaerophorus necrophorus*. It was found in liver abscesses by Bang (1897). Schmorl (1891) encountered apparently the same bacillus in a spontaneous epidemic of labial necrosis among laboratory rabbits. He infected mice with material from the rabbits, and cultivated the mouse organism, which he called *Streptothrix cuniculi* (Orcutt 1930, Beveridge 1934). Hallé (1898) named a bacillus associated with genital infections in man *B. funduliformis*. A similar organism was isolated from foot rot in cattle. The animal and human strains did not differ significantly except that the human strains were usually less pathogenic for laboratory animals (Dack *et al.* 1937) and tended to be more pleomorphic. It continued to be a significant human pathogen in the pre-antibiotic era and was isolated from a variety of putrid infections, but it has been less frequently reported since 1950 as the other gram-negative anaerobes have assumed a more significant role.

F. necrophorum is a filamentous organism with rounded ends. It is often very pleomorphic, with many short bacillary or coccobacillary forms; sphaeroplasts or L-forms are common and the filaments do not branch. In the diphtheritic membranes of calves, the bacilli are long threads arranged in thick heaps or in long wavy rows; in the pleural or pericardial exudates of rabbits dying of labial necrosis and in material from human infections they are usually highly pleomorphic bacilli varying from cocci to bacilli and long threads. The filaments may reach 80–100 μm in length; they are 0.5–1.0 μm thick and one end is often narrow and pointed whereas the other is thicker. In culture the organisms are straight or curved rods or filaments. The ends are usually rounded and the sides generally parallel, although fusiform enlargements and sphaeroplasts up to 5 μm *q* are common, especially in material from infected tissues. Except in young cultures, staining is irregular and beaded forms are common.

Colonies on agar media are pale and semi-translucent with an irregular or dentate edge and a radiate periphery; they are butyrous in consistency and easily emulsified, and the cultures produce a foul, putrid odour. Deep colonies in serum agar are round, minute and white; under magnification they have a thicker centre with an irregular radiate periphery of streaming threads extending into the surrounding medium. Stab growth in serum is filiform with radiate branches but serum is not liquefied.

F. necrophorum is inhibited by bile and sodium taurocholate; it is tolerant of several dyes, including brilliant green (1 in 100 000), gentian violet (1 in 100 000) and crystal violet (1 in 120 000), which may be useful for primary isolation (Lahelle 1947) and in the initial stages of identification. It is killed by moist heat at 55° within an hour and is sensitive to most antibiotics, including penicillin and the high concentrations of neomycin and kanamycin (1000 μg disks) used for preliminary identification, but many strains are resistant to rifampicin. It gives strongly positive tests for indole and H_2S production and negative methyl red, Voges-Proskauer and catalase tests; methylene blue is reduced but nitrate is not. Casein and gelatin are digested. A few sugars are attacked and reactions are weak; gas and a variable amount of acid are produced from glucose, maltose and laevulose but pentoses, lactose and mannitol are not attacked. Lactate is produced from propionate; lysine is not decarboxylated. *n*-Butyric and propionic acids are major metabolic products.

F. necrophorum produces a wide range of exotoxins and exoenzymes that may be related to pathogenicity. Most strains produce an oxygen-stable haemolysin with a wide activity against the red blood cells of many species, including man, horse, cow, sheep and rabbit; they also produce a lipase, a leucocidin, a cytoplasmic toxin and a DNAase. The red cells of several species are agglutinated by strains that are particularly pathogenic for mice and a haemagglutination inhibition test has been developed. The endotoxin is lethal for mice, 11-day chick embryos and rabbits; small quantities are pyrogenic in rabbits and it will induce a local and generalized Shwartzman reaction and tolerance and immunity can be induced in mice. Analysis of the cell-wall lipids shows the absence of sphingosine-based lipid, branched pentadecanoic acid or 3-hydroxy fatty acids (Garcia *et al.* 1975).

The G + C content of the DNA is 33 moles per cent, in keeping with the generally low values for the genus.

F. nucleatum This is a commensal in the human mouth and also a cause of infections related to the mouth. Strains of *F. nucleatum* are identical with strains labelled *F. polymorphum*. *F. nucleatum* is a fastidious anaerobe and nutritionally demanding; it requires adenine or xanthine, pantothenic acid and tryptophan and may be difficult to maintain in repeated subculture on artificial media. It is a small, pleomorphic bacillus 4–6 × 0.5–1 μm with pointed ends; navicular and spindle forms are common and nuclear material can be seen in stained preparations. Elongated chains or filaments may be seen in old cultures.

Surface colonies are 1–2 mm *q*, yellow-grey in colour and with a matt surface (Fig. 26.7); older colonies are surrounded by spreading tufts and they may become rough. Cultures produce an unpleasant but not putrid smell. *F. nucleatum* is sensitive to most antibiotics, including penicillin, rifampicin and the high concentrations of kanamycin and neomycin used for preliminary identification; it is inhibited by bile

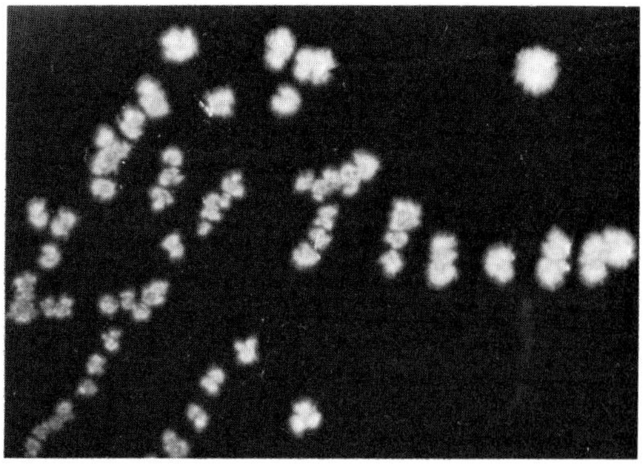

Fig. 26.7 *F. nucleatum*: irregular colonies after incubation for 72 hr on blood agar (× 2).

and sodium taurocholate but is resistant to several dyes, including brilliant green and gentian violet (1 in 100 000). Glucose and, possibly, galactose and fructose may be fermented but only weakly; maltose and lactose are not attacked. Indole and, in general, H_2S are formed but nitrate is not reduced. The G + C content of the DNA is 27 moles per cent. It ferments several amino acids and the major metabolic products are *n*-butyric and acetic acids.

The classification of the remaining species of *Fusobacterium* is unsatisfactory. *F. varium* is less pleomorphic than most fusobacteria and does not show the typical fusiform morphology; it is a small gram-negative bacillus or coccobacillus that does not produce filaments or sphaeroplasts. It has been isolated from the faeces of man, animals and birds and is occasionally implicated in soft-tissue infections. Like many fusobacteria it is sensitive to penicillin and to high concentrations of kanamycin and neomycin but resistant to rifampicin in disk-resistance tests and is tolerant of several dyes; it is also tolerant of sodium taurocholate. It produces indole and H_2S but nitrate is not reduced. Glucose is fermented only weakly and *n*-butyric acid is a major metabolic product. *F. necrogenes* has been isolated from human faeces and from necrotic abscesses in poultry. It is a more typical fusiform organism and produces rough, irregular, opaque colonies on solid media. It gives the same results as *F. varium* in antibiotic-disk resistance tests and in tolerance tests with dyes and sodium taurocholate, and it gives a similar pattern of fatty-acid metabolic products. *F. necrogenes* does not produce indole, but aesculin is hydrolysed and glucose may be fermented weakly. *F. symbiosum* and *F. mortiferum* have been described in human faeces and in soft-tissue infections and necrotic abscesses, in which *F. mortiferum* is associated with the production of large amounts of gas. *F. russi* has been found in similar situations, particularly in perianal abscesses, in infections in cats, and in human and animal faeces. *F. prausnitzi* has been found in faeces and cases of pleurisy, and *F. gonidiaformis* is implicated in infections of the respiratory, genito-urinary and gastro-intestinal tracts; it has also been found in lambs. *F. naviforme*, named for its boat-shaped morphology, and *F. bullosum*, are found in mammalian faeces. *F. bullosum* differs from many fusobacteria in being motile. *F. aquatile* is also a motile fusobacterium found in faeces and river water. *F. plauti* was described as a motile fusobacterium found in the mouth, but no reference strain exists. This also applies to *F. perfoetus*, *F. glutinosum* and *F. stabile*. The validity of these species must await further investigation.

Leptotrichia

The genus *Leptotrichia* contains a single species, *L. buccalis* (Trevisan 1879, Thjötta *et al.* 1939); it is synonymous with Vincent's fusiform organism, *Fusobacterium plauti-vincenti* (Knorr 1922) and *F. fusiforme* (Bergey's Manual 1957). The cells are long, straight or slightly curved rods 1 μm wide × 5–15 μm long and many have pointed ends. Growth in broth culture often produces long, septate filaments that may be 200 μm in length and often become tangled. Bulbous swellings or sphaeroplasts may be found in old cultures. Cells and filaments do not branch; they are non-sporing and non-motile. *L. buccalis* is gram negative and the cell-wall structure is typical of a gram-negative organism (Hofstad 1970); the lipopolysaccharide has strong endotoxic activity; it causes pyrexia and a Shwartzman reaction and there is a serologically-specific O antigen (Gustafson *et al.* 1966). Granules which may be gram positive in young cultures are often present within the cells. This had led to confusion of *L. buccalis* with gram-positive bacilli related to *Lactobacillus* and with *Bacterionema matruchoti*.

L. buccalis is a strict anaerobe that grows as lobate or convoluted colonies on the surface of rich agar media; 5 per cent of CO_2 is essential for primary isolation. It differs from the fusobacteria in its metabolic properties. The only major metabolic product is lactic acid, with a small amount of acetic acid; *n*-butyric acid is not produced. It is a fermentative organism that produces acid but not gas from glucose, fructose, maltose, mannose and sucrose and usually from trehalose and salicin; lactose and starch may be fermented but acid is not produced from rhamnose, mannitol, arabinose, xylose, inulin or sorbitol. It is sensitive to penicillin, rifampicin and to the 1000 μg disks of kanamycin and neomycin; in tolerance tests it is inhibited by bile salts but grows in the presence of Victoria blue 4R, ethyl violet and gentian violet. *L. buccalis* gives negative reactions in most other routine tests: indole and H_2S are not produced; the catalase, oxidase and Voges-Proskaeur reactions are negative; gelatin is not digested and nitrate is not reduced. The G + C content of the DNA is 34 moles per cent.

Cell-wall composition and antigenic properties

The fine structure of the cell wall of members of the Bacteroidaceae resembles that of other gram-negative bacteria (Chapter 2). The ability to form a polysaccharide capsule (p. 119) is important in the pathogenicity of *B. fragilis* and has also been described in *B. asaccharolyticus*. The complex of antigens associated with the outer layer of the cell wall includes the lipopolysaccharide; this has endotoxic activity, and its polysaccharide is responsible for O-antigenic specificity (Hofstad 1975).

The lipopolysaccharide of *B. fragilis* differs from the lipopolysaccharides of aerobic gram-negative bacilli in that it does not contain either 2-keto-3-deoxyoctonate (KDO) or heptoses. It has only weak endotoxic activity; it is not lethal for 11-day-old chick embryos when given intravenously and does not induce a local Shwartzman reaction (Kasper 1976*a*, *b*). It gives a positive reaction in the *Limulus*-lysate test, but only at a much higher concentration than the lipopolysaccharide of *Esch. coli*. Lipopolysaccharides from several other species of *Bacteroides* also appear not to contain KDO and to have only weak endotoxic activity. However, cell-wall extracts of *F. necrophorum* are lethal for mice, 11-day-old chick embryos and rabbits; they are highly pyrogenic for rabbits, induce local and generalized Shwartzman reactions and induce tolerance and immunity in mice; they contain a small amount of KDO (Hofstad and Kristofferson 1971; Garcia *et al.* 1975). The lipopolysaccharide of *L. buccalis* also has strong endotoxic activity (Gustafson *et al.* 1966).

As we have seen (p. 122–3) the cell wall peptidoglycans of closely related members of the melaninogenicus-oralis group may show considerable differences in chemical composition.

The specificity of O antigens in *Bacteroides* is determined by the distribution of oligosaccharides—di- to hexasaccharides—as repeating units in the polysaccharide chain (Hofstad 1977). As with the enterobacteria, O-antigen preparations can be made by boiling suspensions of whole cells. Several workers have attempted to produce a serological typing system for the fragilis group of *Bacteroides* on the basis of agglutination reactions with the O antigens. Weiss and Rettger (1937) showed that groups of *Bacteroides* could be distinguished by serological methods; Werner and Sébald (1968) and Werner (1969*a*) showed that serotypes existed in the species *B. fragilis* and Beerens and his colleagues (1971) divided the fragilis group into three serogroups. Lambe and Moroz (1976) grouped *B. fragilis* by agglutination tests with seven absorbed monospecific antisera but found 21 serogroups and 45 serological patterns amongst 98 test strains. Sheep erythrocytes can be sensitized by phenol-extracted lipolysaccharides from members of the fragilis group and used in direct haemagglutination and haemaglutination-inhibition tests (Hofstad 1977), but the application of this technique is limited because the O antigens are irregularly distributed. Each species is heterogeneous and there are many cross-reactions both within species and within the group, so that specific antisera for many O factors would be needed for reliable serotyping. Elhag and his colleagues (Elhag *et al.* 1977, Elhag and Tabaqchali 1978*a*, *b*) also found cross-reactions in agglutination tests between rabbit O antisera and live cultures of fragilis-group strains, but these were removed by absorption of the sera with O-antigen preparations. Reactions were more reliable with boiled preparations than with whole live cells; this was perhaps attributable to the removal by boiling of cross-reacting substances such as protein and polysaccharide components of the outer membrane (Kasper and Seiler 1975, Kasper 1976*a*, *b*). It is clear that there are many O-antigenic types within the fragilis group and within individual species, and that there is no clear association between O serotype and biotype or species.

Species-specific protein antigens can be extracted from the outer membrane of the cell wall of *Bacteroides* by gentle ultrasonic disintegration and treatment with EDTA. These can be detected in an enzyme-linked immunosorbant assay (ELISA) with rabbit antisera made by the injection of whole live cells (Poxton 1979). This may have applications in the identification of *Bacteroides* strains or in the detection of antibodies in the serum of patients.

Fluorescent-antibody conjugates have been developed for the rapid identification of strains from clinical material. The fragilis group and *B. melaninogenicus* (including *B. asaccharolyticus*) can be identified in smears prepared directly from swabs or pus by indirect immunofluorescence with polyvalent conjugates (Stauffer *et al.* 1975). Lambe (1974) and Lambe and Jerris (1976) found few significant cross-reactions with fluorescein isothiocyanate-labelled antibody conjugates prepared against *B. asaccharolyticus* and *B. melaninogenicus* subsp. *melaninogenicus* and subsp. *intermedius*, although subsp. *intermedius* was divided into two specific serotypes. This suggests that the melaninogenicus-oralis group is less antigenically heterogeneous than the fragilis group. Werner and his colleagues (1977) found that *B. oralis* strains were serologically homogeneous and suggested an antigenic relationship between *B. oralis* and *B. melaninogenicus* subsp. *intermedius*, although conventional biochemical tests suggest a relationship between *B. oralis* and *B. melaninogenicus* subsp. *melaninogenicus*.

Sensitivity to antimicrobial agents

So far we have discussed the susceptibility of gram-negative anaerobic bacilli to antimicrobial agents only inasmuch as it is an aid to the identification of individual species or groups of species. With rare exceptions, members of the family Bacteroidaceae are sensitive to concentrations of metronidazole that are easily attained in the body by conventional doses. Metronidazole is bactericidal and is active only against anaerobic bacteria; it has proved effective in the treatment and prophylaxis of bacteroides infections. Many strains are also sensitive to clindamycin, chloramphenicol and erythromycin but resistance occurs and transferable plasmid-mediated resistance has been described in *B. fragilis* (see p. 130). Although susceptibility to high concentrations of neomycin and kanamycin is used in the identification of some groups of *Bacteroides*, all gram-negative anaerobic bacilli are resistant to the concentrations of aminoglycosides that can be attained in the body.

Members of the fragilis group of *Bacteroides* are almost invariably resistant to most β-lactam antibiotics; the minimal inhibitory concentration of benzyl penicillin usually exceeds 16 mg per l. This resistance is associated with the production of a β-lactamase that has a very broad spectrum of activity on penicillins and cephalosporins. Members of the melaninogenicus-oralis group appeared originally to be sensitive to penicillin, but resistance, mediated by a different penicillinase from that of the fragilis group, has subsequently become quite common (Salyers *et al.* 1977). Similarly, tetracycline resistance, at one time rare in *Bacteroides*, is now found in up to one-half of all strains (Sutter and Finegold 1976). In general, strains of *Fusobacterium* and *Leptotrichia* are sensitive to therapeutically attainable concentrations of penicillin. (For reviews of antimicrobial susceptibility see Finegold 1977, Willis 1977, Garrod *et al.* 1981.)

Genetic mechanisms

There have been few genetic studies of gram-negative anaerobic bacilli and most of the available information relates to *B. fragilis*. The chromosome in *B. fragilis* is a highly coiled circular molecule, like that of *Escherchia coli*, with a molecular weight of *ca* 2.5×10^9. Many strains of *Bacteroides* also contain plasmids; these have been demonstrated in most members of the fragilis group. Stiffler and his colleagues (1974) isolated three distinct plasmids from two strains; one strain contained many copies of a plasmid with a molecular weight of 4×10^6 and a second strain contained many copies of a plasmid with a molecular weight of 2.7×10^6 together with one or a few copies of plasmid of molecular weight 1.6×10^7. Guiney and Davis (1975) and Tinnell and Macrina (1976) found plasmids with a much higher molecular weight (25×10^6 and 31×10^6); these were present as single copies. All of these were described as cryptic plasmids; none was associated with resistance to antibiotics or production of bacteroicines and there was no evidence of any transfer properties. However, the transfer of multiple antibiotic resistance from *B. fragilis* to *Esch. coli* K12 was demonstrated by Mancini and Behme (1977); resistance to ampicillin, amoxycillin, cephalothin, tetracycline, minocycline and chloramphenicol was transferred as a single unit but the resistance markers were unstable during storage of recipient strains. Subsequently plasmid-coded clindamycin and erythromycin resistance from a strain of *B. fragilis* isolated from a blood culture was transferred by conjugation to recipient strains of *B. fragilis* and *B. thetaiotaomicron* (Tally *et al.* 1979). The donor strain contained plasmids of molecular weight 2×10^6 and 20×10^6 and the transcipient strains were always resistant to both antibiotics and contained the two plasmids. Similarly resistance to clindamycin, erythromycin and streptogramins was transferred from *B. distasonis* and *B. fragilis* to a sensitive strain of *B. fragilis* and these resistance markers were curable *en bloc* by exposure to subinhibitory concentrations of acridine orange, proflavine or ethidium bromide (Privatera *et al.* 1979).

It is clear that some plasmids determine resistance to antibiotics and some are transfer- or sex-factors that bring about the transfer of genetic material by conjugation, not only between different bacteroides strains but also between bacteroides strains and members of other genera.

Bacteriocine production

Human faecal strains of the fragilis group of *Bacteroides* may produce bacteriocines active against a wide variety of other bacteroides strains including members of the fragilis group, *B. melaninogenicus* and *B. oralis*, but not inhibitory to strains of any other genera (Beerens and Baron 1965, Beerens *et al.* 1966, Booth *et al.* 1977). Producer strains may be sensitive to the bacteriocines produced by other strains.

The bacteriocines have a high molecular weight and protein is a major component. Bacteriocine activity is stable over the pH range 1–12 and is not affected by RNAase, DNAase or phospholipase but is destroyed by proteolysis; the bacteriocine characterized by Booth and his colleagues was inactivated by trypsin and Pronase but not by proteinase K or pepsin. It was also very resistant to heat; there was no loss of activity after heating at 100° for 1 hr and only a 50 per cent reduction after 15 min at 121°. This heat stability is surprising, because this bacteriocine has a molecular weight of >300 000, although no bacteriophage-like particles or sub-units can be seen on electron microscopy. The bacteriocines described by Beerens and his colleagues were heat labile.

There is wide variation in the pattern of susceptibility to different bacteriocines among strains of the same and closely related species, particularly in the fragilis group. This suggests that it may be possible to develop a bacteriocine-typing scheme. The significance of bacteriocine production *in vivo* is uncertain. Whether bacteriocine production gives strains a competitive advantage in the normal habitat is uncertain; producers and non-producers co-exist in the same gastro-intestinal population and the non-producers greatly outnumber the producers.

Bacteriophages

A virulent bacteriophage active against *B. distasonis* was isolated from sewage by Sabiston and Cohl (1969). Bacteriophages active against other members of the fragilis group have been isolated, mostly from sewage; Booth and his colleagues (1979) described 68 distinct phages but they failed to isolate any directly from human faeces. All the phages except for three in Booth's series are morphologically similar to Bradley's group-B bacteriophages (1967); they have a hexagonal head and a complex tail. They are indistinguishable from each other, regardless of their species specificity; phage heads are 50–90 nm *q* and the flexible, sheathless tails are 12–200 nm long. The atypical phages described by Booth and his colleagues had larger heads and more complex, rigid and sheathed tails. All the phages appear to be virulent and true lysogeny has not been detected, although a phage-carrier or pseudolysogenic state has been described (Keller and Traub 1974, Booth *et al.* 1979). Transduction by *Bacteroides* phage has not been detected. Each phage is specific for a given species of *Bacteroides* and members of all other genera are resistant. Bacteriophages isolated on *B. fragilis* are virulent for some other strains of *B. fragilis*, but not for members of other species in the fragilis group. However, no phage is virulent for all strains of any species; different phages give different patterns of lysis with groups of *B. fragilis* strains and no specific patterns are associated with different sources or different types of infection. A scheme for phage-typing isolates of, for example, *B. fragilis* could be produced, if this would provide useful epidemiological evidence. Similarly, a set of phages could be used to confirm the identity of different species of *Bacteroides*, but a series of phages would be needed for each species to ensure that most isolates were lysed by at least one phage.

Bacteroidaceae in the normal flora

In man

Gram-negative anaerobic bacilli are a major component of the normal commensal flora of the mucous membranes of the mouth, lower gastrointestinal tract and vagina. Eggerth and Gagnon (1933) recognized their importance in the faecal flora and Rosebury (1962) stressed their normal occurrence at all three sites. The species of gram-negative anaerobic bacilli found at the three sites are different, and form distinct populations. (For other organisms in the normal flora, see Chapter 8.)

Oral microflora

The mouth does not have a uniform bacterial population (Chapter 8). The gingival crevice and dental plaque are the principal sites of colonization with gram-negative anaerobic bacilli (Socransky and Manganiello 1971, Hardie and Bowden 1974); the surface of the tongue is almost devoid of Bacteroidaceae and the saliva contains a variable number of gram-negative anaerobes derived from the gingival crevice. Hadi and Russell (1968, 1969) found salivary viable counts of fusobacteria of 2.72×10^5 to 1.79×10^6 per ml; the higher results were from patients with gingivitis. Gram-negative anaerobic bacilli constitute some 16 per cent of the cultivable flora of the gingival crevice but only 4 per cent of that of dental plaque in normal subjects (Gibbons *et al.* 1963, 1964, Loesche *et al.* 1972); in institutionalized subjects, however, they may form 17 per cent of the cultivable flora in plaque. *Bacteroides*—mainly of the melaninogenicus-oralis group—and fusobacteria are present in large numbers in the gingival crevice and subgingival plaque of all healthy adults; in the presence of gingivitis and periodontitis their numbers, particularly those of *B. melaninogenicus* and *B. gingivalis*, increase.

B. melaninogenicus was first described in studies of the microflora of mucous membranes, in particular, of the gingival mucosa (Oliver and Wherry 1921). It is a useful marker-species for the anaerobic flora of the gingival crevice (Holbrook *et al.* 1978). Pigmented strains constitute 4–5 per cent of the cultivable flora of the normal gingival crevice (Gibbons 1974) and 5–6 per cent of the plaque flora from institutionalized subjects (Loesche *et al.* 1972). Non-pigmented oral strains of *Bacteroides* with similar properties are called *B. oralis* (Loesche *et al.* 1964). *B. melaninogenicus* subsp. *intermedius* and subsp. *melaninogenicus* can be isolated from almost all subjects and account for 35–40 per cent of the gram-negative anaerobic bacilli. Similar, but non-pigmented, strains of *B. oralis* account for a similar proportion (*ca* 35 per cent) of the cultivable gram-negative anaerobic bacilli; if the two subspecies of *B. melaninogenicus* are regarded as distinct, *B. oralis* is the commonest more-or-less homogeneous species in the gingival flora (Loesche *et al.* 1972, Gibbons 1974, Duerden 1980*c*). The remainder of the cultivable gram-negative anaerobic bacilli are smaller numbers of other non-pigmented species of the melaninogenicus-oralis group, *B. gingivalis*, fusobacteria and *L. buccalis*. Species of the fragilis group are not part of the normal gingival flora (Hardie 1974). Most studies underestimate grossly the contribution of fusobacteria to the normal gingival flora; many of the strains are difficult to grow and the number cultivated from subgingival plaque falls several thousand-fold below the number seen by direct microscopy. The classification of the fusobacteria from the normal gingival flora is unsatisfactory; strains with different characteristics are allocated to the broad species *F. nucleatum* (*polymorphum*), but many strains do not have the characteristics of any recognized species and cannot be classified by present methods.

Gingival strains of Bacteroidaceae are often nutritionally demanding and may be difficult to maintain in pure culture on solid media; many of these are fusobacteria, but some are pigmented and non-pigmented strains of the melaninogenicus-oralis group. Many species are nutritionally interdependent and will grow only in mixed culture, in which satellitism may be demonstrated. Growth is often improved by the addition of 5–10 per cent of CO_2 in the anaerobic atmosphere (Watt 1973, Stalons *et al.* 1974) and recognized growth factors include metal ions (Caldwell and Arcand 1974) and haemin (Gibbons and MacDonald 1968, Gilmour and Poole 1970); some *B. melaninogenicus* strains also require menadione (vitamin K, Lev 1959) or its precursors (Robins *et al.* 1973) and succinate (Lev *et al.* 1971), but the most demanding strains appear to require other unidentified substances.

Gastro-intestinal microflora

Gram-negative anaerobic bacilli are generally the most numerous bacteria in the normal flora of the gastro-intestinal tract, and the fragilis group of *Bacteroides* forms the majority of them (Drasar and Hill 1974). Obligate anaerobes are seldom found in the normal stomach, duodenum, jejunum or proximal ileum; *Bacteroides* spp. and *Bifidobacterium* spp. begin to appear only in the distal ileum (Finegold 1977) and they form a major part of the caecal and faecal flora. Viable counts yield 10^{11} *Bacteroides* organisims per g wet weight of faeces and caecal contents (Drasar 1967, Finegold *et al.* 1975). The proportion of *Bacteroides* spp. in the faecal flora differs in subjects from different countries and is related to the type of diet. They are found in greater numbers in subjects who consume a mixed Western diet that contains comparatively large amounts of fat and stimulates the production of a large volume of bile; fewer *Bacteroides* spp. are found in subjects from developing countries in Africa and Asia, and from Japan, who consume a mainly vegetarian diet; gram-positive anaerobes and enterococci are the predominant faecal organisms in these subjects. However, subjects in developed countries who change to a vegetarian diet do not convert their faecal flora to a gram-positive predominance and retain their *Bacteroides* organisms (Drasar 1974).

Table 26.5 Strains of *Bacteroides* isolated from the faeces of 20 human subjects (Duerden 1980*a*)

Species	Number of strains
Fragilis group	(165)
B. fragilis	15
B. vulgatus	37
B. distasonis	30
B. thetaiotaomicron	36
B. uniformis	8
B. variabilis/eggerthi	24
B. splanchnicus	15
Melaninogenicus-oralis group	(7)
B. melaninogenicus subsp. *melaninogenicus*	2
subsp. *intermedius*	1
B. ruminicola	4
Asaccharolytic group	(21)
B. asaccharolyticus	8
non-pigmented non-saccharolytic spp.	13
Bacteroides spp. (unidentified)	4
Any	197

Some 80 per cent of faecal gram-negative anaerobic bacilli are members of the fragilis group. The commonest species in most subjects are *B. vulgatus* and *B. thetaiotaomicron*. *B. distasonis* is also common, but in general is isolated from fewer subjects and in smaller numbers than the two commonest species. *B. eggerthi* or *B. variabilis* are found regularly, but *B. fragilis* forms only a small proportion of the fragilis group in the normal faecal flora (see Table 5; Werner 1974, Finegold *et al.* 1975, Moore and Holdeman 1975, Duerden 1980*a*). Asaccharolytic species of *Bacteroides* form a consistent part of the normal faecal flora but are present in much smaller numbers than the fragilis group (Werner *et al.* 1971); the most commonly recognized species is *B. asaccharolyticus* and the lower gastro-intestinal tract is probably the primary habitat of this species. Non-pigmented asaccharolytic species are also present in quite small numbers; they grow slowly to form only small colonies and are difficult to detect. In general, the melaninogenicus-oralis group of *Bacteroides* and the fusobacteria are not a significant part of the cultivable faecal flora; *B. ruminicola* and *B. melaninogenicus* strains may be isolated in small numbers but fusobacteria are rarely found, although this may reflect the difficulty of isolating them from a mixed flora in which they are greatly outnumbered by less demanding anaerobes.

The gram-negative anaerobic bacilli play an important part in the physiological activities of the large intestine. They are responsible for much of the metabolism of bile salts and bile acids. *Bacteroides* of the fragilis group and several clostridia produce enzymes that deconjugate bile salts under conditions of low Eh, and they reduce cholic acid to deoxycholic acid. This is important in the enterohepatic circulation of bile salts. They may also be implicated in the production of carcinogens from bile acids. The fragilis group of *Bacteroides* produce dehydrogenases and dehydroxylases that convert bile acids into aromatic compounds and eventually to substituted cyclopentaphenanthrenes which are known to be carcinogens. Differences in the incidence of colonic carcinoma in various countries are associated with the type of diet consumed and with the relative numbers of gram-negative anaerobic bacilli in the faeces (Drasar and Hill 1974; see also Chapter 8).

Vaginal microflora

The role of the Bacteroidaceae in the normal vaginal flora has been the subject of controversy. The vagina is not a single environment; the flora of the lower vagina is a mixture of organisms from the vagina with others from the perineum and introitus, whereas the flora of the cervix and fornices more closely represents the true vaginal flora. Moreover, the environment is not constant; the state of the mucosa and secretions

Table 26.6 Strains of *Bacteroides* isolated from the vagina of 13 healthy adults (Duerden 1980*b*)

Species	Number of strains
Fragilis group	(6)
B. distasonis	1
B. vulgatus	3
B. thetaiotaomicron	1
B. splanchnicus	1
Melaninogenicus-oralis group	(88)
B. melaninogenicus subsp. *melaninogenicus*	14
subsp. *intermedius*	19
B. bivius	16
B. bivius/disiens	21
B. oralis group (unidentified)	1
B. oralis	10
B. ruminicola	7
Asaccharolytic group	(16)
B. asaccharolyticus	9
non-pigmented non-saccharolytic spp.	7
Bacteroides spp. (unidentified)	3
Any	113

changes with age, with each menstrual cycle and with pregnancy (Hurley *et al.* 1974 and Chapter 8).

Gram-negative anaerobic bacilli are common but not universal commensals of the cervix and vaginal fornices. The presence of *Bacteroides* in the vagina was reported first by Burdon (1928); he isolated *B. melaninogenicus* from 28 out of 35 normal women. Other workers also found that *Bacteroides* were part of the normal vaginal flora (Mead and Louria 1969, Suzuki and Ueno 1971).

Bacteroides spp. were isolated from cervical cultures in 57 per cent of 30 (Gorbach *et al.* 1973) and 65 per cent of 26 (Sanders *et al.* 1975) normal women and from the vaginal fornices in 65 per cent of 20 normal women (Duerden 1980*b*). However, other workers have found them considerably less often: 8.6 per cent of 246 pre-operative gynaecological patients (Neary *et al.* 1973); 4.6 per cent of 500 women attending a family-planning clinic; 5 per cent of 200 patients attending a gynaecological out-patient clinic (Leigh 1974); 5.4 per cent of 280 pregnant women (Hurley *et al.* 1974); and four of 100 pregnant women (Werner *et al.* 1978). In a quantitative study, Lindner and his colleagues (1978) found 10^{7-8} viable *Bacteroides* organisms per ml of vaginal secretions but they isolated *Bacteroides* spp. from only 4 per cent of normal women, 1 per cent of pregnant women and 28 per cent of women with cervicitis. Some of these differences may reflect differences between populations, but most are probably attributable to different methods of investigation and, in particular, to differences in sampling methods and in anaerobic technique. However, *Bacteroides* spp. generally disappear during pregnancy and return soon after delivery (Hite *et al.* 1947, Willis 1977).

The species of *Bacteroides* isolated in most studies of the vaginal flora have not been identified. Duerden (1980*b*) found that members of the melaninogenicus-oralis group predominated. *B. bivius* was the commonest species; *B. melaninogenicus* subsp. *intermedius* and subsp. *melaninogenicus* were also common. *B. asaccharolyticus* can be isolated in smaller numbers but the fragilis group and the fusobacteria are not part of the normal vaginal flora (Table 26.6). The presence of significant numbers of fragilis-group organisms, and in particular of *B. fragilis*, indicates some pathological change in the genital tract. The predominant species in the vagina and the mouth are similar.

In animals

Gram-negative anaerobic bacilli are part of the normal flora of intestinal contents and faeces in many species of animals from invertebrates to man. They are an important component of the flora of two specialized gastro-intestinal structures, the rumen of ruminant animals and the caecum of poultry.

Three species of *Bacteroides* are found in large numbers in the *rumen*—*B. succinogenes*, *B. amylophilus* and *B. ruminicola*. *B. succinogenes* is found in most ruminants and attains a viable count of *ca* 0.5×10^8 per ml, about 8.5 per cent of the total viable count; it is actively cellulolytic and is important in the digestion of cellulose; it also ferments glucose, cellobiose and several other carbohydrates with the production of succinic acid (Hungate 1950). *B. amylophilus* and *B. ruminicola* are not cellulolytic; they are active in the digestion of starch and other carbohydrates. *B. amylophilus* occurs sporadically but when present may be the predominant starch digester and constitute 10 per cent of the total bacteria (Hamlin and Hungate 1956). The fermentative activity of *B. ruminicola* is more varied; it is regularly present and constitutes 6–19 per cent of the total viable count.

The *rumen* bacteria are essential for the nutrition of ruminants. The energy content of the feedstuff is made available to the animal in the form of volatile fatty acids, particularly acetic, butyric and propionic acids, produced by bacterial fermentation of cellulose and other carbohydrates; the acids are absorbed through the rumen wall and metabolized by the animal. The removal of these acid products is essential for the continuation of fermentation. In addition, microbial protein constitutes a major part of the ruminants' protein source. Nitrogenous components of the feed are converted to microbial protein during growth and multiplication of the rumen bacteria, which contain *ca* 65 per cent of protein. In the distal parts of the intestine, dead bacterial cells are digested and provide protein for the animal (Hungate 1966).

In poultry, the *caecum* is highly developed and has an important role in digestion and absorption. Gram-negative anaerobic bacilli are found in large numbers (*ca* 10^{8-9} per g) in the caecal contents. The commonest species is unique to this site—*B. hypermegas*; it is a much larger organism than most *Bacteroides* spp. and ferments a variety of carbohydrates. Fusobacteria are also present, in particular *F. varium* (Barnes and Goldberg 1965, 1968*a, b*).

Bacteroidaceae in clinical infections

Infections caused by gram-negative anaerobic bacilli and the pathogenicity of particular species are discussed in Chapter 62.

References

Bang, B. (1897) *Z. Tiermed.* **1,** 241.

Barnes, E. M. and Goldberg, H. S. (1965) *Ernährungsforschung* **10,** 289; (1968*a*) *J. gen. Microbiol.* **51,** 313; (1968*b*) *J. appl. Bact.* **31,** 530.

Beerens, H. (1970) *Int. J. syst. Bact.* **20,** 297.

Beerens, H. and Baron, G. (1965) *Ann. Inst. Pasteur* **108,** 255.

Beerens, H., Baron, G. and Tahon, M. M. (1966) *Ann. Inst. Pasteur, Lille* **17,** 1.

Beerens, H. and Tahon-Castel, M. (1965) *Infections Humaines à Bactéries Anaérobies non Toxigènes.* Presse Acad. Européenes, Bruxelles.

Beerens, H., Wattre, P., Shinjo, T. and Romand, C. (1971) *Ann. Inst. Pasteur* **121,** 187.

Berger, U. (1956) *Zbl. Bakt.* **166,** 484.

Bergey's Manual of Determinative Bacteriology (1948) 6th edn, p. 564. Ed. by R. S. Breed *et al.* Williams and Wilkins, Baltimore.

Bergey's Manual of Determinative Bacteriology (1957) 7th edn, p. 423. Ed. by R. S. Breed *et al.* Williams and Wilkins, Baltimore.

Beveridge, W. I. B. (1934) *J. Path. Bact.* **38,** 467.

Bibby, B. G. and Berry, G. P. (1939) *J. Bact.* **38,** 263.

Bøe, J. (1941) *Fusobacterium.* Dybwad, Oslo.

Booth, S. J., Johnson, J. L. and Wilkins,, T. D. (1977) *Antimicrob. Agents Chemother.* **11,** 718.

Booth, S. J., van Tassell, R. L., Johnson, J. L. and Wilkins, T. D. (1979) *Rev. infect. Dis.* **1,** 325.

Bradley, D. E. (1967) *Bact. Rev.* **31,** 230.

Bryant, M. P., Small, N., Bouma, C. and Chu, H. (1958) *J. Bact.* **76,** 15.

Burdon, K. L. (1928) *J. infect. dis.* **42,** 161.

Caldwell, D. R. and Arcand, C. (1974) *J. Bact.* **120,** 322.

Carlsson, J. (1973) *Appl. Microbiol.* **25,** 287.

Castellani, A. and Chalmers, A. J. (1919) *Manual of Tropical Medicine*, 3rd edn. Baillière, Tindall and Cox, London.

Cato, E. P. *et al.* (1970) *Outline of Clinical Methods in Anaerobic Bacteriology.* Virginia Polytechnic Institute and State University, Blacksburg, Va.

Cato, E. P. and Johnson, J. L. (1976) *Int. J. syst. Bact.* **26,** 230.

Coykendall, A. L., Kaczmarek, F. S. and Slots, J. (1980) *Int. J. syst. Bact.* **30,** 559.

Dack, G. M., Dragstedt, L. R. and Heinz, T. E. (1936) *J. Amer. med. Ass.* **106,** 7; (1937) *J. infect. Dis.* **40,** 335.

Distaso, A. (1912) *Zbl. Bakt.* **62,** 433.

Dowell, V. R. Jr. (1972) *Amer. J. clin. Nutr.* **25,** 1335.

Drasar, B. S. (1967) *J. Path. Bact.* **94,** 417; (1974) In: *The Normal Microbial Flora of Man*, p. 187. Ed. by F. A. Skinner and J. G. Carr. Academic Press, London.

Drasar, B. S. and Hill, M. J. (1974) *Human Intestinal Flora.* Academic Press, London.

Drasar, B. S., Shiner, M. and Macleod, G. M. (1969) *Gastroenterology* **56,** 71.

Duerden, B. I. (1975) *J. med. Microbiol.* **8,** 113; (1980*a*) *Ibid.* **13,** 69; (1980*b*) *Ibid.* **13,** 79; (1980*c*) *Ibid.* **13,** 89; (1980*d*) *J. Hyg., Camb.* **84,** 301.

Duerden, B. I., Collee, J. G., Brown, R., Deacon, A. G. and Holbrook, W. P. (1980) *J. med. Microbiol.* **13,** 231.

Duerden, B. I., Holbrook, W. P., Collee, J. G. and Watt, B. (1976) *J. appl. Bact.* **40,** 163.

Egerton, J. R. and Parsonson, I. M. (1966) *Aust. vet. J.* **42,** 425.

Eggerth, A. H. and Gagnon, B. H. (1933) *J. Bact.* **25,** 389.

Eiken, M. (1958) *Acta path. microbiol. scand.* **43,** 404.

Elhag, K. M., Bettelheim, K. A. and Tabaqchali, S. (1977) *J. Hyg., Camb.* **79,** 233.

Elhag, K. M. and Tabaqchali, S. (1978*a*) *J. Hyg., Camb.* **80,** 439; (1978*b*) *Ibid.* **81,** 89.

Finegold, S. M. (1977) *Anaerobic Bacteria in Human Disease.* Academic Press, New York.

Finegold, S. M. and Barnes, E. M. (1977) *Int. J. syst. Bact.* **27,** 388.

Finegold, S. M., Flora, D. J., Attebery, H. R. and Sutter, V. L. (1975) *Cancer Res.* **35,** 3407.

Finegold, S. M., Harada, N. E. and Miller, L. G. (1967). *J. Bact.* **94,** 1443.

Garcia, M. M., Charlton, K. M. and McKay, K. A. (1975) *Infect. Immun.* **11,** 371.

Garrod, L. P., Lambert, H. P. and O'Grady, F. (1981) *Antibiotic and Chemotherapy*, 5th edn. Churchill Livingstone, London.

Gesner, B. M. and Jenkin, C. R. (1961) *J. Bact.* **81,** 595.

Gibbons, R. J. (1974) In: *Anaerobic Bacteria: Role in Disease.* Ed. by A. Balows *et al.* Charles C. Thomas, Springfield, Ill.

Gibbons, R. J. and MacDonald, J. B. (1968) *J. Bact.* **80,** 164.

Gibbons, R. J., Socransky, S. S., de Aranjo, W. C. and van Houte, J. (1964) *Arch. oral Biol.* **9,** 365.

Gibbons, R. J., Socransky, S. S., Sawyer, S., Kapsimalis, B. and MacDonald, J. B. (1963) *Arch. oral Biol.* **8,** 281.

Gilmour, M. N., Howell, A. and Bibby, B. G. (1961*a*) *Bact. Rev.* **25,** 131; (1961*b*) *Int. Bull. bact. Nomencl.* **11,** 161.

Gilmour, M. and Poole, A. E. (1970) *Arch. oral Biol.* **15,** 1343.

Gorbach, S. L., Menda, K. B., Thadepalli, H. and Keith, L. (1973) *Amer. J. Obstet. Gynec.* **117,** 1053.

Gregory, E. M., Kowalski, J. B. and Holdeman, L. V. (1977) *J. Bact.* **129,** 1298.

Guillaumie, J., Beerens, H. and Osteaux, R. (1956) *Ann. Inst. Pasteur* **90,** 229.

Guiney, D. G. and Davis, C. E. (1975) *J. Bact.* **124,** 503.

Gustafson, K. L., Kraeger, A. V. and Vaichulis, E. M. K. (1966). *Nature, Lond.* **212,** 301.

Hadi, A. W. and Russell, C. (1968) *Arch. oral Biol.* **13,** 1371; (1969) *Brit. dent. J.* **126,** 82.

Hallé, J. (1898) *Recherches sur la Bacteriologie du Canal Génital de la Femme.* Thèse de Paris.

Hamlin, L. J. and Hungate, R. E. (1956) *J. Bact.* **27,** 548.

Hardie, J. M. (1974) In: *Infection with Non-sporing Anaerobic Bacteria*, p. 99. Ed. by I. Phillips and M. Sussman. Churchill Livingstone, Edinburgh.

Hardie, J. M. and Bowden, G. H. (1974) In: *The Normal Microbial Flora of Man*, p. 47. Ed. by F. A. Skinner and J. G. Carr. Academic Press, London.

Harding, G. K. M., Sutter, V. L., Finegold, S. M. and Bricknell, K. S. (1976) *J. clin. Microbiol.* **4,** 354.

Harrison, A. P. and Hansen, P. A. (1950) *J. Bact.* **59,** 197; (1963) *Leeuwenhoek ned Tijdschr.* **29,** 22.

Henthorne, J. C., Thompson, L. and Beaver, D. C. (1936) *J. Bact.* **31,** 255.

Heyde, M. (1911) *Beitr. klin. Chirurg.* **76,** 1.
Hite, K. E., Hesseltine, H. C. and Goldstein, L. (1947) *Amer. J. Obstet. Gynec.* **53,** 233.
Hoelling, A. (1910) *Arch. Protistenk.* **19,** 239.
Hofstad, T. (1970) *Int. J. syst. Bact.* **20,** 175, (1975) *Acta path. microbiol. scand.*, **B83,** 477; (1977) *Ibid.* **B85,** 9.
Hofstad, T. and Kristofferson, T. (1971) *Acta path. microbiol. scand.* **B79,** 385.
Holbrook, W. P. and Duerden, B. I. (1974) *Arch. oral Biol.* **19,** 1231.
Holbrook, W. P., Ogston, S. A. and Ross, P. W. (1978) *J. med. Microbiol.* **11,** 203.
Holdeman, L. V., Good, I. J. and Moore, W. E. C. (1976) *Appl. environ. Microbiol.* **31,** 359.
Holdeman, L. V. and Johnson, J. L. (1977) *Int. J. syst. Bact.* **27,** 337.
Holdeman, L. V. and Moore, W. E. C. (1974*a*) In: *Bergey's Manual of Determinative Bacteriology*, 8th edn, p. 231. Ed. by R. E. Buchanan and N. E. Gibbons. Williams and Wilkins, Baltimore; (1974*b*) *Int. J. syst. Bact.* **24,** 260.
Hungate, R. E. (1950) *Bact. Rev.* **14,** 1; (1966) *The Rumen and its Microbes.* Academic Press, New York.
Hurley, R., Stanley, V. C., Leask, B. G. S. and de Louvois, J. (1974) In: *The Normal Microbial Flora of Man*, p. 155. Ed. by F. A. Skinner and J. G. Carr. Academic Press, London.
Jackson, F. L. and Goodman, Y. E. (1978) *Int. J. syst. Bact.* **28,** 197.
Johnson, J. L. (1973) *Int. J. syst. Bact.* **23,** 308.
Kaczmarek, F. S. and Coykendall, A. L. (1980) *J. clin. Microbiol.* **12,** 288.
Kasper, D. L. (1976*a*) *J. infect. Dis.*, **133,** 79; (1976*b*) *Ibid.* **134,** 59.
Kasper, D. L. and Seiler, M. W. (1975) *J. infect. Dis.* **132,** 440.
Keller, R. and Traub, N. (1974) *J. gen. Virol.* **24,** 179.
Khairat, O. (1966) *J. dent. Res.* **45,** 1191; (1967) *J. Path. Bact.* **94,** 29.
Kligler, I. J. (1915) *J. allied dent. Soc.* **10,** 282.
Knorr, C. (1922) *Zbl. Bakt.* **89,** 4.
Lahelle, O. (1947) *Acta path. microbiol. scand.* Suppl. No. 67.
Lahelle, O. and Thjötta, T. (1945) *Acta path. microbiol. scand.* **22,** 310.
Lambe, D. W. Jr. (1974) *Appl. Microbiol.* **28,** 561.
Lambe, D. W. Jr. and Jerris, R. C. (1976) *J. clin. Microbiol.* **3,** 506.
Lambe, D. W. Jr. and Moroz, D. A. (1976) *J. clin. Microbiol.* **3,** 586.
Leigh, D. A. (1974) In: *Selected Topics in Clinical Bacteriology*, p. 129. Ed. by J. de Louvois. Baillière Tindall, London.
Lev, M. (1959) *J. gen. Microbiol.* **20,** 697.
Lev, M., Keudall, K. L. and Milford, D. F. (1971) *J. Bact.* **108,** 175.
Lindner, J. G. E. M., Plantema, F. H. F. and Hoogkamp-Korstanje, J. A. A. (1978) *J. med. Microbiol.* **11,** 233.
Loeffler, F. (1884) *Mitt. ReichsgesundhAmt.* **2,** 421.
Loesche, W. J., Hockett, R. N. and Syed, S. A. (1972) *Arch. oral Biol.* **17,** 1311.
Loesche, W. J., Socransky, S. S. and Gibbons, R. J. (1964) *J. Bact.* **88,** 1329.
Mancini, C. and Behme, R. J. (1977) *J. infect. Dis.* **136,** 597.
Mayrand, D. (1979) *Canad. J. Microbiol.* **25,** 927.
Mead, P. B. and Louria, D. B. (1969) *Clin. Obstet. Gynec.* **12,** 219.
Mendel, J. (1919) *C.R. Soc. Biol., Paris* **82,** 583.
Mitsuoka, T., Terada, A., Watanabe, K. and Uchida, K. (1974). *Int. J. syst. Bact.* **24,** 35.
Moore, W. E. C. (1970) *Int. J. syst. Bact.* **20,** 535.
Moore, W. E. C. and Holdeman, L. V. (1973) *Int. J. syst. Bact.* **23,** 69; (1975) *Amer. J. med. Technol.*, **41,** 427.
Müller, H. E. and Werner, H. (1970) *Path. et Microbiol., Basel* **36,** 135.
Neary, M. P., Allen, J., Okubadejo, O. A., and Payne, D. J. H. (1973) *Lancet* **i,** 1291.
Newman, M. G. *et al.* (1979) *J. clin. Microbiol.* **10,** 557.
Olitsky, P. K. and Gates, F. L. (1921) *J. exp. Med.* **33,** 125, 361, 713, (1922) *Ibid.* **36,** 501.
Oliver, W. W. and Wherry, W. B. (1921) *J. infect. Dis.* **28,** 341.
Orcutt, M. L. (1930) *J. Bact.* **20,** 343.
Poxton, I. R. (1979) *J. clin. Path.* **32,** 294.
Poxton, I. R. and Brown, R. (1979) *J. gen. Microbiol.* **112,** 211.
Prévot, A. R. (1938) *Ann. Inst. Pasteur* **60,** 285.
Privatera, G., Dublanchet, A. and Sébald, M. (1979) *J. infect. Dis.* **130,** 97.
Reddy, C. A. and Bryant, M. P. (1977) *Canad. J. Microbiol.* **23,** 1252.
Reinhold, L. (1966) *Zbl. Bakt.* **201,** 49.
Robins, D. J., Yee, R. B. and Bentley, R. (1973) *J. Bact.* **116,** 965.
Rosebury, T. (1962) *Micro-organisms Indigenous to Man.* McGraw-Hill, New York.
Rudek, W. and Haque, R. U. (1976) *J. clin. Microbiol.* **4,** 458.
Sabiston, C. B. Jr. and Cohl, M. E. (1969) *J. dent. Res.* **48,** 599.
Salyers, A. A., Wong, J. and Wilkins, T. D. (1977) *Antimicrob. Agents Chemother.* **11,** 143.
Sanders, C. V., Mickal, A., Lewis, A. C. and Torres, J. (1975) *Clin. Res.* **23,** 30A.
Sawyer, J., MacDonald, J. B. and Gibbons, R. J. (1962) *Arch. oral. Biol.* 7, 685.
Schmorl, G. (1891) *Dtsch. Z. Thiermed.* **17,** 375.
Schröter, G. and Stawru, J. (1970) *Z. med. Microbiol.* **155,** 241.
Schwabacher, H., Lucas, D. R. and Rimington, C. (1947) *J. gen. Microbiol.* **1,** 108.
Sébald, M. (1962) *Étude sur les Bactéries Anaérobies Gram-négatives asporalées.* Thèses de Paris.
Shah, H. N. and Collins, M. D. (1980) *J. appl. Bact.* **48,** 75.
Shah, H. N., Hardie, J. M., van Steenbergen, T. J. M. and de Graaf, J. (1980) *J. dent. Res.* **59D,** 1823.
Shah, H. N., Williams, R. A. D., Bowden, G. H. and Hardie, J. M. (1976) *J. appl. Bact.* **41,** 473.
Shimada, K., Sutter, V. L. and Finegold, S. M. (1970) *Appl. Microbiol.* **20,** 737.
Slots, J. and Genco, R. J. (1979) *J. clin. Microbiol.* **10,** 371.
Socransky, S. S. and Manganiello, S. D. (1971) *J. Periodont.* **42,** 485.
Spiers, M. (1971) *Med. Lab. Tech.* **28,** 360.
Stalons, D., Thornsberry, C. and Dowell, V. R. Jr. (1974) *Appl. Microbiol.* **27,** 1098.
Stauffer, L. R., Hill, E. O., Holland, J. W. and Altemeier, W. A. (1975) *J. clin. Microbiol.* **2,** 337.
Steenbergen, T. J. M. van, De Soet, J. J. and de Graaf, J. (1979) *FEMS Microbiol. Lett.* **5,** 127.

Steenbergen, T. J. M. van, Shah, H. N., Hardie, J. M. and de Graaff, J. (1980) *Leeuwenhoek ned. Tijdschr.* **46,** 231.
Stiffler, P. W., Keller, R. and Traub, N. (1974) *J. infect. Dis.* **130,** 544.
Strom, A., Dyer, J. K., Connell, M. and Tribble, J. L. (1976) *J. dent. Res.* **55,** 252.
Sutter, V. L. and Finegold, S. M. (1971) *Appl. Microbiol.*, **21,** 13; (1976) *Antimicrob. Agents Chemother.* **10,** 736.
Suzuki, S. and Ueno, K. (1971) In: *First Symposium on Anaerobic Bacteria and their Infectious Diseases*, p. 81. Ed. by S. Ishiyama *et al.*, Eisai Co., Tokyo.
Suzuki, S., Ushijama, T. and Ichinose, H. (1966) *Jap. J. Microbiol.* **10,** 193.
Tally, F. P., Snydman, D. R., Gorbach, S. L. and Malamy, M. H. (1979) *J. infect. Dis.* **139,** 83.
Thjötta, T., Hartmann, O. and Bœ, J. (1939) *A Study of Leptotrichia Trevisan.* Dybwad, Oslo.
Tinnell, W. H. and Macrina, F. L. (1976) *Infect. Immun.* **14,** 955.
Trevisan, V. (1879). *R.C. Inst. Lombardo* **12,** 133.
Veillon, A. and Zuber, A. (1897) *C.R. Soc. Biol., Paris* **49,** 253; (1898) *Arch. Med. exp. Anat. path.* **10,** 517.
Watt, B. (1973) *J. med. Microbiol.* **6,** 307.
Weinberg, M., Nativelle, R. and Prévot, A. R. (1937) *Les Microbes Anaérobies.* Masson, Paris.
Weiss, J. E. and Rettger, L. F. (1937). *J. Bact.* **33,** 423.
Werner, H. (1969*a*) *Zbl. Bakt.* **210,** 192; (1969*b*) *Ibid.* **211,** 344; (1974) *Arzneimittel-Forsch.* **24,** 340.
Werner, H., Kunstek-Santos, H., Lohner, C., Schöps, W., Schmitz, H. J. and Karbe, H. H. (1977) *Zbl. Bakt.* **237,** 536.
Werner, H., Lang, N., Kraseman, C., Tolkmitt, G. and Feddern, R. (1978) *Curr. med. Res. Opin.* **5,** Suppl. 2, 52.
Werner, H., Pulverer, G. and Reichertz, C. (1971) *Med. Microbiol. Immunol. (Berl.)* **157,** 3.
Werner, H., Rintelen, G. and Kunstek-Santos, H. (1975) *Zbl. Bakt.* **231,** 133.
Werner, H. and Sébald, M. (1968) *Ann. Inst. Pasteur* **115,** 350.
Wilkinson, F. C., Egerton, J. R. and Dickson, J. (1970) *Aust. vet. J.* **46,** 382.
Williams, B. L., Hollis, D. and Holdeman, L. V. (1979) *J. clin. Microbiol.* **10,** 550.
Williams, R. A. D., Bowden, G. H., Hardie, J. M. and Shah, H. N. (1974) *Proc. Soc. gen. Microbiol.* **1,** 70; (1975) *Int. J. syst. Bact.* **25,** 298.
Williams, R. A. D. and Shah, H. N. (1980) In: *Microbial Classification and Identification*, p. 299. Ed. by M. Goodfellow and R. G. Board, Academic Press, London.
Willis, A. T. (1977) *Anaerobic Bacteriology: Clinical and Laboratory Practice*, 3rd edn, p. 212. Butterworths, London.

27

Vibrio, *Aeromonas*, *Plesiomonas*, *Campylobacter*, and *Spirillum*

M. T. Parker and Geoffrey Smith

Introductory

The credit of first describing a member of this group is attributed by Hugh (1964) to Pacini, who in 1854 in Florence observed in the denuded epithelium and contents of the intestine of patients who had died from cholera tremendous numbers of curved bacilli to which he gave the name *Vibrio cholera* (*sic*). In ignorance of Pacini's observations, Robert Koch (1886) in 1883 cultivated the cholera vibrio and brought strong evidence in support of its pathogenic role. In 1888 Gamaléia (1888*a*) isolated a vibrio from the blood and intestinal contents of chickens dying from a cholera-like disease at Odessa; to this organism he gave the name of *V. metschnikovi*; the alternative spelling *metchnikovi* subsequently gained wide currency and will be used in this book. During the next few years a large number of other vibrios, more or less resembling the cholera vibrio, were isolated from sources such as fresh and sea-water, and the faeces of man and animals. Separation of the cholera vibrios from these organisms was aided by the observation of Pfeiffer (1893) that, when living cholera vibrios were injected intraperitoneally into guinea-pigs previously inoculated with killed cholera vibrios, they underwent bacterilysis.

In 1906, Gotschlich isolated organisms closely resembling but not identical with cholera vibrios from pilgrims at El Tor in Sinai. For many years their pathogenicity was disputed, but this is no longer in doubt. In previous editions of this book, we described these organisms as *V. eltor*, but now accept that they do not differ sufficiently from *V. cholerae* to justify a separate specific name; we shall therefore refer to the two organisms as the classical and the El Tor varieties of *V. cholerae*. Other organisms that differ consistently from *V. cholerae* only in O serotype are commonly found in surface waters; they are usually non-pathogenic but sometimes cause diarrhoea in man. Their designation causes real difficulties; they have been referred to as 'non-cholera' or 'non-agglutinating' vibrios. Both terms are misnomers; these vibrios do occasionally cause a cholera-like disease; and they are non-agglutinating only when tested with antiserum against the O antigen of *V. cholerae*. Some authors include them in the species *V. cholerae* (see, for example, Shewan and Véron, 1974), but it would at present be difficult to define such a broadened species, and the change would confuse non-microbiologists. Reluctantly, therefore, we shall use the term 'non-agglutinating' in the abbreviated form (NAG). (For an alternative view, see Report 1980 and Chapter 70.)

The vibrios found in sea-water are usually halophilic but not markedly curved. One of them, *V. parahaemolyticus*, is an important cause of food poisoning associated with the consumption of seafoods (Sakazaki *et al.* 1963). Many other species have been described but we shall discuss only those that have pathogenic properties.

Three other genera will be described in this chapter. (1) *Aeromonas* and *Plesiomonas* are predominantly straight rods that resemble vibrios in cultural and biochemical characters. (2) *Campylobacter* are curved rods but differ from vibrios in being microaerophilic and non-proteolytic and in failing to attack sugars; they include the organisms described by M'Fadyean and Stockman (Report 1913) as a cause of vibrionic abortion in sheep and cattle and originally named *V. fetus* by Smith (1918), and several other closely related organisms from animals and man.

A brief mention will be made of *Bdellovibrio* (Stolp and Petzold 1962), a group of curved rods characterized by predatory action on other bacteria. The genus *Spirillum* will be referred to, but its only pathogenic member is dealt with in Chapter 44. The strictly anaerobic vibrio-like or spiral bacilli found in the human mouth and the bovine rumen belong to the genus *Selenomonas*.

Vibrio

Definition

Short rods, often curved. Motile by a single polar flagellum, which is usually surrounded by a sheath; some species also have lateral flagella. Non-sporing. Gram negative. Aerobic and facultatively anaerobic. Nutritional requirements simple. Ferment glucose with the production of acid but not gas. Indole positive. Catalase formed; nearly all are oxidase positive, and reduce nitrate to nitrite. Active producers of protease, diastase and DNAase. Some species halophilic. Sensitive to the pteridine compound O/129. Mole percentage G+C in DNA 40–45. Found in fresh and sea-water; some species pathogenic for man.

In this genus we include: (1) *non-halophilic vibrios*, *V. cholerae* and the closely related NAG vibrios, and other vibrios with the ability to grow in media without added salt; and (2) *halophilic vibrios*, which cannot grow in these media.

Morphology

The cholera vibrios are typically but not invariably curved rods. In size they vary considerably, from 1–5 μm in length and 0.3–0.6 μm in breadth. They may be long, thin and delicate or short and thick. They are arranged singly or in short chains; when curved they may form s-shaped or semicircular pairs, or spirals. The halophilic vibrios are usually straight or only slightly curved. Sphaeroplasts are often present in culture. Pleomorphic forms are common in old cultures. The organisms are actively motile by a single polar flagellum, which is surrounded by a sheath. Under certain cultural conditions, many of the halophilic vibrios also form unsheathed lateral flagella (Baumann *et al.* 1971). Fimbriae have been described, but are not numerous.

Growth requirements

Growth occurs readily on ordinary media when individual requirements for electrolytes and temperature are met. Vibrios are strongly aerobic; they grow best in the presence of abundant oxygen. Anaerobically, some fail to grow and the majority grow slowly. One

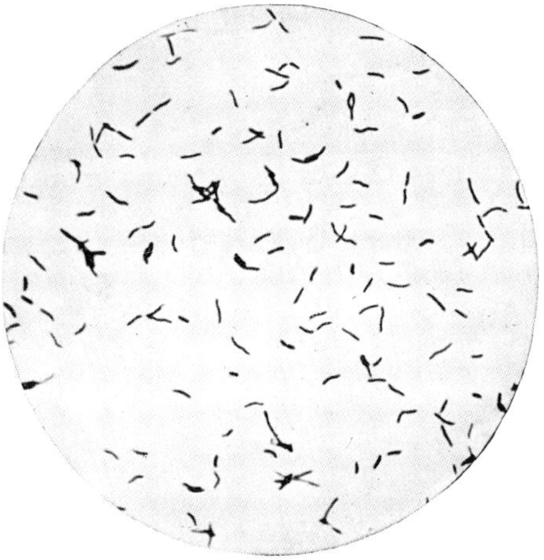

Fig. 27.1 *Vibrio cholerae.* From an agar culture, 24 hr, 37° (× 1000).

of the most characteristic properties of the cholera vibrios is rapid growth in peptone water. Multiplication occurs chiefly at the surface where, after 6–9 hr, a delicate pellicle is formed. There is little turbidity as a rule; the deposit that forms appears to be derived from the surface growth. El Tor strains are said to render broth more turbid, to form a denser surface pellicle and to overgrow classical cholera vibrio strains in mixed culture (Han and Khie 1963, Barua and Mukherjee 1964, Mukherjee and De 1966). Practically all vibrios grow at 30°, but *V. anguillarum* and some other halophilic vibrios grow poorly or not at all at 37°. Among the common species, only the cholera and NAG vibrios, and *V. parahaemolyticus* and *V. alginolyticus* grow at 42°. The cholera vibrios do not grow below 16°.

Vibrios have a high alkali and a low acid tolerance. Growth occurs freely between pH 7.4 and 9.6; the limits are about pH 6.8 and 10.2. Cultures containing a fermentable sugar are sterile in a day or so (Nobechi 1925). The nutritional requirements of vibrios are simple: mineral salts, ammonium ions and usually a single organic source of carbon. Growth on mineral-salts medium with various carbon sources is sometimes helpful in the classification of halophilic vibrios (see Furniss *et al.* 1978).

Halophilism and salt tolerance

Halophilism is defined as the inability to grow on media to which NaCl has not been added, such as an aqueous solution of peptone powder or in cystine lactose electrolyte-deficient (CLED) agar (Furniss *et al.* 1978). It should be noted, however, that the halophilism of *V. anguillarum* is minimal, and that a few strains show slight growth on CLED agar. Non-halophilic vibrios grow on these media, though many, including the cholera vibrios, grow optimally with 3 per cent of added salt. If *salt tolerance* is defined as the ability to grow in peptone water with 3 per cent of added NaCl, all vibrios except *V. albensis* are to some extent tolerant. Nearly all of the halophiles and some cholera vibrios will grow with 6 per cent of added salt. A minority of the halophiles, including some strains of *V. alginolyticus*, tolerate 10 per cent of NaCl, and strains isolated from cured bacon may even tolerate 23 per cent. In addition to NaCl, some of the marine vibrios require other mineral salts (MacLeod 1965).

Cultural characters

Colonies on nutrient agar are typically round, 2–5 mm in diameter, low convex and translucent; but considerable variation occurs, and large and small, flat and domed, and opaque and translucent forms may be seen, even in the same culture. Rugose variants also occur; these are firmly adherent to the medium and almost impossible to emulsify, but may revert to the more usual form.

Numerous *selective media* have been devised for improving the isolation of cholera vibrios from faeces and water. These will be discussed fully when considering the bacteriological diagnosis of the disease (Chapter 70). Most of them make use of such properties of the cholera vibrio as its alkali tolerance, its ability to ferment sucrose and starch, and its growth in the presence of tellurite, thiosulphate and bile salts. Mention must be made, however, of the colonial appearances of the more important species of vibrio on the widely used thiosulphate citrate bile-salt (TCBS) agar of Kobayashi *et al.* (1963). Nearly all vibrios of medical importance, but few other organisms with which they could be confused, form colonies at least 2 mm in diameter after incubation for 24 hr at 37° on this medium. Sucrose-fermenting organisms, such as the cholera vibrios, most of the NAG vibrios, *V. alginolyticus* and *V. anguillarum* give yellow colonies; *V. parahaemolyticus*, which does not ferment sucrose, generally gives blue colonies. However, not all of the other vibrios grow under these conditions.

On MacConkey's bile-salt agar, cholera vibrios form colourless colonies at first; these become pinkish-red on further incubation. Many other vibrios grow poorly on this medium. Not all strains, even of the cholera vibrio, grow on deoxycholate medium (Carpenter 1966); and SS medium is said to be completely inhibitory (Feeley 1962). Some halophiles give spreading growth on agar.

Action on blood

On horse-blood agar, the colonies of most cholera vibrios are surrounded by a 2–3 mm zone of clearing, but this is not true haemolysis. However, the El Tor vibrios, but not the classical cholera vibrios, form dialysable, heat-labile haemolysin for sheep and goat red cells. This difference has long been used to distinguish between the biotypes, but the results obtained depend upon the technique used. The method of Greig (1914), especially as modified by Feeley and Pittman (1963), gave fairly reliable results with El Tor strains isolated before 1963. After this date, strains that gave negative or feebly positive results predominated, and technical modifications were introduced to restore the specificity of the test (Sakazaki *et al.* 1971). It seems, therefore, that haemolysin production is an uncertain means of separating the biotypes.

Haemagglutination of chicken red cells by El Tor but not by classical strains (Finkelstein and Mukerjee 1963) is another means of distinguishing between the biotypes. Sheep, rabbit and human, but not guinea-pig red cells, give similar results. The haemagglutinin is mannitol sensitive (Barua and Chatterjee 1964) but, according to these authors, is not associated with the fimbriae.

The NAG vibrios vary in their production of haemolysins and haemagglutinins.

Many halophilic vibrios are haemolytic on ordinary blood agar, but only certain strains of *V. parahaemolyticus* show haemolysis on a special blood-agar medium containing mannitol, 7 per cent NaCl and rabbit blood (see Sakazaki *et al.* 1968); this is known as the **'Kanagawa phenomenon'** and is attributable to the formation of a heat-stable haemolysin (see p. 144).

Resistance

None of the vibrios forms spores. Their resistance to heat and disinfectants is low, and they are easily destroyed by drying. They are killed by heat at 55° in 15 min or less (Kitasato 1889) and by 0.5 per cent phenol in a few minutes. They are said to be lysed by thymol (Bhaskaran 1957). In gastric juice of normal acidity they are killed in a few minutes, but they may survive for 24 hr in achlorhydric gastric juice (Napier and Gupta 1942). Dried on cover slips they perish in about 3 hr. In faeces they are dead within a day or so at high atmospheric temperatures, but may survive for a few weeks near freezing point. They survive in clean tap water for up to 30 days but perish in 24 hr in cesspool water. (For further information about survival in the environment see Pollitzer 1959, Felsenfeld 1965, Miyaki *et al.* 1967 and Chapter 70.)

All vibrios are sensitive to compound O/129 (2,4 diamino 6,7 diisopropyl pteridine; Shewan and Hodgkiss 1954). In this respect they differ from *Pseudomonas* and most of the other organisms with which they might be confused, except *Plesiomonas*, *Photobacterium*, group-F organisms (p. 145), and possibly *Actinobacillus* (see Chapter 22). However, not all vibrios are equally sensitive (see Furniss *et al.* 1978 and Table 27.1); some, including all the non-halophilic vibrios, are very sensitive and show zones of inhibition around a 10-μg disk (minimal inhibitory concentration: 10 μg/ml); others, including most of the halophilic vibrios, are somewhat less sensitive and show zones only around a 150-μg disk (minimal inhibitory concentration: 10–50 μg/ml).

Classical cholera vibrios are more sensitive than El Tor vibrios to polymyxin B (Han and Khie 1963), but the results of disk-diffusion tests depend upon the medium used. Gangarosa and his colleagues (1967) recommend Difco Mueller-Hinton Medium, on which *V. cholerae* regularly gives a zone of inhibition 6–9 mm

Table 27.1 Usual cultural and biochemical characters of the vibrios (after Furniss *et al.* 1978)

	Non-halophilic				Halophilic		
Character	*V. cholerae*	Non-agglutinating vibrios*	*V. albensis*	*V. metchnikovi*	*V. parahaemolyticus*†	*V. alginolyticus*	*V. anguillarum*
Growth on CLED agar	+	+	+	+	–	–	–‡
Growth with NaCl§:							
3 per cent	+	+	–	+	+	+	+
6 per cent	v	v	–	+	+	+	+
10 per cent	–	–	–	–	–	v	–
Growth at:							
37°	+	+	+	+	+	+	v
42°	+	+	v	–	+	+	–
Growth on TCBS agar‖	Y	Y(G)	NG	Y(NG)	G	Y	Y
Sensitivity to O/129:							
150-μg disk	S	S	S	S	S	S	S
10-μg disk	S	S	S	S	R	R	S
Acid from:							
mannose	+	v	–	v	+	+	+
sucrose	+	v	+	+	–	+	+
arabinose	–	–	–	–	v	–	v
VP reaction	v	v	–	+	–	+	+
Reduction of NO_3	+	+	+	–	+	+	+
Oxidase reaction	+	+	+	–	+	+	+
Attack on:							
arginine	–	–	–	+	–	–	+
lysine	+	+	+	v	+	+	–
ornithine	+	+	+	–	+	+	–

v = Some positive, some negative.
* For *V. mimicus* see p. 146.
† Acid from lactose; *V. vulnificus*, no acid from lactose.
‡ A few show scanty growth.
§ In peptone water with the stated salt concentration.
‖ At 24 hr, 37°: Y = yellow colonies, G = green colonies, NG = no growth.

in width around a 50-μg disk and El Tor strains are not inhibited.

Until early in the 1970s, cholera vibrios were sensitive to many antibiotics, notably tetracycline, chloramphenicol and ampicillin. Sporadic isolations of resistant strains were then reported. More recently, strains with plasmid-mediated resistance to a wide range of antimicrobial agents, including ampicillin, tetracycline, aminoglycosides, sulphonamides, and sometimes chloramphenicol and trimethoprim, have become locally prevalent in Tanzania and Bangladesh (Mhalu *et al.* 1979, Threlfall *et al.* 1980). Gentamicin-resistant strains have since been reported (Threlfall and Rowe 1982).

Most strains of *V. parahaemolyticus* and *V. alginolyticus* are sensitive to chloramphenicol and tetracycline, but ampicillin resistance—and β-lactamase production—are frequent; a few strains are resistant to all three antibiotics (Joseph *et al.* 1978).

Biochemical reactions

The cholera vibrios form acid without gas in glucose, mannose, maltose, sucrose and mannitol in 1–2 days and in lactose in 2–6 days, but do not ferment arabinose. Heiberg (1935, 1936) placed great emphasis on tests with mannose, sucrose and arabinose for the classification of vibrios. He defined six groups according to these reactions, and found that all the cholera vibrios fell into his group I (mannose +, sucrose +, arabinose −), but so do other vibrios of human and marine origin (see Table 27.1). These include most of the NAG vibrios, the rest of which appear in group II (mannose −, sucrose +, arabinose −) or group V (mannose +, sucrose −, arabinose −). Few vibrios, and none belonging to species listed in Table 27.1, attack inositol.

Reactions in the Voges-Proskauer test are variable; El Tor strains tend to give a positive and classical cholera strains a negative result, but the difference is more quantitative than qualitative, and depends somewhat on the method used. (For reactions of the other species see Table 27.1.) All of the common species except *V. metchnikovi* reduce nitrate to nitrite. Many vibrios give the *cholera red reaction*—the development of a red colour when concentrated sulphuric acid is added to an overnight broth culture. This is simply a consequence of nitrate reduction and the production of indole, and is not specific for cholera vibrios. The only oxidase-negative vibrio is *V. metchnikovi*.

Three main patterns of attack on amino acids are recognized: (1) *V. cholerae*, NAG vibrios, *V. parahaemolyticus* and *V. alginolyticus* decarboxylate lysine and ornithine but do not attack arginine; (2) *V. metchnikovi* and some of the halophiles form arginine dihydrolase; and (3) other halophiles (not shown in Table 27.1) are without action on any of these amino acids.

Nearly all of the vibrios form a number of exoenzymes including amylase, gelatinase, DNAase, lecithinase, and a lipase for Tween 80. Burnet and Stone (1947) described the formation of a mucinase by the cholera vibrio that caused desquamation of the intestinal mucosa of the guinea-pig. This enzyme is different from the receptor-destroying enzyme (RDE), which, according to Freter (1955), is not a mucinase and is not antigenic. The mucinase causes panagglutinability of red cells (Felsenfeld 1967). A few of the halophiles form alginase, and a few others are luminescent.

(For further information about the biochemical characters of vibrios, see Baumann *et al.* 1971, Shewan and Véron 1974, Reichelt *et al.* 1976, and Furniss *et al.* 1978.)

Phage-typing

Mukerjee (1961, 1962) developed a phage-typing system for classical strains of cholera vibrios in which five patterns of lysis were recognized by means of three phages (nos. I–III), allowing five types (1–5) to be defined. A fourth phage (no. IV) lysed all the classical but none of the El Tor strains and is now used to aid the separation of these biotypes. Soon afterwards, he described a separate typing system for El Tor strains, but later modified this (Basu and Mukerjee 1968). In the revised system, four phages (nos 1–4), which are best designated by Arabic numbers to distinguish them from the phages used to type classical strains, were used to define six types, and a fifth phage (no. 5) proved valuable as a marker in that it lysed all El Tor but no classical strains. The El Tor typing system was later modified by Lee and Furniss (1976) and has since been extended (Furniss *et al.* 1978). Fifteen distinct phage-typing patterns could be recognized by means of 10 phages. These workers observed, however, that some of the strains that were lysed by the El Tor phages were also lysed by phage IV, and that a small percentage of NAG vibrios were sensitive to phage IV or to phage 5. Tayeka and his colleagues (1981) describe a phage (FK) with a spectrum of lysis rather similar to that of phage IV that is said to distinguish more accurately between classical and El Tor strains. At present there is no phage-typing system for NAG vibrios, but a number of them belong to types in Basu and Mukerjee's (1968) system (Furniss *et al.* 1978).

In an earlier attempt to produce a phage-typing system for cholera vibrios, Nicolle and his colleagues (1960, 1962, 1971) studied the lysogenic state of the organisms. Tayeka and Shimadori (1963) observed that El Tor strains isolated in the current pandemic, but not strains isolated earlier, carried phages with a narrow and characteristic host range. No comprehensive lysogenicity-typing system has yet been developed.

Vibriocine typing

Wahba (1965) observed that El Tor and classical cholera vibrios produced similar bacteriocines and showed similar patterns of susceptibility to bacteriocines. Attempts to produce a vibriocine-typing system have been hindered by difficulties in obtaining reproducible results (see Mukerjee and Tayeka 1974). Mitra and co-workers (1980) describe a single vibriocine-typing system for *V. cholerae* and the NAG vibrios.

Antigenic structure

The cholera and NAG vibrios

Balteanu (1926) described heat-labile H and heat-stable O antigens in vibrios, and Gardner and Venkatraman (1935) established the broad outlines of their distribution among the cholera vibrios (see Fig. 27.2). They showed that the classical and El Tor cholera strains all belonged to the same O group (group I) and that a number of other strains with rather similar biochemical characters that we now call NAG vibrios had different O antigens but the same H antigen as the cholera vibrios.

There are, however, some difficulties in detecting the H antigen; agglutination reactions with live suspensions—particularly of recently isolated strains—are often weak and variable. This may be attributed to masking by the flagellar sheath (Bhattacharyya 1977), which contains O determinants. Treatment of suspensions with 1.5 per cent phenol in saline, which increases H but reduces O sensitivity, is recommended for H-agglutination tests (Sil and Bhattacharyya 1979). These workers advocate the use of the H-agglutination test as a means of defining the extent of the NAG-vibrio group; the possibility of cross-reactions with flagellar components of less closely related vibrios has not been entirely excluded (but see Pastoris *et al.* 1980).

Gardner and Venkatraman (1935) found at least six O groups among the NAG vibrios, and more than 70 have now been described (Sakazaki and Shimada 1977). However, the members of each O group are not always antigenically identical; in addition to the O factors that characterize the group, there are others of

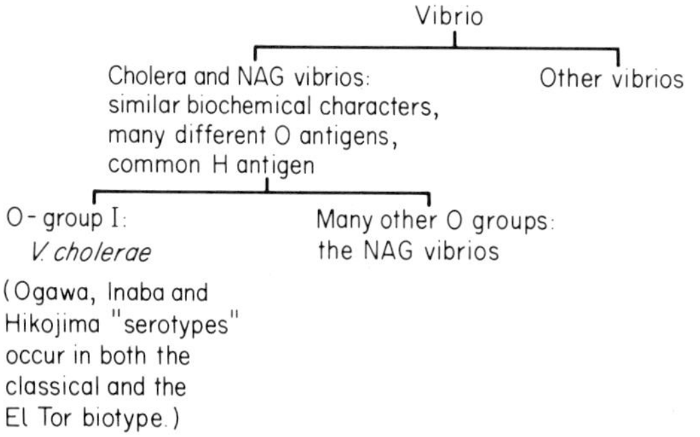

Fig. 27.2 Antigenic classification of the cholera and NAG vibrios.

narrower specificity. Thus, in O-group I, Nobechi (1933) recognized three serotypes: Ogawa, Inaba and Hikojima. Their antigenic constitution, according to White (1937), Burrows and his colleagues (1946), Kauffmann (1950) and others, is considered to be as follows: Ogawa, AB; Inaba, AC; Hikojima, AB(C). However, the antigenic composition of strains is subject to variation *in vitro* and *in vivo*, and it would be unwise to look upon them as stable serotypes. It should be emphasized that all three serological variants occur among both classical and El Tor cholera vibrios.

Early studies of the chemical composition of the cell walls of cholera vibrios were summarized in the last edition of this book. The group and type specificities of the O antigens of smooth strains are, as expected, determined by the polysaccharide moiety of the cell-wall lipopolysaccharide. Serological roughness is not closely related to colonial form or to autoagglutinability. Its development proceeds through successive stages, with a progressive broadening of antigenic specificity (White 1935*a*, 1935*b*, 1940*b*; see also Shimada and Sakazaki, 1977). Most rough strains cannot be serogrouped, but some are agglutinated by both O1 and R antisera. White (1940*a*) described a 'rugose' hapten he believed to be derived from an intercellular secretion of rugose variants. Vibriocidal antibodies were found by Neoh and Rowley (1970) to be directed partly against the polysaccharide determinants of the lipopolysaccharide and partly against cell-wall proteins. One protein antigen common to all vibrios (Okada *et al.* 1977) and another cross-reacting with a cell-wall protein of *Ps. aeruginosa* (Hirao and Homma 1978) have been described.

The halophilic vibrios

Numerous serotypes of *V. parahaemolyticus* have been recognized. Miwatani and co-workers (1969) identified 12 O types and 59 distinct K antigens, but many strains remain untypable. All strains in the species have identical H antigens. It appears that the polar and the lateral flagella differ antigenically. Several workers who used agglutination reactions concluded that the flagellar antigens of the halophilic vibrios were distinct from those of the cholera and NAG vibrios. Subsequently, Shinoda and his colleagues (1976), who studied the properties of dissociated flagellins, concluded that the polar flagella of *V. parahaemolyticus* were antigenically similar to those of a number of other halophilic vibrios and related to those of *V. cholerae* and the NAG vibrios; but that the flagellins of the lateral flagella of *V. parahaemolyticus* and *V. alginolyticus*, though antigenically similar, were different from those of the other halophilic vibrios.

Pathogenicity and toxin formation

In man, the cholera vibrios and some of the NAG vibrios cause a painless diarrhoea of varying severity, the spectrum of which extends from a mild and transient illness to typical Asiatic cholera with severe dehydration and death (Chapter 70). *V. parahaemolyticus* causes diarrhoea, usually with abdominal pain, but this is seldom fatal (Chapter 72). Metchnikoff's vibrio is responsible for a choleraic disease of chickens (Gamaléia 1888*a*) in which the organism may be present in the blood stream as well as in the gut contents. This disease must be distinguished from true chicken cholera, which is due to a *Pasteurella* (Chapter 55). *V. parahaemolyticus* causes food-borne diarrhoea in man (Chapter 72). A closely related organism (Reichelt *et al.* 1976), for which the name *V. vulnificus* has been proposed, causes septic and invasive infections in man (Chapter 61), including wound infections, cellulitis and myositis in persons recently exposed to seawater, and a highly fatal septicaemia in patients with liver insufficiency. *V. vulnificus* differs from *V. parahaemolyticus* principally in its ability to ferment lactose (Hollis *et al.* 1976). *V. alginolyticus* is occasionally responsible for local septic infections, notably of the external ear (Chapter 61). *V. anguillarum* is a common pathogen of eels and other fish.

Non-halophilic vibrios

The cholera and NAG vibrios

Experimental pathogenicity Koch (1886) succeeded in infecting guinea-pigs with cholera vibrios by the oral route, but only after neutralizing the gastric acidity with sodium bicarbonate and inhibiting intestinal motility by the intraperitoneal injection of opium. A cholera-like disease develops in newborn rabbits (Metchnikoff 1894, Sanarelli 1921, Dutta *et al.* 1959) and in newborn mice (Ujiye *et al.* 1968) when cholera vibrios are given by the mouth, and in adult rabbits and dogs when they are injected into the lumen of the duodenum or ileum (Nicati and Rietsch 1884). Injection into ligated sections of small intestine ('gut loops') in adult rabbits (De and Chatterjee 1953) leads to distension of the loops with fluid and hyperaemia of the gut wall.

Toxin production Modifications of these experimental models have been used to detect and measure toxin production. (1) In adult rabbits, culture fluids and control materials were injected into a series of ligated sections of the gut (De 1959) and the volume of fluid per cm of gut was measured (Burrows and Musteikis 1966). (2) In infant rabbits, material was injected into the lumen of the gut (Dutta and Habbu 1955) or given orally (Dutta *et al.* 1959). (3) In the dog, it was given by intraluminal injection or into a gut loop or externally draining fistula (Carpenter *et al.* 1969, Sack and Carpenter 1969).

By these means it was established that cell-free preparations of virulent cultures caused outpouring of fluid into the gut lumen, and that the agent responsible for this was destroyed either by heat for 30 min at 56° or by Pronase, but not by trypsin. This heat-labile toxin is the *enterotoxin* of the cholera vibrio. It can be distinguished from other toxic materials found in culture fluids (Burrows 1968, Burrows and Kaur 1974), which include (1) the cell-wall lipopolysaccharide, which is heat-stable and lethal to the mouse but does not cause fluid outpouring into the gut, and (2) a heat-stable, diffusible agent which inhibits active sodium transport in frog skin, at one time considered to be a possible cause of the symptoms of cholera but now thought to be a laboratory artefact. The haemolysin formed by some El Tor strains is thermolabile, thus differing from the Kanagawa-type haemolysin of food-poisoning strains of *V. parahaemolyticus* (p. 144). It is cytotoxic and kills experimental animals but is thought not to play any part in the pathogenesis of cholera (Honda and Finkelstein 1979).

The enterotoxin is formed intracellularly and diffuses out during exponential growth. The amount produced in culture, and the optimal conditions for its production, vary from strain to strain (see Burrows and Kaur 1974). Enterotoxin is antigenic, and antibody affords both active and passive protection against challenge with toxin in the gut lumen. The enterotoxins of all cholera and NAG vibrios appear to be identical in antigenic specificity (Holmgren *et al.* 1971, Burrows and Kaur 1974). Formalin-treated enterotoxin is antigenic; it is initially non-toxic but tends to revert to toxicity. However, non-toxic but antigenic material, termed *choleragenoid*, can be isolated from preparations of enterotoxin (Finkelstein and LoSpalluto 1969, Kaur *et al.* 1969); under certain circumstances, cultures can be induced to form this product only.

Permeability factor Craig (1965) made the observation that cholera stools and enterotoxic culture fluids, when injected intradermally into rabbits, gave rise to an indurated erythematous lesion. The agent responsible for this, termed permeability factor, is purified along with enterotoxin. The rabbit intradermal test, and the rather similar test for oedema of the rat foot (Finkelstein *et al.* 1969), have been widely used for the assay of enterotoxin, but it is still unsettled whether permeability factor is separable from enterotoxin (see Burrows and Kaur 1974).

Action of enterotoxin In natural and experimental cholera, only minimal histological changes can be seen in the gut epithelium. Enterotoxin becomes rapidly and irreversibly bound to the epithelium; the exudation of fluid begins within 30 min of the initial contact between toxin and epithelium and continues for 12–24 hr. The fluid is isotonic with the blood and contains little protein (Pierce *et al.* 1971). The failure of large molecules to pass through the gut wall affords evidence that the permeability factor is not responsible for the exudation of cholera fluid (see Carpenter *et al.* 1974). It is now widely accepted that the toxin causes the exudation of fluid by activating adenylcyclase in the gut epithelium, thus causing the accumulation of cyclic adenosine monophosphate (cAMP), which provides the energy for the transport of electrolytes across the epithelial membrane (Hendrix 1971). The electrolyte changes induced by the toxin are identical with those caused by cAMP or by substances that lead to the accumulation of endogenous cAMP, and gut preparations that have responded maximally to the toxin are unaffected by these substances (Al-Awqati *et al.* 1972, Field *et al.* 1972). Also, the toxin affects certain functions of other tissues that are known to be mediated by cAMP, for example, activation of lipolysis in rat testicular tissue (Vaughan *et al.* 1970), elongation of Chinese hamster ovarian cells (Donta *et al.* 1973), and histological changes in adrenal-tumour cells (Donta *et al.* 1976), all of which can be used for the assay of toxin, as can chemical estimation of cAMP in tumour cells that have been treated with toxin preparations (Ruch *et al.* 1978).

Chemical examination of purified enterotoxin reveals that it is a protein with a molecular weight of 80 000–90 000 composed of two major polypeptide components: A is responsible for toxicity but is inactive in the absence of the second component, B (Holmgren *et al.* 1974), which binds the A component to the monosialoganglioside of the membrane of the target cell (van Heyningen 1974). The A component comprises two separate portions, the α polypeptide (molecular weight 20 000–24 000) which is biologically active, and a smaller γ polypeptide of unknown function; the B component consists of four to six subunits each of molecular weight *ca* 12 000 (see Holmgren *et al.* 1978 for further references).

The properties of the cholera enterotoxin closely resemble those of the heat-labile enterotoxin of *Esch. coli.* Both activate adenylcyclase, and both cause changes in the permeability of skin capillaries. In precipitation tests they show lines of partial identity, and under certain experimental conditions cross-neutralization of toxic action can be demonstrated (Gyles and Barnum 1969, Evans *et al.* 1973, Pierce 1977, Nalin and McLaughlin 1978). Both the A and B subunits of cholera toxin cross-react with the *Esch. coli* heat-labile toxin, but both toxins appear to possess unshared antigenic determinants (Holmgren and Svennerholm 1979). A heat-stable enterotoxin similar to that of *Esch. coli* has not been demonstrated in the cholera vibrio.

Pathogenesis of cholera To reach the site of infection, the vibrios must first pass the acid barrier of the stomach; in human volunteers, an oral dose of 10^{11} organisms only occasionally causes mild diarrhoea, but if sodium bicarbonate is first given, a dose of 10^6 organisms leads to a moderately severe attack of cholera in about one-quarter of the subjects (Hornick *et al.* 1971). Next, the organism must multiply, and finally it must produce toxin. Not much is known about the conditions under which toxin is formed in the gut. It appears, however that the vibrios become adherent to the epithelium (Patniak and Ghosh 1966, Nelson *et al.* 1976), and that this is essential for the production of disease. It has been suggested that cholera vibrios possess a specific adhesive factor (Finkelstein *et al.* 1978), and also that they are chemotactically attracted to the epithelial surface (Allweiss *et al.* 1977). There appears to be an association between motility and virulence (Guentzel and Berry 1975, Jones *et al.* 1976, Srivastava *et al.* 1980), and between loss of motility and failure to adhere to the epithelium. Very little toxin can be detected in the stools or jejunal aspirates of cholera patients (Aziz and Mosley 1972); there is some evidence that toxin is liberated only in close proximity to the epithelium or after adherence to it (Peterson *et al.* 1972, Srivastava *et al.* 1980).

Sinha and Srivastava (1979) describe two plasmids (P and V) which, when introduced by conjugation into a fully virulent strain, suppress enterotoxin production but do not affect the ability of the strain to multiply in the gut or to adhere to its epithelium. (For a fuller discussion of the pathogenesis of cholera, see Ch. 70.)

(For general reviews of the action of the enterotoxins of *V. cholerae* and *Esch. coli*, see Richards and Douglas 1978, and Field 1979.)

Vibrio metchnikovi

Metchnikoff's vibrio is more invasive than the cholera vibrio. Even after a small dose given intraperitoneally to guinea-pigs, the vibrios can be recovered from the heart's blood. It is fatal to guinea-pigs even when given subcutaneously; under these conditions the cholera vibrio gives rise merely to a local abscess. Both guinea-pigs and chickens can be infected by feeding with *V. metchnikovi*. Moreover this organism is pathogenic to pigeons on intramuscular injection; the cholera vibrio is not, except occasionally in large doses (Wherry 1905). Pigeons given a large intramuscular dose of *V. metchnikovi* die in about 8 hr with general septicaemia (Metchnikoff 1893). Intratracheal injection appears to be even more fatal, since not only guinea-pigs, pigeons and fowls, but also rabbits may be infected by this means (Gamaléia 1888*b*).

The halophilic vibrios

Most but not all of the strains of *V. parahaemolyticus* that cause food poisoning form the Kanagawa-type haemolysin (p. 140) but strains from sea-water seldom do this. The haemolysin is a heat-stable, trypsin-sensitive protein that is lethal for mice and guinea-pigs but causes diarrhoea only when given in very large dose (Obara 1971, Miwatani and Takeda 1976, Miyamoto *et al.* 1980). Whether this agent is responsible for diarrhoea in man is uncertain; culture filtrates from food-poisoning strains cause fluid accumulation when injected into ligated gut loops in rabbits (Sakazaki *et al.* 1974). According to Honda and co-workers

(1976) they cause morphological changes in Chinese-hamster ovarian cells but this effect is not attributable to the Kanagawa agent. Sochard and Colwell (1977) studied a strain that did not produce this haemolysin and obtained from it a heat-labile protein that caused diarrhoea in mice; in rabbits that substance increased skin permeability but did not cause the accumulation of fluid in ligated gut loops. Food-poisoning strains are said to adhere strongly *in vitro* to human fetal intestinal-epithelial cells irrespective of their Kanagawa reaction; strains from sea-water generally do not do this (Hackney *et al.* 1980).

Kanagawa-positive strains of *V. parahaemolyticus* given intraperitoneally or intravenously to mice in doses of *ca* 10^7 organisms give rise to a fatal infection. According to Poole and Oliver (1978), *V. vulnificus* (p. 143) differs from *V. parahaemolyticus* in being rapidly lethal for mice when given subcutaneously.

Classification

Relation to other genera

Before discussing the internal classification of the genus *Vibrio*, its relation to other groups of gram-negative organisms must be considered, particularly *Aeromonas*, *Plesiomonas*, *Campylobacter* and *Pseudomonas* (see Table 27.2).

Aeromonas (p. 146) is closely related to *Vibrio*, but differs from it in its lack of sensitivity to compound O/129, its much higher G+C content, and its frequent but by no means invariable production of gas from glucose.

Plesiomonas (p. 147) resembles *Vibrio* even more closely in that it is sensitive to compound O/129, but it regularly ferments inositol, a character rare among vibrios, and it does not produce exoenzymes such as protease, amylase and DNAase. Both *Aeromonas* and *Plesiomonas* are straight rods that do not exhibit halophilism.

Group-F organisms Furniss and his colleagues (Furniss *et al.* 1977, Lee *et al.* 1978, 1981) have described an organism that has characters intermediate between those of *Vibrio* and *Aeromonas*. It is widely distributed in coastal waters.

Group-F organisms form yellow colonies on TCBS agar; they are marginally halophilic, and grow in 6 per cent and sometimes in 10 per cent NaCl, and at 37° but not at 42°. They are sensitive to compound O/129 when tested with a 150-μg but not with a 10-μg disk. They do not swarm on agar or luminesce, are oxidase positive, produce indole, reduce nitrate, acidify mannose, sucrose and arabinose, and form arginine dihydrolase but do not attack lysine or ornithine. However, they sometimes form gas from glucose and the G+C content of the DNA is 51–52 moles per cent. According to Lee and his colleagues (1981), there are two biotypes of group-F vibrios. Biotype I organisms are nearly always anaerogenic, usually hydrolyse aesculin, and may be isolated from the faeces of persons suffering from diarrhoea; biotype II organisms usually form gas from glucose, never hydrolyse aesculin, and are rarely isolated from human faeces. The name *V. fluviatilis* has been proposed for the group-F organisms; if this is accepted, the only easily applied means of distinguishing *Vibrio* from *Aeromonas* will be a susceptibility test with a disk containing 150 μg of compound O/129.

Photobacterium (Hendrie and Shewan 1974) is a genus of polar-flagellated marine coccobacilli; many are luminescent and halophilic, and most form gas from glucose. They are sensitive to compound O/129; the G+C content of the DNA is 39–42 moles per cent.

Pseudomonas (Chapter 31) does not ferment glucose, is resistant to compound O/129, does not form indole, and has a much higher G+C content in its DNA than *Vibrio*.

Campylobacter (p. 148), though morphologically a vibrio, differs from *Vibrio* in numerous respects: it is microaerophilic, does not attack any sugars, form indole or produce hydrolytic exoenzymes, and has a much lower G+C content in its DNA.

(For *Spirillum* and *Bdellovibrio* see p. 151.)

The non-halophilic vibrios

The cholera vibrios Our decision to consider the classical and El Tor cholera vibrios as biotypes of *V. cholerae* is in conformity with the views of Shewan and Véron (1974). The characteristics used to distinguish

Table 27.2 Usual characters of *Vibrio* and other genera of polar-flagellate, oxidase-positive, gram-negative bacilli*

Character	*Vibrio*	*Aeromonas*	*Plesiomonas*	*Pseudomonas*	*Campylobacter*
Gaseous requirement	Aerobic	Aerobic	Aerobic	Aerobic	Microaerophilic
Number of flagella per pole	1†	1†	>1	1 or >1	1
Attack on glucose	F	F	F	O	None
Gas from glucose	–	v	–	–	...
Acid from: mannitol	+	+	–	–	...
inositol	–	–	+	–	...
Sensitivity to O/129‡	+	–	+	–	–
Hydrolysis of gelatin, starch, etc.	+	+	–	v	–
Indole production	+	+§	+	–	–
Moles per cent G+C	40–45	57–63	51	58–70	30–35

F = fermentative; O = oxidative; v = some positive, some negative.
* For group-F organisms, intermediate between *Vibrio* and *Aeromonas*, see this page (above).
† Some halophilic vibrios and *Aeromonas* strains also have lateral flagella.
‡ Zone of inhibition around a 150-μg disk. § Except *A. salmonicida*.

Table 27.3 Distinguishing features of the classical and El Tor cholera vibrios

Feature	Classical	El Tor
Haemolysis of sheep RBC	−	+
Haemagglutination of chick RBC	−	+
Susceptibility to polymyxin B	S	R
VP reaction	usually weak or −	usually strong
Lysis by phage IV	+	−
Lysis by phage 5	−	+

S = sensitive; R = resistant.

between the biotypes are summarized in Table 27.3. However, the differences are not clear cut, and the results of the various tests are often discrepant. Indeed, Feeley (1965) suggested that up to five biotypes could be recognized by the use of the same tests. Despite the close resemblance between the two biotypes in the laboratory, they behave differently in the human host. The El Tor strains that are now predominant in the world differ from the classical strains in producing a greater proportion of mild and of symptomless infections, in surviving longer in the environment, and in being more easily spread from person to person and in food (see Chapter 70).

The NAG vibrios As shown in Table 27.1, it is very difficult if not impossible to distinguish most of the NAG vibrios from *V. cholerae* by means of cultural and biochemical tests (see also Sakazaki *et al.* 1967); the two groups also show close DNA homology (Citarella and Colwell 1970). The case for including the NAG vibrios in the species *V. cholerae* is therefore very strong and may have to be accepted eventually. It would however be unwise to do this until clear limits can be set to the enlarged species in terms of phenetic characters that can be detected by means of readily applicable cultural and biochemical tests. Although NAG vibrios are sometimes responsible for sporadic diarrhoeal infections and occasionally for local epidemics, it is still of great practical importance to the epidemiologist not to confuse them with 'true' cholera vibrios. We therefore adhere to our decision to distinguish *V. cholerae* from the NAG vibrios by its possession of the O1 antigen.

A minority of NAG vibrios do show biochemical differences from typical *V. cholerae* strains. A number fail to ferment mannitol and a few do not attack sucrose. The sucrose non-fermenting strains give a negative Voges-Proskauer reaction and do not attack starch (Furniss *et al.* 1978). According to Davis and her colleagues (1981), sucrose-negative strains differ significantly from *V. cholerae* and the rest of the NAG vibrios in DNA base-sequence and should be placed in a separate species, *V. mimicus*. Characters that help in distinguishing it from the other NAG vibrios are inability to attack corn oil and ferment tartrate in Jordan's medium, and sensitivity to polymyxins. The O antigens of *V. mimicus* do not include O1; a small proportion of strains form a heat-labile enterotoxin.

Vibrio metchnikovi Gamaléia (1888*a*) isolated this organism from the blood and intestinal contents of chickens dying of a cholera-like disease and named it *V. metchnikovi*. When injected into experimental animals it shows considerable powers of invasion. Lee and his colleagues (1978) re-examined early cultures given this name, and others labelled *V. proteus* (see Finkler and Prior 1884), together with a number of recent isolates from water. They were all oxidase negative and failed to reduce nitrate; but in other respects they were non-halophilic vibrios and were sensitive to compound O/129. Their other characters are given in Table 27.1. They appeared to form a single species, *V. metchnikovi*, which Lee and his colleagues (1978) re-defined.

Vibrio albensis Dunbar (1893) isolated this organism from Elbe water, and Lehmann and Neumann (1896) gave it its present name. The one early strain available for re-examination is a vibrio with distinctive characters. These include luminescence, complete lack of tolerance for NaCl, and failure to grow on TCBS medium (see Table 27.1, Shewan and Véron 1974, and Furniss *et al.* 1978).

The halophilic vibrios

At least nine species of halophilic vibrios other than the three described in this chapter have been recognized. The suggestion has been made that the halophilic vibrios should be placed in a separate genus *Beneckea*, because they form lateral as well as polar flagella (Baumann *et al.* 1971, Reichelt *et al.* 1976).

V. parahaemolyticus forms a fairly distinct species with the characters shown in Table 27.1. *V. vulnificus* can be distinguished from it by its failure to ferment lactose. *V. alginolyticus*—perhaps the most widely distributed halophilic vibrio in sea water—differs from *V. parahaemolyticus* in fermenting sucrose, and in consequence forming yellow colonies on TCBS agar, and in giving a positive Voges-Proskauer reaction. *V. anguillarum* also forms yellow colonies on TCBS agar; it is only marginally halophilic but salt tolerant. Many strains will not grow at 37°. It spreads readily on the surface of agar.

(For further information about the halophilic vibrios see Baumann *et al.* 1971, Reichelt *et al.* 1976, Furniss *et al.* 1978, Love *et al.* 1981, and Hickman *et al.* 1982.)

Aeromonas

Definition *Aeromonas*, Kluyver and van Niel 1936. Short, gram-negative rods. When motile have a single polar flagellum and sometimes also lateral flagella. Non-sporing and not capsulated. Aerobic and facultatively anaerobic. Ferment carbohydrates with the production of acid and gas, or occasionally of acid only. Oxidase and catalase positive.

Active producers of protease, diastase, lipase, lecithinase and DNAase. Not halophilic. Resistant to the vibriostatic compound O/129. Form arginine dihydrolase. Commonly found in water; cause disease in cold-blooded animals and occasionally in man. Mole percentage G+C 57–63.

We shall describe one motile species, *A. hydrophila*, and one non-motile species, *A. salmonicida*.

Aeromonas hydrophila

This is a straight rod (1–4 μm × 0.6 μm) with rounded ends, which occurs singly or in pairs and is motile by means of a single polar flagellum. Leifson and Hugh (1953), however, saw some lateral flagella in young cultures. All strains grow well at 30°, most at 37°, and a few at 37–41°. Growth occurs slowly at 5°. Colonies on plain agar after 24 hours' incubation are 1–3 mm in diameter, circular, smooth, convex, whitish and translucent, becoming light beige on further incubation. On blood agar there is usually a wide zone of β-haemolysis after 1 day, and the growth becomes dark green after 2–3 days. The biochemical reactions are most characteristic at 30°. Acid, and generally gas, are formed in glucose, maltose, mannitol and trehalose, and usually also in sorbitol, arabinose, sucrose and salicin. The production of acid, nearly always with gas, from starch in 24 hr is a characteristic feature. Lactose may be acidified later. Gas production from sugars is irregular or absent at 37°. Aesculin is hydrolysed. All strains are indole positive, and most are Voges-Proskauer positive. They form abundant H_2S in cysteine broth and are usually KCN positive. Most but not all are citrate positive, and all produce ammonia from arginine. Some are gluconate positive but all are malonate negative. Nitrate is reduced to nitrite. Strains that produce good β-haemolysis in blood agar form a powerful soluble haemolysin. Starch and casein are hydrolysed, gelatin is liquefied, and DNAase is formed.

Variable reactions thus include haemolysis, gas formation, and the KCN and VP reactions. Also, some strains but not others form butanediol dehydrogenase (Schubert 1964), hydrolyse aesculin and form elastin (Popoff and Véron 1976). Various combinations of these characters have been used to define species and sub-species (see, for example, Schubert 1968, 1974*a*, Popoff and Véron 1976), but there is little agreement about the validity of these or the names to be given to them.

Motile aeromonads are commonly found in soil and water, and cause *black rot in eggs* (Miles and Halnan 1937). They are responsible for a number of diseases in cold-blooded animals, such as *red leg in frogs*. They have been isolated on a number of occasions from human faeces, sputum and pus, and are occasionally the cause of severe disease in man (Chapter 61). They are said to exhibit enterotoxic activity in the rabbit ileal loop (Annapurna and Sanyal 1977, Ljungh and Kronevi 1982) and to form one or more cytolytic toxins (Bernheimer and Avigad 1974, Donta and Haddow 1978, Ljungh and Kronevi 1982).

Aeromonas salmonicida

This organism appears to be a member of the *Aeromonas* group (Griffin *et al.* 1953*a*), although it is non-motile. It is a gram-negative rod, 1–4 μm × 0.8–1.0 μm. Grows at 5–32°, optimum temperature about 20°. Grows on ordinary media; growth improved by blood or serum. On agar at 15–20° in 24 hr it forms small transparent circular colonies which in 7 days reach a diameter of 1 mm. Both the colonies and the medium become dark brown owing to the production of a diffusible melanin-like pigment which is not formed in the absence of tyrosine and phenylalanine (Griffin *et al.* 1953*b*). Gelatin and Loeffler's serum are liquefied. β-Haemolysis on blood agar in 2 days. Grows but forms no pigment anaerobically. Oxidase positive. Produces acid, and as a rule gas, in glucose, maltose, mannitol and starch in 24–48 hr, and usually also in arabinose and salicin later. It is VP negative, citrate negative and indole negative. Appears to be antigenically homogeneous. Gives rise to *furunculosis of fish*, particularly salmon and trout (see Scott 1968). Experimentally the disease may be produced in fish either by feeding or by inoculation. The organism is believed to be an obligate parasite, living in the intestine but persisting in the kidneys of chronic carriers (Report 1930, 1933, 1935). Two non-pigmented species have also been described (see Schubert 1974*a*).

Plesiomonas

We have summarized the main reasons for placing this organism in a separate genus closely related to *Vibrio* and *Aeromonas* (p. 145 and Table 27.3). Its characters are fairly uniform, and only one species is recognized, *P. shigelloides*.

It is a straight rod with lophotrichous flagella; is not halophilic but tolerates 3 per cent and sometimes 6 per cent of NaCl, and is inhibited by a 10 μg disk of compound O/129. It acidifies glucose, maltose, trehalose and inositol promptly and lactose in 1–4 days but does not attack mannitol or sucrose. It gives a negative Voges-Proskauer reaction, decarboxylates lysine and ornithine, and forms arginine dihydrolase. Protease, amylase, DNAase and haemolytic activity are absent.

In 1947, Ferguson and Henderson described organisms that resembled *Sh. sonnei* in colonial appearance and in the possession of an O antigen identical with that of phase 1 of this organism, but were motile; in 1954, Bader showed that they had polar flagella. Sakazaki and his colleagues (1959) examined a large collection of organisms with similar biochemical characters and found that they were serologically heterogeneous, only a minority having the *Shigella* antigen. Habs and Schubert (1962) created the new genus *Plesiomonas* and distinguished it from *Aeromonas*, and Hendrie and coworkers (1971) drew attention to its close relation to the *Vibrio* genus. (See also Schubert 1974*b*.)

P. shigelloides is often isolated from the faeces of man and animals, but there is no definite evidence that it is a pathogen.

Campylobacter

Definition

Slim, curved, gram-negative rods, motile by a polar flagellum. Microaerophilic. No carbohydrates attacked. Non-proteolytic. Reduce nitrate to nitrite. Oxidase positive. Indole and urease negative. G+C content 30–35 moles per cent. Some species are pathogenic, causing such diseases as abortion in cattle and sheep, and enteritis and septicaemia in man.

Campylobacter fetus

The first member of this group was isolated by M'Fadyean and Stockman from the uterine exudate of aborting sheep (see Report 1913). Smith (1918) cultivated a similar organism in the United States from fetuses of aborting cows (see Chapter 56), and named it *Vibrio fetus* (Smith and Taylor 1919). Because this organism differs in many respects from members of the *Vibrio* group, particularly in its much lower G+C content, it has been transferred to the genus *Campylobacter* (Sebald and Véron 1963, Basden *et al.* 1968, Report 1972).

The complex question of classification and nomenclature is dealt with later in this chapter, but unless stated otherwise the nomenclature used is that of Smibert (1974, 1978).

Morphology

The organism consists of comma or s-shaped forms arranged singly, in pairs, or in short chains; s-shaped threads are also seen stretching nearly across the field of the microscope. In length it is 1.5 to 5 μm or more, and in breadth about 0.2 to 0.3 μm. A single organism has one or two spirals; the length of each spiral is about 2 μm, and the amplitude about 0.5 μm. In the long forms the spirals are drawn out, so that their length is far greater than their breadth. The short forms are sharply curved; the spirals often have an obtuse-angled curve. In young cultures the bacteria are motile by a single polar flagellum about 18 nm in width; in cultures a week old very few are motile. The organism stains well with alkaline methylene blue applied overnight. It is gram negative. In old cultures many of the organisms show granular degeneration, and may become coccoid.

Cultural characters

Growth occurs best in an atmosphere containing a partial pressure of oxygen of 2.5–6.0 per cent and of CO_2 of 10 per cent (Kiggins and Plastridge 1956, Reich *et al.* 1957, Fletcher and Plastridge 1964, Smibert 1978). Little or no growth occurs anaerobically. Optimal temperature for development is between 25° and 37°. Members of subspecies (subsp.) *jejuni* will grow at 43°. When first isolated, campylobacters will not grow on agar without the addition of blood or some other animal fluid. Good growth occurs in soft agar containing 0.005 per cent of glutathione and adjusted before sterilization to pH 6.8 (see Plastridge and Williams 1943, Huddleson 1948, Plastridge *et al.* 1949), in blood broth containing 1.0 per cent of agar (Vinzent *et al.* 1950), and in a modification of Brewer's sodium thioglycollate medium (Hansen *et al.* 1952). To isolate them from human faeces, Butzler and his colleagues (1973) made use of the fact that they passed through a 0.65-μm Millipore filter and could then be cultivated on semisolid thioglycollate blood agar containing appropriate concentrations of bacitracin, colistin, novobiocin and cycloheximide. Skirrow (1977) omitted the filtration, and used a blood agar medium containing (per l) vancomycin 10 mg, colistin 2500 units, and trimethoprim 5 mg; for the isolation of subsp. *jejuni* from faeces, this was incubated at 43° in O_2:CO_2:H_2 1:2:17. Convenient methods of obtaining optimal gas concentrations are discussed by Simmons (1977). The tolerance of campylobacters for oxygen can be increased by the addition of ferrous sulphate, sodium bisulphite and sodium pyruvate to media (see Smibert 1978).

On blood agar incubated in 5–10 per cent CO_2 fine colonies appear within 36 hr and reach their maximum size in 2–4 days. They are circular, domed, rather cloudy, glistening, with an entire edge, a slightly bluish-grey tinge, and a butyrous consistency. Growth is said to occur on MacConkey's agar (King 1957). Smibert (1963) described a chemically defined semisolid medium containing 18 amino acids, B vitamins, and mineral salts.

Resistance

In cultures the organism lives from 2 to 20 weeks at room temperature, but dies rapidly in the ice chest. Dried on thread, it survives for less than 3 hr. It is killed at 56° in 5 min, and is destroyed by pasteurization (Waterman 1982). *C. fetus* is resistant to penicillins, bacitracin and colistin. It is usually sensitive to chloramphenicol, gentamicin, erythromycin and tetracycline, but there are a few exceptions (see Butzler *et al.* 1973, Chow *et al.* 1978, Vanhoof *et al.* 1978, 1980). It is said to be susceptible to dimetridazole (Fernie *et al.* 1977), a substance whose antibacterial activity was formerly thought to be confined to obligate anaerobes. Bacteriophages from *C. fetus* have been described; some can transduce streptomycin resistance and glycine tolerance (Chang and Ogg 1970, 1971) and antigenic characters (Ogg and Chang 1972).

Table 27.4 Intraspecific classification of *Campylobacter fetus* (after Smibert 1978)

Classification*	Subspecies, serotype, or biotype				
S (Marsh and Firehammer 1953)†	3	3,5	5,3	2	1
S (Mitscherlich and Liess 1958)	1	1	...	2	13
S (Morgan 1959)	A	A	...	B	...
B (Florent 1959)	*V. foetus var. venerialis*		*V. foetus var. intestinalis*		...
B (Bryner *et al.* 1962)	1	Subtype 1	2	2	...
B (Mohanty *et al.* 1962)	I	III	II	II	...
S+B (Berg *et al.* 1971)	A-1	A-biosubtype 1	A-2	B	C
B (Véron and Chatelain 1973)	*C. fetus* subsp. *venerealis*	*C. fetus* subsp. *venerealis*, biotype *intermedius*	*C. fetus* subsp. *fetus*		*C. jejuni* *C. coli*
B (Smibert 1974, 1978)	*C. fetus* subsp. *fetus*		*C. fetus* subsp. *intestinalis*		*C. fetus* subsp. *jejuni*‡

* S = serological; B = biochemical.
† 'Montana serotypes'.
‡ 'Related vibrios' of King (1962); 'approved' name *C. jejuni* (Skerman *et al.* 1980).

Biochemical characters

In addition to the properties given in the definition of the genus (p. 148), *C. fetus* forms catalase, and produces only a trace of H_2S or none at all; it does not liquefy gelatin, grow on Simmons's citrate agar or in the presence of 3 per cent NaCl.

Immunochemical characters

Gallut (1952) showed the presence of a polysaccharide fraction corresponding to the type-specific O antigen (Blakemore and Gledhill 1946), and a protein fraction common to the species (see also Gallut *et al.* 1957). Ristic and Brandly (1959*a*, 1959*b*) isolated a heat-stable polysaccharide with immunological reactivity. An endotoxin, present in culture fluids, was described by Osborne and Smibert (1962, 1964); it possessed abortifacient properties and produced a generalized Shwartzman reaction. A strain of *C. fetus* subsp. *intestinalis* was shown to possess an antigenic, antiphagocytic microcapsule that resembled the outer structural protein of the taxonomically related *Spirillum serpens* (McCoy *et al.* 1975, Winter *et al.* 1978). Gel electrophoresis of cell proteins revealed differences between strains that were related to the host of origin (Morris and Park 1973, Fernie and Park 1977). Smibert (1970) found that the cell walls often contained galactose and mannose. Blaser and his colleagues (1980) showed that *C. fetus* subsp. *jejuni* could be distinguished from the subspp. *fetus* and *intestinalis* by gas-liquid chromatographic analysis of cellular fatty acids.

Classification

Serotyping Classification within the species *C. fetus* is based on serotype and biotype, and is related to pathogenicity (see Tables 27.4 and 27.5).

Blakemore and Gledhill (1946) obtained evidence for the existence of type-specific O antigens and for considerable overlapping among the H antigens; subsequent workers have generally confirmed their findings. Marsh and Firehammer (1953) used an agglutination test in which the agglutinating suspension consisted of formolized or heat-killed cells. The majority of 23 sheep strains belonged to a single type (type 1); the remainder formed types 2, 4 and 5, and three bovine strains formed type 3. Berg and his colleagues (1971) found that, because of the occurrence of up to seven heat-labile antigens in any one strain, clear differentiation between groups was not possible by means of agglutination tests with formolized cell suspensions. Other workers have used heat-stable antigen preparations (Mitscherlich and Liess 1958, Morgan 1959). The typing scheme of Mitscherlich and Liess was stated by Smibert (1978) to contain 13 serotypes, of which nos. 1, 2, 7 and 13 belonged to *C. fetus*. (See Table 27.4.)

Biotyping (Florent 1959, Bryner *et al.* 1962, King 1962, Mohanty *et al.* 1962, Véron and Chatelain 1973) is based mainly on ability to grow in media containing 1.0 per cent glycine (Lecce 1958), to grow at 25°, and to produce small amounts of H_2S detectable by a sensitive method, as in a cysteine-supplemented medium over which a lead acetate paper is suspended. The three tests combined are capable of subdividing *C. fetus* strains into four groups (see Table 27.5). Berg and his co-workers (1971) found that clear differentiation between five groups could be obtained by the combined

Table 27.5 Properties of *Campylobacter fetus*

Subspecies (Smibert 1978) and type (Berg *et al.* 1971)*		Slight H_2S production†	Glycine (1 per cent) tolerance	Growth at 25°	Growth at 42°	Venereally transmitted bovine reproductive infections	Orally transmitted reproductive infections (ovine or bovine, or both)
subsp. *fetus*	A-1	−	−	+	−	+	−
	A-biosubtype 1	+	−	+	−	+	−
subsp. *intestinalis*	A-2	+	+	+	−	−	+
	B	+	+	+	−	−	+
subsp. *jejuni*	C	+	+	−	+	−	+

* Other classifications are shown in Table 27.4.
† As shown by lead-acetate paper suspended over medium containing cysteine.
C. fetus subspp. *jejuni* (the 'related vibrios' of King 1962) and *intestinalis* produce human infections, e.g. enteritis and septicaemia, but the former predominates.

use of biotyping and serotyping with heat-stable agglutinogens. Different systems of nomenclature were devised by Florent (1959), Véron and Chatelain (1973), and Smibert (1974, 1978), to take account of existing knowledge of serotyping, biotyping, and pathogenicity (see Table 27.5).

Smibert (1974, 1978) recognizes three subspecies of *C. fetus*: *fetus*, *intestinalis* and *jejuni*, though he agrees that the first two are more closely related to each other than to subsp. *jejuni*. Skerman and his colleagues (1980) give *C. jejuni* as the 'approved' name for Smibert's *C. fetus* subsp. *jejuni*; this nomenclatural decision carries taxonomic implications that require further investigation.

C. fetus susp. *fetus* comprises the organisms whose only pathogenic rôle is in venereally transmitted abortion and infertility of cattle. They grow at 25° but not usually at 42° or in the presence of 1.0 per cent glycine. Those that produce no H_2S correspond to the group A-1 of Berg and his co-workers (1971); those that produce small amounts correspond to group A-biosubtype 1. *C. fetus* subsp. *intestinalis* is transmitted orally; it produces abortion of sheep, abortion and infertility of cattle, and occasional infections in man. It grows at 25° but not usually at 42°, is tolerant to 1.0 per cent glycine, and produces small amounts of H_2S. This subspecies includes the groups A-2 and B of Berg and his colleagues (1971).

C. fetus subsp. *jejuni* produces ovine abortion and many human infections including enteritis and septicaemia; it occurs in the intestinal tract of certain mammals and birds, often in the absence of disease. This subspecies grows at 42° but not 25° and, unlike the two other subspecies, is sensitive to nalidixic acid. It is tolerant of 1.0 per cent glycine, and produces H_2S in amounts that may exceed those produced by the two other subspecies but are nonetheless small. According to Razi and his colleagues (1981) it differs from subsp. *fetus* and subsp. *intestinalis* in its inability to grow anaerobically in a semisolid yeast extract agar medium supplemented with nitrate and aspartate. *C. fetus* subsp. *jejuni* represents the serotype 1 of Marsh and Firehammer (1953), the *related vibrios* of King (1962), and the two species *C. jejuni* and *C. coli* of Véron and Chatelain (1973). It probably includes *Vibrio jejuni*, isolated from calves with diarrhoea (Jones and Little 1931, Jones *et al.* 1931); cultures of this organism no longer exist. It also includes, with one exception, the microaerophilic vibrio-like organisms from swine dysentery (Davis 1961, Lussier 1962, Söderlind 1965), a disease caused primarily by *Treponema hyodysenteriae* (see Chapter 44); the exception is the *Vibrio coli* of Doyle (1944, 1948) which, unlike the other organisms from swine dysentery, did not reduce nitrate. Until very recently, subsp. *jejuni* infections of man were diagnosed only on the few occasions on which invasion of the bloodstream or meninges occurred (King 1957, Bokkenheuser 1970, Ullmann 1975). Since the introduction of suitable methods for its isolation from faeces, it has been recognized as one of the more important causes of enteritis in man (Skirrow 1977; see also Chapter 71).

Skirrow and Benjamin (1980) described strains—mainly from seagulls—that had the thermophilic characters of *C. fetus* subsp. *jejuni* but were resistant to nalidixic acid. Neill and his colleagues (1978, 1979) described microaerophilic organisms from cattle and pigs that resembled *C. fetus* subsp. *fetus* (Smibert 1974) in some but not in all respects. They could be isolated at 30° but not 37°, and became aerobic on subculture. Tests for the accumulation of intracellular poly-β-hydroxybutyric acid (see Krieg and Smibert 1974) were negative, and the G + C content of the DNA was 29–34 moles per cent.

Practical methods for sub-dividing strains of *C. fetus* subsp. *jejuni* would greatly assist in epidemiological studies of disease. Strain-specific antigens have been demonstrated by bactericidal tests, agglutination, co-agglutination, passive haemagglutination, and immunoelectrophoresis (Abbot *et al.* 1980, Kosunen *et al.* 1980, Lauwers *et al.* 1981).

Pathogenicity for laboratory animals Intraperitoneal inoculation of pregnant guinea-pigs causes abortion in 2–3 days (Vinzent *et al.* 1950). Bryner and his colleagues (1978, 1979) tested vaccines in pregnant guinea-pigs by challenging intraperitoneally with *C. fetus* subsp. *jejuni* and subsp. *intestinalis*. It had been found earlier (Bryner *et al.* 1971) that large intraperitoneal doses of subsp. *intestinalis* but not subsp. *fetus* (Smibert 1974) caused infection of the gall-bladder and intestine of mice. Injection into 12-day chick embryos proves fatal in 5 days (Plastridge and Williams 1943). Male hamsters are moderately susceptible, sometimes showing testicular lesions after inoculation by the intraperitoneal route (Ristic *et al.* 1955).

Campylobacter fecalis

This organism, formerly named *Vibrio fecalis*, was frequently isolated by Firehammer (1965) from sheep faeces, but was not associated with obvious disease. It resembled an organism isolated from the genital tract of cattle (Plastridge *et al.* 1964). Most of its properties, including catalase production, were those of *C. fetus*, but it differed in being able to grow in the presence of 2–4 per cent NaCl and in being a strong producer of H_2S, as demonstrated by stab-inoculation of an iron-containing medium. Growth occurred at 37° and 42° but was slight or absent at 25°.

Campylobacter sputorum

This species differs from *C. fetus* and *C. fecalis* in being catalase negative. Like *C. fecalis*, it differs from *C. fetus* in producing H_2S in quantities large enough to be detectable by stab-inoculation in an iron-containing medium. Three subspecies are recognized.

C. sputorum subsp. *sputorum* (Véron and Chatelain 1973, Smibert 1974), formerly known as *Vibrio sputorum* (Prévot 1940), is a non-pathogenic organism commonly isolated from the gingival margins of man (Loesche *et al.* 1965). It grows at 25° and in the presence of 1.0 per cent glycine and 1.0 per cent bile, but not in that of 3.5 per cent NaCl.

C. sputorum subsp. *bubulus* (Véron and Chatelain 1973, Smibert 1974) is the non-pathogenic organism formerly known as *Vibrio bubulus* (Florent 1953). It is found as a saprophyte in the preputial sac of normal bulls and rams, and has been isolated from semen and from the vaginal mucosa of cattle and sheep. It resembles subsp. *sputorum* in its ability to grow at 25° and in the presence of 1.0 per cent glycine; it differs in tolerating 3.5 per cent NaCl but not usually 1.0 per cent bile.

C. sputorum subsp. *mucosalis* (Lawson *et al.* 1975, 1981) is associated with porcine intestinal adenomatosis and related diseases of pigs (Rowland and Lawson 1975, Lawson *et al.* 1976, Love *et al.* 1977, Lawson *et al.* 1979). It differs from the two other subspecies in failing to tolerate 1.0 per cent glycine, and in forming colonies that are yellow instead of whitish. It does not grow in the presence of 3.5 per cent NaCl or 1.0 per cent bile.

(For reviews of *Campylobacter* see Véron and Chatelain 1973, and Smibert 1978.)

Spirillum

Definition

Rigid spiral cells, motile by tufts of usually bipolar flagella. Gram negative. Aerobes or obligate microaerophiles. Oxidase positive. Catalase weakly positive. Can use inorganic sources of nitrogen, but require organic sources of carbon. Yellow or brown pigment formed by some members. Little or no action on carbohydrates. G+C content of DNA 36–65 moles per cent. Non-pathogenic to man and animals. Found mainly in fresh or sea-water. Type species *Spirillum volutans*.

This definition excludes the only known pathogenic species, *Spirillum minus*, which has not been cultivated on artificial media. Because of its resemblance to the pathogenic spirochaetes, this organism is described in Chapter 44. For a general account of the spirilla, see Krieg and Smibert (1974); the very wide range of values reported for the DNA base composition of its members indicates that the group is heterogeneous.

Strictly anaerobic gram-negative spiral organisms of the genus *Selenomonas* (Bryant 1974) have a non-polar tuft of flagella, ferment sugars, and are catalase negative. *Sel. sputigena* is found in the human mouth and *Sel. ruminantium* in the bovine rumen.

Bdellovibrio

This genus comprises vibrios that become attached to bacteria, penetrate their cell walls, and multiply in them (Stolp and Petzold 1962; Greek *bdella* = a leech). They are gram-negative curved bacilli (1–3 μm × 0.3 μm) with a polar sheathed flagellum that move by a series of jumps at a speed of about 100 μm per sec, rotating at the same time about 100 times per sec. When attacked, the host cells are soon immobilized. The vibrios penetrate the periplasmic space within *ca* 20 min. In the bacterial cell they form a large unflagellated spiral structure that later divides into vibrio-shaped motile daughter cells (Stolp 1979) the number of which depends upon the species of *Bdellovibrio*. Within 1–2 hr the host cell disintegrates and the vibrios continue to multiply, presumably feeding on the liberated nutrients. Grown in the presence of a host bacterium, the vibrios cause clearing of a fluid medium; on a solid medium they form plaques in 2–4 days. Most of the bdellovibrios parasitize gram-negative bacilli, notably pseudomonads and enterobacteria.

Parasitism may be obligate or facultative, but obligate parasites may give rise to host-independent variants. Facultative parasites and host-independent variants grow aerobically on complex laboratory media, best at 30°. They are all more or less proteolytic but do not attack sugars fermentatively. Their sensitivity to compound O/129 and their catalase reactions vary; oxidase reactions, when recorded, are positive. The type species is *Bdellovibrio bacteriovorans*; several other species have been described. (See also Burnham and Robinson 1974.)

References

Abbott, J. D., Dale, B., Eldridge, J., Jones, D. M. and Sutcliffe, E. M. (1980) *J. clin. Path.* **33,** 762.

Al-Awqati, Q., Cameron, J. L., Field, M. and Greenough, W. B. (1972) *J. clin. Invest.* **49,** 2a.

Allweiss, B., Dostal, J., Carey, K. E., Edwards, T. F. and Freter, R. (1977) *Nature, Lond.* **266,** 448.

Annapurna, E. and Sanyal, S. C. (1977) *J. med. Microbiol.* **10,** 317.

Aziz, K. M. S. and Mosley, W. H. (1972) *J. infect. Dis.* **125,** 36.

Bader, R. E. (1954) *Z. Hyg. InfektKr.* **140,** 450.

Balteanu, I. (1926) *J. Path. Bact.* **29,** 251.

Barua, D. and Chatterjee, S. N. (1964) *Ind. J. med. Res.* **52,** 828.
Barua, D. and Mukherjee, A. C. (1964) *Bull. Calcutta Sch. trop. Med.* **12,** 147.
Basden, E. H., Tourtellotte, M. E., Plastridge, W. N. and Tucker, J. S. (1968) *J. Bact.* **95,** 439.
Basu, S. and Mukerjee, S. (1968) *Experientia* **24,** 299.
Baumann, P., Baumann, L. and Mandel, M. (1971) *J. Bact.* **107,** 268.
Berg, R. L., Jutila, J. W. and Firehammer, B. D. (1971) *Amer. J. vet. Res.* **32,** 11.
Bernheimer, A. W. and Avigad, L. S. (1974) *Infect. Immun.* **9,** 1016.
Bhaskaran, K. (1957) *Nature, Lond.*, **180,** 43.
Bhattacharyya, F. K. (1977) *Jap. J. med. Sci. Biol.* **30,** 259.
Blakemore, F. and Gledhill, A. W. (1946) *J. comp. Path.* **56,** 69.
Blaser, M. J., Moss, C. W. and Weaver, R. E. (1980) *J. clin. Microbiol.* **11,** 448.
Bokkenheuser, V. (1970) *Amer. J. Epidem.* **91,** 400.
Bryant, M. P. (1974) In: *Bergey's Manual of Determinative Bacteriology*, 8th edn, p. 424. Ed. by R. E. Buchanan and N. E. Gibbons. Williams and Wilkins, Baltimore.
Bryner, J. H., Estes, P. C., Foley, J. W. and O'Berry (1971) *Amer. J. vet. Res.* **32,** 465.
Bryner, J. H., Foley, J. W., Hubbert, W. T. and Matthews, P. J. (1978) *Amer. J. vet. Res.* **39,** 119.
Bryner, J. H., Foley, J. W. and Thompson, K. (1979) *Amer. J. vet. Res.* **40,** 433.
Bryner, J. H., Frank, A. H. and O'Berry, P. A. (1962) *Amer. J. vet. Res.* **23,** 32.
Burnet, F. M. and Stone, J. D. (1947) *Aust. J. exp. Biol. med. Sci.* **25,** 219.
Burnham, J. C. and Robinson, J. (1974) In: *Bergey's Manual of Determinative Bacteriology*, 8th edn, p. 212. Ed. by R. E. Buchanan and N. E. Gibbons. Williams and Wilkins, Baltimore.
Burrows, W. (1968) *Annu. Rev. Microbiol.* **22,** 245.
Burrows, W. and Kaur, J. (1974) In: *Cholera*, p. 143. Ed. by D. Barua and W. Burrows. Saunders, Philadelphia.
Burrows, W., Mather, A. N., McGann, V. G. and Wagner, S. M. (1946) *J. infect. Dis.* **79,** 168.
Burrows, W. and Musteikis, G. M. (1966) *J. infect. Dis.* **116,** 183.
Butzler, J. P., Dekeyser, P., Detrain, M. and Dehain F. (1973) *J. Pediat.* **82,** 493.
Carpenter, C. C. J., Curlin, G. T. and Greenough, W. B. (1969) *J. infect. Dis.* **120,** 332.
Carpenter, C. C. J., Greenough, W. B. and Gordon, R. S. (1974) In: *Cholera*, p. 129. Ed. by D. Barua and W. Burrows. Saunders, Philadelphia.
Carpenter, K. P. (1966) *Mon. Bull. Minist. Hlth Lab. Serv.* **25,** 58.
Chang, W. and Ogg, J. E. (1970) *Amer. J. vet. Res.* **31,** 919; (1971) *Ibid.* **32,** 649.
Chow, A. W., Patten, V. and Bednorz, D. (1978) *Antimicrob. Agents Chemother.* **13,** 416.
Citarella, R. V. and Colwell, R. R. (1970) *J. Bact.* **104,** 434.
Craig, J. P. (1965) *Nature, Lond.* **207,** 614.
Davis, B. R. *et al.* (1981) *J. clin. Microbiol.* **14,** 631.
Davis, J. W. (1961) *J. Amer. vet. med. Ass.* **138,** 471.
De, S. N. (1959) *Nature, Lond.* **183,** 1533.
De, S. N. and Chatterje, D. N. (1953) *J. Path. Bact.* **66,** 559.
Donta, S. T. and Haddow, A. D. (1978) *Infect. Immun.* **21,** 989.
Donta, S. T., King, M. and Sloper, K. (1973) *Nature, New Biol., Lond.* **243,** 246.
Donta, S. T., Kreiter, S. R. and Wendelschafer-Crabb, G. (1976) *Infect. Immun.* **13,** 1479.
Doyle, L. P. (1944) *Amer. J. vet. Res.* **5,** 3; (1948) *Ibid.* **9,** 50.
Dunbar. (1893) *Dtsch. med. Wschr.* **19,** 799.
Dutta, N. K. and Habbu, M. K. (1955) *Brit. J. Pharmacol.* **10,** 153.
Dutta, N. K., Panse, M. V. and Kulkarni, D. R. (1959) *J. Bact.* **78,** 594.
Evans, D. G., Evans, D. J. and Gorbach, S. L. (1973) *Infect. Immun.* **8,** 731.
Feeley, J. C. (1962) *J. Bact.* **84,** 866; (1965) *Ibid.* **89,** 665.
Feeley, J. C. and Pittman, M. (1963) *Bull. Wld Hlth Org.* **28,** 347.
Felsenfeld, O. (1965) *Bull. Wld Hlth Org.* **33,** 725; (1967) *The Cholera Problem.* Warren H. Green Inc., St. Louis.
Ferguson, W. W. and Henderson, N. D. (1947) *J. Bact.* **54,** 179.
Fernie, D. S. and Park, R. W. A. (1977) *J. med. Microbiol.* **10,** 325.
Fernie, D. S., Ware, D. A. and Park, R. W. A. (1977) *J. med. Microbiol.* **10,** 233.
Field, D. M., Fromm, D., Al-Awquati, Q. and Greenough, W. B. (1972) *J. clin. Invest.* **51,** 796.
Field, M. (1979) *Amer. J. clin. Nutr.* **32,** 189.
Finkelstein, R. A., Arita, M., Clements, J. D. and Nelson, E. T. (1978). *Proc. 13th Jt Conf. Cholera*, p. 137. Atlanta, Ga.
Finkelstein, R. A., Jehl, J. J. and Goth, A. (1969) *Proc. Soc. exp. Biol. Med.* **132,** 835.
Finkelstein, R. A. and LoSpalluto, J. J. (1969) *J. exp. Med.* **130,** 185.
Finkelstein, R. A. and Mukerjee, S. (1963) *Proc. Soc. exp. Biol., N.Y.* **112,** 355.
Finkler, D. and Prior, H. (1884) *Dtsch. med. Wschr.* **10,** 632.
Firehammer, B. D. (1965) *Cornell Vet.* **55,** 482.
Fletcher, R. D. and Plastridge, W. N. (1964) *J. Bact.* **87,** 352.
Florent, A. (1953) *C. R. Soc. Biol., Paris* **147,** 2066; (1959) *Proc. 16th int. vet. Congr., Madrid*, vol. 2, p. 953.
Freter, R. (1955) *J. infect. Dis.* **97,** 238.
Furniss, A. L., Lee, J. V. and Donovan, T. J. (1977) *Lancet* **ii,** 565; (1978) *The Vibrios.* Publ. Hlth Monogr. Ser. No. 11. Her Majesty's Stationery Office, London.
Gallut, J. (1952) *Ann. Inst. Pasteur* **83,** 449, 455.
Gallut, J., Chevé, J. and Gauthier, J. (1957) *Ann. Inst. Pasteur* **93,** 683.
Gamaléia, M. N. (1888*a*) *Ann. Inst. Pasteur*, **2,** 482; (1888*b*) *Ibid.* **2,** 552.
Gangarosa, E. J., Bennett, J. V. and Boring, J. R. (1967) *Bull. World Hlth Org.* **36,** 987.
Gardner, A. D. and Venkatraman, K. V. (1935) *J. Hyg., Camb.* **35,** 262.
Gotschlich, F. (1906) *Z. Hyg. InfektKr.* **53,** 281.
Greig, E. D. W. (1914) *Indian J. med. Res.* **2,** 623.
Griffin, P. J., Snieszko, S. F. and Friddle, S. B. (1953*a*) *Trans. Amer. Fish. Soc.* **82,** 129; (1953*b*) *J. Bact.* **65,** 652.
Guentzel, M. N. and Berry, L. J. (1975) *Infect. Immun.* **11,** 890.
Gyles, C. L. and Barnum, D. A. (1969) *J. infect. Dis.* **120,** 419.
Habs, H. and Schubert, R. H. W. (1962) *Zbl. Bakt.* **186,** 316.

Hackney, C. R., Kleeman, E. G., Ray, B. and Speck, M. L. (1980) *Appl. environm. Microbiol.* **40,** 652.
Han, G. K. and Khie, T. S. (1963) *Amer. J. Hyg.* **77,** 184.
Hansen, P. A., Price, K. E. and Clements, M. F. (1952) *J. Bact.* **64,** 772.
Heiberg, B. (1935) *On the Classification of Vibrio cholerae and the Cholera-like Vibrios.* Arnold Busck, Copenhagen; (1936) *J. Hyg., Camb.* **36,** 114.
Hendrie, M. S. and Shewan, J. M. (1974) In: *Bergey's Manual of Determinative Bacteriology*, 8th edn, p. 349. Ed. by R. E. Buchanan and N. E. Gibbons. Williams and Wilkins, Baltimore.
Hendrie, M. S., Shewan, J. M. and Véron, M. (1971) *Int. J. syst. Bact.* **21,** 25.
Hendrix, T. R. (1971) *Bull. N.Y. Acad. Med.* **47,** 1169.
Heyningen, W. E. van (1974) *Science* **183,** 656.
Hickman, F. W. *et al.* (1982) *J. clin. Microbiol.* **15,** 395.
Hirao, Y. and Homma, J. Y. (1978) *Infect. Immun.* **19,** 373.
Hollis, D. G., Weaver, R. E., Baker, C. N. and Thornsberry, C. (1976) *J. clin. Microbiol.* **3,** 425.
Holmgren, J. Lindholm, L. and Lönnroth, I. (1974) *J. exp. Med.* **139,** 801.
Holmgren, J., Lönnroth, I. and Ouchterlony, Ö. (1971) *Infect. Immun.* **3,** 747.
Holmgren and Svennerholm, A.-M. (1979) *Curr. Microbiol.* **2,** 55.
Holmgren, J., Svennerholm, A.-M., Lönnroth, I., Fall-Persson, M., Markman, B. and Lundbäck, H. (1978). *Proc. 13th Jt Conf. Cholerae*, 1977 p. 272. Atlanta, Ga.
Honda, T. and Finkelstein, R. A. (1979) *Infect. Immun.* **26,** 1020.
Honda, T., Taga, S., Takeda, T., Hasibuan, M. A., Takeda, Y. and Miwatani, T. (1976) *Infect. Immun.* **13,** 133.
Hornick, R. B., Music, S. I., Wenzel, R., Cash, R., Libonati, J. P., Snyder, M. J. and Woodward, T. E. (1971) *Bull. N.Y. Acad. Med.* **47,** 1181.
Huddleson, I. F. (1948) *J. Bact.* **56,** 508.
Hugh, R. (1964) *Int. Bull. bact. Nomencl.* **14,** 87.
Jones, F. S. and Little, R. B. (1931) *J. exp. Med.* **53,** 835, 845.
Jones, F. S., Orcutt, M. and Little, R. B. (1931) *J. exp. Med.* **53,** 853.
Jones, G. W., Abrams, G. D. and Freter, R. (1976) *Infect. Immun.* **14,** 232.
Joseph, S. W., DeBell, R. M. and Brown, W. P. (1978) *Antimicriob. Agents Chemother.* **13,** 244.
Kauffmann, F. (1950) *Acta path. microbiol. scand.* **27,** 283.
Kaur, J., Burrows, W. and Cercarski, L. (1969) *J. Bact.* **100,** 985.
Kiggins, E. M. and Plastridge, W. N. (1956) *J. Bact.* **72,** 397.
King, E. O. (1957) *J. infect. Dis.* **101,** 119; (1962) *Ann. N.Y. Acad. Sci.* **98,** 700.
Kitasato, S. (1889) *Z. Hyg. InfektKr.* **5,** 134.
Kluyver, A. J. and Niel, C. B. van (1936) *Zbl. Bakt.* (IIte Abt.) **94,** 369.
Kobayashi, T., Enomoto, S., Sakazaki, R. and Kuwahara, S. (1963) *Jap. J. Bact.* **18,** 387.
Koch, R. (1886) *The Etiology of Cholera*, New Sydenham Soc., **115,** 327.
Kosunen, T. U., Danielsson, D. and Kjellander, J. (1980) *Acta path. microbiol. scand.* **B88,** 207.
Krieg, N. R. and Smibert, R. M. (1974) In: *Bergey's Manual of Determinative Bacteriology*, 8th edn, p. 196. Ed. by R. E. Buchanan and N. E. Gibbons. Williams and Wilkins, Baltimore.
Lauwers, S., Vlaes, L. and Butzler, J. P. (1981) *Lancet* **i,** 158.
Lawson, G. H. K., Leaver, J. L., Pettigrew, G. W. and Rowland, A. C. (1981) *Int. J. syst. Bact.* **31,** 385.
Lawson, G. H. K., Rowland, A. C. and Roberts, L. (1976) *J. med. Micriobiol.* **9,** 163.
Lawson, G. H. K., Rowland, A. C., Roberts, L., Fraser, G. and McCartney, E. (1979) *Res. vet. Sci.* **27,** 46.
Lawson, G. H. K., Rowland, A. C. and Wooding, P. (1975) *Res. vet. Sci.* **18,** 121.
Lecce, J. G. (1958) *J. Bact.* **76,** 312.
Lee, J. V., Donovan, T. J. and Furniss, A. L. (1978) *Int. J. syst. Bact.* **28,** 99.
Lee, J. V. and Furniss, A. L. (1976) *Folia Microbiol., Praha* **21,** 321.
Lee, J. V., Shread, P., Furniss, A. L. and Bryant, T. N. (1981) *J. appl. Bact.* **50,** 73.
Lehmann, K. B. and Neumann, R. (1896) In: *Atlas und Grundriss der Bakteriologie und Lehrbuch der speciellen bakteriologeschen Diagnostik*, 1st edn, p. 340. J. F. Lehmann, Munich.
Leifson, E. and Hugh, R. (1953) *J. Bact.* **65,** 263.
Ljungh, Å. and Kronevi, T. (1982) *Toxicon* **20,** 397.
Loesche, W. J., Gibbons, R. J. and Socransky, S. S. (1965) *J. Bact.* **89,** 1109.
Love, D. N., Love, R. J. and Bailey, M. (1977) *Vet. Rec.* **101,** 407.
Love, M. *et al.* (1981) *Science* **214,** 1139.
Lussier, G. (1962) *Canad. vet. J.* **3,** 267.
McCoy, E. C., Doyle, D., Burda, K., Corbeil, L. B. and Winter, A. J. (1975) *Infect. Immun.* **11,** 517.
MacLeod, R. A. (1965) *Bact. Rev.* **29,** 9.
Marsh, H. and Firehammer, B. D. (1953) *Amer. J. vet. Res.* **14,** 396.
Metchnikoff, E. (1893) *Ann. Inst. Pasteur* **7,** 562; (1894) *Ibid.* **8,** 529.
Mhalu, F. S., Mmari, P. W. and Ijumba, J. (1979) *Lancet* **i,** 345.
Miles, A. A. and Halnan, E. T. (1937) *J. Hyg., Camb.* **37,** 79.
Mitra, S., Balganesh, T. S., Dastidar, S. G. and Chakrabarty, A. N. (1980) *Infect. Immun.* **30,** 74.
Mitscherlich, E. and Liess, B. (1958) *Dtsch. tierärztl. Wschr.* **65,** 36.
Miwatani, T. *et al.* (1969) *Biken J.*, **12,** 9.
Miwatani, T. and Takeda, Y. (1976) *Vibrio parahaemolyticus: a Causative Bacterium of Food Poisoning.* Saikon Publ. Co., Tokyo.
Miyaki, K., Iwahara, S., Sato, K., Fujimoto, S. and Aibara, K. (1967) *Bull. Wld Hlth Org.* **37,** 773.
Miyamoto, Y. *et al.* (1980) *Infect. Immun.* **28,** 567.
Mohanty, S. B., Plumer, G. J. and Faber, J. E. (1962) *Amer. J. vet. Res.* **23,** 554.
Morgan, W. J. B. (1959) *J. comp. Path.* **69,** 125.
Morris, J. A. and Park, R. W. A. (1973) *J. gen. Microbiol.* **78,** 165.
Mukerjee, S. (1961) *J. Hyg., Camb.* **59,** 109; (1962) *Ann. Biochem. exp. Med., Calcutta* **22,** 9.
Mukerjee, S. and Tayeka, K. (1974) In: *Cholera* p. 61. Ed. by D. Barua and W. Burrows. Saunders, Philadelphia.
Mukherjee, B. and De, S. N. (1966) *J. Path. Bact.* **91,** 256.
Nalin, D. R. and McLaughlin, J. C. (1978) *J. med. Microbiol.* **11,** 177.
Napier, L. E. and Gupta, S. K. (1942) *Indian med. Gaz.* **77,** 717.

Neill, S. D., Ellis, W. A. and O'Brien, J. J. (1978) *Res. vet. Sci.* **25,** 368; (1979) *Ibid.* **27,** 180.

Nelson, E. T., Clements, J. D. and Finkelstein, R. A. (1976) *Infect. Immun.* **14,** 527.

Neoh, S. H. and Rowley, D. (1970) *J. infect. Dis.* **121,** 505.

Nicati, W. and Rietsch, M. (1884) *C. R. Acad. Sci.* **99,** 928.

Nicolle, P., Gallut, J., Ducrest, P. and Quiniou, J. (1962) *Rev. Hyg. Méd. Soc.* **10,** 91.

Nicolle, P., Gallut, J. and le Minor, L. (1960) *Ann. Inst. Pasteur* **99,** 664.

Nicolle, P., Gallut, J., Schraen, M.-F. and Brault, J. (1971) *Bull. Soc. Path. exot.* **64,** 603.

Nobechi, K. (1925) *J. Bact.* **10,** 197; (1933) *Bull. Off. int. Hyg. publ.* **25,** 72.

Obara, Y. (1971) *J. Jap. Ass. infect. Dis.* **45,** 392 (in Japanese, quoted by Miwatani and Takeda, 1976).

Ogg, J. E. and Chang, W. (1972) *Amer. J. vet. Res.* **33,** 1023.

Okada, Y., Miyamoto, S. and Yoneda, M. (1977) *Biken J.* **20,** 91.

Osborne, J. C. and Smibert, R. M. (1962) *Nature, Lond.* **195,** 1106; (1964) *Cornell Vet.* **54,** 561.

Pastoris, M. C., Bhattacharyya, F. K. and Sil, J. (1980) *J. med. Microbiol.* **13,** 363.

Patniak, B. K. and Ghosh, H. K. (1966) *Brit. J. exp. Path.* **47,** 210.

Peterson, J. W., LoSpalluto, J. J. and Finkelstein, R. A. (1972) *J. infect. Dis.* **126,** 617.

Pfeiffer, R. (1893) *Z. Hyg. InfektKr.* **11,** 393.

Pierce, N. F. (1977) *Infect. Immun.* **18,** 338.

Pierce, N. F., Greenough, W. B. and Carpenter, C. C. J. (1971) *Bact. Rev.* **35,** 1.

Plastridge, W. N. and Williams, L. F. (1943) *J. Amer. vet. med. Ass.* **102,** 89.

Plastridge, W. N., Williams, L. F. and Roman, J. (1949) *J. Bact.* **57,** 657.

Plastridge, W. N., Williams, L. F. and Trowbridge, D. G. (1964) *Amer. J. vet. Res.* **25,** 1295.

Pollitzer, R. (1959) *Cholera*, World Hlth Org. Monograph Ser., No. 43, p. 97.

Poole, M. D. and Oliver, J. D. (1978) *Infect. Immun.* **20,** 126.

Popoff, M. and Véron, M. (1976) *J. gen. Microbiol.* **94,** 11.

Prévot, A. R. (1940) *Ann. Inst. Pasteur* **64,** 117.

Razi, M. H. H., Park, R. W. A. and Skirrow, M. B. (1981) *J. appl. Bact.* **50,** 55.

Reich, C. V., Dunne, H. W., Bortree, A. L. and Hokanson, J. F. (1957) *J. Bact.* **74,** 246.

Reichelt, J. L., Baumann, P. and Baumann, L. (1976) *Arch. Microbiol.* **110,** 101.

Report. (1913) Rep. Dep. Comm. Epizootic Abortion, Part III. London; (1930) 1st Interim Report Furunculsis Committee, H.M.S.O., Edinburgh; (1933) 2nd Interim Report; (1935) Final Report; (1972) *Int. J. syst. Bact.* **22,** 123; (1980) *Bull. World Hlth. Org.* **58,** 353.

Richards, K. L. and Douglas, S. D. (1978) *Microbiol. Rev.* **42,** 592.

Ristic, M. and Brandly, C. A. (1959*a*) *Amer. J. vet. Res.* **20,** 148; (1959*b*) *Ibid.* **20,** 154.

Ristic, M., Sanders, D. A. and Young, F. (1955) *Amer. J. vet. Res.* **16,** 189.

Rowland, A. C. and Lawson, G. H. K. (1975) *Vet. Rec.* **97,** 178.

Ruch, F. E., Murphy, J. R., Graf, L. H. and Field, M. (1978) *J. infect. Dis.* **137,** 747.

Sack, R. B. and Carpenter, C. C. J. (1969) *J. infect. Dis.* **119,** 138.

Sakazaki, R., Gomez, C. Z. and Sebald, M. (1967) *Jap. J. med. Sci. Biol.* **20,** 265.

Sakazaki, R., Iwanami, S. and Fukumi, H. (1963) *Jap. J. med. Sci. Biol.* **16,** 161.

Sakazaki, R., Namioka, S., Nakaya, R. and Fukumi, H. (1959) *Jap. J. med. Sci. Biol.* **12,** 355.

Sakazaki, R. and Shimada, T. (1977) *Jap. J. med. Sci. Biol.* **30,** 279.

Sakazaki, R., Tamura, K., Kato, T., Obara, Y., Yamai, S. and Hobo, K. (1968) *Jap. J. med. Sci. Biol.* **21,** 325.

Sakazaki, R., Tamura, K. and Murase, M. (1971) *Jap. J. med. Sci. Biol.* **24,** 83.

Sakazaki, R., Tamura, K., Nakamura, A., Kurata, T., Ghoda, A. and Kazumo, Y. (1974) *Jap. J. med. Sci. Biol.* **27,** 35.

Sanarelli, G. (1921) *Ann. Inst. Pasteur* **35,** 745.

Schubert, R. H. W. (1964) *Zbl. Bakt.* **193,** 482; (1968) *Int. J. syst. Bact.* **18,** 1; (1974*a*) In: *Bergey's Manual of Determinative Bacteriology*, 8th edn, p. 345. Ed. by R. E. Buchanan and N. E. Gibbons. Williams and Wilkins, Baltimore; (1974*b*) *Ibid.*, p. 348.

Scott, M. (1968) *J. gen. Microbiol.* **50,** 321.

Sebald, M. and Véron, M. (1963) *Ann. Inst. Pasteur* **105,** 897.

Shewan, J. M. and Hodgkiss, W. (1954) *Nature, Lond.* **173,** 208.

Shewan, J. M. and Véron, M. (1974) In: *Bergey's Manual of Determinative Bacteriology*, 8th edn, p. 340. Ed. by R. E. Buchanan and N. E. Gibbons. Williams and Wilkins, Baltimore.

Shimada, T. and Sakazaki, R. (1977) *Jap. J. med. Sci. Biol.* **30,** 275.

Shinoda, S., Kariyama, R., Ogawa, M., Takeda, Y. and Miwatani, T. (1976) *Int. J. syst. Bact.* **26,** 97.

Sil, J. and Bhattacharyya, F. K. (1979) *J. med. Microbiol.* **12,** 63.

Simmons, N. A. (1977) *Brit. med. J.* **iii,** 707.

Sinha, V. B. and Srivastava, B. S. (1979) *Bull. Wld Hlth Org.* **57,** 643.

Skerman, V. B. D., McGowan, V. and Sneath, P. H. A. (1980) *Int. J. syst. Bact.* **30,** 225.

Skirrow, M. B. (1977) *Brit. med. J.* **iii,** 9.

Skirrow, M. B. and Benjamin, J. (1980) *J. Hyg., Camb.* **85,** 427.

Smibert, R. M. (1963) *J. Bact.* **85,** 394; (1970) *Int. J. syst. Bact.* **20,** 407; (1974) In: *Bergey's Manual of Determinative Bacteriology*, 8th edn, p. 207. Ed. by R. E. Buchanan and N. E. Gibbons. Williams and Wilkins, Baltimore; (1978) *Annu. Rev. Microbiol.* **32,** 673.

Smith, T. (1918) *J. exp. Med.* **28,** 701.

Smith, T. and Taylor, M. S. (1919) *J. exp. Med.* **30,** 299.

Sochard, M. R. and Colwell, R. R. (1977) *Microbiol. Immunol.* **21,** 243.

Söderlind, O. (1965) *Vet. Rec.* **77,** 193.

Srivastava, R., Sinha, V. B. and Srivastava, B. S. (1980) *J. med. Microbiol.* **13,** 1.

Stolp, H. (1979) *Proc. roy. Soc. B.* **204,** 211.

Stolp, H. and Petzold, H. (1962) *Phytopathol. Z.* **45,** 364.

Tayeka, K., Otohuji, T. and Tokiwa, H. (1981) *J. clin. Microbiol.* **14,** 222.

Tayeka, K. and Shimadori, S. (1963) *J. Bact.* **85,** 957.

Threlfall, E. J. and Rowe, B. (1982) *Lancet* **i,** 42.

Threlfall, E. J., Rowe, B. and Huq, I. (1980) *Lancet* **i,** 1247.

Ujiye, A. *et al.* (1968) *Trop. Med., Nagasaki* **10,** 65.
Ullmann, U. (1975) *Zbl. Bakt.* **A230,** 480.
Vanhoof, R., Gordts, B., Dierickx, R., Coignau, H. and Butzler, J. P. (1980) *Antimicrob. Agents Chemother.* **18,** 118.
Vanhoof, R., Vanderlinden, M. P., Dierickx, R., Lauwers, S., Yourassowsky, E. and Butzler, J. P. (1978) *Antimicrob. Agents Chemother.* **14,** 553.
Vaughan, M., Pierce, N. F. and Greenough, W. B. (1970) *Nature, Lond.* **226,** 658.
Véron, M. and Chatelain, R. (1973) *Int. J. syst. Bact.* **23,** 122.
Vinzent, R. Delaune, J. and Hébert, H. (1950) *Ann. Méd.* **51,** 23.
Wahba, A. H. (1965) *Bull. Wld Hlth Org.* **33,** 661.
Waterman, S. C. (1982) *J. Hyg., Camb.* **88,** 529.
Wherry, W. B. (1905) *J. infect. Dis.* **2,** 309.
White, P. B. (1935*a*) *J. Hyg., Camb.* **35,** 347; (1935*b*) *Ibid.* **35,** 498; (1937) *J. Path. Bact.* **44,** 706; (1940*a*) *Ibid.* **50,** 160; (1940*b*) *Ibid.* **51,** 447.
Winter, A. J., McCoy, E. C., Fullmer, C. S., Burda, K. and Bier, P. J. (1978) *Infect. Immun.* **22,** 963.

28

Neisseria, Branhamella and *Moraxella*

Graham Wilson and A. E. Wilkinson

Neisseria

Definition

Gram-negative cocci, usually arranged in pairs with adjacent sides flattened; non-motile. Strict parasites, often growing poorly on ordinary media, but growing well on serum media. Weak fermentative ability. Some species form yellowish pigment. Little or no growth anaerobically. Catalase positive. Oxidase positive. Indole negative. Do not reduce nitrate but may reduce nitrite. Frequently pathogenic. G + C content of DNA 47–52 moles per cent.

Type species is *N. gonorrhoeae*.

The first member of this group to be described was the gonococcus; it was observed by Albert Neisser in 1879 in the pus cells of patients with gonorrhoea, and was successfully cultivated by Bumm (1885*a*, *b*) and by Leistikow and Loeffler (Leistikow 1882) in 1882. Weichselbaum isolated the meningococcus from the cerebrospinal fluid of patients with cerebrospinal meningitis in the year 1887. In 1895 Jaeger described a similar organism which he regarded as the same as the meningococcus; it came to be known as *Diplococcus crassus*. For many years it caused great confusion, and its identity is still in doubt. R. Pfeiffer (see Flügge 1896) described the *Micrococcus catarrhalis* in 1896; he found it in the bronchioles and alveoli of children with bronchopneumonia; it was carefully studied in 1902 by Ghon and H. Pfeiffer. In 1906 von Lingelsheim described a number of gram-negative cocci in the nasopharynx of healthy and diseased persons; these included the *Micrococcus pharyngis siccus*, the *Micrococcus pharyngis cinereus*, the *Diplococcus mucosus*, and the *Micrococcus pharyngis flavus* i, ii, and iii. Two new members of the group, both responsible for occasional cases of meningitis, were described in the United States—one, *N. flavescens*, by Sarah Branham (1930), and one, *N. lactamica*, by Hollis, Wiggins and Weaver (1969). With the exception of the gonococcus, whose habitat is the genito-urinary system, all these organisms are parasites of the human nasopharynx.

Morphology

The members of the group are gram-negative cocci, but they differ somewhat in their morphology and arrangement, and in the ease with which they are decolorized by alcohol. Moreover, the same organism may vary according to environmental conditions; thus, in the body, the meningococcus and the gonococcus present an almost typical arrangement in the form of diplococci with flattened or slightly concave adjacent sides, but in culture they appear as oval or spherical cocci without the typical diplococcal arrangement. Most members of the group are arranged in pairs, tetrads, or small groups; but some such as *N. subflava*, appear frequently as dense clumps with occasional isolated organisms. One difference in arrangement that serves to distinguish gram-negative from gram-positive diplococci is the way in which the main axis of the oval is directed; with gram-negative cocci this axis is always at right angles to the axis joining the two cocci; with gram-positive cocci it is often coincident with it. Moreover gram-negative cocci divide at right angles to the axis joining them, so that the formation of tetrads is commoner than with gram-positive cocci. Most neisseriae are decolorized without difficulty, but some tend to retain the violet stain. This occurs particularly when the organisms are arranged in groups or dense clumps. Films should therefore be made as thin and uniform as possible, and should be examined by daylight rather than by artificial light to avoid the confusion that results from judging the gram reaction of an indeterminate colour. In young cultures the cocci stain fairly evenly, but after about 24 hr autolytic changes set in, with the result that so-called involution forms appear; these are generally large, swollen cocci, which stain poorly. Both the meningococcus and the gonococcus are characterized, as compared with most other neisseriae, by the frequency with which such forms appear; and it may be noted that many of the large swollen forms, in cultures of these species, stain deeply and uniformly. Some workers have described the presence of Babes-Ernst bodies or metachromatic granules in members of this group. Elser and Huntoon (1909) state that the meningococcus, when stained with Loeffler's methylene blue, often shows a brightly stained central spot, the remainder of the cell being scarcely coloured; with Neisser's stain the granules stain bluish-black, the cell body brown. Marx and Woithe (1900) found these granules in gonococci, but only in organisms taken from the florid stage of gonorrhoea; they state that the whole cell may appear filled with granules. Capsules are demonstrable in some freshly isolated strains of meningococci (Clapp *et al.* 1935), and gonococci (Richardson and Sadoff 1977, James and Swanson 1977), in *N. mucosa*, and occasionally in *N. subflava*. Pili are present in gonococci, in *N. subflava*, and in small numbers on primary culture of meningococci (Wishtreich and Baker 1971, DeVoe and Gilchrist 1975). The G + C content of DNA varies in *Neisseria* from 47 to 52 moles per cent (Catlin and Cunningham 1961). The production of L forms of the meningococcus by the penicillin gradient technique was described by Roberts and Wittler (1966).

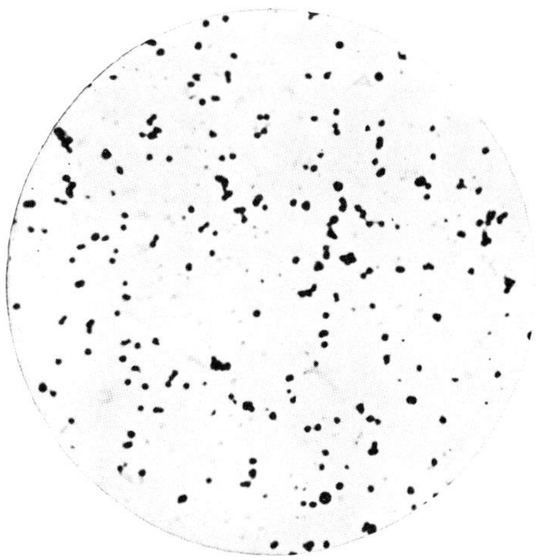

Fig. 28.1 *Neisseria meningitidis*. From a serum agar slope culture, 4 days, 37° (× 1000).

Growth requirements

Though most of the nasopharyngeal cocci will grow on nutrient agar, the meningococcus on primary isolation must be provided with such accessory growth factors as are present in blood, serum, milk and other animal fluids, and in certain vegetable extracts (Lloyd 1916–17). After a few generations on such an enriched medium it may sometimes be brought to grow on ordinary media, but its vitality under these conditions is uncertain (Murray 1929). The gonococcus is even more fastidious. Leistikow and Loeffler (Leistikow 1882) cultivated it first on blood serum gelatin, and Bumm (1885*b*) on coagulated human serum. Since then a host of other media have been introduced (for references to the older ones, see 4th edition, p. 634). During more recent years the media mainly used have been made up with a chocolate agar or proteose peptone base to which various supplements, such as haemoglobin, yeast extract, starch, glucose, cystine, trypticase or serum, have been added (Thayer and Martin 1964, Storck and Rinderknecht 1966, Amies and Garabedian 1967, Brookes and Hedén 1967). Good results can usually be obtained by the use of a simple medium, such as that recommended by McLeod and his colleagues (1934), made up with 10 per cent heated blood agar of pH 7.4 prepared from broth in which extraction of the meat has been carried out by Wright's (1933) method. The tubes or plates should be moist and the air in the incubator kept saturated with moisture and containing 10 per cent CO_2. Stock cultures are best kept in a similar medium containing 0.75 per cent agar, put up in the form of stabs; the tubes should be corked and kept in the incubator. On ordinary slope cultures meningococci and gonococci should be subcultured every two or three days. A number of workers have described chemically defined media on which the more exacting members of the group could be grown (see Mueller and Hinton 1941, Frantz 1942, Grossowicz 1945, Ley and Mueller 1946, Kenny *et al.* 1967, Hunter and McVeigh 1970, Catlin 1973, La Scolea and Young 1974).

All members of the group are aerobic; little or no growth occurs under strictly anaerobic conditions.

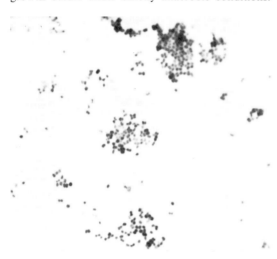

Fig. 28.2 *Neisseria gonorrhoeae*. From an overnight culture at 37° on hydrocele agar (× 1000 *ca*).

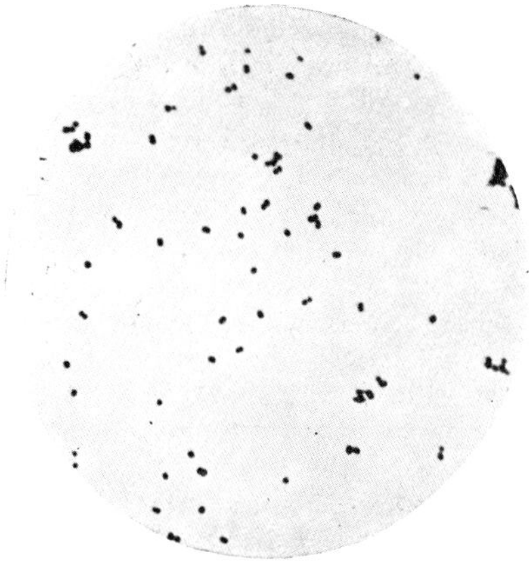

Fig. 28.3 *Neisseria subflava*. From a blood agar slope culture, 24 hr, 37° (× 1000).

Growth of the gonococcus is improved by the addition of cystine or other source of —SH bodies (McLeod *et al.* 1927, Boor 1942).

Workers are at variance on the effect on growth of lowered oxygen pressure or increased CO_2 in the atmosphere. This is not surprising, because of the complexity of the factors concerned. The technique used may, for instance, change the moisture content of the atmosphere and the rate of evaporation from the medium. The presence of 10 per cent CO_2 alters the H-ion concentration, and will interfere with the change in the reaction of the medium that normally occurs during growth; thus CO_2 may be beneficial if the medium has been made too alkaline, and it may by its buffering action prevent the accumulation of acid. The failure to standardize these secondary factors is probably sufficient to explain the diverse results obtained by different workers.

Like many previous workers, McLeod and his colleagues (1934), found that 10 per cent CO_2 improved growth, particularly that of freshly isolated strains (see also James-Holmquest *et al.* 1973). Glucose and glycerol have little or no beneficial effect; peptone in a 1–3 per cent concentration seems to be favourable. The optimum H-ion concentration for growth is about pH 7.4–7.6; the limits within which growth will occur are comparatively narrow, but they depend largely on the constitution of the medium. Brookes and Sikyta (1967) found that in an aerated continuous culture in a proteose peptone medium containing glucose the gonococcus grew best at pH 6.75. The optimum temperature for all the members is 37°; some of them, including the meningococcus will not grow below 30°. Gonococci grow well at 28°, and some strains grow slowly at a temperature as low as 25° (Annear and Grubb 1982). Many of the nasopharyngeal gram-negative diplococci will grow at 22°, but not always on first isolation. The meningococcus forms a weak haemolysin, reaching its maximum in trypagar cultures in about 4 days.

Some of the neisseriae produce a greenish-yellow pigment on solid media. The meningococcus and gonococcus in culture undergo rapid autolysis leading to swelling and loss of staining properties in cultures more than a few hours old. The autolytic system is destroyed by heating at 65° for half an hour.

Cultural characters

The colonial appearances of the neisseriae vary considerably. The meningococcus gives rise to a smooth typically lenticular colony, which tends to be mucoid when the organisms are capsulated. On incubation for some days the edge may become crenated or dentate and secondary papillae may appear on the surface. Organisms of group B, which are non-capsulated, tend to form rather smaller colonies, sometimes with a yellowish tint. Rough colonies often appear under continued laboratory cultivation, and are liable to be confused with the dull, low convex, translucent colonies that are antigenically smooth (Branham 1958). Colonies of the gonococcus are smaller than those of the meningococcus, are slightly more viscous and emulsify less readily. Two different colonial forms of the meningococcus and gonococcus were described by the earlier workers (Wassermann 1898, Lipschütz 1904, Atkin, 1923, 1925, Cohn 1923, Rake 1933); and more recently Kellogg and his colleagues (1963) recognized four colonial forms of the gonococcus. S. P. Wilson (1928) and G. S. Wilson and Smith (1928) observed and studied rough and smooth forms of numerous nasopharyngeal cocci and found a wide variety of colonial forms. The neisseriae in fluid media grow rather poorly. Turbidity is usually slight, and there is a finely granular or less often flocculent deposit. A surface ring growth may be apparent and occasionally a thin pellicle. There is no growth on MacConkey's medium.

Resistance

The resistance of neisseriae to inimical agencies is very low. In culture most of them die out in a few days; though if the organisms are seeded into ascitic agar stab tubes—preferably made up with 0.75 per cent agar—prevented from drying, and kept in the incubator at 37°, they may live for weeks or even months. Though it is not known with certainty why the gram-negative cocci die out in culture so quickly, it appears probable that they are killed by the amount of alkali produced; the production of NH_3 and of alkaline carbonates of organic acids may lower the H-ion concentration of the medium to pH 8.6–9.0, and thus bring about the death of the organisms (Phelon *et al.* 1927). The meningococcus and the gonococcus are killed by heating at 55° in 5 min or less; they are very susceptible to desiccation, death occurring usually within an hour or two. A high proportion of gonococcal cells die on freeze-drying. Ward and Watt (1971) find that, if they are suspended in 1 per cent proteose peptone containing 8 per cent glycerol and frozen at once in liquid nitrogen, all of them survive and can be recovered by rapid thawing in a 1 per cent proteose peptone medium at 37°. Weak disinfectants such as 1 per cent phenol or 0.1 per cent mercuric chloride prove fatal in 1 or 2 min.

When the sulphonamides were introduced, both the meningococcus and the gonococcus proved highly sensitive, but as a result of the widespread treatment of gonorrhoea with these drugs an increasing proportion of gonococci became resistant. With the exception of groups B and C, some members of which may be resistant, the meningococci have remained sensitive (Feldman 1965). Both meningococci and gonococci are normally sensitive to penicillin; a minor degree of resistance is manifested by gonococci in areas where this drug is used extensively. (For penicillinase production by gonococci, see p. 163). Other members of the *Neisseria* group are resistant to penicillin in the range of 0.16–4.0 units per ml (Storck *et al.* 1951).

Biochemical reactions

Biochemically the members of the group are not very active; the production of acid in glucose, maltose, and sucrose is used as a means of classification. Attack on sugars is by the oxidative method (Cowan and Steel 1965). According to Morse and Hebeler (1978), the metabolism of glucose by gonococci is influenced by the pH of the growth medium. At pH 7.2 about 80 per cent is metabolized by the Entner-Doudoroff pathway and 20 per cent by the pentose phosphate pathway. For testing fermentative ability a medium made up with ascitic fluid or with human or rabbit serum is needed. Horse, sheep and ox serum, which contain maltase, should be avoided (Rosher 1936, Hendry 1938). Thompson and Knudsen (1958) recommend Hartley's tryptic digest broth containing 5 per cent sterile rabbit serum, 0.15 per cent agar powder, and 0.2 per cent aqueous solution of phenol red; and Reyn (1965) recommends a solid medium prepared from placenta broth containing 1 per cent peptone, ascitic fluid and co-carboxylase. In our own experience the serum-free medium described by Flynn and Waitkins (1972) has proved most reliable. Rapid sugar utilization tests depend on the presence of preformed enzyme in the inoculum (Kellogg and Turner 1973, Brown 1974, Slifkin and Pouchet 1977). Some neisseriae have no fermentative ability, others attack only monosaccharides, others both monosaccharides and disaccharides, and others disaccharides but not monosaccharides. A polysaccharide is formed by some strains from 5 per cent sucrose (Hehre and Hamilton 1948, Berger 1962*a*). Litmus milk is unaltered. Nitrate is not usually reduced, but most species reduce nitrite, with or without the evolution of gaseous nitrogen (Berger 1961*a*, 1967). The methyl-red test is weakly positive or frankly negative, according to whether or not the organism acidifies glucose; since the pH seldom falls below 6.0, the red colour developed with methyl red is usually faint. The Voges-Proskauer reaction is sometimes positive when Barritt's method is used for testing. All true members produce catalase and give the oxidase reaction described by Gordon and McLeod (1928).

Antigenic structure

Most attention has been concentrated on the meningococcus and the gonococcus. The results of the early workers were rather confused and need not be described (see 6th Edition, p. 689). The classification of *meningococci* now used is the one recommended by a subcommittee of the Nomenclature Committee of the International Association of Microbiologists (see Branham 1953). It recognizes four groups—A, B, C and D. Using the double diffusion method in agar, Slaterus (1961) found that strains of groups A and C gave specific precipitation lines, but not group B—presumably because strains of group B are not capsulated. Strains of group D were never met with, but three new groups, referred to as types X, Y and Z, were isolated from cases or carriers. A new group, called group E, was encountered by Vedros and Culver (1968), forming 237 out of 625 strains isolated from cases and carriers. Group B meningococci appear to be heterogeneous; Frasch and Chapman (1972) found that, by extracting the organisms with hot acid or saline, over ten serotype antigens could be distinguished by microbacterial precipitin and agar gel diffusion techniques. It may be noted that Grados and Ewing (1970) observed an antigenic relation between members of this group and *Esch. coli.*

Serogrouping of meningococci can be carried out on plates containing group-specific antiserum, and serotyping by agar gel double diffusion with specific antisera (Craven *et al.* 1979). Grouping depends on capsular polysaccharide, typing on specific proteins in the outer membrane. The serogroups now recognized are A, B, C, D, X, Y, Z, W135 and 29E(ż). Rough strains, which have lost their capsular substance, react non-specifically by virtue of their common somatic antigen. Polysaccharide and nucleoprotein substances in the cell wall of the meningococcus are shared in part by other members of the *Neisseria* group (Prost *et al.* 1970). Zollinger and Mandrell (1980), who examined the non-capsular surface antigens of group A by the radioimmunoassay inhibition method, found three serologically distinct lipopolysaccharide antigens of which two were specific for group A strains, and five outer-membrane protein antigens of which one was specific.

For demonstration of the capsule either the tube precipitation, the capsular swelling (*Quellung*) (Clapp *et al.* 1935), or the gel-diffusion reaction (Petrie 1932) may be used. In this last method haloes appear around colonies of capsulated organisms grown on agar plates containing the homologous antiserum, owing to the interaction of the specific polysaccharide with antibody. The halo reaction is of value in eliminating the rougher, less specific, and less antigenic strains (Branham 1958).

It should be noted that the type specificity of the meningococcus can be changed by transformation (Alexander and Redman 1953) but there is as yet no evidence that this occurs in the body.

The antigenic structure of the *gonococcus* is even more difficult to unravel than that of the meningococcus. By agglutination and absorption of agglutinin tests, gonococci were tentatively classified into three or four groups (Bruckner and Cristéanu 1906 *a*, *b*, Torrey and Buckle 1922, Tulloch 1923). Atkin (1925) observed that most strains isolated from cases of acute urethritis fell into what he called type I, whereas strains from chronic infections generally fell into type II. He suggested that during residence in the body or in subculture type I might change into type II.

Danielsson (1965) made a fresh study of the subject, using agglutination, complement-fixation, fluorescent antibody, and double diffusion gel precipitation methods. Both heat-labile and heat-stable antigens were found to take part in the fluorescent antibody reaction; and good correlation was observed between the titres in this reaction and in those of the agglutination test. Gel diffusion revealed at least 15 antigenic factors, mostly distributed among four groups, A to D. Though nearly all antigens were present in each group, absorption tests indicated the existence of strain-specific factors (see also p. 164). A close antigenic relation was demonstrated between the gonococcus and the meningococcus, but only a very weak one with other species of *Neisseria.* Maeland (1967) found that antisera prepared against whole gonococci agglutinated both live and boiled cocci, but that sera prepared by the injection into rabbits of erythrocytes on to which a gonococcal antigen was adsorbed agglutinated boiled cocci only. The antibodies contained in this serum belonged to the IgM and the IgG classes, suggesting a superficial location of the sensitizing antigen. An indirect haemolysis test yielded far higher titres than the indirect haemagglutination test, probably owing to the greater sensitivity of the lysis test to the IgM class of antibodies.

Heckels (1978) established the presence of three outer-membrane proteins, of which one was a major component of the surface. Pili, which are present on Kellogg's T1 and T2 colonial types (see Brown and Kraus 1974), are composed of protein, and are antigenically heterogeneous (Novotny and Turner 1975). They facilitate the attachment of gonococci to cells and mucosal surfaces. Some workers have suggested that they also have an antiphagocytic role. Others think that a different surface factor, the leucocyte-association factor, is responsible for interactions between gonococci and neutrophils (Swanson *et al.* 1974).

Chemical fractionation shows that the outer membrane complex of the gonococcus is composed of lipid, lipopolysaccharide and protein (see Sadoff *et al.* 1978, Hendley *et al.* 1978) The proteins differ in their molecular weight and their antigenicity. Sixteen serotypes are distinguishable; these are distributed in different geographical areas. By gas-liquid chromatography, Jantzen, Bryn and Bøvre (1976) found that all strains contained ribose, glucose, glucosamine, 2-keto-3-deoxyoctonate, and heptose (see also Poolman *et al.* 1980).

Pathogenicity

The meningococcus gives rise to rhinopharyngitis, to epidemic cerebrospinal meningitis, and to post-basic meningitis in children; by intraspinal injection of monkeys it is possible to produce a meningitis with pure cultures of the organism. The gonococcus gives rise in human beings to gonorrhoea, with all its complications.

Towards laboratory animals all the gram-negative cocci behave in much the same way. Injected intraperitoneally in large doses into mice or guinea-pigs, they cause death in 1 to 3 days. *Post mortem*, there is a small amount of peritoneal exudate, and sometimes a little fibrin deposit on the organs; the spleen is slightly enlarged, and there is hyperaemia and degeneration of the viscera. The organisms can be cultivated from the peritoneal exudate, but rarely from the heart's blood. There is little multiplication of organisms inside the body; no true infection is set up, and death occurs from toxaemia. A similar result follows the injection of heat-killed organisms, though generally a rather larger dose is needed than of living cocci. It seems probable that the toxicity is due to some constituent of the nucleoprotein, since 'nucleoprotein' extracted from meningococci and gonococci is almost as toxic to mice as are the dead organisms themselves (Boor and Miller 1934). Rabbits are less susceptible than guinea-pigs or mice, and intraperitoneal inoculation generally leaves them unaffected.

The *meningococcus* is a specifically human parasite, giving rise, as a rule, to cerebrospinal meningitis (see Chapter 65). Intracisternal infection by the sub-occipital route with virulent *meningococci* is said to produce cerebrospinal meningitis in young rabbits, but the results are inconstant (Branham and Lillie 1932, Zdrodowski and Voronine 1932). Greater success in this respect can be obtained with *monkeys* (von Lingelsheim 1905), but not all of these animals are susceptible. By intraspinal injection of $\frac{1}{2}$ to 1 agar slope Flexner (1970*b*) succeeded in setting up meningitis in *Macacus rhesus* proving fatal, as a rule, in 18 hours to 4 days. *Post mortem*, the chief lesions were leptomeningitis, particularly at the base of the brain, encephalitis and abscesses, haemorrhages into the pia, inflammation of the dorsal root ganglia, and acute endarteritis. By giving repeated small doses a chronic meningitis could be produced. M'Donald (1908) partly confirmed Flexner's findings. In his hands rhesus monkeys proved insusceptible, but he obtained success with *Callithrix*. Strains of meningococci vary greatly in virulence (see Murray and Ayrton 1924). Some strains, when freshly isolated, will kill mice inoculated intraperitoneally in a dose of about 100 000 organisms, but others require far more than this. Miller (1933, 1934–35) showed, however, that if the culture was suspended in a solution of gastric mucin, as few as 2–10 organisms of a highly virulent strain might suffice to kill a mouse. For the maintenance of virulence, freeze-drying is recommended.

The *gonococcus*, like the meningococcus, is a specifically human parasite; it gives rise to gonorrhoea and ophthalmia neonatorum in both sexes and to vulvovaginitis in females. Except for chimpanzees, in which urethritis has followed the instillation of gonococci into the urethra, the disease cannot be reproduced in any laboratory animal, but the intra-urethral injection of man with pure cultures may set up a typical attack of gonorrhoea. When freshly isolated from the acute disease the organisms belong to colonial type 1, but are rapidly replaced in laboratory subculture by types 2, 3 and 4. Only type 1 is fully virulent, and only with organisms of this type can the typical disease be reproduced in man (Kellogg *et al.* 1963). Injection of a few drops of culture into the anterior chamber of the rabbit's eye gives rise to keratitis and hypopyon (Maslovski 1900); the organisms may invade the intra-ocular tissues, particularly the lens and ciliary body (Miller *et al.* 1945, Drell *et al.* 1947). Injection of 0.000 01 mg of a freshly isolated strain into the yolk sac of an 8-day chick embryo causes death in 3–4 days (Tani and Tashiro 1955). The organisms can be recovered from the blood and tissues. The virulence varies with colonial type (Bumgarner and Finkelstein 1973). Intraperitoneal injection with mucin enhances the pathogenicity of the gonococcus for guinea-pigs (for a review, see Hill 1944).

In the causation of gonorrhoea the surface components of the gonococci—pili, capsules, lipopolysaccharides and certain proteins—are thought to play a critical role, enabling the organisms to attach themselves to the human epithelium, resist the bacteristatic action of antibodies and host cells, and possibly too of antibiotics (Heckels 1978; but see Novotny *et al.* 1977). Cultivation of laboratory strains in perforated plastic chambers implanted in the subcutaneous tissue of guinea-pigs is said to go some way to restoring them to their original pathogenic form (Penn *et al.* 1976, Goldner *et al.* 1979). (For further information on the antigenic structure and virulence of the gonococcus, see Watt and Lambden 1978 and Chapter 66.)

Toxin production

It has already been stated that dead organisms of the *Neisseria* group injected intraperitoneally into mice or guinea-pigs cause death in almost the same dosage as living organisms (see Albrecht and Ghon 1901, Neill and Taft 1920, Report 1920), indicating that the effect is caused by some toxic constituent of the cells. Under suitable conditions the toxin is partly liberated into culture media after incubation as the result of autolysis (Ferry *et al.* 1931), but it can be readily extracted from the bodies of the ground-up organisms with N/20 NaOH solution. With meningococci the MLD of the extract is about 0.1 to 0.15 ml, corresponding to 2 mg of the dried cocci (M. H. Gordon, see Report 1920, Petrie 1937). The toxin appears to be responsible in both animals and man for the frequent occurrence of haemorrhages on the serous membranes, of sterile transudates in the cavities of the body, and for adrenal haemorrhages (Maclagan and Cooke 1917, Petrie 1937). It belongs to a group of non-specific, non-antigenic, thermostable bacterial poisons (Petrie 1937). De Christmas (1897, 1900) claimed to have demonstrated the production of a true exotoxin by the gonococcus, but his work has never been confirmed. The experimental observations of Wassermann (1898), of Schäffer (1897), and of Maslovski (1900) all indicate that the toxin is of the same thermostable non-antigenic type as that formed by other members of the *Neisseria* group. Maeland (1968), who extracted the endotoxin with various solvents, found that it provoked an epinephrine skin reaction in rabbits; it sen-

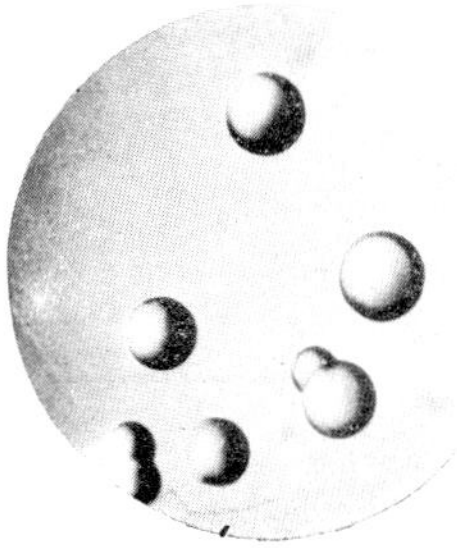

Fig. 28.4. *Neisseria meningitidis*. Surface colonies on serum agar, 24 hr, 37° (×8).

sitized sheep red cells so that they were agglutinated by an anti-gonococcal serum; and by means of haemagglutinin-inhibition and absorption tests it was shown to contain two antigenic determinants, one of which was a protein and the other a carbohydrate.

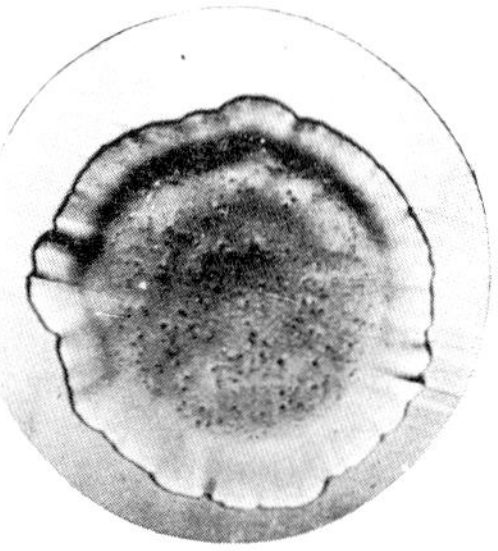

Fig. 28.5. *Neisseria meningitidis*. Surface colony on serum agar, 7 days, 37° (×8).

Classification

The gram-negative cocci, as a group, are difficult to classify satisfactorily. Apart from the meningococcus and the gonococcus, the definition of the different species is far from clear. This is mainly because the colonial appearances are subject to such great variation that the descriptions given of apparently the same species by different workers are often quite contradictory. The cultural descriptions, for example of *'N.' catarrhalis* are most varied (Ghon and Pfeiffer 1902, Dunn and Gordon, M. H. 1905, von Lingelsheim 1906, Arkwright 1907, Gurd 1908, Elser and Huntoon 1909, Martin 1911, Netter and Debré 1911, Dopter 1921, Gordon, J. E. 1921), though all workers agree on its failure to ferment sugars. The nasopharyngeal cocci present a miscellany of characters, forming smooth or rough colonies, fermenting or not fermenting a small range of mono- or disaccharides, and producing a greenish-yellow pigment or no pigment at all. Colonial formation is particularly confusing, owing to its inconstancy, and in our experience cannot be relied upon for the separation of species (S. P. Wilson 1928, G. S. Wilson and Smith 1928).

Study of the base composition of the DNA of the gram-negative cocci and of their genetic transformation shows that they can be divided into two groups: one possessing a G + C content of about 50 moles per cent, and the other, comprising *'N.' catarrhalis*, *'N.' ovis* and *'N.' caviae*, having a G + C content of 40 to 44 moles per cent (Catlin and Cunningham 1961). The organisms in this second group reduce nitrate to nitrite, fail to ferment sugars, and show not only intraspecific but also interspecific genetic transformation with members of the *Moraxella* group (Henriksen and Bøvre 1968*b*). In these three respects, as well as in their low G + C content, they resemble the moraxellae, and Henriksen and Bøvre (1968*b*) would for this reason transfer them to the *Moraxella* genus. This proposal has much to be said for it, but as it would mean classifying cocci and rods in the same genus, it would probably be unacceptable to morphologists. They are clearly distinct from the first group, and we shall therefore follow the recommendation in the 8th edition of *Bergey's Manual of Determinative Bacteriology* (Buchanan and Gibbons 1974) in classifying them in a separate genus, *Branhamella*. The only definite member of this genus is *Branhamella catarrhalis*, but mainly because of their similar DNA base constitution *'N'. caviae* and *'N'. ovis* may be tentatively included in it. Hoke and Vedros, (1982) however, regard *'N'. caviae*, *'N'. cuniculi*, and *'N'. ovis* as sufficiently distinct as to justify their separation into a distinct genus.

Members of the genus *Neisseria* have a G + C content of about 50 moles per cent and show both intra- and interspecific genetic transformation (Catlin and Cunningham 1961). Their classification must depend on fermentative ability, supplemented to some extent by antigenic structure and pigment formation (see Table 28.1).

The subflava group of cocci forming yellowish-green colonies comprises organisms variously referred to as *N. pharyngis flava* i, ii, and iii, *N. flava*, *N. subflava*, and *N. perflava*. The specific individuality of these organisms is very doubtful. There is at present no sure means of distinguishing between them, though Berger and Wulf (1961) would divide them on the basis of a number of minor characters into a perflava and a flava-subflava group. Many years ago (Wilson and Smith 1928) we suggested that they should all be combined into one species, for which some such name as *N. pharyngis* would be appropriate. With this general conclusion Véron, Thibault and Second (1961), Catlin and Cunningham (1961), and Henriksen and Bøvre (1968*b*) agree, but the specific designation *subflava* is preferred to *pharyngis*. We may therefore designate these organisms as *N. subflava*. The taxonomic position of other organisms that have been regarded as neisseriae is doubtful. They will be described briefly in a separate section on p. 165.

For distinguishing pathogenic from non-pathogenic neisseriae, Mulks and Plaut (1978) find that meningococci and gonococci form an IgA protease, whereas the harmless neisseriae do not. For distinguishing the

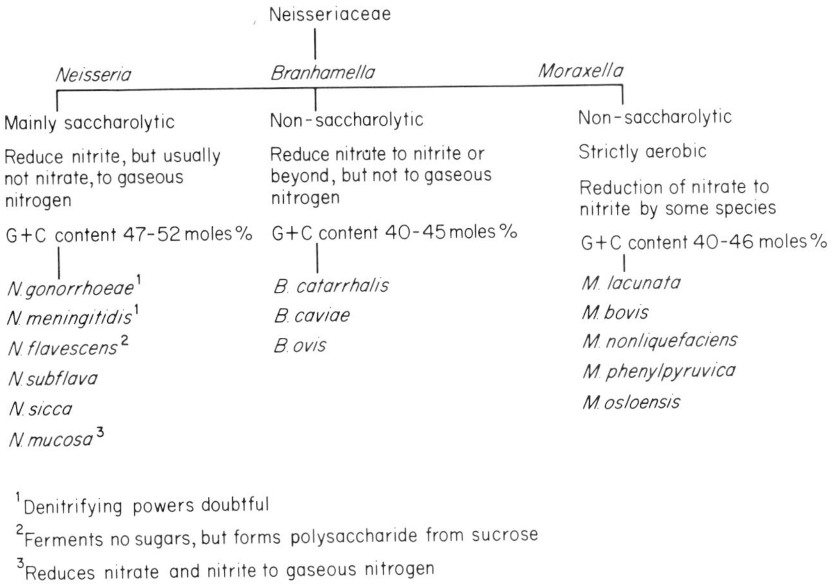

Fig. 28.6 Differentation of *Neisseria*, *Branhamella* and *Moraxella*.

meningococcus from the gonococcus, Odugbemi, McEntegart and Hafiz (1978) recommend growing the organisms on a specially defined medium and testing their ability to grow up to the edge of a disc impregnated with manganous chloride and Congo red; meningococci are able to do so, whereas gonococci are inhibited.

The meningococcus and gonococcus closely resemble each other. Culturally, the gonococcus is more exacting in its nutritive requirements, grows more slowly, and forms smaller colonies, which are slightly viscous and are not as readily emulsified as the butyrous colonies of the meningococcus. It ferments glucose only, and is less toxic than the meningococcus to mice and guinea-pigs injected intraperitoneally.

Fig. 28.6 sets out a provisional classification of the organisms which can for the present be regarded as belonging to the Neisseriaceae.

Description of the species of Neisseria

The following descriptions are intended mainly to supplement the general characters of the genus already given, but some overlapping is inevitable (see also Table 28.1).

Neisseria gonorrhoeae

Oval or spherical cocci 0.8 μm × 0.6 μm, often arranged in pairs with adjacent sides flattened or slightly concave, resembling a pair of kidney beans. In cultures great variation in size and depth of staining occurs, owing to autolysis; in the body the cocci are more regular, and are generally intracellular. Non-motile; may have capsules on first isolation. Three antigenically distinct sorts of pili observed in culture (Novotny and Turner 1975). G+C content of the DNA: 49–50 moles per cent.

Nutritionally exacting, but grows quite well on chocolate agar prepared from broth in which extraction of the meat has been carried out by Wright's method (1933). Incubation should be in 10 per cent CO_2 in a moist atmosphere. Optimum growth occurs at 37°; no growth below 25° or above 38.5°. In culture, four colonial types recognizable, 1, 2, 3 and 4 (Kellogg *et al.* 1963). Types 1 and 2 are found in primary cultures from clinical material; types 3 and 4 appear in subcultures. Brown and Kraus (1974) describe their appearance on GC base medium supplemented with Isovitalex after 20 hours of growth in a candle extinction jar at 36°, and illustrate them in colour. They can also be distinguished by use of a dissecting microscope with a concave mirror illuminated by a fluorescent lamp (Juni and Heym 1977). By this method lenticular colonies of types 1 and 2 refract the light; the flatter, rather larger, colonies of types 3 and 4 do not.

The organisms are very susceptible to inimical agencies. Killed by moist heat at 55° in less than 5 min, in serum cultures by 1 in 4000 $AgNO_3$ in 7 min, and in pus in 2 min. Most strains are sensitive to penicillin, but a few acquire a minor degree of resistance. Strains producing β-lactamase, which are completely resistant to penicillin, appeared in the Philippines and West Africa in 1976, and have now been reported from several countries (see Chapter 66). Produces slight amount of acid in glucose. For testing this, a number of workers have described rapid fermentation methods (Kellogg and Turner 1973, Brown 1974, Tringali and Intonazzo 1978). Does not reduce nitrite. H_2S negative. Antigenically complex. Gel diffusion discloses 15 antigenic factors, mostly

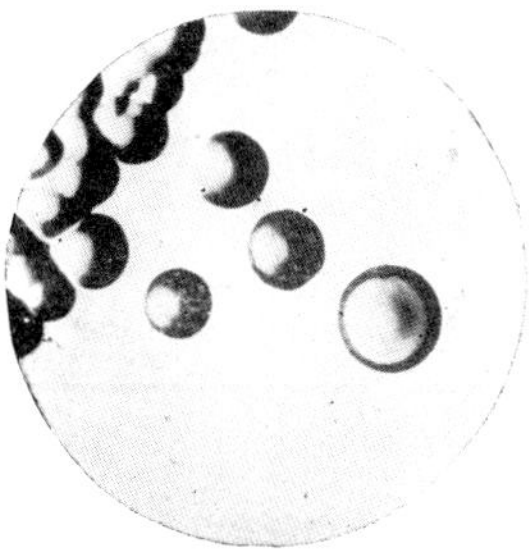

Fig. 28.7 *Neisseria gonorrhoeae*. Surface colonies on serum agar, 24 hrs, 37° (×8).

distributed among four groups (Danielsson 1965). Apicella (1976) designates four distinct acidic polysaccharide antigens, and has devised a typing system using haemagglutination inhibition with appropriately absorbed sera. On the other hand, Johnston, Holmes and Gotschlich (1976) recognize 16 serological types based on the antigenically specific proteins (polypeptides) in the outer membrane of the organisms. In a study of four strains of gonococci, Danielsson and Sandström (1979) demonstrated both common and strain specific antigens. Pathogenically, it is responsible for gonorrhoea and ophthalmia in man, and sometimes arthritis; it is a strict parasite of man. (For reviews of the gonococcus, see Roberts 1977, Morse 1979.)

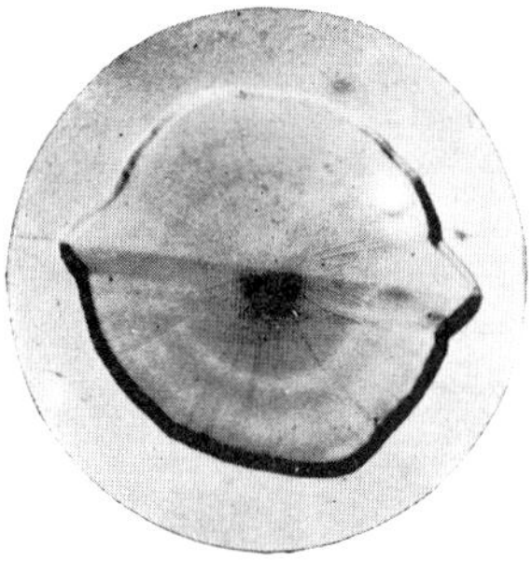

Fig. 28.8 *Neisseria gonorrhoeae*. Surface colony on serum agar, 7 days, 37° (×8).

Neisseria meningitidis

Synonym: Diplococcus intracellularis meningitidis of Weichselbaum, commonly known as the meningococcus.

Resembles the gonococcus morphologically.

Members of groups A, C and D may form capsules. G+C content of DNA: 50–51.5 moles per cent. Grows well on blood or serum agar, but not at first on plain nutrient agar. Forms lenticular colonies, rather larger than those of the gonococcus. Growth favoured by 5–10 per cent CO_2. No growth below 30°. Forms acid in glucose and maltose. Reduces nitrite, but not nitrate. H_2S negative. Killed by moist heat in less than 5 min. Sealed cultures kept at 37° may live for 4 weeks or so. Susceptible to penicillin. Divided by agglutination and absorption of agglutinins into four main groups, A to D. Group B is less homogeneous than group A or C. Group-specificity determined mainly by a polysaccharide antigen. Causes cerebrospinal meningitis and septicaemia in man. Has been isolated from the human genital tract (see Givan *et al.* 1977). Is a strict parasite of man, and is found in the nasopharynx. (For review, see Branham, 1940, 1958.)

Neisseria flavescens

Isolated by Branham (1930) from spinal fluid of patients with meningitis. Differs from the meningococcus chiefly in its production of a yellow pigment, its lack of fermentative activity, and its antigenic constitution. G+C content given variously as 46.5 to 50.1 moles per cent (Catlin and Cunningham 1961, Marmur and Doty 1962, Schildkraut *et al.* 1962, Bøvre 1967). Grows well on blood agar, but poorly on glucose agar. Produces polysaccharide from sucrose, which must therefore be broken down to monosaccharides (Berger 1967). Antigenically homogeneous, different from the meningococcus. Occasionally causes meningitis in man.

Neisseria subflava

A rather heterogeneous species incorporating *Diplococcus pharyngis flavus*, *Neisseria flava*, and *Neisseria perflava*. Isolated by von Lingelsheim in 1906 and studied notably by Elser and Huntoon (1909), J. E. Gordon (1921) and G. S. Wilson and Smith (1928); (see also Martin 1911, Report 1916, Dopter 1921). Diplococcus, often arranged in tetrads or dense clumps. Usually grows at 22° on plain agar, forming a greenish-yellow pigment. Colonies are smooth or rough, generally coherent, tenacious, membranous, friable, auto-agglutinable, and difficult to emulsify. Ferments glucose and maltose, and often sucrose. Antigenically the constituent species appear to be closely related to each other (Warner *et al.* 1952), to possess the same G+C content, 48.0–50.5 moles per cent, and to show interspecific genetic transformation, pointing to the essential unity of the species (Véron *et al* 1961, Catlin and Cunningham 1961, Henriksen and Bøvre 1968*b*). Sheep and rabbit red blood corpuscles may be lysed in blood broth (Berger and Wulf 1961). One case of purpuric septicaemia ascribed to it in man (Muchmore and Venters 1968).

Neisseria sicca

Isolated by von Lingelsheim in 1906. On agar forms an irregularly round, raised, opaque colony up to 3 mm in diameter, with a dull, dry, deeply furrowed surface and a crenated edge; the colony is very firm, often adherent to the medium, difficult to disintegrate, and impossible to emulsify. Grows, as a rule, at 22° on plain agar. Ferments glucose, maltose and usually sucrose. G+C content 49.0–51.5 moles per cent. Non-pathogenic. The claim to specific rank is based on its absence of pigment formation and its antigenic difference from *N. subflava* (Warner *et al.* 1952). This is contested by Véron, Thibault and Second (1961), who regard it as a variant of *N. mucosa*, differing from it in reducing nitrite but not nitrate to gaseous nitrogen, though sharing with it a common O antigen. Our own experience suggests that it is merely a rough variant of one of the other nasopharyngeal diplococci. Thus, we have observed initially smooth colonies changing in later cultures to rough colonies indistinguishable from those described as typical of *N. sicca*. Is a very rare cause of bacterial endocarditis (Gay and Sevier 1978).

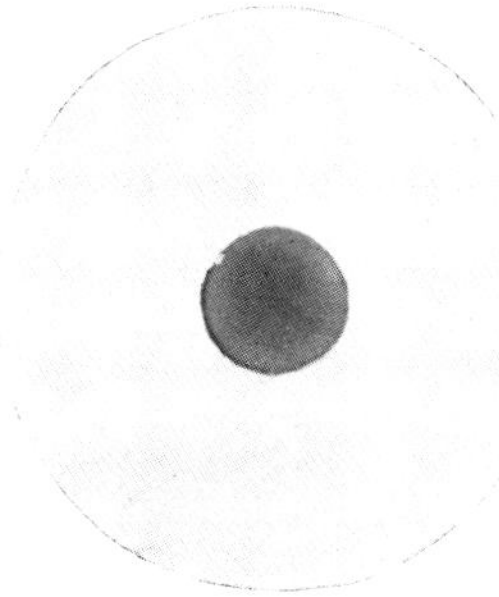

Fig. 28.9 *Neisseria subflava*, Surface colony on agar, 24 hr, 37° (× 8). Smooth type.

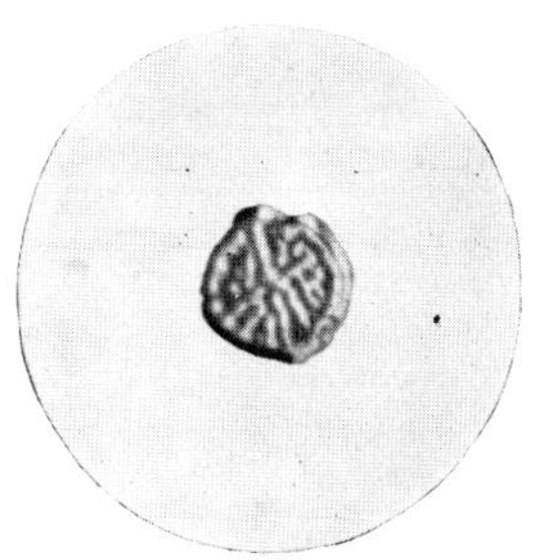

Fig. 28.10 *Neisseria subflava*. Surface colony on agar, 24 hr, 37° showing primary rough type of colonial variant (× 8).

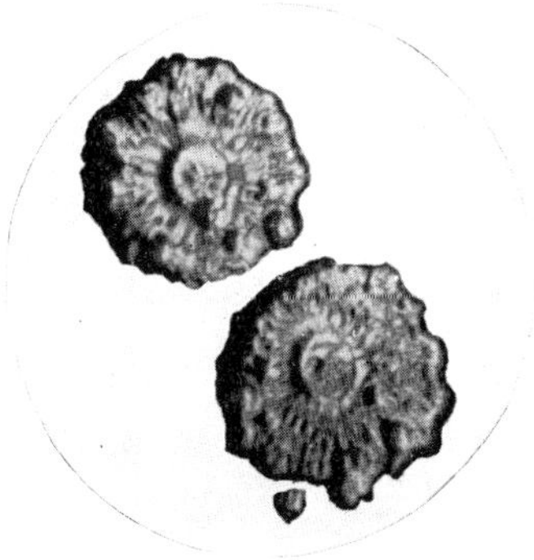

Fig. 28.11 *Neisseria subflava*. Surface colonies on agar, 5 days, 37°, showing differentiation. Secondary rough type (× 8).

Fig. 28.12 *Neisseria subflava*. Surface colonies on agar, 5 days, 37°, showing formation of secondary papillae (× 8).

Neisseria mucosa

Isolated by von Lingelsheim (1908) from the nasopharynx and cerebrospinal fluid. Distinguished by its capsulation, growth on plain agar at room temperature, formation of mucinous colonies, and its reduction of nitrate. Growth is more abundant than that of the meningococcus. Colonies on serum agar are convex, yellowish-grey, mucinous, opaque, and readily emulsified. Ferments glucose, maltose and sucrose, and reduces both nitrate and nitrite to gaseous nitrogen. G+C content of DNA 51.0 moles per cent. Is resistant to penicillin but very sensitive to chloramphenicol. Injected intraperitoneally into mice, it proves fatal in 24–48 hr; *post mortem* there is haemorrhagic inflammation of the abdominal viscera with a fibrino-purulent exudate. Antigenically different from *N. subflava*, but is said to show some affinity with *N. sicca* (Véron *et al* 1961). Not to be confused with *Diplococcus mucosus* (Cowan 1938, Véron *et al.* 1959, 1961), which Seeliger (1953) regards as identical with *Acinetobacter anitratus*.

Organisms whose taxonomic position is still uncertain

'Neisseria' cinerea Von Lingelsheim (1906) described this organism (*N. pharyngis cinerea*) as consisting of plump cocci, arranged in pairs or more usually loose heaps, forming grey or greyish-white granular colonies on agar, 1–1.5 mm in diameter, with an entire edge. Differs from *N. flavescens* in growing less abundantly, in producing no pigment, and in forming no polysaccharide from sucrose (Berger 1967). G+C content of DNA: 49.0 moles per cent.

'Neisseria' canis Isolated by Berger (1962*a*) from the throat of dogs. Grows well at 22°, forms abundant yellowish pigment, ferments no sugars, reduces nitrate to nitrite, is proteolytic, liquefies gelatin, but does not hydrolyse tributyrin. Colonies are tough and friable.

'Neisseria' animalis This name was given by Berger (1960*b*) to an organism isolated from the throat of a guinea-pig. Ferments sucrose, forms a polysaccharide in 5 per cent sucrose agar, reduces nitrite but not nitrate, does not hydrolyse tributyrin or produce pigment.

'Neisseria' cuniculi Isolated by Berger (1962*a*) from rabbits. Has no distinguishing features. Colonies are raised whitish and generally wrinkled. Reduces neither nitrate nor nitrite, ferments no sugars, and forms no pigment. Appears to be antigenically homogeneous.

'Neisseria' crassa Originally described by Jaeger (1895). Said to occur in pairs or tetrads slightly larger and rounder than the meningococcus, but forming rather smaller colonies, greyish-white, smooth, with an entire edge. Grows, though poorly, on agar at 22°; produces acid in glucose, maltose, sucrose, and lactose; reduces nitrate and nitrite to gaseous nitrogen; withstands drying better than the meningococcus,

Table 28.1 Some differential characters of *Neisseria*, *Branhamella* and *Moraxella*

	Pigment formation	Growth at 22°	Acid from glucose	Acid from maltose	Acid from sucrose	H_2S	Polysaccharide from sucrose	G+C content moles %	Pathogenicity for man	Remarks
N. gonorrhoeae	–	–	+	–	–	–	–	49.5–49.6	+	CO_2 necessary, or almost so, for growth. Colonies smooth
N. meningitidis	–	–	+	+	–	–	–	50.0–51.6	+	Growth improved by CO_2. Colonies smooth, sometimes capsulated
N. flavescens	+	+	–	–	–	+	+	46.5–50.1	+	Colonies smooth & golden yellow
N. subflava	+	V	+	+	V	+	V	48.0–50.5	Rarely	Forms capsules. Colonies greenish-yellow
N. sicca	–	V	+	+	+	+	+	49.0–51.5	–	Colonies coarse, dry and wrinkled
N. mucosa	–	+	+	+	+	+	+	50.5–52.0	–	Colonies mucoid, Forms capsules Reduces nitrate Pathogenic to mice when given *iv*.
B. catarrhalis	–	+	–	–	–	–	–	40–44	Seldom	Reduces nitrate
B. caviae	–	S	–	–	–	–	–	43.9–44.5	–	Antigenically distinct
B. ovis	–	S	–	–	–	–	–	43.9–44.5	–	Injected *iv* into mice, it causes haemoglobinuria & death in 18 hr
M. lacunata	–	S	–	–	–	–	–	41.5–43	+	Requires complex media. Liquefies coagulated horse serum. Nitrate usually reduced
M. bovis	–	S	–	–	–	–	–	42.5–43	–	Nitrate not reduced. Liquefies coagulated horse serum. Haemolysis on blood agar. Produces pink eye in cattle
M. nonliquefaciens	–	S	–	–	–	–	–	40.2–42	?	Nitrate reduced. No coagulation of horse serum. No haemolysis. Found in ear, nose and throat
M. phenylpyruvica	–	S	–	–	–	–	–	43–43.5	?	Nitrate reduction variable. Strong urease activity. Deaminates phenylalanine
M. osloensis	–	S	–	–	–	–	–	43–43.5	?	Grows on plain media. No deamination of phenylalanine. Nitrate usually reduced

V = Variable; S = Scanty.

is antigenically homogeneous, but is agglutinated by an anti-meningococcal serum (Kutscher 1912, Berger 1962*a*). Its identity has never been satisfactorily established.

'Neisseria' lactamica Described by Hollis, Wiggins and Weaver (1969). On heart infusion agar colonies resemble those of the meningococcus. May form a yellowish pigment on Loeffler's serum. Some strains grow on nutrient agar at 25° and 31°, some only at 37°, and some at neither temperature. Produces acid from glucose, maltose and lactose, and forms β-D-galactosidase (ONPG reaction). Strongly oxidase positive. Distinguished from the meningococcus only on biochemical and serological grounds, but 70 per cent of strains are auto-agglutinable. Found most frequently in the nasopharynx of infants and young children (Kuzemenská *et al* 1976, Lauer and Fisher 1976, Gold *et al.* 1978, Holten *et al.* 1978). Occasionally gives rise to meningitis. Has been isolated from the human genital tract (Jephcott and Morton 1972).

Branhamella

Definition

Gram-negative cocci, usually arranged in pairs. Grow on nutrient agar at 37°, often at 22°. Ferment no carbohydrates and form no pigment. Produce catalase and oxidase. Reduce nitrate as a rule. G+C content of DNA: 40–45 moles per cent. Seldom pathogenic to man.

Type species is *Branhamella catarrhalis*.

Branhamella catarrhalis

This organism has been variously described by different observers (Ghon and Pfeiffer 1902, von Lingelsheim 1906, Elser and Huntoon 1909, Kutscher 1912, and J. E. Gordon 1921). Ghon and Pfeiffer, who first studied it fully, stated that in sputum the organisms occurred in pairs, tetrads, or small groups and were shaped like coffee-beans. Grew on agar at

22°, forming small convex, whitish-grey colonies which after 3 or 4 days reached a diameter of 3–4 mm and were differentiated into a prominent opaque centre and a thin translucent wave-like periphery with a crenated edge; of friable consistency and auto-agglutinable. Other workers give descriptions that differ in various ways from this, but on the whole agree that the colonies are small, with a smooth glistening surface. Kutscher (1912), however, describes them as tough, dry and friable with an uneven surface, a clearer periphery, radial striation and an irregularly indented edge. Organisms are capsulated. Growth occurs at 18° to 42°, best at 37°, and is improved by blood and serum. Ferments no sugars, but hydrolyses tributyrin (Berger 1962*b*). Resists heating to 65° for 30 min. Antigenically distinct from the neisseriae (see Warner *et al.* 1952). According to Eliasson (1980), it contains a protein antigen that distinguishes it from other members of the *Branhamella* genus. Precipitating antibodies to this so-called P antigen were found in 69 per cent of normal human sera in Sweden. Guinea-pigs given large intraperitoneal doses die of toxaemia in about 24 hr. In man it gives rise not only to upper respiratory infections, but according to Leinonen *et al.* (1981) to a proportion of cases of middle ear disease. Said to differ in its fatty acid composition from typical *Neisseria* (Lewis *et al.* 1968). G+C content of DNA: 40–44 moles per cent.

Branhamella caviae

Isolated from the throat of guinea-pigs, described by Pelczar, Hajek and Faber (1949), and named *N. caviae* by Pelczar (1953). Forms a light caramel to dirty brown growth, ferments no sugars, reduces nitrate and nitrite though not to gaseous nitrogen, hydrolyses tributyrin, and is antigenically distinct. G+C content of DNA: 43.9 moles per cent.

Branhamella ovis

Isolated by Lindqvist (1960) from keratoconjunctivitis in sheep and, later, by Pedersen (1972) from cattle. Grows well on ordinary media, but only scantily at 22°. On bovine blood agar forms glossy, translucent, non-pigmented colonies movable on the surface of the medium and surrounded by a zone of haemolysis. Ferments no sugars; forms catalase and oxidase; and reduces nitrate to nitrite. Injected intravenously into mice, it causes haemoglobinuria in 2 hr and death in 18 hr. G+C content of DNA: 43.9 moles per cent.

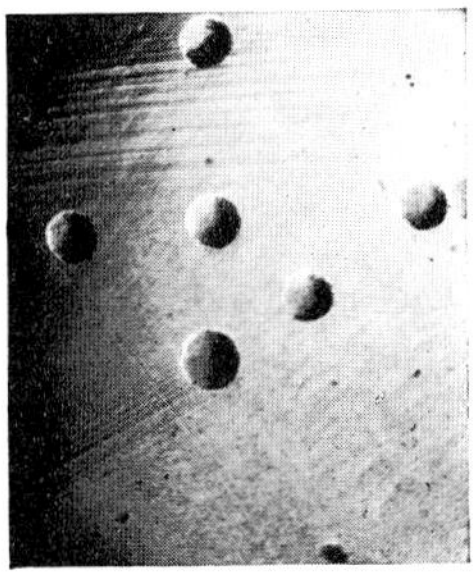

Fig. 28.13 *Morax-Axenfeld Bacillus*. Surface colonies on Fildes's agar plate, 24 hr, 37° (×8).

Moraxella

Definition

Short, plump, gram-negative bacilli arranged in pairs; often pleomorphic in old cultures. Non-flagellated, but said to be sluggishly motile. Non-pigmented. Moderate growth on culture media; at least one species requires natural animal protein. Strictly aerobic. Oxidase and usually catalase positive. Grow best at 32–37°. No acidification of sugars. May or may not liquefy gelatin. Nitrate reduction variable. Indole and H_2S production negative. Sensitive to penicillin and most other antibiotics. Strict parasites of mammals. Some species are pathogenic for man and animals. G+C content of DNA: 40–45 moles per cent.

Type species is *Moraxella lacunata*.

General properties

The first member of this group—*Moraxella lacunata*—was described independently by Morax (1896) and Axenfeld (1897), who isolated it from cases of angular conjunctivitis. The name *'Bacillus lacunatus'* was suggested for it by Eyre (1900). Several other species have since been described. The organisms are parasitic, and have their main habitat on the mucosae of the nasopharyngeal, conjunctival, and genital tracts.

Morphologically, they are plump rod-shaped organisms, 2–3 μm long and 1 μm broad, with parallel or slightly convex sides and rounded ends. They are disposed in pairs end-to-end and sometimes in short chains. They have no flagella, and are usually described as non-motile; Piéchaud (1963), however, examining them on nutrient agar in a special oil chamber, observed sluggish movements of the swinging, longitudinal, and group-translation variety, reminiscent of the myxobacteria; but Halvorsen (1963) was unable to confirm this. They are non-sporing, gram negative, and except for *M. bovis*, show little evidence of capsule formation; in most respects they resemble Friedlaender's pneumobacillus in appearance. In old cultures pleomorphic forms are common, ranging in size and shape from short stunted diplococci to long jointed or filamentous, sometimes fusiform, threads (Eyre 1900).

Culturally they vary in their nutritive requirements. One species, *M. lacunata*, requires natural animal protein such as serum, blood, or ascitic fluid. Others thrive on plain nutrient agar; and *M. osloensis* can use citrate or ethanol as the sole source of carbon. Development of most species is limited to a temperature range of 30–40° and occurs only under aerobic conditions.

On serum agar at 37° colonies have a diameter of about 1 mm after 24 hr. They are circular, raised, greyish, and translucent. On further incubation they increase in size, reaching a diameter of 2–5 mm, and become differentiated into a slightly raised opaque whitish centre and a thin translucent periphery with a lobate edge. The medium is pitted owing to liquefaction of the serum. Loeffler's serum is similarly pitted, and may be digested within 3 or 4 days. Some species are weakly haemolytic, colonies on horse or ox blood agar being surrounded by a narrow zone of lysis. No growth occurs on MacConkey's medium. In serum broth after 24 hr at 37° there is a uniform turbidity; later a greyish-white deposit forms, and increases for 6 to 10 days, after which the medium clears and the organisms sink to the bottom of the tube.

Cultures on solid media usually remain alive at room temperature for not more than 2 to 4 weeks. In liquid media cultures are sterilized by exposure to a temperature of 56° for 15–30 min. No sugars are fermented. Gelatin, serum, and casein are digested by most species. Positive results are obtained in the oxidase, and usually in the catalase and nitrate reduction tests; and negative results in the indole, urease, and phosphatase tests. All strains are sensitive to penicillin and to most other antibiotics. Little is known of their antigenic structure. According to Haug and Henriksen (1969*b*), *M. lacunata*, *M. nonliquefaciens*, *M. bovis*, and *'M.' kingae* agglutinate better after being boiled than before. On the other hand, *M. phenylpyruvica* and *M. oslensis* agglutinate as well or better when unheated than when boiled. *M. lacunata*, *M. nonliquefaciens*, and *M. bovis* have some antigens in common. A few species give rise to conjunctivitis in man or animals, and occasionally to more severe infections in man, such as septic arthritis and endocarditis (Silberfarb and Lawe 1968, Feigin *et al.* 1969) and in cattle, such as infectious bovine keratoconjunctivitis. To laboratory animals most are harmless. The G + C content of their DNA is 40–45 moles per cent (Bøvre 1967, Henriksen and Bøvre 1968*b*).

Classification

The generic name *Moraxella* was suggested by Lwoff (1939). In this genus Lwoff (1964) would include both oxidase-positive and oxidase-negative species. Though not all workers are agreed (see Henderson 1965), there is a strong tendency to reserve the genus *Moraxella* for the oxidase-positive organisms, and to transfer the organisms known variously as *M. lwoffi* and *M. glucidolytica* to the *Achromobacter* or the *Acinetobacter* genus (Gilardi 1967, Henriksen and Bøvre 1968*b*, Baumann *et al.* 1968; Chapter 32). These latter organisms have simpler nutritive requirements, growing, for example, on a synthetic medium containing citrate and ethanol, and failing to react with the oxidase-positive organisms in transformation experiments (Henderson 1965).

Henriksen (1952) drew attention to the close resemblance of *Moraxella* to *Neisseria*. The main differences were the bacillary form of the moraxellae and their division in one plane instead of in two. Subsequent work, however, has shown that the resemblance is to *Branhamella* rather than to *Neisseria* as at present defined. *Moraxella* may therefore be included in the family Neisseriaceae, but not as part of the *Neisseria* genus (see the review by Henriksen 1973). Five species are generally recognized: *M. lacunata* including the species described as *M. duplex* and *M. liquefaciens* (Petit 1900), *M. bovis*, *M. non-liquefaciens*, *M. phenylpyruvica* and *M. osloensis*.

Moraxella lacunata

This organism has the general characters of the genus. In cases of angular conjunctivitis it is found free in the secretion or within the polymorphonuclear and desquamated epithelial cells. Growth occurs best on Fildes's agar and on serum or egg medium; it is poor on blood agar, very poor on chocolate agar, and fails to occur in the absence of natural animal protein. At 37° growth is improved by incubation in a moist atmosphere. On serum agar the medium is pitted beneath the colonies. Liquefies coagulated horse serum. On Fildes's agar young colonies are water-clear, amorphous, low convex with a smooth glistening surface and entire edge; later, they become differentiated and take on a frosted-glass appearance. Catalase is produced, nitrate reduced, and litmus milk is rendered slowly alkaline. The organism is very susceptible to zinc salts, which have an almost specific effect in the treatment of the conjunctival disease. Said to be antigenically related to *Branhamella catarrhalis* and to *Acinetobacter glucidolytica* (see Estevenon *et al.* 1978). Though non-pathogenic to laboratory animals, *M. lacunata*, when instilled into the conjunctival sac of a healthy human volunteer, gives rise in about 5 days to angular conjunctivitis. The organism has a G + C content of 41.5–43 moles per cent. The *liquefaciens* variant differs in growing at room temperature, in doing without the addition of natural animal protein, and in its rapid liquefaction of gelatin. These differences are so slight that Henriksen and Bøvre (1968*b*) consider that it should be included in the species *M. lacunata*. A closely allied organism was isolated by Ryan (1964) from the conjunctiva of healthy guinea-pigs.

Moraxella nonliquefaciens

An organism distinguished by its failure to liquefy gelatin or digest serum was described under this name by Kaffka (1955), who isolated it from the sputum of a woman suffering from bronchopneumonia and from the throat of two normal persons. Subsequently Henriksen (1958; see also Bøvre 1970) found it in 11.3 per cent of 875 cultures from the nose of patients in an ear, nose and throat clinic at Oslo, and Berencsi and Mészáros (1960) in 23 per cent of secretions collected during bronchoscopy from 91 patients suffering from bronchiectasis. It is probably identical with the strain isolated by De Bord (1942) and called *Mima polymorpha* var. *oxidans* (Flamm 1957, Catlin and Cunningham 1964). Bøvre (1967) and Bøvre and Henriksen (1967*a*) observed great variation among strains labelled *M. nonliquefaciens* and concluded that the species was a composite one. They therefore split off *M. osloensis* and *M. phenylpyruvica* on the basis of nutritive requirements, DNA composition, and failure of genetic transformation, and defined *M. nonliquefaciens* afresh. They found it to be a comparatively fastidious organism, failing to grow in Hugh and Leifson's sugar medium, in Koser's or Simmons's citrate media, or in Audureau's (1940) medium. It reduces nitrate to nitrite. Its G+C content is 40–42 moles per cent.

Whether the organism isolated by De Bord (1942) and called *Mima polymorpha* var. *oxidans* (Catlin and Cunningham 1964) is identical with that described by Flamm (1957) as *Moraxella polymorpha*, and whether this latter organism is the same as *M. nonliquefaciens* as the Judicial Commission (Report 1971) thinks, must remain doubtful for the present.

Moraxella osloensis

According to Bøvre and Henriksen (1967*a*) who described this organism, *M. osloensis* differs from *M. nonliquefaciens* mainly in being less fastidious in its nutritive requirements, in growing on plain media, on Hugh and Leifson's medium, in Koser's and Simmons's citrate media, and in Audureau's medium with ethanol as the sole source of carbon; in its greater resistance to heat; in its higher G+C content—43–43.5 moles per cent; and in its genetic incompatibility with *M. nonliquefaciens*.

Moraxella phenylpyruvica

As described by Bøvre and Henriksen (1967*b*), this organism is distinguished from *M. nonliquefaciens* and *M. osloensis* by its strong urease activity, and its deamination of phenylalanine and tryptophan. Nitrate reduction is variable. Grows poorly, even on complex media. No growth occurs in citrate. Its G+C content is 43–43.5 moles per cent. It has been isolated from the genito-urinary tract, blood, cerebrospinal fluid, and pus, but its pathogenicity is unknown.

Moraxella bovis

An organism closely resembling the Morax-Axenfeld bacillus was isolated by Jones and Little (1923) from cattle suffering from infectious ophthalmia or keratitis—a disease characterized by acute conjunctivitis and occasionally corneal ulceration. According to Watt (1951), *M. bovis* differs from *M. lacunata* in growing, though poorly, on ordinary nutrient agar and in broth, in its more rapid digestion of serum, and in its failure to reduce nitrate. Gelatin is liquefied in a stratiform manner, litmus milk is slowly peptonized (see Pugh *et al.* 1966), and colonies on horse blood agar are surrounded by a narrow zone of lysis. On blood or serum agar it forms rather large, somewhat mucoid, easily emulsifiable colonies, greyish and translucent with a smooth convex glistening surface, reaching 4–5 mm in diameter within 48 hours (Ryan 1964). The disease can be reproduced in cattle by conjunctival inoculation with the eye secretion from affected animals (Farley *et al.* 1950) or with pure cultures of the organism (Barner 1952, Henson and Grumbles 1960); and also in mice, but not in rats, rabbits or guinea-pigs (Pugh *et al.* 1968). Its G+C content is 42.5–43 moles per cent (Bøvre 1967), similar to that of *M. lacunata*.

'Moraxella' kingae

Henriksen and Bøvre (1968*a*) studied 8 strains of an organism from Dr Elizabeth King's collection at Atlanta. This organism, at first called *M. kingii*, but later *M. kingae*, was characterized by slight β-haemolysis on ox blood agar plates, by the production of acid from glucose and maltose, by failure to form catalase, by a G+C content of 44.5 moles per cent, and by genetic incompatibility with known species of *Moraxella*. No growth occurred in Hugh and Leifson's medium. Nitrate reduction was slight or absent, and urease production negative. Some strains grew slightly at room temperature. The organism was highly sensitive to penicillin and other antibiotics. Henriksen (1969*a*) later isolated 5 strains from nose and throat swabs. He observed that some colonies on blood agar were pitted and had a central papilla. By agglutination and absorption tests made with boiled suspensions Haug and Henriksen (1969*a*) recognized 3 serotypes. The organism appears to be a parasite of the human nasopharynx; its pathogenicity is uncertain. Its saccharolytic properties and its failure to form catalase exclude it from the genus *Moraxella* as at present defined. Though Henriksen and Bøvre (1968*a*) at one time pleaded to the contrary, they now (1976) propose to transfer it to a new genus, *Kingella*. A full description of this organism is given by Snell and Lapage (1976). Thus the cells are coccoid to medium-sized rods with square ends, ocurring in pairs and short chains. Two types of colony are formed on blood agar: (1) a spreading corroding type associated with twitching motility, fimbriation and competence in transformation, and (2) a smooth stationary type unassociated with the characters of type 1. It forms no catalase, but is strongly oxidase positive. Ferments glucose and maltose. Fails to liquefy gelatin or serum, or to produce indole. Snell and Lapage also describe two further species of *Kingella*, namely *K. denitrificans* and *K. indologenes*. (For a fuller description of *K. kingae*, see Ødum and Frederiksen 1981). The classification of this organism presents difficulties. It certainly differs from the organism described by Flamm (1956) as *M. saccharolytica*, which grows in synthetic media, ferments several sugars, is insensitive to most antibiotics, and has little claim to be classified as a moraxella.

Two new strains of so-called moraxellae, both saccharolytic, catalase negative, and indole positive with a G+C content of 49–50 moles per cent, have been described, one by Bijsterveld (1970) from angular conjunctivitis, the other by Sutton and his colleagues (1972) from a corneal abscess. Their different saccharolytic properties, production of indole, and high G+C content would seem to exclude them from the *Neisseria* and *Moraxella* genera.

References

Albrecht, H. and Ghon, A. (1901) *Wien. klin. Wschr.* **14,** 984.
Alexander, H. E. and Redman, W. (1953) *J. exp. Med.* **97,** 797.
Amies, C. R. and Garabedian, M. (1967) *Brit. J. vener. Dis.* **43,** 137.
Annear, D. I. and Grubb, W. B. (1982) *J. clin. Path.* **35,** 118.
Apicella, M. A. (1976) *J. infect. Dis.* **134,** 377.

Arkwright, J. A. (1907) *J. Hyg., Camb.* **7,** 1945; (1909) *Ibid.* **9,** 104; (1915) *Brit. med. J.* **ii,** 885.
Atkin, E. E. (1923) *Brit. J. exp. Path.* **4,** 325; (1925) *Ibid.* **6,** 235.
Audureau, A. (1940) *Ann. Inst. Pasteur* **64,** 126.
Axenfeld, T. (1897) *Zbl. Bakt.* **21,** 340.
Barner, R. D. (1952) *Amer. J. vet. Res.* **13,** 132.
Baumann, P., Doudoroff, M. and Stainier, R. Y. (1968) *J. Bact.* **95,** 59.
Berencsi, G. and Mészáros, G. (1960) *Zbl. Bakt.* **178,** 406.
Berger, U. (1960*a*) *Z. Hyg. InfektKr.* **146,** 253; (1960*b*) *Ibid.* **147,** 158; (1961*a*) *Ibid.* **148,** 45; (1961*b*) *Int. Bull. bact. Nomencl.* **11,** 17; (1962*a*) *Z. Hyg. InfektKr.* **148,** 445; (1962*b*) *Arch. Hyg. Bakt.* **146,** 388; (1967) *Zbl. Bakt.* **205,** 241; (1971) *Z. med. Mikrobiol. Immunol.* **156,** 154.
Berger, U. and Wezel, M. (1960) *Z. Hyg. InfektKr.* **146,** 244.
Berger, U. and Wulf, B. (1961) *Z. Hyg. InfektKr.* **147,** 257.
Bijsterveld, O. P. van. (1970) *Appl. Microbiol.* **20,** 405.
Boor, A. K. (1942) *Proc. Soc. exp. Biol., N.Y.* **50,** 22.
Boor, A. K. and Miller, C. P. (1934) *J. exp. Med.* **59,** 63; (1944) *J. infect. Dis.* **75,** 47.
Bord, G. G., De. (1942) *Iowa St. Coll. J. Sci.* **16,** 471.
Bøvre, K. (1967) *Acta. path. microbiol. scand.* **69,** 123; (1970) *Ibid. B,* **78,** 780.
Bøvre, K., Fiandt, M. and Szybalski, W. (1969) *Canad. J. Microbiol.* **15,** 335.
Bøvre, K. and Henriksen, S. D. (1967*a*) *Int. J. syst. Bact.* **17,** 127; (1967*b*) *Ibid.* **17,** 343.
Branham, S. E. (1930) *Publ. Hlth. Rep., Wash.* **45,** 845; (1940) *Bact. Rev.* **4,** 59; (1953) *Ibid.* **17,** 175; (1958) *Int. Bull. bact. Nomencl.* **8,** 1.
Branham, S. E. and Lillie, R. D. (1932) *Publ. Hlth. Rep., Wash.* **47,** 2137.
Brookes, R. and Hedén, C. G. (1967) *Appl. Microbiol.* **15,** 219.
Brookes, R. and Sikyta, B. (1967) *Appl. Microbiol.* **15,** 224.
Brown, W. J. (1974) *Appl. Microbiol.* **27,** 1027; (1976) *Hlth. Lab. Sci.* **13,** 54.
Brown, W. J. and Kraus, S. J. (1974) *J. Amer. med. Ass.* **228,** 862.
Bruckner, J. and Cristéanu, C. (1906*a*) *C. R. Soc. Biol.* **60,** 846; (1906*b*) *Ibid.* **60,** 907.
Buchanan, R. E. and Gibbons, N. E. (1974) *Bergey's Manual of Determinative Bacteriology*, 8th edn. Williams and Wilkins, Baltimore.
Bumgarner, L. and Finkelstein, R. A. (1973) *Infect. Immun.* **8,** 919.
Bumm, E. (1885*a*) *Dtsch. med. Wschr.* **11,** 508; (1885*b*) *Ibid.* **11,** 910.
Catlin, B. W. (1973) *J. infect. Dis.* **128,** 178.
Catlin, B. W. and Cunningham, L. S. (1961) *J. gen. Microbiol.* **26,** 303; (1964) *Ibid.* **37,** 353.
Christmas, J. de. (1897) *Ann. Inst. Pasteur.* **11,** 609; (1900) *Ibid.* **14,** 331.
Clapp, F. L., Phillips, S. W. and Stahl, H. J. (1935) *Proc. Soc. exp. Biol. N.Y.* **33,** 302.
Cohn, A. (1923) *Klin. Wschr.* **2,** 873.
Cowan, S.T. (1938) *Lancet* **ii,** 1052.
Cowan, S. T. and Steel, K. J. (1965) *Manual for Identification of Medical Bacteria.* Cambridge University Press, Cambridge.
Craven, D. E., Frasch, C. E., Mocea, L. F., Rose, F. B. and Gonzalez, R. (1979) *J. clin. Microbiol.* **10,** 302.
Danielsson, D. (1965) *Acta path. microbiol. scand.* **64,** 243, 267.
Danielsson, D. and Sandström, E. (1979) *Acta path. microbiol. scand.* 13, **87,** 55.
DeVoe, I. W. and Gilchrist, J. E. (1975) *J. exp. Med.* **141,** 297.
Dopter, C. (1909) *C. R. Soc. Biol.* **67,** 74; (1921) *L'Infection Méningococcique.* Paris.
Drell, M. J., Miller, C. P. and Bohnhoff, M. (1947) *Arch. Ophthal., Chicago* **38,** 221.
Dunn, R. A. and Gordon, M. H. (1905) *Brit. med. J.* ii, 421.
Eliasson, I. (1980) *Acta path. microbiol. scand.* B, **88,** 281.
Elser, W. J. and Huntoon, F. M. (1909) *J. med. Res.* **20,** 371.
Estevenon, A. M., Duriez, T., Breuillaud, J. and Lemoine, A. (1978) *Microbia* **4,** 7.
Eyre, J. W. (1900) *J. Path. Bact.* **6,** 1.
Farley, H., Kliewer, I. O., Pearson, C. C. and Foote, L. E. (1950) *Amer. J. vet. Res.* **11,** 17.
Feigin, R. D., San Joaquin, V. and Middlekamp, J. A. (1969) *J. Pediat.* **75,** 116.
Feldman, H. A. (1965) *J. Amer. med. Ass.* **196,** 391.
Ferry, N. S., Norton, J. F., and Steele, A. H. (1931) *J. Immunol.* **21,** 293.
Flamm, H. (1956) *Zbl. Bakt.* **166,** 498; (1957) *Ibid.* **168,** 261.
Flexner, S. (1907*a*) *J. exp. Med.* **9,** 105; (1907*b*) *Ibid.* **9,** 142.
Flügge, C. (1896) *Die Mikroorganismen.* Leipzig.
Flynn, J. and Waitkins, S. A. (1972) *J. clin. Path.* **25,** 525.
Frantz, I. D. (1942) *J. Bact.* **43,** 757.
Frasch, C. E. and Chapman, S. S. (1972) *Infect. Immun.* **6,** 127.
Gay, R. M. and Sevier, R. E. (1978) *J. clin. Microbiol.* **8,** 729.
Ghon, A. and Pfeiffer, H. (1902) *Z. klin. Med.* **4,** 262.
Gilardi, G. L. (1967) *Amer. J. med. Technol.* **33,** 201.
Givan, K. F., Thomas, B. W. and Johnston, A. G. (1977) *Brit. J. vener. Dis.* **53,** 109.
Gold, R., Goldscheider, I., Lepow, M. L., Draper, T. F. and Randolph, M. (1978) *J. infect. Dis.* **137,** 112.
Goldner, M., Penn, C. W., Sanyal, S. C., Veale, D. R. and Smith, H. (1979) *J. gen. Microbiol.* **114,** 495.
Gordon, J. and McLeod, J. W. (1926) *J. Path. Bact.* **29,** 13; (1928) *Ibid.* **31,** 185.
Gordon, J. E. (1921) *J. infect. Dis.* **29,** 462.
Gotschlich, E. C., Liu, T. Y. and Artenstein, M. S. (1969) *J. exp.* Med. **129,** 1349.
Grados, O. and Ewing, W. H. (1970) *J. infect. Dis.* **122,** 100.
Graham, R. K. and May, J. W. (1965) *J. gen. Microbiol.* **41,** 243.
Grossowicz, N. (1945) *J. Bact.* **50,** 109.
Gurd, F. B. (1908) *J. med. Res.* **18,** 291.
Halvorsen, J. F. (1963) *Acta path. microbiol. scand.* **59,** 200.
Haug, R. H. and Henriksen, S. D. (1969*a*) *Acta path. microbiol. scand.* **75,** 641; (1969*b*) *Ibid.* **75,** 648.
Heckels, J. E. (1978) *J. gen. Microbiol.* **108,** 213.
Hehre, E. J. and Hamilton, D. M. (1948) *J. Bact.* **55,** 197.
Henderson, A. (1965) *Antonie van Leeuwenhoek* **31,** 395.
Hendley, J. O., Powell, K. R., Jordan, J. R., Rodewald, R. D. and Volk, W. A. (1978) In: *Immunobiology of Neisseria gonorrhoeae*, p. 116. Amer. Soc. Microbiology, Washington D.C.
Hendry, C. B. (1938) *J. Path. Bact.* **46,** 383.
Henriksen, S. D. (1952) *J. gen. Microbiol.* **6,** 318; (1958) *Acta path. microbiol. scand.* **43,** 157; (1969*a*) *Ibid.* **75,** 85; (1969*b*) *Ibid.* **75,** 91; (1973) *Bact. Rev.* **37,** 522.
Henriksen, S. D. and Bøvre, K. (1968*a*) *J. gen. Microbiol.* **51,** 377; (1968*b*) *Ibid.* **51,** 387; (1976) *Int. J. syst. Bact.* **26,** 447.

Henson, J. B. and Grumbles, L. C. (1960) *Amer. J. vet. Res.* **21,** 761.
Hill, J. H. (1944) *Amer. J. Syph.* **28,** 471.
Hoke, C. and Vedros, N. A. (1982) *Int. J. syst. Bact.* **32,** 57.
Hollis, D. G., Wiggins, G. L. and Schubert, J. H. (1968) *J. Bact.* **95,** 1.
Hollis, D. G., Wiggins, G. L. and Weaver, R. E. (1969) *J. appl. Microbiol.* **17,** 71.
Holten, E., Bratlid, D. and Bøvre, K. (1978) *Scand. J. infect. Dis.* **10,** 36.
Hunter, K. M. and McVeigh, I. (1970) *Antonie van Leeuwenhoek* **36,** 305.
Jaeger, H. (1895) *Z. Hyg. Infektkr.* **19,** 351.
James, J. F. and Swanson, J. J. (1977) *J. exp. Med.* **145,** 1082.
James-Holmquest, A. N., Wende, R. D., Mudd, R. L. and Williams R. P. (1973) *Appl. Microbiol.* **26,** 466.
Jantzen, E., Bryn, K. and Bøvre, K. (1976) *Acta path. microbiol. scand.* B, **84,** 177.
Jephcott, A. E. and Morton, R. S. (1972) *Lancet* **ii,** 739.
Johnston, K. H., Holmes, K. K. and Gotschlich, E. C. (1976) *J. exp. Med.* **143,** 741.
Jones, F. S. and Little, R. B. (1923) *J. exp. Med.* **38,** 139.
Juni, E. and Heym, C. A. (1977) *J. clin. Microbiol.* **6,** 511.
Kaffka, A. (1955) *Zbl. Bakt.* **164,** 451.
Kellogg, D. S., Peacock, W. L., Deacon, W. E., Brown, L. and Pirkle, C. I. (1963) *J. Bact.* **85,** 1274.
Kellogg, D. S. and Turner, E. M. (1973) *Appl. Microbiol.* **25,** 550.
Kenny, C. P., Ashton, F. E., Diena, B. B. and Greenberg, L. (1967) WHO/VDT/RES/GON/67.21.
Kutscher, K. H. (1912) *Kolle and Wassermann's Handbuch der pathogenen Mikro-organismen, 2te. Auft.*, **4,** 589.
Kuzemenská, P. *et al.* (1976) *Zbl. Bakt.* 1. *Abt.* **A 236,** 559.
La Scolea, L. J. and Young, F. E. (1974) *Appl. Microbiol.* **28,** 70.
Lauer, B. A. and Fisher, C. E. (1976) *Amer. J. Dis. Child.* **130,** 198.
Leinonen, M., Luotonen, J., Herva, E., Valkonen, K. and Mäkelä, P. H. (1981) *J. infect. Dis.* **144,** 570.
Leistikow. (1882) *Berl. klin. Wschr.* **19,** 500.
Lewis, V. J., Weaver, R. E. and Hollis, D. G. (1968) *J. Bact.* **96,** 1.
Ley, H. L. and Mueller, J. H. (1946) *J. Bact.* **52,** 453.
Lindqvist, K. (1960) *J. infect. Dis.* **106,** 162.
Lingelsheim, W. von. (1905) *Dtsch. med. Wschr.* **31,** 1217; (1906) *Klin. Jb.* **15,** 373; (1908) *Z. Hyg. InfektKr.* **59,** 457.
Lipschütz, B. (1904) *Zbl. Bakt.* **36,** 743.
Lloyd, D. J. (1916–17) *J. Path. Bact.* **21,** 113.
Lwoff, A. (1939) *Ann. Inst. Pasteur* **62,** 168; (1964) *Ibid.* **106,** 483.
M'Donald, S. (1908) *J. Path. Bact.* **12,** 442.
Maclagan, P. W. and Cooke, W. E. (1917) *J. R. Army med. Cps.* **29,** 228.
McLeod, J. W., Coates, J. C., Happold, F. C., Priestley, D. P. and Wheatley, B. (1934) *J. Path. Bact.* **39,** 221.
McLeod, J. W., Wheatley, B. and Phelon, H. V. (1927) *Brit. J. exp. Path.* **8,** 25.
Maeland, J. A. (1967) *Acta path. microbiol. scand.* **69,** 145; (1968) *Ibid.* **73,** 413.
Marmur, T. and Doty, P. (1962) *J. molec. Biol.* **5,** 109.
Martin, W. B. M. (1911) *J. Path. Bact.* **15,** 76.
Marx, H. and Woithe, F. (1900) *Zbl. Bakt.* **28,** 1, 33, 65, 97.
Maslovski. (1900) *Zbl. Bakt.* **27,** 541.
Miller, C. P. (1933) *Science* **78,** 340; (1934–35) *Proc. Soc. exp. Biol. N.Y.* **32,** 1136, 1138, 1140.
Miller, C. P., Drell, M. J., Moeller, V. and Bohnhoff, M. (1945) *J. infect. Dis.* **77,** 193.
Morax, V. (1896) *Ann. Inst. Pasteur* **10,** 337.
Morse, S. A. (1979) *CRC Critical Reviews in Microbiology* **7,** 93–189.
Morse, S. A. and Hebeler, B. H. (1978) In: *Immunobiology of Neisseria gonorrhoeae*, p. 9. *Amer. Soc. Microbiology, Washington D.C.*
Muchmore, H. G. and Venters, H. D. (1968) *New Engl. J. Med.* **278,** 1166.
Mueller, J. H. and Hinton, J. (1941) *Proc. Soc. exp. Biol., N.Y.* **48,** 330.
Mulks, M. H. and Plaut, A. G. (1978) *New Engl. J. Med.* **299,** 973.
Murray, E. G. D. (1929) *Spec. Rep. Ser. med. Res. Coun., Lond.* No. 124, p. 15.
Murray, E. G. D. and Ayrton, R. (1924) *J. Hyg., Camb.* **23,** 23.
Neill, M. H. and Taft, C. E. (1920) *Bull. U.S. hyg. Lab.* No. 124, p. 93.
Neisser, A. (1879) *Zbl. med. Wiss.* **17,** 497; (1882) *Dtsch. med. Wschr.* **8,** 279.
Netter, A. and Debré, R. (1911) *La Méningite Cérébro-spinale.* Paris.
Novotny, P., Short, J. A. and Walker, P. D. (1975) *J. med. Microbiol.* **8,** 413.
Novotny, P. and Turner, W. H. (1975) *J. gen. Microbiol.* **89,** 87.
Novotny, P. *et al.* (1977) *J. med. Microbiol.* **10,** 347.
Odugbemi, T. O., McEntegart, M. G. and Hafiz, S. (1978) *J. clin. Path.* **31,** 936.
Ødum, L. and Frederiksen, W. (1981) *Acta path. microbiol. scand.* B, **89,** 311.
Pearce, L. (1915) *J. exp. Med.* **21,** 289.
Pedersen, K. B. (1972) *Acta. path. microbiol. scand.* **B80,** 135.
Pelczar, M. J. (1953) *J. Bact.* **65,** 744.
Pelczar, M. J., Hajek, J. P. and Faber, J. E. (1949) *J. infect. Dis.* **85,** 239.
Penn, C. W., Sen, D., Veale, D. R., Parsons, N. J. and Smith, H. (1976) *J. gen. Microbiol.* **97,** 35.
Petit, P. (1900) *Recherches cliniques et bactériologiques sur les infections aiguës de la cornée.* Thése de la Faculté de Médecine de Paris. G. Steinheil, Paris.
Petrie, G. F. (1932) *Brit. J. exp. Path.* **13,** 380; (1937) *J. Hyg., Camb.* **37,** 42.
Phelon, H. V., Duthie, G. M. and M'Leod, J. W. (1927) *J. Path. Bact.* **30,** 133.
Piéchaud, M. (1963) *Ann. Inst. Pasteur* **104,** 291.
Poolman, J. T., Hopman, C. T. P. and Zanen, H. C. (1980) *J. gen. Microbiol.* **116,** 465.
Prost, B., Vandekerkove, M. and Nicoli, J. (1970) *Bull. World Hlth Org.* **42,** 751.
Pugh, G. W., Hughes, D. E. and McDonald, J. J. (1966) *Amer. J. vet. Res.* **27,** 957; (1968) *Ibid.* **29,** 2057.
Rake, G. (1933) *J. exp. Med.* **57,** 549.
Report. (1916) *Spec. Ser. med. Res. Coun., Lond.* No. 2; (1920) *Ibid.* No. 50; (1971) *Int. J. syst. Bact.* **21,** 100.
Reyn, A. (1965) *Bull. Wld Hlth Org.* **32,** 449.
Richardson, W. P. and Sadoff, J. C. (1977) *Infect. Immun.* **15,** 663.
Roberts, R. B. (1977) *The Gonococcus.* John Wiley & Sons, New York.

Roberts, R. B. and Wittler, R. G. (1966) *J. gen. Microbiol.* **44,** 139.
Rosher, A. B. (1936) *Pers. Comm.*
Ryan, W. J. (1964) *J. gen. Microbiol.* **35,** 361.
Sadoff, J. C., Zollinger, W. D. and Sidberry, H. (Eds) (1978) In: *Immunobiology of Neisseria gonorrhoeae*, p. 93, Amer. Soc. Microbiology, Washington D.C.
Schäffer, J. (1897) *Fortschr. Med.* **15,** 813.
Schildkraut, C. L., Marmur, J. and Doty, P. (1962) *J. molec. Biol.* **4,** 430.
Seeliger, H. (1953) *Zbl. Bakt.* **159,** 173.
Silberfarb, P. M. and Lawe, J. E. (1968) *Arch. intern. Med.* **122,** 512.
Slaterus, K. W. (1961) *Antonie van Leeuwenhoek* **27,** 305.
Slifkin, M. and Pouchet, G. R. (1977) *J. clin. Microbiol.* **5,** 15.
Snell, J. J. S. and Lapage, S. P. (1976) *Int. J. syst. Bact.* **26,** 451.
Sparling, P. F. and Yobs, A. R. (1967) *J. Bact.* **93,** 513.
Storck, H. and Rinderknecht, P. (1966) *Arch. klin. exp. Derm.* **227,** 623.
Storck, H., Rinderknecht, P. and Flury, E. (1951) *Dermatologica, Basel* **103,** 243.
Sutton, R. G. A., O'Keefe, M. F., Bundock, M. A., Jeboult, J. and Tester, M. P. (1972) *J. med. Microbiol.* **5,** 148.
Swanson, J. *et al.* (1974) *Infect. Immun.* **10,** 633.
Tani, T. and Tashiro, M. (1955) *Jap. J. med. Sci. Biol.* **8,** 295.
Thayer, J. D. and Martin, J. E. (1964) *Publ. Hlth Rep., Wash.* **79,** 49.
Thompson, R. E. M. and Knudsen, A. (1958) *J. Path. Bact.* **76,** 501.
Torrey, J. C. and Buckell, G. T. (1922) *J. Immunol.* **7,** 305.
Tringali, G. and Intonazzo, V. (1978) *Ann. Sclavo.* **20,** 390.
Tulloch, W. J. (1923*a*) *J. State Med.* **31,** 501.
Vedros, N. A., Ng, J. and Culver, G. (1968) *J. Bact.* **95,** 1300.
Véron, M., Thibault, P. and Second, L. (1959) *Ann. Inst. Pasteur.* **97,** 497; (1961) *Ibid.* **100,** 166.
Ward, M. E. and Watt, P. J. (1971) *J. clin. Path.* **24,** 122.
Warner, G. S., Faber, J. E. and Pelczar, M. J. (1952) *J. infect. Dis.* **90,** 97.
Wassermann, A. (1898) *Z. Hyg. InfektKr.* **27,** 298.
Watt, J. A. (1951) *Vet. Rec.* **63,** 98.
Watt, P. J. and Lambden, P. R. (1978) In: *Modern Topics in Infection*, p. 134. Ed. by J. D. Williams. Heinemann Medical Books, London.
Weichselbaum, A. (1887) *Fortschr. Med.* **5,** 573, 620.
Wherry, W. B. and Oliver, W. W. (1916) *J. infect. Dis.* **19,** 288.
Wilson, G. S. and Smith, M. M. (1928) *J. Path. Bact.* **31,** 597.
Wilson, S. P. (1928) *J. Path. Bact.* **31,** 477.
Wistreich, G. A. and Baker, R. F. (1971) *J. gen. Microbiol.* **65,** 167.
Wright, H. D. (1933) *J. Path. Bact.* **37,** 257.
Zdrodowski, P. and Voronine, E. (1932) *Ann. Inst. Pasteur.* **48,** 617.
Zollinger, W. D. and Mandrell, R. E. (1980) *Infect. Immun.* **28,** 451.

29

Streptococcus and Lactobacillus

M. T. Parker

Streptococcus

Definition

Spherical or ovoid cells, arranged in chains or pairs. Usually non-motile. Non-sporing. Catalase negative. Many species grow poorly in the absence of fresh meat extract. All are aerobic and facultatively anaerobic, but some require additional CO_2 for growth. With rare exceptions they fail to reduce nitrate. Ferment carbohydrates with the production mainly of lactic acid but never of gas. Many species are parasitic on man or animals, and some are highly pathogenic. A few are saprophytes. G + C content of DNA; 34–41 moles per cent.

Type species: *Streptococcus pyogenes.*

The term *Streptococcus* was first applied by Billroth and Ehrlich (1877) to a chain-forming coccus that they saw in infected wounds. Fehleisen (1883) described a similar coccus as the causative organism of erysipelas. Rosenbach (1884) gave the name *Str. pyogenes* to cocci that grew in chains and had been isolated from suppurative lesions in man. The account by Pasteur, Chamberland and Roux (1881) of a septicaemic infection of rabbits inoculated with human saliva is probably the earliest reference to the pneumococcus, though no clearly identifiable description of the organism was published before the independent studies of Fraenkel and Weichselbaum in 1886. In 1887, Nocard and Mollereau reported the production of mastitis in the cow and goat by inoculation into the udder of a streptococcus from the milk of a cow with this disease. Schütz (1887, 1888) described streptococci isolated from the lesions of equine pneumonia and strangles.

The streptococci are gram-positive cocci that divide in one plane to form pairs or chains. Their metabolism is fermentative; with a few possible exceptions, they are aerobic and facultatively anaerobic; they attack sugars with the production mainly of lactic acid, but do not form any gas. Their carbohydrate metabolism thus resembles closely that of the homofermentative lactobacilli. Certain other gram-positive cocci that resemble streptococci differ from them in being heterofermentative or in forming tetrads. The homofermentative tetrads (*Pediococcus* and *Aerococcus*), and the heterofermentative chain-forming organisms (*Leuconostoc*) will be described briefly at the end of this chapter. We shall also refer to a group of gram-negative or weakly gram-positive cocci (*Gemella*) that appear to be closely related to the streptococci.

Strictly anaerobic chain-forming cocci are often referred to as 'anaerobic streptococci'. Most of these organisms differ so greatly from *Streptococcus* in their metabolic characters that they must be excluded from this genus. However, they form a heterogeneous group (Rogosa 1974*a*), and the possibility cannot be excluded that a few of them, when more adequately studied, may be eligible for inclusion in *Streptococcus*. For the present, however, we shall consider them along with the anaerobic clump-forming gram-positive cocci at the end of Chapter 30.

Classification

Certain of the streptococci—nearly all of them strains that frequently cause disease in man or his domestic animals—have been extensively studied and can be characterized accurately. Others that form part of the normal body flora or live a saprophytic existence are still largely unclassified. Thus, though there is little difficulty in giving a clear account of several well defined species of streptococcus, it is at present very difficult to provide a logical subdivision of the group as a whole. Three general characters have in the past been used for this purpose, but none has proved entirely satisfactory. *Firstly*, the changes induced by the growth of streptococci in blood have proved useful for the recognition of certain pathogenic species. The fortunate circumstance that most of the streptococci that cause the common septic infections produce true or β-haemolysis on blood agar plates provides us with a valuable screening test. Other streptococci produce greening or α-haemolysis of blood-containing media, and some produce no change at all. *Secondly*, biochemical and other physiological tests have been used. In general, the ability to cause specific biochemical changes, such as the fermentation of particular sugars, is of little value for the primary subdivision of streptococci, though individual tests may be useful to characterize particular strains or species. On the other hand, tests of the ability to grow under extreme conditions of temperature or pH, or in the presence of certain chemical substances, are often of more general significance. *Finally*, the presence of group-specific antigens, though of great practical importance for the identification of some of the pathogenic species, is of much less value for the classification of other classes of streptococci.

The most widely accepted general classification of streptococci is that of Sherman (1937), who recognized four main divisions: (1) the *pyogenic streptococci*, which are usually β-haemolytic, have a polysaccharide group antigen, are not heat resistant and do not grow at extremes of temperature or pH, do not have strong reducing activities, and usually hydrolyse arginine; (2) the *enterococci*, which are variable in haemolysis, have the group D antigen, are somewhat heat resistant,

Table 29.1 Classification of the streptococci

Specific name	Other description	Lancefield group[1]	Haemolysis[1]	Main habitat
Pyogenic streptococci				
pyogenes		A	β	Man
equi		C	β	Horse
zooepidemicus		C	β	Many animals
equisimilis		C	β	Many animals; man
dysgalactiae		C	α	Cattle
sp.	'Large-colony' group G	G	β	Many animals, man
agalactiae		B	β(NH, α)	Cattle; man
lentus		E, P or U	β	Pig; cattle
sp.		L	β	Dog, pig
suis		D[2]	β	Pig
milleri		—(A, C, G, F)	NH (α, β)	Man
Enterococci and related organisms				
faecalis	Enterococci *sensu strictu*	D	NH(β)	Faeces of mammals
faecium	Enterococci *sensu strictu*	D	α(β, NH)	Faeces of mammals
faecium var. *durans*	Enterococci *sensu strictu*	D	NH(α)	Faeces of mammals
avium	Enterococci *sensu strictu*	D	NH(α)	Faeces of birds
sp.	Enterococci *sensu strictu*	D	α(NH)	Epiphytic
bovis		D	NH(α)	Faeces of mammals
equinus		D	NH(α)	Faeces of horse
Lactic streptococci				
lactis		N	α(NH)	Milk products
cremoris		N	α(NH)	Milk products
Pneumococci				
pneumoniae		—	α	Man, other animals
Other streptococci				
salivarius		—(K)	NH	Man
sanguis		—(H)	α(NH, β)	Man
mitior	*mitis*	—(O, K, M)	α(NH, β)	Man
mutans		—(E)	NH	Man, other animals
uberis		—(E)	α	Cattle, soil
acidominimus		—	α	Cattle
thermophilus		—	NH(α)	Milk products

NH = non-haemolytic.
[1] Less common reactions in parentheses.
[2] Group-D reaction difficult to obtain; 'group' R, S and T reactions believed to be type specific.

grow over a wide range of temperature and pH, are strongly reducing, and hydrolyse arginine; (3) the *lactic streptococci*, which grow at a low temperature but are rather less tolerant of other extreme environmental conditions; and (4) the '*viridans*' *streptococci*, which are seldom β-haemolytic and grow at 45°, but do not hydrolyse arginine.

We accept the general outline of this classification, though we would place rather different organisms in some of the groups (Table 29.1). It is now clear that the group polysaccharides characteristic of many of the pyogenic streptococci may occasionally be found in otherwise unrelated organisms. We also now recognize that many of the group D streptococci do not have the physiological properties of enterococci. And we can no longer accept the existence of a 'viridans' group as defined by Sherman, but have replaced it by a category of 'other streptococci' that have little in common except a series of negative characters. Finally, we have no hesitation in including the *pneumococci* as a separate and well defined group of streptococci. (For reviews of the streptococci, see Wannamaker and Matsen 1972, Skinner and Quesnel 1978, Parker 1979, and Holm and Christensen 1982.)

In the following descriptions we shall whenever possible use the specific names given in Table 29.1, but sometimes we shall refer to serological groups. When members of one serological group include different sorts of streptococci the reference will be suitably qualified.

Morphology

Streptococci are more or less spherical in shape and are arranged in chains. Growth occurs by elongation on the axis parallel to the chain, and division is at right angles to this, often giving rise to an appearance of pairing within the chain. In some streptococci, the longer dimension of the cell is at right angles to the axis of the chain, but in others—such as the enterococci and the pneumococci—it is coincident with it. A few rod-like forms may be found in many streptococcal cultures. The length of the chain varies; it depends to some extent on the medium in which the organism is grown, but some streptococci characteristically form long chains, whereas others are mainly diplococcal. The cells forming the chain are connected by a bridge of cell-wall material and cannot be separated by shaking, but chains may be partly disrupted without killing many of the cocci by sonic oscillation for short periods (Slade and Slamp 1956). Chains do not elongate indefinitely; in some cases this is because the streptococci produce a 'de-chaining' enzyme; but this may be inhibited by the union of surface antigen with specific antibody (Ekstedt and Stollerman 1960). Some streptococci, such as *Str. pyogenes* and the pneumococcus, give rise to filamentous mutants that form extremely long tangled chains. All the streptococci are non-motile, except for some of the enterococci.

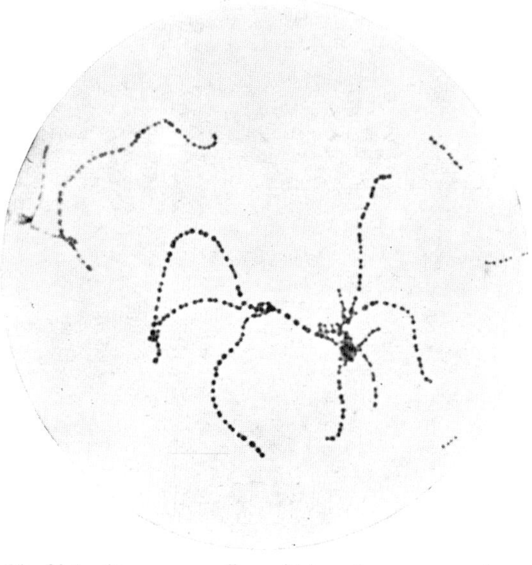

Fig. 29.1 *Str. pyogenes*. From 24-hr culture on agar, showing long chains (× 1000).

Capsulation is not a regular character of streptococci, but some form a capsule of hyaluronic acid in the early phase of growth, and pneumococci have capsules composed of type-specific polysaccharide. Streptococci stain readily with the ordinary dyes and are almost always frankly gram positive. Intracellular cylindrical 'cores', visible by electronmicroscopy, are confined to the enterococci (McCandless *et al.* 1968, 1971).

Various types of extracellular filament have been described in streptococci. Those seen by Henriksen and Henrichsen (1975) in *Str. sanguis* appear to be true fimbriae. Handley and Carter (1979) observed three morphological types of filamentous appendage in *Str. mitior*. The whole cell surface of *Str. pyogenes* is covered by a mat of short, fine filaments (Swanson *et al.* 1969).

Cultural characters

Growth on ordinary nutrient media is generally poor; on media enriched with blood, serum, or glucose it is more rapid but soon ceases. Therefore the amount of growth on solid media is considerably less than that of, for example, staphylococci. Colonies on blood agar seldom exceed 1 mm in diameter after 24 hr at 37°, and further incubation results in no increase. Pigment is not usually formed, except by some strains of *Str. agalactiae* and of the enterococci, which may form yellowish colonies. In broth, strains that form long chains generally give a granular growth with a powdery or floccular deposit and a clear supernatant fluid. Growth in liquid media is greatly favoured by the addition of glucose, but the pH falls rapidly and growth ceases. In buffered glucose media such as Todd-Hewitt broth (Todd and Hewitt 1932), or in media with a high glucose content to which alkali is added continuously during growth, a heavy yield of streptococci is obtained.

Changes on blood media

Marmorek (1895) first noted the ability of some streptococci to lyse red blood corpuscles, and Schottmüller (1903) proposed that the ability to cause haemolysis *in vitro* should be used as a means of classifying streptococci. The various types of change produced in blood agar by streptococci were well described by Smith and Brown (1915; see also Brown 1919). Emphasis was laid on the importance of using a standard technique and of studying deep colonies in pour plates rather than surface colones on streak plates. Most bacteriologists, however, find that they are able to rely on the appearances around surface colonies.

The size and character of the haemolytic zone is influenced not only by the composition of the basal medium but also by the type of blood used. In Britain it is customary to use 5 per cent horse blood agar, on which good haemolytic zones are obtained. The blood agar is poured as a thin layer on top of a layer of nutrient agar, thus making it easier to see small zones of haemolysis. Workers in the United States favour the use of 5 per cent sheep blood agar because this inhibits the growth of haemolytic colonies of *Haem. parainfluenzae.* Sheep blood agar has the disadvantage that it is somewhat less susceptible to lysis by some streptococci of the pyogenic group than is horse blood agar when incubated in air. It is therefore necessary to ensure some sub-surface growth by making stabs into the medium in the area of the primary inoculum. Human blood is inferior to both horse and sheep blood.

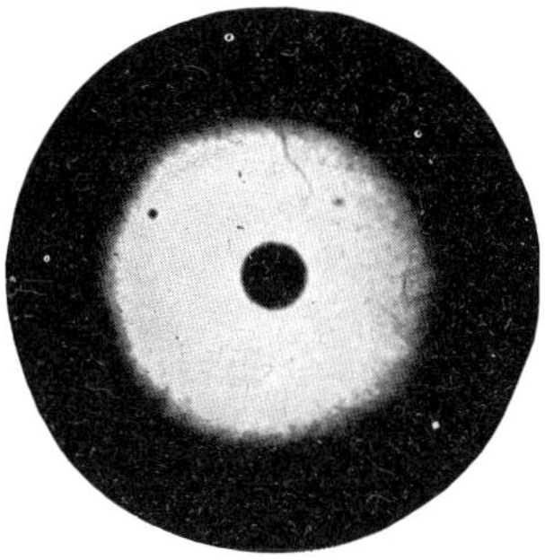

Fig. 29.2 *β-haemolysis.* Deep colony in blood agar plate, showing wide zone of complete haemolysis, and the sharply differentiated margin of the colony (× 8).

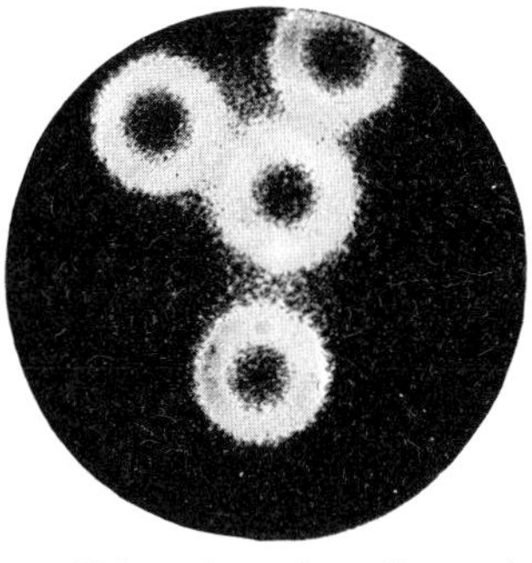

Fig. 29.3 *α-haemolysis.* Deep colonies in blood agar plate, showing zone of discoloured cells round colony, obscuring margin, and zone of incomplete lysis beyond (× 8).

The changes produced in blood agar may be briefly described as follows.

β-haemolysis—The colony is surrounded by a sharply defined, clear and colourless or slightly pink zone in which red blood corpuscles cannot be seen.

α-haemolysis—A greenish discoloration of the blood, 1-3 mm in width, surrounds the colony. Clumps of intact erythrocytes can be seen within it. The margin of the zone is indistinct, and usually consists of a narrow zone of clearer haemolysis visible only on magnification.

α′-(α prime-)haemolysis—Sometimes the zone of clearing around the zone of discoloured erythrocytes is wider and visible to the naked eye. On cursory inspection, then, the appearance may be one of hazy clearing that may be mistaken for β-haemolysis. The width of the clear zone may extend considerably on further incubation or in the refrigerator. Careful examination under the plate microscope shows, however, that the clear zone contains clumps of unlysed erythrocytes, particularly near the colony or immediately beneath it.

The term 'haemolytic' unqualified by any adjective is commonly applied to streptococci only when they are β-haemolytic, and we shall adhere to this convention. The term 'non-haemolytic' is used not only for streptococci that have no action on blood but also for those that produce α- or α′-'haemolysis.'

The β-haemolysis produced by pyogenic streptococci on blood agar plates incubated in air is usually due to the action of an oxygen-stable haemolysin of the S type (p. 184). Improvement in haemolysis by streptococci of groups A, C and G by anaerobic incubation is due to additional action of the oxygen-labile O-lysin. Haemolytic enterococci give clear β-haemolytic zones due to the action of a substance that is also a bacteriocine. It must be emphasized, however, that streptococci may produce a haemolysin but not cause β-haemolysis; thus the pneumococcus, which forms an oxygen-sensitive haemolysin (p. 197), causes greening on aerobic incubation. The addition of glucose to horse-blood agar inhibits β-haemolysis around colonies of *Str. pyogenes* but not of other haemolytic streptococci (Kodama 1936).

The way in which streptococci cause green discoloration of erythrocytes is not well understood. What information we have is derived from the study of pneumococci and may not be applicable to other α-haemolytic streptococci. McLeod and Gordon (1922) suggested that H_2O_2 production by pneumococci led to the formation of methaemoglobin or some related substance, but these organisms cause green discoloration on unheated blood agar, which contains a considerable amount of catalase. Hart and Anderson (1933) found that broth cultures of pneumococci produced an olive-green precipitate when added to haemoglobin or methaemoglobin, and that the green substance was an iron-containing substance resulting from the reduction of haemoglobulin (see also Anderson and Hart 1934). The ability to bring about this change is not, however, confined to organisms that cause greening on blood agar, so factors other than the possession of the necessary enzyme system must determine whether the change occurs in a plate culture.

All strains of *Str. agalactiae*, whether haemolytic or not, produce a substance which completes the lysis of sheep or ox red cells by staphylococcal β-lysin (p. 192). This forms the basis of the CAMP test (Christie *et al.* 1944), which may be performed by streaking a β-lysin forming staphylococcus up to the inoculum of the streptococcus under test, or by dropping a preparation of β-lysin on to isolated streptococcal colonies, preferably in the hot room. *Str. lentus* also gives a positive CAMP test.

Metabolism

All the streptococci of species described in this chapter are aerobic and facultatively anaerobic, but some of the strictly anaerobic chain-forming cocci (Chapter 30) may prove to belong to the genus (Holdeman and Moore 1974). The growth of some streptococci, including most strains of *Str. mutans*, many strains of *Str. milleri*, and some pneumococci, is poor or absent unless some 5 per cent of CO_2 is added to the atmosphere. The range of temperature over which growth occurs is a useful means of classifying streptococci, but accurate temperature control is essential if the tests are to give reliable results. Briefly, the pyogenic streptococci grow between 20° and 42°, and have their optimum at about 37°. Nearly all the group D streptococci grow at 45°, and *Str. faecium* at 50°; the 'strict' enterococci, the lactic streptococci, and *Str. uberis* grow at 10°. The enterococci are also able to grow at pH 9.6; this character is almost unique among streptococci, but the test for it requires careful standardization (Shattock and Hirsch 1947). Enterococci also grow in the presence of 6.5 per cent NaCl, a character shared with a minority of members of several other species.

The nutritional requirements of most of the streptococci are complex, and include amino acids, peptides, purines, pyramidines and vitamins. These requirements are usually met by fresh meat extract, though growth is scanty unless a fermentable sugar is also provided. A few strains will grow only on media supplemented with pyridoxine, and exhibit satellitism around colonies of other bacteria on ordinary nutrient media (George 1974). *Str. bovis* is the least nutritionally exacting of the streptococci; many strains can use ammonium salts as nitrogen source (Wolin *et al.* 1959).

Resistance

With the exception of the enterococci, some of the lactic streptococci, and *Str. thermophilus*, which may resist heating at 60° for 30 min, streptococci are killed in this time at 55°. They are also destroyed by the usual strengths of disinfectant but may survive for months in dry dust in buildings. All group D streptococci (except *Str. suis*) and most strains of *Str. agalactiae* and *Str. mutans* grow on 40 per cent bile agar, but so do some strains of many other streptococci. The ability to grow in 0.1 per cent tellurite broth is an almost unique character of *Str. faecalis*; its growth is accompanied by reduction of the tellurite; black colonies are formed on solid tellurite media.

Most streptococci, with the exception of enterococci, are sensitive to a wide range of therapeutically useful antimicrobial agents; more detailed information will be found at appropriate points in this chapter. The pyogenic streptococci and all but a few pneumococci are highly sensitive to benzylpenicillin. Streptococci are more or less resistant to all the aminoglycoside antibiotics, but the addition of some aminoglycosides will enhance the bactericidal action of penicillins on them.

Differences in susceptibility to certain agents is useful for the presumptive identification of particular streptococci; sensitivity to bacitracin (p. 183) and optochin (p. 196), detected by a standardized disk-diffusion test, are examples of this.

Selective and enrichment media

Streptococci, including pneumococci, are uniformly resistant to several chemicals, including crystal violet at a concentration of 1 in 500 000 (Haxthausen 1927), sodium azide (Edwards 1938) and thallium salts (Rantasalo 1947). These have been used, alone or in various combinations, in a variety of selective and enrichment media. Nowadays, however, more use is made of media containing mixtures of antibacterial agents such as fusidic acid, nalidixic acid, polymyxin and aminoglycosides (Lowbury *et al.* 1964, Beerens and Tahon-Castel 1966) in solid media for the isolation of streptococci from heavily contaminated material. Enrichment broths containing gentamicin have been advocated, especially for the isolation of group B streptococci, but in our experience neomycin 40 mg per l, alone or in combination with nalidixic acid 15 mg per l, is more effective for the isolation of this organism when it is present in small numbers.

Biochemical reactions

Tests for fermentation of carbohydrates may be carried out in meat-extract broth or in serum 'sugars'. All the streptococci ferment glucose and maltose and most of them ferment lactose and sucrose. Other sugar reactions may be of value in distinguishing between closely related streptococci, but departures from the modal fermentation pattern are not infrequent.

All the streptococci give a negative catalase reaction

with the possible exception of a few members of group D (Jones *et al.* 1964). These, like a number of pediococci and lactobacilli, may under certain conditions produce a little gas from hydrogen peroxide; this is not due to a true catalase reaction, because the organisms do not possess a cytochrome system and therefore give a negative reaction in the benzidine test of Deibel and Evans (1960). Tests for the production of H_2O_2 are of value in the classification of the α-haemolytic and non-haemolytic species. Nitrate reduction occurs only in a few group D streptococci. The production of acetylmethylcarbinol from glucose is confined mainly to group D streptococci, *Str. milleri*, and a few other non-haemolytic streptococci. Other tests of value in the classification of streptococci include the production of ammonia from arginine and the hydrolysis of hippurate and aesculin (Chapter 20). A combined test for bile tolerance and aesculin hydrolysis is often used (Rochaix 1924). Evidence of strong reducing activity, e.g. very early reduction in litmus milk, black colonies on tellurite agar, and red colonies on tetrazolium agar (Barnes 1956*b*), is of value for the identification of *Str. faecalis*.

The formation of polysaccharides from sucrose is an important means of identifying several streptococci. After hydrolysis of sucrose to its component monosaccharides, some strains form dextran from glucose and some form laevan from fructose. Tests for the polysaccharides can be made in buffered sucrose broth (Niven and White 1946, Bailey and Oxford 1958). The dextran of *Str. sanguis* when produced in large amount may form a gel; that of *Str. mutans* usually forms adherent deposits on the wall of the tube; but that of *Str. bovis* remains in solution. Dextran and laevan may be detected chemically. A tenfold dilution of the supernatant broth is made in 10 per cent sodium acetate, and 1.2 and 2.5 volumes of ethanol are added to portions of it; dextran gives a precipitate in both tubes and laevan only in the second (Hehre and Neill 1946). Guthof (1970) recommends, for the demonstration of dextran, precipitation with 0.8 volume of acetone and of laevan production with 3.0 volumes of methanol. Dextran may also be demonstrated by its ability to cross-react with type 2 pneumococcus antiserum in a precipitin ring test (Hehre and Neill 1946).

The production of dextran and laevan may be inferred from the type of colony formed on sucrose agar; *Str. salivarius*, which forms laevan, gives a soft, fleshy, non-adherent growth of domed colonies, especially on anaerobic incubation; *Str. bovis*, which forms a soluble dextran, gives a watery growth seen best on plates incubated in air with 5 per cent added CO_2; *Str. mutans* and *Str. sanguis*, which form insoluble dextran, give adherent colonies that may extend down into the agar and on continued incubation be characteristically glass-like. (For the relation of dextran production to the formation of dental plaque, see Chapter 57.)

Liquefaction of gelatin occurs only in the *liquefaciens* variety of *Str. faecalis*, but several other streptococci have some proteolytic activity. On agar plates containing various proteins and observed for clearing around individual colonies, about one-half of *Str. pyogenes* strains hydrolyse casein; enterococci hydrolyse albumin and globulin as well. Group B and C streptococci are inactive in these respects (Sherwood *et al.* 1954).

Bacteriophages and bacteriocines

Phages active on streptococci are of two classes (Maxted 1964): virulent phages, for example those found in sewage, with a wide range of lytic action on streptococci; and temperate phages, which lysogenize streptococcal strains, and have a narrower range. Certain virulent phages produce a *phage-associated lysin.* In 1934, Evans described the lysis of group A streptococci when grown with a group C phage and a group C streptococcus, and gave the name 'nascent lysis' to this phenomenon. Maxted (1957) showed that it was due to an enzyme associated with the phage but separable from it, which attacked the cell wall of streptococci of groups A, C and E. Phage-associated lysin has proved useful in the study of the streptococcal cell wall (Krause 1958), and for the induction of L-forms (Gooder and Maxted 1961). Since the absorption of virulent phages is prevented by a hyaluronic acid capsule, this may have a selective action on the streptococcal population, and sometimes render a non-mucoid culture mucoid or even increase the M-antigen content and mouse virulence (Maxted 1955). Many streptococci are lysogenic; Zabriskie (1964) showed that the ability to form erythrogenic toxin could be transferred from one strain of *Str. pyogenes* to another by lysogenization with a temperate phage.

Inhibition of the growth of one streptococcus by another on solid medium is in many cases attributable to the production of H_2O_2 (Malke *et al.* 1974), but many streptococci also form bacteriocines (Tagg *et al.* 1976). Although some of these agents inhibit a wide range of other gram-positive organisms, the spectrum of activity of a number of them on other streptococci is sufficiently specific for use as a means of typing streptococci (Tagg and Bannister 1979). Among the pyogenic streptococci, patterns of bacteriocine production and of susceptibility to bacteriocines show a number of interesting correlations with antigenic structure (Tagg and Bannister 1979, Johnson *et al.* 1979).

Cell-wall composition and antigenic structure

The cell wall is composed principally of peptidoglycan (for the chemical structure of this, see Schleifer and Kandler 1972, Krause 1975). The *group-specific antigens*—known as C substances—that form the basis of the Lancefield grouping system may be either polysaccharides or teichoic acids (Table 29.2). Polysaccharide group antigens are attached to the peptidoglycan, but teichoic-acid group antigens are found in greater amount beneath the cell wall than in it. Many streptococci also possess type-specific polysaccharides that may be part of the cell wall, or may form a surface

Table 29.2 Group and type antigens of streptococci

Group antigens		
Designation (and occurrence)	Chemical nature	Type antigens
A (in *Str. pyogenes*)	PS	Pr (M, T, R)
C (in *Str. equi*, *Str. zooepidemicus*, *Str. equisimilis* and *Str. dysgalactiae*)	PS	Pr (M-like, T, R)
G (in 'large-colony' group G streptococci)	PS	Pr (M-like, M, T, R)
A, C, G, F (in *Str. milleri*)[1]	PS	PS (Ottens antigens)
B (in *Str. agalactiae*)	PS	PS + Pr (I*c*, R, X)
E, P, U (in *Str. lentus*)	PS	Pr
L	PS	Pr (T, R)
D (in enterococci)	TA	PS
D (in *Str. bovis* and *Str. equinus*)	TA	PS
D (in *Str. suis*)	TA	PS ('group' R, S and T antigens)
N (in *Str. lactis* and *Str. cremoris*)	TA	
Pneumococcus	TA[2]	PS (capsular) + Pr(M)[3]

PS = polysaccharide; Pr = protein; TA = teichoic acid.

[1] Also 'group-like' polysaccharides (see text).
[2] The so-called pneumococcal C substance.
[3] Unrelated to the M antigens of *Str. pyogenes*.

layer or even a capsule. The group D and group N antigens are glycerol teichoic acids, and the C substance of the pneumococci is a choline-containing teichoic acid. In common with many other gram-positive bacteria, streptococci also possess a simple teichoic acid that is a polymer of glycerophosphate with ester-linked D-alanine (McCarty 1959, 1964). This is associated with a lipid component to form a *lipoteichoic acid*, part of thich appears on the cell surface and is responsible for the adherence of streptococci to erythrocytes, epithelial cells and phagocytes (Stewart and Martin 1962, Ofek *et al.* 1975, Ofek and Beachey 1979). Finally, many streptococci have type-specific surface proteins, such as the M, T and R antigens of *Str. pyogenes*.

The cell wall should not be looked upon as a series of concentrically arranged antigens (Wagner *et al.* 1978), nor does the cell surface appear to be uniform in composition; several cell-wall components, including lipoteichoic acid, group antigens, and even peptidoglycan, may participate in surface reactions.

Information about the antigenic constitution of individual streptococci will be given later in this chapter. Here we shall describe the group-specific substances only.

Group antigens

Most of the pyogenic streptococci form an antigen—either polysaccharide or teichoic acid—that can be extracted with hot acid and gives a specific precipitate with antiserum prepared by injecting killed whole cells into rabbits (Hitchcock 1924, Lancefield 1928, 1933, 1941). These antigens serve to define the Lancefield groups. Members of these groups are biochemically uniform or, as in the case of group C, contain recognizable biotypes (Table 29.3). The groups, or biotypes and occasionally serotypes within them, have a characteristic host-range (Table 29.1) and ability to cause particular diseases.

Other acid-extractable polysaccharides and teichoic acids are less regularly distributed. Thus, streptococci with the H and K substances cannot otherwise be distinguished from a number of others that do not form these antigens. Moreover, group antigens that characterize certain pyogenic streptococci are occasionally found in a few members of other quite distinct streptococcal species, for example, antigen E in *Str. mutans* and *Str. uberis*. *Str. milleri*, which has many of the biological characters of a pyogenic streptococcus, does not posess a single group antigen; some strains have the antigen F, which is found only in this species; a few have the antigen A, C or G, but most are ungroupable. The group D teichoic acid is shared by three otherwise distinct sorts of streptococcus: (1) the enterococci, (2) *Str. bovis* and *Str. equinus*, and (3) *Str. suis*.

It must be emphasized that acid extracts contain a number of other polysaccharide and protein antigens. Indeed, whether an antigen present in such an extract should be looked upon as a group antigen depends upon its distribution. A group antigen is therefore an antigen that defines a group of streptococci that can also be recognized by other means. Thus, the 'group' Q polysaccharide is best looked upon as an enterococcal type antigen, and the 'group' polysaccharides R and T are now considered to be type antigens in *Str. suis*.

The group polysaccharides are completely non-toxic. They are haptens, and are antigenic only when attached to the streptococcal cell wall. Individual rabbits vary in their antibody response to the injection of whole streptococci or cell walls. In-breeding of rabbits that give a strong antibody response yields animals that are capable of producing relatively enormous amounts of antibody to group A or C polysaccharide. This antibody is an electrophoretically homogeneous immunoglobulin that in many ways resembles myeloma protein (Osterland *et al.* 1966, Braun *et al.* 1969, 1980), i.e. monoclonal antibody.

The chemical basis of the antigenic specificity of some of the group polysaccharides is known. The group A polysaccharide is composed of *N*-acetylglucosamine and rhamnose, and its antigenic determinant is the terminal *N*-acetylglucosamine attached through a β-linkage to a series of rhamnose molecules (Schmidt 1952, McCarty 1958, Krause 1963). That of the group C polysaccharide has a terminal *N*-acetylgalactosamine attached to the rhamnose. When passaged in mice, group A strains sometimes lose their previous antigenic specificity and acquire another (Wilson 1945, McCarty and Lancefield 1955); these 'variant' strains are antigenically similar to the corresponding variants of group C strains. In each case the terminal amino sugar is lost and the rhamnose exposed (McCarty 1956, Krause 1963, Coligan *et al.* 1980). Rhamnose is the determinant sugar in group B and G polysaccharides, which show some cross-reactions; but the absence of cross-reaction between group B and A-variant polysaccharide suggests a difference in the role of rhamnose in the two antigens (Curtis and Krause 1964). The determinant of the group L polysaccharide is also *N*-acetylglucosamine attached to a rhamnose oligosaccharide, which accounts for the partial cross-reaction between groups A and L, but the linkage of the terminal sugar to the rhamnose sugar appears to be different in the two groups (Karakawa *et al.* 1971). Removal of the *N*-acetylglucosamine results in a polysaccharide that reacts with group G but not with group L polysaccharide.

Grouping techniques

Group antigen may be detected simply by extracting it from the cells in boiling acid, careful neutralization, centrifugation, and layering the supernatant over the serum in a capillary tube; a ring of precipitate will appear within a few minutes (Lancefield 1933). Hydrochloric acid at a strength of N/5 is often used, but this may occasionally destroy the polysaccharide; in such an event the antigen may be extracted in serologically active form by N/15 acid. The group B antigen may sometimes be destroyed by acid at high temperature; it can be extracted with N/5 acid at 50° in 2 hr. The antigen may also be extracted by autoclaving (Rantz and Randall 1955) or with nitrous acid (El Kholy *et al.* 1974). Extraction of group polysaccharides with formamide (Fuller 1938) leaves only a 'ghost' of the cell wall composed of peptidoglycan (Krause and McCarty 1961) and destroys protein antigens. Subsequent purification of the polysaccharide by adding 95 per cent ethanol in which it is soluble, and then precipitating it with acetone, is a useful method of removing cross-reacting material from the preparation.

Two enzymes attack the cell-wall mucopeptide of whole streptococci and release soluble complexes composed of fragments of mucopeptide with attached group polysaccharide (Krause and McCarty 1961). These are (1) the enzyme formed by *Streptomyces albus* (Maxted 1948, McCarty 1952, Schmidt 1965) and (2) the so-called phage-associated lysin (p. 179). They are both active on streptococci of groups A, C and G, but not of group D, and are rapid and convenient alternative methods of obtaining extracts for grouping.

In addition to the capillary-tube precipitation test, grouping may be performed by (1) a double-diffusion test in gel (Rotta *et al.* 1971), which has the advantage that a reaction of identity with a standard preparation of antigen may be obtained, (2) counter-current immunoelectrophoresis (Dajani 1973), or (3) immunofluorescence with labelled $F(ab')_2$ fragments of anti-group IgG (Cars *et al.* 1975). Streptococcal grouping sera give reactions in precipitation tests only when used neat or at low dilution. Preparations of monoclonal antibody give stronger and more specific results (Nahm *et al.* 1980).

Agglutination methods of grouping These are labour-saving and very sensitive, but specially absorbed grouping sera are necessary; if whole cells are used, these must first be trypsinized. A conventional slide-agglutination method may be used (Rosendal 1956), or a co-agglutination method in which the streptococcal suspension is mixed with a series of suspensions of a protein-A-forming strain of *Staph. aureus* to which the various group antibodies have been adsorbed (Christensen *et al.* 1973). The co-agglutination method may be simplified by mixing the staphylococcal suspension, a trypsinized streptococcal suspension and the antiserum simultaneously. A further development of this is to mix the staphylococcus with a formamide extract of the streptococcus and antiserum. Thus the need for trypsinization is eliminated, and streptococcal growth from a primary plate can be used (Efstratiou and Maxted 1979). Agglutination of antibody-coated erythrocytes by formamide extracts has also been advocated.

Toxic properties of streptococcal cell walls

The cell walls of streptococci are toxic for experimental animals, whether or not these are susceptible to infection by the corresponding streptococcus. This toxicity is attributable to the peptidoglycan (Rotta and Raška 1963, Raška and Rotta 1963, Rotta 1967), which has a variety of biological activities (see Schleifer and Heymer 1975); it causes fever, lyses blood platelets, acts as both preparative and provoking agent for the localized Shwartzman reaction, is immunoadjuvant, and gives rise to focal necrosis and diffuse cellular infiltration of the myocardium.

Intradermal injection of soluble peptidoglycan produces an acute necrotic lesion in the skin of rabbits (Abdulla and Schwab 1966); however, if sonically disrupted streptococci are given, multiple skin nodules develop in a circle around the site of injection and may persist, with alternate remission and exacerbation, for up to 2 months (Schwab and Cromartie 1957). This effect is due to a complex of group polysaccharide and peptidoglycan, neither of which alone will cause it. Abdulla and Schwab (1966) concluded that the toxicity of the preparation was due to the peptidoglycan, but that the effect was slowed down by its attachment to the polysaccharide, and possibly other substances, in cell fragments that were resistant to digestion by inflammatory cells (Smialowicz and Schwab 1977, 1978). Under certain circumstances, whole cells may induce similar lesions (Cromartie *et al.* 1977). Cell-wall fragments injected intraperitoneally may give rise to lesions, in which streptococcal group antigen can be detected, in distant parts of the body: in the myocardium, joints and pinnae (Ohanian *et al.* 1969, Cromartie *et al.* 1979), presumably as a reaction to translocated inflammatory cells containing streptococcal material.

The pyogenic streptococci

The streptococci we include in this group are listed in Table 29.3. They are, with certain exceptions, β-haemolytic; they are not specially resistant to heat, do not grow at extremes of temperature or an alkaline pH, and do not have strong reducing properties. All of them are animal parasites; most of them cause septic or respiratory-tract infections and form extracellular toxins. They comprise the following organisms.

1. *Str. pyogenes*, the four species of group-C streptococci, and the un-named species of 'large-colony' group G streptococci have many characters in common, and differ from each other mainly in host-specificity.
2. *Str. agalactiae*—the group B streptococcus—is a distinct organism with several characters that are unique among the pyogenic streptococci.
3. *Str. lentus* strains are biochemically uniform but may have one of three group antigens, E, P and U; they are found in pigs.
4. Group L streptococci have not been fully studied, but appear to be distinct from other pyogenic streptococci.
5. *Str. suis* is a biochemically uniform species found in pigs. The fact that it forms the group D teichoic acid has been recognized only recently. Before this these streptococci were described as members of Lancefield's groups R, S and T.
6. Our reasons for considering *Str. milleri* as one of the pyogenic streptococci will be set out later in this chapter.

Before describing the individual members of the group, we shall discuss certain characters that are common to several of them. When, in this section, we use the terms 'group A', 'group C' and 'group G' streptococcus without qualification we intend to exclude *Str. milleri* strains that form these group antigens.

Morphology

The cell surface of *Str. pyogenes* is covered by short, fine filamentous appendages composed of M protein (Swanson *et al.* 1969) and lipoteichoic acid (Beachey and Ofek 1976). Many members of groups A and C

Table 29.3 Usual characters of the pyogenic streptococci

The following are common characters: do not survive at 60° for 30 min; do not grow at 10° or 45°, at pH 9.6, in 0.1 per cent tellurite or on 0.01 per cent tetrazolium agar; hydrolyse arginine; optochin resistant. Growth in 6.5 per cent NaCl unusual.

		Fermentation of							Hydrolysis of			
Species	Group	mannitol	lactose	sorbitol	trehalose	salicin	inulin	raffinose	hippurate	aesculin	Growth on 40 per cent bile agar	Voges-Proskauer reaction
Str. pyogenes	A	−	+	−	+	+	−	−	−	−	−	−
Str. equi	C	−	−	−	−	+	−	−	−	−	−	−
Str. zooepidemicus	C	−	v	+	−	+	−	−	−	−	−	−
Str. equisimilis	C	−	v	−	+	+	−	−	−	−	−	−
Str. dysgalactiae	C	−	+	v	+	v	−	−	−	−	−	−
Str. sp.	G	−	v	−	+	v	−	−	−	−	−	−
Str. agalactiae[1]	B	−	v	−	+	v	−	−	+	−	+	−
Str. lentus[1]	E, P, U		v	+	+	+	−	−	−	+	v	v
Str. sp.	L	−	+	v	+	v	−	−	v	−	v	−
Str. suis	D	−	+	−	+	+	+	v[2]	−	+	v	−
Str. milleri	−[3]	−	+	−	+	+	−	−	−	+	v	+

v = Some positive, others negative.
[1] CAMP positive.
[2] 'Groups' R and T positive, 'group' S negative.
[3] A minority posess group antigen A, C, F, or G.

form capsules that are composed of hyaluronic acid (Kendall *et al.* 1937). *Str. equi* and *Str. zooepidemicus* are regularly capsulated (Schütz 1888, Frost and Engelbrecht 1940) but *Str. equisimilis* is not. Some strains of *Str. pyogenes* are obviously capsulated and form a mucoid growth on blood agar, but hyaluronic acid formation can often be detected in broth cultures of strains that do not give mucoid colonies. Seastone (1934, 1943) showed that capsules were formed early in the growth cycle and tended to disappear later. This is due to the subsequent production of hyaluronidase.

Seastone (1939) observed that variants of group C streptococci that were highly virulent for guinea-pigs produced more hyaluronic acid and larger capsules than did less virulent variants. Hirst (1941) showed that the addition of hyaluronidase to the inoculum reduced the virulence of group C streptococci for both mice and guinea-pigs by the intraperitoneal route; on the other hand, its addition to group A streptococci had little effect on their virulence for mice by this route (Hirst 1941, Rothbard 1948). According to Johnson and Furrer (1958) the presence of the capsule plays a determining part in virulence for mice infected by aerosols. The significance of the capsule in natural infection is still uncertain, because little is known about the balance between the formation of hyaluronic acid and of hyaluronidase by streptococci in the tissues. Hyaluronic acid is not antigenic.

Resistance to antimicrobial agents

The pyogenic streptococci are still uniformly sensitive to penicillin, but many strains have become resistant to other antimicrobial agents. Sulphonamide resistance became locally prevalent during the Second World War after the use of the drug for mass prophylaxis (Report 1945); nowadays it is found in well under 10 per cent of strains. Tetracycline resistance appeared in many parts of the world after 1959 (Kuharic *et al.* 1960, Parker *et al.* 1962); in Britain it had reached a frequency of about 50 per cent by 1965, but this has since declined considerably. Resistance to chloramphenicol, erythromycin and lincomycin is infrequent in Britain and the USA, but local epidemics caused by erythromycin- and lincomycin-resistant strains have occurred in burns wards in Britain (Lowbury and Hurst 1959) and in isolated communities in Canada (Dixon and Lipinski 1974). Resistance to both antibiotics may appear rapidly in a previously sensitive strain when lincomycin is used for mass prophylaxis (Kohn *et al.* 1968); it may be lost discontinuously when the organism is subcultured serially (Lowbury and Kidson 1968). In Japan, a considerable proportion of all *Str. pyogenes* strains have become resistant to several antibiotics; resistance to tetracycline appeared in 1964, to chloramphenicol in 1969, and to erythromycin in 1971, so that by 1972–74 nearly 40 per cent of all strains isolated were resistant to all three antibiotics (Miyamoto *et al.* 1978).

Several patterns of resistance to erythromycin and lincomycin have been observed. Erythromycin resistance may occur alone, but this is now unusual. Strains may exhibit high resistance to both antibiotics, or low resistance to erythromycin and high resistance to lincomycin. Among the latter, there are strains that show 'paradoxical' lincomycin resistance, that is to say, they are resistant to low and high concentrations but sensitive to intermediate concentrations (Dixon and Lipinski 1972); in the presence of erythromycin, these strains become resistant to intermediate concentrations of lincomycin and to high concentrations of erythromycin. This type of resistance is determined by a plasmid that can be transduced to other streptococci of groups A, C and G (Malke *et al.* 1975).

Group B streptococci are usually resistant to tetracycline (Baker *et al.* 1976). Resistance to chloramphenicol, erythromycin and lincomycin is infrequent, but Lütticken and Laufs (1979*b*) have described single plasmids that determine resistance to all three antibiotics. Streptococci of groups C and G are not often antibiotic resistant.

In groups A, C, G and B, resistance plasmids may be transferred between strains of the same group, and at lesser frequencies to strains of another group, by transduction (Malke *et al.* 1975), or by growth in mixed culture (van Embden *et al.* 1978).

Sensitivity to bacitracin The disk test for sensitivity to bacitracin (Maxted 1953*b*) is a valuable screening test for *Str. pyogenes*. However, the difference in sensitivity between this organism and other haemolytic streptococci is not great, so the potency of the disks must be carefully standardized. If the end-point is to be taken as absence of any zone of inhibition, the disk should contain 0.04 units of bacitracin (Coleman *et al.* 1977). Tests should never be performed on primary-culture plates. Bacitracin sensitivity is not confined to *Str. pyogenes*; it occurs in a number of non-haemolytic streptococci, in many members of group L, in some *Str. lentus* strains, and in some group C and group G strains. In cultures from the human upper respiratory tract, however, there are less than 5 per cent of discrepancies between the results of the bacitracin-sensitivity test and of serological grouping. In material in which group C streptococci predominate this figure may be somewhat higher (Rotta and Jelinková 1967).

The toxins of the pyogenic streptococci

Most of our knowledge concerns the extracellular toxins of *Str. pyogenes*, which have been extensively studied, but several of these substances are formed also by other pyogenic streptococci.

Haemolysins

Marmorek (1895) showed that some streptococci lysed red blood corpuscles and that the lytic agent was filtrable. Todd (1932, 1938*a*) and Weld (1935) established that many haemolytic streptococci form two dinstinct lysins: one is haemolytic in the reduced form (Neill and Mallory 1926) and the other is oxygen stable and apparently soluble in serum. Todd (1938*a*) named them respectively streptolysin O (oxygen labile) and streptolysin S (serum soluble).

Streptolysin O is formed by most group A streptococci and by many strains of groups C and G, but not by members of other groups. It is haemolytic and lethal only when reduced, but is antigenic in both the oxidized and the reduced form. Antibody to it can be detected by inhibition of haemolysis by reduced toxin (Todd 1932), and this is the basis of the anti-streptolysin O test (Chapter 59). The O-lysins of group A, C, and G streptococci are immunologically identical (Todd 1939). The lysin is in the oxidized form when in contact with air, and is rapidly reduced by sodium hydrosulphite or cysteine. It is stable when frozen, but rapidly and irreversibly inactivated at room temperature, probably by the action of streptococcal proteinase. It is a protein (Bernheimer 1948), and some progress has been made in its purification (Halbert and Auerbach 1961, Alouf and Raynaud 1967; see also Bernheimer 1972). When highly potent toxin is injected intravenously into mice, rabbits, and guinea-pigs, death occurs within seconds. Titrations of lethal activity have a sharp end-point and at around the LD50 the animal either dies or survives apparently unharmed. Death is from acute toxic action on the heart (Thompson *et al.* 1970). Haemolysis begins after a short latent period—as little as one minute with a high concentration of lysin—and is almost instantaneous. Reduced lysin rapidly becomes bound to erythrocytes at 0°, but lysis occurs only when the temperature is raised. The oxidized form does not become attached to the cells. Streptolysin O also has a cytotoxic action on leucocytes (Todd 1942, Bernheimer and Schwartz 1960) and lyses platelets (Bernheimer and Schwartz 1965*a*). It lyses parasitic but not saprophytic mycoplasmas (Bernheimer and Davidson 1965). Hewitt and Todd (1939) showed that cholesterol inhibited both the haemolytic and the lethal activity of streptolysin O. This and other evidence (see Halbert 1970) supports the view that the site of attachment of the lysin to the cell membrane is the cholesterol molecule. Some sera contain non-specific inhibitors of streptolysin O (Chapter 59), but serum cholesterol does not act in this way.

There are many similarities between streptolysin O and other oxygen-labile haemolytic toxins, such as pneumococcal haemolysin, tetanolysin, and *Cl. perfringens* θ-toxin, all of which are neutralized by serum from a horse hyperimmunized against streptolysin O (Todd 1934, 1941).

Streptolysin S is responsible for the β-haemolysis around colonies of many streptococci on aerobically incubated blood agar plates. Oxygen-stable streptolysins of the S type are formed by most streptococci of groups A, C and G and also by members of a number of other groups of pyogenic streptococci. Early workers found that the addition of serum to broth stimulated the production of haemolysin, and Todd (1928*b*) showed that this haemolysin was not inactivated by oxygen. Later (Todd 1934) he found that, when serum was added to washed suspensions of streptococci, haemolysin appeared rapidly and could be detected in the filtrate. Several agents other than serum cause the formation of extracellular lysin by washed streptococci; these include yeast ribonucleic acid (Okamoto 1940), egg-yolk lecithovitellin or heated milk (Herbert and Todd 1944), α-lipoprotein, and certain detergents (Ginsburg *et al.* 1963). The lysins are all similar in their general character, but differ somewhat in their physico-chemical properties and in their susceptibility to inactivation by various substances (Ginsburg and Harris 1963). In general, however, these differences correspond to differences between the respective inducers of lysin formation; and, when a haemolysin formed by the action of one inducer is incubated with another inducer, it is possible to detect haemolytic activity in fractions containing the second inducer. These considerations led Ginsburg and Harris (1963, 1964) to conclude that streptolysin S was a complex of a small haemolytic molecule derived from the streptococcus with one of several carriers, which are also the inducers of cell-free lysin. When erythrocytes are mixed with washed streptococci and then separated from them after a short time they subsequently lyse (Ginsburg and Harris 1965). The haemolytic moiety may therefore become attached to the erythrocyte in the same way as it does to the inducer molecule. Bernheimer (1967) concluded that the haemolytic agent was a polypeptide composed of as few as 28 amino-acid molecules. This may explain its apparent lack of antigenicity. (See also Hryniewicz *et al.* 1978.)

Streptolysin S is sensitive to heat and acid, and can be preserved only by storage at very low temperatures. It lyses erythrocytes more slowly than does streptolysin O. Serum extracts of streptococci kill mice in a dose of 0.1 ml (Weld 1934, 1935); and the injection of lysin intravenously into rabbits causes rapid death with evidence of intravascular haemolysis (Todd 1938*b*). On intraperitoneal injection into mice it also causes necrosis of the renal tubules (Tan and Kaplan 1962). *In vitro*, it has a cytopathic action on leucocytes and on a variety of cultured tissue cells, including tumour cells, and it lyses platelets; its action appears to be on the phospholipid of mammalian cell membranes or of their subcellular organelles (see Ginsburg 1970). It has no action on intact bacteria, but lyses cell-wall-deficient forms of various gram-positive organisms, but not those of group A streptococci (Bernheimer and Schwartz 1965*b*). Like streptolysin O, it lyses parasitic but not saprophytic mycoplasmas (Bernheimer and Davidson 1965). Ofek and his colleagues (1970) have suggested that the rapid death and disintegration of leucocytes that have ingested group-A streptococci (Levaditi 1918), once attributed to the action of streptococcal NADase (Bernheimer *et al.* 1957), may be due to cell-bound streptolysin S.

It may be added that an *intracellular haemolysin* extracted from sonically disrupted group A streptococci by Schwab (1956) appears to be different from streptolysin S, because it is antigenic.

Erythrogenic toxin

This was first detected when filtrates of cultures of group A streptococci were injected intradermally into certain healthy adults. It was concluded that the resulting erythematous reaction was a local 'model' for the skin rash of scarlet fever, which was widely held to be a manifestation of the direct toxicity of the erythrogenic substance. There is considerable evidence for the alternative view (Chapter 59) that the rash is a hypersensitivity reaction. Here we shall consider only the properties of the substances in streptococcal filtrates that cause the erythema and associated clinical manifestations.

Skin reactivity in laboratory animals is variable; young guinea-pigs and rabbits are usually insusceptible (Dochez and Sherman 1925, Parish and Okell 1930) but become reactive after the injection of a small dose of toxin (see Watson and Kim 1970, Schlievert *et al.* 1979). The material in crude filtrates that causes the skin reaction in man is heat-stable, requiring 96° for 45 min for complete inactivation. Purified preparations show degrees of heat lability that depend to some extent on the treatment they have received. The toxin is antigenic, and antisera prepared in animals neutralize the skin reaction (Todd *et al.* 1933). Highly purified toxin has poor antigenicity (Watson and Kim 1970). Three immunologically distinct erythrogenic toxins are now recognized, A, B and C (Hooker and Follensby 1934, Watson 1960); they are proteins with molecular weights respectively of 8000, 17 000 and 13 000 (Watson 1979). They are formed only by group A streptococci.

Pyrogenic exotoxin

In 1953, Schwab and his colleagues identified a heat-labile factor in streptococcal skin lesions in rabbits that had the property of enhancing the lethal properties of the endotoxin of gram-negative bacilli. Further investigation established that this property was associated with erythrogenic toxin (Watson 1960), and that pyrogenicity in the rabbit was the most reliable test for it. The later work of Watson and his colleagues (see Kim and Watson 1970, Watson and Kim 1970) revealed that purified erythrogenic toxin—which they prefer to call *streptococcal pyrogenic exotoxin*—had the following actions: pyrogenicity and lethality in rabbits (minimum pyrogenic dose 0.07 μg per kg and LD50 3500 μg per kg, both by the intravenous route); enhancement, sometimes in excess of 100 000-fold, of susceptibility to fatal endotoxic shock in rabbits; toxicity for cultured spleen macrophages, and immunosuppression (see Cunningham and Watson 1978). It causes coagulation of *Limulus* amoebocyte-lysate, but this action is very much more feeble than that of the endotoxin of gram-negative bacilli (Brunson and Watson 1976). According to Watson (1979), pyrogenic exotoxin alters the permeability of the blood-brain barrier and causes fever by direct action on the hypothalamus rather than by causing the production of endogenous pyrogen by granulocytes.

According to Watson and his colleagues (Kim and Watson 1972, Schlievert *et al.* 1979), the pyrogenic exotoxins have two distinct activities, possibly caused by different parts of the same molecule: (1) *primary toxicity*, which is neutralized only by antiserum against the homologous exotoxin and may play a part in the pathogenesis of streptococcal infection quite independent of the ability to give rise to a skin rash, and (2) *secondary toxicity*, which arises from the induction of hypersensitivity to streptococcal products—and also to other antigens—and shows cross-reactivity between the types of pyrogenic exotoxin. Secondary toxicity is responsible for the scarlatiniform rash by inducing skin hypersensitivity of the delayed type and enhancing the local Arthus reaction.

Streptokinase

Filtrates of some streptococci cause the dissolution of fibrin clot by means of a fibrinolysin or streptokinase (Tillett and Garner 1933). This acts not on the fibrin but on a factor present in normal plasma (plasminogen) which is converted into a protease (plasmin), which in turn lyses the fibrin (Kaplan 1944, Christensen 1945). It is formed by most group A streptococci, but in amounts that vary from strain to strain (Report 1947), and also by some members of groups C and G. The streptokinase of group A streptococci acts on the plasminogen of human and bovine but not of rabbit blood. Group A streptococci produce at least two immunologically different streptokinases (Dillon and Wannamaker 1965). Antibody to streptokinase appears in man after streptococcal infections, but its measurement is often complicated by the presence of other factors in serum that render fibrin clots resistant to lysis, including a non-specific anti-proteinase that appears during the acute phase of certain fevers (Kaplan 1946). Experiments on laboratory animals suggest that streptokinase in the presence of human plasminogen may enhance the severity of streptococcal lesions (Krasner and Jannach 1963). According to Ward (1967) plasmin formed by the action of streptokinase acts on the C3 component of complement to give a product with chemotactic activity.

Nucleases

All strains of *Str. pyogenes* form both *deoxyribonuclease* (DNAase) and *ribonuclease* (McCarty 1948, Tillett *et al.* 1949); DNAase is also produced by group B, C, G and L streptococci (Deibel 1963) but usually in smaller amount. Wannamaker (1958) described three immunologically distinct DNAases—A, B and C—any combination of which might be formed by the same strain. The B enzyme, however, is the one most regularly formed, and antibody to it appears in the blood after most infections with group A streptococci (Wannamaker 1959; see also Chapter 59). A fourth DNAase (D) has since been recognized (Wannamaker *et al.* 1967). The B and D enzymes, but not the A and C, also have ribonuclease activity. Taylor (1971) describes a ribonuclease with no action on DNA that is formed by many strains of group A streptococci.

Proteinase

Under suitable conditions most strains of *Str. pyogenes* form a proteinase that attacks casein, fibrin, and gelatin, though it is not responsible for fibrinolysis or gelatin liquefaction (Elliott 1945, Deibel 1963). It destroys several proteins formed by the streptococcus itself, including streptolysin O, streptokinase, hyaluronidase, and M antigen, but it is formed at a rather more acid pH (5.5–6.5) than are the other extracellular products. Destruction by proteinase is an important reason for failing to detect M antigen in a culture; and the value of Neopeptone in Todd-Hewitt broth is that it inhibits proteinase formation (Elliott and Dole 1947). Elliott and Dole (1947) showed that proteinase appeared in cultures as an inactive precursor, which was converted into the active form only under reducing conditions. Conversion is stimulated by the initial presence of a trace of active proteinase; a low concentration of trypsin has the same effect. According to Liu and Elliott (1965) the reducing activity of streptococcal cell walls initiates conversion of precursor to active enzyme. Both proteinase and its precursor have been crystallized; they are both antigenic, and behave as if the precursor had two antigenic components, only one of which is shared with the active enzyme (Elliott 1950).

Hyaluronidase

Many streptococci, including a number that have hyaluronic acid capsules, form hyaluronidase (Meyer *et al.* 1940). Many strains form hyaluronic acid during the early stage of growth *in vitro* and subsequently form hyaluronidase which destroys it (Pike 1948). Hyaluronidase production is stimulated by the addition of hyaluronic acid to a growing culture (McClean 1941). Certain M types of *Str. pyogenes* (e.g. types 4 and 22), in which capsulation is rarely seen, form large amounts of hyaluronidase (Crowley 1944). The formation of hyaluronidase occurs not only in groups A and C, but also in group B, in the large-colony form of group G, in *Str. suis* and *β*-haemolytic strains of *Str. milleri*, and in pneumococci. Hyaluronidase is antigenic, but the enzyme produced by group A strains is immunologically distinct from that of streptococci of groups C and G; antibodies are formed after most group A infections in man. Although streptococcal hyaluronidase increases the permeability of rabbit skin to India ink and to bacterial cells, there is little evidence that it favours invasion of tissues by streptococci under natural conditions.

Nicotinamide adenine dinucleotidase (NADase)

This extracellular enzyme (Carlson *et al.* 1957) is formed by all members of some M types of group A streptococci but not by members of a number of other types (Lazarides and Bernheimer 1957, Lütticken *et al.* 1976); it is also formed by some streptococci of groups C and G. NADase-producing strains inhibit the growth of *Haem. parainfluenzae* on plates of NAD-containing medium (Green 1979). It is doubtful whether NADase plays any part in pathogenesis, but antibody to it may be formed in man as a result of streptococcal infection.

Serum opacity factor

Many group A streptococci give rise to opacity in serum broth (Keogh and Simmons 1940) or around colonies on serum agar. This is probably attributable to the enzymic release of lipids from serum lipoprotein (Krumwiede 1954, Hill and Wannamaker 1968). The opacity factor is closely related to M protein. It is formed only by *Str. pyogenes* and by a few group C and group G streptococci that possess M-like antigens. Among the group A streptococci, all members of some M types but no members of other M types form it (Top and Wannamaker 1968*a*, Widdowson *et al.* 1970; see Table 29.5). The opacity factor is neutralized by streptococcal antisera, and the specificity of this reaction almost exactly parallels that of the corresponding M antigens (Top and Wannamaker 1968*a*, *b*, Maxted *et al.* 1973*b*); thus opacity-neutralization tests provide a useful additional method of typing group A streptococci (p. 189).

The opacity factor is extractable with hot acid from M-positive strains but not from M-negative variants of them; it is also present extracellularly in broth cultures. Purification of M antigen from opacity-forming strains yields material progressively richer in opacity factor (Widdowson *et al.* 1971*a*), but the two substances have been partly separated (Hallas and Widdowson 1979). Opacity factor is trypsin sensitive, but rather more labile to heat and acid than M protein. In rabbits, the antibody response to opacity factor is irregular and shows little correspondence in time or magnitude with the response to M protein; guinea-pigs regularly form antibody to opacity factor in considerable amount (Fraser 1982). In man, opacity-neutralizing antibody is formed as a result of natural infection (Maxted *et al.* 1973*a*).

Other extracellular products

Neuraminidase activity has been detected not only in members of Lancefield groups A, B, C, G and L but also in pneumococci and various other α-haemolytic streptococci (Hayano and Tanaka 1969). The enzymic activity of most strains from culture collections appears to be weak, but that of strains freshly isolated from systemic infections may be strong. The diseases in which neuraminidase may possibly contribute to pathogenesis include neonatal infections due to group B streptococci of type III (p. 192) and pneumococcal infections (p. 197); to these Müller (1974) would add endocarditis associated with *Str. sanguis* and other α-haemolytic streptococci.

Some group A streptococci form starch from maltose, but under other cultural conditions they form an amylase (Crowley 1950, 1959). The ability to form *β*-glucuronidase is said to characterize certain serotypes of group A streptococci (Williams 1954), but it is also found in members of groups C and G. Many extracellular and cellular fractions of group A streptococci cause blastogenic transformation of lymphocytes (see Ginsburg 1972), but the agents responsible for this effect cannot yet be clearly specified.

Streptococcus pyogenes

The terms group A streptococcus and *Str. pyogenes* are almost but not quite synonymous. With the exception of a few strains of *Str. milleri*, the group A polysaccharide is found only in *Str. pyogenes*.

Cultural and biochemical characters

Str. pyogenes is almost invariably β-haemolytic. Members of some serotypes give large clear zones of haemolysis but members of others give smaller zones of incomplete clearing with a fuzzy edge. According to Pinney and her colleagues (1977), poor haemolysis is associated with the production of serum-opacity factor (p. 186). They produce evidence that opacity factor inhibits the liberation of streptolysin S by serum, and state that the incorporation of RNAase-digested yeast RNA into blood agar enhances haemolysis by opacity-producing strains.

Non-haemolytic variants of several types have been described. One type formed α-haemolytic colonies aerobically and β-haemolytic colonies anaerobically; Fuller and Maxted (1939) attributed this, at least in part, to the early production of peroxide. Another strain that was β-haemolytic only on anaerobic growth appeared to inactivate its own haemolysin in the presence of oxygen (Todd 1928*b*). Coburn and Pauli (1941) described a variant that was β-haemolytic only at 22°. Another variant that was completely non-haemolytic aerobically and anaerobically, and did not form streptolysin S, appeared in a hospital ward during an epidemic of infection caused by a β-haemolytic strain of the same serotype (Colebrook *et al.* 1942).

Group A streptococci vary considerably in *colonial appearance*. *Mucoid colonies* are formed by strains synthesizing large amounts of hyaluronic acid. They may be as large as 3–7 mm in diameter, round, oval or running together, with a sharply defined edge, a smooth mirror-like surface and a viscous consistency. *Post-mucoid colonies* apparently result from the collapse of a mucoid colony due to the action of hyaluronidase or drying, or both, and show folds, ridges and papillae (Wilson 1959). *Matt colonies* were described by Todd and Lancefield (1928) as typical of strains producing M antigen; they are small (0.5–1.5 mm in diameter) and domed with a dull matt surface. *Glossy colonies* (Todd 1928*a*) are formed by variants lacking the M antigen, are even smaller than matt colonies, and have a smooth or finely stippled glistening surface. Matt colonies tend to be opaque and glossy colonies to have a bluish translucence. It must be emphasized, however, that it is often difficult to distinguish between M-positive and M-negative variants by means of the colonial appearances. Certain very opaque, flat and irregular colonies (Griffith 1934) are composed of tangled masses of very long chains; according to McCarty (1966) they may be M positive or M negative. A massive intercellular bridge of mucopeptide unites adjacent cocci in the chain (Swanson and McCarty 1969). Most strains of *Str. pyogenes* form chains of moderate length (*ca* 20–40 organisms per chain) and give a granular growth in broth.

The *growth requirements* of *Str. pyogenes* are complex; minimal growth occurs in chemically defined media, but the addition of peptides in the form of a dialysate of meat infusion is necessary for full growth and for the production of extracellular products and M protein (Wannamaker 1958). Alternatively, a digest broth medium freed from substances of high molecular weight by gel filtration and suitably supplemented with pure chemicals can be used (Holm and Falsen 1967).

The *biochemical characters* of *Str. pyogenes* are fairly uniform but not distinctive (Table 29.3). In addition to the properties common to all pyogenic streptococci, they are bile sensitive, many of them failing to grow on 10 per cent bile agar; they do not hydrolyse hippurate, and rarely hydrolyse aesculin. They usually ferment lactose, trehalose, sucrose, and salicin, but not mannitol, sorbitol, inulin, or raffinose. With the exception of type 3 strains, peroxide is generally produced.

Type antigens

In 1919, Dochez, Avery and Lancefield established the existence of types among the haemolytic streptococci pathogenic for man by means of agglutination tests, and showed that the results corresponded with those of mouse-protection tests. Griffith (1926, 1927, 1928, 1934, 1935) distinguished 27 types by means of agglutination, but later it was found that four of them were types of group C or group G streptococci. Lancefield (1928) found that hot-acid extracts of group A streptococci contained not only the group polysaccharide but also a type-specific protein that could be detected in precipitation tests with antiserum from which group antibody had been removed by absorption with a streptococcus of heterologous type. The types so defined corresponded to those of Dochez and his colleagues. Lancefield used many of Griffith's type strains for the production of typing sera, and the type numbers of the Griffith and Lancefield systems were often the same.

Todd (1927*a*, *b*) had found that streptococci freshly isolated from patients could multiply in defibrinated blood from normal persons; some of them were virulent for mice. On subculture in the laboratory these properties were sometimes lost; this was accompanied by a change from the matt to the glossy colonial form (Todd 1928*a*), and by loss of the type-specific protein found in hot-acid extracts (Lancefield and Todd 1928). Hare (1928, 1932) showed that the addition to defibrinated human blood of serum from a patient who had recovered from a streptococcal infection prevented the multiplication of the streptococcus that had caused the infection, but not of other streptococci. This is the basis of the 'bactericidal test' for type-specific antibody (see also Maxted 1956, Lancefield 1957).

Table 29.4 Comparison of properties of M, R and T antigens of *Streptococcus pyogenes*

Property	M antigens	R antigens	T antigens
Effect of:			
boiling at pH 7	Stable	Stable	Destroyed in some media
boiling at pH 2–3	Stable; extracted	Stable; extracted	Destroyed
trypsin	Destroyed	R28: stable R3: destroyed	Stable[1]
pepsin	Destroyed	Destroyed	Stable
Specificity	Distinct M antigen for each type[2]	R28: in several types, and in groups B, C & G. R3: associated with M3	Complex patterns common; for association with M antigens, see Table 29.5
Antigenicity	Moderately or poorly antigenic in cell; poorly antigenic when extracted from cell	Moderately antigenic in cell	Highly antigenic in both intact cell and when extracted from cell
Virulence	An essential virulence factor; antiphagocytic action	Not related to virulence	Not related to virulence
Protection	Antibody confers type-specific protection	Antibody not protective	Antibody not protective

[1] Resist trypsinization of the living cell; but trypsinization of heated suspensions (80° 30 min) releases intact T antigen (Pakula 1951). T-antigen 2 is extracted by autoclaving in buffer, pH 7.8, and is then trypsin sensitive (Maxted 1953*a*).
[2] For exceptions see text.

M and T antigens Further observations by Lancefield (1940, 1943) showed that two classes of protein antigen were responsible for type-specificity in group A streptococci (see Table 29.4). The *M antigens* are resistant to heat and acid, soluble in alcohol and destroyed by trypsin; they are concerned with virulence for mice and the ability to multipy in human blood; antibody to them can be detected by passive protection tests in mice and by bactericidal tests. The *T antigens* are less resistant to heat and acid, insoluble in alcohol, and resistant to trypsin; they play no part in virulence (see also Lancefield 1962). The M antigen is usually detected in hot-acid extracts by a capillary precipitation test with absorbed antisera (Swift *et al.* 1943). A more convenient alternative is an Ouchterlony double-diffusion test with unabsorbed antisera, including purified M antigens as controls. The M antigen is also responsible in part for the agglutination reactions obtained with untrypsinized suspensions of matt streptococci; but with glossy variants or with trypsinized suspensions agglutination is due to T antigen. From about 1940, workers who used the Griffith agglutination typing method made a practice of trypsinizing their suspensions (Elliott 1943), and the method was then a T-typing system.

As a general rule, each streptococcus has only a single M antigen, but there are a few exceptions. The M antigens 14 and 51 sometimes occur together in the same strain, and the relative proportion of the two may vary on subculture or in successive isolations of a strain from the same patient (Wiley and Wilson 1961). Organisms with the M antigen 12 may give rise to variants with the M antigen 62 or 22, without any change in T antigen but with a concomitant gain in the ability to cause opacity in horse serum (Maxted and Valkenburg 1969).

Serological typing The M antigens are sufficiently stable and immunologically distinct to provide a valuable means of typing; partial cross-reactions do occur, but their effects can be eliminated by absorption of sera. The main disadvantage of the M typing method are that, (1) even with a complete set of sera for all presently recognized types, a considerable percentage of strains cannot be typed, and (2) it is not easy to produce good precipitating sera for a number of the types. On the other hand, over 90 per cent of strains can be T typed, and T-agglutinating sera are more easy to prepare. Unfortunately, T typing gives a much less clear differentiation between strains. It is a useful marker for the identification of strains in which no particular M antigen can be detected, and it also gives some indication of the M type to which a strain belongs. Williams and Maxted (1955) therefore advocated a combined typing system in which T antigens were first identified by slide agglutination, and precipitation tests were then performed with antisera for the M types known to carry the T antigens detected. This procedure saves time and conserves sera, and is now widely used (see Parker 1967).

The usual associations between M and T antigens are shown in Table 29.5. Occasionally, as in T-types 6 and 18, a single T antigen occurs only in strains with the M antigen of corresponding number, and this M antigen is never found in association with any other T antigen. Much more often, a series of M types have the same or a similar T-agglutination pattern, though this may be complex. Thus, strains agglutinated by one or more of the T antisera 3, 13, and B3264 include members of 14 recognized M types, and certainly many more that are still unidentified. In a very few instances, the same M antigen may occur in strains with different T antigens (M antigens 2, 22 and 53), but these anomalies nearly always follow a predictable pattern. Whether or not a strain forms opacity factor (Table 29.5) also serves to elimi-

Table 29.5 Association between M and T antigens, and the opacity reaction (OR) of M types, in *Streptococcus pyogenes* (Information provided by Dr C. A. M. Fraser.)

T pattern	M antigen	OR
1	1	−
,,	68	+
2	2	+
(3, 13, B3264)	3	−
,,	13	+
,,	33	−
,,	39*	−
,,	41	−
,,	52	−
,,	53	−
,,	56	−
,,	67	−
,,	69	−
,,	72	−
,,	73	+
,,	77	+
,,	81	+
4	4	+
,,	24	−
,,	26	−
,,	29	−
,,	46*	−
,,	48	+
,,	60	+
,,	63	+
(5, 27, 44)	5	−
6	6	−
9	9	+
,,	74	−
11	11	+
,,	61	+
,,	78	+
12	22	+
,,	↑ 12 ↓	−
,,	62	+
,,	66	+
,,	76	+
14	14⇆51	−
,,	49	+
,,	80	−
(15, 17, 19, 23, 47)	15	−
,,	17	−
,,	19	−
,,	23	−
,,	30*	−
,,	47	−
,,	54	−
18	18*	−
22	22	−
(8, 25, Imp. 19)	2	+
,,	8	−
,,	25	+
,,	31	−
,,	53	−
,,	55	−
,,	57	−
,,	58	+
,,	59	+
,,	65	−
,,	75	+
,,	79	+

M types 32, 34, 36, 37, 38, 40, 42, 43 and 50; T reaction absent or inadequately studied.
T patterns in parentheses: one or more of the reactions present. Arrows indicate known variations in M antigens.
*T antigen often absent.

nate many M types from consideration, and the use of the opacity reaction leads to a further saving of M antisera (Maxted *et al.* 1973*b*).

The inability to M type a considerable proportion of strains—about 30 per cent isolated in Britain and considerably more in some other countries—cannot be attributed to the absence of an M antigen, because nearly all freshly isolated strains will grow in human blood. It merely indicates that current sets of typing sera are incomplete, though the number of new types officially recognized in recent years is considerable (Facklam and Edwards 1979). The opacity-neutralization test (Maxted *et al.* 1973*b*) has been of considerable use in defining new M types among strains that form opacity factor (p. 186), which are usually poor producers of precipitating antibody. The intensive study of strains from a particular locality, in which the sera of naturally infected persons from the same area are used as typing reagents in the opacity-neutralization test (Fraser and Maxted 1979), resulted in a dramatic increase in the typability rate and the provisional recognition of many new types.

The M antigen and related proteins M antigen can be detected, along with lipoteichoic acid, in a series of short, hair-like filaments that cover the surface of the streptococcus (Swanson *et al.* 1969). It is also present in variable amount extracellularly (Pinney and Widdowson 1977). When living streptococci are treated with trypsin the M antigen is destroyed but the cells are not killed; if the trypsin is removed, the M antigen rapidly reappears (Fox and Krampitz 1956). Streptococcal L-forms produce M antigen and liberate it into the medium (Freimer *et al.* 1959). It appears, therefore, that it is formed beneath the cell wall. Hirst and Lancefield (1939) demonstrated that acid-extracted M protein was antigenic (see also Lancefield and Perlman 1952*a*) but its antigenicity was poor. A better antibody response was obtained with M antigen released by the action of phage-associated lysin (Kantor and Cole 1960) or by the use of various adjuvants (Fox and Wittner 1966). Considerable progress has been made in the purification of M antigen. Fox and Wittner

(1965) showed that antigen obtained by hot-acid treatment comprises a mixture of molecules varying in molecular weight but with identical antigenicity. Milder extraction methods (Fox and Wittner 1969, Fischetti *et al.* 1976) yield more homogeneous material of high molecular weight, which on heating in acid can be degraded into at least four molecules of the low-molecular-weight material. The undegraded material is the better antigen. (For a general account of the M proteins, see Fox 1974; and for recent studies of their molecular structure Fischetti 1980, and Beachey *et al.* 1980.)

Other protein antigens are closely associated with the M protein and tend to be progressively purified along with it. One is the *serum-opacity factor* (p. 186) which has parallel antigenic specificity to that of the M protein. Another is the *M-associated protein* (Widdowson *et al.* 1971*b*), present in the M-positive form of all group A streptococci but not in M-negative variants of them. It does not form a precipitate with antisera and does not have antiphagocytic activity; it may be detected by means of a complement-fixation reaction between acid extracts of streptococci and rabbit antisera or selected human sera. (For antibody to M-associated protein in human serum, see Chapter 59). In some M types it can be separated from M protein by careful treatment of intact organisms with streptococcal proteinase followed by acid extraction (Maxted and Widdowson 1972). Although devoid of M-type-specificity, it is of two main antigenic types (Widdowson *et al.* 1976).

The M antigens are not uniform in their antigenic constitution and biological activities, and differences are correlated with the presence or absence of opacity factor and the antigenic type of the M-associated protein (Table 29.6). According to Maxted (1978), M types fall into two broad categories: category-1 types have M and M-associated proteins that give a good antibody response in man and experimental animals, do not form opacity factor, and include most of the types commonly associated with tonsillitis; category-2 types have poorly antigenic M and M-associated proteins, form opacity factor, and include many types responsible for impetiginous skin lesions. It must be emphasized, however, that a few types occupy an intermediate position.

Detection of M antibody As we have seen, type-specific antibody can be detected by means of the bactericidal test and the mouse-protection test. The result of the former is difficult to measure, and the latter can be performed only if a mouse-virulent strain of the correct M type is available. A search has therefore been made for alternative tests for M antibody. Bergner-Rabinowitz and her colleagues (1969) advocate an *in vitro* phagocytosis test with mouse peritoneal leucocytes. Stollerman and Ekstedt (1957) showed that there was an increase in chain length when streptococci were grown in broth in a rotating tube with homologous antiserum. Agglutination tests with tanned erythrocytes (Denny and Thomas 1953, Vosti and Rantz 1964) or inert particles to which streptococcal extracts have been adsorbed, complement-fixation tests (Beachey *et al.* 1974), and tests with radioactively labelled 'M antigen' (Anthony 1970) give numerous cross-reactions, but these can be eliminated by absorption of the serum with streptococci of heterologous type; these cross-reactions are probably due to the presence of antibody to M-associated protein.

Other protein antigens; R antigens Not all antigens that are extractable with hot acid and give precipitation reactions are M antigens. Lancefield (1943) described as R antigen 28 a trypsin-resistant but pepsin-sensitive protein which was apparently not related to virulence (Lancefield and Perlman 1952*b*); it occurs irregularly in various types of *Str. pyogenes* and also in streptococci of groups B, C and G (Maxted 1949). A second R antigen (R3) is found in M type 3 strains but is trypsin sensitive (Lancefield 1958). Several other protein antigens responsible for precipitation reactions have been reported (Wilson and Wiley 1963). The ability to give a precipitate with a hot-acid extract of the homologous streptococcus does not prove, therefore, that a serum contains an M antibody, even if the reaction is abolished by trypsinizing the extract. Confirmation by means of the bactericidal test is essential.

Non-specific binding of immunoglobulins Streptococci of groups A, C and G have surface components, rather similar to the protein A of *Staph. aureus* (Chapter 30), that act as receptors for the non-specific (Fc) portion of immunoglobulins. Unlike protein A, however, the streptococcal components bind all classes of human IgG. Group A streptococci differ from streptococci of groups C and G (and protein-A forming strains of *Staph. aureus*) in binding human IgA; and animal strains of group G differ from human strains of this group in binding bovine IgG (Myhre and Kronvall 1977, Myhre *et al.* 1979).

Pathogenicity

Str. pyogenes is the most common cause of streptococcal disease in man. It is responsible for a variety of

Table 29.6 Association of characters of M antigen and M-associated protein (MAP), and presence of opacity factor (OF), in serotypes of *Str. pyogenes*

	M antigen						
Category of M type	Molecular-weight distribution[1]	Stability after purification	Presence extracellularly	Type of MAP	Antigenicity of M antigen and MAP	OF	Representative M types
1	Wide	Good	Little	I	Good	Absent	1, 5, 6, 12, 19, 30
2	Wide, but peak at 150 000	Poor	Much	II	Poor	Present[2]	4, 22, 25, 49, 60, 63

[1] Hot-acid extracts; by elution from Sephadex G-200.
[2] Associated with M antigen in the high-molecular-weight fraction, both extracellularly and in hot-acid extracts.

respiratory and septic infections, and also for important late sequelae, rheumatic fever and acute glomerulonephritis (Chapter 59). It seldom causes disease in animals, but has been responsible for occasional epizootics in wild or captive mice (Lancefield 1972), or in monkeys (Gourlay 1960). One persistent outbreak of pharyngitis, cervical adenitis and pneumonia among mice in a breeding establishment was due to a strain of M type 50 (Hook *et al.* 1960). This type has never been found in natural infections in man but has given rise to at least one accidental infection in the laboratory (Kurl 1981). (For infections of the bovine udder, see Chapter 59.)

Rabbits and mice are susceptible to experimental infection, but guinea-pigs are more resistant. Freshly isolated strains are seldom very pathogenic for mice; to obtain a highly virulent strain, serial passage should be performed, but this may fail. Given intravenously, a virulent strain causes a fatal septicaemia; intraperitoneally it causes suppurative peritonitis, and intranasally in mice cervical adenitis, in both cases with septicaemia (Glaser *et al.* 1953). Subcutaneous injection leads to a localized abscess, sometimes with septicaemia. Some strains given intradermally to rabbits give rise to a spreading erysipelatous lesion or to severe dermal necrosis with septicaemia; most strains given intradermally to hamsters cause a superficial lesion resembling impetigo in man (Dajani and Wannamaker 1970). In the monkey, nasal instillation of large doses may lead to colonization and the subsequent appearance of antibodies, but the clinical effects are minimal (Watson *et al.* 1946).

Group C streptococci

There are three fairly easily distinguishable species, each with a fairly characteristic distribution among animals: *Str. equi*, *Str. zooepidemicus* and *Str. equisimilis* (see Frost and Engelbrecht 1940). A less well defined, α-haemolytic biotype has been referred to as *Str. dysgalactiae*. The cultural and biochemical characters of these organisms resemble those of *Str. pyogenes*, and they form many similar extracellular substances. A few strains are sensitive to bacitracin. Edwards (1934) subdivided group C streptococci on the basis of their action on sorbitol and trehalose.

1. ***Str. equi*** Most sorbitol-negative, trehalose-negative strains have been isolated from horses, but a few have also been obtained from other animals. Colonies are large, honey-coloured and very viscous, and the organism has a prominent capsule composed of hyaluronic acid (see Woolcock 1974*b*). It ferments salicin but not lactose. According to Bazeley and Battle (1940), all strains fall into a single serotype; they form an M-like protein antigen that has antiphagocytic properties (Woolcock 1974*a*).

2. ***Str. zooepidemicus*** Members of this species ferment sorbitol but not trehalose and are capsulated. They comprise two serotypes—2 and 3. Type 2 is the commoner, and ferments lactose; it has been isolated from respiratory catarrh of horses, and from a variety of other lesions in horses, cattle, guinea-pigs, rabbits and so on, and appears occasionally to cause mastitis in cattle (Buxton 1949). Type 3 does not ferment lactose and was found by Bazeley and Battle only in equine respiratory catarrh.

3. ***Str. equisimilis*** Like *Str. pyogenes*, these organisms ferment trehalose but not sorbitol. They are rather more resistant to bile than *Str. pyogenes*; most grow on 10 per cent and a few on 40 per cent bile agar. They are said not to be capsulated, and have been isolated from both human and animal sources. Many strains form T antigens similar to those of group A streptococci, and a number of other protein antigens that have not been fully investigated (see Woods and Ross 1975).

Group C streptococci isolated from man usually belong to the *equisimilis* species; they may be found in the throat and in skin lesions, and are particularly common in parts of Africa; they are infrequent causes of systemic infections. They produce a streptokinase active against human fibrin, but animal strains do not (Evans 1944).

The term ***Str. dysgalactiae*** is used by veterinary pathologists to denote a non-haemolytic streptococcus responsible for some cases of mastitis in cattle (see Minett 1936, Stableforth 1942). This organism forms the group C polysaccharide. It ferments lactose, trehalose, and sometimes sorbitol and salicin, and hydrolyses arginine but not sodium hippurate. It digests bovine fibrin. At least 3 serological types are recognized.

With the exception of these four organisms, the group C polysaccharide has been found only in a few strains of *Str. milleri*.

Group G streptococci

The large-colony form of group G streptococcus (Simmons and Keogh 1940) almost certainly deserves specific status though it has not yet been accorded this. Like strains of *Str. pyogenes* that do not form opacity factor, it gives a wide zone of β-haemolysis on blood agar. It resembles this organism in a number of other respects: it ferments trehalose but not sorbitol; it forms streptolysin O, a S-type haemolysin, and streptokinase; it possesses T antigens; a few strains have the M antigen 12 of *Str. pyogenes*, and others that do not have a recognized M antigen will grow in human blood and form M-associated protein (Widdowson *et al.* 1971*b*). It may occasionally give rise to tonsillitis, endocarditis and urinary infections in man (MacDonald 1939, Rantz 1942), to respiratory and genital lesions in dogs (Hare and Fry 1938, Laughton 1948), and perhaps to pneumonia in the monkey.

Wensinck and Renaud (1957) described a quite different β-haemolytic group G streptococcus from the respiratory tract of irradiated mice. It fermented mannitol, aesculin, sorbitol and raffinose, sugars not attacked by the large-colony group G streptococcus, and hydrolysed arginine and sometimes hippurate.

In addition to these organisms, a small number of *Str. milleri* strains form the group G polysaccharide.

Streptococcus agalactiae (group B streptococci)

The group B streptococcus forms a well defined species for which the name *Str. agalactiae* has been conserved (Kitt 1893, Report 1954). Most of the strains are β-haemolytic, but zones of haemolysis are usually narrow and indistinct. Some strains produce a haemolysin of the S type (Todd 1934).

The colonies are rather large and soft in consistency, resembling those of enterococci. When incubated anaerobically in the presence of CO_2 on a starch-containing medium, some 90 per cent of strains form a yellow to orange pigment (Orla-Jensen 1919, Fallon 1974, Islam 1977). Most strains will grow on 40 per cent bile agar. Practically all strains form a diffusible substance that completes the lysis of sheep erythrocytes by staphylococcal β-lysin, and thus give a positive CAMP reaction (Christie *et al.* 1944). The CAMP factor is an antigenic, thermolabile protein (Brown *et al.* 1974, Bernheimer *et al.* 1979). The test is usually performed by inoculating perpendicular streaks of the streptococci up to a central streak of the staphylococcus on a sheep blood agar plate; an arrow-shaped area of complete haemolysis indicates the production of CAMP factor. Group B streptococci give a positive result in 5–6 hr if the plate is incubated aerobically with added CO_2 (Darling 1975); spot or disk tests may be made with partially purified β-lysin (Wilkinson 1977). Hippurate is hydrolysed, but not aesculin. Trehalose is fermented, but not sorbitol. Reactions in lactose and salicin are variable, and, together with the presence or absence of haemolysis, have been used to classify the organism into biotypes (Brown 1939*a*). Streptokinase active on human or bovine fibrin is not formed (Hare 1935). Most strains form hyaluronidase. Neuraminidase is formed in variable amount; serotype-III strains, particularly when isolated from serious neonatal infections, produce much more than other strains (Milligan *et al.* 1978, Mattingly *et al.* 1980). Practically all strains form β-glucuronidase (Röd *et al.* 1974).

Type antigens

Lancefield (1934, 1938) defined four serotypes of group B streptococci by means of precipitation reactions between hot-acid extracts and the sera of rabbits given whole-cell vaccines. The antigens responsible for type specificity were polysaccharides, I*a*, I*b*, II and III, but I*a* and I*b* had a minor factor in common. Antisera for types Ia, Ib and II conferred type-specific immunity on mice. Type-III strains are not sufficiently virulent for mice to be used in protection tests, but type-specific immunity against them can be demonstrated in the chick embryo (Tieffenberg *et al.* 1978). The type polysaccharide is often described as capsular, but there is little morphological evidence to support this.

Wilkinson and Eagon (1971) described a protein antigen I*c*, and Lancefield and her colleagues (1975) showed that antibody to this also conferred immunity on mice. The I*c* protein was found in nearly all strains with the I*b* polysaccharide, but in only some with the I*a* polysaccharide. Unfortunately, the serotypes were then redefined as follows.

Type no.	*Polysaccharide Major*	*Polysaccharide Minor*	*Protein*
Ia	I*a*	I*abc*	—
Ib	I*b*	I*abc*	I*c*
Ic	I*a*	I*abc*	I*c*
II	II	—	—
III	III	—	—

This tended to confuse the true relationship between serotypes Ia, Ib and Ic, and the matter was further complicated when it was shown (Jelinková 1977, Bevanger and Maeland 1977, Stringer 1980) that the I*c* protein was sometimes present with polysaccharide II or III, or in the absence of a polysaccharide type antigen. Two other protein antigens occur commonly in group B streptococci; R, which is identical with the R28 antigen of *Str. pyogenes*, and X. Antibody to neither of these confers protection on experimental animals.

Lancefield and Freimer (1966) found that the type-II antigen gave rise to two immunologically distinct antibodies; extracts with hot acid reacted with one of these, and extracts with trichloracetic acid with both. The component present in both extracts consisted of galactose, glucose and glucosamine, but the trichloracetic acid extract contained in addition a heat-labile component. Similar heat-labile components were found in the other type polysaccharides, and were shown to be sialic acid (Wilkinson 1975, Baker *et al.* 1976*a*). Washing the cells with neutral buffer (Baker *et al.* 1976*a*), or digesting them with *Streptomyces* enzyme (Tai and Gotschlich 1979), releases the antigen almost entirely in its 'native' form, which contains sialic acid and is of high molecular weight; this is degraded by hot acid to give the 'core' antigen, which is devoid of sialic acid and of lower molecular weight. Indirect evidence (Chapter 59) suggests that antibody against the native but not the core antigen gives protection against serious infection in man. The core antigen of type III cross-reacts with the pneumococcus type-14 capsular polysaccharide.

The type polysaccharides inhibit phagocytosis. This effect is reversed in the presence of antibody and complement (Mathews *et al.* 1974, Hemming *et al.* 1976).

Pathogenicity

Group B streptococci cause mastitis in cattle (Chapter 57) and miscellaneous infections in other animals. In man they give rise to septicaemia and meningitis (Chapter 59).

Freshly isolated strains vary considerably in their pathogenicity for mice when given in doses of 10^6–10^7 colony-forming units intraperitoneally or intravenously; after serial passage, many strains have quite low LD50 values, but type-III strains cannot be so adapted. Many strains, including members of type III, kill infant rats when given intraperitoneally in doses of 10^2–10^4 (Ferrieri 1979). The intravenous injection of as few as 10–10^2 colony-forming units into 11-day chick embryos usually gives rise to a fatal infection (Tieffenberg *et al.* 1978).

Detection or type-specific antibody In addition to protection tests in mice and chick embryos, type-specific antibody may be assayed by means of an indirect bactericidal or opsonocytophagic test (Mathews *et al.* 1974), by a 'long-chaining' test (Stewardson-Krieger *et al.* 1977), and by binding of radioactively labelled polysaccharide (Wilkinson and Jones 1976).

Human and animal strains Although group B streptococci of human and animal origin show considerable overlap in characters, there is little doubt that they are mainly distinct populations. Thus, types Ia, Ib and Ic are more common in man than in animals; and strains without a type polysaccharide, and strains with the R and X protein antigens, are much more common in cattle than in man (Pattison *et al.* 1955, Butter and de Moor 1967). Human and bovine strains tend to fall into different biotypes (Brown 1939*a*, Butter and de Moor 1967), and human strains are more often bacitracin sensitive than bovine strains.

Phage-typing Stringer (1980) has described a phage-typing method which makes fine subdivisions within the serotypes of both human and animal strains.

For a general account of the group B streptococci, see Jelinková (1977).

Streptococcus lentus (group E, P, and U streptococci)

The group antigen E was identified by Lancefield (1933) in haemolytic streptococci from cows' milk. Similar organisms are a common cause of cervical abscesses, pneumonia and septicaemia in pigs (Thal and Moberg 1953, Collier 1956). Others have confirmed the occasional presence of the organism in milk, but no specific disease of cows is attributed to it. It is not pathogenic for rabbits, guinea-pigs or mice (Deibel *et al.* 1964).

Streptococci with similar cultural and biochemical properties but without the group E antigen have been isolated from cervical abscesses in pigs; these include some with the group antigens P (Moberg and Thal 1954) and U (Thal and Söderlind 1966). De Moor and Thal (1968) examined 86 such strains of groups E, P and U and concluded that they belonged to a single species. Haemolytic streptococci of group E have in the past been named *Str. infrequens*, but the name *Str. lentus* (Brown 1939*b*) is more appropriate. Most strains are weakly β-haemolytic at 24 hr, but give good zones of haemolysis after 48 hr, or when incubated anaerobically. The main biochemical reactions are shown in Table 29.3. Moreira-Jacob (1956) described similar organisms from milk that failed to ferment mannitol and salicin or to hydrolyse aesculin, and named them *Str. subacidus*. Some strains of *Str. lentus* are bacitracin sensitive.

Moreira-Jacob (1956) described three serotypes characterized by polysaccharide antigens among group E strains, and Yao and his colleagues (1964) added a fourth. According to de Moor and Thal (1968), the group E polysaccharide is unrelated to that of groups P and U; the latter have a common antigenic factor that is present in formamide but not in acid extracts, and this may be the 'true' group antigen. Several unrelated streptococci may react with group E antisera (Table 29.1), but it is uncertain whether all of them possess the E polysaccharide. According to Daynes and Armstrong (1973), group E strains of *Str. lentus* form an M-like protein with antiphagocytic properties, but only when grown in the presence of animal serum; under these conditions, the cell surface is covered with hair-like filaments similar to those formed by M-positive strains of *Str. pyogenes*.

Group L streptococci

Fry (1941) first identified group L streptococci from dogs and pigs. They are pathogenic for the dog (Laughton 1948). They have also been isolated from cattle but rarely from man. (See, however, Bevanger and Stammes 1979.)

They give wide zones of β-haemolysis and produce a haemolysin of the S type. They hydrolyse arginine but not aesculin; other biochemical reactions are shown in Table 29.3. Several of the tests gave variable reactions in the hands of different workers. Non-haemolytic group L strains with even more aberrant reactions have been described (McLean 1955). Many group L strains are sensitive to bacitracin (Rotta and Jelinková 1967); many have the R28 antigen and the T-antigen 8 or 20 (Perch and Olsen 1964).

Streptococcus suis

Members of Lancefield 'groups' R, S and T have been isolated mainly from pigs (Field *et al.* 1954), and were characterized by de Moor (1963). They usually give a narrow zone of β-haemolysis rather like that given by *Str. agalactiae* but are occasionally α-haemolytic. Their usual biochemical reactions are shown in Table 29.3. Some strains are bacitracin sensitive.

The R, S and T antigens are cell-wall polysaccharides, but the strains in which they are found also give reactions with group D antiserum. Elliott (1966) examined the antigen responsible for these, and found that, like the group D antigen of the enterococci, it was a glycerol teichoic acid, but it was present only in minimal amount in hot-acid extracts. Later, Elliott and his colleagues (1977) reported that phenol extracts contained much more of the teichoic acid, which was bound to lipid; the antigen released by hot acid was mainly lipid-free. They concluded that most of the teichoic acid in the porcine strains was attached to the cell membrane.

Elliott (1966) proposed the name *Str. suis* for these organisms, and this has been widely accepted. He renamed the 'group' S strains serotype 1 and the 'group' R strains serotype 2 of *Str. suis*. The corresponding type polysaccharides are antiphagocytic in pig and human blood, and antisera specifically reverse this effect (Agarwal *et al.* 1969). Like the type polysaccharides of *Str. agalactiae*, they have high molecular weights, are degraded by hot acid, and contain sialic acid (Elliott and Tai 1978).

Str. suis serotype 1 causes septicaemia, meningitis and arthritis in newborn pigs, and in some countries also in older pigs; serotype 2 causes meningitis in weaned pigs and occasionally in man (Perch *et al.* 1968; see also Chapter 59).

Streptococcus milleri

The existence of this species has only recently gained wide acceptance, though its β-haemolytic members were first described over 40 years ago. Reasons for this belated recognition include (1) variable performance of members of the species in tests usually employed for the primary classification of streptococci: changes produced in blood agar and presence of Lancefield antigens; and (2) failure to recognize that many strains require CO_2 for growth, so that they were described as microaerophilic or even anaerobic streptococci.

Long and Bliss (1934) described β-haemolytic streptococci that formed minute colonies on blood agar, and the group antigens F and G were soon identified in some of them (Lancefield and Hare 1935, Hare 1935, Bliss 1937). Bliss recognized four type antigens among the group F strains, and showed that the group G strains all had the type 1 antigen. Mirick and his colleagues (1944) described a number of non-haemolytic streptococci from the human upper-respiratory tract under the name *Streptococcus* MG, and observed that they were all sulphonamide resistant. Guthof (1956) isolated from dental abscesses a number of non-haemolytic streptococci with characteristic biochemical reactions but without Lancefield-group antigens, and named them *Str. milleri*. Ottens and Winkler (1962) studied non-haemolytic streptococci from the dental root-canal and observed that a number of them had the Lancefield antigens C, F and G. Colman and Williams (1965, 1972, Colman 1968) compared representative strains described by all of these authors and concluded that they should be included in a single species, *Str. milleri*, with an identical pattern of cell-wall constituents, the frequent possession of one of a series of cell-wall type antigens described by Ottens and Winkler (1962), and common biochemical properties; members of the species also cause highly characteristic lesions in man (Chapter 59).

Cultural and biochemical characters

When grown without added CO_2, the colonies of many strains are pinpoint in size after 24 hr, and when β-haemolytic are usually surrounded by a narrow zone of clearing 1–2 mm in diameter, which extends considerably on further incubation. In air with added CO_2, most strains form small but easily recognizable colonies (>0.5 mm in diameter) after incubation for 24 hr at 37°, and β-haemolytic strains show a zone of clearing as wide as or wider than that given by *Str. pyogenes*. The growth of nearly one-half of the strains is enhanced by the addition of CO_2, and nearly one-quarter will not grow in its absence. The cultures have a characteristic honey-like odour.

Over one-half of the cultures are non-haemolytic, about 15 per cent are α-haemolytic and rather more are β-haemolytic. Some three-quarters are ungroupable, and the rest form the antigen F, C, A or G usually in that order of descending frequency, but the proportions tend to vary with the source from which the strains were isolated (Ball and Parker 1979). The presence of a Lancefield antigen and β-haemolysis show some association; thus, some 10 per cent of ungroupable strains, 40 per cent of members of groups A, C and G, and 80 per cent of group-F strains are β-haemolytic.

Other characters are summarized in Table 29.3. The Voges-Proskauer reaction is nearly always strongly positive. On the whole, requirement for CO_2, bile tolerance, aesculin fermentation and the acidification of lactose are rather less frequent in haemolytic than in non-haemolytic strains; and strains from the human vagina tend to have a wider pattern of sugar fermentation than other strains, often acidifying raffinose, melibiose and mannitol, and are usually non-haemolytic and ungroupable (Ball and Parker 1979, Poole and Wilson 1979). The only extracellular enzymes known to be formed are deoxyribonuclease and hyaluronidase (Colman and Williams 1972).

Antigenic structure

In addition to the Lancefield antigens present in a minority, most strains of *Str. milleri* possess one of the four type-specific polysaccharides of Ottens and Winkler (1962), two of which are known to correspond to types described by Bliss (1937). These antigens are also found in a few quite unrelated streptococci, notably in strains of *Str. salivarius*. A number of strains of *Str. milleri* that do not have a recognized Lancefield group antigen are said to form 'group-like' substances, polysaccharides with the physicochemical properties of a group antigen to which precipitating antibody is not formed when the organism is injected into rabbits (Michel *et al.* 1967, Michel and Krause 1967). Many strains also form a heat- and acid-resistant protein antigen (Lütticken *et al.* 1978).

Resistance *Str. milleri* is resistant to bacitracin and sulphonamides, but nearly always sensitive to antibiotics used for the treatment of streptococcal infections.

The pneumococci

The pneumococci conform to our definition of the genus *Streptococcus*, and constitute an easily distinguishable species, *Str. pneumoniae*. They have a characteristic morphology, are soluble in bile, sensitive to

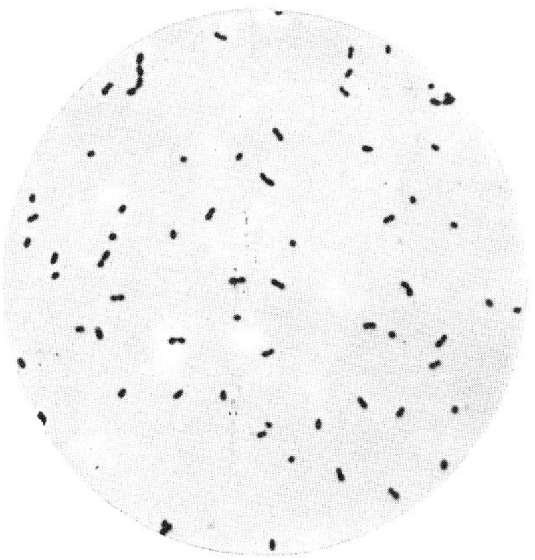

Fig. 29.4 *Str. pneumoniae.* From 24-hr culture on agar, showing diplococci and short chains (× 1000).

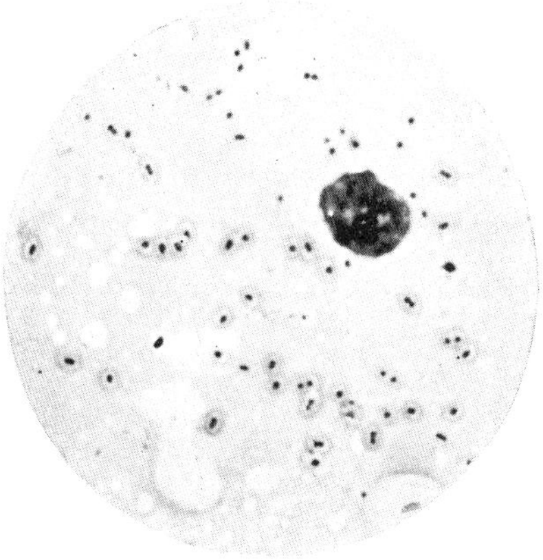

Fig. 29.5 *Str. pneumoniae.* In peritoneal exudate of mouse, showing capsulation (× 1000).

optochin, and possess one of a series of antigenically specific capsular polysaccharides.

Morphology

They are normally ovoid or lanceolate cocci, arranged in pairs or short chains; when in pairs the opposite ends are the more acutely pointed. The capsule surrounds each pair (Fig. 29.5). Filamentous variants, which form very long chains of diplococci, sometimes occur naturally, and can be obtained from many diplococcal cultures after prolonged incubation on the surface of agar (Dawson 1934).

Cultural characters

Pneumococci grow well on unenriched nutrient medium provided that it contains fresh meat infusion and a good peptone, but they often grow poorly on dehydrated media. A large inoculum may be necessary to initiate growth on solid media that have been exposed to the air unless a reducing agent, such as thio-

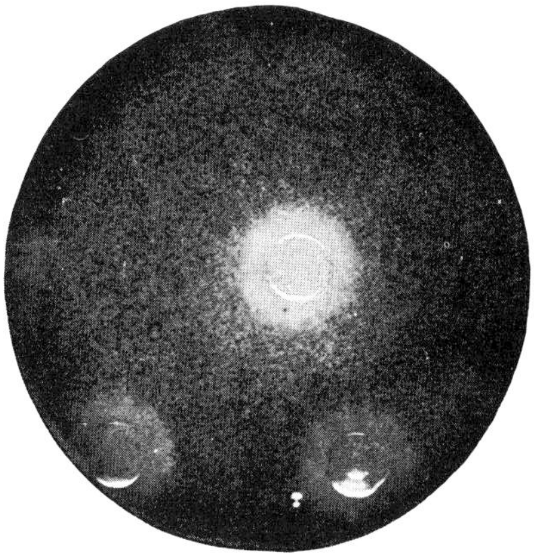

Fig. 29.6 *Str. pneumoniae.* Surface colonies on blood agar plate, showing zones of discoloration around colonies (× 8).

glycollate, is added. Glucose stimulates growth by acting as a source of energy, but the formation of acid from it leads to early cessation of growth unless the medium is well buffered. Blood and serum also improve growth, probably by virtue of their buffering effect. The presence of blood also increases the survival of pneumococci by protecting them from the H_2O_2 formed during growth. About 8 per cent of strains fail to grow aerobically unless 5–15 per cent of CO_2 is added (Austrian and Collins 1966).

Colonies of capsulated pneumococci on blood agar are usually raised, circular, about 1 mm in diameter, with steeply shelving sides, a smooth surface and an entire edge. When autolysis occurs, the centre becomes depressed, giving the characteristic 'draughtsman' appearance. Members of type 3 are usually larger than the rest, and have a watery or mucoid appearance. Non-capsulated variants form colonies with a dull or finely granular surface and a more coherent consistency. They can be induced to appear in capsulated strains by growth in the presence of anti-capsular serum (Stryker 1916). Small irregular outgrowths may develop around colonies after incubation for 5–7 days; subculture from the outgrowths gives rise to flatter colonies with an irregular surface and fimbriate margin (Dawson 1934, Austrian 1953). These are composed of organisms in the filamentous form.

Pneumococci produce α-haemolysis when incubated aerobically on blood agar, but they form a haemolysin in liquid media. After anaerobic incubation on blood agar at 37° and a subsequent period of 24–48 hr at 6°, zones of β-haemolysis appear around colonies near antibiotic-containing disks (Lorian *et al.* 1973). These are most evident when the antibiotic is one that acts on the cell wall, and do not appear on ethanolamine blood agar. The release of haemolysin thus appears to be associated with autolysis.

In fluid media, the diplococcal form gives rise to diffuse turbidity, but the filamentous form produces a cotton-wool-like deposit. Pneumococci are killed by exposure to 55° for 20 min or less, and will not grow at 10° or at 45°, at pH 9.6, or in the presence of 6 per cent NaCl or 0.01 per cent tellurite. All pneumococci produce H_2O_2. They are bacitracin resistant.

Bile solubility and autolysis Pneumococci are the only streptococci that are soluble in bile (Neufeld 1900). They also readily undergo autolysis, which is brought about by an enzyme that lyses the cell walls of pneumococci but not of other streptococci (Avery and Cullen 1923). There is little doubt (see Mair 1929) that the effect of bile salts is to accelerate natural autolysis. The most active constituent of bile salts is sodium deoxycholate (Mair 1917). According to Mosser and Tomasz (1970) the autolytic enzyme acts on the muramic acid of the cell wall only when the normal choline-containing teichoic acid (p. 197) is present; pneumococci in which the choline has been artificially replaced by ethanolamine in the teichoic acid are bile resistant and do not undergo autolysis (Tomasz 1968). In the *bile-solubility test*, pH and salt concentration must be carefully controlled (for a suitable method, see Cowan 1974).

Sensitivity to optochin Morgenroth and Levy (1911) noted the sensitivity of pneumococci to *optochin* (ethylhydrocupreine) and Moore (1915) made use of this property to distinguish them from other streptococci. The test can be performed conveniently with a paper disk containing 5 μg of optochin (Bowers and Jeffries 1955). Pneumococci are killed by a concentration of 1/500 000–1/100 000, whereas most other streptococci require one of 1/5000 or stronger (see Mørch 1943). Lund (1959) concluded that the optochin test was more reliable than the bile-solubility test, on the ground that all pneumococci were sensitive to optochin, whereas some rough pneumococci gave equivocal results in the bile-solubility test.

Antibiotic resistance

Some 10 per cent of pneumococci are resistant to tetracycline (Report 1977), but few are resistant to erythromycin, lincomycin or chloramphenicol. Pneumococci with a diminished sensitivity to benzylpenicillin (minimum inhibitory concentration 0.1–1.0 mg per l) were first reported by Hansman and Bullen in 1967, and have since been found in many countries. They appear to be rather infrequent almost everywhere, but Gratten and his colleagues (1980) found that one-third of strains isolated in New Guinea had this degree of resistance. This is not easily detected by means of a disk-sensitivity test with benzylpenicillin, but is shown by the absence of inhibition around a disk containing 1 μg of oxacillin (Dixon *et al.* 1977). In 1977, strains with a considerably greater resistance to benzylpenicillin (minimum inhibitory concentration 4–8 mg per l) were identified in two districts in South Africa (Appelbaum *et al.* 1977, Jacobs *et al.* 1978), where they appear to have been quite prevalent; in addition, these strains were fully resistant to chloramphenicol, erythromycin, lincomycin, tetracycline and cotrimoxazole. Penicillin-resistant pneumococci do not form penicillinase; for a study of the mechanism of their resistance, see Zighelboim and Tomasz (1980). A few penicillin-resistant pneumococei that are resistant to other antibiotics have been encountered in other countries.

Antigenic structure

Type antigens Neufeld and Händel (1909) first established the existence of antigenically different types of pneumococci by showing that pneumococcal antisera conferred type-specific immunity on mice. Dochez, Avery and their colleagues (see Avery *et al.* 1917) studied the antigenic structure of a large collection of strains by agglutination, and defined the types now numbered 1, 2 and 3. Many other types were identified subsequently, in particular by Lister (1916), Cooper and Walter (1935) and Lund (1970). The number now recognized is 83 (Henrichsen 1979), and few more are likely to be discovered.

Unfortunately two systems of numbering have been used: the American system (Eddy 1944) numbers all distinguishable types, but the Danish system (Mørch 1943) places in a type all antigenically related strains and, when necessary, subdivides the type further by means of letters, small letters being used for antigenic factors and large letters for types. Thus, the Danish types numbered 7 are 7F, 7A, 7B and 7C; the corresponding antigenic formulae are: 7a, 7b; 7a, 7b, 7c; 7a, 7d, 7e, 7h; 7a, 7d, 7f, 7g, 7h; and their respective numbers in the American system are 51, 7, 48 and 50 (Lund 1970). In all, 27 Danish types (including types 1–5) are distinct and unrelated to any others, and the remaining 56 fall into 19 cross-reacting groups. These differences in numbering may cause confusion when accounts of type distribution in different areas are compared, because workers who use sera for the 46 Danish types only may fail to mention this fact.

Neufeld (1902) noted that capsules of pneumococci, when acted upon by specific antiserum, appeared to become greatly swollen, or at least more easily visible microscopically. This phenomenon (*Quellung*) can be made use of for the identification of pneumococcal types (Neufeld and Etinger-Tulczynska 1931) and is the most widely practised typing method. Agglutination, either on slides or in plates, and co-agglutination (Kronvall 1973) are also acceptable methods.

An 'omni-serum' (Lund 1963) containing antibodies to all recognized types may be used for the rapid identification of pneumococci. Its main use is for the diagnosis of pneumococcal infection by detecting type antigen by counter-current immuno-electrophoresis in blood and cerebrospinal fluid (Dorff *et al.* 1971), and in sputum (El-Rafaie and Dulake 1975, Kelsey and Reed 1979); in addition to rapidity in performance, this test has the advantage that positive reactions may be obtained after up to 2 days' antibiotic treatment.

Heidelberger and Avery (1923) showed that type-specificity depended on the formation of capsular polysaccharides and later studies, mainly by these workers and Goebel (see Heidelberger 1938), established the sugar composition of many and the chemical structure of a few of these antigens.

The pneumococcal type polysaccharides show numerous cross-reactions with other streptococcal antigens (Austrian *et al.* 1972), and with the capsular polysaccharides of *Klebsiella* (Heidelberger and Nimmich 1976), as well as cross-reactions with human bloodgroup substances and other tissue antigens, and vegetable gums. The capsular polysaccharides of pneumococci are non-toxic but inhibit phagocytosis of the organism; this takes place, however, in the presence of specific antibody, complement and possibly other serum factors (Johnston *et al.* 1972). The use of vaccines of capsular polysaccharide to protect against pneumococcal infection will be discussed in Chapter 67.

Serum antibody to type polysaccharide can be measured by quantitative precipitation; more sensitive methods are radioimmunoassay and enzyme-linked immunoabsorbent assay (Berntsson *et al.* 1978, Koskela and Leinonen 1981).

C substance The cell walls of pneumococci contain a species-specific antigen often referred to as the C substance (Tillett *et al.* 1930). This is, or includes, the substance that reacts with the so-called C-reactive protein that appears in the blood of patients with certain febrile illnesses. It is a teichoic acid containing choline and galactosamine, and pneumococci require choline for its production (Tomasz 1967, Mosser and Tomasz 1970). When grown in a medium in which choline is replaced by ethanolamine, cell-pairs do not separate after division, and the organism is bile resistant and does not undergo autolysis (Tomasz 1968). Most of this teichoic acid is present in the cell wall, but a minority is linked to a lipid and associated with the cell membrane; it is the latter that reacts wih C-reactive protein (Briles and Tomasz 1973).

M antigens Austrian and MacLeod (1949*a*) extracted from both capsulated and non-capsulated pneumococci a series of type-specific proteins, soluble in acid but sensitive to trypsin, resembling the M proteins of *Str. pyogenes*. Antisera against them agglutinate non-capsulated but not capsulated variants of the homologous strain. Members of the same capsular type may have different M antigens. Variation in capsular type and M protein occur independently in transformation experiments (Austrian and MacLeod 1949*b*). The pneumococcal M antigens differ from those of group A streptococci in playing no part in the virulence of the organism.

Pathogenicity

The pneumococcus is an important pathogen of man, giving rise to pneumonia, particularly the lobar form, sinusitis, otitis media, less frequently meningitis, suppurative arthritis, or peritonitis (see Chapters 57, 65 and 67). It also causes pneumonia in laboratory monkeys, pneumonia and uterine infection in guinea-pigs (Petrie 1933) often said to be associated with type 19, and pneumonia in laboratory rats (Mirick *et al.* 1950, Baer 1967). Pneumococcal carriage without illness is also common in these species. Sporadic cases and epizoötics of pneumococcal mastitis occur in cattle (see Stableforth 1959).

The pneumococcus is highly pathogenic when injected into mice and rabbits, but rather less so for the guinea-pig. The cat, dog and chicken are relatively resistant. It is a characteristically invasive organism, causing a fatal bacteraemic infection when injected intravenously, an acute peritonitis followed by bacteraemia when injected intraperitoneally, localized suppuration followed by generalization when injected subcutaneously, and a spreading inflammatory lesion followed by generalization when injected intradermally in the rabbit. Virulence is readily enhanced by passage, so that death may follow the intraperitoneal injection of less than ten organisms. Different serotypes vary in their virulence for different laboratory animals. Transformation experiments indicate that the type of capsular polysaccharide is the determining factor. Fine (1975) points out that members of several capsular types that are common causes of serious disease in man lack the ability to activate the alternative complement pathway, and thus tend to evade phagocytosis in the absence of type-specific antibody. However, different strains of type 3 pneumococci vary widely in virulence for the rabbit (Tillett 1927, Schaffer *et al.* 1936). Differences in mouse virulence of strains of the same serotype can be related to the amount of capsular polysaccharide formed (MacLeod and Krauss 1950).

Toxin production

The factors we have discussed in the previous section may account for the invasiveness of the pneumococcus but offer little explanation of how it causes death.

Pneumococci produce an intracellular haemolysin which is liberated by autolysis (Cole 1914). This *pneumolysin* (Neill 1926, Cohen *et al.* 1942) is reversibly inactivated by oxygen and is heat-labile. It is said to be related serologically to streptolysin O and to the haemolysins of *Cl. perfringens* and *Cl. tetani*, and to be an acidic protein with lethal and dermonecrotic properties (Kreger and Bernheimer 1969). Oram (1934) described a pneumococcal leucocidin. Pneumococcal neuraminidase (Chu 1948) is formed in large amount by strains freshly isolated from patients, but the activity decreases on serial subculture (Kelly *et al.* 1967). Evidence has been advanced that it contributes to the severe consequences of pneumococcal pneumonia (Fischer *et al.* 1971, Seger *et al.* 1980) and meningitis (Müller 1969). Pneumococci produce hyaluronidase, but there appears to be no relation between the amount formed and their virulence. The experiments of Dick and Gemmell (1971) indicate that uncontrolled multiplication of pneumococci in the mouse results within a few hours in damage that cannot be reversed by subsequent clearance of the blood by penicillin treatment.

The enterococci and related group D streptococci

The term 'enterococcus' was first used by Thiercelin (1899) to describe organisms seen in pairs or short chains in faeces. It was later used in a more restricted sense by Sherman (1937) for streptococci that had an exceptional capacity to survive or multiply under a series of extreme conditions; his criteria for this were: (1) growth at 10°, (2) at 45°, (3) at pH 9.6, and (4) in 6.5 per cent NaCl, and (5) survival at 60° for 30 min. This is the sense in which we shall use the term.

Andrewes and Horder (1906) applied the name *Str. faecalis* to a common faecal streptococcus that formed short chains, clotted milk, and acidified mannitol and lactose but not raffinose. Other characters were added by later authors; resistance to heat, alkali, bile and tellurite, and the ability to hydrolyse aesculin in bile medium. Orla-Jensen (1919) described as *Str. faecium* a second faecal organism differing from *Str. faecalis* in its sugar-fermentation reactions, and later (Orla-Jensen 1943) substantiated the claim that it be considered an enterococcus. *Str. durans* (Sherman and Wing 1935, 1937) had been described earlier, but is now looked upon as a variety of *Str. faecium*. More recently, *Str. avium* has been added to the group (Nowlan and Deibel 1967). In addition, organisms closely resembling enterococci are common saprophytes on the surface of plants. Unlike members of the established enterococcal species, some of them are motile. These, and a number of other atypical enterococci, are difficult to classify.

These organisms all form the group D antigen, first identified by Lancefield (1938) in β-haemolytic strains of *Str. faecalis*; but so do members of two other species of streptococci from animal faeces, *Str. equinus* (Andrews and Horder 1906) and *Str. bovis* (Orla-Jensen 1919), which do not have the physiological characters of enterococci.

For reviews of the classification of the enterococci and other group D streptococci, see Deibel (1964) and Hartman *et al.* (1966).

The following general description applies to the enterococci and to *Str. equinus* and *Str. bovis*, but not to *Str. suis* (p. 193), which is also a group D streptococcus.

Morphology

Enterococci are oval cocci in pairs or short chains that possess thin intracellular 'cores'. Some of the saprophytic strains are motile and have 1-5 flagella (Graudal 1955). Few strains are capsulated.

Cultural characters

The colonies of enterococci are rather larger than those of most streptococci (1-2 mm in diameter), and of a butyrous consistency; those of *Str. equinus* and *Str. bovis* are usually considerably smaller. *Str. faecalis* may be β-haemolytic, but is usually non-haemolytic and rarely α-haemolytic; *Str. faecium*, however, is often α-haemolytic; the remainder are predominantly non-haemolytic. The haemolytic agent of *Str. faecalis* appears to be identical with the type 1 bacteriocine (Brock and Davie 1963). The enterococci produce a number of bacteriocines (Stark 1960, Brock *et al.* 1963), some of which have a wide range of activity on other gram-positive organisms.

Metabolic and biochemical characters

The reactions given by the main species and varieties are shown in Table 29.7. Some of the tests used require careful standardization (Shattock and Mattick 1943, Skadhauge 1950, Facklam 1972). Fermentation of aesculin on 40 per cent bile aesculin agar is probably the most useful means of distinguishing these organisms as a whole from other streptococci (Facklam and Moody 1970). The ability to liquefy gelatin in a stab culture is confined to some strains of *Str. faecalis*, but many other enterococci can be shown by means of a plate test to hydrolyse it (Deibel 1964).

Antigenic structure

The group D antigen is a glycerol teichoic acid (Elliott 1962) which is found in greatest amount between the cell membrane and the cell wall (Smith and Shattock 1964). Better yields of antigen are obtained in unbuffered glucose broth than in Todd-Hewitt medium (Medrek and Barnes 1962*a*). In order to demonstrate the antigen in extracts of *Str. bovis* and *Str. equinus* it may be necessary to concentrate it by precipitation with alcohol (Shattock 1949). The so-called group Q antigen (Guthof 1955) is found in streptococci that also have the group D antigen (Smith and Shattock 1964); unlike the latter, it is a cell-wall polysaccharide. The type antigens of group D streptococci are also cell-wall polysaccharides, and can be demonstrated in acid extracts by precipitation with absorbed antisera; 38 have so far been demonstrated; of these 11 are found only in *Str. faecalis* and the remainder in strains of *Str. faecium*, *Str. bovis* and unclassified strains (Sharpe and Shattock 1952, Sharpe and Fewins 1960, Medrek and Barnes 1962*b*, Sharpe 1964).

Classification

Str. faecalis This is a more or less homogeneous species, recognizable by its strong reducing activities. It differs from nearly all other enterococci in that it decolorizes litmus milk before acidification or clotting have occurred, grows in and reduces 0.1 per cent tellurite broth, forms black colonies on Hoyle's tellurite medium, and forms red colonies on a medium containing tetrazolium 0.01 per cent and glucose 1 per cent at pH 6.0 (Barnes 1956*a*, *b*). However, batch

Table 29.7 Usual characters of the enterococci, and of *Str. bovis* and *Str. equinus*. The following are common characters: grow on 40 per cent bile agar; grow at 45°; ferment salicin; Voges-Proskauer reaction positive; hydrolyse aesculin but not hippurate; resistant to optochin

	Enterococci					
	Str. faecalis	*Str. faecium*	*Str. faecium* var. *durans*	*Str. avium*	*Str. bovis*[1]	*Str. equinus*
Haemolysis	NH(β)	α(β, NH)	NH(α)	NH(α)	NH(α)	NH(α)
Survival at 60° for 30 min	+	+	+	+	−	−
Growth at 10°	+	+	+	+	−	−
Growth in 6.5 per cent NaCl	+	+	+	+	−	−
Growth at pH 9.6	+	+	+	+	−	−
Early reduction of litmus milk	+	−	−	+	−	−
Growth and reduction in 0.1 per cent tellurite	+	−	−	−	−	−
Red colonies on 0.01 per cent tetrazolium agar	+	−	−	−	v	−
Fermentation of pyruvate	+	−	−	−	−	−
Hydrolysis of arginine	+	+	+	−	−	−
Hydrolysis of starch	−	−	−	−	+	+
Polysaccharide from sucrose	−	−	−	−	Dx	−
Acid from:						
mannitol	+	+	−	+	−	−
lactose	+	+	+	+	+	−
sucrose	+	+	−	+	+	+
arabinose	−	+	−	+	−	−
sorbitol	+	−	−	+	−	−
trehalose	+	+	+	+	v	v
raffinose	−	−	−	?	+	−
inulin	−	−	−	+	+	−
glycerol (aerobic)	+	v	−	+	−	−
glycerol (anaerobic)	+	−	−	+	−	−
melibiose	−	+	+	?	v	−
melezitose	+	−	−	+	−	−

NH = non-haemolytic; v = some positive, others negative; Dx = dextran. Less common forms in parentheses.
[1] In bovine faeces; see Table 29.8 for characters of strains found in man.

variation in this medium led Waitkins (1978) to prefer the test for fermentation of pyruvate described by Gross and his colleagues (1975) as a means of identifying *Str. faecalis*. The fermentation reactions given in Table 29.7 are reasonably uniform. Additional points of difference are that *Str. faecalis* utilizes citrate, gluconate, glycerol and uric acid anaerobically as sources of energy but does not require folic acid, and that *Str. faecium* has the opposite characters (Deibel *et al.* 1963, Deibel and Niven 1964, Mead 1974). Strains that produce β-haemolysis on blood agar have been referred to as var. *zymogenes*, and strains that liquefy gelatin as var. *liquefaciens*.

Str. faecium* and *Str. faecium* var. *durans The *durans* organism is now considered to be a variety differing from *Str. faecium* only in its failure to ferment mannitol, sucrose and arabinose. The main points of difference between *Str. faecalis* and *Str. faecium* have already been mentioned (see also Table 29.7). Unlike other enterococci, these organisms grow at 50°.

Str. avium Nowlan and Deibel (1967) isolated an enterococcus with distinctive characters from the faeces of chickens, and sometimes from other sources. In addition to the group D antigen, over half of the strains had the 'group' Q antigen. They resemble *Str. faecalis* in bringing about early reduction of litmus milk, but, unlike *Str. faecalis* and *Str. faecium*, they rarely hydrolyse arginine. They have a wide range of fermentation reactions; in addition to the sugars shown in Table 29.7, they form acid from a number of polyols, including dulcitol and ribitol, that are rarely attacked by enterococci of other species.

Other enterococci Many atypical enterococci live on the surface of plants (Mundt and Johnson 1959). A number of them are motile (Langston *et al.* 1960), and most of them form yellowish colonies (Mundt and Graham 1968). They tend to be less heat tolerant than most other enterococci, a considerable proportion failing to grow at 45° and a few not surviving 60° for 30 min. In other respects they conform to the 'Sherman criteria'. According to Mundt and Graham (1968), they form a distinct group of enterococci. They are usually α-haemolytic and acidify the same sugars as *Str. faecium*, except that raffinose is nearly always attacked. However, they cause early reduction of litmus milk, form grey colonies on tellurite agar and often red colonies on tetrazolium agar. Arginine is seldom hydrolysed, the Voges-Proskauer reaction is usually negative, and polysaccharide is not formed from sucrose.

A number of motile enterococci have been isolated from human faeces. Many of them form yellowish colonies, and most of them resemble the epiphytic strains fairly closely

(Graudal 1957). Langston and his colleagues (1960) proposed the name *Str. faecium* var. *mobilis* for motile strains from plants, and Mundt and Graham (1968) the name *Str. faecium* var. *casseliflavus* for yellow-pigmented epiphytic enterococci as a whole.

Str. bovis* and *Str. equinus These two species will be described together, because many authors find difficulty in distinguishing between them, and there are clearly many intermediate strains. They differ from enterococci in failing to grow at 10° and in not being resistant to heat or alkali; they do not attack arginine. *Str. equinus* does not ferment lactose, raffinose, inulin or melibiose, and does not form dextran from sucrose; *Str. bovis* nearly always ferments lactose and raffinose, often ferments inulin and melibiose, and may form dextran (Smith and Shattock 1962). When a strain of *Str. bovis* produces dextran it forms typical watery colonies on sucrose agar. Smith and Shattock consider that lactose fermentation is the best means of distinguishing between the two species, but recognize that the fermentation pattern is not always distinctive (see also Medrek and Barnes 1962*b*).

Str. bovis is not a homogeneous species (Garvie and Bramley 1979). The characters shown in Table 29.7 are those of the predominant form found in cattle. Most of the strains responsible for endocarditis in man have the following characters: dextran formed; mannitol, inulin and melibiose fermented; starch hydrolysed (Friedberg 1941, Facklam 1972, Kiel and Skadhauge 1973, Parker and Ball 1976); a smaller group of dextran-negative strains from the same source attack these carbohydrates much less often (see Table 29.8).

Distribution

Str. faecalis and *Str. faecium* are present in the faeces of most mammals and birds, but in ruminants and some non-ruminant mammals they are outnumbered by *Str. bovis*. This organism is constantly found in the rumen (see Latham and Jayne-Williams 1978), and some of the strains are said to be strict anaerobes (Chapter 30). *Str. equinus* predominates in the faeces of horses, and some workers but not others have found *Str. bovis* in most samples of pig faeces. In man, *Str. faecalis* and *Str. faecium* are each present in one-half or more of faecal samples, though their relative proportions appear to show some geographical variation (see Mead 1978). *Str. bovis* is found in some 10–20 per cent of normal persons but in a larger proportion of sufferers from colonic cancer (Chapter 57). According to van der Wiel-Korstanje and Winkler (1975), counts of enterococci are increased 100-fold in patients with ulcerative colitis, and many of the strains present are atypical, including a number with the Q antigen and many motile strains.

Pathogenicity

Nearly all enterococcal infections in man are caused by *Str. faecalis*, which gives rise to urinary-tract and wound sepsis, and sometimes to endocarditis (Chapter 57). In sheep, it is said to cause endocarditis in lambs and pneumonia in adults (Jamieson 1950). *Str. bovis* is an important cause of endocarditis in man (Chapter 57), and an occasional cause of bovine mastitis (Garvie and Bramley 1979). Overgrowth of *Str. bovis* in the rumen may play a part in causing digestive disturbances (Latham and Jayne-Williams 1978).

Antibiotic susceptibility

Enterococci are much less sensitive to penicillins than most other streptococci, requiring 1–5 mg per l of benzylpenicillin for inhibition; considerably higher concentrations are seldom bactericidal. Ampicillin and amoxycillin are rather more active, but not the penicillinase-resistant penicillins. Indeed, growth up to a 5 μg methicillin disk is a useful test for *Str. faecalis* and *Str. faecium*, but some other enterococci are rather more sensitive. *Str. bovis* is nearly always very sensitive to benzylpenicillin. Mixtures of amoxycillin and an aminoglycoside are often bactericidal for *Str. faecalis*, but the efficacy of individual aminoglycosides in these combinations is variable. Some strains of *Str. faecalis* are sensitive to tetracycline, chloramphenicol and erythromycin *in vitro*, but these drugs will seldom eliminate systemic infections. Resistance patterns in *Str. faecalis* have tended to become wider in recent years (Toala *et al.* 1969). Several of the resistances are determined by plasmids that are transferable, probably by conjugation; this may be facilitated by the excretion by certain recipient strains of a substance that causes aggregation of donors and recipients into clumps (Dunny *et al.* 1978).

The lactic streptococci

These are found chiefly in milk and cheese, and resemble the enterococci in several respects; they grow at 10° and in 40 per cent bile, and decolorize litmus milk before clotting it. They do not grow at 45°, in 6.5 per cent NaCl or at pH 9.6, and they have the group antigen N (Sherman *et al.* 1940, Shattock and Mattick 1943), which is a glycerol teichoic acid but distinct from the group D antigen (Elliott 1963). They usually survive 60° for 30 min and are non-haemolytic. Two species are recognized, but there are also intermediate strains; *Str. lactis* forms elongated diplococci and *Str. cremoris* forms long tangled chains in milk. The latter is distinguished from the former by its failure to grow at 40°, to ferment maltose and dextrin, and to hydrolyse arginine (Yawger and Sherman 1937). A third species, *Str. raffinolactis*, may merit recognition (see Garvey 1978).

Other streptococci

The remaining streptococci include several species that are reasonably well defined and a number of strains that are difficult to classify. The species that will be described differ from those considered elsewhere in this Chapter in the absence of (1) a clear zone of β-haemolysis, (2) a Lancefield-group antigen that defines the species, (3) tolerance for heat, alkali and NaCl, and (4) sensitivity to optochin. They may for convenience be considered in two groups.

Species found in the mouth and upper respiratory tract

These organisms are often referred to as 'viridans' streptococci, but members of several of the species are usually non-haemolytic. Individual strains may exhibit β-haemolysis, but this is seldom complete and clear-cut. Their colonies are on the whole small for streptococci, and certainly smaller than those of enterococci. The species to be described are *Str. salivarius*, *Str. sanguis*, *Str. mitior* and *Str. mutans*. When encountered by the medical bacteriologist, they may be difficult to distinguish from *Str. bovis* and from α-haemolytic or non-haemolytic strains of *Str. milleri*, which we have already described. In Table 29.8, therefore, the characters of these organisms in the form in which they are likely to be encountered in human clinical material are included for comparison.

These organisms are important constituents of the flora of the human mouth; they also often appear in the faeces, but it is not customary to look upon them as bowel organisms. They rarely give rise to purulent infections, but they are collectively the most common cause of endocarditis (Chapter 57). Rather similar organisms are found in animals, but these have been less intensively studied.

For general accounts of these organisms, see Colman and Williams (1965, 1972, 1973), Colman (1968, 1969, 1970, 1976), Carlsson (1968), Hardie and Bowden (1976), and Facklam (1977).

Streptococcus salivarius

This name was applied by early workers to the short-chained, non-haemolytic streptococcus that is commonly found in the human mouth and intestine. As re-defined by Sherman and his colleagues (Safford *et al.* 1937, Niven *et al.* 1941*a*, *b*, Sherman *et al.* 1943), it does not hydrolyse arginine and acidifies the sugars shown in Table 29.8; but their statement that it grows at 45° is no longer accepted. It forms laevan from sucrose, and characteristic fleshy but not adherent domed colonies on sucrose agar. H_2O_2 is not formed; some strains hydrolyse urea, an uncommon character among streptococci.

Table 29.8 Usual characters of *Str. salivarius*, *Str. sanguis*, *Str. mitior* and *Str. mutans*; and, for comparison, of *Str. bovis* and *Str. milleri*

The following are common characters: do not survive at 60° for 30 min; do not grow at 10°, in 6.5 per cent NaCl or 0.1 per cent tellurite broth; do not hydrolyse hippurate; do not ferment glycerol; with a few exceptions, ferment lactose and sucrose; optochin resistant.

	Str. salivarius	*Str. sanguis*	*Str. mitior*[1]	*Str. mutans*[2]	*Str. bovis*[2]	*Str. milleri*
Haemolysis	NH	α(NH, β)	α(NH, β)	NH	NH(α)	NH(α, β)
Group antigens	−(K)	−(H)	−(O, K, M)	−(E)	D	−(F, G, A, C)
Growth on 40 per cent bile agar	v[3]	v[3]	v[3]	+	+	v[3]
CO_2 required for growth	−	−	−	v[4]	−	v[3]
H_2O_2 production	−	+	+	−	−	−
Voges-Proskauer reaction	v	−	−	+	+	+
Polysaccharide from sucrose	L	Dx	−(Dx)[5]	Dx	Dx(−)[6]	−
Hydrolysis of:						
arginine	−	+	−	−	−	+
aesculin	+	+	−	+	+	+
starch	−	−	−	−	v[4]	−
Acid from:						
mannitol	−	−	−	+	v[4]	−
salicin	+	+	v[3]	+	+	+
sorbitol	−	−	−	+	−	−
trehalose	+	+	v[3]	+	v[4]	+
raffinose	v[4]	v[3]	v[3]	+	+	−
melibiose	−	−	v[3]	+	v[4]	−
inulin	v[4]	+	v[3]	+	v[4]	−

NH = non-haemolytic; v = some positive, others negative; L = laevan; Dx = dextran.
Less common forms in parentheses.

[1] Forms very long chains in broth culture.
[2] Most common form in material of human origin.
[3] Most reactions negative.
[4] Most reactions positive.
[5] Dextran-positive strains usually acidify raffinose and melibiose; dextran-negative strains usually do not.
[6] Dextran-positive strains hydrolyse starch, and acidify mannitol and inulin; dextran-negative strains usually do not.

There are two serotypes (Sherman *et al.* 1943); members of type I give a precipitate with antisera for Lancefield group K (Hare 1935) and for *Str. milleri* of Ottens and Winkler's (1962) type 3 (see Colman 1976). There is an inverse relationship between the formation of the K-like antigen and acetoin production.

Str. salivarius is found in the throat and faeces, but its main habitat appears to be the mouth (Gibbons *et al.* 1964); it is not specially associated with the teeth. It rarely causes disease.

Streptococcus mitior and *Str. sanguis*

The name *Str. mitior* (Schottmüller 1903) may properly be applied to the common α-haemolytic streptococcus from the throat that forms very long chains in broth culture (see Colman and Williams 1972); it has also been called *Str. mitis*. In 1946, White and Niven gave the name *Str. sanguis* to a group of α-haemolytic streptococci from cases of endocarditis, noting particularly their ability to form dextran and to hydrolyse arginine and aesculin, which distinguished them from *Str. mitior*. Others (Porterfield 1950, Carlsson 1968, Colman 1968) recognized a group of strains that form dextran but do not hydrolyse arginine or aesculin; these have been described as *Str. sanguis* II, but we agree with Colman (1968; see also Coykendall and Specht 1975) that they are more closely related to *Str. mitior* than to *Str. sanguis*, and have included them in this species in Table 29.8.

Cultural and biochemical characters Most members of both species are α-haemolytic. H_2O_2 is formed. Dextran, when produced, is of the insoluble type, and colonies on sucrose agar adhere to the medium. If we include in *Str. mitior* strains without action on arginine and aesculin, these can be distinguished from typical *Str. sanguis* strains by their infrequent acidification of salicin, trehalose and inulin. As a whole, *Str. mitior* attacks raffinose more often than does *Str. sanguis*, but this is only because dextran-positive *mitior* strains usually acidify this sugar (Parker and Ball 1976). *Str. mitior*, including the dextran-positive strains, forms a ribitol teichoic acid (Colman and Williams 1972); the cell walls of *Str. sanguis* contain rhamnose but not ribitol. An unusual character of *Str. sanguis* is the formation of thin spreading outgrowths from colonies after prolonged incubation in a moist atmosphere; this is associated with twitching 'motility' and the presence of polar fimbriae (Henriksen and Henrichsen 1975).

Not all members of the *mitior-sanguis* group can be accommodated easily in the two species. According to Parker (1978), 24 per cent of 763 such strains were dextran negative, and hydrolysed either arginine or aesculin but not both.

Antigenic structure Most *Str. mitior* strains are ungroupable by Lancefield's method, but a few have the group-antigen O (Boissard and Wormald 1950), M, or K. The significance of these antigens is doubtful; the O and M antigens are rather heat-sensitive and are destroyed by trypsin and formamide. The M antigen was first identified by Fry (1941) in streptococci from dogs that were β-haemolytic and bore little resemblance to *Str. mitior* (Skadhauge and Perch 1959, Rifkind and Cole 1962). Rather less than one-half of strains with the biochemical characters of *Str. sanguis* form the group-antigen H, first described by Hare (1935). This antigen is rarely if ever present in *Str. mitior*, whether or not dextran is formed, or in unclassified *mitior-sanguis* strains. According to Rosan and Appelbaum (1979), it is a glycerol teichoic acid. Henriksen and Eriksen (1978) describe another group antigen in strains that do not form the H substance.

Streptococcus mutans

Clarke (1924) described as *Str. mutans* a non-haemolytic streptococcus commonly found in carious teeth. When grown in acid media the cocci tended to lengthen into rods. They fermented mannitol, inulin and raffinose. These findings were soon confirmed, but interest in the organism waned, only to be revived when it was shown that under experimental conditions it could cause dental caries in rats and hamsters, and was an important cause of endocarditis in man (Chapter 57).

Cultural and biochemical characters Growth is nearly always improved by the addition of CO_2, and many strains require this for growth. The organism grows in the presence of 40 per cent bile, but this should not be tested for on bile-aesculin medium, on which growth often fails to appear. It forms a 'hard' and insoluble form of dextran. Colonies on sucrose agar are adherent and glass-like; in sucrose broth, much of the dextran is deposited in aggregates on the wall of the vessel. The dextran consists of highly branched $1 \rightarrow 3$ chains that result from the successive action of a glucosyltransferase which forms the soluble $1 \rightarrow 6$ polymer and a dextranase which further modifies this (Guggenheim 1975).

The biochemical characters summarized in Table 29.8 are those of the form of *Str. mutans* usually found in man. This is fairly easily distinguished from other streptococci found in the mouth, but when isolated from human blood or faeces it has to be distinguished from *Str. bovis*. This also is a non-haemolytic streptococcus that forms dextran and ferments mannitol, and its group D antigen may not be easy to demonstrate by routine methods; but *Str. mutans* nearly always acidifies sorbitol and fails to hydrolyse starch.

Str. mutans is highly sensitive to penicillin. Most strains are sulphonamide resistant, and most strains of human origin are bacitracin resistant.

Antigenic structure and classification Strains from human and animal sources are diverse in their characters. Their cultural and biochemical properties show a general similarity, but the G + C content of the DNA ranges from 36 to 46 moles per cent, and strains have little genetic homology and differ in cell-wall composition. Seven serological types have been demonstrated by means of precipitation reactions with hot-acid or formamide extracts (Brathall 1970, Perch *et al.* 1974). These types have some host specificity and are each biochemically uniform. Coykendall (1974, see also Coykendall *et al.* 1976) described five genetically homogeneous varieties, which he later proposed should be considered separate species (Coykendall 1977). The *mutans* variety has the biochemical characters given in Table 29.8, and the Brathall-type antigen *c*; some strains cross-react with Lancefield-group E serum. Var. *rattus* attacks arginine, grows at 45°, and has the antigen *b*; var. *sobrinus* fails to acidify raffinose, forms H_2O_2, and has the antigen *d*; var. *cricetus* is bacitracin sensitive and has antigen *a*; var. *ferus* does not attack raffinose, is bacitracin sensitive, and shares the antigen *c* with var. *mutans*, from which it is however genetically distinct.

Antibody to one of the cell-wall proteins of *Str. mutans* appears to be responsible for protecting monkeys against experimental dental caries. This antigen has to be distinguished and separated from another cell-wall protein that cross-reacts with human heart muscle (van der Rijn *et al.* 1976, Lehner *et al.* 1980, Russell 1980).

Habitat *Str. mutans* has a rather limited distribution in the mouth, being largely confined to the gingival sulcus, dental plaque, and carious areas of teeth.

Species isolated from cows and dairy products

Str. uberis This organism was first described by Ayers and Mudge (1922). It is found in cows' milk, on the skin and in the throat and faeces of cows, and in the soil (Cullen and Little 1969), and is responsible for a form of mastitis of cattle that is particularly common in the winter. It forms oval cocci in pairs or short chains and bears a superficial resemblance to the group D streptococci. It grows at 10° but not at 45°, and seldom survives 60° for 30 min; it will not grow in 6.5 per cent NaCl or at pH 9.6. It grows on 40 per cent bile agar and hydrolyses aesculin, hippurate and arginine. It acidifies mannitol, lactose, sucrose, salicin, sorbitol, trehalose and inulin but not raffinose. About one-fifth of *Str. uberis* strains react with group E antiserum (Cullen 1967), but other strains may react with sera for groups E, P and U (Roguinsky 1971). Some strains are CAMP positive. The species appears to be far from homogeneous (Garvie and Bramley 1979). For further information see Cullen (1969).

Str. acidominimus Ayers and Mudge (1922) also described *Str. acidominimus* from cows' milk and faeces. It ferments a range of carbohydrates rather similar to that fermented by *Str. uberis*, but produces only small amounts of acid. It is said to differ from this organism in giving a negative arginine test, but some strains react with group E antiserum. It may possibly be a variant of *Str. uberis*.

Str. thermophilus This organism is also found in milk (Orla-Jensen 1919, Abd-el-Malek and Gibson 1948). It grows at 50° and is said to survive 63° for 30 min. However, it does not grow at 10°, pH 9.6 or on 40 per cent bile agar, and does not hydrolyse arginine or hippurate. It differs from *Str. bovis* in its narrow range of fermentations; lactose and sucrose, but few other sugars, are attacked and it is one of the very few streptococci that does not acidify maltose.

Other genera closely related to *Streptococcus*

Leuconostoc

This genus comprises chain-forming, heterofermentative, catalase-negative, gram-positive cocci that are found in slimy sugar solutions, fermenting vegetables, and dairy products. Some form detectable gas from sugars when tested in the ordinary way, but others only in special media such as that of Langston and Bouma (1960). They form D(-) lactic acid from glucose, and members of some species form dextran from sucrose.

Pediococcus and *Aerococcus*

The *pediococci* are homofermentative, gram-positive cocci that form tetrads or clusters. They are strictly catalase negative, though some may form a few small bubbles of gas from H_2O_2; but the enzyme responsible for this reaction differs from a true catalase and the organisms do not possess a cytochrome system (Deibel and Evans 1960). They form DL lactic acid from glucose and are microaerophilic. Their main habitat is fermenting vegetable materials, and they are often found in dairy products.

The *aerococci* are rather similar to the pediococci, but differ from them in failing to grow, or growing very sparsely, under strictly anaerobic conditions. One species has been described, *Aerococcus viridans* (Williams *et al.* 1953). This organism is frequently found in the air of occupied places, in dust, and in foodstuffs; little is known about its presence in the human body flora, but a number of strains have been isolated from the blood of patients with endocarditis (Colman 1967). It forms tetrads or packets of gram-positive cocci and grows on blood agar as α-haemolytic colonies that closely resemble faecal streptococci. The catalase reaction is negative, hydrogen peroxide is formed during aerobic growth, and nitrate is not reduced. Aerococci grow on 40 per cent bile agar, at pH 9.6, and in 6.5 per cent NaCl; on the other hand, they do not grow at 45°, do not hydrolyse arginine, are not tellurite resistant and do not reduce litmus milk strongly. Opinions differ about their ability to survive 60° for 30 min. All strains acidify sucrose and most acidify mannitol, lactose, salicin and raffinose. According to Deibel and Niven (1960), the organism described as *Gaffkya homari*, which causes septicaemia in lobsters, is indistinguishable from *Aerococcus viridans*. Aerococci differ from the micrococci in catalase reaction and in DNA base ratio. It is probable, therefore, that they form a distinct group (see also Whittenbury 1965, Evans and Schultes 1969),

Gemella

In 1960, Berger re-investigated an organism described by Thjøtta and Bøe (1938) as *Neisseria haemolysans*. Although a gram-negative coccus, it was oxidase negative, formed H_2O_2 and attacked sugars fermentatively. He removed it from *Neisseria* to a separate genus *Gemella* (Berger 1961). Reyn (1970) states that its cell-wall structure resembles closely that of the gram-positive cocci, and that its DNA base composition (33.5 moles per cent G+C) differs significantly from that of the neisseriae.

The organism is a gram-negative coccus, 0.6–1.0 μm in diameter, which occurs singly or in pairs with flattened adjacent surfaces. On rabbit blood agar it forms smooth circular non-pigmented colonies, surrounded by a clear zone of haemolysis after 48 hr. It grows poorly on nutrient agar, does not reduce nitrate, is aerobic and facultatively anaerobic, and attacks sugars fermentatively. It is a parasite of human mucous membranes.

Lactobacillus

Definition

Straight or curved rods of varying length and thickness, with parallel sides, arranged singly or in chains, sometimes filamentous or pleomorphic, without branching, clubbing, or bifid formation. Gram positive and non-sporing. Usually non-motile. Growth on surface media poor. Complex nutritive requirements. Energy obtained by anaerobic fermentation of sugars. Growth favoured by microaerophilic or anaerobic conditions and by CO_2. Glucose is fermented, and either lactic acid alone or lactic acid along with volatile acids, CO_2, and other by-products is formed. Little or no proteolytic activity. No production of catalase, oxidase, or indole, and no reduction of nitrate. Readily killed by heat, but unusually resistant to acid. Non-pathogenic to man and animals. Widely distributed in fermenting vegetable and animal products, and in the alimentary tract of man and animals. G + C content of DNA: 37–53 moles per cent.

Type species is *Lactobacillus delbruecki*.

Introductory

Our decision to describe *Streptococcus* and *Lactobacillus* in the same chapter was taken on grounds of convenience, and must not be taken to imply a very close relation between the two genera. The lactobacilli are gram-positive rods that have a number of metabolic similarities to the streptococci, particularly in their mode of attack on sugars, but they include both heterofermentative and homofermentative species and grow best under microaerophilic or anaerobic conditions. The first member of the *Lactobacillus* group was isolated by Kern (1881) from kefir, the fermented milk of the Caucasus. He called it *Dispora Kaukasica*, but later its name was changed to *L. caucasicus*. The original description of this organism was incomplete, and its identity is uncertain. The present type species is *L. delbruecki* (Report 1971), an organism originally isolated from milk by Leichmann in 1896 for which a neotype strain has now been established (see Rogosa 1974*b*).

A similar bacillus was observed by Döderlein in 1892 in the vaginal secretion of women; the identity of this organism is likewise in doubt (Rogosa and Sharpe 1959). A slender gram-positive bacillus was observed in the stomach contents of patients suffering from gastric carcinoma by Oppler in 1895; this organism, which was not cultivated, is referred to as the Boas-Oppler bacillus. Moro (1900*a*, *b*) cultivated a similar bacillus, *L. acidophilus*, from the faeces of breast-fed babies. In 1905 Grigoroff isolated from yoghurt the organism now known as *L. bulgaricus*. Similar bacilli were soon found in other forms of fermented milk, and in raw cows' milk and human milk. Lactobacilli from cheese were named *L. casei* by Orla-Jensen (1904). Similar organisms were isolated from the faeces of mammals, fishes and invertebrates (Mereshkowsky 1905, 1906, Petrow 1907), from human saliva and gastric juice, from soil, and from a variety of foods (Heinemann and Hefferan 1909). An organism isolated from carious teeth and named *L. odontolyticus* (McIntosh *et al.* 1922, 1924) is probably the same as *L. plantarum*, which was described by Pederson (1936).

In the following account, we shall divide the species of lactobacilli into three groups corresponding to those originally described by Orla-Jensen (1919, 1943).

1. **Thermobacteria.** *L. lactis*, *L. bulgaricus*, *L. delbruecki*, *L. leichmanni*, *L. helveticus*, *L. acidophilus* and *L. salivarius*.
2. **Streptobacteria.** *L. casei* and *L. plantarum*.
3. **Betabacteria.** *L. fermentum*, *L. cellobiosus*, *L. brevis*, *L. buchneri* and *L. viridescens*.

The thermobacteria and the streptobacteria are described as *homofermentative* because they convert 95 per cent or more of glucose into lactic acid; the betabacteria are described as *heterofermentative* because, in addition to lactic acid, they form considerable amounts of other organic acids, and CO_2.

Habitat

Lactobacilli, are found (1) in the body flora of man and animals, (2) on the surface of plants, and (3) in certain foodstuffs and agricultural products. In the body flora (Chapter 8), they are present in moderately large numbers in the gut, mouth and vagina, but they are seldom the predominant organisms except in the

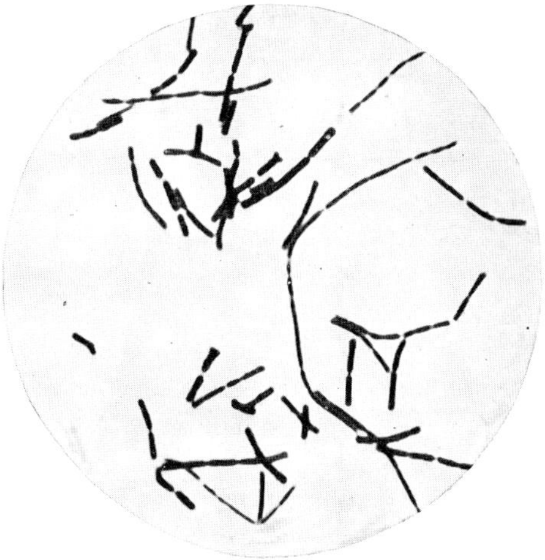

Fig. 29.7 *Lactobacillus acidophilus*. From an agar culture, 48 hr, 37° (× 1000).

small intestine. Members of a number of different species of lactobacilli are found at each of these sites. In general, those most often present in man are: in the lower bowel, *L. acidophilus*, *L. salivarius* and *L. fermentum*; in the vagina, *L. acidophilus*; and in the mouth, *L. casei* and *L. plantarum*.

Lactobacilli are widespread but not numerous on the surface of plants. A number of stored foods and other plant products provide a favourable environment for the multiplication of lactobacilli, which then become the predominant organisms. In some of these the lactobacilli cause spoilage, but in others, for example, certain milk products and silage, they bring about desirable changes. The lactobacilli found in food and similar products include members of a number of widely distributed species as well as a number of highly specialized organisms.

(For a review of the distribution of lactobacilli, see Sharpe 1981.)

Morphology

Members of this group are in general fairly large, non-sporing, gram-positive rods, but they vary a good deal in length and breadth and in old cultures tend to be gram negative. A few strains are motile by peritrichate flagella. Metachromatic granules are prominent in some species, notably *L. lactis*, *L. leichmanni* and *L. bulgaricus* (Sharpe 1962). Some members form long chains with the cells lying parallel to one another and coiled as in a skein of wool or twisted in corkscrew fashion (Kludas and Dobberstein 1959). In others the cells are arranged singly or in pairs, either end-to-end or side by side. Degenerative forms may be visible in cultures more than 3 days old. The thermobacteria are large, thick, and often filamentous. Among the streptobacteria *L. casei* is a short square-ended rod forming chains of varying length; *L. plantarum* varies in length from coccoid to short filamentous forms, arranged singly or in pairs end to end.

Colonial form

Surface colonies vary in appearance, but on the whole conform to one or other of the two types described by Mereshkowsky (1905, 1906) as seen on 2 per cent

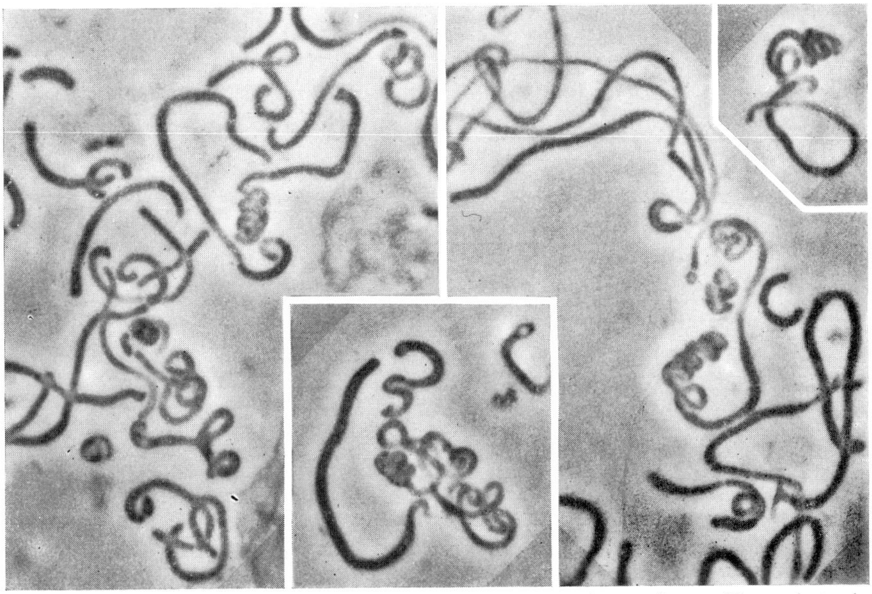

Fig. 29.8 *Lactobacillus bulgaricus*. Showing the characteristic corkscrew forms. (From photomicrograph kindly supplied by Dr Martin Kludas.)

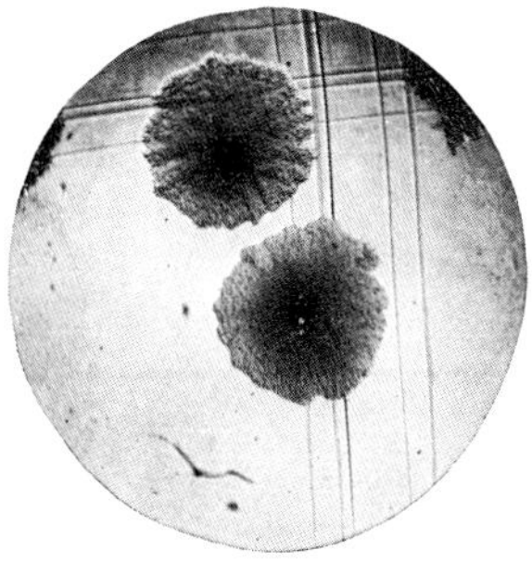

Fig. 29.9 *Lactobacillus acidophilus*. Surface colony on agar, 4 days, 37°, showing differentiation (× 8).

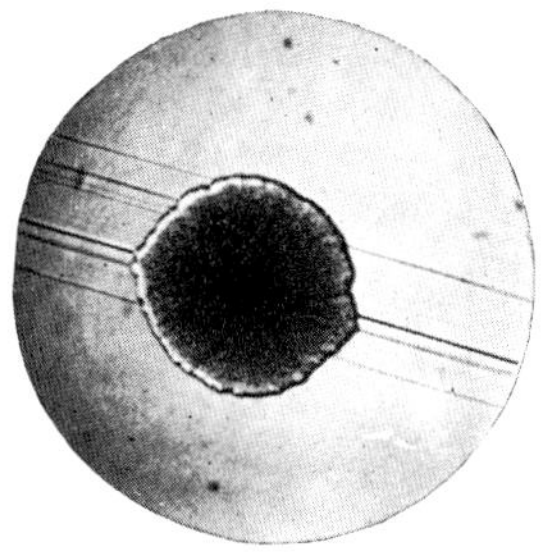

Fig. 29.10 *Lactobacillus acidophilus*. Surface colony on agar, 4 days, 37° (× 8).

glucose agar: (a) round or navicular, amorphous, pinhead, opaque, whitish, encircled by an areola of turbid agar; (b) round or irregularly round, less than pinhead in size, greyish, translucent, having a finely erose edge devoid of an areola; microscopically these colonies have a rhizoid structure. The colonial appearance depends, however, on the nature of the medium, the gaseous conditions of cultivation, the species studied, and other factors. The compact and feathery types of colony are referred to by some writers as smooth and rough respectively; but the colonial form is so subject to environmental conditions that a strain may be recorded by one observer as smooth and by another as rough. Rogosa and Sharpe (1959), for example, include smooth colony formation among the characters of the streptobacterium group and rough colony formation among those of the thermobacterium group; but that this is no more than a general statement is evident from the observations of Lerche and Reuter (1962) and others.

Resistance

The lactobacilli have no particular resistance to heat, and are destroyed by exposure to 60° or 65° for half an hour. They are, however, specially resistant to acid, and are able to survive and even to grow in concentrations of acid that are fatal to most other non-sporing bacteria. Their tolerance to bile salts varies. The ability to grow in the presence of 4 per cent sodium taurocholate (Wheater 1955) or 0.1 per cent Teepol (Naylor and Sharpe 1958) has been used to distinguish between species. In general, the betabacteria except *L. fermentum* and *L. plantarum* exhibit such a resistance. In surface cultures exposed to air lactobacilli die out fairly rapidly. Even in milk some members, such as *L. acidophilus* and *L. salivarius* are dead within 3–5 days, though others survive for as long as 8 weeks. Lactobacilli are resistant to 0.1 per cent thallous acetate.

Nutritional requirements and metabolism

The members of this group obtain their energy by the anaerobic fermentation of sugars. The homofermentative species break down glucose with the production of at least 95 per cent of lactic acid, and only minimal amounts of CO_2 and acetic acid. The heterofermentative species, on the other hand, produce only about 50 per cent lactic acid, the remainder of the sugar being broken down to alcohol, acetic and formic acids, and CO_2. However, the homofermentative streptobacteria form gas from gluconate (Table 29.9). When ribose is attacked by heterofermentative lactobacilli, lactic and acetic acids are formed without gas.

The nutritional requirements of the lactobacilli are complex and varied. Nearly all the amino acids are needed, but there is some variation between species, and the response to the D-form often differs from that to the L-form (Koser and Thomas 1957). Vitamin requirements likewise vary from species to species and so are useful in classification (Rogosa *et al.* 1961; see Table 29.9). Thiamine is necessary for the growth of heterofermentative but not of homofermentative lactobacilli. Unsaturated fatty acids, particularly in the form of Tween 80, have a stimulating effect on growth. (For reviews of nutritional requirements, see Snell 1952, Sharpe 1962.)

Growth on protein media without carbohydrates is very poor, and proteolytic activity is absent or at most slight. No indole, scatole or histamine is formed; some strains produce acetylcholine (see Stephenson and Rowatt 1947). Most of the heterofermentative lactobacilli and a few others form ammonia from arginine (Table 29.9); tests for this are best performed in medium containing 2 per cent glucose. The catalase test is nearly always negative; occasional weakly positive reactions must be attributed to pseudocatalase action, because uniformly negative benzidine tests indicate the absence of a cytochrome system (Sharpe 1962). Peroxide is formed.

Most of the lactobacilli are microaerophilic but grow best in the absence of oxygen. A few are strict anaerobes. A mixture of 95 per cent hydrogen or nitrogen and 5 per cent CO_2 is generally favourable (Kulp 1926, Rogosa and Sharpe 1959).

The temperatures at which growth will occur vary with different species. Thermobacteria grow best at 37–40°; none grow at 15°, and most but not all grow

Table 29.9 General characters of species of lactobacilli

Group	Species	Gas from: glucose	Gas from: gluconate	Growth at 15°	Requirement for: thiamine	Requirement for: riboflavin	Requirement for: pyridoxal	Requirement for: folic acid	NH_3 from arginine	DNA: moles % G+C[1]
Thermobacteria	*L. lactis*	–	–	–	–	+	–	–	–	50.3
	L. bulgaricus	–	–	–	–	+	–	–	–	50.3
	L. delbruecki	–	–	–	–	+	–	–	v	50.0
	L. leichmanni	–	–	–	–	–	–	+	v	50.8
	L. helveticus	–	–	–	–	+	+	–	–	39.3
	L. acidophilus	–	–	–	–	+	–	+	–	36.7
	L. salivarius	–	–	–	–	+	–	+	–	34.7
Streptobacteria	*L. casei*	–	+	+	–	+	+	+	–	46.4
	L. plantarum	–	+	+	–	v	–	–	–	45.0
Betabacteria	*L. fermentum*	+	+	–	+	–	–	–	+	53.4
	L. cellobiosus	+	+	v	+	–	–	–	+	53.1
	L. brevis	+	+	+	+	–	–	+	+	42.7–46.4
	L. buchneri	+	+	+	+	v	v	v	+	44.8
	L. viridescens	+	+	+	+	+	+	+	–	37.5 and 42.3

v = Some positive, others negative.
[1] Data from Rogosa (1974*b*).

at 45°. The optimum for streptobacteria is about 30°; all grow at 15°, but growth at 45° is variable. Thus, ability to grow at 15° is a better means of distinguishing streptobacteria from thermobacteria than is absence of growth at 45° (Rogosa and Sharpe 1959; see Table 29.9). Among the betabacteria, *L. brevis*, *L. buchneri* and *L. viridescens* resemble streptobacteria, *L. fermentum* resembles thermobacteria, and *L. cellobiosus* is variable in temperature-range for growth.

The lactobacilli are not only aciduric; they are acidophilic and grow best in the neighbourhood of pH 6.0. Their range is about pH 4.0–6.8, though some strains will grow at as low as pH 3.5. For testing the fermentation of carbohydrates, Rogosa and Sharpe (1959) recommend a pH of 5.5–6.0.

Media for the isolation of lactobacilli

All media for the growth of lactobacilli are complex and must contain various growth factors. The constituents usually include yeast extract, peptone, citrate, manganese and magnesium; some media also include meat extract; Tween 80 is usually added because it stimulates the growth of many lactobacilli. Two of the more widely used non-selective media are the APT medium of Evans and Niven (1951) and the MRS medium of de Man, Rogosa and Sharpe (1960). A useful selective medium (SL medium, Rogosa *et al.* 1951) has a high content of acetate ions and a pH of 5.4; this is the medium of choice for the growth of lactobacilli from most human sources, particularly when prepared as described by Sharpe (1981). However, not all lactobacilli will grow on it, and various special media have been recommended for the examination of certain foods and beverages (Sharpe 1981).

Action on sugars

The fermentative activity of the different species of lactobacilli remains remarkably constant, but if reliable results are to be obtained great care must be exercised over the technique used in testing. Orla-Jensen (1919) stressed the importance of an adequate nutritional environment. Only vigorously growing cultures in a well balanced medium can be trusted to manifest their full fermentative powers. Strains freshly isolated from an adverse environment may fail to show some of the characters that are apparent after cultivation for a time in a favourable medium (Pederson and Albury 1955).

A good basal medium should contain acetate and oleic acid esters, especially Tween 80, and have a pH value low enough to inhibit the growth of spore bearers and so reduce the sterilization of the medium by heat to a minimum (Rogosa and Sharpe 1959, Sharpe 1962). Heat-susceptible sugars must be Seitz-filtered and added to the medium after sterilization. An initial low pH, such as 5.5–6.0, is to be recommended; in the absence of a suitable indicator, the reaction may have to be determined electrometrically. The concentration of carbohydrates, such as galactose and lactose, should be high enough—1.4 to 2.0 per cent—to permit the adaptive fermentation that sometimes occurs only in the presence of sufficient substrate (Rogosa *et al.* 1953).

In testing for gas production abundant carbohydrate must be available, and large inocula from well growing cultures must be used. As CO_2 is soluble in water, the conventional Durham tube method is inapplicable; gas production can, however, be rendered evident by special devices, of which the simplest is putting a layer of Vaseline over Briggs's medium in a tube (Lerche and Reuter 1960); gas formation is usually evident within a week. In reading fermentation tests titration of the acid formed is usually better than trusting to pH indicators (Orla-Jensen 1943).

Table 29.10 Action of lactobacilli on sugars

Species	Lactic acid formed	Acid formed from: ribose	arabinose	maltose	lactose	sucrose	raffinose	cellobiose	trehalose	salicin
L. lactis	D(−)	−	−	+	+	+	−	−	+	+
L. bulgaricus	D(−)	−	−	−	+	−	−	−	−	−
L. delbruecki	D(−)	−	−	v	−	+	−	−	−	−
L. leichmanni[1]	D(−)	−	−	+	+	+	−	+	+	+
L. helveticus[2]	DL	−	−	+	+	−	−	−	v	−
L. acidophilus	DL	−	−	+	+	+	−	+	+	+
L. salivarius[3]	L(+) or DL	−	−	+	+	+	+	−	+	v
L. casei[4]	L(+)	+	−	+	+	+	−	+	+	+
L. plantarum[5]	DL	+	v	+	+	+	+	+	+	+
L. fermentum	DL	+	v	+	+	+	+	−	v	−
L. cellobiosus	DL	+	+	+	w	+	+	+	+	w
L. brevis	DL	+	+	+	w	v	w	−	−	−
L. buchneri[6]	DL	+	+	+	w	v	w	−	−	−
L. viridescens	DL	+	−	+	−	v	−	−	v	−

v = Some positive, others negative; w = weak reaction.

[1] Differs from *L. acidophilus* in failing to ferment galactose.
[2] Differs from *L. jogurti* in fermenting maltose.
[3] Differs from other thermobacteria in fermenting mannitol.
[4] Several biotypes (see text).
[5] Differs from *L. casei* in fermenting melibiose.
[6] Differs from *L. brevis* in fermenting melezitose.

The type of lactic acid formed by members of the species of lactobacilli is shown in Table 29.10; in this the terms D and L are used to denote the molecular configuration of the acids, L(+) indicating the dextrorotatory and D(−) the laevorotatory form.

Glucose is always fermented. Table 29.10 gives a general indication of the pattern of acidification of other sugars, and is based principally on the findings of Rogosa and Sharpe (1959), Sharpe (1962, 1981) and Rogosa (1974*b*), which do not differ significantly from those of Kundrat (1958), Lerche and Reuter (1962) and Mitsuoka (1969).

Practically all streptobacteria and betabacteria, but none of the thermobacteria, acidify ribose. The species of thermobacteria can usually be distinguished by means of their sugar reactions alone. It should be noted that *L. jogurti* is probably a maltose non-fermenting variety of *L. helveticus*. The streptobacteria tend to give rather similar sugar reactions, but *L. plantarum* differs from *L. casei* in fermenting raffinose and melibiose. *L. casei* comprises several biotypes (p. 210); it is said to be unique among lactobacilli in fermenting pyruvate. In the heterofermentative group, *L. fermentum* is better distinguished from the rest by its failure to grow at 15° than by its sugar reactions. *L. buchneri* ferments melezitose but *L. brevis* does not.

Antigenic structure and cellular constituents

Early work, which we reviewed in the last edition of this book, established that the lactobacilli were antigenically heterogeneous. Sharpe (1955*a*) identified several group antigens by precipitation reactions between hot-acid extracts and antisera prepared by the injection of heat-killed whole-cell vaccines into rabbits. Some cross-reactions were removed by suitable cross-absorption of sera; others could be avoided by using media of different composition for the growth of the vaccine and the absorbing suspension (Sharpe 1955*b*). She was able to group about three-quarters of all strains, and found that members of a single species usually but not invariably belonged to the same serological group. The method was slightly modified by Sharpe and Wheater (1957), and letters were allocated to the groups (see Rogosa and Sharpe 1959, Sharpe 1962). The distribution of group antigens is as follows.

Group A *L. helveticus*
Group B *L. casei*, var. *casei* and var. *alactosus*
Group C *L. casei*, var. *casei*, var. *alactosus* and var. *rhamnosus*
Group D *L. plantarum*
Group E *L. lactis*, *L. bulgaricus*, *L. buchneri* and *L. brevis*
Group F *L. fermentum*
Group G *L. salivarius*

Group antigens have not been identified in other species, including *L. acidophilus*. According to Sharpe (1970), this species is probably heterogeneous antigenically and biochemically; Shimohashi and Mutai (1977) have described four serotypes that correspond to biotypes within the species. It will be noted that *L. casei* strains, except those of var. *rhamnosus*, may possess one of two group antigens, and that the group E antigen occurs in members of four species, two of which are thermobacteria and two betabacteria.

As in the streptococci, the group antigens are chemically diverse. Only antigens B and C are cell-wall polysaccharides. Antigen D is a cell-wall ribitol teichoic aced and antigen E a cell-wall glycerol teichoic acid. Antigen A is a glycerol teichoic acid found in both cell-wall and membrane fractions, and antigen F is a glycerol teichoic acid present only in the cell membrane. Cross-reactions between lactobacilli of different serological groups and a number of other gram-positive organisms are attributable to antibody against the common glycerophosphate component of the lipoteichoic acid of the cell membrane (Sharpe *et al.* 1973*b*).

The cell-wall peptidoglycans of the lactobacilli are also diverse (Schleifer and Kandler 1972). Those of all the thermobacteria we listed at the beginning of this chapter have interpeptide bridges of the L-lysine—D-aspartate type, but two other species that are obligate anaerobes (*L. ruminus* and *L. vitulinus*) have interpeptide bridges of the *meso*-diaminopimelic acid type. Among the streptobacteria, *L. casei* has a peptidoglycan of the L-lysine—D-aspartate type and *L. plantarum* one of the *meso*-diaminopimelic acid type. Three types of peptidoglycan are found in species of betabacteria that we have described: (1) L-lysine—D-aspartate in *L. brevis* and *L. buchneri*; (2) L-ornithine—D-aspartate in *L. fermentum* and *L. cellobiosus*; and (3) L-lysine—L-alanine—L-serine in *L. viridescens*.

Pathogenicity

For all practical purposes the lactobacilli may be regarded as non-pathogenic. Sharpe, Hill and Lapage (1973*a*) reported on 10 strains that had been isolated from cardiac, septicaemic and suppurative lesions in man, but how far they were primarily responsible for the lesions is open to question. The relation between lactobacilli and dental caries is discussed in Chapter 57.

Classification

The lactobacilli form a fairly distinct genus. They cannot be said to have a specially close relation to the streptococci, though they resemble them in their general mode of attack on carbohydrates, the absence of a cytochrome system, and their inability to reduce nitrate. The numerical taxonomic study of Wilkinson and Jones (1977) allocated *Lactobacillus* to a broad group of genera including *Erysipelothrix*, *Listeria* and *Propionibacterium* as well as *Streptococcus*, and distinguished all of these from another group that included the coryneform bacteria. According to Rogosa (1974*b*), gram-positive spore-bearing bacilli that are homofermentative and catalase negative belong to the genus *Sporolactobacillus*, closely related to *Bacillus*. The organism that used to be called *Lactobacillus bifidus* differs from the lactobacilli in so many ways that it is now assigned to the genus *Bifidobacterium* (Chapter 43), as originally proposed by Orla-Jensen (1919; see also Sharpe 1962, Werner and Seeliger 1964).

Orla-Jensen (1919, 1943), whose two monographs have influenced all subsequent workers, laid particular stress on fermentative activity and the type of lactic acid produced from glucose. He divided the lactobacilli into three groups for which he proposed the generic names *Thermobacterium*, *Streptobacterium*, and *Betabacterium*. Breed, Murray and Smith (1957) pointed out that two of these names—*Streptobacterium* and *Betabacterium*—have been used before and are therefore no longer legitimate. We see no reason, however, why they should not be used in a colloquial sense, and printed in roman, not italic, type.

Orla-Jensen's primary division was between the homofermentative thermobacteria and streptobacteria on the one hand and the heterofermentative betabacteria on the other. A secondary division was on the temperature of growth. The thermobacteria had a higher optimum temperature than the streptobacteria or the betabacteria. As we have seen, inability to grow at 15° has proved to be a more reliable distinguishing feature than ability to grow at 45°.

The distribution of several other biochemical and nutritional characters is in conformity with Orla-Jensen's primary subdivision of the lactobacilli into three groups (Rogosa 1970). Thus, the production of acid from ribose and the requirement for thiamine characterize the betabacteria, and the formation of gas from gluconate but not from glucose characterizes the streptobacteria. Also, thermobacteria and streptobacteria, but not betabacteria, form fructose-1,6-diphosphate aldolase (Buyze *et al.* 1957), and only betabacteria form mannitol from fructose (Eltz and Vandemark 1960). The numerical taxonomic studies of Wilkinson and Jones (1977) also distinguished clearly between the three groups.

The characters most commonly used to identify members of individual species are the pattern of acidification of various sugars and the requirement for vitamins. By these means it is usually possible to allocate most of the lactobacilli found in the human body flora and in milk products to a species mentioned in this chapter, though, as we have seen, some pairs of species are separated only by a single character. In certain other environments, notably some vegetable products, different species, and some groups of less well characterized strains, may predominate (see Sharpe 1981).

Numerous other properties have been studied for the light that they may throw on the classification of lactobacilli. In general, members of the well recognized species behave uniformly in these other tests, but it must be conceded that the results obtained show only a partial correlation with Orla-Jensen's classification. This is true of the type of lactic acid formed, the DNA base content (Table 29.9), the presence of group antigens, and the peptidoglycan type of the cell wall. Studies of the electrophoretic mobility and the serological specificity of the lactic dehydrogenases (see London 1976), and of DNA hybridization (see Sharpe 1981), have provided a good deal of additional information about relationships between individual species.

We append a brief note about the species included in the Orla-Jensen groups.

Thermobacterium

Sharpe (1962) recognized eight species, the seven shown in Table 29.9 and *L. jogurti*. This differs from *L. helveticus* only in not fermenting maltose. *L. bulgaricus* is closely related to *L. lactis*, differing from it mainly in fermenting fewer sugars. The two species can be rapidly distinguished from *L. helveticus* by the presence of purple metachromatic granules in films stained with methylene blue. Similar granules are formed by *L. leichmanni*. *L. jenseni* (Gasser *et al.* 1970) is almost identical with *L. leichmanni* and can be distinguished from it only by special tests. Two species of thermobacteria—*L. ruminis* and *L. vitulinus*—are strict anaerobes found in the bovine rumen.

Streptobacterium

The two species that we have described, *L. casei* and *L. plantarum*, can in their typical forms be distinguished by sugar reactions and nutritional requirements. However, *L. casei* is a composite species in which Rogosa and his colleagues (1953) recognized three biotypes: var. *casei*; var. *alactosus*, which differs from it in not acidifying lactose; and var. *rhamnosus*, which acidifies lactose and rhamnose, and is the only variety to grow at 45°. A number of other streptobacteria are found in milk, foods and silage; some have been given separate specific names, others are considered to be further varieties of *L. casei* (see Rogosa 1974*b*), and many remain unclassified.

Betabacterium

The five species we have considered are all active fermentatively; *L. fermentum* and *L. cellobiosus* are considered to be closely related, as are *L. brevis* and *L. buchneri*. In addition to these, there are a number of less well known species that have a much narrower range of fermentations and often grow slowly. These organisms are highly resistant to acid and to ethanol; Rogosa (1974*b*) recognizes five species among them, and allocates these to a separate sub-section of the heterofermentative lactobacilli.

For recent reviews of the classification of the lactobacilli, see Rogosa and Sharpe (1959), Rogosa (1970, 1974*b*), London (1976), Holdeman *et al.* (1977) and Sharpe (1981); and for lactic acid bacteria met with in the dairy industry, see Sharpe (1979).

References

Abd-el-Malek, Y. and Gibson, T. (1948) *J. Dairy Res.* **15,** 233.

Abdulla, E. M. and Schwab, J. H. (1966) *J. Bact.* **91,** 374.

Agarwal, K. K., Elliott, S. D. and Lachman, P. J. (1969) *J. Hyg., Camb.* **67,** 491.

Alouf, J. E. and Raynaud, M. (1967) *C. R. Acad. Sci., Paris* **264,** 2524.

Anderson, A. B. and Hart, P. D'A. (1934) *J. Path. Bact.* **39,** 465.

Andrewes, F. and Horder, T. (1906) *Lancet* **ii,** 708, 775, 852.

Anthony, B. F. (1970) *J. Immunol.* **105,** 379.

Appelbaum, P. C. *et al.* (1977) *Morbid. Mortal. Weekly Rep.* **26,** 285.

Austrian, R. (1953) *J. exp. Med.* **98,** 21, 35.

Austrian, R., Buettger, C. and Dole, M. (1972) In: *Streptococci and Streptococcal Diseases*, p. 355. Ed. by L. W. Wannamaker and J. M. Matsen. Academic Press, New York.

Austrian, R. and Collins, P. (1966) *J. Bact.* **92,** 1281.

Austrian, R. and MacLeod, C. M. (1949*a*) *J. exp. Med.* **89,** 439; (1949*b*) *Ibid.* **89,** 451.

Avery, O. T., Chickering, H. T., Cole, R. and Dochez, A. R. (1917) *Monogr. Rockefeller Inst. med. Res.* No. 7.

Avery, O. T. and Cullen, G. E. (1923) *J. exp. Med.* **38,** 199.

Ayers, S. H. and Mudge, C. S. (1922) *J. infect. Dis.* **31,** 40.

Baer, H. (1967) *Canad. J. comp. Med.* **31,** 216.

Bailey, R. W. and Oxford, A. E. (1958) *Nature, Lond.* **182,** 185.

Baker, C. J., Kasper, D. L. and Davis, C. E. (1976*a*) *J. exp. Med.* **143,** 258.

Baker, C. J., Webb, B. J. and Barrett, F. F. (1976*b*) *Antimicrob. Agents Chemother*, **10,** 128.

Ball, L. C. and Parker, M. T. (1979) *J. Hyg., Camb.* **82,** 63.

Barnes, E. M. (1956*a*) *J. appl. Bact.*, **19,** 193; (1956*b*) *J. gen. Microbiol.* **14,** 57.

Bazeley, P. L. and Battle, J. (1940) *Aust. vet. J.* **16,** 140.

Beachey, E. H. and Ofek, I. (1976) *J. exp. Med.* **143,** 759.

Beachey, E. H., Ofek, I., Cunningham, M. and Bisno, A. (1974) *Appl. Microbiol.* **27,** 1.

Beachey, E. H., Seyer, J. M. and Kang, A. H. (1980). In: *Streptococcal Disease and the Immune Response*, p. 149. Ed. by S. E. Read and J. B. Zabriskie. Academic Press, New York.

Beerens, H. and Tahon-Castel, M. M. (1966) *Ann. Inst. Pasteur* **111,** 90.

Berger, U. (1960) *Z. Hyg. InfektKr.* **146,** 253; (1961) *Int. Bull. bact. Nomencl.* **11,** 17.

Bergner-Rabinowitz, S., Beck, A., Ofek, I. and Davies, A. M. (1969) *Israel J. med. Sci.* **5,** 285.

Bernheimer, A. W. (1948) *Bact. Rev.* **12,** 195; (1967) *J. Bact.* **93,** 2024; (1972) In: *Streptococci and Streptococcal Diseases*, p. 19. Ed. by L. W. Wannamaker and J. M. Matsen. Academic Press, New York.

Bernheimer, A. W. and Davidson, M. (1965) *Science* **148,** 1229.

Bernheimer, A. W., Lazarides, P. D. and Wilson, A. T. (1957) *J. exp. Med.* **106,** 27.

Bernheimer, A. W., Linder, R. and Avigad, L. S. (1979) *Infect. Immun.* **23,** 838.

Bernheimer, A. W. and Schwartz, L. L. (1960) *J. Path. Bact.* **79,** 37; (1965*a*) *Ibid.* **89,** 209; (1965*b*) *J. Bact.* **89,** 1387.

Berntsson, E., Broholm, K.-A. and Kaijser, B. (1978) *Scand. J. infect. Dis.* **10,** 177.

Bevanger, L. and Maeland, J. A. (1977) *Acta path. microbiol. scand.* **B85,** 357.

Bevanger, L. and Stammes, T. I. (1979) *Acta path. microbiol. scand* **B87,** 301.

Billroth and Ehrlich. (1877) *Arch. klin. Chir.* **20,** 403.

Bliss, E. A. (1937) *J. Bact.* **33,** 625.

Boissard, J. M. and Wormald, P. J. (1950) *J. Path. Bact.* **62,** 37.

Bowers, E. F. and Jeffries, L. R. (1955) *J. clin. Path.* **8,** 58.

Brathall, D. (1970) *Odont. Revy* **21,** 143.

Braun, D. G., Eichmann, K. and Krause, R. M. (1969) *J. exp. Med.* **129,** 809.

Braun, D. G., Schalch, W. and Schmid, I. (1980) In: *Streptococcal Diseases and the Immune Response*, p. 317. Ed. by S. E. Read and J. B. Zabriskie. Academic Press, New York.

Breed, R. S., Murray, E. G. D. and Smith, N. R. (1957) *Bergey's Manual of Determinative Bacteriology*, 7th edn., p. 542. Ballière, Tindall & Cox Ltd, Lond.

Briles, E. B. and Tomasz, A. (1973) *J. biol. Chem.* **248,** 6394.

Brock, T. D. and Davie, J. M. (1963) *J. Bact.* **86,** 708.

Brock, T. D., Peacher, B. and Pierson, D. (1963) *J. Bact.* **86,** 702.

Brown, J., Farnsworth, R., Wannamaker, L. W. and Johnson, D. W. (1974) *Infect. Immun.* **9,** 377.

Brown, J. H. (1919) *Monogr. Rockefeller Inst. med. Res.* No. 9; (1939*a*) *J. Bact.* **37,** 133; (1939*b*) *Rep. 3rd Int. Congr. Microbiol. New York*, p. 172.

Brunson, K. W. and Watson, D. W. (1976) *Infect. Immun.* **14,** 1256.

Butter, M. N. W. and de Moor, C. E. (1967) *Leeuwenhoek ned. Tijdschr.* **33,** 439.

Buxton, J. C. (1949) *Brit. vet. J.* **105,** 107.

Buyze, G., van den Hamer, C. J. A. and de Haan, P. G. (1957) *Leeuwenhoek ned. Tijdschr.* **23,** 345.

Carlson, A. S., Kellner, A., Bernheimer, A. W. and Freeman, E. B. (1957) *J. exp. Med.* **106,** 15.

Carlsson, J. (1968) *Odont. Revy* **19,** 137.

Cars, O., Forsum, U. and Hjelm, E. (1975) *Acta path. microbiol. scand.* **B83,** 145.

Christensen, L. R. (1945) *J. gen. Physiol.*, **28,** 363.

Christensen, P., Kahlmeter, G., Jonsson, S. and Kronvall, G. (1973) *Infect. Immun.* **7,** 881.

Christie, R., Atkins, N. E. and Munch-Petersen, E. (1944) *Aust. J. exp. Biol. med. Sci.* **22,** 197.

Chu, C. M. (1948) *Nature, Lond.* **161,** 606.

Clarke, J. K. (1924) *Brit. J. exp. Path.* **5,** 141.

Coburn, A. F. and Pauli, R. H. (1941) *J. exp. Med.* **73,** 551.

Cohen, B., Halbert, S. P. and Perkins, M. E. (1942) *J. Bact.* **43,** 607.

Cole, R. (1914) *J. exp. Med.* **20,** 346.

Colebrook, L., Elliott, S. D., Maxted, W. R., Morley, C. W. and Mortell, M. (1942) *Lancet.* **ii,** 30.

Coleman, D. J., McGhie, D. and Tebbutt, G. M. (1977) *J. clin. Path.* **30,** 421.

Coligan, J. E., Krause, R. M. and Kindt, T. J. (1980) In: *Streptococcal Diseases and the Immune Response*, p. 303. Ed. by S. E. Read and J. B. Zabriskie. Academic Press, New York.

Collier, J. R. (1956) *Amer. J. vet. Res.* **17,** 640.

Colman, G. (1967). *J. clin. Path.* **20,** 294; (1968) *J. gen. Microbiol.* **50,** 149; (1969) *Ibid.* **57,** 247; (1970) Ph.D. Thesis, University of London; (1976) In *Selected Topics in Medical Bacteriology*, p. 179. Ed. by J. de Louvois. Baillière Tindall, London.

Colman, G. and Williams, R. E. O. (1965) *J. gen. Microbiol.*, **41,** 375; (1972) In: *Streptococci and Streptococcal Diseases*, p. 281. Ed. by L. W. Wannamaker and J. M. Matsen. Academic Press, New York; (1973) In: *Recent Advances in Clinical Pathology*, p. 293. Ed. by S. C. Dyke. Churchill, Edinburgh.

Cooper, G. and Walter, A. W. (1935) *Amer. J. publ. Hlth* **25,** 469.

Cowan, S. T. (1974) In *Cowan and Steel's Manual for the Identification of Medical Bacteria*, 2nd ed. Cambridge University Press, London.

Coykendall, A. L. (1974) *J. gen. Microbiol.* **83,** 327; (1977) *Int. J. syst. Bact.* **27,** 26.

Coykendall, A. L., Brathall, D., O'Connor, K. and Dvarskas, R. A. (1976) *Infect. Immun.*, **14,** 667.

Coykendall, A. L. and Specht, P. A. (1975) *J. gen. Microbiol.* **91,** 92.

Cromartie, W. J., Anderle, S. K., Schwab, J. H. and Dalldorf, F. G. (1979) In: *Pathogenic Streptococci*, p. 50. Ed. by M. T. Parker. Reedbooks, Chertsey, Surrey.

Cromartie, W. J., Craddock, J. G., Schwab, J. H., Anderle, S. K. and Yang, C.-H. (1977) *J. exp. Med.* **146,** 1585.

Crowley, N. (1944) *J. Path. Bact.*, **56,** 27; (1950) *J. gen. Microbiol.* **4,** 156; (1959) *J. Hyg., Camb.* **57,** 235.

Cullen, G. A. (1967) *Res. vet. Sci.* **8,** 83; (1969) *Vet. Bull.* **39,** 155.

Cullen, G. A. and Little, W. A. (1969) *Vet. Rec.* **85,** 115.

Cunningham, C. M. and Watson, D. W. (1978) *Infect. Immun.* **19,** 470.

Curtis, S. N. and Krause, R. M. (1964) *J. exp. Med.* **120,** 629.

Dajani, A. S. (1973) *J. Immunol.* **110,** 1702.

Dajani, A. S. and Wannamaker, L. W. (1970) *J. infect. Dis.* **122,** 196.

Darling, C. L. (1975) *J. clin. Microbiol.* **1,** 171.

Dawson, M. H. (1934) *J. Path. Bact.* **39,** 323.

Daynes, R. A. and Armstrong, C. H. (1973) *Infect. Immun.* **7,** 298.

Deibel, R. H. (1963) *J. Bact.*, **86,** 1270; (1964) *Bact. Rev.* **28,** 330.

Deibel, R. H. and Evans, J. B. (1960) *J. Bact.* **79,** 356.

Deibel, R. H., Lake, D. E. and Niven, C. F. (1963) *J. Bact.* **86,** 1275.

Deibel, R. H. and Niven, C. F. Jr. (1960) *J. Bact.* **79,** 175; (1964) *Ibid.* **88,** 4.

Deibel, R. H., Yao, J., Jacobs, R. H. and Niven, C. F. (1964) *J. infect. Dis.* **114,** 327.

Denny, F. W. and Thomas, L. (1953) *J. clin. Invest.* **32,** 1085.

Dick, T. B. and Gemmell, C. G. (1971) *J. med. Microbiol.* **4,** 153.

Dillon, H. C. and Wannamaker, L. W. (1965) *J. exp. Med.* **121,** 351.

Dixon, J. M. S. and Lipinski, A. E. (1972) *Antimicrob. Agents Chemother.* **1,** 333; (1974) *J. infect. Dis.* **130,** 351.

Dixon, J. M. S., Lipinski, A. E. and Graham, M. E. P. (1977) *Canad. med. Ass. J.* **117,** 1159.

Dochez, A. R., Avery, O. T. and Lancefield, R. C. (1919) *J. exp. Med.* **30,** 179.

Dochez, A. R. and Sherman, L. (1925) *Proc. Soc. exp. Biol., N.Y.* **22,** 282.

Döderlein. (1892) *Das Scheidensekret und seine Bedeutung für das Puerperalfieber*. Leipzig.

Dorff, G. J., Coonrod, J. D. and Rytel, M. W. (1971) *Lancet* **i,** 578.

Dunny, G. M., Brown, B. L. and Clewell, D. B. (1978) *Proc. nat. Acad. Sci., Wash.* **75,** 3479.

Eddy, B. E. (1944) *Publ. Hlth Rep., Wash.* **59,** 449, 451, 1041.

Edwards, P. R. (1934) *J. Bact.* **27,** 527.

Edwards, S. J. (1938) *J. comp. Path.* **51,** 250.

Efstratiou, A. and Maxted, W. R. (1979) *J. clin. Path.* **32,** 1228.

Ekstedt, R. D. and Stollerman, G. H. (1960) *J. exp. Med.* **112,** 671, 687.

El Kholy, A., Wannamaker, L. W. and Krause, R. M. (1974) *Appl. Microbiol.* **28,** 836.
Elliott, S. D. (1943) *Brit. J. exp. Path.* **24,** 159; (1945) *J. exp. Med.* **81,** 573; (1950) *Ibid.* **92,** 201; (1962) *Nature, Lond.* **193,** 1105; (1963) *Ibid.* **200,** 1184; (1966) *J. Hyg., Camb.* **64,** 205.
Elliott, S. D. and Dole, V. P. (1947) *J. exp. Med.* **85,** 305.
Elliott, S. D., McCarty, M. and Lancefield, R. C. (1977) *J. exp. Med.* **145,** 490.
Elliott, S. D. and Tai, J. Y. (1978) *J. exp. Med.* **148,** 1699.
El-Refaie, M. and Dulake, C. (1975) *J. clin. Path.* **28,** 801.
Eltz, R. W. and Vandemark, P. J. (1960) *J. Bact.* **79,** 763.
Embden, J. D. A. van, Soedirman, N. and Engel, H. W. B. (1978) *Lancet* **i,** 655.
Evans, A. C. (1934) *Publ. Hlth Rep., Wash.,* **49,** 1386; (1944) *J. Bact.* **48,** 263, 267.
Evans, J. B. and Niven, C. F. (1951) *J. Bact.* **62,** 599.
Evans, J. B. and Schultes, L. M. (1969) *Int. J. syst. Bact.* **19,** 159.
Facklam, R. R. (1972) *Appl. Microbiol.* **23,** 1131; (1977) *J. clin. Microbiol.* **5,** 184.
Facklam, R. R. and Edwards, L. R. (1979) In: *Pathogenic Streptococci,* p. 251. Ed. by M. T. Parker, Reedbooks, Chertsey, Surrey.
Facklam, R. R. and Moody, M. D. (1970) *Appl. Microbiol.* **20,** 245.
Fallon, R. J. (1974) *J. clin. Path.* **27,** 902.
Fehleisen. (1883) *Aetiologie des Erysipels.* Berlin.
Ferrieri, P. (1979) In: *Pathogenic Streptococci,* p. 164. Ed. by M. T. Parker. Reedbooks, Chertsey, Surrey.
Field, H. I., Buntain, D. and Done, J. T. (1954) *Vet. Rec.* **66,** 453.
Fine, D. P. (1975) *Infect. Immun.* **12,** 772.
Fischer, K., Poschmann, A. and Oster, H. (1971) *Mschr. Kinderheilk.* **119,** 2.
Fischetti, V. A. (1980) In: *Streptococcal disease and the immune response,* p. 111. Ed. by S. E. Read and J. B. Zabriskie. Academic Press, New York.
Fischetti, V. A., Gotschlich, E. C., Siviglia, G. and Zabriskie, J. B. (1976) *J. exp. Med.* **144,** 32.
Fox, E. N. (1974) *Bact. Rev.* **38,** 57.
Fox, E. N. and Krampitz, L. O. (1956) *J. Bact.* **71,** 454.
Fox, E. N. and Wittner, M. K. (1965) *Proc. nat. Acad. Sci., Wash.* **54,** 1118; (1966) *J. Immunol.* **97,** 86; (1969) *Immunochemistry* **6,** 11.
Fraenkel, A. (1886) *Z. klin. Med.* **10,** 401; (1886) *Ibid.* **11,** 437.
Fraser, C. A. M. (1982) *J. med. Microbiol.* **15,** 153.
Fraser, C. A. M. and Maxted, W. R. (1979) In: *Pathogenic Streptococci,* p. 259. Ed. by: M. T. Parker. Reedbooks, Chertsey, Surrey.
Friedberg, R. (1941) *Studier over Ikke-haemolytiske Streptokokker.* Einar Munksgaard, Copenhagen.
Freimer, E. H., Krause, R. M. and McCarty, M. (1959) *J. exp. Med.* **110,** 853.
Frost, W. D. and Engelbrecht, M. A. (1940) *The Streptococci.* Wildorf Book Co., Madison, Wis.
Fry, R. M. (1941) Unpublished.
Fuller, A. T. (1938) *Brit. J. exp. Path.* **19,** 130.
Fuller, A. T. and Maxted, W. R. (1939) *J. Path. Bact.* **49,** 83.
Garvie, E. I. (1967) *J. Dairy Res.* **34,** 31; (1978) *Int. J. syst. Bact.* **28,** 190.
Garvie, E. I. and Bramley, A. J. (1979) *J. appl. Bact.* **46,** 295.
Gasser, F., Mandel, M. and Rogosa, M. (1970) *J. gen. Microbiol.,* **62,** 219.
George, R. H. (1974) *J. med. Microbiol.* **7,** 77.
Gibbons, R. J., Kapsimalis, B. and Socransky, S. S. (1964) *Arch. oral Biol.* **9,** 101.
Ginsburg, I. (1970) In: *Microbial Toxins,* **3,** 100. Ed. by T. C. Montie, S. Kadis. and S. J. Ajl Academic Press, New York; (1972) *J. infect. Dis.* **126,** 294.
Ginsburg, I. and Harris, T. N. (1963) *J. exp. Med.,* **118,** 919; (1964) *Ergebn. Mikrobiol.* **38,** 198; (1965) *J. exp. Med.* **121,** 647.
Ginsburg, I., Harris, T. N. and Grossowicz, N. (1963) *J. exp. Med.* **118,** 905.
Glaser, J., Berry, J. W. and Loeb, L. H. (1953) *Proc. Soc. exp. Biol., N.Y.* **82,** 87.
Gooder, H. and Maxted, W. R. (1961) *Soc. gen. Microbiol., Symposium* No. XI, p. 151.
Gourlay, R. N. (1960) *J. comp. Path.* **70,** 339.
Gratten, M., Naraqi, S. and Hansman, D. (1980) *Lancet,* **ii,** 192.
Graudal, H. (1955) *Undersøgelser over Bevægliger Streptokokker.* Christtreus Bogtrykkeri, Copenhagen; (1957) *Acta path. microbiol. scand.* **41,** 403.
Green, N. E. (1979) *J. clin. Path.* **32,** 556.
Grigoroff, S. (1905) *Rev. méd. Suisse rom.* **25,** 714.
Griffith, F. (1926) *J. Hyg. Camb* **25,** 385; (1927) *Ibid.* **26,** 363; (1928) *Ibid.* **27,** 113; (1934) *Ibid.* **34,** 542; (1935) *Ibid.* **35,** 23.
Gross, K. C., Houghton, M. P. and Senterfit, L. B. (1975) *J. clin. Microbiol.* **1,** 54
Guggenheim, B. (1975) *Caries Res.* **9,** 309.
Guthof, O. (1955) *Zbl. Bakt.* **164,** 60; (1956) *Ibid.* **166,** 553; (1970) *Ibid.* **215,** 435.
Halbert, S. P. (1970) In: *Microbial Toxins,* **3,** 60. Ed. by T. C. Montie, S. Kadis and S. J. Ajl. Academic Press, New York.
Halbert, S. P. and Auerbach, T. (1961) *J. exp. Med.* **113,** 131.
Hallas, G. and Widdowson, J. P. (1979) In: *Pathogenic Streptococci,* p. 36. Ed. by M. T. Parker. Reedbooks, Chertsey, Surrey.
Handley, P. S. and Carter, P. (1979) In: *Pathogenic streptococci,* p. 241. Ed. by M. T. Parker. Reedbooks, Chertsey, Surrey.
Hansman, D. and Bullen, M. M. (1967) *Lancet* **ii,** 264.
Hardie, J. M. and Bowden, G. H. (1976) *J. dent. Res.* **55,** suppl. A, 166.
Hare, R. (1928) *Brit. J. exp. Path.* **9,** 337; (1932) *J. Path. Bact.* **35,** 701; (1935) *Ibid.* **41,** 499.
Hare, R. and Fry, R. M. (1938) *Vet. Rec.* **50,** 213.
Hart, P. D'A. and Anderson, A. B. (1933) *J. Path. Bact.* **37,** 91.
Hartman, P. A., Reinbold, G. W. and Saraswat, D. S. (1966) *Int. J. bact. Nomencl.* **16,** 197.
Haxthausen, H. (1927) *Ann. Derm. Syph., Paris* **8,** 201.
Hayano, S. and Tanaka, A. (1969) *J. Bact.* **97,** 1328.
Hehre, E. J. and Neill, J. M. (1946) *J. exp. Med.* **83,** 147.
Heidelberger, M. (1938) *Rev. Immunol.* **4,** 293.
Heidelberger, M. and Avery, O. T. (1923) *J. exp. Med.* **38,** 73.
Heidelberger, M. and Nimmich, N. (1976) *Immunochemistry* **13,** 67.
Heinemann, P. G. and Hefferan, M. (1909) *J. infect. Dis.* **6,** 304.
Hemming, V. G., Hall, R. T., Rhodes, P. G., Shigeoka, A. O. and Hill, H. R. (1976) *J. clin. Invest.* **58,** 1379.

Henrichsen, J. (1979) *J. Infect.* **1** (suppl. 2), 31.
Henriksen, S. D. and Eriksen, J. (1978) *Immunochemistry* **15,** 761.
Henriksen, S. D. and Henrichsen, J. (1975) *Acta path. microbiol. scand.* **B83,** 133.
Herbert, D. and Todd, E. W. (1944) *Brit. J. exp. Path.* **25,** 242.
Hewitt, L. F. and Todd, E. W. (1939) *J. Path. Bact.* **49,** 45.
Hill, M. J. and Wannamaker, L. W. (1968) *J. Hyg., Camb.* **66,** 37.
Hirst, G. K. (1941) *J. exp. Med.* **73,** 493.
Hirst, G. K. and Lancefield, R. C. (1939) *J. exp. Med.* **69,** 425.
Hitchcock, C. H. (1924) *J. exp. Med.* **40,** 445.
Holdeman, L. V., Cato, E. P. and Moore, W. E. C. (1977) *Anaerobe Laboratory Manual,* 4th ed. Virginia Polytechnic Institute, Blacksberg, Virginia.
Holdeman, L. V. and Moore, W. E. C. (1974) *Int. J. syst. Bact.* **24,** 260.
Holm, S. E. and Christensen, P. [Eds] (1982) *Basic Concepts of Streptococci and Streptococcal Diseases.* Reedbooks, Chertsey, Surrey.
Holm, S. E. and Falsen, E. (1967) *Acta path. microbiol. scand.* **69,** 264.
Hook, E. W., Wagner, R. R. and Lancefield, R. C. (1960) *Amer. J. Hyg.* **72,** 111.
Hooker, S. B. and Follensby, E. M. (1934) *J. Immunol.* **27.** 177.
Hryniewicz, W., Gray, E. D., Tagg, J. R., Wannamaker, L. W., Kanclerski, K. and Laible, N. (1978) *Zbl. Bakt.,* I Abt. Orig. **A242,** 327.
Islam, A. K. M. S. (1977) *Lancet,* **i,** 256.
Jacobs, M. R. *et al.* (1978) *New Engl. J. Med.* **299,** 735.
Jamieson, S. (1950) *Vet. Rec.* **62,** 772.
Jelinková, J. (1977) *Curr. Top. Microbiol. Immunol.* **76,** 126.
Johnson, B. H. and Furrer, W. (1958) *J. infect. Dis.* **103, 135.**
Johnson, D. W., Tagg, J. R. and Wannamaker, L. W. (1979) *J. med. Microbiol.* **12,** 413.
Johnston, R. B. Jr., Klemperer, M. R. Alper, C. A. and Rosen, F. S. (1972) *J. exp. Med.* **129,** 1275.
Jones, D., Deibel, R. H. and Niven, C. F. (1964) *J. Bact.* **88,** 602.
Kantor, F. S. and Cole, R. M. (1960) *J. exp. Med.* **112,** 77.
Kaplan, M. H. (1944) *Proc. Soc. exp. Biol., N.Y.* **57,** 40; (1946) *J. clin. Invest.* **25,** 337.
Karakawa, W. W., Wagner, J. E. and Pazur, J. H. (1971) *J. Immunol.* **107,** 554.
Kelly, R. T., Farmer, S. and Greiff, D. (1967) *J. Bact.* **94,** 272.
Kelsey, M. C. and Reed, C. S. (1979) *J. clin. Path.* **32,** 960.
Kendall, F. E., Heidelberger, M. and Dawson, M. H. (1937) *J. biol. Chem.* **118,** 61.
Keogh, E. V. and Simmons, R. T. (1940) *J. Path. Bact.* **50,** 137.
Kern. (1881) *Bull. Soc. Nat. Mosou.* No. 3.
Kiel, P. and Skadhauge, K. (1973) *Acta path. microbiol. scand.* **B81,** 10.
Kim, Y. B. and Watson, D. W. (1970) *J. exp. Med.,* **131,** 611; (1972) In: *Streptococci and Streptococcal Diseases,* p. 33. Ed. by L. W. Wannamaker and J. M. Matsen. Academic Press, New York.
Kitt. (1893) see Minett (1935).
Kludas, M. and Dobberstein, H. (1959) *Zbl. Bakt.* **175,** 520.
Kodama, T. (1936) *Kitasato Arch.* **13,** 217.
Kohn, J., Hewitt, J. H. and Fraser, C. A. M. (1968) *Brit. med. J.* **i,** 703.
Koser, S. A. and Thomas, J. L. (1957) *J. Bact.* **73,** 477.
Koskela, M. and Leinonen, M. (1981) *J. clin. Path.* **34,** 93.
Krasner, R. I. and Jannach, J. R. (1963) *J. infect. Dis.* **112,** 134.
Krause, R. M. (1958) *J. exp. Med.* **108,** 803; (1963) *Bact. Rev.* **27,** 369; (1975) *Z. ImmunForsch.* **149,** 136.
Krause, R. M. and McCarty, M. (1961) *J. exp. Med.* **114,** 127.
Kreger, A. S. and Bernheimer, A. W. (1969) *J. Bact.* **98,** 306.
Kronvall, G. (1973) *J. med. Microbiol.* **6,** 187.
Krumwiede, E. (1954) *J. exp. Med.* **100,** 629.
Kuharic, H. A., Roberts, C. E. and Kirby, W. M. (1960) *J. Amer. med. Ass.* **174,** 1779.
Kulp, W. L. (1926) *Science,* **64,** 204.
Kundrat, W. (1958) *Zentbl. Bakt., IIte Abt.* **111,** 249.
Kurl, D. N. (1981) *Lancet,* **ii,** 752.
Lancefield, R. C. (1928) *J. exp. Med.* **47,** 91, 469, 481, 843, 857; (1933) *Ibid.* **57,** 571; (1934) *Ibid.* **59,** 441; (1938) *Ibid.* **67,** 25; (1940) *Ibid.* **71,** 521, 539; (1941) *Harvey Lectures,* Ser. 36, 251; (1943) *J. exp. Med.* **78,** 465; (1957) *Ibid.* **106,** 525; (1958) *Ibid.* **108,** 329; (1962) *J. Immunol.* **89,** 307; (1972) In: *Streptococci and Streptococcal Diseases,* p. 313. Ed. by L. W. Wannamaker and J. M. Matsen. Academic Press, New York.
Lancefield, R. C. and Freimer, E. H. (1966) *J. Hyg. Camb.* **64,** 191.
Lancefield, R. C. and Hare, R. (1935) *J. exp. Med.* **61,** 335.
Lancefield, R. C., McCarty, M. and Everly, W. N. (1975) *J. exp. Med.* **142,** 165.
Lancefield, R. C. and Perlman, G. E. (1952*a*) *J. exp. Med.* **96,** 71; (1952*b*) *Ibid.* **96,** 83.
Lancefield, R. C. and Todd, E. W. (1928) *J. exp. Med.* **48,** 769.
Langston, C. W. and Bouma, C. (1960) *Appl. Microbiol..* **8,** 212.
Langston, C. W., Gutierrez, J. and Bouma, C. (1960) *J. Bact.* **80,** 714.
Latham, M. J. and Jayne-Williams, D. J. (1978) In: *Streptococci,* p. 207. Ed by F. A. Skinner and L. B. Quesnel. Academic Press, London.
Laughton, N. (1948) *J. Path. Bact.* **60,** 471.
Lazarides, P. D. and Bernheimer, A. W. (1957) *J. Bact.* **74,** 412.
Lehner, T, Russell, M. W. and Caldwell, J. (1980) *Lancet,* **1,** 995.
Leichmann, G. (1896) *Zbl. Bakt.,* II Abt. **2,** 281.
Lerche, M. and Reuter, G. (1960) *Zbl. Bakt.* **179,** 354; (1962) *Ibid.* **185,** 466.
Levaditi, C. (1918) *C. R. Soc. Biol.* **81,** 1064.
Lister, F. S. (1916) *Publ. S. Afr. Inst. med. Res.* No. 8.
Liu, T. Y. and Elliott, S. D. (1965) *Nature, Lond.* **206,** 33.
London, J. (1976) *Ann. Rev. Micriobiol.* **30,** 279.
Long, P. H. and Bliss, E. A. (1934) *J. exp. Med.* **60,** 619.
Lorian, V., Waluschka, A. and Popoola, B. (1973) *Appl. Microbiol.* **25,** 290.
Lowbury, E. J. L. and Hurst, L. (1959) *J. clin. Path.* **12,** 163.
Lowbury, E. J. L. and Kidson, A. (1968) *Brit. med. J.* **ii,** 490.
Lowbury, E. J. L., Kidson, A. and Lilly, H. A. (1964) *J. clin. Path.* **17,** 231.
Lund, E. (1959) *Acta path. microbiol. scand.* **47,** 308; (1963) *Ibid.* **59,** 533; (1970) *Int. J. syst. Bact.* **20,** 321.

Lütticken, R. and Laufs, R. (1979) In: *Pathogenic streptococci*, p. 279. Ed. by M. T. Parker. Reedbooks, Chertsey, Surrey.
Lütticken, R., Lütticken, D., Johnson, D. R. and Wannamaker, L. W. (1976) *J. clin. Microbiol.* **3,** 533.
Lütticken, R., Wendorff, U., Lütticken, D., Johnson, E. A. and Wannamaker, L. W. (1978) *J. med. Microbiol.* **11,** 419.
McCandless, R. G., Cohen, M., Kalmanson, G. M. and Guze, L. B. (1968) *J. Bact.* **96,** 1400.
McCandless, R. G., Hensley, I. J., Cohen, M., Kalmanson, G. M. and Guze, L. B. (1971) *J. gen. Microbiol.* **68,** 357.
McCarty, M. (1948) *J. exp. Med.* **88,** 181; (1952) *Ibid.* **96,** 555, 569; (1956) *Ibid.* **104,** 629; (1958) *Ibid.* **108,** 311; (1959) *Ibid.* **109,** 361; (1964) *Proc. nat. Acad. Sci., Wash.* **52,** 259; (1966) *J. Hyg., Camb.* **64,** 185.
McCarty, M. and Lancefield, R. C. (1955) *J. exp. Med.* **102,** 11.
McClean, D. (1941) *J. Path. Bact.* **53,** 13; (1942) *Ibid.* **54,** 284
MacDonald, I. (1939) *Med. J. Aust.* **ii,** 471.
McIntosh, J., James, W. W. and Lazarus-Barlow, P. (1922) *Brit. J. exp. Path.* **3,** 138; (1924) *Ibid.* **5,** 175.
McLean, S. J. (1955) *Aust. J. exp. Biol. Med.* **33,** 275.
MacLeod, C. M. and Krauss, M. R. (1950) *J. exp. Med.* **92,** 1.
McLeod, J. W. and Gordon, J. (1922) *Biochem. J.* **16,** 499.
Mair, W. (1917) *J. Path. Bact.* **21,** 305; (1929) *A System of Bacteriology, Med. Res. Coun.*, London, **2,** 168.
Malke, H., Starke, R., Jacob, H. E. and Köhler, W. (1974) *J. med. Microbiol.* **7,** 367.
Malke, H., Starke, R., Köhler, W., Kolesnichenko, G. and Totolian, A. A. (1975) *Zbl. Bakt.* **A233,** 24.
Man, J. C. de, Rogosa, M. and Sharpe, M. E. (1960) *J. appl. Bact.* **23,** 130.
Marmorek, A. (1895) *Ann. Inst. Pasteur* **9,** 593.
Mathews, J. H., Klesius, P. H. and Zimmerman, R. A. (1974) *Infect. Immun.* **10,** 1315.
Mattingly, S. J., Milligan, T. W., Pierpont, A. A. and Straus, D. C. (1980) *J. clin. Microbiol.* **12,** 633.
Maxted, W. R. (1948) *Lancet* **ii,** 255; (1949) *J. gen. Microbiol.* **3,** 1; (1953*a*) *J. Path. Bact.* **65,** 345; (1953*b*) *J. clin. Path.* **6,** 224, (1955) *J. gen. Microbiol.* **12,** 484; (1956) *Brit. J. exp. Path.* **37,** 415; (1957) *J. gen. Microbiol.* **16,** 584; (1964) In: *Rheumatic Fever with Glomerulonephritis*, p. 25. Ed. by J. W. Uhr. Williams and Wilkins, Baltimore; (1978) In: *Streptococci*, p. 107. Ed. by F. A. Skinner and L. B. Quesnel. Academic Press, London.
Maxted, W. R. and Valkenburg, H. A. (1969) *J. med. Microbiol.* **2,** 199
Maxted, W. R. and Widdowson, J. P. (1972) In: *Streptococci and Streptococcal Diseases*, p. 251. Ed. by L. W. Wannamaker and J. M. Matsen. Academic Press, New York.
Maxted, W. R., Widdowson, J. P. and Fraser, C. A. M. (1973*a*) *J. Hyg., Camb.* **71,** 35.
Maxted, W. R., Widdowson, J. P., Fraser, C. A. M., Ball, L. C. and Bassett, D. C. J. (1973*b*) *J. med. Microbiol.* **6,** 83.
Mead, G. C. (1974) *J. gen. Microbiol.*, **82,** 241; (1978) In: *Streptococci*, p. 245. Ed. by F. A. Skinner and L. B. Quesnel. Academic Press, London.
Medrek, T. F. and Barnes, E. M. (1962*a*) *J. gen. Microbiol.* **28,** 701; (1962*b*) *J. appl. Bact.* **25,** 169.
Mereshkowsky, S. S. (1905) *Zbl. Bakt.* **39,** 380, 584, 696; (1906) *Ibid.* **40,** 118.
Meyer, K., Hobby, G. L., Chaffee, E. and Dawson, M. H. (1940) *J. exp. Med.* **71,** 137.
Michel, M. F. and Krause, R. M. (1967) *J. exp. Med.* **125,** 1075.
Michel, M. F., Moor, C. E. de, Ottens, H., Willers, J. M. N. and Winkler, K. C. (1967) *J. gen. Microbiol.* **49,** 49.
Milligan, T. W., Baker, C. J., Straus, D. C. and Mattingly, S. J. (1978) *Infect. Immun.* **21,** 738.
Minett, F. C. (1935) *Proc. Twelfth Int. Vet. Congr.* p. 511; (1936) *J. Hyg., Camb.* **35,** 504.
Mirick, G. S. *et al.* (1950) *Amer. J. Hyg.* **52,** 48.
Mirick, G. S., Thomas, L., Curnen, E. C. and Horsfall, F. L. (1944) *J. exp. Med.* **80,** 391, 407, 431.
Mitsuoka, T. (1969) *Zbl. Bakt.*, **210,** 32.
Miyamoto, Y., Takizawa, K., Matsushima, A., Asai, Y. and Nakatsuka, S. (1978) *Antimicrob. Agents Chemother.* **13,** 399.
Moberg, K. and Thal, E. (1954) *Nord. VetMed.* **6,** 69.
Moor, C. E. de (1963) *Leeuwenhoek ned. Tijdschr.* **29,** 272.
Moor, C. E. de and Thal, E. (1968) *Leeuwenhoek ned. Tijdschr.* **34,** 377.
Moore, H. F. (1915) *J. exp. Med.* **22,** 269.
Mørch, E. (1943) *Serological Studies on the Pneumococci.* Humphrey Milford, London.
Moreira-Jacob, M. (1956) *J. gen. Microbiol.* **14,** 268.
Morgenroth, J. and Levy, R. (1911) *Berl. klin. Wschr.* **48,** 1560, 1979.
Moro, E. (1900*a*) *Jb. Kinderheilk.* **52,** 38; (1900*b*) *Wien. klin. Wschr.* **13,** 114.
Mosser, J. L. and Tomasz, A. (1970) *J. biol. Chem.* **245,** 287.
Müller, H. E. (1969) *Dtsch. med. Wschr.* **94,** 2149; (1974) *Infect. Immun.* **9,** 323.
Mundt, J. O. and Graham, W. F. (1968) *J. Bact.* **95,** 2005.
Mundt, J. O. and Johnson, A. H. (1959) *Food Res.* **24,** 218.
Myhre, E. B., Holmberg, O. and Kronvall, G. (1979) *Infect. Immun.* **23,** 1.
Myhre, E. B. and Kronvall, G. (1977) *Infect. Immun.* **17,** 475.
Nahm, N. H., Murray, P. R., Clevinger, B. L. and Davie, J. M. (1980) *J. clin. Microbiol.* **12,** 506.
Naylor, J. and Sharpe, M. E. (1958) *J. Dairy Res.* **25,** 92.
Neill, J. M. (1926) *J. exp. Med.* **44,** 199.
Neill, J. M. and Mallory, T. B. (1926) *J. exp. Med.* **44,** 241.
Neufeld, F. (1900) *Z. Hyg. InfektKr.* **34,** 454; (1902) *Ibid.* **40,** 54.
Neufeld, F. and Etinger-Tulczynska, R. (1931) *Z. Hyg. InfektKr.* **112,** 492.
Neufeld, F. and Händel, L. (1909) *Arb. ReichsgesundhAmt.* **34,** 293.
Niven, C. F., Smiley, K. L. and Sherman, J. M. (1941*a*) *J. Bact.* **41,** 479; (1941*b*) *J. biol. Chem.* **140,** 105.
Niven, C. F. Jr. and White, J. C. (1946) *J. Bact.* **51,** 790.
Nocard and Mollereau. (1887) *Ann. Inst. Pasteur* **1,** 109.
Nowlan, S. S. and Deibel, R. H. (1967) *J. Bact.* **94,** 291.
Ofek, I. and Beachey, E. H. (1979) In: *Pathogenic Streptococci*, p. 44. Ed. by M. T. Parker. Reedbooks, Chertsey, Surrey.
Ofek, I., Beachey, E. H., Jefferson, W. and Campbell, G. L. (1975) *J. exp. Med.*, **141,** 990.
Ofek, I., Bergner-Rabinowitz, S. and Ginsburg, I. (1970) *J. infect. Dis.* **122,** 517.
Ohanian, S. H., Schwab, J. H. and Cromartie, W. J. (1969) *J. exp. Med.* **129,** 37.
Okamoto, H. (1940) *Jap. J. med. Sci.* Sect. 4, **12,** 167.
Oppler, B. (1895) *Dtsch. med. Wschr.* **21,** 73.
Oram, F. (1934) *J. Immunol.* **26,** 233.

Orla-Jensen. (1904) *Zbl. Bakt.*, IIte Abt. **13**, 161, 291, 428, 514, 604, 687, 753.
Orla-Jensen, S. (1919) *The Lactic Acid Bacteria.* Copenhagen; (1943) *The Lactic Acid Bacteria.* Einar Munksgaard, Copenhagen.
Osterland, C. K., Miller, E. J., Karakawa, W. W. and Krause, R. M. (1966) *J. exp. Med.* **123,** 599.
Ottens, H. and Winkler, K. C. (1962) *J. gen. Microbiol.* **28,** 181.
Pakula, R. (1951) *J. gen. Microbiol.* **5,** 640.
Parish, H. J. and Okell, C. C. (1930) *J. Path. Bact.* **33,** 527.
Parker, M. T. (1967) *Bull. Wld Hlth Org.*, **37,** 513; (1978) In: *Streptococi*, p. 71. Ed. by F. A. Skinner and L. B. Quesnel. Academic Press, London; [Ed.] (1979) *Pathogenic Streptococci.* Reedbooks, Chertsey, Surrey.
Parker, M. T. and Ball, L. C. (1976) *J. med. Microbiol.* **9,** 275; (1979) In: *Pathogenic Streptococci*, p. 234. Ed. by M. T. Parker. Reedbooks, Chertsey, Surrey.
Parker, M. T., Maxted, W. R. and Fraser, C. A. M. (1962) *Lancet*, **i,** 1550.
Pasteur, L., Chamberland, C. and Roux, E. (1881) *C. R. Acad. Sci.* **92,** 159.
Pattison, I. H., Matthews, P. R. J. and Maxted, W. R. (1955) *J. Path. Bact.* **69,** 43.
Pederson, C. S. (1936) *J. Bact.* **31,** 217.
Pederson, C. S. and Albury, M. N. (1955) *J. Bact.* **70,** 702.
Perch, B., Kjems, E. and Ravn, T. (1974) *Acta path. microbiol. scand.* **B82,** 357.
Perch, B., Kristjansen, P. and Skadhauge, K. (1968) *Acta path. microbiol. scand.* **74,** 69.
Perch, B. and Olsen, S. J. (1964) *Nord. Vet. Med.* **16,** 241.
Petrie, G. F. (1933) *Vet. J.* **89,** 25.
Petrow, N. P. (1907) *Zbl. Bakt.* **43,** 349.
Pike, R. M. (1948) *J. infect. Dis.* **83,** 1, 12, 19.
Pinney, A. M. and Widdowson, J. P. (1977) *J. med. Microbiol.* **10,** 415.
Pinney, A. M., Widdowson, J. P. and Maxted, W. R. (1977) *J. Hyg., Camb.* **78,** 355.
Poole, P. M. and Wilson, G. (1979) *J. clin. Path.* **32,** 764.
Porterfield, J. S. (1950) *J. gen. Microbiol.* **4,** 92.
Rantasalo, I. (1947) *Ann. Med. intern. Fenn.* **36,** 341.
Rantz, L. A. (1942) *J. infect. Dis.* **71,** 61.
Rantz, L. A. and Randall, E. (1955) *Stanford med. Bull.* **13,** 290.
Raška, K. and Rotta, J. (1963) *J. Hyg. Epidem., Praha* **7,** 319.
Report. (1945) *J. Amer. med. Ass.* **129,** 921; (1947) *J. exp. Med.* **85,** 441; (1954) *Int. Bull. bact. Nomencl.* **4,** 145; (1971) *Int. J. syst. Bact.* **21,** 100; (1977) *Brit. med. J.* **i,** 131.
Reyn, A. (1970) *Int. J. syst. Bact.* **20,** 19.
Rifkind, D. and Cole, R. M. (1962) *J. Bact.* **84,** 163.
Rijn, I. van de, Bleiweis, A. S. and Zabriskie, J. B. (1976) *J. dental Res.* **55,** suppl C, p. 59.
Rochaix, A. (1924) *C. R. Soc. Biol.* **90,** 771.
Röd, T. O., Haug, R. H. and Midtvedt, T. (1974) *Acta path. microbiol. scand.* **B82,** 533.
Rogosa, M. (1970) *Int. J. syst. Bact.*, **20,** 519; (1974*a*) In: *Bergey's Manual of determinative bacteriology*, 8th edn, p. 517. Ed. by R. E. Buchanan and N. E. Gibbons, Williams and Wilkins, Baltimore; (1974*b*) *Ibid.* p. 576.
Rogosa, M., Franklin, J. G. and Perry, K. D. (1961) *J. gen. Microbiol.* **25,** 473.
Rogosa, M., Mitchell, J. A. and Wiseman, R. F. (1951) *J. Bact.* **62,** 132.
Rogosa, M. and Sharpe, M. E. (1959) *J. appl. Bact.* **22,** 329.
Rogosa, M., Wiseman, R. F., Mitchell, J. A., Disraely, M. N. and Beaman, A. J. (1953) *J. Bact.* **65,** 681.
Roguinsky, M. (1971) *Ann. Inst. Pasteur* **120,** 154.
Rosan, B. and Appelbaum, B. (1979) In: *Pathogenic Streptococci*, p. 208. Ed. by M. T. Parker. Reedbooks, Chertsey, Surrey.
Rosenbach, F. J. (1884) *Microorganismen bei den Wundinfektionskrankheiten.* Wiesbaden.
Rosendal, K. (1956) *Acta path. microbiol. scand.* **39,** 127.
Rothbard, S. (1948) *J. exp. Med.* **88,** 325.
Rotta, J. (1967) *Folia microbiol., Praha* **12,** 255.
Rotta, J. and Jelinková, J. (1967) *Int. J. syst. Bact.* **17,** 297.
Rotta, J., Krause, R. M., Lancefield, R. C., Everly, W. and Lackland, H. (1971) *J. exp. Med.* **134,** 1298.
Rotta, J. and Raška, K. (1963) *J. Hyg. Epidem., Praha* **7,** 16.
Russell, R. R. B. (1980) *J. gen. Microbiol.* **118,** 383.
Safford, C. E., Sherman, J. M. and Hodge, H. M. (1937) *J. Bact.* **33,** 263.
Schaffer, M. F., Enders, J. F. and Wu, C.-J. (1936) *J. exp. Med.* **64,** 281.
Schleifer, K. H. and Heymer, B. (Eds.) (1975) *Z. Immun. Forsch.* **149,** 103.
Schleifer, K. H. and Kandler, O. (1972) *Bact. Rev.* **36,** 407.
Schlievert, P. M., Bettin, K. M. and Watson, D. W. (1979) *Infect. Immun.* **26,** 467.
Schmidt, W. C. (1952) *J. exp. Med.* **95,** 105; (1965) *Ibid.* **121,** 771.
Schottmüller, H. (1903) *Münch. med. Wschr.* **50,** 849, 909.
Schütz, W. (1887) *Arch. wiss. prakt. Tierheilk.* **13,** 27; (1888) *Ibid.* **14,** 456.
Schwab, J. H. (1956) *J. Bact.* **71,** 94, 100.
Schwab, J. H. and Cromartie, W. J. (1957) *J. Bact.* **74,** 673.
Schwab, J. H., Watson, D. W. and Cromartie, W. J. (1953). *Proc. Soc. exp. Biol., N.Y.* **82,** 754.
Seastone, C. V. (1934) *J. Bact.*, **28,** 481; (1939) *J. exp. Med.* **70,** 361; (1943) *Ibid.* **77,** 21.
Seger, R., Joller, P., Baerlocher, K. and Hitzig, W. H. (1980) *Schweiz. med. Wschr.* **110,** 1454.
Sharpe, M. E. (1955*a*) *J. gen. Microbiol.* **12,** 107; (1955*b*) *Ibid.* **13,** 198; (1962) *Dairy Sci Abst.* **24,** 109; (1964) *J. gen. Microbiol.* **36,** 151; (1970) *Int. J. syst. Bact.*, **20,** 509; (1979) *J. Soc. Dairy Technol.*, **32,** 9; (1981) In: *The Prokaryotes*, **2,** 1653. Ed. by M. P. Starr *et al.* Springer, Berlin, Heidelberg and New York.
Sharpe, M. E., Brock, J. H., Knox, K. W. and Wicken, A. J. (1973*b*) *J. gen. Microbiol.* **74,** 119.
Sharpe, M. E. and Fewins, B. G. (1960) *J. gen. Microbiol.* **23,** 621.
Sharpe, M. E., Hill, L. R. and Lapage, S. P. (1973*a*) *J. med. Microbiol.* **6,** 281.
Sharpe, M. E. and Shattock, P. M. F. (1952) *J. gen. Microbiol.* **6,** 150.
Sharpe, M. E. and Wheater, D. M. (1957) *J. gen. Microbiol.* **16,** 676.
Shattock, P. M. F. (1949) *J. gen. Microbiol.* **3,** 80.
Shattock, P. M. F. and Hirsch, A. (1947) *J. Path. Bact.* **59,** 495.
Shattock, P. M. F. and Mattick, A. T. R. (1943) *J. Hyg., Camb.* **43,** 173.
Sherman, J. M. (1937) *Bact. Rev.* **1,** 3.
Sherman, J. M., Niven, C. F. and Smiley, K. L. (1943) *J. Bact.* **45,** 249.
Sherman, J. M., Smiley, K. L. and Niven, C. F. (1940) *J. Dairy Sci.* **23,** 529.

Sherman, J. M. and Wing, H. V. (1935) *J. Dairy Sci.* **18,** 657; (1937) *Ibid.* **20,** 165.
Sherwood, N. P., Paretsky, D., Nachtigall, A., McLain, A. R. and Trufelli, G. T. (1954) *J. infect. Dis.* **95,** 1.
Shimohashi, H. and Mutai, M. (1977) *J. gen. Microbiol.* **103,** 337.
Simmons, R. T. and Keogh, E. V. (1940) *Aust. J. exp. Biol. med. Sci.* **18,** 151.
Skadhauge, K. (1950) *Studies on Enterococci.* Einar Munksgaard, Copenhagen.
Skadhauge, K. and Perch, B. (1959) *Acta path. microbiol. scand.* **46,** 239.
Skinner, F. A. and Quesnel, L. B. [Eds.] (1978) '*Streptococci*' (*Soc. appl. Bact. Symp. Ser. no. 7*). Academic Press, London.
Slade, H. D. and Slamp, W. C. (1956) *J. Bact.* **71,** 624.
Smialowicz, R. J. and Schwab, J. H. (1977) *Infect. Immun.* **17,** 591; (1978) *Ibid.* **20,** 258.
Smith, D. G. and Shattock, P. M. F. (1962) *J. gen. Microbiol.* **29,** 731; (1964) *Ibid.* **34,** 165.
Smith, T. and Brown, J. H. (1915) *J. med. Res.* **31,** 455.
Snell, E. E. (1952) *Bact. Rev.* **16,** 235.
Stableforth, A. W. (1942) *Proc. R. Soc. Med.* **35,** 625; (1959) In: *Infectious Diseases of Animals due to Bacteria,* Vol. 2, p. 589. Ed. A. W. Stableforth and I. A. Galloway. Butterworth, London.
Stark, J. M. (1960) *Lancet* **i,** 733.
Stephenson, M. and Rowatt, E. (1947) *J. gen. Microbiol.* **1,** 279.
Stewardson-Krieger, P., Allbrandt, K., Kretschmer, R. R. and Gotoff, S. P. (1977) *Infect. Immun.* **18,** 666.
Stewart, F. S. and Martin, W. T. (1962) *J. Path. Bact.* **84,** 251.
Stollerman, G. H. and Ekstedt, R. (1957) *J. exp Med.* **106,** 345.
Stringer, J. (1980) *J. med. Microbiol.* **13,** 133.
Stryker, L. M. (1916) *J. exp. Med.* **24,** 49.
Swanson, J., Hsu, K. C. and Gotschlich, E. C. (1969) *J. exp. Med.* **130,** 1063.
Swanson, J. and McCarty, M. (1969) *J. Bact* **100,** 505.
Swift, H. F., Wilson, A. T. and Lancefield, R. C. (1943) *J. exp. Med.* **78,** 127.
Tagg, J. R. and Bannister, L. V. (1979) *J. med. Microbiol.* **12,** 397.
Tagg, J. R., Dajani, A. S. and Wannamaker, L. W. (1976) *Bact. Rev.* **40,** 722.
Tai, J. Y. and Gotschlich, E. C. (1979) *J. exp. Med.* **149,** 58.
Tan, E. M. and Kaplan, M. H. (1962) *J. infect. Dis.* **110,** 55.
Taylor, A. G. (1971) *J. gen. Microbiol.* **65,** 193.
Thal, E. and Moberg, K. (1953) *Nord. VetMed.* **5,** 835.
Thal, E. and Söderlind, O. (1966) *Proc. 10th Nord. vet. Congr., Stockh.* p. 336.
Thiercelin. (1899) *C. R. Soc. Biol.* **5,** 269.
Thjøtta, T. and Bøe, J. (1938) *Acta path. microbiol. scand.* Suppl. **37,** 527.
Thompson, A., Halbert, S. P. and Smith, U. (1970) *J. exp. Med.* **131,** 745.
Tieffenberg, J., Vogel, L., Kretschmer, R. R., Padnos, D. and Gotoff, S. P. (1978) *Infect. Immun.* **19,** 481.
Tillett, W. S. (1927) *J. exp. Med.* **45,** 1093.
Tillett, W. S. and Garner, R. L. (1933) *J. exp. Med.* **58,** 485.
Tillett, W. S., Goebel, W. F. and Avery, O. T. (1930) *J. exp. Med.* **52,** 95.
Tillett, W. S., Sherry, S. and Christensen, L. R. (1949) *Proc. Soc. exp. Biol., N.Y.* **68,** 184.
Toala, P., MacDonald, A. Wilcox, C. and Finland, M. (1969) *Amer. J. med. Sci.* **258,** 416.
Todd, E. W. (1927*a*) *Brit. J. exp. Path.* **8,** 1; (1927*b*) *Ibid.* **8,** 289; (1928*a*) *Ibid.* **9,** 1; (1928*b*) *J. exp. Med.* **48,** 493; (1932) *Ibid.* **55,** 267; (1934) *J. Path. Bact.* **39,** 299; (1938*a*) *Ibid,* **47,** 423; (1938*b*) *Brit. J. exp. Path.* **19,** 367; (1939) *J. Hyg., Camb.* **39,** 1; (1941) *Brit. J. exp. Path.* **22,** 172; (1942) *Ibid.* **23,** 136.
Todd, E. W. and Hewitt, L. F. (1932) *J. Path. Bact.* **35,** 973.
Todd, E. W. and Lancefield, R. C. (1928) *J. exp. Med.* **48,** 751.
Todd, E. W., Laurent, L. J. M. and Hill, N. G. (1933) *J. Path. Bact.* **36,** 201.
Tomasz, A. (1967) *Science,* **157,** 694; (1968) *Proc. nat. Acad. Sci., Wash.* **59,** 86.
Top, F. H. Jr and Wannamaker, L. W. (1968*a*) *J. Hyg., Camb.* **66,** 49; (1968*b*) *J. exp. Med.* **127,** 1013.
Vosti, K. L. and Rantz, L. A. (1964) *J. Immunol.* **92,** 185.
Wagner, M., Wagner, B. and Rýc, M. (1978) *J. gen. Microbiol.* **108,** 283.
Waitkins, S. A. (1978) *J. clin. Path.* **31,** 692.
Wannamaker, L. W. (1958) *J. exp. Med.* **107,** 797; (1959) *Amer. J. Med.* **27,** 567.
Wannamaker, L. W., Hayes, B. and Yasmineh, W. (1967) *J. exp. Med.* **126,** 497.
Wannamaker, L. W. and Matsen, J. M. (Eds.) (1972) *Streptococci and Streptococcal Diseases: Recognition, Understanding, and Management.* Academic Press, New York.
Ward, P. A. (1967) *J. exp. Med.* **126,** 189.
Watson, D. W. (1960) *J. exp. Med.,* **111,** 255; (1979) In: *Pathogenic Streptococci,* p. 62. Ed. by M. T. Parker. Reedbooks, Chertsey, Surrey.
Watson, D. W. and Kim, Y. B. (1970) In: *Microbial Toxins,* **3,** 69. Ed. by Montie, T. C., Kadis, S. and Ajl, S. J. Academic Press, New York.
Watson, R. F., Rothbard, S. and Swift, H. F. (1946) *J. exp. Med.* **84,** 127.
Weichselbaum, A. (1886) *Med. Jb.* **i,** 483.
Weld, J. T. (1934) *J. exp. Med.* **59,** 83, (1935) *Ibid.* **61,** 473.
Wensinck, F. and Renaud, H. (1957) *Brit. J. exp. Path.* **38,** 489.
Werner, H. and Seeliger, H. P. R. (1964) *Path. et Microbiol., Basel,* **27,** 202.
Wheater, D. M. (1955) *J. gen. Microbiol.* **12,** 123, 133.
White, J. C. and Niven, C. F. (1946) *J. Bact* **51,** 717.
Whittenbury, R. (1965) *J. gen. Microbiol.* **40,** 97.
Widdowson, J. P., Maxted, W. R. and Grant, D. L. (1970) *J. gen. Microbiol.* **61,** 343.
Widdowson, J. P., Maxted, W. R., Grant, D. L. and Pinney, A. M. (1971*a*) *J. gen. Microbiol.* **65,** 69.
Widdowson, J. P., Maxted, W. R. and Pinney, A. M. (1971*b*) *J. Hyg., Camb.* **69,** 553; (1976) *J. med. Microbiol.* **9,** 73.
Wiel-Korstanje, J. A. A. van der and Winkler, K. C. (1975) *J. med. Microbiol.* **8,** 491.
Wiley, G. G. and Wilson, A. T. (1961) *J. exp. Med.* **113,** 451.
Wilkinson, B. J. and Jones, D. (1977) *J. gen. Microbiol.* **98,** 399.
Wilkinson, H. W. (1975) *Infect. Immun.* **11,** 845; (1977) *J. clin. Microbiol.* **6,** 43.
Wilkinson, H. W. and Eagon, R. G. (1971) *Infect. Immun.* **4,** 596.

Wilkinson, H. W. and Jones, W. L. (1976) *J. clin. Microbiol.* **3,** 480.

Williams, R. E. O. (1954) *J. gen. Microbiol.* **10,** 337.

Williams, R. E. O., Hirch, A. and Cowan, S. T. (1953) *J. gen. Microbiol.* **8,** 475.

Williams, R. E. O. and Maxted, W. R. (1955) *Atti VI Congr. int. Microbiol., Roma,* 1953, vol. 1, p. 46.

Wilson, A. T. (1945) *J. exp. Med.* **81,** 593; (1959) *Ibid.* **109,** 257.

Wilson, A. T. and Wiley, G. G. (1963) *J. exp. Med.* **118,** 527.

Wolin, M. J., Manning, G. B. and Nelson, W. O. (1959) *J. Bact.* **78,** 147.

Woods, R. D. and Ross, R. F. (1975) *Infect. Immun.* **12,** 881.

Woolcock, J. B. (1974*a*) *Infect. Immun.* **10,** 116, (1974*b*) *J. gen. Microbiol.* **85,** 372.

Yao, J., Jacobs, N. J., Deibel, R. H. and Niven, C. F. (1964) *J. infect. Dis.* **114,** 327.

Yawger, E. S. and Sherman, J. M. (1937) *J. Dairy Sci.* **20,** 205.

Zabriskie, J. B. (1964) In: *Rheumatic Fever and Glomerulonephritis,* p. 53. Ed. by J. W. Uhr. Williams and Wilkins, Baltimore.

Zanen, H. C. and Engel, H. W. B. (1979) In: *Pathogenic Streptococci,* p. 232. Ed. by M. T. Parker, Reedbooks, Chertsey, Surrey.

Zighelboim, S. and Tomasz, A. (1980) *Antimicrob. agents Chemother.* **17,** 434.

30

Staphylococcus and *Micrococcus*; the anaerobic gram-positive cocci

M. T. Parker

Introductory

Staphylococci were seen in pus by Koch in 1878, were cultivated in liquid media by Pasteur in 1880, and were shown by Ogston (1881) to be frequent in acute and chronic abscesses. Rosenbach (1884) made a thorough study of the staphylococci, obtained them in pure culture, and, adopting the generic name *Staphylococcus* proposed by Ogston (1881), divided them into two species—*Staph. pyogenes aureus* and *Staph. pyogenes albus*. Winslow and his colleagues (Winslow and Rogers 1906, Winslow and Winslow 1908) made a broader classification of the cluster-forming cocci of the animal body and of the natural environment. (For early references, see Hill 1981). In this Chapter we shall recognize among the aerobic, catalase-positive, gram-positive cocci a genus *Staphylococcus* of animal parasites that form irregular clusters, are facultative anaerobes and ferment glucose, and shall distinguish it from a genus *Micrococcus* composed mainly of saprophytic organisms, which may form irregular clusters or tetrads, are strictly aerobic and do not ferment glucose. We do not suggest, however, that it is always easy to allocate a strain to one or other genus by means of tests that can be readily performed in routine laboratories. Brief mention will also be made of a genus of motile marine cocci, *Planococcus*.

Among the staphylococci, Rosenbach's *Staph. pyogenes aureus*, now officially named *Staph. aureus* (Report 1958), is a reasonably well defined species, which includes most of the staphylococci that are pathogenic for man and animals. Pigmentation is now recognized as an inconstant character of the species, and for most practical purposes we may define *Staph. aureus* as the 'coagulase-positive staphylococcus', though Hájek (1976) proposes the creation of a second species of coagulase-positive staphylococci, *Staph. intermedius*, to accommodate certain aberrant strains found in animals.

The nomenclature of the coagulase-negative staphylococci is still in a state of confusion. For the most part we shall follow Baird-Parker (1974) in recognizing two species, *Staph. epidermidis* and *Staph. saprophyticus*, each composed of several biotypes. Kloos, Schleifer and their colleagues (Schleifer and Kloos 1975*b*, Kloos and Schleifer 1975*a*, *b*, Kloos *et al.* 1976) have since described a number of other species that show partial correspondence with the Baird-Parker biotypes of *Staph. epidermidis* and *Staph. saprophyticus*. Unfortunately they use both these specific names but with a much narrower connotation. When we use them without qualification, it will be in the sense of Baird-Parker (1974).

Staphylococcus

Definition

Spherical or ovoid, non-motile, gram-positive cells, arranged in grape-like clusters on solid media. On agar the growth is opaque and of a golden, beige or white colour. Catalase positive. Oxidase negative. Aerobic and facultatively anaerobic organisms that require complex media for growth and attack glucose fermentatively. Parasites of animals; one species an important pathogen. G+C content of DNA: 30–40 moles per cent.

The type species is *Staphylococcus aureus*.

Habitat

Staphylococci occur on the body surface of many species of mammals and birds, and this appears to be their normal habitat. They are also present in varying numbers in the air and dust of occupied buildings, and in milk, food and sewage. *Staph. aureus* is carried in the nose of some 30 per cent of persons, many of whom are also skin carriers. Carriage of this organism is frequent also in a number of other animals, but evidence is accumulating that the *Staph. aureus* strains of several animal species form distinct populations. Staphylococcal disease is mainly but not exclusively due to *Staph. aureus*, and is of importance in man and in several of his domestic animals.

Morphology

The staphylococci are round or somewhat oval cells with an average diameter of 0·8–1·0 μm. The size varies from strain to strain and also depends partly on the age of the culture and the medium on which it is grown. All staphylococci are non-motile, non-flagellated, and non-sporing.

Staphylococci are arranged in grape-like clusters in which the members are disposed in three planes of space without any definite configuration. This is more evident on solid than in liquid media; in broth, pairs and short chains may occur and be mistaken for streptococci. On solid media streptococci tend to lose their capacity for forming chains and may develop in small clusters. The appearance of a coccus should therefore always be studied in both liquid and solid media. The

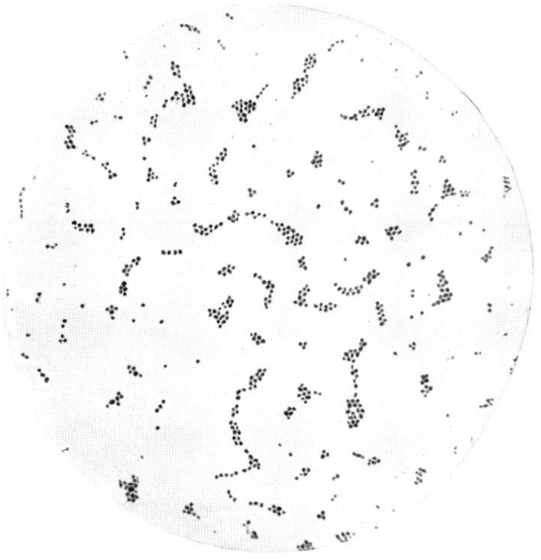

Fig. 30.1 *Staphylococcus aureus*. From an agar culture, 24 hr, 37° (× 1000).

staphylococci stain well with most aniline dyes and are uniformly gram positive in young cultures. Most staphylococci are unencapsulated, but a few have capsules that can be seen microscopically. Heavily capsulated strains, such as the Smith strain, are unusually virulent for mice when given intraperitoneally (Hunt and Moses 1958) because the polysaccharide capsular material inhibits phagocytosis (Chapter 60). They also give a negative slide-coagulase reaction and form diffuse colonies in semisolid agar containing serum (p. 225).

The observations of Price and Kneeland (1954, 1956) suggested that many non-capsulated strains have similar material on the cell surface but in smaller amount, and this was later confirmed. Four distinct capsular antigens are now recognized (Yoshida 1971, 1972). Although only some 4 per cent of *Staph. aureus* strains are evidently capsulated, nearly all strains react with antisera against one or more of the capsular antigens (Yoshida *et al.* 1979). According to Yoshida and Minegishi (1979), *Staph. epidermidis* strains may occasionally be capsulated and possess an antiphagocytic surface polysaccharide (see Yoshida *et al.* 1976).

Cultural characters

Nearly all staphylococci grow abundantly in unenriched nutrient media. On *nutrient agar* they form smooth, circular, opaque, often yellow-pigmented colonies, 1–2 mm in diameter after overnight incubation at 37°. The growth of *Staph. aureus* is butyrous and easy to emulsify, but *Staph. epidermidis* is sometimes glutinous, adherent to the medium and difficult to emulsify. Heavily capsulated strains may form mucoid colonies. In *nutrient broth* most strains give a moderate to dense turbidity with a powdery deposit, but some coagulase-negative strains form a thick sticky deposit, leaving the supernatant fluid clear. No pigment is formed in liquid media.

Nearly all strains of *Staph. aureus* liquefy *gelatin* within a few days, and many strains of *Staph. epidermidis* do so rather more slowly, but the difference between the species is not clear cut. On *MacConkey's agar*, the colonies of *Staph. aureus* are small and pale pink after 24 hr and deep pink after 48 hr.

Variant colonial forms

Dwarf or G forms of *Staph. aureus* can be isolated from most strains by growth under conditions in which the normal forms are inhibited, including growth in the presence of lithium chloride, barium chloride, gentian violet, acridines, and various antibiotics (Hale 1947, Browning and Adamson 1950, Wise and Spink 1954). They are occasionally isolated in pure culture from septic lesions in man (Hale 1947, Quie 1969) and from bovine mastitis (Lernau and Sompolinsky 1962). Most dwarf forms show a tendency to reversion on continued subculture, or by 'sectoring'. The ability to form catalase, coagulase, haemolysins or staphylokinase may be absent, and mouse virulence may be low. Various metabolic defects have been described in dwarf forms. Some grow normally in the presence of yeast extract and others on nutrient agar in an atmosphere containing one per cent CO_2 (Hale 1947, 1951). Two separate defects in the synthesis of thiamine (Sompolinsky *et al.* 1967), a requirement for pantothenic acid (Sompolinsky *et al.* 1969) or haemin (Borderon and Horodniceanu 1976), and the absence of vitamin K due to a variety of blocks in menoquinone synthesis (Sásărman *et al.* 1971), have been found in different dwarf strains. Lacey (1969) described intensely pigmented dwarf-colony variants selected by exposure to linoleic acid that were resistant to all aminoglycoside antibiotics, and identical variants selected with paromomycin that were resistant to linoleic acid. Similar strains are occasionally isolated from antibiotic-treated patients (Lacey and Mitchell 1969).

Pigmentation

Colonies of *Staph. aureus* are usually golden-yellow but may be pale yellow or fawn, or even completely white. Those of *Staph. intermedius* and the coagulase-negative staphylococci are usually white, but less often may be pale yellow. The development of pigment, and its actual tint, depend on several factors. Carbon dioxide favours pigmentation and oxygen is essential; if a culture that has been incubated anaerobically is exposed to air it soon develops its characteristic colour (Lubinski 1894). On nutrient agar, maximal pigmentation occurs after overnight incubation in the dark at 37°, followed by 48 hr in the light at room temperature (Brown 1965). Pigmentation on nutrient or blood agar is, however, only moderate, and growth on potato medium, 30 per cent milk agar (Christie and Keogh 1940) or 10 per cent cream agar (Willis *et al.* 1966) enhances this. Apparently the fatty acid in milk or cream stimulates pigmentation and, according to Steuer (1956), glycerol monoacetate or monophosphate has a similar effect. However, O'Connor and her colleagues (1966) state that glycerol monoacetate media give optimal pigmentation only if some nutrient bases are used, but that cream agar made with most bases gives good results. Two distinct types of pigmentation—orange and lemon-yellow—occur among *Staph. aureus* strains grown on glycerol monoacetate agar (Willis and Turner 1962).

The ability to form yellow pigment is often lost irreversibly, especially after repeated subculture anaerobically or prolonged culture in broth (Barber 1955). According to Grinsted and Lacey (1973), non-pigmented variants are more susceptible to drying and to linoleic acid than are the original strains, and so would be less likely to survive at carriage sites; the ability to form pigment could not be transduced, but its loss was associated with a reduction in the cellular DNA content. The yellow pigments are carotene-like, and at least seven types can be distinguished (Sandvik and Brown 1965). Some coagulase-negative staphylococci, described as *Micro. violagabriellae* by Castellani (1955), produce a violet pigment. This occurs constantly on potato but irregularly on other media (Steel 1964, Marples 1969).

Selective and enrichment media for *Staph. aureus*

Use may be made of the fact that staphylococci have a high salt tolerance. Ten per cent salt broth and salt agar permit the growth of *Staph. aureus* and inhibit most gram-negative bacilli (Hill and White 1929). Enrichment in Robertson's cooked meat medium with 10 per cent of added sodium chloride (Maitland and Martyn 1948), followed by plating on a suitable selective medium, permits the detection of quite small numbers of staphylococci in heavily contaminated material. Advantage may also be taken of the fact that *Staph. aureus* liberates phenolphthalein rapidly from phenolphthalein phosphate at 37°. Overnight cultures on nutrient agar containing 0·01 per cent of this reagent, when exposed to ammonia vapour, become bright pink (Barber and Kuper 1951). In addition to salt resistance, other characters made use of in selective and indicator media include acid production from mannitol or lactose, resistance to 2-phenyl ethanol, and the reduction of potassium tellurite. Tellurite glycine agar (Zebovitz *et al.* 1955) is somewhat inhibitory for *Staph. aureus*, but Baird-Parker (1962) increased the amount of glycine and added pyruvate and egg yolk. On this medium, *Staph. aureus* gives characteristic black-centred colonies surrounded by an area of clearing, sometimes with an area of opacity within it; many other organisms are inhibited, and isolation of *Staph. aureus* is usually quantitative (see also Holbrook *et al.* 1969). Finegold and Sweeney (1961) showed that the addition of 75 mg per l of polymyxin B to nutrient agar inhibited coagulase-negative staphylococci but permitted the growth of *Staph. aureus*; this antibiotic may be addded to Baird-Parker's medium. Staphylococci in the air and in dehydrated foods are often poorly viable and so cannot be isolated on selective media. For air-sampling, phenolphthalein phosphate agar containing 5 per cent serum is recommended. If this medium is incubated anaerobically at 41°, the growth of many of the other cocci is inhibited (Harding and Williams 1969).

Many different media have been proposed for the enumeration of *Staph. aureus* in foods (see Report 1972, Idziak and Mossel 1980, Devriese 1981).

Metabolism

Staphylococci are facultative anaerobes but their growth is improved by oxygen. They attack sugars fermentatively, though this activity may be very weak. Anaerobically they form mainly lactic acid from glucose. In air the chief product is acetic acid, and a little CO_2 is formed; this, however, is insufficient to be detected by means of a Durham's tube. They have a cytochrome system, and catalase is formed. Growth occurs at 10° (or lower in some strains) and 45°, but is usually most rapid between 30° and 37°; it occurs over a wide range of pH (4–9) and is optimal at pH 7.0–7.5.

Nutritive requirements are complex. *Staph. aureus* will grow aerobically with glucose, salts, up to 14 amino acids, thiamine and nicotinic acid (Fildes *et al.* 1936, Knight 1937; see also Mah *et al.* 1967); an organic source of sulphur is necessary (Fildes and Richardson 1937, Gladstone 1937). Growth does not occur in the complete absence of CO_2 (Gladstone *et al.* 1935). Strains of *Staph. aureus* show some differences in their requirements for amino acids and vitamins; according to Tschäpe and Rische (1972) these differences are of value in the definition of host-specific varieties of the organism. *Staph. epidermidis* has an additional requirement for biotin (Evans 1948) shared only with some strains of *Staph. intermedius*.

Resistance to heat and disinfectants

The staphylococci are among the more resistant of the non-sporing organisms. In broth or agar tubes sealed with paraffin and kept in the ice-chest, cultures may remain alive for months. Dried on threads they retain their vitality for 3 to 6 months, and from dried pus they have been cultivated after 2 to 3 months. Unlike some of the micrococci, true staphylococci show no particular resistance to heat and are usually killed by a temperature of 60° in half an hour. (For survival in the natural environment see Chapter 60.) Staphylococci will grow in media containing high concentrations of salt or sucrose, though they perish in similar concentrations of NaCl in distilled water. They grow well in media containing 12 per cent sodium chloride, and some growth occurs in all attainable concentrations (Hucker and Haynes 1937). In pure culture they resist a concentration of 1 per cent phenol for 15 minutes, but are killed by 2 per cent. Mercuric chloride is a poor disinfectant for staphylococci; to kill them in 10 minutes a 1 per cent solution is required.

Staphylococci are very sensitive to some aniline dyes. This is made use of by incorporating crystal violet in media for the isolation of *Br. abortus* from milk and of streptococci from skin lesions contaminated with staphylococci.

Strains of *Staph. aureus* and *Staph. intermedius* show various appearances on media containing subinhibitory concentrations of crystal violet (Epstein 1935, Meyer 1967*b*). When seeded heavily on to plates containing crystal violet 1 in 500 000, the more sensitive strains give a thin streak of purple-stained growth, and the more resistant give definite colonies that may be yellow or white according to the pigment-forming ability of the strain (see p. 236). Fatty acids inhibit the growth of staphylococci, the highly unsaturated acids having a more powerful action on coagulase-positive than on coagulase-negative strains. The minimal inhibitory concentration of linolenic acid for *Staph. aureus* is usually 0.05 per cent or less and for other staphylococci 1.0 per cent or more (Lacey and Lord 1981).

Resistance to heavy metals is discussed below.

Resistance to enzymes that attack the cell wall

Staphylococci are uniformly resistant to lysozyme, but some micrococci are sensitive to it. On the other hand, staphylococci are generally sensitive to lysostaphin, but most micrococci are resistant to it (Lachica *et al.* 1971). Lysostaphin (Schindler and Schuhardt 1964) is a mixture of enzymes produced by a particular strain of *Staph. epidermidis*; among these are (1) a peptidase that acts specifically on the interpeptide bridges of staphylococcal but not of micrococcal peptidoglycan (p. 224) and (2) lysozyme. Thus, according to Schleifer and Kloos (1975*a*), if an organism is sensitive to lysostaphin (200 mg per l) and resistant to lysozyme (25 mg per l) it is probably a staphylococcus.

Resistance to antimicrobial agents

Strains of *Staph. aureus* isolated before 1942 were sensitive to a wide range of antibiotics, but a few of them were resistant to benzylpenicillin by virtue of the production of penicillinase. From 1945 onwards, penicillinase-forming strains became increasingly common, first in hospitals and subsequently in the general population and among animals.

At least four immunological types of staphylococcal penicillinase are known (Richmond 1965*a*, Rosdahl 1973); they differ in their specific activity and in the extent to which the enzyme appears extracellularly. The A-type penicillinase is usually formed in large amount and much of it is extracellular (Novick and Richmond 1965, Richmond 1965*b*); it is the type formed by many of the strains that became widely endemic in hospitals and acquired resistance to other antibiotics. Strains that form A-type penicillinase may also be resistant to several heavy-metal ions, including mercury, cadmium, arsenic and lead (Moore 1960, Richmond and John 1964, Novick and Roth 1968). The genes determining penicillinase production are usually on a plasmid (Novick 1963) but are occasionally chromosomal (Asheshov 1966). The type-A plasmid has separate loci determining production of the enzyme, inducibility, 'extracellularity', and, in some strains, also resistance to the various heavy metals. Although resistance to mercury is nearly always determined by the A-type plasmid, resistance to cadmium and arsenic may also be chromosomally determined and occur in strains that form other types of penicillinase (Dyke *et al.* 1970). Heavy-metal resistance appears to be due to the exclusion of the metal from the interior of the cell (Váczi *et al.* 1962, Chopra 1971).

A second form of penicillin resistance, often referred to as 'methicillin resistance', was detected soon after methicillin came into use (Jevons 1961). This is a broad-spectrum resistance to all penicillins and to cephalosporins that is fully expressed only at temperatures below 37° (Annear 1968, Parker and Hewitt 1970) and in which the antibiotic is not destroyed enzymically. Because methicillin resistance is found almost exclusively in strains that also form A-type penicillinase, it was not detected before methicillin became available. However, methicillin-resistant strains proved to be locally prevalent in some geographical areas in which this antibiotic had not been used (see, for example, Parnas and Jabłonski 1961); there is thus no reason to believe that it arose in response to its use (Chapter 60).

Resistance to almost every other widely used antimicrobial drug has been detected in *Staph. aureus*. With certain agents, for example, novobiocin and fusidic acid, spontaneous mutation to resistance occurs regularly. Resistance to other drugs did not appear until they had been in use for some time, in some cases not for several years. When these resistances developed, they tended to be concentrated in a minority of penicillinase-forming strains, which then spread widely in hospitals (Chapter 60). These 'multiple-antibiotic-resistant' strains often carry resistance determinants for four or five chemically unrelated antibiotics, and exceptionally for eight or more. Several of these resistances are mediated by antibiotic-destroying enzymes but others are non-enzymic (Chapter 5). Staphylococci often possess more than one mechanism of resistance to the same group of antibiotics, and these may co-exist in the same strain. Many of the resistance determinants are on plasmids, which may be present in the cell as multiple copies (Chopra *et al.* 1973). Nearly all of them are carried on distinct genetic elements and are separately transferable. There are a few exceptions: plasmids determining penicillinase production together with resistance to erythromycin (Mitsuhashi *et al.* 1965), fusidic acid (Evans and Waterworth 1966, Lacey and Grinsted 1972, Lacey and Rosdahl 1974) or kanamycin (Annear and Grubb 1972) have been observed, but these appear to be rare and tend to fragment.

Transfer of both plasmid and chromosomal resistance by transduction occurs at low frequency when the transducing agent is a single phage (see, for example, Asheshov 1966), but transfer may occur at much higher frequency in mixed culture or when crude lysates are used as the source of phage. This may take place in the presence of a 'helper' phage, usually of serological group B, which is thought to confer the ability for high-frequency transduction (Novick and Morse 1967, Lacey 1971*a*, *b*) or competence to receive exogenous DNA by transformation (Thompson and Pattee 1977). According to Meijers and his colleagues (1981), transfer in mixed culture occurs by general transduction when the donor strain carries a transducing phage of serological group B and a transducible plasmid, and the acceptor exerts little restriction and does not carry an incompatible plasmid; if the acceptor is immune to the transducing phage, transfer in the reverse direction will not occur. A mathematical model based on these assumptions gives rates of transfer that correspond well with those observed experimentally.

Coagulase-negative staphylococci found in hospitals are often resistant to several antibiotics. Resistance plasmids may be transferred between strains of *Staph. aureus* and *Staph. epidermidis* in mixed culture. Most strains of *Staph. saprophyticus* are naturally resistant to novobiocin 2 mg per l, whereas other staphylococci are sensitive, unless a resistant variant has been selected by contact with the antibiotic. According to Curry and Borovian (1976) micrococci grow on a soya digest medium containing nitrofuran 50 mg per l but staphylococci do not. In the absence of a specific resistance mechanism, staphylococci are slightly more resistant to erythromycin than are micrococci, and will grow on solid media containing 0.4 mg per l. (For sensitivity to polymyxin B, see p. 221.)

Biochemical characters

The biochemical activities of staphylococci are not easy to determine with consistency, even when care is taken to use standard techniques, yet these are the characters upon which we have to rely in routine practice to distinguish between staphylococci and micrococci. In general, the methods now used are either those of Baird-Parker (1963) or those of Kloos and Schleifer (Kloos *et al.* 1974, Kloos and Schleifer 1975*a*, Schleifer and Kloos 1975*a*, *b*), which differ considerably in detail (see Marples 1981).

All staphylococci and many micrococci form acid from glucose aerobically; the ability to do this anaerobically is generally considered to be the most distinctive character of staphylococci. The results of tests for the fermentation of glucose depend very much on the exact composition of the medium (Cowan and Steel 1964), and the standard technique should be used (Report 1965). Even when performed in this way, however, the test may give a negative or equivocal result with some organisms that must now be considered to be staphylococci by virtue of the base composition of their DNA, because they form very little acid from glucose either aerobically or anaerobically. Evans and Kloos (1972) therefore proposed that a test of growth in the depths of a tube of semisolid thioglycollate agar should be used to identify staphylococci, but this too does not always give clear results.

Staph. aureus usually forms acid from mannitol *anaerobically* and is the only staphylococcus to do so, but a standard medium must be used to detect this character (Evans and Pate 1980). Aerobically, a number of other staphylococci acidify mannitol, and most staphylococci, including *Staph. aureus*, acidify maltose, but, according to Hájek (1976), *Staph. intermedius* attacks neither mannitol nor maltose. Few staphylococci acidify raffinose, starch, dulcitol or salicin. Schleifer and Kloos (1975*a*) say that staphylococci acidify glycerol aerobically but micrococci do this rarely.

Many staphylococci, and all *Staph. aureus*, are both methyl red positive and Voges-Proskauer positive. The latter reaction is often weak but is of value in the classification of coagulase-negative staphylococci. Indole is not produced. Most staphylococci produce ammonia from arginine, hydrolyse urea, and reduce nitrate to nitrite; some staphylococci produce a trace of hydrogen sulphide. They are all oxidase negative, and with rare exceptions catalase positive.

Phosphatase production is useful in the classification of staphylococci. In *Staph. aureus* it is almost invariably detectable after 18 hours' incubation at 37°. The detection of a weaker reaction after 2 to 5 days' incubation at 30° is of value in the classification of coagulase-negative staphylococci (Baird-Parker 1963).

Lipolysis is common in staphylococci. The ability to hydrolyse tributyrin is a character of all strains of *Staph. aureus* and of many other staphylococci. The range of activity of the lipases of different staphylococci varies, and more strains attack the short-chained Tween 20 than the longer-chained Tween 80. Hydrolysis of Tween 80 is a property of some strains of *Staph. aureus* (Sierra 1957, Jessen *et al.* 1959). It can be repressed by lysogenization with certain phages (Rosendal and Bülow 1965). The production of opacity in egg yolk is due to a lipase (Gillespie and Alder 1952, Shah and Wilson 1963, 1965) and is confined to strains that attack Tween 80; it is an almost invariable character of strains that cause boils (Gillespie and Alder 1952). The formation of a dense red band around colonies on human blood agar (Weld 1962) and of fatty acid plaques in plasma agar (Weld *et al.* 1963) is also due to lipolytic action (see O'Leary and Weld 1964).

Staphylococci produce various *proteolytic enzymes*. Nearly all strains of *Staph. aureus*, and many other staphylococci, form a gelatinase, and most strains of *Staph. aureus* digest casein rapidly in the presence of serum (Fisk and Mordvin 1943). The coagulase-negative staphylococci form a number of proteinases that can be distinguished serologically or by their electrophoretic mobility (Sandvik and Fossum 1965, Scherer and Brown 1974). Some strains of *Staph. epidermidis* have stronger elastolytic activity than *Staph. aureus* (see Hartman and Murphy 1977).

All *Staph. aureus* strains produce a powerful *deoxyribonuclease* that resists boiling (Weckman and Catlin 1957), but other staphylococci may form a heat-labile enzyme. If a test for the nuclease is to be used for identifying *Staph. aureus*, the culture must be heated before testing (for suitable methods, see Lachica 1976, Kamman and Tatini 1977). *Staph. intermedius* forms a heat-stable DNAase and coagulase (Hájek 1976) as do some animal strains of *Staph. epidermidis* biotype 2 (Devriese and van de Kerckhove 1979; see

p. 235); others, and certain similar strains of human origin (Dornbusch *et al.* 1976) form heat-stable DNAase but not coagulase. *Lysozyme* is produced by all Staph. *aureus* strains, and also by some other staphylococci and micrococci (Kashiba *et al.* 1959, Holt 1971).

Cellular composition and antigenic structure

Several classes of substance present in the cell wall or outside it are responsible for immunological reactions with staphylococcal antisera. The main structural unit of the cell wall is the *peptidoglycan*, to which, in staphylococci but not in micrococci, is linked a *teichoic acid.* More superficially, a number of other antigenic substances may be present: surface polysaccharides, protein A, clumping factor, and the so-called type antigens that can be detected in agglutination tests.

Peptidoglycan

The peptidoglycans of staphylococci and micrococci have a common glycan but differ in the amino-acid composition of the linking peptides. Seidl and Schleifer (1978) have devised a test for distinguishing staphylococci from micrococci that makes use of this fact; the respective peptides are coupled to albumin and used to produce sera that give a specific agglutination reaction with cocci that have been extracted with trichloracetic acid. Another difference is that the interpeptide bridge in staphylococci is composed mainly of glycine but in micrococci does not contain this amino acid. Since the endopeptidase of lysostaphin attacks glycyl-glycine linkages, staphylococci are lysostaphin sensitive and micrococci are resistant (p. 222). However, *Staph. aureus* is rather more sensitive than many other staphylococci because its interpeptide bridge is composed entirely of glycine and that of the others may contain only four to five glycine molecules and on average one molecule of serine. However, the amino-acid content of the growth medium may affect the composition of the interpeptide bridge (Schleifer *et al.* 1976), so it is not easy to use differences in lysostaphin sensitivity as a means of classifying staphylococci (Heddaeus *et al.* 1979).

Teichoic acid

The antigens extracted by Julianelle and Wieghard (1934, 1935, Wieghard and Julianelle 1935) with trichloracetic acid are teichoic acids (Haukenes *et al.* 1961) that give precipitation reactions with antisera against whole-cell vaccines. Those present in *Staph. aureus* are ribitol teichoic acids and are of two forms, containing respectively β- and α-linked *N*-acetylglucosamine, that occur either separately or together (Hofstad 1964). *Staph. epidermidis* strains form glycerol teichoic acids, which may contain α-linked or β-linked glucose or α-linked glucosamine. *Staph. saprophyticus* forms a ribitol teichoic acid different from those of *Staph. aureus* (Oeding and Hasselgren 1972). The teichoic acids of the animal strains now included in *Staph. intermedius* are various, and are distinct from those of *Staph. aureus* (Oeding 1974). Removal of teichoic acid renders staphylococci susceptible to lysozyme. Although forming part of the cell-wall structure, and usually demonstrated by precipitation reactions with extracted material, there is evidence that some teichoic acid appears at the cell surface (James and Brewer 1968) and that antibody to it may be detectable by agglutination (Sanderson *et al.* 1961).

Red blood cells can be sensitized with extracts from staphylococcal cultures, and exhibit haemagglutination in the presence of staphylococcal antisera (Rountree and Barbour 1952, Oeding 1957, Weld and Rogers 1960). Antibodies to the erythrocyte-coating antigen are absent at birth but are common in young adults (Rountree and Barbour 1952). The haemagglutinating antibody does not, however, appear to be specific for staphylococci, and the corresponding antigen is probably identical with the so-called 'non-species-specific antigen' (Rantz *et al.* 1952). Similar antigens in streptococci and lactobacilli are lipoteichoic acids.

Surface polysaccharide

As we have seen (p. 220), a few *Staph. aureus* strains have polysaccharide capsules, and similar polysaccharides can be detected serologically on the surface of a large number of apparently unencapsulated strains. Karakawa and his colleagues (1974) characterized an acidic polysaccharide from a *Staph. aureus* strain that showed no evidence of capsulation.

Protein A

This is present on the surface of nearly all *Staph. aureus* strains of human origin, but is much less common on animal strains. It was first described by Verwey (1940) and rediscovered by Jensen (1958), who showed that it was responsible for a precipitin line in agar gel between crude acid extracts of *Staph. aureus* and normal human serum. The protein nature of this antigen was demonstrated later (Löfqvist and Sjöquist 1962, Yoshida *et al.* 1963). It has a molecular weight of *ca* 13 000 and sensitizes tanned but not untreated sheep erythrocytes (Grov 1967, 1968).

Protein A is also responsible for the agglutination of most strains of *Staph. aureus* by all normal human sera (Lenhart *et al.* 1963). This, however, is not because the sera contain specific antibody, but because protein A combines non-specifically with the Fc-portion of human IgG (Forsgren and Sjöquist 1966, Forsgren and Forsum 1970). Protein A reacts similarly with the IgG of guinea-pig and mouse serum, but gives only trace reactions with rabbit serum (Grov 1973). In human sera, it is bound to only subclasses 1, 2 and 4 of IgG, and it is bound to subclass 2 of IgA and IgM, a pattern of reactivity different from that of the Fc-binding components of streptococci (Myhre and Kronvall 1979; see also Chapter 29).

Rabbits develop a general Arthus reaction if protein A is given after an intravenous dose of human IgG, or a local Arthus reaction if preformed aggregates of protein A and IgG are injected intramuscularly (Gustafson *et al.* 1967). In guinea-pigs (Gustafson *et al.* 1968) and in man (Martin *et al.* 1967), an injection of protein A alone gives rise to an immediate hypersensitivity reaction; a second injection elicits a delayed hypersensitivity reaction in both guinea-pigs and rabbits (Heczko *et al.* 1973).

Protein A can be demonstrated at the cell surface by electronmicroscopy (Nickerson *et al.* 1970). According to Dossett and his colleagues (1969), it has antiphagocytic activity. The amount of protein A formed by human strains varies widely, and is to some extent correlated with phage group (Lind 1972). Kronvall and his colleagues (1972) found that bovine strains from acute mastitis formed much more protein A than did strains from chronic mastitis. According to Forsgren (1970) occasional strains of *Staph. epidermidis* form protein A.

The non-specific adsorption of immunoglobulin to protein A has been widely exploited as a means of detecting antigen-antibody reactions. Co-agglutination reactions, in which protein-A forming staphylococci coated with immunoglobulin are mixed with bacterial suspensions, cellular extracts and samples of pathological material are now used for the rapid identification of many different micro-organisms (Chapter 20).

Clumping factor ('bound coagulase')

As early as 1908, Much observed that staphylococci that coagulated plasma were also clumped by it. Birch-Hirschfeld (1934) found a close relation between the tube test for coagulase and what came to be known as the 'slide-coagulase' reaction. Williams and Harper (1946) showed that the slide test, when performed carefully and with a negative control, very rarely gave a positive result if the tube test gave a negative result; on the other hand, some 12 per cent of tube-coagulase-positive strains were not clumped in the slide test.

Duthie (1954*a*, 1955) produced evidence that the two reactions were due to different substances. Clotting was caused by an extracellular substance ('free coagulase', p. 228) and required the presence not only of fibrinogen but also of a serum factor; clumping occurred when fibrinogen alone was mixed with a washed suspension of cells. Duthie found that the cellular component was released from the cells by autolysis; he purified it partily and injected it into rabbits, the sera from which inhibited the clumping but not the clotting reaction. Animal plasmas that are not clotted by most *Staph. aureus* strains, such as mouse plasma, are suitable for the clumping test. The later observation that *Staph. aureus* was clumped in a slide test by serum that had as far as possible been freed of fibrinogen (Brown and Faruque 1963) suggested that the reaction was due to a serum factor, but Lipiński and his colleagues (1967) later showed that staphylococci were clumped by a lysate of fibrin containing complexes of fibrin monomer that were not clotted by thrombin. They concluded, therefore, that similar substances in serum were responsible for the clumping reaction.

Clumping factor, the cellular component responsible for the slide-coagulase reaction, is found almost exclusively in strains that form the 'true' or extracellular coagulase. As we have seen (p. 220) capsulated strains give a negative clumping reaction, presumably because the clumping factor is covered by extracellular polysaccharide, but a few other strains that clot plasma appear to lack this factor.

Colony-compacting factor Finkelstein and Sulkin (1958) observed that staphylococci that clotted plasma could be distinguished from coagulase-negative staphylococci in a soft agar medium containing human or rabbit plasma. Without plasma, both types produced elongated diffuse colonies; with plasma the coagulase-positive but not the coagulase-negative staphylococci formed compact spherical colonies. They also observed that the test worked equally well with serum, and concluded that it was due to an antigen-antibody reaction. Alami and Kelly (1959) showed that the positive correlation was between this 'compacting factor' and clumping factor, and that the reaction was with fibrinogen. This suggested that the compacting factor might be identical with the clumping factor. It is certainly a surface component, because capsulated strains grow diffusely in serum soft agar.

Forsum and his colleagues (1972) took the view that the compacting factor is protein A, but this has been challenged. According to Yoshida, Ohtomo and Minegishi (1977), compacting factor is not clumping factor, protein A or teichoic acid, but a distinct surface polysaccharide. This view is, however, difficult to reconcile with the observation (Yoshida 1971) that capsulated strains of *Staph. aureus* formed compact colonies when grown in the presence of specific anticapsular serum.

Type antigens

Because all *Staph. aureus* strains possess common antigenic factors at the cell surface, they are agglutinated by any staphylococcal antiserum. Absorption of the sera with heterologous strains removes the common agglutinins and reveals a number of type-specific antibodies (Julianelle 1922, Hine 1922). These form the basis of the two distinct systems that have been used for typing by slide agglutination. In the first, sera are absorbed so as to divide the species into a series of distinct types. Thus Cowan (1938, 1939) distinguished three main types; Christie and Keogh (1940) increased this number to nine, and Hobbs (1948) to 15. This system has since been extended and refined by Pillet and his co-workers (see Pillet 1966). The second system (Oeding 1952) is more comprehensive in that an attempt is made to identify all the type-specific antigens that are detectable by slide agglutination even when they give overlapping reactions. Some 30 antigenic factors can be recognized, but many of them have not yet been properly characterized. Both the Oeding and the Pillet typing systems are now widely used and give consistent results in the hands of experts, but few comparisons have been made of the results obtained with them (see, however, Modjadedy and Fleurette 1974, Flandrois *et al.* 1975).

Toxin production

Among the many extracellular substances produced by *Staph. aureus*, there are (1) general toxic substances, such as the haemolytic toxins, that may contribute to the clinical picture of local or systemic septic infections, and (2) specialized toxins, such as the enterotoxins and the epidermolytic toxins, that are responsible for specific, non-suppurative clinical manifestations. The other staphylococci, even those now considered to be pathogenic, produce few if any substances that have claims to be considered to be important toxins.

The haemolytic toxins

Staphylococci produce four substances that damage the cell membranes of erythrocytes, two of which also have important actions on other tissues.

Table 30.1 Properties of staphylococcal α-, β-, λ- and δ-toxins

	α	β	λ	δ
Action on red blood cells of:				
sheep	++	+++	+++	++
rabbit	+++	+	+++	++
horse	−	−	−	+++
man	±	+	+++	+++
Lethal activity	+++	±	±	+
Dermonecrosis	Yes	No	No	Yes
Leucocidal activity	Yes	No	Yes	Yes

α-lysin This toxin is responsible for haemolysis and for the killing of leucocytes; it causes necrosis on intradermal injection into rabbits and guinea-pigs, and acute and fatal toxaemia when given intravenously to rabbits and mice. It is antigenic, and treatment with formalin at 37° leads to the rapid disappearance of its haemolytic and toxic but not of its antigenic activity (see Burnet 1931).

The toxin lyses rabbit erythrocytes rapidly at 37°; sheep cells are less sensitive to haemolysis, and horse and human cells are almost completely resistant. The rate of haemolysis of rabbit cells is depicted by an S-shaped curve in which a short pre-lytic lag phase is followed by a period of linear release of haemoglobin, after which the process slows down (Lominski and Arbuthnott 1962). The maximum rate of haemolysis is directly proportional to the concentration of toxin. Haemolysis is arrested by the addition of antitoxin only during the early part of the lag phase, and loss of K^+ begins well before haemolysis (Madoff *et al.* 1964).

The resistance of α-lysin to heat is peculiar in that inactivation occurs more readily at lower than at higher temperatures, and that lysin heated at 70° loses most of its activity but regains it when heated for a further 30 min at 100° (Arrhenius 1907). When the toxin is heated at 60° for 30 min, an inactive precipitate is formed; if this is separated, resuspended and heated at 80° it regains its haemolytic activity, but the supernatant fluid remains inactive on further heating (Cooper *et al.* 1966). This effect is due to the aggregation of toxin molecules (Arbuthnott *et al.* 1967). Purified toxin contains a soluble active 3S component and a soluble inactive 12S component. Brief heating at 60° results in the formation of an insoluble inactive aggregate; this can be dissociated with 8M urea, when biological activity reappears.

Highly purified lysin (see Arbuthnott 1970) appears to be a protein; it loses activity at temperatures above 20°, probably as a result of molecular aggregation. Its active (3S) component is not homogeneous, and consists of molecules with molecular weights of between 21 000 and 50 000 (see McNiven *et al.* 1972)

Haemolytic, dermonecrotic, lethal, and leucocidal properties are all present in highly purified material (Goshi *et al.* 1963); the dermonecrotic and lethal activities show an Arrhenius effect on heating (Manohar *et al.* 1966). There is little doubt therefore, that they are all properties of the α-lysin. The haemolytic unit of purified toxin contains about 0.01 μg of toxin protein; 0.5–2.4 μg causes necrosis on intradermal injection into a rabbit, and 2 μg given intravenously will kill the rabbit (Arbuthnott 1970). Large doses kill within seconds with few distinctive histological changes. Small doses may take 1–4 days to kill, and there is usually severe necrosis of the kidneys.

The toxin acts widely on animal cells; a variety of cultured cells are killed and intracellular material may be released (for references see Arbuthnott 1970). Rabbit and human leucocytes are killed, but without gross morphological changes, and platelets are lysed (Gengou 1935, Siegel and Cohen 1964, Bernheimer and Schwartz 1965). Wiseman and Caird (1970; see also Wiseman *et al.* 1975) believe that the effect of α-toxin is to activate proteolysis of the cell membrane. Arbuthnott and his colleagues (Freer *et al.* 1973, Arbuthnott *et al.* 1973) do not accept this, and consider that the toxin becomes attached to the cell-wall lipids and brings about their structural disorganization; at the same time, the lipids cause the polymerization of the toxin to its inactive form, but the relation of these two processes is not yet clear.

(For further information about the action of α-toxin, see Harshman 1979.)

***β*-lysin** This acts on sheep and ox, but not on rabbit or human, red corpuscles, and causes lysis only after the tubes have stood at room temperature or in the ice-chest overnight—the so-called *hot-cold lysis* (Walbum 1921, Glenny and Stevens 1935).

Staphylococci that produce β-lysin give rise on sheep blood agar plates to zones of darkening around the bacterial growth. The red blood cells are so modified that they are lysed not only by cooling, but by small alterations of pH and osmotic pressure (Pulsford

1954), by substances produced by certain lipolytic and proteolytic organisms (Christie and Graydon 1941), and by practically all strains of *Str. agalactiae* (Munch-Petersen and Christie 1947). This last cause of lysis forms the basis of the CAMP test for this organism (Chapter 29).

Staphylococci that form β-lysin possess a phospholipase active on sphingomyelin (Doery *et al.* 1963, 1965) and purified β-lysin releases water-soluble organic phosphorus compounds from sheep erythrocyte ghosts, which are rich in sphingomyelin (Maheswaran and Lindorfer 1967, Wiseman and Caird 1967). β-lysin is formed in the absence of CO_2, though its production is enhanced by this gas, and under anaerobic as well as aerobic conditions. It is more resistant to formalin and is somewhat more resistant to heating at 60° than is α-lysin, but it is destroyed more rapidly at 100° than at 60° (Wiseman 1965). It is antigenically distinct from α-lysin, kills laboratory animals only in large doses (Bryce and Rountree 1936, Wadström and Möllby 1971), and is not dermonecrotic (Smith and Price 1938*a*). It lyses platelets but not polymorphonuclear leucocytes (Bernheimer and Schwartz 1965, Wadström and Möllby 1971). Staphylococcal β-lysin gives rise to opacity in horse serum, but not in the serum of most other animals, apparently by action on the albumen fraction (Christie and North 1941). The appearance is rather like that of the opacity produced by the α-toxin of *Cl. perfringens* but the staphylococcal toxin does not render egg yolk opaque (see also Gulasekharam *et al.* 1963). For conditions of production and activity of β-lysin see Haque and Baldwin (1964), and for methods of purification see Wiseman (1970).

γ-lysin Morgan and Graydon (1936) suggested that there were two serologically distinct haemolysins for rabbit erythrocytes and referred to them as α_1 and α_2. Smith and Price (1938*b*) described a γ-toxin, which caused rapid lysis of the red corpuscles of the rabbit and sheep but not of the horse, and was distinct from the α- and β-haemolysins. This toxin has now been purified and characterized (Guyonnet and Plommet 1970, Möllby and Wadström 1971, Taylor and Bernheimer 1974, Fackrell and Wiseman 1976); it is a protein that is antigenically distinct from other haemolysins; it acts strongly on sheep, rabbit and human erythrocytes; it is non-toxic for mice but large doses kill guinea-pigs; it has some leucocidal action but is not dermonecrotic.

δ-lysin This haemolysin was observed by Williams and Harper (1947) on blood agar plates, and was active on the red cells of most animals. Marks and Vaughan (1950) found that it was dermonecrotic and resisted boiling. It rapidly lyses leucocytes and a variety of other tissues (Hallander and Bengtsson 1967, Gladstone and Yoshida 1967), and also disrupts lysosomes, and bacterial protoplasts and sphaeroplasts (Kreger *et al.* 1971). It is said to have antibiotic activity on certain gram-positive organisms (see Kreger *et al.* 1971). Kapral and his colleagues (1976) showed that δ-toxin inhibited the absorption of water through the rabbit and guinea-pig intestine; the level of cyclic adenosine monophosphate in the gut contents was increased, but other tests for the stimulation of adenylcyclase production (see Chapter 27) were negative (O'Brien and Kapral 1977).

This lysin has proved difficult to purify and assay. According to Murphy and Haque (1980), preparations uncontaminated with other extracellular products have not yet been obtained; the toxic activities attributable to δ-lysin are therefore still in some doubt. Haemolysis by δ-lysin is non-specifically inhibited by serum lipoproteins, which must be removed before antibody can be detected (Fackrell and Wiseman 1974, Birkbeck and Whitelaw 1980). Antigenicity is poor and is more regularly domonstrable in rabbits than in mice. Chao and Birkbeck (1978) have shown that the erythrocytes of some marine fish, notably cod, are highly susceptible to δ-lysin but not to any other staphylococcal haemolysin. According to Turner (1978*a*, *b*; see Turner and Pickard 1980), certain canine strains—presumably of *Staph. intermedius*—form a δ-lysin that gives only partial reactions of identity with the δ-lysin of *Staph. aureus* strain Wood 46, and differs from it in other respects.

(For a review of staphylococcal haemolysins, see Wiseman 1975.)

The detection of staphylococcal haemolytic toxins

Attempts to determine the distribution of haemolysins among staphylococci have been hampered by the absence of pure antisera against the various lytic antigens. Tests may be made on broth cultures, but are rather laborious. Therefore, tests on blood agar plates are preferred (Bryce and Rountree 1936). Gillespie and Simpson (1948) placed a strip of filter paper soaked in antitoxic serum on a rabbit blood agar plate and identified α-lysin by observing inhibition of haemolysis. Elek and Levy (1950) modified this method to detect α-, β-, and δ-lysins, alone or in any combination. Two plates, one of rabbit and one of sheep blood agar were taken; a strip of paper soaked in a serum known to contain much anti-α, less anti-β, and little anti-δ lysin was placed across the diameter, and staphylococci were streaked up to the strip at right angles. When α-lysin was present there was on both plates a wide zone of complete laking with a hazy edge, which was inhibited at some distance from the paper strip. The effect of β-lysin was to cause darkening, or even partial haemolysis, on the sheep but not on the rabbit blood plate, which was inhibited only in a V-shaped area close to the paper strip. The δ-lysin was indicated by narrow, sharply demarcated areas of laking on both plates, which were little inhibited by the serum. There are a number of discrepancies between the results of tube and plate tests (Elek and Levy

1954), and γ-lysin cannot be detected in plate tests because it is inhibited by agar. Another way to identify haemolysins is to separate them electrophoretically in agar gel and then localize them by running agar suspensions of the appropriate erythrocytes into trenches cut in the agar (Haque 1967). If the appropriate substrates are used, the method can be modified to detect a number of other extracellular products (Ali and Haque 1974, Madler *et al.* 1976). Use may also be made of the ability of sodium azide to inhibit the production of α- and δ-lysins, but not of β-lysin, in sheep blood agar (Binder and Blobel 1970).

Leucocidins

The leucocidal activity of toxic preparations has been studied by two different methods. In the Neisser–Wechsberg (NW) method, their ability to inhibit reduction of methylene blue by rabbit leucocytes is measured (Neisser and Wechsberg 1901); in the Panton–Valentine (PV) method, destructive action on human leucocytes is observed microscopically (Panton and Valentine 1932). According to Valentine (1936) and Wright (1936), the NW leucocidin is identical with α-haemolysin and PV leucocidin is different from it; but Proom (1937) showed that both leucocidins impaired the respiratory activity of rabbit leucocytes, though α-lysin did not destroy the cell and its nucleus. Jackson and Little (1957) found that δ-lysin also had leucocidal activity; and Gladstone and van Heyningen (1957) distinguished clearly between the three leucocidins by their effect on the morphology of human polymorphonuclear leucocytes: α-lysin caused no gross changes; δ-lysin caused nuclear swelling with loss of lobulation and subsequent lysis of many of the cells; PV leucocidin caused loss of motility and progressive disappearance of the granules but the cell was not disrupted. Woodin (1959, 1960, 1961) showed that PV leucocidin could be separated chromatographically into a 'fast' (F) and a 'slow' (S) component, and that neither component alone had any leucocidal effect. Both are proteins which have since been purified and crystallized. The effect of PV leucocidin on leucocytes has been studied exhaustively, and the evidence suggests that the F and S components act successively on the triphosphoinositide of the leucocyte cell membrane (see Woodin 1970). The two components form separate antibodies on injection into animals. For a further account of PV leucocidin see Woodin (1970), and for methods of detecting antibody to it see Chapter 60.

Coagulase ('free coagulase')

The ability of certain staphylococci to coagulate plasma was first described by Loeb (1903–04) and confirmed by Much in 1908, but its significance as a means of defining *Staph. aureus* was not generally recognized until nearly 30 years later (for references, see Elek 1959). Coagulation is brought about by an extracellular substance often called 'free coagulase' to distinguish it from 'bound coagulase' or clumping factor (p. 225). Coagulase is formed by *Staph. aureus*, *Staph. intermedius*, and by a few animal strains that are at present thought to belong to *Staph. epidermidis* biotype 2 (see p. 237).

It is thermostable; considerable activity is retained even after heating at 100° for 30 min (Vanbreuseghem 1934) or after autoclaving (Walker *et al.* 1948). It is formed abundantly in infusion or digest broth during the logarithmic phase of growth, and its release from the cell is enhanced by the presence of albumin (Duthie 1954*b*, Altenbern 1966). It may be concentrated and partly purified by chemical means (see Murray and Gohdes 1960).

Coagulase does not clot purified fibrinogen, and does not require the presence of calcium to clot plasma. Smith and Hale (1944) showed that for coagulation a third substance, in addition to coagulase and fibrinogen, must be present. They considered that the interaction of coagulase with this activator, generally referred to as the coagulase-reacting factor, resulted in the production of a thermostable thrombin-like substance which converted fibrinogen into fibrin. Coagulase-reacting factor is found in the globulin fraction of serum (Tager 1948, Miale 1949), closely resembles prothrombin, and may even be identical with it (Tager and Lodge 1951, Duthie and Lorenz 1952, Tager 1956, Rubinstein 1958). The existence of the coagulase-reacting factor offers a partial explanation of the behaviour of plasmas from different animal species in the coagulase test. Mouse and fowl plasmas are ordinarily not coagulated by staphylococci, and this has been attributed to the absence of coagulase-reacting factor. Guinea-pig plasma is as a rule slowly clotted at 20° but not at 37°. However, some strains bring about rapid clotting of guinea-pig plasma, and a few even clot mouse plasma. Moreover most strains from sheep and cattle, but very few from man, cause rapid clotting of sheep and bovine plasma (Stamatin *et al.* 1949, G. S. Smith 1954, Meyer 1966*a*); strains from sheep and cattle also clot human and rabbit plasma, though a few bovine strains clot rabbit plasma only at room temperature overnight (Edwards and Rippon 1957). Dog and pigeon staphylococci (now classified as *Staph. intermedius*) coagulate bovine plasma strongly, but have a weak and uncertain action on human plasma (Haughton 1966, Hájek and Marsálek 1969); dog strains clot dog plasma but human strains do not. Pig strains coagulate human plasma but not bovine plasma (Hájek and Marsálek 1970). Illès (1963) observed that *Staph. aureus* strains that did not coagulate bovine plasma alone did so when human serum was added. This suggests that certain animal species have different coagulase-reacting factors, and that there is specificity in the reaction with them of the varieties of *Staph. aureus* from different animal hosts.

Coagulase may be assayed by a clotting-time method (see Wegrzynowicz *et al.* 1969), or by measuring the increase of light-scattering that occurs when it is incubated with fibrinogen and coagulase-reacting factor (Stutzenberger and San Clemente 1967). Antibody to coagulase can be produced in rabbits. It is also demonstrable in some human sera, especially in those from patients suffering from chronic staphylococcal infections (Gross 1933, Lominski and Roberts 1946, Rammelkamp *et al.* 1950, Duthie and Lorenz 1952). There is evidence of several antigenically distinct coagulases (Rammelkamp *et al.* 1950, Duthie 1952). Barber and Wildy (1958) showed that members of each of the three main phage groups (I, II and III) produced antigenically distinguishable coagulases, as did strains lysed by phages 3A, 42E and 187.

Coagulase is non-toxic, but it has been suggested that it may play some part in initiating staphylococcal lesions by causing the deposition of masses of fibrin around the cocci and so protecting them from phagocytosis (Hale and Smith 1945, Smith *et al.* 1947). Pathogenic staphylococci multiply in undiluted normal human serum but non-pathogenic staphylococci do not (Spink and Paine 1940). Ekstedt and Nungester (1955) concluded that coagulase protected *Staph. aureus* from a bacteristatic factor in the serum, but Cybulska and Jeljaszewicz (1966) could not confirm this. Nevertheless, free coagulase production and the ability to grow in serum are highly correlated (Cybulska and Jeljaszewicz 1965).

The tube-coagulase test This may be performed in a variety of ways (see Williams and Harper 1946), but the following can be recommended. About 0·1 ml of an overnight broth culture, or of a broth suspension of an agar slope culture made up to the same density as a broth culture, is mixed with 1·0 ml of a freshly prepared 1 in 10 dilution of plasma in saline or broth. The mixture is incubated at 37° for 3–6 hr; if no clot has appeared in this time it should be left overnight at room temperature and again examined. Since plasma may undergo spontaneous coagulation, and since some specimens of plasma are more suitable for the test than others, control tubes containing diluted plasma alone, as well as others inoculated with known coagulase-positive and coagulase-negative cultures, should always be put up. Media containing fermentable carbohydrates should be avoided. The plasma may be stored undiluted in the ice-chest for several months, or can be freeze-dried. Human, rabbit or pig plasma may be used, and may be oxalated, heparinized or treated with EDTA, but merthiolate and liquoid inhibit the reaction. It is important to ensure that the cultures tested are, in fact, staphylococci. Many gram-negative bacilli (Harper and Conway 1948) and enterococci (Evans *et al.* 1952) utilize citrate and so bring about the coagulation of citrated plasma. Several organisms, such as *Ps. aeruginosa*, *Ser. marcescens*, and some strains of *Actinomyces* and *Y. pestis*, also cause the clotting of plasma by mechanisms not connected with the utilization of citrate.

Plate tests for coagulase When coagulase-positive staphylococci are grown on plasma agar, a zone of turbidity due to the deposition of insoluble fibrin appears around the colonies (Penfold 1944); but zones of opacity may be produced in serum and in plasma agar plates by a variety of agents other than coagulase, and the opacity due to coagulase is often dispersed by proteolytic enzyme. Media in which plasma is replaced by purified fibrinogen and prothrombin are not open to this objection (van der Vijver *et al.* 1972).

Staphylokinase

Staph. aureus digests fibrin (Minett 1936) by forming staphylokinase which, like streptokinase (Chapter 29), activates plasminogen (Lack 1948). It acts on human, dog, guinea-pig, and rabbit plasma, but not on ox, sheep, horse, rat, or chicken plasma (Gerheim and Ferguson 1949). It can conveniently be demonstrated on plates by the clearing of heat-precipitated fibrin (Christie and Wilson 1941). The *Müller phenomenon*—the appearance of discrete spots of clearing in blood agar plates at some distance from the growth of *Staph. aureus*—is now believed to be due to staphylokinase diffusing from the staphylococcus and acting on particulate plasminogen; and the heat-stable inhibitor of the Müller phenomenon found in the serum of normal adults is probably anti-staphylokinase (Quie and Wannamaker 1961, 1962).

The genetic determinant for staphylokinase production may be chromosomal or part of a phage genome; converting phages may determine staphylokinase production only, or may also cause the suppression of β-lysin formation (Winkler *et al.* 1965, Kondo and Fujise 1977). The strong negative correlation between β-lysin and staphylokinase production suggests that the third determinant may be widely distributed.

Hyaluronidase

After the discovery that the 'spreading factor' of Duran-Reynals (1933) was hyaluronidase (Chain and Duthie 1940), the production of this enzyme by staphylococci, and its possible role in the production of staphylococcal lesions, was widely studied (see Elek 1959). Nearly all strains of *Staph. aureus* form hyaluronidase, but in varying amount; a clear relation between the amount of the enzyme formed and potential pathogenicity has not been established.

Enterotoxins

Some coagulase-positive staphylococci produce toxins that cause acute vomiting and diarrhoea in man, and so give rise to staphylococcal food poisoning (Chapter 72) and probably also some cases of post-antibiotic diarrhoea (Chapter 60). In the former the toxin is ingested in the food; in the latter the staphylococcus is assumed to form it in the lumen of the gut, but no direct evidence of this has yet been obtained. Enterotoxins can be produced in the laboratory by a variety of methods, including growth in semi-solid agar medium with added CO_2 or in a gently agitated deep liquid culture. For the production of small quantities of concentrated toxin, growth in a cellophan sac lying on the surface of fluid medium in a Roux bottle (Casman and Bennett 1963), or on a cellophan sheet on an agar plate (Hallander 1965), is suitable. To obtain high yields, certain protein hydrolysate media supplemented with thiamin and nicotinic acid are recommended (see Bergdoll 1970).

Early tests for enterotoxin were made by feeding filtrates to human volunteers (Dack *et al.* 1931), and much subsequent work was done with rhesus monkeys or kittens. The monkey is susceptible by the oral route (Jordan and McBroom 1931); the toxin is best given by stomach tube (Surgalla *et al.* 1953). Vomiting is the only trustworthy sign of a positive reaction. Unfortunately the monkey is only about one-tenth as susceptible as is man, and rapidly becomes tolerant on repeated feeding. At least six monkeys are needed for each test, and two or more must vomit if the result is to be considered positive. The chimpanzee is apparently more susceptible than the rhesus monkey (B. J. Wilson 1959).

Enterotoxin can also be demonstrated by its ability to cause vomiting and diarrhoea in kittens when injected intraperitoneally (Dolman *et al.* 1936), but this test proved difficult to standardize, and for a time it fell into disrepute. It was re-examined by Matheson and Thatcher (1955) and found to be reliable if the experimental conditions were strictly controlled (see also Thatcher and Matheson 1955). The intravenous injection of filtrates into kittens has also given reliable results (Davison *et al.* 1938, Casman 1958). Kittens are, however, relatively refractory to the C-type enterotoxin.

Attempts to concentrate and purify the enterotoxins were at first very laborious because the process had to be monitored by assay of the toxin in monkeys or kittens. It soon appeared, however, that purified enterotoxin was antigenic for rabbits and monkeys (Bergdoll *et al.* 1959); and once reliable antisera had been made, immunological tests could be substituted for assay *in vivo*. Treatment with formalin inactivates much of the toxic effect of the inoculum, including its enterotoxic action on the monkey.

Detection of enterotoxins Several types of precipitation test may be used: the single-diffusion tube method of Oudin is useful for quantitative determination of toxin; the double-diffusion tube method of Oakley can be used to detect very small amounts of toxin; the Ouchterlony plate, or its micro-slide modification, is a sensitive method of detecting enterotoxin in food extracts or culture fluids (see Hall *et al.* 1963, 1965, Casman *et al.* 1969, Bergdoll 1970). Single radial-immunodiffusion tests on slides are useful for screening cultures for enterotoxin production (Meyer and Palmieri 1980). Enterotoxin may also be adsorbed on to tanned or glutaraldehyde-treated erythrocytes and used in a haemagglutination-inhibition test (Morse and Mah 1967, Shibata *et al.* 1977); or tanned cells (Silverman *et al.* 1968) or latex particles (Salomon and Tew 1968) may be coated with antitoxin and used in a reversed passive haemagglutination test. Radioimmunoassay (Johnson *et al.* 1971, Miller *et al.* 1978) or enzyme-linked immunosorbent assay (Stiffler-Rosenburg and Fey 1978) may be used.

Properties of enterotoxins Early in the immunological studies it became clear that there were at least two antigenic types of enterotoxin (Bergdoll *et al.* 1959). Enterotoxin B, which is easy to produce *in vitro*, was the first to be studied in detail. It appears, however, to be an uncommon cause of food-poisoning, but is the type most often associated with staphylococcal post-antibiotic diarrhoea (Chapter 60). Enterotoxin A, on the other hand, is formed by some three-quarters of the *Staph. aureus* strains responsible for food-poisoning outbreaks (Wieneke 1974). Four other types of enterotoxin were described subsequently, C, D, E and F (see Bergdoll 1972), and each appears to have been responsible for occasional outbreaks of food-poisoning. A number of strains form more than one type of enterotoxin. Some 40 per cent of strains of *Staph. aureus* of human origin form one or more enterotoxins; the frequency of production of individual toxins is: A, nearly 20 per cent; B, C and D, 7–11 per cent; E, 2 per cent. Enterotoxin production is much less common among animal strains, and enterotoxins A and B are seldom formed (Olson *et al.* 1970, Hájek and Maršálek 1973).

(See Chapter 60 for the possible role of enterotoxin F in toxic shock syndrome.)

Enterotoxins are trypsin-resistant proteins that resist boiling and have molecular weights of 28 000–35 000. Their mode of action is uncertain. The site of emetic action in the monkey is in the abdominal viscera; the sensory stimulus travels from the abdomen to the vomiting centre via the sympathetic nerves and vagus (Sugiyama and Hayama 1965). Enterotoxin B causes enteritis when given by the mouth to monkeys (Kent 1966), dogs, and chinchillas, but in monkeys the emetic effect and the enteritis do not always appear together (Bergdoll 1970). There are unexplained similarities between the effects on experimental animals of staphylococcal enterotoxin and the endotoxin of gram-negative bacteria (Sugiyama 1966). Repeated feeding of enterotoxin to monkeys leads to the rapid development of a temporary tolerance (Sugiyama *et al.* 1962). Injected intravenously into rabbits and cats it causes pyrexia (Martin and Marcus 1964) to which tolerance also develops rapidly, but cross-tolerance with endotoxin has not been observed. Animals given enterotoxin parenterally become more susceptible to the lethal action of endotoxin (Sugiyama *et al.* 1964) and to necrosis at the site of a subsequent intradermal injection of adrenaline (Bokkenheuser *et al.* 1963). Enterotoxins do not in general cause dilatation of ligated segments of rabbit ileum, but Koupal and Deibel (1977) studied one staphylococcal strain that had this property. The material responsible was a trypsin-resistant protein that had the serological specificity of enterotoxin A, but its effect on the ileal loop was neutralized by antisera against the A and B toxins.

For further information, see Bergdoll (1970, 1972) and Bergdoll *et al.* (1974).

Epidermolytic toxins

Staph. aureus strains from vesicular and exfoliative diseases in man form toxins that bring about separation of the epidermis when injected into baby mice (p. 235). These toxins are proteins of molecular weight *ca* 25 000. Two serological types have been described: A, which is heat stable, and B, which is heat labile (Kondo *et al.* 1974). In Europe and North America, the predominant epidermolytic-toxin forming strains belong to phage-group II (p. 232) and most of them form both toxins or only the A toxin; in Japan, many of the strains have other phage-typing patterns and form only the A toxin (Kondo *et al.* 1975). Many of the phage-group II strains that cause exfoliative diseases form a particular bacteriocine (p. 232). The ability to produce exfoliative toxin and bacteriocine is often eliminated by growth at high temperature with ethidium bromide, and this is associated with the loss of a plasmid that determines both characters (Warren *et al.* 1974, 1975). However, some strains continue to form toxin after the plasmid has been eliminated, suggesting that they also have a chromosomal determinant for toxin production (Rogolsky *et al.* 1976). According to Arbuthnott (1981), production of A toxin is determined chromosomally and of B toxin by a plasmid that also carries the gene for bacteriocine production; but some strains that form the A toxin have plasmids that determine only bacteriocine production.

The action of epidermolytic toxins is to cause separation of epidermal cells along the line of cleavage just below the stratum granulosum, but the cells are apparently not killed. Purified toxin causes epidermal splitting in baby mice, and in monkeys and man but not in other species of animals. In man, natural disease caused by epidermolytic strains of *Staph. aureus* usually leads to the formation of vesicles or more extensive acute exfoliation of the skin. The evidence that they are also sometimes responsible for 'staphylococcal scarlet fever', an erythematous disease in which exfoliation occurs late and is limited in extent, may need re-evaluation (Chapter 60).

For further information about the epidermolytic toxins, see Rogolsky (1979) and Arbuthnott (1981).

Pyrogenic exotoxins

Schlievert and his colleagues (1979*a*, *b*, Schlievert 1981) have isolated and characterized substances with toxic properties similar to those of streptococcal pyrogenic exotoxins (or erythrogenic toxins; see Chapter 29). They are proteins of rather low molecular weight (*ca* 20 000) that are pyrogenic in rabbits and mice, enhance susceptibility to endotoxic shock, are potent non-specific mitogens for lymphocytes, and are immunosuppressive. They are antigenically distinct from all other staphylococcal extracellular products. Three antigenic types of pyrogenic exotoxin have been described (Schlievert *et al.* 1981). Types A and B are formed by most strains of *Staph. aureus* but type C appears to be more limited in distribution. Pyrogenic exotoxins have not been identified in *Staph. epidermidis.* Rabbits sensitized to one of these toxins develop erythematous or oedematous rashes when subsequently given an injection of the toxin of homologous type (Schlievert 1981).

In Chapter 60 we shall describe staphylococcal *toxic shock syndrome*, which is associated with the local multiplication of *Staph. aureus* at a carrier site or in a minor lesion and in which a scarlatiniform rash almost invariably appears. According to Schlievert and his colleagues (1981), this is attributable to the type C pyrogenic exotoxin; all strains of *Staph. aureus* isolated from patients suffering from this disease but only 16 per cent of strains from control subjects formed this toxin. For further discussion of the aetiology of toxic shock syndrome, see Chapter 60.

Production of toxins and other extracellular products

Most strains of *Staph. aureus* form several of the substances mentioned in the last section, but only the production of coagulase and heat-stable nuclease are constant characters of the whole species as at present defined, and these are shared with *Staph. intermedius* and occasional animal strains of *Staph. epidermidis.* There is, however, little to suggest that either coagulase or nuclease production are important virulence factors.

Staph. aureus and *Staph. intermedius* strains mostly form one or more of the haemolytic toxins α, β, or δ. Many other staphylococci are haemolytic on blood agar; they may form a broad-spectrum haemolysin resembling δ-lysin (Kleck and Donahue 1968), but these have not been studied in detail. The α-lysin, which has some claim to be considered an important virulence factor (Chapter 60), is formed by some 85–95 per cent of *Staph. aureus* strains of human origin, but non-producers occasionally cause severe infections (Lack and Wailling 1954); α-lysin is much less often formed by animal strains. Most human strains form δ-lysin, but only a few—less than one-quarter according to most observers—form β-lysin. Most animal strains of *Staph. aureus* form β-lysin and many form δ-lysin. Strains devoid of any lysin may cause pyaemic infections in chickens (H.W. Smith 1954). There is a strong negative correlation between the production of β-lysin and staphylokinase (Christie and Wilson 1941), so that staphylokinase is seldom formed by animal strains.

Quantitative comparisons of toxin formation by *Staph. aureus* strains from septic lesions and from healthy carriers are seldom informative. Strains from human carriers are occasionally poor formers of one or more of these substances, but most of them are indistinguishable in this respect from strains that have caused severe infections. Individual cells in a culture are far from uniform in their ability to produce haemolysins (Elek and Levy 1954), coagulase (Smith *et al.* 1952) and hyaluronidase (Rogers 1953), so studies of single-colony isolations may be misleading. Toxin production in the test-tube may be a poor guide to what happens in the body. Gladstone and Glencross (1960), working with staphylococci in cellophan sacs in the peritoneal cavity of animals, found that the production of α-lysin and of PV leucocidin, but not of coagulase, was much increased *in vivo.* Both lysin and leucocidin could be detected in this manner in a number of strains that did not produce them under optimal conditions *in vitro.*

None of the toxins or other extracellular products of *Staph. aureus* appears to be a 'general' determinant for pathogenicity in the species as a whole. The only agents to which a definite role can be assigned are the enterotoxins, the epidermolytic toxins, and perhaps the α-lysin.

Production of antibiotic substances

Staphylococci produce a number of **bacteriocines,** often referred to as staphylococcines (Fredericq 1946), that have a wide spectrum of activity, mainly on gram-positive organisms. In the early part of this century, it was their action on diphtheria bacilli that attracted most interest (see Abraham and Florey 1949). Their distribution among staphylococci has not been investigated in detail, but they are formed by both coagulase-positive and -negative staphylococci (Halbert *et al.* 1953, Jones and Edwards 1966). Within *Staph. aureus*, many different patterns of inhibition of one strain by another are seen (Pulverer and Sieg 1972). Preparations of staphylococcal δ-lysin are said to have antibiotic activity on a rather similar range of bacteria to those inhibited by staphylococcines.

The bacteriocine that has received most recent attention is that formed by many of the *Staph. aureus* strains that produce exfoliative toxin (p. 230). This was first recognized by its powerful action on *C. diphtheriae* (Parker *et al.* 1955), but was later shown to inhibit all strains of *Staph. aureus* that did not produce it, as well as pneumococci and many streptococci (Parker and Simmons 1959, Dajani and Wannamaker 1969), and on *Neisseria gonorrhoeae* (Morriss *et al.* 1978). It is a heat-resistant but trypsin-sensitive protein of high molecular weight that kills but does not lyse bacteria (Dajani *et al.* 1970*a*); the RNA is disintegrated and the cytoplasm undergoes dissolution leaving cell-wall 'ghosts' (Dajani *et al.* 1970*b*, Clawson and Dajani 1970). The ability to form this bacteriocine is genetically determined by a plasmid that usually also carries the genes for the production of epidermolytic toxin B.

(For lysostaphin, see p. 222).

Bacteriophage typing

The temperate phages of *Staph. aureus* have a narrow host range and lyse only some other strains of the same species (Fisk 1942); they are therefore suitable for use as typing phages. The present phage-typing system is a direct descendant of that of Wilson and Atkinson (1945), as modified by Williams and Rippon (1952). These workers showed that a staphylococcus could seldom be characterized by lysis by a single phage, but that many different patterns of lysis could be obtained with a set of phages. The differences between these patterns could be used to make fine distinctions between staphylococcal strains. But the patterns given by strains of common origin were not always identical. It was therefore necessary to record the strength of the reactions, and to take account of the known range of variability in interpreting the results. In general, a pair of strains are probably unrelated if their patterns differ by two or more 'strong reactions' (i.e. >50 plaques with one culture and none with the other) when tested with the phages at the routine test dilution (RTD). The phage-typing pattern is expressed as a list of the phages by which the staphylococcus is lysed strongly, e.g. 3C/55/71.

Some 70 per cent of all *Staph. aureus* strains are lysed by one or more of the phages when tested at RTD. This percentage is increased to *ca* 90 per cent by re-testing the untypable cultures with the same phages at 1000 times the RTD strength (RTD $\times$ 1000). But typing with the stronger phages is not so reproducible as typing at RTD. It is now recommended, therefore, that the strength of phage for secondary typing should be RTD $\times$ 100 (Report 1971). This results in a slight reduction in the percentage of typable cultures but is justified by increased accuracy.

A basic set of typing phages was established by international agreement in 1953 and has since undergone several modifications (see Report 1959, 1963, 1967, 1971, 1975 and Fig. 30.2). A number of new phages were introduced, either to characterize pre-existent strains (e.g. phage 71) or because a general fall in the typability-rate indicated the appearance of new strains either in hospitals or in the general population (e.g. phages 80, 81, 83A, 84, 85, 94, 95 and 96). When new phages were introduced, others were removed, with the object of keeping the number within the limits dictated by the use of a single plate for typing. With the present 23 phages, several hundred different phage-typing patterns are possible, but certain combinations of reactions occur much more frequently than others. Thus, lysis by phage 52 is much more likely to be accompanied by lysis by phage 29, 52A, 79 or 80, or a combination of these, than is lysis by other phages. Patterns including 3C often also contain 3A, 55 or 71, but rarely members of the 29/52 etc. group. Thus, it is possible to subdivide the typing phages into a small number of *lytic groups*, members of which appear together in typing patterns, and to divide staphylococci into corresponding *phage groups* (Williams and Rippon 1952). Most human strains fall into one or other of the phage groups I, II, III and V. Strains lysed by phage 42D are rarely of human origin. Two phages, 81 and 95, cannot be allocated to any of the lytic groups.

Susceptibility to particular phages or combinations of phages is in several instances associated with other biological characters in the staphylococcus. In general, members of phage groups I, II and III correspond to the similarly numbered serotypes in Cowan's system and possess antigenically distinct coagulases. Enterotoxic food poisoning is caused mainly by members of group III, but this cannot be entirely explained by the distribution of enterotoxins in the phage groups (Chapter 60). Antibiotic resistances appeared very much more slowly in members of phage group II than in groups I and III. Certain other characters are associated more with particular patterns than with the phage group. In group I, the 52, 52A, 80, 81 complex of strains, which were prevalent in hospitals between 1954 and 1964, had a remarkable ability to cause widespread outbreaks of severe sepsis. Strains that cause exfoliative diseases and produce epidermolytic toxin (p. 230) usually belong to phage-group II and most of them are lysed only by phage 71 (Parker *et al.* 1955); they always give a negative egg-yolk reaction.

The preparation of the typing phages is best undertaken by a single laboratory in each country. The host range of the

Lytic group	Human 1953	Human 1974	Bovine 1970
I	29 52 52A 79	29 52 52A 79 (80)	29 52A
II	3A (3B) 3C 55	3A 3C 55 (71)	3A 116
III	6 7 42E 47 53 54 (70) (73) 75 77	6 42E 47 53 54 75 77 (83A) (84) (85)	6 42E 53 75 84
IV	(42D)		42D 102 107 117
V		(94) (96)	
...		(81)	78
...		(95)	118 119

( ) = subsequently removed ◯ = added since 1953 —— = also in human typing set

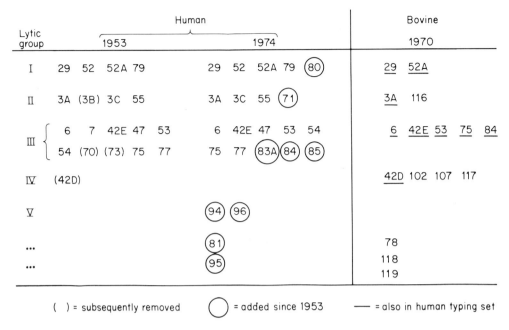

Fig. 30.2 Constitution of international basic set of phages for typing *Staphylococcus aureus* strains of human origin at its inception (1953) and at present (1974); constitution of international basic set of phages for typing strains of bovine origin (1970).

propagated phage should conform closely to that of a standard preparation in parallel tests. Before typing, the strength of each phage must be measured by titration on the propagating strain and adjustment made if necessary. Many of the phages are calcium dependent and the typing medium must contain at least 400 mg of ionized calcium per l. Plates are flooded with log-phase culture; after drying, a standard drop of each phage is placed on the plate by means of a mechanical applicator. The plate is examined after overnight incubation at 30°. (For further technical details see Blair and Williams 1961, Parker 1972.)

The typing system we have described was developed for the study of staphylococcal disease in man, and should now be looked upon as a means of typing members of the human biotype A of *Staph. aureus* (p. 236). Some members of the bovine biotype C are lysed by phage 42D or by other phages from bovine staphylococci that have a similar host range; these phages form the lytic group IV (Meyer 1967*a*). Other bovine strains give wide and inconstant patterns with phages of lytic groups I and III. A separate set of phages for bovine staphylococci was agreed upon after an international collaborative study (Davidson 1972; Fig. 30.2). Porcine strains of biotype B are lysed by phages of the human set, usually of groups I and III, and strains from hares (biotype D) only by group II phages. *Staph. intermedius* strains are generally not lysed by the human or bovine phages, but may be typable by means of phages derived from staphylococci of homologous species.

The temperate phages of *Staph. aureus* are absorbed by all members of the species. In general, however, the phages carried by a particular strain will lyse only staphylococci belonging to the same phage group (Rosenblum and Dowell 1960). There are at least six serological groups of phages, but phages of different serological groups may belong to the same lytic group (Rountree 1949, Rippon 1956). The insusceptibility of staphylococci to lysis by the typing phages appears to be due partly to the activity of the restriction and modification systems of the various strains, which appear to be the main determinants of phage group (van Boven *et al.* 1974, Stobberingh and Winkler 1976). Brief heating of a staphylococcus at 56° often broadens its pattern of lysis by inactivating the restriction endonucleases. However, lysogenic immunity is also of great importance in determining the pattern of lysis within the phage group and sometimes outside it. Lysogenization of a staphylococcus sometimes narrows its pattern of lysis; less often it widens it, and this may be due to prophage substitution. Both of these processes are concerned in changes that can be brought about in the 80/81 staphylococcus. Certain converting phages change the pattern from 80/81 to 80, or 81 or to untypability; others, however, that replace a defective phage present in the 80/81 staphylococcus widen the pattern to 52/52A/80/81 (Rountree 1959, Asheshov and Rippon 1959, Rountree and Asheshov 1961). But the relatively narrow 80/81 pattern is itself due to temperate phages, and when two of these are lost the organism has a broad pattern of lysis by phages of groups I and III (Asheshov and Winkler 1966).

Lysogenization by phage may change other characters of the staphylococcus. The acquisition of certain phages which change the typing pattern from 80/81 to 80 (Rosendal *et al.* 1964), and of others which change it from 83A/84/85 to 84/85 (Jevons *et al.* 1966), results in a simultaneous loss of Tween-splitting activity; sometimes the latter change is accompanied by the appearance of lemon-yellow pigmentation. Loss of phage reverses the process. Certain phages of lytic group II cause a simultaneous conversion of a staphylococcus that is β-lysin negative and staphylokinase positive to one having the opposite characters (Winkler *et al.* 1965). It appears, therefore, that the determinants for several phenotypic characters or their repressors may be part of a phage genome.

A number of different phage-typing systems for coagulase-negative staphylococci have been proposed, but none of them has yet received international recognition (see Pulverer *et al.* 1979).

Pathogenicity

Disease in man and animals

In man, *Staph. aureus* causes a variety of septic diseases that are described in Chapter 60. Here it is necessary to note only that primary skin lesions are of two types: subepidermal (the boil), and intraepidermal (the various epidermolytic diseases). In addition, it may also cause vomiting and diarrhoea by virtue of the production of enterotoxin (Chapter 72).

Staph. epidermidis, usually of Baird-Parker's (1974) biotype 1, may cause minor superficial sepsis; occasionally it is responsible for severe septicaemic disease in certain classes of susceptible persons. *Staph. saprophyticus* of Baird-Parker's biotype 3 causes a specific type of urinary-tract disease in young women (Chapter 60).

Staph. aureus is responsible for mastitis in cattle, sheep and goats (Chapter 57), tick pyaemia in lambs, and suppurative arthritis in domestic fowl, as well as a variety of sporadic septic lesions in other domestic and wild animals. Little is known about the natural pathogenicity of *Staph. intermedius*, but it certainly appears to cause local sepsis in dogs. Some *Staph. epidermidis* strains, often called *Staph. hyicus*, cause skin lesions in pigs.

Experimental infection

Man Garré in 1885 (see Neisser 1912) found that by rubbing staphylococci into the skin of his arm he was able to produce boils. Several other workers have since established that the intradermal injection of strains of *Staph. aureus* from boils and abscesses, and some also from the nose of normal persons, results in the appearance of small boils, but it is usually necessary to inject at least a million organisms. If, however, a small number of organisms is introduced into the skin on a suture, a large abscess may result; lesions have been produced with as few as 300 organisms (Elek and Conen 1957). Removing the outer layer of the epidermis, placing as few as 100 *Staph. aureus* on the damaged skin, and sealing the area with an occlusive dressing leads to an acute cellulitis (Marples and Kligman 1972).

Soon after Garré's original experiment, Almquist (1891) injected a staphylococcus from a case of neonatal pemphigus into himself; this gave rise to a superficial blister 1 cm in diameter. Similar results were obtained with strains from cases of tropical pemphigus by Clegg and Wherry (1906) and of staphylococcal impetigo by Bigger and Hodgson (1943). These vesicle-forming staphylococci also occasionally cause lesions when rubbed into the scarified skin (Sheehan and Fergusson 1943, Bigger and Hodgson 1943).

Rabbits The *subcutaneous* injection of 1 ml of a 24-hr broth culture of *Staph. aureus* gives rise to a local abscess, from which the organisms can be recovered. If the culture is mixed with melted agar and inoculated intracutaneously into the rabbit's back, a spreading necrotic lesion occurs (Jackson *et al.* 1940). An extensive oedematous and haemorrhagic lesion results from the injection of as few as 10 000 organisms into the area of a necrotic burn (Johnson *et al.* 1961). Susceptibility to intradermal inoculation of *Staph. aureus* is increased by the local or intravenous injection of adrenaline (Miles and Niven 1950), or of the endotoxin of certain gram-negative bacteria (Conti *et al.* 1961). *Intravenous* injection of large doses of *Staph. aureus*, *e.g.* 0·1–0·5 ml of a broth culture, usually causes death in 24–48 hr. Smaller doses (e.g. 0·01–0·05 ml) give rise to a pyaemia proving fatal in 1–6 wk. *Post mortem*, abscesses are found in internal organs, particularly the kidneys. Endocarditis may develop without previous artificial wounding of the heart valves. Continuous passage by the intrathoracic route greatly enhances the virulence of the organism (Adlam *et al.* 1968).

Mice A large dose of *Staph. aureus* given *intravenously* causes rapid death; the LD50 for death within 24 hr is usually over 10^8 organisms (Gorrill and McNeil 1963). Death also occurs after the injection of somewhat smaller doses, but it is usually delayed. The LD50 for death in up to 14 days is about 4×10^6 organisms; the later deaths are usually associated with the presence of multiple microscopic abscesses in the kidneys (Gorrill 1951). The intracerebral LD50 is *ca* 5×10^5 organisms. The mouse is less susceptible to *intraperitoneal* injections of ordinary strains of *Staph. aureus* and usually survives doses of at least 10^8 organisms; capsulated strains, however, cause death within a few hours in much smaller doses. *Subcutaneous* injections of washed suspensions into baby mice regularly causes a lethal infection (McKay and Arbuthnott 1979). In adult mice it is necessary to give 10^6–10^7 organisms subcutaneously to induce the formation of a local abscess, but the implantation of as few as 10–100 organisms on a tied suture results in local suppuration (James and MacLeod 1961). A more convenient means of producing a skin lesion with a small measured dose is to introduce subcutaneously a small plug of cotton dust impregnated with staphylococci through a needle by means of a trocar (Noble 1965). The result is a characteristic abscess, at least 3 mm in diameter, surrounded by a white or necrotic zone. Intramammary injection in lactating mice cause mastitis and often death (Anderson 1974). Strains that form epidermolytic toxin give rise to generalized exfoliation when given intracutaneously or intraperitoneally to baby mice (Melish and Glasgow 1970).

Mice have been used to compare the virulence of staphylococcal strains by measuring the death-rates or the percentage of kidneys containing abscesses (Gorrill 1958) after intravenous injection; the death-rate after intraperitoneal injection with or without an adjuvant; the diameter of the thigh after intramuscular injection (Selbie and Simon 1952); and the production of a skin abscess with (Noble 1966) or without (Fisher 1965) an adjuvant. In 3-day-old mice, the death rate or the inhibition of normal weight-gain after subcutaneous injection may be used (Kinsman and Arbuthnott 1980). Cutler (1979) compared the virulence of strains by measuring the diameter of skin lesions after the subcutaneous injection of washed suspensions into guinea-pigs.

The coagulase-negative staphylococci are non-pathogenic when tested in laboratory animals under the usual conditions, but, when given intracerebrally to 2-day-old mice, *Staph. saprophyticus* strains from urinary-tract infections are nearly as virulent as *Staph. aureus* (Namavar *et al.* 1978*a*).

Classification

Staphylococcus, *Micrococcus* and *Planococcus*

These three genera comprise the catalase-positive cocci that are arranged in clumps or packets. They must be distinguished from packet-forming cocci, such as aerococci and pediococci, that are devoid of a cytochrome system and do not give a true catalase reaction (Chapter 29).

It is now evident that *Staphylococcus* and *Micrococcus* are sharply demarcated genera, differing not only in the composition of their DNA and cell wall (Table 30.2) but also in their cytochromes, menoquinones and respiratory enzymes. However, members of the two genera cannot easily be distinguished by means of tests that can be used in the routine laboratory. For some time, opinion favoured the use of fermentation of glucose (Evans *et al.* 1955) or the ability to grow anaerobically (Evans and Kloos 1972) as differentiating tests, but *Staph. saprophyticus* strains often gave equivocal or misleading results in one or both of these. In practice they can be relied upon when they indicate active fermentation or unequivocal anaerobic growth, which will be the case with *Staph. aureus*, *Staph. intermedius* and most strains of *Staph. epidermidis*.

Table 30.2 Usual characters of staphylococci and micrococci

	Staphylococcus	*Micrococcus*
Morphology: form cubical packets	−	v
Growth anaerobically	+†	−
Acid from glucose anaerobically	+†	−
Susceptibility to:		
lysostaphin*	S	R
erythromycin (0.4 mg per l)	R	S
furoxone (50 mg per l)	S	R
NaCl (15 per cent)	R	S
Acid from glycerol aerobically	+	−
Peptidoglycan: glycine in interpeptide bridge	+	−
Teichoic acid	+	−
DNA: moles per cent G+C	30–40	66–75

v = Some positive, others negative; S = sensitive, R = resistant.

*Unless sensitive also to lysozyme (see text).

†Many strains give negative or equivocal results.

At present, therefore, we are forced to rely on a series of other tests that so far have been evaluated only by testing quite small numbers of strains of known DNA-base content and cell-wall composition. Only one of these tests, for lysostaphin sensitivity, has even an indirect relation to the characters on which the two genera have been separated by the taxonomists. It is therefore unwise to rely upon the results of a single test. Schleifer and Kloos (1975*a*) favour the use in combination of two tests to recognize staphylococci: (1) growth and acid production aerobically on 1 per cent glycerol agar containing erythromycin 0·4 mg per l; and (2) inhibition of growth on an agar plate by one drop of lysostaphin solution (200 mg per l) but not by one drop of lysozyme solution (25 mg per l). Sensitivity to furoxone (p. 223) and growth in 15 per cent NaCl broth are useful additional characters.

The genus *Planococcus* (p. 238) is composed of marine cocci intermediate in DNA-base composition between *Staphylococcus* and *Micrococcus*.

Staph. aureus and *Staph. intermedius*

Until recently the view was held that the staphylococci that clotted plasma and formed a heat-resistant nuclease constituted a single species. This was easy to maintain as long as attention was confined to strains of human origin, but is now questioned by workers who have studied staphylococci from animals. Hájek (1976) proposes that certain animal strains formerly considered to be aberrant strains of *Staph. aureus* should be placed in a separate species, *Staph. intermedius*, because they share several characters with *Staph. epidermidis*. The boundary between *Staph. aureus* and *Staph. epidermidis* has been further eroded by the observation that other animal strains generally considered to belong to *Staph. epidermidis* biotype 2 (Baird-Parker 1974) may form DNAase and occasionally coagulase (Devriese and Oeding 1975).

Some of the characters of *Staph. aureus* and *Staph. intermedius* are given in Table 30.3. Although both coagulate rabbit plasma and form heat-resistant DNAase, some of the *Staph. intermedius* strains do not give a coagulase reaction in

Table 30.3 Usual reactions of *Staph. aureus*, *Staph. intermedius* and *Staph. epidermidis*

	Staph.		
	aureus	*intermedius*	*epidermidis*
Yellow or orange pigment	+	−	−
Coagulase*	+	+	−
Heat-stable DNAase	+	+	−
Acid from:			
mannitol (AnO_2)	+	−	−
mannitol (O_2)	+	−	v†
maltose (O_2)	v	−	v†
lactose (O_2)	+	+	v†
VP reaction	+	−	v†
Phosphatase	+	+	v†
Lysis by human and bovine phages	+	−	−
Requirement for biotin	−	+	+

AnO_2 = anaerobically; O_2 = aerobically; v = some positive, others negative.

*Clots rabbit plasma.
†See also Table 30.4.

human plasma and most of them give a weaker DNAase reaction than *Staph. aureus* strains. Both form haemolytic toxins, but *Staph. intermedius* appears not to form α-lysin; both are phosphatase positive. Characters in which *Staph. intermedius* resembles *Staph. epidermidis* rather than *Staph. aureus* are: failure to form an orange or yellow pigment, to acidify mannitol anaerobically, and to be lysed by phages of the human or bovine basic sets; requiring biotin for growth; and having some serine in the interpeptide bridges of the peptidoglycan. *Staph. intermedius* does not form the teichoic acids characteristic of either of *Staph. aureus* or of *Staph. epidermidis*. It does not acidify maltose or mannitol aerobically and it gives a negative Voges-Proskauer reaction, thus differing from *Staph. aureus* and many strains of *Staph. epidermidis*.

Host-adapted varieties of *Staph. aureus* and *Staph. intermedius*

The differences between *Staph. aureus* strains found in various species of mammals and birds are so striking that Meyer (1966*b*) proposed the recognition of three host-adapted varieties ('Standortvarianten') defined by differences in phage-typing pattern (p. 233), ability to clot plasma from various species of animals (p. 228), and type of growth on crystal-violet agar (p. 222): var. *humanus*, var. *bovinus*, and var. *caninus*. After a more comprehensive comparison of strains, Hájek and Maršálek (1971) recognized six biotypes A–F, of which two (E and F) were subsequently transferred to *Staph. intermedius* (Hájek 1976).

***Staph. aureus*, biotype A: human staphylococci** These clot human but not bovine plasma, usually give a purple growth on crystal violet agar, produce α-lysin more often than β-lysin and form staphylokinase, and are lysed by phages of the human set.

***Staph. aureus*, biotype B: porcine staphylococci** Like human staphylococci, they fail to clot bovine plasma and are lysed by human phages; they usually form β-lysin, rarely form staphylokinase, and give yellow colonies on crystal violet agar (Oeding *et al.* 1972).

***Staph. aureus*, biotype C: bovine staphylococci** Meyer's (1966*a*, *b*) var. *bovinus* clots bovine as well as human plasma, forms β-lysin, is generally lysed by group IV phages, and gives yellow colonies on crystal violet agar. However, several other biotypes probably occur in cattle; Meyer (1966*a*) described a group of strains lysed by phages of groups I and III that did not clot bovine plasma, and Oeding and his colleagues (1970) found many strains in the bovine respiratory tract that gave a slow coagulase reaction with bovine and human plasma, a purple growth on crystal violet agar, and reactions only with group-I phages.

***Staph. aureus*, biotype D: staphylococci of hares** This characteristic group of strains resembles human staphylococci in coagulase reaction and growth on crystal violet agar but forms β-lysin and not staphylokinase. The strains are lysed by phages of group II (Pulverer and Lucas 1966, Hájek and Maršálek 1971); members of this phage group isolated from man nearly always form α-lysin.

Staph. intermedius These organisms, formerly described as biotypes E and F of *Staph. aureus*, include var. *caninus* of Meyer (1966*a*). In addition to dogs, they are found in pigeons (Hájek and Maršálek 1969), mink (Oeding *et al.* 1973) and horses (Hájek *et al.* 1974). They have the characters of *Staph. intermedius* shown in Table 30.3. Many of them fail to clot human plasma, but they clot rabbit and bovine plasma. Unlike *Staph. aureus*, they rarely reduce tellurite. They grow well on crystal violet agar and, because they are non-pigmented, form white colonies on it.

This picture of host-specifity is rather over-simplified. A single host species may harbour several cultural varieties of staphylococci. This is so in cattle, and probably also in chickens; chicken strains include some apparently of the human biotype, and others resembling biotype B (Gibbs *et al.* 1978) that may however belong to a separate var. *gallinarum* (Witte *et al.* 1977).

Staphylococcus epidermidis and *Staph. saprophyticus*

Modern schemes for the classification of the coagulase-negative staphylococci stem from the work of Baird-Parker (1963, 1965), who divided staphylococci, which fermented glucose, into six biotypes SI–SVI, and micrococci, which did not, into eight biotypes M1–M8. His SI was *Staph. aureus*, and biotypes SII and SV were subsequently combined, so that he was left with four biotypes of coagulase-negative staphylococci, which he renamed *Staph. epidermidis* biotypes 1–4 (Baird-Parker 1974). It later appeared that the first four biotypes of his genus *Micrococcus* resembled staphylococci in G + C content of the DNA and in cell-wall composition; in 1974, therefore, he reclassified them as *Staph. saprophyticus* biotypes 1–4 (Table 30.4).

The tests relied upon for the recognition of the Baird-Parker biotypes are the phosphatase and Voges-Proskauer

Table 30.4 Biotypes of *Staph. epidermidis* and *Staph. saprophyticus*

	Staph. epidermidis biotype*				*Staph. saprophyticus* biotype			
	1	2	3	4	1	2	3	4
Earlier biotype designation†	SII,SV	SIII	SIV	SVI	M1	M2	M3	M4
Phosphatase	+	+	–	–	–	–	–	–
VP reaction	+	–	+	+	+	+	+	+
Acid from:								
mannitol	–	–	–	+	–	–	+	+
lactose	+‡	v	–	v	–	+	v	+
maltose	+	–	v	v	v	+	+‡	+§

v = Some strains positive, others negative.

*Baird-Parker (1974).
†Baird-Parker (1963).
‡Usual reaction.
§Also acidifies arabinose.

tests and the acidification of certain sugars. There is generally no difficulty in recognizing the two phosphatase-positive biotypes, *Staph. epidermidis* 1 and 2. The real difficulty is in distinguishing *Staph. epidermidis* 3 from *Staph. saprophyticus* 1 and 2, and *Staph. epidermidis* 4 from *Staph. saprophyticus* 3 and 4. Nearly all *Staph. epidermidis* strains are sensitive to novobiocin 2 mg per l; and many but by no means all *Staph. saprophyticus* strains are resistant (Namavar *et al.* 1978*b*). Thus, the novobiocin-resistance test is of value when it gives a positive result.

Kloos, Schleifer and their colleagues (Schleifer and Kloos, 1975*b*, Kloos and Schleifer 1975*a*,*b*, Kloos *et al.* 1976*a*) reclassified the coagulase-negative staphylococci; they used rather different testing methods and recognized eight species in addition to *Staph. epidermidis* and *Staph. saprophyticus*. A comparison of the descriptions with those of Baird-Parker (1974) suggests that (1) Kloos and Schleifer's species *epidermidis* corresponds closely to *Staph. epidermidis* biotype 1 (Baird-Parker); (2) that their species *capitis*, *haemolyticus*, *hominis*, *warneri*, and possibly *simulans* and *sciuri*, correspond generally to *Staph. epidermidis* biotypes 3 and 4, and their species *cohni* and *xylosus* to *Staph. saprophyticus* biotypes 1, 2 and 4 (Baird-Parker); and their species *saprophyticus* falls within *Staph. saprophyticus* biotype 3 (Baird-Parker). When the same set of cultures is classified by the two methods, however, many discrepancies have been reported. Marples (1981) found a wide scatter of *warneri*, *hominis* and *haemolyticus* strains in Baird-Parker's *Staph. epidermidis* and *Staph. saprophyticus* biotypes, and Namavar and his colleagues (1978*b*) found many strains of the species *haemolyticus*, *capitis*, *cohni* and *warneri* in Baird-Parker's *Staph. saprophyticus* biotype 3. Varaldo and his colleagues (Varaldo *et al.* 1978, 1980, Varaldo and Satta 1978) describe a method of classification of staphylococci by their bacterilytic activity under different cultural conditions on micrococci. They define six 'lyogroups' that are said to correspond closely to species or groups of species in the Kloos and Schleifer classification system.

***Staph. epidermidis* biotype 1 (Baird-Parker)** This is the predominant coagulase-negative staphylococcus of the human nose and axilla, but tends to be outnumbered by other staphylococci, and to a lesser extent micrococci, on the arms and legs (Kloos and Musselwhite 1975; for its role in human disease, see Chapter 60).

***Staph. epidermidis* biotype 2 (Baird-Parker)** Members of this biotype include the organism described by Sompolinsky (1953) as a cause of exudative dermatitis in pigs. Similar organisms are found in cows and chickens, in whom they do not appear to cause disease (Devriese and Oeding 1975). In addition to the characters given in Table 30.4, these strains may form heat-stable DNAase and clot rabbit plasma; they also form hyaluronidase. They, and other members of biotype 2, resemble *Staph. intermedius* in giving a negative Voges-Proskauer reaction, failing to ferment maltose, and giving white colonies on crystal violet agar. Devriese and his colleagues (1978) propose the name *Staph. hyicus* for biotype 2 and recognize two subspecies: var. *hyicus*, producing heat-stable DNAase and sometimes clotting plasma, and var. *chromogenes*, doing neither.

***Staph. saprophyticus* biotype 3 (Baird-Parker)** The members of this biotype that cause urinary-tract infection in young women are almost always novobiocin resistant and belong to Kloos and Schleifer's species *saprophyticus*. However, biotype-3 strains isolated from the skin of the arms and legs are less uniform in character. According to Namavar and his colleagues (1978*b*), they may be novobiocin resistant or sensitive, and may fall into one of several Kloos and Schleifer species.

For the distribution of staphylococci and micrococci on the skin, see Kloos and Musselwhite (1975) for human adults and children, Carr and Kloos (1977) for newborn babies, and Kloos *et al.* (1976*b*) for animals; see also Kloos (1980). Hill (1981) gives a comprehensive review of the classification of staphylococci (See also Feltham and Sneath 1982).

Micrococcus

Definition

Spherical or ovoid, non-sporing, gram-positive cocci that on solid media may be arranged in grape-like clusters or may form tetrads or cubical packets. Usually non-motile. On agar form opaque colonies that may be bright yellow or pink. Catalase positive. May give a faintly positive oxidase reaction. All aerobic and some strictly aerobic. Some acidify glucose oxidatively but none ferment it. Widespread in nature. DNA base composition: 66–75 moles per cent G + C.

The type species is *Micrococcus luteus*.

This definition excludes the anaerobic cocci (p. 238) and the spore-forming aerobic cocci (*Sporosarcina*, p. 238). As we have seen, the distinction between *Staphylococcus* and *Micrococcus* cannot always be made by tests for the fermentation of glucose, and on p. 235 we discussed ways of doing this in practice.

Morphology Some of the micrococci are a little larger than staphylococci, and, unlike staphylococci, some form tetrads or cubical packets. Some of the pink micrococci are motile.

Cultural characters Colonies are opaque and round, with a smooth or rough surface. Some strains form bright yellow and others pink colonies; many, however, have only a slight lemon-yellow tinge that is not intensified by growth on milk agar. Some yellow micrococci produce a violet pigment on potato but not on nutrient agar.

Micrococci are less tolerant of salt than staphylococci; none grow in the presence of 15 per cent and some not in 10 per cent NaCl. Many have an optimum temperature for growth of 30° or lower, and most grow at 10°. A few grow at 45°, and some, including strains from dairy equipment, survive heating at 60° or even 70° for 30 min.

Resistance *Micro. varians* is novobiocin resistant, but most of the others are sensitive. *Micro. luteus* is generally sensitive to lysozyme. All micrococci are resistant to the endopeptidase of lysostaphin (p. 222).

Table 30.5 Characters of the species of *Micrococcus*

	Micrococcus		
	luteus	*roseus*	*varians*
Pigmentation*	Yellow	Pink	Yellow
Acid from glucose and xylose	–	v	+
Lipolysis	+	–	v
NH_3 from arginine	–	–	v†
Susceptibility to;			
novobiocin (2 mg per l)	S	S	R
lysozyme	S	R	R
DNA: moles per cent G+C	71–75	66–75	66–72

R = resistant; S = sensitive; v = some positive, others negative.
* Many strains not definitely pigmented.
† Usually +.

Biochemical characters Some strains, mainly of *Micro. luteus*, grow on a medium in which ammonia is the sole source of nitrogen, but others have complex nutritive requirements. *Micro. varians* forms acid from several sugars aerobically; some strains of *Micro. roseus* may acidify sugars, but less strongly; *Micro. luteus* rarely acidifies any sugars. Members of this last species are, however, regularly lipolytic and often give a weak oxidase reaction.

The three species shown in Table 30.5 are those recognized by Baird-Parker (1974), who gives detailed information about their properties (see also Ogasawara-Fugita and Sakaguchi 1976). Kloos and his colleagues (1974) describe two additional species, *Micro. lylae* and *Micro. kristinae*, from human skin.

Micro. tetragenus A brief mention must be made of the organism described by Gaffky (1883) as *Micrococcus tetragenus*. In the human and animal body a capsule is formed. The growth on agar is glutinous, often adherent to the medium, and difficult to emulsify. Colonies are whitish. Gelatin is not liquefied. Nitrate is not reduced. Acid is said to be produced in glucose, maltose, lactose, and sucrose. The organism is occasionally pathogenic to man, and may even give rise to septicaemia. One of its most notable characteristics is its ability to give rise to a fatal septicaemia when injected subcutaneously or intraperitoneally into mice. According to Pike (1962), similar organisms are commonly found in the mouth. Bergan and his colleagues (1970) described a somewhat different micrococcus which is a frequent inhabitant of the pharynx. It, too, forms mucoid, adherent colonies but is diplococcal with flattened adjacent sides, and is not pathogenic for mice. They consider that it represents the organism described by earlier workers as *Staph. salivarius* and that its correct name is *Micro. mucilaginosus*.

Planococcus

This is a group of motile marine cocci that form tetrads. They produce a yellow-orange pigment and are salt tolerant but grow in the absence of added salt. Glucose is attacked oxidatively. The mole percentage of G+C (39–52) is intermediate between those of staphylococci and micrococci.

Sporosarcina

These organisms are motile, aerobic gram-positive cocci that form cubical packets and are common saprophytes. They form heat-resistant spores and extremely long flagella. The organism formerly called *Sarcina ureae* is motile only in broth containing urea, grows well on unenriched media, acidifies no sugars, and produces ammonia from urea. The DNA base ratio is close to that of the aerobic spore-bearing bacilli (MacDonald and MacDonald 1962), and these organisms have now been placed in the genus *Sporosarcina* (Fam. Bacillaceae).

The anaerobic gram-positive cocci

These organisms have been less well studied that many other groups of anaerobes. It is customary to consider that anaerobic gram-positive cocci of medical importance belong to two genera, *Peptostreptococcus* and *Peptococcus*, comprising respectively chain-forming and clump-forming cocci, but this morphological difference is not clear cut, and few if any other characters are of assistance in separating the two groups. A common feature of members of both genera is that they use the products of protein decomposition, such as peptones and amino acids, as their main source of energy and nitrogen.

Several other sorts of anaerobic gram-positive cocci may be encountered in the animal body. These include organisms that derive their energy from the fermentation of carbohydrates: *Ruminococcus* (see Rogosa 1974), *Coprococcus* (Holdeman and Moore 1974) and *Sarcina* (p. 239). Some anaerobic chain-forming cocci share with streptococci the property of forming mainly lactic acid from glucose. These include several microaerophilic organisms, such as those given the specific names *intermedius*, *morbillorum*, and *constellatus*, that have in the past been classified as peptostreptococci or peptococci. Their relation to the CO_2-requiring streptococci has yet to be established. Evidence is also accumulating, however, that a few streptococci are true obligate anaerobes, including *Str. hanseni* (Holdeman and Moore 1974) from human faeces, *Str. pleomorphus* (Barnes *et al.* 1977) from the caecum of birds, and certain aberrant strains of *Str. bovis* found in the rumen of cattle (Latham *et al.* 1979; see Chapter 29).

Peptostreptococci and peptococci

These organisms are non-motile, non-sporing cocci of varying diameter (0.4–2.5 μm) that tend to occur in clusters, though some form short or long chains. Many retain the gram stain poorly. They are strict anaerobes but are not particularly oxygen sensitive. They grow on blood agar, forming low convex, butyrous greyish-white non-haemolytic colonies 0.7–1.5 mm in diameter after 1–2 days. Some strains form black colonies. Neither cellular nor colonial morphology is characteristic of individual species. The optimal temperature is usually 35–37°. All have complex nutritive requirements.

Practically all are sensitive to metronidazole, and give a zone of inhibition exceeding 15 mm in width around a 5-μg disk (Watt and Jack 1977). All are sensitive to penicillin. According to Wren and his colleagues (1977), peptostreptococci are sensitive to novobiocin 1.6 mg per l but peptococci resist concentrations of 25 mg per l or greater. Sensitivity to polyanethol sulphonate is said to characterize *Peptostreptococcus anaerobius*, at least on some media (Wideman *et al.* 1976). The G+C content of the DNA is 33–37 moles per cent.

Metabolic activities Sugars and organic acids may or may not be fermented; they are not an essential source of energy. Some strains attack sugars poorly or not at all unless a fatty acid is added to the basal medium, and others form gas from sugars only when provided with a source of sulphur, such as thiosulphate (Wildy and Hare 1953). Action on glucose is heterofermentative. The organisms ferment nitrogenous compounds, usually with the formation of NH_3 and the evolution of gas. In complex media without added carbohydrate, the products of fermentation include a variety of organic acids and, in some cases, a mixture of CO_2 and H_2. Proteolysis is variable; catalase reactions are negative or weak.

Classification

This is at present unsatisfactory. In the following brief account, we shall follow mainly the descriptions and nomenclature of Rogosa (1974). For other views, and different names, see Hare (1967), Smith (1975) and Holdeman *et al.* (1977). Only strict anaerobes will be included.

Chain-forming cocci One group of species has the following characters: (1) production of a foetid odour, (2) formation of gas from complex media without added sugar, and from pyruvate, and (3) having higher fatty acids ($C3$–C_6) among the products of fermentation. Individual species are recognized by their patterns of sugar fermentation. *Peptostreptococcus anaerobius* (or *putridus*) was described as a cause of puerperal fever by Schottmüller (1910, 1928) and Colebrook (1930). It ferments glucose, fructose and maltose weakly in ordinary media, but forms acid and gas in the presence of thiosulphate. *Peptostreptococcus productus* ferments a wide range of sugars in ordinary media, and *Peptostreptococcus lanceolatus* attacks glucose and sucrose only, even when sulphur is provided.

The species *parvulus* and *micros* are biochemically less active; they do not give a foetid odour or form gas from complex media or pyruvate, and they have little or no action on sugars.

Cluster-forming cocci The more biochemically active species of peptococci form gas from complex media and from pyruvate; they ferment a wide range of amino acids—serine, threonine, glutamic acid and histidine but not glycine; they reduce nitrate and form indole. Three species can be recognized by their action on sugars, organic acids and gelatin. *Peptococcus activus* forms acid and gas from a number of sugars on ordinary media, and is the only one of the three to liquefy gelatin; *Peptococcus aerogenes* forms acid and gas from glucose and fructose, but only in media supplemented with oleic acid; and *Peptococcus asaccharolyticus* has no action on sugars, and differs from the others in forming gas from citrate and tartrate.

A fourth species, *Peptococcus magnus* (sometimes considered to be a peptostreptococcus) is much less active. It fails to form gas in complex media or pyruvate; it ferments glycine, but no other amino acids, with gas production; it does not reduce nitrate, form indole or acidify any sugars. Whether *Peptococcus niger*, which is said not to attack sugars or to reduce nitrate, but to form black colonies regularly, is a separate species is uncertain (but see Smith 1975). Mention must also be made of peptococci that cause abscesses and lymphadenitis in sheep and goats (Morel 1911, Joubert 1958, Shirlaw and Ashford 1962). They form gas from pyruvate, and acid but not gas from sugars.

Many different anaerobic cocci are present in the flora of the upper-respiratory, intestinal and female genital tracts. *Peptostreptococcus anaerobius* is often present at all of these sites, and is commonly associated with puerperal infection. *Peptococcus asaccharolyticus* has a similar wide distribution, and *Peptostreptococcus productus* is frequent in the gut contents. *Peptococcus magnus*, though sparse at carrier sites, is often isolated from pathological material, sometimes in pure culture. (For clinical infections with anaerobic cocci, see Lambe *et al.* 1974 and Finegold 1977.)

Sarcina

This name, which was at one time used to describe aerobic cocci that formed packets of eight, is now reserved for strict anaerobes that are similarly arranged. *Sarcina* grows on simple media and can grow at very low pH; it derives its energy from the fermentation of sugars. It is said to form heat-resistant spores. The mole percentage of G+C is 28–31. It is a soil organism and is regularly present in the faeces of vegetarians (Crowther 1971). For further information, see Canale-Parola (1970).

References

Abraham, E. P. and Florey, H. W. (1949) In: *Antibiotics* **1,** 493. Ed. by Florey, H. W. *et al.* Oxford University Press, London.

Adlam, C., Pearce, J. H. and Smith, H. (1968) *Nature, Lond.* **219,** 641.

Alami, S. Y. and Kelly, F. C. (1959) *J. Bact.* **78,** 539.

Ali, M. I. and Haque, R. (1974) *J. med. Microbiol.* **7,** 375.

Almquist, E. (1891) *Z. Hyg. InfektKr.* **10,** 253.

Altenbern, R. A. (1966) *J. infect. Dis.* **116,** 593.

Anderson, J. C. (1974) *J. med. Microbiol.* **7,** 205.

Annear, D. I. (1968) *Med. J. Aust.* **i,** 444.

Annear, D. I. and Grubb, W. B. (1972) *J. med. Microbiol.* **5,** 109.

Arbuthnott, J. P. (1970) In: '*Microbial Toxins*', **3,** 189. Ed. by T. C. Montie, S. Kadis, and S. J. Ajl. Academic Press, New York; (1981) In *The Staphylococci*, p. 109. Ed. by A. Macdonald and G. Smith. Aberdeen University Press, Aberdeen.

Arbuthnott, J. P., Freer, J. H. and Billcliffe, B. (1973) *J. gen. Microbiol.* **75,** 309.

Arbuthnott, J. P., Freer, J. H. and Bernheimer, A. W. (1967) *J. Bact.* **94,** 1170.

Arrhenius, S. (1907) *Immunochemie.* Akad. Verlagsgesell., Leipzig.

Asheshov, E. H. (1966) *Nature, Lond.* **210,** 804.

Asheshov, E. H. and Rippon, J. E. (1959) *J. gen. Microbiol.* **20,** 634.

Asheshov, E. H. and Winkler, K. C. (1966) *Nature, Lond.* **209,** 638.

Baddiley, J., Brock, J. H., Davison, A. L. and Partridge, M. D. (1968) *J. gen. Microbiol.* **54,** 393.

Baddiley, J., Buchanan, J. G., Rajbhandary, U. L. and Sanderson, A. R. (1962) *Biochem. J.* **82,** 439.

Baird-Parker, A. C. (1962) *J. appl. Bact.* **25,** 12; (1963) *J. gen. Microbiol.* **30,** 409; (1965) *Ibid.*, **38,** 363; (1974) In: *Bergey's Manual of Determinative Bacteriology*, Ed. by R. E. Buchanan and N. E. Gibbons, 8th edn., p. 478. Williams and Wilkins, Baltimore.

Barber, M. (1955) *J. gen. Microbiol.* **13,** 338.

Barber, M. and Kuper, S. W. A. (1951) *J. Path. Bact.* **63,** 65.

Barber, M. and Wildy, P. (1958) *J. gen. Microbiol.* **18,** 92.

Barnes, E. M., Impey, C. S., Stevens, B. J. H. and Peel, J. L. (1977) *J. gen. Microbiol.* **102,** 45.

Bergan, T., Bøvre, K. and Hovig, B. (1970) *Acta path. microbiol. scand.* **B 78,** 85.

Bergdoll, M. S. (1970) In: *Microbial Toxins*, **3,** 266. Ed. by T. C. Montie, S. Kadis, and S. J. Ajl, Academic Press, New York; (1972) In: *The Staphylococci*, p. 301. Ed by J. O. Cohen. Wiley-Interscience, New York.

Bergdoll, M. S., Huang, I. Y. and Schantz, E. J. (1974) *J. agric. food Chem.* **22,** 9.

Bergdoll, M. S., Surgalla, M. J. and Dack, G. M. (1959) *J. Immunol.* **83,** 334.

Bernheimer, A. W. and Schwartz, L. L. (1965) *J. Path. Bact.* **89,** 209.

Bigger, J. W. and Hodgson, G. A. (1943) *Lancet* **i,** 544.

Binder, N. and Blobel, H. (1970) *Zbl. Bakt.* **215,** 511.

Birch-Hirschfeld, L. (1934) *Klin. Wschr.*, **13,** 331.

Birkbeck, T. H. and Whitelaw, D. D. (1980) *J. med. Microbiol.* **13,** 213.

Blair, J. E. and Williams, R. E. O. (1961) *Bull. World Hlth Org.* **24,** 771.

Bokkenheuser, V., Cardella, M. A., Gorzynski, E. A., Wright, G. G. and Neter, E. (1963) *Proc. Soc. exp. Biol., N.Y.* **112,** 18.

Borderon, E. and Horodniceanu, T. (1976) *Ann. Microbiol., Inst. Pasteur* **A127,** 503.

Boven, C. P. A. van, Stobberingh, E. E., Verhoef, J. and Winkler, K. C. (1974) *Ann. N.Y. Acad. Sci.* **236,** 376.

Brown, A. E. (1965) *J. med. Lab. Technol.* **22,** 121.

Brown, A. E. and Faruque, A. A. (1963) *J. clin. Path.* **16,** 457.

Browning, C. H. and Adamson, H. S. (1950) *J. Path. Bact.* **62,** 499.

Bryce, L. M. and Rountree, P. M. (1936) *J. Path. Bact.* **43,** 173.

Burnet, F. M. (1931) *J. Path. Bact.* **34,** 759.

Canale-Parola, E. (1970) *Bact. Rev.* **34,** 82.

Carr, D. L. and Kloos, W. E. (1977) *Appl. envir. Microbiol.* **34,** 673.

Casman, E. P. (1958) *Publ. Hlth Rep., Wash.* **73,** 599.

Casman, E. P. and Bennett, R. W. (1963) *J. Bact.* **86,** 18.

Casman, E. P., Bennett, R. W., Dorsey, A. E. and Stone, J. E. (1969) *Hlth. Lab. Sci.* **6,** 185.

Castellani, A. (1955) *Ann. Inst. Pasteur* **89,** 475.

Chain, E. and Duthie, E. S. (1940) *Brit. J. exp. Path.* **21,** 324.

Chao, L.-P. and Birkbeck, T. H. (1978) *J. med. Microbiol.* **11,** 303.

Chopra, I. (1971) *J. gen. Microbiol.* **63,** 265.

Chopra, I., Bennett, P. M. and Lacey, R. W. (1973) *J. gen. Microbiol.* **79,** 343.

Christie, R. and Graydon, J. J. (1941) *Aust. J. exp. Biol. med. Sci.* **19,** 9.

Christie, R. and Keogh, E. V. (1940) *J. Path. Bact.* **51,** 189.

Christie, R. and North, E. A. (1941) *Aust. J. exp. Biol. med. Sci.* **19,** 323.

Christie, R. and Wilson, H. (1941) *Aust. J. exp. Biol. med. Sci.* **19,** 329.

Clawson, C. C. and Dajani, A. S. (1970) *Infect. Immun.* **1,** 491.

Clegg, M. T. and Wherry, W. B. (1906) *J. infect. Dis.* **3,** 165.

Colebrook, L. (1930) *Brit. med. J.* **ii,** 134.

Conti, C. R., Cluff, L. E. and Scheder, E. P. (1961) *J. exp. Med.* **113,** 845.

Cooper, L. Z., Madoff, M. A. and Weinstein, L. (1966) *J. Bact.* **91,** 1686.

Cowan, S. T. (1938) *J. Path. Bact.* **46,** 31; (1939) *Ibid.* **48,** 169.

Cowan, S. T. and Steel, K. J. (1964) *J. Bact.* **88,** 804.

Crowther, J. S. (1971) *J. med. Microbiol.* **4,** 343.

Curry, J. C. and Borovian, G. E. (1976) *J. clin. Microbiol.* **4,** 455.

Cutler, R. R. (1979) *J. med. Microbiol.* **12,** 55.

Cybulska, J. and Jeljaszewicz, J. (1965) *J. clin. Path.* **18,** 759; (1966) *J. Bact.* **91,** 953.

Dack, G. M., Jordan, E. O. and Woolpert, O. (1931) *J. prev. Med.* **4,** 167.

Dajani, A. S., Gray, E. D. and Wannamaker, L. W. (1970*a*) *J. exp. Med.* **131,** 1004; (1970*b*) *Infect. Immun.* **1,** 485.

Dajani, A. S. and Wannamaker, L. W. (1969) *J. Bact.* **97,** 985.

Davidson, I. (1972) *Bull. World Hlth Org.* **46,** 81.

Davison, E., Dack, G. M. and Cary, W. E. (1938) *J. infect. Dis.* **62,** 219.

Devriese, L. A. (1981) *J. appl. Bact.* **50,** 351.

Devriese, L. A., Hájek, V., Oeding, P., Meyer, S. A. and Schleifer, K. H. (1978) *Int. J. syst. Bact.* **28,** 482.

Devriese, L. A. and van de Kerckhove, A. (1979) *J. appl. Bact.* **46,** 385.
Devriese, L. A. and Oeding, P. (1975) *J. appl. Bact.* **39,** 197.
Doery, H. M., Magnusson, B. J., Cheyne, I. M. and Gulasekharam, J. (1963) *Nature, London* **198,** 1091.
Doery, H. M., Magnusson, B. J., Gulasekharam, J. and Pearson, J. E. (1965) *J. gen. Microbiol.* **40,** 283.
Dolman, C. E., Wilson, R. J. and Cockcroft, W. H. (1936) *Canad. publ. Hlth J.* **27,** 489.
Dornbusch, K., Nord, C.-E., Olsson, B. and Wadström, T. (1976) *Med. Microbiol. Immunol.* **162,** 143.
Dossett, J. H., Kronvall, G., Williams, R. C. Jr and Quie, P. G. (1969) *J. Immunol.* **103,** 1405.
Duran-Reynals, F. (1933) *J. exp. Med.* **58,** 161.
Duthie, E. S. (1952) *J. gen. Microbiol.* **7,** 320; (1954*a*) *Ibid.* **10,** 427; (1954*b*) *Ibid.* **10,** 437; (1955) *Ibid.* **13,** 383.
Duthie, E. S. and Lorenz, L. L. (1952) *J. gen. Microbiol.* **6,** 95.
Dyke, K. G. H., Parker, M. T. and Richmond, M. H. (1970) *J. med. Microbiol.* **3,** 125.
Edwards, S. J. and Rippon, J. E. (1957) *J. comp. Path.* **67,** 111.
Ekstedt, R. D. and Nungester, W. J. (1955) *Proc. Soc. exp. Biol., N.Y.* **89,** 90.
Elek, S. D. (1959) *Staphylococcus pyogenes and its Relation to Disease.* Livingstone, Edinburgh and London.
Elek, S. D. and Conen, P. E. (1957) *Brit. J. exp. Path.* **38,** 573.
Elek, S. D. and Levy, E. (1950) *J. Path. Bact.* **62,** 541; (1954) *J. Path. Bact.* **68,** 31.
Epstein, S. (1935) *Derm. Ztschr.* **70,** 328.
Evans, J. B. (1948) *J. Bact.* **55,** 793.
Evans, J. B., Bradford, W. L. and Niven, C. F. (1955) *Int. Bull. bact. Nomencl.* **5,** 61.
Evans, J. B., Buettner, L. G. and Niven, C. F. (1952) *J. Bact.* **64,** 433.
Evans, J. B. and Kloos, W. E. (1972) *Appl Microbiol.* **23,** 326.
Evans, J. B. and Pate, C. A. (1980) *Int. J. syst. Bact.* **30,** 557.
Evans, R. J. and Waterworth, P. M. (1966) *J. clin. Path.* **19,** 555.
Fackrell, H. B. and Wiseman, G. M. (1974) *J. med. Microbiol.* **7,** 411; (1097) *J. gen. Microbiol.* **92,** 1, 11.
Feltham, R. K. A. and Sneath, P. H. (1982) *J. gen. Microbiol.,* **128,** 713.
Fildes, P. and Richardson, G. M. (1937) *Brit. J. exp. Path.* **18,** 292.
Fildes, P., Richardson, G. M., Knight, B. C. J. G. and Gladstone, G. P. (1936) *Brit. J. exp. Path.* **17,** 481.
Finegold, S. M. and Sweeney, E. E. (1961) *J. Bact.* **81,** 636.
Finkelstein, R. A. and Sulkin, S. E. (1958) *J. Bact.* **75,** 339.
Fisher, S. (1965) *J. infect. Dis.* **115,** 285.
Fisk, R. T. (1942) *J. infect. Dis.* **71,** 153, 161.
Fisk, R. T. and Mordvin, O. E. (1943) *J. Bact.* **46,** 392.
Flandrois, J. P., Fleurette, J. and Modjadedy, A. (1975) *Ann. Microbiol., Inst. Pasteur* **B126,** 333.
Forsgren, A. (1970) *Infect. Immun.* **2,** 672.
Forsgren, A. and Forsum, U. (1970) *Infect. Immun.* **2,** 387.
Forsgren, A. and Sjöquist, J. (1966) *J. Immunol.* **97,** 882.
Forsum, U., Forsgren, A. and Hjelm, E. (1972) *Infect. Immun.* **6,** 582.
Fredericq, P. (1946) *C. R. Soc. Biol., Paris* **140,** 1167.
Freer, J. H., Arbuthnott, J. P. and Billcliffe, B. (1973) *J. gen. Microbiol.* **75,** 321.
Gaffky. (1883) *Arch. klin. Chir.* **28,** 495.
Gengou, O. (1935) *Ann. Inst. Pasteur* **54,** 428.
Gerheim, E. B. and Ferguson, J. H. (1949) *Proc. Soc. exp. Biol., N.Y.* **71,** 261.
Gibbs, P. A., Patterson, J. T. and Harvey, J. (1978) *J. appl. Bact.* **44,** 57.
Gillespie, W. A. and Alder, V. G. (1952) *J. Path. Bact.* **64,** 187.
Gillespie, W. A. and Simpson, P. M. (1948) *Brit. med. J.* **ii,** 902.
Gladstone, G. P. (1937) *Brit. J. exp. Path.* **18,** 322.
Gladstone, G. P., Fildes, P. and Richardson, G. M. (1935) *Brit. J. exp. Path.* **16,** 335.
Gladstone, G. P. and Glencross, E. J. G. (1960) *Brit. J. exp. Path.* **41,** 313.
Gladstone, G. P. and Heyningen, W. E. van. (1957) *Brit. J. exp. Path.* **38,** 123.
Gladstone, G. P. and Yoshida, A. (1967) *Brit. J. exp. Path.* **48,** 11.
Glenny, A. T. and Stevens, M. F. (1935) *J. Path. Bact.* **40,** 201.
Gorrill, R. H. (1951) *Brit. J. exp. Path.* **32,** 151; (1958) *Ibid.* **39,** 203.
Gorrill, R. H. and McNeil, E. M. (1963) *Brit. J. exp. Path.* **44,** 404.
Goshi, K., Cluff, L. E. and Norman, P. S. (1963) *Bull. Johns Hopk. Hosp.* **112,** 15.
Grinsted, J. and Lacey, R. W. (1973) *J. gen. Microbiol.* **75,** 259.
Gross, H. (1931) *Z. ImmunForsch.,* **73,** 14; (1933) *Klin. Wschr.* **12,** 304.
Grov. A. (1967) *Acta path. microbiol. scand.* **69,** 567; (1968) *Ibid.* **73,** 400; (1973) *Ibid.* **A236,** suppl., 77.
Gulasekharam, J., Cheyne, I., Doery, H. M. and Magnusson, B. J. (1963) *Nature, Lond.* **198,** 1114.
Gustafson, G. T., Sjöquist, J. and Stålenheim, G. (1967) *J. Immunol.* **98,** 1178.
Gustafson, G. T., Stålenheim, G., Forsgren, A. and Sjöquist, J. (1968) *J. Immunol.* **100,** 530.
Guyonnet, F. and Plommet, M. (1970) *Ann. Inst. Pasteur* **118,** 19.
Hájek, V. (1976) *Int. J. syst. Bact.* **26,** 401.
Hájek, V. and Maršálek, E. (1969) *Zbl. Bakt.,* **212,** 60; (1970) *Ibid.* **214,** 68; (1971) *Ibid.* **A217,** 176; (1973) *Ibid.* **A223,** 63.
Hájek, V., Maršálek, E. and Harna, V. (1974) *Zbl Bakt.* **A229,** 429.
Halbert, S. P., Swick, L. and Sonn, C. (1953) *J. Immunol.* **70,** 400.
Hale, J. H. (1947) *Brit. J. exp. Path.* **28,** 202; (1951) *Ibid.* **32,** 307.
Hale, J. H. and Smith, W. (1945) *Brit. J. exp. Path.* **26,** 209.
Hall, H. E., Angelotti, R. and Lewis, K. H. (1963) *Amer. J. publ. Hlth* **78,** 1089; (1965) *Hlth Lab. Sci.* **2,** 179.
Hallander, H. O. (1965) *Acta path. microbiol. scand.* **63,** 299.
Hallander, H. O. and Bengtsson, S. (1967) *Acta path. microbiol. scand.* **70,** 107.
Haque, R. (1967) *J. Bact.* **93,** 525.
Haque, R. and Baldwin, J. N. (1964) *J. Bact.,* **88,** 1304.
Harding, L. and Williams, R. E. O. (1969) *J. Hyg., Camb.* **67,** 35.
Hare, R. (1967) In: *Recent Advances in Medical Microbiology,* p. 284. Ed. by A. P. Waterson. Churchill, London.
Harper, E. M. and Conway, N. S. (1948) *J. Path. Bact.* **60,** 247.

Harshman, S. (1979) *Molec. cell. Biochem.* **23,** 143.
Hartman, D. P. and Murphy, R. A. (1977) *Infect. Immun.* **15,** 59.
Haughton, G. (1966) *Ann. Med. exp. Fenn.* **44,** 307.
Haukenes, G., Ellwood, D. C., Baddiley, J. and Oeding, P. (1961) *Biochim. biophys. acta* **53,** 425.
Heczko, P. B., Grov, A. and Oeding, P. (1973) *Acta path. microbiol. scand.* **B81,** 731.
Heddaeus, H., Heczko, P. B. and Pulverer, G. (1979) *J. med. Microbiol.* **12,** 9.
Hill, J. H. and White, E. D. (1929) *J. Bact.* **18,** 43.
Hill, L. R. (1981) In: *The Staphylococci*, p. 33. Ed. by A. Macdonald and G. Smith. Aberdeen University Press, Aberdeen.
Hine, T. G. M. (1922) *Lancet* **ii,** 1380.
Hobbs, B. C. (1948) *J. Hyg., Camb.* **46,** 222.
Hofstad, T. (1964) *Acta path. microbiol. scand.* **61,** 558.
Holbrook, R., Anderson, J. M. and Baird-Parker, A. C. (1969) *J. appl. Bact.* **32,** 187.
Holdeman, L. V., Cato, E. P. and Moore, W. E. C. (1977) *Anaerobe Laboratory Manual*, 4th edn. Virginia Polytechnic Institute Anaerobe Laboratory, Blacksburg, Va.
Holdeman, L. V. and Moore, W. E. C. (1974) *Int. J. syst. Bact.* **24,** 260.
Holt, R. J. (1971) *J. med. Microbiol.* **4,** 375.
Hucker, G. J. and Haynes, W. C. (1937) *Amer. J. publ. Hlth* **27,** 590.
Hunt, G. A. and Moses, A. J. (1958) *Science* **128,** 1574.
Idziak, E. S. and Mossel, D. A. A. (1980) *J. appl. Bact.* **48,** 101.
Illès, J. (1963) *Path. et Microbiol., Basel* **26,** 225.
Jackson, A. W. and Little, R. M. (1957) *Canad. J. Microbiol.* **3,** 101.
Jackson, S. H., Nicholson, T. F. and Holman, W. L. (1940) *J. Path. Bact.* **50,** 1.
James, A. M. and Brewer, J. E. (1968) *Biochem. J.* **107,** 817.
James, R. C. and MacLeod, C. M. (1961) *Brit. J. exp. Path.* **42,** 266.
Jensen, K. (1958) *Acta path. microbiol. scand.* **44,** 421.
Jessen, O., Faber, V., Rosendal, K. and Eriksen, K. R. (1959) *Acta path. microbiol. scand.* **47,** 316.
Jevons, M. P. (1961) *Brit. med. J.* **i,** 124.
Jevons, M. P., John, M. and Parker, M. T. (1966) *J. clin. Path.* **19,** 305.
Johnson, H. M., Bucovik, J. A., Kauffman, P. E. and Peeler, J. T. (1971) *Appl. Microbiol.* **22,** 837.
Johnson, J. E., Cluff, L. E. and Goshi, K. (1961) *J. exp. Med.* **113,** 259.
Jones, G. W. and Edwards, S. J. (1966) *J. Dairy Res.* **33,** 271.
Jordan, E. O. and McBroom, J. (1931) *Proc. Soc. exp. Biol. N.Y.* **29,** 161.
Joubert, L. (1958) *Ann. Inst. Pasteur* **95,** 215.
Julianelle, L. A. (1922) *J. infect. Dis.* **31,** 256.
Julianelle, L. A. and Wieghard, C. W. (1934) *Proc. Soc. exp. Biol., N.Y.* **31,** 947; (1935) *J. exp. Med.* **62,** 11, 31.
Kamman, J. F. and Tatini, S. R. (1977) *J. Food Sci.* **42,** 421.
Kapral, F. A., O'Brien, A. D., Ruff, P. D. and Drugan, W. I. jr. (1976) *Infect. Immun.* **13,** 140.
Karakawa, W. W., Kane, J. A. and Smith, M. R. (1974) *Infect. Immun.* **9,** 511.
Kashiba, S., Niizu, K., Tanaka, S., Nozu, H. and Amano, T. (1959) *Biken's J.* **2,** 50.
Kent, T. H. (1966) *Amer. J. Path.* **48,** 387.
Kinsman, O. S. and Arbuthnott, J. P. (1980) *J. med. Microbiol.* **13,** 281.
Knight, B. C. J. G. (1937) *Biochem. J.* **31,** 731.
Kleck, J. L. and Donahue, J. A. (1968) *J. infect. Dis.* **118,** 317.
Kloos, W. E. (1980) *Annu. Rev. Microbiol.* **34,** 559.
Kloos, W. E. and Musselwhite, M. S. (1975) *Appl. Microbiol.* **30,** 381.
Kloos, W. E. and Schleifer, K. H. (1975*a*) *J. clin. Microbiol.* **1,** 82; (1975*b*) *Int. J. syst. Bact.* **25,** 62.
Kloos, W. E., Schleifer, K. H. and Smith, R. F. (1976*a*) *Int. J. syst. Bact.* **26,** 22.
Kloos, W. E., Tornabene, T. G. and Schleifer, K. H. (1974) *Int. J. Syst. Bact.* **24,** 79.
Kloos, W. E., Zimmerman, R. J. and Smith, R. F. (1976*b*) *Appl. envir. Microbiol.* **31,** 53.
Koch, R. (1878) *Untersuchungen über die Aetiologie der Wundinfektionskrankheiten.* Vogel, Leipzig.
Kondo, I. and Fujise, K. (1977) *Infect. Immun.* **18,** 266.
Kondo, I., Sakurai, S. and Sarai, Y. (1974) *Infect. Immun.* **10,** 851.
Kondo, I., Sakurai, S., Sarai, Y. and Futaki, S. (1975) *J. clin. Microbiol.* **1,** 397.
Koupal, A. and Deibel, R. H. (1977) *Infect. Immun.* **18,** 298.
Kreger, A. S., Kim, K-S., Zaboretzky, F. and Bernheimer, A. W. (1971) *Infect. Immun.* **3,** 449.
Kronvall, G., Holmberg, O. and Ripa, T. (1972) *Acta path. microbiol. scand.* **B80,** 735.
Lacey, R. W. (1969) *J. med. Microbiol.* **2,** 187; (1971*a*) *Ibid.* **4,** 73; (1971*b*) *J. gen. Microbiol.* **69,** 229.
Lacey, R. W. and Grinsted, J. (1972) *J. gen. Microbiol.* **73,** 501.
Lacey, R. W. and Lord, V. L. (1981) *J. med. Microbiol.* **14,** 41.
Lacey, R. W. and Mitchell, A. A. B. (1969) *Lancet* **ii,** 1425.
Lacey, R. W. and Rosdahl, V. T. (1974) *J. med. Microbiol.* **7,** 1.
Lachica, R. V. F. (1976) *Appl. envir. Microbiol.* **32,** 633.
Lachica, R. V. F., Hoeprich, P. D. and Genigeorgis, C. (1971) *Appl. Microbiol.* **21,** 823.
Lack, C. H. (1948) *Nature, Lond.* **161,** 559.
Lack, C. H. and Wailling, D. G. (1955) *J. Path. Bact.* **68,** 431.
Latham, M. J., Sharpe, M. E. and Weiss, N. (1979) *J. appl. Bact.* **47,** 209.
Lenhart, N. A., Mudd, S., Yoshida, A. and Li, I. W. (1963) *J. Immunol.* **91,** 771.
Lernau, H. and Sompolinsky, D. (1962) *Cornell Vet.* **52,** 445.
Lind, I. (1972) *Acta path. microbiol. scand.* **B80,** 702.
Lipiński, B., Hawiger, J. and Jeljaszewicz, J. (1967) *J. exp. Med.* **126,** 979.
Loeb, L. (1903–4) *J. med. Res.* **10,** 407.
Löfkvist, T. and Sjöquist, J. (1962) *Acta path. microbiol. scand.* **56,** 295.
Lominski, I. and Arbuthnott, J. P. (1962) *J. Path. Bact.* **83,** 515.
Lominski, I. and Roberts, G. B. S. (1946) *J. Path. Bact.* **58,** 187.
Lubinski, W. (1894) *Zbl. Bakt.* **16,** 769.
MacDonald, R. E. and MacDonald, S. W. (1962) *Canad. J. Microbiol.* **8,** 795.
McKay, S. E. and Arbuthnott, J. P. (1979) *J. med. Microbiol.* **12,** 99.
McNiven, A. C., Owen, P. and Arbuthnott, J. P. (1972) *J. med. Microbiol.* **5,** 113.

Madler, J. J., Lee, S.-H. and Haque, R. (1976) *Appl. envir. Microbiol.* **32,** 575.
Madoff, M. A., Cooper, L. Z. and Weinstein, L. (1964) *J. Bact.* **87,** 145.
Mah, R. A., Fung, D. Y. C. and Morse, S. A. (1967) *Appl. Microbiol.* **15,** 866.
Maheswaran, S. K. and Lindorfer, R. K. (1967) *J. Bact.* **94,** 1313.
Maitland, H. B. and Martyn, G. (1948) *J. Path. Bact.* **60,** 553.
Manohar, M., Kumar, S. and Lindorfer, R. K. (1966) *J. Bact.* **91,** 1681.
Marks, J. and Vaughan, A. C. T. (1950) *J. Path. Bact.* **62,** 597.
Marples, R. R. (1969) *J. Bact.* **100,** 47; (1981) *Zbl Bakt.* I Abt. Orig., suppl. 10, p. 9.
Marples, R. R. and Kligman, A. M. (1972) In: *Epidermal Wound Healing*, p. 24. Ed. by H. I. Maibach and D. T. Rovee. Year Book Medical Publishers, Chicago.
Martin, R. R., Crowder, J. G. and White, A. (1967) *J. Immunol.* **99,** 269.
Martin, W. J. and Marcus, S. (1964) *J. Bact.* **87,** 1019.
Matheson, B. H. and Thatcher, F. S. (1955) *Canad. J. Microbiol.* **1,** 372.
Meijers, J. A., Winkler, K. C., Stobberingh, E. E. and Schrauwen, P. P. (1981) *J. med. Microbiol.* **14,** 21.
Melish, M. E. and Glasgow, L. A. (1970) *New. Engl. J. Med.* **282,** 1114.
Meyer, R. F. and Palmieri, M. J. (1980) *Appl. envir. Microbiol.* **40,** 1080.
Meyer, W. (1966*a*) *Zbl. Bakt.* **201,** 331; (1966*b*) *Ibid.* **201,** 465; (1967*a*) *J. Hyg., Camb.* **65,** 439; (1967*b*) *Z. med. Mikrobiol.* **153,** 158.
Miale, J. B. (1949) *Blood* **4,** 1039.
Miles, A. A. and Niven, J. S. F. (1950) *Brit. J. exp. Path.* **31,** 73.
Miller, B. A., Reiser, R. F. and Bergdoll. M. S. (1978) *Appl. envir. Microbiol.* **36,** 421.
Minett, F. C. (1936) *J. Path. Bact.* **42,** 247.
Mitsuhashi, S., Hashimoto, H., Kono, M. and Morimura, M. (1965) *J. Bact.* **89,** 988.
Modjadedy, A. and Fleurette, J. (1974) *Ann. Microbiol., Inst. Pasteur* **B125,** 367.
Möllby, R. and Wadström, T. (1971) *Infect. Immun.* **3,** 633.
Moore, B. (1960) *Lancet* **ii,** 453.
Morel, M. G. (1911) *J. Méd. vét. Zootechnie* **72,** 513.
Morgan, F. G. and Graydon, J. J. (1936) *J. Path. Bact.* **43,** 385.
Morriss, D. M., Lawson, J. W. and Rogolsky, M. (1978) *Antimicrob. Agents Chemother.* **14,** 218.
Morse, S. A. and Mah, R. A. (1967) *Appl. Microbiol.* **15,** 58.
Murphy, R. A. and Haque, R. (1980) *J. med. microbiol.* **13,** 193.
Much, H. (1908) *Biochem. Z.* **14,** 143.
Munch-Petersen, E. and Christie, R. (1947) *J. Path. Bact.* **59,** 367.
Murray, M. and Gohdes, P. (1960) *Biochim. biophys. acta* **40,** 518.
Myhre, E. B. and Kronvall, G. (1979) In: *Pathogenic Streptococci*, p. 76. Ed. by M. T. Parker, Reedbooks, Chertsey, Surrey.
Namavar, F., de Graaff, J., de With, C. and MacLaren, D. M. (1978*a*) *J. med. Microbiol.* **11,** 243.
Namavar, F., de Graaff, J. and MacLaren, D. M. (1978*b*) *Leeuwenhoek ned. Tidschr.* **44,** 425.
Neisser, M. (1912) See Kolle and Wassermann, *Hdb. der path. Mikroorg.*, IIte Aufl. (1912–13), **4,** 356.
Neisser, M. and Wechsberg, F. (1901) *Z. Hyg. InfektKr.* **36,** 299.
Nickerson, D. S., White, J. G., Kronvall, G., Williams, R. C. and Quie, P. G. (1970) *J. exp. Med.* **131,** 1039.
Noble, W. C. (1965) *Brit. J. exp. Path.* **46,** 254; (1966) *J. Path. Bact.* **91,** 181.
Novick, R. P. (1963) *J. gen. Microbiol.* **33,** 121.
Novick, R. P. and Morse, S. I. (1967) *J. exp. Med.* **125,** 45.
Novick, R. P. and Richmond, M. H. (1965) *J. Bact.* **90,** 467.
Novick, R. P. and Roth, C. (1968) *J. Bact.* **95,** 1335.
O'Brien, A. D. and Kapral, F. A. (1977) *Infect. Immun.* **16,** 812.
O'Connor, J. J., Willis, A. T. and Smith, J. A. (1966) *J. Path. Bact.* **92,** 585.
Oeding, P. (1952) *Acta path. microbiol. scand.*, Suppl. 93, 356; (1957) *Ibid.*, **41,** 435; (1974) *Ann. N.Y. Acad. Sci.*, **236,** 15.
Oeding, P., Hájek, V. and Maršálek, E. (1973) *Acta path. microbiol. scand.* **B81,** 567.
Oeding, P. and Hasselgren, I. L. (1972) *Acta path. microbiol. scand.* **B80,** 265.
Oeding, P., Marandon, J.-L., Hájek, V. and Maršálek, E. (1970) *Acta path. microbiol. scand.* **B78,** 414.
Oeding, P., Marandon, J.-L., Meyer, W., Hájek, V. and Maršálek, E. (1972) *Acta path. microbiol. scand.* **B80,** 525.
Ogasawara-Fujita, N. and Sakaguchi, K. (1976) *J. gen. Microbiol.* **94,** 97.
Ogston, A. (1881) *Brit. med. J.* **i,** 369.
O'Leary, W. M. and Weld, J. T. (1964) *J. Bact.* **88,** 1356.
Olson, J. C., Casman, E. P., Baer, E. F. and Stone, J. F. (1970) *Appl. Microbiol.* **20,** 605.
Panton, P. N. and Valentine, F. C. O. (1932) *Lancet* **i,** 506.
Parker, M. T. (1972) In: *Methods of Microbiology* Vol. 7B, p. 1. Ed. by J. R. Norris and D. W. Ribbons. Academic Press, London.
Parker, M. T. and Hewitt, J. H. (1970) *Lancet* **i,** 800.
Parker, M. T. and Simmons, L. E. (1959) *J. gen. Microbiol.* **21,** 457.
Parker, M. T., Tomlinson, A. J. H. and Williams, R. E. O. (1955) *J. Hyg., Camb.* **53,** 458.
Parnas, J. and Jabłonski, L. (1961) *Polski Tygod. lek.* **16,** 1341.
Pasteur, L. (1880) *C.R. Acad. Sci., Paris* **90,** 1033.
Penfold, J. B. (1944) *J. Path. Bact.* **56,** 247.
Pike, E. B. (1962) *J. appl. Bact.* **25,** 448.
Pillet, J. (1966) *Postepy Mikrobiol.* **5,** 231.
Price, K. M. and Kneeland, Y. (1954) *J. Bact.* **67,** 472; (1956) *Ibid.* **71,** 229.
Proom, H. (1937) *J. Path. Bact.* **44,** 425.
Pulsford, M. F. (1954) *Aust. J. exp. Biol. med. Sci.* **32,** 347.
Pulverer, G., Heczko, P. B. and Peters, G. (1979) *Phage-typing of Coagulase-negative Staphylococci.* Fischer, Stuttgart and New York.
Pulverer, G. and Lucas, A. (1966) *Z. med. Mikrobiol. Immunol.* **152,** 353.
Pulverer, G. and Sieg, J. F. (1972) *Zbl Bakt.* **A222,** 446.
Quie, P. G. (1969) *Yale J. Biol. Med.* **41,** 394.
Quie, P. G. and Wannamaker, L. W. (1961) *J. Bact.* **82,** 770; (1962) *J. clin. Invest.* **41,** 1962.
Rammelkamp, C. H., Hezebicks, M. M. and Dingle, J. H. (1950) *J. exp. Med.* **91,** 295.

Rantz, L. A., Zuckerman, A. and Randall, E. (1952) *J. Lab. clin. Med.* **39,** 443.
Report. (1958) *Int. Bull. bact., Nomencl.* **8,** 153; (1959) *Ibid.* **9,** 115; (1963) *Ibid.* **13,** 119; (1965) *Ibid.* **15,** 107; (1967) *Ibid.* **17,** 113; (1971) *Int. J. syst. Bact.* **21,** 165, 167, 171; (1972) *J. appl. Bact.* **35,** 673; (1975) *Int. J. syst. Bact.* **25,** 233.
Richmond, M. H. (1965*a*) *Biochem. J.* **94,** 584; (1965*b*) *Brit. med. Bull.* **21,** 260.
Richmond, M. H. and John, M. (1964) *Nature, Lond.* **202,** 1360.
Rippon, J. E. (1956) *J. Hyg., Camb.* **54,** 213.
Rogers, H. J. (1953) *J. Path. Bact.* **66,** 545.
Rogolsky, M. (1979) *Microbiol. Rev.* **43,** 320.
Rogolsky, M., Wiley, B. B. and Glasgow, L. A. (1976) *Infect. Immun.* **13,** 44.
Rogosa, M. (1974) In: *Bergey's Manual of Determinative Bacteriology*, 8th edn. p. 517. Ed. by R. E. Buchanan and N. E. Gibbons. Williams and Wilkins, Baltimore.
Rosdahl, V. T. (1973) *J. gen. Microbiol.* **77,** 229.
Rosenbach, F. J. (1884). *Mikroorganismen bei den Wundinfektionskrankheiten des Menschen.* Wiesbaden.
Rosenblum, E. D. and Dowell, C. E. (1960) *J. infect. Dis.* **106,** 297.
Rosendal, K. and Bülow, P. (1965) *J. gen. Microbiol.* **41,** 349.
Rosendal, K., Bülow, P. and Jessen, O. (1964) *Nature, Lond.* **204,** 1222.
Rountree, P. M. (1949) *J. gen. Microbiol.* **3,** 164; (1959) *Ibid.* **20,** 620.
Rountree, P. M. and Asheshov, E. H. (1961) *J. gen. Microbiol.* **26,** 111.
Rountree, P. M. and Barbour, R. G. H. (1952) *Aust. Ann. Med.* **1,** 80.
Rubinstein, H. M. (1958) *Brit. J. Haematol.* **4,** 326.
Salomon, L. L. and Tew, R. W. (1968) *Proc. Soc. exp. Biol., N.Y.* **129,** 539.
Sanderson, A. R., Juergens, W. G. and Strominger, J. L. (1961) *Biochem. biophys. Res. Comm.* **5,** 472.
Sandvik, O. and Brown, R. W. (1965) *J. Bact.* **89,** 1201.
Sandvik, O. and Fossum, K. (1965) *Amer. J. vet. Res.* **26,** 357.
Sásărman, A., Surdeanu, M., Portelance, V., Dobardzic, R. and Sonea, S. (1971) *J. gen. Microbiol.* **65,** 125.
Scherer, R. K. and Brown, R. W. (1974) *Appl. Microbiol.* **28,** 768.
Schindler, C. A. and Schuhardt, V. T. (1964) *Proc. nat. Acad. Sci., Wash.* **51,** 415.
Schleifer, K. H., Hammes, W. P. and Kandler, O. (1976) *Adv. microb. Physiol.* **13,** 245.
Schleifer, K. H. and Kloos, W. E. (1975*a*) *J. clin. Microbiol.* **1,** 377; (1975*b*) *Int. J. syst. Bact.* **25,** 50.
Schlievert, P. M. (1981) *Infect. Immun.* **31,** 732.
Schlievert, P. M., Schoettle, D. J. and Watson, D. W. (1979*a*) *Infect. Immun.* **23,** 609; (1979*b*) *Ibid.* **25,** 1075.
Schlievert, P. M., Shands, K. N., Dan, B. B., Schmid, G. P. and Nishimura, R. D. (1981) *J. infect. Dis.* **143,** 509.
Schottmüller, H. (1910) *Mitt. Grenzgeb. Med. Chir.* **21,** 450; (1928) *Münch. med. Wschr.* **75,** 1580, 1634.
Seidl, P. H. and Schleifer, K. H. (1978) *Appl. envir. Microbiol.* **35,** 479.
Selbie, F. R. and Simon, R. D. (1952) *Brit. J. exp. Path* **33,** 315.
Shah, D. B. and Wilson, J. B. (1963) *J. Bact.* **85,** 516; (1965) *Ibid.* **89,** 949.
Sheehan, H. L. and Fergusson, A. G. (1943) *Lancet* **i,** 547.
Shibata, Y., Morita, M., Amano, Y. and Ishida, N. (1977) *Microbiol. Immunol.* **21,** 45.
Shirlaw, J. F. and Ashford, W. A. (1962) *Vet. Rec.* **74,** 1025.
Siegel, I. and Cohen, S. (1964) *J. infect. Dis.* **114,** 488.
Sierra, G. (1957) *Leeuwenhoek ned. Tijdschr.* **23,** 15.
Silverman, S. J., Knott, A. R. and Howard, M. (1968) *Appl. Microbiol.* **16,** 1019.
Smith. D. D., Morrison, R. B. and Lominski, I. (1952) *J. Path. Bact.* **64,** 567.
Smith, G. S. (1954) *J. med. Lab. Technol.* **12,** 98.
Smith, H. W. (1954) *J. Path. Bact.* 67, 73.
Smith, L. DS. (1975) *The Pathogenic Anaerobic Bacteria*, 2nd edn. Thomas, Springfield.
Smith, M. L. and Price, S. A. (1938*a*) *J. Path. Bact.* **47,** 361; (1938*b*) *Ibid.* **47,** 379.
Smith, W., and Hale, J. H. (1944) *Brit. J. exp. Path.* **25,** 101.
Smith, W., Hale, J. H. and Smith, M. M. (1947) *Brit. J. exp. Path.* **28,** 57.
Sompolinsky, D. (1953) *Schweiz. Arch. Tierheilk.* **95,** 302.
Sompolinsky, D., Ernst-Geller, Z. and Segal, S. (1967) *J. gen. Microbiol.* **48,** 205.
Sompolinsky, D., Gluskin, I. and Ziv, G. (1969) *J. Hyg., Camb.* **67,** 511.
Spink, W. W. and Paine, J. R. (1940) *J. Immunol.* **38,** 383.
Stamatin, N., Tacu, A. and Marica, D. (1949) *Ann. Inst. Pasteur* **76,** 178.
Steel, K. G. (1964) *J. gen. Microbiol.* **36,** 133.
Steuer, W. (1956) *Zbl. Bakt.* **167,** 210.
Stiffler-Rosenberg, G. and Fey, H. (1978) *J. clin. Microbiol.* **8,** 473.
Stobberingh, E. E. and Winkler, K. C. (1976) *Zbl. Bakt.*, I Abt Orig. **A** suppl. 5, 313.
Stutzenberger, F. J. and San Clemente, C. L. (1967) *J. Bact.* **94,** 821.
Sugiyama, H. (1966) *J. infect. Dis.* **116,** 162.
Sugiyama, H., Bergdoll, M. S. and Dack, G. M. (1962) *J. infect. Dis.* **111,** 233.
Sugiyama, H. and Hayama, T. (1965) *J. infect. Dis.* **115,** 330.
Sugiyama, H., McKissic, E. M., Bergdoll, M. S. and Heller, B. (1964) *J. infect. Dis.* **114,** 111.
Surgalla, M. J., Bergdoll, M. S. and Dack, G. M. (1953) *J. Lab. clin. Med.* **41,** 782.
Tager, M. (1948) *Yale J. Biol. Med.* **20,** 369; (1956) *J. exp. Med.* **104,** 675.
Tager, M. and Lodge, A. L. (1951) *J. exp. Med.* **94,** 73.
Taylor, A. G. and Bernheimer, A. W. (1974) *Infect. Immun.* **10,** 54.
Thatcher, F. S. and Matheson, B. H. (1955) *Canad. J. Microbiol.* **1,** 382.
Thompson, N. E. and Pattee, P. A. (1977) *J. Bact.* **129,** 778.
Tschäpe, H. and Rische, H. (1972) *Z. allg. Mikrobiol.* **12,** 59.
Turner, W. H. (1978*a*) *Infect. Immun.* **20,** 485; (1978*b*) *J. appl. Bact.* **45,** 291.
Turner, W. H. and Pickard, D. J. (1980) *J. med. Microbiol.* **13,** 151.
Váczi, L., Fodor, M., Milch, H. and Réthy, A. (1962) *Acta microbiol hung.* **9,** 81.
Valentine, F. C. O. (1936) *Lancet*, **i,** 526.
Vanbreuseghem, R. (1934) *C. R. Soc. Biol.* **116,** 650.
Varaldo, P. E., Grazi, G., Soro, O., Cisani, G. and Satta, G. (1980) *J. clin. Microbiol.* **12,** 63.
Varaldo, P. E. and Satta, G. (1978) *Int. J. syst. Bact,* **28,** 148.

Varaldo, P. E., Satta, G., Grazi, G. and Romanzi, C. A. (1978) *Int. J. syst. Bact.* **28,** 141.
Verwey, W. F. (1940) *J. exp. Med.* **71,** 635.
Vijver, J. C. M. van der, Kraayeveld, C. A. and Michel, M. F. (1972) *J. clin. Path.* **25,** 450.
Wadström, T. and Möllby, R. (1971) *Biochim. biophys. acta* **242,** 288, 308.
Walbum, L. E. (1921) *C. R. Soc. Biol., Paris* **85,** 1205.
Walker, B. S., Derow, M. A. and Schaffer, N. K. (1948) *J. Bact.* **56,** 191.
Warren, R., Rogolsky, M., Wiley, B. B. and Glasgow, L. A. (1974) *J. Bact.* **118,** 980; (1975) *Ibid.* **122,** 99.
Watt, B. and Jack, E. P. (1977) *J. med. Microbiol.* **10,** 461.
Weckman, B. G. and Catlin, B. W. (1957) *J. Bact.* **73,** 747.
Wegrzynowicz, Z., Jeljaszewicz, J. and Lipinski, B. (1969) *Appl. Micriobiol.* **17,** 556.
Weld, J. T. (1962) *Proc. Soc. exp. Biol., N.Y.* **109,** 693.
Weld, J. T., Kean, B. H. and O'Leary, W. M. (1963) *Proc. Soc. exp. Biol., N.Y.* **112,** 448.
Weld, J. T. and Rogers, D. E. (1960) *Proc. Soc. exp. Biol. N.Y.* **103,** 311.
Wideman, P. A., Vargo, V. L., Citronbaum, D. and Finegold, S. M. (1976) *J. clin. Microbiol.,* **4,** 330.
Wieghard, C. W. and Julianelle, L. A. (1935) *J. exp. Med.* **62,** 23.
Wieneke, A. A. (1974) *J. Hyg., Camb.* **73,** 255.
Wildy, P. and Hare, R. (1953) *J. gen. Microbiol.* **9,** 216.
Williams, R. E. O. and Harper, G. J. (1946) *Brit. J. exp. Path.* **27,** 72; (1947) *J. Path. Bact.* **59,** 69.
Williams, R. E. O. and Rippon, J. E. (1952) *J. Hyg., Camb.* **50,** 320.
Williams, R. E. O., Rippon, J. E. and Dowsett, L. M. (1953) *Lancet* **i,** 510.
Willis, A. T., O'Connor, J. J. and Smith, J. A. (1966) *J. Path. Bact.* **92,** 97.
Willis, A. T. and Turner, G. C. (1962) *J. Path. Bact.* **84,** 337.
Wilson, B. J. (1959) *J. Bact.* **78,** 240.
Wilson, G. S. and Atkinson, J. D. (1945) *Lancet* **i,** 647.
Winkler, K. C., de Waart, J., Grootsen, C., Zegers, B. J. M., Tellier, N. F. and Vertregt, C. D. (1965) *J. gen. Microbiol.* **39,** 321.
Winslow, C.-E. A. and Rogers, A. F. (1906) *J. infect. Dis.* **3,** 485.
Winslow, C.-E. A. and Winslow, A. R. (1908) *The Systematic Relationship of the Coccaceae.* New York.
Wise, R. I. and Spink, W. W. (1954) *J. clin. Invest.* **33,** 1611.
Wiseman, G. M. (1965) *J. Path. Bact.* **89,** 187; (1970) In: *Microbial Toxins,* **3,** 237. Ed. by T. C. Montie, S. Kadis and S. J. Ajl. Academic Press, New York; (1975) *Bact. Rev.* **39,** 317.
Wiseman, G. M. and Caird, J. D. (1967) *Canad. J. Microbiol.* **13,** 369; (1970) *Ibid.* **16,** 47.
Wiseman, G. M., Caird, J. D. and Fackrell, H. B. (1975) *J. med. Microbiol.* **8,** 29.
Witte, W., Hummel, R., Meyer, W., Exner, H. and Wundrak, R. (1977) *Z. allg. Mikrobiol.* **17,** 639.
Woodin, A. M. (1959) *Biochem. J.* **73,** 225; (1960) *Ibid.* **75,** 158; (1961) *J. Path. Bact.* **81,** 63; (1970) In: *Microbial Toxins,* **3,** 327. Ed. by T. C. Montie, S. Kadis and S. J. Ajl, Academic Press, New York.
Wren, M. W. D., Eldon, C. P. and Dakin, G. H. (1977) *J. clin. Path.* **30,** 620.
Wright, J. (1936) *Lancet* **i,** 1002.
Yoshida, A., Mudd, S. and Lenhart, N. A. (1963) *J. Immunol.* **91,** 777.
Yoshida, K. (1971) *Infect. Immun.* **3,** 535; (1972) *Ibid.* **5,** 833.
Yoshida, K., Ichiman, Y. and Ohtomo, T. (1976) *Zbl. Bakt.* **A,** suppl. 5, 819.
Yoshida, K. and Minegishi, Y. (1979) *J. appl. Bact.* **47,** 299.
Yoshida, K., Ohtomo, T. and Minegishi, Y. (1977) *J. gen. Microbiol.* **98,** 67.
Yoshida, K. *et al.* (1979) *J. appl. Bact.* **46,** 147.
Zebovitz, E., Evans, J. B. and Niven, C. F. (1955) *J. Bact.* **70,** 686.

31

Pseudomonas

M. T. Parker

Introductory

Definition

Gram-negative straight or slightly curved rods. Motile by means of one or more polar flagella, though some strains also have lateral flagella of different wavelength, and one species is non-motile. Non-sporing and not acid-fast. Strict aerobes, but some grow anaerobically in the presence of nitrate. Catalase positive. Attack sugars oxidatively; none is fermentative and none photosynthetic. Nutritionally unexacting; nearly all grow with ammonium salts and a single carbon source. Some strains are pathogenic for man and animals. Mole percentage G+C:57–70.
Type species: *Ps. aeruginosa*.

The pseudomonads most familiar to medical microbiologists are those that form blue or green pigments. *Pseudomonas aeruginosa* was first named by Schroeter in 1872. Gessard (1882, 1890, 1891, 1892) showed that it formed two pigments: one, *pyocyanin*, is bluish-green, non-fluorescent, and soluble in chloroform and water; the other, *fluorescin*, is greenish-yellow, fluorescent and insoluble in chloroform but soluble in water. In 1886, Flügge recognized two other fluorescent pseudomonads, one of which, now called *Ps. fluorescens*, liquefies gelatin, and the other of which, *Ps.*

putida, does not. Since then, numerous species have been added to the genus, many of them non-fluorescent organisms previously described as *Alkaligenes* or *Vibrio*. A broad definition of the genus *Pseudomonas*, such as the one given at the head of this Chapter, then became necessary. Its boundaries are, however, rather ill defined, and it is not easy to distinguish it from other genera in the family Pseudomonadaceae, such as *Xanthomonas* (see Palleroni 1978).

We recognize an 'inner' group of pseudomonads that form the greenish-yellow fluorescent pigment or are closely related to strains that do (the 'fluorescent pseudomonads'), and a more heterogeneous collection of organisms that do not form this pigment, though some of them form coloured colonies. Among the fluorescent pseudomonads, the species of chief medical interest is *Ps. aeruginosa*; among the non-fluorescent species, Whitmore's bacillus (*Ps. pseudomallei*) and the glanders bacillus (*Ps. mallei*) have long been recognized as causes of disease in man and animals, and several others are now considered to be pathogens.

General description of *Pseudomonas*

Morphology

The organisms of this group are rod-shaped, somewhat variable in length, and have parallel sides and rounded ends. They are arranged singly, in small bundles, or short chains. All except the glanders bacillus are motile by means of polar flagella at one or both ends. The number of flagella varies from one cell to another in the same culture, but some species are predominantly monotrichate and others predominantly lophotrichate. The *index of flagellation* (percentage of flagellated cells with more than one flagellum per pole; see Lautrop and Jessen, 1964) of some species, such as *Ps. aeruginosa*, is 3 per cent or less, and of others, such as *Ps. fluorescens* and *Ps. pseudomallei*, exceeds 50 per cent. As in certain other genera characterized by polar flagella, such as *Vibrio, Aeromonas* and *Chromobacterium*, some strains also have a few morphologically distinct lateral flagella with a different wavelength (Palleroni *et al.* 1970). The presence of molecular oxygen is necessary for motility (Shoesmith and Sherris 1960); to test for this property, migration through agar is therefore an unsuitable method. Most strains possess fimbriae; these are usually polar but occasionally peritrichate (Fuerst and Hayward 1969).

Some pseudomonads, such as *Ps. aeruginosa*, stain evenly, but others—mainly those that accumulate poly-β-hydroxybutryate in the cells (Forsyth *et al.* 1958)—stain irregularly, deeply staining areas alternating with poorly staining or even unstained areas, particularly in films prepared by Wright's or Giemsa's method. Granules, which correspond to the unstained areas, may be seen by phase-contrast microscopy or in films stained with Sudan black (Burdon 1946). They are characteristic of *Ps. mallei* and *Ps. pseudomallei* and certain other non-fluorescent pseudomonads.

Cultural and biochemical characters

Growth occurs readily on ordinary bacteriological media, but *Ps. mallei* grows rather less profusely than the rest. Colonial forms are highly variable even within a species, and will be described under the separate organisms. Many pseudomonads form *pigments*, which include the yellow-green fluorescent pigment characteristic of *Ps. aeruginosa* and *Ps. fluorescens*, and the various phenazine pigments. Of the latter the blue-green diffusible pigment pyocyanin is found only in *Ps. aeruginosa*, but several other species form different phenazines; thus two yellow and one blue phenazine are found in strains resembling *Ps. fluorescens* and another yellow phenazine is found in *Ps. cepacia*. Several other pseudomonads form yellow, brownish or pink pigments, including a few that, like *Xanthomonas*, form carotenoids.

Growth occurs over a wide range of temperatures; differences in this respect are of value in classification. Thus, *Ps. aeruginosa* and a number of the non-fluorescent pseudomonads grow at 41°. On the other hand, *Ps. putida* and nearly all *Ps. fluorescens* strains fail to grow at 41°—and indeed a number of them fail to grow even at 37°; all *Ps. fluorescens* and a number of *Ps. putida* strains will grow at 4°.

All are obligate aerobes, but some can use denitrification as an alternative mechanism for respiration under anaerobic conditions (Jessen 1965) and will grow in the depths of a tube of nitrate agar. Denitrification, defined by Stanier and his colleagues (1966) as the ability to grow anaerobically in the presence of nitrate, is a useful character for the classification of pseudomonads. Most but not all denitrifying pseudomonads form gaseous nitrogen in nitrate broth. Other pseudomonads form nitrite from nitrate or have no action on it.

The nutritional requirements of pseudomonads are very simple. With the exception of a few species, all grow in a solution of mineral salts with ammonia as nitrogen source and a single organic compound as source of carbon and energy. For this purpose citrate will serve, but a wide variety of other compounds can be substituted. The pattern of utilization of organic compounds as sources of energy is of great value in taxonomy (Stanier *et al.* 1966).

The oxidation of sugars by pseudomonads yields rather small amounts of acid. Peptone-water sugars are unsuitable for detecting this, because it is often neutralized by alkali produced from attack on the peptone. Some produce acid from sugars in Hugh and Leifson's medium (1953), but this, too, may be alkalinized. Acidification is more easily detected in ammonium-salts medium, in which the sugar is the only carbon source (Snell and Lapage 1971).

All pseudomonads are catalase positive, and all except *Ps. maltophilia* and some of the fluorescent plant pathogens are oxidase positive. The fluorescent pseudomonads form ammonia from arginine, but this character is irregularly distributed among the other species. Decarboxylation of lysine and ornithine, and hydrolysis of aesculin, are infrequent characters. Indole is not formed.

Numerous extracellular enzymes, including proteases, lipases, lecithinases and esterases are formed. Gelatinase production is of taxonomic significance. Some but not all of the strains that accumulate poly-β-hydroxybutyrate also form an enzyme that destroys it. Amylase activity is found in *Ps. stutzeri*, *Ps. pseudomallei* and some strains of *Ps. mallei*.

Antigenic structure

Limited studies (Munoz *et al.* 1949) show that the somatic antigens of most species are heterogeneous. Few cross reactions occur between the different species, except *Ps. pseudomallei* and *Ps. mallei*. Gel-precipitation studies of the extracellular products of *Ps. aeruginosa*, *Ps. fluorescens* and *Ps. pseudomallei* suggest that they are species-specific (Liu 1961). (For cell-wall antigens, see Tunstall *et al.* 1975).

Classification

The widespread distribution and diverse activities of the pseudomonads have led to confusion in their classification. Strains from a variety of environments have been studied in isolation, and by different methods. In 1966, Stanier and his colleagues subjected a large collection of strains of diverse origin to detailed study, placing most emphasis on the general similarities of their metabolic activities. Their testing system was much too elaborate for routine use, but the species so defined can usually be recognized by more simple means, though not by the biochemical tests used for the classification of the enterobacteria. Since then, a number of further species have been defined by similar methods (Redfearn *et al.* 1966, Ballard *et al.* 1968, Davis *et al.* 1969, Ballard *et al.* 1970, Palleroni *et al.* 1970; see also Doudoroff and Palleroni 1974, Palleroni 1978). Well over 30 species are now recognized, and it is not practicable to give accounts of all of them in this Chapter.

In Table 31.1 we give the main characters of a number of species of interest to medical bacteriologists; a few more will be mentioned later in this chapter. Many of these characters are far from invariable, and it has proved difficult to design a dichotomous

Table 31.1 Characters of some species of *Pseudomonas*

	Fluorescent group			Pseudomallei group								
	aeruginosa	*putida*	*fluoresscens*	*pseudomallei*	*mallei*	*cepacia*	*acidovorans*	*stutzeri*	*alkaligenes* and *pseudoalkaligenes*	*maltophilia*	*diminuta* and *vesicularis*	*paucimobilis*
Flagella per pole	1	>1	>1	>1	0	>1	>1	1	1	>1	1	1
Accumulation of poly-β-hydroxybutyrate	–	–	–	+	+	+	+	–	v§	–	+	+
Pigment: fluorescent	+†	+	+	–	–	–	–	–	–	–	–	–
Growth with single carbon source	+	+	+	+	+	+	+	+	+	–	–	+
Denitrification*	+	–	v	+	v	–	–	+	–	–	–	–
Growth at 4°	–	v	+	–	–	–	–	–	–	–	–	–
Growth at 41°	+†	–	–‡	+	+	+	–	v	+	–	v	+
Oxidase	+	+	+	+	+	+	+	+	+	–	+‖	+
NH_3 from arginine	+	+	+	+	+	–	–	–	–	–	–	–
Gelatinase	+	–	+	+	+	+	–	–	v§	+	–	–
Amylase	–	–	–	+	v	–	–	+	–	–	–	–
G+C (moles per cent)	67	60–63	59–61	69	69	67–68	65–66	61–66	§	67	66–67	64–67

v: some positive; others negative.

* Growth in the depths of a nitrate-agar stab.

† Only *Ps. aeruginosa* forms pyocyanin and grows at 42° through three successive subcultures.

‡ A few strains of *Ps. fluorescens* grow at 41°.

§ *Ps. alkaligenes* does not accumulate poly-β-hydroxybutyrate, liquefies gelatin and has 66 moles of G+C per cent; *Ps. pseudoalkaligenes* accumulates poly-β-hydroxybutyrate, does not liquefy gelatin and has 62–63 moles of G+C per cent.

‖ *Ps. vesicularis* gives a very weak oxidase reaction.

key for the identification of pseudomonads. King and Phillips (1978) recommend the use of a multi-stage system in which blocks of tests are applied successively; they give considerable weight to the pattern of acidification of selected sugars in ammonium-salts medium.

The fluorescent pseudomonads These include two fairly homogeneous species, *Ps. aeruginosa* and *Ps. putida*, and a third, less clearly defined, *Ps. fluorescens*. Most but not all of them form the fluorescent pigment; all split arginine but none accumulates poly-β-hydroxybutyrate. *Ps. aeruginosa* includes all the producers of pyocyanin, but is best defined by its ability to grow at 42°. *Ps. fluorescens* is distinguished from *Ps. putida* by its liquefaction of gelatin (Flügge 1886). A very dissimilar group of plant-pathogenic fluorescent pseudomonads that are slow growing and do not attack arginine (Sands *et al*. 1970) have been described under the name *Ps. syringae*.

The pseudomallei group *Ps. pseudomallei* is now firmly established as a multitrichate non-pigmented pseudomonad that accumulates poly-β-hydroxybutyrate, grows anaerobically in the presence of nitrate, and has a high optimum temperature. The general characters of *Ps. mallei* are similar to those of *Ps. pseudomallei*, except for its poor growth and absence of motility. The two organisms are antigenically related and produce similar lesions in experimental infections. *Ps. cepacia* (syn. *Ps. multivorans*) is one of a group of soil organisms and plant pathogens that somewhat resemble *Ps. pseudomallei* in morphology and cultural characters and grow over a similar range of temperature. *Ps. cepacia* differs from *Ps. pseudomallei* in that it does not grow anaerobically in the presence of nitrate or produce ammonia from arginine, but other members of the group resemble *Ps. pseudomallei* more closely. Another named species, *Ps. picketti*, not shown in the Table, is rather like *Ps. cepacia*.

The acidovorans group Doudoroff and Palleroni (1974) describe 10 species of soil organisms that accumulate poly-β-hydroxybutyrate, do not form ammonia from arginine and—with rare exceptions—fail to grow at 41°. Of these we include only *Ps. acidovorans*. Some species in this group grow autotrophically with hydrogen.

Pseudomonas stutzeri This is a well defined species of monotrichate, actively denitrifying soil organisms that do not accumulate poly-β-hydroxybutyrate.

Pseudomonas alkaligenes and ***Ps. pseudoalkaligenes*** These are representative of the polar-flagellated, non-pigmented, oxidase-positive organisms originally included in the genus *Alkaligenes*. The two species differ as shown in Table 31.1.

Other species *Ps. maltophilia* is a fairly homogeneous species that requires methionine for growth; it usually acidifies maltose, but this character is not unique among pseudomonads. *Ps. diminuta* and *Ps. vesicularis* have even more complicated growth requirements. *Ps. paucimobilis* is a yellow-pigmented organism easily confused with *Flavobacterium*.

The fluorescent pseudomonads

Morphology

The microscopic appearances of these organisms are not characteristic. They are rods (1.5–3.0 μm × 0.5 μm) with a straight axis, rounded ends and parallel sides, arranged singly, in pairs or short chains. Nearly all are motile; *Ps. aeruginosa* is monotrichate and the rest are multitrichate. All stain evenly. Some strains form abundant extracellular slime but none forms a definite capsule. The normal fimbriae of *Ps. aeruginosa* are long, thin and polar (Bradley 1966), but some strains have thicker non-polar fimbriae associated with the carriage of drug-resistance plasmids (Bradley 1974).

Cultural appearances

Many strains of *Ps. aeruginosa* can be identified immediately by their colonial appearance on nutrient agar; they form irregularly round, effuse colonies 2–3 mm in diameter with a matt surface, a floccular internal structure and a butyrous consistency (Table 31.2: colonial type 1). Others, however, are smaller, raised, and coliform-like (type 2). *Rough forms* (type 3) may be raised and umbonate or frankly rugose

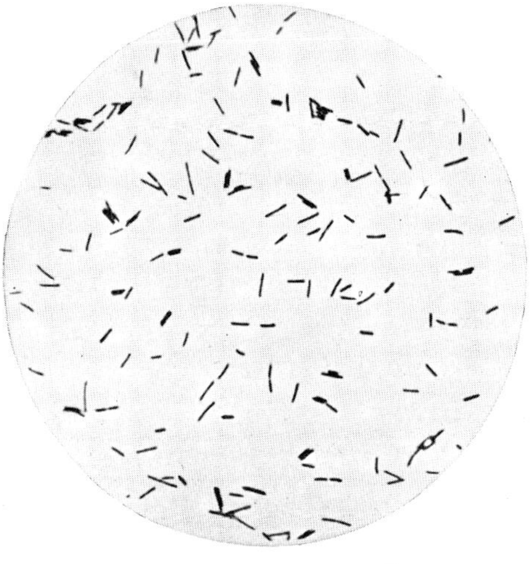

Fig. 31.1 *Pseudomonas fluorescens*. From an agar culture, 24 hr, 37° (× 1000).

Table 31.2 Colonial forms of *Pseudomonas aeruginosa*. Appearance on nutrient agar after 18 hours' incubation at 37°

Colonial type no.	Description	Size (mm)	Shape	Elevation	Surface	Edge	Opacity	Consistency	Emulsifiability	Suspension
1	Typical ('S')	2–3	irregular	effuse	matt	fimbriate	semi-opaque	butyrous	easy	uniform
2	Coliform-like ('SR')	1–2	circular	convex	smooth, shiny	entire	translucent	slightly viscid	easy	uniform
3	Rough									
	(a)	1–2	circular	raised, umbonate	rough	entire	opaque	butyrous	difficult	granular
	(b)	2–3	irregular	edge flattened, centre raised	rugose	undulate	opaque centre, transluscent periphery	butyrous	difficult	granular
4	Mucoid	2–3	circular, tend to coalesce	raised, with higher centre	smooth, shiny	entire	almost transparent	viscid	difficult	streaky
5	Dwarf	*	circular	convex	smooth, shiny	entire	translucent	slightly viscid	easy	uniform

Modified from Wahba and Darrell (1965) and Phillips (1969).
*Colonies not visible after 18 hr; $\leqslant$0.5 mm after 48 hr.

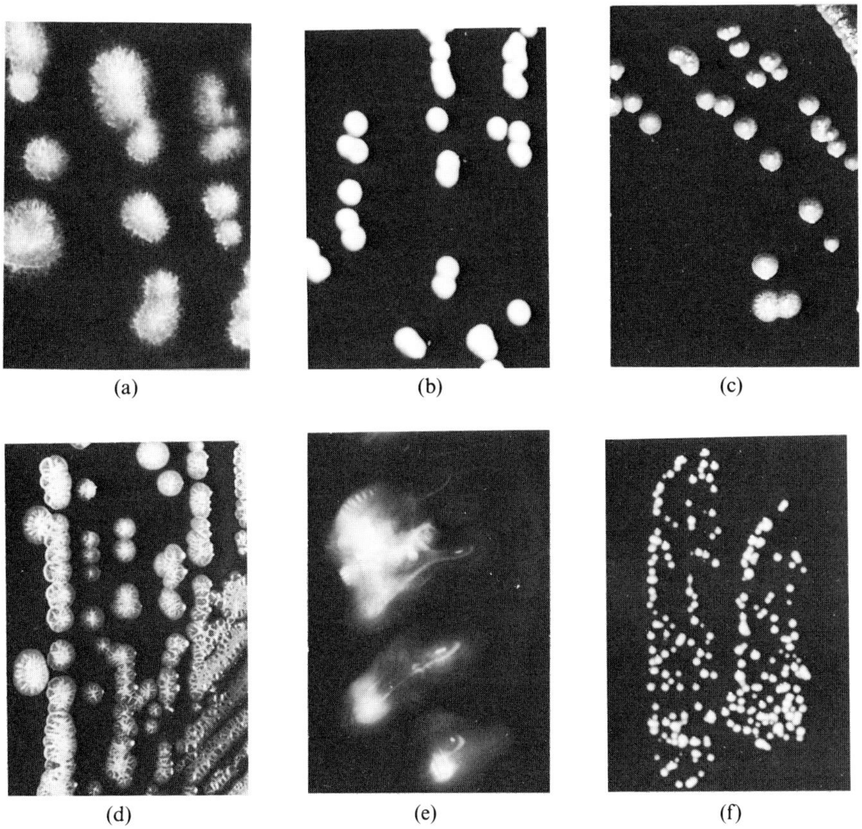

Fig. 31.2 Colonial variants of *Pseudomonas aeruginosa*: (a) Type 1: normal form. (b) Type 2: coliform-like. (c) Type 3: rough, raised, umbonate. (d) Type 3: rough, rugose. (e) Type 4: mucoid. (f) Type 5: dwarf. Nutrient agar, 37°, a–e 24 hr, f 48 hr (× 3). (From photographs kindly supplied by Professor Ian Phillips.)

(Phillips 1969). Colonial types 1 and 2 form even suspensions in saline but type 3 forms a granular suspension. Any combination of these colonial forms may appear in the same culture. *Mucoid forms* (Sonnenschein 1927) are at first convex and almost transparent and tend to coalesce; later they become flatter and contoured. Many of them do not produce pyocyanin. *Dwarf forms* do not form visible colonies unless incubation is prolonged beyond 24 hr (Wahba and Darrell 1965). Colonies of *Ps. fluorescens* and *Ps. putida* are generally round, with a moist, glistening surface, and of a pale yellowish-green colour. Some, strain resembling *Ps. fluorescens* form yellow, green or blue non-fluorescent pigments.

The production of a blue-green pigment which diffuses into the surrounding medium confirms the identity of *Ps. aeruginosa*, but many cultures do not form pyocyanin except on special media and some do not form it at all. The ability to form pyocyanin may be irreversibly lost in culture. Most but not all *Ps. aeruginosa* cultures have a characteristic fruity odour due to the production of *o*-aminoacetophenone from tryptophan (Habs and Mann 1967).

The mucoid form of *Ps. aeruginosa* produces large amounts of an extracellular polysaccharide somewhat similar to alginic acid (Eagon 1956, Linker and Jones 1964, Doggett *et al.* 1964). *In vitro*, most mucoid forms tend to give rise to non-mucoid variants; but they remain stable when subcultured in the presence of surface-active agents such as sodium deoxycholate and lecithin (Govan 1975). Non-mucoid variants tend to outgrow the parent mucoid strain, particularly in stationary culture (Govan *et al.* 1979). Mucoid growth often appears immediately around plaques of phage lysis (Martin 1973), but Govan and Fyfe (1978) succeeded in isolating mucoid variants from all of a series of non-mucoid strains without adding phage. Mucoid variants predominate among *Ps. aeruginosa* strains isolated from the sputum of cases of cystic fibrosis (Chapter 61) and are occasionally found in other parts of the body.

On nutrient agar, many cultures of *Ps. aeruginosa*, particularly of colonial types 1 and 3, form iridescent patches with a metallic sheen. Beneath these patches crystals are visible in the medium, and within the patches the organisms are lysed. They are not the result of phage action (Warner 1950, Don and van den Ende 1950) and are relatively constant characters of individual strains (Wahba 1964). According to Sierra and Zagt (1960), the autolysis is caused by extracellular proteolytic enzymes that digest dead cells, and the crystals are salts of fatty acids liberated by autolysis. (For early references, see Zierdt 1971.)

The fluorescent pseudomonads grow on MacConkey's agar but not as well as on nutrient agar. On blood agar, *Ps. aeruginosa* gives rise to diffuse clearing and later browning of the medium. Growth in broth is abundant after 1 day, with dense turbidity, a thick white ring and fine surface pellicle. *Ps. aeruginosa* often shows green pigmentation near the surface.

For the isolation of *Ps. aeruginosa* from heavily contaminated material, use is often made of its resistance to quaternary ammonium compounds (Harper and Cawston 1945). A suitable selective medium contains 0.03 per cent of cetrimide (see Lowbury 1951, Lowbury and Collins 1955). For further information about the isolation and identification of *Ps. aeruginosa* see Chapter 61.

Pigment formation

Jordan (1899) found that both phosphate and sulphate were required for the formation of fluorescin but not of pyocyanin; and Turfitt (1936) showed that the production of both pyocyanin and fluorescin was favoured by glycerol and of fluorescin by asparagine. Burton, Campbell and Eagles (1948) concluded that the ions Mg,SO_4,K,PO_4 and Fe were essential for the formation of pyocyanin. According to Garibaldi (1967), the main requirement for fluorescin production is that the content of free iron in the medium should be restricted; the increased production of the pigment in media containing egg-white is due to the binding of iron (see also Meyer and Abdallah 1978). King, Ward and Raney (1954) devised two media, A and B, respectively for the optimal production of pyocyanin and fluorescin, and these are widely used.

The proportion of *Ps. aeruginosa* cultures that do not form pyocyanin depends on the medium used for its detection, and under the most favourable conditions for pigmentation is probably 10 per cent or less. The presence of pyocyanin in cultures may be confirmed by the addition of 1 ml of water and 1 drop of N/1 H_2SO_4 to a chloroform extract; on shaking a red colour appears in the watery layer. Colours other than blue and green may be seen in cultures of *Ps. aeruginosa*, particularly after prolonged incubation. Black or brown pigmentation occurs, and this may be due to the formation of melanin (Yabuuchi and Ohyama 1972). Red coloration is attributable to the formation of pyorubrin, which is related to pyocyanin. Variants producing only the red or the black pigment have been described. Wrede (1930) showed that pyocyanin was a phenazine derivative; it is now considered to be a resonance hybrid of the mesomeric forms of *n*-methyl l-hydroxyphenazine (Jensen and Holten 1949). Other phenazines are formed by pseudomonads: chlororaphine and phenazine α-carboxylic acid are found in strains closely related to *Ps. fluorescens*; another yellow phenazine is found in *Ps. cepacia*.

Fluorescin imparts a faintly yellowish tinge to cultures, but this is sometimes not easy to detect; under ultraviolet light, however, intense greenish fluorescence is seen. Examination by this method is useful for detecting fluorescent pseudomonads in mixed culture (Lowbury *et al.* 1962). Practically all strains of *Ps. aeruginosa* form fluorescin, but some 10 per cent of the other organisms assigned to the fluorescent group on the basis of their biochemical activities do not do so (Stanier *et al.* 1966).

Resistance

The fluorescent pseudomonads are not particularly resistant to heat; they succumb when exposed to a temperature of 55° for 1 hr. They survive for many months in water at ambient temperature. When drops of fluid dry in air there is a heavy initial mortality, but the survivors die more slowly thereafter (Lowbury and Fox 1953). Although as susceptible to most disinfectants as other gram-negative bacteria, they are unusually resistant to quaternary ammonium compounds. Their simple nutrient requirements and their ability to metabolize a variety of organic chemical substances often enables them to multiply in weak disinfectant solutions and in a variety of other fluids found in hospital wards (see Chapter 61). The sensitivity of *Ps. aeruginosa* to acid (Phillips *et al.* 1968), to silver salts (Moyer *et al.* 1965, Ricketts *et al.* 1970), and to *p*-aminomethylbenzene sulphonamide (Sulfamylon) (Lindberg *et al.* 1965) has been successfully exploited in the prevention and treatment of superficial infections; but silver-resistant strains have appeared in some hospitals (Bridges *et al.* 1979). The action of a number of disinfectants and antibiotics on *Ps. aeruginosa* is strongly potentiated by the addition of ethylenediamine tetraacetate (Brown and Richards 1965). *Ps. aeruginosa* is more resistant to cadmium salts than are the other fluorescent pseudomonads; nearly all strains grow at 37° on agar containing 0.2 per cent of $CdSO_4$ (Wahba and Darrell 1965).

Susceptibility to antibiotics Of the antibiotics active on other gram-negative bacilli, *Ps. aeruginosa* can be expected to be sensitive only to colistin, carbenicillin and some of the aminoglycosides. It is resistant to most penicillins and cephalosporins, partly by virtue of possessing a β-lactamase. Small amounts of methicillin inhibit this enzyme and therefore act synergistically with benzylpenicillin or cephalosporins against *Ps. aeruginosa* (Sabath and Abraham 1964). Carbenicillin (α-carboxybenzylpenicillin) is resistant to the β-lactamase normally present in *Ps. aeruginosa*. Carbenicillin-resistant strains, which destroy the antibiotic rapidly by means of a different β-lactamase, have become locally prevalent in hospitals in which the drug has been extensively used (Lowbury *et al.* 1969). This β-lactamase is mediated by a resistance factor that is transmissible between *Ps. aeruginosa* and various enterobacteria (Sykes and Richmond 1970, Richmond 1976).

Although resistant to streptomycin and kanamycin, *Ps. aeruginosa* was until recently sensitive to gentamicin and several other aminoglycosides. Gentamicin-resistant strains, which became increasingly prevalent after 1970, are of two sorts: (1) those with non-enzymic resistance, usually slight or moderate in degree (minimal inhibitory concentration 4–32 mg per l), and (2) those that form aminoglycoside-inactivating enzymes and are usually considerably more resistant. Several different enzymes have been described, which mediate various patterns of cross-resistance between gentamicin, tobramycin, amikacin and other aminoglycosides (see Phillips *et al.* 1978).

Ps. fluorescens and *Ps. putida* are usually resistant to gentamicin; *Ps. putida* is unique among fluorescent pseudomonads in being sulphonamide sensitive (King and Phillips 1978).

(For further information about the resistance of *Ps. aeruginosa* see Brown 1975).

Metabolic and biochemical characters

Fluorescent pseudomonads differ in the range of temperatures over which they will grow. Ability to grow at a temperature of 41° is widely quoted as a character of *Ps. aeruginosa*, but a few strains of *Ps. fluorescens* will also do this. Growth over three successive subcultures at an accurately controlled temperature of 42° is, however, one of the most constant characters of *Ps. aeruginosa* (Haynes 1951). The maximum temperature for growth of other fluorescent pseudomonads is variable; some just fail to grow at 42°, but others will not grow even at 37°. On the other hand, no strain of *Ps. aeruginosa* will grow at 4°, but some strains of *Ps. putida* and all *Ps. fluorescens* strains grow slowly at this temperature.

All are strict aerobes when tested in medium with a sufficiently low nitrate content. *Ps. aeruginosa* and some strains of *Ps. fluorescens* grow anaerobically in the depths of a nitrate agar stab. In aerobic culture most of these denitrifying cultures will form gas in nitrate broth.

Fluorescent pseudomonads are oxidase positive, produce ammonia from arginine, grow in Koser's medium and are KCN resistant, but do not accumulate poly-β-hydroxybutyrate intracellularly. *Ps. aeruginosa* has a number of distinctive biochemical characters in tests that are carried out strictly at 37°. Thus, at this temperature it oxidizes gluconate and produces slime; other fluorescent pseudomonads may oxidize gluconate at a lower temperature, and a few even at 37°, but none of them also produces slime at the higher temperature (Haynes 1951). It reduces tetrazolium salts and forms red colonies on plates containing 1 per cent triphenyl tetrazolium chloride incubated overnight at 37° (Selenka 1958); almost no other fluorescent pseudomonad does this, though some form red colonies at lower temperatures. It also reduces selenite, but *Ps. fluorescens* does not (Lapage and Bascomb 1968). It deaminates acetamide (Buhlmann *et al.* 1961), a property shared with various enterobacteria but not not with other fluorescent pseudomonads. An interesting feature of *Ps. aeruginosa* is its ability both in culture and in the animal body to form hydrocyanic acid (Patty 1921).

Growth with single organic carbon compounds may be used for the classification of fluorescent pseudomonads: only *Ps. aeruginosa* grows with geraniol and only *Ps. putida* with trehalose or meso-inositol (Doudoroff and Palleroni 1974). In ammonium-salts medium, *Ps. fluorescens* does not form acid from mannitol (King and Phillips 1978). Many strains of *Ps. fluorescens*, but no other fluorescent pseudomonads, form laevan from sucrose.

Extracellular enzymes

Ps. aeruginosa and *Ps. fluorescens* produce a variety of extracellular enzymes, including proteases, lipases and lecithinase—a property that distinguishes them from *Ps. putida*.

Three different proteinases are formed (Morihara 1964) which attack gelatin, casein, fibrin and elastin (Liu 1974). The alkaline phosphatase of *Ps. aeruginosa*, unlike that of the other fluorescent pseudomonads, is not inactivated by heating at 70° for 20 min (Liu 1966*c*). Lipolytic action is manifest on a variety of fats and on Tween 80. Opalescence occurs in lecithinase agar (Willis and Gowland 1960); the production of lecithinases C and D can be detected chemically (Klinge 1957). Whether or not opacity is produced in egg-yolk agar depends on the composition of the medium; in phosphate-containing medium, only *Ps. fluorescens* causes opacity (Klinge 1957), but *Ps. aeruginosa* produces phospholipase C, and hence opacity, only in a phosphate-free medium (Esselmann and Liu 1961, Liu 1964*a*).

Two haemolytic substances are formed by *Ps. aeruginosa*: one is heat-labile and appears to be the phospholipase C (Esselmann and Liu 1961); the other is a heat-stable glycolipid (Jarvis and Johnson 1949, Sierra 1960). Some strains of *Ps. fluorescens* appear to possess a somewhat weaker haemolytic activity (Liu 1957).

(For a discussion of the toxicity of the extracellular products see p. 254).

Other extracellular products Emmerich and Löw (1899) and Emmerich, Löw and Korschun (1902) found that old cultures of *Ps. aeruginosa* were highly bactericidal to many organisms. Schoental (1941) brought evidence to show that this was due to the pigments, particularly to α-oxyphenazine which has a wide action on gram-positive and gram-negative bacteria. She also isolated an oily substance from very old cultures that was active against *Vibrio cholerae*. Young (1947) identified four fractions with antibiotic activity. They included, in addition to pyocyanin and α-oxyphenazine, a substance present in ether extracts of acidified cultures and a fluorescent residue, both of which were active mainly on gram-positive organisms. The haemolytic glycolipid from *Ps. aeruginosa* described by Jarvis and Johnson (1949) has antibiotic action on the tubercle bacillus.

Bacteriocines

Many strains of *Ps. aeruginosa* and *Ps. fluorescens* form bacteriocines active against other strains of the respective species (Hamon 1956, Klinge 1959). The *pyocines* formed by *Ps. aeruginosa* are of three types: S-type or soluble bacteriocines that resemble colicines (Ito *et al.* 1970); R-type, which are contractile, sheathed rods resembling headless phages (Kageyama

1964, Govan 1974); and F-type, which are longer, thinner, non-contractile, unsheathed rods (Govan 1974). The bacteriocines of *Ps. aeruginosa* are active against other members of this species and occasionally against other fluorescent pseudomonads but not against non-fluorescent pseudomonads (Jones *et al.* 1974). Pyocines also give various patterns of inhibition of gonococci (Morse *et al.* 1976, Sidberry and Sadoff 1977) and serologically ungroupable meningococci (Blackwell and Law 1981). Those active in this way are said to be of the R type (Blackwell *et al.* 1979). (Pyocine typing of *Ps. aeruginosa* is discussed in Chapter 61.)

Phages Most strains of *Ps. aeruginosa* are lysogenic, and many strains carry several phages. Various sets of these phages are used for typing (see Bergan 1978 and Chapter 61).

Antigenic structure

Boivin and Mesrobeanu (1937) extracted from *Ps. aeruginosa* with cold trichloracetic acid a lipolysaccharide complex containing a polysaccharide hapten rather like the haptens of enterobacteria. This complex has a general structure like that of enterobacterial lipopolysaccharide but with different chemical constituents both in the polysaccharide and lipid portions (Fensom and Gray 1969, Michaels and Eagon 1969, Hancock *et al.* 1970). The polysaccharide has a common core, and the side chains are composed predominantly of amino sugars that determine O-antigenic specificity (Chester *et al.* 1973, Koval and Meadow 1975).

The O antigens can be recognized by precipitation reactions with acid or formamide extracts (Christie 1948, van der Ende 1952, Köhler 1957), or by tube- or slide-agglutination reactions with boiled or live bacterial suspensions as long as the sera are prepared with autoclaved or boiled vaccines (Habs 1957, Kleinmaier 1957). A short course of immunization with large and repeated doses of boiled suspension is recommended (Mikkelsen 1968); if the course is prolonged beyond 21 days the titre falls.

Habs (1957) described 12 O types, only two pairs of which showed cross-reactions. Unfortunately at least six other type-numbering systems were subsequently proposed by other workers and confusion ensued. The correspondence between types in the various systems has now been established (see Lányi and Bergan 1978). There is fairly general agreement that the original Habs-type numbers, with a few additions, should form the basis for an agreed system of type designation. But the question whether all serologically distinguishable organisms, or only serologically unrelated organisms, should be given type numbers is unsettled. Lányi and Bergan (1978) propose the latter, and recognize 13 serologically unrelated O groups in *Ps. aeruginosa*: 1, 2, 3, 4, 6, 7, 9, 10, 11, 12, 13, 14 and 15, nine of which can, if desired, be further subdivided, giving a total of 27 serologically distinguishable O types.

The H antigens of *Ps. aeruginosa* are difficult to distinguish from other heat-labile antigens except by demonstrating specific inhibition of motility by the corresponding antibody (Lányi 1970). Several different *fimbrial antibodies* can be detected in sera prepared with formalinized or live vaccines (Bradley and Pitt 1975), but these are usually present in fairly low titre unless the vaccine strain has been treated with osmium tetroxide to inhibit fimbrial retraction. However, high titres of agglutinating antibody to other heat-labile components may be present; these may conveniently be distinguished from H antibody by their failure to inhibit migration of the organism through semisolid nitrate agar (Pitt and Bradley 1975). For the preparation of H antisera, Pitt (1981) recommends the use of mutants without fimbriae prepared by selection with fimbria-specific phages (Bradley and Pitt 1974); washed suspensions of flagella prepared from these mutants are used to immunize rabbits. Lányi (1970) described two unrelated H-antigen complexes and a total of eight flagellar factors, and Pitt (1981) six flagellar factors present in strains in various combinations.

Various cellular proteins are common to all *Ps. aeruginosa* strains; these include the 'common protective antigen' (p. 255) and the antigens responsible for the multiple lines of precipitation between autolysates of any strain of the organism and the sera of infected cystic-fibrosis patients (Chapter 61). One protein antigen gives cross-reactions with a wide range of gram-negative bacteria, including *Bord. pertussis* and *Neiss. meningitidis* (Sompolinsky *et al.* 1980); it is said to be distinct from the 'common protective antigen'.

In agglutination tests, most strains of *Ps. fluorescens* are antigenically distinct from *Ps. aeruginosa*, though occasional cross-reactions have been reported (Holl and Kleinmaier 1961).

Toxicity

Extracellular products

Dialyzable substances It is unlikely that the phenazine pigments cause significant damage to animal tissues (but see Armstrong *et al.* 1971). *Ps. aeruginosa* and related plant pathogens form a heat-stable phytotoxin (Liu 1974).

Haemolysins and extracellular enzymes Of the haemolysins of *Ps. aeruginosa*, the glycolipid is only mildly toxic but the phospholipase C causes redness and induration on intradermal injection (Liu 1966*a*). It appears, however, that little phospholipase is formed in the tissues of infected animals, in culture in animal serum, or in media without added glucose; it can be detected in considerable amount in cultures grown in the serum of uncontrolled diabetics (Liu 1966*b*). Protease causes haemorrhage and necrosis on intradermal injection (Liu 1966*a*), and there is little doubt that it is responsible for damage to the cornea by virtue of its action on elastin (Kawaharjaro and Homma 1975, Kreger and Gray 1978). It has been suggested (Johnson *et al.* 1967) that it is responsible in a similar manner for the necrotizing vasculitis seen in severe infections (Chapter 61). It is unlikely that protease has a significant lethal effect (Liu 1974, Wretlind and Kronevi 1978).

Exotoxin A A heat-labile lethal agent can be demonstrated in the serum of rabbits given an overwhelming infection with protease-negative strains of *Ps. aeruginosa*, and in cultures of these strains grown in rabbit-serum semisolid agar (Liu 1966*b*). This agent, named exotoxin A, is a protein. It causes leucopenia and necrotic and haemorrhagic lesions in the internal organs of mice, and fatal shock in dogs (Liu 1974); it is toxic for human blood macrophages (Pollack and Anderson 1978). Exotoxin A inhibits intracellular protein synthesis (Inglewski and Kabat 1975), and is very similar in its mode of action to diphtheria toxin (Vasil *et al.* 1977). Like diphtheria toxin, it is composed of an enzyme (ADP-ribosyl transferase) which inhibits this synthesis and a carrier or binding component (see Vasil and Inglewski 1978). Two other antigenically distinct exotoxins, B and C, have been recognized.

Other exotoxins Scharmann (1976) has described a leucocidin, distinct from other toxins and present in a minority of *Ps. aeruginosa* strains. It acts on granulocytes, lymphocytes and various tissue cells, but not on erythrocytes (see also Nonoyama *et al.* 1979*a*, *b*). According to Kubota and Liu (1971), filtrates of *Ps. aeruginosa* cause accumulation of fluid in ligated loops of rabbit ileum.

Extracellular slime One of the main components of this is polysaccharide (Brown *et al.* 1969) which may be similar to that produced in larger amount by mucoid variants. This has antiphagocytic activity, but the highly purified polysaccharide appears not to be toxic. Some preparations of slime may have lethal activity (Liu *et al.* 1961), but they probably contain other substances derived from the cell wall or the outer cell membrane.

Endotoxins

The lipopolysaccharide of *Ps. aeruginosa* has biological activities similar to those of enterobacterial lipopolysaccharide (Greer and Milazzo 1976). In some experimental models at least, protection against lethal infection by whole-cell vaccines is O-group specific (Chapter 61). This suggests that the lipopolysaccharide may play some part in the pathogenesis of severe infection.

Homma and his colleagues (1958) obtained from cell autolysates an endotoxin fraction composed of lipopolysaccharide and protein; from this and from whole cells they separated a component consisting mainly of protein (Homma and Suzuki 1964). The protein components from *Ps. aeruginosa* strains of all O serogroups were antigenically similar and were cross-protective in mice that were challenged intraperitoneally (Abe *et al.* 1975, 1977). This '*common protective antigen*' does not have the biological properties usually associated with lipopolysaccharide, but has anti-tumour activity and induces the formation of interferon (Tanamoto *et al.* 1978). An antigen cross-reactive with it is found in other pseudomonads and in cholera vibrios (Hirao and Homma 1978).

Pathogenicity

Ps. aeruginosa is widely distributed in soil, water and sewage and in the mammalian gut. It is pathogenic for man and for certain animals; and in plants it is responsible for rots and wilts in sugar-cane, tobacco and lettuce (Elrod and Braun 1942). *Ps. fluorescens* and *Ps. putida* are soil and water organisms that appear to cause disease in animals only when injected directly into the tissues. (For diseases due to pseudomonads see Chapter 61).

Experimental infection Different strains of *Ps. aeruginosa* vary in their virulence for animals. In *guinea-pigs* and *rabbits* a large dose (10^9 organisms) of some strains given intravenously or even subcutaneously will cause death in 24–48 hours with evidence of septicaemia, but other strains do not kill for weeks or not at all. Large doses of organisms are necessary to cause infection by most routes, but intraocular injection of as few as 50 organisms in rabbits causes severe infection and destruction of the eye (Crompton *et al.* 1962).

In *mice*, some strains given intravenously will kill in a dose of *ca* 4×10^6, but at least 100 times this dose of other strains is needed (Klyhn and Gorrill 1967). With doses well above the MLD, death usually occurs in 24–48 hr, but with smaller doses it may be delayed for 10–14 days, when multiple abscesses are frequently seen in the kidneys (Gorrill 1952).

Fatal septicaemic infections occur in rats and mice after the heavy contamination of a relatively small burned area of skin with *Ps. aeruginosa* (Lindberg *et al.* 1965, Jones and Lowbury 1965). Strains differ widely in their ability to cause fatal infections in burned mice (Jones *et al.* 1966).

Ps. fluorescens is generally considered to be non-pathogenic for animals, but, according to Liu (1964*b*), a small burn at the base of a rat's tail contaminated with some strains becomes acutely inflamed and necrotic; there is often a short lived bacteraemia. He attributes the failure of these strains to cause lesions in enclosed tissues to the fact that they grow suboptimally at the temperature of the animal body.

Whitmore's bacillus and the glanders bacillus

History and classification

The glanders bacillus was isolated by Loeffler and Schütz in 1882 (see Loeffler 1886) from a horse dying of glanders, and was named *B. mallei* by Zopf in 1885 (see Buchanan *et al.* 1966). An organism in many respects resembling it was isolated by Whitmore and Krishnaswami (1912) from a glanders-like disease of man. It was designated *B. pseudomallei* by Whitmore (1913). Subsequently Stanton and Fletcher (1921, 1925) gave the name of melioidosis to the disease and *B. whitmori* to the causative organism. Both organisms were at first thought to be strict animal parasites; this appears to be true of the glanders bacillus, but Whitmore's bacillus is now recognized as a natural inhabitant of the soil of certain tropical areas (see Chapter 61).

The view that the two organisms were closely related—held at first on the grounds that they caused rather similar diseases and had some antigenic affinity with each other (Stanton and Fletcher 1925)—led to their being classified together, but uncertainty about their taxonomic position led to their being placed successively in the genera *Actinobacillus*, *Pfeifferella*, *Malleomyces*, *Loefflerella* and *Acinetobacter*. The resemblance of Whitmore's bacillus to *Ps. aeruginosa* was recognized at least 40 years ago (Legroux and Djemil 1931), and was strengthened by the observation that it was oxidase positive (Miller *et al.* 1948*a*) and had polar flagella (Brindle and Cowan 1951). Numerous other biochemical characters common to Whitmore's bacillus and other pseudomonads were subsequently discovered. But the glanders bacillus grew poorly on ordinary laboratory media, was non-motile, and was a strict animal parasite. It seemed almost unthinkable to include it in a genus of culturally unexacting, motile, free-living organisms. Re-examination of the evidence by Redfearn and his colleagues (1966) left little doubt, however, that the glanders bacillus closely resembles Whitmore's bacillus in metabolic activities and biochemical characters, and that both are pseudomonads. Additional evidence (1) that these two organisms are closely related is furnished by the successful hybridization of their DNA (Rogul *et al.* 1970); and (2) that they are pseudomonads by their G + C content of 68 per cent, which is similar to that of other members of the group. (For review of the *pseudomallei* group see Howe *et al.* 1971.)

Morphology

Both are slender rods, 0.3–0.5 μm in width; the glanders bacillus is 1–3 μm in length, but Whitmore's bacillus is usually rather shorter (< 2 μm), particularly in its rugose form. The glanders bacillus is straight or slightly curved, with rounded ends and irregularly parallel or wavy sides; the cells are arranged singly, in pairs end to end, in parallel bundles or in Chinese-letter form. The 'ordinary' smooth form of Whitmore's bacillus often appears as long parallel bundles of 'filaments'—really chains of closely associated bacilli (Miller *et al.* 1948*a*)—embedded in interstitial substance. In the rugose form, the shorter, oval cells are more irregularly arranged and there is no interstitial substance. Staining is irregular, and bipolar staining is seen, particularly of Whitmore's bacillus in films of pus. When stained with Sudan black, both organisms show dark granules of poly-β-hydroxybutyrate.

Cultural characters

Whitmore's bacillus grows well on unenriched media. After overnight incubation on nutrient agar at 37°, the colonies are 1–2 mm in diameter. In the smooth form, they are round, low convex, amorphous, translucent and greyish-yellow, with a smooth, glistening surface and an entire edge; although mucoid in consistency, they are easily emulsified. After several days they become opaque, yellowish-brown, uneven and often umbonate. Chambon and Fournier (1956) described numerous different colonial forms, and Stanton and Fletcher (1927) illustrated an ultra-rugose form, with an extremely corrugated surface and a tenacious consistency. The smooth colonial form of Whitmore's bacillus tends not only to die out rapidly in culture but

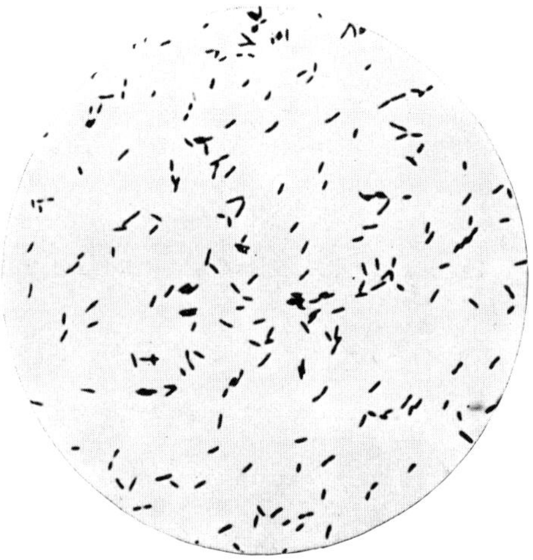

Fig. 31.3 *Pseudomonas mallei*. From an agar culture, 48 hr, 37° (× 1000).

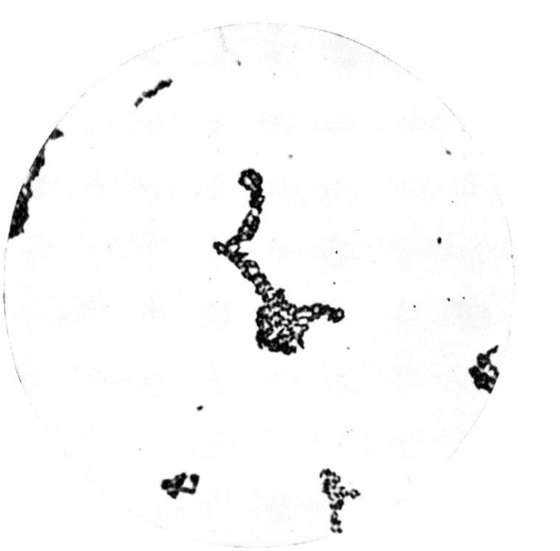

Fig. 31.4 *Pseudomonas pseudomallei*. From an agar culture, 24 hr, 37° (× 1000).

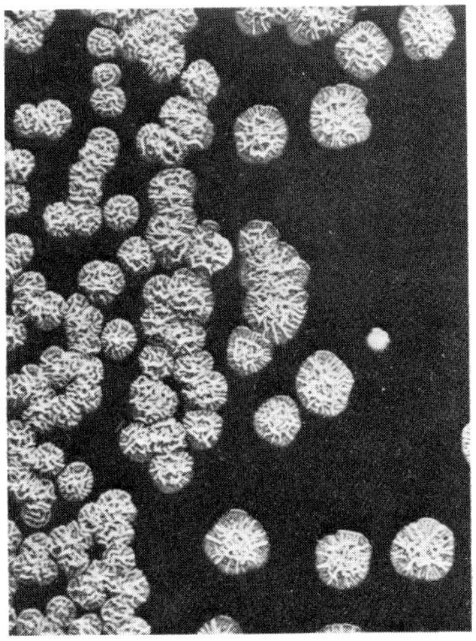

Fig. 31.5 *Pseudomonas pseudomallei.* Surface colonies on glycerol agar plate. Rugose form. (After Stanton and Fletcher.)

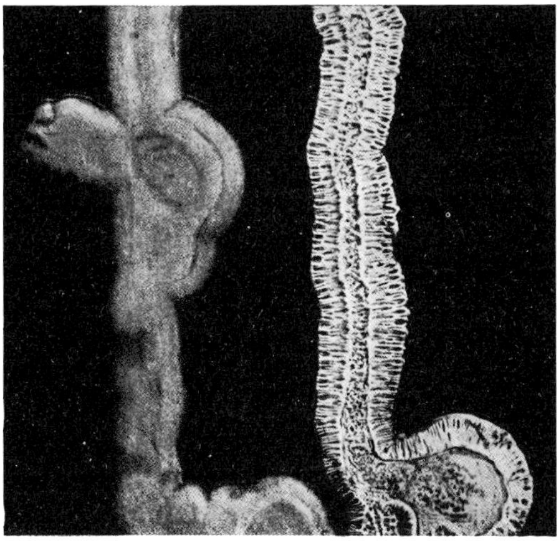

Fig. 31.6 *Pseudomonas pseudomallei.* Growth on glycerol agar. Left: smooth form. Right: rugose form. (After Stanton and Fletcher.)

to kill other pseudomonads, including rough variants, in its vicinity; Rogul and Carr (1972) attribute this to the production of ammonia. The *glanders bacillus* grows less well on plain agar and forms smooth, grey, translucent colonies 0.5–1 mm in diameter in 18 hr at 37°. After 48 hr they resemble those of *Ps. pseudomallei* at 24 hr, and later they become opaque, granular and slightly brownish. (For descriptions of colonial variants see Nigg *et all.* 1956). Neither organism forms any diffusible pigment of the phenazine or fluorescent type.

Whitmore's bacillus forms large red, opaque colonies on MacConkey's medium; the glanders bacillus grows poorly or not at all. Neither organism grows on deoxycholate citrate or SS agar. Growth of both is increased by glycerol but not by serum. On glycerol agar, Whitmore's bacillus may give either a profuse mucoid growth or dull, wrinkled growth with a corrugated or honeycombed appearance (Fig. 31.6).

Metabolic and biochemical characters

Both organisms grow at 41°, but the range of temperature over which they will grow is rather narrower than that of *Ps. aeruginosa*, and growth is often poor at temperatures as high as 21°. Both will grow anaerobically in the presence of nitrate. However, they do not form gaseous nitrogen in nitrate broth incubated aerobically; this character distinguishes *Ps. pseudomallei* from *Ps. stutzeri* (Lapage *et al.* 1968). Despite the relatively poor growth of the glanders bacillus, its nutritional requirements are simple; both it and Whitmore's bacillus can grow with ammonium salts and a single organic compound.

Both organisms form acid from various sugars in Hugh and Leifson's or ammonium salts media, but the glanders bacillus does this rather slowly. They produce ammonia from arginine and have extracellular hydrolytic enzymes that attack poly-β-hydroxybutyrate and starch—three characters that distinguish them from *Ps. cepacia.* They hydrolyse gelatin and Tween 80, and are usually KCN resistant and do not attack malonate. Like *Ps. aeruginosa*, *Ps. pseudomallei* forms a heat-stable alkaline phosphatase (Liu 1966*c*) and a haemolysin (Liu 1957).

The following characters of Whitmore's bacillus help to distinguish it from *Ps. cepacia*: production of ammonia from arginine, hydrolysis of starch, formation of heat-stable phosphatase and negative lysine decarboxylase reaction. The patterns of acidification of sugars are rather similar. Atypical strains occur in both species, and final confirmation of identity by animal-pathogenicity tests may occasionally be necessary (Bremmelgaard 1975).

Resistance Unlike *Ps. aeruginosa*, neither organism will grow on 0.1 per cent cetrimide agar (Wetmore and Gochenour 1956). Both are resistant to various dyes, and 1 in 200 000 crystal violet is a useful selective agent, especially in glycerol agar (Miller *et al.* 1948*a*). Ashdown (1979) recommends a selective medium containing glycerol, crystal violet, neutral red and gentamicin on which *Ps. pseudomallei* forms pink, wrinkled colonies. *Ps. pseudomallei* survives at least a month in water, faeces and dried soil and for a week in putrefying carcasses; *Ps. mallei* survives 4 weeks in water and then rapidly disappears (Miller *et al.* 1948*a*).

Ps. pseudomallei is sensitive to a rather different range of antibiotics than is *Ps. aeruginosa*, but the behaviour of individual strains is far from uniform. Most strains appear to be sensitive to therapeutically attainable concentrations of sulphonamides, tetracyclines, chloramphenicol, novobiocin and rifampicin; on the other hand, polymyxin B, colistin, gentamicin and carbenicillin have little or no activity (Eikhoff *et al.* 1970). Tetracycline and novobiocin act synergically against most strains (Calabi 1973). Information about the antibiotic sensitivity of *Ps. mallei* is scanty. The few strains reported on appear to be sensitive to sulphonamides and usually also to streptomycin, tetracycline and novobiocin; some were sensitive to chloramphenicol (see Miller *et al.* 1948*c*, Sîrmon and Marica 1956, Nagy and Zalay 1967).

Antigenic relationships It is generally agreed that strains of *Ps. pseudomallei* form an antigenically homogeneous group and that *Ps. mallei* strains are heterogeneous, and form two or three groups, only one of which is antigenically related to *Ps. pseudomallei* (Stanton and Fletcher 1927, de Moor *et al.* 1932, Cravitz and Miller 1950, Dodin and Fournier 1970).

Phages Several temperate phages from *Ps. pseudomallei* will lyse *Ps. mallei* but not other pseudomonads (Smith and Cherry 1957).

Toxins Liu (1957) showed that *Ps. pseudomallei* produced extracellular toxic material that was both lethal and dermonecrotic; and Heckly and Nigg (1958) separated two heat-labile toxic constituents, both of which killed mice when injected intraperitoneally but only one of which was dermonecrotic. Levine and his colleagues (1959) identified a polypeptide from *Ps. pseudomallei* which, though not toxic, enhanced mortality in experimental melioidosis and plague.

Pathogenicity

Glanders is a natural disease which spreads among equine animals, and man is occasionally infected from them. Melioidosis occurs sporadicallly in a wide range of mammalian species including man, and the horse, pig, sheep, goat, cat, dog, and several rodents (see Chapter 61).

Experimental infection

Small laboratory animals may be infected with either organism by feeding; inhalation, scarification of the skin or injection into the tissues. Individual strains of the organisms undergo frequent changes in virulence in the laboratory. Virulence often falls on prolonged culture but may be restored by animal passage (Miller *et al.* 1948*b*, Nigg *et al.* 1956). Strains of low virulence may produce non-fatal subacute or chronic infection.

Judged by the LD50 for the subcutaneous or intraperitoneal injection of virulent strains of either species (Miller *et al.* 1948*b*), the hamster is the most susceptible animal (LD50 *ca* 10), followed by the ferret (LD50:50–100). Somewhat larger doses are required to kill guinea-pigs; there are, moreover, individual differences between animals, so that a few will survive infection with as many as 10^6 organisms. Rabbits are resistant to *Ps. mallei* and moderately resistant to *Ps. pseudomallei*; and laboratory mice, rats and monkeys have little susceptibility to either organism unless it is given in large doses. In the more susceptible animal species, the LD50 by the subcutaneous and intraperitoneal routes is similar; infection by the respiratory route occurs with even lower doses than these, but the response is somewhat more variable; and the response to oral dosing is very variable.

Guinea-pigs Subcutaneous inoculation of a virulent strain of either organism results in local lesion which ulcerates in a few days. In the second week widespread lesions develop; at necropsy, abscesses or greyish-yellow nodules may be seen in bones, soft tissues, glands, viscera and serous membranes. With *Ps. mallei*, nodules may appear on the nasal mucous membrane.

Intraperitoneal injection of either organism usually causes death within 1–2 weeks. There is a fibrinous and later nodular peritonitis together with the characteristic *Straus reaction* (Straus 1889) in male animals. The testicles swell in 2–3 days and later become greatly enlarged. The lesion begins in the tunica vaginalis, which is at first covered with tiny yellowish-white granules. Later the layers are united by thick purulent and eventually caseous exudate. The scrotal skin becomes inflamed and may ulcerate later. If very large doses are given, death may occur before the Straus reaction has developed. Organisms with lowered virulence may cause a Straus reaction but not kill the animal.

Other pseudomonads

The main cultural and biochemical characters of a selection of these are given in Table 31.1. We append a few additional notes on them, together with brief descriptions of others of possible medical importance.

Pseudomonas cepacia

This is one of a group of non-fluorescent pseudomonads found in soil, some of which cause disease in plants. It was first described by Burkholder (1950) as a pathogen of onion bulbs, and given its present name. Later, Stanier and his colleagues (1966) referred to it as *Ps. multivorans*, and gave a full account of its properties (Ballard *et al.* 1970). It is an important cause of sepsis in hospital patients (Chapter 61).

Ps. cepacia is a slender multitrichous rod that accumulates poly-β-hydroxybutyrate and so stains irregularly. It grows well on nutrient agar and forms opaque colonies; in some strains these are greyish-white, but in others they are at first yellowish and later take on an intense reddish-purple colour owing to the formation of a non-diffusible phenazine. Agar cultures tend to be self-sterilizing in 2–3 days, but the organism survives for up to a year in distilled water. It resembles *Ps. pseudomallei* in growing at 41°, liquefying gelatin and forming acid in ammonium salts medium from dulcitol but not from ethanol (King and Phillips 1978), but differs from it in not attacking arginine or starch and in forming lysine and ornithine decarboxylase (see also King *et al.* 1979). Unlike *Ps. aeruginosa*, it is resistant to gentamicin, colistin and usually carbenicillin, but is sulphonamide sensitive.

Ps. picketti

This organism, first described by Ralston and his colleagues (1973), is a non-pigmented pseudomonad that accumulates poly-β-hydroxybutyrate, is oxidase positive and grows at 41°, but does not attack arginine. It thus resembles *Ps. cep-*

acia, and causes sepsis in hospital patients in rather similar circumstances. However, it differs from this organism in its failure to acidify several sugars, notably dulcitol, inositol and sorbitol, and to decarboxylate lysine and ornithine. According to King and co-workers (1979), the organism earlier described as *Ps. thomasi* (Phillips *et al.* 1972) is indistinguishable from *Ps. picketti*.

Ps. acidovorans

Although normally found in the soil, this organism has been isolated on a number of occasions from hospital patients and their environment. In addition to the characters shown in Table 31.1, it acidifies mannitol but not ethanol, and glucose or maltose in ammonium salts media. It is usually sensitive to colistin and sulphonamides but resistant to gentamicin; it is variable in its susceptibility to carbenicillin. (See King and Phillips 1978.)

Ps. stutzeri

This is a very actively denitrifying soil organism that is occasionally found in clinical material. It is an evenly staining rod with a single polar flagellum, but some strains have lateral flagella of shorter wavelength (Palleroni *et al.* 1970). The colonial appearances are highly characteristic (van Niel and Allen 1952). Most cultures include 'rough', 'smooth' and intermediate forms, so giving an appearance of impurity (Fig. 31.7). 'Rough' colonies are dry, with branching and merging

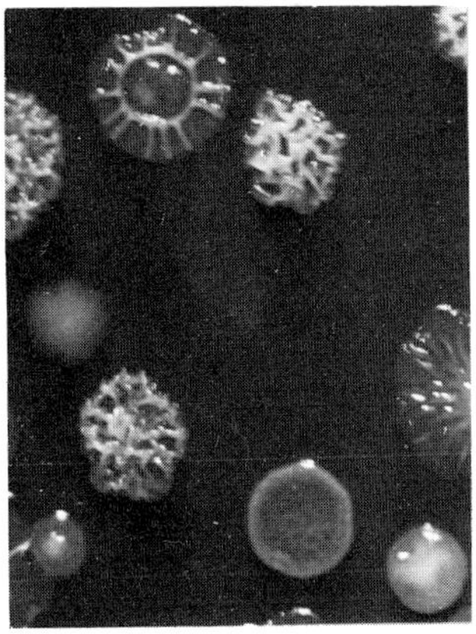

Fig. 31.7 *Pseudomonas stutzeri*. Grown for 48 hr at 37° and 3 days at room temperature. Oblique light, ×5. (From a photograph kindly supplied by Dr S. P. Lapage.)

ridges, and often with concentric spreading edges. They can be removed entire from the medium. 'Smooth' colonies often mucoid in consistency. Growth flesh-coloured or light brown in older cultures, but no diffusible pigment is formed. The 'rough' colonies may be confused with those of *Ps. pseudomallei*, but the colonies of *Ps. pseudomallei* are usually highly uniform (Lapage *et al.* 1968).

The main characters of *Ps. stutzeri* are given in Table 31.1. Freshly isolated strains usually form large volumes of gas from anaerobically incubated nitrate broth. Growth at 41° is somewhat variable, but all strains grow at 40°. It differs from *Ps. pseudomallei* in the absence of the following: gelatin liquefaction; accumulation of poly-β-hydroxybutyrate; attack on arginine; KCN resistance; hydrolysis of casein; opalescence on lecithin agar; formation of an acid-stable alkaline phosphatase (Stanier *et al.* 1966; Lapage *et al.* 1968). The closely related soil organism *Ps. mendocina* forms ammonia from arginine, does not hydrolyse starch, and forms a brownish pigment (Palleroni *et al.*1970).

Ps. stutzeri has a rather wider pattern of susceptibility to antibiotics than other pseudomonads (Russell and Mills 1974); it is usually sensitive to clinically attainable concentrations of ampicillin, as well as to carbenicillin, colistin and gentamicin. It is much more sensitive than *Ps. aeruginosa* to quaternary ammonium compounds.

Ps. alkaligenes and *Ps. pseudoalkaligenes*

These monotrichate, non-pigmented organisms do not denitrify or attack arginine; they produce little or no acid from any sugars in ammonium salts sugars (see also Table 31.1). *Ps. pseudoalkaligenes* is occasionally isolated from patients; it is sensitive to carbenicillin and sulphonamides (King and Phillips 1978).

Ps. maltophilia

This organism is found in water and raw milk, and has been isolated from patients on a number of occasions, at times under conditions that suggest some degree of pathogenicity (Hugh and Ryschenko 1961, Gilardi 1969, Holmes *et al.* 1979). It stains evenly, and on nutrient agar at 37° forms opaque greyish colonies with a yellowish tinge. Because of its requirement for methionine it will not grow in Koser's citrate medium. The oxidase reaction is negative or equivocal. It usually acidifies maltose and may acidify glucose, but it does not attack other sugars in ammonium salts media. It decarboxylates lysine but not ornithine and splits aesculin (see Holmes *et al.* 1979). It is usually sensitive to carbenicillin, colistin and sulphonamides but is variable in its susceptibility to gentamicin (King and Phillips 1978).

Ps. diminuta and *Ps. vesicularis*

These are closely related species of monotrichate organisms with flagella of very short wavelength. They require pantothenate, biotin and cyanocobalamin for growth; *Ps. diminuta* also requires cysteine or methionine (Ballard *et al.* 1968). They grow rather slowly on ordinary nutrient media. Features other than growth requirements that distinguish *Ps. vesicularis* from *Ps. diminuta* are that the former gives only a weak oxidase reaction, forms a carotenoid pigment and thus has yellow colonies, acidifies glucose and maltose in ammonium salts media, and splits aesculin. *Ps. vesicularis* is occasionally isolated from clinical material.

Ps. paucimobilis

This is a yellow-pigmented, non-fermentative rod that is likely to be confused with flavobacteria because its motility is difficult to demonstrate (Holmes *et al.* 1977); in a hanging-drop preparation only a small proportion of the cells move actively. It has the characters shown in Table 31.1; in addition, it forms acid in ammonium salts media from ethanol, glucose, maltose and a number of other sugars, and hydrolyses aesculin.

'Pseudomonas' putrefaciens

This organism was first described as *Achromobacter putrefaciens* but later transferred to the genus *Pseudomonas*. It is widely distributed in nature and is a well recognized cause of spoilage of fish. Organisms that resemble pseudomonads but form large amounts of H_2S have been isolated from human sources on a number of occasions (see Holmes *et al.* 1975, Debois *et al.* 1975).

The salient features of this organism are as follows: polar monotrichate; forms salmon-pink or reddish-brown colonies; oxidizes sugars; grows with a single organic source of carbon; does not form poly-β-hydroxybutyrate; is oxidase positive; does not form ammonia from arginine but decarboxylates ornithine; gelatinase and amylase negative; forms abundant H_2S in triple-sugar iron agar; some strains reduce nitrate beyond nitrite. It is sensitive to chloramphenicol, tetracycline, neomycin and gentamicin, and usually to carbenicillin and streptomycin; colistin sensitivity is variable.

The G + C content of the DNA varies over a considerable range—43–53 moles per cent—and is much lower than that of pseudomonads. Several phenotypic characters vary from strain to strain, including tolerance for 6–7 per cent NaCl, ability to grow at 5° and 42°, and the pattern of acidification of ammonium salts sugars. Owen and co-workers (1978) recognize four DNA homology groups, with uniform phenotypic characters and DNA base ratios, among the strains. It is therefore unlikely that *'Pseudomonas' putrefaciens* is a single species, or a pseudomonad.

References

Abe, C., Shionoya, H., Hirao, Y., Okada, K. and Homma, J. Y. (1975) *Jap. J. exp. Med.* **45,** 355.

Abe, C., Tanamoto, K. and Homma, J. Y. (1977) *Jap. J. exp. Med.* **47,** 393.

Armstrong, A. V., Stewart-Tull, D. E. S. and Roberts, J. S. (1971) *J. med. Microbiol.* **4,** 249.

Ashdown, L. R. (1979) *Rev. infect. Dis.* **1,** 891.

Ballard, R. W., Doudoroff, M., Stanier, R. Y. and Mandel, M. (1968) *J. gen. Microbiol.* **53,** 349.

Ballard, R. W., Palleroni, N. J., Doudoroff, M., Stanier, R. Y. and Mandel, M. (1970) *J. gen. Microbiol.* **60,** 199.

Bergan, T. (1978) In: *Methods in Microbiology*, vol. 10, p. 169. Ed by T. Bergan and J. R. Norris. Academic Press, London.

Blackwell, C. C. and Law, J. A. (1981) *J. Infect.* **3,** 370.

Blackwell, C. C., Young, H. and Anderson, I. (1979) *J. med. Microbiol.* **12,** 321.

Boivin, A. and Mesrobeanu, L. (1937) *C. R. Soc. Biol.* **125,** 273.

Bradley, D. E. (1966) *J. gen. Microbiol.* **45,** 83; (1974) *Virology* **58,** 149.

Bradley, D. E. and Pitt, T. L. (1974) *J. gen. Virol.* **23,** 1; (1975) *J. Hyg., Camb.* **74,** 419.

Bremmelgaard, A. (1975) *Acta path. microbiol. scand.* **B83,** 65.

Bridges, K., Kidson, A., Lowbury, E. J. L. and Wilkins, M. D. (1979) *Brit. med. J.* **i,** 446.

Brindle, C. S. and Cowan, S. T. (1951) *J. Path. Bact.* **63,** 571.

Brown, M. R. W. (Ed.) (1975) *Resistance of Pseudomonas aeruginosa.* John Wiley and Sons, Chichester.

Brown, M. R. W., Foster, J. H. S. and Clamp, J. R. (1969) *Biochem. J.* **112,** 521.

Brown, M. R. W. and Richards, R. M. E. (1965) *Nature, Lond.* **207,** 1391.

Buchanan, R. E., Holt, J. G. and Lessel, E. F. (1966) *Index Bergeyana.* Williams and Wilkins, Baltimore.

Buhlmann, X., Vischer, W. A. and Bruhin, H. (1961) *J. Bact.* **82,** 787.

Burdon, K. L. (1946) *J. Bact.* **52,** 665.

Burkholder, W. H. (1950) *Phytopathology* **40,** 115.

Burton, M. O., Campbell, J. J. R. and Eagles, B. A. (1948) *Canad. J. Res.* **26,** C 15.

Calabi, O. (1973) *J. med. Microbiol.* **6,** 293.

Chambon, L. and Fournier, J. (1956) *Ann. Inst. Pasteur* **91,** 355, 472.

Chester, I. R., Meadow, P. M. and Pitt, T. L. (1973) *J. gen. Microbiol.* **78,** 305.

Christie, R. (1948) *Aust. J. exp. Biol. med. Sci.* **26,** 425.

Cravitz, L. and Miller, W. R. (1950) *J. infect. Dis.* **86,** 46.

Crompton, D. O., Anderson, K. F. and Kennare, M. A. (1962) *Trans. ophth. Soc. Austral.* **22,** 81.

Davis, D. H., Doudoroff, M. and Stanier, R. Y. (1969) *Int. J. syst. Bact.* **19,** 375.

Debois, J., Degreef, H., Vandepitte, J. and Spaepen, J. (1975) *J. clin. Path.* **28,** 993.

Dodin, A. and Fournier, J. (1970) *Ann. Inst. Pasteur* **119,** 221.

Doggett, R. G., Harrison, G. M. and Wallis, E. S. (1964) *J. Bact.* **87,** 427.

Don, P. A. and Ende, M. van den. (1950) *J. Hyg., Camb.* **48,** 196.

Doudoroff, M. and Palleroni, N. J. (1974) In: *Bergey's Manual of Determinative Bacteriology*, 8th edn., p. 217. Ed. by R. E. Buchanan and N. E. Gibbons. Williams and Wilkins, Baltimore.

Eagon, R. G. (1956) *Canad. J. Microbiol.* **2,** 673.

Eikhoff, T. C., Bennett, J. V., Hayes, P. S. and Feeley, J. (1970) *J. infect. Dis.* **121,** 95.

Elrod, R. P. and Braun, A. C. (1942) *J. Bact.* **44,** 633.

Emmerich, R. and Löw, O. (1899) *Z. Hyg. InfektKr.* **31,** 1.

Emmerich, R., Löw, O. and Korschun, A. (1902) *Zbl. Bakt.* **31,** 1.

Ende, M. van den. (1952) *J. Hyg., Camb.* **50,** 405.

Esselmann, M. T. and Liu, P. V. (1961) *J. Bact.* **81,** 939.

Fensom, A. H. and Gray, G. W. (1969) *Biochem. J.* **114,** 185.

Flügge, C. G. F. W. (1886) *Die Mikro-organismen.* 2te Auflage. Vogel, Leipzig.

Forsyth, W. G. S., Hayward, A. C. and Roberts, J. B. (1958) *Nature, Lond.* **182,** 800.

Fuerst, J. A. and Hayward, A. C. (1969) *J. gen. Microbiol.* **58,** 227.

Garibaldi, J. A. (1967) *J. Bact.* **94,** 1296.

Gessard, C. (1882) *C. R. Acad. Sci.* **94,** 563; (1890) *Ann. Inst. Pasteur* **4,** 88; (1891) *Ibid.* **5,** 65; (1892) *Ibid.* **6,** 801.

Gilardi, G. I. (1969) *Amer. J. clin. Path.* **51,** 58.

Gorrill, R. H. (1952) *J. Path. Bact.* **64,** 857.

Govan, J. R. W. (1974) *J. gen. Microbiol.* **80,** 1, 17; (1975) *J. med. Microbiol.* **8,** 513.

Govan, J. R. W. and Fyfe, J. A. M. (1978) *J. antimicrob. Chemother.* **4,** 233.

Govan, J. R. W., Fyfe, J. A. M. and McMillan, C. (1979) *J. gen. Microbiol.* **110,** 229.

Greer, G. G. and Milazzo, F. H. (1976) *Canad. J. Microbiol.* **22,** 800.

Habs, H. and Mann, S. (1967) *Zbl. Bakt.*, I. Abt. Orig. **203,** 473.

Habs, I. (1957) *Z. Hyg. InfektKr.* **144,** 218.

Hamon, Y. (1956) *Ann. Inst. Pasteur* **91,** 82.

Hancock, L., Humphreys, G. and Meadow, P. (1970) *Biochim. biophys. acta* **202,** 389.

Harper, G. J. and Cawston, W. C. (1945) *Bull. Inst. med. Lab. Tech.* **11,** 40.

Haynes, W. C. (1951) *J. gen. Microbiol.* **5,** 939.

Heckly, R. J. and Nigg, C. (1958) *J. Bact.* **76,** 427.

Hirao, Y. and Homma, J. Y. (1978) *Infect. Immun.* **19,** 373.

Holl, K. and Kleinmaier, H. (1961) *Z. ImmunForsch.* **121,** 170.

Holmes, B., Lapage, S. P. and Easterling, B. G. (1979) *J. clin. Path.* **32,** 66.

Holmes, B., Lapage, S. P. and Malnick, H. (1975) *J. clin. Path.* **28,** 149.

Holmes, B., Owen, R. J., Evans, A., Malnick, H. and Willcox, W. R. (1977) *Int. J. syst. Bact.* **27,** 133.

Homma, J. Y., Hamamura, N., Naoi, M. and Egami, F. (1958) *Bull. Soc. Chim. biol., Paris* **40,** 647.

Homma, J. Y. and Suzuki, N. (1964) *J. Bact.* **87,** 630.

Howe, C., Sampath, A. and Spotnitz, M. (1971) *J. infect. Dis.* **124,** 598.

Hugh, R. and Leifson, E. (1953) *J. Bact.* **66,** 24.

Hugh, R. and Ryschenkow, E. (1961) *J. gen. Microbiol.* **26,** 123.

Inglewski, B. H. and Kabat, D. (1975) *Proc. nat. Acad. Sci., Wash.* **72,** 2284.

Ito, S., Kageyama, M. and Egami, F. (1970) *J. gen. appl. Microbiol., Tokyo* **16,** 205.

Jarvis, F. G. and Johnson, M. J. (1949) *J. Amer. chem. Soc.* **71,** 4124.

Jensen, K. A. and Holten, C. H. (1949) *Acta chem. scand.* **3,** 1446.

Jessen, O. (1965) *Pseudomonas aeruginosa and other Green Fluorescent Pseudomonads.* Copenhagen.

Johnson, G. G., Morris, J. M. and Berk, R. S. (1967) *Canad. J. Microbiol.* **13,** 711.

Jones, L. F., Thomas, E. T., Stinnett, J. D., Gilardi, G. L. and Farmer, J. J. (1974) *Appl. Microbiol.* **27,** 288.

Jones, R. J., Jackson, D. McG. and Lowbury, E. J. K. (1966) *Brit. J. plastic Surg.* **19,** 43.

Jones, R. J. and Lowbury, E. J. L. (1965) *Lancet* **ii,** 623.

Jordan, E. O. (1899) *J. exp. Med.* **4,** 627.

Kageyama, M. (1964) *J. Biochem., Tokyo* **55,** 49.

Kawaharajo, K. and Homma, J. Y. (1975) *Jap. J. exp. Med.* **45,** 515.

King, A., Holmes, B., Phillips, I. and Lapage, S. P. (1979) *J. gen. Microbiol.* **114,** 137.

King, A. and Phillips, I. (1978) *J. med. Microbiol.* **11,** 165.

King, E. O., Ward, M. K. and Raney, D. E. (1954) *J. Lab. clin. Med.* **44,** 301.

Kleinmaier, H. (1957) *Zbl. Bakt.* **170,** 570.

Klinge, K. (1957) *Arch. Hyg., Berl.* **141,** 348; (1959) *Arch. Mikrobiol.* **33,** 406.

Klyhn, K. M. and Gorrill, R. H. (1967) *J. gen. Microbiol.* **47,** 227.

Köhler, W. (1957) *Z. ImmunForsch.* **114,** 282.

Koval, S. F. and Meadow, P. M. (1975) *J. gen. Microbiol.* **91,** 437.

Kreger, A. S. and Gray, L. D. (1978) *Infect. Immun.* **19,** 630.

Kubota, Y. and Liu, P. V. (1971) *J. infect. Dis.* **123,** 97.

Lányi, B. (1970) *Acta microbiol. hung.* **17,** 35.

Lányi, B. and Bergan, T. (1978) In: *Methods in Microbiology*, vol. 10, p. 93. Ed. by T. Bergan, and J. R. Norris. Academic Press, London.

Lapage, S. P. and Bascomb, S. (1968) *J. appl. Bact.* **31,** 568.

Lapage, S. P., Hill, L. R. and Reeve, J. D. (1968) *J. med. Microbiol.* **1,** 195.

Lautrop, H. and Jessen, O. (1964) *Acta path. microbiol. scand.* **60,** 588.

Legroux, R. and Djemil, K. (1931) *C. R. Acad. Sci.* **193,** 1117.

Levine, H. B., Lein, O. G. and Maurer, R. L. (1959) *J. Immunol.* **83,** 468.

Lindberg. R. B., Moncrief, J. A., Switzer, W. E., Order, S. E. and Mills, W. (1965) *J. Trauma* **5,** 601.

Linker, A. and Jones, R. S. (1964) *Nature, Lond.* **204,** 187.

Liu, P. V. (1957) *J. Bact.* **74,** 718; (1961) *Ibid.* **81,** 28; (1964*a*) *Ibid.* **88,** 1421; (1964*b*) *Amer. J. clin. Path.* **41,** 150; (1966*a*) *J. infect. Dis.* **116,** 112; (1966*b*) *Ibid.* **116,** 481; (1966*c*) *Amer. J. clin. Path.* **45,** 639; (1974) *J. infect. Dis.* **130,** S94.

Liu, P. V., Abe, Y. and Bates, J. L. (1961) *J. infect. Dis.* **108,** 218.

Loeffler. (1886) *Arb. ReichsgesundhAmt.* **1,** 141.

Lowbury, E. J. L. (1951) *J. clin. Path.* **4,** 66.

Lowbury, E. J. L. and Collins, A. G. (1955) *J. clin. Path.* **8,** 47.

Lowbury, E. J. L. and Fox, J. E. (1953) *J. Hyg., Camb.* **51,** 203.

Lowbury, E. J. L., Kidson, A., Lilly, H. A., Ayliffe, G. A. J. and Jones, R. J. (1969) *Lancet* **ii,** 448.

Lowbury, E. J. K., Lilly, H. A. and Wilkins, M. D. (1962) *J. clin. Path.* **15,** 339.

Martin, D. F. (1973) *J. med. Microbiol.* **6,** 111.

Meyer, J. M. and Abdallah, M. A. (1978) *J. gen. Microbiol.* **107,** 319.

Michaels, G. B. and Eagon, R. G. (1969) *Proc. Soc. exp. Biol., N.Y.* **131,** 1346.

Mikkelsen, O. S. (1968) *Acta path. microbiol. scand.* **73,** 373.

Miller, W. R., Pannell, L., Cravitz, L., Tanner, W. A. and Ingalls, M. S. (1948*a*) *J. Bact.* **55,** 115.

Miller, W. R., Pannell, L., Cravitz, L., Tanner, W. A. and Rosebury, T. (1948*b*) *J. Bact.* **55,** 127.

Miller, W. R., Pannell, L. and Ingalls, M. S. (1948*c*) *Amer. J. Hyg.* **47,** 205.

Moor, C. E. de, Soekarnen and Walle, N. van de. (1932) *Geneesk. Tijdsch. Ned.-Ind.* **24,** 1618.

Morihara, K. (1964) *J. Bact.* **88,** 745.

Morse, S. A., Vaughan, P., Johnson, D. and Inglewski, B. H. (1976) *Antimicrob. Agents Chemother.* **10,** 354.

Moyer, C. A., Brentano, L., Gravens, D. L., Margraf, H. W. and Monafo, W. W. (1965) *Arch. Surg., Chicago* **90,** 812.

Munoz, J., Scherago, M. and Weaver, R. H. (1949) *J. Bact.* **57,** 269.

Nagy, G. and Zalay, L. (1967) *Acta vet. Hung.* **17,** 285.

Niel, C. B. van and Allen, M. B. (1952) *J. Bact.* **64,** 413.

Nigg, C., Ruch, J., Scott, E. and Noble, K. (1956) *J. Bact.* **71,** 530.

Nonoyama, S., Kojo, H., Mine, Y., Nishida, M., Goto, S. and Kuwahara, S. (1979*a*) *Infect. Immun.* **24,** 394; (1979*b*) *Ibid.* **24,** 399.

Owen, R. J., Legros, R. M. and Lapage, S. P. (1978) *J. gen. Microbiol.* **104,** 127.
Palleroni, N. J. (1978) The *Pseudomonas* group. (*Patterns of Progress*, no. 15). Meadowfield Press, Durham, England.
Palleroni, N. J., Doudoroff, M., Stanier, R. Y., Solanes, R. E. and Mandel, M. (1970) *J. gen. Microbiol.* **60,** 215.
Patty, F. A. (1921) *J. infect. Dis.* **29,** 73.
Phillips, I. (1969) *J. med. Microbiol.* **2,** 9.
Phillips, I., Eikyn, S. and Laker, M. (1972) *Lancet* **i,** 1258.
Phillips, I., King, B. A. and Shannon, K. P. (1978) *J. antimicrob. Chemother.* **4,** 121.
Phillips, I., Lobo, A. Z., Fernandes, R. and Gundara, N. S. (1968) *Lancet* **i,** 11.
Pitt, T. L. (1981) *J. med. Microbiol.* **14,** 251.
Pitt, T. L. and Bradley, D. E. (1975) *J. med. Microbiol.* **8,** 97.
Pollack, M. and Anderson, S. E. Jr. (1978) *Infect. Immun.* **19,** 1092.
Ralston, E., Palleroni, N. J. and Doudoroff, M. (1973) *Int. J. syst. Bact.* **23,** 15.
Redfearn, M. S., Palleroni, N. J. and Stanier, R. Y. (1966) *J. gen. Microbiol.* **43,** 293.
Richmond, M. H. (1976) *J. med. Microbiol.* **9,** 363.
Ricketts, C. R., Lowbury, E. J. L., Lawrence, J. C., Hale, M. and Wilkins, M. D. (1970) *Brit. med. J.* **ii,** 444.
Rogul, M., Brendle, J. J., Haapala, D. K. and Alexander, A. D. (1970) *J. Bact.* **101,** 827.
Rogul, M. and Carr, S. R. (1972) *J. Bact.* **112,** 372.
Russell, A. D. and Mills, A. P. (1974) *J. clin. Path.* **27,** 463.
Sabath, L. D. and Abraham, E. P. (1964) *Nature, Lond.* **204,** 1066.
Sands, D. C., Schroth, M. N. and Hildebrand, D. C. (1970) *J. Bact.* **101,** 9.
Scharmann, W. (1976) *J. gen. Microbiol.* **93,** 283, 292.
Schoental, R. (1941) *Brit. J. exp. Path.* **22,** 137.
Selenka, F. (1958) *Arch. Hyg., Berl.* **142,** 569.
Shoesmith, J. G. and Sherris, J. C. (1960) *J. gen. Microbiol.* **22,** 10.
Sidberry, H. D. and Sadoff, J. C. (1977) *Infect. Immun.* **15,** 628.
Sierra, G. (1960) *Leeuwenhoek ned. Tijdschr.* **26,** 189.
Sierra, G. and Zagt, R. (1960) *Leeuwenhoek ned. Tijdschr.* **26,** 193.
Sîrmon, E. and Marica, D. (1956) *Probl. Epiz. Microbiol. Inst. Pat. Igien. Anim., Bucaresti* No. 5, p. 51.
Smith, P. B. and Cherry, W. B. (1957) *J. Bact.* **74,** 668.
Snell, J. J. S. and Lapage, S. P. (1971) *J. gen. Microbiol.* **68,** 221.
Sompolinsky, D., Hertz, J. B., Høiby, N., Jensen, K., Mansa, B. and Samra, Z. (1980) *Acta path. microbiol. scand.* **B88,** 143.
Sonnenschein, C. (1927) *Zbl. Bakt.* **104,** 365.
Stanier, R. Y., Palleroni, N. J. and Doudoroff, M. (1966) *J. gen. Microbiol.* **43,** 159.
Stanton, A. T. and Fletcher, W. (1921) *Proc. 4th Congr. Far Eastern Ass. trop. Med. Hyg.* **2,** 196; (1925) *J. Hyg., Camb.* **23,** 347; (1927) *Ibid.*, **26,** 31.
Straus, I. (1889) *Arch. Méd. exp.* **1,** 460.
Sykes, R. B. and Richmond, M. H. (1970) *Nature, Lond.* **226,** 952.
Tanamoto, K., Abe, C., Homma, Y., Kuretani, K., Hoshi, A. and Kojima, Y. (1978) *J. Biochem., Tokyô* **83,** 711.
Tunstall, A. M., Gowland, G. and Hobbs, G. (1975) *J. appl. Bact.* **38,** 159.
Turfitt, G. E. (1936) *Biochem. J.* **30,** 1323.
Vasil, M. L. and Inglewski, B. H. (1978) *J. gen. Microbiol.* **108,** 333.
Vasil, M. L., Kabat, D. and Inglewski, B. H. (1977) *Infect. Immun.* **16,** 353.
Wahba, A. H. (1964) *Nature, Lond.* **31,** 242.
Wahba, A. H. and Darrell, J. H. (1965) *J. gen. Microbiol.* **38,** 329.
Warner, P. T. J. C. P. (1950) *Brit. J. exp. Path.* **31,** 242.
Wetmore, P. W. and Gochenour, W. S., Jr. (1956) *J. Bact.* **72,** 79.
Wetmore, P. W., Thiel, J. F., Herman, Y. F. and Harr, J. R. (1963) *J. infect. Dis.* **113,** 186.
Whitmore, A. (1913) *J. Hyg., Camb.* **13,** 1.
Whitmore, A. and Krishnaswami, C. S. (1912) *Indian med. Gaz.* **47,** 262.
Willis, A. T. and Gowland, G. (1960) *Nature, Lond.* **187,** 432.
Wrede, F. (1930) *Z. Hyg. InfektKr.* **111,** 90.
Wretland, B. and Kronevi, T. (1978) *J. med. Microbiol.* **11,** 145.
Yabuuchi, E. and Ohyama, A. (1972) *Int. J. syst. Bact.* **22,** 53.
Young, G. (1947) *J. Bact.* **54,** 109.
Zierdt, C. H. (1971) *Leeuwenhoek ned. Tijdschr.* **37,** 319.

32

Chromobacterium, Flavobacterium, Acinetobacter, and Alkaligenes

M. T. Parker

Chromobacterium

Definition

Rod-shaped, gram-negative, non-sporing bacilli that are motile and possess both polar and lateral flagella; often show bipolar or barred staining. Grow in air on ordinary media; some facultative anaerobes that ferment sugars and others strict aerobes that oxidize sugars. Some mesophilic and others psychrotrophic. Oxidase positive. Usually produce a non-diffusible pigment. Saprophytic organisms that occasionally cause disease in man and animals. G+C content of DNA: 63–72 moles per cent.

Type species: *Chromobacterium violaceum*.

Chromobacterium violaceum was described by Bergonzini in 1881 (see Sneath 1956*c*). Cruess-Callaghan and Gorman (1935) drew attention to the fact that the violet-pigmented chromobacteria could be divided into those that grow at 37° but not at 4°, and those that grow at 4° but not at 37°. These two groups differ in a number of cultural and biochemical characters; though usually called respectively 'mesophilic' and 'psychrophilic' chromobacteria, the term 'psychrotrophic' might be more appropriate for the latter group.

Morphology

The members of this group are rods with rounded ends. Most mesophils are small, but the psychrophils are usually larger and may be slightly curved. The organisms are usually arranged singly or in pairs. The mesophils typically show bipolar staining and many of the longer psychrophils are barred. Mesophils accumulate poly-β-hydroxybutyrate, and contain granules that stain with Sudan black. All are motile and the type of flagellation is characteristic (Figs. 32.1 and 32.2). There is a single short polar flagellum which stains with difficulty and has a relatively long wavelength, and one or more longer subpolar or lateral flagella which stain easily and have a shorter wavelength (Leifson 1956, Sneath 1956*b*). Lateral flagella

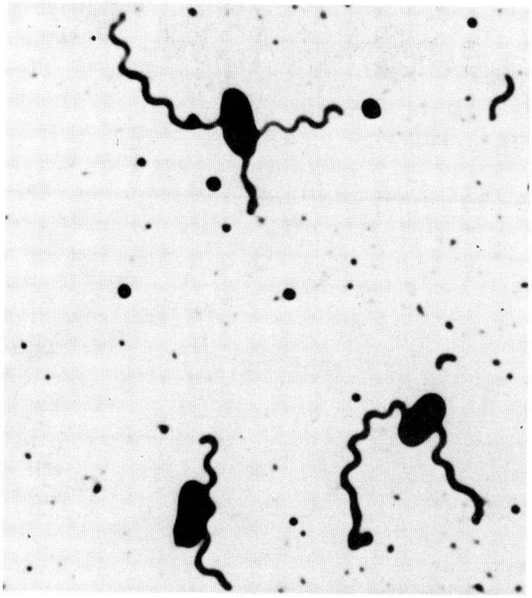

Fig. 32.1 *Chromobacterium violaceum.* Stained preparation to show arrangement of flagella. One organism shows both polar and lateral flagella (×2500). (From photographs kindly supplied by Professor P. H. A. Sneath.)

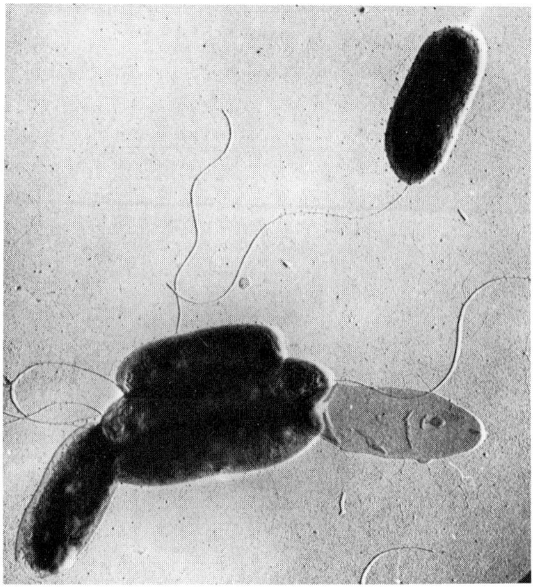

Fig. 32.2 *Chromobacterium violaceum.* Electronmicrograph showing the different wavelengths of polar and lateral flagella (×8000). (From photographs kindly supplied by Professor P. H. A. Sneath.)

are formed most abundantly in young agar cultures and are rare in broth cultures. Neither spores nor capsules are formed.

Cultural characters

Growth occurs readily on nutrient agar when the inoculum is heavy, but single, well separated organisms often fail to form colonies. This inhibition is abolished by catalase (Sneath 1955). Growth is sometimes poor on over-dried plates. Colonies of mesophils are usually round and low convex, 1.0–1.5 mm in diameter after 2 days, with an entire edge and a smooth surface, butyrous and easily emulsified; a few strains form rough, irregular, granular colonies. In 1–3 days, the colonies become opaque and deeply pigmented at the centre, and may emit an odour of cyanide. Many of the psychrophils form a viscous extracellular polysaccharide (Corpe 1960, 1964) and so give colonies of a rubbery consistency; pigmentation is often less intense and slower to develop. On blood agar, mesophils, but not psychrophils, give a wide zone of partial clearing with a diffuse edge. In broth, the mesophils give a densely turbid growth, generally with a pigmented pellicle that sinks to the bottom leaving a characteristic violet collar; the psychrophils also have a pigmented pellicle and collar, but give rise to little turbidity. In gelatin stabs, the mesophils produce rapid liquefaction but the psychrophils seldom cause liquefaction in less than 14 days.

Resistance

The organisms of this group show no particular resistance to heat or disinfectants. They are killed at a temperature of 55° in 10 min (Sneath 1956*a*). They die out rapidly in culture and are difficult to preserve by drying. Mesophilic strains responsible for infections are resistant to penicillin and sulphonamides, but are sensitive to streptomycin and other aminoglycosides, and to chloramphenicol and tetracycline (Sneath 1960). They form an inducible penicillinase that destroys all penicillins except carbenicillin and related compounds, and most cephalosporins (Farrar and O'Dell 1976).

Metabolism and biochemical characters

All are aerobic. The mesophils are usually also facultative anaerobes, but do not form pigment in the absence of oxygen. The psychrophils are strict aerobes. All the mesophils grow at temperatures between 16° and 37°, and some at higher temperatures; all the psychrophils grow between 4° and 25°. No strain will grow both at 4° and 37° but all grow moderately well at 25°. According to Sneath (1956*a*), growth occurs slowly in Koser's medium and citrate is utilized. Acid but not gas may be produced from carbohydrates in peptone media. Most of the mesophils acidify glucose and trehalose promptly, both aerobically and anaerobically. Acid may be formed late from sucrose, glycogen, dextrin and starch. The psychrophils may give small amounts of acid in glucose aerobically, but never in the absence of oxygen, and never acidify trehalose. The mesophils produce HCN, often in considerable quantity. Aesculin is hydrolysed by psychrophils but not by mesophils. Saline extracts of bacteria and broth supernatants of mesophilic strains will lyse human red blood cells (Han and King 1959. In general, the mesophils are proteolytic, digesting gelatin, casein and serum. All chromobacteria are VP negative, and give negative or doubtful MR reactions. None forms indole, but cultures of some gelatinous psychrophils give a pink colour with Ehrlich's reagent owing to the formation of hydroxyindoles, presumably as precursors of violacein (Corpe 1961). Chromobacteria are oxidase positive; the reaction can be detected in pigmented growth by a modification of Kovács's method (Dhar and Johnson 1973). All are urease negative and produce little or no H_2S. Arginine is usually attacked, but often without the formation of NH_3. All are catalase positive, but the reaction is often weak, and the organisms are very sensitive to H_2O_2. They give a negative phenylpyruvic acid test and are usually malonate negative. A few form a reducing substance other than 2-keto-D-gluconic acid from gluconate (Sneath 1956*a*). There is some disagreement about their action on nitrate, and it appears that the result obtained depends to some extent on the medium used. Eltinge (1956) states that all psychrophils reduce nitrate completely, with or without the formation of gas, but that some mesophils, including those from lesions in animals, produce nitrite only from nitrate.

Pigment formation

Pigment is formed only in the presence of oxygen. Poorly pigmented and non-pigmented variants occur in many cultures and entirely non-pigmented mesophils have been isolated from surface water in the tropics (Sivendra *et al.* 1975). All chromobacteria appear to produce the same pigment, violacein. It is non-diffusing, soluble in ethanol and acetone, but not in water, chloroform or benzene. It has some antibiotic activity and is a complex indole derivative (Sneath 1960).

Antigenic structure

The chromobacteria exhibit considerable antigenic heterogeneity. The somatic antigens of the mesophils and the psychrophils appear to be quite distinct. The mesophilic strains cross-agglutinate extensively and the psychrophils less. No clear cut antigenic groups have been established (Sneath and Buckland 1959). The polar and lateral flagella of the same strain are antigenically distinct (Sneath 1956*b*).

Pathogenicity

Chromobacteria are common inhabitants of water and, according to Corpe (1951), are also widespread in soil, though special techniques are required for their isolation. Infections with *Chr. violaceum* have been recorded in man and various other mammals (for references, see Farrar and O'Dell 1976) but are uncommon; in man it causes a highly fatal septicaemic illness with the formation of internal abscesses (Chapter 61). Only the mesophils cause infections.

There are wide differences in the virulence of chromobacteria for laboratory animals. A few mesophils, including those from human infections, are quite virulent (Sneath and Buckland 1959), having an LD50 of about 5×10^6 for the guinea-pig on intraperitoneal injection. Large doses result in an acute and fatal septicaemia, but smaller doses often give rise to a pyaemia more closely resembling the natural disease of man.

Classification

The foregoing description of the mesophilic and psychrophilic chromobacteria is largely based on the papers of Sneath (1956*a*, *b*, *c*, 1960, 1974). He advocates the acceptance of the epithet *Chr. violaceum* for the mesophilic strains and would apply the name *Chr. lividum* to the psychrophils (Table 32.1). The Judicial Commission of the International Committee on Bacterial Nomenclature (Report 1958) ruled that the type species of the genus should be called *Chr. violaceum* and that it should be a mesophil. Leifson (1956), while agreeing that most mesophils are fermenters of carbohydrates, also describes a small group of oxidative mesophils under the name of *Chr. laurentium*. Eltinge (1956, 1957) considers that nitrate reduction is the most significant criterion for the subdivision of the genus. She recognizes a psychrophilic group that reduces nitrate to gaseous nitrogen, and a group of mesophils that reduce nitrate to nitrite or are nitrate negative.

According to Moss and his colleagues (1978), psychrophilic but fermentative chromobacteria that form spreading colonies are common in water; they would allocate these organisms to a separate species, *Chr. fluviatilis*. The G+C content of their DNA (50–52 moles per cent) differs considerably from that of other chromobacteria.

It must be conceded that members of the genus *Chromobacterium* have few common characters except the production of violacein. Members of the two main species show differences that among other groups of organisms would result in their being placed in different genera.

Table 32.1 Differences between *Chr. violaceum* and *Chr. lividum*

	Chr. violaceum	*Chr. lividum*
Growth at 37°	+	–
Growth at 4°	–	+
Anaerobic growth	+	–
Uusual size	0.75 × 2.0 μm	1.0 × 3.7 μm
Intracellular granules*	+	–
Staining	Bipolar	Barred
Gelatin stab	Liquefaction 3–6 days	No liquefaction in 14 days
Blood agar	Clearing in 2 days	No change in 2 days
Attack on glucose	Usually fermentative	Oxidative
Acid from trehalose	+	–
Hydrolysis of aesculin	–	+
Production of HCN	+	–
Egg yolk reaction	+	–

* Stainable with Sudan black.

Flavobacterium

Introduction

This name has been applied rather indiscriminately to gram-negative bacilli that grow on ordinary nutrient media and form a yellow non-diffusible pigment, but nowadays organisms that ferment sugars, have polar flagella, or exhibit gliding movement are specifically excluded (Weeks 1974, McMeekin and Shewan 1978). Despite this, it remains a rather heterogeneous group and includes at least two sorts of organism: (1) those with a low G+C content (30–45 moles per cent), which are non-motile and, according to McMeekin and Shewan (1978), catalase and oxidase positive, and (2) those with a high G+C content (55–70 moles per cent), which may be non-motile or motile by means of peritrichate flagella and are catalase and oxidase negative. According to Holmes and Owen (1979), *Flavobacterium* is a genus of non-motile, pigmented organisms that are strict aerobes, oxidize sugars, and form protease, DNAase and phosphatase; indole is often produced but in small amount; the G+C content of the DNA is 31–40 moles per cent.

Flavobacteria are commonly present in fresh or sea water, and in soil and foodstuffs. Members of one species (*Fl. meningosepticum*) are undoubtedly pathogenic for man. Other flavobacteria may be found in pathological material from time to time, but their ability to cause disease is much less certain. The wide patterns of resistance to antibiotics exhibited by these strains may account for their presence in the flora of hospital patients.

Fl. meningosepticum

In 1959, King defined a homogeneous group of flavobacteria from cases of hospital-acquired neonatal meningitis in various parts of the world; similar strains have since been isolated from the blood stream of adults (Chapter 61). Indistinguishable organisms have been found on the body surface of patients and in the inanimate environment in hospitals.

Morphology In the cerebrospinal fluid, *Fl. meningosepticum* appears as a small rod (*ca* 1.5 × 0.4 μm) extracellularly and intracellularly; in culture it tends to form filaments. Although usually described as non-motile, Webster and Hugh (1979) state that it spreads on semisolid medium and forms flagella. These are generally short and straight, one per bacilus, and attached at the pole or laterally. It is non-sporing and usually unencapsulated, but capsules have been seen after mouse passage.

Cultural characters On plain agar after 24 hr at 37° it forms smooth, circular glistening colonies 1–2 mm in diameter. A pale, yellowish iridescent pigment is formed after 4–7 days, but only at temperatures well below 37°; some strains are unpigmented. It grows poorly on MacConkey's agar and not at all on SS agar. On blood agar it does not give rise to β-haemolysis but green discoloration may appear.

Metabolism It is aerobic and facultatively anaerobic, and grows well between 25° and 37°, but there is little growth at 40°. According to Olsen (1966), strains from the meninges grew at 39° but strains from the blood stream did not.

Susceptibility to antimicrobial agents It is resistant to a wide range of agents, including penicillins and aminoglycosides, but is usually sensitive to cotrimoxazole, novobiocin, rifampicin, clindamycin and cefoxitin: disk-sensitivity tests are said to give unreliable results (see Olsen 1967, Altmann and Bogokovsky 1971, Aber *et al.* 1978).

Biochemical reactions Sugars are oxidized, and acid is formed after 7 days in glucose, fructose, lactose, maltose, mannitol, trehalose and xylose, but not sucrose or salicin; members of serotype F acidify arabinose. Gelatin and casein are hydrolysed. Catalase and oxidase are formed. A weak indole reaction is given in Tryptone broth after 2 days at 30°. Nitrate is not reduced and urease not formed. The H_2S reaction is weak and variable.

Antigenic structure Six serotypes (A–F) were originally described; strains from neonatal meningitis usually belong to type C, and members of type F have been isolated from the blood stream of adults. Richard and his colleagues (1979) have since added other serotypes.

The G+C content of the DNA is 36–38 moles per cent.

(For further information, see King 1959, Buttiaux and Vandepitte 1960, Olsen 1966, 1969, Lapage and Owen 1973, Owen and Snell 1976.)

Other flavobacteria of possible medical importance

Saccharolytic flavobacteria other than *Fl. meningosepticum* are occasionally found in clinical material (Pickett and Manclark 1970). According to Holmes and his colleagues (1978), they include *Fl. breve*, an organism normally found in water and resistant to many antibiotics.

Fl. odoratum This is a non-saccharolytic, non-motile flavobacterium that has been isolated from human urine and wound swabs and occasionally from the blood stream (Holmes *et al.* 1979), but its pathogenicity is in doubt. It is non-motile, forms catalase and oxidase, reduces nitrate, is urease positive, and hydrolyses gelatin. According to these authors, it can be distinguished from *Alk. odorans* (p. 269) by being non-motile, pigmented, proteolytic, lipolytic, and forming urease. *Fl. odoratum* is highly resistant to many antibiotics, including ampicillin, carbenicillin and all the aminoglycosides, but may be moderately sensitive to cotrimoxazole, cephaloridine and nalidixic acid.

Acinetobacter

Definition

Short, stout, non-motile gram-negative bacilli that become almost coccoid in stationary-phase culture. Frequently capsulated. Strict aerobes. Do not form pigment or reduce nitrate. Catalase positive. Oxidase negative. Grow in mineral salts medium with ammonium salts and a single carbon source, such as acetate, but not glucose. Some form acid from sugars oxidatively but others do not. G+C content of DNA: 39–47 moles per cent. Saprophytes; parasites and occasional pathogens of man and animals.

The name *Acinetobacter* was proposed by Brisou and Prévot (1954) for gram-negative non-fermenting bacilli with a characteristic cellular morphology: they were short, plump 'bag-like' rods that under certain cultural conditions resembled diplococci. It is now customary to include in the genus only strict aerobes that give a negative oxidase reaction and are penicillin resistant (Baumann *et al.* 1968, Lautrop 1974).

The first recognizable member of the group to be described was a soil organism isolated by Beijerinck in 1911, and named by him *Micrococcus calco-aceticus* (see Baumann *et al.* 1968). Subsequently, strains isolated by medical bacteriologists were given a variety of other names: those that acidified sugars, and so resembled Beijerinck's organism, were called successively *Herellea vaginicola*, 'B5W', *Bact. anitratum*, *Achromobacter anitratus*, and *Acinetobacter anitratus*; and asaccharolytic strains *Mima polymorpha*, *Moraxella lwoffi*, *Achromobacter lwoffi* and *Acinetobacter lwoffi* (see Henriksen 1973). Despite the difference in action on sugars, these two groups of organisms are rather similar. If only one species is recognized, its name would be *Aci. calcoaceticus*; if the asaccharolytic strains are given separate specific status, the name *Aci. lwoffi* would be appropriate for them.

Habitat Members of the genus are widespread in nature, being common in soil, water and foodstuffs. They are found on the skin of the axilla and groin in some 25 per cent of normal persons (Taplin *et al.* 1963) and rather less often in the upper respiratory tract. Numerous isolations have been made from pus, cerebrospinal fluid and blood in man (Chapter 61).

Morphology After overnight culture on solid medium, apparently diplococcal forms predominate, though a prolonged search will usually reveal an occasional solid bacillary form. The 'diplococci' are in fact oval coccobacilli, usually about 1.0–1.5 × 0.6–0.8 μm in size, that show bipolar staining. In exponentially growing broth cultures, on the other hand, they are typically rod shaped and somewhat larger (1.5–2.5 × 0.9–1.6 μm), and are often in pairs or chains (Baumann *et al.* 1968). A proportion of the cells retain the gram stain. The organisms are non-flagellated and non-motile in the conventional sense, though a twitching movement has been described (Lautrop 1961). Fimbriae are often present, and many strains have capsules.

Cultural characters They grow well on ordinary media, and form white or cream, glistening, smooth and often rather viscid colonies about 1 mm in diameter; those of encapsulated strains are raised and opaque. Surface spreading associated with twitching motility can be demonstrated only on certain media (Lautrop 1974; see also Barker and Maxted 1975). They grow on MacConkey's medium, on which saccharolytic strains form pink colonies; they occasionally grow on deoxycholate citrate agar. A few strains are haemolytic on blood agar. In broth there is uniform turbidity with a variable amount of powdery, granular or viscous deposit. Many strains liquefy gelatin slowly.

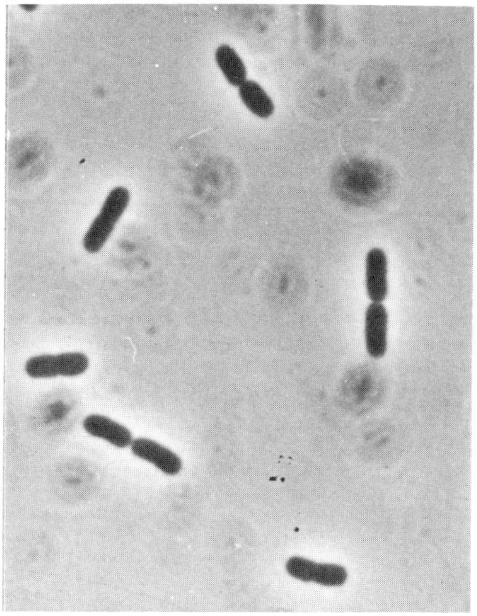

Fig. 32.3 *Acinetobacter anitratus.* Logarithmic phase of growth in liquid medium (phase contrast ×2000). (From photographs kindly supplied by Dr N. J. Palleroni.)

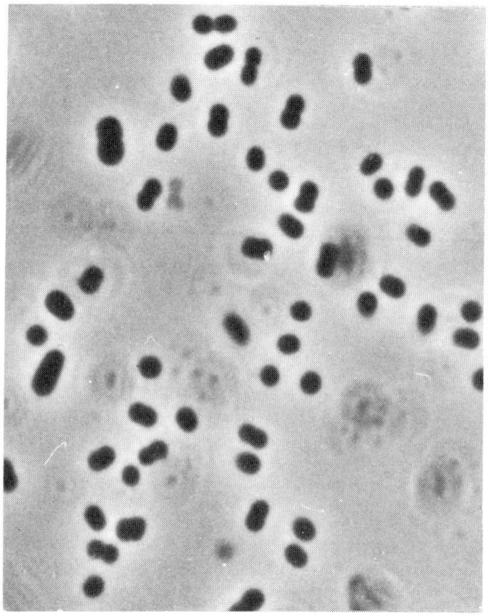

Fig. 32.4 *Acinetobacter anitratus.* Stationary phase of growth in liquid medium (phase contrast ×2000). (From photographs kindly supplied by Dr N. J. Palleroni.)

Metabolic activities Many strains from food and water have an optimum temperature of 25–30° and grow at 0°, but not at 37°. Most of those isolated from human sources will grow even at 42°. All are strict aerobes and will not grow anaerobically in the presence of nitrate. They grow in salt solutions with ammonia as nitrogen source and a number of organic compounds as single source of carbon and energy; these compounds vary from strain to strain but always include acetate. On the other hand, they will not use carbohydrates as carbon source (Baumann *et al.* 1968).

Susceptibility to antimicrobial agents All strains are resistant to penicillins other than carbenicillin, and a penicillinase is said to be formed. Resistance to sulphonamides, streptomycin, chlorampenicol and erythromycin is usual, at least in saccharolytic strains, and tetracycline resistance is now common. In general, saccharolytic strains are more often resistant to non-penicillin antibiotics than are asaccharolytic strains (Hugh and Reese 1969, Pedersen *et al.* 1970), but they are usually sensitive to certain of the newer cephalosporins. Minocycline is more active than tetracycline against *Acinetobacter* (Kuck 1976). High-level amino-glycoside resistance of enzymic origin has been observed in some strains (Murray and Moellering 1979).

Biochemical characters Some strains have no action on sugars; others attack them by means of a single enzyme, an aldolase dehydrogenase, that converts glucose, galactose, mannose, ribose, xylose, arabinose, maltose, lactose and cellobiose to the corresponding acid (Villecourt and Blachère 1955, Baumann *et al.* 1968). Strains that possess this enzyme will acidify the corresponding Hugh and Leifson sugars in the presence of oxygen; acid is formed in peptone-water sugars containing glucose, arabinose, xylose and occasionally rhamnose, but reactions may be delayed. A few strains form acid in 1 per cent lactose peptone water, but all saccharolytic strains acidify a 10 per cent lactose agar slope in 1–2 days. All grow in Koser's citrate medium; none decarboxylates amino acids or attacks phenylalanine or gluconate; some use malonate, some form urease, and some are KCN resistant. Some strains cause 'ropy' milk (Morton and Barrett 1982).

Antigenic structure At least 28 capsular types have been identified by means of capsular reactions or immunofluorescence (Marcus *et al.* 1969). Agglutination tests are unsatisfactory.

Classification Opinions are divided on whether one or two species should be recognized. The non-saccharolytic strains are said to be less often capsulated, but otherwise resemble the saccharolytic strains in most characters other than their inability to produce the single enzyme responsible for the attack on sugars. The DNA base-composition of the group as a whole covers a fairly wide range; saccharolytic strains give uniform and non-saccharolytic strains rather more scattered values (De Ley 1968).

The DNA-base composition of *Acinetobacter* and *Moraxella* covers the same range. As at present defined, the two genera are distinguished by the positive oxidase reaction and sensitivity to penicillin of *Moraxella*. According to Thornley

(1967), neither is a good taxonomic criterion, and strains that are difficult to allocate to either genus are quite commonly seen. The relation to *Acinetobacter* of some saccharolytic, oxidase-positive strains at present classified with *Moraxella* is uncertain.

(For further information about *Acinetobacter*, see Gilardi 1969, and Juni 1978.)

Alkaligenes

Petruschky (1896) isolated from human faeces a gram-negative bacillus actively motile by peritrichate flagella that produced an alkaline reaction in milk; he named it *Bact. faecalis alkaligenes*. In the following years, many workers applied the terms *Bact. alkaligenes* and *Alkaligenes faecalis* rather loosely to gram-negative bacilli that failed to ferment any of the common sugars in peptone water.

Introduction

Such organisms were often found in human faeces, in the blood stream and in various lesions. The evidence that these organisms were pathogenic was sometimes good, but many of them would probably not have conformed to the present definition of *Alkaligenes*. Nyberg (1935) carried out a thorough study of strains from human sources and from sewage; he recognized a short thick bacillus, usually non-motile or feebly motile, which possessed as a rule poorly formed peritrichate flagella, did not liquefy gelatin, form indole or produce any change in litmus milk. This he called *Bact. alkaligenes*, and distinguished it from the longer, thinner, slightly curved rods with polar flagella that alkalinizes milk and would nowadays be considered to be pseudomonads.

The genus *Alkaligenes* is now reserved for motile organisms with peritrichate flagella that are strict aerobes and do not oxidize glucose. Most of these organisms are oxidase positive and grow in Koser's citrate medium. Their action on nitrate is variable, but some of those found in the soil are active denitrifiers. Their DNA contains 56–67 moles per cent G+C (Thornley 1967). According to Pickett and Manclark (1970), the alkaligenes strain most often found in or on the human body is *Alk. odorans*, which is common in the gut and sometimes isolated from urine and pus.

Alk. odorans This organism was described fully by Málek and his colleagues (1963). It is a small rod, 0.5–0.8 × 0.8–1–2 μm, and is not capsulated. Young cultures have a characteristic aromatic or fruity odour. Colonies on nutrient agar are greyish-white and either (1) umbonate with a smooth central plateau and a veil-like spreading edge and irregular outline or (2) high convex and circular. Both types of colony appear in the same culture. The organism was said to be non-haemolytic on blood agar, but Mitchell and Clarke (1965) described a variant (*Alk. odorans* var. *viridans*) that gave bright green zones of α-haemolysis.

Alk. odorans is a strict aerobe, with variable action on nitrate, but the *viridans* variety is said to be able to reduce nitrite to gaseous nitrogen though it does not reduce nitrate (Gilardi 1967, Chatelain 1969). It is catalase and oxidase positive and grows in 6.6 per cent NaCl, but gives negative results in tests for indole, H_2S, urease, gluconate oxidation, deamination of phenylalanine and decarboxylation of arginine. It is sensitive to KCN but forms ammonia in peptone water. Gelatin is sometimes liquefied, and litmus milk is alkalinized and peptonized. It is resistant to penicillin and streptomycin but sensitive to most other antibiotics (Pedersen *et al.* 1970).

Alkaligenes must be distinguished from *Bordetella bronchiseptica*. It differs from this organism in growing in Koser's citrate medium, in not requiring niacin, in being urease negative, and not causing haemagglutination of mammalian erythrocytes (Szturm and Bourdon 1948, Ulrich and Needham 1953, Thibault *et al.* 1955).

(For a general account of *Alkaligenes*, see Pichinoty *et al.* 1978.)

'*Achromobacter*' *xylosoxidans*

The genus *Achromobacter* should probably no longer be recognized (Hendrie *et al.* 1974) because of the close correspondence between former descriptions of it and the present definition of *Alkaligenes*. However, the organism named *Achromobacter xylosoxidans* by Yabuuchi and his colleagues (1974), which appears to be of some medical importance, could not be accommodated in *Alkaligenes*, and judgement must be reserved on its generic status.

This organism was first isolated from ear swabs of patients with chronic otitis media (Yabuuchi and Ohyama 1971) and has since been found in a variety of other clinical specimens (Yabuuchi *et al.* 1974, Holmes *et al.* 1977). It has been responsible for a number of cases of postoperative meningitis (Shigeota *et al.* 1978). It is a short, thick gram-negative bacillus that is motile by means of peritrichate flagella and forms intracellular granules of poly-β-hydroxybutyrate. It grows on ordinary nutrient media at room temperature and usually at 37°. It does not form pigment. Catalase and oxidase reactions are positive. Glucose, xylose, and sometimes ethanol, are acidified in ammonium salts sugars, but the reaction in Hugh and Leifson's glucose medium is neutral or alkaline. Nitrate is reduced to nitrite and sometimes beyond this. H_2S is not formed. It is not proteolytic or lipolytic. The G+C content of the DNA is 67 moles per cent. Most strains are resistant to several antimicrobial agents, but patterns of susceptibility are far from uniform (Yabuuchi *et al.* 1974, Holmes *et al.* 1977). Shigeota and his colleagues (1978) report that its resistance to chlorhexidine is remarkably high, nearly all strains surviving for 10 min in concentrations of 1 per cent or more.

It should be noted that peritrichate non-fermenting rods that oxidize sugars are commonly isolated from the soil. Many of them are oxidase positive, produce H_2S in Kliger's medium, reduce nitrate and are penicillin resistant (Thornley 1967); they are therefore difficult to distinguish from *Agrobacterium* (Fam. Rhizobiaceae).

We conclude this chapter by presenting a tentative classification of the non-fermenting gram-negative rods that grow on plain agar aerobically and may be encountered in the medical and veterinary laboratory (Table 32.2). (See also Gilardi 1971, Snell 1973.)

Table 32.2 Some non-fermenting aerobic gram-negative bacilli

Organism	Morphology	Flagella	Oxidase reaction	Action on carbo-hydrates	Reduction of nitrate	Other characters
Acinetobacter	Short, stout rods and coccal forms:	None	–	Oxidative or –	–	Reistant to >1 unit penicillin
Moraxella	chains and filaments	None	+	–*	Variable	Sensitive to 0.1 unit penicillin
Alkaligenes†	Short rods	Peritrichous	+	–	Variable	Resistant to >1 unit penicillin
Bordetella bronchiseptica	Small rods	Peritrichous	weak +	–	To NO_2	Urease+; more nutritionally exacting than *Alkaligenes*. See Chapter 39
Flavobacterium meningosepticum‡	Small rods and filaments	None	+	Oxidative	–	Weak indole producer. Usually yellow colonies
Pseudomonas	Slender rods or shorter, irregularly staining rods	Polar (or mixed)	+§	Oxidative or –	Variable	See Chapter 31
Chromobacterium lividum‖	Relatively large rods: barred staining	Mixed	+	Oxidative	Usually to N_2	Violet colonies Grows at 4°

+ = Positive reaction. – = Negative reaction.

* Except '*Moraxella*' *kingae*.

† For peritrichous, saccharolytic organisms, see p. 269.

‡ For other flavobacteria, see p. 267.

§ Except *Ps. maltophilia*.

‖ *Chr. violaceum* is fermentative.

References

Aber, R. C., Wennersten, C. and Moellering, R. C. Jr. (1978) *Antimicrob. agents Chemother.* **14,** 483.

Altmann, G. and Bogokovsky, B. (1971) *J. med. Microbiol.* **4,** 296.

Barker, J. and Maxted, H. (1975) *J. med. Microbiol.* **8,** 443.

Baumann, P., Doudoroff, M. and Stanier, R. Y. (1968) *J. Bact.* **95,** 1520.

Brisou, J. and Prévot, A.-R. (1954) *Ann. Inst. Pasteur* **86,** 722.

Buttiaux, R. and Vandepitte, J. (1960) *Ann. Inst. Pasteur* **98,** 398.

Chatelain, R. (1969) *Ann. Inst. Pasteur* **116,** 498.

Corpe, W. A. (1951) *J. Bact.*, **62,** 515; (1960) *Canad. J. Microbiol.* **6,** 153; (1961) *Nature, Lond.* **190,** 191; (1964) *J. Bact.* **88,** 1433.

Cruess-Callaghan, G. and Gorman, M. J. (1935) *Sci. Proc. R. Dublin Soc.* **21,** N.S., 213.

De Ley, J. (1968) *Leeuwenhoek ned. Tijdschr.* **34,** 109.

Dhar, S. K. and Johnson, R. (1973) *J. clin. Path.* **26,** 304.

Eltinge, E. T. (1956) *Leeuwenhoek ned. Tijdschr.* **22,** 139; (1957) *Int. Bull. bact. Nomencl.* **7,** 37.

Farrar, W. E. Jr. and O'Dell, N. M. (1976) *J. infect. Dis.* **134,** 290.

Gilardi, G. L. (1967) *Canad. J. Microbiol.*, **13,** 895; (1969) *Leeuwenhoek ned. Tijdschr.* **35,** 421; (1971) *J. appl. Bact.* **34,** 623.

Han, E. A. and King, J. W. (1959) *Amer. J. clin. Path.* **31,** 133.

Hendrie, M. S., Holding, A. J. and Shewan, J. M. (1979) *Int. J. syst. Bact.* **24,** 534.

Henricksen, S. D. (1973) *Bact. Rev.* **37,** 522.

Holmes, B. and Owen, R. J. (1979) *Int. J. syst. Bact.* **29,** 416.

Holmes, B., Snell, J. J. S. and Lapage, S. P. (1977) *J. clin. Path.* **30,** 595; (1978) *Int. J. syst. Bact.* **28,** 201; (1979) *J. clin. Path.* **32,** 73.

Hugh, R. and Reese, R. (1969) *Publ. Hlth Lab.* **27,** 174.

Juni, E. (1978) *Ann. Rev. Microbiol.* **32,** 349.

King, E. O. (1959) *Amer. J. clin. Path.* **31,** 341.

Kuck, N. A. (1976) *Antimicrob. Agents Chemother.* **9,** 493.

Lapage, S. P. and Owen, R. J. (1973) *J. clin. Path.* **26,** 747.
Lautrop, H. (1961) *Int. Bull. bact. Nomencl.* **11,** 107; (1974) In: *Bergey's Manual of Determinative Bacteriology*, 8th edn, p. 436. Ed. by R. E. Buchanan and N. E. Gibbons. Williams and Wilkins, Baltimore.
Leifson, E. (1956) *J. Bact.* **71,** 393.
McMeekin, T. A. and Shewan, J. M. (1978) *J. appl. Bact.* **45,** 321.
Málek, I., Radochová, M. and Lysenko, O. (1963) *J. gen. Microbiol.* **33,** 349.
Marcus, B. B., Samuels, S. B., Pittman, B. and Cherry, W. B. (1969) *Amer. J. clin. Path.* **52,** 309.
Mitchell, R. G. and Clarke, S. K. R. (1965) *J. gen. Microbiol.* **40,** 343.
Morton, D. J. and Barrett, E. L. (1982) *Curr. Microbiol.* **7,** 107.
Moss, M. O., Ryall, C. and Logan, N. A. (1978) *J. gen. Microbiol.* **105,** 11.
Murray, B. E. and Moellering, R. C. Jr. (1979) *Antimicrob. Agents Chemother.* **15,** 190.
Nyberg. C. (1935) *Zbl. Bakt.* **133,** 443.
Olsen, H. (1966) *Acta path. microbiol. scand.* **67,** 291; (1967) *Ibid.* **70,** 601; (1969) *Ibid.* **75,** 313.
Owen, R. J. and Snell, J. J. S. (1976) *J. gen. Microbiol.* **93,** 89.
Pedersen, M. M., Marso, E. and Pickett, M. J. (1970) *Amer. J. clin. Path.* **54,** 178.
Petruschky, J. (1896) *Zbl. Bakt.* **19,** 187.
Pichinoty, F., Véron, M., Mandel, M., Durand, M., Job, C. and Garcia, J. L. (1978) *Canad. J. Microbiol.* **24,** 743.
Pickett, M. J. and Manclark, C. R. (1970) *Amer. J. clin. Path.* **54,** 155.
Report. (1958) *Int. Bull. bact. Nomencl.* **8,** 151.
Richard, C., Monteil, H. and Laurent, B. (1979) *Ann. Microbiol., Inst. Pasteur* **B130,** 141.
Shigeota, S., Yasunaga, Y., Honzumi, K., Okamura, H., Kumata, R. and Endo, S. (1978) *J. clin. Path.* **31,** 156.
Sivendra, R., Lo, H. S. and Lim, K. T. (1975) *J. gen. Microbiol.* **90,** 21.
Sneath, P. H. A. (1955) *J. gen. Microbiol.* **13,** i; (1956*a*) *Ibid.* **15,** 70; (1956*b*) *Ibid* **15,** 99; (1956*c*) *Int. Bull. bact. Nomencl.* **6,** 65; (1960) *Iowa St. J. Sci.* **34,** 243; (1974) In: *Bergey's Manual of Determinative Bacteriology*, 8th edn, p. 354. Ed. by R. E. Buchanan and N. E. Gibbons. Williams and Wilkins, Baltimore.
Sneath, P. H. A. and Buckland, F. E. (1959) *J. gen. Microbiol.* **20,** 414.
Snell, J. J. S. (1973) *The Distribution and Identification of Non-Fermenting Bacteria. PHLS Mono. Ser.*, No. 4. HMSO, London.
Szturm, S. and Bourdon, D. (1948) *Ann. Inst. Pasteur* **75,** 65.
Taplin, D., Rebell, G. and Zaias, N. (1963) *J. Amer. med. Ass.* **186,** 952.
Thibault, P., Szturm-Rubinsten, S. and Piéchaud-Bourbon, D. (1955) *Ann. Inst. Pasteur* **88,** 246.
Thornley, M. J. (1967) *J. gen. Microbiol.* **49,** 211.
Ulrich, J. A. and Needham, G. M. (1953) *J. Bact.* **65,** 210.
Villecourt, P. and Blachère, H. (1955) *Ann. Inst. Pasteur* **88,** 523.
Webster, J. A. and Hugh, R. (1979) *Int. J. syst. Bact.* **29,** 333.
Weeks, O. B. (1974) In: *Bergey's Manual of Determinative Bacteriology*, 8th edn, p. 357. Ed. by R. E. Buchanan and N. E. Gibbons. Williams and Wilkins, Baltimore.
Yabuuchi, E. and Ohyama, A. (1971) *Jap. J. Microbiol.* **15,** 477.
Yabuuchi, E., Yano, I., Goto, S., Tanimura, E., Ito, T. and Ohyama, A. (1974) *Int. J. syst. Bact.* **24,** 470.

33

The Enterobacteriaceae

Roger J. Gross and Barry Holmes

Introductory

Definition

Gram-negative, non-sporing rods; often motile, usually by means of peritrichate flagella. Capsulated or non-capsulated. Easily cultivable on ordinary laboratory media. Aerobic and facultatively anaerobic. All species ferment glucose with the formation of acid or of acid and gas. Reduce nitrate to nitrite with the exception of some strains of *Erwinia* and *Yersinia*. Oxidase negative; catalase positive, except *Shigella dysenteriae* 1. Typically intestinal parasites of man and animals, though some species may occur in other parts of the body, on plants and in the soil. Many species are pathogenic. The G + C content of the DNA in most of the Enterobacteriaceae is in the range 49–59 moles per cent, but in *Proteus* and *Providencia* it is in the range 37–42 moles per cent and in *Yersinia* 46–47 moles per cent.

In the past the term *Bacterium* was used to comprise a broad group of gram-negative, non-sporing rods occurring in the intestinal canal of man and animals, on plants and in the soil, and leading either a saprophytic, commensal or pathogenic existence. These organisms have been studied for many years by groups of workers with widely differing objectives. The medical and veterinary bacteriologists concentrated their attention on those organisms which caused illness in man and animals. The plant pathologists were most interested in those which possessed enzymes capable of digesting vegetable tissues. Industrial bacteriologists studied in detail a few strains that carried out useful fermentations or interfered with manufacturing processes. Sanitary bacteriologists were concerned in distinguishing the organisms found in the intestine from those living saprophytically outside the animal body. This diversity of approach led to much confusion, and the same organism was often studied independently, and under different names, by several groups of workers who used methods that were not comparable. Of recent years, and largely under the stimulus of the Enterobacteriaceae Sub-Committee of the International Committee on Systematic Bacteriology, an attempt has been made to co-ordinate the experience of workers in the different fields, and there is now general agreement on the definition of the family Enterobacteriaceae and the main lines on which it should be subdivided.

As long ago as 1893 it was pointed out by Theobald Smith that certain important groups of organisms pathogenic to man and animals differed from most of the non-pathogenic forms in failing to ferment lactose, and the genera *Salmonella* and *Shigella* were created for these non-lactose fermenters. The lactose fermenters were, for the most part, thought to be normal inhabitants of the intestinal tract of man and the higher animals, or to exist on plants or in the soil. However, we now know that some genera contain members which acidify lactose promptly, others which do so late and irregularly, and yet others which are non-lactose-fermenting. Furthermore, it is now known that *Esch. coli*, usually a prompt lactose fermenter, is an important cause of acute enteritis in persons of all age groups, and in animals (see Chapter 71). For convenience we shall continue to use the term 'coliform bacteria' to refer to those members of the Enterobacteriaceae that generally, though by no means invariably, ferment lactose.

We have chosen to consider the genus *Yersinia* separately (Chapter 38) although it conforms to the definition of the Enterobacteriaceae. The inclusion of *Yersinia* in the family Enterobacteriaceae was proposed by Thal (1954) and supported by Sneath and Cowan (1958) and Talbot and Sneath (1960). Most workers now accept this classification, although *Yersinia*, particularly *Y. enterocolitica* and related organisms, are the subject of intensive study at present.

There is now general agreement that the genera and species of the Enterobacteriaceae should be distinguished by their biochemical characters (Reports 1954*a*, 1954*b*, 1958, 1963), and that it is only within these genera and species that serological methods can be used for further subdivision.

After the early separation of *Salmonella* and *Shigella* from the rest of the Enterobacteriaceae on the grounds of their pathogenicity for animals, their antigenic structure was studied intensively. Serological methods proved of great value for subdividing these genera, but yielded little information about their relation to other genera. In fact, before it was realized how often cross reactions occurred between otherwise dissimilar organisms, undue reliance on serological methods led to the inclusion of a number of coliform bacteria in the genus *Salmonella*.

Early attempts to subdivide the coliform bacteria according to their ability to ferment a large number of carbohydrates also failed to produce a rational system of classification (see Winslow *et al.* 1919). The water bacteriologists, however, who were mainly interested in lactose-fermenting organisms, arrived at a broad classification into three groups, based on a few arbitrarily selected tests, of which the Voges-Proskauer test (VP), the methyl-red test (MR), and the Koser citrate test were the most important, and recognized a coli group (MR+, VP−, citrate−), an aerogenes group (MR−, VP+, citrate+) and an intermediate group (MR+, VP−, citrate+). Attempts were later made to subdivide the late-lactose-fermenting and lactose non-fermenting 'paracolons' in the same way (Borman *et al.* 1944). Although nobody would now accept their suggestion of a genus *Paracolobactrum*, divided into a coli-, an aerogenes-, and an intermediate-like species, there is no doubt that most enterobacteria do fall into one of these three divisions, whether or not they ferment lactose. The MR+, VP−, citrate− division includes the faecal coliform and dysentery bacteria; the MR+, VP−, citrate+ division contains the intermediate coliform bacteria and the salmonellae; and the MR−, VP+, citrate+ division includes all the aerogenes-like organisms. It must be emphasized, however, that some species, subspecies and biochemical varieties within these broad divisions may be aberrant in their MR, VP and citrate reactions. *Proteus*, *Providencia* and *Morganella* are generally believed to fall outside this classification, and form a fourth division. Ewing and Edwards (1960) advocated a similar classification into four groups: (1) Shigella-Escherichia, (2) Salmonella-Arizona-Citrobacter, (3) Klebsiella-Cloaca-Serratia, and (4) Proteus-Providence, though the biochemical criteria they used were different.

The Enterobacteriaceae can thus be divided into a number of biochemical groups, each of which may be further subdivided by serological, biochemical and other methods. There has been disagreement on the taxonomic rank to be given to these subdivisions, though most workers would now accord generic status to the main biochemical groups. Cowan (1956, 1957) recognized six genera: *Salmonella*, *Escherichia*, *Shigella*, *Citrobacter*, *Klebsiella* and *Proteus*. Kauffmann (1959) divided the Enterobacteriaceae into three tribes and 14 genera, including in addition to those given by Cowan, *Arizona*, *Cloaca*, *Hafnia*, *Erwinia*, *Serratia*, *Morganella*, *Rettgerella* and *Providencia*. Ewing (1966) revised the earlier classification of Ewing and Edwards (1960) to comprise four tribes and 10 genera: (1) Escherichieae (*Escherichia*, *Shigella*); (2) Salmonelleae (*Salmonella*, *Arizona*, *Citrobacter*); (3) Klebsielleae (*Klebsiella*, *Enterobacter*, *Serratia*); (4) Proteeae (*Proteus*, *Providencia*), and envisaged the addition in future of a fifth tribe Edwardsielleae for the genus *Edwardsiella*. We propose to use the generic names *Citrobacter*, *Edwardsiella*, *Enterobacter*, *Erwinia*, *Escherichia*, *Hafnia*, *Klebsiella*, *Kluyvera*, *Morganella*, *Proteus*, *Providencia*, *Salmonella*, *Serratia*, *Shigella* and *Yersinia*. In addition we shall mention briefly the recently proposed genera *Cedecea* (Grimont *et al.* 1981) and *Tatumella* (Hollis *et al.* 1981). We shall include the Arizona group in the genus *Salmonella*.

The definition of species within these genera also presents difficulties. For example, in the genus *Salmonella* it has been customary to give a specific epithet to each of the hundreds of serotypes or 'serovars'. In contrast, *Esch. coli* is usually considered to be a single species divisible into a large number of serotypes, each designated by a numerical formula. Several proposals have been made to introduce some uniformity (Kauffmann 1959, Ewing 1966, Le Minor *et al.* 1970). Those relating to the genus *Salmonella* are summarized in Chapter 37.

Table 33.1 An alphabetical key to the genera and species of Enterobacteriaceae

Genera	Species and other subgeneric groups	Synonyms
Cedecea	*Ced. davisae*	
	Ced. lapagei	
Citrobacter	*Citro. amalonaticus*	*Levinea amalonatica*
	Citro. freundi	
	Citro. koseri	*Citro. diversus, Levinea malonatica*
Edwardsiella	*Ed. tarda*	*Ed. anguillimortifera*
Enterobacter	*Ent. aerogenes*	*K. mobilis*
	Ent. cloacae	
	Ent. gergoviae	
	Ent. sakazakii	
Erwinia	*Erw. herbicola*	*Ent. agglomerans*
Escherichia	*Esch. adecarboxylata*	
	Esch. blattae	
	Esch. coli	
Hafnia	*Haf. alvei*	
Klebsiella	*K. oxytoca*	
	K. pneumoniae	
	subsp. *aerogenes*	*K. aerogenes*
	subsp. *ozaenae*	*K. ozaenae*
	subsp. *pneumoniae*	*K. pneumoniae*
	subsp. *rhinoscleromatis*	*K. rhinoscleromatis*
Kluyvera	*Kluy. ascorbata*	
	Kluy. cryocrescens	
Morganella	*Morg. morgani*	
Proteus	*Prot. mirabilis*	
	Prot. vulgaris	
Providencia	*Prov. alcalifaciens*	
	Prov. rettgeri	
	Prov. stuarti	
Salmonella	*Salm.* subgenus I	
	Salm. subgenus II	
	Salm. subgenus III	*Arizona*
	Salm. subgenus IV	
	Salm. choleraesuis	
	Salm. gallinarum	
	Salm. paratyphi A	
	Salm. pullorum	*Salm. gallinarum* var. *pullorum*
	Salm. typhi	
Serratia	*Ser. liquefaciens*	*Ent. liquefaciens*
	Ser. marcescens	
	Ser. marinorubra	*Ser. rubidaea*
	Ser. odorifera	
	Ser. plymuthica	
Shigella	*Sh. boydi*	
	Sh. dysenteriae	
	Sh. flexneri	
	Sh. sonnei	
Tatumella	*Tat. ptyseos*	
Yersinia	*Y. enterocolitica*	
	Y. frederikseni	
	Y. intermedia	
	Y. kristenseni	
	Y. pestis	
	Y. pseudotuberculosis	

The generic and specific names used in this book are listed alphabetically in Table 33.1, along with a few of their common synonyms. In this chapter we shall consider the general characters of the Enterobacteriaceae; in Chapter 34 we shall give a more detailed account of the individual genera other than the following, which have separate chapters or parts of chapters devoted to them: *Proteus*, *Morganella* and *Providencia* (Chapter 35), *Shigella* (Chapter 36), *Salmonella* (Chapter 37) and *Yersinia* (Chapter 38).

Morphology

The modal form of the individual cell is that of a rod, 2 to 3 μm in length and 0.6 μm in breadth, with parallel sides and rounded ends. By the usual methods of examination the cell appears to be almost devoid of internal structure. It stains evenly, forms no spores, and shows no granules. It is gram negative and non-acid-fast. This modal form is, however, widely departed from. Some strains are almost coccal in form, others show long, sometimes filamentous rods. There is a tendency for the coccobacillary, or the elongated, form to predominate in any single strain, but some cultures show a wide diversity in this respect.

Motile strains are found in all genera of the Enterobacteriaceae except *Shigella* and *Klebsiella* (although some authorities have proposed that *Enterobacter aerogenes* should be reclassified as *Klebsiella mobilis*). *Y. pestis* is non-motile. Some 80 per cent of strains of *Esch. coli* are motile. In the genus *Salmonella* there is only one constantly non-motile serotype, comprising *Salm. gallinarum* and *Salm. pullorum*. In the other genera, absence of motility is uncommon. Flagella are always peritrichate—except in *Tatumella*, in which the flagella observed tend to be polar, subpolar or lateral.

Peluffo (1953) found the flagellar wave-lengths to vary between 1.95 and 2.5 μm. Leifson (1960), who made measurements on stained films, gave somewhat higher figures for normal flagella (2.13–3.02 μm). He also described several variations in the shape of the flagella, of which the 'curly' was the most frequent. 'Curly' flagella have about half the normal wave-length, and function inefficiently. They occur as genetically stable mutants, and can also be induced phenotypically, in *Proteus* but not in other of the Enterobacteriaceae, by growth at pH 5.5 (Hoeniger 1965). Non-motile variants may be either non-flagellated or flagellated but 'paralysed'. In the salmonellae there is a unique type of variation—phase variation (Chapter 37)—in which the production of flagella of different antigenic constitution is rapidly alternated. (For reviews of the genetic control of flagellation see Joys 1968 and Iino 1969).

The upper limit of temperature for motility is often considerably lower than that for growth. Raising the temperature from 37 to 44° inhibits the motility of salmonellae, and the production of flagella, while having no effect on the growth rate (Quadling and Stocker 1962). Most of the Enterobacteriaceae are motile at 37°, but some—notably *Enterobacter*, *Hafnia*, *Tatumella* and *Yersinia*—should be tested at a lower temperature. 'Swarming' over the surface of solid media is a character only of certain strains of *Proteus* (Chapter 35).

Many of the Enterobacteriaceae are fimbriate (Duguid *et al.* 1955, Constable 1956, Duguid and Gillies 1957, 1958, Duguid 1959). Duguid distinguishes between *common fimbriae*, which are usually present in large numbers (50–100) per cell, often influence the physico-chemical characters of the organism, are chromosomally determined, and do not have an essential role in bacterial conjugation, and *sex fimbriae*, which are present in smaller numbers, are episomally determined, and appear to be organs of conjugation (Duguid *et al.* 1966, Duguid 1968). Strains may carry both common fimbriae and sex fimbriae, and more than one type of common fimbriae. Within the strain, there are individual cells with many common fimbriae and others with none, and there is reversible variation between the fimbriate and the non-fimbriate phase.

Five of the six types of common fimbriae are found among the Enterobacteriaceae (Duguid *et al.* 1966), and all are peritrichate. Type 1 fimbriae are strongly hydrophobic and render the organism adhesive to plant, fungus and animal cells, cause agglutination of untreated erythrocytes, and lead to the formation of a pellicle on the surface of liquid media. Adhesiveness and the ability to cause haemagglutination are reversed by D-mannose and so are described as mannose sensitive (MS). Type 2 fimbriae do not confer adhesiveness. Type 3 fimbriae cause adhesiveness to plant and fungal cells and to glass surfaces, but not to animal tissue cells or erythrocytes unless these are first treated with tannic acid or heated to 70°; these properties are uninfluenced by mannose (MR). Type 4 fimbriae also cause MR adhesiveness, but this is active against untreated red cells. Type 6 fimbriae are rather scanty in number (4–40 per cell), longer than fimbriae of other types and do not cause adhesiveness. Fimbriae of type 1 are found in *Escherichia*, *Sh. flexneri*, most salmonellae, *Klebsiella*, *Enterobacter* and *Serratia*; of type 2 in a few of the salmonellae that do not form type 1 fimbriae; of type 3 in klebsiella and serratia strains together with fimbriae of type 1; of type 4 only in proteus strains, and of type 6 only in otherwise non-fimbriate strains of *K. pneumoniae* subsp. *ozaenae*.

Little is known about the function of the common fimbriae. Adhesion to animal cells might favour pathogenicity, but there are few examples of this (Duguid 1968, Duguid and Old 1980). On the other hand, the ability of bacteria with type 1 fimbriae to form a pellicle on the surface of static fluids increases their access to oxygen and leads to additional multiplication of the organism (Old *et al.* 1968, Old and Duguid 1970). This might be an advantage to saprophytic enterobacteria in their natural environment; and adhesiveness of type 3 fimbriae to plant root hairs and fungal mycelium might serve a similar function (Duguid 1968).

More recently a number of filamentous protein structures resembling fimbriae have been described in *Esch. coli*. These cause MR haemagglutination, and there is good evidence that they play an important part in the pathogenesis of diarrhoeal disease and in urinary-tract infection. They include the K88 antigen found in strains causing enteritis of pigs, the K99 antigen found in strains causing enteritis of calves and lambs, and the colonization-factor antigens CFA/I and

CFA/II found in enterotoxigenic *Esch. coli* of human origin. These will be described in more detail in Chapter 71.

Some species are normally *capsulated* and form mucoid colonies when first isolated, though the ability to form capsules may be lost on prolonged subculture on artificial media. The amount of capsular material varies enormously from strain to strain. Sometimes it forms a well circumscribed capsule around the organisms, but at others it becomes free in the surrounding medium as 'loose slime' (Duguid 1951). Capsulation is the rule in *Klebsiella*, and also occurs in a minority of *Enterobacter* and *Esch. coli* strains. Klebsiella capsules are composed mainly of acid polysaccharide, and many antigenically distinct capsular polysaccharides have been recognized. The capsules of *Esch. coli* appear to contain one of two substances (1) type-specific polysaccharides (A antigens) and (2) a more widely distributed mucoid (M) substance formed by many different strains (Kauffmann 1944). Rare strains that produce extracellular slime of the latter type occur in *Salmonella*, *Shigella* and most other genera. Slime production is the rule among D-tartrate-negative forms of *Salm. paratyphi B*, and is favoured by continued incubation at low temperature. When grown on medium with high osmolarity, most salmonella and *Esch. coli* strains produce extracellular slime (Anderson 1961, Anderson and Rogers 1963). This is a colanic acid (Goebel 1963, Grant *et al.* 1969) which appears to be identical with Kauffmann's M substance (Chapter 37).

Conditions for growth

Members of the Enterobacteriaceae grow readily on nutrient media without the addition of any accessory substances. They are aerobic, and facultatively anaerobic, though the growth is usually far less copious under the latter conditions. The optimum temperature is, for most species isolated by medical bacteriologists, in the neighbourhood of 37°, and the range over which growth occurs is fairly wide, extending from about 42° to 18° or lower. Many *Enterobacter* and *Serratia* strains grow poorly, or not at all, at a temperature of 44°, and differ in this respect from *Esch. coli.* Moreover, many coliform bacteria have a lower temperature optimum than 37°, and others, which appear to grow as well at 37° as at 30°, are unable to carry out some of their biochemical activities at the higher temperature. These organisms frequently grow slowly at temperatures as low as −1.5°. They form an important part of the coliform flora of water and soil, and are commonly found in chilled meat (Eddy and Kitchell 1959). In our experience, many of the coliform bacteria of water which can grow at refrigerator temperatures are *Hafnia alvei* or *Serratia* strains.

Type of growth

The type of growth given by the various members of the Enterobacteriaceae is very similar. When normal smooth strains are grown in broth a uniform turbidity develops, increasing rapidly during the first 12 to 18 hr of growth, and then more slowly up to 48 to 72 hr. Pellicle formation is rare except when the fimbriate phase has been selected by serial subculture in liquid medium. A slight deposit forms as growth increases, and this is easily dispersed on shaking the tube.

On agar, the colonies are relatively large, with an average diameter of 2–3 mm, but vary considerably in size. They may be circular, raised and low convex, with an entire edge and smooth surface; they may be flatter with a more irregular surface, and a more effuse and irregular edge, or they may assume the well known vine-leaf form which is commonly described as characteristic of *Salm. typhi.* Even with freshly isolated strains the range of variation is wide; and when old laboratory strains are under examination the most varied colonial forms may be seen. Apart from the possible appearance of rough variants, a single strain may show several different types of colony, if successive subcultures in broth are interspersed with platings and subculture of individual colonies. A few strains produce dwarf colonies on nutrient agar, but form colonies of normal size when provided with a source of reduced sulphur, such as sulphide or cysteine (Gillespie 1952).

Many members of the genus *Klebsiella*, when freshly isolated, form typically mucoid colonies. The differential value of this characteristic is diminished by the fact that certain other organisms, including some strains of *Esch. coli*, may similarly give rise to this type of growth.

All members of the Enterobacteriaceae, with the exception of a few klebsiella strains from the respiratory tract, grow in the presence of bile salts and give rise to colonies on MacConkey's agar nearly as large as those on nutrient agar. Occasional strains produce yellow or orange pigments (Parr 1937, Thomas and Elson 1957), and many *Serratia* strains form a red pigment called *prodigiosin*.

Resistance to heat and to various chemical substances

Most members of Enterobacteriaceae are killed by exposure to a temperature of 55° for about 1 hr, or of 60° for 15–20 min. On the whole, the *Esch. coli* strains tend to have a slightly higher resistance to heat than most other coliform bacteria. A small proportion of them are not completely destroyed by exposure to 60° for 30 min in broth, or to pasteurization at 62.8° for the same time in milk (Henneberg and Wendt 1935, Wilson *et al.* 1935). Towards chlorine in water the aerogenes type tends to be slightly more resistant than the coli or intermediate types (Bardsley 1938). In raw water stored under atmospheric conditions coliform bacteria may remain alive for weeks or months. *K. pneumoniae* subsp. *aerogenes* tends to survive rather longer than *Esch. coli*, but the results are influenced, among other things by the temperature (see Platt 1935, Raghavachari and Iyer 1939*b*). In faeces stored at 0° coliform organisms may be demonstrated for a year or more, the coli type being often gradually supplanted by the intermediate and aerogenes types (Parr 1938).

Members of the Enterobacteriaceae are susceptible to all classes of chemical disinfectants (Chapter 4) and, like other gram-negative rods, to sodium azide (Snyder and Lichstein 1940). Differences in susceptibility to dyes and other chemicals that are of use in enrichment and selective media for the isolation of pathogenic members of the family are mentioned in Chapters 37, 68 and 69. The resistance of individual organisms to antibiotics and other antimicrobial agents is highly various and will be described at appropriate points in Chapters 34–37.

Biochemical differentiation

It was early recognized that the fermentation of sugars provided a useful means of distinguishing certain pathogenic enterobacteria from the coliform bacteria. Theobald Smith, for instance, observed that *Esch. coli* fermented lactose, whereas *Salm. typhi* did not (Smith 1890); and the production of acid and gas from glucose by *Esch. coli*, but of acid only by *Salm. typhi*, was pointed out by Chantemesse and Widal in 1891. Shigellae were also distinguished from the coliform bacteria and divided into a few broad groups by similar methods (Chapter 36). For many years afterwards, interest was concentrated on the serological identification of salmonellae and shigellae, and the biochemical classification of the remaining enterobacteria was left mainly in the hands of the water bacteriologists, who had a profound influence on it between 1900 and 1950 (Chapter 9).

In his original description of the coliform bacteria, Escherich (1885) noted the occurrence of two types: one, '*Bact. coli*', formed fairly large rods, was motile and clotted milk slowly; the other, '*Bact. lactis aerogenes*', formed shorter, plumper rods, was non-motile and clotted milk more actively. In the next few years, the introduction of suitable indicators of acidity (Wurtz 1892) and of means of detecting gas production (Smith 1890, 1893, Durham 1898) added to the ease with which fermentation tests could be carried out, and various elaborate classifications of the coliform bacteria were made according to the range of their ability to attack sugars but none of these proved satisfactory (Bergey and Deehan 1908, MacConkey 1905, 1909, Jackson 1911).

An important correlation between biochemical activity and natural habitat was observed at about this time; it was found that '*Bact. lactis aerogenes*' (later called '*Bact. aerogenes*') was an infrequent inhabitant of the intestine, but was often found on vegetation and in the soil, whereas '*Bact. coli*' was a constant intestinal parasite (Winslow and Walker 1907). This was of practical as well as theoretical importance, and the presence or absence of '*Bact. coli*' in water supplies came to be regarded as a valuable indication of faecal pollution. The ability to distinguish between those that were of intestinal origin and those that might occur in unpolluted waters was aided by the discovery that there were differences in kind between the results of fermentation of one and the same sugar by different coliform bacteria. Thus there were differences in the amount of gas produced from glucose and the ratio of CO_2 to H_2 in the gas (Smith 1895, Harden and Walpole 1906), in whether or not acetylmethylcarbinol was produced, and in the final pH (Chapter 20). Levine (1916*a*, *b*) established that there was a high negative correlation between the results of the Voges-Proskauer (VP) test for acetylmethylcarbinol and the methyl-red (MR) test for final acidity in glucose-containing medium. In a series of intensive studies he placed on a firm foundation the conclusion that the lactose-fermenting coliform bacteria could be divided into two primary divisions: the first comprised strains that produced a moderate amount of gas with a $CO_2:H_2$ ratio of about 1:1, were VP negative and MR positive, included most of the strains isolated from the faeces of man and animals, and corresponded to '*Bact. coli*'; the second comprised strains that produced a greater amount of gas with a $CO_2:H_2$ ratio of about 2:1, were VP positive and MR negative, and were found often on plants and in unpolluted water. These could be further subdivided into '*Bact. aerogenes*' which did not liquefy gelatin and '*Bact. cloacae*' which did.

Later, Brown (1921) drew attention to the usefulness of a medium containing citrate for distinguishing '*Bact. aerogenes*' from '*Bact. coli*', and Koser (1923, 1924, 1926*a*, *b*) devised a synthetic medium in which citrate was the sole source of carbon. He was able to subdivide coliform bacteria into an MR+, VP−, citrate− coli type, an MR−, VP+, citrate+ aerogenes type, and an MR+, VP−, citrate+ intermediate type; only the first of these was commonly of faecal origin.

Considerable help was also afforded by the test introduced in 1904 by Eijkman who found that coli but not aerogenes strains were able to form gas in a glucose broth medium incubated at 46°. After a chequered career, this test was improved and standardized (Levine *et al.* 1934, Wilson *et al.* 1935) by the institution of accurate temperature control at 44° and the replacement of glucose broth by MacConkey's lactose bile salt broth. It was then shown to be a highly specific test for faecal coliform bacteria in the natural waters of many countries. However, as shown by Raghavachari and Iyer (1939*a*) and by Boizot (1941), organisms that were MR−, VP+, citrate+ but produced gas from lactose at 44° were common in the surface waters in some tropical countries. These organisms—referred to by British water bacteriologists as 'Irregular VI'—can be distinguished from faecal coliform bacteria by their inability to form indole at 44° (Mackenzie *et al.* 1948). These tests, with the addition of indole production and gelatin liquefaction, have for many years formed the basis of the classification of coliform bacteria used by British water bacteriologists. In addition to their usefulness in practice, they have the advantage of conforming in broad outline with the biochemical classification of the Enterobacteriaceae as a whole that gained general acceptance in the 1950s (Report 1954*a*, *b*, Kauffmann 1956*c*, 1959, Kauffmann *et al.* 1956).

For the detailed biochemical classification of the Enterobacteriaceae, however, a number of additional tests are required. The fermentation of sugars is of little value for the establishment of broad groupings, but action on individual substrates is often of use to distinguish between closely related groups. For example, the production of acid, with or without gas, from glycerol, adonitol, inositol, cellobiose and starch is useful in making subdivisions within *Klebsiella* and *Enterobacter* (see Kauffmann 1956*a*, *b*). In several genera, lactose fermentation is late or irregular. The introduction of the ONPG test for β-galactosidase (Chapter 20), which quickly reveals potential ability to attack this sugar, has added uniformity to the characters of *Citrobacter* and some biochemical varieties of *Klebsiella*, *Enterobacter* and *Shigella* (Bülow 1964, Lapage and Jayaraman 1964). Other useful tests also described in Chapter 20 include those for the utilization of various organic acids as sole carbon sources or their fermentation in peptone media, for the oxidation of gluconate, the decarboxylation or deamination of amino acids and for the hydrolysis of hippurate. It must be pointed out that virtually all members of the Enterobacteriaceae produce some hydrogen sulphide under optimal conditions, and that the term 'H_2S production' is used in this group to indicate the ability to produce blackening in triple sugar iron agar, in Kligler's medium, or in ferrous chloride gelatin. The test for urea hydrolysis was originally introduced for the recognition of *Proteus*, but many other members of the Enterobacteriaceae give positive reactions that are of value in classification when a suitable medium is used (see Chapter 20). Tetrathionate reductase is produced by nearly all strains of *Citrobacter*, *Proteus*, *Salmonella* and *Serratia* and by a few strains of *Esch. coli*, *Enterobacter* and *Klebsiella*, but is not produced by strains of *Shigella* and *Edwardsiella* (Le Minor 1967). In general, the Enterobacteriaceae produce few exoenzymes. Rapid liquefaction of gelatin (in 1–2 days) occurs only in some *Proteus* species and in *Serratia*, slower liquefaction (in 3–7 days) in salmonellae of subgenera II–IV, and very slow liquefaction (in 15–30 days or more) in *Enterobacter* species and *K. oxytoca*. Deoxyribonuclease production is confined to *Serratia* (Martin and Ewing 1967), and lipolysis to these organisms and to *Pr. vulgaris* and *Pr. mirabilis* (Davis and Ewing 1964). According to Mulczyk and Szewczuk (1970), nearly all *Citrobacter*, *Klebsiella*, *Enterobacter* and *Serratia* strains, but no other enterobacteria, produce pyrrolidonyl peptidase. (For further information about the biochemical characters of enterobacteria see Kauffmann 1969, Edwards and Ewing 1972 and Lapage *et al.* 1979).

In Table 33.2 we summarize the usual behaviour of the genera of the Enterobacteriaceae in some commonly used biochemical tests. Details of the biochemical characters which differentiate between the species within these genera will be given in Chapters 34–37. In some genera exceptions occur so frequently that it is difficult to describe adequately the biochemical reactions in a single table. In other genera there are distinct differences between different sub-genera, species or biochemical varieties ('biovars'). For simplicity, we have chosen to indicate the reaction of the majority of strains of each genus by using + and − signs. Where reactions vary too much to show a majority reaction we have used a V, regardless of whether the variation is between sub-genera, species, 'biovars' or strains. For example, we have shown the reaction of *Salmonella* in the test for β-galactosidase as a V, since strains of *Salmonella* sub-genus III (*Arizona*) are usually positive. Nevertheless, the great majority of *Salmonella* strains encountered belong to sub-genus I or II and are usually β-galactosidase negative. Similarly, we have shown the production of gas from glucose by *Salmonella* as variable (V), since although most strains produce gas the clinically important *Salm. typhi* strains are anaerogenic. On the other hand we have shown *Salmonella* as motile although strains of *Salm. gallinarum* and its *pullorum* variety are non-motile, and non-motile strains occasionally occur in other serogroups. It is therefore important to consider the biochemical characters shown in Table 33.2 together with those in Chapters 34–37.

Antigenic structure

The arrangement of the various antigenic components in the bodies and flagella of the enterobacteria has already been discussed in Chapter 13, and further information about the antigenic constitution of individual groups and species will be given in Chapters 34–38. The main structural components of the cell envelope are (1) a thin inner layer of mucopeptide, (2) a thicker outer layer composed of a polymolecular complex of lipopolysaccharide with protein and lipid, and (3) probably also a lipoprotein element linked to the mucopeptide (Freer and Salton 1971). The lipopolysaccharide complex not only determines the specificity of the O antigens, but is the main endotoxic element in the bacterial cell. Members of the Enterobacteriaceae possess a so-called common antigen (see p. 281) which may be distinct from the lipopolysaccharide. They often have surface antigens; these may be present in a definite capsule or as loose extracellular slime, or they may form a thin envelope that is not recognizable microscopically. In addition, many strains form flagella and fimbriae, which are also antigenic.

Table 33.2 Differential reactions of the genera of Enterobacteriaceae

Property	*Cedecea*	*Citrobacter*	*Edwardsiella tarda*	*Enterobacter*	*Erwinia herbicola*	*Escherichia coli*	*Hafnia alvei*	*Klebsiella**	*Kluyvera*	*Morganella morgani*	*Proteus*	*Providencia*	*Salmonella*	*Serratia*	*Shigella*	*Tatumella ptyseos*	*Yersinia*
β-Galactosidase	+	+	–	+	+	+	+	+	+	–	–	–	V	+	V	–	+
Growth on Simmons's citrate medium	+	+	–	+	V	–	+†	+	+	–	V	V	V	+	–	+‡	–
Decarboxylase:																	
arginine	+	+	–	V	–	V	–	–	–	–	–	–	+	–	–	V	–
lysine	–	–	+	V	–	+	+	+	+	–	–	–	+	+	–	–	–
ornithine	V	V	+	+	–	V	+	–	+	+	V	–	+	V	V	–	V
Gelatin liquefaction	–	–	–	+	+	–	–	–	–	–	+	–	V	+	–	–	–
Gluconate oxidation	V	–	–	+	V	–	+	V	–	–	V	–	–	+	–	–	–
H_2S production§	–	V	+	–	–	–	–	–	–	–	+	–	+	–	–	–	–
Indole production	–	V	+	–	–	+	–	V	+	+	V	+	–	–	V	–	V
Growth in KCN medium	+	V	–	+	–	–	+	V	+	+	+	+	–	+	–	–	–
Malonate fermentation	+	V	–	+	+	–	V	V	+	–	–	–	V	–	–	–	–
Motility	+	+	+	+	+	V	+	–	+	+	+	+	+	+	–	V‡	V
Methyl-red test	V	+	+	–	+	+	–†	V	+	+	+	+	+	V	+	–	+
PPA reaction‖	–	–	–	–	–	–	–	–	–	+	+	+	–	–	–	+	–
Urease production	–	V	–	V	–	–	–	+	–	+	+	V	–	–	–	–	V
Voges-Proskauer test	V	–	–	+	V	–	+†	V	–	–	–	–	–	+	–	–	V
Gas in glucose	V	+	+	+	–	V	+	V	+	+	+	V	V	V	–	–	–
Acid in:																	
adonitol	–	V	–	V	–	–	–	+	–	–	–	V	–	V	–	–	–
dulcitol	–	V	–	V	–	V	–	V	V	–	–	–	V	–	V	–	–
inositol	–	–	–	V	–	–	–	+	–	–	–	V	–	+	–	–	–
lactose	V	+	–	+	V	+	–	+	+	–	–	–	V	V	–	–	–
mannitol	+	+	–	+	+	+	+	+	+	–	–	V	+	+	V	–	+
salicin	+	V	–	+	+	V	–	+	+	–	V	V	–	+	–	+	V
sucrose	V	V	–	+	+	V	–	+	+	–	V	V	–	+	–	+	V
xylose	V	+	–	+	+	+	+	+	+	–	+	–	+	V	V	V	V

+ = Most strains positive; – = most strains negative; V = some strains positive, others negative.

* Excluding *K. pneumoniae* subsp. *rhinoscleromatis* (see Table 34.2 for the biochemical reactions of this subspecies).

† Results are those given at 30°; some strains may give positive MR test and negative VP test results and fail to grow on Simmons's citrate medium when incubated at 37°.

‡ At 25°; strains usually give negative results in these tests when incubated at 36°.

§ In triple sugar iron agar.

‖ Phenylpyruvic acid from phenylalanine.

O antigens

Most of the early observations on the toxicity of dead enterobacteria were made with *Salm. typhi* (Pfeiffer 1894, Sanarelli 1894, Chantemesse 1897, Brieger 1902), but it later became clear that most gram-negative bacteria behaved similarly. The behaviour of *Salmonella* and *Proteus* in agglutination tests (Smith and Reagh 1903, Weil and Felix 1917, 1920) led to the recognition of a series of heat-stable somatic O antigens, and chemical studies (Furth and Landsteiner 1928, 1929, White 1929*a*, *b*, 1931, Kurauchi 1929, Meyer 1930, 1931, Morgan 1931) revealed the presence of specific polysaccharide antigens in *Salmonella* and *Shigella*. The independent work of Boivin and his colleagues (1933*a*, *b*, 1934*a*, *b*, 1935), and of Raistrick and Topley (1934), Delafield (1934) and Martin (1934), established that the somatic antigens of salmonellae were themselves toxic and contained a polysaccharide hapten that was responsible for O-antigenic specificity and was linked to lipid and protein constituents. Delafield (1934) showed that material with similar properties was produced by a variety of other gram-negative bacteria, including members of the genera *Escherichia*, *Klebsiella*, *Proteus*, *Shigella*, *Haemophilus* and *Pasteurella*.

Boivin and his colleagues obtained their 'glycolipid' antigen, which was both toxic and antigenic, by extracting the bacteria with trichloracetic acid in the cold. Morgan (1936, 1937) extracted the complete antigen of *Sh. dysenteriae* 1 with diethylene glycol, and found that it contained a specific polysaccharide hapten, and protein and phospholipid fractions that were without O-specificity. Goebel and his colleagues (Goebel *et al.* 1945, Binkley *et al.* 1945) studied the degradation products of the complete antigen and found that toxicity was sometimes associated with the polysaccharide and sometimes with the protein fraction. They therefore concluded that the antigen contained a separate toxic substance. Westphal and Lüderitz (1954) were able to separate two lipid components from the antigen: lipid B was easily detachable and biologically inactive, and lipid A was more strongly bound to the polysaccharide and was responsible for toxicity. Lipid A in watery solution was almost inactive, but when linked artificially to a protein of non-bacterial origin it had all the endotoxic properties of the naturally occurring complex (Westphal 1960; see also Lüderitz *et al.* 1973, Rietschel *et al.* 1975).

When the bacteria are extracted at 65° with a mixture of phenol and water (Westphal *et al.* 1952), the phenol layer contains lipid B and protein and the overlying watery layer contains lipopolysaccharide (i.e. polysaccharide with attached lipid A). The lipolysaccharide has O-specificity and is a potent endotoxin, but is a poor antigen (Table 33.3). It forms an opalescent solution, and has a high molecular weight (1–20 × 10^6) and a linear structure. Extraction of cells, or treatment of lipopolysaccharide, with acetic acid yields polysaccharide with a molecular weight of 20 000–30 000; this has O-specificity but no endotoxic action, and is not antigenic for the rabbit. Material extracted with alkali, on the other hand, consists of chemically altered polysaccharide and lipid A with an intermediate molecular weight (*ca* 200 000) and little endotoxic action. It is readily adsorbed to red cells and so is particularly suitable for use in the passive haemagglutination test (Thomas and Mennie 1950); although it is not antigenic when given alone, specific antibody is formed if sensitized erythrocytes are injected into rabbits (see Lüderitz *et al.* 1966). Treatment of lipopolysaccharide with deoxycholate results in its break-up into non-toxic subunits with a molecular weight of *ca* 20 000, but reaggregation occurs and toxicity is restored when the deoxycholate is removed (Ribi *et al.* 1966). Toxic preparations contain varying amounts of lipid A, and the presence of some lipid is clearly necessary for toxicity. Possibly, however, the state of molecular aggregation rather than the lipid itself is responsible for the biological activity of the endotoxin (Tarmina *et al.* 1968). Rietschel and his colleagues (1975) have shown that lipid A is itself immunogenic and induces the production of antibodies which cross-react with lipid A from a variety of bacteria. Such antibodies may, under certain circumstances, protect against the toxic effects of lipid A.

Table 33.3 Properties of lipopolysaccharide fractions extracted by various means (modified from Lüderitz *et al.* 1966)

Method of extraction	Components	Antigenicity for rabbit	Specificity	Endotoxin	Sensitization of RBC	Molecular weight
Trichloracetic acid or diethylene glycol	Polysaccharide—lipid A—protein—lipid B	+ + +	O	+ + +	+	> 10^6
Phenol-water	Polysaccharide—lipid A	±	O	+ + +	±	> 10^6
Alkali	Polysaccharide (de-acetylated)—lipid A	–*	O‡	±	+ + +	200 000
Acetic acid	Polysaccharide	–†	O	–	–	20 000–30 000

* Unless adsorbed to RBC.
† Unless coupled to protein.
‡ Except *Salmonella* O-factors 5 and 10, which are destroyed.

The effects of endotoxin on the animal body are numerous; when administered parenterally to a suitable animal it gives rise to fever, to leucopenia followed by leucocytosis, to hyperglycaemia with a subsequent fall in the blood sugar far below the normal level, and to lethal shock after a latent period; it provokes the localized and the generalized Shwartzman reaction and causes haemorrhagic necrosis in tumours and at the site of an intradermal injection of adrenalin; it gives rise to non-specific resistance to the intraperitoneal injection of a variety of gram-negative bacteria and to damage by irradiation; and it has a powerful adjuvant effect on the antibody response to unrelated protein antigens but not to polysaccharide antigens. No other biological material gives rise to such a wide variety of reactions, though some other bacterial products, for example, staphylococcal enterotoxin and the pyrogenic exotoxin of *Str. pyogenes*, cause a number of similar effects. The mode of action of endotoxin is not understood. There is reason to believe that its action may be responsible for many of the consequences of established infections with gram-negative rods. The fact that it is produced by many non-pathogenic bacteria, and by rough variants of pathogenic species, suggests that it plays no part in virulence. (For further information about endotoxin see Landy and Braun 1964, Nowotny 1969, Milner *et al.* 1971 and Chapter 13.)

Each of the main enterobacterial genera can be subdivided by means of agglutination tests into a number of O-antigenic groups, some characterized by a single antigen and others by a combination of antigenic 'factors', which may appear in different combinations in other members of the genus. There is also extensive sharing of O antigens between otherwise quite unrelated organisms. Nearly all the shigella serotypes are antigenically related to one or more O groups of *Esch. coli*; the relation is sometimes unilateral, sometimes reciprocal, and sometimes one of complete identity. For example, the O antigen of *Sh. dysenteriae* 3 is identical with that of *Esch. coli* O124. Such a relation between salmonella and *Esch. coli* serotypes is less common, but by no means infrequent. There is also considerable sharing between the O antigens of *Esch. coli* and *Providencia*.

Chemical examination of the products of hydrolysis of the O polysaccharides has thrown some light on their antigenic relationships. A wide variety of sugars has been identified in the polysaccharides of smooth strains of *Salmonella* and *Esch. coli*. In addition to various hexoses, hexosamines, 6-deoxyhexoses and deoxyhexosamines that are widely distributed in bacterial polysaccharides, there are others that appear to be found only among the enterobacteria, such as the 3–6 dideoxyhexoses (colitose, abequose, paratose and tyvelose) and a heptose (see Pon and Staub 1952, Staub *et al.* 1959, Westphal *et al.* 1959). An invariable constituent of the O antigens not only of enterobacteria but of many other gram-negative bacteria is 2-keto-3-deoxyoctonate or KDO (Ellwood 1970). Kauffmann and his colleagues (1960) analysed the sugars obtained from the polysaccharides of salmonellae belonging to 36 O groups. Members of the same O group always had polysaccharides with an identical qualitative composition, but so did many organisms with different O antigens. The O groups could therefore be classified into a smaller number of chemical types ('chemotypes') each containing organisms with polysaccharides composed of the same 'building blocks'. Five sugars (D-glucose, D-galactose, D-galactosamine, L-glycero-D-mannoheptose, and KDO) were found in all the polysaccharides, and formed the 'basal core'. Other sugars were attached to this to form side-chains; the qualitative composition of the side-chains determined the chemotype, but the arrangement of the sugars in the side-chains, in repeating units of oligosaccharides, and often showing branching, was responsible for O-antigenic specificity. The more 'complex' chemotypes had three side-chain sugars, one of which was a 3–6 dideoxyhexose. In *Esch. coli* (Ørskov *et al.* 1967, 1977) the basal-core sugars were generally the same, except that a few strains lacked galactose. Twelve of the 17 *Salmonella* chemotypes were also found in *Esch. coli*, though salmonella and *Esch. coli* strains of the same chemotype rarely showed serological cross-reactions. However, some salmonella chemotypes did not contain any *Esch. coli* strains and a considerable number of *Esch. coli* chemotypes did not contain any salmonella strains.

Rough variants of the Enterobacteriaceae form lipopolysaccharide, but this does not contain any of the side-chain sugars of the corresponding smooth form (Lüderitz *et al.* 1960). Rough variants are of two classes: RII (or Ra), which contains all the basal sugars, and RI, which includes a series of mutants (Rb-Re) with various defects in core structure (Lüderitz *et al.* 1965). A class of 'semi-rough' mutants (Naide *et al.* 1965) intermediate in characters between smooth and RII organisms has basal core sugars but may form only part of the side-chain or a small number of side-chains. The so-called T-forms of salmonellae (Chapter 37) have serological characters resembling those of rough mutants and have a normal core structure, but form side-chains composed of ribose and additional galactose in place of the sugars of the corresponding smooth form (Wheat *et al.* 1967).

There is uncertainty about the part played by the O antigen in the virulence of the Enterobacteriaceae. If by virulence we mean the ability to multiply in the tissues of experimental animals after intravenous or intraperitoneal injection, and to cause a fatal infection, there is little doubt that most smooth enterobacteria are more virulent than their rough variants. Semi-rough variants are usually of intermediate virulence. Mäkelä and her colleagues (1973) have shown that even quite subtle changes in the O antigens of otherwise isogenic *Salm. typhimurium* derivatives can affect their virulence for mice when injected intraperitoneally. Heat-killed vaccines made from smooth organisms, or lipopolysaccharide extracted from them, give protection against injections of the homologous organisms, and also sometimes against related organisms sharing the same O antigen; for example, vaccines of smooth *Salm. paratyphi B* will protect against infection with *Salm. typhimurium* (Schütze 1930). Vaccines made from rough variants, or 'rough' polysaccharide, usually do not have this effect. On the other hand, protection by living vaccines, whether of smooth or rough strains, is often superior to that obtained with killed vaccines of smooth strains, and may be effective against organisms with unrelated O antigens. We know very little about the significance of the polysaccharide portion of the O antigen in infections in the natural host. (For further information about the virulence of enterobacteria and the part played in it by O antigens see Chapters 34, 36 and 37, and Roantree, 1967.)

The enterobacterial common antigen (Kunin *et al.* 1962, Whang and Neter 1962) was first observed in indirect haemagglutination tests between antisera against *Esch. coli* of O-group 14 and erythrocytes coated with supernatants or extracts of cultures of other enterobacteria. The antigen-antibody reaction can also be detected by haemolysis in the presence of complement, and by immunofluorescence (Aoki *et al.* 1966). The common antigen has been detected in strains of *Esch. coli*, *Edwardsiella*, *Citrobacter*, *Salmonella*, *Shigella*, *Klebsiella*, *Enterobacter*, *Serratia*, *Proteus*, *Yersinia*, *Erwinia*, *Plesiomonas* and *Aeromonas* (Mäkelä and Mayer 1976). When vaccines of intact cells were injected into animals, only *Esch. coli* O14 and O56 gave rise to antibody against common antigen, but ethanol-soluble fractions of lipopolysaccharides from other members of the Enterobacteriaceae also did so (Suzuki *et al.* 1964). Hammarström and his colleagues (1971) were unable to separate the common antigen of *Esch. coli* O14 from the O polysaccharide and found only core constituents in their preparations. They pointed out that *Esch. coli* O14 and O56 had O polysaccharides that were devoid of side-chains; some rough strains formed much antibody to common antigen, and members of O groups with complicated side-chains formed little. Antigenicity was thus related to the accessibility of core constituents. Chemical analysis has shown that the common antigen is an amino-sugar polymer consisting of *N*-acetyl-D-glucosamine and *N*-acetyl-D-mannosaminuronic acid, partly esterified by acetic and palmitic acids (Mayer and Schmidt, 1979).

(For further information about the chemistry of the O polysaccharides see Lüderitz *et al.* 1966, 1971, Stocker *et al.* 1966, Lüderitz 1970 and Ørskov *et al.* 1977; for general reviews of the structure and function of O antigens see the monograph of Weinbaum, Kadis and Ajl 1971, and Chapter 13.)

Surface antigens

Many of the Enterobacteriaceae possess somatic antigens that are thought to be superficial to the O antigens on the ground that they are responsible for 'O inagglutinability', that is to say, absence of agglutination of living bacteria by antiserum made by injecting heated suspensions of the homologous organism. Many of these antigens have been described as heat-labile on the evidence of agglutination tests, but the main effect of heating appears to be to cause separation of the antigen from the bacterial cell. The best known representative of this group is the Vi antigen of *Salm. typhi* (Chapter 37). Kauffmann (1947), recognized several classes of surface antigens in *Esch. coli* that caused O inagglutinability, and classified them according to the effect of heat on the agglutinability, antigenicity and antibody-binding power of bacteria that possessed them. He proposed the term K antigen to include all such surface antigens, whether present at or near the surface of unencapsulated organisms or forming part of a capsule, and it has since been widely used in this sense. Increasing knowledge of these antigens obtained by methods other than agglutination (Ørskov *et al.* 1971), and growing doubt about the assumption that O inagglutinability always indicates the presence of a surface antigen distinct from the lipopolysaccharide, make it likely that the term K antigen will in future be used less and less or will be redefined. A number of the surface antigens appear to be acid polysaccharides; these include the Vi antigen, some of the K antigens of *Esch. coli* and the capsular polysaccharides of *Klebsiella*. We have also referred (p. 276) to the mucoid substance that appears to be a colanic acid. Two K antigens of *Esch. coli*, K88 and K99, are filamentous protein structures. In the future these may be placed in a separate class of antigens together with the colonization factor antigens CFA/I and CFA/II. These are discussed in more detail in Chapter 71.

H antigens

The strains included in each genus of the Enterobacteriaceae possess many different H antigens. In *Salmonella*, which is unique in showing diphasic variation of the H antigen, there is a close mosaic-like interrelation between the flagellar antigens of the different serotypes. In other groups, however, the flagellar antigens are usually distinct; and H antigens common to different enterobacterial groups are rare. The H antigens are almost certainly the fibrous proteins—flagellins—that make up the bulk of the flagella. Antigenically distinct flagella sometimes differ in amino-acid composition, but specificity appears to be determined by the amino-acid sequence in the flagellin molecule, and possibly also by the organization of the flagellins within the flagellum. (See Joys 1968, Iino 1969).

Fimbrial antigens

Antibodies to fimbriae cause the agglutination of homologous organisms that are in the fimbrial phase; loose bulky floccules closely resembling H agglutination appear rapidly at 37° (Gillies and Duguid 1958). Heating at 100° detaches the fimbriae and renders the organism inagglutinable by fimbrial antiserum, but the fimbrial antigen is not destroyed, and detached fimbriae retain their antibody-binding power. Organisms heated at 60° for $\frac{1}{2}$ hr and treated with formalin remain fimbriate. The properties of the fimbrial antigen are thus similar to those of the X antigen described by Topley and Ayrton (1924) in salmonella suspensions that had been grown in broth for several days. They may cause confusion in serum agglutination tests if the suspension is fimbriate (Cruickshank 1939), because many human sera contain fimbrial antibody (Gillies and Duguid 1958). To obtain a suitable H suspension for use in agglutination tests, a small inoculum should be taken from an agar plate or from a swarm through semi-solid agar and should be grown for not more than 6 hr in broth; for O suspensions, the organism should be grown for 24 hr on a well dried agar plate. Confusion may also result if diagnostic antisera contain fimbrial antibody. The best way to avoid this is to use non-fimbrial variants for immunization.

Several antigens have been recognized among the type 1 fimbriae. All fimbriate strains of *Sh. flexneri* share a common antigen, and this is distinct from the three main fimbrial antigens found in different strains of *Esch. coli* (Gillies and Duguid 1958); but all fimbrial strains of *Sh. flexneri* and *Esch. coli*, and all *Klebsiella* strains that have type 1 fimbriae, share a common component that is absent from the fimbriae of *Salmonella*. On the other hand, all salmonellae with type 1 fimbriae share a major common fimbrial antigen, which is also found in *Citrobacter* strains (Duguid and Campbell 1969). Some minor salmonella fimbrial antigens are less widely distributed. As we noted above, several other *Esch. coli* surface antigens are filamentous proteins that resemble fimbriae.

Many coliform bacteria and *Providencia* strains contain a common antigen referred to by Stamp and Stone (1944) as the α antigen. It is more thermostable than the H antigen, withstanding a temperature of 75° for 1 hr, but is inactivated at 100° in 15 min and by exposure for 1 hr to ethanol at 65°. It is most common in recently isolated strains and tends to be lost on subculture. Agglutinins to the α antigen are sometimes present in rabbit sera and may lead to confusion in the identification of suspected pathogenic organisms. It appears that the α antigen is not a single substance; Emslie-Smith (1948) distinguished between seven separate agglutinins in rabbit sera to antigens of this description. A thermolabile β antigen was reported by Mushin (1949) in strains of coliform bacteria, *Proteus*, and *Sh. flexneri*. It is inactivated by boiling, gives floccular agglutination, and has an optimal agglutinating temperature of 52°. According to Gillies and Duguid (1958) the general properties of the α- and β-antigens distinguish them from fimbrial antigens.

References

Anderson, E. S. (1961) *Nature, Lond.* **190,** 284.
Anderson, E. S. and Rogers, A. H. (1963) *Nature, Lond.* **198,** 714.
Aoki, S., Merkel, M. and McCabe, W. R. (1966) *Proc. Soc. exp. Biol., N.Y.* **121,** 230.
Bardsley, D. A. (1938) *J. Hyg., Camb.* **38,** 721.
Bergey, D. H. and Deehan, S. J. (1908) *J. med. Res.* **19,** 175.
Binkley, F., Goebel, W. F. and Perlman, E. (1945) *J. exp. Med.* **81,** 331.
Boivin, A., Mesrobeanu, I. and Mesrobeanu, L. (1933*a*) *C. R. Soc. Biol.* **113,** 490; (1933*b*) *Ibid.* **114,** 307; (1934*a*) *Ibid.* **115,** 306; (1934*b*) *Ibid.* **117,** 271; (1935) *Arch. roum. Path. exp. Microbiol.* **8,** 45.
Boizot, G. E. (1941) *J. Hyg., Camb.* **41,** 566.
Borman, E. K., Stuart, C. A. and Wheeler, K. M. (1944) *J. Bact.* **48,** 351.
Brieger, L. (1902) *Dtsch. med. Wschr.* **28,** 477.
Brown, H. C. (1921) *Lancet*, **i,** 22.
Bülow, P. (1964) *Acta path. microbiol. scand.* **60,** 376.
Chantemesse, A. (1897) *C. R. Soc. Biol.* **49,** 96, 101.
Chantemesse, A. and Widal, F. (1891) *Bull. Méd.* **5,** 935.
Constable, F. L. (1956) *J. Path. Bact.* **72,** 133.
Cowan, S. T. (1956) *J. gen. Microbiol.* **15,** 345; (1957) *Bull. Hyg.* **32,** 101.
Cruickshank, J. C. (1939) *J. Hyg., Camb.* **39,** 224.
Davis, B. R. and Ewing, W. H. (1964) *J. Bact.* **88,** 16.
Delafield, M. E. (1934) *Brit. J. exp. Path.* **15,** 130.
Duguid, J. P. (1951) *J. Path. Bact.* **63,** 673; (1959) *J. gen. Microbiol.* **21,** 271; (1968) *Arch. Immunol. Ther. exp.* **16,** 173.
Duguid, J. P., Anderson, E. S. and Campbell, I. (1966) *J. Path. Bact.* **92,** 107.
Duguid J. P. and Campbell, I. (1969) *J. med. Microbiol.* **2,** 535.
Duguid, J. P. and Gillies, R. R. (1957) *J. Path. Bact.* **74,** 397; (1958) *Ibid.* **75,** 519.
Duguid, J. P., Smith, P. N. Dempster, G. and Edmunds, P. N. (1955) *J. Path. Bact.* **70,** 335.
Duguid, J. P. and Old, D. C. (1980) In: *Bacterial Adherence*, p. 185. Ed. by E. H. Beachey. Chapman and Hall, London.
Durham, H. E. (1898) *Brit. med. J.* **i,** 1387.
Eddy, B. P. and Kitchell, A. G. (1959) *J. appl. Bact.* **22,** 57.
Edwards, P. R. and Ewing, W. H. (1972) *Identification of Enterobacteriaceae*, 3rd edn. Burgess, Minneapolis.
Eijkman, C. (1904) *Zbl. Bakt.* **37,** 436, 742.
Ellwood, D. C. (1970) *J. gen. Microbiol.* **60,** 373.
Emslie-Smith, A. H. (1948) *J. Path. Bact.* **60,** 307.
Escherich, T. (1885) *Fortschr. Med.* **3,** 515, 547.
Ewing, W. H. (1966) *Enterobacteriaceae Taxonomy and Nomenclature*. U.S. Dept. Hlth, Educ. Welfare, Atlanta.
Ewing, W. H. and Edwards, P. R. (1960) *Int. Bull. bact. Nomencl.* **10,** 1.
Freer, J. H. and Salton, M. R. J. (1971) In: *Microbial Toxins*, **4,** p. 67. Ed. by G. Weinbaum, S. Kadis and S. J. Ajl. Academic Press, N.Y.
Furth, J. and Landsteiner, K. (1928) *J. exp. Med.* **47,** 171; (1929) *Ibid.* **49,** 727.
Gillespie, W. A. (1952) *J. Path. Bact.* **64,** 551.
Gillies, R. R. and Duguid, J. P. (1958) *J. Hyg., Camb.* **56,** 303.
Goebel, W. F. (1963) *Proc. nat. Acad. Sci., Wash.* **49,** 464.
Goebel, W. F., Binkley, F. and Perlman, E. (1945) *J. exp. Med.* **81,** 315.
Grant, W. D., Sutherland, I. W. and Wilkinson, J. F. (1969) *J. Bact.* **100,** 1187.
Grimont, P. A. D., Grimont, F., Farmer, J. J. III and Asbury, M. A. (1981) *Int. J. syst. Bact.* **31,** 317.
Hammarström, S., Carlsson, H. E., Perlmann, P. and Svensson, S. (1971) *J. exp. Med.* **134,** 565.
Harden, A. and Walpole, G. S. (1906) *Proc. roy. Soc. B*, **77,** 399.
Henneberg, W. and Wendt, H. (1935) *Zbl. Bakt.*, IIte Abt. **93,** 39.
Hoeniger, J. F. M. (1965) *J. Bact.* **90,** 275.
Hollis, D. G., Hickman, F. W., Fanning, G. R., Farmer, J. J. III, Weaver, R. E. and Brenner, D. J. (1981) *J. clin. Microbiol* **14,** 79.
Iino, T. (1969) *Bact. Rev.* **33,** 454.
Jackson, D. D. (1911) *J. infect. Dis.* **8,** 241.
Joys, T. M. (1968) *Leeuwenhoek ned. Tijdschr.* **34,** 205.
Kauffmann, F. (1944) *Acta path. microbiol. scand.* **21,** 20; (1947) *J. Immunol.* **57,** 71; (1956*a*) *Acta path. microbiol. scand.* **39,** 85; (1956*b*) *Ibid.* **39,** 103; (1956*c*) *Zbl. Bakt.* **165,** 344; (1959) *Int. Bull. bact. Nomencl.* **9,** 1; (1969) *The Bacteriology of the Enterobacteriaceae*, 2nd edn. Munksgaard, Copenhagen.
Kauffmann, F., Edwards, P. R. and Ewing, W. H. (1956) *Int. Bull. bact. Nomencl.* **6,** 29.
Kauffmann, F., Lüderitz, O., Stierlin, H. and Westphal, O. (1960) *Zbl. Bakt.* **178,** 442.
Koser, S. A. (1923) *J. Bact.* **8,** 493; (1924) *Ibid.* **9,** 59; (1926*a*) *J. Amer. Wat. Wks Ass.* **15,** 641; (1926*b*) *J. Bact.* **11,** 409.
Kunin, C. M., Beard, M. V. and Halmagyi, N. E. (1962) *Proc. Soc. exp. Biol. N.Y.* **111,** 160.
Kurauchi, K. (1929). See Ando, K. (1929) *J. Immunol.* **17,** 555.
Landy, M. and Braun, W. [Eds] (1964) *Bacterial Endotoxins*. Rutgers Univ. Press, New Brunswick, N.J.
Lapage, S. P. and Jayaraman, M. S. (1964) *J. clin. Path.* **17,** 117.
Lapage, S. P., Rowe, B., Holmes, B. and Gross, R. J. (1979) In: *Identification Methods for Microbiology*, 2nd edn., p. 123. Ed. by F. A. Skinner and D. W. Lovelock. Academic Press, London.
Leifson, E. (1960) *Atlas of Bacterial Flagellation*. Academic Press, N.Y.
Le Minor, L. (1967) *Ann. Inst. Pasteur* **113,** 117.
Le Minor, L., Rohde, R. and Taylor, J. (1970) *Ann. Inst. Pasteur* **119,** 206.
Levine, M. (1916*a*) *J. Bact.* **1,** 87; (1916*b*) **1,** 153.
Levine, M., Epstein, S. S. and Vaughn, R. H. (1934) *Amer. J. publ. Hlth.* **24,** 505.
Lüderitz, O. (1970) *Angew. Chem.* **82,** 708.
Lüderitz, O., Galanos, C., Lehmann, V., Nurminen, M., Rietschel, E. T., Rosenfelder, G., Simon, M. and Westphal, O. (1973) *J. infect. Dis.* **128,** (Suppl.) S17.
Lüderitz, O., Kauffmann, F., Stierlin, H. and Westphal, O. (1960) *Zbl. Bakt.* **179,** 180.

Lüderitz, O., Risse, H. J., Schultze-Holthausen, H., Strominger, J. L., Sutherland, I. W. and Westphal, O. (1965) *J. Bact.* **89,** 343.

Lüderitz, O., Staub, A. M. and Westphal, O. (1966) *Bact. Rev.* **30,** 192.

Lüderitz, O., Westphal, O., Staub, A. M. and Nikaido, H. (1971) In: *Microbial Toxins,* **4,** p. 145. Ed. by G. Weinbaum, S. Kadis and S. J. Ajl, Academic Press, N.Y.

MacConkey, A. (1905) *J. Hyg., Camb.* **5,** 333; (1909) *Ibid.* **9,** 86.

Mackenzie, E. F. W., Taylor, E. W. and Gilbert, W. E. (1948) *J. gen. Microbiol.* **2,** 197.

Mäkelä, P. H. and Mayer, H. (1976) *Bact. Rev.* **40,** 591.

Mäkelä, P. H., Valtonen, V. V. and Valtonen, M. (1973) *J. infect. Dis.* **128,** (Suppl.) S81.

Martin, A. R. (1934) *Brit. J. exp. Path.* **15,** 137.

Martin, W. J. and Ewing, W. H. (1967) *Canad. J. Microbiol.* **13,** 616.

Mayer, H. and Schmidt, G. (1979) *Curr. Topics Microbiol. Immunol.* **85,** 99.

Meyer, K. (1930) *Z. ImmunForsch.* **68,** 98; (1931) *Ibid.* **69,** 134, 499.

Milner, K. C., Rudbach, J. A. and Ribi, E. (1971) In: *Microbial Toxins,* **4,** p. 1. Ed. by G. Weinbaum, S. Kadis and S. J. Ajl. Academic Press, N.Y.

Morgan, W. T. J. (1931) *Brit. J. exp. Path.* **12,** 62; (1936) *Biochem. J.* **30,** 909; (1937) *Ibid.* **31,** 2003.

Mulczyk, M. and Szewczuk, A. (1970) *J. gen. Microbiol.* **61,** 9.

Mushin, R. (1949) *J. Hyg., Camb.* **47,** 227.

Naide, Y., Nikaido, H., Mäkelä, P. H., Wilkinson, R. G. and Stocker, B. A. D. (1965) *Proc. nat. Acad. Sci., Wash.* **53,** 147.

Nowotny, A. (1969) *Bact. Rev.* **33,** 72.

Ørskov, F., Ørskov, I., Jann, B. and Jann K. (1971) *Acta path. microbiol., scand.* **B79,** 142.

Ørskov, F., Ørskov, I., Jann, B., Jann, K., Müller-Seitz, E. and Westphal. O. (1967) *Acta path. microbiol. scand.* **71,** 339.

Ørskov, I., Ørskov, F., Jann, B. and Jann, K. (1977) *Bact. Rev.* **41,** 667.

Old, D. C., Corneil, I., Gibson, L. F., Thomson, A. D. and Duguid, J. P. (1968) *J. gen. Microbiol.* **51,** 1.

Old, D. C. and Duguid, J. P. (1970) *J. Bact.* **103,** 447.

Parr, L. W. (1937) *Proc. Soc. exp. Biol., N.Y.* **35,** 563; (1938) *Amer. J. publ. Hlth.* **28,** 445.

Peluffo, C. A. (1953) *Proc. VIth int. Congr. Microbiol.* **1,** 370.

Pfeiffer, R. (1894) *Dtsch. med. Wschr.* **20,** 898.

Platt, A. E. (1935) *J. Hyg., Camb.* **35,** 437.

Pon, G. and Staub, A. M. (1952) *Bull. Soc. Chim. biol.* **34,** 1132.

Quadling, C. and Stocker, B. A. D. (1962) *J. gen. Microbiol.* **28,** 257.

Raghavachari, T. N. S. and Iyer, P. V. S. (1939*a*) *Indian J. med. Res.* **26,** 867; (1939*b*) *Ibid.,* **26,** 877.

Raistrick, H. and Topley, W. W. G. (1934) *Brit. J. exp. Path.* **15,** 113.

Report (1954*a*) *Int. Bull. bact. Nomencl.* **4,** 1; (1954*b*) *Ibid.* **4,** 47; (1958) *Ibid.* **8,** 25; (1963) *Ibid.* **13,** 69.

Ribi, E., Anacker, P. L., Brown, R., Haskins, W. T., Malmgren, B., Milner, K. C. and Rudbach, J. A. (1966) *J. Bact.* **92,** 1493.

Rietschel, E. T., Galanos, C. and Lüderitz, O. (1975) In: *Microbiology 1975,* p. 307. Ed. by D. Schlessinger. American Society for Microbiology, Washington.

Roantree, R. J. (1967) *Annu. Rev. Micriobiol.* **21,** 443.

Sanarelli, J. (1894) *Ann Inst. Pasteur* **8,** 193, 353.

Schütze, H. (1930) *Brit. J. exp. Path.* **11,** 34.

Smith, T. (1890) *Zbl. Bakt.* **7,** 502; (1893) *Misc. Invest. infect. parasit. Dis. domest. Animals,* 8° Washington, 53; (1895) *Amer. J. med. Sci.* **110,** 283.

Smith, T. and Reagh, A. L. (1903) *J. med. Res.* **10,** 89.

Sneath, P. H. A. and Cowan, S. T. (1958) *J. gen. Microbiol.* **19,** 551.

Snyder, M. L. and Lichstein, H. C. (1940) *J. infect. Dis.* **67,** 113.

Stamp (Lord) and Stone, D. M. (1944) *J. Hyg. Camb.* **43,** 266.

Staub, A. M., Tinelli, R., Lüderitz, O. and Westphal, O. (1959) *Ann. Inst. Pasteur* **96,** 303.

Stocker, B. A. D., Wilkinson, R. G. and Mäkelä, P. H. (1966) *Ann. N.Y. Acad. Sci* **133,** 334.

Suzuki, T., Gorzynski, E. A. and Neter, E. (1964) *J. Bact.* **88,** 1240.

Talbot, J. M. and Sneath, P. H. A. (1960) *J. gen. Microbiol.* **22,** 303.

Tarmina, D. F., Milner, K. C., Ribi, E. and Rudbach, J. A. (1968) *J. Bact.* **96,** 1611.

Thal, E. (1954) *Untersuchungen über Pasteurella pseudotuberculosis.* Berlingska Boktryckeriet, Lund.

Thomas, J. C. and Mennie, A. T. (1950) *S. Afr. med. J.* **24,** 897.

Thomas, S. B. and Elson, K. (1957) *J. appl. Bact.* **20,** 50.

Topley, W. W. C. and Ayrton, J. (1924) *J. Hyg., Camb.* **23,** 198.

Weil, E. and Felix, A. (1917) *Wein. klin. Wschr.* **30,** 1509; (1920) *Z. ImmunForsch.* **29,** 24.

Weinbaum, G., Kadis, S. and Ajl, S. J. [Eds.] (1971) *Microbial Toxins,* **4.** Academic Press, N.Y.

Westphal, O. (1960) *Ann. Inst. Pasteur* **98,** 789.

Westphal, O. and Lüderitz, O. (1954) *Angew. Chemie.* **66,** 407.

Westphal, O., Lüderitz, O. and Bister, F. (1952) *Z. Naturf.* **7*b*,** 148.

Westphal, O., Lüderitz, O., Staub, A.-M. and Tinelli, R. (1959) *Zbl. Bakt.* **174,** 307.

Whang, H. Y. and Neter, E. (1962) *J. Bact.* **84,** 1245.

Wheat, R. W., Berst, M., Ruschmann, E., Lüderitz, O. and Westphal, O. (1967) *J. Bact.* **94,** 1366.

White, P. B. (1929*a*) *J. Path. Bact.* **32,** 85; (1929*b*) *Med. Res. Coun., System of Bacteriology,* **4,** 86; (1931) *J. Path. Bact.* **34,** 325.

Wilson, G. S., Twigg, R. S., Wright, R. C., Hendry, C. B., Cowell, M. P. and Maier, I. (1935) *Spec. Rep. Ser. med. Res. Coun. London,* No. 206.

Winslow, C.-E. A., Kligler, I. J. and Rothberg, W. (1919) *J. Bact.* **4,** 429.

Winslow, C.-E. A. and Walker, L. T. (1907) *Science* **26,** 797.

Wurtz, R. (1892) *Arch. Méd. exp.* **4,** 85.

34

Coliform bacteria; various other members of the Enterobacteriaceae

Barry Holmes and Roger J. Gross

Introductory

In this chapter we shall describe important members of the Enterobacteriaceae that are commonly referred to as 'coliform bacteria' (Chapter 33), that is to say, members of genera in which lactose fermentation is usual, such as *Escherichia*, *Klebsiella* and *Enterobacter*. In considering these organisms together we do not imply that lactose fermentation is of any great taxonomic significance. Later in the chapter we shall give shorter descriptions of other genera in which lactose fermentation is variable or absent. Separate chapters or parts of chapters are devoted to (1) *Proteus*, *Morganella* and *Providencia* (Chapter 35), (2) *Shigella* (Chapter 36), (3) *Salmonella* (Chapter 37), and (4) *Yersinia* (Chapter 38).

With a few exceptions, the genera described in this chapter conform to the general definition of the Enterobacteriaceae given on Chapter 33: they are aerobes and facultative anaerobes that ferment glucose and give a positive catalase and a negative oxidase reaction; all of them except a few strains of *Erwinia* reduce nitrate to nitrite; and only in *Tatumella* are there motile strains that do not exhibit peritrichous flagellation.

Escherichia

Definition

Usually motile and usually produce gas from fermentable carbohydrates. Mannitol is fermented and most strains acidify lactose promptly. Indole is usually produced. Give a positive methyl-red reaction and a negative Voges-Proskauer reaction. Do not grow on Simmons's citrate medium. Most strains fail to hydrolyse urea or to produce H_2S detectable in triple sugar iron agar. Phenylalanine is not deaminated and gluconate is not oxidized. There is no growth in Møller's KCN medium and gelatin is not liquefied. Most strains decarboxylate lysine.

Inhabit the intestine of man and animals; cause both suppurative and diarrhoeal diseases. G + C content of DNA: 50–51 moles per cent.

Type species: *Esch. coli*

The above definition applies to *Esch. coli.* Two other species have been described, *Esch. adecarboxylata* and *Esch. blattae.* We do not accept that these are correctly placed in the genus *Escherichia* but shall describe them briefly at the end of this section.

In 1885 Escherich described the isolation from the faeces of infants of an organism that he named *Bakterium coli commune.* The name *Bacterium coli* was subsequently widely used for many years. However, in 1920 Castellani and Chalmers described the genus *Escherichia* and gave an account of the biochemical reactions of the genus which was very similar to modern descriptions of it.

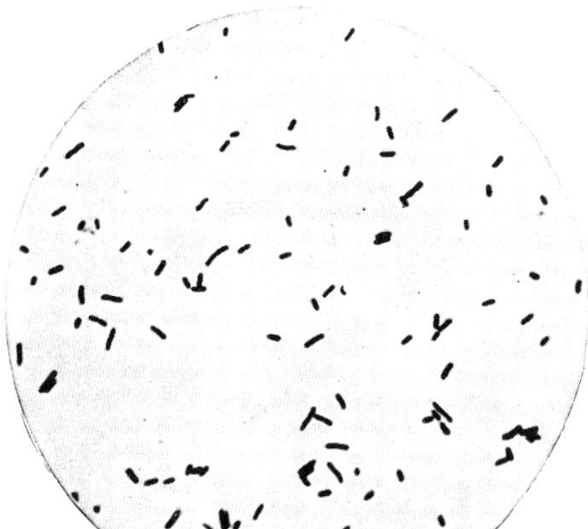

Fig. 34.1 *Escherichia coli.* From 24 hr culture on nutrient agar, 37° (× 1000).

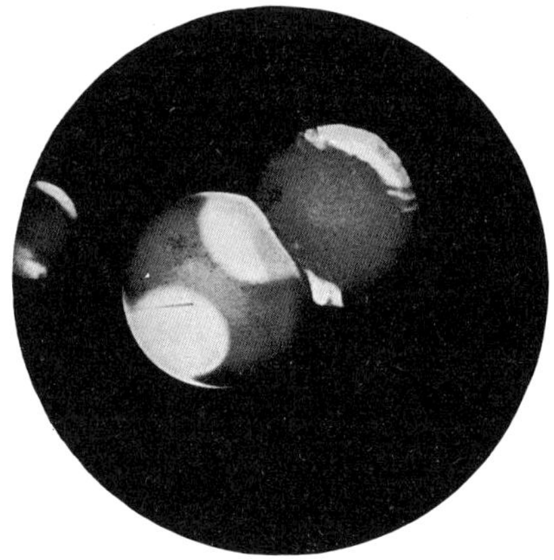

Fig. 34.2 *Escherichia coli.* Colonies on nutrient agar after 24 hr, 37° (× 8).

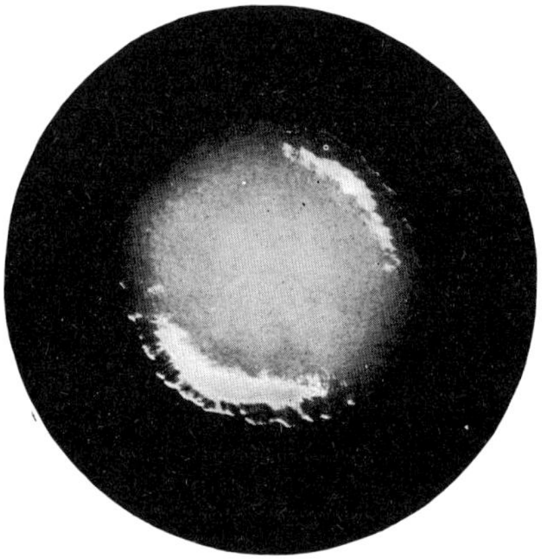

Fig. 34.3 *Escherichia coli.* Larger and flatter type of colony on nutrient agar after 24 hr, 37° (× 8).

Morphological and cultural characters

Strains of *Esch. coli* are usually motile and seldom capsulated. They grow well on ordinary media, forming large (2–3 mm diameter) circular, low convex, smooth and colourless colonies in 18 hr on nutrient agar, and large red colonies on MacConkey's lactose agar; blood agar is discoloured around the growth and there may be haemolysis. They grow over a wide range of temperature (15–45°). Some strains are more heat resistant than other members of the Enterobacteriaceae and will survive 60° for 15 min or 55° for 60 min. On the other hand, they are more sensitive than members of the *Salmonella* genus to certain dyes, particularly brilliant green.

Biochemical activities

The characteristic biochemical reactions of *Esch. coli* were given in Table 33.2. Acid, usually with gas, is formed from glucose, and usually from maltose, mannitol, xylose, rhamnose, arabinose, sorbitol, trehalose and glycerol. Although most strains ferment lactose with the production of acid and gas within 24 to 48 hr, some otherwise typical strains acidify it late or not at all. Action on sucrose, salicin, dulcitol and raffinose varies from strain to strain. Adonitol, inositol and cellobiose are rarely fermented. Gas is not formed from starch. Many characteristic biochemical reactions, such as indole production and the formation of acid and gas from lactose and other carbohydrates, take place at 44° as well as at 37°. All strains give a positive methyl-red and a negative Voges-Proskauer reaction. Although there is usually no growth on Simmons's citrate medium, *Esch. coli* will grow in the presence of inorganic salts with acetate as sole carbon source (Trabulsi and Ewing 1962); unlike *Shigella* it will also ferment citrate in a medium with a simple nutrient base, such as Christensen's citrate medium. Variants of *Esch. coli* which utilize citrate have been isolated and the citrate-utilizing ability has been shown to be controlled by a transferable plasmid (Sato *et al.* 1978). *Esch. coli* is usually described as urease negative but certain groups of strains, e.g. members of O group 26 from cases of infantile enteritis (Le Minor *et al.* 1954) and strains from oedema disease of piglets (Quinchon *et al.* 1959), contain a high proportion of strains that give an alkaline reaction in Chistensen's urea medium. Although H_2S is produced in cysteine broth it is not detected by blackening of triple sugar iron agar and *Esch. coli* is regarded as H_2S negative for identification purposes. Nevertheless, H_2S positive variants have been described in which the ability to produce H_2S is controlled by a transferable plasmid (Lautrop *et al.* 1971, Ørskov and Ørskov 1973). Almost all strains give a negative KCN test and fail to oxidize gluconate or utilize malonate. Most strains of *Esch. coli* decarboxylate lysine and differ in this respect from members of the genus *Shigella*. Action on arginine is weak, late or absent, and may or may not be accompanied by the production of ammonia. (For further information see the monograph of Ewing *et al.* 1972*a*.)

Resistance to antimicrobial agents

Esch. coli is ordinarily sensitive to many antibiotics; although moderately resistant to benzylpenicillin, it is sensitive to ampicillin and the cephalosporins, and usually to tetracycline, streptomycin, chloramphenicol, kanamycin, gentamicin, trimethoprim, sulphonamides and the polymyxins. Many strains, however, have acquired plasmids conferring resistance to one or more of these drugs. Strains resistant to streptomycin, sulphonamides, tetracyclines and, to a lesser extent, ampicillin are particularly common and the resistances are usually transferable. Strains with transferable resistance to one or more drugs can be isolated from the faeces of most members of the general population (Smith and Armour 1966, Moorhouse and McKay 1968, Lewis 1968, Moorhouse 1969, Datta 1969, Datta *et al.* 1971), and some twenty per cent of strains responsible for urinary-tract infection in patients outside hospitals have transferable drug resistance (Brumfitt *et al.* 1971). Strains with transferable drug resistance may also be isolated from intensively reared food animals such as calves, pigs and poultry (Linton 1977). (For further information about antibiotic resistance in *Esch. coli* see Chapter 61.)

Antigenic structure

Early attempts to study the antigenic structure of *Esch. coli* by means of agglutination revealed a great heterogeneity of antigenic factors (Mackie 1913). Theobald Smith (1928) demonstrated the presence of a capsular antigen in certain bovine strains, and Lovell (1937), using a precipitation test, was able to recognize two antigens, a soluble specific carbohydrate substance associated with the capsule (see also D. E. Smith 1927) and a somatic antigen. It was left, however, to workers in Scandinavia to reveal the full antigenic complexity of members of this species and to reduce it to some sort of order (see Knipschildt 1945, Vahlne 1945, Kauffmann 1944, 1947).

Kauffmann (1947) published a diagnostic scheme based on the distribution of H, O and K antigens, as detected by the reactions obtained in agglutination tests (see also Edwards and Ewing 1972, Ørskov and Ørskov 1977).

(1) So far over 50 H antigens are known, all of them monophasic. There are few significant cross-reactions between them, or with the H antigens of other members of the Enterobacteriaceae. Before use for H-antigen determination it is often necessary to grow cultures in semi-solid agar.

(2) Over 160 different O antigens have been described. Agglutination tests must be carried out on boiled or autoclaved cultures to overcome inagglutinability caused by K antigens. Numerous cross-reactions occur between individual *Esch. coli* O antigens, and between these and the O antigens of *Salmonella*, *Shigella*, *Citrobacter* and *Providencia*. In some instances the antigens appearing in the different genera are identical.

(3) The term K antigen is used collectively for sur-

face or capsular antigens that cause 'O inagglutinability' (Chapter 33). These antigens have been divided into three classes according to the effect of heat on the agglutinability, antigenicity and antibody-binding power of bacterial strains that carry them. (a) The *L antigens* were considered to be thermolabile surface antigens; after 1 hr at 100° the bacterial suspension is agglutinated by O antiserum but not by L antiserum, and has lost its L antigenicity and its ability to remove L antibody from serum. To prepare an L antiserum, living organisms are injected and the serum is absorbed with boiled organisms of the homologous strain. (b) The *A antigens* are capsular and are unaffected by boiling. If the organism is heated for 2 hr at 121°, O inagglutinability, agglutination by A antiserum and A antigenicity are abolished, but ability to bind A antibody remains. Therefore, to prepare a pure A antiserum it is necessary to absorb OA antiserum with a non-capsulated variant or with an organism of the same O group that possesses a different K antigen. (c) The characters attributed to *B antigens* are as follows: O inagglutinability, agglutination by B antiserum and antigenicity are destroyed in 1 hr at 100°, but antibody-binding power is unaffected after 2 hr at 121°. Pure B antisera have rarely been prepared, and the presence of B antigen is inferred from a rise in the titre of agglutination by OB antiserum when the suspension is boiled. With a few exceptions, only one K antigen, either L, A or B, is present in any one strain.

In Kauffmann's original scheme, the primary subdivision was into a number of O groups, members of which might contain subgroups possessing different K antigens. Each of these might include strains with different H antigens. Consecutive numbers were allotted to the antigens of each class, but the B antigens were numbered separately from the rest of the K antigens. Later (Kauffmann *et al.* 1956), the numbers of the B antigens were included in the same consecutive series as the rest of the K antigens. In the antigenic formula, the nature of the K antigen is indicated after the serial number, e.g. O18:K76 (B):H14. Where the number of the B antigen was changed in the reorganization, the old number is often also included in the formula, e.g. O111:K58 (B4):H12.

Two antigens (K88 and K99) originally placed in the L class (Ørskov *et al.* 1961, Ørskov *et al.* 1975) are heat-labile proteins. By electron microscopy they can be seen as filaments resembling fimbriae on the bacterial surface. Their genetic determinants are carried on transferable plasmids (Stirm *et al.* 1966, 1967; Smith and Linggood, 1972). They are similar to the plasmid-determined colonization factor antigens CFA/I and CFA/II found in some enterotoxigenic strains that cause human diarrhoea (Evans *et al.* 1975, Evans and Evans, 1978). We consider that these antigens would more appropriately be placed in a separate series together with the fimbrial antigens (see also Chapter 71).

The antigens present in mucoid strains of *Esch. coli* are of two classes (Kauffmann 1944): (1) acid polysaccharides that are type-specific and probably correspond to the A antigens (Ørskov *et al.* 1963), and (2) a more widely distributed M substance that is a poor antigen and may be present in many strains of *Esch. coli* and of other enterobacteria (Henriksen 1954*b*, *c*, Rees 1957) and is a colanic acid (Goebel 1963; see also Chapter 33).

We can have little hesitation in accepting the evidence for the existence of a number of Kauffmann's L and A antigens, which appear to be two distinct classes of polysaccharides. The chemistry and genetics of these polysaccharide K antigens are summarized in the review of Ørskov and her colleagues (1977).

The evidence for the existence of the B antigens is less conclusive. Knipschildt (1945), who described the first B antigens, found that they caused complete O inagglutinability in their respective O antisera. Further, he was able to produce pure B sera by absorption with strains belonging to the same O group but possessing a different K antigen, or no K antigen. F. Ørskov (1956) pointed out that the O inagglutinability exhibited by many of the gastro-enteritis strains reported to possess B antigen was very slight, and that there was little difference in the serological behaviour between many strains to which L and B antigens had been allotted. A small difference in the titre of agglutination of heated and unheated suspensions by the same serum constitutes inadequate evidence on which to postulate the existence of a K antigen. Such a result might equally well be due to the presence in the organism of a heat-labile non-antigenic substance which inhibits agglutination. It has been reported (Braun *et al.* 1956, Jann *et al.* 1970) that purified lipopolysaccharide from strains said to possess B antigens has both O and B specificity.

New light was thrown on the constitution of the somatic antigens of *Esch. coli* by Ørskov and Ørskov (1970, 1972; see also Ørskov *et al.* 1971), who carried out precipitation reactions after electrophoresis of bacterial extracts in gel. The extracts were made by heating cultures in saline at 60° or at 100°, and were tested against homologous O and OK antisera. Precipitation arcs corresponding to the O antigens could be identified, and were of two main types, those migrating towards the anode and those migrating towards the cathode; in certain serological types in which the main arc was on the cathode side, a small additional arc appeared on the anode side, but only with extracts made at the lower temperature. Arcs considered to correspond to K antigens—because they were obtained only with antiserum against unheated bacteria (OK serum)—always appeared on the anode side. They appeared when the extracts had been made either at 60° or at 100°, thus confirming the conclusion that the K antigens survived boiling but then acted as haptens.

The serotypes of *Esch. coli* fell into a few clear cut groups according to the electrophoretic characters of the O and K antigens (Table 34.1). Group 1 comprised strains with an O antigen that migrated towards the cathode, and could be subdivided into 1A, in which the additional heat-labile anodic arc was present, and

Table 34.1 Immunoelectrophoretic properties of O and K antigens of *Escherichia coli* (modified from Ørskov and Ørskov 1972)

Group			Immunoelectrophoretic properties: O-antigen arc moves towards anode	O-antigen arc moves towards cathode	K-antigen arc*	Characters of O groups
1	A	*a*	(+)†	+	+‡	Many have L-type K antigens (e.g. O3, O4, O5). Cause parenteral disease.
1	A	*b*	(+)†	+	−	K antigens not detectable
1	B	*a*	−	+	+§	Have A-type K antigens (O8, O9). Cause parenteral disease.
1	B	*b*	−	+	−	Many associated with infantile gastro-enteritis (e.g. O55, O111).
2		*a*	+	−	+‡	Cause parenteral disease (e.g. O14, O22).
2		*b*	+	−	−	Associated with dysentery-like disease (e.g. O124).

* Moves towards anode.
† Weak arc, absent when extract made at 100°.
‡ Fast.
§ Slow.

IB, in which it was absent. These in their turn could be subdivided into (*a*) strains with and (*b*) strains without a K-antigen arc. The K antigens in subgroup 1A*a* migrated further towards the anode than did the K antigens in subgroup 1B*a*. Group 2, with O antigens migrating towards the anode, could be subdivided into subgroup 2*a* with a K antigen and 2*b* without; the few serotypes that fell into subgroup 2*a* all have K antigens resembling those of subgroup 1A*a* strains.

Serotypes classified in this way had other characters in common. Subgroup 1A*a*, with an electrophoretically mobile K antigen, contained many strains with so-called L antigens, and included most of the serotypes responsible for parenteral septic infections. In subgroup 1B*a* were the strains with less electrophoretically mobile K antigens of the A type, some of which were also associated with parenteral disease. Subgroup 1B*b* comprised strains with no detectable K antigen, and included many serotypes that have caused infantile gastro-enteritis and were believed to possess B antigens. Members of subgroup 2*b*, in which no K antigen could be found, included a number of serotypes associated with dysentery-like disease in adults and children. Many of these are biochemically aberrant and their O antigens show frequent cross-reactions with those of shigella subgroups B and C. They differ from subgroup 1B*b* mainly in the direction of migration of the O antigen. (For further information on the serology, chemistry and genetics of *Esch. coli* O & K antigens see the review by Ørskov *et al.* 1977.)

Other typing methods

Nicolle and his colleagues (1952) devised a phage-typing system for O groups 111, 55 and 26 using unadapted phages. Twenty-eight phage types were recognized and a great deal of information about their distribution throughout the world has been obtained (see Buttiaux *et al.* 1956). Smith and Crabb (1956) used an *ad hoc* method of classifying previously unselected strains of bovine origin by their susceptibility to a large collection of O phages, and found it suitable for studying the acquisition of faecal coliform bacteria by newborn calves. For further information on the phage typing of *Esch. coli* see the review by Milch (1978).

The subdivision of O groups, or of phage types within an O group, by colicine-typing methods also proved to be practicable (Fredericq *et al.* 1956, Shannon 1957, Hamon 1958, 1959). 'Resistogram typing' (Chapter 20) is said to be helpful in distinguishing between *Esch. coli* strains (Elek and Higney 1970). Wilson and her colleagues (1981) advocate the use of this method in combination with selected biochemical and metabolic tests for the study of strains isolated sequentially from the urinary tract.

Habitat and pathogenicity

Esch. coli is a widespread intestinal parasite of mammals and birds, but appears not to lead an independent existence outside the animal body. It is a pathogen in man and animals, and causes both septic infection and diarrhoea.

In *man*, the most common and important septic infections due to *Esch. coli* are of the urinary tract (Chapter 57). The organism also causes sepsis in operation wounds and abscesses in a variety of organs, and neonatal meningitis and septicaemia (Chapter 61). In *calves*, it causes septicaemia particularly when they have been deprived of colostrum or for some other reason have too little immune globulin in the serum (Chapter 61).

In Chapter 71 we shall discuss the diarrhoeal diseases caused by *Esch. coli*. These are of several sorts, and include (1) acute enteritis of young animals, including human infants, piglets, calves and lambs, (2) 'oedema disease' of piglets, in which there is sudden death with swelling of various abdominal organs after multiplication of certain *Esch. coli* strains in the gut, (3) acute enteritis in human subjects of all ages occurring mainly in the tropics and including travellers' diarrhoea, and (4) a dysentery-like disease affecting man at all ages.

The pathogencity of *Esch. coli* for laboratory animals is not great when injections are made by the usual routes. In the mouse, intraperitoneal injection of the organism causes a fatal infection, and the LD50 for individual strains varies considerably. Kauffmann (1948) observed that strains isolated from septic infections of man were virulent for the mouse, caused necrosis when injected into the rabbit's skin, and usually had an antigen of the L type. Medearis and Kenny (1968) also found striking differences between the mouse virulence of strains, and associated this with their ability to resist phagocytosis; mutants with incomplete O side-chains had greatly reduced virulence and a poor ability to persist in the peritoneal cavity (Medearis *et al.* 1968). Wolberg and de Witt (1969), like Kauffmann, found a relation between mouse virulence and the presence of L antigen; L— variants underwent phagocytosis *in vitro*, but L+ variants resisted phagocytosis except in the presence of specific antibody. Glynn and Howard (1970) showed that there was a direct relation between the presence of the K polysaccharide and resistance of the organism to killing by complement. In members of the same serotype, resistance to complement and to phagocytosis, and mouse virulence by the intracerebral route, were each proportional to the amount of K antigen formed (Howard and Glynn 1971*a*, *b*). Strains of *Esch. coli* that caused renal infection in women had significantly more K antigen than those that caused infection only of the bladder, or strains from the faeces of a control population (Glynn *et al.* 1971).

According to Ørskov and his colleagues (1971), the strains responsible for parenteral disease in man usually have an O antigen that migrates towards the cathode and an acid-polysaccharide K antigen of the L or A class. Strains of *Esch. coli* from cases of bacteraemia in calves will invade the blood stream when given orally to calves that are deficient in immune globulins, but not when given to normal calves; correspondingly, they will grow *in vitro* in globulin-deficient calf serum but not in normal calf serum (Smith and Halls 1968).

In recent years it has been shown that as many as 80 per cent of *Esch. coli* strains causing neonatal meningitis and 40 per cent of those isolated from infants with septicaemia but without meningitis possess the K1 antigen (Schiffer *et al.* 1976). Clearly the presence of this antigen enhances the invasiveness of *Esch. coli* strains, although the mechanism of this enhancement remains uncertain. Gemski and his colleagues (1980) have shown that the presence of K1 confers resistance to the bactericidal effect of serum, while Welch and his colleagues (1979) suggest that the presence of K1 on the bacterial surface exerts an antiphagocytic effect.

Many strains of *Esch. coli* form a *haemolysin*. According to Kauffmann (1948), mouse-virulent strains from septic lesions are often haemolytic and agglutinate red blood cells, but haemagglutinating activity also occurs in non-haemolytic strains. Bamforth and Dudgeon (1952) could not prepare a filtrable haemolysin from strains of human origin, but Lovell and Rees (1960), working with strains from pigs, obtained haemolytic filtrates from young broth cultures. Smith (1963) found that most strains from the gut of cattle, pigs and sheep, but less than 20 per cent of strains of human origin, were haemolytic on sheep blood agar. He recognized two different haemolysins: the filtrable agent described by Lovell and Rees, which he called α-haemolysin, and a non-filtrable β-haemolysin. The ability to form α-lysin is transferable from one strain to another by conjugation (Smith and Halls 1967). Potent preparations of α-lysin are toxic for mice, rabbits and guinea-pigs when given intravenously, but the ability to form the lysin has no effect on the pathogenicity of the strain by the intraperitoneal route.

Two other plasmid-determined characters appear to contribute to the virulence of invasive strains of *Esch. coli*. (1) The *vir* plasmid (Smith 1978) controls the synthesis of a heat-labile, non-dialysable toxin lethal for rabbits, mice and chickens when administered intravenously. It was found in only six of 247 *Esch. coli* strains from unrelated cases of generalized infection; two were from calves and four from lambs. (2) Smith (1978) also found that the Col V plasmid was widely distributed among strains from generalized infections in animals. The presence of the Col V plasmid increased the ability of *Esch. coli* strains to proliferate in the tissues of chickens after intramuscular injection. Nevertheless, the production of colicine V itself is not responsible for the increased virulence (Quackenbush and Falkow 1979). Williams and Warner (1980) have shown that the Col V plasmid also carries the genetic determinants for an iron uptake system; this may be responsible for the enhanced virulence.

There is evidence that the ability of *Esch. coli* to infect the urinary tract is associated with surface components that specifically mediate adherence to the epithelial cells of the tract (Chapter 57).

The pathogenesis of diarrhoea due to *Esch. coli*, and the experimental models that have been used to study it, will be discussed in Chapter 71. In some strains, particularly those causing infantile enteritis, the pathogenic mechanism remains largely unknown. Other strains depend on the ability to colonize the small intestine and there to produce enterotoxins that act on the intestinal epithelium. These enterotoxins can be detected in the laboratory by tests on animals and in tissue culture. It should be noted that the heat-labile toxin of these *Esch. coli* strains resembles very closely the toxin of cholera vibrios (Chapter 27). Strains that cause dysentery-like disease in man share with *Shigella* strains the ability to penetrate the cells of the intestinal epithelium and to multiply intracellularly; experimentally they also cause keratoconjunctivitis in the guinea-pig and penetrate and multiply within HeLa cells in tissue culture (see Chapter 36). Strains causing oedema disease in pigs produce a toxin (Clugston and Nielsen 1974) which may act as a vasotoxin (Clugston *et al.* 1974).

The strains of *Esch. coli* responsible for the main groups of disease are generally distinct, and belong to different O groups. Thus O-groups 1, 2, 4, 6, 7, 8, 9, 11, 18ab, 22 and 75 predominate in parenteral septic infections; infantile gastroenteritis is associated with members of O-groups 18ac, 20, 25, 26, 44, 55, 86, 111, 114, 119, 125, 126, 127, 128 and 142; and the dysenteric type of diarrhoea in man has been attributed to strains in O-groups 28ac, 112ac, 124, 136, 143, 144, 152 and 164 (Rowe 1979). Enterotoxigenic *Esch. coli* from human sources often belong to O-groups 6, 8, 15, 25, 27, 63, 78, 148 and 159. In pigs, however, in which oedema disease and diarrhoea of older piglets appear to be due to members of a few O groups (138, 139 and 141), diarrhoea of very young piglets appears to be associated with a variety of O groups, but most strains have the transferable protein antigen K88. Similarly, many strains causing diarrhoea of calves and lambs possess the K99 antigen (Smith and Linggood 1971, 1972).

Biochemically atypical *Escherichia coli* strains

Studies of DNA-DNA recombination suggest that *Esch. coli* and *Shigella* should be looked upon as members of a single species (Brenner *et al.* 1973); it is therefore to be expected that intermediate strains will occur. Certain of these, which are non-motile and anaerogenic, and often ferment lactose late or not at all, have caused difficulties in classification (Chapter 36). They were at one time included in the so-called 'alkalescens-dispar' group, among which a number of O serogroups were recognized (Frantzen 1950, 1951; see also Ewing *et al.* 1950). Such strains are now considered to be atypical *Esch. coli* and their O antigens are included in the O-serogrouping scheme for *Esch. coli*.

Having decided that these aberrant strains should be classified as *Esch. coli*, the main difficulty is to distinguish them from members of the genus *Shigella*. For this purpose reliance is placed on the production of lysine decarboxylase, the fermentation of citrate in Christensen's medium or of mucate, and the utilization of sodium acetate as sole carbon source, any of which properties would suggest that an organism is not a shigella. Nevertheless, some *Esch. coli* strains give negative results in each of these tests.

Some workers believe that an intermediate group between *Escherichia* and *Shigella* should be established and that an important criterion for inclusion in it should be the ability to cause dysentery-like disease (Trifonova, 1968). Shmilovitz and his colleagues (1974) have suggested that the group should be known as Intermediate Shigella Coli Alkalescens Dispar (ISCAD). Alternatively, Stenzel (1978) proposed that these organisms should be included in *Shigella* subgroup D under the name *Sh. metadysenteriae*. There is no doubt that a number of biochemically atypical *Esch. coli* can cause dysentery-like disease. Moreover, like true dystentery bacteria, they cause keratoconjunctivitis in the guinea-pig (Sakazaki *et al.* 1974). However, recent studies by Silva and her colleagues (1980) show that many strains of *Esch. coli* isolated from patients with dysentery-like disease and able to cause keratoconjunctivitis in the guinea-pig are motile, aerogenic, and ferment lactose promptly. The only atypical biochemical character shared by these enteroinvasive strains is their failure to produce lysine decarboxylase. We therefore support the view of the Enterobacteriaceae Sub-Committee of the International Committee on Bacteriological Nomenclature (Report 1963) that pathogenicity should not be considered in the classification of Enterobacteriaceae and that strains with biochemical reactions that do not conform strictly to those of *Shigella* should be classified as atypical *Esch. coli*.

The curious organism described by Massini (1907) as *Bact. coli mutabile* is itself a non-lactose fermenter but has the property of giving rise to lactose-fermenting mutants that show no tendency to revert to the parent form. Dulaney and Michelson (1935) described a severe outbreak of diarrhoea in infants in which this organism was present in the faeces, and Ewing, Tanner and Tatum (1956) showed subsequently that their strains had the antigenic formula O18ac:K77 (B21):H7. However, strains able to ferment lactose mutatively have been observed in a number of other O groups of *Escherichia*.

Other proposed species of *Escherichia*

The species *Esch. adecarboxylata* was described by Leclerc (1962) from a collection of strains isolated from water. Strains of the species are easily distinguished from *Esch. coli* in that they acidify adonitol, ferment malonate and are KCN tolerant, and that they fail to form lysine decarboxylase or produce alkali when grown on Christensen's citrate medium. Ewing and Fife (1972) included *Esch. adecarboxylata* in *En-*

terobacter agglomerans (syn. *Erwinia herbicola*), but it forms gas from glucose, produces indole and acidifies dulcitol and inositol, all characters that distinguish it from typical strains of *Erw. herbicola*. However, the genus *Erwinia* as at present constituted is biochemically heterogeneous (p. 302), and further study will be needed before the relation of *Esch. adecarboxylata* to it is clarified.

Esch. blattae was described by Burgess and co-workers (1973) from bacteria isolated from the hind-gut of the cockroach. Strains of it differ from *Esch. coli* in oxidizing gluconate and failing to form indole, acidify mannitol or produce β-galactosidase. It would perhaps be better placed in another genus. (For atypical biogroups of *Esch. coli* found in clinical specimens, see Brenner *et al.* 1982.)

Klebsiella

Definition

Non-motile and usually capsulated. Ferment mannitol, salicin, adonitol, inositol and sorbitol, and often also lactose and sucrose. Most strains form gas from sugars; gas production from starch is an important diagnostic feature. Characteristically give a negative methyl-red and a positive Voges-Proskauer reaction, but strains from the respiratory tract often give the opposite reactions and may have other atypical biochemical reactions. Nearly all grow on Simmons's citrate medium and in Møller's KCN medium. Do not deaminate phenylalanine. Non-pigmented. With a few exceptions fail to liquefy gelatin. Habitat: found in the bowel and respiratory tract of man and animals; and in soil and water. Cause a variety of septic infections in man and animals. G + C content of the DNA: 52–56 moles per cent.

Type species: *Klebsiella pneumoniae*.

Classification

In the last two editions of this book, we used the name *K. aerogenes* for the non-motile, capsulated, gas-producing, acetylmethylcarbinol-forming strains commonly found in human faeces and in water. They were thought to correspond to the non-motile gas-producing organism from faeces described by Escherich (1885) as *Bakterium lactis aerogenes*, referred to by later workers as *Bact. aerogenes*, and subsequently transferred to a reconstituted genus *Klebsiella* (Report 1954). Unfortunately the term *Bact.* (later *Aerobacter*) *aerogenes* was also used by water bacteriologists to refer to organisms that were later shown to be motile and are now placed in the genus *Enterobacter*. In an attempt to resolve the resultant confusion (see p. 296), some taxonomists adopted *pneumoniae* as the specific epithet for the non-motile *aerogenes*-like organisms although it had earlier been used to designate certain biochemically atypical strains of *Klebsiella* (see below). The view of these workers has now prevailed, and the name *K. pneumoniae* appears in the Approved Lists of Bacterial Names (Skerman *et al.* 1980) while *K. aerogenes* is omitted. In the interests of uniformity we shall use the name *K. pneumoniae* for the species as a whole, but propose to refer to the most frequently encountered, biochemically typical form of it as *K. pneumoniae* subsp. *aerogenes*.

A number of other less easily classifiable non-motile and capsulated organisms occur in the respiratory tract of man and other animals. Von Frisch in 1882 cultivated a capsulated organism from patients with rhinoscleroma; in 1883 Friedländer isolated a similar organism—generally known as Friedländer's bacillus—from the lungs of patients who had died of pneumonia; and Abel (1896) described the ozaena bacillus, which must not be confused with the *Coccobacillus foetidus ozaenae* of Perez (Chapter 35). These three organisms, and others from the respiratory tract, form a biochemically heterogeneous group but are antigenically related to some of the *K. pneumoniae* subsp. *aerogenes* strains. Various specific names, including *K. atlantae*, *K. edwardsi*, *K. ozaenae*, *K. pneumoniae* and *K. rhinoscleromatis* have been attached to biochemical varieties in this group. Both American and British authors have used the name *K. pneumoniae* for respiratory strains believed to correspond to Friedländer's bacillus. Studies of DNA-DNA hybridization (Brenner *et al.* 1972) indicate, however, that these respiratory strains are so closely related genetically to *K. pneumoniae* subsp. *aerogenes* that they should all form part of a single species, for which *K. pneumoniae* is now the accepted name. For medical purposes, however, it is important to distinguish between the various biochemical varieties within this species. Although the name *K. aerogenes* does not appear on the Approved Lists of Bacterial Names (Skerman *et al.* 1980) and is at present without official standing in nomenclature we shall recognize it as a subspecies of *K. pneumoniae* along with the subspecies *ozaenae*, *pneumoniae* and *rhinoscleromatis* (see Table 34.2).

Two other members of the genus *Klebsiella*, though biochemically similar to *K. pneumoniae* subsp. *aerogenes*, are accorded separate specific status. An organism that formed indole and liquefied gelatin slowly was described by Flügge (1886) as *Bact. oxytocum* (see Lautrop 1956). The decision to consider it a separate species—*K. oxytoca*—is supported by evidence from DNA-DNA hybridization tests, which suggest that assignment to a separate genus may even be appropriate (Jain *et al.* 1974). Another organism named *K. terrigena* shows similar genetic differences from *K. pneumoniae* but is indistinguishable from it in routinely used tests (Izard *et al.* 1981*a*). It has been found in soil and water and will not be discussed further.

Table 34.2 Differentiation of species and subspecies of *Klebsiella*

		K. pneumoniae			
Property	*K. oxytoca*	subsp. *aerogenes*	subsp. *ozaenae*	subsp. *pneumoniae*	subsp. *rhinoscleromatis*
β-galactosidase	+	+	+	+	−
Lysine decarboxylase	+	+	V	+	−
Gas from glucose	+	+	V	+	−
Gelatin liquefaction	V	−	−	−	−
Gluconate oxidation	+	+	−	V	−
Indole formation	+	−	−	−	−
Growth in KCN medium	+	+	+	−	+
Malonate fermentation	+	+	−	+	+
Methyl-red test	V	−	+	+	+
Growth on Simmons's citrate	+	+	V	+	−
Urease formation	+	+	−	+	−
Voges-Proskauer test	V	+	−	−	−
Dulcitol: acid	V	V	−	+	−
Lactose: acid	+	+	+	+	−

+ = Most strains positive; − most strains negative; V = some strains positive, others negative.

Morphological and cultural characters

Members of the genus *Klebsiella* tend to be somewhat shorter and thicker than the other enterobacteria and are straight rods about 1–2 μm long and 0.5–0.8 μm wide, with parallel or bulging sides and rounded or slightly pointed ends (Fig. 34.4). The cells are either in pairs end-to-end or are arranged singly. In the body diplobacilli, very like pneumococci, are commonly seen. They are non-motile. When the capsule is pronounced, it can be demonstrated even by Gram's stain. The India-ink method is to be preferred to the ordinary capsule stains, since distortions due to drying are avoided and the capsular material present as 'loose slime' can be seen as well as that around individual cells (Duguid 1951).

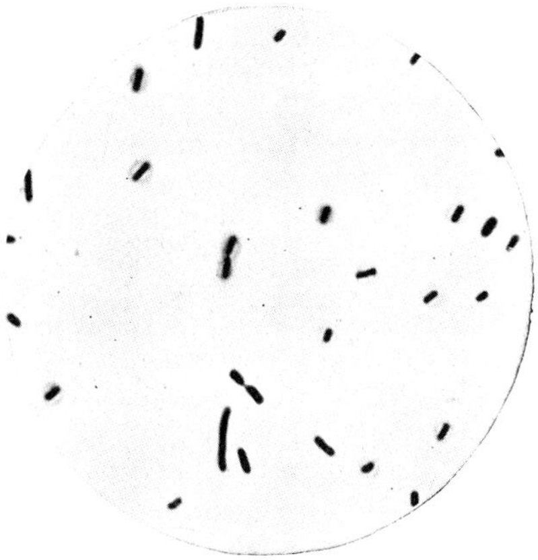

Fig. 34.4 *Klebsiella pneumoniae* subsp *aerogenes*. From a nutrient agar culture, 24 hr, 37° (× 1000).

Capsular material is produced in greater amount in media containing a relative excess of carbohydrate (Duguid and Wilkinson 1953) and is a nitrogen-free polysaccharide (Toenniessen 1921, Heidelberger *et al.* 1925). Wilkinson, Dudman and Aspinall (1954) purified the capsular polysaccharides of a number of strains belonging to capsular type 54, and showed that they contained about 50 per cent of D-glucose, 10 per cent of L-fucose and 29 per cent of a uronic acid which was subsequently shown to be glucuronic acid (Aspinall *et al.* 1956). Mucoid strains produced the largest amount of polysaccharide, and most of it was present as capsule or slime: smooth strains produced less and nearly all of it was intracellular. Rough strains produced least (Wilkinson, Duguid and Edmunds 1954). Wherever the polysaccharide material was situated, its chemical composition was the same (Dudman and Wilkinson 1956). There were, however, differences in physical properties and in molecular weight between capsular material and loose slime (see Wilkinson 1958). The polysaccharides of the different capsular types are all complex acid polysaccharides (Dudman and Wilkinson 1956), which contain a uronic acid—usually glucuronic acid but occasionally galacturonic acid—as well as a considerable amount of pyruvic acid (Wheat *et al.* 1965). In addition to the type-specific acid polysaccharide, a neutral polysaccharide, which is serologically active but not type-specific, has also been described (Gormus and Wheat 1971).

Most strains of *Klebsiella* are fimbriate but some of the respiratory strains form an exception (Duguid 1959, 1968). Biochemically typical strains of whatever capsular type usually form fimbriae of both type 1 and type 3. *K. pneumoniae* subsp. *pneumoniae* forms only type 1 fimbriae. *K. pneumoniae* subspp. *ozaenae* and *rhinoscleromatis* are invariably non-fimbriate, as are certain biochemically aberrant respiratory strains belonging to capsular types 1 and 2.

When much capsular material is produced, the growth on agar is luxuriant, greyish-white, mucoid and almost diffluent. This is due, no doubt, to the

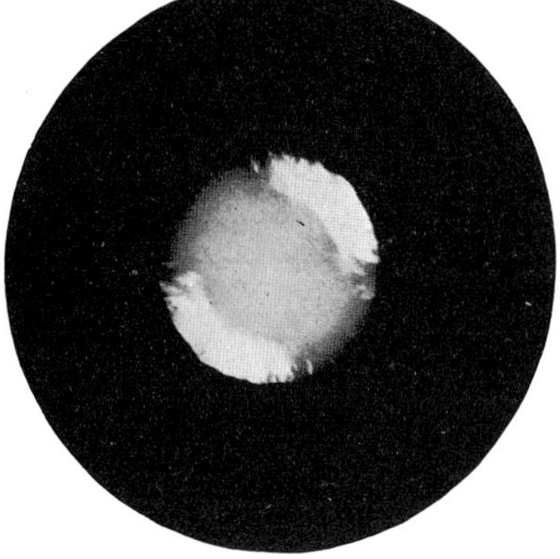

Fig. 34.5 *Klebsiella pneumoniae* subsp *aerogenes*. Surface colony on nutrient agar, 24 hr, 37° (×8).

high proportion of water—92 per cent (Toenniessen 1921)—in the capsular material. The condensation water on an agar slope is converted into a greyish-white mucoid mass. In broth the organism grows freely, giving rise after a few days to a pronounced viscosity, so that the medium takes on the consistency of melted gelatin. Great stress used to be laid on the nail-headed growth in stab gelatin cultures; a circular, convex growth may occur on the surface, with a filiform growth in the stab, but this appearance is given only by some strains. Usually there is no liquefaction of the gelatin, but often a large napiform bubble of gas accumulates just beneath the surface, giving on first sight the appearance of liquefaction.

The cultural appearances of capsulated strains are subject to considerable variation. Non-mucoid variants appear on serial subculture on solid medium, particularly when the plates are incubated for several days (Toenniessen 1913), either as translucent peripheral outgrowths or as secondary colonies in the substance or on the surface of the original colonies. Sometimes the whole colony may dry up and wither away, leaving an effuse translucent layer looking like ground glass—aptly called by Collins (1924) 'suicide colonies'. Non-mucoid variants form colonies closely resembling those of *Esch. coli*. Reversion to the mucoid form may occur—often suddenly. (For further information about colonial variation see Toenniessen 1914, Hadley 1927, Goslings 1935, Read *et al.* 1957.) Some of the respiratory strains grow poorly or not at all on MacConkey's agar.

The organisms are killed by moist heat at 55° in half an hour. They may survive drying for months (Loewenberg 1894). When kept at room temperature, cultures remain viable as a rule for weeks or months. They are facultatively anaerobic: but growth under strictly anaerobic conditions is poor. There is no haemolysis of horse or sheep red cells. The optimum temperature for growth is 37°, the limits are 12° and 43°. Some strains form a slightly brownish pigment.

Biochemical activities

Members of the genus ferment a wide range of sugars, but their behaviour in this and other respects is far from uniform (Table 34.2). *K. pneumoniae* subspp. *aerogenes* and *pneumoniae* are the most active fermenters of the genus, producing acid and gas from all or nearly all of the sugars usually employed, with the possible exception of dulcitol in the case of the former subspecies. Attention should be drawn to their ability to form gas from starch. Hormaeche and Munilla (1957) advocate a medium containing non-soluble starch, and practically all strains of *K. pneumoniae* subspp. *aerogenes* and *pneumoniae*, and practically no other members of the Enterobacteriaceae, are able to form gas within 4 days. The other biochemical reactions of *K. pneumoniae* subspp. *aerogenes* and *pneumoniae* are fairly uniform. They are indole negative, do not liquefy gelatin, and grow on Simmons's citrate medium. They are urease positive, utilize malonate and have a lysine decarboxylase. The two subspecies differ from each other in that strains of subsp. *aerogenes* usually give a positive Voges-Proskauer reaction, grow in Møller's KCN medium and oxidize gluconate whereas strains of subsp. *pneumoniae* give the opposite results. Strains of subsp. *pneumoniae* were found by Cowan *et al.* (1960) to belong always to capsular type 3 and to be fimbriate. *K. oxytoca* resembles *K. pneumoniae* subsp. *aerogenes* closely, except that it liquefies gelatin slowly, forms indole, and nearly always acidifies dulcitol (Malcolm 1938, Ørskov 1955*c*, Lautrop 1956).

The respiratory klebsiellas other than *K. pneumoniae* subsp. *pneumoniae* are somewhat less active biochemically. Among them *K. pneumoniae* subsp. *rhinoscleromatis* is fairly well defined. It is anaerogenic, does not acidify lactose, gives a positive methyl-red reaction and a negative Voges-Proskauer reaction, does not grow on Simmons's citrate medium, does not form lysine decarboxylase, and is urease and gluconate negative. Some strains fail to grow in Møller's KCN medium. *K. pneumoniae* subsp. *ozaenae* forms a less clear-cut group, including strains that are late lactose and sucrose fermenters, are sometimes anaerogenic, and also give a positive methyl-red and a negative Voges-Proskauer reaction. They are malonate and gluconate negative. Few strains form lysine decarboxylase, but some attack arginine slowly.

Cowan and his colleagues (1960) described a further species, *K. edwardsi*, of dulcitol-negative, late-lactose-fermenting non-fimbriate members of capsular types 1 and 2 with two varieties: subsp. *edwardsi*, which is anaerogenic, sometimes fails to grow on Simmons's medium but is always Voges-Proskauer positive; and subsp. *atlantae* which forms gas in sugars but is malonate negative and sometimes Voges-Proskauer negative. Cowan (1974) elevated these two varieties to specific status but Bascomb and her colleagues (1971) were unable to confirm the existence of *K. edwardsi* by numerical taxonomic analysis. (See also Brooke 1953, Report

1954, Henriksen 1954*a*, Edwards and Fife 1955, Ørskov 1955*c*, Hormaeche and Munilla 1957, Epstein 1959, Fife *et al.* 1965, Ślopek and Durlakowa 1967, Dubay 1968.)

Resistance to antimicrobial agents

Biochemically typical strains of *K. pneumoniae* subsp. *aerogenes* are resistant to a wider range of antibiotics than are most *Esch. coli* strains. They are nearly always naturally resistant to ampicillin; they were at one time usually sensitive to cephalosporins, thus differing from strains of *Enterobacter* which are almost invariably cephalosporin resistant (Benner *et al.* 1965). Their resistance to streptomycin, chloramphenicol and tetracycline varied from strain to strain but they were usually sensitive to gentamicin and the polymyxins. Since that time, resistance to a number of antibiotics, notably gentamicin and various cephalosporins, have become common in strains found in hospitals (Chapter 61). Many of the respiratory strains, on the other hand, are more sensitive to antibiotics. A few of them are sensitive to as little as 2 μg per ml of benzylpenicillin, and most of them are sensitive to ampicillin, streptomycin, chloramphenicol and tetracycline.

Antigenic structure

In studies of the antigenic structure of this group, most attention has been paid to the capsular antigens. A great advance was made when American workers showed the presence in the group of capsular polysaccharides (Avery *et al.* 1925, Heidelberger *et al.* 1925, Julianelle 1926*a*, *b*, *c*). Julianelle (1926*a*) recognized three serological types A, B and C among the respiratory klebsiellas, distinguishable by agglutination, absorption, precipitation and protection tests, and an undifferentiated group. The type specificity depended on a polysaccharide in the capsule. Goslings and Snijders (1936) described three further capsular antigens (D, E and F) among *K. pneumoniae* subsp. *ozaenae* strains, and Kauffmann (1949) eight more. He proposed that the capsular types should be referred to by arabic numerals in place of the capital letters used previously. The former types A, B, C and D became types 1, 2, 3 and 4, and the new types were numbered consecutively. Since then, many further types have been recognized (Brooke 1951, Edwards and Fife 1952, Edmunds 1954, Ørskov 1955*a*, Durlakowa *et al.* 1967) and the present total is more than 80.

In addition to the capsular antigens there are smooth somatic antigens which occur in various combinations with the capsular antigens (Goslings 1933, Goslings and Snijders 1936). Kauffmann (1949) studied spontaneous non-capsulated variants of capsulated strains and recognized three of these O antigens among members of the first 14 capsular types. I. Ørskov (1954) examined a much larger selection of strains, and was able to add only two more O antigens. Four of the five *Klebsiella* O antigens were identical with or related to *Escherichia* O antigens. It is thus possible to divide *Klebsiella* strains into a small number of O groups which may be further subdivided into capsular types. Organisms with the same capsular antigen may, however, occur in several O groups. It is unlikely, therefore, that the identification of the O antigen, which in any case presents considerable technical difficulties, will be of much practical value in the classification of the capsulated members of the genus *Klebsiella*.

There is some association between antigenic structure, biochemical activities and habitat. Thus, members of the first six capsular types occur most frequently in the human respiratory tract, though occasionally they may be found in other parts of the body or even in other animals (Edwards and Fife 1955). They are not, however, by any means the only types found in the respiratory tract. Ørskov (1955*b*), for example, found that only 16 per cent of strains from sputum belonged to capsular types 1–6. Nearly all the biochemically aberrant respiratory klebsiellas are to be found in capsular types 1–6, though a number of strains in types 1–3 are biochemically typical strains of *K. pneumoniae* subsp. *aerogenes*. Capsular type 3 contains nearly all the subsp. *rhinoscleromatis* strains, as well as most of the strains of subsp. *pneumoniae*. The strains described by Cowan and his colleagues (1960) as *K. edwardsi* belong to types 1 and 2. Members of subsp. *ozaenae* belong to capsular types 4–6, and make up all, or nearly all, of these types.

There is considerable overlapping between the antigens of the genus *Klebsiella* and certain quite unrelated organisms. Capsular type 2, for example, is similar immunologically to the type 2 pneumococcus (Avery *et al.* 1925). Capsular antigens are usually detected by means of the capsular 'swelling' reaction, but agglutination, complement-fixation, indirect immunofluorescence (Riser *et al.* 1976) and counter-current immunoelectrophoresis (Palfreyman 1978) have also been employed.

Other typing methods

Many *Klebsiella* strains produce bacteriocines, which appear to be distinct from colicines because they have no action on *Esch. coli* (Stouthamer and Tieze 1966). Many of these so-called *pneumocines* have a narrow range of activitiy on other klebsiella strains, often mainly on members of the same biochemical variety. Bacteriocine typing can be carried out by a traditional cross-streak method (Hall 1971) or by means of liquid preparations of bacteriocines after induction with mitomycin C (Edmondson and Cooke 1979). *Phage-typing* by means of a set of temperate phages derived from other klebsiella strains has also been advocated (Ślopek *et al.* 1967). Humphries (1948) described an enzyme present in lysates of phage-infected organisms capable of dissolving the capsule of type 1 organism. It did not inhibit growth itself nor was it toxic to the cells.

Pathogenicity

In man, strains of *Klebsiella* occasionally give rise in members of the general population to cases of severe bronchopneumonia, and also to more chronic destructive lesions with multiple abscess formation in the lungs (Limson *et al.* 1956). Such lesions are usually but not always associated with members of capsular types 1–5, which are frequently biochemically atypical. However, the main importance of *Klebsiella* as a pathogen for man is in causing infections in hospital patients; the strains responsible are nearly always biochemically typical members of *K. pneumoniae* subsp. *aerogenes* most of which belong to higher-numbered capsular types. These strains cause widespread colonization of hospital patients. Clinical sepsis develops in surgical wounds and in the urinary tract; a number of the patients have bacteraemic infections and some of them die. Colonization of the respiratory tract is very common, but its clinical significance is often difficult to assess in patients with serious underlying diseases. Some of them develop bronchopneumonia in which the klebsiella appears to be the primary infecting agent. (See also Chapters 57, 58 and 61.) Evidence for the pathogenic relationship of the *ozaenae* and *rhino-scleromatis* subspecies of *K. pneumoniae* to the diseases from which they take their names is considered in Chapter 61. *K. pneumoniae* subsp. *aerogenes* is also responsible for a variety of septic diseases in animals (Chapter 61).

Experimentally the virulence of *K. pneumoniae* is subject to considerable variation. Apart from the fact that smooth variants tend to be pathogenic for laboratory animals and rough variants non-pathogenic, there is a great difference in the virulence of individual smooth strains. Some strains will kill mice in a dose of 0.2 ml of a 24-hour broth culture diluted one in a million times; others fail to kill even in a dose of 0.2 ml of the undiluted culture. Capsular types 1 and 2 are usually very virulent to mice when injected intraperitoneally; other types are generally non-virulent or of only low virulence to mice. Virulence does not appear to depend on capsule formation. The types just quoted as being of low virulence possess capsules in the same way as the highly virulent types 1 and 2. Moreover, non-capsulated highly virulent variants have been described (Toenniessen 1914).

After subcutaneous injection of a very small dose—about 0.000 000 1 ml of a 24-hr broth culture of a virulent strain—into mice, the animals die in 12 to 72 hr. *Post mortem*, there is a local exudate, the focal glands are swollen, and the spleen is enlarged. Capsulated rods are found in the blood and viscera.

Guinea-pigs are refractory to subcutaneous, but succumb to intraperitoneal injection, death occurring in 12 to 72 hr. The fatal dose is about 0.01 ml of a 24-hr broth culture. *Post mortem*, there is a viscous exudate in the peritoneum; the spleen may be enlarged, and the adrenals haemorrhagic. The rods are found in large numbers in the blood and viscera.

Rabbits appear to be more resistant, but they succumb after intravenous or intraperitoneal injection with a dose of about 0.1 ml of a broth culture. Intraperitoneal inoculation is likewise fatal to pigeons.

Enterobacter

Definition

Motile; less often and less heavily capsulated than *Klebsiella*. Ferment mannitol and form gas from some sugars, including cellobiose but not starch. Generally methyl-red negative, Voges-Proskauer positive, and grow on Simmons's citrate and in Møller's KCN medium, if tested at 30°. No H_2S production in triple sugar iron agar. Do not deaminate phenylalanine. Gluconate is generally oxidized and ornithine decarboxylase is produced. Non-pigmented or yellow-pigmented. Many liquefy gelatin. Habitat: soil and water, but occasionally found in the human bowel flora. Occasionally cause septic infection in man. G + C content of the DNA: 52–60 moles per cent.

Type species: *Enterobacter cloacae*.

The organism that Escherich (1885) described as *Bakterium lactis aerogenes* was non-motile (see p. 292); it was subsequently re-named *Bact. aerogenes*, and in 1900 was transferred to a separate genus *Aerobacter* by Beijerinck. The terms *Bacterium*, *Aerobacter* or *Klebsiella aerogenes* have been applied by successive generations of water bacteriologists to all Voges-Proskauer-positive coliform bacteria, whether motile or non-motile, except for those that liquefy gelatin rapidly. Later, some bacteriologists defined *Aerobacter aerogenes* as a motile organism and was so distinguished from *Klebsiella*. Motile, gelatin-liquefying *Aerobacter aerogenes*-like organisms were usually described by water bacteriologists as *Bact. cloacae*. When it was discovered that many of the motile organisms from water and faeces liquefied gelatin slowly they were transferred to *Cloaca*, taking the specific epithet *aerogenes* with them. Hormaeche and Munilla (1957), and Hormaeche and Edwards (1958), recognized as 'Cloaca A' and 'Cloaca B' the organisms now called *Ent. cloacae* and *Ent. aerogenes* respectively (Report 1963). The genus has been re-named successively *Aerobacter* (Hormaeche and Edwards 1958) and *Enterobacter* (Hormaeche and Edwards 1960). The latter name has now gained general acceptance (Report 1963), and the genus *Enterobacter* is considered to include all the motile *Aerobacter aerogenes*-like organisms, excepting those that form a red pigment and strains related to them, which are placed in the genus *Serratia*.

As well as *Ent. aerogenes* and *Ent. cloacae*, other species have been recently recognized in the genus. In 1976 Richard and colleagues described a new group of organisms that was most similar to *Ent. aerogenes*. Subsequently the name *Ent. gergoviae* was proposed for this group, strains of which are found in various clinical specimens and in the natural environment (Brenner *et al.* 1980). Strains previously regarded as yellow-pigmented non-sorbitol-fermenting variants of *Ent. cloacae* have been shown by DNA-DNA hybridization to warrant recognition as a separate species for which the name *Ent. sakazakii* has been proposed (Farmer *et al.* 1980). Two other recently described species will not be considered further because they have so far been isolated only from water; *Ent. amnigenus* (Izard *et al.* 1981*b*) and *Ent. intermedium* (Izard *et al.* 1980). The name *Ent. agglomerans* is commonly used in American literature for the organism we describe under the name *Erwinia herbicola*.

Morphological and cultural characters

Although motility is the main distinguishing character between *Enterobacter* and *Klebsiella*, a few strains that in all other respects resemble *Ent. cloacae* are non-motile. The colonies of *Enterobacter* strains may be somewhat mucoid, but the amount of extracellular material formed is usually not great.

Biochemical activities

In general, the fermentative activity of *Enterobacter* strains is more limited than that of typical strains of *Klebsiella*; the most significant restriction is of the ability to form gas from starch, and often also from inositol and glycerol; but *Enterobacter* strains do form gas from cellobiose and ferment rhamnose, characters which distinguish them from most species of *Serratia*. Some *Enterobacter* strains give a weak urease reaction in Christensen's medium, but in the modified medium advocated by Kauffmann (1954) *Klebsiella* strains are urease positive and *Enterobacter* strains usually negative. Although growth at 37° is usually good, many *Enterobacter* strains give atypical biochemical reactions unless tested at a lower temperature. The main cultural and biochemical characters of the genus were given in Table 33.2 and the characters of four of its species are given in Table 34.3.

Ent. aerogenes resembles *Klebsiella* in its wide range of fermentative activity and in its ability to form gas from inositol, glycerol and cellobiose, but it does not form gas from starch. It also differs from *Klebsiella* in being motile and in forming ornithine as well as lysine decarboxylase. Liquefaction of gelatin in 7–60 days distinguishes it from all of the subspecies of *K. pneumoniae* but not from *K. oxytoca*. Nevertheless, Bascomb and her colleagues (1971) believe that it should be transferred to a reconstituted *Klebsiella* genus as *Klebsiella mobilis* and this is supported by DNA–DNA hybridization data. This proposal has, however, not gained wide acceptance. *Ent. cloacae* differs from *Ent. aerogenes* in attacking arginine but not lysine and in its inability to form gas from inositol and glycerol. It sometimes fails to liquefy gelatin. *Ent. gergoviae* strains differ from *Ent. aerogenes* in producing urease, and in giving negative results in the following tests: KCN tolerance, and fermentation of adonitol, *i*-inositol, D-sorbitol and mucate. *Ent. sakazakii* strains differ from *Ent. cloacae* in being yellow-pigmented, producing extracellular deoxyribonuclease, and failing to ferment D-sorbitol.

Resistance to antimicrobial agents

Strains of *Enterobacter* are usually resistant to ampicillin and nearly always highly resistant to cephalosporins (Benner *et al.* 1965) by virtue of producing a β-lactamase with predominantly cephalosporinase activity (Jack and Richmond 1970). Some are highly resistant to carbenicillin and many are also resistant to tetracycline, chloramphenicol and streptomycin. Until quite recently most were sensitive to kanamycin and nearly all to gentamicin (Toala *et al.* 1970). *Enterobacter* strains differ from *Serratia* strains in being sensitive to the polymyxins (Greenup and Blazevic 1971).

Antigenic structure According to Sakazaki and Namioka (1960), there is no significant relation between the capsular antigens of *Klebsiella* and *Enterobacter*. In their opinion, the capsules of *Enterobacter* strains contain only M or slime antigen.

Habitat and pathogenicity The normal habitat of

Table 34.3 Differentiation of species of *Enterobacter*

Property	*Ent. aerogenes*	*Ent. cloacae*	*Ent. gergoviae*	*Ent. sakazakii*
Decarboxylase:				
arginine	–	+	–	+
lysine	+	–	V	–
Deoxyribonuclease*	–	–	–	+
Growth in KCN medium	+	+	–	+
Methyl-red test	–	–	–	V
Urease formation	–	V	+	–
Yellow pigmentation*	–	–	–	+
Adonitol: acid	+	V	–	–
i-Inositol: acid	+	V	–	V
D-sorbitol: acid	+	+	–	–

For symbols, see Table 34.2.
* Tested at 25°.

Enterobacter is believed to be soil and water, but the organism is occasionally found in the faeces and the respiratory tract of man. In recent years, infection of hospital patients with *Ent. cloacae* and *Ent. aerogenes* has been reported more often than formerly, but *Enterobacter* is a much less important cause of hospital infection than is *Klebsiella*. Most infections are of the urinary tract (Eikhoff *et al.* 1966).

Hafnia

Definition

Usually motile and usually produce gas from glucose. Acid formed from arabinose, glycerol, maltose, mannitol, rhamnose, trehalose and xylose. Indole is not produced. Most strains utilize citrate (if tested at 30°). Do not hydrolyse urea or produce H_2S detectable in triple sugar iron agar. Phenylalanine is not deaminated and gelatin is not liquefied. Most strains grow in Møller's KCN medium. Arginine dihydrolase is not produced but both lysine and ornithine are decarboxylated. Occurs in the faeces of man and other animals; also widespread in the natural environment. G+C content of the DNA: 52–57 moles per cent.

Type species: *Hafnia alvei.*

This organism has at times been placed in the genus *Enterobacter* under the names *Ent. alvei* and *Ent. hafniae*. However, DNA–DNA hybridization studies reveal that the organism should not be placed in *Enterobacter* and deserves separate generic status (Brenner 1978). Three DNA-related groups have been recognized in *H. alvei* and these may be elevated to separate species at some future date. At present *Hafnia* contains no other species, but it should be noted that the organism known as *Obesumbacterium proteus* (syn. *Hafnia protea*) biogroup 1 is a brewery-adapted biochemical variety of *H. alvei* (see also p. 305).

Morphological and cultural characters Strains grow well on ordinary media producing greyish convex colonies 1.0–2.0 mm in diameter on nutrient agar at 37° after 24 hours' incubation.

Biochemical activities *H. alvei* ferments a much narrower range of sugars than do members of the genus *Enterobacter*: it acidifies sucrose and salicin late or not at all and lactose and sorbitol only rarely, and does not attack dulcitol, adonitol or inositol. It is not capsulated and does not liquefy gelatin. Møller (1954) greatly advanced our knowledge of this species when he showed that, although strains were able to grow at 37°, many of their biochemical activities at this temperature were irregular. The effect of temperature on motility was also critical: many were non-motile at 37°, but nearly all were motile at 30°. At 22° they all gave a positive Voges-Proskauer reaction, grew on Simmons's citrate medium, and formed gas in sugars, though any one or all of these reactions might be inhibited at 37°. According to Guinée and Valkenburg (1968) all *H. alvei* strains are lysed by a single phage, which is without action on any other members of the Enterobacteriaceae.

Antigenic structure Forty-nine serotypes were recognized by Sakazaki (1961) and the total was extended to 197 by Matsumoto (1963, 1964).

Habitat and pathogenicity Isolated from the faeces of man and other animals; also found in sewage, soil, water and dairy products. Rarely if ever pathogenic.

Serratia

Definition

Small, motile, gram-negative rods. Ferment mannitol, sucrose and salicin with the production of acid and sometimes of a small bubble of gas. Indole negative, generally give a negative methyl-red and a positive Voges-Proskauer reaction, and grow on Simmons's citrate medium. Do not deaminate phenylalanine. All liquefy gelatin rapidly, usually within 2–3 days and all produce lecithinase, lipase and deoxyribonuclease. Some strains produce a red non-diffusible pigment. Typically found in soil and water, but some strains occur in the animal body. May cause septic infection in man. G+C content of the DNA: 53–59 moles per cent.

Type species: *Serratia marcescens.*

Serratia marcescens was first described by Bartolomeo Bizio in 1823 as a cause of 'bleeding polenta' (Breed and Breed 1924). Similar organisms have been isolated at various times from water, soil and sewage, from contaminated food-stuffs and, less frequently, from the animal body. It has in recent times become obvious that the pigmented strains form only a small part of a group of mainly non-pigmented strains (Davis *et al.* 1957), to which this name is now applied.

Numerous other species of *Serratia* have been described, but until about 10 years ago only *Ser. marcescens* was generally recognized. Since then, several others have been added. (1) *Ser. liquefaciens*; this species was formerly known as *Ent. liquefaciens*, and was transferred to *Serratia* as a result of a numerical-taxonomy study (Bascomb *et al.* 1971). (2) *Ser. marinorubra* (see Grimont *et al.* 1977); the name *Ser. rubidaea* was used for this organism by Ewing and his colleagues (1972*b*, 1973), but their description of it differs in several respects from that given originally

Table 34.4 Differentiation of species of *Serratia*

Property	*Ser. liquefaciens*	*Ser. marcescens*	*Ser. marinorubra*	*Ser. odorifera* biovar 1	*Ser. odorifera* biovar 2	*Ser. plymuthica*
Decarboxylase:						
lysine	+	+	V	+	+	−
ornithine	+	+	−	+	−	−
Gas from glucose	V	V	−	−	−	V
Red pigmentation	−	V	V	−	−	V
Acid from:						
adonitol	−	V	+	V	V	−
arabinose	+	−	+	+	+	+
lactose	V	−	+	+	+	V
raffinose	+	−	+	+	−	V
sorbitol	+	+	−	+	+	V
sucrose	+	+	+	+	−	+
xylose	+	V	+	+	+	+

For symbols, see Table 34.2.

for *Bact. rubidaeum*. We accept the reasons given by Grimont and co-workers (1977) for preferring the name *Ser. marinorubra* for this organism, but would point out that it frequently appears in the American literature as *Ser. rubidaea*. (3) *Ser. plymuthica* (see Grimont *et al.* 1977) somewhat resembles *Ser. liquefaciens* in biochemical characters. (4) *Ser. odorifera* (see Grimont *et al.* 1978*a*) comprises two distinct biochemical varieties, but DNA-DNA hybridization tests indicate that it forms a single species. The differential characters of these species are given in Table 34.4.

There are several other proposed species that we shall not describe in detail. *Ser. ficaria* is associated with figs and a species of fig wasp (Grimont *et al.* 1979) but it has recently been isolated from a human source (Gill *et al.* 1981). *Ser. proteamaculans*, another plant-associated species, was originally thought to be the same as *Ser. liquefaciens* (Grimont *et al.* 1978*b*) but more recent work suggests that it is a distinct species. *Ser. fonticola* (Gavini *et al.* 1979), a new species from water, is possibly misplaced in *Serratia*.

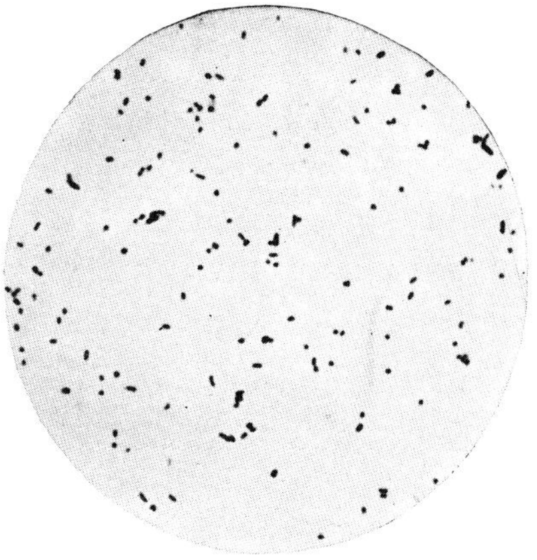

Fig. 34.6 *Serratia marcescens*, coccobacillary forms, from a nutrient agar culture, 24 hr, 37° (× 1000).

Morphological and cultural characters

Ser. marcescens is described in detail because it is the species most commonly encountered in clinical specimens. It is on the whole smaller than the average coliform bacterium. Its size is, however, subject to considerable variation; even on the same type of medium a single strain may at one time give rise to coccobacilli and at another to rods indistinguishable from other coliform bacteria (Figs. 34.6 and 34.7). Capsules are not ordinarily formed, but Bunting, Robinow and Bunting (1949) found that capsular material was formed on a well aerated medium poor in nitrogen and phosphate. Most *Serratia* strains are motile, and flagellation is peritrichate. The flagella are usually best seen in cultures grown at temperatures below 37° (Fulton *et al.* 1959).

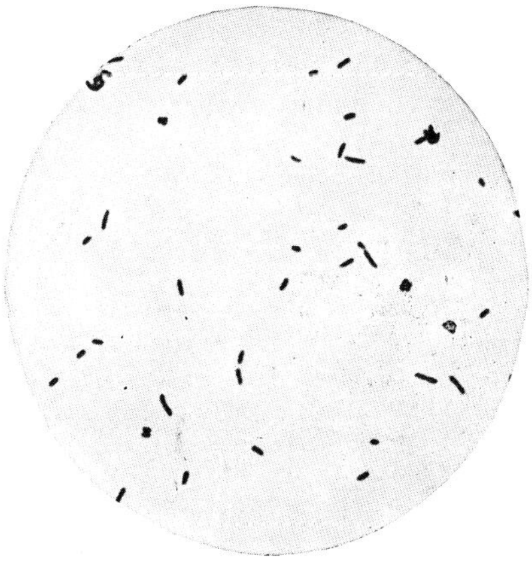

Fig. 34.7 *Serratia marcescens*, rod forms, from a nutrient agar culture, 24 hr, 37° (× 1000).

Colonies of *Ser. marcescens* on agar are usually

homogeneous for the first day or two, and then may become differentiated into a convex, pigmented and relatively opaque centre and an effuse, colourless, almost transparent periphery with an irregular crenated edge. However, the colonial characters of a strain are liable to considerable variation; colonies may vary in size, shape, opacity, surface and consistency. In their ability to produce the red pigment there is likewise great variation. Pigment is formed only in the presence of oxygen and at a suitable temperature. The optimum temperature for pigment formation is not necessarily the same as that for growth. Thus many strains grow best at 30–37° but form little or no pigment, whereas at lower temperatures growth is poorer and pigment formation is abundant. The ability to form pigment is commonly diminished and often irreversibly lost after repeated subculture. *Ser. liquefaciens* and *Ser. odorifera* are non-pigmented but red pigment is formed by many strains of *Ser. marinorubra* and *Ser. plymuthica*.

The red pigment, *prodigiosin*, is soluble in absolute alcohol, ether, chloroform, acetone, benzol and carbon disulphide, but is insoluble in water. Wrede and Rothhaas (1934) purified it and assigned to it a tripyrrylmethene structure. This was confirmed by Hubbard and Rimington (1949, 1950). Williams, Green and Rappoport (1956*a*, *b*; see also Green *et al.* 1956) consider that prodigiosin is not a single substance. They separated it into three red fractions and one blue fraction with different absorption spectra. Prodigiosin is also formed by certain organisms unrelated to *Serratia*, including an actinomycete (Perry 1961), and certain gram-negative rods from sea water that attack sugars oxidatively, are oxidase positive and have polar flagella (Lewis and Corpe 1964).

Except for the possibly misplaced species *Ser. fonticola*, all serratia strains are actively proteolytic. Liquefaction of gelatin is usually obvious within 2–3 days. On blood agar plates many produce a narrow zone of haemolysis in 24 or 48 hr. This does not appear to be due to the production of a soluble haemolysin. Most strains give opacity in egg-yolk media, owing to the formation of a lecithinase C (Monsour and Colmer 1952, Klinge 1957); they also produce deoxyribonuclease (Schreier 1969), except for *Ser. fonticola*.

The optimum temperature for growth varies; some strains grow as well at 37° as at 30°, many have an optimum in between 30 and 37°, and a few fail to grow at all at 37°. Many grow at 1–5°.

Biochemical activities

The biochemical reactions of the genus were given in Table 33.2, and those of the various species of *Serratia* are shown in Table 34.4. Acid is usually formed promptly in glucose, maltose, mannitol, sucrose, sorbitol, trehalose and glycerol. Inositol and cellobiose may be acidified in a few days. Lactose is not acidified, except by strains of *Ser. fonticola* and *Ser. marinorubra*, or is acidified late, but β-galactosidase is always formed. Dulcitol is fermented only by *Ser. fonticola* and rhamnose only by strains of *Ser. ficaria*, *Ser. fonticola* and *Ser. odorifera*. The strains are either completely anaerogenic, or form small bubbles of gas filling no more than 5 per cent of the Durham's tubes. Gas is not formed in inositol, glycerol, cellobiose or starch.

Resistance to antimicrobial agents

Serratia strains, like enterobacter strains, are nearly always highly resistant to cephalosporins. Resistance to ampicillin and gentamicin varies from strain to strain, but many strains destroy these antibiotics enzymically (see also Chapter 61). Serratia strains, unlike enterobacter strains, are resistant *in vitro* to polymyxins (Ramirez 1968).

Antigenic structure

Davis and Woodward (1957) studied the O antigens of 16 cultures of *Ser. marcescens* and identified six O groups. There was no apparent relation between serological grouping and pigment production. Ewing, Davis and Reavis (1959), who examined 115 *Serratia* cultures, established an antigenic scheme including nine O groups, 13 separate H antigens, and 46 serotypes. This scheme has since been extended to 15 O groups (Ewing *et al.* 1962). Seven additional H antigens have also been described by Traub and Kleber (1977) and LeMinor and Pigache (1977); these may conveniently be detected by motility-inhibition tests (Traub and Kleber 1977).

Other typing methods

Serratia strains produce two different groups of *bacteriocines* or 'marcescines' (Prinsloo 1966, Eichenlaub and Winkler 1974). Group A bacteriocines are resistant to heat, chloroform and proteolytic enzymes; they are active against other serratia strains and sometimes also against less closely related enterobacteria. Group B bacteriocines are sensitive to these agents; they are active against other enterobacteria but not against serratia strains. Bacteriocine-susceptibility tests (Traub and Raymond 1971) have been used as a means of typing *Ser. marcescens*. According to Traub (1980), 73 patterns of susceptibility can be detected; unfortunately, however, small differences in pattern may occur on replicate typing. (For an assessment of typing methods for *Serratia*, see Chapter 61.)

Habitat and pathogenicity

Ser. marcescens is widely distributed in nature. Pigmented strains have at times caused alarm by giving rise to red colours in various foods (e.g. Klein 1894), by simulating the appearance of blood in the sputum (Woodward and Clarke 1913), or by producing stains on babies' napkins (Waisman and Stone 1958). Pigmented and non-pigmented strains are found from time to time in the human respiratory tract and in faeces. Accounts of human disease due to *Ser. marcescens* have become increasingly frequent in recent years. Most of the infections occur in hospital patients; they include sepsis of the urinary and respiratory tract, of the meninges, and of wounds. Septicaemia is common, and endotoxic shock and endocarditis have been reported; some strains become endemically established in hospitals (see Wilfert *et al.* 1970 and Chapter 61). Only a small proportion—usually less than 10 per cent—of the strains responsible for infection are pigmented.

On *experimental inoculation* into laboratory animals, organisms of this group prove harmless except in very large doses. Culture filtrates of *Ser. marcescens* have been used at times in the treatment of cancer, but there is no reason to believe that the substance producing haemorrhage and necrosis in the tumour is other than the endotoxin (see Wharton and Creech 1949). The inhalation of aerosols of *Ser. marcescens* is said to cause an acute, short-lived illness with respiratory and general constitutional symptoms (Paine 1946).

The other *Serratia* species also occur commonly in the natural environment, especially in water. *Ser. liquefaciens* and *Ser. odorifera* occur quite often in human clinical specimens whilst strains of *Ser. marinorubra* and *Ser. plymuthica* may only occasionally do so.

Citrobacter

Definition

Motile. Ferment mannitol, usually with gas production. May or may not ferment lactose but nearly always produce β-galactosidase. Methyl-red test positive and Voges-Proskauer test negative. Grow in Simmons's citrate medium. May or may not hydrolyse urea. Do not decarboxylate lysine. Many strains produce a dihydrolase for arginine and most strains decarboxylate ornithine. Found in human clinical specimens, soil and water, and in the intestinal canal of animals. G+C content of the DNA: 52–54 moles per cent.

Type species: *Citrobacter freundi.*

Werkman and Gillen (1932) proposed the establishment of the genus *Citrobacter* to include certain 'intermediate' coliform bacteria (Chapter 33) which produced trimethylene glycol from glycerol. The name did not find immediate acceptance and these organisms were described as *Escherichia freundi* by Yale (1939). A number of lactose-negative or late lactose-fermenting organisms described some years later shared certain somatic antigens with salmonellae and many possessed the Vi antigen of *Salm. typhi.* These organisms were known as the Ballerup-Bethesda group until West and Edwards (1954) showed their similarity to strains of *Esch. freundi.* Kauffmann (1956) subsequently revived the name *Citrobacter* for them and the species became known as *Citro. freundi.*

An outstanding feature of *Citro. freundi* is its ability to blacken triple sugar iron agar owing to the production of H_2S. In 1971, Young and her colleagues described a group of organisms that were H_2S negative in this medium and placed them in a new genus, *Levinea*, in which there were two species, *L. malonatica* and *L. amalonatica.* Ewing and Davis (1972) suggested that *L. malonatica* was a later synonym of *Citro. diversum* (see Werkman and Gillen 1932) and proposed the revival of the name *Citro. diversus*, with the change of spelling for grammatical reasons. However, in 1970 Frederiksen had also described a group of organisms for which he proposed the name *Citro. koseri.* Subsequent studies (Crosa *et al.* 1974, Sakazaki *et al.* 1976) showed that *Citro. diversus* and *Citro. koseri* are synonyms, and agreement has yet to be reached on a single name for this species. We prefer the name *Citro. koseri* because the description of *Citro. diversus* given by Ewing and Davis (1972) differs in several respects from the original description of the species (Holmes *et al.* 1974). Brenner and co-workers (1977) re-examined strains of *L. amalonatica* and suggested that these might also be included in the genus *Citrobacter* as a third species, *Citro. amalonaticus.* We agree with the view that the three species *Citro. freundi*, *Citro. amalonaticus* and *Citro. koseri* (*Citro. diversus*) should be included in the genus.

Morphological and cultural characters In these respects strains of *Citrobacter* resemble most other members of the Enterobacteriaceae. They grow well on ordinary media producing smooth, convex colonies 2–4 mm in diameter on nutrient agar. They are not pigmented. Rough or mucoid forms sometimes occur.

Biochemical activities The chief biochemical reactions of the genus were shown in Table 33.2 and the distinguishing characters of the three species are set out in Table 34.5.

Resistance to antimicrobial agents *Citro. freundi* is usually sensitive to aminoglycosides and chloramphenicol and sensitivity to ampicillin, tetracycline and cephalosporins varies. Holmes and his colleagues (1974) showed that 79 per cent of *Citro. freundi* strains were resistant to cephaloridine and sensitive to carbenicillin, while 96 per cent of *Citro. koseri* strains were sensitive to cephaloridine and resistant to carbenicillin. Lund and his colleagues (1974) found that both *Citro. koseri* and *Citro. amalonaticus* were generally resistant to ampicillin and carbenicillin.

Typing methods Serotyping schemes have been proposed for *Citro. freundi* (West and Edwards 1954, Sedlák and Slajsová 1966), *Citro. koseri* (Gross and

Table 34.5 Differentiation of species of *Citrobacter*

Property	*Citro. amalonaticus*	*Citro. freundi*	*Citro. koseri*
H_2S production*	−	+	−
Indole production	+	−	+
Growth in KCN medium	+	+	−
Malonate fermentation	−	−	+
Ornithine decarboxylase	+	V	+
Adonitol: acid	−	−	+
Dulcitol: acid	−	V	V
Raffinose: acid	−	V	−
Sucrose: acid	−	V	V

For symbols, see Table 34.2.
* In triple sugar iron agar.

Rowe 1975, Gross *et al.* 1981), and *Citro. amalonaticus* (Sourek and Aldova 1976).

Habitat and pathogenicity Members of the genus *Citrobacter* are often found in the faeces of man and may be isolated from a variety of clinical specimens. They do not often give rise to serious infections, except *Citro. koseri*, which has been responsible for a number of outbreaks of neonatal meningitis (Gross *et al.* 1973, Gwynn and George 1973, Duhamel *et al.* 1975, Graham and Band 1981; see also Chapter 61).

Erwinia

Definition

Usually motile and usually fail to produce gas from glucose. Often form a yellow pigment. Acid produced from arabinose, maltose, mannitol, rhamnose, salicin, sucrose, trehalose and xylose. Indole is produced only by a few strains associated with plants. Do not hydrolyse urea or produce H_2S detectable in triple sugar iron agar. Phenylalanine is not deaminated but gelatin is liquefied by most strains on prolonged incubation. Few strains grow in Møller's KCN medium. There is no decarboxylase activity. Predominantly associated with plants and occasionally responsible for disease in man. G + C content of the DNA: 50–58 moles per cent.

Type species: *Erwinia amylovora.*

Bacteria with the biochemical characters of the Enterobacteriaceae that form yellow non-diffusible pigments are found often in water and soil (Thomas and Elson 1957), occasionally in milk and other foods, and from time to time in material from human sources. An anaerogenic, yellow-pigmented organism, which was described as *Bacterium typhiflavum* by Dresel and Stickl (1928), and as *Chromobacterium typhiflavum* in some of the earlier editions of this book, deserves special mention. This organism is not uncommon in the human upper respiratory tract and in swabs from superficial lesions (see Slotnick and Tulman 1967, Gilardi *et al.* 1970, Bottone and Schneierson 1972, Meyers *et al.* 1972); it has also been isolated from various animals (Muraschi *et al.* 1965, Lev *et al.* 1969). It resembles closely organisms that are frequently isolated from plants and soil, some of which are plant pathogens. These have been placed in the genus *Erwinia*, a diverse group, some members of which form gas from sugars and form pectinase. The 'typhiflavum' strains are indistinguishable from *Erw. herbicola*, which is anaerogenic and does not liquefy pectin (Graham and Hodgkiss 1967). Although *Erwinia herbicola* is the name accepted by many microbiologists the name *Enterobacter agglomerans* has been commonly used in the American literature. We do not accept the name *Ent. agglomerans* proposed by Ewing and Fife (1972) because the description of *Ent. agglomerans* differs from the original description of *Bacillus agglomerans* in several important respects; in particular, the latter was described as having polar flagella. Ewing and Fife (1972) recognized seven anaerogenic and four aerogenic biochemical varieties within the species. However, this classification showed little correlation with the results of DNA-DNA hybridization, by which ten or more DNA-relatedness groups can be identified within the species. Further study is necessary before definite proposals can be made about the classification and nomenclature of these organisms. At present we shall regard this heterogeneous group as a single species named *Erwinia herbicola*.

Morphological and cultural characters Strains are usually slender rods (1–3 μm × 0.5–0.7 μm), but occasionally form filaments up to 20 μm in length. The axis is straight, the sides are parallel and the ends rounded. They are motile, and the flagella, though sparse, can generally be shown to be peritrichate. In fluid media, sausage-shaped aggregations or *symplasmata* may occur; these appear to be chains of spherical bodies, 7–13 μm in diameter, consisting of organisms within a matrix (Lev *et al.* 1969).

Colonies on nutrient agar are 1–2 mm in diameter after 24 hr at 37°, and are round, convex, amorphous, smooth, glistening and opaque with an entire edge. The pigment is of an ochre or rusty yellow colour. On further incubation, granular structures corresponding to the symplasmata appear, as well as biconvex bodies with a clear cut margin. The latter probably represent downgrowths into the medium (Cruickshank 1935, Graham and Hodgkiss 1967). The organism grows poorly on MacConkey's agar and is not haemolytic on blood agar. In a gelatin stab, liquefaction usually begins in 6–10 days. Growth occurs between 20° and 37°, but the optimum is near 37°.

Biochemical activities The characteristics of the species were given in Table 33.2. It is catalase positive and oxidase negative, and usually but not invariably reduces nitrate to nitrite. Sugars are attacked fermentatively, without gas production. Acid is formed promptly from glucose and mannitol, and from maltose after 2–3 days; arabinose, rhamnose, salicin, sucrose, trehalose and xylose are also fermented, but not dulcitol or inositol. Only a few strains ferment lactose, though the organism produces β-galactosidase. When tested at 37°, the methyl-red reaction is positive and the Voges-Proskauer reaction negative, though the latter is often positive at 30°. Most strains are indole negative, but some plant strains are indole positive. Most strains grow on Simmons's citrate medium. Urease is not produced, and reactions in the malonate and gluconate tests vary from strain to strain. No strains whether from plant or animal sources decarboxylate arginine, lysine or ornithine. (See also Graham and Hodgkiss 1967, Bascomb *et al.* 1971, Ewing and Fife 1971, 1972).

Habitat and pathogenicity As well as being an occasional pathogen in predisposed human patients this organism has caused bacteraemia when administered in contaminated intravenous fluids (see Lapage *et al.* 1973).

Edwardsiella

Definition

Motile. Glucose is fermented, often with the production of some gas. Maltose and D-mannose are also fermented, but fewer other sugars than by members of many other taxa of Enterobacteriaceae. Lysine and ornithine are decarboxylated but arginine dihydrolase is not produced. Pathogenic for eels, catfish and other animals; occasional human infections have been reported. G+C content of the DNA:55–59 moles per cent.

Type species: *Edwardsiella tarda*.

In 1962, Sakazaki and Murata described under the name 'Asakusa' a group of enterobacteria that appeared to be quite distinct from all others (see Sakazaki 1965). The main habitat of this organism appears to be the gastro-intestinal canal of lower animals but a number of isolations have been made from human faeces, and a few cases of febrile diarrhoea and systemic infection in man have been reported (Sonnenworth and Kallus 1968, Okubadejo and Alausa 1968, Jordan and Hadley 1969, Bockemühl *et al.* 1971). According to Gilman and his colleagues (1971) the presence of this organism in the faeces of jungle-dwelling Malaysians is closely associated with *Entamoeba histolytica* infection; however, the organism does not necessarily play a role in the pathogenesis of amoebic dysentery and may have multiplied in the gut as a result of a change in the local micro-environment.

The organism has been named *Edwardsiella tarda* (Ewing *et al.* 1965). Sakazaki and Tamura (1975) proposed the new combination *Ed. anguillimortifera* for the organism previously known as *Paracolobactrum anguillimortiferum* and considered this an earlier synonym of *Ed. tarda*. This issue remains to be settled, and we prefer to use the name *Ed. tarda* because it has been in common use for so long and because the characters of the type strain of *Ed. anguillimortifera* differ from those described for *Paracolobactrum anguillimortiferum*. Grimont and co-workers (1980) described a distinct biochemical variety of *E. tarda* in which strains differed from those typical of the species in being able to ferment L-arabinose, mannitol and sucrose. Since these atypical strains are closely related to *Ed. tarda* by DNA-DNA hybridization they do not constitute a separate species and are likely to be referred to as '*Ed. tarda* biogroup 1'. Two other species have been described: *Ed. hoshinae* (Grimont *et al.* 1980) has been isolated only from animals and will not be described further; the name *Ed. ictaluri* (Hawke *et al.* 1981) has been proposed for strains earlier called *Edwardsiella* group GA7752 (Hawke 1979) that have caused several outbreaks of enteric septicaemia in catfish.

Cultural characters *Ed. tarda* grows well on ordinary media but produces only small colonies of 0.5–1 mm diameter after 24 hr.

Biochemical activities The chief biochemical reactions of *Ed. tarda* were shown in Table 33.2 (for further information, see Sakazaki 1967, Ewing *et al.* 1969, Bascomb *et al.* 1971).

Resistance to antimicrobial agents Strains of *Edwardsiella* are usually resistant to colistin but sensitive to most other antibiotics (including penicillin, which is unusual in the Enterobacteriaceae). Occasional *Ed. tarda* strains are resistant to sulphonamides.

Antigenic structure The H and O antigens of *Ed. tarda* appear to be unrelated to those of other enterobacteria (Sakazaki 1965). Two different serotyping schemes have been described for this species. Sakazaki

(1967) recognized 17 O antigens, 11 H antigens and 18 O-H combinations. Edwards and Ewing (1972) recognized 49 O antigens, 37 H antigens and 148 O-H combinations.

Other typing methods Some *Ed. tarda* strains produce bacteriocines, which are active against other *Ed. tarda* strains and against *Yersinia*, but not against *Proteus* (Hamon *et al.* 1969). *Ed. tarda* strains are not sensitive to colicines.

Habitat and pathogenicity *Edwardsiella* strains are frequently isolated from faeces or other specimens from healthy cold-blooded animals and their environment, particularly fresh water. They are pathogenic for eels, catfish and other animals, sometimes causing economic losses. *Ed. tarda* is an occasional pathogen of man and is quite often isolated from wounds (Jordan and Hadley 1969). *Ed. tarda* is rarely found in the faeces of healthy people. Although a higher isolation rate has been invariably found in patients with diarrhoea (Ewing *et al.* 1965), the role of the species in the aetiology of diarrhoea requires further study. The natural reservoir of *Edwardsiella* appears to be the gastro-intestinal tract of cold-blooded animals. Human infection with *Ed. tarda* probably results from contact with materials contaminated from this source.

Kluyvera

Definition

Motile. Usually form gas from glucose. Acid produced from arabinose, cellobiose, glycerol, lactose, maltose, mannitol, raffinose, rhamnose, salicin, sucrose, trehalose, and xylose. Indole is produced by most strains. Most strains grow in Simmons's citrate medium. Do not hydrolyse urea or form H_2S detectable in triple sugar iron agar. Phenylalanine is not deaminated and gelatin not liquefied. Most strains grow in Møller's KCN medium. Arginine dihydrolase is not produced; most strains decarboxylate lysine and all decarboxylate ornithine. Occur in human clinical specimens and the natural environment. G+C content of the DNA: 55–57 moles per cent.

Type species: *Kluyvera ascorbata*.

The name *Kluyvera* was originally proposed in 1956 by Japanese workers; they recognized two species: *Kluy. citrophila* in which citrate was utilized and *Kluy. noncitrophila* in which it was not. These names were little used in succeeding years and do not appear in the Approved Lists of Bacterial Names (Skerman *et al.* 1980). In 1981, Farmer and co-workers re-proposed the name *Kluyvera*; their studies of DNA-DNA hybridization led them to recognize three groups, two of which they named *Kluy. ascorbata* and *Kluy. cryocrescens*.

Cultural characters Strains grow well on ordinary media and form colonies 1–2 mm in diameter on nutrient agar after 24 hr at 37°.

Biochemical activities The characteristic biochemical reactions of the genus were given in Table 33.2. The tests used to distinguish *Kluy. ascorbata* from *Kluy. cryocrescens* are given in Table 34.6 (for test methods see Farmer *et al.* 1981), but most workers will attempt to identify strains only to the generic level. Most of the biochemical characters of the genus were given in the definition at the head of this section. Neither adonitol nor *i*-(*myo*)inositol is acidified. The methyl-red test is positive and the Voges-Proskauer test negative. Most strains ferment malonate.

Table 34.6 Differentiation of species of *Kluyvera*

Property	*Kluy. ascorbata*	*Kluy. cryocrescens*
Ascorbate test*	+	–
Lysine decarboxylase	+	V
Gas from glucose	+	V
D-Glucose†: acid	–	+
Dulcitol: acid	V	–

For symbols, see Table 34.2
*For method see Farmer *et al.* (1981).
† Incubated at 5° for 21 days.

Resistance to antimicrobial agents *Kluyvera* strains are generally susceptible to antimicrobial agents (see Farmer *et al.* 1981). It may be noted that strains of *Kluy. cryocrescens* often display large zones of inhibition around discs containing carbenicillin and cephalothin whereas strains of *Kluy. ascorbata* do not.

Habitat and pathogenicity Strains of *Kluyvera* have been isolated from water, sewage, soil and milk, but most of these so far identified have come from human sources, especially sputum and less commonly urine; a number were from faeces and a few from blood cultures. There is, however, as yet little definite evidence that the organisms are of clinical significance in man.

Cedecea

Definition

Usually motile and usually produce gas from glucose. Acid produced from cellobiose, maltose, D-mannitol, salicin and trehalose. Indole is not produced. Most strains utilize citrate. Do not hydrolyse urea or produce H_2S detectable in triple sugar iron agar. Phenylalanine is not deaminated and gelatin not liquefied. Most strains grow in Møller's KCN medium. Lysine is not decarboxylated. Isolated from the human respiratory tract but clinical significance uncertain. G+C content of the DNA:48–52 moles per cent.

Type species: *Cedecea davisae*.

Cedecea and its two named species have been described only recently but three more species may eventually be recognized (Grimont *et al.* 1981).

Cultural characters Members of the genus grow well on ordinary media producing convex colonies about 1.5 mm diameter on nutrient agar at 37°.

Biochemical activities The usual biochemical characters of *Cedecea* were summarized in Table 33.2. Properties not mentioned in the definition of the genus (above) include failure to form acid from adonitol, L-arabinose, dulcitol, *i*-inositol or L-rhamnose; most strains tolerate KCN, ferment malonate, and form arginine dihydrolase but not lysine decarboxylase. The differential characters of *Ced. davisae* and *Ced. lapagei* are given in Table 34.7.

Table 34.7 Differentiation of species of *Cedecea*

Property	*Ced. davisae*	*Ced. lapagei*
Methyl-red test	+	V
Voges-Proskauer test	–	+
Ornithine decarboxylase	+	–
Lactose: acid	–	+
Sucrose: acid	+	–
D-xylose: acid	+	–

For symbols, see Table 34.2.

Resistance to antimicrobial agents *Cedecea* strains are susceptible to carbenicillin, sulphonamides, trimethoprim, streptomycin, kanamycin, tobramycin, gentamicin, amikacin, chloramphenicol, tetracycline, minocycline, nalidixic acid and furantoin. Strains are resistant to ampicillin, cephalothin, colistin and polymyxin B.

Habitat and pathogenicity All *Cedecea* strains so far described have been isolated from human clinical specimens, predominantly from the respiratory tract. There is not yet any clear evidence that they play a pathogenic role.

Tatumella

Definition

Differ from other members of the Enterobacteriaceae in that, when grown at 25°, over one-half of the strains are motile by means of polar, subpolar or lateral flagella. Gas is not formed from glucose; acid is produced from glycerol, salicin, sucrose and trehalose. Indole is not formed. Most strains grow on Simmons's citrate medium at 25°. Do not hydrolyse urea or produce H_2S detectable in triple sugar iron agar. Phenylalanine is deaminated and gelatin is not liquefied. Growth does not occur in Møller's KCN medium. Lysine and ornithine are not decarboxylated but some strains produce a dihydrolase for arginine. Isolated predominantly from the human respiratory tract but clinical significance as yet uncertain. G+C content of the DNA:53–54 moles per cent.

Type species: *Tatumella ptyseos*.

This genus and its single species were described by Hollis and co-workers (1981). Studies of DNA-DNA hybridization show that, despite the absence of peritrichous flagellation, the species is correctly placed in the Enterobacteriaceae. Its members also differ from others in this family in growing less well and in surviving less readily on laboratory media.

Cultural characters Colonies are 0.5–1.0 mm in diameter, low convex with entire edges, semitranslucent, smooth and glossy after incubation for 24 hr on blood agar. In general, colonies are not as large as those of other species of Enterobacteriaceae, but growth occurs on MacConkey's agar.

Biochemical activities The characteristic biochemical reactions of the species were given in Table 33.2. In addition to the properties mentioned in the definition (above), strains fail to form acid from adonitol, L-arabinose, dulcitol, *i*-inositol, lactose, maltose, D-mannitol, raffinose, L-rhamnose, D-sorbitol, and starch; the methyl-red and Voges-Proskauer reactions are negative, and malonate is not fermented. A few strains form arginine dihydrolase, particularly at 25°. It should be noted that strains of *Tatumella*, unlike members of all the other genera described in this chapter, deaminate phenylalanine in a suitable medium.

Resistance to antimicrobial agents Unlike most of the other members of the Enterobacteriaceae, strains of *Tatumella* are sensitive to penicillins and cephalosporins, and to chloramphenicol, tetracycline and aminoglycoside antibiotics.

Habitat and pathogenicity The organism is commonly associated with the human respiratory tract; most isolations of it have been from sputum, but it has also been found in faeces and urine and in blood cultures. Its pathogenicity is at present uncertain.

Other genera

The following genera have been proposed, but will not be discussed in detail: *Obesumbacterium* (see Shimwell 1963, Priest *et al.* 1973) from breweries; *Rahnella* (see Izard *et al.* 1979) from water; and *Xenorhabdus* (see Thomas and Poinar 1979) associated with certain nematodes and with insect larvae parasitized by them.

References

Abel, R. (1896) *Z. Hyg. InfektKr.* **21,** 89.

Aspinall, G. O., Jamieson, R. S. P. and Wilkinson, J. F. (1956) *J. chem. Soc.* 3483.

Avery, O. T., Heidelberger, M. and Goebel, W. F. (1925) *J. exp. Med.* **42,** 709.

Bamforth, J. and Dudgeon, J. A. (1952) *J. Path. Bact.* **64,** 751.

Bascomb, S., Lapage, S. P., Willcox, W. R. and Curtis, M. A. (1971) *J. gen. Microbiol.* **66,** 279.

Beijerinck, M. W. (1900) *Zbl. Bakt.*, IIte Abt. **6,** 193.

Benner, E. J., Micklewait, J. S., Brodie, J. L. and Kirby, W. M. M. (1965) *Proc. Soc. exp. Biol., N.Y.* **119,** 536.

Bockemühl, J., Pan-Urai, R. and Burkhardt, F. (1971) *Path. et Microbiol., Basel* **37,** 393.

Bottone, E. and Schneierson, S. S. (1972) *Amer. J. clin. Path.* **57,** 400.

Braun, O. H., Lüderitz, O., Schäfer, E. and Westphal, O. (1956) *Z. Hyg. InfektKr.* **142,** 552.

Breed, R. S. and Breed, M. E. (1924) *J. Bact.* **9,** 545.

Brenner, D. J. (1978) *Progr. clin. Path.* **7,** 71.

Brenner, D. J., Fanning, G. R., Miklos, G. V. and Steigerwalt, A. G. (1973) *Int. J. syst. Bact.* **23,** 1.

Brenner, D. J., Farmer, J. J., Hickman, F. W., Asbury, M. A. and Steigerwalt, A. G. (1977) *Taxonomic and Nomenclature Changes in Enterobacteriaceae.* Publication No. 79-8356, Center for Disease Control, Atlanta.

Brenner, D. J., Richard, C., Steigerwalt, A. G., Asbury, M. A. and Mandel, M. (1980) *Int. J. syst. Bact.* **30,** 1.

Brenner, D. J., Steigerwalt, A. G. and Fanning, G. R. (1972) *Int. J. syst. Bact.* **22,** 193.

Brenner, D. J. *et al.* (1982) *J. clin. Microbiol.* **15,** 703.

Brooke, M. S. (1951) *Acta path. microbiol. scand.*, **28,** 313; (1953) *J. Bact.* **66,** 721.

Brumfitt, W., Faiers, M. C., Reeves, D. S. and Datta, N. (1971) *Lancet* **i,** 315.

Bunting, M. I., Robinow, C. F. and Bunting, H. (1949) *J. Bact.* **58,** 114.

Burgess, N. R. H., McDermott, S. N. and Whiting, J. (1973) *J. Hyg., Camb.* **71,** 1.

Buttiaux, R., Nicolle, P., Le Minor, L., Le Minor, S. and Gaudier, B. (1956) *Ann. Inst. Pasteur* **91,** 799.

Castellani, A. and Chalmers, A. J. (1920) *Ann. Inst. Pasteur* **34,** 600.

Clugston, R. E. and Nielsen, N. O. (1974) *Canad. J. comp. Med.* **38,** 22.

Clugston, R. E., Nielsen, N. O. and Smith, D. L. T. (1974) *Canad. J. comp. Med.* **38,** 34.

Collins, G. (1924) See Hadley, P. (1927) *J. infect. Dis.* **40,** 1.

Cowan, S. T. (1974) *Cowan and Steel's Manual for the Identification of Medical Bacteria*, 2nd edn. Cambridge University Press, London.

Cowan, S. T., Steel, K. J., Shaw, C. and Duguid, J. P. (1960) *J. gen. Microbiol* **23,** 601.

Crosa, J. H., Steigerwalt, A. G., Fanning, G. R. and Brenner, D. J. (1974) *J. gen. Microbiol.* **83,** 271.

Cruickshank, J. C. (1935) *J. Hyg., Camb.* **35,** 354.

Datta, N. (1969) *Brit. med. J.* **ii,** 407.

Datta, N., Faiers, F., Reeves, D. S., Brumfitt, W., Ørskov, F. and Ørskov, I. (1971) *Lancet* **i,** 312.

Davis, B. R., Ewing, W. H. and Reavis, R. W. (1957) *Int. Bull. bact. Nomencl.* **7,** 151.

Davis, B. R. and Woodward, J. M. (1957) *Canad. J. Microbiol.* **3,** 591.

Dresel, E. G. and Stickl, O. (1928) *Dtsch. med Wschr.* **54,** 517.

Dubay, L. (1968) *Arch. Immunol. Ther. exp.* **16,** 486.

Dudman, W. F. and Wilkinson, J. F. (1956) *Biochem. J.* **62,** 289.

Duguid, J. P. (1951) *J. Path. Bact.* **63,** 673; (1959) *J. gen. Microbiol.* **21,** 271; (1968) *Arch. Immunol. Ther. exp.* **16,** 173.

Duguid, J. P. and Wilkinson, J. F. (1953) *J. gen. Microbiol.* **9,** 174.

Duhamel, M., Cuvelier, A., Cousin, J. and Fournier, A. (1975) *Nouv. Pr. méd.* **4,** 428.

Dulaney, A. D. and Michelson, I. D. (1935) *Amer. J. publ. Hlth* **25,** 1241.

Durlakowa, I., Lachowicz, Z. and Ślopek, S. (1967) *Arch. Immunol. Ther. exp.* **15,** 497.

Edmondson, A. S. and Cooke, E. M. (1979) *J. Hyg., Camb.* **82,** 207.

Edmunds, P. N. (1954) *J. infect. Dis.* **94,** 65.

Edwards, P. R. and Ewing, W. H. (1972) *Identification of Enterobacteriaceae.* 3rd edn. Burgess, Minneapolis.

Edwards, P. R. and Fife, M. A. (1952) *J. infect. Dis.* **91,** 92; (1955) *J. Bact.* **70,** 382.

Eichenlaub, R. and Winkler, U. (1974) *J. gen. Microbiol.* **83,** 83.

Eikhoff, T. C., Steinhauer, B. W. and Finland, M. (1966) *Arch. intern. Med.* **65,** 1163.

Elek, S. D. and Higney, L. (1970) *J. med. Microbiol.* **3,** 103.

Epstein, S. S. (1959) *J. clin. Path.* **12,** 52.

Escherich, T. (1885) *Fortschr. Med.* **3,** 515, 547.

Evans, D. G. and Evans, D. J. (1978) *Infect. Immun.* **21,** 638.

Evans, D. G., Silver, R. P., Evans, D. J., Chase, D. G. and Gorbach, S. L. (1975) *Infect. Immun.* **12,** 656.

Ewing, W. H. and Davis, B. R. (1972) *Int. J. syst. Bact.* **22,** 12.

Ewing, W. H., Davis, B. R. and Fife, M. A. (1972*b*). *Biochemical Characterization of Serratia liquefaciens and Serratia rubidaea.* U.S. Dept. of Health, Education and Welfare, Atlanta.

Ewing, W. H., Davis, B. R., Fife, M. A. and Lessel, E. F. (1973) *Int. J. syst. Bact.* **23,** 217.

Ewing, W. H., Davis, B. R. and Johnson, J. G. (1962) *Int. Bull. bact. Nomencl.*, **12,** 47.

Ewing, W. H., Davis, B. R. and Martin, W. J. (1972*a*) *Biochemical Characterization Escherichia coli.* U.S. Dept. of Health, Education and Welfare, Atlanta.

Ewing, W. H., Davis, B. R. and Reavis, R. W. (1959) *Studies on the Serratia Group.* U.S. Dept. of Health, Education and Welfare, Atlanta.

Ewing, W. H. and Fife, M. A. (1971) *Enterobacter agglomerans. The Herbicola-Lathyri bacteria.* U.S. Dept. Hlth., Educ. and Welfare, Atlanta; (1972) *Int. J. syst. Bact.* **22,** 4.

Ewing, W. H., McWhorter, A. C. and Bartes, S. F. (1969) *Publ. Hlth. Lab.* **27,** 129.

Ewing, W. H., McWhorter, A. C., Escobar, M. R. and Lubin, A. H. (1965) *Int Bull. bact. Nomencl.* **15,** 33.

Ewing, W. H., Tanner, K. E. and Tatum, H. W. (1956) *Publ. Hlth Lab.* **14,** 106.

Ewing, W. H., Taylor, M. W. and Hucks, M. C. (1950) *Publ. Hlth Rep., Wash.* **65,** 1474.

Farmer, J. J. III, Asbury, M. A., Hickman, F. W., Brenner,

D. J. and The Enterobacteriaceae Study Group (1980) *Int. J. syst. Bact.* **30,** 569.
Farmer, J. J. III, Fanning, G. R., Huntley-Carter, G. P., Holmes, B., Hickman, F. W., Richard, C. and Brenner, D. J. (1981) *J. clin. Microbiol.* **13,** 919.
Fife, M. A., Ewing, W. H. and Davis, B. R. (1965) *The Biochemical Reactions of the Tribe Klebsielleae.* U.S. Dept. of Health, Education and Welfare, Atlanta.
Flügge, C. (1886) *Die Mikroorganismen.* 1st edn. Leipzig.
Frantzen, E. (1950) *Acta path. microbiol. scand.* **27,** 236; (1951) *Ibid.,* **28,** 103.
Frederiksen, W. (1970) *Spišy přir. Fak. Univ. Brne* **47,** 89.
Fredericq, P., Betz-Bareau, M. and Nicolle, P. (1956) *C. R. Soc. Biol.* **150,** 2039.
Friedländer, C. (1883) *Fortschr. Med.* **1,** 715.
Frisch, A. von. (1882) *Wien. med. Wschr.* **32,** 969.
Fulton, M., Forney, C. E. and Leifson, E. (1959) *Canad. J. Microbiol.* **5,** 269.
Gavini, F. *et al.* (1979) *Int. J. syst. Bact.* **29,** 92.
Gemski, P., Cross, A. S. and Sadoff, J. C. (1980) *FEMS Microbiol. Lett.* **9,** 193.
Gilardi, G. L., Bottone, E. and Birnbaum, M. (1970) *Appl. Microbiol.* **20,** 151.
Gill, V. J., Farmer, J. J. III, Grimont, P. A. D., Asbury, M. A. and McIntosh, C. L. (1981) *J. clin. Microbiol.* **14,** 234.
Gilman, R. H., Madasamy, M., Gan, E., Mariappan, M., Davis, C. E. and Kyser, K. A. (1971) *Sth E. Asian J. trop. Med. publ. Hlth* **2,** 186.
Glynn, A. A., Brumfitt, W. and Howard, C. J. (1971) *Lancet* **i,** 514.
Glynn, A. A. and Howard, C. J. (1970) *Immunology* **18,** 331.
Goebel, W. F. (1963) *Proc. nat. Acad. Sci., Wash.* **49,** 464.
Gormus, R. J. and Wheat, R. W. (1971) *J. Bact.* **108,** 1304.
Goslings, W. R. O. (1933) *Onderzoekingen over de Bacteriologie en de Epidemiologie van het Scleroma Respiratorium.* A. van Straelen, Amsterdam; (1935) *Zbl. Bakt.* **134,** 195.
Goslings, W. R. O. and Snijders, E. P. (1936) *Zbl. Bakt.* **136,** 1.
Graham, D. C. and Hodgkiss, W. (1967) *J. appl. Bact.* **30,** 175.
Graham, D. R. and Band, J. D. (1981) *J. Amer. med. Ass.* **245,** 1923.
Green, J. A., Rappoport, D. A. and Williams, R. P. (1956) *J. Bact.* **72,** 483.
Greenup, P. and Blazevic, D. J. (1971) *Appl. Microbiol.* **22,** 309.
Grimont, P. A. D., Grimont, F., Dulong De Rosnay, H. L. C. and Sneath, P. H. A. (1977) *J. gen. Microbiol.* **98,** 39.
Grimont, P. A. D., Grimont, F., Farmer, J. J. III and Asbury, M. A. (1981) *Int. J. syst. Bact.* **31,** 317.
Grimont, P. A. D., Grimont, F., Richard, C., Davis, B. R., Steigerwalt, A. G. and Brenner, D. J. (1978*a*) *Int. J. syst. Bact.* **28,** 453.
Grimont, P. A. D., Grimont, F., Richard, C. and Sakazaki, R. (1980) *Curr. Microbiol.* 4, 347.
Grimont, P. A. D., Grimont, F. and Starr, M. P. (1978*b*) *Int. J. syst. Bact.* **28,** 503; (1979) *Curr. Microbiol.* **2,** 277.
Gross, R. J. and Rowe, B. (1975) *J. Hyg., Camb.* **75,** 121.
Gross, R. J., Rowe, B. and Easton, J. A. (1973) *J. clin. Path.* **26,** 138.
Gross, R. J., Rowe, B., Sechter, I., Cahan, D. and Altman, G. (1981) *J. Hyg., Camb.* **86,** 111.
Guinée, P. A. M. and Valkenburg, J. J. (1968) *J. Bact.* **96,** 564.
Gwynn, C. M. and George, R. H. (1973) *Arch. Dis. Childh.* **48,** 455.
Hadley, P. (1927) *J. infect. Dis.* **40,** 1.
Hall, F. A. (1971) *J. clin. Path.* **24,** 712.
Hamon, Y. (1958) *Ann. Inst. Pasteur* **95,** 117; (1959) *Ibid.* **96,** 614.
Hamon, Y., Kayser, A., Le Minor, L. and Maresz, J. (1969) *C. R. Acad. Sci., Paris* **268,** 2517.
Hawke, J. P. (1979) *J. Fish. res. Board Canad.* **36,** 1508.
Hawke, J. P., McWhorter, A. C., Steigerwalt, A. G. and Brenner, D. J. (1981) *Int. J. syst. Bact.* **31,** 396.
Heidelberger, M., Goebel, W. F. and Avery, O. T. (1925) *J. exp. Med.* **42,** 701.
Henriksen, S. D. (1954*a*) *Acta. path. microbiol. scand.* **34,** 259; (1954*b*) *Ibid.* **34,** 266; (1954*c*) *Ibid.,* **34,** 271.
Hollis, D. G., Hickman, F. W., Fanning, G. R., Farmer, J. J. III, Weaver, R. E. and Brenner, D. J. (1981) *J. clin. Microbiol.* **14,** 79.
Holmes, B., King, A., Phillips, I., and Lapage, S. P. (1974) *J. clin. Path.* **27,** 729.
Hormaeche, E. and Edwards, P. R. (1958) *Int. Bull. bact. Nomencl.* **8,** 111; (1960) *Ibid.* **10,** 71.
Hormaeche, E. and Munilla, M. (1957) *Int. Bull. bact. Nomencl.* **7,** 1.
Howard, C. J. and Glynn, A. A. (1971*a*) *Immunology,* **20,** 767; (1971*b*) *Infect. Immun.* **4,** 6.
Hubbard, R. and Rimington, C. (1949) *Biochem. J.* **44,** 1; (1950) *Ibid.* **46,** 220.
Humphries, J. C. (1948) *J. Bact.* **56,** 68.
Izard, D., Ferragut, C., Gavini, F., Kersters, K., De Ley, J. and Leclerc, H. (1981*a*) *Int. J. syst. Bact.* **31,** 116.
Izard, D., Gavini, F. and Leclerc, H. (1980) *Zbl. Bakt.,* I Abt. Orig., **C1,** 51.
Izard, D., Gavini, F., Trinel, P. A. and Leclerc, H. (1979) *Ann. Microbiol., Inst. Pasteur* **A130,** 163. (1981*b*) *Int. J. syst. Bact.* **31,** 35.
Jack, G. W. and Richmond, M. H. (1970) *J. gen. Microbiol.* **61,** 43.
Jain, K., Radsak, K. and Mannheim, W. (1974) *Int. J. syst. Bact.* **24,** 402.
Jann, B., Jann, K., Schmidt, G., Ørskov, I. and Ørskov, F. (1970) *Europ. J. Biochem.* **15,** 29.
Jordan, G. W. and Hadley, W. K. (1969) *Ann. intern. Med.* **70,** 283.
Julianelle, L. A. (1926*a*) *J. exp. Med.* **44,** 113; (1926*b*) *Ibid.* **44,** 683; (1926*c*) *Ibid.* **44,** 735.
Kauffmann, F. (1944) *Acta path. microbiol. scand.* **21,** 20; (1947) *J. Immunol.* **57,** 71; (1948) *Acta path. microbiol. scand.* **25,** 502; (1949) *Ibid.* **26,** 381; (1954) *Enterobacteriacae.,* 2nd edn. Einar Munksgaard, Copenhagen; (1956) *Zbl. Bakt.,* I Abt. Orig., **A165,** 344.
Kauffmann, F., Ørskov, F. and Ewing, W. H. (1956) *Int. Bull. bact Nomencl.* **6,** 63.
Klein, E. (1894) *J. Path. Bact.* **2,** 217.
Klinge, K. (1957) *Arch. Hyg., Berl.,* **141,** 334.
Knipschildt, H. E. (1945) *Undersøgelser over Coligruppens Serologi med saerligt Henblik paa Kapselformerne.* Arnold Busck, Copenhagen.
Lapage, S. P., Johnson, R. and Holmes, B. (1973) *Lancet* **ii,** 284.

Lautrop, H. (1956) *Acta path. microbiol. scand.* **39,** 375.
Lautrop, H., Ørskov, I. and Gaarslev, K. (1971) *Acta path. microbiol. scand,* **B79,** 641.
Leclerc, H. (1962) *Ann. Inst. Pasteur* **102,** 726.
Le Minor, S. and Pigache, F. (1977) *Ann. Microbiol., Inst. Pasteur* **B128,** 207.
Le Minor, S., Le Minor, L., Nicolle, P. and Buttiaux, R. (1954) *Ann. Inst. Pasteur* **86,** 204.
Lev, M., Alexander, R. H. and Sobel, H. J. (1969) *J. appl. Bact.* **32,** 429.
Lewis, M. J. (1968) *Lancet* **i,** 1389.
Lewis, S. M. and Corpe, W. A. (1964) *Appl. Microbiol* **12,** 13.
Limson, B. M., Romansky, M. J. and Shea, J. G. (1956) *Ann. intern. Med.* **44,** 1070.
Linton, A. H. (1977) *Vet. Rec.* **100,** 354.
Loewenberg. (1894) *Ann. Inst. Pasteur* **8,** 292.
Lovell, R. (1937) *J. Path. Bact.* **44,** 125.
Lovell, R. and Rees, T. A. (1960) *Nature, Lond.* **188,** 755.
Lund, M. E., Matsen, J. M. and Blazevic, D. J. (1974) *Appl. Microbiol.* **28,** 22.
Mackie, T. J. (1913) *J. Path. Bact.* **18,** 137.
Malcolm, J. F. (1938) *J. Hyg., Camb.* **38,** 395.
Massini, R. (1907) *Arch. Hyg., Berl.* **61,** 250.
Matsumoto, H. (1963) *Jap. J. Microbiol.,* **7,** 105; (1964) *Ibid.* **8,** 139.
Medearis, D. N., Camitta, B. M. and Heath, E. C. (1968) *J. exp. Med.* **128,** 399.
Medearis, D. N. Jr and Kenny, J. F. (1968) *J. Immunol.* **101,** 534.
Meyers, B. R., Bottone, E., Hirschman, S. Z. and Schneierson, S. S. (1972) *Ann. intern. Med.* **76,** 9.
Milch, H. (1978) In: *Methods in Microbiology,* vol. 11, p. 87. Ed. by T. Bergan and J. R. Norris. Academic Press, London.
Møller, V. (1954) *Acta path. microbiol. scand.* **35,** 259.
Monsour, V. and Colmer, A. R. (1952) *J. Bact.* **63,** 597.
Moorhouse, E. C. (1969) *Brit. med. J.* **ii,** 405.
Moorhouse, E. C. and McKay, L. (1968) *Brit. med. J.* **ii,** 741.
Muraschi, T. E., Friend, M. and Bolles, D. (1965) *Appl. Microbiol.* **13,** 128.
Nicolle, P., Le Minor, L., Buttiaux, R. and Ducrest, P. (1952) *Bull. Acad. nat. Méd.* **136,** 480.
Okubadejo, O. A. and Alausa, K. O. (1968) *Brit. med. J.* **iii,** 357.
Ørskov, F. (1956) *Acta path. microbiol. scand.* **39,** 147.
Ørskov, F. and Ørskov, I. (1970) *Acta path. microbiol. scand.* **B78,** 593; (1972) *Ibid.* **B80,** 905.
Ørskov, F., Ørskov, I., Jann, B. and Jann, K. (1971) *Acta path. microbiol. scand.* **B79,** 142.
Ørskov, I. (1954) *Acta path. microbiol. scand.* **34,** 145; (1955*a*) *Ibid.* **36,** 449; (1955*b*) *Ibid.* **36,** 454; (1955*c*) *Ibid.* **37,** 353.
Ørskov, I. and Ørskov, F. (1973) *J. gen. Microbiol.* **77,** 487; (1977) *Med. Microbiol. Immunol.* **163,** 99.
Ørskov, I., Ørskov, F., Jann, B. and Jann, K. (1963) *Nature, Lond.* **200,** 144; (1977) *Bact. Rev.* **41,** 667.
Ørskov, I., Ørskov, F., Smith, H. W. and Sojka, W. J. (1975) *Acta path. microbiol. scand.* **B83,** 31.
Ørskov, I., Ørskov, F., Sojka, W. J. and Leach, J. M. (1961) *Acta path. microbiol. scand.* **53,** 404.
Paine, T. F. (1946) *J. infect. Dis.* **79,** 226.
Palfreyman, J. M. (1978) *J. Hyg., Camb.* **81,** 219.
Perry, J. J. (1961) *Nature, Lond.* **191,** 77.
Priest, F. G., Somerville, H. J., Cole, J. A. and Hough, J. S. (1973) *J. gen. Microbiol.* **75,** 295.
Prinsloo, H. E. (1966) *J. gen. Microbiol.* **45,** 205.
Quackenbush, R. L. and Falkow, S. (1979) *Infect. Immun.* **24,** 562.
Quinchon, C., Henry, M. and Henry, G. (1959) *Ann. Inst. Pasteur* **96,** 765.
Ramirez, M. L. (1968) *Appl. Microbiol.* **16,** 1548.
Read, B. E., Keller, R. and Cabelli, V. J. (1957) *J. Bact.* **73,** 765.
Rees, T. A. (1957) *Brit. vet. J.* **113,** 171.
Report. (1954) *Int. Bull. bact. Nomencl.* **4,** 47; (1963) *Ibid.* **13,** 69.
Richard, C., Joly, B., Sirot, J., Stoleru, G. H. and Popoff, M. (1976) *Ann. Microbiol., Inst. Pasteur* **A127,** 545.
Riser, E., Noone, P. and Poulton, T. A. (1976) *J. clin. Path.* **29,** 296.
Rowe, B. (1979) In: *Clinics in Gastroenterology,* vol. 8, p. 625. Ed. by H. P. Lambert. W. B. Saunders, London.
Sakazaki, R. (1961) *Jap. J. med. Sci. Biol.* **14,** 223; (1965) *Int. Bull. bact. Nomencl.* **15,** 45; (1967) *Jap. J. med. Sci. Biol.* **20,** 205.
Sakazaki, R. and Murata, Y. (1962) *Jap. J. Bact.* **17,** 616. Cited in Sakazaki R. (1965) *Int. Bull. bact. Nomencl.* **15,** 45.
Sakazaki, R. and Namioka, S. (1960) *Jap. J. med. Sci. Biol.* **13,** 1.
Sakazaki, R. and Tamura, K. (1975) *Int. J. syst. Bact.* **25,** 219.
Sakazaki, R., Tamura, K., Johnson, R. and Colwell, R. R. (1976) *Int. J. syst. Bact.* **26,** 158.
Sakazaki, R., Tamura, K., Nakamura, A., Kurata, T., Gohda, A. and Takeuchi, S. (1974) *Jap. J. med. Sci. Biol.* **27,** 19.
Sato, G., Asagi, M., Oka, C., Ishiguro, N. and Nobuyuki, N. (1978) *Microbiol. Immunol.* **22,** 357.
Schiffer, M. S., Oliveira, E., Glode, M. P., McCracken, G. H., Sarff, L. M. and Robbins, J. B. (1976) *Pediat. Res.* **10,** 82.
Schreier, J. B. (1969) *Amer. J. clin. Path.* **51,** 711.
Sedlák, J. and Slajsová, M. (1966) *Zbl. Bakt.* **200,** 369.
Shannon, R. (1957) *J. med. Lab. Technol.* **14,** 199.
Shimwell, J. L. (1963) *Brewers' J.* **99,** 759.
Shmilovitz, M., Kretzer, O. and Levy, E. (1974) *Israel. J. med. Sci.* **10,** 1425.
Silva, R. M., Toledo, M. R. F. and Trabulsi, L. R. (1980) *J. clin. Microbiol.* **11,** 441.
Skerman, V. B. D., McGowan, V. and Sneath, P. H. A. (1980) *Int. J. syst. Bact.* **30,** 225.
Ślopek, S. and Durlakowa, I. (1967) *Arch. Immunol. Ther. exp.* **15,** 481.
Ślopek, S., Przondo-Hessek, A., Milch, H. and Deák, S. (1967) *Arch. Immunol. Ther. exp.* **15,** 589.
Slotnick, I. J. and Tulman, L. (1967) *Amer. J. Med.* **43,** 147.
Smith, D. E. (1927) *J. exp. Med.* **46,** 155.
Smith, D. H. and Armour, S. E. (1966) *Lancet* **ii,** 15.
Smith, H. W. (1963) *J. Path. Bact.* **85,** 197; (1978) *J. Amer. vet. med. Ass.* **173,** 601.
Smith, H. W. and Crabb, W. E. (1956) *J. gen. Microbiol.* **15,** 556.
Smith, H. W. and Halls, S. (1967) *J. gen. Microbiol.* **47,** 153; (1968) *J. med. Microbiol.* **1,** 61.
Smith, H. W. and Linggood, M. A. (1971) *J. med. Microbiol.* **4,** 467; (1972) *J. med. Microbiol.* **5,** 243.
Smith, T. (1928) *J. exp. Med.* **48,** 351.

Sonnenworth, A. C. and Kallus, B. A. (1968) *Amer. J. clin. Path.* **49,** 92.
Sourek, J. and Aldova, E. (1976) *Zbl. Bakt.* Abt. Orig. **A234,** 480.
Stenzel, W. (1978) *Int. J. syst. Bact.* **28,** 597.
Stirm, S., Ørskov, I. and Ørskov, F. (1966) *Nature, Lond.* **209,** 507.
Stirm, S., Ørskov, F., Ørskov, I. and Mansa, B. (1967) *J. Bact.* **93,** 731.
Stouthamer, A. H. and Tieze, G. A. (1966) *Leeuwenhoek ned. Tijdschr.* **32,** 171.
Thomas, G. M. and Poinar, G. O. Jr. (1979) *Int. J. syst. Bact.* **29,** 352.
Thomas, S. B. and Elson, K. (1957) *J. appl. Bact.* **20,** 50.
Toala, P., Lee, Y. H., Wilcox, C. and Finland, M. (1970) *Amer. J. med. Sci.* **260,** 41.
Toenniessen, E. (1913) *Zbl. Bakt.* **69,** 391; (1914) *Ibid.* **73,** 241; (1921) *Ibid.* **85,** 225.
Trabulsi, L. R. and Ewing, W. H. (1962) *Publ. Hlth Lab.* **20,** 137.
Traub, W. H. (1980) In: *The Genus Serratia*, p. 79. Ed. by A. von Graevenitz and S. J. Rubin. CRC Press, Florida.
Traub, W. H. and Kleber, I. (1977) *J. clin. Microbiol.* **5,** 115.
Traub, W. H. and Raymond, E. A. (1971) *Appl. Microbiol.* **22,** 1058.
Trifonova, A. (1968) *Zbl. Bakt.* **208,** 399.
Vahlne, G. (1945) *Serological Typing of the Colon Bacteria.* Gleerupska Univ.-Bokhandeln, Lund.
Waisman, H. A. and Stone, W. H. (1958) *Pediatrics, Springfield* **21,** 8.
Welch, W. D., Martin, W. J., Stevens, P. and Young, L. S. (1979) *Scand. J. infect. Dis.* **11,** 291.
Werkman, C. H. and Gillen, G. F. (1932) *J. Bact.* **23,** 167.
West, M. G. and Edwards, P. R. (1954) *The Bethesda-Ballerup Group of Paracolon Bacteria.* U.S. Dept of Health, Education and Welfare, Atlanta.
Wharton, D. R. A. and Creech, H. J. (1949) *J. Immunol.* **62,** 135.
Wheat, R. W., Dorsch, C. and Godoy, G. (1965) *J. Bact.* **89,** 539.
Wilfert, J. N., Barrett, F. F., Ewing, W. H., Finland, M. and Kass, E. H. (1970) *Appl. Microbiol.* **19,** 345.
Wilkinson, J. F. (1958) *Bact. Rev.* **22,** 46.
Wilkinson, J. F., Dudman, W. F. and Aspinall, G. O. (1954) *Biochem. J.* **59,** 446.
Wilkinson, J. F., Duguid, J. P. and Edmunds, P. N. (1954) *J. gen. Microbiol.* **11,** 59.
Williams, P. H. and Warner, P. J. (1980) *Infect. Immun.* **29,** 411.
Williams, R. P., Green, J. A. and Rappoport, D. A. (1956*a*) *J. Bact.* **71,** 115; (1956*b*) *Science* **123,** 1176.
Wilson, M. I., Crichton, P. B. and Old, D. C. (1981) *J. clin. Path.* **34,** 424.
Wolberg, G. and Witt, C. W. de. (1969) *J. Bact.* **100,** 730.
Woodward, H. M. M. and Clarke, K. B. (1913) *Lancet* **i,** 314.
Wrede, F. and Rothhaas, A. (1934) *Hoppe-Seyl. Z.* **226,** 95.
Yale, M. W. (1939) In: *Bergey's Manual of Determinative Bacteriology*, 5th edn., p. 389. Ed. by D. H. Bergey, R. S. Breed, R. G. E. Murray and A. P. Hitchens. Williams and Wilkins, Baltimore.
Young, V. M., Kenton, D. M., Hobbs, B. J., and Moody, M. R. (1971) *Int. J. syst. Bact.* **21,** 58.

35

Proteus, *Morganella* and *Providencia*

M. T. Parker

Introductory

The organisms described in this chapter are members of the Enterobacteriaceae and thus conform to the definition given in Chapter 33. They have one almost unique character—the oxidative deamination of amino acids; this is tested for by growing the organisms in a medium containing phenylalanine, from which they form phenylpyruvic acid (the PPA reaction; Henriksen and Closs 1938, Henriksen 1950). The only other members of the Enterobacteriaceae to give a positive PPA reaction belong to the recently described genus *Tatumella*. *Proteus*, *Morganella* and *Providencia* also share the following common but not unique characters: motility, resistance to KCN, and inability to acidify dulcitol, to form arginine or lysine decarboxylase, or ferment malonate. With a few exceptions in *Proteus*, they are all methyl-red positive and Voges-Proskauer negative, and do not acidify lactose or form β-galactosidase.

Proteus

Definition

Many strains give a characteristic spreading growth ('swarming') on agar media. Usually form gas from glucose but in small amount. Acidify xylose but not mannose, mannitol, adonitol or inositol, and very rarely lactose. Decompose urea and give a positive PPA reaction. Liquefy gelatin and form lipase. G+C content of the DNA: 37–42 moles per cent.

Type species: *Prot. vulgaris*.

The two species in this genus, *Prot. vulgaris* and *Prot. mirabilis*, were first described by Hauser in 1885 and were redefined by Wenner and Rettger in 1919.

Habitat

These organisms are widely distributed in nature, and constitute an important part of the flora of decomposing organic matter of animal origin. They are constantly present in rotten meat and in sewage, and very frequently in the faeces of man and animals. They are common in garden soil and on vegetables, but their access to these materials probably results largely from contamination with sewage or manure. Besides their wide saprophytic existence, proteus bacilli are able under certain conditions to grow in the animal body and cause septic infections. Certain strains, referred to as *Proteus X* strains, were isolated from the urine, faeces and blood of patients suffering from typhus fever, and are agglutinated by the sera of typhus patients (p. 315).

Morphology

In agar cultures after 24–48 hr, the majority are rods of the coliform type, 1–3 μm long by 0.4–0.6 μm wide, though short, fat cocco-bacillary forms are not uncommon (Figs. 35.1 and 35.2). In young cultures on solid media, in which swarming (see below) is apparent, many of the organisms are long, curved, and filamentous, reaching 10, 20 or even 30 μm in length (Fig. 35.3). In older cultures there is no very charcteristic arrangement; they are distributed singly, in pairs, or in short chains. In young swarming cultures the filamentous forms tend to be arranged concentrically, more or less simulating the isobars in a diagram of a cyclone. There is some variation in depth of staining. Except for non-flagellated O variants, all members are actively motile by peritrichate flagella in young cultures.

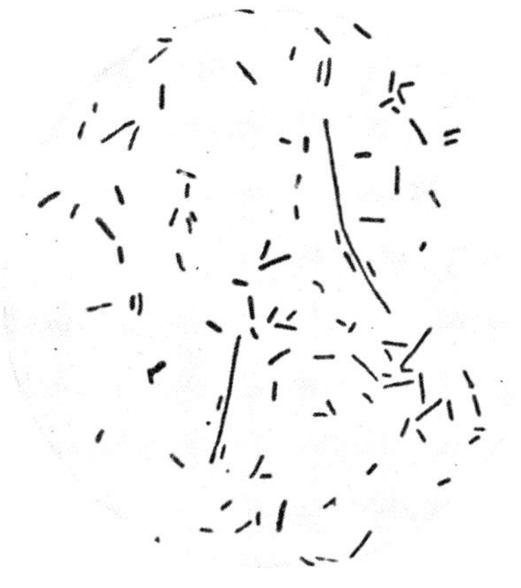

Fig. 35.1 *Proteus vulgaris*. From an agar culture, 24 hr, 37°, showing chiefly rod forms (× 1000).

The flagella of proteus bacilli are much more variable in shape than those of most other enterobacteria (Leifson *et al* 1955). The normal and 'curly' forms, which in other groups are characters of particular strains, may occur together in the same organism, or even in the same flagellum. The form of the flagellum is also influenced by the pH of the medium. All proteus strains are fimbriate (Duguid and Gillies 1958). Their fimbriae belong to Duguid's type 3. Neither spores not capsules are formed. The reaction to Gram's stain is uniformly negative.

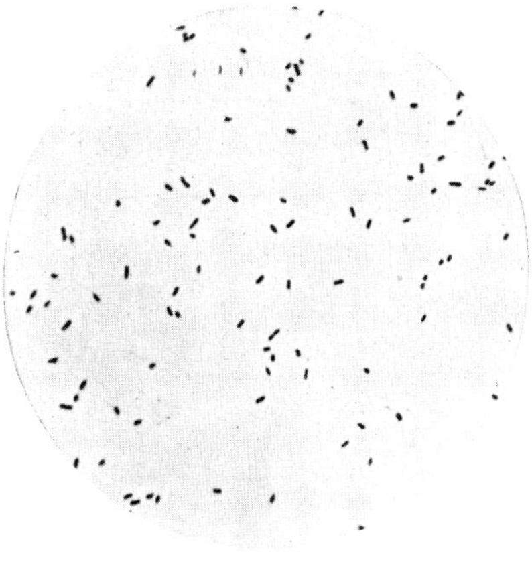

Fig. 35.2 *Proteus vulgaris*. From an agar culture, 48 hr, 37°, showing short rods only (× 1000).

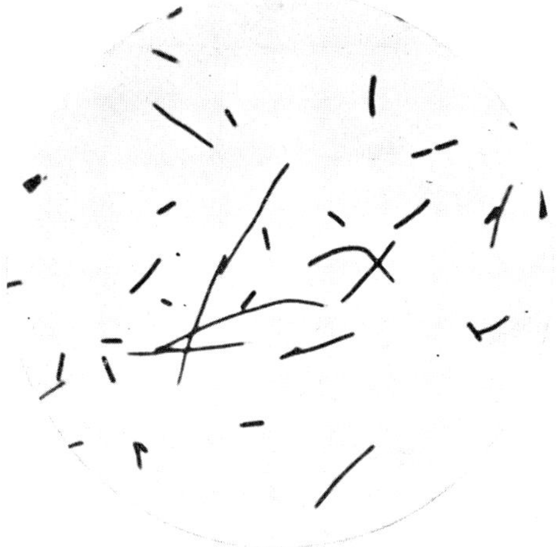

Fig. 35.3 *Proteus vulgaris*. From an agar culture, 6 hr, 37°, showing long filaments (× 1000).

Cultural characters

Growth occurs freely on ordinary nutrient media. One of the most characteristic properties of *Prot. vulgaris* and *Prot. mirabilis* is their ability to *swarm* on solid media (Cantu 1911, Moltke 1927, 1929, Russ-Münzer 1935). Swarming may be defined as 'progressive surface spreading by the microbes from the edge of the parent colony'. Its most characteristic form is *discontinuous swarming*; this begins after about 4 hours' incubation, when a thin, ground-glass layer of growth appears around the site of inoculation and extends progressively for about 2 hr. Further progress outwards then ceases, but the layer of growth thickens. At about the 8th hour, swarming starts again and a fresh ring of growth appears. The alternation of swarming and rest occurs regularly about every 4 hr till the plate is covered. It is owing to this periodic

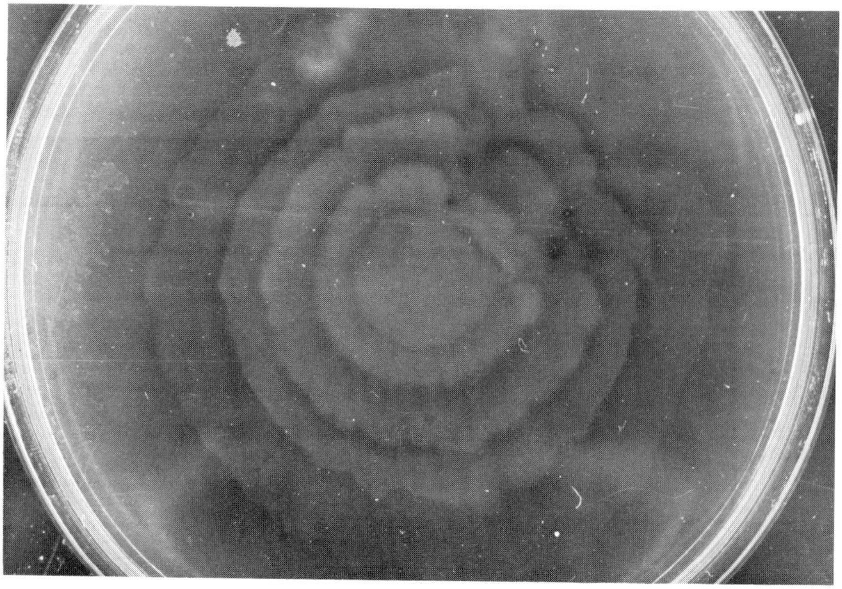

Fig. 35.4 Discontinuous swarming of *Proteus* on nutrient agar (× 1.5). (From a photograph kindly supplied by Dr J. Alwen.)

extension that the surface of the growth often appears rippled or contoured (Fig. 35.4).

Microscopic examination shows that, when swarming begins, the edge of the growth consists of long slender rods which are in continuous motion. Bundles of them break away from the edge, and, after travelling some distance outwards, join neighbouring lateral offshoots to form arches which are rapidly filled with other rods from within. Whole rafts of rods tear loose from the peninsula so formed and work across the agar, so that in a short time the colony is surrounded by an archipelago of islands and solitary organisms, all constantly in motion. The very long rods are the predominant feature of the picture; they form arches, whorls, spirals and question-mark forms. When swarming ceases, the long rods are replaced by quite short forms. Other strains or variants of *Prot. vulgaris* and *Prot. mirabilis* show *continuous swarming*, either as a uniform sheet of growth extending over the whole surface of the plate or as a compact ring of growth of limited extent. In these cases, long forms are not seen.

The cause of swarming has given rise to much speculation. Lominski and Lendrum (1947) concluded that it was determined by the negative chemotactic effect of metabolites formed at the site of stationary growth (see also Grabow 1972). This view is not accepted by Williams and his colleagues (1976). Douglas (1979) attributes cessation of swarming to reduction in the velocity of the long forms at the edge of the growth.

Russ-Münzer (1935) attributed swarming to the depletion of nutrients, and several observations suggest that its occurrence is dependent on the composition of the medium. Swarming is inhibited in peptone meat extract agar with a low electrolyte content (Sandys 1960), but this is not due to the low osmolarity of the medium, because the addition of further nutrients permits swarming to occur (Naylor 1964). Similarly, Jones and Park (1967*b*) found that swarming did not take place on a minimal medium with normal electrolyte content on which growth was active, but it was stimulated by the addition of nutrients. A high-molecular weight fraction of yeast extract is a potent stimulator of swarming on minimal medium (Brogan *et al.* 1971). Jeffries and Rogers (1968) showed that, on a defined medium, swarming did not occur unless the content of ordinary agar exceeded 0.8 per cent. A higher percentage of purified agar was necessary for swarming, and extracts of agar contained a substance that stimulated swarming. The addition of activated charcoal to agar inhibits swarming and the appearance of long forms (Smith and Alwen 1966).

According to Jones and Park (1967*a*), long forms arise at the edge of surface growth and are never seen in liquid medium. They contain many nuclear units and have no cross walls. They have more flagella per unit length, and longer flagella, than do short forms (Hoeniger 1965). Morrison and Scott (1966) suggested that long forms move continuously only when in contact with each other and obtain thrust by sliding along one another.

When two different proteus strains are inoculated at different points on the same agar plate, and both of them swarm, the two sheets of growth never quite coalesce (Dienes 1946). This is thought to be due to the production of mutually inhibitory substances, which can be detected by re-inoculation of the cleared medium (Hughes 1957). Lines of demarcation are formed only when both strains are in the active stage of discontinuous swarming; if one is in the stationary phase it is overrun by the other; and strains that swarm continuously always appear to be compatible (Skirrow 1969).

Swarming occurs readily on ordinary agar at 37°. It is not seen in *Morganella* or *Providencia*; these organisms, like many other enterobacteria, may give spreading growth on 1 per cent agar at 20–25°, but this is probably of a different nature from 'true' swarming.

Several methods have been devised for the inhibition of swarming, mainly to help in the isolation of organisms of the streptococcal and enteric groups. (*a*) Addition of narcotic drugs, such as chloral hydrate, morphine, and sodium phenylethylbarbiturate (Krämer and Koch 1931, Lode and Howard 1932). Not all strains are equally sensitive to these drugs. (*b*) Incorporation of 5–6 per cent alcohol in the medium is effective (Floyd and Dack 1939), but may cause inhibition of, for example, streptococci. (*c*) Addition of sodium azide in a final concentration of 1/5000–1/10 000 (Snyder and Lichstein 1940, Lichstein and Snyder 1941); this does not inhibit streptococcal growth, though the blood around β-lytic colonies appears green and around α-lytic colonies brown. (*d*) Incorporation of 0.1 per cent (w/v) boric acid in heated blood agar (Sykes and Reed 1949). (*e*) A polyvalent rabbit serum may be added to the medium to immobilize the motile organisms (Beattie 1945). (*f*) Shake cultures or poured plates may be used, as the deep colonies of proteus are compact (Fry 1932). (*g*) Use of 6 per cent agar in the medium (Hayward and Miles 1943, Hayward *et al.* 1978). (*h*) Incorporation of sulphonamides. Holman (1957) advocates the use of 1/8000 sulphamezathine in heated blood agar for the isolation of fastidious organisms. (*i*) Swarming rarely occurs on a plain Lab-Lemco medium without added salt (Sandys 1960). This is a very simple and useful method. (*j*) The addition of 0.2 mM *p*-nitrophenyl glycerol to medium inhibits swarming without affecting growth or motility (Kopp *et al.* 1966). Swarming is also inhibited by some surface-active agents, such as the alkyl sulphates (Lominski and Lendrum 1942), by bile salts, and by the bismuth sulphite medium of Wilson and Blair.

The colonial appearances of *Proteus* vary in culture. On MacConkey's agar Belyavin (1951) recognized three varieties which he refered to as phases.

Phase A forms smooth colonies, 3–4 mm across, with a beaten-copper surface and an entire or finely radially striated edge; morphologically it consists of regular bacillary forms 5–6 μm × 0.5 μm. *Phase B* forms smaller smooth conical colonies with a shiny surface; morphologically it consists of highly pleomorphic cells—coccobacilli, filaments and giant cells. *Phase C* forms flat colonies, larger than those of A, with a rough surface, granular structure, and a dentate or fimbriate edge; morphologically it consists of long tangled filaments. Phase A swarms intermittently and phase C in a continuous film; phase B is usually non-swarming. Antigenically phases A and C contain a dominant type-specific O antigen which B mainly lacks. Phase variations A→B and A→C are said to be reversible. Coetzee and Sacks (1960) described five colonial variants, three of which were recognizable as phase A, B and C. Cultures initiated from a single cell of one particular variant eventually gave rise to all the others. The ability to swarm can be transduced by bacteriophage. Three genetic factors have been identified which in various combinations are responsible for discontinuous (phase A), continuous (phase C) and limited swarming (Coetzee 1963*b*). A determinant for swarming may be present in a non-motile strain.

In broth *Proteus* gives rise to a uniform turbidity accompanied by a slight to moderate powdery deposit and a faint ammoniacal smell. A thin fragile pellicle may develop in older cultures.

Resistance

Proteus bacilli are readily destroyed by moist heat at 55° for 1 hr, and by common disinfectants such as phenolics or halogens. However, some strains isolated in hospitals, particularly of *Prot. mirabilis*, have a significant resistance to chlorhexidine, of which as much as 800 mg per l may be required to prevent multiplication (Gillespie *et al.* 1967, Stickler 1974).

Resistance to antimicrobial agents The pattern of resistance to penicillins and cephalosporins is complicated (see Garrod *et al.* 1981) and is determined in part by mechanisms of intrinsic resistance and in part by the production of β-lactamase. Broadly speaking, *Prot. mirabilis* strains that do not form this enzyme are sensitive to amounts of benzylpenicillin that are attainable in the urine and are fully sensitive to ampicillin, carbenicillin and cephalosporins. β-Lactamase producers are highly resistant to benzylpenicillin, ampicillin and carbenicillin but often remain sensitive to cephalosporins. *Prot. vulgaris* strains are resistant to ampicillin but often sensitive to carbenicillin; they are generally resistant to cephaloridine but may be sensitive to newer cephalosporins such as cefuroxime and cefoxitin. Susceptibility to chloramphenicol and tetracyclines is variable but all strains are resistant to polymyxins. Originally, proteus strains were generally sensitive to aminoglycosides, but this is no longer so. Strains with transferable resistance to various aminoglycosides, including gentamicin, to trimethoprim, and to carbenicillin have lately become prevalent in hospitals (see, for example, Shafi and Datta 1975, Yoshikawa *et al.* 1978); their resistance plasmids may be transferred between *Proteus*, other enterobacteria and *Ps. aeruginosa*.

Metabolism

The optimum temperature for growth is about 34–37°, though rapid multiplication occurs above 20°. The limits of growth are between about 10° and 43°.

Early reports on the haemolytic activity of *Proteus* are discrepant and it seems probable that the nature of the blood was partly responsible. Taylor (1928), using human blood, observed haemolysis regularly within 24 hours in 1 per cent blood broth, but not on 10 per cent blood agar plates. Yacob (1932) used 5 per cent rabbit blood agar plates, and found that all strains produced β-haemolysis in 18–24 hr. Phillips (1955*a*) found that broth supernatants, but not filtrates, were often haemolytic. Sheep cells were more sensitive to haemolysis than horse cells, and *Prot. vulgaris* was more often haemolytic than *Prot. mirabilis*.

Biochemical reactions

All proteus strains form acid and gas from glucose. The volume of gas produced is, however, rather small and seldom fills more than 15 per cent of the Durham's tube. Occasional strains are anaerogenic on first isolation (Edwards 1942). No strain forms acid from mannitol, adonitol or inositol. It is also generally true that proteus strains rarely ferment lactose, but a few lactose-fermenting strains have been seen (Sutter and Foecking 1962). They possess a plasmid determinant for lactose fermentation that is readily transferable to other enterobacteria and appears to have arisen outside the *Proteus* group (Falkow *et al.* 1964). By definition (Wenner and Rettger 1919, Rustigian and Stuart 1945) *Prot. vulgaris* acidifies maltose, sucrose and salicin promptly. *Prot. mirabilis* does not attack maltose and usually gives a late positive reaction in sucrose; most strains are without action on salicin, but some show late acidification. Unlike *Morganella* and *Providencia*, *Proteus* acidifies xylose but not mannose.

Gelatin is liquefied, and most strains are lipolytic. It is generally stated that *Prot. vulgaris* is indole positive, but indole-negative, maltose-fermenting strains were at one time prevalent in some hospitals (Keating 1956). The ability to grow in Koser's or Simmons's citrate medium is variable, but, according to Proom (1955), nicotinic acid is required for growth by all strains. H_2S is formed in triple sugar iron (TSI) medium. All strains are catalase positive, oxidase negative.

Proteus forms urease (Moltke 1927) in common with *Morganella* and some strains of *Providencia*; members of all three genera deaminate phenylalanine. *Prot. mirabilis* forms ornithine decarboxylase, a property it shares with *Morg. morgani* but not with *Prot. vulgaris*. Proteus strains decarboxylate glutamic acid only weakly and at 25° (Møller 1955), and do not ferment sodium citrate (Kauffmann and Petersen 1956). (For further information, see Ewing *et al.* 1960, Brenner *et al.* 1978, and Table 35.1.)

According to Brenner and co-workers (1978), a third species of *Proteus* exists, comprising strains that resemble *Prot. vulgaris* except in being non-motile and indole negative, and in failing to produce H_2S or acidify xylose; this has been named *Prot. myxofaciens*.

Antigenic structure

The early workers reported much antigenic heterogeneity among strains of *Pr. vulgaris* and *Pr. mirabilis* (Cantu 1911, Wenner and Rettger 1919, Moltke 1927, Taylor 1928). The presence of flagellar and somatic antigens was demonstrated by Weil and Felix (1917, 1918). Many different O and H antigens have since been described (Winkle 1945, Perch 1948, Belyavin *et al.* 1951). In the scheme of Kauffmann (1966), O groups are further subdivided according to their H-antigenic structure, and 110 distinct serotypes can be recognized. The antigens of *Proteus* are in general different from those of *Morganella* and *Providencia*.

Table 35.1 Usual biochemical reactions of species of *Proteus*, *Morganella* and *Providencia*

The following properties are common to all: motile; acid in glucose; no acid in dulcitol, sorbitol, raffinose and arabinose; nitrate reduced to nitrite; KCN resistant; phenylalanine deaminated; lysine and arginine decarboxylase –; malonate –; mucate –. Except for a few strains of *Proteus*, all MR +, VP –, lactose –.

	Prot.		*Morg.*	*Prov.*		
	vulgaris	*mirabilis*	*morgani*	*rettgeri*	*alcalifaciens*	*stuarti*
Gas from glucose	+	+	+	–*	+	–*
Acid from maltose	+	–	–	–	–	–
sucrose	+	(+) or –	–	(+)	(+)	(+)
salicin	+	– or (+)	–	V	–	–
mannitol	–	–	–	+	–	–*
mannose	–	–	+	+	+	+
xylose	+	+	–	–*	–	–
adonitol	–	–	–	+	+	–
inositol	–	–	–	+	–	+
trehalose	(+)	+	–	–	–	+
Indole	+	–	+	+	+	+
Gelatin liquefaction	+	+	–	–	–	–
H_2S (TSI medium)	+	+	–	–	–	–
Urease	+	+	+	+	–	–*
Ornithine decarboxylase	–	+	+	–	–	–
Growth in Simmons's citrate	V	V	–	+	+	+
Require nicotinic acid	+	+	+	–	–	–
Lipase production	+	+	–	–	–	–

* A minority (> 10 per cent) give a positive result.
+ = Positive reaction within 48 hr; (+) = delayed positive reaction; V = some strains positive, others negative; – = negative reaction.

Although earlier workers had reported considerable sharing of antigens between the two species of *Proteus*, Penner and Hennessy (1980) describe separate O-serogrouping schemes for them.

The *Proteus* X strains—X19, X2 and XK—which react with the sera of typhus patients (Chapter 77), each possess a distinctive O antigen (Kauffman and Perch 1947). According to White (1933), the X19 strain has two antigenic receptors, one of which is alkali-labile and is mainly responsible for agglutination of the organism by antisera prepared against it, and the other of which is alkali-stable and is responsible for the so-called Weil-Felix reaction with the sera of typhus patients. Meisel and Mikulaszek (1933) and Castañeda (1934, 1935) extracted polysaccharides from the X strains; Castañeda's results agreed closely with those of White, and showed that the alkali-stable polysaccharide was common to *Proteus* X19 and *Rickettsia prowazeki* (Chapter 46), whereas the alkali-labile polysaccharide was specific for *Proteus* X19.

The O antigens of *Proteus* are resistant to heating at 95° in ethanol and to dilute HCl, both of which treatments destroy the H antigens. Agglutination tests for the determination of O antigens should be carried out by the tube method with cultures boiled for 1 hr, not with living cultures by slide agglutination, because some living and some formolized cultures are O-inagglutinable. O agglutination tests are best read after 20 hours' and H agglutination tests after 4 hours' incubation in a water-bath at 50°. There are many cross reactions between the O antigens of the *Proteus*, *Escherichia* and *Salmonella* groups (Kauffmann 1966, Rauss and Vörös 1967).

Phage typing Several different phage-typing schemes have been proposed (Vieu 1958, Pavlatou *et al.* 1965, France and Markham 1968, Schmidt and Jeffries 1974). Numerous patterns of lysis can be recognized; those of *Pr. vulgaris* and *Pr. mirabilis* strains are generally similar. The typing phages rarely lyse other enterobacteria.

Bacteriocine typing Cradock-Watson (1965) described a method of typing *Pr. vulgaris* and *Pr. mirabilis* by determining the range of activity of their bacteriocines on a set of indicator strains. Later, Senior (1977) developed an improved method in which both production of and sensitivity to proticines were used to characterize the strains.

Other typing methods The *Dienes phenomenon* forms the basis of a very discriminating method of distinguishing between strains that exhibit discontinuous swarming (Story 1954). Reactions of identity (Fig. 35.5) occur in only about 1 in 70 pairs of epidemiologically unrelated strains (Skirrow 1969). If used with care it is a valuable means of comparing small numbers of cultures from a single outbreak of infection (Skirrow 1969, de Louvois 1969). A 'resistotyping' method (see Chapter 20) has also been used (Kashbur *et al.* 1974).

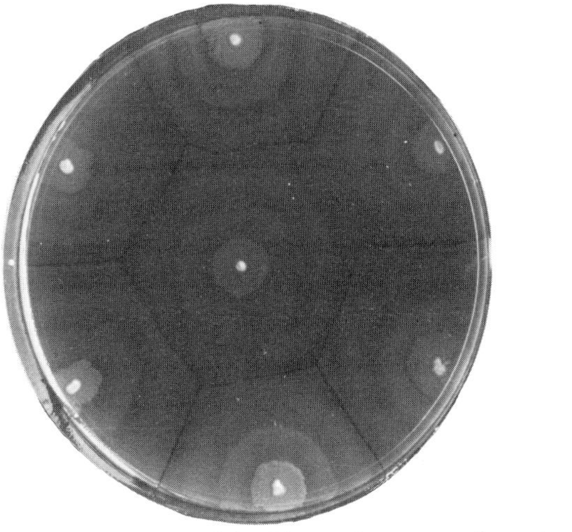

Fig. 35.5 The use of the Dienes phenomenon for typing *Proteus mirabilis*. A line of demarcation is seen between the strain at the centre and five of the six strains at the periphery. The strain at 2 o'clock, on the other hand, gives a reaction of identity with the strain at the centre. (From a photograph kindly supplied by Dr M. B. Skirrow.)

Pathogenicity

Prot. mirabilis and *Prot. vulgaris* are pathogenic for man. They occur in a variety of different types of septic lesion, either alone or together with other organisms. Their presence in mixed culture in wounds and burns is believed to favour the development of pathogenic anaerobes. From time to time they are isolated in pure culture from abscesses, from the meninges or from the blood.

Proteus organisms are an important cause of urinary-tract infection. Such infections tend to occur in hospital patients with obstructive lesions of the urinary tract, after instrumentation, or in association with prolonged catheterization (Chapter 58), but they may develop in healthy male children (Chapter 57). (For further information about human infections, see Chapter 61, and for the relation of *Proteus* X strains to typhus fever, Chapter 46.)

Pathogenicity for laboratory animals is variable; virulent strains on introduction into the tissues are able to proliferate and invade the blood stream. Strains of lower virulence cause chronic inflammation, either of the suppurative or of the infective granuloma type (Larson and Bell 1913, Wenner and Rettger 1919); the latter are best seen after intraperitoneal injection.

The inoculation intraperitoneally of 0.5–1.0 ml of a 24-hour broth culture of a virulent strain generally proves fatal to rats and mice in 18–48 hours, and to guinea-pigs and rabbits in 1–7 days. The LD50 for mice of 21 strains ranged from 2 to 600 million living organisms (Miles 1951). Large doses given intravenously to mice result in a fatal septicaemic illness, and smaller doses of some strains produce necrotic lesions in the kidneys of survivors (Phillips 1955*b*).

Morganella

Definition

Do not show true swarming on agar media. Usually form a small amount of gas from glucose. Acidify mannose but not lactose, sucrose, salicin, mannitol, xylose, adonitol or inositol. Decompose urea and convert phenylalanine to phenylpyruvic acid. Form indole but do not produce H_2S in TSI medium, liquefy gelatin or form lipase. G+C content of the DNA: 51-53 moles per cent.

Type species: *Morg. morgani.*

Morgan (1906) first described this organism, which he had isolated from human faeces. In 1936, Rauss suggested that Morgan's bacillus was closely related to *Proteus*. Though it failed to liquefy gelatin, it was able to swarm at 20-28° on 1 per cent agar, to decompose urea, and to produce indole. His conclusions were supported by the observation of Henriksen and Closs (1938) that both *Proteus* and Morgan's bacillus were able to break down phenylalanine to phenylpyruvic acid. For a number of years Morgan's bacillus was known as *Prot. morgani*. The proposal of Kaufmann (1956) that it should be placed in a separate genus *Morganella* did not at first receive general acceptance, but the evidence in favour of this view has now become overwhelming: notably, the considerably higher G+C content of the DNA in *Morganella* than in *Proteus*, the low level of genetic homology between the two genera (Brenner *et al.* 1978), and serological differences between certain of their enzymes.

General properties

Morg. morgani is motile; it spreads on soft agar at low temperatures, but without the characteristic changes in cellular morphology associated with true swarming. It is resistant to ampicillin and cephaloridine but usually sensitive to carbenicillin. Susceptibility to tetracycline and chloramphenical is variable; most strains continue to be sensitive to aminoglycosides. It is resistant to polymyxins. The biochemical characters of *Morg. morgani* are fairly uniform (Table 35.1) and serve to distinguish it from *Proteus* and *Providencia*. It is antigenically heterogeneous; Rauss and Vörös (1959, 1967) recognized 66 serotypes. These are distinct from the serotypes of *Proteus* and *Providencia*. Phages that lyse *Morg. morgani* are generally different from those that lyse proteus strains.

Morg. morgani is found in human faeces, but not very frequently. It is occasionally responsible for septic infections, notably of the urinary tract. Despite its wide spectrum of antibiotic resistance it appears rather seldom to give rise to large outbreaks of infection in hospitals.

Perez's bacillus In 1899 Perez described, under the name *Cocco-bacillus foetidus ozoenae*, a small organism that he had isolated from the nose of patients suffering from ozaena, but there is insufficient evidence to show that it is causally related to this disease. Perez (1901, 1913) found his bacillus also in the saliva and nasal mucus of apparently healthy dogs.

Morphologically, Perez's bacillus is a small gram-negative coccobacillus, which was originally described as non-motile, but which is said by Ferry and Noble (1918) to be sluggishly motile, and by Ward (1917) occasionally to acquire motility in culture. Biochemically, it produces acid and gas in glucose only; indole is produced; and urea is hydrolysed. Ferry and Noble (1918) drew attention to its similarity to *Bordetella bronchiseptica*, but the more recent observations of Blăškovič (1944) suggest that it is identical with Morgan's bacillus.

An organism resembling Perez's bacillus was isolated by Shiga (1922) from the nose of ozaena patients, but had rather more fermentative and proteolytic ability. Shiga's four strains had the same biochemical reactions as Morgan's bacillus; the two motile strains swarmed on 1 per cent agar at 22° and were aggutinated by Morgan H antisera; and three of the four strains were agglutinated by Morgan O antisera. Their pathogenicity to mice and rabbits also resembled that of *Morg. morgani*.

Providencia

Definition

Do not swarm on agar media. Gas production from glucose variable. Mannose acidified promptly and sucrose usually after some delay; no action on maltose or lactose. Other sugar reactions variable. Ferment citrate, and grow consistently on Simmons's citrate medium. Convert phenylalanine to phenylpyruvic acid, but action on urea variable. Form indole. Do not produce H_2S in TSI medium, liquefy gelatin or form lipase. G+C content of the DNA: 37-42 moles per cent.

Stuart and his colleagues (1943, 1946) described a group of organisms that they considered to be intermediate in character between *Proteus* and *Shigella*; these fermented glucose with or without the formation of a small bubble of gas but did not decompose urea. Kauffmann (1951) named them the 'Providence group' and later established a new genus *Providencia* for them. At about the same time as the Providence group was being recognized, Rustigian and Stuart (1943) drew attention to organisms, first described by Hadley and co-workers (1918), that were anaerogenic, fermented mannitol, and formed urease; these were added to the genus *Proteus* as *Prot. rettgeri*. Neither the Providence nor the *rettgeri* strains formed a homogenous group and their classification presented difficulties.

Ewing and his colleagues (1954) found that the Providence strains could be separated into two main biochemical varieties: (1) organisms that formed gas from glucose and usually fermented adonitol but not inositol, and (2) anaerogenic organisms that usually fermented inositol but not adonitol. These were subsequently given specific names, respectively *Prov. alcalifaciens* and *Prov. stuarti* (Ewing 1962). Later, Ewing, Davis and Sykes (1972) described six biochemical varieties in these two species, and Penner, Hinton and Hennesy (1975) an even greater number of varieties among the *rettgeri* strains. Ewing (1958) had drawn attention to biochemical similarities between some *Providencia* and *rettgeri* strains, and Coetzee (1963*a*) had shown that members of both groups had similar phenylalanine-attacking enzymes and phage receptors. Penner and his colleagues (1976*b*) described strains resembling one of their *rettgeri* biotypes that had *Providencia* O antigens and were variable in urease production when isolated from different patients in the same outbreak of hospital-acquired infection. Studies of genetic homology (Brenner *et al.* 1978) strongly supported the addition of the *rettgeri* strains to *Providencia*, with consequent changes in the definition of this genus. The three species, *alcalifaciens*, *stuarti* and *rettgeri* could be recognized, but only by accepting that a simple distinction between them on the basis of gas production, mannitol fermentation and presence of urease was no longer possible (see Table 35.1). There were also indications for the existence of additional species that could not be clearly defined by existing biochemical tests.

General properties

Strains of *Providencia* are generally motile, though not very actively. They can be trained to spread over a 1 per cent agar plate, usually as a continuous film but occasionally as branching filaments (Stuart *et al.* 1946). They form characteristic orange-centred colonies on deoxycholate citrate agar (Buttiaux *et al.* 1953–54). They are generally resistant to ampicillin, often to carbenicillin and usually to cephaloridine, but may be sensitive to some of the newer cephalosporins; resistance to tetracycline and chloramphenical are variable. Strains of *Prov. stuarti* are on the whole resistant to a wider range of antibiotics that are those of *Prov. alcalifaciens* (Perch 1954). Strains of *Prov. stuarti* and *Prov. rettgeri* that are now endemic in some hospitals have additional resistances, e.g. to aminoglycosides (Iannini *et al.* 1976, Prentice and Robinson 1979). Resistance to chlorhexidine may be even greater than that seen in *Prot. mirabilis* (Stickler and Thomas 1976).

The biochemical reactions characteristic of the genus are summarized in the definition at the head of this section, and those useful for the recognition of species are shown in Table 35.1. Gas production from glucose occurs regularly in *Prov. alcalifaciens* but also in a minority of members of the other two species. A considerable minority of the strains that must be classified as *Prov. stuarti* on grounds of genetic homology (Brenner *et al.* 1978) give atypical reactions in mannitol-fermentation and urease tests and so are difficult to distinguish from *Prov. rettgeri.* Action on adonitol, inositol and trehalose are probably better criteria for the identification of species.

Providencia is antigenically heterogeneous. Edwards and Ewing (1955) described 62 O groups and 30 different H antigens in *Prov. alcalifaciens* and *Prov. stuarti*, but their typing system was not much used for epidemiological studies. Penner and his colleagues (1976*a*) produced a different typing scheme in which only O serogroups were determined and used this to study infections in hospitals (Penner *et al.* 1979) that were caused mainly by *Prov. stuarti.* They found that 97 per cent of the strains fell into one of 14 serogroups. Earlier, Penner and Hinton (1973) had found extensive cross-reactions between strains of *Prov. rettgeri* attributable to the presence of a common heat-labile antigen. When the antibody to this were removed from typing sera. 18 O and 12 H antigens could be recognized. In general, the O antigens of the three species of *Providencia* seem to belong to separate series. Phage typing and bacteriocine typing both appear to be practicable in this genus (see Coetzee 1967).

Providence strains occur not infrequently in human faeces. They cause clinical infections in hospital patients, mainly in the urinary tract of catheterized patients. These are difficult to eradicate and are quite often complicated by septicaemic episodes (Chapter 61). Long-continued endemic prevalences of infection with particular strains are common (Iannini *et al.* 1976, Wenzel *et al.* 1976).

Classification of *Proteus*, *Morganella* and *Providencia*

The most striking features of the two original members of the genus *Proteus*, the species *vulgaris* and *mirabilis*, was that they swarmed on nutrient agar, formed gas from glucose, did not form acid from mannitol, liquefied gelatin, split urea and deaminated phenylalanine (Hauser 1885, Wenner and Rettger 1919, Moltke 1927, Henriksen and Closs 1938). *Prot. morgani* was added to the genus (Rauss 1936), despite the fact that it did not swarm or liquefy gelatin; and later *Prot. rettgeri* (Rustigian and Stuart 1943), although it was anaerogenic and fermented mannitol. Eventually there were those who would have included the Providence group, most members of which were urease negative. Had this been done, the deamination of phenylalanine would have been the only important common character of members of the genus. The contrary process of subdividing the genus was begun by Kauffmann (1956), who recognized four genera in a Tribe Proteae: *Proteus*, *Morganella*, *Rettgerella* and *Providencia* (see also Rauss 1963).

There is little doubt that *Prot. vulgaris* and *Prot. mirabilis* are closely related; indeed some workers would place them in a single species, *Prot. hauseri*. They are the only true swarmers, share a number of O antigens and phage receptors (Coetzee 1963*a*) and have antigenically similar phenylalanine deaminases (Smit and Coetzee 1967). Brenner and his colleagues (1978) found that *Prot. mirabilis* was a genetically homogeneous species; *Prot. vulgaris* was rather less uniform but included one large group that merited specific status. They would also recognize a third species, *Prot. myxofaciens* (see p. 314), isolated from insects.

Morganella is in many ways the most distinct of the genera we have recognized in this chapter. It is very uniform in its biochemical characters, has a higher G+C content in its DNA than *Proteus* or *Providencia* and is genetically homologous (Brenner *et al.* 1978). It has a series of O antigens and phage receptors of its own, and an antigenically distinct urease (Guo and Liu 1965) and phenylalanine deaminase (Smit and Coetzee (1967).

Reasons for transferring *Prot. rettgeri* to *Providencia* have already been given (p. 317). To these may be added possession of a common phenylalanine deaminase (Smit and Coetzee 1967). It must be conceded, however, that the DNA homology groups in this genus do not correspond entirely with biochemically recognizable species.

References

Beattie, M. K. (1945) *J Path. Bact.* **57,** 388.
Belyavin, G. (1951) *J. gen. Microbiol.* **5,** 197.
Belyavin, G., Miles, E. M. and Miles, A. A. (1951) *J. gen. Microbiol.* **5,** 178.
Blaškovič, D. (1944) *Z. Hyg. InfektKr.* **126,** 190.
Brenner, D. J. *et al.* (1978) *Int. J. syst. Bact.* **28,** 269.
Brogan, T. D., Nettleton, J. and Reid, C. (1971) *J. med. Microbiol.* **4,** 1.
Buttiaux, R., Fresnoy, R. and Moriamez, J. (1953–4) *Ann. Inst. Pasteur, Lille* **6,** 62.
Cantu, C. (1911) *Ann. Inst. Pasteur* **25,** 852.
Castañeda, M. R. (1934) *J. exp. Med.*, **60,** 119; (1935) *Ibid.* **62,** 289.
Coetzee, J. N. (1963*a*) *J. gen. Microbiol.* **31,** 219; (1963*b*) *Ibid.* **33,** 1; (1967) *Nature, Lond.* **213,** 614.
Coetzee, J. N. and Sacks, T. G. (1960) *J. gen. Microbiol.* **23,** 209.
Cradock-Watson, J. E. (1965) *Zbl. Bakt.* **196,** 385.
Dienes, L. (1946) *Proc. Soc. exp. Biol., N.Y.* **63,** 265.
Douglas, C. W. I. (1979) *J. med. Microbiol.* **12,** 195.
Duguid, J. P. and Gillies, R. R. (1958) *J. Path. Bact.* **75,** 519.
Edwards, J. L. (1942) *J. Hyg. Camb.* **42,** 238.
Edwards, P. R. and Ewing, W. H. (1955) *Identification of Enterobacteriaceae.* Burgess, Minneapolis.
Ewing, W. H. (1958) *Int. Bull. bact. Nomencl.* **8,** 17; (1962) *Ibid.* **12,** 93.
Ewing, W. H., Davis, B. R. and Sykes, J. V. (1972) *Publ. Hlth Lab.* **30,** 25.
Ewing, W. H., Suassuna, I. and Suassuna, I. R. (1960) *The Biochemical Reactions of the Genus Proteus.* U.S. Dept. of Health, Education and Welfare, Atlanta.
Ewing, W. H., Tanner, K. E. and Dennard, D. A. (1954) *J. infect. Dis.* **94,** 134.
Falkow, S., Wohlheiter, J. A., Citarella, R. V. and Baron, L. S. (1964) *J. Bact.* **88,** 1598.
Ferry, N. S. and Noble, A. (1918) *J. Bact.* **3,** 499.
Floyd, T. M. and Dack, G. M. (1939) *J. infect. Dis.* **64,** 269.
France, D. R. and Markham, N. P. (1968) *J. clin. Path.* **21,** 97.
Fry, R. M. (1932) *Brit. J. exp. Path.* **13,** 456.
Garrod, L. P., Lambert, H. P. and O'Grady, F. (1981) *Antibiotic and Chemotherapy.* Churchill Livingstone, Edinburgh.
Gillespie, W. A., Lennon, G. G., Linton, K. B. and Phippen, G. A. (1967) *Brit. med. J.* **iii,** 90.
Grabow, W. O. K. (1972) *J. med. Microbiol.* **5,** 191.
Guo, M. M. S. and Liu, P. V. (1965) *J. gen. Microbiol.* **38,** 417.
Hadley, P., Elkins, M. W. and Caldwell, D. W. (1918) *Agr. Exp. Sta. Rhode Island Coll.* Bull. No. 174.
Hauser, G. (1885) *Ueber Fäulnisbakterien.* Leipzig.
Hayward, N. J., Incledon, G. M. and Spragg, J. E. (1978) *J. med. Microbiol.* **11,** 155.
Hayward, N. J. and Miles, A. A. (1943) *Lancet* **ii,** 116.
Henriksen, S. D. (1950) *J. Bact.* **60,** 225.
Henriksen, S. D. and Closs, K. (1938) *Acta. path. microbiol. scand.* **15,** 101.
Hoeniger, J. M. F. (1965) *J. gen. Microbiol.* **40,** 29.
Holman, R. A. (1957) *J. Path. Bact.* **73,** 91.
Hughes, W. H. (1957) *J. gen. Microbiol.* **17,** 49.
Iannini, P. B., Eickhoff, T. C. and La Force, F. M. (1976) *Ann. intern. Med.* **85,** 161.
Jeffries, C. D. and Rogers, H. E. (1968) *J. Bact.* **95,** 732.
Jones, H. E. and Park, R. W. A. (1967*a*) *J. gen. Microbiol.* **47,** 359; (1967*b*) *Ibid.* **47,** 369.
Kashbur, I. M., George, R. H. and Ayliffe, G. A. J. (1974) *J. clin. Path.* **27,** 572.
Kauffmann, F. (1951) *Enterobacteriaceae*, 1st edn. Munksgaard, Copenhagen; (1956) *Zbl. Bakt.*, **165,** 344; (1966) *The Bacteriology of the Enterobacteriaceae.* Munksgaard, Copenhagen.
Kauffmann, F. and Perch, B. (1947) *Acta path. microbiol. scand.* **24,** 135.
Kauffmann, F. and Petersen, A. (1956) *Acta path. microbiol. scand.* **38,** 481.
Keating, S. (1956) *Med. J. Aust.* **ii,** 168.
Kopp, R., Müller, J. and Lemme, R. (1966) *Appl. Microbiol.* **14,** 873.
Krämer, E. and Koch, F. E. (1931) *Zbl. Bakt.* **120,** 452.
Larson, W. P. and Bell, E. T. (1913) *J. infect. Dis* **13,** 510.
Leifson, E., Carhart, S. R. and Fulton, M. (1955) *J. Bact.* **69,** 73.
Lichstein, H. C. and Snyder, M. L. (1941) *J. Bact.* **42,** 653.
Lode, A. and Howard, A. (1932) *Zbl. Bakt.* **124,** 538.
Lominski, I and Lendrum, A. C. (1942) *J. Path. Bact.* **54,** 421; (1947) *Ibid.* **59,** 688.
Louvois, J. de. (1969) *J. clin. Path.* **22,** 263.
Meisel, H. and Mikulaszek, E. (1933) *C. R. Soc. Biol.* **114,** 364.
Miles, A. A. (1951) *J. gen. Microbiol.* **5,** 307.
Møller, V. (1955) *Acta path. microbiol. scand.* **36,** 158.
Moltke, O. (1927) *Contributions to the Characterization and Systematic Classification of Bac. proteus vulgaris (Hauser).* Levin and Munksgaard, Copenhagen; (1929) *Zbl. Bakt.* **111,** 399.
Morgan, H. de R. (1906) *Brit. med. J.* **i,** 908.
Morrison, R. B. and Scott, A. (1966) *Nature, Lond.* **211,** 255.
Naylor, P. G. D. (1964) *J. appl. Bact.* **27,** 422.
Pavlatou, M., Hassikou-Kaklamani, E. and Zantioti, M. (1965) *Ann. Inst. Pasteur* **108,** 402.

Penner, J. L. and Hennessy, J. N. (1980) *J. clin. Microbiol.* **12,** 304.

Penner, J. L. and Hinton, N. A. (1973) *Canad. J. Microbiol.* **19,** 271.

Penner, J. L., Hinton, N. A., Duncan, I. B. R., Hennessy, J. N. and Whiteley, G. R. (1979) *J. clin. Microbiol.* **9,** 11.

Penner, J. L., Hinton, N. A. and Hennessy, J. (1975) *J. clin. Microbiol.* **1,** 136.

Penner, J. L., Hinton, N. A., Hennessy, J. N. and Whiteley, G. R. (1976*a*) *J. infect. Dis.* **133,** 283.

Penner, J. L., Hinton, N. A., Whiteley, G. R. and Hennessy, J. N. (1976*b*) *J. infect. Dis.* **134,** 370.

Perch, B. (1948) *Acta. path. microbiol. scand.* **25,** 703; (1954) *Ibid.* **35,** 278.

Perez, F. (1899) *Ann. Inst. Pasteur* **13,** 937; (1901) *Ibid* **15,** 409; (1913) *Berl. klin. Wschr.* **50,** 2411.

Phillips, J. E. (1955*a*) *J. Hyg. Camb.* **53,** 26; (1955*b*) *Ibid,* **53,** 212.

Prentice, B. and Robinson, B. L. (1979) *Canad. med. Ass. J.* **121,** 745.

Proom, H. (1955) *J. gen. Microbiol.* **13,** 170.

Rauss, K. F. (1936) *J. Path. Bact.*, **42,** 183; (1963) *Int. Bull. bact. Nomencl.* **13,** 85.

Rauss, K. and Vörös, S. (1959) *Acta microbiol., Budapest* **6,** 233; (1967) *Ibid.* **14,** 195, 199.

Russ-Münzer, A. (1935) *Zbl. Bakt.* **133,** 214.

Rustigian, R. and Stuart, C. A. (1943) *Proc. Soc. exp. Biol., N.Y.* **53,** 241; (1945) *J. Bact.* **49,** 419.

Sandys, G. H. (1960) *J. med. Lab. Technol.* **17,** 224.

Schmidt, W. C. and Jeffries, C. D. (1974) *Appl. Microbiol.* **27,** 47.

Senior, B. W. (1977) *J. med. Microbiol.* **10,** 7.

Shafi, M. S. and Datta, N. (1975) *Lancet* **i,** 1355.

Shiga, M. (1922) *Zbl. Bakt.* **88,** 521.

Skirrow, M. B. (1969) *J. med. Microbiol.* **2,** 471.

Smit, J. A. and Coetzee, J. N. (1967) *Nature, Lond.* **214,** 1238.

Smith, D. G. and Alwen, J. (1966) *Nature, Lond.* **212,** 941.

Snyder, M. L. and Lichstein, H. C. (1940) *J. infect. Dis.* **67,** 113.

Stickler, D. J. (1974) *J. clin. Path.* **27,** 284.

Stickler, D. J. and Thomas, B. (1976) *J. clin. Path.* **29,** 815.

Story, P. (1954) *J. Path. Bact.* **68,** 55.

Stuart, C. A., Wheeler, K. M. and McCann, V. (1946) *J. Bact.* **52,** 431.

Stuart, C. A., Wheeler, K. M., Rustigian, R. and Zimmerman, A. (1943) *J. Bact.* **45,** 101.

Sutter, V. L. and Foecking, F. J. (1962) *J. Bact.* **83,** 933.

Sykes, J. A. and Reed, R. (1949) *J. gen. Microbiol.* **3,** 117.

Taylor, J. F. (1928) *J. Path. Bact.* **31,** 897.

Vieu, J.-F. (1958) *Zbl. Bact.* **171,** 612.

Ward, H. C. (1917) *J. infect. Dis.* **21,** 338.

Weil, E. and Felix, A. (1917) *Wien. klin. Wschr.* **30,** 1509; (1918) *Ibid.* **31,** 637.

Wenner, J. J. and Rettger, L. F. (1919) *J. Bact.* **4,** 331.

Wenzel, R. P., Hunting, K. J., Osterman, C. A. and Sande, M. A. (1976) *Amer. J. Epidem.* **104,** 170.

White, P. B. (1933) *Brit. J. exp. Path.* **14,** 145.

Williams, F. D., Anderson, D. M., Hoffman, P. S., Schwarzhoff, R. N. and Leonard, S. (1976) *J. Bact.* **127,** 237.

Winkle, S. (1945) *Zbl. Bakt.* **151,** 494.

Yacob, M. (1932) *Indian J. med. Res.* **19,** 787.

Yoshikawa, T. T., Shibata, S. A., Chow, A. W. and Guze, L. B. (1978) *Antimicrob. agents Chemother.* **13,** 177.

36

Shigella

Bernard Rowe

Introductory

Definition

Non-motile organisms conforming to the definition of the Enterobacteriaceae (Chapter 33). Acid produced from a number of carbohydrates. Gas production almost entirely restricted to members of one serotype. Lactose usually not fermented, except by *Sh. sonnei* and occasional strains in other serotypes. Adonitol and inositol never and salicin very rarely fermented. Indole production variable. Give a positive methyl-red and a negative Voges-Proskauer reaction, and do not grow on Simmons's citrate medium. Do not hydrolyse urea, deaminate pheylalanine, form H_2S on triple sugar iron agar, grow in KCN medium, ferment malonate, oxidize gluconate, or liquefy gelatin. Do not form lysine decarboxylase. Serotypes are characterized by their somatic antigens. Cause bacillary dysentery or acute enteritis in man and the higher apes. Found, as a rule, in the intestinal tract of these animals. G + C content of the DNA: 50–52 moles per cent.

Type species: *Shigella dysenteriae*.

The shigellae form a group that, like the salmonellae, was created by medical bacteriologists to include organisms sharing a common habitat and the ability to cause a particular disease. Their cultural and biochemical characters indicate that they are enterobacteria and that they are most closely related to *Esch. coli* (Chapter 34).

The *Shigella* group is generally considered to consist of four main subgroups that are differentiated by a combination of biochemical and serological characters.

Subgroup A. These are organisms that typically do not ferment mannitol. They include Shiga's bacillus (Shiga 1898*a*, *b*, 1901, Kruse, 1900, 1901), Schmitz's bacillus (Schmitz 1917), and a number of less common types described by Andrewes (1918), Dudgeon and Urquhart (1919), Large and Sankaran (1934), Sachs (1943), and others. The members of subgroup A are nowadays described as serotypes of *Sh. dysenteriae*, a name originally applied to Shiga's bacillus alone.

Subgroup B. This is a group of antigenically interrelated organisms that usually ferment mannitol and are referred to as *Sh. flexneri*. The first members were isolated by Flexner (1900*a*, *b*) and by Strong and Musgrave (1900).

Subgroup C. Boyd (1931, 1932, 1936, 1938, 1940) described a group of mannitol fermenters resembling *Sh. flexneri* biochemically but differing from it antigenically; they are now generally referred to as *Sh. boydi*.

Subgroup D. The only member of this subgroup is *Sh. sonnei*, a late-lactose-fermenting organism defined by Sonne (1915), though almost certainly recognized by earlier workers (Duval 1904, Castellani 1907) and probably identical with Kruse's type E bacillus (Kruse *et al.* 1907).

The organisms described by Andrewes (1918) as *B. alkalescens* and *B. dispar* were for a long time included among the dysentery bacilli. They are non-motile, anaerogenic organisms that may acidify lactose after some delay, and are now considered to be biochemically atypical strains of *Esch. coli* (see Chapter 34). They often share antigens with shigella serotypes and this may lead to difficulties in identification (see p. 329).

Table 36.1 gives a list of the organisms included in the genus *Shigella* and of some of the names under which they were described previously.

Morphological and cultural characters

Dysentery bacilli are non-motile gram-negative organisms indistinguishable morphologically and culturally from other enterobacteria. Generally, they grow rather less profusely than most coliform bacteria, but the colonies of Sonne's bacillus tend to be larger and more opaque than those of other dysentery bacilli. Shigellae grow on deoxycholate citrate agar (Leifson 1935) forming characteristic, low convex colonies which are translucent to transmitted light but appear more opaque against a dark background. Sonne's bacillus frequently undergoes colonial variation. On primary isolation it usually forms circular convex colonies with a smooth surface and an entire edge, but a

Table 36.1 Classification of *Shigella* (modified from Report 1958*b*)

Subgroup	Species and serotype	Synonyms
A	*Sh. dysenteriae*	
Mannitol not fermented	Type 1	*Sh. shigae*; Shiga-Kruse bacillus
	2	*Sh. schmitzi*; possibly some strains of *B. ambiguus* (Andrewes)
Types antigenically distinct	3	Q771 — *Sh. arabinotarda A*
	4	Q1167 } Large-Sachs group (possibly includes some strains of *B. ambiguus*)
	5	Q1030 }
	6	Q454 }
	7	Q902 }
	8	559–52
	9	58
	10	2050
B	*Sh. flexneri*	*Sh. paradysenteriae*. *B. paradysenteriae*
Mannitol usually fermented	Type 1 Subtype 1a	V } (Andrewes and Inman)
Types antigenically interrelated	" 1b	VZ }
	2 " 2a	W }
	" 2b	WX }
	3 " 3a	Z }
	" 3b	
	" 3c	
	4 " 4a	103 (Boyd)
	" 4b	103Z (Rewell and Bridges)
	5	P119
	6	Boyd 88, Newcastle and Manchester biochemical subtypes
	X variant	X } (Andrewes and Inman)
	Y "	Y }
C	*Sh. boydi*	*Sh. paradysenteriae*
Mannitol usually fermented	Type 1	170 } Boyd
Types antigenically distinct[1]	2	P288 }
	3	D1 }
	4	P274 }
	5	P143 }
	6	D19 }
	7	Lavington type T; *Sh. etousa*
	8	112
	9	1296/7
	10	430
	11	34
	12	123
	13	425
	14	2770–51
	15	703
D	*Sh. sonnei*	Duval's bacillus; *B. ceylonensis* A; Kruse's *B. pseudodysenteriae* type E
Mannitol usually fermented and lactose at 3–8 days		

[1]Except types 10 and 11, which are related.

few colonies may show, at one or more points on the periphery, a small tangled-hair-like projection, giving the appearance of a bursting bomb (Braun and Weil 1928). Sometimes on first isolation, but nearly always in early subculture, a second type of colony is seen, which is larger and less opaque, has a coarsely ground-glass surface and irregular undulate or crenated edge. These two types of colony consist of organisms differing antigenically and in certain other respects (Mita 1921, Ørskov and Larsen 1925). They have been regarded as corresponding to smooth and rough forms, but Wheeler and Mickle (1945) consider that the terms phase 1 and phase 2 are preferable. These workers describe a third colonial form, which is even larger and flatter, gives a sediment in broth and is salt sensitive. This they consider the true R form. The change from phase 1 to phase 2 is irreversible. Although phase 2 organisms give a smooth growth in broth and are not sensitive to salt at 37°, they are unstable on boiling in 0.9 per cent NaCl, are agglutinable by acid, sensitive to acriflavine, and give a positive Millon's reaction (Rauss *et al.* 1961). They grow poorly on deoxycholate citrate agar. Flexner's bacillus also undergoes a form of irreversible colonial variation (Boyd 1931, 1932, 1936, Cooper *et al.* 1957), but this does not occur so readily, nor is the difference between the colonial forms so obvious as the change from phase 1 to phase 2 in Sonne's bacillus. As in *Salmonella*, antigenic structure and serological behaviour are of far more importance in defining smooth and rough types than the colonial appearance.

A capsulated variety of Boyd's bacillus was described by Ewing and his colleagues (1951). Among the dysentery bacilli, fimbriation is found only in certain serotypes of *Sh. flexneri* and is always of type 1 (Duguid and Gillies 1956; see also Chapter 33).

Resistance

The members of this group are not specially resistant. They are killed by a temperature of 55° in 1 hr, by 0.5 per cent phenol in 6 hr, and by 1 per cent phenol in about 15–30 min. When dried on linen and kept in the dark at room temperature, they survive for from 5 to 46 days (Vaillard and Dopter 1903, Roelcke 1938). In garden earth at room temperature in the dark they survive for 9 to 12 days (Roelcke 1938). In naturally infected faeces kept alkaline and prevented from drying they may remain alive for some days, but in stools that are allowed to become acid through growth of coliform or other bacteria they perish often in a few hours. On the whole, Sonne's bacillus is more resistant to inimical agents generally than the Shiga or Flexner bacillus. When dried on cotton threads it survives less well at relative humidities of 40–60 per cent than at higher or lower humidities (Spicer 1959). It remains viable for long periods in faecal matter dried on cloth, when kept at a cool room remperature in a dark, moist place (Hutchinson 1956). Observations, however, on the survival of dysentery bacilli in various foods at different temperatures revbealed little difference between Flexner's and Sonne's bacillus (Taylor and Nakamura 1964).

Resistance to antimicrobial agents

When shigellae were first tested with sulphonamides nearly all were found to be sensitive, but within a few years isolated reports of resistant strains began to appear (Cooper and Keller 1942, Hardy, 1945). After the end of the second World War, serious outbreaks of bacillary dysentery requiring the extensive use of sulphonamides and antibiotics occurred in Japan and resistant strains were isolated; many possessed multiple resistance (see Tanaka *et al.* 1977). It was soon demonstrated that this multiple resistance was usually transferable between shigellae and *Escherichia coli* by direct cell-to-cell contact (Mitsuhashi *et al.* 1960, 1961). These transferable resistances were subsequently shown to be determined genetically by plasmids (see Falkow 1975).

Between 1947 and 1950 antibiotic-resistant strains became predominant in many countries (Tateno 1950, 1955, Davies 1954). In Britain, ampicillin resistance was first reported in 1966 (Scrimgeour 1966). By 1969, 94 per cent of strains of *Sh. sonnei* in London were resistant to ampicillin, 83 per cent to sulphonamides, 70 per cent to streptomycin, 42 per cent to tetracycline, and 1.5 per cent to neomycin; 34 per cent of strains showed multiple-antibiotic resistance (Davies *et al.* 1970).

In more recent years a high incidence of plasmid-mediated resistance has been observed in enterobacteria, including shigellae, in many countries (Report 1978). The most common pattern was resistance to sulphonamides and streptomycin which were carried by a single plasmid (Crosa *et al.* 1977). Resistance to ampicillin and tetracyclines also occurred frequently and was usually plasmid-mediated. Resistance to chloramphenicol has been less frequently reported, but serious epidemics due to chloramphenicol-resistant strains have occurred. Between 1968 and 1972 an epidemic of bacillary dysentery due to *Sh. dysenteriae* serotype 1 occurred in Central America and Mexico (Mata *et al.* 1970, Gangarosa *et al.* 1970); the epidemic strain was resistant to chloramphenicol, sulphonamides, streptomycin and tetracyclines; the resistances were encoded on a single plasmid. Later in the outbreak, strains with plasmid-mediated ampicillin resistance were reported (Olarte *et al.* 1976). This outbreak showed a high attack rate with appreciable mortality, especially in children. Similar observations were made during epidemics with multiple-resistant strains of this serotype in Bangladesh (Rahaman *et al.* 1974, 1975 and Report 1974*a*) and Sri Lanka (Report 1979). In Southern India, Paniker *et al.* (1978) reported that chloramphenicol resistance appeared to have become established in shigellae; they reported several localized outbreaks due to *Sh. dysenteriae* serotype 1 with many deaths in children; the chloramphenicol resistance was plasmid-mediated and formed part of a multiple resistance linkage group. In many countries information is lacking on the incidence of antibiotic resistance in shigellae.

Metabolism

The dysentery bacilli are aerobes and facultative anaerobes. Their optimum temperature is about 37°. According to Braun and Weil (1928) Sonne's bacillus grows as readily at 45° as at 37°, and more readily at 10° than most coliform bacteria. Shiga's bacillus is characterized by its inability to form catalase; for this reason it may fail to grow aerobically on the surface of nutrient agar in which peroxide has been formed during heating (Proom *et al.* 1950). All members of the *Shigella* group are oxidase negative.

Shigellae will not grow in a medium containing only salts and a simple carbon source. Most strains grow when glucose and nicotinic acid are added, but some also require one or more amino acids (Davies 1954, Sen 1960*a*). Occasional strains need purines in addition (Sen 1960*b*) and *Sh. flexneri* serotype 6 is said to need vitamins of the B group (T. M. Lachowicz *et al.* 1963).

Biochemical reactions

Shigellae are sensitive to potassium cyanide and fail to utilize malonate, oxidize gluconate, or form large quantities of H_2S. They resemble *Escherichia* in producing glutamic acid decarboxylase. None of the dysentery bacilli forms lysine decarboxylase, liquefies gelatin, splits urea, deaminates phenylalanine or gives a positive Voges-Proskauer reaction. Not only do shigellae fail to grow in Koser's medium, but they are unable to utilize citrate in the presence of readily assimilable sources of carbon and nitrogen. They therefore bring about no change in the pH of Christensen's citrate medium, which is alkalinized by most *Escherichia* strains. A similar test for the utilization of sodium acetate (Trabulsi and Ewing 1962) affords a sharper distinction, in that shigellae, with the exception of certain strains of *Sh. flexneri* 4a, almost invariably give a negative reaction; over 90 per cent of *Esch. coli*, and of the organisms intermediate between it and *Shigella*, give a positive reaction, though this may sometimes be delayed.

All the shigellae produce acid from glucose, but not from adonitol and inositol. Failure to ferment salicin is an almost invariable character, though rare salicin-fermenting strains of *Sh. sonnei* have been described (Serény 1959). Table 36.2 shows some of the more common fermentative reactions seen in the four subgroups.

Subgroup A organisms generally do not ferment mannitol, but a mannitol-positive strain of *Sh. dysenteriae* 3 has been reported. With a few exceptions they also fail to attack lactose, sucrose, dulcitol and raffinose. Shiga's bacillus differs from the other serotypes in that it never ferments sorbitol or arabinose.

Subgroup B organisms generally ferment mannitol, but mannitol non-fermenting strains have been seen in all six serotypes. Flexner bacilli do not ferment lactose, but may occasionally acidify sucrose slowly, even on first isolation. The fermentative reactions of *Sh. flexneri* serotype 6 may be used to recognize three biochemical subtypes: 88 which is an anaerogenic mannitol fermenter like the rest of the Flexner bacilli; Manchester which produces both acid and gas from glucose

Table 36.2 Typical biochemical reactions of the dysentery bacilli

	Gas	Glucose	Mannitol	Lactose	Sucrose	Dulcitol	Sorbitol	Arabinose	Xylose	Raffinose	Indole
Subgroup A											
(*Sh. dysenteriae*)											
Shiga's bacillus	–	+	–	–	–	–	–	–	–	–	–
Schmitz's bacillus	–	+	–	–	–	–	v	v	–	–	+
Serotypes 3–10	–	+	–	–	–	–	v	v	v	–	v
Subgroup B											
(*Sh. flexneri*)											
Serotypes 1–5	–	+	+	–	– (or x)	–	v	v	–	v	v
Serotype 6											
88 variety	–	+	+								
Newcastle variety	+	+	–	–	– (or x)	v	v	v	–	–	–
Manchester ,,	+	+	+								
Subgroup C											
(*Sh. boydi*)											
Serotypes 1–15	–	+	+	–	–	v	v	v	v	–	v
Subgroup D											
(*Sh. sonnei*)	–	+	+	x (or –)	x (or –)	–	–	+	v	x (or –)	–

+ = positive reaction within 48 hr.
v = some strains positive, others negative.
x = negative, or late and irregularly positive reaction.
– = negative reaction.

and mannitol; and Newcastle which does not acidify mannitol but usually forms gas from glucose. It has been repeatedly stated that aerogenic members of this serotype produce small bubbles of gas, but if a suitable peptone is used, they usually produce sufficient to fill a third of the Durham's tube (see J. O. Ewing and Taylor 1945).

Subgroup C (*Sh. boydi*) contains anaerogenic organisms that ferment mannitol. Lactose, sucrose and raffinose are not attacked.

Subgroup D organisms (*Sh. sonnei*) form acid promptly from mannitol, arabinose and rhamnose, but not from dulcitol or sorbitol. Characteristically, acid is formed from lactose, sucrose, and raffinose between the 3rd and the 8th day, but certain strains, which are particularly common in France but have given rise to occasional outbreaks in Great Britain, are non-fermenters of these sugars. These lactose-sucrose negative strains usually acidify xylose rapidly (Braun and Weil 1928, Szturm-Rubinsten and Piechaud 1957). Rarely *Sh. sonnei* fails to ferment mannitol (Graham 1958).

Other biochemical reactions, many of which are useful in strain differentiation, include the following.

1. None of the shigellae ferments mucate, except for a few strains of *Sh. sonnei* which give a delayed reaction. In contrast, *Esch. coli* usually gives a strongly positive reaction with mucate.

2. Only *Sh. sonnei* and *Sh. boydi* serotype 13 form ornithine decarboxylase. The action of shigellae on arginine is variable.

3. The production of indole varies with serotype, but Schmitz's bacillus is always indole positive and Shiga's bacillus, *Sh. flexneri* serotype 6 and *Sh. sonnei* are uniformly indole negative. The remaining types may be indole positive or negative.

4. Shiga's bacillus is always catalase negative. A number of other dysentery bacilli also fail to produce catalase. Many members of *Sh. dysenteriae* serotypes 3, 4, 6 and 9, *Sh. flexneri* serotype 4a and *Sh. boydi* serotype 13 are catalase negative. Schmitz's bacillus and Sonne's bacillus always give a positive reaction (Carpenter and K. Lachowicz 1959, Carpenter 1966).

5. *Sh. sonnei* commonly ferments lactose but the reaction is nearly always delayed; the remaining serotypes are lactose negative but lactose-fermenting strains of *Sh. flexneri* serotype 2a (Trifanova *et al.* 1974) and *Sh. boydi* serotype 9 (Manolov *et al.* 1962) have been described. A number of lactose-negative shigellae form β-D-galactosidase, including almost all strains of Shiga's bacillus and a minority of other members of subgroups A and C (Szturm-Rubinsten and Piéchaud 1963, Lubin and Ewing 1964). The action of shigellae on maltose is variable, and of little value in classification.

6. The production of gas from glucose, which characterizes certain biotypes of *Sh. flexneri* serotype 6, has also been recorded in *Sh. boydi* serotype 14 (Carpenter 1961) and *Sh. boydi* serotype 13 (Rowe *et al.* 1975).

(For further information about the biochemical reactions of the group see Kauffmann 1969, Ewing *et al.* 1971, Edwards and Ewing 1972.)

Antigenic structure

We shall consider separately the antigenic structure of the four subgroups of dysentery bacilli.

Subgroup A According to the nomenclature agreed by the Enterobacteriaceae Subcommittee (Report 1958*a*, *b*, 1968) subgroup A contains ten serotypes of *Sh. dysenteriae*, of which Shiga's bacillus and Schmitz's bacillus are numbered 1 and 2 respectively. Each serotype possesses a type or main antigen by which it can be differentiated and there are only a few minor cross-reactions between serotypes within the subgroup. Many of the serotypes show strong antigenic relationships to *Esch. coli* serotypes; these relationships may be identical or reciprocal (Edwards and Ewing 1972).

According to Schütze (1944) Shiga strains vary in their agglutinability; hypoagglutinable strains may be rendered more agglutinable by growth at 20–26°, by heating in saline to 60° for 1 hr, or by adding 0.5 per cent phenol. It is now known that this inhibition of O agglutination is caused by surface antigens which are inactivated by heat. Similar observations have been made on other members of the genus (Edwards and Ewing 1972).

Subgroup B The members of this subgroup have complex antigenic structures. The work of Gettings (1919), Murray (1918) and Andrewes and Inman (1919) afforded a picture of it as containing at least four antigenic components. According to Andrewes and Inman, each of the four components V, W, X and Z was represented to some extent in every strain, but in any one strain there was usually a preponderance of one antigen over the rest. An exception was the so-called Y race, which appeared to contain more evenly balanced V, W, and Z components, with a small amount of X.

This conception was challenged by the work of Boyd (1931, 1932, 1936, 1938) in India. Studying a strain of dysentery bacillus known as 103 which, when newly isolated, was not agglutinated by antisera to the V, W, Y and Z races, and only very feebly by an X antiserum, Boyd observed that after some time in artificial culture it gave rise to two types of colony. One of them, referred to as 103A, was virtually inagglutinable by V to Z antisera. The other, referred to as 103B, was agglutinated readily by antisera of the V to Z group. This variant differed further from 103A in that it bred true, whereas 103A behaved like the parent strain in consistently giving rise to a variant of the 103B type. Further study showed that a serum prepared against 103A agglutinated both 103A and 103B, but that a serum prepared against 103B had practically no action on 103A. Absorption of agglutinin experiments confirmed the suggestion that 103A contained two antigens—a type and a group—but that 103B contained only the group antigen. Strain 103B grew uniformly in broth and was perfectly stable in saline; it was therefore not a rough variant. Boyd later found other examples of A→B variation among Flexner bacilli. They occurred more readily in some strains than in others and were always irreversible. He also showed that the V, W and Z races of Andrewes and Inman each possessed a type antigen and a group antigen. The X and Y strains had different group antigens and no type antigens. He found evidence of at least six components in the group antigens, which occurred in a variety of combinations in different strains of Flexner bacilli.

Boyd (1940) included in the Flexner group all organisms that possessed the group antigens, and numbered the serotypes according to their specific antigens. He placed strains devoid of group antigen but biochemically similar to Flexner's bacillus in a separate species for which the name *boydi* has since been adopted.

In the terminology now agreed upon by the Enterobacteriaceae Sub-committee there are six main serotypes of Flexner bacilli (1–6), each characterized by a different specific antigen. These may be further subdivided into sub-serotypes (*e.g.* 1a and 1b) according to the exact constitution of the group antigen. The type antigens themselves are now designated by roman and the group antigens by arabic numerals. A simplified version of the antigenic scheme is given in Table 36.3. Not all the known group factors are included, but only those that serve to distinguish the commonly recognized serotypes and sub-serotypes. Varieties lacking one of the group factors, or with an additional group factor, are seen occasionally but are often transitory in their appearance (Ewing and Carpenter 1966). There are numerous cross-reactions between the O antigens of *Sh. flexneri* and the *Esch. coli* serogroups.

Although the antigens of Flexner's bacillus have been studied mainly by the agglutination method, similar results may be obtained by precipitation (González and Otero 1945, Brodie and Green 1951), by haemagglutination (Chun and Park 1956), and by immunofluorescence techniques (La Brec *et al.* 1959). Flexner bacilli also possess thermolabile surface antigens (Madsen 1949, Ślopek *et al.* 1960, Edwards and Ewing 1962).

The O-antigenic specificity of Flexner's bacillus is determined by the polysaccharide portion of the cell-wall lipopolysaccharide, which has the same general structure as that of other enterobacteria (Johnston *et al.* 1968). The O-specific side-chains of the Flexner serotypes, unlike those of salmonella serotypes, are all composed of the same three sugars: D-glucose, *N*-acetyl glucosamine and L-rhamnose. Antigenic specificity depends on the sequence of these sugars in the side-chain (Simmons 1969, 1971). Determinants for type and group factors form part of the same macromolecule. Petrovskaya and Bondarenko (1977) have studied the immunochemical and genetic basis of the *Sh. flexneri* antigens. All serotypes possess a basic structure of group antigens 3, 4; the type-specific antigens I to V and group antigens 7, 8 arise from the 3, 4 antigens by phage conversions. For type-specific antigen III and group antigen 6 the conversions result in the incorporation of acetyl groups in the side chains of the cell-wall lipopolysaccharide. For the other type-specific antigens and group antigens 7, 8, α-glycosyl groups are incorporated. The lipopolysaccharide of *Sh. flexneri* serotype 6 does not control the immunochemical determinants of group antigens 3, 4 and these workers suggest that this serotype resembles *Sh. boydi* immunochemically. Using the information from this study, Petrovskaya and Kohmenko (1979) have proposed an amendment to the classification of *Sh. flexneri*. As well as recommending the transfer of *Sh. flexneri* serotype 6 to *Sh. boydi*, they recommended that *Sh. flexneri* serotype 3c be deleted, re-designated the antigenic structures of *Sh. flexneri* serotype 3a and 3b and suggested antigenic formulae for subtypes 5a and 5b. These recommendations have not yet been confirmed by the Enterobacteriaceae Subcommittee.

The *fimbriae* of all strains of Flexner's bacillus are antigenically identical, but different from those of fimbriate coliform bacteria. A serum made with a fimbriate Flexner strain will agglutinate to high titre a living suspension of a member of another Flexner serotype if it is also in the fimbriate form. The floccules form rapidly at 37° and are loose and bulky. It is therefore important to ensure that diagnostic sera are free from fimbrial antibodies by preparing them from cultures grown on solid medium (Gillies and Duguid 1958).

Subgroup C Members of this subgroup have simpler antigenic structures than *Sh. flexneri*. With the exception of a reciprocal antigenic relation between

Table 36.3 A simplified version of the antigenic scheme for *Sh. flexneri* (modified from Ewing and Carpenter 1966)

Serotype	Sub-serotype	Abbreviated antigenic formula
1	1a	I:4 . . .
	1b	I:6 . . .
2	2a	II:3, 4 . . .
	2b	II:7, 8 . . .
3	3a	III:6, 7, 8 . . .
	3b	III:3, 4, 6 . . .
	3c	III:6 . . .
4	4a	IV:3, 4 . . .
	4b	IV:6 . . .
5	5	V:7, 8 . . .
6		VI:— . . .
X variant		—:7, 8 . . .
Y variant		—:3, 4 . . .

types 10 and 11, all serotypes are antigenically distinct. Many *Sh. boydi* O antigens are related to *Esch. coli* antigens (Edwards and Ewing 1962, Rowe *et al.* 1976). Some strains are not agglutinated by homologous O sera unless heated (Madsen 1949). According to Edwards and Ewing (1962) this is due to the presence of surface antigens with the same general properties as the *Escherichia* B antigens. Metzger (1958) describes a type of heat-labile surface antigen which is destroyed by autoclaving but not by boiling and does not cause O inagglutinability. The capsular antigens of serotypes 1 and 2 (Ewing *et al.* 1951) are not related to each other.

Subgroup D Sonne's bacillus may undergo an antigenic variation that affects the somatic antigens and has been variously referred to as phase or form variation (Wheeler and Mickle 1945). Cultures often contain a mixture of both forms but by selection of freshly isolated smooth cultures it is possible to prepare a pure S or form I serum which does not react with R or form II cultures. Unless careful selection is made of cultures for antigen preparation, the antisera will react with both forms; this is of little concern in routine work, and indeed it is possibly better that such sera for routine use should be a mixture of the two forms. Form I antigen shows no significant relation to the O antigens of the other shigellae or to those of any of the *Esch. coli* of O serogroups 1 to 164. It is however identical with the O antigen of one type (C27) of *Plesiomonas shigelloides* (see Chapter 27). Form II antigen is closely related to an antigen of *Sh. boydi* serotype 6; this reaction is only seen with old cultures of the latter and is thought to be due to common rough antigens (Martin *et al.* 1968). According to Romanowska and Mulczyk (1967) the specific portion of the phase 1 antigen contains both sugars and amino acids, but that of the phase 2 antigen is a polysaccharide. The form I antigen is genetically determined by a plasmid (see Chapter 6).

Provisional serotypes

In addition to the serotypes described in Table 36.3, a number of provisional serotypes have been described (Ewing *et al.* 1958, Gross *et al.* 1980). These have the biochemical reactions of *Shigella* but, because they do not belong to accepted serotypes, they have remained sub-judice. However Petrovskaya and Khomenko (1979) have proposed that two of them (3873–50 and 3341–55) be regarded as *Sh. dysenteriae* 11 and 12, and two others (2710–54 and 2615–53) as *Sh. boydi* 16 and 17 respectively.

Usually the antisera for the identification of these strains are available only at reference laboratories. It follows that when a laboratory isolates a strain which is biochemically a *Shigella*, even though it does not react with any of the complete range of antisera, it requires further identification.

For methods of identifying the individual serotypes, reference should be made to articles by Carpenter (1968), Kauffmann (1969), and Edwards and Ewing (1972). It is particularly important to remember that because of surface antigens some strains of *Shigella* especially when freshly isolated may not agglutinate in diagnostic sera as living suspensions. It is necessary to heat the suspension for 30 min at 100° C.

A further complication in the serological identification is caused by the sharing of antigens with other Enterobacteriaceae, particularly *Esch. coli* (Edwards and Ewing 1972, Rowe *et al.* 1976). For this reason biochemical and serological testing must be done in parallel.

Bacteriophage and colicine typing

Thomen and Frobisher (1945) used phages obtained from chicken faeces, and were able to subdivide Flexner's bacillus into types which showed a close correspondence with the recognized serotypes (see also Ślopek *et al.* 1972). Working with Sonne's bacillus, Hammarström (1949) succeeded by the use of 11 phages acting on the phase 2 organism in dividing 1834 strains into 68 types and subtypes. Several other workers have used this method; if carefully carried out it gives fairly consistent and useful results (Mayr-Harting 1952, Kallings *et al.* 1968), but the range of lytic activity of typing phages may be restricted by resistance-transfer factors (Tschäpe and Rische 1970). An extended bacteriophage-typing scheme for *Sh. sonnei* was described by Pruneda and Farmer (1977).

D'yakov (1957) showed that Sonne and Boyd bacilli frequently produced bacteriocines with a wide range of activity against enterobacteria. He found that Sonne strains could be subdivided into groups according to the range of activity of the substances they produced against other strains of the same species. This method was more extensively applied by Abbott and Shannon (1958), who used 9 indicator strains, which included 7 strains of *Sh. sonnei* and one each of *Esch. coli* and Schmitz's bacillus (see also Abbott and Graham 1961). Horák (1980) has developed an extended scheme in which 33 indicator strains are used and 26 colicine types can be identified. The ability to produce a particular colicine is a fairly stable character, and the results of colicine typing are therefore of value in epidemiological studies.

Toxin formation

The toxicity of the dysentery bacilli, like that of the other enterobacteria, is associated in the main with the cell-wall lipopolysaccharide (Chapter 33). In addition, Shiga's bacillus forms a toxic product which has been described as 'exotoxin' or 'neurotoxin'. Many years ago it was observed that cultures of Shiga's bacillus, when injected intravenously into rabbits, gave rise to diarrhoea and paralysis (Conradi 1903, Todd 1904, Dopter 1905, Flexner and Sweet 1906). *Post mortem*, the intestinal wall was often swollen, haemorrhagic or ulcerated, and histological changes could be found in the central nervous system. The agent responsible could be separated from the bacterial bodies by filtration, and was heat-labile. No similar substance was produced by Flexner's bacillus.

Kraus and Dörr (1905), Bessau (1911), and Olitsky and Kligler (1920) concluded that Shiga's bacillus formed two toxins: (1) a soluble exotoxin which affected rabbits but not guinea-pigs, gave rise to paralysis, and resulted in the production of specific neutralizing antitoxin, and (2) an insoluble endotoxin, present in bacterial bodies, which was fatal to both rabbits and guinea-pigs. The chemical work of Boivin and Mesrobeanu (1937*a–h*, 1938) left little doubt that two toxic substances were formed. They demonstrated in cultures of Shiga's bacillus (1) a thermostable 'glycolipid' (i.e. lipopolysaccharide) antigen and (2) a thermolabile protein which they regarded as an exotoxin having a neurotropic effect. The protein toxin is specific to Shiga's bacillus and may be found in either antigenically smooth or rough strains of this organism.

It is perhaps unfortunate that the terms endotoxin and exotoxin have been used to refer to these two substances. The so-called exotoxin is closely bound up with the cell bodies and, as the observations of Okell and Blake (1930) showed, is not excreted by living bacilli. It is found in filtrates of broth cultures older than one week, in autolysates and phage lysates of younger cultures, and in dried bacterial bodies. Dubos and Geiger (1946) found that the presence of more than 0·15 μg of iron per l in the medium inhibited its production. Van Heyningen and Gladstone (1953) extracted the toxin from heat-killed cultures with alkali and purified it. The amount of toxin produced by the organisms was low but the potency of the purified material was high; its toxicity for the rabbit was of the same order as that of tetanus and botulinum A toxins for the guinea-pig, and considerably greater than that of diphtheria toxin for the guinea-pig. It gives rise to, and is neutralized by, a specific antitoxin which combines with it in constant proportions. There is now little doubt that it is not primarily a neurotoxin. Its main action is on small blood vessels (Penner and Bernheim 1942, Howard 1955, Bridgwater *et al.* 1955). In the mouse and the rabbit most of the effect is on the blood vessels of the central nervous system, but there is some action on the bowel. Lesions in the rat are confined to the gastro-intestinal tract, and in the hamster are chiefly pulmonary. The guinea-pig appears to be insusceptible (Cavanagh *et al.* 1956). The toxin has no action when given to monkeys by the mouth or into isolated pouches of the gut (Branham *et al.* 1949, 1953).
(For further information about this toxin, see van Heyningen 1971.)

In general, the high toxicity of Shiga's bacillus for experimental animals is not rivalled by most of the other dysentery bacilli. Buchwald (1939) described a heat-labile 'neurotoxin' in Schmitz's bacillus, but this appeared to be much less potent than the toxic protein of Shiga's bacillus.

Keusch and his colleagues (1970) found that some strains of Shiga's bacillus produced a heat-labile protein that gave rise to the accumulation of fluid when injected into ligated loops of rabbit ileum. This agent, which is also produced by some strains of *Sh. flexneri* and *Sh. sonnei*, is lethal and paralytic for mice, produces histological changes in the ileal mucosa and is cytotoxic for tissue cultures (Keusch and Jacewicz 1975, McIver *et al.* 1975). When injected into rabbit ileal loops, the fluid secreted is bloody and small ulcers are formed. These changes resemble those seen in the jejunum or colon of the shigella-infected monkey (Rout *et al.* 1975). The cytotoxic and enterotoxic activity of this agent may thus have roles in the pathogenesis of bacillary dysentery. Early observations suggested that the agent formed by *Sh. dysenteriae* serotype 1 did not stimulate adenylcyclase as do the heat-labile enterotoxin of *Esch. coli* and the enterotoxin of *V. cholerae*, but Charney and co-workers (1976) have shown that under optimal conditions mucosal adenylcyclase activity is increased.

Most strains of *Sh. flexneri* serotype 2a and a few X and Y variants produce a *mucinase* (Formal and Lowenthal 1956, Formal *et al.* 1958*a*). Its significance is unknown.

Pathogenicity

Shigellae give rise to bacillary dysentery in man. The disease caused by Sonne's bacillus tends to be mild and of short duration, whilst that caused by *Sh. flexneri* tends to be more severe. *Sh. boydi* and *Sh. dysenteriae* produce disease of varying severity, but Shiga's bacillus has often caused epidemics of severe disease (Cahill *et al.* 1966, Mata *et al.* 1970). Apart from chimpanzees and monkeys, which may become infected in their natural habitat and in captivity (Ruch 1959), bacillary dysentery is a specifically human disease. There have been occasional reports of infections in dogs.

Parenteral injection As we have seen, Shiga's bacillus is toxic for rabbits and mice. The intravenous injection of 0·01 mg of bacteria into the *rabbit* causes diarrhoea, paralysis and collapse, with death in 1–4 days. *Post mortem* there are scattered petechial haemorrhages, and the intestine is congested and oedematous, with mucoid or bloody fluid in the lumen. A similar picture is seen after injection of dead bacilli in larger quantity, or of toxin. When a small dose of bacilli is given, there may be time for necrosis and actual ulceration of the intestine to occur. Very large doses of Shiga's bacillus may cause death in the guinea-pig, and there are usually no macroscopic changes to be seen *post mortem*.

With other dysentery bacilli the death of rabbits, mice and guinea-pigs can be brought about by large doses given intravenously or intraperitoneally, but no specific lesions are seen in the intestine. Phase 1 strains of *Sh. sonnei* are said to be more virulent for mice than phase 2 organisms when injected intraperitoneally (Branham and Carlin 1950, Rauss *et al.* 1961). Variants of *Sh. flexneri* serotypes 2a and 2b that have lost their specific antigen are also said to be of lower virulence (Cooper *et al.* 1957). In both instances, loss of virulence is associated with loss of specific immunogenicity.

Oral infection Most workers have failed completely to reproduce true dysenteric lesions in cats, dogs and rabbits. The feeding of *monkeys* with large doses (*ca* 10^{10} organisms) of Flexner's bacillus gives rise to severe dysentery (Dack and Petran 1934, La Brec *et al.* 1964); acute inflammation of the mucosa occurs in the stomach as well as in the colon (Kent *et al.* 1967). With Sonne's bacillus the disease is less severe (Takasaka *et al.* 1969).

Guinea-pigs are normally very resistant to infection with *Sh. flexneri* by the oral route, but if they are starved for four days and given $CaCO_3$ before and morphine after feeding they suffer from an acute infection with an LD 50 of *ca* 10^6 (Formal *et al.* 1958*b*, 1959). Instead of starvation, a subcutaneous injection of CCl_4 may be given. Death occurs within 24–48 hr, but there is no diarrhoea. The lesions are mainly in the small intestine (Formal *et al.* 1963), and consist of areas of cellular infiltration and ulceration resembling the dysenteric lesions of man. The organisms penetrate the epithelium and appear in the lamina propria within 8 hr of infection (La Brec *et al.* 1964). Young guinea-pigs with a dietary deficiency of folic acid also suffer from a fatal infection when *Sh. flexneri* is administered orally (Haltalin *et al.* 1970).

Mice A more chronic but non-fatal infection of mice can be produced by giving streptomycin-resistant Flexner bacilli with $CaCO_3$ after preliminary treatment with streptomycin and erythromycin (Freter 1956, Cooper 1959, Cooper and Pillow 1959). Infection can be established with as few as 100 organisms, but it is necessary to continue giving antibiotics to prevent the re-establishment of the normal bowel flora. The mice remain well but excrete the organism for at least two weeks. There is gross mucosal destruction and inflammatory exudate in the caecal region.

Other experimental infections In the guinea-pig, the introduction of large numbers of organisms into the urinary bladder (Bingel 1943), the vagina (Piéchaud and Szturm-Rubinsten 1959) or the conjunctival sac (Serény 1955, 1957) causes acute inflammation. After the instillation of 10^5–10^7 organisms into the eye, a muco-purulent conjunctivitis appears in 1–3 days, and is followed by a severe keratitis. Freshly isolated shigellae, of whatever species, give rise to keratoconjunctivitis, as do certain escherichia strains that cause a dysentery-like disease in man; other enterobacteria do not. Rough Flexner strains, *Sh. sonnei* phase 2, and many old laboratory cultures of dysentery bacilli usually fail to cause keratoconjunctivitis.

Virulent shigella strains penetrate cultured HeLa cells, multiply in them, and cause cytopathic changes (La Brec *et al.* 1964, Nakamura 1967); similar changes occur in primary cultures of guinea-pig corneal cells (Ogawa *et al.* 1967).

Repeated subculture of dysentery bacilli in the laboratory often leads to loss of virulence without loss of type-specific O-antigen. Strains that have undergone this change may fail to cause disease in the starved guinea-pig, in the monkey and in man (La Brec *et al.* 1964, Istrati *et al.* 1967), and do not cause keratoconjunctivitis in the guinea-pig or penetrate HeLa cells; but there is no increase in the LD50 for the mouse by the intraperitoneal route. Formal and his colleagues (1965) obtained hybrids by recombination of a virulent strain of *Sh. flexneri* serotype 2a with a non-virulent strain of *Esch. coli* which produced keratoconjunctivitis and penetrated and multiplied within HeLa cells but caused only an abortive infection in the starved guinea-pig and was relatively non-virulent for monkeys. The organisms invaded the epithelial cells and the lamina propria of the guinea-pig ileum but failed to multiply, and most of them were destroyed within 24 hr. Gemski and his colleagues (1972) carried out feeding experiments using Shiga's bacillus in monkeys in parallel with in-vitro studies of epithelial penetration and enterotoxin production. Non-invasive toxigenic variants failed to produce disease in monkeys, whereas invasive, non-toxigenic strains caused disease in both, but of a milder nature than that produced by the fully virulent invasive and toxigenic strain. Levine *et al.* (1973) obtained similar results in human volunteers. It may be concluded that epithelial penetration is the main virulence property and that the role of the enterotoxin has yet to be defined.

If the experimental data are applicable to the natural disease, a virulent shigella must have the ability (1) to penetrate epithelial cells, (2) to multiply there, and then (3) to produce toxic products which affect the physiological functions of the epithelial cells and may cause cell death and ulceration.

Classification

The definition given at the beginning of this chapter cannot be relied upon to exclude all strains of *Esch. coli* from the genus *Shigella*. Typical strains of *Esch. coli* are easily recognised, but an almost continuous series of biochemical types connects them to the shigellae. Non-motile, anaerogenic strains that fail to produce acid from lactose or do so late present difficulties in classification and identification. Such strains (p. 321) have been referred to as the Alkalescens-Dispar Group, but as we have seen (Chapter 34) most workers regard them as biotypes of *Esch. coli*; the proposal to create a separate taxonomic group for them has not been generally accepted. (Report 1974*b*).

The fact that most of these biochemically atypical *Esch. coli* strains belong to a limited number of *Escherichia* O serogroups and that their O antigens are often related to those of *Shigella* adds further difficulties to their identification. In practice, reliance is usually placed on the decarboxylation of lysine, and the fermentation of citrate in Christensen's special medium, and of acetate and mucate, as grounds for excluding a borderline organism from the *Shigella* group, but the result of one test can never be conclusive, and some decisions must inevitably be arbitrary (see Cefalú 1955, Ewing, Reavis and Davis 1958; see Table 36.4).

In this chapter *Shigella* is classified into four subgroups, as proposed by the Shigella Sub-Committee (see Table 36.1). Although this has the advantages of tidiness and simplicity it presents certain dif-

Table 36.4 Differential characters of *Shigella* and *Escherichia* (modified from Ewing 1966)

Character	*Shigella*	*Escherichia*
Motility	–	v
Gas from glucose	–[1]	+[1]
Acid from lactose	– or (+)	+ or (+)[1]
Acid from sucrose	– or (+)	v
Acid from salicin	–[1]	v
Indole	v	+
Lysine decarboxylase	–	v
Fermentation of Christensen's citrate	–	v
Fermentation of sodium acetate	–	+ or (+)
Fermentation of mucate	–[2]	+

+ = positive reaction within 48 hr.
(+) = late positive reaction.
v = some strains positive, others negative.
– = negative reaction.
[1] Some exceptions.
[2] Some *Sh. sonnei* strains(+).

ficulties, particularly in the allocation of new serotypes to the appropriate subgroup. Most trouble is due to the fact that the primary division between the fermenters and non-fermenters of mannitol is not so clear cut as it was thought to be. Thus, if a new organism is mannitol negative, and does not possess the group antigens of *Sh. flexneri* or the type antigen of a known member of subgroups A, C or D, it would be allocated to subgroup A as a new serotype. Some strains that are otherwise identical with *Sh. dysenteriae* serotype 3 ferment mannitol, but have been retained in subgroup A. If the first members of this serotype to be isolated had been mannitol fermenters, they would have been considered as members of subgroup C. Similarly, mannitol-negative variants of several of the Boyd serotypes have been described, but they are retained in subgroup C in the interests of consistency. Since some of these serotypes are rarely seen, it is clear that their classification as serotypes of *Sh. dysenteriae* or *Sh. boydi* may have been fortuitous.

The decision to consider an organism a serotype of subgroup B depends on the possession of one or more of the characteristic *flexneri* group antigens, irrespective of whether or not it ferments mannitol. Thus, the main distinction appears to be between, on the one hand, the 'closed' group of antigenically interrelated Flexner bacilli and, on the other, an 'open' group containing the antigenically heterogeneous serotypes of subgroups A and C. It is unfashionable nowadays to accord much significance to antigenic structure in bacterial classification. In this case, however, it is the peculiar mosaic-like relation between the antigenic constitution of the various Flexner serotypes, rather than the mere possession of antigens in common, that is so characteristic.

Sh. sonnei, the sole member of subgroup D, stands rather on its own. The traditional definition of it as a late fermenter of lactose and sucrose can no longer be maintained, but it differs from the rest of the dysentery bacilli in several of its other biochemical characters. On these grounds alone it must be regarded as a separate species. Cowan (1956) suggested that it should be removed to the *Escherichia* group; admittedly certain of the biochemical characters in which it differs from the rest of the dysentery bacilli are ones in which it agrees with *Escherichia*. It is, however, one of the few dysentery bacilli with no antigenic relation to *Esch. coli*.

The following are reviews on the classification and nomenclature of the *Shigella* group: Boyd (1948), Madsen (1949), Report (1958*a*,*b*, 1968, 1974*b*), Ewing *et al.* (1958, 1971), Ewing and Carpenter (1966), Kauffmann (1969), Edwards and Ewing (1972).

References

Abbott, J. D. and Graham, J. M. (1961) *Mon. Bull. Minist. Hlth Lab. Serv.* **20,** 51.
Abbott, J. D. and Shannon, R. (1958) *J. clin. Path.* **11,** 71.
Andrewes, F. W. (1918) *Lancet* **i,** 560.
Andrewes, F. W. and Inman, A. C. (1919) *Spec. Rep. Ser. med. Res. Coun., Lond.* No. 42.
Bessau, G. (1911) *Zbl. Bakt.* **57,** 27.
Bingel, K. F. (1943) *Z. Hyg. InfektKr.* **125,** 110.
Boivin, A. and Mesrobeanu, L. (1937*a*) *C. R. Acad. Sci.* **204,** 302; (1937*b*) *Ibid.* **204,** 1759; (1937*c*) *C. R. Soc. Biol.* **124,** 442; (1937*d*) *Ibid.* **124,** 1078; (1937*e*) *Ibid.* **125,** 796; (1937*f*) *Ibid.* **126,** 222; (1937*g*) *Ibid.* **126,** 323; (1937*h*) *Ibid.* **126,** 652; (1938) *Ibid.* **128,** 446.
Boyd, J. S. K. (1931) *J. R. Army med. Cps.* **57,** 161; (1932) *Ibid.* **59,** 241, 331; (1936) *Ibid.* **66,** 1; (1938) *J. Hyg., Camb.* **38,** 477; (1940) *Trans. R. Soc. trop. Med. Hyg.* **33,** 553; (1948) *Proc. 4th int. Congr. trop. Med. Malaria, Wash., D.C.,* p. 290.
Branham, S. E. and Carlin, S. A. (1950) *J. Immunol.* **65,** 407.
Branham, S. E., Dack, G. M. and Riggs, D. B. (1953) *J. Immunol.* **70,** 103.
Branham, S. E., Habel, K. and Lillie, R. D. (1949) *J. infect. Dis.* **85,** 295.
Braun, H. and Weil, A. J. (1928) *Zbl. Bakt.* **109,** 16.
Bridgwater, F. A. J., Morgan, R. S., Rowson, K. E. K. and Wright, G. P. (1955) *Brit. J. exp. Path.* **36,** 447.
Brodie, J. and Green, D. M. (1951) *J. gen. Microbiol.* **5,** 1001.
Buchwald, H. (1939) *Z. Immun Forsch.* **96,** 445.
Cahill, K. M., Davies, J. A. and Johnson, R. (1966) *Amer. J. trop. Med. Hyg.* **15,** 52.
Carpenter, K. P. (1961) *J. gen. Microbiol.* **26,** 535; (1966) *Ann. Immunol. Hung.* **9,** 29; (1968) *Identification of Shigella.* (Assoc. Clin. Path. Broadsheet No. 60) London.
Carpenter, K. P. and Lachowicz, K. (1959) *J. Path. Bact.* **77,** 645.
Castellani, A. (1907) *J. Hyg., Camb.* **7,** 1.
Cavanagh, J. B., Howard, J. G. and Whitby, J. L. (1956) *Brit. J. exp. Path.* **37,** 272.
Cefalú, M. (1955) *Z. Hyg. InfecktKr.* **141,** 421.
Charney, A. N., Gots, R. E., Formal, S. B. and Giannella, R. A. (1976) *Gastroenterology* **70,** 1085.
Chun, D. and Park, B. (1956) *J. infect. Dis.* **98,** 82.
Conradi, H. (1903) *Dtsch. med. Wschr.* **29,** 26.
Cooper, G. N. (1959) *Aust. J. exp. Biol. med. Sci.* **37,** 193.

Cooper, G. N. and Pillow, J. A. (1959) *Aust. J. exp. Biol. med. Sci.* **37,** 210.

Cooper, M. L. and Keller, H. M. (1942) *Proc. Soc. exp. Biol., N.Y.* **51,** 238.

Cooper, M. L., Keller, H. M. and Walters, E. W. (1957) *J. Immunol.* **78,** 160.

Cowan, S. T. (1956) *J. gen. Microbiol.* **15,** 345.

Crosa, J. H., Olarte, J., Mata, L. J., Luttropp, L. K. and Peñaranda, M. E. (1977) *Antimicrob. Agents Chemother.* **11,** 553.

Dack, G. M. and Petran, E. (1934) *J. infect. Dis.* **55,** 1.

Davies, J. R. (1954) *Mon. Bull. Minist. Hlth Lab. Serv.* **13,** 114.

Davies, J. R., Farrant, W. N., and Uttley, A. H. C. (1970) *Lancet* **ii,** 1157.

Dopter, C. (1905) *Ann. Inst. Pasteur* **19,** 353.

Dubos, R. J. and Geiger, J. W. (1946) *J. exp. Med.* **84,** 143.

Dudgeon, L. S. and Urquhart, A. L. (1919) *Spec. Rep. Ser. med. Res. Coun., Lond.* No. 40, p. 25.

Duguid, J. P. and Gillies, R. R. (1956) *J. gen. Microbiol.* **15,** vi.

Duval, C. W. (1904) *J. Amer. med. Ass.* **43,** 381.

D'yakov, S. I. (1957) *J. Microbiol., Epidem., Immunobiol.* **28,** 1296.

Edwards, P. R. and Ewing, W. H. (1962) *Identification of Enterobacteriaceae*, 2nd edn. Burgess, Minneapolis; (1972) *Ibid.*, 3rd edn.

Ewing, J. O. and Taylor, J. (1945) *Mon. Bull. Minist. Hlth Lab. Serv.* **4,** 130.

Ewing, W. H. and Carpenter, K. P. (1966) *Int. J. syst. Bact.* **16,** 145.

Ewing, W. H., Edwards, P. R., and Hucks, M. C. (1951) *Proc. Soc. exp. Biol., N.Y.* **78,** 100.

Ewing, W. H., Reavis, R. W. and Davis, B. R. (1958) *Canad. J. Microbiol.* **4,** 89.

Ewing, W. H., Sikes, J. V., Wathen, H. G., Martin, W. H. and Jaugstetter, J. E. (1971) *Biochemical Reactions of Shigella.* U.S. Dept. Hlth, Educat. & Welfare. Atlanta.

Falkow, S. (1975) *Infectious Multiple Drug Resistance.* Pion, London.

Flexner, S. (1900*a*) *Zbl. Bakt.* **28,** 625; (1900*b*) *Bull. Johns Hopk. Hosp.* **11,** 231.

Flexner, S. and Sweet, J. E. (1906) *J. exp. Med.* **8,** 514.

Formal, S. B., Abrams, G. D., Schneider, H. and Sprinz, H. (1963) *J. Bact.* **85,** 119.

Formal, S. B., Dammin, G. J., La Brec, E. H. and Schneider, H. (1958*b*) *J. Bact* **75,** 604.

Formal, S. B., Dammin, G. J., Schneider, H. and La Brec, E. H. (1959) *J. Bact.* **78,** 801.

Formal, S. B., La Brec, E. H., Schneider, H. and Falkow, S. (1965) *J. Bact.* **89,** 835.

Formal, S. B. and Lowenthal, J. P. (1956) *Proc. Soc. exp. Biol., N.Y.* **92,** 10.

Formal, S. B., Lowenthal, J. P. and Galindo, E. (1958*a*) *J. Bact.* **75,** 467.

Freter, R. (1956) *J. exp. Med.* **104,** 411.

Gangarosa, E. J., Perera, D. R., Mata, C. J., Mendizábal-Morris, C., Guzmán, G. and Reller, L. B. (1970) *J. infect. Dis.* **122,** 181.

Gemski, P. Jr, Takeuchi, A., Washington, O. and Formal, S. B. (1972) *J. infect. Dis.* **126,** 523.

Gettings, H. S. (1919) *Spec. Rep. Ser. med. Res. Coun., Lond.* No. 30.

Gillies, R. R. and Duguid, J. P. (1958) *J. Hyg., Camb.* **56,** 303.

González, L. M. and Otero, P. M. (1945) *J. Immunol.* **50,** 373.

Graham, J. M. (1958) *J. Path. Bact.* **76,** 291.

Gross, R. J., Thomas, L. V. and Rowe, B. (1980) *J. clin. Microbiol.* **12,** 167.

Haltalin, K. C., Nelson, J. D., Woodman, E. B. and Allen, A. A. (1970) *J. infect. Dis.* **121,** 275.

Hammarström, E. (1949) *Phage-typing of Shigella sonnei.* Stockholm.

Hardy, A. V. (1945) *Publ. Hlth Rep., Wash.* **60,** 1037.

Heyningen, W. E. van. (1971) In: *Microbial Toxins*, Vol. IIa, p. 255. Ed. by S. Kadis, T. C. Montie and S. J. Ajl, Academic Press, New York.

Heyningen, W. E. van and Gladstone, G. P. (1953) *Brit. J. exp. Path.* **34,** 202.

Horák, V. (1980) *Zbl. Bakt.*, I Abt. Orig. **A246,** 191.

Howard, J. G. (1955) *Brit. J. exp. Path.* **36,** 439.

Hutchinson, R. I. (1956) *Mon. Bull. Minist. Hlth Lab. Serv.* **15,** 110.

Istrati, G., Meitert, T. and Ciufecu, C. (1967) *Zbl. Bakt.* **203,** 295.

Johnston, J. H., Johnston, R. J. and Simmons, D. A. R. (1968) *Immunology* **14,** 657.

Kallings, L. O., Lindberg, A. A. and Sjöberg, L. (1968) *Arch. Immunol. Ther. exp.* **16,** 280.

Kauffmann, F. (1969) *The Bacteriology of Enterobacteriaceae.* 2nd edn. Munksgaard, Copenhagen.

Kent, T. H., Formal, S. B., La Brec, E. H., Sprinz, H. and Maenza, R. M. (1967) *Amer. J. Path.* **51,** 259.

Keusch, G. T. and Jacewicz, M. (1975) *J. infect. Dis.* **131,** S33.

Keusch, G. T., Mata, L. J. and Grady, C. F. (1970) *Clin. Res.* **18,** 442.

Kraus, R. and Dörr, R. (1905) *Wien. klin. Wschr.* **18,** 1077.

Kruse, W. (1900) *Dtsch. med. Wschr.* **26,** 637; (1901) *Ibid.* **27,** 370, 386.

Kruse, Rittershaus, Kemp and Metz. (1907) *Z. Hyg. InfektKr.* **57,** 417.

La Brec, E. H., Formal, S. B. and Schneider, H. (1959) *J. Bact.* **78,** 384.

La Brec, E. H., Schneider, H., Magnani, T. J. and Formal, S. B. (1964) *J. Bact.* **88,** 1503.

Lachowicz, T. M., Mulczyk, M. and Jankowski, S. (1963) *Arch. Immunol. Ther. exp.* **11,** 563.

Large, D. T. M. and Sankaran, O. K. (1934) *J. R. Army med. Cps.* **63,** 231.

Leifson, E. (1935) *J. Path. Bact.* **40,** 581.

Levine, M. M. *et al.* (1973) *J. infect. Dis.* **127,** 261.

Lubin, A. H. and Ewing, W. H. (1964) *Publ. Hlth Lab.* **22,** 83.

MacIver, J., Grady, G. F. and Keusch, G. T. (1975) *J. infect. Dis.* **131,** 559.

Madsen, S. (1949) *On the Classification of the Shigella Types.* Munksgaard, Copenhagen.

Manolov, D. G., Trifanova, A. and Ghinchev, P. (1962) *J. Hyg. Epidem., Praha* **6,** 422.

Martin, W. J., Mock, W. E. and Ewing, W. H. (1968) *Canad. J. Microbiol.* **14,** 737.

Mata, L. J., Gangarosa, E. J., Cáceres, A., Perera, D. R. and Mejicanos, M. L. (1970) *J. infect. Dis.* **122,** 170.

Mayr-Harting, A. (1952) *J. gen. Microbiol.* **7,** 382.

Metzger, M. (1958) *Schweiz. Z. allg. Path. Bakt.* **21,** 645.

Mita, K. (1921) *J. infect. Dis.* **29,** 580.

Mitsuhashi, S., Harada, K. and Hashimoto, H. (1960) *Jap. J. exp. Med.* **30,** 179.

Mitsuhashi, S., Harada, K. and Kameda, M. (1961) *Nature, Lond.* **189,** 947.
Murray, E. G. D. (1918) *J. R. Army med. Cps.* **31,** 257, 353.
Nakamura, A. (1967) *Jap. J. med. Sci. Biol.* **20,** 213.
Ogawa, H., Yoshikura, H., Nakamura, A. and Nakaya, R. (1967) *Jap. J. med. Sci. Biol.* **20,** 329.
Okell, C. C. and Blake, A. V. (1930) *J. Path. Bact.* **33,** 57.
Olarte, J., Filloy, L. and Galindo, E. (1976) *J. infect. Dis.* **133,** 573.
Olitsky, P. K. and Kligler, I. J. (1920) *J. exp. Med.* **31,** 19.
Ørskov, J. and Larsen, A. (1925) *J. Bact.* **10,** 473.
Paniker, C. K., Vimala, K. N., Bhat, P. and Stephen, S. (1978) *Indian J. med. Res.* **68,** 413.
Penner, A. and Bernheim, A. I. (1942) *J. exp. Med.* **76,** 271.
Petrovskaya, V. G. and Bondarenko, V. M. (1977) *Int. J. syst. Bact.* **27,** 171.
Petrovskaya, V. G. and Khomenko, N. A. (1979) *Int. J. syst. Bact.* **29,** 400.
Piéchaud, D. and Szturm-Rubinsten, S. (1959) *Ann. Inst. Pasteur* **97,** 511.
Proom, H., Woiwod, H. J., Barnes, J. M. and Orbell, W. G. (1950) *J. gen. Microbiol.* **4,** 270.
Pruneda, R. C. and Farmer, J. J. III (1977) *J. clin. Microbiol.* **5,** 66.
Rahaman, M. M., Alam, A. K., Islam, M. R., Greenough, W. B. and Lindenbaum, J. (1975) *Johns Hopk. med. J.* **136,** 65.
Rahaman, M. M., Huq, I., Dey, C. R., Kibriya, A. K. M. G. and Curlin, G. (1974) *Lancet* **i,** 406.
Rauss, K., Kétyi, I., Vertenyi, A. and Vörös, S. (1961) *Acta microbiol. hung.* **8,** 53.
Reavis, R. W. and Ewing, W. H. (1958) *Int. Bull. bact. Nomencl.* **8,** 75.
Report. (1958*a*) *Int. Bull. bact. Nomencl.* **8,** 25; (1958*b*) *Ibid.*, **8,** 93; (1968) *Int. J. syst. Bact.* **18,** 191; (1974*a*) *Wkly Epidem. Rec.* **49,** 311; (1974*b*) *Int. J. syst. Bact.* **24,** 385; (1978) *Wkly Epidem. Rec.*, **53,** 208; (1979) *Ibid.* **21,** 167.
Roelcke, K. (1938) *Z. Hyg. InfektKr.* **120,** 307.
Romanowska, E. and Mulczyk, M. (1967) *Biochim. biophys. Acta* **136,** 312.
Rout, W. R., Formal, S. B., Giannella, R. A. and Dammin, G. J. (1975) *Gastroenterology* **68,** 270.
Rowe, B., Gross, R. J. and Guiney, M. (1976) *Int. J. syst. Bact.* **26,** 76.
Rowe, B., Gross, R. J. and van Oye, E. (1975) *Int. J. syst. Bact.* **25,** 301.
Ruch, T. C. (1959) *Diseases of Laboratory Primates.* Saunders, Philadelphia.
Sachs, A. (1943) *J. R. Army med. Cps.* **80,** 92. [Wrongly described as Sachs, H.]
Schmitz, K. E. F. (1917) *Z. Hyg. InfektKr.* **84,** 449.
Schütze, H. (1944) *J. Path. Bact.* **56,** 250.
Scrimgeour, G. (1966) *Mon. Bull. Minist. Hlth Lab. Serv.* **25,** 278.
Sen, R. (1960*a*) *Nature, Lond.*, **185,** 267; (1960*b*) *J. Bact.* **80,** 585.
Serény, B. (1955) *Acta microbiol. hung.* **2,** 299; (1957) *Ibid.* **4,** 367; (1959) *Ibid.* **6,** 217.
Shiga, K. (1898*a*) *Zbl. Bakt.*, **23,** 599; (1898*b*) *Ibid.*, **24,** 817, 870, 913; (1901) *Dtsch. med. Wschr.* **27,** 741, 765, 783.
Simmons, D. A. R. (1969) *Biochem. J.* **114,** 34r; (1971) *Bact. Rev.*, **35,** 117.
Ślopek, S., Mulczyk, M. and Grzybek-Hryncewicz, K. (1960) *Arch. Immunol. Ter. dósw.* **8,** 191.
Ślopek, S. *et al.* (1972) *Arch. Immunol. Ther. exp.* **20,** 1.
Sonne, C. (1915) *Zbl. Bakt.* **75,** 408.
Spicer, C. C. (1959) *J. Hyg., Camb.* **57,** 210.
Strong, R. P. and Musgrave, W. E. (1900) *J. Amer. med. Ass.* **35,** 498.
Szturm-Rubinsten, S. and Piéchaud, D. (1957) *Ann. Inst. Pasteur* **92,** 335; (1963) *Ibid.* **104,** 284.
Takasaka, M., Honjo, S., Imaizumi, K., Ogawa, H., Nakaya, R., Nakamura, A. and Mise, K. (1969) *Jap. J. med. Sci. Biol.* **22,** 389.
Tanaka, T., Tsunoda, M. and Mitsuhashi, S. (1977) In: *R Factor Drug Resistance Plasmids*, p. 187. Ed. by S. Mitsuhashi, and H. Hashimoto. University Park Press, Baltimore.
Tateno, I. (1950) *Jap. J. exp. Med.*, **20,** 795; (1955) *J. Louisiana State med. Soc.* **107,** 116.
Taylor, B. C. and Nakamura, M. (1964) *J. Hyg., Camb.* **62,** 303.
Tee, G. H. (1952) *Mon. Bull. Minist. Hlth Lab. Serv.* **11,** 68.
Thomen, L. F. and Frobisher, M. (1945) *Amer. J. Hyg.* **42,** 225.
Todd, C. (1904) *J. Hyg., Camb.* **4,** 480.
Trabulsi, L. R. and Ewing, W. H. (1962) *Publ. Hlth Lab.* **20,** 137.
Trifanova, A., Bretoeva, M. and Tekelieva, R. (1974) *Zbl. Bakt.*, I Abt. Orig. **A226,** 343.
Tschäpe, H. and Rische, H. (1970) *Zbl. Bakt.* **214,** 91.
Vaillard, L. and Dopter, C. (1903) *Ann. Inst. Pasteur* **17,** 463.
Weil, A. J. (1943) *J. Immunol.* **46,** 13.
Wheeler, K. M. and Mickle, F. L. (1945) *J. Immunol.* **51,** 257.

37

Salmonella

M. T. Parker

Introductory

Definition

Organisms that conform to the definition of the Enterobacteriaceae. Most strains are motile. Apart from a few exceptions that form acid only, they produce acid and gas from glucose and mannitol, and usually also from sorbitol. Rarely ferment sucrose or adonitol, and rarely form indole. Acetylmethylcarbinol not formed. Do not hydrolyse urea or deaminate phenylalanine. Usually form H_2S on triple sugar iron agar and grow on Christensen's citrate medium. Form lysine decarboxylase. The many serotypes in the group are closely related to each other by somatic and flagellar antigens, and most strains show diphasic variation of the flagellar antigen. Primarily intestinal parasites of vertebrates. Pathogenic for many species of animals, giving rise to enteritis and to typhoid-like diseases. G+C content of DNA, 50–52 moles per cent.

Type species: *S. choleraesuis*.

Nomenclature

The genus *Salmonella* was originally created by medical bacteriologists to include organisms that gave rise to a certain type of illness in man and animals, and were related to one another antigenically. It later became clear that these organisms had a large number of biochemical characters in common, and it is now customary to lay more emphasis on biochemical activity than on antigenic structure in defining the group. Individual strains may behave atypically in one or more biochemical tests, but they usually have so many other biochemical characters in common with the group that there is little difficulty in classifying them. An organism which, on biochemical grounds, appears to be a *Salmonella* will nearly always be found to possess one or more antigens in common with other salmonella types, but the possession of salmonella antigens does not automatically qualify an organism for inclusion in the group.

The definition at the head of the chapter conforms to the current conception of a 'greater' *Salmonella* genus (Kauffmann 1960, Rohde 1967, Le Minor *et al.* 1970) which includes not only the familiar serotypes that are pathogens of mammals, but also the Arizona group and certain other organisms formerly considered to be biochemically aberrant salmonella types. *Salmonella* is subdivided into four sub-genera; the first of these includes the typhoid and paratyphoid bacilli and most of the other types that are responsible for widespread disease in mammals; the remaining three sub-genera comprise organisms that are in the main parasites of cold-blooded animals.

Sub-genus I. With a few exceptions, these organisms ferment dulcitol but not lactose or salicin, are malonate negative, do not liquefy gelatin, and are KCN sensitive.

Sub-genus II. These also ferment dulcitol but differ from sub-genus I organisms in that they ferment malonate and liquefy gelatin.

Sub-genus III. (The Arizona group). This comprises organisms that acidify lactose but not dulcitol, ferment malonate and liquefy gelatin.

Sub-genus IV. Members of this sub-genus acidify salicin but not lactose or dulcitol; they do not ferment malonate, but liquefy gelatin and are KCN resistant.

The terminology introduced by White (1929*a*, *b*) and modified by Kauffmann accorded specific rank to each antigenically distinguishable salmonella type, and the convention was established that each new type should be named after the place at which it was first isolated. The first table to be published contained some 20 types, but the number is now about 2000. Nowadays few bacteriologists would accord specific status to most of these serotypes.

Several workers have favoured a drastic reduction in the number of species; Cowan (1957) proposed that only a few important species should be retained, and the rest considered as varieties of *S. kauffmanni*. Ewing (1963) recognized only three species, two serologically homogeneous, *S. choleraesius* and *S. typhi*, and a third, *S. enteritidis* in which all the other serotypes appear as varieties, e.g. *S. enteritidis* var. *enteritidis*, *S. enteritidis* var. *typhimurium*. More recently, Le Minor and his colleagues (1970) have suggested that sub-genera I–IV should be reclassified respectively as *S. kauffmanni*, *S. salamae*, *S. arizonae* and *S. houtenae*. In *S. kauffmanni*, serotypes would be referred to by names, but these would be printed in Roman; in *S. salamae* and *S. houtenae*, antigenic formulae would be used, but where a name had already been allotted this would be given as well and prefixed by * in the former species and by ** in the latter; serotypes of *S. arizonae* would, as at present, be described by their antigenic formulae. The Approved Lists of Bacterial Names (Skerman *et al.* 1980) include only *S. choleraesuis*, *S. enteritidis*, *S. typhi*, *S. typhimurium* and *S. arizonae*.

None of these solutions to the problem of nomenclature has yet received general support, and we shall for the present continue to refer to the serotypes of sub-genus I salmonellae by their Linnaean names, while emphasizing that this is merely a matter of convenience. The serotypes may, in some organisms, be further subdivided into biotypes or phage types. When we use the term 'salmonella types' without qualification, however, we shall be referring to serotypes.

Habitat

There is no reason to doubt that the salmonellae are primarily intestinal parasites, though they may also be isolated from the blood and internal organs of vertebrates. They are frequently to be found in sewage, in river and sea-water, and in certain foods. Here they may survive for varying lengths of time, but it is doubtful whether they can exist indefinitely in any environment outside the animal body.

Members of sub-genus I predominate among mammals. Some serotypes, such as the typhoid bacillus and *S. gallinarum*, are confined to one or a few closely related species. The majority, however, have a wide range of mammalian hosts. The other sub-genera include mainly parsites of cold-blooded vertebrates, but these organisms occasionally cause disease in man and other mammals. Cold-blooded animals also harbour members of sub-genus I.

Morphology

The shape, size, structure, and arrangement of the bacterial cells do not differ materially from those of other enterobacteria. With the exception of *S. gallinarum* and its variant *pullorum*, all salmonellae are motile by peritrichate flagella. Individual non-motile strains may be encountered occasionally in the body, and non-motile variants may be thrown off under cultural conditions in the laboratory.

Most salmonellae are fimbriate (Duguid and Gillies 1958, Duguid *et al.* 1966). Notable exceptions include all *S. sendai*, almost all *S. paratyphi A*, most *S. gallinarum* var. *pullorum*, and a minority of *S. typhi*, *S. paratyphi B* and *S. typhimurium* strains. In most salmonella types the fimbriae are of the mannose-sensitive, haemagglutinating and adhesive type 1 (Chapter 33), but a small proportion of *S. paratyphi B* strains, and all fimbriate strains of *S. gallinarum* and its *pullorum* variety, form type 2 fimbriae, which are not haemagglutinating or adhesive. (See Chapter 33 for the antigenic properties of salmonella fimbriae.) *S. sendai*, though not fimbriate, forms a mannose-resistant haemagglutinin.

Cultural characters

Most members of this group grow readily on ordinary nutrient media and cannot be distinguished from coliform bacteria; a few types such as *S. paratyphi A*, *S. abortusovis*, *S. typhisuis*, *S. sendai*, and *S. gallinarum* var. *pullorum*, grow less abundantly. On brilliant green agar plates the difference is particularly noticeable, the growth being both slower and less abundant than that of other members of the group. *S. typhi* and *S. rostock* likewise grow poorly on brilliant green agar, though developing fairly well on ordinary nutrient agar. Strains that form small colonies on ordinary media are less tolerant of extremes of temperature and pH, and have more exacting nutritional requirements than other strains (Stokes and Bayne 1957). In broth, smooth strains give rise to a uniform turbidity.

On agar, the colonies are fairly large, with an average diameter of 2–3 mm, but they vary considerably in size. They may be circular and low convex with a smooth surface and entire edge; they may be flatter with a less regular surface and a more effuse serrated edge; or they may assume the vine-leaf form, which used to be regarded as characteristic of *S. typhi*. Dwarf colony forms of *S. typhi* are sometimes met with. They were first described by Jacobsen in 1910, and have since been reported on by several workers (Mellon and Jost 1926, W. J. Wilson 1938, Morris *et al.* 1941, 1943). On ordinary agar the colonies after 24 hours' incubation are only about 0.2–0.3 mm in diameter, but colonies of more normal size are formed on media containing assimilable sulphur compounds such as sulphite and thiosulphate. Biochemically dwarf colony forms behave more or less normally. Serologically they react only with O antisera, but when grown on a sulphur-containing medium they form the usual H and Vi antigens.

According to Stokes and Bayne (1958*a*) dwarf colony forms can be obtained from most salmonella strains by growth in broth containing lithium sulphate or sodium selenite. They differ from the parent strains in that they require additional growth factors. Some require cysteine alone, and others cysteine together with other amino acids or with pyridoxine.

Certain strains, notably of *S. paratyphi B*, give rise under favourable conditions to a mucoid growth. Sometimes the colonies are mucoid after 24 hours' incubation; they are about twice the size of normal colonies and resemble large drops of mucilage (Fletcher 1918). More often the mucoid appearance develops as a secondary phenomenon after prolonged incubation. Thus, when an agar plate is inoculated in three or four places with the point of a needle, and after one day's incubation at 37° is left at room temperature for a few days, large colonies are formed characterized by a depressed centre surrounded by a luxuriant mucoid wall. The 'mucoid wall test', or *Schleimwall-Versuch*, described originally by Müller (1910), is positive with most freshly isolated D-tartrate-negative strains of *S. paratyphi B* and generally negative with *S. typhimurium* (Kauffmann 1941). Most *Salmonella* and *Escherichia* strains form mucoid colonies when grown on medium with increased osmolarity (Anderson 1961, Anderson and Rogers 1963).

The mucoid material contains a polysaccharide (Birch-Hirschfeld 1935) that appears to be antigenically similar no matter by which type of *Salmonella* it is formed. Antibody to this so-called M antigen (Kauffmann 1935*a*, 1936*a*) can be demonstrated by agglutination if mixtures of serum and mucoid organisms are incubated for 2 hr at 37° and overnight at room temperature, but sera usually have a low titre, seldom exceeding 1 in 160. Strains with a well developed M antigen are usually inagglutinable by O and H antisera. The salmonella M antigen is closely related to, if not identical with, the M antigen of *Esch. coli* (Henriksen 1950). It appears to be a colanic acid (see also Chapter 33).

Growth requirements

In general, the organisms can make use aerobically of simple carbon compounds, e.g. lactate or citrate, as the source of carbon and energy. Individual strains differ widely in their requirements for nitrogenous compounds, both on first isolation and after cultivation in the laboratory (see Knight 1936, Braun 1948). Individual strains may require up to three amino acids, with thiamine or biotin in addition. In many types only an occasional strain is exacting, but in others, all require one or more additional nutrients (Stokes and Bayne 1957, 1958*b*). Many of these exacting strains grow poorly on ordinary nutrient media.

The typhoid bacillus requires tryptophan, but can be trained to grow without it. *S. gallinarum*, and its *pullorum* variant, grow without tryptophan, but individual strains have a variety of different requirements for amino acids and vitamins (Johnson and Rettger 1943, Kuwahara *et al.* 1958). *S. sendai* resembles the dwarf forms of *S. typhi* in being unable to use sulphate. Fukumi, Sayama and Nakaga (1953) showed that this was due to an inability to form cysteine and methionine from sulphate, though both sulphite and thiosulphate could be utilized (see also Urano *et al.* 1958, and p. 349).

Resistance to heat and to various chemical substances

Most members of this group are killed by exposure to a temperature of 55° for about 1 hr, or of 60° for 15–20 min. Many observations have been made on the resistance of salmonellae to different chemical reagents, chiefly in an endeavour to prepare selective media on which the growth of coliform bacilli would be inhibited. Malachite green, in suitable concentration, kills *Esch. coli* or inhibits its growth without exerting the same effect on *S. typhi* (Loeffler 1903, 1906, Lentz and Tietz 1903, 1905). There are other green dyes that have a similar selective action; but the studies of Browning, Gilmour and Mackie (1913) and of Krumwiede and Pratt (1914) showed that brilliant green gives the best results. To this dye bacilli of the paratyphoid group are most resistant, the typhoid bacillus is somewhat less resistant, whereas the dysentery bacilli, and still more the group of coliform bacilli, are very susceptible. Lithium chloride (Gray 1931, Havens and Mayfield 1933) also inhibits the growth of *Esch. coli* in concentrations that have no effect on the typhoid bacillus. Sodium tetrathionate favours the growth of salmonellae at the expense of the coliform bacilli (Muller 1923)—an action that appears to be due not to any inhibitory action it possesses on coliform bacilli, but to the ability of most salmonellae to reduce this substance and use it as a source of energy (Pollock *et al.* 1942). However, nearly all strains of *Citrobacter*, *Proteus* and *Serratia*, and a few strains of *Enterobacter*, also form tetrathionate reductase (Le Minor 1967) and so possess a similar advantage. Sodium deoxycholate, in the presence of certain other substances, inhibits the growth of coliform but not of dysentery or salmonella bacteria, and is used in the preparation of selective media for the isolation of intestinal pathogens (Leifson 1935, Hynes 1942). Selenium salts are also of value for the same purpose (Guth 1916, Leifson 1936, H. G. Smith 1959). (The use of selective and enrichment media for the isolation of salmonellae is considered in Chapters 68 and 72.)

Resistance to antimicrobial agents

Until about 1960, nearly all salmonellae were sensitive *in vitro* to a wide range of antimicrobial agents, including chloramphenicol, tetracycline, streptomycin, kanamycin, ampicillin (and benzylpenicillin at concentrations of 8–32 μg per ml), the cephalosporins, the sulphonamides, the nitrofurazones and trimethoprim. Tetracycline, streptomycin and kanamycin, however, appear to be without therapeutic action on infections with apparently sensitive strains.

Resistance to tetracycline was first noted by Huey and Edwards in 1958, and in the following three years was observed to increase somewhat in frequency in the USA and the Netherlands. Ampicillin resistance appeared in Britain in 1962 (Anderson and Datta 1965) and rapidly became common. Datta (1962) first observed transferable resistance to streptomycin, tetracycline and sulphonamides in a *S. typhimurium* strain of phage type 29 from a hospital epidemic in London in 1958. Anderson (1965, 1968*a*, see also Anderson and Lewis 1965) described the rapid accumulation of more and more resistances by this organism between 1963 and 1965. Resistant strains, predominantly of phage-type 29, appeared first in calves and soon afterwards in the human population. By 1968, the commonest spectrum of resistance was to ampicillin, streptomycin, sulphonamide, tetracycline and furazolidone, but some strains are also resistant to chloramphenicol and kanamycin. The resistances were nearly always transferable (see also Anderson 1968*b*). The incident came to an end in 1969, but after 1976 there were further epidemics in calves, with subsequent spread to man, caused by resistant strains of *S. typhimurium* of phage types 204 and 193 (Threlfall *et al.* 1978*a*, *b*, 1980). These strains carried plasmids coding for resistance to four to six antibiotics, including chloramphenicol. These events were probably attributable to the therapeutic use of antibiotics in calf-rearing establishments (see Linton 1981). In Britain, infections with multiple-antibiotic resistant salmonellae have been caused mainly by strains of *S. typhimurium* belonging to a few phage types, and have been seen chiefly in cattle and man.

In other countries quite different situations have been described. In the USA, for example, multiple-antibiotic resistance is common in a number of serotypes isolated from farm animals, including *S. saintpaul* and *S. heidelberg* as well as *S. typhimurium* (Neu *et al.* 1975) and has been attributed to the continued use of the corresponding antibiotics for growth promotion (Ryder *et al.* 1980). However, according to Cherubin and his colleagues (1980) the frequency of resistance in salmonellae of human and animal origin in the north-eastern parts of the USA varies independently with time.

In the Netherlands, multiple-antibiotic resistance became prevalent in the mid-1960s in several serotypes, notably in *S. typhimurium* and *S. panama* (Manten *et al.* 1971). In 1972–74 resistant strains of *S. dublin* appeared in large numbers in cattle, and strains isolated from man showed similar resistances (Voogd *et al.* 1977). By 1974, 64 per cent of isolations of *S. typhimurium* were resistant to ampicillin, tetracycline, chloramphenicol and kanamycin, and 25 per cent of *S. dublin* to tetracycline and chloramphenicol. Later, van Leeuwen and co-workers (1979) observed a marked decrease of tetracycline resistance in salmonellae from pigs and man that coincided in time with the institution of a ban on the incorporation of tetracycline into animal feeds. (See also van Leeuwen *et al.* 1982)

During the 1970s there were a number of very extensive epidemics of human infections with multiple-antibiotic resistant salmonellae in which there was little evidence of an animal origin (see Cherubin 1981). Several distinct strains of *S. typhimurium* and a number of members of other serotypes were implicated. Resistance was usually to at least six antibiotics and was specified by plasmids. For example, one strain of *S. typhimurium* widely distributed in the Middle East and India was resistant to ampicillin, chloramphenicol, gentamicin, kanamycin, streptomycin, sulphonamides, tetracycline and trimethoprim; all these resistances were encoded on a single plasmid (Frost *et al.* 1982). In most of these epidemics there were many episodes of hospital-acquired infection, mainly in young children, with a certain amount of spread into the surrounding population. An epidemic of infection with *S. wien* began in paediatric hospitals in Algeria in 1969. At first the strain was antibiotic sensitive, but resistance to several antibiotics, including ampicillin and chloramphenicol, soon appeared. The resistant strain reached France in 1970, where it caused many outbreaks in hospitals and became for a time the second most common salmonella serotype in human infections in the country. Subsequently similar events occurred in Jugoslavia, Italy, Austria and Iraq. (For a review of this epidemic, see Cherubin 1981). The evidence suggests that this and a number of other extensive epidemics resulted from intra-human spread, presumably made possible by extensive antibiotic use in man.

Antibiotic resistance was rarely seen in *S. typhi* before 1972, when a strain with plasmid-encoded resistance to chloramphenicol streptomycin, sulphonamides and tetracycline gave rise to a large epidemic which affected several states in Mexico (see Anderson 1975); this lasted for 2 years and then ceased. At about the same time a strain of *S. typhi* with a similar resistance pattern but belonging to a different Vi-phage type caused a local epidemic in South India (Paniker and Vimala 1972). Several other small foci of infection with chloramphenicol-resistant *S. typhi* have since been reported in India (Sharma *et al.* 1979) but the infections have remained localized. Chloramphenicol-resistant *S. typhi* appeared in Viet Nam in 1972, and by 1975 85 per cent of strains from Saigon were resistant (Ricossé *et al.* 1979). Chloramphenicol resistance is also common in Thailand (Vongsthongsri and Tharavanij 1980) but not so far in several other countries in South-East Asia. Although chloramphenicol-resistant strains of *S. typhi* from different parts of the world belong to a variety of Vi-phage types they nearly all carry resistance plasmids of a single compatibility group, H2 (Report 1978). It should be noted that the resistance plasmid of the Mexican strain of *S. typhi* specified the same resistances as did the plasmid in the strain of *Shig. dysenteriae* serotype 1 that had somewhat earlier been responsible for widespread bacillary dysentery in Central America (Chapter 36), but the two plasmids belonged to different compatibility groups (Datta and Olarte 1974).

A strain of *S. typhimurium* isolated from patients in a burns unit in which silver nitrate prophylaxis was used (see Chapter 61) had resistance to silver salts and to various antibiotics, encoded on a single plasmid (McHugh *et al.* 1975).

Biochemical activities

The common biochemical characters of the salmonellae were shown in Table 33.2. Although most strains conform closely to this pattern, there are a number of exceptions; no organism should therefore be excluded from the group on the result of a single test. The usual reactions include (1) fermentation of glucose, maltose, mannitol and sorbitol with the production of acid and gas, (2) absence of fermentation of sucrose and adonitol, (3) failure to produce indole, to hydrolyse urea, or to deaminate phenylalanine, and (4) a positive methyl-red reaction and a negative Voges-Proskauer reaction. Gas is not formed by *S. typhi*, by *S. gallinarum*, or by occasional strains of a number of other serotypes: maltose is usually not fermented by *pullorum* strains; and *S. typhisuis* does not attack mannitol. Occasional strains that form some acid from sucrose, produce indole or split urea have been recorded. The positive reaction to the methyl-red test, and the negative reactions to the phenylalanine and Voges-Proskauer tests appear to be constant. Other biochemical characters are variable, and some of them are the means by which *Salmonella* is subdivided into the four sub-genera (Table 37.1). Within sub-genus I there are several types that depart from the usual pattern; in Table 37.2 we give the reactions of three of these (*S. typhi*, *S. paratyphi A* and *S. choleraesuis*), and other examples will be found in the descriptions of individual types at the end of this chapter (pp. 347–350).

Table 37.1 Biochemical differentiation of the four sub-genera of *Salmonella*

	Sub-genus			
	I	II	III	IV
Acid in dulcitol	+	+	−	−
,, ,, lactose	−	−	+/x	−
β-galactosidase	−	−/x	+	−
Acid in salicin	−	−	−	+
Fermentation of D-tartrate	+	−/x	−/x	−/x
,, ,, mucate	+	+	v	−
,, ,, malonate	−	+	+	−
Liquefaction of gelatin	−*	+	+	+
Growth in KCN medium	−	−	−	+

* A few exceptions.
x = Late or irregular results. v = Some strains positive, others negative.

Most salmonellae grow with citrate as the sole carbon source, though important exceptions include the typhoid and paratyphoid A bacilli, *S. typhisuis*, *S. sendai* and the *pullorum* variety of *S. gallinarum*. Most salmonellae also give a positive reaction for H_2S in triple sugar iron agar, but exceptions include *S. paratyphi A*, the diphasic variety of *S. choleraesuis*, *S. typhisuis*, *S. sendai*, *S. berta* and a few strains of the typhoid bacillus. Members of sub-genus I nearly all fail to liquefy gelatin, but three serotypes (*S. abortusbovis*, *S. schleissheim* and *S. texas*) liquefy it rapidly. Gelatin liquefaction, which may be slow or rapid, is characteristic of the three remaining sub-genera.

Nearly all salmonellae produce lysine, arginine and ornithine decarboxylase but no glutamic acid decarboxylase (Møller 1955, Kauffmann and Møller 1955). Notable exceptions are the typhoid bacillus and some strains of *S. paratyphi B* var. *odense* which form no decarboxylase for ornithine and *S. paratyphi A* which does not attack lysine. Kauffmann (1941) drew attention to the value of tests for the fermentation of organic acids that were introduced by Brown, Duncan and Henry (1924). Most sub-genus I salmonellae rapidly attack D-tartrate, citrate and mucate, though there are many exceptions. Several types, notably *S. paratyphi A*, *S. sendai* and the 'human' type of *S. paratyphi B*, are D-tartrate negative, and others attack this substance late and irregularly (Kauffmann and Petersen 1956, Kauffmann 1960).

Some other biochemical reactions are variable within the *Salmonella* group, and are at times useful for the identification of individual serotypes, or for sub-dividing them further

Table 37.2 Common biochemical reactions of sub-genus I salmonellae and of three aberrant species*

	Common pattern of sub-genus I	*S. typhi*	*S. paratyphi A*	*S. choleraesuis*†
Gas from sugars	+	−	+	+
Acid from dulcitol	+	v	+	+
,, ,, arabinose	+	−	+	−
Growth on Christensen's citrate medium	+	−	−	+
H_2S‡	+	+	− or (+)	v
Decarboxylase: lysine	+	+	−	+
,, ,, : arginine	+	v	+	+
,, ,, : ornithine	+	−	+	+
Fermentation: D-tartrate	+	+	−	v
,, ,, : mucate	+	−	−	−
,, ,, : citrate	+	+	−	+

* For other aberrant species see text.
† See also Table 37.4. ‡ In triple sugar iron or similar medium.
+ = Positive reaction within 2 days. (+) = Delayed positive reaction. v = Some strains positive, others negative.
− = Negative reaction.

(see p. 344). They include the fermentation of rhamnose, xylose, arabinose, trehalose and inositol. Most strains attack these sugars, and it is the absence of fermentation that is of significance.

In the past considerable attention has been devoted to certain special reactions. One of these, first described by Stern (1916), consists in growing organisms in a fuchsin sulphite glycerol meat-extract medium. Some organisms, known as 'Stern-positive', produce in this medium a deep lilac colour within 3 days. The reaction is apparently due to the formation of an aldehyde. It is certainly not due solely to acid formation.

The second reaction was described by Bitter, Weigmann and Habs (1926). It is essentially a test of the ability of the organism to grow in a synthetic medium containing ammonium salts as the main source of nitrogen. A 1 per cent solution of a sugar, such as glucose, rhamnose, arabinose, or dulcitol, is added to test the fermentative power of the strain under these conditions. The ammonium reactions and the reaction in Stern's medium are not often employed nowadays, except for distinguishing between the fermentative types of *S. enteritidis* (Kauffmann 1935*b*).

Screening media for the simultaneous detection of several biochemical reactions are widely used in the routine identification of salmonellae. On triple sugar iron agar they show H_2S production and absence of fermentation of lactose, sucrose and salicin, and on lysine iron agar (Edwards and Fife 1961) the production of H_2S and lysine decarboxylase. The use of screening media is time-saving but will lead to error unless a purity plate is seeded at the same time and inspected carefully after incubation. Stenzel (1976) advocates early acidification of sorbose as a means of excluding *Citrobacter*.

For further information about the biochemical reactions of salmonellae see Kauffmann (1941, 1956*a*, 1969), Ewing and Ball (1966), Ewing, *et al.* (1970) and Edwards and Ewing (1972).

Antigenic structure

The general characters of the various antigenic components of the cells and appendages of members of the Enterobacteriaceae were described in Chapter 33. We shall now describe in more detail the antigens that are used in the definition of the serological types of salmonellae: (1) the O antigens—heat-stable polysaccharides that form part of the cell-wall lipopolysaccharide, (2) the H antigens—heat-labile proteins of the flagellae that in salmonellae have the almost unique character of *diphasic variation*, and (3) surface polysaccharides that inhibit the agglutinability of the organism by homologous O antisera, of which the Vi antigen of the typhoid bacillus is the most important example.

It will be recalled that the O polysaccharides have a core structure that is common to all enterobacteria to which are attached side chains of sugars that determine O specificity. Salmonellae may be classified into 'chemotypes' each of which contains organisms with the same sugars in their side chains. Chemotypes may include salmonellae with different O antigens, and in many cases also *Esch. coli* strains of several O groups and other enterobacteria, but the organisms are serologically distinct because of differences in the sequence of sugars in the side chains, particularly at the distal end. As might be expected, minor cross-reactions between salmonella O serogroups, and between the O antigens of salmonellae and other members of the Enterobacteriaceae, are quite common. The traditional method of detecting salmonella O antigens is by the agglutination of heated suspensions of organisms by rabbit antisera against boiled organisms; known cross-reactions are removed by suitable absorption of the sera. More specific reactions are claimed with antisera prepared against synthetic antigens composed of the 'type-determining' disaccharide, e.g., paratose 1→3 mannose for antigen O2, covalently linked to bovine serum albumin (Svenungsson and Lindberg 1977, 1978), which may be used in immunofluorescence or coagglutination tests.

The H antigens of *Salmonella* are found only in this genus; they comprise a large number of factors arranged in different combinations in the different serotypes. In most salmonellae the flagellar antigens exist in two alternative phases; in phase 1 there are one or more antigenic factors and in phase 2 two or more. To identify a serotype completely it is necessary to determine the antigenic factors present in both phases. This is done by agglutination tests with live or formolized suspensions and factor sera prepared in rabbits by injecting formolized cells and subsequently performing suitable absorptions.

For information about the practical identification of the various serotypes the reader is referred to publications by Kauffmann (1964, 1969) and Edwards and Ewing (1972). An up-to-date list of salmonella serotypes can be obtained from most national salmonella centres. Kelterborn (1967, 1979) gives a comprehensive account of the first isolation and geographical distribution of nearly 2000 salmonella serotypes. Additions to the list of officially recognized serotypes are published annually (see, for example, Le Minor *et al.* 1981).

Antigenic notation

The identification and labelling of the salmonella antigens was initiated by White (1926, 1929*a*, *b*), and continued and extended by Kauffmann (1929*a*, *b*, 1930, 1931, 1934*a*). In their earlier studies the two investigators used a different system of labelling, so that descriptions given in the English and German papers during the period can be correlated only by the aid of an appropriate key giving the equivalent numbers and letters in the two systems (see Lovell 1932). Later, however, the terminology introduced by Kauffmann was adopted for general use (Report 1934).

The antigenic formula consists of three parts, describing the somatic (O) antigen, the phase 1 H antigen and the phase 2 H antigen, in this order. The three parts are separated by colons, and the components of each part by commas.

The somatic (O) antigens are given arabic numerals. The flagellar (H) antigens of phase 1 are accorded small letters; these antigens have already illustrated the limitations of an alphabetical notation by exceeding twenty-six in number. By convention, those discovered later than the antigen that received the label z have been accorded an additional distinguishing numeral, z_1, z_2, and so on. The flagellar antigens of phase 2 are labelled in two different ways. At first they were accorded arabic numerals, but later it was found that the second phase of certain species uniformly contained the antigens e and n, often associated with x or with one of the z series of antigens. It thus happens that phase 2 may contain antigenic components of the 1, 2, ... series, or of the e, n, ... series. The position is confusing, because both the e and some of the z antigenic components may be found in the first phase. Strains also occur in which phase 2 contains neither the 1, 2 ... nor the e, n ... series, but instead antigenic components that are characteristic of phase 1.

A few examples may be given to illustrate the use of this notation (Table 37.3). The formula for *S. paratyphi A*, which is a monophasic flagellated bacillus existing only in phase 1, is 1, 2, 12 : a : –. The 1, 2 and 12 indicate that there are three components to the somatic antigen. The letter a represents the flagellar antigen and the dash following it shows that there is no second phase. The formula for *S. paratyphi B* is 1, 4, [5], 12 : b : 1, 2. Here b is the flagellar antigen of phase 1 and 1, 2 are the flagellar antigens of phase 2. For *S abony* the formula is 1, 4, 5, 12 : b : e, n, x. Here the somatic antigens and the phase 1 flagellar antigens are the same as those of *S. paratyphi B*, but the second phase contains components of the e, n ... series in place of the more usual 1, 2 ... series. In *S. gloucester* (1, 4, 12, [*27*] : i : 1, w) the phase 2 component contains antigens from neither of these series, but the two components 1 and w that are normally found in phase 1 of other organisms. An organism so far met with only in phase 2 is *S. abortusequi*, which has the formula 4, 12 : – : e, n, x. *S. gallinarum*, which is non-motile, has the formula 1, 9, 12 : – : –, the two dashes showing that there is neither a first nor a second flagellar phase.

Not all of the somatic factors accorded numerals are, however, entirely clear cut in their specificity; it sometimes happens that an organism may possess only part of such a factor. Thus, the round bracket in the antigenic formula of *S. minneapolis*, which is (3), 15, 34 : e, h : 1, 6, indicates that this organism has only part of the factor 3 which is present in *S. anatum*. Certain somatic factors are present only in some members of a particular serotype. For example, the factor 5 is absent from many strains of *S. paratyphi B*; this is indicated by placing the factor within a square bracket in the antigenic formula (1, 4, [5], 12 : b : 1, 2). Finally, there are somatic factors that are determined by the presence of a temperate phage (p. 341). These are underlined, e.g. the factor 1 in *S. typhimurium* and the factor 15 in *S. minneapolis*.

The Kauffmann-White diagnostic scheme

Primary subdivision is into groups, each of which shares a common somatic antigen. Where more than one somatic antigen is present, one of these antigens is regarded as determining the group to which the strain concerned shall be allocated. Thus, group B consists of those organisms that possess the O antigen 4, group D of those organisms that possess the O antigen 9, and so on.

Several of the groups are divided further into subgroups whose members each possess a second antigen in common. Group C, for example, is divided into subgroup C1, characterized by the O antigens 6, 7, and subgroup C2 characterized by the O antigens 6, 8. The situation in group E is even more complicated,

Table 37.3 Illustration of the Kauffmann-White scheme of serological classification

Group	Serotype	O antigen	H antigen Phase 1	H antigen Phase 2
A	*S. paratyphi A*	$\underline{1}$, 2, 12	a	—
B	*S. abortusequi*	4, 12,	—	e, n, x
	S. paratyphi B	$\underline{1}$, 4, [5], 12	b	1, 2
	S. limete	$\underline{1}$, 4, 12, $\underline{27}$	b	1, 5
	S. abony	$\underline{1}$, 4, [5], 12	b	e, n, x
	S. abortusbovis	$\underline{1}$, 4, 12, $\underline{27}$	b	e, n, x
	S. schleissheim	4, 12, $\underline{27}$	b, z_{12}	—
	S. wien	$\underline{1}$, 4, 12, $\underline{[27]}$	b	1, w
	S. abortusovis	4, 12	c	1, 6
	S. stanley	$\underline{1}$, 4, [5], 12, $\underline{27}$	d	1, 2
	S. chester	4, [5], 12	e, h	e, n, x
	S. derby	$\underline{1}$, 4, [5], 12	(f), g	—
	S. typhimurium	$\underline{1}$, 4, [5], 12	i	1, 2
	S. agama	4, 12	i	1, 6
	S. gloucester	$\underline{1}$, 4, 12, $\underline{[27]}$	i	1, w
	S. bredeney	$\underline{1}$, 4, 12, $\underline{27}$	1, v	1, 7
	S. heidelberg	$\underline{1}$, 4, [5], 12	r	1, 2
C 1	*S. paratyphi C*	6, 7, [Vi]	c	1, 5
	S. choleraesuis	6, 7	c	1, 5
	S. montevideo	6, 7	g, m, s	—
	S. oranienburg	6, 7	m, t	—
	S. thompson	6, 7	k	1, 5
	S. bareilly	6, 7	y	1, 5
	S. tennessee	6, 7	z_{29}	—
C 2	*S. manhattan*	6, 8	d	1, 5
	S. newport	6, 8	e, h	1, 2
	S. bovismorbificans	6, 8	r	1, 5
	S. kentucky	8, 20	i	z_6
D	*S. sendai*	$\underline{1}$, 9, 12	a	1, 5
	S. typhi	9, 12, Vi	d	—
	S. enteritidis	$\underline{1}$, 9, 12	g, m	—
	S. dublin	$\underline{1}$, 9, 12, [Vi]	g, p	—
	S. panama	$\underline{1}$, 9, 12	1, v	1, 5
	S. gallinarum	$\underline{1}$, 9, 12	—	—
E 1	*S. anatum*	3, 10	e, h	1, 6
	S. meleagridis	3, 10	e, h	1, w
	S. london	3, 10	1, v	1, 6
E 2	*S. newington*	3, $\underline{15}$	e, h	1, 6
E 3	*S. minneapolis*	(3), $\underline{15}$, 34	e, h	1, 6
E 4	*S. senftenberg*	1, 3, 19	g, s, t	—

Numbers underlined represent phage-determined antigenic factors. Numbers in square brackets represent antigens present only in some strains of the serotype.

and four subgroups are now recognized (see Table 37.3). There are now some 40 groups, the first 26 designated by letters and subsequent ones, after exhaustion of the alphabet, by the numbers of their group-determining antigens (50, 51, 52, etc.).

Within the O groups, twelve flagellar antigens or antigenic complexes occur commonly in phase 1:

a	d	k	y
b	e, h	1, v	z
c	i	r	z_{10}

and seven in phase 2:

1, 2	e, n, x
1, 5	e, n, z_{15}
1, 6	
1, 7	z_6

The complex 1, w may occur in either phase. In addition, most O groups contain types possessing the antigen g as part of a complex with either m, s or t; these types are usually monophasic.

The organisms classified in the Kauffmann-White scheme include members of sub-genera I, II and IV. A different scheme of notation is used in sub-genus III (the Arizona group) in which both somatic and flagellar antigens are represented by arabic numerals. Most of the Arizona antigens correspond closely to antigens already recognized in *Salmonella*, and it will ultimately be possible to reclassify them in the Kauffmann-White scheme.

It must be made clear that the Kauffmann-White scheme is not a record of the complete antigenic structure of each organism. The antigenic constitution of most organisms is far more complex than is suggested by the formulae given in the table, in which only antigens of differential importance are recorded. Many of the antigenic components identified by a single numeral or letter are themselves complex, consisting of two or more fractions. For instance, the 6 somatic antigen comprises two portions labelled 6_1 and 6_2. The 12 somatic antigen contains three portions, 12_1, 12_2, and 12_3. The d flagellar antigen contains five partial antigens, d, d_1, d_2, d_3, and d_4. If small antigenic differences are taken into account, the number of serotypes that can be recognized is almost unlimited.

For most purposes a simplification of the Kauffmann-White scheme proposed by Edwards and Kauffmann (1952) gives sufficient information; in this, only 12 O sera, 18 H sera and the Vi serum are used. All H complexes containing g, 1 and z_4 are known respectively as G, L and Z_4 and are not further differentiated; in phase 2, e,n,x and e,n,z_{15} are combined as e,n; and 1,2, 1,5, 1,6, 1,7 and z_6 are designated simply as 1. From 1963 (Kauffmann and Rohde 1962) an attempt has been made to slow down the rate of increase in the number of serotypes in the following ways: in sub-genus I, a difference within the G or the L complex is no longer considered sufficient for the establishment of a new type, and, in sub-genus II, additions must be recognizable by means of sera used in the simplified antigenic scheme.

About 60 per cent of all known salmonella serotypes belong to sub-genus I, 16 per cent to sub-genus II and 1 per cent to sub-genus IV; the remaining 23 per cent are sub-genus III (Arizona) serotypes (Rohde 1967). The majority of the sub-genus I types belong to the first five O groups (A–E), and the rest are scattered throughout the remaining groups. The first five O groups are composed almost exclusively of sub-genus I strains, together with a few of sub-genus II. Most of the sub-genus II strains, and all sub-genus IV strains belong to O groups from F onwards; and Arizona O antigens also usually correspond to antigens of the 'higher' salmonella O groups.

Diphasic variation of the H antigens occurs in all four sub-genera. Many of the common phase-1 factors (e.g. a, c, i and k) are found both in sub-genus I and in sub-genus II, but a few (e.g. e, h) occur only in sub-genus I; nearly all of them are absent from sub-genus IV. Several of the Arizona H antigens have not yet been found in other salmonellae.

Variation in the O antigens

The smooth → rough variation The chemical basis of this type of variation was discussed in Chapter 33. Rough variants are organisms with defects in the biosynthesis of the O polysaccharide. All of them lack the specific side-chains responsible for O specificity, and some of them have additional abnormalities of the core structure. They can be classified in a series from Ra, which lacks only the side chains, through Rb to Re, which show a progressive loss of sugar constituents from the core. The term 'semi-rough' is applied to organisms that form less than the normal number of side-chains, or form side-chains with less than the normal number of repeating units. Rough mutants have lost their agglutinability by homologous O antiserum; Ra mutants may be agglutinated by antisera against certain other smooth salmonellae, particularly members of O-chemotype I, and Rb-Re mutants are agglutinated by antisera against other rough enterobacteria.

Rough variation is usually associated with changes in colonial morphology, but a clear distinction between rough and smooth colonies can be seen only on some media (Schmidt *et al.* 1968–69). The traditional test for roughness is agglutinability in salt solution; most strains are stable in 0·3 per cent NaCl, but many rough strains undergo spontaneous agglutination in 0·9 per cent NaCl, and most of the remainder in 3·5 per cent NaCl. According to Schmidt and his colleagues (1968–69), however, the most reliable tests are agglutination by heating to 100° for 1 hr or in the presence of certain dyes, such as 0·3 per cent auramine (Kröger 1955). These characters are common to all types of rough variant, but the chemotypes Ra–Re show a steady gradation in their sensitivity to inhibition by certain dyes and antibiotics (Schmidt *et al.* 1969–70, Schlecht and Westphal 1970).

Rough variation occurs rarely in nature, is common in strains maintained through many generations on ordinary laboratory media, and can readily be induced by several different procedures—the prolonged incubation of a broth culture, growth in the presence of antibodies acting on the smooth somatic antigens, subjection to the action of an anti-smooth phage, and so on. To prevent its occurrence, strains should be dried from the frozen state *in vacuo*, or, if this is not practicable, kept on a dry Dorset egg medium in the ice-chest and subcultured as infrequently as possible. Moist solid media and sugar-containing media should be avoided.

T variants Kauffmann (1956*b*) observed that, in certain cultures of *S. paratyphi B* and *S. typhimurium*, some colonies were not agglutinated by O antisera though they had the naked-eye appearances of smooth forms. They were always found in cultures in which rough forms were also appearing. A serum prepared from an inagglutinable variant agglutinated to high titre both the homologous strain and others in the same state, forming small dense aggregates like those seen in O agglutination. Although the T forms of *S. paratyphi B* and *S. typhimurium* were identical, an antigenically distinct T antigen was found in *S. bareilly* (Kauffmann 1957). These two antigens are now referred to respectively as T1 and T2. As we have seen (Chapter 33), the T antigens are O antigens with abnormal side-chains.

Form variation Kauffmann (1954) gave this name to the complete or partial loss of one of the factors from an O antigen. In some cultures individual colonies differ widely in the amount of one factor that is present; the ones most likely to be affected are factors 1 and 12. Factor 1 is present irregularly in many serotypes of O groups A, B and D, and in some other groups the antigen 12 is complex, containing 12_1, 12_2 and 12_3 components, the relative proportions of which may vary (Kauffmann 1941); this has been most carefully studied in the *pullorum* variety of *S. gallinarum* (see p. 348). In a number of strains of salmonellae, such as *S. paratyphi B* and *S. typhimurium*, that normally possess factors 4 and 5, the 5 factor may be missing, not only from certain colonies but from the whole clone.

Lysogenic conversion Phages obtained from members of the subgroup E2, which contains the factors 3, 15, were used to infect organisms from other E subgroups with the structures 3, 10 and 1, 3, 19. In each instance, the antigen 15 appeared after lysogeny had been effected (Iseki and Sakai 1953, Uetake *et al.* 1955, 1958). Similarly, a lysogenic phage from *S. kentucky* transferred the factor 20 to *S. newport* (Baron *et al.* 1957). The mechanism responsible for the addition of the new antigen differs from that of transduction in that the antigenic change occurs in a large proportion of the cells exposed to phage lysates, and the new antigen appears in the cells within a few minutes of its infection with phage.

A somewhat similar situation exists in those members of groups A, B and D which possess factor 1 (Iseki and Kashiwagi 1955, Zinder 1957, Stocker 1958). Strains of *S. typhimurium* lacking the factor 1 acquire it after lysogenization by one of a number of phages, though there is no antigenic relation between the antigen 1 and any constituent of the phage. Lysogenization is associated with the appearance in cell extracts of those polysaccharide components responsible for the antigenic specificity of factor 1. The presence of the antigen 1 in organisms of group E4 with the O-antigenic structure 1, 3, 19 is, however, independent of the presence of bacteriophage (Stocker *et al.* 1960). Numerous other O factors may be acquired by lysogenic conversion (Le Minor 1969). It is thus apparent that some components of the O-antigenic structure of many salmonellae are determined genetically by phage DNA.

Variation in the H antigens

OH → O variation Flagellated strains may occasionally give rise to variants that contain only the O antigen. As a rule, this type of variation is irreversible. Some strains, like *S. gallinarum* and its *pullorum* variety, are permanently non-flagellated. The OH → O variation cannot readily be produced in the laboratory. The production of O forms by growth on phenolized agar cannot be considered a true variation, since there is a rapid reversion to the normal flagellated form when the organisms are inoculated on to ordinary media.

Occasionally, organisms may be non-motile but possess flagella and form H antigen. Non-flagellated strains which possess no H antigens are often difficult to allot to a salmonella serotype. Some of them possess H-determining genes, and after transduction of motility produce their 'own' H antigens (Stocker *et al.* 1953). Similarly, non-motile cultures may be able to act as donors of flagellar antigens. In this way Lederberg and Edwards (1953) showed that a number of strains of *S. gallinarum*, though non-motile, carried genetic factors determining the flagellar antigens g, m.

Diphasic variation Andrewes (1922, 1925) first showed that the flagella of salmonellae may assume two alternative forms.

Any given culture of a particular diphasic strain may consist entirely of bacilli in one or other of the two phases or may contain representatives of both. An organism in one phase, though always capable of giving rise to descendants in the alternative phase, usually maintains a constant phase over a number of generations. If, therefore, we prepare plate cultures of a diphasic organism and make numerous subcultures, each from a single colony, we should as a rule obtain some suspensions in one and others in the alternative phase. If these are killed by the addition of formalin after 18–24 hours' growth, there will usually not have been time for any change in phase to occur. Strains may occur in which the alternative phase is difficult to demonstrate. If they are grown in semi-solid agar to which has been added an antiserum containing agglutinins for the existing phase, and subculture is made from the spreading edge of the growth, the alternative phase may be detected. Alternatively, the serum may be added to semi-solid agar in a Craigie tube (see Chapter 20).

In diphasic strains the rate of mutation from phase 1 to phase 2, and in the reverse direction, is usually constant, so that any culture will tend, as it grows, towards an equilibrium characteristic of the strain (Stocker 1949). The antigens of phases 1 and 2 are transduced separately, and with very rare exceptions replace antigens in the corresponding phase in the recipient (Lederberg and Edwards 1953). The antigenic complexes e,n,x and l,w are always transduced as a unit. The l,w complex may appear in different serotypes in combination with either a phase-1 or a phase-2 antigen; the results of transduction experiments show that in the former case it functions as a phase-2 and in the latter case as a phase-1 antigen (Edwards *et al.* 1955).

According to Lederberg and Iino (1956), phase-1 antigens are determined by a series of genes at one locus (H1) and phase-2 antigens by genes at another locus (H2). Phase variation is the alternative expression of one or other of these genes. In transduction experiments between single-phase cultures of diphasic strains, H1 can be expressed only when transduced to phase-1 cells, regardless of the phase of the donor, but H2 can be expressed in any phase of the recipient, but only when the donor is in phase 2. This indicates that the H2 gene controls the expression of the phases. It can exist in two different states—active and inactive; when it is in the active state, production of phase-1 antigen by H1 is repressed, while H2 produces phase-2 antigen; and when it is inactive, the production of phase-1 antigen, specified by H1, proceeds. Each H locus is linked to a gene which controls its activity. Monophasic serotypes may either be deficient in H2 locus or may be fixed in one phase by a mutation of one of the 'activity loci'. (For further information about the genetics of phase variation see Iino 1969 and Chapter 6.)

Organisms in the two flagellar phases resemble each other culturally. Sertic and Boulgakov (1936*b*), however, observed that organisms in phase 2, but not those in phase 1, were agglutinable by acriflavine. Bernstein and Lederberg (1955) found that, though this was indeed true for organisms suspended in broth, all motile salmonellae were agglutinated by acriflavine when suspended in water, irrespective of their phase.

Artificial phase variation It was observed by Kauffmann (1936*b*) that, if a certain strain of *S. typhi* was cultured in broth containing agglutinating serum to the normal d antigen, this antigen was lost and replaced by a new antigen j, which had never been met with in typhoid bacilli under natural conditions. Similarly, the b antigen of *S. schleissheim* may be replaced by antigen z_5. In both these organisms, which are monophasic, the antigen appearing in the new phase was one commonly found in phase 1. Bruner and Edwards (1941), however, were able to obtain in the monophasic *S. paratyphi A* a second phase containing the 1, 5 ... antigens, which are characteristic of phase 2. An even more remarkable series of imposed variations was recorded by Edwards and Bruner (1939) in *S. abortusequi*. This organism exists normally only in phase 2, in which it possesses the antigens e, n, x. By growth in appropriate sera a phase 1, containing the antigen a, was obtained, and a phase 3, containing the antigen z_5. All three phases were reversible. It is by no means always easy to decide whether a new phase that appears as the result of growth in an antiserum is to be regarded as an artificial phase or as the alternative phase of an organism that is generally, though not invariably, monophasic. For example, *S. choleraesius* var. *kunzendorf* was for many years regarded as a well established monophasic type occuring in phase 2. When it was found, however, by Bruner and Edwards (1939) that a specific phase could be produced containing the same antigen c as *S. choleraesuis*, it was decided to omit the *kunzendorf* type from the Kauffmann-White scheme.

The examples quoted so far concern the acquisition of new antigens. Peso and Edwards (1951) describe examples in which growth in the presence of a serum specific for one factor in the antigenic complex of a monophasic organism resulted in loss of that factor. In this manner they transformed, for example, a strain of *S. rostock* (9, 12 : g, p, u : −) into *S. dublin* (9, 12 : g, p : −). A somewhat different situation exists with certain organisms which have the factor d in both phases. *S. salinatis* has the formula 4, 12 : d, e, h, : d, e, n, z_{15}, and all colonies are agglutinated by *S. typhi* H serum. When grown in the presence of anti-d serum, they are transformed into *S. sandiego* (4, 12 : e, h : e, n, z_{15}), which then exhibits normal phase variation. This is not an example of an organism with three phases, since the loss of the factor d is irreversible (Edwards *et al.* 1957).

The Vi antigen

Grinnell (1932) observed that a laboratory strain of the typhoid bacillus that was weakly virulent afforded little protection to mice against the intraperitoneal injection of a virulent strain, but that a recently isolated virulent strain afforded good protection against a similar test dose. This finding was confirmed by Bensted and Findlay (see Perry *et al.* 1933*a*, *b*, 1934*a*, *b*), who showed that there was a correspondence between high virulence for the mouse by the intraperitoneal route and failure, when tested in living suspension, to be agglutinated by O antiserum. Virulent strains stimulated the formation of bactericidins, and the bactericidal content of an antiserum and its protective power in passive immunity experiments on mice were closely related. Serial passage of a weakly virulent strain in mice resulted in the concomitant appearance of virulence, O-inagglutinability of living organisms, and the ability to protect mice against otherwise lethal doses of virulent typhoid bacilli. Further studies by Felix and Pitt (1934*a*, *b*) rendered it evident that the degree of agglutinability was determined by the presence of a substance which protected the O antigen from the antiserum and was itself antigenic. Because of its association with virulence for mice it was given the name of Vi antigen.

Rabbits immunized with living cultures of highly mouse-virulent typhoid bacilli produce, in addition to the usual O antibodies, a Vi antibody which agglutinates these virulent strains, whereas rabbits immunized with less virulent smooth strains, or with heat-killed virulent strains, produce O antibodies alone.

Agglutination tests to detect the presence of the Vi antibody must be made with suspensions incubated at a temperature of 37°. Either living organisms or organisms treated with 0·2 per cent formol may be used. A pure Vi antiserum may be made by absorbing a serum prepared against a mouse-virulent strain with organisms devoid of Vi antigen, but more conveniently by immunization with a rough strain possessing the Vi antigen. A Vi antiserum gives effective passive protection in mice against living virulent typhoid bacilli; an O antiserum in mice is almost or quite ineffective, though it does protect the animals against lethal doses of endotoxin (see also Felix and Bhatnagar 1935). Mice may be immunized actively, either with a suspension containing Vi antigen, or with a preparation of purified antigen.

The Vi antigen, unlike the O antigen, is not toxic on injection, and organisms with a Vi antigen but no O antigen have little pathogenicity for mice. The most pathogenic strains are those possessing the greatest quantities of both O and Vi antigens. The part the Vi antigen plays in the virulence of the typhoid bacillus for mice appears, as suggested by Felix and Pitt, to be due to the protection of the bacterial cell against phagocytosis and the bactericidal effect of the serum.

The amount of Vi antigen present in different strains of *S. typhi* varies considerably and in a continuous rather than a discontinuous manner. Total loss of the Vi antigen appears to be irreversible (Craigie and Brandon 1936*b*). Strains rich in Vi antigen form more opaque colonies than those with little antigen (Giovanardi 1938) and can readily be recognized by examination under oblique illumination (Landy 1950, Nicolle *et al.* 1950). According to Sertic and Boulgakov (1936*a*) smooth forms containing Vi antigen are agglutinated by acriflavine.

In their original description of the Vi antigen, Felix and Pitt described it as heat-labile, but this conclusion was based on the reactions of heated cells in the agglutination reaction. Other workers found the Vi antigen to be highly resistant to heat (Peluffo 1941, Chu and Hoyt 1954, Chu *et al.* 1956). After watery suspensions had been heated, the supernatant fluid contained material which reacted with Vi antibody, and was detectable by precipitation, complement fixation and indirect haemagglutination methods. It did not, however, stimulate the formation of antibodies unless first adsorbed on to red blood cells (Chu and Hoyt 1954). It seems, therefore, that the material detached from the cells by heating aqueous suspensions is not the complete antigen. Milder methods of extraction yielded materials with some antigenic properties (Webster *et al.* 1952, Landy and Webster 1952). The material obtained and purified in different ways varies widely in its ability to cause the formation of antibody in particular animal species; thus some preparations are more antigenic in mice than in rabbits, and others are more antigenic in rabbits than in mice (Landy *et al.* 1961, 1963). Treatment with weak alkali causes deacetylation, which abolishes antigenicity for the mouse but not for man. Deacetylation is accompanied by loss of viscosity and reduction of the molecular weight from 10^6 to 4×10^4 (Jarvis *et al.* 1967). Acetone suspensions dried *in vacuo* provide a rich and stable source of Vi antigen, both as a standard material for serological tests and as a vaccine.

The Vi antibody may be detected by agglutination or with more precision by indirect haemagglutination tests with purified antigen adsorbed to red blood cells. The use of these methods for the detection of typhoid carriers is discussed in Chapter 68. The injection of purified Vi antigen into some strains of mice leads to the formation of incomplete (albumin-potentiated) antibody, but this may have protective action (Gaines *et al.* 1960). In persons given TAB vaccine, considerably more incomplete than complete antibody may be formed (Gaines *et al.* 1965).

The Vi antigen is a highly polymerized acidic polysaccharide consisting of repeating units of *O*- and *N*-acetyl-*d*-galactosaminuric acid. The determinant groups responsible for its specific reactions in serological tests are stable to heat; but the ability to form antibody appears to depend on the macromolecular structure of the polysaccharide and is easily destroyed by heat and some mild chemical treatments.

Vi antigens identical with or closely related to that of the typhoid bacillus have since been discovered in *S. paratyphi C* (Kauffmann 1935*a*, 1936*a*), in occasional strains of *S. dublin* (Le Minor and Nicolle 1964) and in number of strains of *Escherichia* and *Citrobacter*. Active immunity to the typhoid bacillus can be produced in mice by the injection of organisms containing the Vi antigen but otherwise unrelated to the typhoid bacillus. Many of these coliform bacilli contain more Vi antigen than the average strain of the typhoid bacillus (Luippold 1944, Landy 1952). The Vi antigen of *S. paratyphi C*, on the other hand, is present in only small amount, and appears to play little part in the pathogenicity of this organism for mice (Kauffmann 1936*a*). Whiteside and Baker (1959) were unable to detect any antigenic differences between highly purified Vi antigen from *Citrobacter*, *Escherichia*, *S. typhi* and *S. paratyphi C* (see also Baker and Whiteside 1960).

Felix and Pitt (1936) also described a Vi antigen in *S. paratyphi A* which was antigenically different from that of the typhoid bacillus. Although causing only slight inhibition of O agglutination it was associated with an increase of virulence for the mouse. Kauffmann (1953) believes that this antigen is identical with the somatic factor 2, and that the so-called Vi antigen described by Felix and Pitt (1936) in *S. paratyphi B* and *S. typhimurium* (see Felix 1952) is the same as the somatic factor 5, though this is rather more heat-labile than most other O antigens.

Other antigens

The possibility that constituents of the cell body other than the lipopolysaccharide and the Vi antigen might contribute to the virulence or pathogenicity of salmonellae has received insufficient attention. According to Barber and Eylan (1975, 1976) preparations of somatic proteins are toxic for laboratory animals; some of these are common to salmonellae of different O groups. Immunization of mice with protein vaccines prepared from *S. paratyphi B* are said to afford protection against a lethal dose of *S. paratyphi C*. (For the antigenic properties of the fimbrial antigens, see Chapter 33.)

Bacteriophage-typing methods

S. typhi The Vi-phage typing method for *S. typhi* (Craigie and Brandon 1936*a*, *b*, Craigie and Yen 1938) was the first phage-typing method to be developed and is still in many ways unique. Soon after the reports of Felix and Pitt (1934*a*, *b*) on the Vi antigen of the typhoid bacillus, several workers established the existence of bacteriophages acting specifically on bacilli containing this antigen. The special contribution of Craigie and his co-workers was their observation of a peculiar adaptability possessed by one particular anti-Vi phage. When this phage was grown on typhoid strains of different origin, races of bacteriophage were obtained that had developed a high degree of specificity for the particular strain on which they had been propagated. By means of a series of bacteriophages prepared in this way, Craigie and Yen were able to classify nearly all of 592 strains of typhoid bacilli into eleven bacteriophage types. A study of the origin of the strains revealed a high degree of correlation between the bacteriophage type and the epidemic source. These observations have since been abundantly confirmed (Felix 1943, 1951, 1955) so that the bacteriophage method of typing is well established as a routine procedure in the investigation of outbreaks of typhoid fever. The technical methods used in Vi-phage typing have been internationally standardized (Craigie and Felix 1947) and a series of national and regional reference laboratories has been set up. More than 70 types have been recognized (Bernstein and Wilson 1963, Anderson 1972). Type A strains are sensitive to all adaptations of the Vi-phage, but the rest of the types possess a fairly high degree of specificity for one or a few related adaptations.

The method has its limitations. About a quarter of the strains isolated in Great Britain are untypable. They include some strains that are completely devoid of Vi antigen, some basically resistant to the Vi phage, and probably also some representatives of as yet unidentified types for which an adapted phage has not yet been produced. Most untypable strains, however, belong to the group of so-called 'degraded' strains which are sensitive to a large number of the adapted phages. A more serious difficulty is that, in many countries, one Vi-phage type—usually either type A or type E1—is so common as to limit the epidemiological information obtainable from typing. Attempts have been made to subdivide the common types further into fermentative types (Pavlatou and Nicolle 1953, Brandis and Maurer 1954). Nicolle, Pavlatou and Diverneau (1953, 1954) advocated the use of a battery of unadapted O phages as a complementary typing system for the further subdivision of the common Vi-phage types.

We have no space for a full discussion of the mechanisms underlying the Vi-type specificity of *S. typhi*. Briefly, it may be said that the extraordinary adaptability of the Vi phage is partly due to the selection of spontaneously occurring host-range mutants of the phage by the bacterium, and partly to a non-mutational, phenotypic modification of the phage by the host. As a rule, susceptibility to typing phages appears to be determined partly by the presence or absence of certain temperate phages ('type-determining phages') and partly by some unknown character of the non-lysogenic precursor. The former is responsible for the selection of specific host-range mutants of the Vi phage and the latter for changes in its phenotype (Anderson 1951, 1955, 1957, Felix and Anderson 1951, Anderson and Felix 1953*a*, *b*, Anderson and Fraser 1955, 1956, Anderson and Wiliams 1956, Bernstein and Wilson 1963).

S. paratyphi B Felix and Callow (1943) produced a series of apparent adaptations of a phage specific for the so-called Vi antigen of *S. paratyphi B*, and were able to define several epidemiologically valid phage types. It proved difficult, however, to adapt the phage to the remaining untypable strains, and they therefore introduced a series of unrelated phages into the system (Felix and Callow 1951). They observed that some of the supposedly adapted phages were antigenically different from the original strain, and concluded that they were unlikely to have been adaptations. Probably the typing phages are a heterogeneous collection of O phages, some of them identical with phages carried naturally by strains of *S. paratyphi B*. Many of these phages are 'type-determining', and numerous examples of transformation of type by lysogenization have been recorded. For a more recent account of the phage-typing of *S. paratyphi B* see Anderson (1964).

S. typhimurium Felix (1956) and Callow (1959) developed a system of typing this organism making use of a series of adaptations of one of the *S. paratyphi B* phages supplemented by other unrelated O phages. It originally distinguished 34 phage types but this number has now been increased to 207 (Anderson 1964, Anderson *et al.* 1977). Discrimination between strains is very fine. In a number of cases the acquisition of a particular resistance plasmid narrows the pattern of susceptibility of the strain to the typing phages (Anderson 1968*b*). Two strains that were prevalent in Britain in the late 1970s (see p. 335) belonged respectively to phage-types 204 and 193; they were derived from phage-type 49 by the acquisition of different resistance plasmids. In the Netherlands a quite different system is used (Scholtens 1962, 1969) in which typing by another set of bacteriophages is supplemented by biotyping.

Other salmonellae Phage-typing systems for other salmonellae include the following: *S. enteritidis* (Lilleengen 1950, Lalko 1977); *S. dublin* (Lilleengen 1950, Smith 1951*c*); *S. thompson* (Smith 1951*a*, *b*); *S. gallinarum* (Lilleengen 1952); *S. paratyphi A* (Banker 1955, Sanborn *et al.* 1977); *S. virchow* (Velaudapillai 1959); *S. panama* (Guinée and Scholtens 1967); *S. newport* (Petrow *et al.* 1974). The typing systems for *S. paratyphi A* and *S. panama* are said to make use of adapted phages.

The methods so far decribed are based on the susceptibility of the salmonellae to phages. An alternative approach, to define the characters of the phages carried by the salmonellae, has been shown to be practicable (see Boyd 1950, Boyd and Bidwell 1957, Atkinson *et al.* 1952) but is now seldom employed in salmonella typing.

Biotyping

Subdividing common salmonella serotypes according to their biochemical characters is sometimes of value in epidemiological investigations. In many serotypes, however, there are few biochemical tests in which significant numbers of strains behave differently, so the number of identifiable biotypes is small. For this reason, biotyping has proved most useful as an adjunct to phage typing, when it may serve to subdivide a large group of untypable strains or members of common phage types (as in *S. typhi*). Its potential usefulness is probably greatest in studying the epidemiology of sporadic human infections with *S. typhimurium*, but the results obtained are at present difficult to interpret. In a comprehensive investigation by biotyping and phage typing, Duguid, Anderson and their colleagues examined over 2000 strains of *S. typhimurium* (Duguid *et al.* 1975, Anderson *et al.* 1978); they were able to identify 204 phage types, 19 'primary' and 147 'full' biotypes. When the results obtained by both methods were considered together, 507 different 'types' could be recognized, but the study of long-term prevalences of infection yielded numerous examples of variation in both phage type and biotype.

There are several examples of stable biochemical varieties in serotypes that exhibit marked differences in biological characters such as host range, disease spectrum and geographical distribution. These are seen in *S. enteritidis*, *S. paratyphi B* and *S. gallinarum* (pp. 348 and 349). Four salmonellae that are traditionally accorded specific status because of wide differences in their natural pathogenicity (*S. paratyphi C*, *S. choleraesuis*, *S. decatur* and *S. typhisuis*) are serologically identical and distinguishable only by their biochemical characters (see Table 37.4).

Pathogenicity

Natural pathogenicity

Salmonellae cause disease in a wide range of species of vertebrates. In mammals and birds most of the pathogenic serotypes belong to sub-genus I (for exceptions see pp. 350–351). Broadly speaking, the disease is either an acute enteritis or a systemic infection. Salmonellae of most serotypes usually cause a diarrhoeal disease in which the incubation period is short (1–2 days). A few salmonellae habitually cause systemic disease in which symptoms referable to the intestine are not prominent and may be absent; and the incubation is longer (7–14 days). When the manifestations of systemic disease are mainly septicaemic, the clinical picture is one of enteric fever, as is usually the case in infections with *S. typhi* and *S. paratyphi A*, *B* and *C* in man; but certain other salmonellae, for example, *S. choleraesuis*, *S. blegdam*, *S. enteritidis* var. *chaco* and to a lesser extent *S. dublin* in man, and *S. gallinarum* in adult chickens, tend to cause pyaemic infections and to localize in the viscera, the meninges, the bones or joints or the serous cavities. Nevertheless, these distinctions are not clear cut. Local abscesses and even pyaemia may develop occasionally as a late complication of a diarrhoeal episode, and bacteraemia can be detected in all but the mildest attacks of enteritis. Even though diarrhoea is not a prominent symptom in the first two weeks of an attack of typhoid fever, there are important lesions in the lymphoid tissue of the intestine. The age of the patient also influences the type of disease. Salmonellae that usually cause enteritis in healthy adults cause septicaemic or pyaemic infections with much greater frequency in young children and in adults predisposed to this by other diseases. The one exception to this rule is the typhoid bacillus, which causes mild and atypical infections in children.

Convalescents from salmonella disease continue to excrete the organism, usually in the faeces and occasionally after systemic infections in the urine or in a purulent discharge. Many infected subjects harbour the organism in the intestine for a time without showing symptoms of disease. Both convalescent and symptomless excreters remain infectious for a variable period of time, but with some salmonellae in certain hosts often do so for life.

A few salmonella serotypes exhibit strong host specificity, for example, *S. typhi*, *S. paratyphi A* and *C* and *S. sendai* in man, *S. abortusequi* in horses, *S. abortusovis* in sheep and *S. gallinarum* in birds. With some exceptions the serotypes with a restricted range of hosts cause systemic disease and are nutritionally exacting and biochemically aberrant. Most other salmonellae are found in a number of animal species and tend to cause mild infections confined to the intestine. *S. typhimurium* is to some extent an exception in that it has a very wide host range but nevertheless gives rise to severe disease in many species of animals.

Nothing is known about the characters of salmonellae that are responsible for host specifity. Indeed, information about the mechanisms of pathogenicity in systemic salmonella disease is scanty. It is questionable whether the lipopolysaccharide is an important toxic agent in enteric fever, or whether the antiphagocytic activity of the Vi antigen is of significance in natural typhoid infection (Chapter 68). Some evidence suggests that the ability to cause diarrhoea is a specific attribute of some strains that is absent from others. Living salmonellae may cause the accumulation of fluid in ligated loops of rabbit intestine (Taylor and Wilkins 1961). According to Giannella and co-workers (1973, 1975) this occurs only when the salmonellae invade the gut wall and cause inflammation, but not all invasive strains cause fluid accumulation. Deibel and his colleagues, on the other hand, report the presence of an enterotoxin in sterile filtrates of salmonella cultures that cause diarrhoea in suckling mice and fluid accumulation in rabbit ileal loops (Koupal and Deibel 1975, Sedlock *et al.* 1978). According to Sedlock and Deibel (1978) the action of the agent is enhanced by first washing out the gut loops with a mucolytic substance or by giving it along with a live salmonella that does not form the agent. Sandefur and Peterson (1976) studied the production by some strains of salmonellae of substances that cause erythema and induration in rabbit skin, and identified two permeability factors (see Chapter 27), a heat-stable and rapidly acting substance formed by all salmonellae and a heat-labile substance with delayed action formed only by some strains and in variable amount (see also Kühn *et al.* 1978, Jiwa 1981). Duguid and co-workers (1976) advance evidence that fimbriation is of importance in establishing and maintaining salmonellae in the mouse gut. Fimbriate and non-fimbriate variants of the same strain of *S. typhimurium* were of similar virulence by the intraperitoneal and ocular routes, but the fimbriate form caused more infections and deaths than the non-fimbriate when given orally. Fimbriate organisms were excreted in the faeces by survivors for a longer time than non-fimbriate organisms.

Most attention has been paid to the oral route of infection with salmonellae, but it would be incorrect to assume that it is the only one. Infection may occur by inhalation in babies, and possibly in other animals, and by the conjunctiva (see p. 346). The transmission of certain salmonellae to man by means of arthropod bites is fairly well established (see p. 346 and Chapter 68).

It is impossible here to describe in detail the natural pathogenicity of all serotypes of salmonellae for man and animals, but some information will be found in the descriptions of the individual species on pp. 347–351. Enteric fever and certain other generalized salmonella infections are discussed in Chapter 68, salmonella enteritis in Chapter 72 and hospital-acquired salmonella infections in Chapter 58.

Experimental infection

S. typhi If massive doses of living typhoid bacilli are administered by the mouth to chimpanzees, it is possible to produce a disease that is very similar to typhoid fever in man (Metchnikoff and Besredka 1911, Edsall *et al.* 1960, Gaines *et al.* 1968*a*, *b*), though the incubation period is shorter and ulceration of the lymphoid tissue of the small intestine does not occur. (For experimental typhoid infection in man see

Chapter 68.) Administration of typhoid bacilli by the mouth to small laboratory animals (rabbit, guinea-pig, rat or mouse) does not give rise to an infection of this type, or usually to any harmful result. The intraperitoneal or intravenous injection of living typhoid bacilli in adequate doses induces a fatal infection, and the bacilli can be recovered from the blood and tissues *post mortem*. There is nothing characteristic in the findings at necropsy. With some strains of typhoid bacilli it may be necessary to inject 10^9 living bacilli or more into the peritoneum of a mouse to produce a fatal result; five to ten times this dose of heat-killed organisms will usually cause a purely toxaemic death. A highly virulent strain given intraperitoneally will kill mice in a dose of 5×10^7 or less. There is no evidence that the typhoid bacillus possesses an ability to multiply freely in the tissues of small laboratory animals when injected in small doses, as is the case with some strains in other species of salmonellae (see below). Fatal infection can, however, be produced in the mouse by the intraperitoneal injection of small numbers of organisms together with such adjuvants as mucin (Olitzki and Koch 1945). Small numbers of organisms injected intracerebrally into mice without adjuvant give rise to a fatal infection (Landy *et al.* 1957). In rabbits given an intravenous dose of living bacilli too small to produce a rapidly fatal infection, the organisms tend to localize in certain tissues, particularly the gall-bladder (see Chapter 68.

Though milk-borne typhoid fever is not uncommon, there is little or no evidence to suggest that cows suffer from natural infection with typhoid bacilli (Scott and Minett 1947); on the other hand they may occasionally acquire and excrete paratyphoid B bacilli (see Chapter 68).

S. typhimurium* and *S. enteritidis The effects produced by the administration of living cultures of these organisms to the small laboratory animals are entirely different from those produced by the typhoid bacillus. We are dealing with a natural pathogen of these animals which gives rise in them to a characteristic disease, usually known as mouse typhoid. This disease is produced when living cultures of *S. typhimurium* are given by the mouth as well as when they are administered by injection, though the time to death is longer in the former case than in the latter. The organism has a very definite invasive power for the tissues of mice and of other laboratory rodents. A virulent strain will kill 50 per cent or more of mice when injected intraperitoneally in a dose of 100 organisms. However, there are wide differences in the virulence of different strains of *S. typhimurium* for the mouse when given intraperitoneally. Rough variants are avirulent, but the LD50 of culturally smooth strains may also vary between 10^2 and 10^6 organisms.

Mice dying within 2 to 3 days after the injection of a moderate dose of a virulent strain will be found to have succumbed to an acute septicaemia with few obvious lesions; but mice dying after the more usual period of 5 to 10 days often show characteristic lesions, including a varying degree of splenic enlargement, often associated with the presence of small necrotic foci, larger and very characteristic necrotic lesions in the liver, and sometimes scattered pneumonic patches in the lungs, accompanied by a scanty pleural exudate. These lesions have been described in some detail by many observers (see, for instance, Seiffert *et al.* 1928). The spread of infection from the intestine, and the subsequent involvement of the various tissues, were studied and described by Müller (1912) and by Ørskov and his colleagues (Ørskov *et al.* 1928, Ørskov and Moltke 1928, Ørskov and Lassen 1930). *S. enteritidis* behaves in much the same way as *S. typhimurium*.

One other point should be emphasized. The disease produced in the mouse by *S. typhi* shows no tendency to spread by contact from mouse to mouse but that produced by *S. typhimurium* and *S. enteritidis* is highly contagious. It is surprising, therefore, that these animals often appear to be quite refractory to feeding with organisms unless large doses are given. This has led some workers to suggest that natural spread may not take place by the oral route. Moore (1957) working with two strains of *S. enteritidis* derived from natural epizoötics, found that guinea-pigs could be infected much more easily by the conjunctival route than by the mouth. Infection could be established by the instillation into the eye of about 100 organisms, whereas doses of 10^8 by the mouth were ineffective. The spread of infection could be prevented by protecting the conjunctiva from contamination with salmonellae. Respiratory carriage of *S. typhimurium* can be induced in mice by intranasal instillation of the organism (Tannock and Smith 1971).

There is evidence that the resistance of mice to infection by feeding is due to other organisms normally present in the flora of the gut. Bohnhoff, Drake and Miller (1954) found that the number of organisms necessary to establish infection by the oral route could be reduced 100 000-fold by a single dose of streptomycin. Infection was then initiated by less than 10 organisms. The same increase of susceptibility occurred with both *S. typhimurium* and *S. enteritidis*. Penicillin was as effective as streptomycin. Resistance to infection could be re-established by feeding normal mouse faeces. The significant change brought about by the antibiotics appeared to be a destruction of the anaerobic gram-negative flora of the gut (Miller *et al.* 1956, 1957, Miller 1959). The caecum of the normal mouse contains amounts of fatty acids—notably acetic, butyric and propionic—that will inhibit *S. typhimurium in vitro*, but these acids are not formed when streptomycin is given by the mouth (Meynell and Subbaiah 1963). Considerable quantities of acetic and butyric acid are formed in anaerobic culture by some of the Bacteroidaceae; steam distillates of the gut contents of untreated mice inhibit infection by the oral route, and contain the same fatty acids (Bohnhoff *et al.* 1964). On the other hand, lowering intestinal motility by injecting opium intraperitoneally (Kent *et al.* 1966), or ligating the ileum (Abrams and Bishop 1966), increases penetration of the bowel wall by salmonellae, and it has been suggested that the normal gut flora also protects against salmonella infection by its effect on the rate of intestinal emptying.

A generalized infection with *S. enteritidis* can be produced in guinea-pigs by swabbing the shaven skin with broth cultures (Liu *et al.* 1937, Liu 1938*a*). Messerlin and Couzi (1942) demonstrated the transmission of *S. choleraesuis* between guinea-pigs by the bite of the human flea. *S. enteritidis* can also be transmitted to guinea-pigs by the bite of *Dermacentor andersoni* and *Ornithodorus moubata* (Parker and Steinhaus 1943, Jadin 1951) and to mice by rat fleas (Eskey *et al.* 1949).

S. paratyphi B This species occupies a position in some ways intermediate between that of *S. typhi* and *S. typhimurium*. It is a natural pathogen of man but not of rodents. When injected into mice it kills them in far smaller doses than does *S. typhi*, though it is less virulent than *S. typhimurium*. When administered by the mouth it has a limited ability to invade the tissues and multiply in them; but it rarely gives rise to fatal infections.

S. paratyphi C The intraperitoneal injection of large doses of *S. paratyphi C* into mice gives rise to an acute intoxication like that following a fatal dose of *S. typhi*, but doses 100 times smaller cause a true infection, which often results in death. Small doses by the intragastric route sometimes give rise to generalized infection (Archer and Whitby 1957).

Many other species of *Salmonella* will probably be found to possess high or moderate virulence for laboratory animals, and in some instances we already know this to be true. *S. choleraesuis*, for instance, is highly virulent for the rabbit. According to Milner and Shaffer (1952) 1-day-old chicks can readily be infected by the mouth with different types of *Salmonella*; the death-rate is low and the organism can be recovered from the faeces for days and sometimes weeks (see also Shaffer *et al.* 1957). Day-old chicks may also be infected by the inhalation of small numbers of organisms, but with most salmonellae the mortality is low (Clemmer *et al.* 1960). Infection with *S. gallinarum* var. *pullorum* is more easily established by spraying than by feeding (J. E. Wilson 1955).

We append a few notes on the members of *Salmonella* sub-genus I that are the more important causes of disease in man and animals, or have unusual characters, and we follow these with brief accounts of sub-genera II, III and IV.

Notes on selected species or serotypes of salmonellae

Sub-genus I

S. abortusequi (4, 12: —: e, n. x). A natural pathogen of the horse in many countries, causing abortion in mares (Kilbourne 1893, Buxton and Field 1959) and, occasionally, orchitis in stallions (Henning and McIntosh 1946). Surviving foals may suffer from a fatal septicaemia or from abscesses in the joints. There have been few isolations of the organism from other animals. One food-borne outbreak of infection has been recorded in man (Bruner 1946).

S. abortusovis (4, 12: c: 1, 6). A natural pathogen only in sheep, in which it gives rise to abortion (Schermer and Ehrlich 1921). Compared with most other organisms of the *Salmonella* group, this organism grows poorly. Many of the usual sugar reactions are negative, late or irregular.

S. choleraesuis (6, 7: c: 1, 5). The American hog-cholera bacillus, isolated by Salmon and Smith (1885, 1886), from the former of whom the name of *Salmonella* is derived. Though hog cholera, or swine fever, is now known to be a virus disease, *S. choleraesuis* is a common secondary invader. It also gives rise to generalized infections in pigs not suffering from hog cholera. Occasional infections have been observed in many other animal species. *S. choleraesuis* is an important human pathogen, not because it gives rise to many infections, except in China (Huang and Lo 1967), but because of the severity of its effects. About 20 per cent of recorded infections are fatal, nearly half are associated with prolonged pyrexia, usually of the septic type, and over one-third result in local pus formation (Saphra 1950, Saphra and Wassermann 1954, Saphra and Winter 1957). The common pyaemic manifestations include pneumonia, septic arthritis and osteomyelitis, purulent meningitis and bacterial endocarditis. Most of the severe cases are sporadic and difficult to trace to their source. A few food-borne epidemics have been described (Clauberg 1931, Boecker and Silberstein 1932, Kauffmann 1934*a*, 1941, Buczowski 1961) but these were of diarrhoeal disease and seldom included septicaemic infections. Symptomless faecal excretion is said to be rare. The organism is grown without difficulty from blood and pus on ordinary nutrient media, but is difficult to isolate from faeces. It grows poorly on MacConkey's medium and not very well on deoxycholate citrate agar. It is not enriched by growth in selenite broth. Selective media containing brilliant green should be used (Gitter 1959, Smith 1959), but must not contain sulphadiazine (Chung and Frost 1971). H_2S-negative, diphasic strains were at one time referred to as the American type and H_2S-positive strains in phase 2 as the European or *kunzendorf* type. It is doubtful whether these two varieties differ significantly in geographical distribution or natural pathogenicity; differences in H_2S production are only quantitative, and phase 1 can usually be demonstrated if suitable methods are used (p. 341). *S. choleraesuis* is closely related to *S. paratyphi C* but is distinguished from it by its failure to ferment arabinose and trehalose, and to some extent by its lack of the Vi antigen. Two other organisms, *S. typhisuis* and *S. decatur*, also have the same antigenic formula in the Kauffmann-White scheme as *S. choleraesuis*, though the H antigenic factor c is slightly different in each of them (Kauffmann *et al.* 1955). *S. typhisuis* (see p. 350) is usually distinguishable by its poor growth and weak gas production, and by its inability to acidify mannitol or to attack the organic acids (see Table 37.4). *S. decatur*, which was isolated from a child with diarrhoea (Kauffmann *et al.* 1955), differs from all the others in giving a positive Stern reaction. It is customary to consider these organisms as separate species, rather than as biochemical types, because of the differences between them in habitat and pathogenicity.

S. dublin (1, 9, 12, [Vi]: g, p: —). Sometimes known as the *kiel* variety of *S. enteritidis*. In Europe its principal habitat appears to be cattle. Before the recent increase in *S. typhimurium* infection in calves it was the commonest salmonella in bovine faeces in Britain, although its geographical distribution within the country was irregular (Field 1948, Smith and Buxton 1951, Murdoch and Gordon 1953). In the early 1970s there was a considerable increase in the frequency of *S. dublin* infections in the Netherlands and neighbouring countries, but not in Britain, associated with the presence of multiple-antibiotic resistant strains (p. 335). In the United States it has been reported infrequently. Was reported by Henning (1953) to be the principal cuasc of calf 'paratyphoid' in South Africa. Though Lütje (1939) in Germany and Henning (1953) observed the disease mainly in calves, in Britain cows are more often affected. The disease often causes abortion, and has a mortality of 70 per cent. Most of the survivors become permanent carriers (Field 1948, 1949). Pohl (1954) found that transient faecal excretion also occurred among contact animals and that the organism persisted for many months on contaminated pasture. *S. dublin* may cause gastro-enteritis or abortion also in sheep and goats (Levi 1949, Watson 1960), and occasional infections have been described in other animals. Human infections are quite frequent in England, and in western Europe. Numerous milk-borne outbreaks of gastro-enteritis have been described (for instance, see Conybeare and Thornton 1938, Tulloch 1939). Sporadic infections often give rise to an enteric or septic type of fever (Smith and Scott 1930), sometimes associated with abscess formation in bones and joints. Five fermentative varieties, other than the usual one, have been recognized, *accra, koeln, teheran, hessarek* and *dawa* (Neel *et al.* 1953, Hughes 1954). Anaerogenic variants are common in Britain (Walton and Lewis 1971).

S. enteritidis (1, 9, 12:g, m:—). The *gaertner* or *jena* variety was first recognized as a cause of food-borne enteritis (Gaertner 1888). There are three other biochemical varieties: the *danysz* variety, sometimes known as the Ratin bacillus, the *essen* variety, and the *chaco* variety. These four biochemical subtypes can be distinguished from each other only by the use of special tests, such as those of Bitter and Stern (see Kauffmann 1941), yet they differ greatly from each other in host specificity and geographical distribution. The *jena* variety appears to be responsible for most of the outbreaks of food poisoning due to *S. enteritidis* that follow the consumption of meat. The *danysz* variety was isolated by Danysz (1900) from an epidemic of mouse typhoid in field mice. It is a natural pathogen of the brown rat in Britain (Ludlam 1954, Brown and Parker 1957), The Ratin and Liverpool 'viruses' are strains of this variety that have been used in attempts to reduce the wild rodent population by establishing natural epizoötics; and these have sometimes led to outbreaks of illness in man (Leslie 1942, Dathan *et al.* 1947, Kokko 1947, Taylor 1956, Brown and Parker 1957). The *essen* variety was isolated from cases of gastro-enteritis, from ducks, and from ducks' eggs by Hohn and Herrmann (1935*a*, *b*) and described as *S. moskau*; it is not to be confused with *S. essen* which belongs to the B group, nor with *S. moscow* (see below). The *chaco* variety was isolated by Savino and Menéndez (1934) from cases of continued fever in South America. It usually gives rise to a severe septicaemic or pyaemic disease without diarrhoea. It has since been met with in India (Hayes and Freeman 1945), in Java (Gispen and Stibbe 1947), and in West Africa (Cosgrove and Reid 1953), where it was isolated only from man.

It is appropriate to mention here two other salmonella serotypes—*S. moscow* (9, 12:g, q:—) and *S. blegdam* (9, 12:h, m, q:—)—which closely resemble *S. enteritidis* in antigenic structure, and may give rise to severe pyaemic illness. *S. moscow* was isolated in Russia from a typhoid-like illness (Kulescha and Titowa 1923, Sütterlein 1923, Weigmann 1925*a*, *b*, Iwaschenzoff 1926, Hicks 1930); most of the patients were also suffering from relapsing fever, and it was concluded that the salmonella was a secondary invader. Evidence obtained from a similar disease in Peking (Huang *et al.* 1937, Liu *et al.* 1937, Liu 1938*b*) suggests, however, that both the salmonella and the spirochaete were louse-borne (see also Chapter 68). *S. blegdam* first attracted attention as a cause of a severe pyaemic disease among Australian troops in New Guinea and neighbouring islands (Fenner and Jackson 1946, Baker *et al.* 1949, Atkinson *et al.* 1949). Infection with this serotype was frequently seen in Shanghai, sometimes in association with typhus fever (Fournier 1949, Fournier *et al.* 1950, Wu and Fournier 1951). It was seen among prisoners of war in Korea, often associated with relapsing fever (Zimmerman 1953). Raynal and Fournier (1947; see also Fournier 1949) reported that it was the most common salmonella in wild rats in Shanghai.

S. gallinarum (1, 9, 12:—:—). Isolated from fowls suffering from fowl typhoid (Klein 1889). Differs from other salmonellae in being non-flagellated. Forms little or no gas from glucose; fermentation of other sugars is relatively slow and feeble. Sugar fermentation tests should be read after 8 days. Is D-tartrate positive and H_2S positive, though the reaction may be weak or delayed. When grown in nutrient gelatin containing cysteine produces a characteristic cloudiness (Hinshaw 1941). *S. gallinarum* appears to be seldom pathogenic for man, but a biochemical subtype, known as the Duisberg variety, which fails to attack D-tartrate or to produce H_2S, may be responsible for cases of acute gastro-enteritis (see Müller 1933, Kauffmann 1934*b*). This variant should not be confused with *S. duisberg* (see Fulton and Fulton 1966). A more important biochemical subtype, often regarded as a separate species and called *S. pullorum*, is responsible for the widespread disease of chicks known as *bacillary white diarrhoea* or pullorum disease (Rettger 1900). Most strains of the *pullorum* variety grow more slowly than *S. gallinarum* and form small translucent colonies; they also usually differ from it in producing gas from glucose, and in failing to ferment maltose, dulcitol and D-tatrate, but the sugar reactions are not a reliable means of distinguishing between the two organisms. Maltose-fermenting strains of the *pullorum* variety are fairly often seen, and as many as 10 per cent of strains may be anaerogenic (Hinshaw *et al.* 1943). Both varieties are H_2S positive, but the *pullorum* strains produce no cloudiness in the cysteine gelatin medium. Blaxland and his colleagues (1956) state that acidification of dulcitol, fermentation of D-tartrate, the cysteine reaction and the colonial appearance are the most reliable differential tests. According to Trabulsi and Edwards (1962), *S. gallinarum* always gives a negative ornithine and the *pullorum* variety a positive ornithine decarboxylase reaction (see also Costin and Costin 1966). The *pullorum* variety grows poorly on the usual media for isolating salmonellae. It does not produce characteristic colonies on Wilson and Blair's medium, and as a rule forms rather small colonies on deoxycholate citrate agar after 24 hr. Brilliant green agar is to be recommended. As we mentioned on page 341, there is

Table 37.4 Biochemical reactions of *S. paratyphi C*, *S. choleraesuis*, *S. typhisuis* and *S. decatur*

	Acid in						Fermentation of	
	dulcitol	arabinose	trehalose	mannitol	H_2S	Stern glycerol	tartrate	citrate
S. paratyphi C	+	v	+	+	+	—	+	+
S. choleraesuis	x	—	—	+	v	—	+	+
S. typhisuis	(+)	+	+	— or x	—	—	—	—
S. decatur	+	+	+	+	+	+	+	+

+ = Positive within 2 days. (+) = Delayed positive. x = Late and irregularly positive, or negative. v = Some strains positive, some negative. — = Negative reaction.

considerable variation from strain to strain in the relative proportion of the factors 12_2 and 12_3 in the somatic antigen of the *pullorum* variety. Normally, factor 12_3 is present in excess of 12_2, but in some instances factor 12_2 predominates, and there is little or no 12_3. Thus, an antigen made from a normal strain may fail to detect antibodies in chicks suffering from infection with a 'variant' strain with little 12_3 (Younie 1941, Edwards and Bruner 1946). *S. gallinarum* causes fowl typhoid, usually in adult chickens and much less often in chicks.The *pullorum* variety gives rise to acute diarrhoeal disease in chicks. Adult birds are often chronically affected, but the disease does not give rise to epidemic illness among them. Neither disease is common in other domestic birds in Great Britain (Blaxland *et al.* 1958). Occasional strains have been isolated from wild birds and from mammals. Although clinical illness due to these organisms is rare in man, the *pullorum* variety apparently was responsible for one large outbreak of food poisoning (Mitchell *et al.* 1946).

S. paratyphi A (1, 2, 12 : a : —). Isolated from enteric fever in man (Gwyn 1898). Is an important cause of enteric fever in Asia, the Middle East and Africa (Zimmerman *et al.* 1952). Causes a continued fever in which the organism is regularly present in the blood and faeces. Is a natural pathogen only in man, but has been isolated occasionally from other animals. Does not ferment xylose. Occasional strains fail to form gas. Different strains vary in their production of H_2S. Cultures are not enriched by growth in tetrathionate broth. (For other reactions see Fig. 37.2.) Strains lacking the O factor 1 are described as the *durazzo* variety. There are many resemblances between *S. paratyphi A* and *S. sendai* (see below).

S. paratyphi B (1, 4, [5], 12 : b : 1, 2). Isolated from cases of enteric fever in man (Achard and Bensaude 1896, Schottmüller 1900, 1901). The 1 antigen is present in only some strains. The 5 antigen is generally present, but may be missing, particularly in strains from carriers (Kauffmann 1934*c*); such strains have been named *S. paratyphi B* var. *odense*. Some strains cause an enteric type of fever in which the incubation period is long and there is little diarrhoea. Others cause an acute, but mild, enteritis after a short incubation period. Typical strains from cases of enteric fever form a mucoid wall when grown on solid media at room temperature. Brown, Duncan and Henry (1924) showed that they differed from *S. typhimurium* in their inability to ferment D-tartrate. Kristensen and Kauffmann (1937) observed that strains of *S. paratyphi B* from cases of enteritis were D-tartrate positive and failed to form a mucoid wall. Most of them were diphasic, but some, which resembled the strains isolated by de Moor (1935) from cases of enteritis in Java, were monophasic. Diphasic, D-tartrate-positive strains were isolated from animals as well as from man, but the monophasic, D-tartrate-positive *java* strains were restricted almost entirely to man (see also Brandis 1948, Petersen 1949). Diphasic, D-tartrate-negative strains which formed mucoid walls were also rarely isolated from animals other than man. Cherry, Davis and Edwards (1953) confirmed these findings in the main, but suggested that the three groups were not entirely clear cut. A considerable minority of animal strains were D-tartrate negative and some of the D-tartrate-positive animal strains were monophasic. Kauffmann (1955) proposed that differences in antigenic structure should be disregarded, and two separate species recognized on the grounds of mucoid wall formation and pathogenicity: (1) *S. paratyphi B*, which forms a mucoid wall, is usually but not always D-tartrate negative and disphasic, and gives rise to enteric fever, and (2) *S. java*, which does not form a slime wall, is usually D-tartrate positive, may be either monophasic or diphasic, and causes enteritis. The use of phage-typing has revelead a variety of types which give rise to diarrhoea. In one outbreak (McDonald *et al.* 1955) the organism was monophasic, did not ferment dulcitol or produce a mucoid wall, and could not be typed. Other strains responsible for diarrhoea belonged to the phage type Worksop (Bernstein 1958) or to type 1 var. 6. Unfortunately there are still few published accounts of well defined outbreaks in which both the clinical manifestations and the relevant characters of the causative organism are recorded. Until more information of this sort is available it will not be possible to decide which of these cultural and antigenic characters is most related to pathogenicity and host specificity. According to Costin (1967), *java* strains differ from *S. paratyphi B* in being able to grow on an agar medium with sodium acetate as the sole source of carbon.

S. paratyphi C (6, 7, [Vi] : c : 1, 5). Isolated from cases of enteric fever in man, at first mainly in Eastern Europe during and immediately after the First World War (see Weil 1917, Neukirch 1918, MacAdam 1919, Mackie and Bowen 1919, Hirschfeld 1919, Schütze 1920, Dudgeon and Urquhart 1920, Andrewes and Neave 1921, Weigmann 1925*a*, *b*, Iwaschenzoff 1926, White 1926). Many of the patients, like those infected with *S. moscow* (see p. 348), also had relapsing fever. The organism received many other names and, in particular, was often referred to as 'Hirschfeld's Bacillus' or the 'Eastern European type of *Bact. paratyphosum C*'. It was common in British Guiana in 1927–30 (Giglioli 1930, 1958), occurred among troops in south-east Asia in the Second World War (Hayes and Freeman 1945), and was said to be common in China (Raynal and Fornier 1947). The organism is an important pathogen of man, giving rise to enteric fever that is often associated with septic lesions. It has rarely been isolated from animals. It is closely related antigenically to *S. cholerаesuis*, *S. typhisuis* and *S. decatur*, and is best distinguished from them by biochemical means (see Table 37.4).

S. sendai (1, 9, 12 : a : 1, 5). Isolated from cases of enteric fever in Japan (Aoki and Sakai 1925), but few reports of its occurrence there or elsewhere have appeared in recent years (Dunlop 1950). Resembles *S. paratyphi A* in growing more poorly than most other salmonellae, in being a poor producer of H_2S, in failing to ferment tartrate or grow in citrate media, and in causing enteric fever in man. Also closely related antigenically to *S. paratyphi A*, from which it differs in containing the O antigen 9 and in possessing a second phase, which can, however, be artificially induced in *S. paratyphi A* by growth in the presence of antiserum. *S. sendai* can also be distinguished from *S. paratyphi A* by its ability to ferment xylose, and its frequent failure to form gas. Some light has been thrown on the cultural abnormalities of *S. sendai* by comparing it with *S. miami*, which also has the antigenic formula 1, 9 12 : a : 1, 5. This organism is quite common in Florida (Galton and Hardy 1948); it causes gastro-enteritis in man, grows vigorously and is biochemically a typical salmonella. An important difference between the two organisms is the inability of *S. sendai* to use sulphate as a source of sulphur. In media containing sulphite some, but not all, of the biochemical differences between *S. sendai* and *S. miami* disappear (Fukumi *et al.* 1953, Le Minor 1955, 1956).

S. typhi (9, 12, Vi : d : —). The cause of typhoid fever in man (Chapter 68). Its general characters have been referred to in previous sections. The majority of freshly isolated strains contain the Vi antigen. Though many strains of *S.*

typhisuis, *S. sendai* and *S. gallinarum* fail to produce gas from glucose, *S. typhi* is unique among the *Salmonella* group in never forming gas. Does not appear to infect animals under natural conditions. According to Weiner and Price (1956), has some antigenic relation to the larvae of *Trichinella spiralis* (see also Weiner and Neely 1964).

S. typhimurium (1, 4, [5], 12 : i : 1, 2). This organism was isolated in 1892 by Loeffler from rodents suffering from a typhoid-like disease. Other specific names given to it include *psittacosis* (Nocard 1893), *aertrycke* (de Nobele (1898) and *pestis caviae* (Wherry 1908); it was frequently referred to in German literature as the 'Breslau bacillus'. Strains in which the O antigen 5 is missing are sometimes referred to as the *copenhagen* (Kauffmann 1934*c*) or the *storrs* (Edwards 1935) variety; this type is common in American and Dutch pigeons (van Dorrsen 1955). The disease in rodents was described on p. 346; in man, *S. typhimurium* usually gives rise to an acute enteritis, but occasionally causes a prolonged fever of the enteric type. It is probably the salmonella with the widest distribution in the animal kingdom, including man. Buxton (1957) lists 37 animal species from which it has been isolated. The part it plays in causing disease in the various species varies from country to country. It was for many years common in cattle in Germany (Lütje 1937, Bartel 1938, Diageler and Kotter 1956), and was the predominant salmonella in cattle, sheep and horses in New Zealand (Josland 1952). Although it has been recorded as causing epidemic disease in mice and rats, it is not one of the types most commonly isolated from these animals in the wild state. In the 1950's, the biggest single reservoir of *S. typhimurium* in Britain was believed to be domestic poultry, and many epidemics occurred in chicks, turkey poults and ducklings (Blaxland *et al.* 1958). After 1964, antibiotic-resistant strains of *S. typhimurium* belonging to several phage types caused successive outbreaks of disease in calves in Britain and subsequently spread to the human population. In other countries, however, antibiotic-resistant strains for which an animal source could not be established have caused extensive epidemics in man (see p. 335). Mention must be made of two other salmonellae with antigenic structures resembling those of *S. typhimurium*. One of these, *S. gloucester* (1, 4, 12, 27 : i : l, w), is easily distinguished from *S. typhimurium* in non-specialized laboratories if both flagellar phases are examined, but the other, *S. agama* (4, 12 : i : 1, 6), can be identified only if single-factor sera are used.

S. typhisuis (6, 7 : c : 1, 5). This organism was isolated from young pigs suffering from a typhoid-like disease by Glässer (1909, 1910). *S. typhisuis* is almost antigenically identical with *S. choleraesuis*. It differs, however, in growing poorly on ordinary media and in its fermentation reactions. It forms gas very slowly and sparsely; it usually fails to ferment mannitol or does so late; it does not attack citrate, mucate, or the tartrates (see Table 37.4). It also ferments arabinose and trehalose, which *S. choleraesuis* does not. So far as is known, *S. typhisuis* does not naturally infect animals other than the pig, nor does it give rise to disease in man. It seems to be very uncommon in the United States (Bruner and Edwards 1940).

Sub-genus II

In 1960, Kauffmann observed that many serotypes of African origin liquefied gelatin, fermented malonate and were slow and irregular in their attack on D-tartrate. In these characters they resembled the Arizona group, from which they could be distinguished by their ability to ferment dulcitol but not lactose (Kauffman 1972), though nearly half of them gave a weak β-galactosidase reaction (Kauffmann 1963*a*). Kauffmann's proposal (1960) that they should form a separate sub-genus has been generally accepted, though Le Minor and his colleagues (1970) would place them in one species (*S. salamae*). Their other biochemical characters are given in Table 37.1).

Over 100 serotypes have been placed in sub-genus II (Kauffmann 1963*b*). In general, their antigenic characters are more like those of sub-genus I than of the Arizona group. A few belong to salmonella O-groups B–E, but most belong to the 'higher' groups. Their phase-1 H antigens are usually like those of sub-genus I, but some of their phase-II antigens, though bearing a superficial resemblance to sub-genus I antigens, differ in detail. (For further information see Rohde 1965, 1967).

The main habitat of sub-genus II strains is the faeces of cold-blooded animals, mainly but not exclusively in the tropics. Occasionally they appear to have caused disease in man.

Sub-genus III (the Arizona group)

The first member of this group was isolated by Caldwell and Ryerson (1939) from reptiles in the United States of America. Numerous further strains were isolated subsequently, mainly from reptiles and turkeys (see Edwards *et al.* 1943, Hinshaw and McNeil 1946), less often from mammals and man.

Sub-genus III conforms to the general definition of *Salmonella* given at the beginning of this chapter. The characters by which it can be distinguished from the other sub-genera are shown in Table 37.1. (See also Edwards *et al.* 1947*a*, *b*, Kauffmann 1956*a*, Shaw 1956, Ewing and Fife 1966).

Many different somatic and flagellar antigens have been identified (Edwards *et al.* 1947*a*, *b*, 1965, Fife and Martin 1967, Kauffmann 1969). The O and H antigens are designated by arabic numerals and the two flagellar phases are shown separately. Most of the O antigens show a strong reciprocal relation to or are identical with factors in salmonella O groups from F onwards. The H antigen of monophasic types consists of several factors that appear in different combinations in various serotypes. The diphasic H antigens, however, usually consist of single factors. Many of the common H antigens and H complexes of sub-group I salmonella also occur in sub-genus III (see Rohde 1967).

Soon after the first isolation of Arizona organisms from reptiles in the western United States it became apparent that they were an important cause of diarrhoeal illness in turkey poults in the same area (Hinshaw and McNeil 1946). Spread could be traced from farm to farm and infection appeared to be transmitted by hatchery eggs (Edwards *et al.* 1947*a*; see also Greenfield *et al.* 1971). Occasional isolations have been made from many other animals, but the main reservoirs of infection appear to be lizards, snakes, tortoises, chickens and turkeys (Edwards *et al.* 1956). Organisms of the Arizona group occasionally cause serious illness in man in some parts of the world. Edwards, Fife and Ramsey (1959) collected information about 229 individual human infections. The organism was isolated from the blood or from a localized abscess in one-third of them. About half of the patients had an acute diarrhoeal disease, and 10 per cent were symptomless excreters. The proportion of infections that gave rise to a serious septicaemic and pyaemic illness was thus higher than in infections with most salmonella types (see also Johnson *et al.* 1976). Outbreaks of infection in man appear to be predominantly food-borne (Murphy and Morris 1950, Edwards *et al.* 1959). Most of the strains responsible for human infections and for outbreaks among poultry, as well as the majority from lizards and tortoises, are monophasic. The diphasic ones appear to be very common in snakes in the wild state (Edwards and Boycott 1955, Le Minor *et al.* 1958), and have caused diarrhoeal disease in monkeys and sheep (Greenfield *et al.* 1971).

Sub-genus IV

It has now become clear that salicin-fermenting salmonella-like organisms are not uncommon in cold-blooded animals, particularly in the tropics. The characters by which they are distinguished from members of the other sub-genera are shown in Table 37.1. According to Rohde (1967), 1.5 per cent of recognized salmonella types have these characters, and would thus be placed in sub-genus IV. Most of them have O antigens in the 'higher' groups, but few of their H antigens correspond to antigens found in sub-genus I.

Not all salmonellae can be fitted neatly into the four sub-genera, and examples are found of strains with some of the biochemical characters of one sub-genus and the antigens usually associated with another. Such organisms may have arisen by hybridization between salmonellae or of salmonellae with other enterobacteria (Baron *et al.* 1959, Schneider *et al.* 1961, Mäkelä 1964).

References

Abrams, G. D. and Bishop, J. E. (1966) *J. Bact.* **92,** 1604.

Achard, C. and Bensaude, R. (1896) *Bull. Mém. Soc. méd Hôp.* **13,** 820.

Anderson, E. S. (1951) *J. Hyg., Camb.* **49,** 458; (1955) *Nature, Lond.* **175,** 111; (1957) *Zbl. Bakt.,* **168,** 489; (1961) *Nature, Lond.* **190,** 284; (1964) In: *The World Problem of Salmonellosis.* Ed. by E. van Oye, p. 89. Dr W. Junk Publishers, The Hague; (1965) *Brit. med. J.* **ii,** 1289; (1968*a*) *Ibid.* **ii,** 333; (1968*b*) *Annu. Rev. Microbiol.* **22,** 131; (1972) *Pers. Comm.* (1975) *J. Hyg., Camb.* **74,** 289.

Anderson, E. S. and Datta, N. (1965) *Lancet* **i,** 407.

Anderson, E. S. and Felix, A. (1953*a*) *J. gen. Microbiol.* **8,** 408; (1953*b*) *Ibid.* **9,** 65.

Anderson, E. S. and Fraser, A. (1955) *J. gen. Microbiol.* **13,** 519; (1956) *Ibid.* **15,** 225.

Anderson, E. S. and Lewis, M. J. (1965) *Nature, Lond.* **206,** 579.

Anderson, E. S. and Rogers, A. H. (1963) *Nature, Lond.* **198,** 714.

Anderson, E. S., Ward, L. R., de Saxe, M. J. and de Sa, J. D. H. (1977) *J. Hyg., Camb.* **78,** 297.

Anderson, E. S., Ward, L. R., de Saxe, M. J., Old, D. C., Barker, R. and Duguid, J. P. (1978) *J. Hyg., Camb.* **81,** 203.

Anderson, E. S. and Williams, R. E. O. (1956) *J. clin. Path.* **9,** 94.

Andrewes, F. W. (1922) *J. Path. Bact.* **25,** 505; (1925) *Ibid.* **28,** 345.

Andrewes, F. W. and Neave, S. (1921) *Brit. J. exp. Path.* **2,** 157.

Aoki, K. and Sakai, K. (1925) *Zbl. Bakt.* **95,** 152.

Archer, G. T. L. and Whitby, J. L. (1957) *J. Hyg., Camb* **55,** 513.

Atkinson, N., Geytenbeek, H., Swann, M. C. and Wallaston, J. M. (1952) *Aust. J. exp. Biol. med. Sci.* **30,** 333.

Atkinson, N., Woodroffe, G. M., Macbeth, A. M., Chibnall, H. and Mander, S. (1949) *Aust. J. exp. Biol. med. Sci.* **27,** 597.

Baker, E. E. and Whiteside, R. E. (1960) *Proc. Soc. exp. Biol., N.Y.* **105,** 328.

Baker, M. P., Bragdon, J. H., Fenner, F. and Jackson, A. V. (1949) *J. Amer. med. Ass.* **141,** 330.

Banker, D. D. (1955) *Nature, Lond.* **175,** 309.

Barber, C. and Eylan, E. (1975) *Zbl. Bakt.,* I Abt. Orig. **A230,** 451, 461; (1976) *Ibid.* **A234,** 53.

Baron, L. S., Formal, S. B. and Washington, O. (1957) *Virology* **3,** 417.

Baron, L. S., Spilman, W. M. and Carey, W. F. (1959) *Science* **130,** 566.

Bartel. H. (1938) *Tierärztl. Rdsch.* **44,** 601.

Bernstein, A. (1958) *Mon. Bull. Minist. Hlth Lab. Serv.* **17,** 92.

Bernstein, A. and Lederberg, J. (1955) *J. Bact.* **69,** 152.

Bernstein, A. and Wilson, E. M. J. (1963) *J. gen. Microbiol.* **32,** 349.

Birch-Hirschfeld, L. (1935) *Z. Hyg. InfectKr.* **117,** 626.

Bitter, L., Weigmann, F. and Habs, H. (1926) *Münch. med. Wschr.* **73,** 940.

Blaxland, J. D., Sojka, W. J. and Smither, A. M. (1956) *J. comp. Path.* **66,** 270; (1958) *Vet. Rec.* **70,** 374.

Boecker, E. and Silberstein, W. (1932) *Zbl. Bakt.* **125,** 256.

Bohnhoff, M., Drake, B. L. and Miller, C. P. (1954) *Proc Soc. exp. Biol., N.Y.* **86,** 132.

Bohnhoff, M., Miller, C. P. and Martin, W. R. (1964) *J. exp. Med.* **120,** 805, 817.

Boyd, J. S. K. (1950) *J. Path. Bact.* **62,** 501.

Boyd, J. S. K. and Bidwell, D. E. (1957) *J. gen. Microbiol.* **16,** 217.

Brandis, H. (1948) *Z. Hyg. InfektKr.* **140,** 688.

Brandis, H. and Maurer, H. (1954) *Z. Hyg. InfektKr.* **140,** 138.

Braun, H. (1948) *Progr. Med., Istanbul* **1,** 81.
Brown, C. M. and Parker, M. T. (1957) *Lancet* **ii,** 1277.
Brown, H. C., Duncan, J. T. and Henry, T. A. (1924) *J. Hyg., Camb.* **23,** 1.
Browning, C. H., Gilmour, W. and Mackie, T. J. (1913) *J. Hyg., Camb.* **13,** 335.
Bruner, D. W. (1946) *J. Bact.* **52,** 147.
Bruner, D. W. and Edwards, P. R. (1939) *J. Bact.* **37,** 365; (1940) *Agric. Exp. Sta. Univ. Kentucky*, Bull. No. 404; (1941) *J. Bact.*, **42,** 467.
Buczowski, Z. (1961) *Bull. Inst. mar. trop. Med., Gdańsk* **12,** 51.
Buxton, A. (1957) *Salmonellosis in Animals.* Commonwealth Agricultural Bureau, Farnham Royal.
Buxton, A. and Field, H. I. (1959) In: *Infectious Diseases of Animals,* **2.** Ed. by A. W. Stableforth and L. A. Galloway. Butterworth, London.
Caldwell, M. E. and Ryerson, D. L. (1939) *J. infect. Dis.* **65,** 242.
Callow, B. R. (1959) *J. Hyg., Camb.* **57,** 346.
Cherry, W. B., Davis, B. R. and Edwards, P. R. (1953) *Amer. J. publ. Hlth* **43,** 1280.
Cherubin, C. E. (1981) *Revs. infect. Dis.* **3,** 1105.
Cherubin, C. E., Timoney, J. F., Sierra, M. F., Ma, P., Marr, J. and Shin, S. (1980) *J. Amer. med. Ass.* **243,** 439.
Chu, D. C.-Y. and Hoyt, R. E. (1954) *J. Hyg., Camb.* **52,** 100.
Chu, D. C.-Y., Hoyt, R. E. and Pickett, M. J. (1956) *J. Hyg., Camb.* **54,** 585.
Chung, G. T. and Frost, A. J. (1971) *Res. vet. Sci.* **12,** 287.
Clauberg, K. W. (1931) *Klin. Wschr.* **10,** 540.
Clemmer, D. I., Hickey, J. L. S., Bridges, J. F., Schliessmann, D. J. and Shaffer, M. F. (1960) *J. infect. Dis.* **106,** 197.
Conybeare, E. T. and Thornton, L. H. D. (1938) *Rep. publ. Hlth. med. Subj., Min. Hlth., Lond.* No. 82.
Cosgrove, P. C. and Reid, J. (1953) *Trans. R. Soc. trop. Med. Hyg.* **47,** 154.
Costin, I. D. (1967) *Zbl. Bakt.* **204,** 485.
Costin, I. D. and Costin, E. (1966) *Ann. Inst. Pasteur* **111,** 592.
Cowan, S. T. (1957) *Bull. Hyg., Lond.* **32,** 101.
Craigie, J. and Brandon, K. F. (1936*a*) *J. Path. Bact.* **43,** 233; (1936*b*) *Ibid.* **43,** 249.
Craigie, J. and Felix, A. (1947) *Lancet* **i,** 823.
Craigie, J. and Yen, C. H. (1938) *Canad. publ. Hlth J.* **29,** 448, 494.
Danysz, J. (1900) *Ann. Inst. Pasteur* **14,** 193.
Dathan, J. C., McCall, A. J., Orr-Ewing, J. and Taylor, J. (1947) *Lancet* **i,** 711.
Datta, N. (1962) *J. Hyg., Camb.* **60,** 301.
Datta, N. and Olarte, J. (1974) *Antimicrobial Agents Chemother.* **5,** 310.
Diageler, A. and Kotter, L. (1956) *Berl. Münch. tierärztl. Wschr.* **69,** 281.
Dorrsen, C. A. van (1955) *Tijdschr. Diergeneesk.* **80,** 1188.
Dudgeon, L. S. and Urquhart, A. L. (1920) *Lancet* **ii,** 15.
Duguid, J. P., Anderson, E. S., Alfredsson, G. A., Barker, R. and Old, D. C. (1975) *J. med. Microbiol.* **8,** 149.
Duguid, J. P., Anderson, E. S. and Campbell, I. (1966) *J. Path. Bact.* **92,** 107.
Duguid, J. P., Darekar, M. R. and Wheater, D. W. F. (1976) *J. med. Microbiol.* **9,** 459.
Duguid, J. P. and Gillies, R. R. (1958) *J. Path. Bact.* **75,** 519.
Dunlop, S. J. C. (1950) *Docum. neerl. indones. Morb. trop.* **2,** 21.
Edsall, G. *et al.* (1960) *J. exp. Med.* **112,** 143.
Edwards, P. R. (1935) *J. Bact.* **30,** 465.
Edwards, P. R. and Boycott, J. A. (1955) *J. gen. Microbiol.* **13,** 569.
Edwards, P. R. and Bruner, D. W. (1939) *J. Bact.* **38,** 63; (1946) *Cornell Vet.* **36,** 318.
Edwards, P. R., Cherry, W. B. and Bruner, D. W. (1943) *J. infect. Dis.* **73,** 229.
Edwards, P. R., Davis, B. R. and Cherry, W. B. (1955) *J. Bact.* **70,** 279.
Edwards, P. R. and Ewing, W. H. (1972) *Identification of Enterobacteriaceae.* 3rd edn. Burgess, Minneapolis.
Edwards, P. R. and Fife, M. A. (1961) *Appl. Microbiol.* **9,** 478.
Edwards, P. R., Fife, M. A. and Ewing, W. H. (1965) *Antigenic Scheme for the Genus Arizona.* U.S. Dept Hlth, Educ. and Welfare, Atlanta.
Edwards, P. R., Fife, M. A. and Ramsey, C. H. (1959) *Bact. Rev.* **23,** 155.
Edwards, P. R. and Kauffmann, F. (1952) *Amer. J. clin. Path.* **22,** 692.
Edwards, P. R., Kauffmann, F. and Huey, C. (1957) *Acta path. microbiol. scand.* **41,** 517.
Edwards, P. R., McWhorter, A. C. and Fife, M. A. (1956) *Bull. World Hlth Org.* **14,** 511.
Edwards, P. R., West, M. G. and Bruner, D. W. (1947*a*) *Kentucky agric. Exp. Sta. Bull.* No. 499; (1947*b*) *J. infect. Dis.* **81,** 19.
Eskey, C. R., Prince, F. M. and Fuller, F. B. (1949) *Publ. Hlth Rep., Wash.* **64,** 933.
Ewing, W. H. (1963) *Int. Bull. bact. Nomencl.* **13,** 95.
Ewing, W. H. and Ball, M. M. (1966) *The Biochemical Reactions of Members of the Genus Salmonella.* U.S. Dept. Hlth, Educ. and Welfare, Atlanta.
Ewing, W. H., Ball, M. M., Bartes, S. F. and McWhorter, A. C. (1970) *J. infect. Dis.* **121,** 288.
Ewing, W. H. and Fife, M. A. (1966) *Int. J. syst. Bact.* **16,** 427.
Felix, A. (1943) *Brit. med. J.*, **i,** 435; (1951) *Brit. med. Bull.* **7,** 153; (1952) *J. Hyg., Camb.* **50,** 550;(1955) *Bull. World Hlth Org.*, **13,** 109; (1956) *J. gen. Microbiol.* **14,** 208.
Felix, A. and Anderson, E. S. (1951) *Nature, Lond.* **167,** 603.
Felix, A. and Bhatnagar, S. S. (1935) *Brit. med. J.* **ii,** 127.
Felix, A. and Callow, B. R. (1943) *Brit. med. J.* **ii,** 127; (1951) *Lancet*, **ii,** 10.
Felix, A. and Pitt, R. M. (1934*a*) *J. Path. Bact.* **38,** 409; (1934*b*) *Lancet*, **ii,** 186; (1936) *Brit. J. exp. Path.* **17,** 81.
Fenner, F. and Jackson, A. V. (1946) *Med. J. Aust.* **i,** 313.
Field, H. I. (1948) *Vet. J.* **104,** 251, 294, 323; (1949) *Vet. Rec.* **61,** 275.
Fife, M. A. and Martin, W. J. (1967) *Antigenic Scheme for the Genus Arizona.* U.S. Dept. Hlth, Educ. and Welfare, Atlanta.
Fletcher, W. (1918) *Lancet* **ii,** 102.
Fournier, J. (1949) *Ann. Inst. Pasteur* **77,** 269.
Fournier, J., Sung, C., Koang, N. K. and Velliot, G. (1950) *Chin. med. J.* **68,** 63.
Frost, J. A., Rowe, B., Ward, L. R. and Threlfall, E. J. (1982) *J. Hyg., Camb.* **88,** 193.
Fukumi, H., Sayama, E. and Nakaga, R. (1953) *Jap. J. med. Sci.* **6,** 327.

Fulton, M. and Fulton, A. W. (1966) *Int. J. syst. Bact.* **16,** 417.

Gaertner. (1888) *KorrespBl. ärztl. Ver. Thüringen* **17,** 573.

Gaines, S., Currie, J. A. and Tully, J. G. (1960) *Proc. Soc. exp. Biol., N.Y.* **104,** 602; (1965) *Amer. J. Epidem.* **81,** 350.

Gaines, S., Sprinz, H., Tully, J. G. and Tigertt, W. D. (1968*a*) *J. infect. Dis.* **118,** 293.

Gaines, S., Tully, J. G. and Tigertt, W. D. (1968*b*) *J. infect. Dis.* **118,** 393.

Galton, M. M. and Hardy, A. V. (1948) *Publ. Hlth Rep., Wash.* **63,** 847.

Giannella, R. A., Formal, S. B., Dammin, G. J. and Collins, H: (1973) *J. clin. Invest.* **52,** 441.

Giannella, R. A., Gots, R. E., Charney, A. N., Greenough, W. B. III and Formal, S. B. (1975) *Gastroenterology* **69,** 1238.

Giglioli, G. (1930) *J. Hyg., Camb.* **29,** 273; (1958) *W. Ind. med. J.* **7,** 29.

Giovanardi, A. (1938) *Zbl. Bakt.* **141,** 341.

Gispen, R. and Stibbe, W. K. M. (1947) *Med. Maanbdbl.* **1,** 247.

Gitter, M. (1959) *Vet. Rec.* **71,** 234.

Glässer, K. (1909) *Dtsch. tierärztl. Wschr.* **16,** 513; (1910) *Zbl. Bakt. Ref.* **45,** 612.

Gray, J. D. A. (1931) *J. Path. Bact.* **34,** 335.

Greenfield, J., Bigland, C. H. and Dukes, T. W. (1971) *Vet. Bull.* **41,** 605.

Grinnell, F. B. (1932) *J. exp. Med.* **56,** 907.

Guinée, P. A. M. and Scholtens, R. T. (1967) *Leeuw. ned. Tijdschr.* **33,** 25.

Guth, F. (1916) *Zbl. Bakt.* **77,** 487.

Gwyn, N. B. (1898) *Johns Hopk. Hosp. Bull.* **9,** 54.

Havens, L. C. and Mayfield, C. R. (1933) *J. infect. Dis.* **52,** 157.

Hayes, W. and Freeman, J. F. (1945) *Indian J. med. Res.* **33,** 177.

Henning, M. W. (1953) *Onderstepoort J. vet. Sci.* **26,** 3.

Henning, M. W. and McIntosh, B. M. (1946) *J. S. Afr. vet. med. Ass.* **17,** 88.

Henricksen, S. D. (1950) *Acta path. microbiol. scand.* **27,** 107.

Hicks, E. P. (1930) *J. Hyg., Camb.* **29,** 446.

Hinshaw, W. R. (1941) *Hilgardia* **13,** 583.

Hinshaw, W. R., Browne, A. S. and Taylor, T. J. (1943) *J. infect. Dis.* **72,** 197.

Hinshaw, W. R. and McNeil, E. (1946) *J. Bact.* **51,** 281.

Hirschfeld, L. (1919) *Lancet* **i,** 296.

Hohn, J. and Herrmann, W. (1935*a*) *Zbl. Bakt.* **133,** 183; (1935*b*) *Ibid.* **134,** 177.

Huang, C. H., Chang, H. C. and Lieu, V. T. (1937) *Chin. med. J.* **52,** 345.

Huang, C. T. and Lo, C. B. (1967) *J. Hyg., Camb.* **65,** 149.

Huey, C. R. and Edwards, P. R. (1958) *Proc. Soc. exp. Biol., N.Y.* **97,** 550.

Hughes, M. H. (1954) *W. Afr. med. J.* **3,** 57.

Hynes, M. (1942) *J. Path. Bact.* **54,** 193.

Iino, T. (1969) *Bact. Rev.* **33,** 454.

Iseki, S. and Kashiwagi, K. (1955) *Proc. Japan Acad.* **31,** 558.

Iseki, S. and Sakai, T. (1953) *Proc. Japan Acad.* **29,** 121, 127.

Iwaschenzoff, G. (1926) *Arch. Schiffs- u. Tropenhyg.* **30,** 1.

Jacobsen, K. A. (1910) *Zbl. Bakt.* **56,** 208.

Jadin, J. (1951) *Rev. belge pathol.* **21,** 8.

Jarvis, F. G., Mesenko, M. T., Martin, D. G. and Perrine, T. D. (1967) *J. Bact.* **94,** 1406.

Jiwa, S. F. H. (1981) *J. clin. Microbiol.* **14,** 463.

Johnson, E. A. and Rettger, L. F. (1943) *J. Bact.* **45,** 127.

Johnson, R. H., Lutwick, L. I., Huntley, G. A. and Vosti, K. L. (1976) *Ann. intern. Med.* **85,** 587.

Josland, S. W. (1950) *Aust. vet. J.,* **26,** 249; (1952) *N.Z. J. Med.* **51,** 180.

Kauffmann, F. (1929*a*) *Zbl. Bakt. Ref.* **94,** 282; (1929*b*) *Z. Hyg. InfectKr.* **110,** 537; (1930) *Ibid.* **111,** 221, 233, 247; (1931) *Zbl. ges. Hyg.* **25,** 273; (1934*a*) *Ergebn. Hyg. Bakt.* **15,** 219; (1934*b*) *Zbl. Bakt.* **132,** 337; (1934*c*) *Z. Hyg. Infekt Kr.* **116,** 368; (1935*a*) *Ibid.* **116,** 617; (1935*b*) *Ibid.* **117,** 431; (1936*a*) *Ibid.* **117,** 778; (1936*b*) *Ibid.* **119,** 103; (1941) *Die Bakteriologie der Salmonella-Gruppe.* Munksgaard, Copenhagen; (1953) *Acta path. microbiol. scand.* **32,** 574; (1954) *Enterobacteriaceae,* 2nd edn. Munksgaard, Copenhagen; (1955) *Z. Hyg. InfektKr.* **141,** 546; (1956*a*) *Acta path. microbiol. scand.* **39,** 85; (1956*b*) *Ibid.,* **39,** 299; (1957) *Ibid.* **40,** 343; (1960) *Ibid.* **49,** 393; (1963*a*) *Ibid.* **58,** 109; (1963*b*) *Ibid.* **58,** 348; (1964) In: *The World Problem of Salmonellosis.* p. 21. Ed. by E. van Oye. Dr. W. Junk Publisher, The Hague; (1969) *The Bacteriology of the Enterobacteriaceae.* 2nd edn. Munksgaard, Copenhagen; (1972) *Zbl. Bakt.* **A222,** 40.

Kauffmann, F., Edwards, P. R. and McWhorter, A. C. (1955) *Acta path. microbiol. scand.***36,** 568.

Kauffmann, F. and Møller, V. (1955) *Acta path. microbiol. scand.* **36,** 173.

Kauffmann, F. and Petersen, A. (1956) *Acta path. microbiol. scand.* **38,** 481.

Kauffmann, F. and Rohde, R. (1962) *Acta path. microbiol. scand.* **56,** 341.

Kelterborn, E. (1967) *Salmonella species. First isolations, names and occurence.* Dr. W. Junk Publisher, The Hague; (1979) *Zbl. Bakt.* I Abt. Orig. **A243,** 289.

Kent, T. H., Formal, S. B. and Labrec, E. H. (1966) *Arch. Path.* **81,** 501.

Kilbourne, F. L. (1893) *Misc. Invest. infect. parasit. Dis. dom. Anim.,* 8° Washington, 49.

Klein, E. (1889) *Zbl. Bakt,* **5,** 689.

Knight, B. C. J. G. (1936) *Spec Rep. Ser. med. Res. Coun., Lond.* No. 210.

Kokko, U. P. (1947) *Nord. Med.* **36,** 2325.

Koupal, L. R. and Deibel, R. H. (1975) *Infect. Immun.* **11,** 14.

Kristensen, M. and Kauffmann, F. (1937) *Z. Hyg. InfectKr.* **120,** 149.

Kröger, E. (1955) *Ergebn. Hyg. Bakt.* **29,** 475.

Krumwiede, C. and Pratt, J. S. (1914) *J. exp. Med.* **19,** 501.

Kühn, H., Tschäpe, H. and Rische, H. (1978) *Zbl. Bakt.,* I Abt. Orig. **A240,** 171.

Kulescha, G. S. and Titowa, N. A. (1923) *Virchows Arch.* **241,** 319.

Kuwahara, S., Yasushi, S., Owa, T. and Shinoda, M. (1958) *Jap. J. Microbiol.* **2,** 343.

Lalko, J. (1977) *Bull. Inst. mar. trop. Med. Gdańsk* **28,** 187.

Landy, M. (1950) *Publ. Hlth Rep., Wash.* **65,** 950; (1952) *Proc Soc. exp. Biol. N.Y.* **80,** 55.

Landy, M., Gaines, S. and Sprinz, H. (1957) *Brit. J. exp. Path.* **38,** 15.

Landy, M., Johnson, A. G. and Webster, M. E. (1961) *Amer. J. Hyg.* **73,** 55.

Landy, M., Trapani, R. J., Webster, M. E. and Jarvis, F. G. (1963) *Tex. Rep. Biol. Med.* **21,** 214.

Landy, M. and Webster, M. E. (1952) *J. Immunol.* **69,** 143.

Lederberg, J. and Edwards, P. R. (1953) *J. Immunol.* **71,** 232.

Lederberg, J. and Iino, T. (1956) *Genetics.* **41,** 743.
Leeuwen, W. J. van. *et al.* (1979) *Antimicrob. Agents Chemother.* **16,** 237.
Leeuwen, W. J. van, Voogd, C. E., Guineé, P. A. M. and Manten, A. (1982) *Leeuwenhoek ned. Tydschr.* **48,** 85.
Leifson, E. (1935) *J. Path. Bact.* **40,** 581; (1936) *Amer. J. Hyg.* **24,** 423.
Le Minor, L. (1955) *Ann. Inst. Pasteur* **88,** 76; (1956) *Ibid.* **91,** 664; (1967) *Ibid.* **113,** 117; (1969) *Int. J. syst. Bact.* **18,** 197.
Le Minor, L., Bockemühl, J. and Rowe, B. (1981) *Ann. Microbiol., Inst. Pasteur* **B132,** 85.
Le Minor, L., Fife, M. A. and Edwards, P. R. (1958) *Ann. Inst. Pasteur* **95,** 326.
Le Minor, L. and Nicolle, P. (1964) *Ann. Inst. Pasteur* **107,** 550.
Le Minor, L. Rohde, R. and Taylor, J. (1970) *Ann. Inst. Pasteur* **119,** 206.
Lentz, O. and Tietz, J. (1903) *Münch. med. Wschr.* **50,** 2139; (1905) *Klin. Jb.* **14,** 495.
Leslie, P. H. (1942) *J. Hyg., Camb.* **42,** 552.
Levi, M. L. (1949) *Vet. Rec.* **61,** 555.
Lilleengen, K. (1950) *Acta path. microbiol. scand.* **27,** 625; (1952) *Ibid.* **30,** 194.
Linton, A. H. (1981) *Vet. Rec.* **108,** 328.
Liu, P. Y. (1938*a*) *Chin. med. J.* Suppl. No. 2, p. 227; (1938*b*) *Ibid.* p. 279.
Liu, P. Y., Zia, S. H. and Chung, H. L. (1937) *Proc. Soc. exp. Biol., N.Y.* **37,** 17.
Loeffler, F. (1892) *Zbl. Bakt.,* **11,** 129; (1903) *Dtsch. med. Wschr.* **29,** 36; (1906) *Ibid.* **32,** 289.
Lovell, R. (1932) *Bull. Hyg.* **7,** 405.
Ludlam, G. B. (1954) *Mon. Bull. Minist. Hlth. Lab. Serv.* **13,** 196.
Luippold, G. F. (1944) *Science* **99,** 497.
Lütje. (1937) *Dtsch. tierärztl. Wschr.* **45,** 242; (1939) *Ibid.* **47,** 227, 246, 257.
MacAdam, W. (1919) *Lancet* **ii,** 189.
McDonald, J. C., Price, A. and Taylor, J. (1955) *Mon. Bull. Minist. Hlth Lab. Serv.***14,** 83.
McHugh, G. L., Moellering, R. C., Hopkins, C. C. and Swartz, M. N. (1975) *Lancet,* **i,** 235.
Mackie, F. P. and Bowen, C. J. (1919) *J. R. Army med. Cps.* **33,** 154.
Mäkelä, P. H. (1964) *J. gen. Microbiol.* **35,** 503.
Manten, A., Guinée, P. A. M., Kampelmacher, E. H. and Voogd, C. E. (1971) *Bull. World Hlth Org.* **45,** 85.
Mellon, R. R. and Jost, E. L. (1926) *J. Immunol.* **12,** 331.
Messerlin, A. and Couzi, G. (1942) *Bull. Inst. Hyg. Maroc,* N.S. **2,** 15.
Metchnikoff, E. and Besredka, A. (1911) *Ann. Inst. Pasteur* **25,** 193.
Meynell, G. G. and Subbaiah, T. V. (1963) *Brit. J. exp. Path.* **44,** 197, 209.
Miller, C. P. (1959) *Univ. Michigan med. Bull.* **25,** 272.
Miller, C. P., Bohnhoff, M. and Drake, B. L. (1956) *Antibiot. Annual,* p. 453.
Miller, C. P., Bohnhoff, M. and Rifkind, D. (1957) *Science* **125,** 749.
Milner, K. C. and Shaffer, M. F. (1952) *J. infect. Dis.* **90,** 81.
Mitchell, R. B., Garlock, F. C. and Broh-Kahn, R. H. (1946) *J. infect. Dis.* **79,** 57.
Møller, V. (1955) *Acta path. microbiol. scand.* **36,** 158.
Moor, C. E. de. (1935) *Geneesk. Tijdschr. Ned.-Ind.* **75,** 743.
Moore, B. (1957) *J. Hyg., Camb.* **55,** 414.
Morris, J. F., Barnes, C. G. and Sellers, T. F. (1943) *Amer. J. publ. Hlth* **33,** 246.
Morris, J. F., Sellers, T. F. and Brown, A. W. (1941) *J. infect. Dis.* **68,** 117.
Muller, L. (1923) *C. R. Soc. Biol.* **89,** 434.
Müller, M. (1912) *Zbl. Bakt.* **62,** 335.
Müller, R. (1910) *Dtsch. med. Wschr.* **36,** 2387; (1933) *Münch. med. Wschr* **80,** 1771.
Murdoch, C. R. and Gordon, W. A. M. (1953) *Mon. Bull. Minist. Hlth Lab. Serv.* **12,** 72.
Murphy, W. J. and Morris, J. F. (1950) *J. infect. Dis.* **86,** 255.
Neel, R., Jorgensen, K., Le Minor, L. and Machoun, A. (1953) *Ann. Inst. Pasteur* **84,** 400.
Neu, H. C., Cherubin, C. E., Longo, E. D., Flouton, B. and Winter, J. (1975) *J. infect. Dis.* **132,** 617.
Neukirch, P. (1918) *Z. Hyg. InfektKr.* **85,** 103.
Nicolle, P., Jude, A. and Le Minor, L. (1950) *Ann. Inst. Pasteur* **78,** 572.
Nicolle, P., Pavlatou, M. and Diverneau, G. (1953) *C. R. Acad. Sci.* **236,** 2453; (1954) *Ann. Inst. Pasteur* **87,** 493.
Nobele, de. (1898) *Ann. Soc. Méd. Gand.* **72,** 281.
Nocard. (1893) *Cons. Publ. Sal. Dept. Seine,* Séance Mar 24.
Olitzki, L. and Koch, P. K. (1945) *J. Immunol.* **50,** 229.
Ørskov, J., Jensen, K. and Kobayashi, K. (1928) *Z. ImmunForsch.* **55,** 34.
Ørskov, J. and Lassen, H. C. A. (1930) *Z. ImmunForsch.* **67,** 137.
Ørskov, J. and Moltke, O. (1928) *Z. ImmunForsch.* **59,** 357.
Paniker, C. K. J. and Vimala, K. N. (1972) *Nature, Lond.,* **239,** 109.
Parker, R. R. and Steinhaus, E. A. (1943) *Publ. Hlth Rep., Wash.* **58,** 27.
Pavlatou, M. and Nicolle, P. (1953) *Ann. Inst. Pasteur* **85,** 185.
Peluffo, C. A. (1941) *Proc. Soc. exp. Biol., N.Y.* **48,** 340.
Perry, H. M., Findlay, H. T. and Bensted, H. J. (1933*a*) *J. Roy. Army med. Cps.* **60,** 241; (1933*b*) *Ibid.,* **61,** 81; (1934*a*) *Ibid.,* **62,** 161; (1934*b*) *Ibid.* **63,** 1.
Peso, O. A. and Edwards, P. R. (1951) *Publ. Hlth Rep., Wash.* **66,** 1694.
Petersen, K. F. (1949) *Z. Hyg. InfektKr.* **129,** 634.
Petrow, S., Kasatiya, S. S., Pelletier, J., Ackermann, H.-W. and Peloquin, S. (1974) *Ann. Microbiol., Inst. Pasteur* **A125,** 433.
Pohl, G. (1954) *Mh. Veterinärmed.* **20,** 449.
Pollock, M. R., Knox, R. and Gell, P. G. H. (1942) *Nature, Lond.* **150,** 94.
Raynal, J. H. and Fournier, J. (1947) *Méd. trop.* **7,** 199.
Report (1934) Salmonella Sub-Committee of Nomenclature Comm., Int. Soc. Microbiol., *J. Hyg., Camb.* **34,** 333; (1978) *Wld Hlth Org. techn. Rep. Ser.* no. 624.
Rettger, L. F. (1900) *N.Y. med. J.* **71,** 803.
Ricossé, J.-H., Goudineau, J.-A., Doury, J.-C., Vieu, J.-F., Meyruey, M.-H. and Pelloux, H. (1979) *Méd. trop, Marseille* **39,** 415.
Rohde, R. (1965) *J. appl. Bact.* **28,** 368; (1967) *Zbl. Bakt.* **205,** 404.
Ryder, R. W. *et al.* (1980) *J. infect. Dis.* **142,** 485.
Salmon, E. and Smith, T. (1885) *Ann. Rep. Bureau Animal Industry*; (1886) *Amer. mon. micr. J.* **7,** 204.
Sanborn, W. R. *et al.* (1977) *J. Hyg., Camb.* **79,** 1.

Sandefur, P. D. and Peterson, J. W. (1976) *Infect. Immun.* **14,** 671.
Saphra, I. (1950) *Amer. J. med. Sci.* **220,** 74.
Saphra, I. and Wassermann, M. (1954) *Amer. J. med. Sci.* **228,** 525.
Saphra, I. and Winter, J. W. (1957) *New Engl. J. Med.* **256,** 1128.
Savino, E. and Menéndez, P. E. (1934) *Rev. Inst. Bact., Buenos Aires* **6,** 347.
Schermer and Ehrlich (1921) *Berl. tierärztl. Wschr.* **37,** 469.
Schlecht, S. and Westphal, O. (1970) *Zbl. Bakt.* **213,** 356.
Schmidt, G., Schlecht, S., Lüderitz, O. and Westphal, O. (1968–69) *Zbl. Bakt.* **209,** 483.
Schmidt, G., Schlecht, S. and Westphal, O. (1969–70) *Zbl. Bakt.* **212,** 88.
Schneider, H., Formal, S. B. and Baron, L. S. (1961) *J. exp. Med.* **114,** 141.
Scholtens, R. T. (1962) *Leeuwenhoek. ned. Tijdschr.* **28,** 373; (1969) *Arch. roum. Path. exp. Microbiol.* **28,** 984.
Schottmüller, H. (1900) *Dtsch. med. Wschr.* **26,** 511; (1901) *Z. Hyg. InfektKr.* **36,** 368.
Schütze, H. (1920) *Lancet,* **i,** 93.
Scott, W. M. and Minett, F. C. (1947) *J. Hyg., Camb.* **45,** 159.
Sedlock, D. M. and Deibel, R. H. (1978) *Canad. J. Microbiol.* **24,** 268.
Sedlock, D. M., Koupal, L. R. and Deibel, R. H. (1978) *Infect. Immun.* **20,** 375.
Seiffert, G., Jahncke, A. and Arnold, A. (1928) *Zbl. Bakt.* **109,** 193.
Sertic, V. and Boulgakov, N.-A. (1936*a*) *C. R. Soc. Biol.* **122,** 35; (1936*b*) *Ibid.* **123,** 951.
Shaffer, M. F., Milner, K. C., Clemmer, D. I. and Bridges, J. F. (1957) *J. infect. Dis.* **100,** 17.
Sharma, K. B., Bhat, B. M., Pasricha, A. and Vaze, S. (1979) *J. antimicrob. Chemother.* **5,** 15.
Shaw, C. (1956) *Int. Bull. bact. Nomencl.* **6,** 1.
Skerman, V. D. B., McGowan, V. and Sneath, P. H. A. (1980) *Int. J. syst. Bact.* **30,** 225.
Smith, H. G. (1959) *J. gen. Microbiol.* **21,** 61.
Smith, H. W. (1951*a*) *J. gen. Microbiol.,* **5,** 458; (1951*b*) *Ibid.* **5,** 472; (1951*c*) *Ibid.* **5,** 919.
Smith, H. W. and Buxton, A. (1951) *J. Path. Bact.* **63,** 459.
Smith, J. and Scott, W. M. (1930) *J. Hyg., Camb.* **30,** 32.
Stenzel, W. (1976) *Zbl. Bakt,* I Abt. Orig. **A236,** 269.
Stern, W. (1916) *Zbl. Bakt.* **78,** 481.
Stocker, B. A. D. (1949) *J. Hyg., Camb.* **47,** 398; (1958) *Proc. VII Int. Congr. Microbiol., Stockh.,* p. 65.
Stocker, B. A. D., Staub, A. M., Tinelli, R. and Kopacka, B. (1960) *Ann. Inst. Pasteur* **98,** 505.
Stocker, B. A. D., Zinder, N. D. and Lederberg, J. (1953) *J. gen. Microbiol.* **9,** 410.
Stokes, J. L. and Bayne, H. G. (1957) *J. Bact.* **74,** 200; (1958*a*) *Ibid.* **76,** 136; (1958*b*) *Ibid.* **76,** 417.
Sütterlein, T. (1923) *Zbl. Bakt.* **90,** 419.
Svenungsson, B. and Lindberg, A. A. (1977) *Med. Microbiol. Immunol.* **163,** 1; (1978) *Acta path. microbiol. scand.* **B86,** 35, 283.
Tannock, G. W. and Smith, J. M. B. (1971) *J. infect. Dis.* **123,** 502.
Taylor, J. (1956) *Lancet* **i,** 630.
Taylor, J. and Wilkins, M. P. (1961) *Ind. J. med. Res.* **49,** 544.
Threlfall, E. J., Ward, L. R., Ashley, A. S. and Rowe, B. (1980) *Brit. med. J.* **i,** 1210.
Threlfall, E. J., Ward, L. R. and Rowe, B. (1978*a*) *Vet. Rec.* **103,** 438; (1978*b*) *Brit. med. J.* **ii,** 997.
Trabulsi, L. R. and Edwards, P. R. (1962) *Cornell Vet.* **52,** 563.
Tulloch, W. J. (1939) *J. Hyg., Camb.* **39,** 324.
Uetake, H., Luria, S. E. and Burrous, J. W. (1958) *Virology* **5,** 68.
Uetake, H., Nakagawa, T. and Akiba, T. (1955) *J. Bact.* **69,** 571.
Urano, T., Kawakami, M., Koganesawa, S. and Mitsuhashi, S. (1958) *Jap. J. Microbiol.* **2,** 241.
Velaudapillai, T. (1959) *Z. Hyg. InfektKr.* **146,** 84.
Vongsthongsri, U. and Tharavanij, S. (1980) *S. E. A. J. trop Med. publ. Hlth* **11,** 256.
Voogd, C. E., van Leeuwen, W. J., Guinée, P. A. M., Manten, A. and Valkenburg, J. J. (1977) *Leeuwenhoek ned. Tijdschr.* **43,** 269.
Walton, J. R. and Lewis, L. E. (1971) *Vet. Rec.* **89,** 112.
Watson, W. A. (1960) *Vet. Rec.* **72,** 62.
Webster, M. E., Landy, M. and Freeman, M. E. (1952) *J. Immunol.* **69,** 135.
Weigmann, F. (1925*a*) *Zbl. Bakt.* **95,** 396; (1925*b*) *Ibid.* **97,** Beiheft, 299.
Weil, E. (1917) *Wien. klin. Wschr.* **30,** 1061.
Weiner, L. M. and Neely, J. (1964) *J. Immunol.* **92,** 908.
Weiner, L. M. and Price, S. (1956) *J. Immunol.* **77,** 111.
Wherry, W. B. (1908) *J. infect. Dis.* **5,** 519.
White, P. B. (1926) *Spec. Rep. Ser. med. Res. Coun., Lond.* No. 103; (1929*a*) *J. Path. Bact.* **32,** 85; (1929*b*) Med. Res. Coun. *System of Bacteriology,* **4,** 86.
Whiteside, R. E. and Baker, E. E. (1959) *J. Immunol.* **83,** 687.
Wilson, J. E. (1955) *Vet. Rec.* **67,** 849.
Wilson, W. J. (1938) *J. Hyg., Camb.* **38,** 507.
Wu, M. and Fournier, J. (1951) *Amer. J. trop. Med.* **31,** 42.
Younie, A. R. (1941) *Canad. J. comp. Med.* **5,** 164.
Zimmerman, L. E. (1953) *Amer. J. publ. Hlth* **43,** 279.
Zimmerman, L. E., Cooper, M. and Graber, C. D. (1952) *Amer. J. Hyg.* **56,** 252.
Zinder, N. D. (1957) *Science* **126,** 1237.

38

Pasteurella, *Francisella* and *Yersinia*

Geoffrey Smith and Graham Wilson

Introductory

Perroncito (1878) appears to have been the first to isolate and describe the organism known as the bacillus of fowl cholera (see Kitt 1903). Similar organisms were cultured by subsequent workers from cattle, pigs, sheep and other animals suffering from a disease characterized by septicaemia with widespread haemorrhages. Trevisan (1887) suggested for these various organisms the generic name *Pasteurella*, and Lignières (1900) added the specific name for each organism according to the animal it attacked; thus the name of the organism from fowls was *P. aviseptica*, from pigs *P. suiseptica*, from cattle *P. boviseptica*, from sheep *P. oviseptica*, and from rabbits *P. lepiseptica*. As these organisms behaved as if they belonged to a single species we suggested in the first edition of this book in 1929 that they should all be referred to by the name of *Pasteurella septica*. Later, Rosenbusch and Merchant (1939) proposed the alternative name of *Pasteurella multocida*. The authors of *Bergey's Manual* (Breed *et al.* 1957) regarded the term *P. multocida* as claiming priority in accordance with the rules of nomenclature, but on their own showing this was certainly not the first specific name given. Because the name *P. multocida* has achieved almost universal acceptance we shall now use it. Another organism belonging to the *Pasteurella* group but differing mainly in its haemolytic activity was isolated by F. S. Jones (1921), Tweed and Edington (1930), Newsom and Cross (1932), and Rosenbusch and Merchant (1939) from diseases of calves and sheep, and named *P. haemolytica*. What was thought to be a variant of this organism was described by Henriksen and Jyssum (1961*a*, *b*) under the name *P. haemolytica* var. *ureae*. A very similar, if not identical, organism was isolated from the human respiratory tract by D. M. Jones (1962), who called it *P. ureae*. This name seems more appropriate, as the organism has little relation to *P. haemolytica* (Henriksen 1961).

Malassez and Vignal (1883) were apparently the first to describe pseudotuberculosis in the guinea-pig. Several workers recorded the finding of a bacillus in this disease (see Chapter 55), chief amongst whom were Eberth (1886) and Pfeiffer (1890), who named it *Bacillus pseudotuberculosis*. Eberth considered it probable that pseudotuberculosis of rabbits was caused by the same organism as that responsible for the *Tuberculose zoogléique* of the French authors. It is not to be confused with *Corynebacterium pseudotuberculosis ovis*, described by Preisz and Nocard as the cause of pseudotuberculosis in sheep, or with *Corynebacterium pseudotuberculosis murium*, described by Kutscher (1894) and Bongert (1901) as the cause of pseudotuberculosis in mice, which are described in Chapter 25 under their currently accepted names, respectively *C. ovis* and *C. kutscheri*. More recently a closely allied organism, referred to as *Pasteurella X* by Knapp and Thal (1963) and called *Yersinia enterocolitica* by Frederiksen in 1964, was isolated from cases of a disease often resembling pseudotuberculosis in man and animals. The plague bacillus, formerly known as *Pasteurella pestis*, was isolated almost simultaneously by Kitasato (1894) and by Yersin (1894) from human patients suffering from plague. The claim of Kitasato to have discovered the plague bacillus is often challenged (see Ogata 1955), but no one who reads Kitasato's original description can have any serious doubt that what he described was the genuine plague bacillus and not a contaminant, though of course his cultures may have become contaminated later. (See Bibel and Chen 1976.)

The plague and pseudotuberculosis organisms were at one time regarded as belonging to the *Pasteurella* group, but it was decided (Report 1971*a*) to create for them the new genus *Yersinia*. According to Buchanan and Gibbons (1974) this genus belongs to the family Enterobacteriaceae (see Chapter 33).

American workers formerly included in the plague group the causative organism of tularaemia; for reasons given on p. 363 we regarded it as being closely related to the *Brucella* group. Work on its DNA base composition has revealed a closer affinity with *Pasteurella multocida*. However, as it differs from this organism in many respects, it has now been accorded a genus of its own and is referred to as *Francisella tularensis* (Report 1971*a, b*).

Pasteurella

Definition

Very small gram-negative, non-motile, coccoid, ovoid, or rod-shaped organisms, often showing bipolar staining. Aerobic and facultatively anaerobic. Do not grow, or grow only scantily, in the presence of bile salts. Attack carbohydrates weakly by the fermentative method, forming small amounts of acid but no gas. Ferment sucrose but rarely lactose. Catalase and oxidase positive. Some species produce indole and some urease. Do not liquefy gelatin. Are parasites of man and animals, causing characteristic diseases. G + C content of DNA 37–41 moles per cent.

The type species is *Pasteurella multocida* (*septica*).

Habitat

P. multocida is found in a wide variety of animals and probably has its main habitat in the respiratory tract. J. E. Smith (1955) found it in 50 per cent of tonsils of normal dogs. It is occasionally isolated from the nasopharynx of healthy human beings. *P. haemolytica* is met with particularly in sheep, cattle and goats; it occurs in the nasopharynx and tonsils of healthy animals. *P. ureae* occurs in human sputum. So far as we know all pasteurellae lead a parasitic existence.

Morphology and staining

The usual appearance is that of small ovoid non-motile gram-negative bacilli with convex sides and rounded ends showing bipolar staining. There is no special arrangement; the organisms are disposed singly, in pairs, short chains or small groups. Coccoid and short rod forms, often staining irregularly, are not uncommon. In general, the bacilli tend to be ovoid and to show bipolar staining when taken from animal tissues or smooth colonies, and to be more bacillary without bipolar staining when taken from rough colonies, but many exceptions occur. An indefinite capsule, sometimes referred to as an envelope substance, is often formed in the animal body, and in mucoid and smooth iridescent colonies of *P. multocida*. It can be revealed by the India ink method and by Giemsa's stain. According to Priestley (1936*a, c*) it is seen only in virulent strains. It reaches its maximum development after 24 hr at 37° and then gradually disappears. It is not formed below 20° or above 40°. It is composed either of hyaluronic acid or of polysaccharide (J. E. Smith 1958), is heat-labile, and is antigenically distinct from the somatic substance. Most freshly isolated strains of *P. haemolytica* are capsulated (Carter 1967). Henriksen and Frøholm (1975) described a fimbriated strain of *P. multocida* showing twitching movement and forming spreading colonies.

Cultural characters

P. multocida grows fairly well on nutrient agar, forming circular colonies about 1 mm in diameter after 24 hr at 37°. The work of de Kruif (1921, 1922*a*, *b*, 1923) and of Webster and Burn (1926) showed the existence of three different colonial forms: (*a*) a smooth form, virulent for rabbits, growing diffusely in broth, and forming smooth, moderately opaque iridescent colonies on serum agar; (*b*) a rough form, completely avirulent for rabbits, giving a granular deposit in broth, and forming translucent bluish colonies; (*c*) a mucoid form of intermediate virulence. (See also Anderson *et al.* 1929, Mørch and Krogh-Lund 1930, Hughes 1930.) The highly virulent smooth form contains a type-specific polysaccharide capsular antigen; the mucoid form is rich in hyaluronic acid, and may or may not possess a polysaccharide capsular antigen in addition. The rough form has neither a capsular nor a mucoid antigen (Carter 1955, J. E. Smith 1958, Namioka and Murata 1961). Carter (1967) recognizes two forms of smooth colony—an S and an S^R. The S form, like the mucoid form, is capsulated, does not flocculate in acriflavine, and is typable by indirect haemagglutination; the S^R form, like the rough form, behaves in the opposite manner.

The colonial form varies to some extent with the animal source.

According to J. E. Smith (1958, 1959), nearly all strains from pigs form on blood agar translucent mucoid iridescent colonies, up to 1.8 mm in diameter, tending to become confluent. Cattle strains are non-mucoid. Pig strains form hyaluronic acid: cattle strains do not. Strains from cats and dogs are greyish, smooth, less translucent, non-iridescent, and not over 1.2 mm in diameter; they cause a slight browning of the blood. The property of *iridescence*, best exhibited by colonies grown for 24 hr on 10 per cent horse serum agar, is appreciated most readily by examination in oblique strong sunlight against a darkish background. It depends apparently on the regular mosaic of opaque coccobacilli and transparent capsules in the colony; this acts as a diffraction grating. The colour effects are seen only by transmitted light, and differ from those of fluorescence, which is caused by absorption of light of a specific wavelength, usually ultraviolet, with emission of part of the absorbed energy in the form of light of greater wavelength (J. E. Smith 1958).

G. R. Smith (1959, 1960, 1961) recognized two forms of *P. haemolytica*, now usually referred to as biotypes. Biotype T, derived mainly from septicaemia in lambs aged 5–12 months, had largish colonies with dark brown centres. Biotype A, derived from sheep pneumonia and from septicaemia in lambs within a few weeks of birth, had slightly smaller colonies with only a little central thickening; unlike biotype T, it rapidly lost its viability in ageing broth cultures.

In *sheep* blood agar, colonies of *P. haemolytica* (biotypes A and T) are surrounded by a single narrow zone of haemolysis, but in *lamb* blood agar a double zone may be seen—a narrow inner zone and an outer wider incomplete zone which increases in size at room temperature (G. R. Smith 1962).

Colonies of *P. ureae* on blood agar are mucoid and are usually accompanied by some greening of the medium.

In broth the S form of *P. multocida* causes uniform turbidity with a light powdery deposit later tending to become viscous. The R form produces a granular deposit without turbidity. There is little or no growth on potato, and none at all on MacConkey's bile salt medium. *P. haemolytica* produces slight growth on MacConkey's medium.

Metabolism and growth requirements

All the members are aerobes and facultative anaerobes. Webster (1924*a*, 1925) and Webster and Baudisch (1925) found that the smooth variant of *P. multocida* would not grow aerobically in plain broth unless large numbers of organisms were introduced—about 100 000 per ml; the rough variant grew from an inoculum of only a few organisms. If a trace of rabbit blood, or an iron compound with strongly catalytic properties, was added to the medium, or the partial pressure of oxygen was lowered mechanically, the smooth variant grew from only a small inoculum. Jordan (1952) showed that H_2O_2 was produced during growth, and that haematin, or better still catalase, promoted aerobic growth.

The *nutritional requirements* have not been worked out completely. McKenzie and her colleagues (1948; see also Wessman and Wessman 1970) grew *P. multocida* successfully on a defined medium containing a mixture of salts, glucose, asparagine, tryptophan, glutamic acid, glycine, thiamine, cystine, biotin, nicotinamide, and calcium pantothenate. In a slightly modified medium Watko (1966) found that an inoculum of only 2 or 3 organisms was sufficient to start growth. Namioka and Murata (1961) prepared a medium (YPC agar) containing yeast extract, proteose peptone, L-cystine, glucose, sucrose, sodium sulphite, potassium diphosphate and agar on which the organisms grew well. For *P. haemolytica* Wessman (1966), finding that the organisms flourished in a casein hydrolysate medium, devised a medium containing 15 individual amino acids from casein, a mixture of salts and vitamins, and galactose and glucose as sources of carbon. Selective media for the isolation of *P. multocida* were described by Das (1958) and Morris (1958).

Pasteurellae grow between about 12° and 43°; their optimal temperature is 37°; but, according to Burrows and Gillett (1966), their nutritive requirements are more exacting at 37° than at lower temperatures.

None of the pasteurellae forms a soluble haemolysin, but *P. haemolytica* forms colonies surrounded by a narrow zone of weak β-haemolysis on nutrient agar containing horse, sheep, rabbit, or ox blood.

Resistance

None of the pasteurellae is specially resistant to adverse agencies. The organisms are killed by heat at 55°, and by 0.5 per cent phenol within 15 min. In cultures or infected organs kept in the ice-chest the bacilli may survive for months. In infected blood kept in the dried state *P. multocida* may retain its viability and its virulence for about three weeks; in blood allowed to putrefy in a glass tube for 100 days it may still be capable of causing infection (Ostertag 1908). Pasteurellae are susceptible *in vitro* to sulphadiazine, penicillin, streptomycin, and chloramphenicol. The A form of *P. haemolytica* is said to be highly susceptible to penicillin, but the T form to be fairly resistant (G. R. Smith 1961, Biberstein and Francis 1968). About half the strains of *P. multocida* are able to grow in the presence of 0.4 per cent selenite, but *P. haemolytica* and *P. ureae* are invariably unable to do so (Lapage and Bascomb 1968).

Biochemical reactions

For testing the sugar reactions of *P. haemolytica*, Bosworth and Lovell (1944) recommended a peptone water medium containing 10 per cent broth, 1 per cent sugar, and bromothymol blue as an indicator. Growth is light, acid production is weak, and the results obtained by different observers are not always in agreement. There is also a good deal of variation between different strains, depending to some extent on the animal source and the nature of the medium.

All strains of *P. multocida* produce acid but not gas in glucose and sucrose; most strains ferment galactose, mannitol, mannose, sorbitol and xylose. Dextrin, dulcitol, lactose, maltose and raffinose are occasionally attacked; inulin, inositol, rhamnose and salicin are usually not attacked. Strains from certain sources tend to possess particular biochemical characteristics (Magnusson 1914, Tanaka 1926, Khalifa 1934, Gourdon *et al.* 1957, J. E. Smith 1958, Heddleston 1976). Avian strains not infrequently ferment arabinose. Bovine strains often ferment trehalose. Canine and feline strains often attack maltose but may fail to attack mannitol and sorbitol. Dog strains, it may be noted, are usually non-capsulated, non-mucoid, non-iridescent, acid agglutinable, and only moderately pathogenic for mice; cat strains are similar, but are highly pathogenic for mice (J. E. Smith 1959). Strains of *P. multocida* are without action on litmus milk and gelatin. They produce indole, reduce nitrate, and form a small quantity of H_2S as detected by lead acetate paper. The MR and VP reactions are both negative; the catalase and oxidase reactions are both positive, though rather weakly so. Methylene blue is reduced. Citrate cannot be used as the sole source of carbon. Urea is seldom decomposed except by the strains often referred to as *P. pneumotropica* (see p. 362), and by *P. ureae*. The β-galactosidase and malonate tests are negative. Most strains are positive for ornithine decarboxylase but negative for lysine and glutamic acid decarboxylases and for arginine dihydrolase; most are resistant to potassium cyanide.

P. haemolytica differs from *P. multocida* in not forming indole; in addition it regularly attacks maltose, dextrin and, after several days, inositol. Strains of biotype T but not biotype A ferment trehalose within about two days; strains of biotype A but not T produce small amounts of acid from arabinose within about seven days. Certain strains of biotype T are said to be catalase negative (Biberstein *et al.* 1960). *P. ureae* differs from *P. multocida* in not forming indole or H_2S, and in decomposing urea strongly.

Antigenic structure

The early observations of Lal (1927), Cornelius (1929), Yusef (1935) and Roberts (1947) showed that multiple antigenic types of *P. multocida* could be distinguished by use of agglutination and absorption, or of mouse-protection tests; by means of protection tests, strains were classified by Roberts into four types (I, II, III and IV). The serological behaviour is determined by capsular and by somatic antigens. Both smooth iridescent and mucoid colonies possess a capsule; the blue granular rough colonies do not. Capsular antigens can be typed either by slide agglutination (Namioka and Murata 1961), or by indirect haemagglutination in which the soluble capsular antigen is adsorbed on to human group O red cells and then mixed with various antisera (Carter 1955, 1957, 1961, 1962, 1967). To reveal the somatic antigen Namioka and Murata (1961) removed the capsular material by treatment with N/1 HCl, suspended the washed organisms in saline at pH 7.0 and tested them against antisera prepared in rabbits by the injection of similarly decapsulated organisms. Carter (1972) treated fowl cholera strains with hyaluronidase to make them agglutinable for the identification of their somatic antigens. Organisms from rough colonies autoagglutinate when 1 or 2 drops of a 1 in 1000 solution of acriflavine are added to a fairly thick saline suspension; mucoid colonies form a slimy precipitate, and smooth colonies remain homogeneously suspended (Carter 1957). The rough variants are devoid of both capsular and smooth somatic antigens. Cat and dog strains are frequently unencapsulated and untypable.

On the basis of polysaccharide capsular antigens, Carter (1955) established among smooth iridescent strains four serotypes A, B, C, and D. Later a further type—type E—was recognized (Carter 1961), and type C was shown not to be a true capsular type. There are therefore only four capsular types, namely A, B, D and E (Carter 1957). Type B is generally agreed to be the equivalent of Roberts's type I (Roberts 1947); an opinion on the relationship between all Roberts's types and the capsular types was given by Prodjoharjono and his colleagues (1974). Mucoid strains are rich in hyaluronic acid and have a common mucoid antigen. Carter (1955) stated that they did not fall into the four capsular types that he defined among smooth strains; but it would appear that some mucoid strains also contain a polysaccharide antigen that enables them to be typed in the A, B, D, E series (J. E. Smith 1958, Carter 1967). Namioka and Murata (1961), by the use of agglutination and absorption tests, established six somatic (O) groups. Further studies, reviewed by Namioka (1973), established eleven O groups based on antigens that consisted of combinations of several factors; the capsular groups A, B, D and E contained six, two, six and one O group respectively (Namioka 1973). Thus 15 serotypes were recognized. But Bhasin and Lapointe-Shaw (1980) described even greater antigenic complexity. By crossed immunoelectrophoresis they found in an envelope-cytoplasm preparation of serotype 1 at least 19 cell envelope and 55 cytoplasmic antigens. Moreover by bacteriocine typing Mushin (1980) divided avian strains of this organism into 23 types. It is now usual to describe the serotype of a strain by stating both the somatic and capsular groups, e.g. serotype 5:A. Capsular type A strains can be rapidly identified by a staphylococcal hyaluronidase test (Carter and Rundell 1975) and type D strains by an acriflavine flocculation test (Carter 1973). Heddleston and his colleagues (1972) described an agar-gel precipitation test in which the antigen preparations were boiled for an hour; the test was useful for serotyping avian strains but necessitated a further system of serotype nomenclature. Brogden and Packer (1979) compared *P. multocida* typing systems; a single serotype in one system often represented more than one serotype in another.

According to Carter (1967), capsular type A strains are associated on the North American continent with fowl cholera and with pneumonia in cattle and pigs, as well as with primary and secondary infections in a wide variety of animals; type B strains with haemorrhagic septicaemia of cattle and buffaloes in Asia and Australia; type D strains with primary and secondary infections in a wide range of animals; and type E strains with haemorrhagic septicaemia in central Africa. Strains from human beings belong mainly to type A and to a less extent to type D. (See also Baxi *et al.* 1970.)

By means of the trichloracetic acid technique Pirosky (1938*a, b, c*) extracted from smooth and rough variants of *P. multocida* four different glycolipid antigens. Judged by cross-precipitation tests one of these antigens was related to the Vi antigen of the typhoid bacillus, and another to the O antigen shared by *Salmonella typhi* and *Salmonella enteritidis*.

Like *P. multocida*, *P. haemolytica* possesses capsular and somatic antigens. By means of an indirect haemagglutination test, Biberstein and co-workers (1960) and Biberstein and Gills (1962) distinguished 11 numbered capsular serotypes; by means of agglutination tests with autoclaved bacteria, they distinguished several less clearly defined somatic serotypes, identified by letters. Fifteen capsular serotypes are now known. Serotypes 1, 2, 5, 6, 7, 8, 9, 11, 12, 13 and 14 correspond to G. R. Smith's (1959, 1960, 1961) colonial type A, which ferments arabinose but not trehalose, is sensitive to penicillin, and rapidly loses viability in ageing broth cultures. Serotypes 3, 4, 10 and 15 correspond to Smith's colonial type T, which ferments trehalose but not arabinose, is resistant to penicillin, and does not rapidly lose viability in ageing broth cultures. Later, the A type was found to contain major somatic antigens A and B; the T type contained major somatic antigens C and D (Biberstein and Francis 1968).

It will be realized that the nomenclature employed here is very confusing. In *P. multocida*, capsular antigens are designated by letters and somatic antigens by numbers; in *P. haemolytica* the reverse procedure is employed. Moreover the A somatic antigen of *P. haemolytica* is quite distinct from Smith's A type, which was defined on cultural, biochemical, and pathological grounds. It may be best to regard Smith's A and T types as two separate species, each with its own series of capsular and somatic antigens (J. E. Smith and Thal 1965, Biberstein and Francis 1968, Thompson and Mould 1975). Not all strains of *P. haemolytica* are typable. Åarsleff and his colleagues (1970) found that up to 90 per cent of strains from nasal swabs of sheep failed to react in the indirect haemagglutination test; such strains have carbohydrate fermenting properties that resemble those of G. R. Smith's type A (Gilmour 1978). By means of a plate-agglutination test, Frank (1980) divided 10 so-called 'untypable' strains into three groups which were distinct from established serotypes.

P. ureae, according to D. M. Jones (1962), is antigenically homogeneous, but according to Henriksen (1961) has at least two serotypes, of which one contains a type-specific capsular antigen. Both observers agree, however, that it has no antigenic relation to other members of the *Pasteurella* group.

Little is known of the nature of the protective substance associated with the capsule of pasteurellae. Rebers and co-workers (1967), however, extracted from a capsulated virulent strain a heat-stable lipopolysaccharide protein complex that was toxic to calves, mice and rabbits, but immunized the survivors. The gross chemical composition of this substance was similar to that of the enterobacterial endotoxins (see Chapter 33). The immunizing ability of *P. multocida* is associated with the capsular antigen (Priestley 1936*b*). It is still doubtful what degree of specificity this possesses. The older workers held that a strain from one animal could be used to vaccinate an animal of a different species (Chamberland and Jouan 1906, Magnusson 1914); but Roberts (1947), who prepared antisera in horses and rabbits and tested their ability to protect mice inoculated intraperitoneally with living organisms, classified 37 strains of *P. multocida* into four different types, suggesting the presence of some heterogeneity in immunizing ability as well as in antigenic structure. Nagy and Penn (1976) found that endotoxin-free capsular antigens of *P. multocida* types B and E protected cattle against intravenous challenge with a virulent type E strain.

Pathogenicity

P. multocida, as already mentioned, is pathogenic for a wide variety of animals, producing anything from a harmless latent infection to a rapidly fatal haemorrhagic septicaemia (see Chapter 55); man is an occasional victim.

Strains of *P. multocida* from different sources vary in their virulence for mice and other animals (Carter 1967; see also Chapter 55). Most capsulated strains from acute infections are virulent for mice and rabbits. Cat strains as a rule are highly virulent for mice, dog strains only moderately so (J. E. Smith 1958). The resistance of passively immunized mice is said to be broken down by the intraperitoneal injection of a solution of haemoglobin or of ferric ammonium citrate. This kataphylactic effect is thought to be due to the provision of iron to the bacteria (Bullen and Rogers 1968). For testing virulence, mice, rabbits and pigeons are the most suitable experimental animals. Though they may occasionally suffer from spontaneous disease, guinea-pigs are not so susceptible (J. Wright 1936).

Mice Subcutaneous inoculation of a small quantity of a 24-hr broth culture proves fatal in 24 to 48 hr. *Post mortem*, there may be local oedema and congestion, with practically no other signs. Microscopically the bacilli are found in large numbers in the blood and viscera. When very few organisms are injected, or a culture of relatively low virulence is used, the mouse does not die for 2 to 8 days, or even longer; at necropsy there is a fibrino-purulent pericarditis, a layer of fibrin over the pleura, partial consolidation of the lungs, and not infrequently a purulent exudate in the peritoneum. Bacilli are plentiful in the blood and organs. Intraperitoneal inoculation is more rapidly fatal.

Rabbits can be infected by subcutaneous, intraperitoneal, intravenous, intratracheal, or intranasal inoculation. Death occurs in 2 to 5 days as a rule after intraperitoneal injection, with lesions similar to those in mice. In addition there may be a haemorrhagic tracheitis (Magnusson 1914), and hyperaemia of the kidneys and intestine (Poels 1886). Intranasal insufflation of the bacilli is often followed by snuffles or pleuro-pneumonia (Beck 1893, Webster 1924*b*, 1926), and sometimes by purulent otitis media (Smith and Webster 1925).

Pigeons are very susceptible to intravenous or intraperitoneal, less so to intramuscular, injection. Death occurs in 24 to 48 hr. Bacilli are abundant in the blood.

P. haemolytica is naturally pathogenic for sheep, cattle, and goats (see Chapter 55), but has little effect on laboratory animals by the usual methods of infection, except in massive doses.

Mice can be infected by the intraperitoneal route if the organisms are mixed with mucin; the animals die within 48 hr (G. R. Smith 1958). Mice can also be infected intracerebrally. G. R. Smith's type A strains, when injected intraperitoneally into young lambs—not more than 3 weeks old—in a dose of about 5000 organisms, give rise to acute fibrinous peritonitis proving fatal within 36 hr; adult sheep are not affected by this treatment (G. R. Smith 1960). Type T strains produce disease in young lambs only when injected in large doses.

P. ureae is only slightly pathogenic for mice, guinea-pigs, and rabbits (Henriksen and Jyssum 1961*b*), but strains vary in virulence. Some strains injected intraperitoneally kill mice in two days; others do so only when injected with mucin. It is probably part of the flora of the human respiratory tract; and it is the only member of the *Pasteurella* group that has no animal host (Kolyvas *et al.* 1978).

Classification

Organisms of the *Pasteurella* group differ in many respects from those of the *Yersinia* group, which contains organisms that cause plague, pseudotuberculosis, and enteritis. The main differences between the two genera were pointed out by van Loghem (1944). Briefly, pasteurellae are smaller, often fail to grow on potato or in the presence of bile salts, ferment sucrose and sorbitol but not as a rule rhamnose or salicin, form indole and H_2S, give a negative methyl-red reaction, are oxidase positive, and have a distinctive antigenic structure. In addition, strains of *P. multocida* and some of *P. haemolytica* produce a decarboxylase for ornithine and grow in a medium containing KCN (Steel and Midgley 1962). Many yersiniae have the opposite characters, and some strains are motile. The distinction between the two genera receives strong support from the numerical method of analysis as carried out by Talbot and Sneath (1960) and Smith and Thal (1965), and is supported by studies on DNA homology (Ritter and Gerloff 1966).

The classification of members of the *Pasteurella* group itself has already been dealt with in this chapter. Mannheim and his colleagues (1980) consider that the *Actinobacillus-Haemophilus-Pasteurella* group appears to rank as a family. Briefly it may be said that the genus *Pasteurella* contains three main species—*P. multocida*, *P. haemolytica* and *P. ureae*. *P. haemolytica* differs from *P. multocida* chiefly in its formation of haemolytic colonies on blood agar, its growth, though poor, on MacConkey's agar, its greater range of fermentation of sugars, its failure to form indole, its non-pathogenicity for laboratory animals, its lysis of lambs' blood, and the different G + C content of its DNA (40.75 moles per cent). Mráz (1969*a*) is of the opinion that it should be classified with *Actinobacillus* rather than with *Pasteurella*. Comparing 46 strains of *P. haemolytica* with 99 strains of *A. lignieresi* he noted only three major differences between them, namely the double zone of lysis on lambs' blood agar, the negative urease test, and the slightly lower G + C content of the DNA of *P. haemolytica*. The biotypes A and T may eventually prove sufficiently different from each other to justify the establishment of two separate species (J. E. Smith and Thal 1965, Biberstein and Francis 1968); if so, the ability of both to produce an unusual haemolytic effect on lambs' blood (G. R. Smith 1962) must be regarded as a remarkable coincidence.

P. ureae is pleomorphic, grows poorly at 22°, fails to grow on MacConkey's agar or to form H_2S, ferments maltose but not xylose, is indole negative, hydrolyses urea strongly, is serologically distinct, and is usually non-pathogenic for laboratory animals. Jawetz (1950) described under the name of *P. pneumotropica* an organism causing broncho-pneumonia in mice. It ferments maltose but not mannitol, and hydrolyses urea.

The taxonomic status of *P. anatipestifer*, *P. gallinarum* and *P. aerogenes* (see McAllister and Carter 1974, Chladek and Ellis 1979) and of *P. piscicida* (Koike *et al.* 1975) is uncertain. (For reviews of pasteurella taxonomy and nomenclature, see Frederiksen 1973, Hussaini 1975; and of the relation of *Pasteurella* to *Actinobacillus* and *Haemophilus*, see Zinnemann 1981.)

Pasteurella multocida

Isolation.—First member of the *Pasteurella* group isolated by Perroncito in 1878. Isolated from birds by Pasteur in 1880.
Habitat.—Parasites of domestic and wild animals and birds.
Morphology.—Very small, 0.7 μm × 0.3–0.6 μm, ovoid bacilli, with straight axis, slightly convex sides, and rounded ends; arranged singly, in pairs, or in small bundles. In smears from the animal body the organisms are regular, ovoid, and evenly distributed; on agar cultures they are more rod-shaped and often show pleomorphism. Non-motile, non-sporing. Virulent strains form a capsule in the animal body and in culture media at 37°. Show bipolar staining. Gram negative and non-acid-fast.

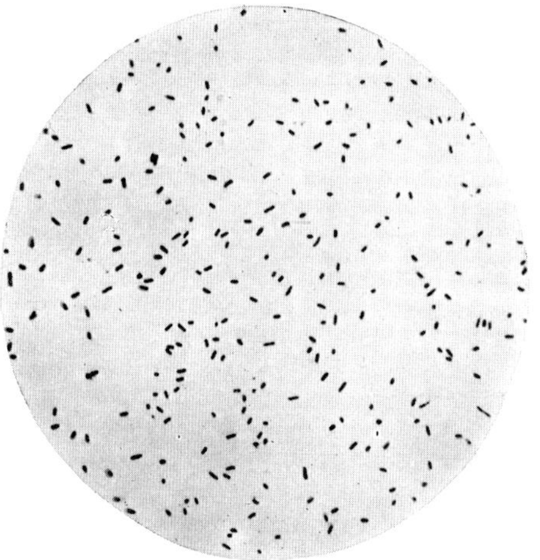

Fig. 38.1 *Pasteurella multocida*. From an agar culture, 24 hr 37° (× 1000).

Agar plate.—24 hr at 37°. Round, 0.5–1.0 mm in diameter, low convex, amorphous, greyish-yellow, translucent colonies, with smooth, glistening surface and entire edge; consistency butyrous; emulsifiability easy.
5 days at 37°. Up to 6 mm in diameter, differentiated into a brownish, finely granular, sometimes ringed or striated, nearly opaque centre and a clearer, smooth, homogeneous, greyish-yellow translucent periphery.
Deep agar shake.—5 days at 37°. Thick surface growth; numerous, punctiform, undifferentiated colonies scattered throughout medium.
Agar stroke.—24 hr at 37°. Moderate, confluent, raised, greyish-yellow, translucent growth, with glistening, wavy or beaten-copper surface and finely lobate edge.
Gelatin stab.—7 days at 22°. Good, filiform growth, confluent at top, discrete below, extending to bottom; raised surface growth, 5 mm in diameter, with crenated edge, no liquefaction.
Broth.—24 hr at 37°. Moderate growth with slight turbidity, and a slight powdery or viscous deposit. Later the turbidity increases, and a heavy, viscous deposit forms, disintegrating partly on shaking but leaving irregular sized wisp-like masses of growth in suspension. An incomplete surface pellicle forms with an inconspicuous ring growth.
Loeffler's serum.—24 hr at 37°. Good confluent growth, similar to that on agar.
Horse blood agar plate.—2 days at 37°. Good growth similar to that on agar; no haemolysis, but blood plate is slightly cleared and browned.
Potato.—7 days at 22°. No visible growth.
MacConkey plate.—5 days at 37°. No visible growth.
Resistance.—Very susceptible to inimical agencies; killed by heat at 60° in a few minutes, by 0.5 per cent phenol in 15 min.
Metabolism.—Aerobe, facultative anaerobe. May require low O–R potential on first isolation. Opt. temp. 37°, limits 12°–43°. Growth improved slightly by serum, uninfluenced by glucose, slightly inhibited by glycerol. No haemolysis. Said to form neuraminidase (Müller and Krasemann 1974).
Biochemical.—Acid, no gas, in glucose, sucrose, mannitol, and usually sorbitol and xylose, within 14 days. Some strains ferment maltose, arabinose, trehalose, or glycerol. LM unchanged; indole+; MR−; VP−; nitrate reduced, citrate−; NH_3 very slightly+; MB reduction+; H_2S+; catalase+; oxidase+. Urea seldom decomposed except by strains referred to as *P. pneumotropica*.
Antigenic structure.—Mucoid and smooth iridescent colonies contain capsulated organisms, the capsule being characterized by hyaluronic acid, a polysaccharide, or both. Four capsular serotypes are recognized—A, B, D and E—and 11 smooth somatic O groups; on the basis of capsular and somatic antigens, 15 serotypes can be established. Blue and granular colonies are non-capsulated and contain rough but not smooth O antigens.
Pathogenicity.—No true exotoxin produced. Virulence subject to alteration. Causes fowl cholera in birds. Members of this group produce septicaemia with widespread haemorrhages in cattle, buffaloes, reindeer, sheep, pigs, rabbits, and other animals; they also produce respiratory infections. Acute forms of disease can be reproduced by experimental inoculation. *Subcutaneous* inoculation of a 24-hr broth culture into a mouse proves fatal in 18 to 72 hr. PM local oedema and congestion; often no other signs; microscopically bacilli present in enormous numbers in blood and viscera. If a small dose is given and the animal does not die for 4 to 7 days,

there is often a fibrinopurulent pericarditis, a layer of fibrin over the pleura, and partial consolidation of the lungs. Bacilli are numerous in blood and organs.

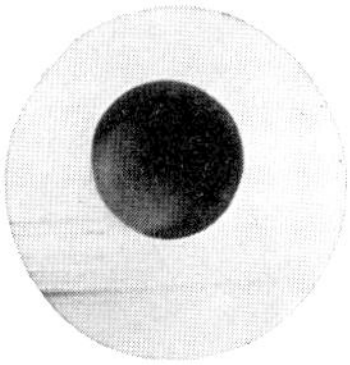

Fig. 38.2 *Pasteurella multocida.* Surface colony of smooth type on agar, 24 hr, 37° (×8).

Pasteurella haemolytica

Isolation.—By F. S. Jones (1921), Tweed and Edington (1930), Newsom and Cross (1932) and others from pneumonia in sheep, cattle, and goats.
Morphology.—Short, non-motile, evenly stained rods. Gram negative.
Agar plate.—24 hr at 37°. Similar to *P. multocida*, but on horse, sheep, or rabbit blood agar forms circular, glistening, convex colonies up to 4 mm in diameter, surrounded by a narrow zone of β-haemolysis. A so-called T variant from lamb septicaemia forms colonies that in young cultures are larger, with dark brown centres (G. R. Smith 1959, 1961). On agar plates made with the blood of very young lambs *P. haemolytica* gives rise to a double zone of haemolysis—an inner complete and a wide outer partial (G. R. Smith 1962).
MacConkey plate.—Small pink colonies.
Broth.—24 hr at 37°. Grows readily producing a turbidity. Later a granular deposit and a pellicle are formed.
Metabolism.—No soluble haemolysin, except that for the erythrocytes of very young lambs.
Biochemical.—Acid, no gas, in glucose, maltose, mannose, mannitol, sucrose, sorbitol, usually in galactose, dextrin, xylose, and inositol, and sometimes in lactose, salicin, rhamnose, raffinose, trehalose, arabinose, glycerol, glycogen, and starch. Dulcitol is rarely attacked. Inulin, adonitol and erythritol are not attacked. MR −; VP −; nitrate +; citrate −; H_2S −; urea −; MB reduction +; catalase almost always +; oxidase −; gelatin liquefaction −. The T type ferments trehalose but not arabinose, as opposed to the A type which is said to ferment arabinose but not trehalose.
Antigenic structure.—Fifteen capsular serotypes, which are numbered, and several less clearly defined somatic serotypes, which are distinguished by letters. G. R. Smith's colonial type A contains eleven of the capsular serotypes and the major somatic antigens A and B; his colonial type T contains four capsular serotypes—3, 4, 10 and 15—and the major somatic antigens C and D.
Pathogenicity.—Gives rise to septicaemia in lambs and to pneumonia in sheep and less often in cattle; also pathogenic for goats. Type A and T strains produce different forms of disease in sheep. Non-pathogenic for laboratory animals unless reinforced with mucin, or injected intracerebrally. Injected intraperitoneally or intravenously into very young lambs it produces a disease fatal within 24 hr or so. (For a review of papers on *P. haemolytica*, see Mráz 1969*b*.)

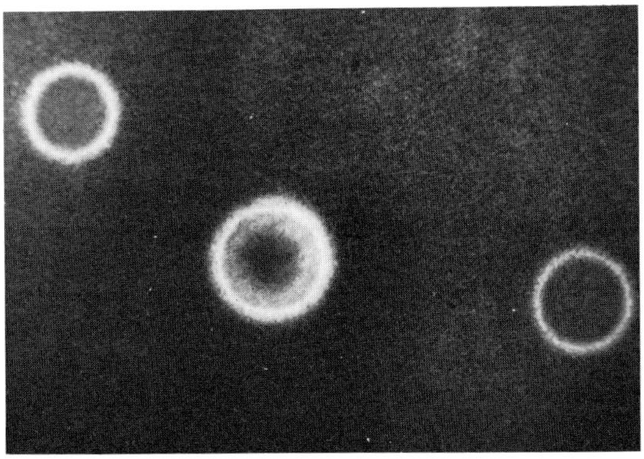

Fig. 38.3 *Pasteurella haemolytica.* One type T colony (centre) and two of type A, on blood agar, 24 hr, 37° (×5.6).

Cardiobacterium The name *Cardiobacterium hominis* was proposed by Slotnick and Dougherty (1964) for a pasteurella-like organism first reported by Tucker and coworkers (1962) as an occasional cause of endocarditis in man (see Chapter 57). The same organism has been found in the nose and throat of 68 per cent of healthy persons (Slotnick *et al.* 1964). On primary isolation small colonies are produced on blood agar in an atmosphere supplemented with 3–5 per cent CO_2. The organism, which ferments a wide range of sugars, is oxidase positive and produces indole; it does not produce catalase or reduce nitrate (Midgley *et al.* 1970). These reactions are helpful in distinguishing cardiobacteria from species of the genera *Pasteurella*, *Haemophilus*, *Moraxella*, *Actinobacillus*, and *Streptobacillus*.

Francisella

Francisella tularensis

This organism is a tiny, non-motile, gram-negative bacillus, which was isolated by McCoy and Chapin in 1912 from rodents suffering from tularaemia (see Chapter 55). Its taxonomic position is doubtful. Reimann (1932), along with several other workers, would assign it to the *Pasteurella* group on account of its bipolar staining, its solubility in 1 in 800 sodium ricinoleate, its transmission by insect vectors, and the nature of the lesions it produces in animals. On the other hand the beneficial effect of CO_2 on its growth, its cytotropism (Buddingh and Womack 1941), its failure to develop anaerobically, its production of H_2S, its very weak fermentative ability, its sensitivity to methylene blue (Yaniv and Avi-Dor 1952), and its high pathogenicity for man in the laboratory qualify it perhaps even better for inclusion in the *Brucella* group. However, as already mentioned, the G + C content of its DNA, namely 33–36 moles per cent, is very much nearer to the figure of 37–41 for *Pasteurella* than to that of 55–58 for *Brucella* or of 46–47 for *Yersinia pestis*. It differs from *Pasteurella* in its pleomorphism, its demand for cystine, and in several other ways. The International Committee on Bacterial Nomenclature therefore recommend that it should be accorded a separate genus—*Francisella* (Report 1971*a*).

In the animal body it occurs as a coccoid or rod-shaped organism surrounded by a clear area, which probably represents a capsule. The diameter of the organism is 0.3–0.7 μm long by 0.2 μm wide; the diameter with the capsule is 0.4–1.0 μm by 0.3–0.5 μm. In culture it tends to be pleomorphic; coccoid, ovoid, bacillary, bean-shaped, dumb-bell and even filamentous forms may be seen (Eigelsbach *et al.* 1946). The organisms stain best with carbol-fuchsin or aniline gentian violet; with methylene blue they stain very poorly and show no capsule. In culture, coccoid forms alone are seen (Wherry and Lamb 1914); a capsule is visible when the organisms are mixed with serum. No growth occurs in unenriched media. It was first cultivated on Dorset's egg, but later it was found that coagulated egg yolk was more satisfactory (McCoy 1912). On this medium the maximum growth is reached in 2 days; it is pale, translucent, slightly mucoid, and pearly in appearance, not easily distinguishable from the medium; it is readily emulsifiable. Growth occurs also on glucose blood agar, glucose serum agar, and blood agar slopes, provided that a piece of rabbit's spleen is rubbed over the surface and then left in the condensation water (Francis and Lake 1922); and on agar to which 0.02 per cent of cystine is added (Francis 1922, 1923). On these media the organism should be subcultured every other day, but on egg yolk it may remain viable for 3 months (Wherry and Lamb 1914). Shaw and Hunnicutt (1930) recommended a medium composed of brain veal infusion agar, pH 7.6, containing 5 per cent rabbit serum, 1 per cent dextrose, and 0.05 per cent cystine; and Ringertz and Dahlstrand (1968) one containing tryptose broth, cysteine-HCl, sodium thioglycollate, glucose, and rabbit blood. Kudo (1934) prepared a mixture of 60 per cent egg yolk and 40 per cent rabbit serum sterilized at 70–75° on 3 successive days. Liquid media can also be used (Tamura and Gibby 1943); Steinhaus and co-workers (1944) recommended for this purpose a medium containing 1 per cent dextrose, 0.15 per cent cystine, and 0.5 per cent haemoglobin. According to Halmann and his colleagues (1967) a growth-initiating factor derived from the organisms themselves is required when less than 100 000 cells are inoculated into an artificial medium. This can be obtained by filtration of a 24–48-hr broth culture through a Millipore membrane. The factor is said to be of low molecular weight.

F. tularensis grows well in the developing chick embryo, multiplying particularly in the ectodermal epithelial cells. When inoculated on to the chorio-allantoic membrane, it leads to the death of the embryo in 3–4 days. In both these respects it resembles *Brucella melitensis* (Buddingh and Womack 1941, Ransmeier 1943). Under suitable conditions the organism is said to produce acid in glucose and glycerol, and usually in maltose, mannose and laevulose (Downs and Bond 1935, Francis 1942). Olsufjev and Emelyanova (1962) found that most American strains fermented glycerol and were highly pathogenic for the rabbit, whereas European and Asiatic strains did not ferment glycerol and were only mildly pathogenic for the rabbit. Fleming and Foshay (1955) found that strains of low virulence did not attack citrulline. Catalase production is weak.

The organism is very susceptible to inimical agencies, including chloramphenicol and the tetracyclines, and is killed by moist heat at 55–60° in 10 min. Antigenically *F. tularensis* appears to be homogeneous. It shows a slight relation to organisms of the *Brucella* group, but no cross-absorption of antibodies occurs from rabbit sera. There is also a slight relation to *Yersinia pestis* (Larson *et al.* 1951, Olsufjev and Emelyanova 1957). In contrast to *Brucella* it is said to have a low cytochrome C content (Report 1971*b*).

Under natural conditions *F. tularensis* gives rise to tularaemia in rodents—especially ground-squirrels and jack-rabbits—and occasionally in man. It causes a latent infection in a wide variety of mammals and birds (Burroughs *et al.* 1945). Experimentally the disease can be reproduced in ground-squirrels, gophers, guinea-pigs, rabbits, mice, hamsters, cotton-rats, and monkeys; rats are more resistant (Dieter and Rhodes 1926); cats, dogs, chickens and pigeons are practically immune (see Downs *et al.* 1947). Feeding, nasal instillation, cutaneous, subcutaneous, intraperitoneal, and conjunctival infection are all successful. After subcutaneous inoculation the guinea-pig dies in 5 to 8 days. *Post mortem* there is a whitish membrane-like area at the site of inoculation; the regional lymphatic glands may be enlarged and caseous. The spleen is enlarged, very dark in colour, and contains discrete, yellowish-white, caseous granules up to 1 mm in diameter, projecting slightly above the surface; there are numerous granules in the liver. Focal necrotic areas are sometimes present in the bone marrow (Lillie and Francis 1933). The lungs are rarely affected. The bacilli are present in large numbers in the blood and organs; as little as 1×10^{-7} ml of the heart's blood may prove infective for other animals. The virulence of the organism may decline in culture, so that instead of causing an acute or subacute disease in guinea-pigs, it gives rise to a chronic disease from which the animal often recovers (McCoy 1912, Foshay 1932). Strains of lowered virulence have also been isolated directly from ticks (Davis *et al.* 1934). Removal of the capsule reduces virulence for guinea-pigs without affecting viability (Hood 1977). The organism is extremely dangerous to handle in the laboratory, and large numbers of workers have contracted the infection.

Larson and co-workers (1955) isolated from a sample of turbid water an organism to which they gave the name *Pasteurella novicida*. Girard and Gallut (1957), who studied this organism, considered it to be more closely related to *F. tularensis* than to other members of the *Pasteurella* group. It differs from *F. tularensis* in fermenting sucrose, in being less dependent on cystine, in its somewhat different antigenic structure, and in its lower degree of pathogenicity for laboratory animals. On the other hand living organisms afford protection against infection with *F. tularensis* (Owen *et al.* 1964), and the two organisms are closely homologous at the molecular level as shown by DNA hybridization (Tyeryar and Lawton 1970). There seems little doubt that it is a member of the *Francisella* group, though whether it should be given specific rank is less certain.

Yersinia

Definition

Small, gram-negative, motile or non-motile, pleomorphic, coccoid or ovoid bacilli, often showing bipolar staining. Aerobic and facultatively anaerobic. Grow in the presence of bile salts. Attack carbohydrates by the fermentative method; gas not formed, except in small amount at 22° by *Y. enterocolitica*. Nitrate usually reduced. Catalase positive; oxidase and phenylalanine deaminase negative. Do not form indole, except for some strains of *Y. enterocolitica*, and do not liquefy gelatin. Are parasites of man and animals, causing characteristic diseases. G + C content of the DNA: 46 moles per cent.

The type species is *Yersinia pestis*.

Habitat

The three species of *Yersinia* are all parasites of animals. *Y. pestis* also infects fleas. In man this organism gives rise to a systemic, often fatal, disease characterized by the formation of buboes. *Y. pseudotuberculosis* is found in a wide variety of animals including rodents and birds, and occasionally man, causing a disease that is often localized in the abdominal lymph glands. *Y. enterocolitica* is the least pathogenic of the three. It leads a commensal existence in normal pigs and rats, and less often in dogs and cats (Yanagawa *et al.* 1978), and in small rodents such as voles and field mice (Kapperud 1975). In man it occasionally produces an enteritis accompanied by diarrhoea. It is also a saprophyte, living and multiplying in water. (See also Kapperud 1981 and Chapter 55.)

Morphology

The members of this group resemble the pasteurellae in being ovoid or coccoid bacilli, but they are larger and more variable in size and shape. Pleomorphism is most apparent with *Y. pestis*, which, particularly on 3 per cent sodium chloride nutrient agar, shows shadow forms, club forms, snake-like forms, large yeast-like globules, and other bizarre forms. *Y. pseudotuberculosis* and *Y. enterocolitica* are motile at 22° but not at 37°, though often only slightly so (Arkwright 1927, Weitzenberg 1935, Knapp and Thal 1963). The pseudotubercle bacillus is said to possess 1–6 peritrichate flagella (Preston and Maitland 1952).

In the animal body and often on 5–10 per cent serum agar or in buffered broth containing 0.2 per cent of Fildes's peptic digest of blood the plague bacillus forms what is variously described as a capsule or an envelope (Kitasato 1894, Rowland 1914). This is best revealed by examining specimens suspended either in India ink under darkground illumination (Amies 1951), or in a mixture of India ink and 6 per cent glucose, stained with equal parts of methyl alcohol and 1 per cent methylene blue (Butt *et al.* 1936). The capsule develops best at 37°, poorly at 26°, and not at all at 20° (Schütze 1932*a*); aerobic conditions favour and blood agar hinders its formation. It is dissolved by alkali but, unlike the pneumococcal capsule, it is soluble in potassium thiocyanate as well. It contains a protein antigen distinct from the somatic antigen. L-forms of *Y. enterocolitica* have been described (Pease 1979).

Cultural characters

On nutrient agar *Y. pseudotuberculosis* and *Y. enterocolitica* grow fairly rapidly giving a confluent growth in 24 hr: *Y. pestis* grows more slowly giving a poorer growth, often barely noticeable after this time. Colonies of *Y. pseudotuberculosis* and *Y. pestis* resemble each other in many respects, and are characterized

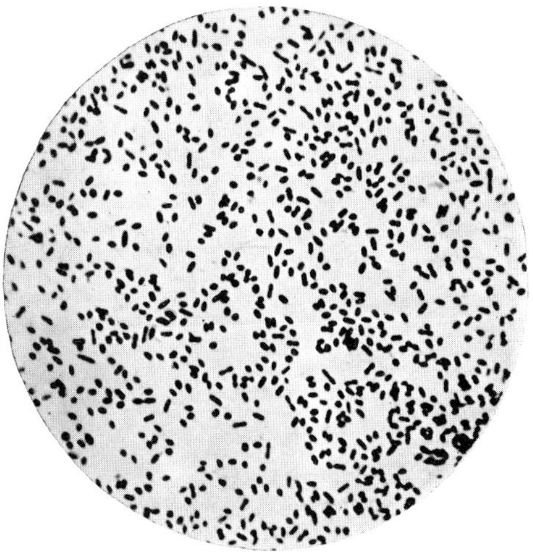

Fig. 38.4 *Yersinia pestis*. From an agar culture, 3 days, 37° (× 1000).

by an effuse clear or slightly granular, peripheral extension after 2 to 4 days' growth (Figs. 38.5 and 38.6); the colonies of *Y. pseudotuberculosis*, however, develop more rapidly and are larger and more granular; later the central raised part of the colony may assume a ringed or draughtsman-like appearance. On moist agar the colonies, especially of *Y. pestis*, are viscous and tend to adhere to the medium (Eastwood and Griffith 1914). When grown on a defined medium rich in haemin, colonies of *Y. pestis* may be dark brown (Jackson and Burrows 1956). The colonies of *Y. enterocolitica* are described on p. 375.

On deoxycholate citrate agar plague bacilli grow scantily and form reddish pin-point colonies in 48 hr, whereas pseudotubercle bacilli grow abundantly and form large opaque colonies which turn the medium

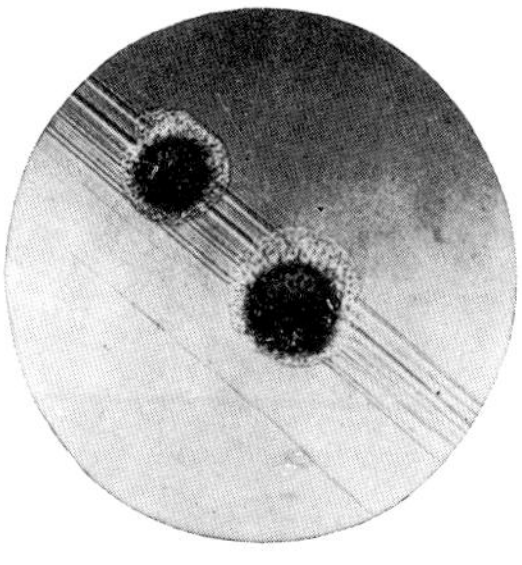

Fig. 38.5 *Yersinia pestis*. Surface colonies on agar, 3 days, 37°, showing differentiation, and effuse edge (× 8).

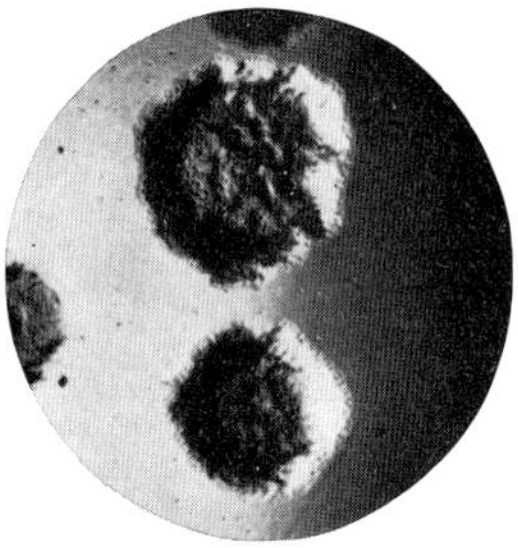

Fig. 38.6 *Yersinia pseudotuberculosis*. Surface colonies on agar, 24 hr, 37°, showing irregular granular surface and effuse edge (× 8).

yellow (Thal and Chen 1955). On Clauberg's medium containing about 0.06 per cent of potassium tellurite the pseudotubercle bacillus forms small greyish-black colonies reaching up to 2 mm in diameter in 48 hr. Apart from some strains of *Proteus* spp., no other gram-negative bacillus, including the plague bacillus, is able to grow under these conditions (Brzin 1963). On media such as deoxycholate or SS agar, *Y. enterocolitica* colonies grow slowly at 37° or 22°, often being little more than pinpoint in size after 24 hr.

Numerous colonial variants of *Y. pestis* have been described—smooth dewdrop, smooth low convex, large flat, sunflower, and other forms (Gotschlich 1912, Pirie 1929, Bessanowa and Lenskaja 1931, Bhatnagar 1940*a*, Eisler *et al.* 1958). On the whole, smooth forms tend to be capsulated and virulent, irregular and rough forms to be non-capsulated and avirulent; but variants are greatly influenced by environmental factors and are often unstable. According to Levine and Garber (1950) colonies of smooth forms after 4 days at 37° on tryptose agar containing 0.005 per cent tetrazolium chloride are circular with a sharply defined carmine centre; colonies of rough forms are irregular and diffuse pink. Smooth and rough colonies of *Y. pseudotuberculosis* have also been described (Kakehi 1915-16, Zlatogoroff and Moghilewskaja 1928*a, b*).

With the exception of *Y. enterocolitica*, yersiniae produce little turbidity in broth, but a deposit of fine or coarse flocculi is formed on the walls and at the bottom of the tube. Growth continues for several days and at room temperature for several weeks; there is then a surface pellicle or ring growth accompanied by a heavy deposit difficult to disintegrate. In broth covered with melted butter or oil and left undisturbed, growth occurs in the form of stalactites depending from the under-surface of the droplets.

In a gelatin stab culture there is a surface layer and a filiform growth extending to the bottom, but no liquefaction. Potato is not a suitable medium, *Y. pestis* grows hardly at all, and *Y. pseudotuberculosis* either not at all or as a thin yellow layer later turning brown (Preisz 1894). Like *Y. enterocolitica*, both organisms grow slightly but definitely in MacConkey's bile salt medium; the growth on MacConkey's agar disappears in 2 or 3 days, owing presumably to autolysis of the bacilli. Many strains of *Y. enterocolitica* when grown at 20° produce adhesins that cause the agglutination of erythrocytes and are associated with the presence of fimbriae (MacLagan and Old 1980).

Metabolism and growth requirements

The bacilli have a wide range of growth. Both *Y. pestis* and *Y. pseudotuberculosis* can multiply to some extent at very low temperatures—according to Tumansky and his colleagues (1935) even at 0°. Their upper limit of growth is 43°, their optimum about 27-30°. Growth occurs best between pH 6.3 and 7.3 (Pirt *et al.* 1961). *Y. enterocolitica* will multiply at 4°.

Small numbers of plague bacilli inoculated on to the surface of an agar plate usually fail to develop unless blood, a peptic digest of blood, haemin, or 0.025 per cent sodium sulphite is added to the medium, or the culture is incubated anaerobically (Schütze and Hassanein 1929, Wright 1934). Herbert (1949), who confirmed and amplified these findings, suggests that H_2O_2 is produced aerobically and that haemin acts by promoting the synthesis of catalase. Its demand for haematin places *Y. pestis* between *Pasteurella* and *Haemophilus*.

Observers differ over the nutritional requirements of the plague bacillus. Rao (1939), who probably used a large inoculum, found that only three amino acids—proline, phenylalanine and cystine—were required in a defined medium. Herbert (1949), on the other hand, obtained anaerobic growth in a defined medium containing glucose, salts, and no fewer than 20 amino acids; and Burrows and Gillett (1966) found that, when incubated aerobically in the presence of 5 per cent CO_2, the plague bacillus grew on media containing cystine, methionine, phenylalanine, glycine, valine, isoleucine, glutamic acid and thiamine. There is general agreement, however, with the conclusion of Hills and Spurr (1952) that the organism is more exacting at 37° than at 28°. This applies also to the pseudotubercle bacillus, which requires either no growth factors or only thiamine or pantothenate at 28°, but glutamic acid, thiamine, cystine and pantothenate at 37° (Burrows and Gillett 1966). Similarly at 37° *Y. enterocolitica* requires thiamine and either cystine or methionine. The addition of iron in the form of ferrous sulphate to a serum-containing medium increases the growth of yersiniae (Jackson and Morris 1961), whereas saturated and unsaturated fatty acids have the opposite effect (Eisler and von Metz 1968).

No haemolysin is produced, but on horse blood agar the whole plate is slightly cleared and browned. Old broth cultures of the plague bacillus are very toxic to animals.

Biochemical reactions

Y. pestis, *Y. pseudotuberculosis*, and typical strains of *Y. enterocolitica* produce acid by the fermentative method from glucose, mannitol, and arabinose, and usually from maltose, xylose, and trehalose. Gas is not produced except by *Y. enterocolitica*, which is said to produce a small amount at 22° (Niléhn 1967). Acid is not produced from raffinose, dulcitol, erythritol, or inositol, and usually not from lactose. Fermentation is sometimes more active at 28° than at 37°. Yersiniae produce catalase in abundance, unlike pasteurellae, which produce only a small amount (Burrows *et al.* 1964). Nitrate is usually reduced, and the MR and β-galactosidase tests are positive. Negative results are given in tests for oxidase, citrate utilization, H_2S production, phenylalanine deaminase, lysine decarboxylase, arginine dihydrolase, and gelatin hydrolysis.

Characters that distinguish *Y. pestis*, *Y. pseudotuberculosis*, and typical strains of *Y. enterocolitica* are given in Table 38.1. *Y. enterocolitica* differs from the other two members of the group mainly in fermenting sucrose, sorbitol, and cellobiose, but not salicin, and in giving a positive ornithine decarboxylase reaction, and a positive VP reaction at 22° (Niléhn 1967). Devignat (1954) considered that differences in their sugar reactions were not constant enough to distinguish between *Y. pestis* and *Y. pseudotuberculosis*; the hydrolysis of urea (Fauconnier 1950), reduction of methylene blue (Savino *et al.* 1939), and fermentation of melibiose and rhamnose by the pseudotubercle bacillus are of help.

Devignat (1951, 1953) divides *Y. pestis* into three varieties on the basis of fermentation of glycerol and the reduction of nitrate to nitrite.

	Fermentation of glycerol	Reduction of nitrate
Y. pestis var. *orientalis*	−	+
Y. pestis var. *antiqua*	+	+
Y. pestis var. *mediaevalis*	+	−

This division is of interest in relation to the distribution of the three variants in different countries and in different rodents (see Chapter 55).

The literature contains many references to atypical strains of *Y. enterocolitica*. Niléhn (1969) and Wauters (1970) each described five biotypes of the species; Wauters (1973) found a relationship between lecithinase activity and biotype. Knapp and Thal (1973) excluded certain strains from the species on the grounds of their reactions in tests for the production of indole, and of acid in xylose, aesculin, and salicin. Stevens and Mair (1973) suggested the exclusion of certain leporine strains that grew in the presence of 4 per cent sodium chloride. On the basis of DNA-relatedness and other characters, Brenner and co-workers (1980) recognized the five classical biotypes of *Y. enterocolitica*, of which one (type 5) contained metabolically inactive strains isolated mainly from hares. Other atypical strains were placed in three separate species: *Y. intermedia* and *Y. frederikseni* attacked rhamnose and *Y. kristenseni* failed to attack sucrose. *Y. intermedia* differed from *Y. frederikseni* by giving

Table 38.1 Some differential characters of *Yersinia* spp.

Test	*Y. pestis*	*Y. pseudotuberculosis*	*Y. enterocolitica*
Motility in 18-hr broth cultures at 25°	−	+	+
Acid in sugars	Salicin, aesculin, usually dextrin	Salicin, rhamnose, melibiose, aesculin, usually adonitol	Sucrose, sorbitol, cellobiose, usually dextrin
Litmus milk	Nil or slight acid	Alkaline	Nil
Indole	−	−	+ or −
VP	−	−	+ (at 22°)
Urease	−	+	+
Ornithine decarboxylase	−	−	+
Lysis by specific plague phage	+	−	−
Pathogenicity for mice	+ (rapid)	+ (slow)	±*

* See Highsmith and her colleagues (1977).

positive reactions with melibiose, α-methyl-D-glucoside and raffinose. *Y. frederikseni* itself contained three DNA-relatedness groups (Ursing *et al.* 1980). With a set of 24 bacteriophages from raw sewage Baker and Farmer (1982) divided over 300 strains belonging to the four species into 106 phage types, only 9 per cent of which occurred in more than one of the species. Further properties of these species and of other yersiniae are given by Bercovier and co-workers (1980). However, numerical analysis does not always bear out genetically defined relationships (Harvey and Pickett 1980, Kapperud *et al.* 1981). Rhamnose-positive strains are derived mainly from the external environment, particularly from water (see Bottone 1977). (For further information see Darland *et al.* 1975, Falcão *et al.* 1978.)

Resistance

None of the members is specially resistant. In broth they are killed by heat at 55° and by 0.5 per cent phenol, within 15 minutes. In dried bubo juice, or dried on threads and kept at room temperature in a desiccator, plague bacilli live for not more than a few days (Kitasato 1894); but in dry flea faeces they may survive for 5 weeks (Eskey and Haas 1940). In the infected guinea-pig spleen kept in pure neutral glycerol at −15° they retain their virulence for years (Francis 1932); and even in corked tubes of nutrient agar they may remain viable, though not necessarily of full virulence, up to 25 years or so (Francis 1949). *Y. pestis* cannot grow in the presence of 0.4 per cent selenite, but about half the strains of *Y. enterocolitica* and most strains of *Y. pseudotuberculosis* can do so (Lapage and Bascomb 1968). All members are susceptible *in vitro* to sulphadiazine, streptomycin, tetracyclines, and chloramphenicol. *Y. pestis* is moderately susceptible *in vitro* to penicillin. Unlike *Y. enterocolitica*, *Y. pseudotuberculosis* is usually susceptible to penicillin and cephalothin.

Antigenic structure

The antigenic structure of members of this group is complex and has not yet been fully worked out. Schütze (1932*a*) first showed that in the plague bacillus there were two main antigens, one corresponding to the envelope and the other to the somatic substance. The envelope antigen is best developed at 37° and is heat-labile: the somatic antigen is formed both at 20° and at 37° and is heat-stable.

Both antigens may be present in avirulent as well as in virulent forms. In fact Schütze (1939) was unable to distinguish serologically between virulent and avirulent or rough and smooth forms. Baker and his colleagues (1947) found the envelope substance to contain two water-soluble fractions—a carbohydrate protein and a carbohydrate-free protein. Further work (see Meyer, Foster *et al.* 1948) showed that the second fraction was immunogenic in rats, mice, and monkeys but not in guinea-pigs. By Ouchterlony's agar diffusion precipitin technique Crumpton and Davies (1956) obtained evidence of at least ten different antigens in the plague bacillus. Three of these—antigens 3, 4 and 5—had their maximal production at 37°. Antigen 3—an envelope antigen—was protective for mice and rats and was absent from *Y. pseudotuberculosis*; antigens 4 and 5 were shared with this organism. Absence of antigen 4 appeared to be associated with the formation of rough colonies. Antigen 8 corresponded to the toxin. The specificity of another antigen was determined by a polysaccharide that was unique in being composed largely of an aldoheptose sugar not previously found in nature. Burrows and Bacon (1956, 1958) brought evidence to show that all virulent strains contained, besides the capsular antigen F1, two other antigens referred to as V and W. Some avirulent strains also contained V and W antigens, but for full virulence the strain had to possess the F1 capsular antigen, the V and W antigens, the ability to form densely pigmented colonies on defined media containing haemin, independence of exogenously supplied purines for growth, and resistance to phagocytosis in the absence of visible capsulation. The conditions, however, determining both virulence and immunogenicity vary according to the species of animal selected for test. Extracts of the tissues of

Table 38.2 Antigens of *Yersinia pestis*

Antigens			
Fraction 1 F1		1A water-soluble CHO protein	Part of envelope antigen, formed at 37°. Low toxicity; immunogenic for mice but less so for guinea-pigs. Serologically indistinguishable from each other. Species-specific
		1B water-soluble CHO-free protein	(as above, 1A and 1B bracketed together)
		1C water-soluble	Toxic to mice
		Water-insoluble	Formed at 37°. Non-toxic and non-immunogenic for mice, but immunogenic for guinea-pigs
Antigen	1	Virulence antigen, specific to *Y. pestis*. Immunogenic for mice and guinea-pigs. Contained in the envelope substance	
,,	3	Formed at 37°. Corresponds to F1; immunogenic for mice. Species-specific	
,,	4	Formed at 37° below pH 6.9. Heat-stable protein, whose absence is associated with roughness of colony. Not immunogenic. Shared with virulent *Y. pseudotuberculosis*	
,,	5	Formed at 37° aerobically but at 28° only in presence of excess oxygen. Shared with *Y. pseudotuberculosis*	
,,	8	Toxin	
,,	—	Polysaccharide containing an aldoheptose sugar	
,,	V	Not present in envelope. Formed best at 37°. Associated with virulence (V and W bracketed together)	
,,	W	Shared with freshly isolated strains of *Y. pseudotuberculosis* (V and W bracketed together)	
,,		Other antigens common to *Y. pestis* and *Y. pseudotuberculosis*	
,,		Rough O. Polysaccharide, heat-stable, shared with *Y. pseudotuberculosis*.	

infected guinea-pigs obtained by ultrasonic vibration were found to have aggressive properties, enhancing the virulence of plague bacilli for guniea-pigs and inhibiting phagocytosis; four precipitinogens—V, W, X and Y—were detected (H. Smith *et al.* 1960). By alkali treatment of the disintegrated organisms an almost non-toxic residue was obtained, apparently consisting of the cell-wall-envelope complex, that was capable of immunizing guinea-pigs and to a less extent mice (Keppie *et al.* 1960).

The production of antigen 4 is dependent on pH; it is completely suppressed above pH 6.9 and has its optimum at pH 5.9. At 28° antigen 5 is formed only in the presence of excess oxygen, but at 37° oxygen pressure is unimportant (Pirt *et al.* 1961). Later Burrows (1968) described a U antigen present in most strains of *Y. pestis* and *Y. pseudotuberculosis* but absent from *Y. enterocolitica*, and also a T antigen; neither of these antigens was necessary for virulence. According to Chen and Meyer (1966) sixteen antigens have so far been distinguished in the plague bacillus, of which thirteen are shared with *Y. pseudotuberculosis*. Included in these is the heat-stable polysaccharide rough somatic antigen (see Table 38.2). A *haemagglutinating* polysaccharide fraction acting on sheep cells was described by Fukumi and co-workers (1954). Gemski and his colleagues (1980*a*, *b*) showed that plasmids may be associated with the production of V and W antigens by *Y. pseudotuberculosis* and *Y. enterocolitica*.

(For review of the antigens of the plague bacillus, see Burrows 1963.)

Y. pseudotuberculosis contains heat-labile flagellar antigens formed at 18 to 26° and heat-stable somatic antigens formed at both 26 and 37° (Arkwright 1927).

Schütze (1928, 1932*b*) defined by agglutination 4 groups based on somatic antigens, all possessing the same flagellar antigens. The somatic antigen of his group II was related to serotypes in salmonella group B (see also Knapp 1955, 1960). Further observations were made by Bhatnagar (1940*b*). Our present concept of the antigenic structure of the pseudotubercle bacillus rests on the more recent observations of Thal (1956, 1966), Davies (1958), Crumpton, Davies and Hutchinson (1958), Itagaki and his colleagues (1968), and Thal and Knapp (1971), who, working mainly with precipitin, haemagglutinin, or haemagglutination-inhibition tests, defined a number of smooth O and H antigens by which they were able to classify the strains into six O groups (Table 38.3). Each of the groups contains a specific polysaccharide.

Y. pestis and *Y. pseudotuberculosis* share many antigens, including the R antigen, which appears to be present in all strains, smooth or rough, virulent or avirulent, toxic or non-toxic (Davies 1958). Whether the pseudotubercle bacillus contains any entirely specific antigens is a little uncertain. Ransom (1956), for example, could not find any, but Crumpton and Davies (1957) speak of a specific smooth O antigen differing from antigen 4; and Winter and Moody (1959) were able by injection of rabbits to prepare agglutinating sera to both organisms that were strictly specific. Antigenic relationships between *Y. pseudotuberculosis* and other members of the Enterobacteriaceae have frequently been noted (see Mair and Fox 1973).

In *Y. enterocolitica* Winblad (1968) identified 8 different O antigenic factors. These were later extended to 34, and together with 19 H factors served to divide the species into a large number of serotypes (Wauters *et al.* 1972). Some of the newer serotypes possess biochemical properties that differ from those of well recognized ones (Knapp and Thal 1973, Winblad 1973). The V and W antigens in biotype 2, serotype O:8 are said to be identical with those in *Y. pestis* and *Y. pseudotuberculosis* (Carter *et al.* 1980). The O:9 serotype contains a lipopolysaccharide factor that is responsible for cross-reactions with *Brucella abortus* and *Vibrio cholerae* (Sandulache and Marx 1978); and the factor in serotype O:12 for cross-reactions with the O47 antigen of *Salmonella* (see Ahvonen 1972).

Bacteriophages

A specific bacteriophage acting on all strains of *Y. pestis* but on none of *Y. pseudotuberculosis* was described by Gunnison and co-workers (1951), and used

Table 38.3 Antigenic structure of *Yersinia pseudotuberculosis* (Thal and Knapp 1971)

O group	O subgroup	R antigen	Thermostable O antigens	Thermolabile H antigens
I	A	(1)	2, 3	a, c
	B	(1)	2, 4	a, c
II	A	(1)	5, 6	a, d
	B	(1)	5, 7	a, d
III	—	(1)	8	a
IV	A	(1)	9, 11	b or a, b
	B	(1)	9, 12	a, b, d
V	A	(1)	10, 14	a; a, e, (b)
	B		10, 15	a
VI	—	(1)	?13	a

Note. By the indirect haemagglutination test a heat-stable somatic antigen common to all types can be demonstrated (Bader 1972).

for purposes of identification by Cavanaugh and Quan (1953) and Chadwick (1963). Numerous phages have been isolated from *Y. enterocolitica* acting on different strains of this organism but not on strains of *Y. pestis* or *Y. pseudotuberculosis* (Niléhn and Ericson 1969). The phage type (Nicolle *et al.* 1973, Niléhn 1973) of *Y. enterocolitica* strains is partly related to their origin and to their serological and biochemical characters.

Virulence, toxicity and immunogenicity

All fully virulent strains of *Y. pestis* are *toxic* (Burrows 1963), and so, according to Ajl and his colleagues (1955), are avirulent strains. The mouse is about 250 times as susceptible as the guinea-pig. According to H. Smith (1968) the toxin affecting the guinea-pig is qualitatively different from that affecting the mouse (see below), though no evidence for this had previously been found (Keppie *et al.* 1957, Cocking *et al.* 1960).

The soluble so-called *murine toxin* can be liberated from the bacilli in various ways, such as by autolysis, by lysis with bile salts, or by ultrasonic disintegration. From acetone-dried organisms it can be extracted and partly purified by chemical and electrophoretic treatment. The final product has an LD50 for mice injected intravenously of about 0.2 μg (Ajl *et al.* 1955, 1958, Spivack and Karler 1958). The toxin, whose formation is affected by temperature of incubation and other factors (Goodner 1955) and is quite distinct from the envelope antigen, partakes of the nature of both an endotoxin and an exotoxin. It differs from the endotoxins of the Enterobacteriaceae in not being a lipopolysaccharide, and from an exotoxin in its predominantly cellular location, its failure to give rise to a true antitoxin, and its insusceptibility to penicillin, which inactivates diphtheria and staphylococcal exotoxins (Ramon *et al.* 1947). It is associated with the V W antigens, and possibly with two other antigens, X and Y, produced during growth *in vivo* in the guinea-pig, and found by H. Smith and his colleagues (1960) to have an aggressive action, enhancing virulence and resistance to phagocytosis in plague cultures. Chemically, this toxin is a protein with no detectable carbohydrate. It is partly inactivated by heating to 65° for 1 hr, and completely destroyed at 80°. Formalin—0.2 per cent for 24 hr at 37°—removes its toxicity without affecting its antigenicity. Biologically, it acts mainly on the peripheral vascular system and on the liver leading to haemoconcentration and fatal shock. It differs from the *guinea-pig toxin*, which has two synergically active protein components unassociated with the murine toxin (H. Smith 1968).

Besides the murine exotoxin, Walker (1967) recognized as an *endotoxin* the toxic factor extracted by Davies (1956) with 45 per cent phenol at 60°. This substance causes fever in rabbits and is lethal to mice in 5–15 mg quantities. After mild hydrolysis the polysaccharide portion yields an unusual 7-carbon sugar, L-glycero-D-mannoheptose. The toxin also contains a phospholipid. The part played by the murine exotoxin-like toxin and Davies's endotoxin in the pathogenesis of plague is doubtful. (For further reference to this toxin see Burrows 1963, Walker 1967.)

Besides these toxins, a substance described as a *pesticine* is formed by the plague bacillus under certain cultural conditions (Ben-Gurion and Hertman 1958). A second pesticine, produced by both *Y. pestis* and *Y. pseudotuberculosis*, was described by Brubaker and Surgalla (1962). Pesticine 1 inhibits the growth of *Y. pseudotuberculosis* (Beesley and Surgalla 1970).

The virulence of plague bacilli for rats, mice, guinea-pigs and other animals depends on several different factors.

Jackson and Burrows (1956) specified six properties of virulent strains: elaboration of the envelope antigen; high toxigenicity; resistance to phagocytosis in the absence of visible capsulation; formation of V and W antigens; ability to synthesize purines; and the production of pigmented colonies on a specially defined medium containing haemin. Later, Burrows and Bacon (1958) found that the first requirement, namely formation of the F1 antigen present in the envelope fraction, was not essential for full virulence to mice, though it was for guinea-pigs. On the other hand, the V W factors were essential; without them the strain was avirulent for both species of animal. The ability to resist phagocytosis, or for a strain to develop the property of resistance, is manifested only at 37°; it may be due either to the antigenic fraction F1 (H. Smith *et al.* 1960), or to the V W antigens (Burrows and Bacon 1956). Virulent strains have a lethal dose for mice of 5–100 cells, as opposed to the 150 million cells of avirulent strains (Burrows 1955). They contain an antigen, detectable by the agar diffusion method, that is absent from avirulent strains (Burrows 1956). Atypical strains of lowered virulence were isolated from rodents, fleas, and man in Indonesia by Williams and his colleagues (1978), who thought that they might be responsible for mild plague—pestis minor—in man. Laird and Cavanaugh (1980) found a relationship between autoagglutination and virulence in *Y. pestis*, *Y. pseudotuberculosis* and *Y. enterocolitica*.

Schütze (1939) found that both smooth and rough, virulent and avirulent strains grown at 37° conferred *protection* on rats; and Bhatnagar (1940*a*) found that avirulent strains were protective only when they formed the envelope antigen. Killed virulent strains are immunogenic, but killed avirulent strains are less effective than living avirulent strains, presumably because the living organisms multiply to some extent in the tissues (Burrows 1963). Killed vaccines to be fully effective for the guinea-pig require an oily or mineral adjuvant (Spivack *et al.* 1958).

Though both the F1 fraction and the V W antigens are immunogenic for mice, no extracted product has yet proved superior to whole organisms. Keppie and his colleagues (1960), however, prepared an effective comparatively non-toxic vaccine by alkali treatment of organisms disintegrated by ultrasonic vibration; it consisted of a complex of envelope and cell wall material. The V W antigens do not appear to contribute much to the immunization of guinea-pigs, but the F1 antigen is of importance. Thal (1956) was able to immunize guinea-pigs against infection with virulent plague bacilli by means of a living avirulent vaccine of *Y. pseudotuberculosis*, though the plague bacillus failed to immunize guinea-pigs against virulent pseudotubercle bacilli (Thal *et al.* 1967). This difference can hardly be explained by the presence or absence of antigen 4 (see Table 38.2). Vaccines for man are dealt with in Chapter 55. (For further references to immunogenicity, see Crumpton and Davies 1957, Burrows and Bacon 1958, and Burrows 1963.)

Phage lysates of some strains of *Y. pseudotuberculosis* were found by Lazarus and Nozawa (1948) to be toxic to mice. The toxin was inactivated by heating to 60° for 30 min; it was thus more heat-labile than the murine toxin of the plague bacillus. The toxicity appears to be associated with the polysaccharide present in each of the serological types of the pseudotubercle bacillus (Davies 1958). Most toxigenic strains belong to group III. They differ from non-toxigenic strains in producing experimental infections that are fatal within 2 days instead of 2 weeks; toxoid but not living avirulent vaccine strains will usually protect against toxigenic strains (Thal 1973). Polysaccharides extracted from rough strains differ from the smooth polysaccharides and are related to the rough polysaccharide of *Y. pestis*. When cultured at 30° or less, certain strains of *Y. enterocolitica* produced a heat-stable enterotoxin demonstrable in infant mice (Pai and Mors 1978, Okamoto *et al.* 1980). Tissue invasiveness in certain human pathogenic strains of *Y. enterocolitica* appears to be plasmid-controlled (Zink *et al.* 1980).

Pathogenicity

Y. pestis and *Y. pseudotuberculosis* both cause natural disease in rodents, and are able to attack man; *Y. pseudotuberculosis* also produces disease in a wide range of mammals and birds. The virulence of the plague bacillus is subject to variation *in vitro* depending on such environmental factors as nature of the medium, temperature of incubation, pH, and partial pressure of oxygen. There is reason to believe that changes may also occur *in vivo*; strains, for example, from chronic lesions in rats, may be of lowered virulence and unable to give rise to the characteristic picture of plague till after a few passages through normal animals (see Chapter 55). *Y. enterocolitica* occurs in man and in numerous animal species, especially chinchillas, hares, pigs, and dogs; its presence is sometimes but not always associated with disease.

Experimental reproduction of plague in animals

Bubonic plague can be reproduced in rodents and monkeys by experimental inoculation. Dogs, cats, pigs, cattle, sheep, goats and horses are difficult to infect; birds, with the exception of sparrows, are completely resistant. The disease is said to occur naturally in camels; but these animals are refractory to experimental inoculation (Zabolotny 1923). Even the rodents show great variation in susceptibility to infection. The less resistant members succumb rapidly, whereas the more resistant ones either fail to develop the disease, or else develop a subacute or chronic type. Spencer (1921) found in America that about 30 per cent of the rats from a plague-free district were resistant to subcutaneous inoculation of *Y. pestis*. According to Sokhey (1939), the white mouse is the most susceptible laboratory animal, but Otten (1938) on the contrary maintains that the guinea-pig is more susceptible than the mouse.

Strong and his colleagues (1912) were able to produce pneumonic plague in monkeys by making them inhale sprayed cultures of *Y. pestis*. Guinea-pigs also developed pneumonic plague under these conditions, but as a rule only secondarily to a septicaemia. The same appears to hold true of rats and marmots (*Spermophilus citellus*) (Eberson and Wu Lien Teh 1917, Wu Lien Teh and Eberson 1917). Pneumonic plague can be produced in mice by intranasal instillation of highly virulent bacilli (Meyer, Quan and Larson 1948) or by exposure to aerosols (P. N. Smith 1959).

Rats *Subcutaneous injection* of a very small number of virulent plague bacilli leads to death in 2 to 8 days. *Post mortem*, there is necrosis and oedema at the site of inoculation; the regional lymph glands are swollen and surrounded by a haemorrhagic infiltration of the subcutaneous tissue. Glands in other parts of the body are often congested and swollen: the spleen may be enlarged and dark red; the liver and lungs are hyperaemic, and sometimes a pleural exudate is seen. Bacilli are found in large numbers in the local lesion and the nearest glands; they are usually irregular in size and shape, exhibit bipolar staining, and may show involution forms. They are often present in the spleen and in the blood. If the animal lives for a week, small, irregular, necrotic foci may be observed in the spleen and liver.

The German Plague Commission (Report 1899) had apparently no difficulty in infecting rats by the mouth. They fed the animals with a drop of plague culture or with the cadaver of a rat that had died of plague. Death followed uniformly in 2 to 3 days. *Post mortem*, three types of lesion were found. (*a*) Most commonly enlargement and congestion of the submaxillary and suprahyoid glands, with the general picture of septicaemia. (*b*) Less often primary infection of the stomach and intestine, numerous punctiform haemorrhages round the pylorus, swelling and haemorrhagic infiltration of the lymph follicles and mesenteric glands, which were often quite large, and contained innumerable plague bacilli. (*c*) Quite frequently an aspiration pneumonia; the lungs showed inflammatory foci of varying size, containing large numbers of plague bacilli; the spleen was enlarged, and the liver hyperaemic.

The English Plague Commission had less success with feeding. Of the wild rats of Bombay only 38 per cent were found to be susceptible when fed on the carcasses of dead plague rats. In the Punjab the proportion was nearly 70 per cent. The lesions found in rats infected by feeding were of a type similar to those in rats infected naturally. Two striking differences, however, were present. (*a*) In naturally infected plague rats the bubo is in the neck; a mesenteric bubo was not encountered in 5000 examinations: in artificially fed rats the bubo is generally in the mesentery. (*b*) In naturally infected rats the stomach and intestines show no obvious pathological changes: in rats infected by feeding numerous lesions are found in the intestines—haemorrhages in the stomach wall 3 per cent, congestion of intestines 27 per cent, enlargement of Peyer's patches 31 per cent (Report 1906).

The *nasal mucosa* and *conjunctiva* are favourable sites for inoculation in rats (Report 1899). A trace of infective material smeared over the conjunctiva proves fatal in 3 to 4 days. *Post mortem*, there is a swelling of the cervical lymph glands, enlargement of the spleen, and frequently numerous petechiae in the stomach and jejunum. The appearances are in fact similar to those of an animal dying after oral infection. Contamination of the nasal mucosa is frequently followed by an inhalation pneumonia.

The English Plague Commission (Report 1912, p. 287) were able to reproduce the lesions of chronic or—as they preferred to call it—of resolving plague in rats by inoculating large numbers of animals with small doses, and examining the survivors 3 weeks after inoculation. The chief lesions found were chronic buboes, necrotic areas in the spleen, and chronic abscesses in the spleen or more rarely the liver.

Mice The mouse reacts to inoculation in much the same way as the rat, but is said to be more susceptible (Sokhey 1939). After subcutaneous inoculation the septicaemia is very striking; the blood and internal organs swarm with bacilli. Death occurs in 3 days. Infection can be accomplished by feeding, if a sufficiently large dose of a virulent culture is used.

Guinea-pigs Guinea-pigs are highly susceptible to plague, dying in 2 to 5 days after *subcutaneous* injection of a pure culture. *Post mortem*, there is a necrotic focus at the site of injection surrounded by intense congestion and oedema; the regional lymph glands are swollen and embedded in a bloody oedema; their interior is soft and necrotic. There is enlargement and congestion of the spleen, which is often studded with miliary, soft, grey nodules up to 1 mm in diameter, sometimes projecting above its surface, and containing large numbers of bacilli. The suprarenals may be congested (Yersin 1894). The liver may be peppered with tiny necrotic foci, and occasional small necrotic nodules are visible in the lungs. Sometimes there is a pleural effusion. Guinea-pigs can be infected by the *cutaneous* route. If the plague material is rubbed on the shaven skin of the abdomen, an inflammatory reaction appears in the neighbourhood, marked by a slight reddening and the formation of umbilicated pustules in which plague bacilli are present (Dieudonné and Otto 1912). After a few days the regional glands swell and death occurs in 4 to 5 days. The post-mortem signs are similar to those after subcutaneous inoculation. Even a single living bacillus injected into the skin may prove fatal (Strong *et al.* 1912).

Intraperitoneal injection is fatal in 24 to 36 hr. *Post mortem*, there is a rich fibrinous exudate, containing enormous quantities of plague bacilli. Infection by the mouth, nasal mucosa, and vaginal mucosa is not constant. The animals are very sensitive to *conjunctival* infection. By *inhalation* or *intratracheal* inoculation it is possible to set up an acute primary pneumonia. Symptoms appear in 48 hr, and death occurs in about 72 hr. *Post mortem* there is a confluent bronchopneumonia with oedema or commencing necrosis of the lung tissue (Bessonowa *et al.* 1927, Bablet and Girard 1933). However infection occurs, the organisms sooner or later gain access to the blood stream. At post mortem they can be isolated from the blood, spleen, liver, lung and bone marrow. After bacteraemia has developed, the organisms may often be found in the bile, urine, and less frequently in other excretions (Sémikoz *et al.* 1927).

Rabbits are less susceptible to plague than rats and guinea-pigs, but they can generally be infected by subcutaneous inoculation (Dieudonné and Otto 1912).

Monkeys vary in susceptibility. The German Commission (Report 1899) infected *Macacus radiatus* by subcutaneous and intraperitoneal inoculation, and by feeding. This species was not nearly so suceptible, however, as *Presbytes entellus* (*Semnopithecus entellus*), which succumbed after minute quantities of plague culture subcutaneously.

Experimental reproduction of pseudotuberculosis in animals

The disease caused by *Y. pseudotuberculosis* can be reproduced in guinea-pigs, rabbits, rats, mice and, according to Pallaske (1933) to whom reference should be made for a detailed account of the lesions produced, in cats, pigeons, canaries and turkeys. For laboratory purposes the guinea-pig is the most suitable animal to study.

Subcutaneous or *intramuscular* inoculation of the guinea-pig is followed by a disease which, depending on the dose and the virulence of the strain, may be acute, subacute, or chronic. The acute disease resembles plague and is fatal in a few days. The differential diagnosis can be made only by cultivation and a thorough study of the organism responsible. Macroscopically, the local lymphatic glands tend to be more affected in plague than in pseudotuberculosis. The subacute disease proves fatal in about 2 weeks, and the chronic in 3 weeks or longer. *Post mortem*, there is a caseous local swelling, the regional lymphatic glands are enlarged, and there are nodules varying in number, size and degree of caseation, in the spleen, liver and lungs. If the animal lives for 3 weeks or so, the nodules are usually very conspicuous. They are more or less spherical, greyish-white in colour and 0.2 to 3.0 mm in diameter; they project above the surface of the organ, and in the liver they show no particular localization at the free border, as do the necrotic areas in rodent typhoid infection. Microscopically the bacilli are generally present in considerable numbers at the site of inoculation and in the regional glands, and they can be readily cultivated from all the lesions. The disease can also be reproduced by *feeding*, which is the natural method of infection. Death occurs in 1 to 3 weeks. Nodules of varying size are found in the intestinal wall, the mesenteric glands are enlarged and often caseous, and nodules may be present in the spleen, liver, and lungs.

Y. enterocolitica, though it gives rise to a disease resembling pseudotuberculosis in animals and to enteritis and mesenteric lymphadenitis in man, was formerly thought to have little or no pathogenicity for laboratory animals (Knapp and Thal 1963).

Further studies (P.B. Carter 1975, see also Highsmith *et al.* 1977) showed that mice and other laboratory animals were susceptible; the results of pathogenicity experiments appeared to depend on virulence, incubation temperature, and route of inoculation. Some strains, particularly those of serotype 8, will produce keratoconjunctivitis in guinea-pigs (Sereny test) and lethal infections in Mongolian gerbils; and many serotype 3 strains that are enterotoxic and invasive for Hela cells will produce diarrhoea in rabbits if given in large doses by stomach tube (Mors and Pai 1980, Pai *et al.* 1980, Schiemann and Devenish 1980).

Classification

The genus *Yersinia* was created as a result of the division of the genus *Pasteurella*; it contains the three species *Y. pestis*, *Y. pseudotuberculosis*, and *Y. enterocolitica* (Buchanan and Gibbons 1974). The name *Y. philomiragia* has been suggested (Jensen *et al.* 1969) for an organism isolated from the lungs and liver of a dead musk rat in Utah; but data based on DNA-relatedness and phenotype (Ursing *et al.* 1980) indicated that the organism did not belong to the genus *Yersinia*. The name *Y. ruckeri* has been suggested (Ewing *et al.* 1978) for an organism causing 'redmouth' in rainbow trout and certain other fish.

There is good reason for classifying the plague bacillus and the pseudotubercle bacillus in the same genus. The two organisms resemble each other closely; the main differences between them are listed in Table 38.1. In addition to these should be mentioned the greater rapidity and luxuriance of growth of *Y. pseudotuberculosis* on plain nutrient agar and on deoxycholate citrate agar (Thal and Chen 1955), and the greater pathogenicity of *Y. pestis* for white rats (Report 1912, p. 350).

Y. enterocolitica has an antigenic structure separate from that of *Y. pestis* and *Y. pseudotuberculosis*. Its taxonomic position is further complicated by the antigenic and biochemical heterogeneity of many of the less well recognized types often included in the species (see Knapp and Thal 1973, Stevens and Mair 1973, Wauters 1973, Brenner *et al.* 1976, Bottone 1977). According to Baker and Farmer (1982) *Y. enterocolitica* can be split by bacteriophage typing into four separate species: *Y. enterocolitica*, *Y. kristenseni*, *Y. intermedia*, and *Y. frederikseni*.

Yersinia pestis

Isolation.—Independently by Kitasato (1894) and by Yersin (1894).
Habitat.—Parasite of rodents and man.
Morphology.—Small, straight, ovoid bacillus, 1.5 μm × 0.7 μm, with rounded ends and convex sides; arranged singly, in short chains, or in small groups. Shows high degree of pleomorphism, especially in buboes and on 3 per cent salt agar; there is every degree of variation in depth of staining, and clubs, shadow forms, snake-like filaments, coccoid forms, yeast-like forms, and numerous others may be seen. Non-motile. Non-sporing. In the animal body and in cultures on serum agar at 37° a true capsule may be formed. Shows bipolar staining, is gram negative, and non-acid-fast.
Agar plate.—24 hr at 37°. Very small, 0.1–0.2 mm in diameter, round, glistening, transparent, colourless, finely granular, umbonate colonies, with smooth or finely granular surface and an entire or delicately notched edge; differentiated into a raised centre and a flat periphery.

5 days at 37°. Larger, up to 4 mm in diameter, with a raised, sometimes ringed, nearly opaque, greyish-yellow centre and a flat or shelving, finely granular, translucent, greyish-white periphery; consistency is butyrous or viscous, emulsifiability easy; sometimes a secondary ring of growth is seen. Variant colonies occur.
Deep agar shake.—5 days at 37°. Maximum growth at the surface; numerous round, transparent, colourless, punctiform colonies, visible with a hand lens, scattered throughout the medium.
Agar stroke.—24 hr at 37°. Poor, slightly raised, translucent, greyish-yellow, glistening growth, with a wavy or frosted-glass surface, and an irregularly lobate edge. Growth increases very little with subsequent incubation.
Gelatin stab.—7 days at 22°. Good, filiform growth, confluent at top, discrete below, extending to bottom of tube, and sometimes sending out little feathery projections into medium. Surface growth is raised, 5 mm in diameter, with a slightly lobate edge. No liquefaction.
Broth.—24 hr at 37°. Moderate growth; little or no turbidity; a floccular or powdery deposit, not disintegrating completely on shaking. Later the flaky deposit increases and may crawl up the sides of the tube; a delicate surface pellicle often forms. If butter or oil is floated on the medium, stalactites grow down from the under-surface of the droplets.
Loeffler's serum.—24 hr at 37°. Fairly good, confluent growth, better than that on agar.
Horse blood agar plate.—2 days at 37°. Colonies are similar to those on agar but show less tendency to differentiation and peripheral spread. No haemolysis; whole plate is slightly cleared and browned.
Potato.—7 days at 22°. Usually a thin layer of growth.
MacConkey plate.—24 hr at 37°. Very slight, effuse, confluent growth, just visible to the naked eye. Colonies disappear after 2 or 3 days, owing presumably to autolysis.
Resistance.—Fairly susceptible to inimical agencies. Killed by drying in a day or two, by heat at 55° in 5 min, by 5 per cent phenol immediately, and by 0.5 per cent phenol in 15 min. Agar plate cultures exposed to sun are sterilized in 1 to 5 hr. Cultures in the ice-chest may survive for months.
Metabolism.—Aerobe, facultative anaerobe. Requires low O-R potential for initiation of growth. Opt. temp. 27–28°; limits −2° to +45°. Opt. pH 7.2; limits pH 5.0–9.6. Forms alkali in broth. Growth favoured slightly by serum, uninfluenced by glucose; partly inhibited by glycerol. No haemolysis.
Biochemical.—Acid, no gas, in glucose, mannitol, arabinose, salicin, aesculin, maltose, xylose, trehalose and usually dextrin; but not in raffinose, dulcitol, erythritol, inositol, rhamnose, glycerol, sucrose, sorbitol, cellobiose, melibiose, lactose and adonitol. Catalase + +; nitrate +; MR +; β-galactosidase +; NH_3 +; LM unaltered or turned slightly acid; indole −; VP −; H_2S −; oxidase −; citrate −; urease −; ornithine decarboxylase −; phenylalanine deaminase −; lysine decarboxylase −; arginine dihydrolase −; gelatin hydrolysis −; methylene blue reduction −.
Antigenic structure.—Complex. Has specific envelope antigens formed at 37°, numerous somatic antigens formed at 26° and 37° many of which are shared with *Y. pseudotuberculosis*, and a rough antigen also shared with this organism (see Table 38.2). Is lysed by a specific bacteriophage. G+C content of DNA is 46–47 moles per cent.
Pathogenicity.—Virulence subject to variation. Toxic substances are formed, but they remain chiefly within the cells. A soluble protein murine toxin having some of the properties of an exotoxin, and a lipopolysaccharide endotoxin have been described. Causes plague in man and rodents. Experimental inoculation reproduces disease in mice, rats, guinea-

pigs, rabbits, marmots, ground-squirrels, and other rodents; also in monkeys. Dogs, cats, pigs, cattle, sheep, goats and horses are difficult to infect. Birds, with the exception of sparrows, are completely resistant.

Subcutaneous inoculation of a 24 hr broth culture into a mouse or guinea-pig is fatal in 2 to 5 days. PM necrotic local lesion surrounded by congestion and oedema. Regional glands enlarged and embedded in bloody oedema; they are soft and necrotic on section. Spleen firm, slightly enlarged and congested; may contain miliary, soft, grey nodules; liver peppered with tiny necrotic foci. Microscopically, bacilli found in abundance in local lesion and bubo; smaller numbers in spleen and heart's blood.

Yersinia pseudotuberculosis

Synonym.—B. pseudotuberculosis rodentium.

Isolation.—First observed by Malassez and Vignal in 1883; named *B. pseudotuberculosis rodentium* by Pfeiffer (1890) in 1889.

Habitat.—Parasite of rodents, particularly guinea-pigs; also affects man, other mammals, and birds.

Morphology.—Small, pleomorphic coccobacillus varying greatly in length and shape. Some strains consist of regular ovoid or coccoid organisms, 0.8–2.0 μm × 0.8 μm, with convex sides, rounded ends, and straight axis; arranged singly. Other strains consist of rod-shaped organisms, 1.5–5.0 μm × 0.6 μm, with parallel sides, rounded ends, and straight or curved axis; arranged singly, in groups, or in short chains. Long curved filaments are not uncommon (Fig. 38.7). Motile in broth cultures at 22°. Non-sporing. Non-capsulated. Ovoid forms show bipolar staining; rod forms show great irregularity of staining; the barred and granular type of staining is very common. Gram negative, and non-acid-fast.

Agar plate.—24 hr at 37°. Round, 0.5–1.0 mm in diameter, umbonate, granular, translucent, greyish-yellow colonies, with dull, finely granular or beaten-copper surface and entire edge; butyrous consistency; easily emulsifiable; differentiated into a raised, more opaque centre and a flat, clearer periphery with radial striation. A rough variant with an irregular surface and a crenated edge also occurs.

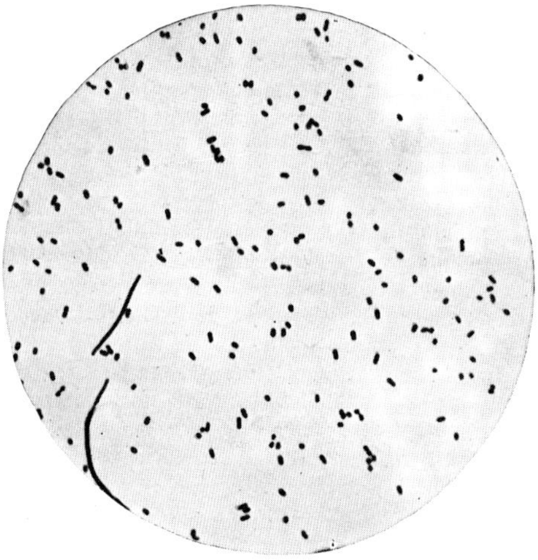

Fig 38.7 *Yersinia pseudotuberculosis.* From an agar culture, 24 hr, showing occasional long rods, 37° (× 1000)

Deep agar shake.—5 days at 37°. Heavy surface growth. No colonies beneath surface.

Agar stroke.—24 hr at 37°. Moderate, confluent, raised, greyish-yellow, translucent growth, with glistening, wavy or beaten-copper surface and an irregularly lobate edge.

Gelatin stab.—7 days at 22°. Good filiform growth, confluent at top, discrete below, extending to bottom of tube. Raised surface growth, 5 mm in diameter, with finely lobate edge. No liquefaction.

Broth.—24 hr at 37°. Moderate growth without turbidity and a viscous deposit disintegrating on shaking. Later the broth clears and a heavy, flocculo-membranous deposit forms, partly disintegrating on shaking. Incomplete surface and ring growth.

Loeffler's serum.—24 hr at 37°. Confluent growth, not so good as on agar.

Horse blood agar plate.—2 days at 37°. Good growth, but colonies are more compact and less differentiated than on agar. No haemolysis; whole plate is slightly cleared and browned.

Potato.—7 days at 22°. A thin yellowish membrane, which subsequently turns brown.

MacConkey plate.—24 hr at 37°. Very slight effuse confluent growth. Colonies disappear after a few days, owing presumably to autolysis.

Resistance.—Fairly susceptible to inimical agencies. Killed by moist heat at 60°, in 10 min. Susceptible to penicillin, 500 I.U./ml.

Metabolism.—Aerobe, facultative anaerobe. Opt. temp. 30°, limits 5–43°. Growth uninfluenced by serum and glucose, slightly inhibited by glycerol. No haemolysis.

Biochemical.—Acid, no gas, in glucose, mannitol, arabinose, salicin, rhamnose, melibiose, aesculin, maltose, xylose, trehalose and usually adonitol, within 14 days. Sucrose, sorbitol, cellobiose, raffinose, dulcitol, inositol and lactose, are not attacked. LM usually slight alkali formation; MR +; nitrate +; NH_3 +; methylene blue reduction + +; catalase + +; urease +; β-galactosidase +; indole −; VP −; H_2S −; oxidase −; citrate −; ornithine decarboxylase −; phenylalanine deaminase −; lysine decarboxylase −; arginine dihydrolase −; gelatin hydrolysis −.

Antigenic structure.—Contains (1) a heat-labile flagellar antigen formed below 25°, (2) numerous heat-stable somatic antigens formed at 22 and 37° some of which are shared with *Y. pestis* and some with members of the *Salmonella* group, and (3) a rough antigen shared with *Y. pestis*. Strains divided into six O groups.

Pathogenicity.—No true exotoxin formed. Virulence subject to alteration. Causes pseudotuberculosis in rodents, especially guinea-pigs, and in other mammals and birds. Causes mesenteric lymphadenitis and other forms of disease in man. Experimental inoculation reproduces the disease in rodents. Subcutaneous injection of a 24-hr broth culture into a guinea-pig proves fatal in 1 to 3 weeks. PM caseous local swelling, enlargement of regional glands, and nodules in spleen, liver and lungs. Microscopically the bacilli are numerous in the local lesion and glands.

Yersinia enterocolitica

Synonym.—Bacterium enterocolitica, Pasteurella X.

Isolation.—Observed by Schleifstein and Coleman (1939) in America, and by Hässig and co-workers (1949) in Switzerland.

Habitat.—Parasite of animals and man. Also occurs in the natural environment, especially in water.
Morphology.—In young cultures incubated at 22° or 37° occurs as gram-negative coccobacilli and short rods. Older cultures show pleomorphism. In cultures incubated at 22–25°, motile by means of peritrichate flagella. No capsule formed in culture, but zones interpreted as capsules may occur in organisms grown *in vivo*.
Agar plate.—24 hr at 37°. Forms circular, smooth, low convex colonies, 1–2 mm in diameter, with glistening surface and entire or slightly crenated edge, easily emulsifiable.
Broth.—24 hr at 37°. Uniform turbidity with a slight powdery deposit.
MacConkey plate.—24 hr. At 22°, pinpoint in size. At 37°, 1.5–2 mm in diameter.
Resistance.—Killed by moist heat at 55° within 15 min. Susceptible *in vitro* to tetracyclines and chloramphenicol, but not usually to penicillin and cephalothin.
Metabolism.—Aerobe, facultative anaerobe. Will grow at 4°. Organisms grown at 22° may differ in certain biochemical respects and in pathogenicity from those grown at 37°.
Biochemical.—Differs from *Y. pseudotuberculosis* chiefly in producing acid and small amounts of gas from carbohydrates including sucrose, sorbitol, and cellobiose, but not salicin, rhamnose, melibiose, aesculin, and adonitol; and in being VP +, indole + or −, ornithine decarboxylase +, methylene blue reduction −, and LM −. The taxonomic position of certain biochemically atypical strains is uncertain. Brenner and his colleagues divided strains into four species (p. 367).
Antigenic structure.—Distinct from that of *Y. pseudotuberculosis.* Thirty-four O serotypes and 19 H factors have been reported. Serological cross reactions between serotype 9 and *Brucella* spp. occur.
Pathogenicity.—Causes enteritis, mesenteric lymphadenitis, septicaemia and secondary immunological complications in man. Isolated from pseudotuberculosis-like lesions in animals, and from apparently healthy animals. Formerly thought to be non-pathogenic for laboratory animals; under certain circumstances mice and other animals can be infected (p. 372). (For reviews see Report 1973, 1979, Sonnenwirth 1974, Bottone 1977 and Highsmith *et al.* 1977.)

References

Åarsleff, B., Biberstein, E. L., Shreeve, B. J. and Thompson, D. A. (1970) *J. comp. Path.* **80,** 493.
Ahvonen, P. (1972) *Ann. clin. Res.* **4,** 30, 39.
Ajl, S. J., Reedal, J. S., Durrum, E. L., and Warren, J. (1955) *J. Bact.* **70,** 158.
Ajl, S. J., Rust, J., Hunter, D., Woebke, J. and Bent, D. F. (1958) *J. Immunol.* **80,** 435.
Amies, C. R. (1951) *Brit. J. exp. Path.* **32,** 259.
Anderson, L. A. P., Coombes, M. G. and Mallick, S. M. K. (1929) *Indian J. med. Res.* **17,** 611.
Arkwright, J. A. (1927) *Lancet* **i,** 13.
Bablet, J. and Girard, G. (1933) *C. R. Soc. Biol.* **114,** 471.
Bader, R. E. (1972) *Zbl. Bakt.* **221,** 327.
Baker, E. E., Sommer, H., Foster, L. E., Meyer, E. and Meyer, K. F. (1947) *Proc. Soc. exp. Biol., N.Y.* **64,** 139.
Baker, P. M. and Farmer, J. J. (1982) *J. clin. Microbiol.* **15,** 491.
Baxi, K. K., Blobel, H. and Brückler, J. (1970) *Zbl. Bakt.* **214,** 101.
Beck, M. (1893) *Z. Hyg. InfektKr.* **15,** 363.
Beesley, E. D. and Surgalla, M. J. (1970) *Appl. Microbiol.* **19,** 915.
Ben-Gurion, R. and Hertman, I. (1958) *J. gen. Microbiol.* **19,** 289.
Bercovier, H. *et al.* (1980) *Curr. Microbiol.* **4,** 201.
Bessonowa, A., Kotelnikow, G. and Sémikoz, F. (1927) *C. R. 1st Congr. antipest. U.R.S.S.* p. 485.
Bessonowa, A. and Lenskaja, G. (1931) *Zbl. Bakt.* **119,** 430.
Bhasin, J. L. and Lapointe-Shaw, L. (1980) *Canad. J. Microbiol.* **26,** 676.
Bhatnagar, S. S. (1940*a*) *Indian J. med. Res.* **28,** 1; (1940*b*) *Ibid.* **28,** 17.
Bibel, D. J. and Chen, T. H. (1976) *Bact. Rev.* **40,** 633.
Biberstein, E. L. and Francis, C. K. (1968) *J. med. Microbiol.* **1,** 105.
Biberstein, E. L. and Gills, M. G. (1962) *J. comp. Path.* **72,** 316.
Biberstein, E. L., Gills, M. and Knight, H. (1960) *Cornell Vet.* **50,** 283.
Bongert (1901) *Z. Hyg. InfektKr.* **37,** 449.
Bosworth, T. J. and Lovell, R. (1944) *J. comp. Path.* **54,** 168.
Bottone, E. J. (1977) *CRC crit. Rev. Microbiol.* **5,** 211.
Breed, R. S., Murray, E. G. D. and Smith, N. R. (1957) *Bergey's Manual of Determinative Bacteriology.* 7th edn. Baillière, Tindall & Cox, London.
Brenner, D. J., Steigerwalt, A. G., Falcão, D. P., Weaver, R. E. and Fanning, G. R. (1976) *Int. J. syst. Bact.* **26,** 180.
Brenner, D. J. *et al.* (1980) *Curr. Microbiol.* **4,** 195.
Brogden, K. A. and Packer, R. A. (1979) *Amer. J. vet Res.* **40,** 1332.
Brubaker, R. R. and Surgalla, M. J. (1962) *J. Bact.* **84,** 539.
Brzin, B. (1963) *Zbl. Bakt.* **189,** 543.
Buchanan, R. E. and Gibbons, N. E. (1974) *Bergey's Manual of Determinative Bacteriology.* 8th edn. Williams and Wilkins, Baltimore.
Buddingh, G. J. and Womack, F. C. (1941) *J. exp. Med.* **74,** 213.
Bullen, J. J. and Rogers, H. J. (1968) *Nature, Lond.* **217,** 86.
Burroughs, A. L., Holdenried, R., Longanecker, D. S. and Meyer, K. F. (1945) *J. infect. Dis.* **76,** 115.
Burrows, T. W. (1955) In: *Mechanisms of Microbial Pathogenicity.* 5th Symp. Soc. gen. Microbiol., Lond., p. 152; (1956) *Nature, Lond.* **177,** 426; (1963) *Ergebn. Mikrobiol.* **37,** 59; (1968) *Int. Sympos. Pseudotuberculosis, Paris 1967,* p. 145.
Burrows, T. W. and Bacon, G. A. (1956) *Brit. J. exp. Path.* **37,** 481; (1958) *Ibid.* **39,** 278.
Burrows, T. W., Farrell, J. M. F. and Gillett, W. A. (1964) *Brit. J. exp. Path.* **45,** 579.
Burrows, T. W. and Gillett, W. A. (1966) *J. gen. Microbiol.* **45,** 333.
Butt, E. M., Bonynge, C. W. and Joyce, R. L. (1936) *J. infect. Dis.* **58,** 5.
Carter, G. R. (1955) *Amer. J. vet. Res.* **16,** 481; (1957) *Ibid.* **18,** 210; (1961) *Vet. Rec.* **73,** 1052; (1962) *Canad. J. publ. Hlth* **53,** 158; (1967) *Advances in Veterinary Science*, p. 321. Academic Press, N.Y.; (1972) *Vet. Rec.* **91,** 150; (1973) *Amer. J. vet. Res.* **34,** 293.
Carter, G. R. and Rundell, S. W. (1975) *Vet. Rec.* **96,** 343.
Carter, P. B. (1975) *Infect. Immun.* **11,** 164.
Carter, P. B., Zahorchak, R. J. and Brubaker, R. R. (1980) *Infect. Immun.* **28,** 638.

Cavanaugh, D. C. and Quan, S. F. (1953) *Amer. J. clin. Path.* **23,** 619.

Chadwick, P. (1963) *Canad. J. Microbiol.* **9,** 829.

Chamberland and Jouan (1906) *Ann. Inst. Pasteur* **20,** 81.

Chen, T. H. and Meyer, K. F. (1966) *Bull. Wld Hlth Org.* **34,** 911.

Chladek, D. W. and Ellis, R. P. (1979) *Amer. J. vet. Res.* **40,** 446.

Cocking, E. C., Keppie, J., Witt, K. and Smith, H. (1960) *Brit. J. exp. Path.* **41,** 460.

Cornelius, J. T. (1929) *J. Path. Bact.* **32,** 355.

Crumpton, M. J. and Davies, D. A. L. (1956) *Proc. roy. Soc. B.* **145,** 109; (1957) *Nature, Lond.* **180,** 863.

Crumpton, M. J., Davies, D. A. L. and Hutchinson, A. M. (1958) *J. gen. Microbiol.* **18,** 129.

Darland, G., Ewing, W. H. and Davis, B. R. (1975) *DHEW Publ. No. (CDC) 75.8294.*

Das, M. S. (1958) *J. comp. Path.* **68,** 288.

Davies, D. A. L. (1956) *Biochem. J.* **63,** 105; (1958) *J. gen. Microbiol.* **18,** 118.

Davis, G. E., Philip, C. B. and Parker, R. R. (1934) *Amer. J. Hyg.* **19,** 449.

Devignat, R. (1951) *Bull. Wld Hlth Org.* **4,** 247; (1953) *Schweiz. Z. allg. Path.* **16,** 509; (1954) *Bull. Wld Hlth Org.* **10,** 463

Dieter, L. V. and Rhodes, B. (1926) *J. infect. Dis.* **38,** 541.

Dieudonné, A. and Otto, R. (1912) See *Kolle and Wassermann's Hdb. path. Mikroorg,* IIte Aufl., 1912–13, **4,** 155.

Downs, C. M. and Bond, G. C. (1935) *J. Bact.* **30,** 485.

Downs, C. M., Coriell, L. L., Pinchot, G. B., Maumenee, E., Klauber, A., Chapman, S. S. and Owen, B. (1947) *J. Immun.* **56,** 217.

Eastwood, A. and Griffith, F. (1914) *J. Hyg., Camb.* **14,** 285.

Eberson, F. and Wu Lien Teh. (1917) *J. infect. Dis.* **20,** 170.

Eberth, C. J. (1886) *Virchow's Arch.* **103,** 488.

Eigelsbach, H. T., Chambers, L. A. and Coriell, L. L. (1946) *J. Bact.* **52,** 179.

Eisler, D. M., Kubik, G. and Preston, H. (1958) *J. Bact.* **76,** 41.

Eisler, D. M. and Metz, E. K. von (1968) *J. Bact.* **95,** 1767.

Eskey, C. R. and Haas, V. H. (1940) *Publ. Hlth Bull., Wash.* No. 254.

Ewing, W. H., Ross, A. J., Brenner, D. J. and Fanning, G. R. (1978) *Int. J. syst. Bact.* **28,** 37.

Falcão, D. P., Ewing, W. H., Daws, B. R. and Hermann, G. J. (1978) *Rev. Microbiol. (S. Paulo)* **9,** 109.

Fauconnier, J. (1950) *Ann. Inst. Pasteur* **79,** 104.

Fleming, D. E. and Foshay, L. (1955) *J. Bact.* **70,** 345.

Foshay, L. (1932) *J. infect. Dis.* **51,** 280.

Francis, E. (1922) *Publ. Hlth Rep., Wash.* **37,** 987; (1923) *Ibid.* **38,** 1391; (1932) *Ibid.* **47,** 1287; (1942) *J. Bact.* **43,** 343; (1949) *Publ. Hlth Rep., Wash.* **64,** 238.

Francis, E. and Lake, G. C. (1922) *Publ. Hlth Rep., Wash.* **37,** No. 3, 83.

Frank, G. H. (1980) *J. clin. Microbiol.* **12,** 579.

Frederiksen, W. (1964) *Proc. 14th scand. Congr. Path. Microbiol., Oslo,* p. 103.

Frederiksen, W. (1973) See *Report* 1973, p. 170.

Fukumi, H., Kaneko, H. and Sayama, E. (1954) *Jap. J. med. Sci., Biol.* **7,** 621.

Gemski, P., Lazere, J. R. and Casey, T. (1980*a*) *Infect. Immun.* **27,** 682.

Gemski, P., Lazere, J. R., Casey, T. and Wohlhieter, J. A. (1980*b*) *Infect. Immun.* **28,** 1044.

Gilmour, N. J. L. (1978) *Vet. Rec.* **102,** 100.

Girard, G. and Gallut, J. (1957) *Ann. Inst. Pasteur, Paris* **92,** 544.

Goodner, K. (1955) *J. infect. Dis.* **97,** 246.

Gotschlich, E. (1912) See *Kolle and Wassermann Hdb. path. Mikroorg.,* IIte Aufl., 1912–13, **1,** 167.

Gourdon, R., Gourdon, J-M., Henry, M. and Quinchon, C. (1957) *Ann. Inst. Pasteur* **93,** 251.

Gunnison, G. B., Larson, A. and Lazarus, A. S. (1951) *J. infect. Dis.* **88,** 254.

Halman, M., Benedict, M. and Mager, J. (1967) *J. gen. Microbiol.* **49,** 451.

Harvey, S. and Pickett, M. J. (1980) *Int. J. syst. Bact.* **30,** 86.

Hässig, A., Karrer, J. and Pusterla, F. (1949) *Schweiz. med. Wschr.* **79,** 971.

Heddleston, K. L. (1976) *Amer. J. vet. Res.* **37,** 745.

Heddleston, K. L., Gallagher, J. E. and Rebers, P. A. (1972) *Avian Dis.* **16,** 925.

Henriksen, S. D. (1961) *Acta path. microbiol. scand.* **53,** 425.

Henriksen, S. D. and Frøholm, L. O. (1975) *Acta path. microbiol. scand.* **B83,** 129.

Henriksen, S. D. and Jyssum, K. (1961*a*) *Acta path. microbiol. scand.* **50,** 443; (1961*b*) *Ibid.* **51,** 354.

Herbert, D. (1949) *Brit. J. exp. Path.* **30,** 509.

Highsmith, A. K., Feeley, J. C. and Morris, G. K. (1977) *Hlth Lab. Sci.* **14,** 253.

Hills, G. M. and Spurr, E. D. (1952) *J. gen. Microbiol.* **6,** 64.

Hood, A. M. (1977) *J. Hyg., Camb.* **79,** 47.

Hughes, T. P. (1930) *J. exp. Med.* **51,** 225.

Hussaini, S. N. (1975) *Vet. Bull., Weybridge* **45,** 403.

Itagaki, K., Tsubokura, M., Sasaki, T. and Nagai, T. (1968) *Symp. Ser. immunobiol. Standard.* **9,** 27. S. Karger, Basel.

Jackson, S. and Burrows, T. W. (1956) *Brit. J. exp. Path.* **37,** 570.

Jackson, S. and Morris, B. C. (1961) *Brit. J. exp. Path.* **42,** 363.

Jawetz, E. (1950) *J. infect. Dis.* **86,** 172.

Jensen, W. I., Owen, C. R. and Jellison, W. L. (1969) *J. Bact.* **100,** 1237.

Jones, D. M. (1962) *J. Path. Bact.* **83,** 143.

Jones, F. S. (1921) *J. exp. Med.* **34,** 561.

Jordan, R. M. M. (1952) *Brit. J. exp. Path.* **33,** 27.

Kakehi, S. (1915–16) *J. Path. Bact.* **20,** 269.

Kapperud, G. (1975) *Acta path. microbiol. scand.* **B83,** 335; (1981) *Ibid.* **B89,** 29.

Kapperud, G., Bergan, T. and Lassen, J. (1981) *Int. J. syst. Bact.* **31,** 401.

Keppie, J., Cocking, E. C., Witt, K. and Smith, H. (1960) *Brit. J. exp. Path.* **41,** 577

Keppie, J., Smith, H. and Cocking, E. C. (1957) *Nature, Lond.* **180,** 1136.

Khalifa, A. I. B. (1934) See *Vet. Bull., Weybridge,* 1936, **6,** 792.

Kitasato, S. (1894) *Lancet* **ii,** 428.

Kitt, T. (1903) *Kolle and Wassermann's Handbuch der pathogenen Mikroorganismen,* Vol. ii, p. 543.

Knapp, W. (1955) *Zbl. Bakt.* **164,** 57; (1960) *Z. Hyg. InfektKr.* **146,** 315.

Knapp, W. and Thal, E. (1963) *Zbl. Bakt.* **190,** 472; (1973) See Report 1973, p. 10.

Koike, Y., Kuwahara, A. and Fujiwara, H. (1975) *Jap. J. Microbiol.* **19,** 241.

Kolyvas, E., Sorger, S., Marks, M. I. and Pai, C. H. (1978) *J. Pediat.* **92,** 81.

Kruif, P. H. de (1921) *J. exp. Med.* **33,** 773; (1922*a*) *Ibid.* **35,** 561; (1922*b*) *Ibid.* **36,** 309; (1923) *Ibid.* **37,** 647.
Kudo, M. (1934) *Jap. J. exp. Med.* **12,** 371.
Kutscher (1894) *Z. Hyg. InfektKr.* **18,** 327.
Laird, W. J. and Cavanaugh, D. C. (1980) *J. clin. Microbiol.* **11,** 430.
Lal, R. B. (1927) *Amer. J. Hyg.* **7,** 561.
Lapage, S. P. and Bascomb, S. (1968) *J. appl. Bact.* **31,** 568.
Larson, C. L., Philip, C. B., Wicht, W. C. and Hughes, L. E. (1951) *J. Immun.* **67,** 289.
Larson, C. L., Wicht, W. and Jellison, W. L. (1955) *Publ. Hlth Rep., Wash.* **70,** 253.
Lazarus, A. S. and Nozawa, M. M. (1948) *J. Bact.* **56,** 187.
Levine, H. B. and Garber, E. D. (1950) *J. Bact.* **60,** 508.
Lignières, J. (1900) *Bull. Soc. cent. Méd. vét.* **54,** 469. Quoted from Med. Res. Coun. *System of Bacteriology*, 1929, **4,** 446.
Lillie, R. D. and Francis, E. (1933) *Publ. Hlth Rep., Wash.* **48,** 1127.
Loghem, J. J. van. (1944) *Antonie van Leeuwenhoek* **10,** 15.
McAllister, H. A. and Carter, G. R. (1974) *Amer. J. vet. Res.* **35,** 917.
McCoy, G. W. (1912) *Publ. Hlth Bull., Wash.* No. 53, p. 17.
McCoy, G. W. and Chapin, C. W. (1912) *J. infect. Dis.* **10,** 61.
McKenzie, D., Stradler, M., Boothe, J., Oleson, J. J. and Subbarow, Y. (1948) *J. Immun.* **60,** 283.
MacLagan, R. M. and Old, D. C. (1980) *J. appl. Bact.* **49,** 353.
Magnusson, H. (1914) *Z. InfektKr. Haustiere* **15,** 61.
Mair, N. S. and Fox, E. (1973) See *Report* 1973, p. 180.
Malassez, L. C. and Vignal, W. (1883) *Arch. Physiol. norm. path., 3rd Series*, **2,** 369.
Mannheim, W., Pohl, S. and Holländer, R. (1980) *Zbl. Bakt.* I. Abt. Orig. **A246,** 512.
Meyer, K. F., Foster, L. E., Baker, E. E., Sommer, H. and Larson, A. (1948) *Proc. 4th int. Congr. trop. Med. Malaria* **1,** 264.
Meyer, K. F., Quan, S. F. and Larson, A. (1948) *Amer. Rev. Tuberc.* **57,** 312.
Midgley, J., Lapage, S. P., Jenkins, B. A. G., Barrow, G. I., Roberts, M. E. and Buck, A. G. (1970) *J. med. Microbiol.* **3,** 91.
Mørch, J. R. and Krogh-Lund, G. (1930) *C. R. Soc. Biol.* **105,** 319.
Morris, E. J. (1958) *J. gen. Microbiol.* **19,** 305.
Mors, V. and Pai, C. H. (1980) *Infect. Immun.* **28,** 292.
Mráz, O. (1969*a*) *Zbl. Bakt.* **209,** 336, 349; (1969*b*) *Zbl. Bakt. Ref.* **215,** 267.
Müller, H. E. and Krasemann, C. (1974) *Zbl. Bact.*, I. Abt. Orig. **A229,** 391.
Mushin, R. (1980) *J. Hyg., Camb.* **85,** 59.
Nagy, L. K. and Penn, C. W. (1976) *Res. vet. Sci.* **20,** 249.
Namioka, S. (1973) See *Report* 1973, p. 177.
Namioka, S. and Murata, M. (1961) *Cornell Vet.* **51,** 498, 507, 522.
Newsom, I. E. and Cross, F. (1932) *J. Amer. vet. med. Ass.* **80,** 711.
Nicolle, P., Mollaret, H. H. and Brault, J. (1973) See Report 1973, p. 54.
Niléhn, B. (1967) *Acta path. microbiol. scand.* **69,** 83; (1969) *Acta path. microbiol. scand.* suppl. 206, 1; (1973) See *Report* 1973, p. 59
Niléhn, B. and Ericson, C. (1969) *Acta path. microbiol. scand.* **75,** 177.
Ogata, N. (1955) *Zbl. Bakt.* **163,** 171.
Okamoto, K., Ichikawa, H., Kawamoto, Y., Miyama, A. and Yoshii, S. (1980) *Microbiol. Immunol.* **24,** 401.
Olsufjev, N. G. and Emelyanova, O. S. (1957) *J. Hyg. Epidem., Praha* **1,** 357; (1962) *Ibid.* **6,** 193.
Ostertag, R. (1908) *Z. InfektKr. Haustiere* **4,** 1.
Otten, L. (1938) *Meded. Dienst. Volksgezondh. Ned.-Ind.* **27,** 111.
Owen, C. R., Buker, E. O., Jellison, W. L., Lackman, D. B. and Bell, J. F. (1964) *J. Bact.* **87,** 676.
Pai, C. H. and Mors, V. (1978) *Infect. Immun.* **19,** 908
Pai, C. H., Mors, V. and Seemayer, T. A. (1980) *Infect. Immun.* **28,** 238.
Pallaske, G. (1933) *Z. InfektKr. Haustiere* **44,** 43.
Pease, P. (1979) *J. med. Microbiol.* **12,** 337.
Perroncito. (1878) *Ann. Rep. Acad. Agric., Torino* **20,** 89.
Pfeiffer, A. (1890) *Zbl. Bakt.* **7,** 219.
Pirie, J. H. H. (1929) *Publ. S. Afr. Inst. med. Res.* **4,** 203.
Pirosky, I. (1938*a*) *C. R. Soc. Biol.* **127,** 98; (1938*b*) *Ibid.* **128,** 346; (1938*c*) *Ibid.* **128,** 347.
Pirt, S. J., Thackeray, E. J. and Harris-Smith, R. (1961) *J. gen. Microbiol.* **25,** 119.
Poels, J. (1886) *Fortschr. Med.* **4,** 388.
Preisz, H. (1894) *Ann. Inst. Pasteur* **8,** 231.
Preston, N. W. and Maitland, H. B. (1952) *J. gen. Microbiol.* **7,** 117.
Priestley, F. W. (1936*a*) *Brit. J. exp. Path.* **17,** 374; (1936*b*) *J. comp. Path.* **49,** 340; (1936*c*) *Ibid.* **49,** 348.
Prodjoharjono, S., Carter, G. R. and Conner, G. H. (1974) *Amer. J. vet. Res.* **35,** 111.
Ramon, G., Girard, G. and Richou, R. (1947) *C. R. Acad. Sci., Paris* **224,** 1259.
Ransmeier, J. C. (1943) *J. infect. Dis.* **72,** 86.
Ransom, J. P. (1956) *Proc. Soc. exp. Biol. Med.* **93,** 551.
Rao, M. S. (1939) *Indian J. med. Res.* **27,** 75, 617, 833.
Rebers, P. A., Heddleston, K. L. and Rhoades, K. R. (1967) *J. Bact.* **93,** 7.
Reimann, H. A. (1932) *Amer. J. Hyg.* **16,** 206.
Reports. (1899) Germ. Plague Comm., *Arb. ReichsgesundhAmt* **16,** 1; (1906) Engl. Plague Comm., *J. Hyg., Camb.* **6,** 421; (1912) Engl. Plague Comm., *J. Hyg., Camb.* **12,** Suppl., p. 287; (1971*a*) *Int. J. syst. Bact.* **21,** 157; (1971*b*) *World Hlth Org. Tech. Rep. Ser.* No. 464; (1973) Contributions to Microbiology and Immunology, vol. 2. *Yersinia, Pasteurella* and *Francisella*. S, Karger, Basel; (1979) *Ibid.* vol. 5. *Yersinia enterocolitica:* Biology, Epidemiology and Pathology.
Ringertz, O. and Dahlstrand, S. (1968) *Acta path. microbiol. scand.* **72,** 464.
Ritter, D. B. and Gerloff, R. K. (1966) *J. Bact.* **92,** 1838.
Roberts, R. S. (1947) *J. comp. Path.* **57,** 261.
Rosenbusch, C. T. and Merchant, I. A. (1939) *J. Bact.* **37,** 69.
Rowland, S. (1914) Engl. Plague Comm., *J. Hyg., Camb.* **13,** Suppl. 418.
Sandulache, R. and Marx, A. (1978) *Ann. Microbiol., Paris* **B129,** 425
Savino, E., Aldao, A. and Anchezar, B. (1939) *Rev. Inst. bact., Buenos Aires* **9,** 110.
Schiemann, D. A. and Devenish, J. A. (1980) *Infect. Immun.* **29,** 500.

Schleifstein, J. and Coleman, M. B. (1939) *N.Y. St. J. Med.* **39,** 1749.
Schütze, H. (1928) *Arch. Hyg.* **100,** 181; (1932*a*) *Brit. J. exp. Path.* **13,** 284; (1932*b*) *Ibid.* **13,** 289; (1939) *Ibid.* **20,** 235.
Schütze, H. and Hassanein, M. A. (1929) *Brit. J. exp. Path.* **10,** 204.
Sémikoz, F., Bessonowa, A. and Kotelnikow, G. (1927) *C. R. 1st Congr. antipest. U.R.S.S.* p. 488.
Shaw, F. W. and Hunnicutt, T. (1930) *J. Lab. clin. Med.* **16,** 46.
Slotnick, I. J. and Dougherty, M. (1964) *Leeuwenhoek J. Microbiol.* **30,** 261.
Slotnick, I. J., Mertz, J. A. and Dougherty, M. (1964) *J. infect. Dis.* **114,** 503.
Smith, D. T. and Webster, L. T. (1925) *J. exp. Med.* **41,** 275.
Smith, G. R. (1958) *J. comp. Path.* **68,** 455; (1959) *Nature, Lond.* **183,** 1132; (1960) *J. comp. Path.* **70,** 326; (1961) *J. Path. Bact.* **81,** 431; (1962) *Ibid.* **83,** 501.
Smith, H. (1968) *Bact. Rev.* **32,** 164.
Smith, H., Keppie, J., Cocking, E. C. and Witt, K. (1960) *Brit. J. exp. Path.* **41,** 452.
Smith, J. E. (1955) *J. comp. Path.* **65,** 239; (1958) *Ibid.* **68,** 315; (1959) *Ibid.* **69,** 231.
Smith, J. E. and Thal, E. (1965) *Acta path. microbiol. scand.* **64,** 213.
Smith, P. N. (1959) *J. infect. Dis.* **104,** 78.
Sokhey, S. S. (1939) *Indian J. med. Res.* **27,** 341.
Sonnenwirth, A. C. (1974) In: *Manual of Clinical Microbiology*, 2nd edn, p. 222. Amer. Soc. Microbiol., Washington, D.C.
Spencer, R. R. (1921) *Publ. Hlth Rep., Wash.* No. 36, p. 2836.
Spivack, M. L., Foster, L., Larson, A., Chen, T. H., Baker, E. E. and Meyer, K. F. (1958) *J. Immunol.* **80,** 132.
Spivack, M. L. and Karler, A. (1958) *J. Immunol.* **80,** 441.
Steel, K. J. and Midgley, J. (1962) *J. gen. Microbiol.* **29,** 171.
Steinhaus, E. A., Parker, R. R. and McKee, M. T. (1944) *Publ. Hlth Rep., Wash.* **59,** 78.
Stevens, M. and Mair, N. S. (1973) See *Report* 1973, p. 17.
Strong, R. P., Teague, O., Barber, M. A. and Crowell, B. C. (1912) *Philipp. J. Sci.*, B. **7,** 129–270.
Talbot, J. M. and Sneath, P. H. A. (1960) *J. gen. Microbiol.* **22,** 303.
Tamura, J. T. and Gibby, I. W. (1943) *J. Bact.* **45,** 361.
Tanaka, A. (1926) *J. infect. Dis.* **38,** 421.
Thal, E. (1956) *Ann. Inst. Pasteur* **91,** 68; (1966) *Zbl. Bakt.* **200,** 56; (1973) See Report 1973, p. 190.
Thal, E. and Chen, T. H. (1955) *J. Bact.* **69,** 103.
Thal, E. and Knapp, W. (1971) *Symp. Ser. immunobiol. Standard* **15,** 219. S. Karger, Basel.
Thal, E., Knapp, W. and Hanko, E. (1967) *Zbl. Bakt.* **204,** 399.
Thompson, D.A. and Mould, D. L. (1975) *Res. vet. Sci.* **18,** 342.
Trevisan, V. (1887) *Rc. Ist Lombardo Sci. Lett., Piso.*, p. 94. Quoted from R. E. Buchanan's *General Systematic Bacteriology* (1925). Williams and Wilkins, Baltimore.
Tucker, D. N., Slotnick, I. J., King, E. O., Tynes, B., Nicholson, J. and Crevasse, L. (1962) *New Engl. J. Med.* **267,** 913.
Tumansky, W., Müller, M., Bokalo, A., Wedistschew, S. and Sabinin, A. (1935) *Rev. Microbiol. Saratov* **14,** 128.
Tweed, W. and Edington, J. W. (1930) *J. comp. Path.* **43,** 234.
Tyeryar, F. J. and Lawton, W. D. (1970) *J. Bact.* **104,** 1312.
Ursing, J., Steigerwalt, A. G. and Brenner, D. J. (1980) *Curr. Microbiol.* **4,** 231.
Walker, R. V. (1967) *Curr. Top. Microbiol. Immunol.* **41,** 23.
Watko, L. P. (1966) *Can. J. Microbiol.* **12,** 933.
Wauters, G. (1970) Thesis. University of Louvain, Belgium; (1973) See Report 1973, p. 38.
Wauters, G., Le Minor, L., Chalon, A. M. and Lassen, J. (1972) *Ann. Inst. Pasteur* **122,** 951.
Webster, L. T. (1924*a*) *Proc. Soc. exp. Biol.* **22,** 139; (1924*b*) *J. exp. Med.* **40,** 109, 117; (1925) *Ibid.* **41,** 571; (1926) *Ibid.* **43,** 555.
Webster, L. T. and Baudisch, O. (1925) *J. exp. Med.* **42,** 473.
Webster, L. T. and Burn, C. G. (1926) *J. exp. Med.* **44,** 343, 359.
Weitzenberg, R. (1935) *Zbl. Bakt.* **133,** 343.
Wessman, G. E. (1966) *Appl. Microbiol.* **14,** 597.
Wessman, G. E. and Wessman, G. (1970) *Canad. J. Microbiol.* **16,** 751.
Wherry, W. B. and Lamb, B. H. (1914) *J. infect. Dis.* **15,** 331.
Williams, J. E. *et al.* (1978) *Amer. J. publ. Hlth* **68,** 262.
Winblad, S. (1968) *Int. Sympos. Pseudotuberculosis, Paris* 1967, p. 337; (1973) See Report 1973, p. 27.
Winter, C. C. and Moody, M. D. (1959) *J. infect. Dis.* **104,** 274.
Wright, H. D. (1934) *J. Path. Bact.* **39,** 381.
Wright, J. (1936) *J. Path. Bact.* **42,** 209.
Wu Lien Teh and Eberson, F. (1917) *J. Hyg., Camb.* **16,** 1.
Yanagawa, Y., Maruyama, T. and Sakai, S. (1978) *Microbiol. Immunol.* **22,** 643.
Yaniv, H. and Avi-Dor, J. (1952) *Nature, Lond.* **169,** 201.
Yersin. (1894) *Ann. Inst. Pasteur* **8,** 662.
Yusef, H. S. (1935) *J. Path. Bact.* **41,** 203.
Zabolotny, D. (1923) *Ann. Inst. Pasteur* **37,** 618.
Zink, D. L. *et al.*, (1980) *Nature, Lond.* **283,** 224.
Zinnemann, K. (1981) In: *Haemophilus, Pasteurella and Actinobacillus*, p. 1. Ed. by M. Kilian and W. Frederiksen Academic Press, London
Zlatogoroff, S. I. and Moghilewskaja, B. I. (1928*a*) *Ann. Inst. Pasteur* **42,** 1615; (1928*b*) *C. R. Soc. Biol.* **99,** 506.

39

Haemophilus and *Bordetella*

J. W. G. Smith

Haemophilus

Definition

Very small rods, sometimes almost coccal, sometimes thread-like; may be pleomorphic. Non-motile, non-sporing, gram negative, non-acid-fast. Dependent for aerobic growth on haemin or other porphyrins (X factor) or on diphosphopyridine nucleotide or other co-enzyme-like substance (V factor) contained in blood or plant tissues. Some species require both X and V factor. Most species grow poorly in the absence of oxygen. All known species appear to be obligatory parasites, inhabiting particularly the upper respiratory tract, and most species are pathogenic. G + C content of DNA 37–44 moles per cent.

Type species: *H. influenzae*.

Since the isolation and description by Pfeiffer (1892, 1893) of the haemophilic bacillus then thought to be associated with influenza, several other small gram-negative bacilli have been described that share certain of its characteristic growth requirements and are parasites, and often pathogens, of man or animals. These organisms have been grouped together under the generic name *Haemophilus* (see Winslow *et al.* 1920). The definition later permitted the inclusion, by other workers, of organisms like the Bordet–Gengou bacillus and Ducrey's bacillus. Such an extension of the term 'haemophilic bacilli' was soon opposed (Kristensen 1922, Fildes 1923), and the definition of the genus and the species it should include have caused controversy ever since.

The haemophilic nature of Pfeiffer's bacillus is determined by its requiring for growth two main factors: X factor, an iron-containing porphyrin, and V factor, a coenzyme (see p. 382). But if the genus is confined to strains that need one or both of these compounds, there is a risk of excluding similar organisms which can synthesize these growth factors either in ineffective or suboptimal amounts. Nevertheless, to most workers a requirement for X or V factor, or their precursors, is necessary for the inclusion of a strain of small gram-negative bacilli in the *Haemophilus* genus (Report 1967, Sneath and Johnson 1973, Kilian 1976*a*, Zinnemann 1980).

Species in the genus

The organisms we have included in this genus are listed in Table 39.1, but a number of the groupings adopted can be disputed, and clear cut differences between the species may be difficult to recognize in individual isolates. *H. aegyptius*, the Koch-Weeks bacillus, is regarded by Kilian and his colleagues (Kilian 1976*a*, Kilian *et al.* 1976) as a biotype of *H. influenzae* possessing a particular association with conjunctival infection. It can certainly be difficult to distinguish *H. aegyptius* from *H. influenzae* (Pittman and Davis 1950, Ingham and Turk 1969), but its failure to form indole, and usually to ferment xylose, its haemagglutinating ability, and the close serological similarity of its strains, perhaps justify the retaining of the older term for the present. We have also retained the *H. parahaemolyticus* group, although Kilian (1976*a*) prefers to dispense with it because haemolysis was an unstable property among the strains he tested. *H. aphrophilus*, although requiring X factor on first isolation, readily loses this property on culture; since it is independent of V factor its place in the genus is disputed (Kilian 1976*a*). The dependence of *H. suis* on X factor is also in doubt (Biberstein *et al.* 1963, Biberstein and White 1969, Kilian 1974, Broom and Sneath 1981), so that *H. suis* and *H. parasuis* may be indistinguishable (Kilian 1976*a*). Haemophili isolated from the respiratory tract of fowls are usually described as requiring V factor only and can be identified as *H. paragallinarum*; strains dependent on both X and V factors appear to be very rare, and the existence of *H. gallinarum* is in doubt (Zinnemann 1980). Other isolates resembling *H. paragallinarum* have been named *H. avium*, being distinguished by a lack of pathogenicity for fowls and a positive alkaline-phosphatase reaction (Hinz and Kunjara 1977).

Table 39.1 Species of *Haemophilus*: growth requirements and haemolytic activity

Species	Requirement for X	Requirement for V	Requirement for CO_2	Haemolysis on horse-blood agar
influenzae	+	+	−	−
aegyptius	+	+	−	−
haemolyticus	+	+	−	+
parainfluenzae	−	+	∓	∓
parahaemolyticus	−	+	−	+
suis	+	+	−	−
parasuis	−	+	−	−
haemoglobinophilus (syn. *canis*)	+	−	−	−
gallinarum	+	−	+	−
paragallinarum	−	+	+	−
aphrophilus	∓	−	+	−
paraphrophilus	−	+	+	−
ducreyi	+	−	∓	−

∓ = Variable, most strains negative.

In a detailed study of 426 *Haemophilus* strains, Kilian (1976*a*) identified four other groups of V-factor-dependent haemophilic bacteria (not shown in Table 39.1). For one of these he proposed the name *H. segnis*; it contained 17 strains isolated from human saliva and dental plaque. Kilian (1976*a*) also suggested, based on an examination of 15 strains, that the name *H. pleuropneumoniae*, originally used by Shope (1964), be reintroduced for a group of V factor-dependent haemophili highly pathogenic to the respiratory tract of pigs. These were all β-haemolytic on first isolation though five of them lost this property. They could be distinguished by biochemical tests from other haemophili which require V factor, including *H. parahaemolyticus*, in which species swine strains are usually placed (Nicolet 1968, Report 1971). The group appears to be related to *H. parasuis*, since transformation can occur between the species (White *et al.* 1964).

Several other organisms that have been named *Haemophilus* cannot at present be accepted as members of this genus. *H. vaginalis* requires neither X nor V factor and is now considered to belong to a separate genus, *Gardnerella* (see Chapter 43). *H. piscium*, associated with ulcer disease in trout (Snieszko *et al.* 1950), does not require X or V factor, but requires diphosphothiamine or adenosine triphosphate for growth and can be cultivated in a medium containing fish extract; its taxonomic position remains undefined (Broom and

Sneath 1981). This is true also of *H. ovis*, isolated from sheep suffering from bronchopneumonia (Mitchell 1925), and of the organism cultured by Kairies (1935) from the respiratory tract of ferrets and called by him *Bacterium influenzae putorium multiforme*. Neither *H. somnus* (Bailie *et al.* 1973), associated with encephalitis and generalized infection in cattle (Bailie 1969), nor *H. agni* (Report 1971), associated with lamb septicaemia (Kennedy *et al.* 1958), can be regarded as haemophili because they do not have a strict requirement for X or V factor (Report 1971, but see Broom and Sneath 1981). An organism isolated from the small intestine of rabbits with mucoid enteritis was provisionally named *H. paracuniculus* by Targowski and Targowski (1979) but only one culture was examined in detail. This did have the characters of a *Haemophilus*, but judgement must be reserved until further strains have been described. The organism described as *H. murium*, which was said to require X factor, can be recognized only from its original description (Kairies and Schwartzer 1956); it was isolated from mice with an influenza-like illness.

The organism known as *H. equigenitalis* is described briefly on p. 391. It cannot be considered a member of the genus *Haemophilus* as at present constituted.

The relation of *Haemophilus* to other genera of small gram-negative rods is also uncertain. When studied by numerical taxonomy it is found to be closely related to *Pasteurella* and *Actinobacillus* (Johnson and Sneath 1973, Sneath and Johnson 1973, Mannheim *et al.* 1980, Broom and Sneath 1981, Kilian *et al.* 1981). The organisms that will be included in the genus by taxonomists in the future may well be different from those given in Table 39.1.

Morphology

H. influenzae, as originally described by Pfeiffer (1893) and as most commonly seen in strains recently isolated from acute infection, such as meningitis, is a very small, almost coccal rod, 1–1.5 μm by 0.3–0.4 μm, with rounded, sometimes rather pointed ends. In some cultures these coccobacillary forms are the only ones seen. More usually, a proportion of longer bacilli and a few long thread forms are found. In some cultures longer and stouter rods may predominate, while other strains may present an entirely different picture, the bacilli being thin, long and wavy, sometimes forming tangled masses. It is not unusual, especially where there are thread forms, to encounter large, spherical or fusiform swollen bodies, often attaining a diameter of 2–3 μm, or even more, which are sometimes attached to the end of a lateral thread, and sometimes to the end of a short lateral stalk (Wade and Manalang 1920, Kristensen 1922). Another form occasionally met with is a long thread with an enormous fusiform swelling, situated centrally or towards one end.

Any of these types may predominate in a single

Fig. 39.1 *H. influenzae*. From 24-hr culture on Fildes's agar; typical coccobacillary forms (× 1000).

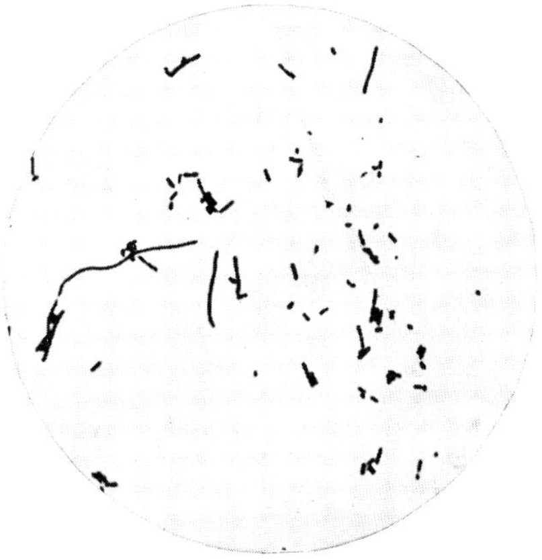

Fig. 39.2 *H. influenzae*. From 24-hr culture on Fildes's agar; short and long bacillary forms (× 1000).

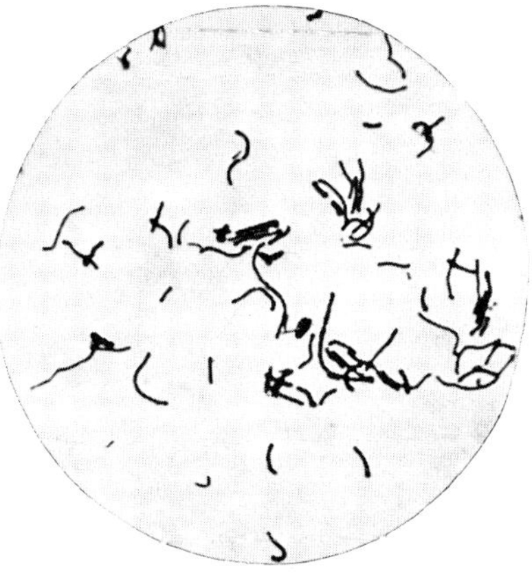

Fig. 39.3 *H. influenzae*. From 24-hr culture on Fildes's agar; atypical form; long, curved bacilli (× 1000).

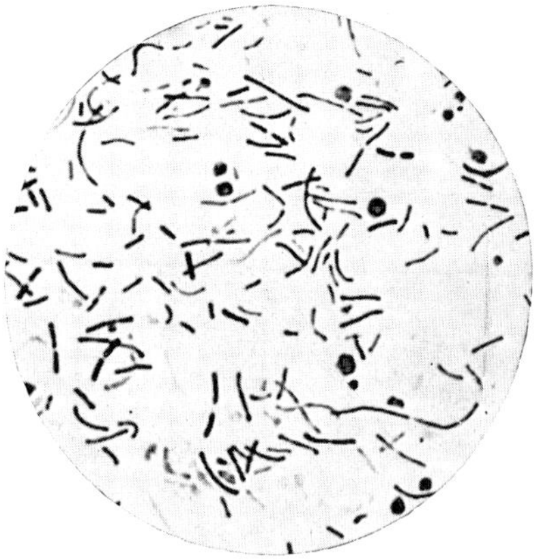

Fig. 39.4 *H. influenzae*. From 24-hr culture on Fildes's agar; thread forms and large spherical bodies (× 1000).

strain, and two strains may differ so much in morphology as to suggest quite different organisms. Dible (1924) regarded the different morphological types as varieties of *H. influenzae*, but others noted that morphology often changed with prolonged subculture, usually from coccobacilli to longer bacilli or to curved filaments (Wollstein 1915, Kristensen 1922, Smith 1931); that is, from 'typical' to 'atypical'. The typical morphology usually predominates in strains isolated from pathological conditions, and these strains may be capsulated; the change to atypical morphology is associated with the S→R type of variation (Pittman 1931).

The haemophilic bacilli are non-motile, non-sporing, non-acid-fast, and gram negative. Strains of *H. influenzae* from acute infections are often capsulated (Pittman 1931). With the exception of *H. parahaemolyticus*, and some strains of *H. parainfluenzae*, other strains are non-capsulated (White *et al.* 1964). The organisms stain with some difficulty, and many of the ordinary bacteriological dyes are unsuitable for this purpose. Dilute carbol fuchsin applied for 5–15 min usually gives satisfactory results.

Cultural characters

Growth requirements

Nutritional requirements provide criteria by which this genus has been defined and by which many of the species included within it are distinguished. The most characteristic feature of the true haemophilic bacilli is their failure to grow in the absence of certain factors which are present in blood. The ability of the influenza bacillus to grow on blood-containing media, and its inability to grow on nutrient agar, with or without the addition of serum or other native protein, has been noted by all workers from Pfeiffer onwards. Grassberger (1897) described another phenomenon—that of *satellitism*—which is highly characteristic of *H. influenzae*. In cultures on blood agar plates, streaked with sputum or bronchial secretion, he noted the appearance of relatively large colonies of the influenza bacillus with a slightly granular central portion. These large colonies (1 mm or more in diameter) always developed in the immediate vicinity of a colony of *Staphylococcus* (Fig. 39.5). *Staph. aureus*, coagulase-negative staphylococci and certain other chromogenic micrococci all produced this effect. These observations have been repeatedly confirmed (see Davis 1921, Kristensen 1922).

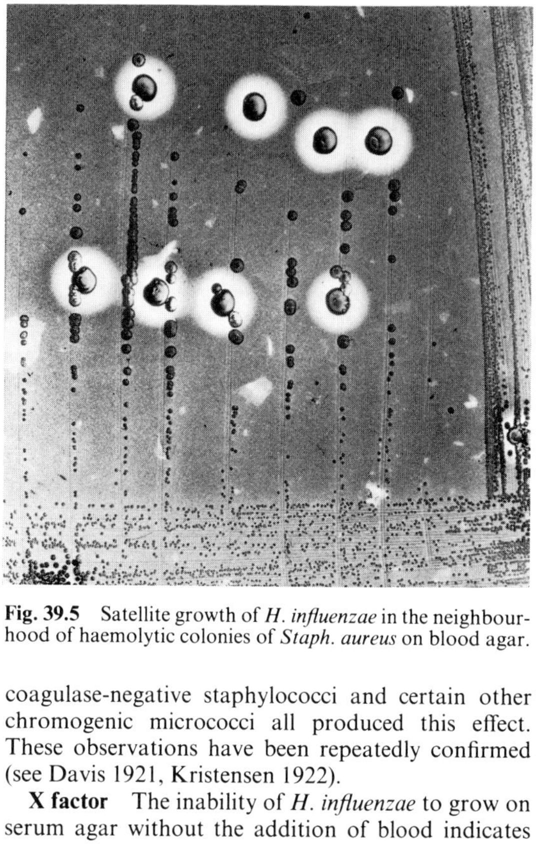

Fig. 39.5 Satellite growth of *H. influenzae* in the neighbourhood of haemolytic colonies of *Staph. aureus* on blood agar.

X factor The inability of *H. influenzae* to grow on serum agar without the addition of blood indicates that some constituent of the red blood cells (RBC) is essential for growth, and the term 'X factor' is generally applied to the growth-promoting derivatives of blood pigments. During anaerobic growth the requirement for X factor is certainly reduced (Anderson 1931) but very small amounts may be needed (Gilder and Granick 1947, Smith *et al.* 1953). Lwoff and Lwoff (1937*a,b,c*) concluded that it was used for the synthesis of the iron-containing respiratory enzymes cytochrome, cytochrome oxidase, catalase and peroxidase.

It was at first assumed that X factor was haemoglobin, but the pure, crystallized substance does not promote growth (Fildes 1921). Haematin was found to be growth promoting, and Fildes (1921) suggested that it owed its activity to its being a peroxidase. But not all peroxidases are growth promoting, and some iron compounds without peroxidase activity promote the growth of *H. influenzae* (Baudisch 1932). X factor may be regarded as the factor or factors necessary to permit the aerobic growth of organisms incapable of synthesizing porphyrins or haem from the precursor δ-aminolaevulinic acid (White and Granick 1963, Biberstein *et al.* 1963), a process which requires five enzyme-mediated steps (see Granick and Mauzerall 1961); presumably any or all of these enzymes may be lacking in X-factor-dependent haemophili. The ability to convert δ-aminolaevulinic acid to porphobilinogen, the first step in this chain, has been used as a convenient means of measuring X factor requirement (Kilian 1974), but this test does not exclude a need for haematin because of lack of one of the remaining enzymes in the pathway. Some haemophilus strains apparently require haemin itself, lacking the ability to incorporate iron into protoporphyrin (White and Granick 1963, Kilian *et al.* 1976). Granick and Gilder (1946) found that protoporphyrin and other iron-free porphyrins could be substituted for haematin provided they contained single side-chains, which were essential for the insertion of iron into the haem enzymes.

V factor Davis (1917) showed that *H. influenzae* also required a factor besides that derived from haemoglobin; it was present in the tissues of various plants and animals, and synthesized by most bacterial species other than *H. influenzae*. The second factor, unlike haematin, is thermolabile, being inactivated by heating to 120° for 30 min. The label 'V factor' is applied to this thermolabile substance (see also Thjøtta and Avery 1921, Davis 1921, Fildes 1922, 1923, 1924, Kristensen 1922, Valentine and Rivers 1927). Both X and V factors are present in blood. V factor is closely concerned in oxidation-reduction processes of the growing cell (Pittman 1935) and is one of the two co-dehydrogenases, di- and tri-phosphopyridine nucleotide (Lwoff and Lwoff 1937*a*, *b*, *c*, see Chapter 3), or precursors to these enzymes (Shifrine and Biberstein 1960).

In satellitism the staphylococci, and many other organisms of greater synthetic ability than the haemophilic bacilli, synthesize the co-dehydrogenase and presumably some haem compounds, which diffuse into the medium and stimulate the growth of bacilli that require them.

The requirements of the different members of the *Haemophilus* group for X and V factors, and for CO_2 which some of them need on primary isolation, are set out in Table 39.1. It should be noted that the need for X or V factor does not necessarily imply complete inability of a bacterium to synthesize them. Thus, certain strains of *H. parainfluenzae* may form suboptimal amounts of V factor (Miles and Gray 1938), and *H. influenzae* suboptimal (Smith *et al.* 1953) or even optimal amounts (Butler 1962) of X factor. But determination of X and V factor requirements is not without difficulty, and Gilder and Granick (1947), for example, found *H. influenzae* strains wrongly classified as *H. parainfluenzae* because there was sufficient X factor in the yeast extract used as a source of V factor to allow for their growth. For this reason Kilian (1974) prefers a test of the ability to synthesize porphyrins from δ-aminolaevulinic acid as a more reliable indicator of strain characteristics, the formation of porphyrin being detected by fluorescence in Wood's light (see also White and Granick 1963).

Preliminary testing for X and V requirements consists in growing the organism on autoclaved chocolate agar (Zinnemann 1960), and in noting the phenomenon of satellitism near a staphylococcus streak on 2 per cent proteose peptone agar previously inoculated with the organism (Zinnemann *et al.* 1968). Alternatively the strain may be seeded into glucose peptone water and into glucose peptone water to which X, V, or X+V factors have been added. The medium, which contains bromothymol blue, changes colour to yellow if growth occurs (Cooper and Attenborough 1968).

In the final test for V requirement 1 μg per ml of a pure preparation of nicotinamide adenine dinucleotide is added to a Difco Proteose Peptone No. 3 medium at a temperature below 45°; for X requirement, a stock solution of haemin is prepared by adding 2.5 mg of crystalline haemin to 1 ml of 0.2M KOH in 47.5 per cent ethanol and diluting to 5 ml with sterile water. The final dilution in the medium is 1 μg per ml (White 1963).

Media for growth

Various media are used for the isolation and study of the haemophilic bacilli. Ordinary blood agar is by no means satisfactory; far better results are obtained with media in which the red cells have been broken up, and their modified contents distributed throughout the medium. The well known 'chocolate' agar, prepared by adding blood to melted agar and raising the mixture to the boiling-point for 3 min is a considerable improvement on the ordinary blood agar plate, but it shares the disadvantage that the medium is opaque.

The medium devised by Levinthal (1918) has the advantage of being colourless and transparent and has proved useful in revealing the iridescent colonies of capsulated strains of *H. influenzae*. It is prepared by adding 5 per cent of defibrinated rabbit or human blood to melted agar in a flask, and raising it to boiling-point over a flame, with several shakings. The precipitate is allowed to settle, and the clear supernatant fluid is decanted or filtered through sterile glass-wool. The medium may, for safety, be sterilized by a further short heating, but must not be subjected to prolonged sterilization in the steamer. The simpler method devised by Pittman (1931) of preparing Levinthal agar may be recommended.

Fildes (1920) introduced a peptic digest of blood, which is preserved with chloroform and may be added to broth or melted agar as required. These media, which are transparent and have the colour of ordinary broth or agar, give copious growths of *H. influenzae* and inhibit the growth of many other organisms. They are admirably suited for the primary isolation of the influenza bacillus.

The ability to support the optimal growth of haemophilic bacilli is not a property of blood from all species of animals. Thus, Krumwiede and Kuttner (1938) describe thermolabile substances inhibiting the growth of *H. influenzae* and *H. parainfluenzae* in sheep, goat, bovine and human blood but not in the blood of the rat, rabbit or guinea-pig. When horse blood is lysed and held at 37° before its addition to the medium, the V factor liberated from the RBC is destroyed by a serum constituent; when the blood is lysed after addition to the medium, the concentration of serum is too low to destroy V factor. Human serum contains no inhibitor of V factor, but the stromata of the RBC are inhibitory, the effect being destroyed by heating (see Waterworth 1955)—as happens, for instance, in chocolate agar.

For primary isolation from such a source as the nasopharynx or the bronchial tract, advantage may be taken of the selective action of penicillin in concentrations inhibiting the growth of gram-positive cocci and diphtheroid bacilli but not *H. influenzae* (Fleming 1929). Hovig and Aandahl (1969) recommended for the same purpose the use of bacitracin (300 μg/ml) incorporated in chocolate agar plates.

Conditions for growth

The optimum temperature for growth is around 37° and the minimum between 20° and 25°. There is general agreement that *H. influenzae* grows better under aerobic than under anaerobic conditions. Statements about its ability to develop anaerobically are somewhat contradictory (see Kristensen 1922). According to Fildes (1921) it grows well initially on a suitable medium but quickly dies out. Some strains on first isolation require an increased partial pressure of CO_2 in the atmosphere. This character is of some assistance in classification (Table 39.1), but many of the strains rapidly become adapted to grow in ordinary air.

Type of growth

The type of colony given by *H. influenzae* on solid media varies widely with the strain and the kind of medium employed. On blood agar it forms tiny, transparent, pinpoint colonies, usually low convex but sometimes flat and tending to become confluent. On a more favourable medium, such as Levinthal's agar, and especially Fildes's agar, the colonies are far larger,

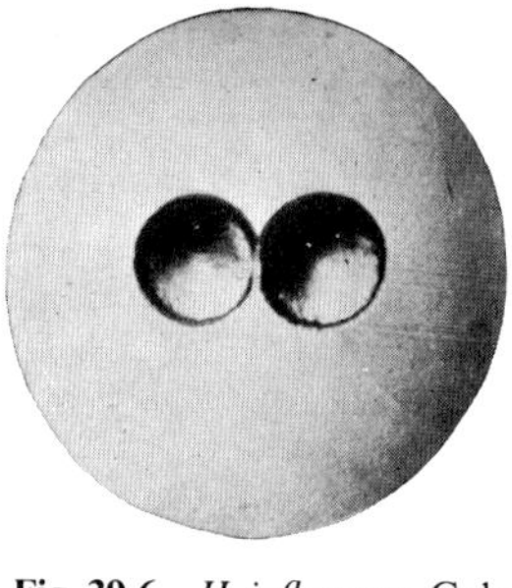

Fig. 39.6 *H. influenzae*. Colonies on Fildes's agar after 24 hr (×8).

0.5–0.8 mm after 24 hours' incubation. They are circular, raised and dome shaped, with a slightly splayed-out, entire edge; the colony is translucent and there is little differentiation (Fig. 39.6); the growth emulsifies easily. Strains isolated from acute infections may form mucoid colonies up to 3 mm diameter, owing to the presence of a capsule, but when non-capsulated they usually give rise to colonies with a smooth surface.

On further incubation, and frequently during the first 24 hr of growth, the colony becomes differentiated into a central portion with a granular or contoured surface, an intermediate flattened portion, and a sharp bevelled periphery with a narrow splayed-out edge. Between 24 and 48 hr there is usually an enlargement of the colonies, which may attain a diameter of 1–1.5 mm. This increase in size results in the formation of a flatter colony, retaining a raised central boss, sometimes smooth, sometimes granular or contoured (Fig. 39.7), and usually more opaque and more friable.

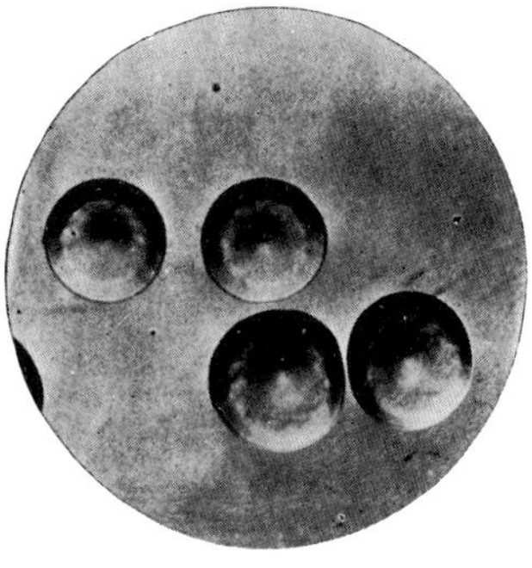

Fig. 39.7 *H. influenzae*. Colonies on Fildes's agar after 48 hr (×8).

Some colonies may be, from the start, flatter and more granular; others may be raised, conical and smooth, with little central differentiation. The morphologically typical strains tend to give, on Fildes's agar, smooth colonies with little differentiation. The morphologically atypical strains, or variants, tend to produce a more granular colony, with earlier and greater differentiation (Kristensen 1922, Smith 1931, Rosher 1947).

Pittman's (1931) 'smooth' strains have the colonial characters of capsulated bacteria. Pittman distinguished smooth and rough strains by their possession or not of a specific antigen, and associated smooth strains with acute infections in man. She regarded the two colonial types as examples of S→R variation, but many strains from infected processes and most of the strains from the normal respiratory tract have the characters of Pittman's rough strains, which would have been regarded by the earlier observers as 'smooth'.

Such difficulties may be resolved if we accept three forms of *H. influenzae*: (1) a mucoid, iridescent, capsulated M form with a characteristic specific soluble substance; (2) a non-capsulated, non-iridescent S form; and (3) a non-capsulated, non-iridescent R form (Chandler *et al.* 1937, 1939). The M form of this classification corresponds to Pittman's S form; the S and R forms together correspond to her R form.

The various colonial forms of the other haemophilus species generally resemble those of non-capsulated *H. influenzae*, with certain exceptions. *H. canis* colonies, at first indistinguishable from those of *H. influenzae*, become larger and more opaque as they grow older. On blood agar in 24 hr, *H. ducreyi* forms round, raised, greyish-white, glistening colonies 0.5–1.0 mm in diameter, which in three days may have a crateriform depression. *H. aphrophilus* forms small, highly convex granular colonies.

In liquid media, such as Fildes's broth, most strains of *H. influenzae* give rise to a uniform turbidity, with or without a slight powdery deposit. Some, on the other hand, give a flocculent deposit with a varying degree of turbidity of the supernatant fluid. Coccobacillary strains give a uniform turbidity. Many of those showing long bacilli or threads form flocculent growths.

H. canis grows diffusely with a slight deposit. The growth of *H. parainfluenzae* is usually flocculent or granular, occurring at the bottom of the tube, on the sides or in the form of a pellicle. *H. aphrophilus* gives a granular growth and a heavy deposit, the granules adhering to the sides of the glass.

Haemolysis

Pritchett and Stillman (1919) and Stillman and Bourn (1920) observed that a small proportion of haemophilic bacilli from patients with influenza and from normal persons gave a well defined zone of haemolysis on blood agar, and were either bacillary or filamentous in shape.

Many haemolytic strains are 'atypical' not only morphologically but in growth requirements (Kristensen 1922), being dependent on V factor but not on X factor (Fildes 1924). Valentine and Rivers (1927) reported that the majority of these haemolytic strains required only V factor for their growth, but that a minority required both X and V factors; a proportion of non-haemolytic strains of *Haemophilus* required only V factor. Later, Pittman (1953) used haemolysis as the primary feature that distinguished *H. haemolyticus* from *H. influenzae*, both of which required X and V factors, and *H. parahaemolyticus* from *H. parainfluenzae*, which both needed only V factor. However, Kilian (1976*a*) found the haemolysis of *H. parahaemolyticus* to be unstable; since loss of this property made the strains indistinguishable from *H. influenzae* he does not admit the species *H. parahaemolyticus* and includes such strains among the biotypes of *H. parainfluenzae*. Kilian (1976*b*) reported that porcine strains of *H. parahaemolyticus* (syn. *H. pleuropneumoniae*) acted synergistically with the β toxin of *Staphylococcus aureus* to give a CAMP-like reaction (see Chapter 30).

Resistance

Resistance to heat of the haemophilic bacilli is low; they are usually killed by exposure for 30 min to a temperature of 55°.

In-vitro susceptibility to chemotherapeutic agents

H. influenzae is commonly sensitive to sulphonamides. It shows some sensitivity to benzylpenicillin *in vitro*, but is considerably more sensitive to ampicillin and related compounds, the minimal inhibitory concentration of which for most strains is *ca* 0.5 mg per l. It is usually sensitive to tetracycline, chloramphenicol, gentamicin and polymyxin, and to trimethoprim and cotrimoxazole. However, resistance to many of these antimicrobial substances has been reported from different countries. Williams and Andrews (1974), working in Britain, tested the sensitivity of 68 isolates of *H. influenzae* from respiratory infections to 15 antimicrobial agents, and interpreted the results in terms of the blood levels to be expected in treated patients. All strains were sensitive to chloramphenicol, ampicillin and amoxycillin, and fairly sensitive to erythromycin. Ten per cent of the isolates were resistant to tetracycline. Sensitivity to cephalosporins was at an intermediate level. Trimethoprim was fairly active, but 28 per cent of strains showed a moderate degree of resistance to sulphamethoxazole, although a synergistic effect of the combination was found, all strains being judged sensitive to co-trimoxazole (but see May and Davies 1972).

Ampicillin resistance (Thomas *et al.* 1974) appears to be increasing in frequency (Nelson 1980); it is mediated by β-lactamases. The most common β-lactamase in *H. influenzae* is identical with the TEM 1 enzyme of the enterobacteria and is encoded by a transposon that may be in either a plasmid or on the chromosome (Saunders *et al.* 1978, Matthew 1979); a second plasmid-borne β-lactamase has been described (Rubin *et al.* 1981). Occasional chloramphenicol-resistant strains of *H. influenzae* (Manten *et al.* 1976) and *H. parainfluenzae* (Cavanagh *et al.* 1975) have been described. Plasmid-borne resistance to chloramphenicol and tetracycline can be transferred between strains of *H. influenzae* as a single unit (van Klingeren *et al.* 1978).

Biochemical activities

The results of different observers who have studied the fermentation reactions of haemophili are, owing to a number of factors, often at variance. The reactions are influenced considerably by the medium used, the time of observation and the other conditions of testing. A medium is required which allows good growth and, unless a pH meter is used, is sufficiently clear to reveal the colour changes of pH indicators. The degree of acidity produced may be low (Stillman and Bourn 1920), and acid may be produced in media devoid of sugar (Kristensen 1922, Tunevall 1951 *a*, *b*). Despite these difficulties, formation of acid and gas have been found to be of value for classification. Table 39.2 represents a consensus view of these and other biochemical reactions, based especially on the work of Stillman and Bourn (1920), Kristensen (1922), Dible (1924), Fildes (1924), Smith (1931), Sneath and Johnson (1973), Kilian (1976*a*), Zinnemann (1980) and Broom and Sneath (1981).

All haemophili, with the exception of *H. ducreyi*, ferment glucose. Gas is formed by *H. aphrophilus*, *H. paraphrophilus*, and some members of several other species (Table 39.2; see

Table 39.2 Biochemical characters of species of *Haemophilus*

Species	Glucose, A	Glucose, G	Sucrose, A	Lactose, A	Maltose, A	Mannitol, A	Xylose, A	Galactose, A	Laevulose, A	Production of indole	Production of oxidase	Production of catalase	Production of urease
influenzae	+	−	−	−	∓	−	±	+	±	±	+	+	±
aegyptius	+	−	−	−	∓	−	∓	+	+	−	+	+	+
haemolyticus	+	∓	−	−	+	−	±	∓	∓	∓	+	+	+
parainfluenzae	+	±	+	−	+	−	−	∓	±	−	+	±	±
parahaemolyticus	+	∓	+	−	+	∓	∓	−	?	−	+	±	±
suis and *parasuis*	+	−	+	−	+	−	−	C	−	−	−	+	−
haemoglobinophilus (syn. *canis*)	+	−	+	−	−	+	+	+	+	+	+	+	−
gallinarum and *paragallinarum*	+	−	+	V	V	+	+	V	V	−	−	+	−
aphrophilus	+	+	+	+	+	−	−	+	+	−	−	−	−
paraphrophilus	+	+	+	+	?	−	−	?	?	−	+	−	−
ducreyi	−	−	−	−	−	−	−	−	−	−	−	−	−

A = acid; G = gas; ± = variable, most strains positive;
∓ = variable, most strains negative; V = variable;
C = reports conflicting.

Kilian 1976*a*). The fermentation of lactose characterizes *H. aphrophilus* and *H. paraphrophilus*, and has been described in strains of *H. gallinarum* and *H. paragallinarum*.

Almost all strains of *Haemophilus* reduce nitrate (Kilian 1976*a*). Catalase and oxidase formation by cultures grown on chocolate agar may be used for distinguishing species of *Haemophilus*, but catalase activity is variable according to the conditions of growth. For example, Biberstein and Gills (1961) noted that catalase-negative strains might become positive when supplied with enough haemin. Kilian (1976*a*) examined 188 strains of *H. influenzae* grown on chocolate agar and found all to be oxidase positive; the reaction of other haemophili was variable, except for *H. aphrophilus* and *H. ducreyi* which were invariably negative. Many strains of *H. influenzae*, *H. parainfluenzae* and *H. parahaemolyticus* (Gibson 1970, Kilian 1976*a*), but few of other species of *Haemophilus*, form urease (Table 39.2).

H. influenzae and *H. parainfluenzae* have been subdivided by Kilian (1976*a*) into biotypes. *H. influenzae* biotype I, isolated mostly from acute infections, is usually capsulated, and indole, urease and ornithine decarboxylase positive. Biotype II is found in the upper respiratory tract and may be isolated from infections; it is distinguished from biotype I by lack of ornithine decarboxylase. Biotype III strains are isolated from the conjunctiva and normal throat; they are non-capsulated and often long and slender morphologically. In contrast to the other biotypes they haemagglutinate RBC and are indole negative; they include strains corresponding to the species *H. aegyptius*. Biotype IV is somewhat heterogeneous, but biotype V is a homogeneous group of non-capsulated, urease negative, indole-positive strains isolated from ear infections. *H. parainfluenzae* was tentatively divided into three biotypes, differing in biochemical features such as production of urease and ornithine decarboxylase, and of gas from glucose. Kilian (1976*a*) found that types II and III of *H. parainfluenzae* were sometimes haemolytic, but since this property was unstable he proposed that such strains should be included among the biotypes of *H. parainfluenzae* rather than accorded specific status as *H. parahaemolyticus*. Kilian (1976*a*) supported the validity of his proposed biotypes by the observation that they also showed differences in pathogenicity. More recent studies by others have shown that the biotypes are sufficiently stable to be of epidemiological value (Bruun and Friis-Møller 1976, Oberhofer and Back 1979).

Antigenic constitution

Serological studies of *H. influenzae* first revealed a wide antigenic heterogeneity.

By direct agglutination with 20 antisera, followed by absorption tests where necessary, Park, Williams and Cooper (1918) could find only four identical pairs among 160 strains. This extreme heterogeneity was amply confirmed by numerous workers. Iizuka (1938), for example, recorded more than 50 different agglutinating types among 249 strains isolated from sick and healthy persons.

The complexity was to some extent resolved by the discoveries of Pittman (1931). Among 97 strains of influenza bacilli, she noted 15 producing colonies of her S type (p. 384), all isolated under conditions suggestive of pathogenicity. When tested by agglutination they fell into two antigenic classes, A and B, one containing 12 strains, the other three. It was possible to separate from these 15 strains a soluble specific substance, apparently carbohydrate, and presumably associated with the capsule, which in precipitin tests indicated the same antigenic grouping as the agglutination tests. Her S strains, in artificial culture, readily gave rise to non-capsulated variants, usually with a bacillary or filamentous morphology, which no longer produced the soluble specific substance. These findings were confirmed, in their essential points at least, by several subsequent workers (see Dochez *et al.* 1932, Wright and Ward 1932, Platt 1937). It will be recalled (see p. 385) that Pittman's S forms correspond to the M forms of the Chandler notation, and we shall refer to them as such.

Characteristic M strains are commonly isolated from infections of the meninges, the bronchi and the pleural cavity, the lung, the middle ear (Bjuggren and Tunevall 1950) and the larynx (Huntington 1935, Jones and Camps 1957). Among strains isolated from the respiratory tract of healthy carriers or persons with mild infections the M types are less common. It should be noted, however, that the non-capsulated S strains, which are more frequently isolated than capsulated strains from various kinds of infection, are almost certainly pathogenic; in, for example, bronchiectasis (Allison *et al.* 1943, Franklin and Garrod 1953, Allibone *et al.* 1956) and chronic bronchitis (Brown *et al.* 1954, Knox *et al.* 1955, Edwards *et al.* 1957). They are also found in the blood after tonsillectomy, in the urine of patients with damaged kidneys, in the inflamed appendix, and in infections of bone and joints (see Zinnemann 1960, Rogers *et al.* 1960).

Pittman divided her capsulated smooth strains into six serological types, a, b, c, d, e, and f. Type b occurs most frequently among capsulated strains from infections (see, e.g. Silverthorne *et al.* 1943, Alexander 1943, Straker 1945), and the type b antigen can be detected in the body fluids, including the urine, of infected patients (Newman *et al.* 1970, Kaldor *et al.* 1979).

As in the pneumococci, type specificity depends on capsular polysaccharides (Platt 1937, Alexander and Heidelberger 1940), which are distinguishable by chemical and physical as well as by serological means (MacPherson *et al.* 1946). The specific substances of types a, b and c are polyribophosphates (Zamenhof *et al.* 1953, Zamenhof and Leidy 1954, Branefors-Helander *et al.* 1977). The capsule of the type e strain is complex; some strains have two distinguishable antigens e_1 and e_2, others e_2 alone (Williamson and Zinnemann 1951, 1954). Tunevall (1952*a*) reports that capsulated strains, but not non-capsulated strains, release hyaluronic acid in the culture medium; none of his capsulated strains produced a hyaluronidase. Whilst the capsulated form of the pneumococcus alone is virulent for mice and pathogenic for human beings, both the capsulated and the S forms of the influenza bacillus are virulent and pathogenic. The capsulated M forms of this bacillus are nevertheless more virulent for mice and more pathogenic for man than the non-capsulated S forms; they are found, for example, in the more severe cases of meningitis; the less severe cases are caused by the non-capsulated S forms (Gordon *et al.* 1944). The type b capsular polysaccharide is a protective antigen and vaccines effective against haemophilus meningitis have been prepared from the purified material (Anderson *et al.* 1972; see Chapter 65).

The serological types may be determined in the whole bacilli by the agglutination reaction and by the capsular swelling reaction (Alexander *et al.* 1946), and in extracts of the bacilli by precipitation (see Zinnemann 1960).

The M→S change associated with loss of a capsule does not always result in a complete loss of the specific polysaccharide; MacPherson (1948), for example, found some type b substance in phenol extracts of non-capsulated S forms derived from M type b. Buckmire (1976) described a 'class I' variant which contained about a quarter of the polyribophosphate capsular material of the mucoid parent and a 'class II' variant with one-tenth the amount.

The parallelism demonstrable between capsulated pneumococci and influenza bacilli extends to the genetic determination of the type capsular substance. As we saw in Chapter 29, type transformation of R forms can be induced, and even hitherto undescribed mixed types obtained, by the addition of DNA from heterologous S forms (Tunevall 1952*b*). The capsular polysaccharides of some influenza bacilli and some pneumococci are antigenically related, and the type b capsular antigen is immunochemically similar to the K1 antigen of *Esch. coli* (Schneerson *et al.* 1972).

Little exact information is available on the antigenic structure of the non-capsulated S forms. Both Platt (1939) and Tunevall (1953) brought evidence for the existence of a species-specific antigen. From the result of agglutination experiments made with antigens and antisera prepared from 6-hr capsulated and 24-hr non-capsulated strains, heated to 56° and unheated, May (1965) concluded that in the species *H. influenzae* three main antigens were present: (1) an M antigen, already described, occurring in 6-hr capsulated cultures, disappearing largely after 24 hr; (2) a thermolabile somatic type-specific S antigen destroyed by heating at 56° within 24 hr; and (3) a thermostable somatic species-specific R antigen. The M strains contain all three antigens, but the S and R antigens are covered over by the capsule in 6-hr strains and cannot therefore take part in agglutination. The S strains contain both S and R antigens, and the R strains R alone. In 24-hr non-capsulated cultures the S antigen can be revealed, but as it is destroyed by heat at 56° within 24 hr, agglutination tests must be carried out at a lower temperature. Agglutination, however, is not a satisfactory method for determining the type specificity of S strains because so many strains are autoagglutinable; for this purpose the precipitation reaction must be used. How many type-specific antigens there are is still uncertain. Omland's (1964) studies support the common view that there is great irregularity in the distribution of these antigens between strains.

H. influenzae possesses endotoxin activity (DeClerq and Merigan 1969); as in other gram-negative bacilli, this is associated with the lipopolysaccharide, which can be extracted with hot phenol-water. In haemagglutination studies with antiserum against lipopolysaccharide from *H. influenzae* type b absorbed with lipopolysaccharides from the other capsular types, three antigenic specificities were found (Flesher and Insel 1978).

The antigenic structure of the parainfluenza bacilli is still undefined; Miles and Gray (1938) found some antigenic relation between non-haemolytic strains of *H. parainfluenzae*. Though Knorr (1924) observed cross-agglutination and precipitation between strains of *H. influenzae* and *H. aegyptius*, Pittman and her colleagues were unable to confirm this. They found a substantial similarity among the strains of *H. aegyptius*, and noted that they contained a haemagglutinin for human RBC not found in *H. influenzae* strains (Pittman and Davis 1950, Davis *et al.* 1950). *H. para-*

haemolyticus (*H. pleuropneumoniae*), which causes pleuropneumonia in swine, is a capsulated organism. Nicolet (1971), who studied strains from different countries, established the existence of at least three serotypes based on the nature of the M antigen. Gunnarsson and his colleagues (1977) recognized five serotypes, and obtained type-specific antigens by means of phenol-water extraction (Gunnarsson 1979). Bakos and co-workers (1952) distinguished S and R forms of *H. suis*, and classified 43 out of 50 S forms into three serological types. Their strains contained haemagglutinins for the red blood cells of various animals which were independent of serological type.

Pathogenicity

The part played by *H. influenzae* in the causation of human disease is by no means easy to define. Turk and May (1967) classify the diseases with which it is associated into three groups. (1) Acute, usually severe, pyogenic diseases undoubtedly caused by *H. influenzae*, almost always of the capsulated type b form (see Greene 1978); the organism is usually present in pure culture and is commonly found in the blood stream (see Farrand 1969). Of this group meningitis is the typical example; epiglottitis in children and suppurative arthritis may likewise be included. (2) Diseases not caused primarily by *H. influenzae*, but in which this organism appears to play an important secondary part. The typical examples of this group are chronic bronchitis and bronchiectasis; the organisms present are usually of the non-capsulated S type (Turk and Holdaway 1968). (3) Diseases in which the occurrence of *H. influenzae* is incidental. In this group there are many examples, such as pneumonia, bronchopneumonia and otitis media, but because this organism is frequently present in the normal upper respiratory tract it is often impossible to distinguish between this group and the second group.

Of other members of the *Haemophilus* group the Koch-Weeks bacillus (*H. aegyptius*) gives rise in many parts of the world to epidemic conjunctivitis, mainly in children. *H. parainfluenzae* is occasionally associated with acute pharyngitis and occasionally causes endocarditis (Jemsek *et al.* 1979). *H. haemolyticus* is likewise found in acute pharyngitis (see Miles and Gray 1938).

H. ducreyi is responsible for soft chancre and for the buboes that are sometimes associated with it (see Chapter 73). *H. suis* is associated with a virus in the causation of swine influenza (see Chapter 96). It appears to be the cause of Glässer's disease of pigs, which is characterized by fibrinous inflammation of serous cavities, meninges, and joints (Hjärre and Wramby 1943). *H. canis* was isolated by Friedberger (1903) from 19 out of 20 dogs suffering from a suppurative inflammation of the prepuce; but he was unable to reproduce the disease with it, and concluded that it was a harmless parasite of the preputial sac. It was isolated from normal dogs by Krage (1910), Kristensen (1922), Rivers (1922*b*), and Kirchenbauer (1934). *H. parahaemolyticus* gives rise to pleuropneumonia of swine—a disease that varies in severity in different countries (Shope 1964, White *et al* 1964). Since its first isolation by Khairat (1940), *H. aphrophilus* has occasionally been cultured from the blood stream of patients suffering from endocarditis, but there is reason to believe that not all these strains were correctly identified (Khairat 1971). According to Kraut and her colleagues (1972) it is frequently present between the gums and the teeth and forms part of the normal flora of the mouth. Among patients in whom it has been identified as a pathogen, 26 per cent have been reported to have had underlying malignant disease (Bieger *et al.* 1978).

Experimental infection

H. influenzae By the intratracheal injection of highly virulent strains Blake and Cecil (1920) reproduced in monkeys a disease bearing many resemblances to influenza.

The injection of large doses of living culture (the growth from ½ to 1 blood agar slope suspended in saline) into the peritoneum of rabbits, guinea-pigs, or mice, often results in death within 24–48 hr. The organisms are isolable from the peritoneal cavity but not often from the heart's blood. The cause of death seems to be a toxaemia rather than an invasive infection (see Pfeiffer 1893, Delius and Kolle 1897, McIntosh 1922). Similar results are obtained with filtrates of cultures in liquid media, which in 0.5–5.0 ml doses may produce death on intravenous injection into rabbits or guinea-pigs (see Parker 1919, Ferry and Houghton 1919, Wollstein 1919, McIntosh 1922). The wide variation in virulence is probably a reflection of the origin of the strains and their state when tested. Strains of *H. influenzae* isolated from cases of meningitis are usually far more virulent for laboratory animals than strains isolated from the respiratory tract, and some of these meningeal strains have definite invasive powers (Cohen 1909, Henry 1912, Wollstein 1915). Capsulated strains have a significantly higher virulence than non-capsulated (Chandler *et al.* 1937, 1939). The incorporation of the intraperitoneal infecting dose in mucin—a technique which was first successfully applied to enhancing the virulence of meningococci (see Chapter 65)—increases the virulence of *H. influenzae*; small doses produce a fatal septicaemia in mice against which anti-*influenzae* horse serum is protective (Fothergill *et al.* 1937). Certain strains of *H. influenzae* of human origin give rise to a fatal infection after intracerebral injection of about 2000 organisms into mice; other strains, and strains of *H. parainfluenzae*, are non-virulent by this route (de Torregrosa and Francis 1941). The newborn rat is susceptible to intranasal challenge with capsulated *H. influenzae* type b strains, the resulting infection giving rise to bacteraemia and meningitis (Moxon *et al.* 1974, 1977). Using a mixture of streptomycin-sensitive and -resistant strains, Moxon and Murphy (1978) deduced that the blood of the rat was often infected by a single organism from the nasopharynx, and the cerebrospinal fluid by a single blood-borne organism.

H. suis In association with a filtrable virus (see Chapter 96) this organism is an important natural pathogen of swine. The disease can be experimentally produced in these animals. In relation to the small animals of the laboratory it appears to behave much in the same way as *H. influenzae*. Large intravenous injections may be fatal for rabbits, and large intraperitoneal injections for guinea-pigs or mice; but the results are very irregular, and there appear to be great differences in the virulence, or toxicity, of different strains (see

Lewis and Shope 1931, Kirchenbauer 1934). Freshly isolated strains, from swine with influenza or Glässer's disease, produce infections in guinea-pigs lethal in 14–20 hr, the organisms being cultivable from the serous cavities, meninges and heart blood; in young pigs they induce an infection resembling naturally occurring Glässer's disease (Bakos *et al.* 1952). *H. parasuis* produces a septicaemic disease in experimentally infected swine (Riley *et al.* 1977).

H. parahaemolyticus This organism which, as already stated, gives rise to porcine contagious pleuropneumonia, is pathogenic to swine when injected intranasally, causing up to a 50 per cent mortality. In this respect it differs from *H. suis* which, even in high dosage, is harmless to swine in the absence of the accompanying virus. The virulence of the organism appears to vary somewhat, but highly virulent strains may cause disease in swine in a dose as low as 100 organisms (White *et al.* 1964).

H. parainfluenzae and ***H. aphrophilus*** Both of these organisms are non-pathogenic to laboratory animals.

H. gallinarum Some strains of this organism kill mice when injected intraperitoneally (McGaughey 1932). Pure cultures given intranasally to chickens regularly cause coryza (Eliot and Lewis 1934), as does *H. paragallinarum* (Hinz and Kunjara 1977).

H. canis Information about the pathogenicity of this organism for laboratory animals is scanty (but see Rivers 1922*b*).

H. ducreyi Tomasczewski (1903) was successful in reproducing soft chancre in human subjects with pure cultures. In man, progressive purulent lesions follow the intradermal injection of cultures; and it was apparently a common practice to separate *H. ducreyi* from contaminating saprophytes in genital material by injecting it intradermally into the patient (see Cunha 1939). Ulcerative lesions have followed the intracutaneous inoculation into monkeys and rabbits of cultures several generations removed from primary isolation (Reenstierna 1921, Nicolle 1923, Feiner and Mortara 1945). A high proportion of successes is obtained in rabbits by the intracutaneous injection of cultures in defibrinated rabbits' blood (Kaplan *et al.* 1956).

Classification

Table 39.1 shows that the members of this group can be classified on their need for X and V factors and on their haemolytic powers. Demand for an increase in the partial pressure of CO_2 in the atmosphere on isolation is also of some help. Below we give a summarized description of *H. influenzae*, and some shorter notes on the other members. (For taxonomic studies, see Sneath and Johnson 1973, Kilian 1976*a*, Zinnemann 1980, Broom and Sneath 1981.)

Haemophilus influenzae

Isolated by Pfeiffer (1892) from cases of influenza in man. A strict parasite, living particularly in the upper respiratory tract of man.

Morphology In its typical form *H. influenzae* is a tiny coccobacillus (1–1.5 by 0.3–0.4 μm). In its virulent mucoid (M) form it is capsulated. Among any large sample of strains, or in any one strain during prolonged subculture in the laboratory, wide departures from the typical morphology will usually be found. These include longer rods, filaments and irregular forms. (For a fuller description see p. 381.)

Growth requirements *H. influenzae* requires both the X factor and V factor for growth. It grows far more readily under aerobic than under anaerobic conditions; it would appear that some strains are incapable of prolonged anaerobic subcultivation. The optimal temperature for growth is in the neighbourhood of 37°. Added CO_2 does not improve growth.

Growth on solid media The usual types of colony seen on Fildes's or Levinthal's medium were described on p. 384. At 24 hr, the colonies of fully capsulated (M) strains are smooth, slightly opaque and slightly mucoid, and are somewhat larger (diameter 1.3 mm) than those of uncapsulated strains (diameter 0.5–0.8 mm), which are usually translucent and butyrous in consistency. Capsulated (M) strains also differ from S and R strains in being iridescent by oblique transmitted light. Minute colonies of L forms develop on agar containing penicillin and glycine (Lapinski and Flakas 1967).

Growth in liquid media Most strains give rise to a uniform turbidity, with or without a slight powdery deposit. Some give a more flocculent deposit. The latter usually show an atypical morphology, and the colonial appearances associated with the more advanced stage of R variation.

Resistance See p. 385.

Biochemical activities See Table 39.2. Five biotypes have been recognized, distinguished by their indole, urease and ornithine decarboxylase activity and other properties. Biotype I is characteristic of acute infection and is usually capsulated, produces indole and is active against both urea and ornithine.

Antigenic structure There appear to be four main sets of antigens: (a) type-specific polysaccharide antigens, of which six well differentiated types are recognized, a, b, c, d, e and f, present in the M forms of the organism; (b) probably type-specific somatic antigens present in M and S forms; of these there appear to be many, but little is known about them; (c) a lipopolysaccharide endotoxin, in which a number of antigenic specificities may occur, differing in their distribution among the capsular serotypes; (d) a species-specific somatic antigen present in M, S and R forms.

Pathogenicity Pathogenic for man (see p. 388). Produces toxic death when injected in large doses into laboratory animals, and infective death when injected in small doses together with mucin. The more pathogenic, capsulated strains cause an invasive infection when given intranasally to infant rats.

(For fuller description of *H. influenzae*, see Turk 1982.)

Haemophilus aegyptius

Isolated by Koch (1887) and Weeks (1887) from cases of acute conjunctivitis and known as the Koch–Weeks bacillus. Morphologically and biochemically it closely resembles non-capsulated *H. influenzae*, but is said to be more rod-shaped (Mazloum *et al.* 1982). It grows rather more slowly, and is consistently rather than occasionally indole negative; few strains acidify xylose. After 24–36 hr in semi-fluid medium (0.15 per cent agar) the colony is fluffy, with comet-like projections, whereas colonies of capsulated *H. influenzae* are fluffy without any projections, and those of non-capsulated *H. influenzae* are granular. Cultures in Fildes's broth agglutinate 0.5 per cent suspensions of human

RBC at room temperature, the haemagglutinin being specifically inhibited by *H. aegyptius* antiserum. Haemagglutinin-formation is pronounced in strains from acute conjunctivitis (Pittman and Davis 1950; but see Smith 1954). There is a high degree of genetic homology between the two organisms (Leidy *et al.* 1965).

Distinguishable by its agglutinogens, which form a separate though heterogeneous group (Pittman and Davis 1950). Contains some antigens in common with *H. influenzae*, and some peculiar to *H. aegyptius* (Olitzki and Sulitzeanu 1959).

Non-pathogenic for mice intravenously in doses that, if of *H. influenzae*, would kill in 4 days (Orfila and Courden 1961).

Both *H. influenzae* and *H. parainfluenzae* also occur in association with acute conjunctivitis (see, e.g. Pittman and Davis 1950, Huet and Benslama 1956).

Haemophilus haemolyticus

Described by Pritchett and Stillman (1919) as a haemolytic form of *H. influenzae*. Has the same general characters as this organism, but lyses RBC in both solid and liquid medium. Minor differences from *H. influenzae* in saccharolytic properties and frequency of indole production have been described; some strains produce gas from glucose (see Table 39.2).

Haemophilus parainfluenzae

Isolated by Rivers (1922*a*) from cases of acute pharyngitis and bacterial endocarditis in man. Differs from *H. influenzae* in requiring the V but not the X factor for growth, and in fermenting maltose, sucrose, and dextrin more frequently (Table 39.2). Both capsulated, antigenically homogeneous and non-capsulated, antigenically heterogeneous strains occur.

Haemophilus parahaemolyticus

A haemolytic variety of *H. parainfluenzae* distinguished by Pittman (1953). A similar organism isolated in the USA from cases of contagious porcine pleuropneumonia was called *H. pleuropneumoniae* (Shope 1964, Shope *et al.* 1964, White *et al.* 1964). Capsulated. Requires V but not X factor. Organisms form long tangled threads staining unevenly. Colonies larger than those of other haemophili, less translucent, and surrounded by a wide zone of haemolysis, but the haemolytic property is unstable (Kilian 1976*a*). Stringy flocculent deposit in broth. Ferments glucose, sucrose and maltose. Three serological types recognized on basis of M antigens. Gives rise to pneumonia on intranasal injection into swine in the absence of a virus. According to Kilian, Nicolet and Biberstein (1978), who studied 47 *Haemophilus* strains isolated mainly from pneumonic lesions in pigs, the correct name for this organism is *H. pleuropneumoniae*. May be related to or identical with the organism isolated in England from a similar but milder disease in swine and designated '*H. parainfluenzae*' (Pattison *et al.* 1957, Matthews and Pattison 1961). An organism resembling *H. parahaemolyticus* but requiring extra CO_2 for good growth and haemolysis was described by Zinnemann and his colleagues (1971) under the name of *H. paraphrohaemolyticus*.

Haemophilus suis and *H. parasuis*

H. suis was isolated by Shope (1931; see also Lewis and Shope 1931) from cases of swine influenza in which it is associated with a virus; also found in Glässer's disease of swine. Called originally *H. influenzae-suis*. Grows rather poorly and requires both X and V factors. Does not produce indole. Ferments glucose, sucrose and maltose. Three serotypes based on capsular antigens are described. *H. parasuis* differs from *H. suis* only in not requiring the X factor for growth (Biberstein and White 1969), although the dependence of *H. suis* on X factor has been debated (Kilian 1976*a*).

Haemophilus haemoglobinophilus (syn. *canis*)

Isolated by Friedberger (1903) from the prepuce of dogs and called *Bacillus haemoglobinophilus canis*. Parasitic but not pathogenic. Requires X but not V factor for growth. Old colonies are larger and more opaque than those of *H. influenzae*. For biochemical characters see Table 39.2.

Haemophilus gallinarum

Isolated by de Blieck (1931, 1932, 1934) in Holland from fowls suffering from coryza and called *Bacillus haemoglobinophilus coryzae gallinarum*; and by McGaughey (1932) in England from a coryzal disease of fowls. Named *H. gallinarum* by Eliot and Lewis (1934). Pure cultures give rise to coryza in chickens and turkeys when given intranasally. Extra CO_2 needed for isolation. Requires V factor for growth; about X factor there is a difference of opinion. Biberstein and White (1969) would regard strains requiring X factor as *H. gallinarum* and those requiring only V factor as *H. paragallinarum*. For sugar reactions see Table 39.2. Does not form urease or oxidase, nor does it produce indole. (See also Sawata *et al.* 1980.)

Haemophilus aphrophilus

Isolated by Khairat (1940) from cases of infective endocarditis. Since reported by several other workers, but not all strains were correctly identified (Khairat 1971). Strains require X factor and extra CO_2 for growth on first isolation but both requirements may be lost on subculture. V factor is not needed. For biochemical reactions, see Table 39.2, and Boyce and his colleagues (1969). Non-pathogenic to guinea-pigs and mice, but may be associated with infective endocarditis in man. An organism isolated from this disease and from various other lesions and found to require the V factor and extra CO_2 for growth but not the X factor was called *H. paraphrophilus* (Zinnemann *et al.* 1968). Similar strains which were haemolytic have been described as *H. paraphrohaemolyticus* (Zinnemann *et al.* 1971; see also Frazer *et al.* 1974).

Haemophilus ducreyi

Isolated by Ducrey (1890) from soft sore.

Morphologically in the purulent discharge from the ulcerated surface of the lesion the organisms appear as small ovoid rods, arranged in pairs, in groups, or in chains lying parallel to one another. Several forms may, however, be assumed. Thus, it may appear as a short rod with parallel sides and rounded ends, staining evenly; or ovoid or navicular in shape with bipolar staining; or in pairs end-to-end, having a dumb-bell appearance. The bacillus is about 1.1–1.5 μm long by 0.6 μm broad (Stein 1929). It may be intra- or extracellular in position. On solid media the organisms appear as isolated individuals, in groups, and in short chains; in fluid media very long chains are frequently formed, and in certain media there is a pellicle with dependent 'stalactites' of growth (Cunha 1939, 1943).

Growth requirements and cultural appearances The organisms may be cultivated by inoculating scrapings from the floor of the ulcer on to a medium consisting of 3 per cent agar containing 20–33 per cent defibrinated rabbit's blood; the medium should be prepared on the day of inoculation, and should be distributed into wide tubes having a large surface exposed to the air (Nicolle 1923, Reenstierna 1923, Nicolle and Durand 1924). Several tubes should be inoculated, and incubated at 35°. Colonies appear in 24 hr. On blood agar after 24 hr the colonies are 0.5–1.0 mm in diameter, low, convex, greyish-white and glistening, with a smooth surface and entire edge; after 2 to 3 days, they may reach a diameter of 2 mm, and the surface may show a crateriform depression. According to Hunt (1935), growth occurs best in sealed tubes, suggesting that it is promoted by an increased partial pressure of CO_2, but stock strains do not require CO_2 for growth (Kilian 1976*a*, Broom and Sneath 1981).

The necessity for blood in the medium, and a low partial oxygen pressure, is stressed by Sanderson and Greenblatt (1937). Watanabe (1939) confirmed the necessity for blood; rabbit blood was best, followed by that of the goat, sheep, ox or man. Deacon and his colleagues (1956) recommend primary culture in fresh clots of human blood, and subculture in media containing defibrinated rabbit blood; and, in conformity with Reymann (1947), excess of CO_2 in the atmosphere.

According to Lwoff and Pirosky (1937) *H. ducreyi* requires X but not V factor for growth. Only small quantities of haemin are required. The growth of some strains in the absence of blood or serum in the medium (Hababou-Sala 1925, de Assis 1926) may be attributed to the presence of small but sufficient quantities of haemin in nutrient broth.

Another medium that is recommended for primary isolation consists of 1 part of 5 per cent glycerine agar and 4 parts of Besredka's egg medium (Hababou-Sala 1925). (For media more recently advocated for primary isolation, see Hammond *et al.* 1979, Hafiz *et al.* 1981, Hannah and Greenwood 1982.) Incubation at 34–36°, with added moisture and CO_2, is generally recommended.

In Martin's broth to which 20 per cent of defibrinated rabbit's blood has been added, the organism develops rapidly, forming granules, which are suspended in the liquid or become attached to the walls of the tube. After a few days, an incomplete film may form on the surface. Cultures in this medium remain viable in the incubator for at least 10 days.

For preservation the organism should be inoculated into a medium consisting of 0.25 per cent of nutrient agar, 1 per cent starch, and 20 per cent of defibrinated rabbit's blood. Cultures on this medium remain alive for a month at incubator temperature, and for a similar period at room temperature, provided they are previously incubated for 5 days.

Resistance *H. ducreyi* is not specially resistant; it is killed by moist heat at 55° within 1 hr, and by 0.5 per cent phenol in a short time.

Biochemical activities Reymann (1949) records reduction of nitrate but no fermentation of carbohydrate or growth in milk; V.P. –, indole –, H_2S –, NH_3 –; no liquefaction of gelatin, coagulated egg albumin or serum, but has phosphatase activity (Hannah and Greenwood 1982).

Antigenic structure Suspensions from blood agar cultures are agglutinated by a specific antiserum; this reaction may be used for identification.

Pathogenicity *H. ducreyi* causes soft chancre in man. Monkeys and rabbits have been successfully infected. The organism has a low pathogenicity for chick embryos (Anderson and Snow 1940). (See also Stein 1929.)

'Haemophilus' equigenitalis

Special mention must be made of the organism responsible for contagious metritis of mares (Chapter 56), tentatively named *H. equigenitalis* by Taylor and his colleagues (1978). This is a small, non-motile, gram-negative rod that forms minute colonies on chocolate Columbia blood agar after incubation for 48 hr in air with 5–10 per cent of added CO_2 or with a reduced partial pressure of O_2. It shows little if any growth on unheated blood agar. Neither X nor V factor is required. The optimum temperature for growth is 37°. Catalase, phosphatase and oxidase are formed, and the porphyrin test is positive (Tainturier *et al.* 1981) but sugars are not attacked, nitrate is not reduced, and nearly all other biochemical tests give negative results. The organism is sensitive to all penicillins, tetracycline, chloramphenicol, erythromycin and all aminoglycosides except streptomycin; it is resistant to lincosamides, sulphamethoxazole and trimethoprim (see Sugimoto *et al.* 1981). The G+C content of the DNA is 36 moles per cent. According to Timoney and coworkers (1979) it exhibits a remarkable degree of heat sensitivity; in vaginal discharge it is reported to be destroyed at 50° in less than 1 min and at 40° in 27 min. (For a study of its ultra-structure, see Swaney and Breese 1980.) This organism does not fit into the *Haemophilus* genus (Brewer and Corbel 1983)

Bordetella

Definition

Minute coccobacilli. Motile and non-motile species occur. Non-sporing, gram-negative, non-acid-fast. On first isolation growth may occur only on complex media. Carbohydrates not fermented. All known species appear to be obligatory parasites, inhabiting the respiratory tract, and are pathogenic. Mole percentage G+C in DNA: 61–66.

Type species: *Bord. pertussis.*

The three organisms included in this genus are *Bord.*

pertussis, described by Bordet and Gengou (1906, 1907, 1909) as the cause of whooping cough; *Bord. parapertussis*, isolated by Bradford and Slavin (1937) and Eldering and Kendrick (1938), also from cases of whooping cough; and *Bord. bronchiseptica*, isolated by Ferry (1911) from the respiratory tract of dogs suffering from distemper and named by him *Bacillus bronchicanis* (see also M'Gowan 1911). Ferry (1912, 1912–13) later found the same bacillus in guinea-pigs, rabbits and monkeys, and changed its name to *Bacillus bronchisepticus*. Though previously grouped with *Brucella*, the conspicuous antigenic similarity of this organism to *Bord. pertussis*, its production in guinea-pigs of lesions similar to those caused by *Bord. pertussis* in rabbits and puppies (Smith 1913, Mallory and Hornor 1912, Mallory *et al.* 1912), and in the rabbit lung of lesions similar to those in human whooping cough (Rhea 1915)—all bring it into closer relation with *Bord. pertussis* than with *Brucella abortus*. A close genetic relationship between the species has been demonstrated by DNA–DNA reassociation; the relative binding of labelled *Bord. pertussis* DNA with *Bord. parapertussis* DNA was 88–94 per cent and with *Bord. bronchiseptica* DNA 72–93 per cent of that of homologous DNA (Kloos *et al.* 1978). Although, in consequence, Kloos and his colleagues questioned the validity of according these organisms separate specific status, we still accept the view of Lopez (1952) that the genus *Bordetella* comprises three species. (See also Johnson and Sneath 1973.)

Morphology

The organisms are rod-shaped, coccoid, or oval, 0.5–1.0 by 0.3–0.5 μm, arranged singly, in pairs, or small groups. They cannot be distinguished with certainty from members of the *Haemophilus* group, but long bacillary and thread forms are far less common. In virulent strains of *Bord. pertussis* rod forms predominate, in avirulent strains coccobacilli (Ungar *et al.* 1954). *Bord. parapertussis* and *Bord. bronchiseptica* are mainly rod-shaped. Electronmicroscopy reveals the usual structure of gram-negative bacilli (Hatasa 1964, Richter and Kress 1967). When freshly isolated, *Bord. pertussis* may possess a poorly defined capsule. *Bord. bronchiseptica* is furnished with peritrichate flagella; the other two members are non-motile. Freshly isolated strains of *Bord. pertussis* (Morse and Morse 1970) and some strains of *Bord. bronchiseptica* (Irons and MacLennan 1978) may possess fimbriae. L forms (Wittler 1951) and sphaeroplasts (Mason 1966) appear in cultures containing glycine.

Growth and cultural characters

For primary isolation of the whooping-cough bacillus Bordet and Gengou (1906) employed an agar medium containing blood, glycerine, and potato extract. This medium, with minor modifications, is still in general use, though supplemented by various selective media for purposes of isolation (see Chapter 67). The three species of *Bordetella* are independent of X and V factors; *Bord. pertussis* grows poorly in the absence of blood or vegetable and tissue extracts.

Blood is not, however, essential, for growth occurs on a medium containing salts, amino acids, nicotinic acid and 0.1 per cent soluble starch (Hornibrook 1940; see also Cohen and Wheeler 1946, Ungar *et al.* 1950). Pollock (1947) found that the serum albumin was the fraction of blood active in promoting the growth of *Bord. pertussis* on a beef-digest medium; growth promotion was indirect, through the absorption of toxic substances—perhaps unsaturated fatty acids—from the medium; and blood could be replaced by charcoal (see Powell *et al.* 1951). Blood can also be replaced by charcoal and catalase (Mazloum and Rowley 1955) and by ion exchange resins (Kuwajima *et al.* 1958*b*).

Proom (1955) studied the nutritional requirements of *Bord. pertussis, Bord. parapertussis* and *Bord. bronchiseptica* in defined media. Eight amino acids were sufficient for growth, and nicotinamide was the only vitamin needed. Starch was necessary only for *Bord. pertussis*. Jebb and Tomlinson (1955, 1957) found that glutamic acid and a sulphur-containing amino acid—either cysteine, cystine or methionine—together with nicotinic acid, were sufficient for a number of strains of *Bord. pertussis*, and concluded that difficulties in growing the organism were probably due to its susceptibility to inhibitors in the medium. These appear to be unsaturated fatty acids, colloidal sulphur or sulphides in commercial peptones (Proom 1955) and an unidentifiable substance, probably an organic peroxide, since it is removed by lysed RBC, haemin or ferrous sulphate (Rowatt 1957*a*). Alkali is formed during growth, necessitating adequate buffering of the medium. Rowatt (1957*b*), in reviewing the growth requirements of *Bord. pertussis*, notes that, besides the minimal requirements, purines, haematin and biotin stimulate growth, and that for rapid profuse growth, constituents of blood, yeast extract and casein hydrolysate are required, together with factors present in large, unwashed inocula. (For selective media for the isolation of *Bord. pertussis*, see Lacey 1954, Rowatt 1957*b*, Turner 1961, Sutcliffe and Abbott 1972, Regan and Lowe 1977; for *Bord. bronchiseptica*, see Füzi 1973). Goldner and his colleagues (1966) confirmed the importance of glutamic acid for both growth and antigenicity, and found that this acid, L-proline, and L-glutamine, were the only amino acids whose absence seriously limited growth. *Bord. parapertussis* and *Bord. bronchiseptica* are not susceptible to the inhibitors of *Bord. pertussis* found in complex media. Stainer and Scholte (1970) developed a chemically defined medium containing proline, glutamic acid and cystine, together with ascorbic acid, niacin, glutathione and salts and buffer.

Bord. pertussis, when grown on Bordet-Gengou medium, forms colonies that, during the early stages of growth, may resemble those of *H. influenzae*. But when incubation is prolonged beyond 24 hr the colonies become larger, more opaque and greyish, a form that is never assumed by *H. influenzae*. They are also smoother, more shiny and more distinctly dome-shaped. The combination of slight opacity, greyness of hue and shining surface, gives them a characteristic 'bisected pearl' appearance. A confluent row of colonies has been compared to an 'aluminium streak'. Colonies of *Bord. parapertussis* on Bordet–Gengou medium tend to be slightly larger than those of *Bord. pertussis*; growth occurs on nutrient agar and is accompanied by a brown discoloration of the medium. As already noted, *Bord. bronchiseptica* grows more readily than the other two members of the group; on nutrient agar it forms round, smooth, butyrous, lenticular, glistening, greyish-yellow colonies, 1 mm in diameter, within 24 hr. In suitable liquid media virulent strains of *Bord. pertussis* form a slight deposit and a surface pellicle within a few days; avirulent strains grow diffusely without a surface pellicle (Ungar *et al.* 1954). *Bord. parapertussis* grows in nutrient broth, forms a viscid deposit, and turns the medium brown within 4–5 days (Eldering and Kendrick 1938). *Bord. bronchiseptica* grows likewise in plain nutrient broth, in which it gives rise to a moderate turbidity within 24 hr and a slight flocculent or viscous deposit.

Resistance

Exposure to a temperature of 56° is usually lethal within 30 min. Outside the body, *Bord. pertussis* in dried droplets is said to survive up to 5 days on glass, 3 days on cloth, and a few hours on paper (Ocklitz and Milleck 1967).

Bord. pertussis is moderately sensitive to the sulphonamides. It is sensitive to the polymyxins, to which *Bord. bronchiseptica* is also highly susceptible, and to erythromycin, chloramphenicol, the tetracyclines and penicillin in a roughly equipotent group, also to co-trimoxazole and streptomycin, which is moderately active (see Brownlee and Bushby 1948, McLean *et al.* 1949, Wells *et al.* 1950, Hegarty *et al.* 1950, Gastal 1958, Albrecht and Husmann 1959). Virulent strains are more susceptible to penicillin than avirulent strains (Ungar and Muggleton 1954); *Bord. parapertussis* is reported to be rather more resistant *in vitro* than *Bord. pertussis* (Day and Bradford 1952). Strains of *Bord. bronchiseptica* isolated from pigs and dogs are not infrequently resistant to sulphonamides, penicillin and streptomycin (Harris and Switzer 1969, Wilkins and Helland 1973).

Biochemical activities

Biochemical activity, as usually studied in the laboratory, is weak.

Bord. pertussis is recorded by Stillman and Bourn (1920) as failing to ferment glucose, laevulose, galactose, maltose, sucrose, dextrin, mannitol, lactose or inulin, as producing no indole and as failing to reduce nitrate. It produces a hazy zone of haemolysis. *Bord. bronchiseptica* resembles *Bord. pertussis* in fermenting none of the commonly used carbohydrates but differs from it in reducing nitrate, and forming urease. *Bord. parapertussis* is also urease positive, and catalase positive more consistently than *Bord. pertussis*; both are coagulase positive, particularly with the plasma of dog and hamster (Bilaudelle 1955). Some strains of *Bord. bronchiseptica* are haemolytic, activity being most evident at acid pH (Petersen 1976). All three organisms render litmus milk alkaline in 5 days.

Antigenic constitution

On first isolation from man on optimal medium, *Bord. pertussis* is invariably in a smooth state. Early workers found that all recent isolates belonged to a single antigenic type, but that laboratory strains were not agglutinated by antiserum to fresh strains (Bordet and Sleeswyck 1910, Kristensen 1922, 1927). Leslie and Gardner (1931), in a careful study of apparently smooth strains, described four different antigenic states, which they named phases I, II, III and IV. Recent isolates were mostly in phase I and laboratory strains in phases II, III or IV. Later workers regarded the phases not as interchangeable states but as stages in the course of an S–R variation with varying amounts of phase I antigen on the bacillary surface (Shibley and Hoelscher 1934, Toomey *et al.* 1935, 1936).

Smooth strains were found to contain a number of antigens (Flosdorf *et al.* 1941), and Andersen (1952*a*, 1953) recognized five heat-labile capsular antigens (a_1, a_2, a_3, a_4, a_5); a_1 was common to all strains, but these were divisible into five types by means of their content of a_2, a_3, a_4 and a_5 antigens. Eldering and her colleagues (1957, 1967) identified six species-specific agglutinogens (1–6) in *Bord. pertussis*, together with two that were shared with *Bord. parapertussis* and *Bord. bronchiseptica*. Agglutinogen 1 was present in all freshly isolated strains of *Bord. pertussis* and is probably the common antigen described by earlier workers. Fresh isolates may also possess agglutinogens 2 or 3 or both, or mutate to reveal these antigens. Moreover, strains with antigens 1, 2 or 1, 3 can change to 1, 2 and 3 (Stanbridge and Preston 1974). Agglutinogens 5 and 6 are uncommon and the serotyping of clinical isolates relies mainly on antigens 1, 2, 3 and also 4. Although the common agglutinogen has been purified and shown to have the properties of a protein (Onoue *et al.* 1961), the chemical nature and the structural relations of the different agglutinogens to features such as the capsule or cell wall remain uncertain. The outer cell membrane of *Bord. pertussis*, isolated in the form of vesicles, was found to contain agglutinogens 1, 2 and 3, as well as lipopolysaccharide and proteins (Novotny and Cownley 1978). Aprile and Wardlaw (1973*a*) found that phase I cultures, heated to 56°, could adsorb anti-lipopolysaccharide antibodies from antisera, suggesting that antigenic determinants of lipopolysaccharide may take part in the agglutination of phase I cells. One of the agglutinogen antigens is water soluble and is liberated from the cell by sonic vibration and by acid extraction (Flosdorf *et al.* 1939, Flosdorf and Kimball 1940 *a*, *b*, Smolens and Mudd 1943). It is non-toxic, induces agglutinins in the rabbit and removes agglutinins from antisera to S bacilli. It is fixed to the cells by treatment with formalin (de Bock and Worst-van Dam 1960).

The S form of *Bord. pertussis* is capsulated (Lawson 1933) but the capsule is ill defined and readily removed by washing (Miller 1937, Klieneberger-Nobel 1948), and it may therefore be distinct from agglutinogens present in washed suspensions. Capsular material by itself stimulates little agglutinin response in the rabbit (Evans and Adams 1952).

The heat-stable somatic O antigen of *Bord. pertussis* is common to all strains (Andersen 1952*a*) and cross-reactions have been found to occur with the O antigens of *Bord. parapertussis* and *Bord. bronchiseptica*, as well as between the somatic antigens of R strains of these organisms (Flosdorf *et al.* 1941, Bondi and Flosdorf 1943, Eldering *et al.* 1957). MacLennan (1960) found that the endotoxins extracted from the three species were antigenically different although related, but Ross and his colleagues (1969) found them to be antigenically distinct from each other. Bilaudelle and his colleagues (1960) found cell-wall preparations from *Bord. pertussis* to be protein-lipopolysaccharide, the polysaccharide consisting mainly of glucose, galactose and glucosamine. Aprile and Wardlaw (1973*b*) studied lipopolysaccharide extracts from three smooth phase I and one rough phase IV strain of *Bord pertussis*, all of which contained hexosamine and heptose. Antigenic analysis with rabbit antisera suggested the presence of six distinct determinants, A–F, of which B was common to all strains, and F, present in two of the phase I strains, was shared with *Brucella melitensis*, type 1. The phase IV strain had only two of the antigenic determinants, suggesting that it had undergone a loss variation, but the lipopolysaccharide of one of the phase I strains also had only the same two determinants. Evidently the antigenic properties of the endotoxin may vary independently of phase variation.

The heat-labile antigens of *Bord. parapertussis* and *Bord. bronchiseptica* are different from those of *Bord. pertussis*. Little if any antigenic relation is noticeable between *Bord. pertussis* and *H. influenzae* (see Schlüter 1936).

Bord. pertussis contains, of course, numerous substances other than the agglutinogens and O antigen that call forth the production of antibodies; they are mostly defined in relation to their biological properties. They include a heat-labile toxin, haemagglutinin, histamine-sensitizing factor, lymphocytosis-promoting factor, islet-activating protein and protective antigen; these are considered below.

Antigenic variation

Like *H. influenzae*, *Bord. pertussis* gives rise in artificial culture to R variants. In contrast to *H. influenzae*, R strains of which are commonly present in the normal nasopharynx, R variants of *Bord. pertussis* have not been cultured from the human body, though they can readily be induced in the laboratory by growing the organism on an unfavourable medium.

In our discussion of *Bord. pertussis* we have assumed the identity of the S form and phase I. The S form of a pathogenic bacterium is by definition the form that possesses a somatic antigen or antigens associated with virulence. Phase I was originally defined serologically; in the absence of any direct measure of vi rulence in man, the virulence of phase I strains is ensured as effectively as possible by their being freshly isolated from human infection. By these criteria the identification of S and phase I strains can readily be made. But in the course of attempts to determine qualities needed for an effective whooping-cough vaccine, the following additional characters have been proposed for phase I strains: inability to grow on blood agar or nutrient agar; high mouse virulence; high immunizing power in the killed state; presence of haemagglutinin; and two properties added by Ungar and Muggleton (1949)—precipitability by aluminium phosphate and solubility in sodium deoxycholate (see also Ungar *et al.* 1954). Standfast (1951*b*) measured all but the last of these qualities in a large number of freshly isolated strains. Each had a wide inter-strain variability, and each varied independently of the others; and in a given strain, the characters were lost at varying rates with prolonged subculture. Standfast concluded that no comprehensive, quantitative definition of phase I was possible, but that it was legitimate to regard the various characters as commonly associated with the S form, provided that no one of them was regarded as a necessary part of the definition of the S state.

The four phase variants, I to IV, described by Leslie and Gardner (1931) are by no means well defined; the characteristics of neither phase I (Standfast 1951*a*) nor phase IV can be described with certainty (Parker 1978); and Aprile (1972) failed to produce a phase IV variant from phase I strains in the laboratory. Phase IV strains lack the mouse virulence, protective antigen and histamine-sensitizing factor, as well as the agglutinogens that are usually found in phase I strains (Kind 1953, Kasuga *et al.* 1954, Aprile 1972); and isolated cell envelopes of phase IV cells are deficient in two of the polypeptide bands that characterize extracts of phase I cells when subjected to polyacrylamide-gel electrophoresis (Wardlaw *et al.* 1976). Phase variation is regarded by Parker (1976, 1978) as the result of a tendency of *Bord. pertussis* readily to undergo loss mutation in a number of different genes independently in response to the selection pressure of artificial culture; she proposes that the irregular changes occurring *in vitro* are better described by the terms 'fresh isolate', 'intermediate strain' and 'degraded strain'.

In addition to loss variation, *Bord. pertussis* was observed by Lacey (1960) to undergo a rapid, apparently phenotypic change which he called *'modulation'*. When cultured in medium in which the normal sodium ions were replaced by magnesium, typical strains, which he termed 'X-mode', became converted to 'C-mode', with change in antigenic specificity and loss of haemolytic activity and of the ability to agglutinate red blood cells. 'C-mode' cells also lack the mouse-protective antigen, histamine-sensitizing and lymphocytosis-promoting factors (Holt and Spasojević 1968, Parton and Wardlaw 1975, Wardlaw *et al.* 1976). Modulation from X- to C-mode can also be induced by lowered temperature of incubation (Lacey 1960), and by a high concentration of nicotinic acid in the culture medium (Pusztai and Joo 1967). The change from X- to C-mode occurs rapidly—within 15 cell divisions—and is reversible by culture on a medium with a low magnesium and high sodium content; modulation must be considered as an example of phenotypic variation. In addition to the properties mentioned above, change from X- to C-mode is associated with loss of the two polypeptide components that are present in phase I cells but not in phase IV cells (Wardlaw *et al.* 1976), as well as a marked loss in intra- and extracellular adenylate cyclase activity, which is also low in phase IV cells (Parton and Durham 1978). Though phase-IV cells and C-mode cells thus show a number of similar differences from the corresponding phase I and X-mode cells, the two terms should not be regarded as interchangeable, phase variation being a loss variation and modulation the result of a phenotypic change.

Natural pathogenicity

Bord. pertussis and *Bord. parapertussis* give rise to whooping cough, and as such are among the more important human pathogens. Healthy carriers appear to be practically non-existent, although subclinical parapertussis may occur.

Bord. bronchiseptica is likewise parasitic, giving rise to lesions in dogs, monkeys, guinea-pigs and other laboratory animals; it is occasionally found in the nasopharynx of man. It causes atrophic rhinitis in swine, a disease of economic importance among animals reared in confined conditions (see Goodnow 1980). The organism also causes acute infectious canine tracheobronchitis (kennel cough) (Wright *et al.* 1973, Thompson *et al.* 1976). It appears to be a secondary invader in dogs suffering from distemper, being frequently responsible for the pulmonary complications of the disease (M'Gowan 1911, Laidlaw and Dunkin 1926). Spooner (1938) found it playing a similar role in a spontaneous distemper-like disease of ferrets.

Toxins of *Bord. pertussis*

Bord. pertussis, or its culture filtrates, causes a variety of different effects in experimental animals, the nature or significance of which has been only partly unravelled. These effects have been attributed to various biologically active factors, including heat-labile toxin, heat-stable endotoxin, lymphocytosis-promoting factor, histamine-sensitizing factor, islet-activating protein, immunological adjuvant, haemagglutinin, and the mouse-protective antigen which immunizes mice against later intracerebral challenge with live *Bord. pertussis*. Some of these factors may represent different activities of the same substance. Of particular interest is their relation to the protective antigen or antigens that must be responsible for the immunity of children who have either recovered from whooping cough or have been vaccinated; but so far it has not been possible to isolate such a protective antigen or to identify it with certainty with any particular product or products of the bacillus.

Heat-labile toxin was described by Evans and Maitland (1937), who extracted it from ground bacilli. It was lethal on intravenous injection in guinea-pigs, and produced areas of necrosis on intradermal injection in the rabbit. Evans and Maitland's preparations also contained the agglutinogen; this appeared to be distinct from the toxin, since antisera to the extract were protective against experimental infection and agglutinated bacillary suspensions but had no antitoxic activity as judged by their ability to modify skin necrosis induced by the toxin. The toxin was easily destroyed by formalin, was unstable at 37° and was rapidly destroyed at 55°. A toxin with very similar properties was obtained from *Bord. bronchiseptica* (Evans and Maitland 1939) and from *Bord. parapertussis* (Breuckner and Evans 1939). Evans (1940, 1942) found that formolized toxin was antigenic, and that antitoxin prepared against it neutralized the toxins of all three *Bordetella* species. Bilaudelle and his colleagues (1960) partly separated the toxin into two relatively labile factors, a toxin lethal for mice and a toxin dermatonecrotic in the rabbit.

Highly active preparations can be made by fractionating weak $CaCl_2$ extracts of freeze-dried cells (Robbins and Pillemer 1950). With further purifications by chromatography on diethylaminoethyl cellulose, Onoue, Kitagawa and Yamamura (1963) obtained a product, protein in nature, with an LD50 for mice of 0.33 μg. Iida and Okonogi (1971) likewise prepared a highly toxic fraction that caused atrophy of the spleen in mice given sub-lethal doses. In rabbits, the heat-labile toxin produces severe oedema of the lungs followed by a characteristic accumulation of macrophages in the alveoli and of lymphocytes around the blood vessels, and to severe necrosis in scattered areas—a histological picture not unlike that found in the lung in whooping cough; antitoxin protected rabbits against this effect (Sprunt and Martin 1943). The protective action of antitoxin in experimental infection appears to be limited to neutralizing the toxin contained in the infecting dose, and thus reducing the likelihood of the organisms establishing a foothold. Thus in mice antitoxin, but not antibacterial serum, protects against challenge by the intraperitoneal route, whereas antibacterial serum, but not antitoxin, protects against intranasal infection (Anderson and North 1943); and only intoxication results from intraperitoneal infection, whereas bacteraemia follows intranasal infection (Proom 1947). However, the toxin appears to have an important role in causing atrophy of the nasal turbinate bones in experimental infection in young pigs and dogs (Hanada *et al.* 1979, and see Goodnow 1980). The toxin may act as an aggressin in infection, since antitoxin instilled into the nose with the bacilli lowered their infectivity in mice (Evans 1944).

Toxin may help in inducing protective antibodies; gently treated whole bacilli are better protective antigens than bacilli heated or formolized sufficiently to destroy contained toxin (North *et al.* 1941, see also North 1946). Most of the evidence, however, is against regarding heat-labile toxin as a protective antigen of *Bord. pertussis*. The serum of convalescents from whooping cough may contain protective antibodies but little or no antitoxin (North *et al.* 1939, Evans 1947); toxin does not induce protective antibodies (Verwey and Thiele 1949, Kuwajima *et al.* 1958*a*); bacilli heated sufficiently to destroy the toxin are protective, and extracted protective antigen is relatively heat-stable (Robbins and Pillemer 1950).

The heat-stable toxin, described by Ehrich and colleagues (1942) and Eldering (1942), is evidently the lipopolysaccharide endotoxin, possessing the typical properties of the endotoxins of gram-negative bacilli, although its toxicity to laboratory animals may be variable between strains of *Bord. pertussis* (MacLennan 1960). Like other endotoxins it possesses adjuvant activities (Farthing and Holt 1962). Endotoxin is able to induce some protection against living organisms (Cruickshank and Freeman 1937), but has only a weak effect against intracerebral challenge in mice (Dolby *et al.* 1975*a*, *b*); it is not regarded as playing an important part in protection in man (Aprile and Wardlaw 1973*a*, Munoz and Bergman 1977, Pittman 1979).

Lymphocytosis-promoting factor (LPF) is that fraction of *Bord. pertussis* cultures that gives rise to a leucocytosis on injection in mice. The effect is detectable within a few hours, and reaches its maximum in 3–5 days. Small lymphocytes predominate, constituting about 60 per cent of the circulating white cells (Morse 1965, Munoz and Bergman 1966), but a polymorphonuclear leucocytosis also occurs. LPF is probably responsible for the same effect in children with whooping cough. Morse and Morse (1976) isolated from the supernatant fluids of *Bord. pertussis* cultures a purified protein fraction which gave a single precipitation line with anti-*Bord. pertussis* antisera. On electronmicroscopy the protein was ring shaped, 75–80 Å in diameter and composed of 4–5 subunits; it could be dissociated into four polypeptide bands (see also Morse 1971). As little as 0.02 μg was active in producing lymphocytosis in the mouse. Purified LPF prepared by Arai and Sato (1976) contained lipid and carbohydrate in addition to protein. Although purified LPF was found by Morse and his colleagues (Morse *et al.* 1977, Ho *et al.* 1977, Kong and Morse 1977) to be a mitogen for peripheral blood lymphocytes both in the mouse and in man, acting in the mouse on T lymphocytes and depending also on a contribution from B cells, the lymphocytosis has mainly been attributed to a reduced homing of circulating lymphocytes (Morse 1965, Taub *et al.* 1972). The purified LPF of Morse and Morse (1976) was also toxic to the mouse, 4 μg proving fatal, and sensitized to histamine and decreased the normal hyperglycaemic response to adrenaline.

The histamine-sensitizing factor (HSF) increases the normally low sensitivity of mice to histamine 50- to 100-fold (Parfentjev and Goodline 1948). Only phase I bacilli contain HSF (Halpern and Roux 1949, Kind 1953) and, together with mouse protective activity and an envelope protein, it is lost in antigenic modulation from X- to C-mode (Wardlaw *et al.* 1976). A similar toxin may be produced by *Bord. bronchiseptica* (Dixon *et al.* 1979). HSF also sensitizes rats to histamine (Malkiel and Hargis 1957), but makes rabbits and guinea-pigs slightly less sensitive (Stronk and Pittman 1955). It is difficult to purify owing to the insolubility of partly purified preparations (Morse 1976). By means of urea extraction and ion exchange chromatography, Lehrer (1978) prepared a fraction with a 320-fold increase in HSF activity, but it contained at least three components when examined by polyacrylamide-gel electrophoresis. HSF is probably a protein, being susceptible to proteolytic enzymes (Pieroni *et al.* 1976).

The means by which HSF acts has been studied. Mice that have been given injections of *Bord. pertussis* organisms not only become hypoglycaemic but also fail to develop the hyperglycaemia that normally occurs in response to histamine or adrenaline, through which this effect of histamine is mediated (Parfentjev and Schleyer 1949, Stronk and Pittman 1955, Szentivani *et al.* 1963). Pertussis-treated mice also have lowered splenic levels of cyclic adenosine monophosphate (cAMP), and adenylate cyclase in the spleen is not stimulated, as it is normally, by adrenaline (Ortez *et al.* 1970, 1975). Chemical β-adrenergic blocking agents have similar effects and also give rise to histamine sensitivity in mice (Fishel *et al.* 1962). HSF activity might therefore be explained by a capacity to block cellular receptors for β-adrenergic agonists—the so-called β-adrenergic blockade effect (Fishel *et al.* 1964)—but for a number of reasons this explanation is unlikely. Different β-adrenergic blocking agents differ in ability to cause histamine sensitization (Bergman and Munoz 1968). Lymphocytes treated *in vitro* with an LPF fraction containing HSF activity become unresponsive, in respect of cAMP accumulation, both to a β-adrenergic agonist and to a prostaglandin, which has a different receptor (Parker and Morse 1973). Sumi and Ui (1975) studied the blood insulin levels in rats as an indicator of α- and β-adrenergic stimulation of islet cells; α- stimulation causes a fall and β- stimulation a rise in blood insulin. They found that in mice treated with pertussis vaccine a markedly increased rise in blood insulin occurred in response to β-adrenergic stimulation. More recently, Hewlett and colleagues (1978) reported that reticulocytes taken from rats which had been treated with *Bord. pertussis* and, as a result, could not develop hyperglycaemia in response to adrenaline, had no detectable reduction in the number of β-adrenergic receptors, measured by iodohydroxybenzylpindolol binding, or in their affinity, measured by adenylcyclase accumulation when stimulated with isoproterenol. Ui and his colleagues later purified and studied the fraction in pertussis vaccine responsible for enhanced insulin secretion in response to insulin secretagogues, which they called *islet-activating protein* (IAP) (Yajima *et al.* 1978*a*, *b*). Purified IAP has a histamine-sensitizing action, so that HSF activity may be accounted for by the release of excess insulin in response to histamine injection, the resulting hypoglycaemia contributing to the toxicity of histamine (Ui *et al.* 1978). (The interrelations of the various *Bord. pertussis* factors are further considered below.) In the presence of an inhibitor of cAMP breakdown, isolated islets from rats previously treated with IAP accumulated more cAMP than normal rat islets when, in the presence of calcium, they were incubated with glucose or other insulin secretagogues; Katada and Ui (1979) suggest that IAP acts on the islet B-cell membrane to cause a movement of calcium into the cell with a consequent stimulation of cAMP formation and of insulin secretion (see also Ui *et al.* 1978, Katada and Ui 1982).

Adjuvant action *Bord. pertussis* has the capacity to modify the immunological response. When suspensions of it are injected into animals along with other antigens they increase the production of all classes of immunoglobulins to these antigens (Greenberg and Fleming 1947, 1948, Dresser *et al.* 1970), and especially of IgE (Suko *et al.* 1977). Part of this effect is due to the endotoxin component of the bacterial cell (Farthing and Holt 1962), but a heat-labile adjuvant is also present (Clausen *et al.* 1970); its mode of action is uncertain (see reviews by Morse 1976, Munoz and Bergman 1977). In addition to stimulating immunoglobulin formation *Bord. pertussis* promotes the development of experimental allergic encephalomyelitis (EAE) when injected with homologous nerve tissue and Freund's adjuvant (Lee and Olitsky 1955). In the production of EAE in guinea-pigs, *Bord. pertussis* can replace the mycobacteria in Freund's complete adjuvant (Weiner *et al.* 1959), or when incorporated in aqueous suspension with guinea-pig spinal cord promote the development of a hyperacute form of encephalomyelitis in suitable strains of rats (Levine *et al.* 1966, Lennon *et al.* 1976). Other auto-immune diseases, such as orchitis and thyroiditis, may also be promoted by *Bord. pertussis* (Hargis *et al.* 1968, Twarog and Rose 1969). The effect has been attributed to an ability of *Bord. pertussis* to stimulate cellular hypersensitivity coupled, in the case of EAE, with an increase in vascular permeability (Bergman *et al.* 1978).

Haemagglutination of a variety of species of RBC, including those of man and many laboratory animals, is induced *in vitro* by freshly isolated *Bord. pertussis* grown in liquid medium. Extracts of *Bord. pertussis* contain two haemagglutinating components. The more active of these is filamentous in structure and may be derived from fimbriae (Arai and Sato 1976, Morse and Morse 1976, Sato *et al.* 1978). Purified filamentous haemagglutinin adheres to mammalian cells other than RBC, the activity being inhibited by cholesterol and by lipoprotein inhibitors in normal serum. The serum inhibitors can be removed by kaolin or acetone to reveal inhibition by antibodies. Electronmicrographic studies of the respiratory ciliated epithelium of suckling mice infected with *Bord. pertussis* by inhalation suggest that adhesion of the bacilli to the cilia is mediated by fimbriae. Sato and his colleagues (1978) found that their purified filamentous haemagglutinin was a protective antigen in mice, inducing immunity to intracerebral challenge, and rabbit antiserum was protective when given passively to suckling mice before respiratory challenge. It seems likely that the fimbriae play an important part in pathogenesis, although the possible protective role of this antigen in vaccination has yet to be established (Irons and MacLennan 1978, Sato *et al.* 1978).

Studies of tracheal organ cultures from the chick (Iida and Ajiki 1974) and hamster (Baseman 1977, Muse *et al.* 1977, 1978), infected *in vitro* with *Bord. pertussis* and examined by a variety of microscopic techniques including scanning electronmicroscopy, have illuminated the early phases of pathogenesis. The bacilli appear to adhere to the cilia of the respiratory epithelium and not to invade the tissues. Ciliary activity is impaired and the ciliated cells appear to become damaged and then extruded from the mucosal surface leaving denuded areas in which only the microvillous cells of the tracheal mucosa remain. Although fimbriae may play a part in the adhesion process in suckling mice (Sato *et al.* 1978), Muse and his colleagues (1978) saw no evidence in their electronmicrographs that fimbriae mediated adhesion to the cilia in hamster tracheal organ cultures.

The second, less active *Bord. pertussis* haemagglutinin is closely associated and possibly identical with LPF; it appears to be specific for a sialic acid-containing receptor (Arai and Sato 1976, Morse and Morse 1976, Irons and MacLennan 1978).

Other toxic effects have been attributed to bacterial adenylate cyclase, which impairs the activity of phagocytes (Confer and Eaton 1982), and to a factor which affects ciliary activity in hamster tracheal cells (Goldman *et al.* 1982).

Protective antigens

The relation of the bacterial components responsible for these various activities to each other and to the protective antigens is not yet clear (see Munoz and Bergman 1977, 1978). The findings of the various workers who have studied isolated components are not easy to interpret or to reconcile, because of differences between the methods used and the uncertain purity of the preparations. In examining the protective effect of bacterial components, much has depended upon the mouse-protection test (Kendrick *et al.* 1947), in which the immunized mouse is protected against subsequent intracerebral challenge with live *Bord. pertussis*, since this laboratory test has been found to reflect the protective effect of vaccines in children (see Chapter 67). Heat-labile toxin and endotoxin are separate identifiable products that are not regarded as protective components of pertussis vaccine. The place of the surface agglutinogens in protection is usually regarded as unimportant, because they have been found to play no part in the mouse-protection test (Andersen and Bentzon 1958, Eldering *et al.* 1967, Holt and Spasojević 1968). However, other workers have noticed an association between agglutinogens and the protective factor (Flosdorf and Kimball 1940*a*, Smolens and Mudd 1943, Evans and Perkins

1953, 1954*a*). Preston (1963, 1965, 1976) associated field evidence of a lowered protective effect in vaccines with failure to include strains with the agglutinogens of prevalent whooping-cough bacilli. The fimbrial antigen might be expected to play a part in protection, but it is certainly not the only factor concerned (Sato *et al.* 1978); it may be unimportant in pertussis vaccine, possibly because its natural role in protection might depend upon the induction of secretory antibodies. Protective activity can be adsorbed to the stromata of RBC (Pillemer *et al.* 1954), but the preparation when used in vaccine field trials caused more reactions than whole-cell vaccines (Report 1959); it is presumably a complex antigen containing one or more toxic factors and haemagglutinin.

Munoz and Bergman (1977) have put forward the concept of *pertussigen*, a single substance in the organism responsible for histamine sensitization, promotion of lymphocytosis, islet activation, heat-labile adjuvancy and mouse protection. Pittman (1979) prefers to use the term *pertussis toxin* for this substance. Highly purified preparations of HSF promote lymphocytosis (Lehrer *et al.* 1974, 1975), just as LPF preparations sensitize to histamine, cause hypoglycaemia and the proliferation of lymphocytes *in vitro* (Morse 1976), and both HSF and LPF induce protection in mice against intracerebral challenge. Purified preparations of pertussigen also act as an adjuvant, stimulate allergic encephalomyelitis in rats and give rise to hypoglycaemia (Munoz and Bergman 1978). The IAP of Ui and his colleagues (1978) possesses histamine-sensitizing and lymphocytosis-promoting activities, although in mice doses of less than 0.1 μg decrease the hypoglycaemic response to adrenaline without affecting histamine sensitivity or leucocytosis (Ui *et al.* 1978). If pertussigen, or pertussis toxin, is one of the major components responsible for protection, the preparation of pertussis vaccine devoid of toxicity will depend upon the practicability of detoxifying it while leaving its antigenicity relatively unimpaired, possibly by treatment with formaldehyde (Pittman 1952, Munoz and Hestekin 1966).

The observations of Standfast (1958), that the immunogenicity in mice of various pertussis vaccines varies with the route—intranasal (i.n.) and intracerebral (i.c.)—by which living bacilli are injected to test immunity, led to the characterization of two protective antigens: one, destroyed by boiling for 1 hr, largely responsible for protection against i.c. challenge; the other, stable to this treatment (see also Fisher 1955), responsible for protection against i.n. challenge. Strains of *Bord. pertussis* may have one or the other, or both antigens. Their corresponding rabbit antisera may contain antibodies to one, the other, or both; so that the two protective effects are also distinguishable by passive protection tests in mice (Dolby and Standfast 1958). The two antigens have also been partly separated by fractionation (Dolby 1958). The effect of the 'i.n.' antibody, given at or soon after infection, is to decrease a lethal to a sublethal dose, which is eliminated from the lung within a few weeks (Dolby *et al.* 1961); 'i.c.' antibody, on the other hand, whether induced actively or given passively, acts only when, on the 4th day of infection, the bacilli induce an inflammatory change that permits exudation of circulating antibody into the infective lesion (Dolby and Standfast 1961, Holt *et al.* 1961). The antigen protecting mice against intracerebral challenge is of particular interest, since laboratory assay by this method has proved to be the most reliable indicator of the protective value of whooping-cough vaccines in man (see Standfast and Dolby 1961, and Chapter 67); Pittman (1979) suggests that Standfast's i.c. antigen may be the filamentous haemagglutinin, which Morse and Morse (1976) and Sato and his colleagues (1978) found to be active in the mouse-protection test. Both Yoshida and his colleagues (1955) and Bilaudelle and his colleagues (Bilaudelle 1960, Bilaudelle *et al.* 1960) locate the antigen protecting mice against i.c. infection in the comparatively non-toxic fraction associated with the cell of the bacillus.

Experimental infection

The effect of the *intraperitoneal* injection of *Bord. pertussis* into rabbits or guinea-pigs is very similar to that of *H. influenzae*. Large doses are required to produce death, and the infection seems to be toxaemic rather than invasive (Bordet and Gengou 1907, 1909, Wollstein 1909). Phase III and IV strains were 20-30 times less toxic than phase I or II strains on intraperitoneal injection into guinea-pigs (Leslie and Gardner 1931).

The *intranasal* instillation of *Bord. pertussis* into anaesthetized mice produces a patchy or diffuse interstitial pneumonia, leucocytic infiltration around vessels and bronchioles, proliferation of the bronchiolar epithelium, and mucous secretion in the bronchioles containing masses of bacteria (Burnet and Timmins 1937, Bradford 1938). Young cultures are more virulent than old (Gray 1947, 1949).

The histological picture in many respects resembles that of the lung in human pertussis. After sublethal doses, the bacilli multiply in the lung up to about the 16th day, after which they slowly decrease in number, being eliminated several weeks later. The lethal dose appears to be one that can multiply in the lung rapidly enough to reach the critical number of about 10^8 after 10–14 days, before the animal has made a specific immune response to the infection (see Fisher 1955, Dolby *et al.* 1961). Toxin appears to promote infectivity by this route, perhaps by reason of its capacity, proved in sheep tissue, to inhibit the action of cilia in the bronchi and trachea (Standfast 1958). (For a fuller description of the pathological response to respiratory-tract infection in the mouse, see Pittman, Furman and Wardlaw 1980.)

For intraperitoneal infection, the normally low virulence of *Bord. pertussis* by the intraperitoneal route, which, as we have seen, results mainly in intoxication, may be enhanced by starch (Powell and Jamieson 1937) or mucin (Silverthorne 1938). Witebsky and Salm (1937) produced by intradermal injection into rabbits inflammatory lesions followed in 2–3 days by necrosis. A progressively fatal respiratory infection may be induced in suckling mice by the inhalation of an aerosol containing *Bord. pertussis* (Sato *et al.* 1978); the infection was inhibited by anti-fimbrial serum, suggesting that fimbriae may play a part in the pathogenesis of respiratory infection.

Kendrick and her colleagues (1947) used the *intracerebral* route as a means of testing the degree of immunization in mice. On intracerebral inoculation, after an immediate overflow of some of the inoculum into the blood stream, the bacteria in the brain increase steadily in number until the brain content of living organisms is about 10^8, when the mouse dies. This critical concentration is largely independent of the size of the inoculum, small doses being as lethal as large. The organisms multiply in between the cilia of the ependymal cells of the ventricles giving rise to choroiditis and meningitis, but do not invade the substance of the brain (Berenbaum *et al.* 1960). In this respect there is a close analogy with the behaviour of the organisms in the human respiratory tract in whooping cough. When the number of bacilli reaches 10^5–10^6, circulating protective antibodies, if present, are thought to pass into the infected tissue, with curative effect (Dolby and Standfast 1961, Holt *et al.* 1961), though the evidence for this is regarded by Wardlaw and Jakus (1968) as unconvincing.

In this connection a peculiarity of *Bord pertussis* described by Evans and Perkins (1954*b*, 1955) must be noted; namely the capacity of killed bacilli to induce resistance to intracerebral infection with living organisms. The effect is so rapid that the response of normal mice to large doses of living culture may be modified by reason of the large number of dead cells present in the inoculum (Fisher 1955). The resistance, as induced by a pertussis vaccine, is detectable within 5 hr, and is maximal at 3 days. It declines at the 14th day, after which there is a secondary rise, associated with demonstrable agglutinins and complement-fixing antibodies in the blood of the mice. The first phase of resistance, up to the 14th day, was regarded as a non-antibody effect, analogous to viral interference.

Virulence tests by the three routes, intraperitoneal, intranasal and intracerebral, may give entirely different answers. Standfast (1951*b*) found the range of LD50s of S strains (see Chapter 67), in millions of bacilli, to be 280–800 for the intraperitoneal, 0.26–15.5 for the intranasal and 0.18–2.5 for the intracerebral; there was no correlation between high and low values for each route among the strains tested. By the intranasal test (Standfast 1951*a*), all grades of virulence were found in S strains, but there was evidence of a discontinuous variation in the possession of the power to kill mice quickly; the strains either had or had not this 'quick-killing factor' (q.k.f.). There was no detectable association of virulence with the presence of q.k.f., toxin or haemagglutinin. The independence of virulence and toxin production was also evident in a non-toxic variant of *Bord. pertussis* isolated by Andersen (1952*b*) from mice after intracerebral injection, which was highly virulent, tending to produce a septicaemic disease in mice.

After *intratracheal* inoculation, Culotta, Harvey and Gordon (1935) produced in three monkeys a disease with a 10-day incubation period, a catarrhal stage, and a febrile coughing stage not unlike human pertussis. Monkey virulence does not run parallel with mouse virulence (Lin 1958). By similar means Sprunt, Martin and Williams (1935) produced an interstitial pneumonia in the monkey, characterized by a mononuclear cell reaction, and accompanied by a lymphocytosis; and North and his colleagues (1940) induced in *Macacus* monkeys an infection which by the seventh day resulted in a sticky tracheal and bronchiolar exudate full of *Bord. pertussis*, and pulmonary congestion with conspicuous fibrinous and cellular infiltration, both interstitially and in the alveoli. In none of the monkeys did a cough develop. Huang and his colleagues (1962) by spraying the nose and pharynx of Taiwan monkeys (*Macaca cyclopsis*) produced, after an incubation period of 7–15 days, a cough lasting for some weeks; the organisms could be recovered in culture for 3–5 weeks after inoculation.

The *chick embryo* is susceptible to *Bord. pertussis* infection, especially when the injections are made into the yolk sac (Shaffer and Shaffer 1946). *Bord. pertussis* has a cytopathic effect on some, but not on all, cultures of tissues from man, cat and chick embryo (see Felton *et al.* 1954, Pozsgi *et al.* 1961).

Bord. parapertussis behaves like *Bord. pertussis*, but is much less virulent for experimental animals and seldom kills mice when injected intracerebrally.

Bord. bronchiseptica produces in experimental animals lesions that have already been described (p. 395; see also p. 400). Injected intracerebrally in mice it multiplies in the cavity of the ventricle and in the brain parenchyma, giving rise to a purulent ventriculitis (Iida *et al.* 1962).

Summarized descriptions

Bordetella pertussis

Morphology *Bord. pertussis* bears a general resemblance to *H. influenzae*. The cell form is more constant, being usually of the short bacillary type. Longer bacillary or thread forms are relatively uncommon.

Growth requirements Not dependent on either V factor or X factor for growth. On first isolation it requires a complex medium (p. 392), but can be trained to grow on nutrient agar. Optimal temperature for growth *ca* 37°. No growth anaerobically.

Growth on solid media Colonial appearances on the Bordet–Gengou medium were described on p. 392. When fully developed the colonies tend to be rather larger than those of *H. influenzae*; but they develop more slowly and the characteristic appearances are often not obvious in less than 48–72 hours' incubation.

Growth in liquid media See p. 393.

Resistance Killed by exposure to a temperature of 55° for 30 min.

Biochemical activities Ferments no sugars. Renders litmus milk slightly alkaline. Forms catalase, but not indole. Does not reduce nitrate. Produces a hazy zone of haemolysis.

Antigenic structure In the normal S state it possesses surface agglutinogens, one common to all strains and the others of more limited distribution, thus permitting the identification of serotypes. In artificial culture, particularly on a relatively unfavourable medium, it gives rise to R, or partly R, variants, with a different antigenic structure. The antigens characterizing the endotoxin of the S and R forms of *Bord. pertussis* are serologically related to the corresponding substances in *Bord. parapertussis* and *Bord. bronchiseptica*.

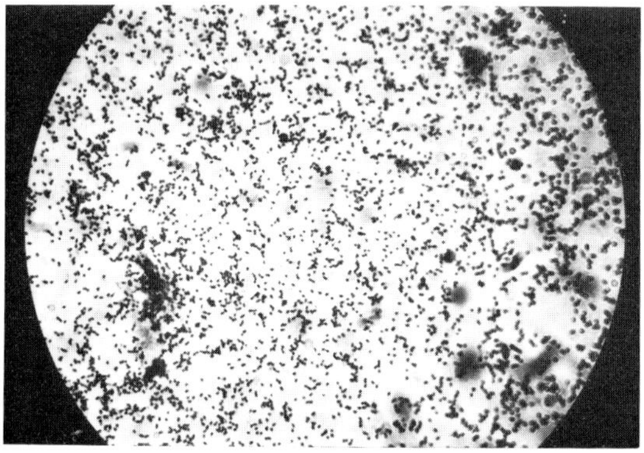

Fig. 39.8 *Bord. pertussis*. (Kindly supplied by Dr A. F. B. Standfast).

Pathogenicity *Bord pertussis* is the cause of whooping cough in man and is found only in clinical cases of the disease. Injected in large doses into laboratory animals it gives rise to a fatal toxaemic infection very similar to that produced by *H. influenzae*. Introduced intranasally or intratracheally, it produces a fatal broncho-pneumonic infection.

Bordetella parapertussus

Found only in man, in certain countries, in whom it causes mild whooping cough. Differs from *Bord. pertussis* (a) in growing readily on nutrient agar and turning the medium brown in 4–5 days; (b) in forming larger colonies on Bordet–Gengou medium; (c) in forming colonies surrounded by a haemolytic zone on Bordet–Gengou medium containing 50 per cent of blood; (d) in forming a viscid sediment in nutrient broth and colouring the medium brown; (e) in having an S endotoxin that is serologically distinct from, though related to, that of *Bord. pertussis*; and (f) in being, as a rule, less virulent to experimental animals. (See also Lautrop 1958.)

Bordetella bronchiseptica

Habitat Strict parasite, occurring in several different species of animals, and sometimes in man.

Morphology Similar to *Bord. pertussis*, but is motile by peritrichate flagella.

Cultural characters Grows fairly well on nutrient agar media, producing small round, convex amorphous colonies, with smooth glistening surface, of butyrous consistency. Grows best under aerobic conditions; no growth under strictly anaerobic conditions. In agar shake cultures, growth is almost entirely on the surface. Some strains are haemolytic.

Resistance Similar to *Bord. pertussis*.

Biochemical activities Haemolysin produced, active on RBC of rabbit, dog and guinea-pig. Grows freely on MacConkey's agar. No carbohydrates fermented. Produces strong alkalinity in litmus milk. Nitrate often reduced. H_2S-, NH_3 very slight production or none at all. Catalase+ + +. Urease+. Oxidase weak+. Grows in Koser's citrate.

Antigenic structure The surface antigen of the S form,

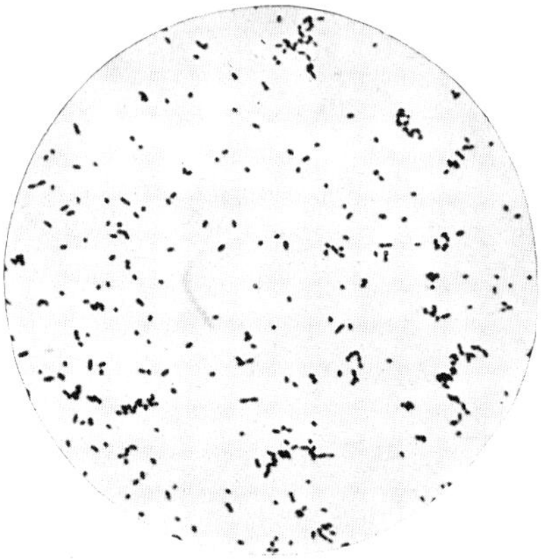

Fig. 39.9 *Bord. bronchiseptica*. From a nutrient agar culture, 24 hr, 37° (× 1000).

and its endotoxin, are serologically homogeneous and related to the corresponding substances in *Bord. pertussis* and *Bord. parapertussis*.

Pathogenicity Frequent cause of broncho-pneumonia in rodents, and of broncho-pneumonia complicating distemper in dogs. Causes atrophic rhinitis in pigs and kennel cough in dogs. Experimentally, intraperitoneal inoculation of guinea-pigs with 0.5 ml of a 24-hr broth culture causes death in 24–48 hr. *Post mortem*, there are small haemorrhages on the peritoneum and a viscid translucent exudate forming pseudomembranes on the liver, spleen and the less mobile parts of the intestine. The bacilli are easily recovered from the peritoneal cavity, but with difficulty from the blood, liver and lungs. Subcutaneous inoculation produces only a local lesion. Feeding and inhalation are without effect. Non-pathogenic to mice except by intracerebral injection, when it gives rise to a purulent ventriculitis. Rapidly loses virulence in culture (see Goodnow 1980).

(See also Table 32.2, p. 270.)

(For a fuller description of these three species, see Pittman and Wardlaw 1981.)

References

Albrecht, J. and Husmann, K. H. (1959) *Arch. Hyg. Bakt.* **143,** 366.

Alexander, H. E. (1943) *Amer. J. Dis. Child.* **66,** 160.

Alexander, H. E. and Heidelberger, M. (1940) *J. exp. Med.* **71,** 1.

Alexander, H. E., Leidy, G. and MacPherson, C. (1946) *J. Immunol.* **54,** 207.

Allibone, E. C., Allison, P. R. and Zinnemann, K. (1956) *Brit. med. J.* **i,** 1457.

Allison, P. R., Gordon, J. and Zinnemann, K. (1943) *J. Path. Bact.* **55,** 465.

Andersen, E. K. (1952*a*) *Acta path. microbiol. scand.* **30,** 54; (1952*b*) *Ibid.* **31,** 546; (1953) *Ibid.* **33,** 202.

Andersen, E. K. and Bentzon, M. W. (1958) *Acta path. microbiol. scand.* **43,** 106.

Anderson, G. and North, E. A. (1943) *Aust. J. exp. Biol. med. Sci.* **21,** 1.

Anderson, K. and Snow, J. S. (1940) *Amer. J. Path.* **16,** 269.

Anderson, L. R. (1931) *Amer. J. Hyg.* **13,** 164.

Anderson, P., Peter, G., Johnston, R. B. Jr., Wetterlow, L. H. and Smith, D. H. (1972) *J. clin. Invest.* **51,** 39.

Aprile, M. A. (1972) *Canad. J. Microbiol.* **18,** 1793.

Aprile, M. A. and Wardlaw, A. C. (1973*a*) *Canad. J. Microbiol.* **19,** 231; (1973*b*) *Ibid.* **19,** 537.

Arai, H. and Sato, Y. (1976) *Biochim. biophys. Acta.* **444,** 765.

Assis, A. de (1926) *C. R. Soc. Biol.* **95,** 1008.

Bailie, W. E. (1969) *Characterization of Haemophilus somnus (new species), a micro-organism isolated from infectious thromboembolic meningencephalomyelitis of cattle.* PhD Thesis, Kansas State University.

Bailie, W. E., Coles, E. H. and Weide, K. D. (1973) *Int. J. syst. Bact.* **23,** 231.

Bakos, K., Nilsson, A. and Thal, E. (1952) *Nord. vet. Med.* **4,** 241.

Baseman, J. B. (1977) *J. infect. Dis.* **136,** S196.

Baudisch, O. (1932) *Biochem. Z.* **245,** 265.

Berenbaum, M. C., Ungar, J. and Stevens, W. K. (1960) *J. gen. Microbiol.* **22,** 313.

Bergman R. K. and Munoz, J. (1968) *Nature, Lond.* **217,** 1173

Bergman, R. K., Munoz, J. J. and Portis, J. L. (1978) *Infect. Immun.* **21,** 627.

Biberstein, E. L. and Gills, M. (1961) *J. Bact.* **81,** 380.

Biberstein, E. L., Mini, P. D. and Gills, M. G. (1963) *J. Bact.* **86,** 814.

Biberstein, E. L. and White, D. C. (1969) *J. med. Microbiol.* **2,** 75.

Bieger, R. C., Brewer, N. S. and Washington, J. A. (1978) *Medicine* **57,** 345.

Bilaudelle, H. (1955) *Acta path. microbiol. scand.* **37,** 434; (1960) *Z. ImmunForsch.* **120,** 173.

Bilaudelle, H. *et al.* (1960) *Acta path. microbiol. scand.* **50,** 208.

Bjuggren, G. and Tunevall, G. (1950) *Acta oto-laryng., Stockh.* **38,** 130.

Blake, F. G. and Cecil, R. L. (1920) *J. exp. Med.* **32,** 691.

Blieck, L. de (1931) *Tijdschr. Diergeneesk.* 15 March; (1932) *Vet. J.* **88,** 9; (1934) XII *int. vet. Congr. N.Y.* **iii,** 161.

Bock, C. A. de and Worst-Van Dam, A. M. (1960) *Leeuwenhoek. ned. Tijdschr.* **26,** 126.

Bondi, A. and Flosdorf, E. W. (1943) *J. Immunol.* **47,** 315.

Bordet, J. and Gengou, O. (1906) *Ann. Inst. Pasteur* **20,** 731; (1907) *Ibid.* **21,** 720; (1909) *Ibid.* **23,** 415.

Bordet, J. and Sleeswyk. (1910) *Ann. Inst. Pasteur* **24,** 476.

Boyce, J. M. H., Frazer, J. and Zinnemann, K. (1969) *J. med. Microbiol.* **2,** 55.

Bradford, W. L. (1938) *Amer. J. Path.* **14,** 377.

Bradford, W. L. and Slavin, B. (1937) *Amer. J. publ. Hlth.* **27,** 1277.

Branefors-Helander, P., Erbing, C., Keure, L. and Lindberg, B. (1977) *Carbohyd. Res.* **56,** 117.

Brewer, R. A. and Corbel, M. J. (1983). *Brit. vet. J.* **139,** 200.

Broom, A. K. and Sneath, P. H. A. (1981) *J. gen. Microbiol.* **126,** 123.

Brown, C. C. *et al.* (1954) *Amer. J. Med.* **17,** 478.

Brownlee, G. and Bushby, S. R. M. (1948) *Lancet*, **i,** 127.

Brueckner, I. E. and Evans, D. G. (1939) *J. Path. Bact.* **49,** 563.

Bruun, B. and Friis-Møller, A. (1976) *Acta. path. microbiol. scand.* **B84,** 201.

Buckmire, F. L. A. (1976) *Infect. Immun.* **13,** 1733.

Burnet, F. W. and Timmins, C. (1937) *Brit. J. exp. Path.* **18,** 83.

Butler, L. O. (1962) *J. gen. Microbiol.* **28,** 189.

Cavanagh, P., Morris, C. A., and Mitchell, N. J. (1975) *Lancet* **ii,** 96.

Chandler, C. A., Fothergill, L. D. and Dingle, J. H. (1937) *J. exp. Med.* **66,** 789; (1939) *J. Bact.* **37,** 415.

Clausen, C. R., Munoz, J. and Bergman, R. K. (1970) *J. Immunol.* **104,** 312.

Cohen, C. (1909) *Ann. Inst. Pasteur* **23,** 273.

Cohen, S. M. and Wheeler, M. W. (1946) *Amer. J. publ. Hlth.* **36,** 371.

Confer, D. L. and Eaton, J. W. (1982) *Science* **217,** 948.

Cooper, R. G. and Attenborough, I. D. (1968) *Aust. J. exp. Biol. med. Sci.* **46,** 803.

Cruickshank, J. C. and Freeman, G. G. (1937) *Lancet* **ii,** 567.

Culotta, C. S., Harvey, D. F. and Gordon, E. F. (1935) *J. Pediat.* **6,** 743.

Cunha, R. (1939) *Zbl. Bakt.*, **144,** 508; (1943) *O Hospital* **23,** 393.

Davis, D. J., Pittman, M. and Griffiths, J. J. (1950) *J. Bact.* **59,** 427.

Davis, J. D. (1917) *J. infect. Dis.* **21,** 392; (1921) *Ibid.* **29,** 178, 187.

Day, E. and Bradford, W. L. (1952) *Pediatrics* **9,** 320.

Deacon, W. E. *et al.* (1956) *J. invest. Derm.* **26,** 399.

DeClerq, E. and Merigan, T. C. (1969) *J. Immunol.* **103,** 899.

Delius, W. and Kolle, W. (1897) *Z. Hyg. InfektKr.* **24,** 327.

Dible, J. H. (1924) *J. Path. Bact.* **27,** 151.

Dixon, M., Jackson, D. M. and Richards, I. M. (1979) *Amer. Rev. resp. Dis.* **120,** 843.

Dochez, A. R., Mills, K. C. and Kneeland, Y. Jr. (1932) *Proc. Soc. exp. Biol., N.Y.* **30,** 314.

Dolby, J. M. (1958) *Immunology* **1,** 326.

Dolby, J. M., Ackers, J. P. and Dolby, D. E. (1975*a*) *J. Hyg., Camb.* **74,** 71.

Dolby, J. M., Dolby, D. E. and Bronne-Shanbury, C. J. (1975*b*) *J. Hyg., Camb.* **74,** 85.

Dolby, J. M. and Standfast, A. F. B. (1958) *Immunology* **1,** 144; (1961) *J. Hyg., Camb.* **59,** 205.

Dolby, J. M., Thow, D. C. W. and Standfast, A. F. B. (1961) *J. Hyg., Camb.* **59,** 191.

Dresser, D. W., Wortis, H. H. and Anderson, H. R. (1970) *Clin. exp. Immunol.* **7,** 817.

Ducrey, A. (1890) *Ann. Derm. Syph., Paris,* III sér. **1,** 56.

Edwards, G. *et al.* (1957) *Brit. med. J.* **ii,** 259.

Ehrich, W. E. *et al.* (1942) *Amer. J. med. Sci.* **204,** 530.

Eldering, G. (1941) *Amer. J. Hyg.* **34,** B., 1; (1942) *Ibid.* **36,** 294.

Eldering, G., Holwerda, J. and Baker, J. (1967) *J. Bact.* **93,** 1758.

Eldering, G., Hornbeck, C. and Baker, J. (1957) *J. Bact.* **74,** 133.

Eldering, G. and Kendrick, P. (1938) *J. Bact.* **35,** 561.

Eliot, C. P. and Lewis, M. R. (1934) *J. Amer. vet. med. Ass.* **84,** 878.

Evans, D. G. (1940) *J. Path. Bact.* **51,** 49; (1942) *Lancet* **i,** 529; (1944) *J. Path. Bact.* **56,** 49; (1947) *Ibid.* **59,** 341.

Evans, D. G. and Adams, M. O. (1952) *J. gen. Microbiol.* **7,** 169.

Evans, D. G. and Maitland, H. B. (1937) *J. Path. Bact.* **45,** 715; (1939) *Ibid.*, **48,** 67.

Evans, D. G. and Perkins, F. T. (1953) *J. Path. Bact.* **66,** 479; (1954*a*) *Ibid.* **68,** 251; (1954*b*) *Brit. J. exp. Path.* **35,** 322, 603; (1955) *Ibid.* **36,** 391.

Farrand, R. J. (1969) *Brit. med. J.* **i,** 150.

Farthing, J. R. and Holt, L. B. (1962) *J. Hyg., Camb.* **60,** 411.
Feiner, R. R. and Mortara, F. (1945) *Amer. J. Syph.* **29,** 71.
Felton, H. M., Gaggero, A. and Pomerat, C. (1954) *Texas Rep. Biol. Med.* **12,** 960.
Ferry, N. S. (1911) *J. infect. Dis.* **8,** 399; (1912) *Vet. J.* **68,** 376; (1912–13) *Amer. vet. Rev.* **41,** 77.
Ferry, N. S. and Houghton, E. M. (1919) *J. Immunol.* **4,** 233.
Fildes, P. (1920) *Brit. J. exp. Path.* **1,** 129; (1921) *Ibid.* **2,** 16; (1922) *Ibid.* **3,** 210; (1923) *Ibid.* **4,** 265; (1924) *Ibid.* **5,** 69.
Fishel, C. W., Szentivanyi, A. and Talmage, D. W. (1962) *J. Immunol.* **89,** 8; (1964) In: *Bacterial Endotoxins*, p. 474. Ed. by M. Landy and W. Braun, Rutgers Univ. Press, N.Y.
Fisher, S. (1955) *Aust. J. exp. Biol. med. Sci.* **33,** 609.
Fleming, A. (1929) *Brit. J. exp. Path.* **10,** 226.
Flesher, A. R. and Insel, R. A. (1978) *J. infect. Dis.* **138,** 719.
Flosdorf, E. W., Bondi, A. and Dozois, T. F. (1941) *J. Immunol.* **42,** 133.
Flosdorf, E. W., Dozois, T. F. and Kimball, A. C. (1941) *J. Bact.* **41,** 457.
Flosdorf, E. W. and Kimball, A. C. (1940*a*) *J. Immunol.* **39,** 287; (1940*b*) *Ibid.* **39,** 475.
Flosdorf, E. W., Kimball, A. C. and Chambers, L. A. (1939) *Proc. Soc. exp. Biol., N.Y.* **41,** 122.
Fothergill, L. D., Dingle, J. H. and Chandler, C. A. (1937) *J. exp. Med.* **65,** 721.
Franklin, A. W. and Garrod, L. P. (1953) *Brit. med. J.* **ii,** 1067.
Frazer, J., Zinnemann, K. and Boyce, J. M. H. (1974) *J. med. Microbiol.* **8,** 89.
Friedberger, E. (1903) *Zbl. Bakt.* **33,** 401.
Füzi, M. (1973) *Zbl. Bakt.* I Abt. Orig. **244,** 270.
Gastal, R. (1958) *Ann. Inst. Pasteur* **94,** 636.
Gibson, L. (1970) *Amer. J. clin. Path.* **54,** 199.
Gilder, H. and Granick, S. (1947) *J. gen. Physiol.* **31,** 103.
Goldman, W. E., Klapper, D. G. and Baseman, J. B. (1982) *Infect. Immun.* **36,** 782.
Goldner, M., Jakus, C. M., Rhodes, H. K. and Wilson, R. J. (1966) *J. gen. Microbiol.* **44,** 439.
Goodnow, R. A. (1980) *Microbiol. Rev.* **44,** 722.
Gordon, J., Woodcock, H. E. de C. and Zinnemann, K. (1944) *Brit. med. J.* **i,** 779.
Granick, S. and Gilder, H. (1946) *J. gen. Physiol.* **30,** 1.
Granick, S. and Mauzerall, D. (1961). In: *Metabolic Pathways*, vol. 3, p. 525. Ed. by D. M. Greenberg, Academic Press, New York.
Grassberger, R. (1897) *Z. Hyg. InfektKr.* **25,** 453.
Gray, D. F. (1947) *J. Path. Bact.* **59,** 235; (1949) *J. Immunol.*, **61,** 35.
Greenberg, L. and Fleming, D. S. (1947) *Canad. J. publ. Hlth* **38,** 279; (1948) *Ibid.* **39,** 131.
Greene, G. R. (1978) *Pediatrics* **62,** 1021.
Gunnarsson, A. (1979) *Amer. J. vet. Res.* **40,** 469.
Gunnarsson, A., Biberstein, E. L. and Horvell, B. (1977) *Amer. J. vet. Res.* **38,** 1111.
Hababou-Sala, J. (1925) *C. R. Soc. Biol.* **92,** 498.
Hafiz, S., Kinghorn, G. R. and McEntegart, M. G. (1981) *Brit. J. vener. Dis.* **57,** 382.
Halpern, B. N. and Roux, J. (1949) *C. R. Soc. Biol.* **143,** 923.
Hammond, G. W., Lian, C. J., Wilt, J. C. and Ronald, A. R. (1978) *J. clin. Microbiol.* **7,** 39.
Hanada, M., Shimoda, K., Tomita, S., Nakase, Y. and Nishiyama, Y. (1979) *Jap. J. vet. Sci.* **41,** 1.
Hannah, P. and Greenwood, J. R. (1982) *J. clin. Microbiol.* **16,** 861.
Hargis, B. J., Malkiel, S. and Berkelhammer, J. (1968) *J. Immunol.* **101,** 374.
Harris, D. L. and Switzer, W. P. (1969) *Amer. J. vet. Res.* **30,** 1161.
Hatasa, K. (1964) *Acta paediat. jap.* **68,** 841, 848.
Hegarty, C. P., Thiele, E. and Verwey, W. F. (1950) *J. Bact.* **50,** 651.
Henry, H. (1912) *J. Path. Bact.* **17,** 174.
Hewlitt, E., Spiegel, A., Wolff, J., Aurback, G. and Manclark, C. R. (1978) *Infect. Immun.* **22,** 430.
Hinz, K. and Kunjara, C. (1977) *Int. J. syst. Bact.* **27,** 324.
Hjärre, A. and Wramby, G. (1943) *Z. InfektKr. Haust.* **60,** 37.
Ho, M.-K., Kong, A. S. and Morse, S. I. (1977) *J. exp. Med.* **149,** 1001.
Holt, L. B. *et al.* (1961) *J. Hyg., Camb.* **59,** 373.
Holt, L. B. and Spasojević, V. (1968) *J. med. Microbiol.* **1,** 119.
Hornibrook, J. W. (1940) *Proc. Soc. exp. Biol., N.Y.* **45,** 598.
Hovig, B. and Aandahl, E. H. (1969) *Acta path. microbiol. scand.* **77,** 676.
Huang, C. C. *et al.* (1962) *New Engl. J. Med.* **266,** 105.
Huet, M. and Benslama, T. (1956) *Arch. Inst. Pasteur, Tunis* **33,** 65.
Hunt, G. A. (1935) *Proc. Soc. exp. Biol., N.Y.* **33,** 293.
Huntington, R. W. (1935) *J. clin. Invest.* **14,** 459.
Iida, T. and Ajiki, Y. (1974) *Jap. J. Microbiol.* **18,** 119.
Iida, T., Kusano, N., Yamamoto, A. and Shiga, H. (1962) *Jap. J. exp. Med.* **32,** 471.
Iida, T. and Okonogi, T. (1971) *J. med. Microbiol.* **4,** 51.
Iizuka, A. (1938) *Z. ImmunForsch.* **94,** 312, 318.
Ingham, H. R. and Turk, D. C. (1969) *J. clin. Path.* **22,** 258.
Irons, L. I. and MacLennan, A. T. (1978). See C. R. Manclark and J. C. Hill, p. 338.
Ishida, S., Asakawa, S. and Kurokawa, M. (1970) *Jap. J. med. Sci. Biol.* **23,** 67.
Jackson, F. L., Goodman, Y. E., Bel. F. R. Wong, P. C. and Whitehouse, R. L. S. (1971) *J. med. Microbiol.* **4,** 171.
Jebb, W. H. H. and Tomlinson, A. H. (1955) *J. gen. Microbiol.* **13,** 1; (1957) *Ibid.*, **17,** 59.
Jemsek, J. G., Greenberg, S. B., Gentry, L. O., Walton, D. E. and Mattox, K. L. (1979) *Amer. J. Med.* **66,** 51.
Johnson, R. and Sneath, P. H. A. (1973) *Int. J. syst. Bact.* **23,** 381.
Jones, H. M. and Camps, F. E. (1957) *Practitioner* **178,** 223.
Kairies, A. (1935) *Z. Hyg. InfektKr.* **117,** 12.
Kairies, A. and Schwartzer, K. (1936) *Zbl. Bakt.* **137,** 351.
Kaldor, J., Asznowicz, R. and Dwyer, B. (1979) *J. clin. Path.* **32,** 538.
Kaplan, W. *et al.* (1956) *J. invest. Derm.* **26,** 407.
Kasuga, T. Y., Nakase, Y., Ukishuma, K. and Takatsu, K. (1954) *Kitasato. Arch. exp. Med.* **27,** 49.
Katada, T., and Ui, M. (1979) *J. biol. Chem.* **254,** 469; (1982) *Ibid.* **257,** 7210.
Kendrick, P. L., Eldering, G., Dixon, M. K. and Misner, J. (1947) *Amer. J. publ. Hlth.* **37,** 803.
Kennedy, P. C., Frazier, L. M., Theilen, G. H. and Biberstein, E. E. (1958) *Amer. J. vet. Res.* **19,** 645.
Khairat, O. (1940) *J. Path. Bact.* **50,** 497; (1971) *Brit. med. J.* **i,** 728.
Kilian, M. (1974) *Acta path. microbiol. scand.* **B82,** 835; (1976*a*) *J. gen. Microbiol.* **93,** 9; (1976*b*) *Acta path. microbiol. scand.* **B84,** 339.
Kilian, M., Frederiksen, W. and Biberstein, E. L. B. (1981) In: *Haemophilus, Pasteurella, and Actinobacillus*, p. vii. Ed. by M. Kilian and W. Frederiksen. Academic Press, London.
Kilian, M., Mordhorst, C.-H., Dawson, C. R. and Lautrop, H. (1976) *Acta path. microbiol. scand.* **B84,** 132.
Kilian, M., Nicolet, J. and Biberstein, E. L. (1978) *Int. J. syst. Bact.* **28,** 20.
Kind, L. S. (1953) *J. Immunol.* **70,** 411; (1955) *J. Allergy* **26,** 507.
Kirchenbauer, H. (1934) *Z. InfektKr. Haustiere* **45,** 273.
Klieneberger-Nobel, E. (1948) *J. Hyg. Camb.* **46,** 345.
Klingeren, B. van, Dessens-Kroon, M. and van Embden, J. D. A. (1978) In: *Current Chemotherapy*, vol. 2, p. 450. Ed. by W. Siegenthaler and R. Lüthy. American Society for Microbiology, Washington, DC.
Kloos, W. E., Dolorogosz, W. J., Ezzell, J. W., Kimbo, B. R. and Manclark, C. R. (1978). See Manclark, C. R. and Hill, J. C., p. 70.

Knorr, M. (1924) *Zbl. Bakt.* **92,** 371, 385.
Knox, K., Elmes, P. C. and Fletcher, C. M. (1955) *Lancet* **i,** 120.
Koch, R. (1887) *Arb. ReichsgesundhAmt.* **3,** 62.
Kong, A. S. and Morse, S. I. (1977) *J. exp. Med.* **145,** 163.
Krage, P. (1910) *Z. InfektKr. Haustiere* **7,** 380.
Kraut, M. S., Attebery, H. R., Finegold, S. M. and Sutter, V. L. (1972) *J. infect. Dis.* **126,** 189.
Kristensen, M. (1922) *Haemoglobinophilic Bacteria*. Copenhagen; (1927) *C. R. Soc. Biol.* **96,** 355.
Krumwiede, E. and Kuttner, A. G. (1938) *J. exp. Med.* **67,** 429.
Kuwajima, Y. *et al.* (1958*a*) *Osaka City Med. J.* **4,** 177.
Kuwajima, Y., Matsui, T. and Kishigami, M. (1958*b*) *Jap. J. Microbiol.* **1,** 375.
Lacey, B. W. (1954) *J. Hyg., Camb.* **52,** 273; (1960) *Ibid.* **58,** 57.
Laidlaw, P. P. and Dunkin, G. W. (1926) *J. comp. Path.* **39,** 222.
Lapinski, E. M. and Flakas, E. D. (1967) *J. Bact.* **93,** 1438.
Lautrop, H. (1958) *Acta path. microbiol, scand.* **43,** 255.
Lawson, G. M. (1933) *Amer. J. Dis. Child.* **47,** 1454; (1939) *Amer. J. Hyg.* **29,** 119; (1940) *J. Lab. clin. Med.* **25,** 435.
Lee, J. M. and Olitsky, P. K. (1955) *Proc. Soc. exp. Biol., N.Y.* **89,** 263.
Lehrer, S. B. (1978). See Manclark, C. R. and Hill, J. C., p. 160.
Lehrer, S. B., Tan, E. M. and Vaughan, J. H. (1974) *J. Immunol.* **113,** 18.
Lehrer, S. B., Vaughan, J. H. and Tan, E. M. (1975) *J. Immunol.* **114,** 34.
Leidy, G., Jaffee, I. and Alexander, H. E. (1965) *Proc. Soc. exp. Biol., N.Y.* **118,** 671.
Lennon, V. A., Westall, F. C., Thompson, M. and Ward, W. (1976) *Europ. J. Immunol.* **6,** 805.
Leslie, P. H. and Gardner, A. D. (1931) *J. Hyg., Camb.* **31,** 423.
Levine, S., Wenk, E. J., Devlin, H. B., Pieroni, R. E. and Levine, L. (1966) *J. Immunol.* **97,** 363.
Levinthal, W. (1918) *Z. Hyg. InfektKr.* **86,** 1.
Lewis, P. A. and Shope, R. F. (1931) *J. exp. Med.* **54,** 361.
Lin, T. (1958) *J. Formosan med. Ass.* **57,** 505.
Linnemann, C. C. and Perry, E. B. (1977) *Amer. J. Dis. Child* **131,** 560.
López, M. M. (1952) *Microbiol. española* **5,** 177.
Lwoff, A. and Lwoff, M. (1937*a*) *Proc. roy. Soc. B* **122,** 352, 360; (1937*b*) *Ann. Inst. Pasteur* **59,** 129; (1937*c*) *C. R. Acad. Sci.* **204,** 1510.
Lwoff, A. and Pirosky, I. (1937) *C. R. Soc. Biol.* **124,** 1169.
McGaughey, C. A. (1932) *J. comp. Path.* **45,** 58.
M'Gowan, J. P. (1911) *J. Path. Bact.* **15,** 372.
McIntosh, J. (1922) *Spec. Rep. Ser. med. Res. Coun., Lond.* No. 63.
McLean, I. W., Schwab, J. L., Hillegas, A. B. and Schlingman, A. S. (1949) *J. clin. Invest.* **28,** 953.
MacLennan, A. P. (1960) *Biochem. J.* **74,** 398.
MacPherson, C. F. C. (1948) *Canad. J. Res.*, Sect. E. **26,** 197.
MacPherson, C. F. C., Heidelberger, M., Alexander, H. E. and Leidy, G. (1946) *J. Immunol.* **52,** 207.
Malkiel, S. and Hargis, B. J. (1957) *Proc. Soc. exp. Biol., N.Y.* **81,** 689.
Mallory, F. B. and Hornor, A. A. (1912) *J. med. Res.* **27,** 115.
Mallory, F. B., Hornor, A. A. and Henderson, F. F. (1912) *J. med. Res.* **27,** 391.
Manclark, C. R. and Hill, J. C. (1978) [Eds] *International Symposium on Pertussis*, US Dept. of Health, Bethesda, Md.
Mannheim, W., Pohl, S. and Holländer, R. (1980) *Zbl.Bakt. 1 Abt.* **A24b,** 512.
Manten, A., van Kingeren, B. and Dessers-Kroon, M. (1976) *Lancet* **i,** 702.
Mason, M. A. (1966) *Canad. J. Microbiol.* **12,** 539.
Matthew, M. (1979) *J. antimicrob. Chemother.* **5,** 349.
Matthews, P. R. J. and Pattison, I. H. (1961) *J. comp. Path.* **71,** 44.
May, J. R. (1965) *J. Path. Bact.* **90,** 379.
May, J. R. and Davies, J. (1972) *Brit. med. J.* **iii,** 376.
Mazloum, H. A. and Rowley, D. (1955) *J. Path. Bact.* **70,** 439.
Mazloum, H. A., Kilian, M., Mohamed, Z. M. and Saw, M. D. (1982) *Acta path. microbiol. immunol. scand.* **B90,** 109.
Miles, A. A. and Gray, J. (1938) *J. Path. Bact.* **47,** 257.
Miller, J. J. (1937) *Proc. Soc. exp. Biol., N.Y.* **37,** 45.
Mitchell, C. A. (1925) *J. Amer. vet. med. Ass.* **68,** 8.
Morse, J. H., Kong, A. L., Lindenbaum, J. and Morse, S. I. (1977) *J. clin. Invest.* **60,** 683.
Morse, J. H. and Morse, S. I. (1970) *J. exp. Med.* **131,** 1342.
Morse, S. I. (1965) *J. exp. Med.* **121,** 49; (1971) *J. infect. Dis.* **136,** 234; (1976) *Adv. appl. Microbiol.* **20,** 9.
Morse, S. I. and Bray, K. K. (1969) *J. exp. Med.* **129,** 523.
Morse, S. I. and Morse, J. H. (1976) *J. exp. Med.* **43,** 1483.
Moxon, E. R., Glode, M. P., Sutton, A. and Robbins, J. B. (1977) *J. infect. Dis.* **136,** 186.
Moxon, E. R. and Murphy, P. A. (1978) *Proc. nat. Acad. Sci., Wash.* **75,** 1534.
Moxon, E. R., Smith, A. L., Averill, D. R. and Smith, D. H. (1974) *J. infect. Dis.* **129,** 154.
Munoz, J. J. and Bergman, R. K. (1966) *J. Immunol.* **97,** 120; (1977) *Bordetella pertussis—Immunological and other biological activities.* Immunology series no. **4**. Marcel Dekker, New York; (1978) See Manclark, C. R. and Hill, J. C., p. 143.
Munoz, J. and Hestekin, B. M. (1966) *J. Bact.* **91,** 2175.
Muse, K. E., Collier, A. M. and Baseman, J. B. (1977) *J. infect. Dis.* **136,** 768.
Muse, K. E., Findley, D., Allen, L. and Collier, A. M. (1978). See Manclark, C. R. and Hill, J. C., p. 41.
Nelson, J. D. (1980) *J. Amer. med. Ass.* **244,** 239.
Newman, R. B., Stevens, R. W. and Gaafar, H. A. (1970) *J. Lab. clin. Med.* **76,** 107.
Nicolet, J. (1968) *Path. et Microbiol., Basel* **31,** 215; (1971) *Zbl. Bakt.* **216,** 487.
Nicolle, C. (1923) *C. R. Soc. Biol.* **88,** 871.
Nicolle and Durand (1924) *Arch. Inst. Pasteur, Tunis* **13,** 243.
North, E. A. (1946) *Aust. J. exp. Biol. med. Sci.* **24,** 253.
North, E. A., Anderson, G. and Graydon, J. J. (1941) *Med. J. Aust.* **ii,** 589.
North, E. A., Keogh, E. V., Anderson, G. and Williams, S. (1939) *Aust. J. exp. Biol. med. Sci.* **17,** 275.
North, E. A., Keogh, E. V., Christie, R. and Anderson, G. (1940) *Aust. J. exp. Biol. med. Sci.* **18,** 125.
Novotny, P. and Cownley, K. (1978). See Manclark, C. R. and Hill, J. C., p. 99.
Oberhofer, T. R. and Back, A. E. (1979) *J. clin. Microbiol.* **10,** 168.
Ocklitz, H. W. and Milleck, J. (1967) *Zbl. Bakt.* **203,** 79.
Olitzki, A. L. and Sulitzeanu, A. (1959) *J. Bact.* **77,** 264.
Omland, T. (1964) *Acta path. microbiol. scand.* **62,** 79, 89.
Onoue, K., Kitagawa, M. and Yamamura, Y. (1961) *J. Bact.* **82,** 648; (1963) *Ibid.*, **86,** 648.
Orfila, J. and Courden, B. (1961) *Ann. Inst. Pasteur* **100,** 252.
Ortez, R., Klein, T. W. and Szentivanyi, A. (1962) *J. Immunol.* **89,** 8; (1970) *J. Allergy clin. Immunol.* **89,** 8.
Ortez, R. A., Seshachalam, D. and Szentivanyi, A. (1975) *Biochem. Pharmacol.* **24,** 1297.
Parfentjev, I. A. and Goodline, M. A. (1948) *J. Pharmacol.* **92,** 411.
Parfentjev, I. A. and Schleyer, W. L. (1949) *Arch. Biochem.* **20,** 341.
Park, W. H., Williams, A. W. and Cooper, G. (1918) *Proc. Soc. exp. Biol., N.Y.* **16,** 120.
Parker, C. (1976) *Adv. appl. Microbiol.*, **20,** 27.
Parker, C. D. (1978). See Manclark, C. R. and Hill, J. C., p. 65.

Parker, C. W. and Morse, S. I. (1973) *J. exp. Med.* **137,** 1078.
Parker, J. T. (1919) *J. Amer. med. Ass.* **72,** 476.
Parton, R. and Durham, J. P. (1978) *FEMS Microbiol. Letters* **4,** 287.
Parton, R. and Wardlaw, A. C. (1975) *J. med. Microbiol.* **8,** 47.
Pattison, I. H., Howell, D. G. and Elliot, J. (1957) *J. comp. Path.* **67,** 320.
Petersen, K. B. (1976) *Acta path. microbiol. scand.* **B84,** 75.
Pfeiffer, R. (1892) *Dtsch. med. Wschr.* **18,** 28; (1893) *Z. Hyg. InfektKr.* **13,** 357.
Pieroni, R. E., Broderick, E. J. and Levine, L. (1976) *J. Bact.* **91,** 2169.
Pillemer, L., Blum, L. and Lepow, I. H. (1954) *Lancet* **i,** 1257.
Pittman, M. (1931) *J. exp. Med.* **53,** 471; (1935) *J. Bact.*, **30,** 149; (1952) *J. Immunol.* **69,** 201; (1953) *J. Bact.* **65,** 750; (1979) *Revs. infect. Dis.* **1,** 401.
Pittman, M. and Davis, D. J. (1950) *J. Bact.* **59,** 413.
Pittman, M., Furman, B. L. and Wardlaw, A. C. (1980) *J. infect. Dis.* **142,** 56.
Pittman, M. and Wardlaw, A. C. (1981). In: *The Prokaryotes.* Ed. by M. P. Starr, H. Stolp, H. G. Trüper, A. Balows and H. G. Schlegel. Springer-Verlag, Berlin.
Platt, A. E. (1937) *J. Hyg., Camb.* **37,** 98; (1939) *Aust. J. exp. Biol. med. Sci.* **17,** 19.
Pollock, M. R. (1947) *Brit. J. exp. Path.* **28,** 295.
Powell, H. M., Culbertson, C. G. and Ensminger, P. W. (1951) *Publ. Hlth Rep., Wash.* **66,** 346.
Powell, H. M. and Jamieson, W. A. (1937) *J. Immunol.* **32,** 153.
Pozsgi, N., Andreesco-Tigoiu, V. and Dona, D. (1961) *Arch. roum. Path. exp. Microbiol.* **20,** 431.
Preston, N. W. (1963) *Brit. med. J.* **ii,** 724; (1965) *Ibid.* **ii,** 11; (1976) *J. Hyg., Camb.* **77,** 85.
Pritchett, I. W. and Stillman, E. G. (1919) *J. exp. Med.* **29,** 259.
Proom, H. (1947) *J. Path. Bact.* **59,** 165; (1955) *J. gen. Microbiol.* **12,** 63.
Pusztai, S. and Joo, I. (1967) *Ann. Immunol. Hung.* **10,** 63.
Raettig, H. (1940) *Zbl. Bakt.* **145,** 386.
Reenstierna, J. (1921) *Acta derm-venereol., Stockh.* **2,** 1; (1923) *Arch. Inst. Pasteur, Tunis* **12,** 273.
Regan, J. and Lowe, F. (1977) *J. clin. Microbiol.* **6,** 303.
Report (1959) *Brit. med. J.* **i,** 994; (1967) *Int. J. syst. Bact.* **17,** 165; (1971) *Ibid.* **21,** 132.
Reymann, F. (1947) *Acta path. microbiol. scand.* **24,** 208; (1949) *Ibid.* **26,** 345.
Rhea, L. J. (1915) *J. med. Res.* **32,** 471.
Richter, G. W. and Kress, Y. (1967) *J. Bact.* **94,** 1216.
Riley, M. G. I., Russell, E. G. and Callinan, R. B. (1977) *J. Amer. vet. med. Ass.* **171,** 649.
Rivers, T. M. (1922*a*) *Johns Hopk. Hosp. Bull.* **33,** 429; (1922*b*) *J. Bact.* **7,** 579.
Robbins, K. S. and Pillemer, L. (1950) *Proc. Soc. exp. Biol., N.Y.* **74,** 75.
Rogers, K. B., Zinnemann, K. and Foster, W. P. (1960) *J. clin. Path.* **13,** 519.
Rosher, A. B. (1947) *Proc. R. Soc. Med.* **40,** 749.
Ross, R., Munoz, J. and Cameron, C. (1969) *J. Bact.* **99,** 57.
Rowatt, E. (1957*a*) *J. gen. Microbiol.* **17,** 279; (1957*b*) *Ibid.* **17,** 297.
Rubin, L. G., Medeiros, A. A., Yolken, R. H. and Moxon, E. R. (1981) *Lancet* **ii,** 1008.
Sanderson, E. S. and Greenblatt, R. B. (1937) *Sth. med. J.* **30,** 147.
Sato, Y., Izumiya, K., Oda, M.-A. and Sato, M. (1978). See Manclark, C. R. and Hill, J. C., p. 51.
Saunders, J. R., Elwell, L. P., Falkow, S., Sykes, R. B. and Richmond, M. H. (1978) *Scand. J. infect. Dis.* **13** (suppl.), 16.
Sawata, A., Kume, K. and Nakase, Y. (1980) *Amer. J. vet. Res.* **41,** 1901.
Schlüter, W. (1936) *Zbl. Bakt.* **136,** 362.
Schneerson, R., Bradshaw, M. W., Whisnant, J. K., Myerowitz, R. L., Parke, J. C. and Robbins, J. B. (1972), *J. Immunol.* **108,** 1551.
Shaffer, M. F. and Shaffer, L. S. (1946) *Proc. Soc. exp. Biol., N.Y.* **62,** 244.
Shibley, G. S. and Hoelscher, H. (1934) *J. exp. Med.* **60,** 403.
Shifrine, M. and Biberstein, E. L. (1960) *Nature, Lond.* **187,** 623.
Shope, R. E. (1931) *J. exp. Med.* **54,** 349; (1964) *Ibid.*, **119,** 357.
Shope, R. E., White, D. C. and Leidy, G. (1964) *J. exp. Med.* **119,** 369.
Silverthorne, N. (1938) *Canad. J. publ. Hlth* **29,** 233.
Silverthorne, N., Cameron, C. and Paterson, M. (1943) *Canad. J. publ. Hlth* **34,** 175.
Smith, C. H. (1954) *J. Path. Bact.* **68,** 284.
Smith, M. M. (1931) *J. Hyg., Camb.* **31,** 321.
Smith, T. (1913) *J. med. Res.* **29,** 291.
Smith, W., Hale, J. H. and O'Callaghan, C. H. (1953) *J. Path. Bact.* **65,** 229.
Smolens, J. and Mudd, S. (1943) *J. Immunol.* **47,** 155.
Sneath, P. H. A. and Johnson, R. (1973) *Int. J. syst. Bact.* **23,** 405.
Snieszko, S. F., Griffin, P. J. and Friddle, S. B. (1950) *J. Bact.* **59,** 699.
Spooner, E. T. C. (1938) *J. Hyg., Camb.* **38,** 79.
Sprunt, D. H. and Martin, D. S. (1943) *Amer. J. Path.* **19,** 255.
Sprunt, D. H., Martin, D. S. and Williams, J. E. (1935) *J. exp. Med.* **62,** 449.
Stainer, D. W. and Scholte, M. J. (1970) *J. gen. Microbiol.* **63,** 211.
Stanbridge, T. N. and Preston, N. W. (1974) *J. Hyg., Camb.* **72,** 213.
Standfast, A. F. B. (1951*a*) *J. gen. Microbiol.* **5,** 250; (1951*b*) *Ibid.* **5,** 531; (1958) *Immunology* **1,** 123, 135.
Standfast, A. F. B. and Dolby, J. M. (1961) *J. Hyg., Camb.* **59,** 217.
Stein, R. O. (1929) See *Kolle and Wassermann Hdb. path. Mikroorg.*, III te Aufl., **6,** 185.
Stillman, E. G. and Bourn, J. M. (1920) *J. exp. Med.* **32,** 665.
Straker, E. A. (1945) *Lancet* **i,** 817.
Stronk, M. G. and Pittman, M. (1955) *J. infect. Dis.* **96,** 152.
Sugimoto, C., Isayame, Y. and Mitani, K. (1981) *Nat. Inst. Anim. Quart.* **21,** No. 4.
Suko, M., Ogita, T., Okendaira, H. and Horiuchi, Y. (1977) *Int. Arch. Allergy, Basel* **54,** 329.
Sumi, T. and Ui, M. (1975) *Endocrinology*, **97,** 352.
Sutcliffe, E. N. and Abbott, J. D. (1972) *J. clin. Path.* **25,** 732.
Swaney, L. M. and Breese, S. S. (1980) *Amer. J. vet. Res.* **41,** 127.
Szentivanyi, A., Fishel, C. W. and Talmage, D. W. (1963) *J. infect. Dis.* **113,** 86.
Tainturier, D. J., Delmas, C. F. and Dabernat, H. J. (1981) *J. clin. Microbiol.* **14,** 355.
Targowski, S. and Targowski, H. (1979) *J. clin. Microbiol.* **9,** 33.
Taub, A. N., Rosett, W., Adler, A. and Morse, S. I. (1972) *J. exp. Med.* **136,** 1581.
Taylor, C. E. D., Rosenthal, R. O., Brown, D. F. J., Lapage, S. P., Hill, L. R. and Legros, R. M. (1978) *Equine vet. J.* **10,** 136.
Thjøtta, T. and Avery, O. T. (1921) *J. exp. Med.* **34,** 97, 455.
Thomas, W. J., McReynolds, J. W., Moch, C. R. and Bailey, D. W. (1974) *Lancet* **i,** 313.
Thompson, H., McCandlish, I. A. P. and Wright, N. G. (1976) *Res. vet. Sci.* **20,** 16.
Timoney, P. J., Ward, J., McArdle, J. F. and Harrington, A. M. (1979) *Vet. Rec.* **104,** 530.
Tomasczewski, E. (1903) *Z. Hyg. InfektKr.* **42,** 327.
Toomey, J. A., Ranta, K., Robey, L. and McClelland, J. E. (1935) *J. infect. Dis.* **57,** 49.

Toomey, J. A., Takacs, W. S. and Ranta, K. (1936) *J. infect. Dis.* **59,** 326.
Torregrosa, M. C. de and Francis, T. (1941) *J. infect. Dis.* **68,** 59.
Tunevall, G. (1951*a*) *Acta path. microbiol. scand.* **29,** 203; (1951*b*) *Ibid.* **29,** 387; (1952*a*) *Ibid.* **30,** 203; (1952*b*) *Ibid.* **31,** 233; (1953) *Ibid.* **32,** 193, 258.
Turk, D. C. (1982) *H. influenzae. Publ. Hlth Lab. Serv. Monogr.* No. 17.
Turk, D. C. and Holdaway, M. D. (1968) *J. med. Microbiol.* **1,** 79.
Turk, D. C. and May J. R. (1967) *Haemophilus influenzae. Its Clinical Importance.* English University Press, London.
Turner, G. C. (1961) *J. Path. Bact.* **81,** 15.
Twarog, F. J. and Rose, N. R. (1969) *Proc. Soc. exp. Biol. N.Y.* **130,** 434.
Ui, M., Katada, T. and Yajima, M. (1978). See Manclark, C. R. and Hill, J. C., p. 166.
Ungar, J., James, A. M., Muggleton, P. W., Pegler, H. F. and Tomich, E. G. (1950) *J. gen. Microbiol.* **4,** 345.
Ungar, J. and Muggleton, P. W. (1949) *J. gen. Microbiol.* **3,** 353; (1954) *J. Path., Bact.* **67,** 285.
Ungar, J. Muggleton, P. W. and Stevens, W. K. (1954) *J. Hyg., Camb.* **52,** 475.
Valentine, F. C. O. and Rivers, T. M. (1927) *J. exp. Med.* **45,** 993.
Verwey, W. F. and Thiele, E. H. (1949) *J. Immunol.* **61,** 27.
Wade, H. W. and Manalang, C. (1920) *J. exp. Med.* **31,** 95.
Wardlaw, A. C. and Jakus, C. M. (1968) *Canad. J. Microbiol.* **14,** 989.
Wardlaw, A. C., Parton, R. and Hooker, M. J. (1976) *J. med. Microbiol.* **9,** 89.
Watanabe, S. (1939) *Kitasato Arch.* **16,** 1.
Waterworth, P. M. (1955) *Brit. J. exp. Path.* **36,** 186.
Weeks, J. E. (1887) *Arch. Augenheilk.* **17,** 318.
Weiner, S. L., Tinker, M. and Bradford, W. L. (1959) *Arch. Path.* **67,** 694.
Wells, E. B., Chang, S. M., Jackson, G. G. and Finland, M. (1950) *J. Pediat.* **36,** 752.
White, D. C. (1963) *J. Bact.* **85,** 84.
White, D. C. and Granick, S. (1963) *J. Bact.* **85,** 842.
White, D. C., Leidy, G., Jamieson, J. D. and Shope, R. E. (1964) *J. exp. Med.* **120,** 1.
Wilkins, J. R. and Helland, D. R. (1973) *J. Amer. vet. med. Ass.* **162,** 47.
Williams, J. D. and Andrews, J. (1974) *Brit. med. J.* **i,** 134.
Williamson, G. M. and Zinnemann, K. (1951) *J. Path. Bact.* **63,** 695; (1954) *Ibid.* **68,** 453.
Winslow, C.-E. A., Broadhurst, J., Buchanan, R. E., Krumwiede, C., Rogers, L. A. and Smith, G. H. (1920) *J. Bact.* **5,** 191.
Witebsky, E. and Salm, H. (1937) *J. exp. Med.* **65,** 43.
Wittler, R. G. (1951) *J. gen. Microbiol.* **5,** 1024.
Wollstein, M. (1909) *J. exp. Med.* **11,** 41; (1915) *Ibid.* **22,** 445; (1919) *Ibid.* **30,** 555.
Wright, J. and Ward, H. K. (1932) *J. exp. Med* **55,** 235.
Wright, N. G., Thompson, H., Taylor, D. and Cornwell, H. J. C. (1973) *Vet. Rec.* **93,** 486.
Yajima, M. *et al.* (1978*a*) *J. Biochem.* **83,** 295; (1978*b*) *Ibid.* **83,** 305.
Yoshida, N. *et al.* (1955) *Tokushima J. exp. Med.* **2,** 11.
Zamenhof, S. *et al.* (1953) *J. biol. Chem.* **203,** 695.
Zamenhof, S. and Leidy, G. (1954) *Fed. Proc.* **13,** 327.
Zinnemann, K. (1960) *Ergebn. Mikrobiol.* **33,** 307; (1980) *Zbl. Bakt.* I Abt. Org. **A247,** 248.
Zinnemann, K., Rogers, K. B., Frazer, J. and Boyce, J. M. H. (1968) *J. Path. Bact.* **96,** 413.
Zinnemann, K., Rogers, K. B., Frazer, J. and Devaraj, S. K. (1971) *J. med. Microbiol.* **4,** 139.

40

Brucella

Graham Wilson

Introductory

Definition

Small, non-motile, non-sporing, gram-negative coccobacilli. Grow rather poorly on ordinary media, or may require special media. Aerobic; no growth under strict anaerobic conditions. Growth often improved by CO_2. Little fermentative action on carbohydrates in usual media. Urea hydrolysed to a variable extent. Usually tend to produce alkali in litmus milk, and a brown pigmentation on potato. Intracellular parasites, occurring in man and animals, and producing characteristic infections. G+C content of DNA: 56–58 moles per cent.

Type species: *Brucella melitensis.*

The first member of the group, *Br. melitensis*, was isolated in 1887 by Bruce from the spleen of patients who had died of Malta fever. At that time, and for a long time afterwards, the bacillary nature of the organism was not recognized; in all the older textbooks it is therefore described as a micrococcus. The organism finds its natural habitat in the goat and the sheep. It may, however, infect other animals. In man it gives rise to undulant fever. It is fairly widely distributed throughout the world.

The discovery of the second member, *Br. abortus*, was made by Bang of Copenhagen in 1897. Working in conjunction with Stribolt, he isolated the organism from cows suffering from infectious abortion, and by a series of experiments demonstrated its specific role in this disease. The organism is parasitic in cattle. To a less extent it infects certain other animals. In man, it gives rise to undulant fever. It is perhaps even more widespread than *Br. melitensis*, having been found in practically every country of the world.

The third member, *Br. suis*, was isolated by Traum (1914) from the fetus of a sow. It is a natural parasite of pigs, in which it gives rise to a disease frequently characterized by inflammatory lesions in the reproductive organs. It may occasionally infect other animals. In man it shares with *Br. melitensis* and *Br. abortus* the ability to produce undulant fever. It appears to be very much less widespread than these two organisms, it chief home being in the large hog-raising districts of the middle western states of North America. In Denmark, *Br. suis* strains were isolated by Thomsen (1931, 1934) which differed in certain respects from those found in the United States; they will be referred to as the Danish porcine type. Their natural hosts are hares and swine. The American type has been found occasionally in Europe (see Thomsen 1934), and has been reported from Brazil (Neiva 1934), the Argentine, and Australia (King 1934).

Three other named organisms may be mentioned here; their claim to specific rank will be discussed in the section on Classification. One is *Br. ovis*, which gives rise to genital disease of sheep in New Zealand and Australia (Buddle and Boyes 1953). The second is *Br. neotomae*, which was isolated by Stoenner and Lackman (1957) from desert wood rats in Utah. The third, *Br. canis*, was described by Carmichael and Bruner (1968) as the cause of widespread abortion in dogs in the United States. (For the distribution of the different species of *Brucella* in various countries of the world, see Abdussalam and Fein 1976.)

The names paramelitensis, para-abortus, and parasuis were at one time used to refer to inagglutinable strains of *Brucella*, corresponding most closely to the melitensis, abortus, and suis types. So long as the so-called para-strains were regarded as distinct species, no objection could be raised to this terminology; but now that they are known to be merely rough variants of the original smooth forms, this practice is no longer justifiable and merely serves to confuse the nomenclature. We shall refer to these, therefore, as rough melitensis, abortus, or suis strains, as the case may be.

Morphology and staining

The bacilli are short and slender; the axis is straight; the ends are rounded; the sides may be parallel or convex outwards. In length they vary from about 0.6 to 1.5 μm, and in breadth from 0.5 to 0.7 μm. The short forms may appear as oval cocci, or, if they are about to divide, as diplococci. As a rule they are arranged singly, in pairs end-to-end, or in small groups; sometimes short chains of 4–6 members may be seen, especially in liquid media. Owing to the frequent coccoid appearance, their bacillary nature may be in doubt, but it may be noted that in size they are smaller than any of the gram-negative cocci. Moreover, when arranged in pairs, their long diameter is in the same axis as that in which they are lying, in distinction to the gram-negative diplococci, whose long axis is generally at right angles to that in which they are lying. By the electron microscope they are seen to have the usual structure of gram-negative bacilli, but in addition they are characterized by peripheral formations occurring singly or in clusters associated with the cytoplasmic membrane (de Petris *et al.* 1964; see also Dubray 1972). Reports on capsulation are discrepant. L-forms may develop on special media containing penicillin (Christophorov and Peschkov 1969), and sphaeroplasts on media containing penicillin and glycine (Hines *et al.* 1964).

Br. melitensis is generally considered to be more coccal in form than *Br. abortus*, and for this there is some justification. The difference in size and shape, however, is so slight as to render it impossible to distinguish with certainty between individual strains. Duncan (1928) pointed out that these organisms, when grown on agar or glucose agar, show no distinctive morphological differences; but if they are cultivated on a relatively rich medium, such as Fildes's peptic digest blood agar, *Br. abortus* strains frequently develop long bacillary forms, reaching 2.0 or 3.0 μm in length, whereas *Br. melitensis* strains usually retain their coccal shape and rarely exceed 1.0 μm in length.

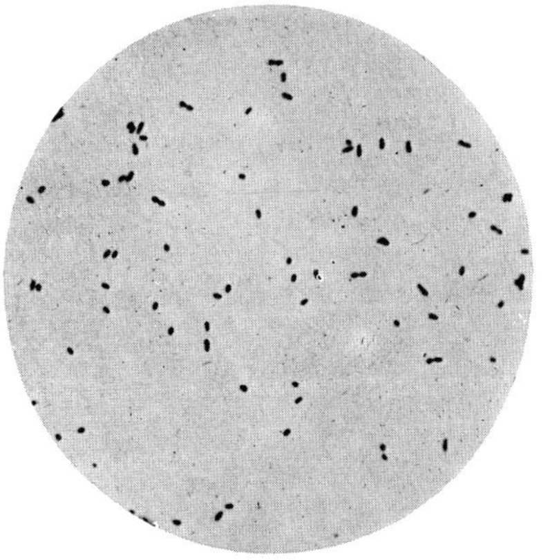

Fig. 40.1 *Brucella abortus*. From an agar culture, 24 hr, 37°, showing very short bacillary forms (× 1000).

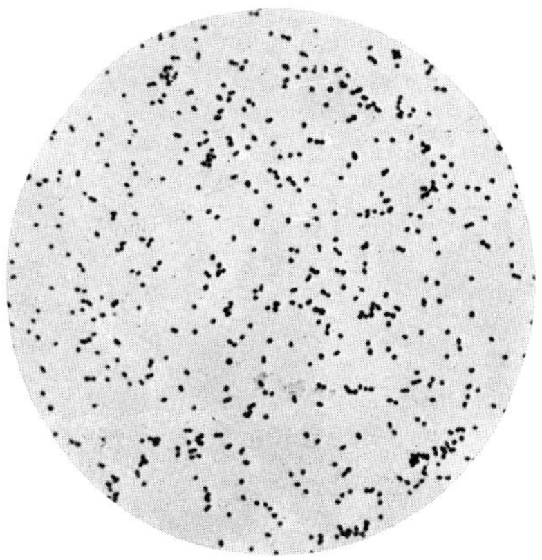

Fig. 40.2 *Brucella melitensis*. From an agar culture, 2 days, 37°, showing mainly coccal forms (× 1000).

The organisms stain fairly well with the ordinary dyes. They are gram negative, non-acid-fast, non-motile, and non-sporing. Bipolar staining is not uncommon, and irregularity in the depth of colour may be seen. In old cultures irregular forms appear.

To demonstrate brucellae in infected tissues advantage may be taken of their resistance to alkalis. Hansen's method uses alkaline methylene blue, and Köster's method alkaline safranin; both methods afford differential staining (Hansen and Köster 1936). Alternatively, labelled antisera may be used for identification by the immunofluorescence technique (Moody *et al.* 1961, Meyer 1966*b*).

Cultural reactions

Apart from their different CO_2 requirements, the members of this group resemble each other closely in their cultural characters. None of them is difficult to grow; none grows profusely. On agar the colonies are small, translucent and undifferentiated. In broth there is a moderate turbidity with a slight powdery or viscous sediment, which disintegrates completely on shaking; after about 2 weeks in the incubator or at room temperature the deposit becomes extremely viscous, and can be disintegrated only with difficulty. According to Thomsen (1933), if the organisms are grown in flasks of broth instead of in tubes, the suis and CO_2-requiring abortus types give rise in 1 to 3 weeks to a mealy or scaly surface pellicle and a heavy deposit that is difficult to disintegrate by shaking. Aerobic abortus strains form no pellicle, but produce a uniform turbidity and a slight deposit that is easily disintegrated. Strains of *Br. melitensis* give rise to a fairly dense turbidity, a moderately heavy deposit, and a granular, usually incomplete, surface growth. Growth in gelatin is poor, and is unaccompanied by liquefaction. On potato a yellow colour develops in 2 to 3 days, deepening to a café-au-lait or chocolate tint in the course of a fortnight. It will be recalled that the glanders bacillus, Whitmore's bacillus, the cholera vibrio, and *Ps. aeruginosa* behave similarly. Crystals of ammonium magnesium phosphate may form in cultures of *Brucella* on liver agar incubated aerobically (Huddleson *et al.* 1927).

Growth is rather slow; and, unless a fairly heavy inoculum is made, colonies are not usually visible for 2 days or even longer. In broth the maximum turbidity is not reached for a week or more. On the whole the American porcine strains probably give the best, and the Danish porcine strains the poorest growth, the melitensis and abortus strains occupying an intermediate position.

Br. abortus, *Br. melitensis* and *Br. suis*, when inoculated on to the chorioallantoic membrane of the developing chick embryo, are able to multiply and to bring about death of the embryo in a few days with lesions in the spleen and liver. All three organisms grow intracellularly—*Br. melitensis* in the ectodermal epithelium, *Br. abortus* and *Br. suis* in cells of mesodermal origin and in the vascular endothelium. Rough strains are non-invasive (Goodpasture and Anderson 1937, Buddingh and Womack 1941, de Ropp 1944).

The colonial appearance of *Brucella* depends on the smoothness or roughness of the strain. The difference is best brought out by examination under a binocular plate microscope with obliquely reflected light (Henry 1933) in 4-day-old cultures on glycerol dextrose agar or trypticase soy agar (Report 1967*a*). Colonies of antigenically smooth strains are small, circular, translucent, bluish, with a glistening surface. The individual cells are uniformly short rods arranged singly. Colonies of antigenically rough strains are of much the same size, but are less convex, more opaque, and have a dull dry yellowish-white granular appearance. The individual cells are rather larger than those of the smooth form, and occasional long slender rods are present. Mucoid colonies are transparent, greyish or slightly orange, and slimy. The distinction between smooth and rough colonies can be increased by flooding plates with a 1 in 2000 aqueous solution of crystal violet and pouring off the excess dye 15 seconds later. Smooth colonies then appear a light bluish-green, rough colonies red to bluish-red (White and Wilson 1951). Another method is to grow the organisms on tryptose agar containing 0.01 per cent of the dye, 2, 3, 5-triphenyltetrazolium chloride (Huddleson 1952). The differential inhibition effect of sodium diethyldithiocarbamate was recommended by Renoux (1952*b*) for distinguishing between colonies of the three main species (see also Cruickshank 1955).

Growth requirements

Growth is improved by the addition of natural animal protein to the medium, particularly by bovine serum in a 7.5 per cent concentration (Painter *et al.* 1966). The most satisfactory media are liver extract agar—first described by Holth (1911), subsequently by Stafseth (1920), and frequently referred to as Huddleson's medium—potato infusion agar, trypticase soy agar, and 2.5 per cent glucose or glycerol agar, all with the addition of serum. *Br. abortus* biotype 2 grows on liver agar only when enriched either with serum, 0.02 per cent crystalline bovine albumin, 0.1 per cent Tween 40, or an extract of disintegrated brucella cells (Huddleson 1956, 1964). Erythritol has a specially stimulating effect on the growth of most strains of *Brucella* (Smith *et al.* 1961, Pearce *et al.* 1962, Williams *et al.* 1964). The nutritional requirements vary with the different species and biotypes, but as a rule several amino acids are needed, together with the accessory factors thiamine, nicotinic acid, pantothenic acid, and biotin. Some laboratory strains grow best in aerated medium (Gerhardt and Gee 1946, Sanders and Huddleson 1950). Toxic substances inhibiting growth may be present in peptone and in absorbent cotton wool through which the medium is filtered (Schuhardt *et al.* 1949, Boyd and Casman 1951). (For detailed information on nutritional requirements and on the composition of defined media, see Gerhardt 1958.)

The range of temperature consistent with growth is 20–40°; the optimum is about 37°. At 20° growth is very slow. The effect of pH is rather difficult to dissociate from that of CO_2. Many strains of *Br. abortus* require for their optimum development a concentration of 5–10 per cent CO_2 in the atmosphere. This has the effect of turning an alkaline medium acid. For the growth of these organisms an initial H-ion concentration of pH 6.6 is desirable. The other members of the group usually grow as well on an alkaline as on a slightly acid medium; but since, as will be pointed out directly, even melitensis strains are often benefited by a small amount of extra CO_2, it is advisable for practical purposes to adjust media to pH 6.6–6.8.

CO_2 requirements

One of the most interesting features of the *Brucella* group is their peculiar respiratory behaviour. Ever since its original isolation by Bang (1897), *Br. abortus* has presented certain difficulties in cultivation. No growth occurs on a solid medium under aerobic conditions. If, however, the tube is suitably sealed (Preisz

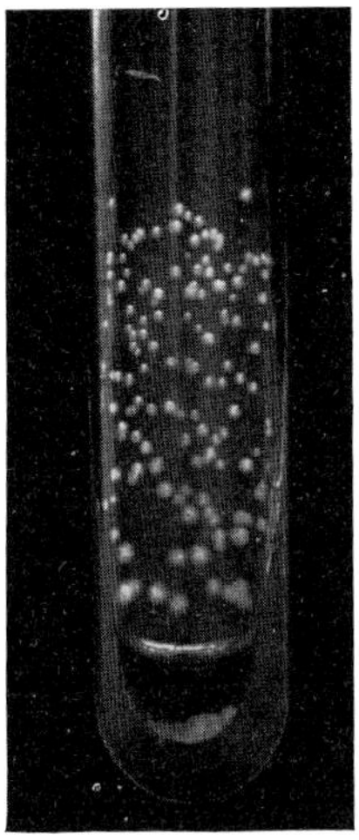

Fig. 40.3 *Brucella abortus*. Glycerine agar slope culture, 3 days, 37°, in a corked tube, showing character of growth on direct isolation from the tissues.

1903) (see Fig. 40.3), or if it is attached by rubber tubing to another tube inoculated with an organism such as *Bac. subtilis* (Nowak 1908), growth occurs after a delay of a few days. These observations were generally interpreted as showing that *Br. abortus* was microaerophilic, and could not grow till the partial pressure of oxygen over the culture had been lowered to a suitable extent. A similar interpretation was also placed on the fact that, when inoculated into a serum agar gelatin shake medium, it grew in the form of a band situated about ½ cm below the surface (Fig. 40.4).

Credit is due to Huddleson (1921) for showing that this organism requires for its development a partial pressure of CO_2 higher than that normally present in the atmosphere (0.03–0.04 per cent). He found that if slopes of *Br. abortus* were incubated in a glass jar containing 10 per cent CO_2, good growth occurred in 24 hr, whereas under aerobic conditions there was no growth at all. Analysis of the gas over a culture of *Bac. subtilis* revealed the presence of CO_2, and it was therefore concluded that the success of Nowak's method depended on the evolution of this gas rather than on a decrease in the partial pressure of oxygen. Further work by Smith (1924) and McAlpine and

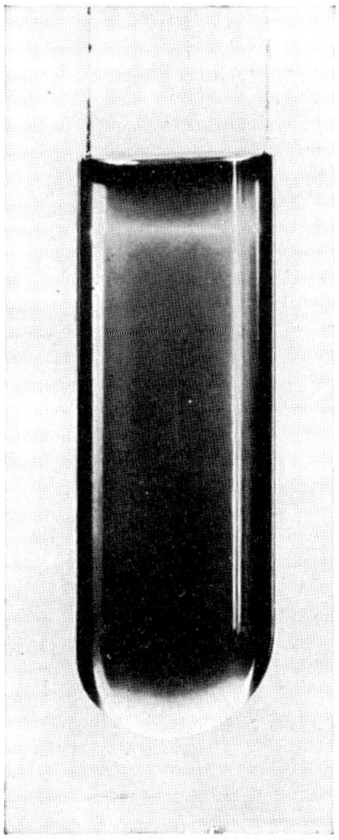

Fig. 40.4 *Brucella abortus*. Growth in form of band situated 6–8 mm below the surface. Bang's gelatin agar serum medium, pH 7.0, incubated aerobically.

Slanetz (1928) confirmed the importance of CO_2. Smith showed that development was much better in an atmosphere of 10 per cent CO_2 than in sealed tubes, and that in agar shake cultures, either sealed or incubated in an atmosphere of 10 per cent CO_2, growth occurred, not in a band below the surface as Bang (1897) and his co-worker Stribolt had found but on the surface itself.

In spite of these observations, it was not easy to understand why growth should occur in sealed tubes, or why, in the absence of added CO_2, growth in shake tubes should occur in a band below the surface.

A fuller study of the gaseous requirements of *Br. abortus* (Wilson 1931*a*) showed (1) that the organism would not grow anaerobically even in the presence of added CO_2, or aerobically in its absence; (2) that growth would occur in partial pressures of oxygen varying from 0.5–99.0 per cent, provided a minimum of 0.5 per cent CO_2 was added, and in partial pressures of CO_2 varying from 0.5–98.0 per cent, provided a minimum of 0.5 per cent oxygen was added; (3) that the optimum partial pressure of oxygen for development was about 21 per cent, i.e., that normally present in the atmosphere, and of CO_2 about 10 per cent. No evidence was obtained to suggest that a partial pressure of oxygen lower than that normally present in the atmosphere was beneficial to growth. It seemed clear, therefore, that neither growth in sealed tubes nor the band phenomenon in shake tubes could be due to a preference of the organism for microaerophilic conditions. Earlier observations (Wilson 1930) had shown that CO_2 was given off by burning cotton-wool plugs, and to a less extent by heated paraffin wax, rubber stoppers, and sealing wax. Analysis of the gas inside sterile sealed tubes revealed the presence of CO_2 in amounts varying from about 1–3 per cent—a proportion ample to initiate growth in inoculated tubes. The larger the number of organisms inoculated, the less need was there for additional CO_2, since the organisms themselves produced a certain amount of this gas. But with inocula of any size, growth was always most rapid and luxuriant when the partial pressures of oxygen and CO_2 approached the optima.

Similarly, evidence was brought (Wilson 1931*b*) to suggest that the band phenomenon in shake tubes was due to the necessity of an adequate concentration of CO_2 (Fig. 40.4). It was found that this gas was given off to a certain extent by the organisms themselves, and to a still greater extent by certain media, particularly those containing serum. Growth could not occur at the surface, because the CO_2 was given off into the atmosphere; nor could it occur in the depths of the medium, because the conditions were anaerobic. It therefore commenced in a zone as near the surface as was consistent with the maintenance of an adequate partial pressure of CO_2. If the tube was sealed, or was incubated in an atmosphere of 10 per cent CO_2, then growth occurred at the surface, where the optimum partial pressure of oxygen existed. This explanation, when slightly amplified, was found to fit the numerous observations on variation in the distance of the band from the surface, and on the so-called double zone phenomenon, in which two bands of growth, separated from each other by apparently unaltered medium, are visible.

The demand for an increased partial pressure of CO_2 is particularly characteristic of *Br. abortus* strains. Not all strains, however, of this organism require it (see Smith 1924, 1926*b*). Strains from Rhodesia (Bevan 1930), for example, grow quite well under ordinary aerobic conditions. Moreover, even strains that need extra CO_2 on isolation frequently become adapted, after a variable time in the laboratory, to do without it. The growth of many strains of *Br. melitensis* is greatly benefited by incubation in an atmosphere of 5–10 per cent CO_2, though some growth always occurs in ordinary air. The porcine strains, both American and Danish, appear to be least dependent upon CO_2. The addition of this gas to the air never improves their growth and sometimes actually inhibits it. How the CO_2 acts is not definitely known. Alteration in the pH of the medium does not seem to be the explanation. It has been suggested that the gas passes through the cell wall and brings about a change in the intracellular pH or oxidation-reduction potential, which is necessary for the initiation of growth. As Gladstone, Fildes, and Richardson (1935) have shown, CO_2 seems to be required in a greater or less degree by many human pathogens, and presumably plays an important part in their metabolism (see Chapter 3).

Cultivation in the presence of dyes

To Huddleson and Abell (1928) and Huddleson (1929, 1931) we owe a valuable method of distinguishing between the melitensis, abortus, and suis types, depending on their ability to grow in the presence of certain aniline dyes. Without entering into the detailed technique of the method, we may say that the general procedure is to prepare plates of liver agar, pH 6.6, containing 1 in 30 000 and 1 in 60 000 thionin, 1 in 25 000 and 1 in 50 000 basic fuchsin, 1 in 50 000 and 1 in 100 000 methyl violet, and 1 in 100 000 and 1 in 200 000 pyronin. The dyes used must be obtained from the Allied Chemical and Dye Company of New York, or standardized against these dyes. The organisms are inoculated rather heavily on to the plates, which are then incubated for 3 days aerobically, or in 10 per cent CO_2, according to the probable nature of the strains under examination.

Alternatively, and for some purposes more conveniently, strips of filter paper impregnated with stronger solutions of these dyes can be incorporated in liver agar plates, which are then streaked transversely with suspensions of the strains to be tested; the degree of sensitivity of the strain is indicated by the amount of growth over and on either side of the underlying strips of paper (Cruickshank 1948). This method is recommended by Wundt (1958*b*), who regards it as preferable to the incorporation of dyes in the medium itself.

Strains of *Br. melitensis* usually grow to some extent in the presence of all four dyes; typical *Br. abortus* strains are inhibited by thionin, but grow freely in the presence of the other three; *Br. suis* strains grow well in the presence of thionin, but may be inhibited by basic fuchsin, methyl violet, and pyronin. Though this is the general behaviour of the three species, there is considerable variation between different strains of the same species, especially those coming from different localities (Meyer and Zobell 1932, Wilson 1933). For example, Rhodesian strains of *Br. abortus* often have a rather greater resistance to thionin than abortus strains from other sources. Strains of *Br. abortus* referred to as biotype 2 are very susceptible to all four dyes. If reliance is placed exclusively on this method of differentiation, confusion will not infrequently result between strains of different types. If, on the other hand, it is used, as we believe it should be used, in conjunction with other methods, it will be found of considerable value. (See Table 40.3.)

Two other dyes are useful for certain purposes—thionin blue (not to be confused with thionin) and safranin O. Thionin blue in a concentration of 1 in 500 000 allows the growth of all three main species, but is inhibitory to the attenuated vaccinal strain *Br. abortus* S19 and to *Br. abortus* biotype 2. Safranin O in a concentration of 1 in 5000–1 in 10 000 inhibits *Br. suis*, but with a few exceptions not other strains of *Brucella* (Moreira-Jacob 1963, Jones 1964).

The vaccinal strain S19 of *Br. abortus*, which grows without added CO_2, is inhibited by 1 in 500 000 thionin blue, by 1 in 5000 safranin O, by the antibiotic mitomycin C in a concentration of 1 μg/ml (Myers and Dobosch 1969), and to some extent by penicillin, to which it is less resistant than other abortus strains (Jones *et al.* 1965).

The vaccinal strain Rev. 1 of *Br. melitensis*, it may be added, grows more slowly and is more sensitive to penicillin and more resistant to streptomycin than other melitensis strains (Crouch and Elberg 1967).

(For the older references to the use of dye tests, see 5th edition, pp. 998 and 999.)

Resistance

The members of this group exhibit the usual susceptibility of vegetative bacteria to heat and disinfectants. In aqueous suspensions of moderate density they are destroyed by heating for about 10 min at 60°, and by exposure for about 15 min to 1.0 per cent phenol. In milk they are readily destroyed by holder and by high-temperature short-time pasteurization. In agar cultures kept sealed at 0° they generally live for at least 1 month, and often longer. Considerable attention has been paid to their resistance under natural conditions, and much information on this subject will be found in the Report of the Mediterranean Fever Commission (1905–7). So many factors determine the exact outcome of any given observation under natural conditions that it is dangerous to draw general conclusions. In favourable circumstances, however, *Br. melitensis* may remain alive for 6 days in urine, 6 weeks in dust, and 10 weeks in water or soil. In pickled hams from naturally infected pigs it may live for as long as 3 weeks, but is apparently destroyed by smoking (Hutchings *et al.* 1951). *Br. abortus* may survive for 7 months in infected uterine exudate kept at about freezing-point (Bang 1897). In raw milk at room temperature it seems to die out fairly rapidly with the production of acid. Acid production also seems to be the cause of its rapid death in butter and cheese; the organisms can rarely be found in these articles for more than a few days (Smith 1934, Pullinger 1935). *Br. suis* may live on sacking for 4 weeks and in sterile faeces for 100 days in the dark (Cameron 1932, 1933).

King (1957) gives the following figures, collected from various authors, for the survival of *Br. abortus*: bovine urine 4 days; bovine faeces in a dark cupboard 120 days; aborted fetus in cold weather 75 days; manure and soil at 77° F 29 days; manure and soil at freezing point 2½ years; wet soil at room temperature 66 days; sacking in an unheated cellar 30 days; water at 77° F 10 days; milk at 50° F less than 10 days; ice-cream kept at 0° 30 days; butter stored at 46° F 142 days; Roquefort cheese 2 months. Both of these last figures, it may be noted, differ considerably from those given by Smith (1934) and Pullinger (1935).

In vitro the organisms of the *Brucella* group are moderately susceptible to the sulphonamides, very susceptible to streptomycin, the tetracyclines, rifampicin, and gentamicin, but resistant to penicillin, with the exception of strain 19, the vaccinal strain of *Br. abortus*. Apart from biotype 2, strains of *Br. abortus* are moderately resistant to polymyxin, amphotericin, and bacitracin (Robertson *et al.* 1973).

Metabolism and biochemical properties

The effect of temperature and pH on growth has already been considered. All the members require the presence of oxygen; most strains of *Br. abortus* require in addition a partial pressure of CO_2 considerably higher than that found in atmospheric air. Under ordinary aerobic conditions of incubation broth cultures become alkaline, owing to the production of ammonia. Litmus milk is turned weakly alkaline. Occasional haemolytic strains of *Br. melitensis* have been described (Forni 1927), but usually neither the melitensis, abortus, nor suis strains have any lytic action on blood. The effect of bile salt on growth has not been studied fully; on MacConkey's medium strains of *Br. abortus*, *Br. melitensis*, and *Br. suis* generally give rise to small non-lactose-fermenting colonies after 3 or 4 days.

In ordinary sugar media no fermentation is observable. Under special conditions, however, it can be shown that they do possess some oxidative activity. McCullough and Beal (1951) demonstrated this by manometric methods; and Pickett and Nelson (1955), using a plain phosphate buffer solution containing 1 per cent sugar, 0.2 per cent agar with cresol red as an indicator, incubating at 35°, and reading the results after 2–4 days, established qualitative differences between the three main species of *Brucella*. The study of amino acid requirements was begun by Gerhardt and Wilson (1948) and Rode, Oglesby, and Schuhardt (1950), and developed by Meyer and Cameron (1958, 1959, 1961) who, using the Warburg manometric technique on resting suspensions, tested the oxidative ability of the three main species on a number of amino acids and carbohydrates. A modified version of their results is given in Table 40.1.

It will be seen that *Br. abortus* differs from *Br. melitensis* in its capacity to oxidize certain sugars, and that *Br. suis* differs from the other two in its oxidation of the four amino acids of the urea cycle. This method is of value also in the classification of irregular strains. For example, intermediate strains that behave like *Br. melitensis* in the CO_2, H_2S, and dye tests but like *Br. abortus* serologically can be shown to have the metabolic character of *Br. melitensis* biotype 2; and strains that behave like *Br. melitensis* in the CO_2, H_2S, dye, and serological tests may have the metabolic character of *Br. abortus* biotype 5. Oxidative metabolism is the principal means by which the organisms obtain their energy (Morgan and Corbel 1976). The rate at which oxygen is taken up varies with the species of organism and the nature of the substrate. Verger and Grayon (1977), who studied the metabolic pattern of 370 strains of *Brucella* in the presence of 12 selected amino acid and carbohydrate substances, recognized three

Table 40.1 Oxidative metabolism of species of *Brucella*

Species	Amino acids							Carbohydrates					
	L-alanine	L-asparagine	L-glutamic acid	L-arginine	DL-citrulline	L-lysine	DL-ornithine	L-arabinose	D-galactose	D-ribose	D-xylose	D-glucose	*i*-erythritol
melitensis	+	+	+	−	−	−	−	−	−	−	−	+	+
abortus	+	+	+	−	−	−	−	+	+	+	v	+	+
suis	v*	v*	v*	+	+	v*	+	v*	v*	+	+	+	+
neotomae	v	+	+	−	−	−	−	+	+	v	−	+	+
canis	v	−	+	+	+	+	+	v	v	+	−	+	−
ovis	v	+	+	−	−	−	−	−	−	−	−	−	−

v = Variation between strains; v* = variation between biotypes of some assistance in classification.

oxidative groups, namely those with low, moderate, and high rates of uptake, corresponding with the three principal species and their subtypes.

The methyl-red and Voges-Proskauer tests are negative. No indole is formed. According to Zobell and Meyer (1932), all types reduce nitrate to nitrite. Nitrite is also rapidly reduced, so that the Griess-Ilosvay test on nitrate broth cultures may be negative. American suis strains are more active than the other types in reducing nitrite. Ammonia is produced to a variable extent from peptone, urea, and asparagin.

Püschel (1936) drew attention to the ability of certain strains of *Brucella* to split urea, and several workers have amplified her general findings and made the quantitative production of urease the basis of a useful differential test (Pacheco and Mello 1950, Renoux and Quatrefages 1951, Sanders and Warner 1951). Wundt (1958*a*) recommends carrying out the test as described by Hoyer (1950). A slope of growth is washed off with 3 ml of saline; 0.3 ml of this suspension is added to 1 ml of a buffer solution containing urea and phenol red, and incubated at 37°. *Br. suis* turns the medium red in 5–15 min, *Br. melitensis* in 15–30 min, and *Br. abortus* not for 2–24 hr. Not all strains of *Br. melitensis*, however, are as active as this and some cannot be distinguished from *Br. abortus*. The conditions under which the test is carried out have to be carefully standardized, and the results, though often helpful, are of limited value.

Catalase formation is strongest with *Br. suis* and weakest with *Br. abortus*; it is said to be quantitatively associated with virulence (Huddleson and Stahl 1943). The oxidase reaction is negative. The reducing action of these organisms is comparatively weak (Habs 1930), and in broth cultures methylene blue is often not decolorized. It may be noted that some strains of *Br. abortus* reduce basic fuchsin. Huddleson (1931) thought that this was a property of non-pathogenic strains, but our observations do not bear this out.

H_2S production

One of the most important differential criteria, to which attention was first drawn by Huddleson and Abell (1927), and Huddleson (1929), is the production of H_2S. This test should be carried out on liver agar with lead acetate papers (see Chapter 20). Freshly isolated strains of *Br. abortus* and *Br. suis* (American variety) give off H_2S for at least the first 4 days, whereas strains of *Br. melitensis* produce either none at all, or only during the first 24 hr of incubation. The Danish variety of *Br. suis* forms no H_2S. In the laboratory, abortus and American suis strains sometimes lose their ability to produce H_2S; the test should therefore be made as soon after isolation as possible. The interpretation of this test can be summed up by saying that, although failure of a given strain to produce H_2S beyond the first day does not exclude its being of abortus or American suis type, the continued production of H_2S after the first day affords a strong presumption that it is not of melitensis or Danish suis type.

Antigenic structure

It would be idle to recapitulate here the confusion that reigned for so long over the antigenic relationship of members of this group. A brief summary of it, with references to some of the main papers, was given in the 5th edition (pp. 1001–3). Before 1918, when Evans demonstrated an antigenic affinity between *Br. melitensis* and *Br. abortus*, most workers had concerned themselves with comparison of melitensis and so-called paramelitensis strains. There was a failure to realize that these organisms belonged not to separate species, but were manifestations of smooth→rough variation. It was some time, in fact, before the readiness with which the transition from S to R forms occurred was appreciated (for references see Pandit and Wilson 1932). Rough strains are agglutinable by non-specific agents, particularly acid, salt, peptone, and certain aniline dyes. Varying degrees of roughness occur, from strains that agglutinate to titre with an S antiserum but are slightly susceptible to non-specific agents to those that are unaffected by such a serum and are incapable of remaining homogeneously distributed even in cold saline. Moreover the degree of roughness of a given strain appears to vary from one culture to another. Once roughness has appeared, it persists, or recurs after an intervening period of apparent smoothness.

Roughness may be tested for by boiling in saline for 2 hr (thermo-agglutination test), by incubation at 37° with 1 in 500 or 1 in 1000 acriflavine (Alessandrini and Sabatucci 1931, Pampana 1931), by a study of the colonial appearance (see p. 408), by agglutination with an antiserum prepared against a completely rough strain of the corresponding type, or by the ease of phagocytosis by normal leucocytes (Munger and Huddleson 1938).

If, as Wilson and Miles (1932) showed, care is taken to exclude all but fully S strains, it is possible by quantitative agglutinin-absorption tests to distinguish between *Br. melitensis* on the one hand and *Br. abortus* and *Br. suis* on the other. The hypothetical antigenic picture so obtained is represented in Fig. 40.5. It will be seen that all three types contain the same two antigens, but with a different quantitative distribution, the M antigen being in excess in the melitensis, the A antigen in the abortus and suis types. By adjusting the absorbing dose, it is generally possible to absorb out all the minor agglutinins without removing more than a fraction of the major agglutinins. The resulting serum is therefore monospecific, and will agglutinate only those strains in which the corresponding antigen is predominant. Usually an absorbing dose standardized by opacity to match 3000×10^6 *Esch. coli* per ml is satisfactory for absorbing a serum diluted to 1/32–1/64 of its titre. With such a serum, strains of unknown type can be identified by direct agglutination.

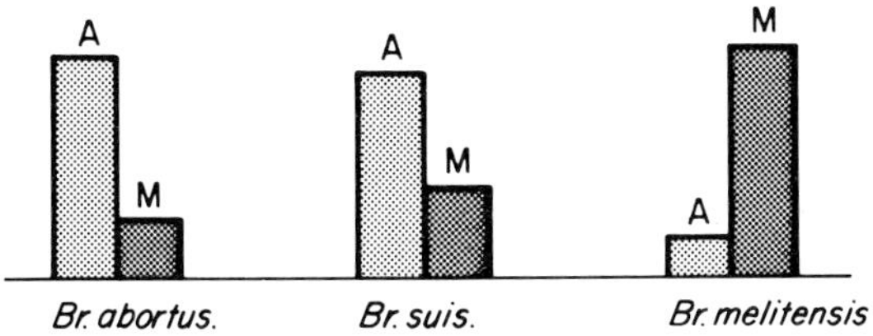

Fig. 40.5 Schematic representation of antigenic structure of strains of *Brucella*.

The inagglutinability of the heterologous strain by a monospecific serum, in spite of its having on its surface, as its minor antigen, receptors corresponding to those of the homologous strain, is explicable on the assumption that the minor antigen is sparsely distributed in discrete patches over the surface; so that when, after absorbing the monospecific agglutinin, several such bacilli approach each other in suspension, the number of sensitized patches on each that happen to coincide with those on another cell is insufficient to hold the bacilli together as an agglutinated mass (Miles 1939); this explanation is borne out by the observations of Miles (1939), who found that the antigenic A : M ratio in *Br. abortus* is about 20 : 1, and in *Br. melitensis* about 1 : 20.

The conclusions reached on the antigenic structure of these organisms received confirmation from the work of Miles (1933), who showed that the melitensis may be separated from the abortus-suis types by the optimal proportion agglutination method, and from that of Habs (1933), Habs and Sievert (1935), Sievert (1936), Olin and Lindström (1934), Olin (1935), and Veazie and Meyer (1936), using the method of quantitative absorption. It must not be supposed, of course, that the two antigens depicted in Fig. 40.5 are simple. They are probably composite. For example, using the agar-gel precipitation technique, Olitzki and Sulitzeanu (1958) obtained evidence that in *Br. suis* there were at least six different antigenic components, and Nagy (1968) that in the vaccinal S19 strain of *Br. abortus* there were thirteen, some of which were shared by biotypes 1, 2, 4 and 5. Joshi and Prakash (1971) found that ultrasonic extracts of *Br. melitensis* gave six precipitin lines by the Ouchterlony technique and nine by immunoelectrophoresis; *Br. abortus* behaved similarly.

As a means of typing unknown strains, direct agglutination with monospecific serum is one of the simplest and most rapid methods. Jones (1958), who describes the mode of preparation of monospecific sera, found that only 2 of 180 strains isolated in different parts of the world failed to agglutinate to titre with one or other of the two sera. Like other tests, however, for differentiation of this group, it cannot be relied upon entirely, because, as Wilson (1933) showed, certain strains, particularly from the south-east of France, may possess the antigenic structure of abortus, while having the biochemical and pathogenic characteristics of melitensis.

Several workers have described an antigenic affinity between members of the *Brucella* group and those of other groups. Many of their conclusions were based on the observation that a given serum agglutinated strains of both *Brucella* and some other group. Absorption and other experiments failed to confirm these conclusions (see Priestley 1933, Wilson 1934). Serological cross-reactions, however, have now been demonstrated between *Brucella* species and *Escherichia coli* O : 116 and O : 157; *Francisella tularensis*; *Salmonella* serotypes of Kauffman-White group N (O : 30 antigen); *Pseudomonas maltophilia*; *Vibrio cholerae* (McCullough *et al.* 1948, Gallut 1950, 1953, Feeley 1969); and *Yersinia enterocolitica* serogroup O : 9 (Ahvonen *et al.* 1969, Corbel and Cullen 1970, Hurvell and Lindberg 1973). (For further references, see Corbel 1982.)

Little work has been done on the antigenic structure of the R forms of *Brucella*. Our own observations suggest that, though the R forms of the three main species may share a common antigen, they are not identical. Members of the species *Br. canis* and *Br. ovis* form neither of the 'smooth' agglutinogens but are agglutinable by antisera prepared against rough brucellae.

Cellular composition

The reported results of chemical fractionation of brucella strains are difficult to summarize. Different methods of extraction have been used, and the results obtained are somewhat conflicting.

The early attempts at fractionation were summarized in the 5th edition (pp. 1004–5) and need not be repeated here. The later work has focused mainly on the antigens present in the cell wall. By digestion of acetone-dried organisms Pennell and Huddleson (1937, 1938) extracted toxic antigens from the three main species of *Brucella* and observed the occurrence of cross-precipitation between them and their respective antisera. Miles and Pirie (1939*a*, *b*) attempted to separate from *Br. melitensis* the A and M antigens postulated by Wilson and Miles (1932). Mild treatment resulted in their obtaining a 'native' antigen composed of large particles that corresponded to the protein-lipopolysaccharide endotoxins of the enterobacteria. From this, among other products, they extracted a lipopolysaccharide similar to the endotoxin of Pennell and Huddleson. Though failing to separate the A and M antigens, Miles and Pirie did obtain evidence that the antigens of the S forms of *Br. abortus* and *Br. melitensis* showed a qualitative similarity but a quantitative difference in distribution.

By means of various techniques numerous subsequent workers prepared specimens of cell walls and extracted from them lipopolysaccharide protein substances that stimulated an antibody and an immune response (Smith *et al.* 1962, Markenson *et al.* 1962, Börger 1963, Ellwood *et al.* 1967, Diaz *et al.* 1968*b*, Dranovskaya and Vershilova 1977). The soluble cytoplasmic constituents, on the other hand, were without immunogenic activity.

The lipid content of the cell walls appears to be 10–18 per cent (Bobo and Eagon 1968, Dranovskaya and Vershilova 1977). Ellwood and his colleagues showed that the immunogenic substance had a high phospholipid content, devoid of heptose but containing formyl groups; and brought some evidence to suggest that the protective antigen was identical with the agglutinogen.

Börger by paper chromatography found the lipopolysaccharide to contain three sugars, of which two were identified as glucose and mannose. Diaz and his colleagues obtained a lipopolysaccharide protein component from *Br. abortus* corresponding to the A antigen and one from *Br. melitensis* corresponding to the M antigen. They agreed with Miles and Pirie that these antigens were present together on a single antigenic complex. Other workers (Börger 1963, Serre *et al.* 1969) likewise agreed that the antigens in these two organisms were qualitatively alike but different in their quantitative distribution. (See also Renoux *et al.* 1976.)

Hoyer and McCullough (1968), who examined the polynucleotide homologies of the deoxyribonucleic acids of *Brucella*, found that the mole percentages of guanine and cytosine were in all species 57–58; but whereas *Br. abortus*, *Br. melitensis*, *Br. suis*, and *Br. neotomae* had very similar polynucleotide sequences, *Br. ovis* differed to some extent, in that it was only 93–94 per cent as effective as a competitor in DNA interactions.

Bacteriophages

Phages active against brucellae were first isolated from lysogenic strains and from manure by the Russian workers Drozewkina and Veršilova (see Parnas *et al.* 1958). These phages had a similar host range, typified by that of phage Tbilisi (Tb), which was shown by Meyer (1961) to lyse only strains of *Br. abortus*. This phage was later shown to give partial lysis of *Br. neotomae* when applied at routine test dilution (RTD); stronger preparations of phage (RTD × 1000) caused 'lysis from without' of *Br. suis*, but the phage could not be propagated on *Br. suis* or *Br. neotomae* (Jones *et al.* 1968*a*). Several other phages have since been obtained, notably those named Weybridge (Wb), Firenze (Fi) and Bk_2. When these are used at RTD, it is possible to distinguish between strains of *Br. melitensis*, *Br. abortus*, *Br. suis* and *Br. neotomae* if these are in the smooth or the intermediate phase. *Br. canis* and *Br. ovis*, which do not possess the specific antigens M or A of smooth brucellae, are untypable with these phages, but can be recognized if the two 'rough' phages R/O and R/C are added to the typing set (see Table 40.2, Corbel 1977, and Corbel and Thomas 1980).

Table 40.2 Differentiation between species of *Brucella* by phages at routine test dilution

	Lysis by phage*					
Species	Tb	Wb	Fi	Bk_2	R/O	R/C
melitensis	–	–	–	+	–	–
abortus	+	+	+	+	P	–
suis	–	+	P(or +)	+	–	–
neotomae	P	+	+	+	–	–
canis	–	–	–	–	–	+
ovis	–	–	–	–	+	+

+ = Lysis; P = partial lysis; – = no lysis.
* See text.

Virulence

Brucella resembles the tubercle bacillus and *Listeria monocytogenes* in growing intracellularly. The ability of virulent strains of *Br. abortus* to resist the destructive action of phagocytes, and even to multiply within them, is ascribed to the presence of a substance in the cell wall of the bacterium that can be neutralized by a specific antiserum (Smith and Fitzgeorge 1964). Also present in the cell wall is an immunogenic substance that interferes with the intensely bactericidal action of bovine serum on *Br. abortus*. According to Smith (1968) these two substances are distinct.

Under artificial conditions of cultivation brucella strains tend to undergo a change that is characterized by a gradual loss of specific and increase of non-specific agglutinability, together with a decrease in virulence to animals. This change appears to be an example of the smooth→rough variation. It occurs more readily with some strains than with others, is commoner with *Br. melitensis* than with *Br. abortus*, and is almost always irreversible. It is said to be prevented by addition to the culture medium of serum from susceptible animals; the substance in the serum responsible for this is a protein and is sometimes referred to as the SS or smooth selecting factor (see Braun 1950).

Pathogenicity

Br. melitensis, *Br. abortus*, and *Br. suis* are all infective for man and animals. Though undulant fever is the most characteristic result of infection in man, numerous other clinical manifestations occur. Often the infection remains latent, causing no recognizable symptoms of disease. Since we cannot carry out large-scale experiments on man under similar conditions, it is impossible to make any definite statement on the comparative virulence of these three organisms for man, but the limited observations on human volunteers of Morales-Otero (1929, 1930, 1933), a careful study of the available epidemiological records, and the frequency with which laboratory infections occur, suggest that *Br. melitensis* is the most pathogenic and *Br. abortus* the least pathogenic of the three types. The American suis type appears to occupy an intermediate position. The Danish suis type, on the other hand, is probably even less pathogenic than *Br. abortus*, since there is no record of its ever having been responsible for disease in man.

Under natural conditions *Br. melitensis* is pathogenic for goats and sheep, *Br. abortus* for cattle, and *Br. suis* for pigs. None of the three main species is readily transmitted to other hosts, and when transmission to an unusual host does occur the organisms tend to localize in the mammary gland and the reticuloendothelial system rather than in the uterus and fetal membranes. Thus, in cattle, abortion due to *Br. melitensis* is rare and to *Br. suis* unknown; in sheep and goats *Br. abortus* seldom gives rise to abortion and *Br. suis* never; nor in pigs is abortion ever caused by *Br. abortus* or *Br. melitensis* (Meyer 1964). Horses, dogs, cats, reindeer, and many small wild mammals (Thorpe *et al.* 1967) are subject to infection, particularly with *Br. abortus* (see Chapter 56).

All three species are infective to a variable extent for laboratory animals. On the whole the guinea-pig appears to be the most susceptible, but rabbits, rats, and mice can often be infected. In the guinea-pig the disease produced by parenteral inoculation with moderate doses is usually chronic and retrogressive. The brunt of the infection is borne by the reticulo-endothelial system. The resulting lesions are relatively inconspicuous, and consist mainly of a non-hyperaemic enlargement of the lymphatic glands, some degree of enlargement of the spleen, and the presence of a variable number of circular necrotic foci in the spleen and liver. In male guinea-pigs abscess formation is not uncommon in the testicle or epididymis, and intraperitoneal inoculation is sometimes followed by a Straus reaction. Occasionally the bones, joints, or other organs may be affected. The lesions are extremely variable in size and number, and may be completely absent on naked-eye inspection. In infections with *Br. suis* (American variety) the necrotic lesions tend to be few, large and purulent. The lesions in melitensis and abortus infections, on the other hand, are smaller, more numerous, and generally non-purulent, except in the testicle.

Distinction between the different brucella types cannot be made by intramuscular injection of guinea-pigs. This is, however, said to be possible by use of the Schultz-Dale technique, the reaction in sensitized animals being type-specific (Jadassohn, Riedmüller and Schaaf 1934). Guinea-pigs are of value for distinguishing broadly between strains of different degrees of virulence. Either a quantitative estimate can be made of the number of organisms in the spleen some weeks after intramuscular inoculation (see de Ropp 1945*a*), or the duration of bacteraemia after the injection of large doses subcutaneously may be observed (Jacotot and Vallée 1956, Cruickshank 1957).

Little work has been done on monkeys, but the observations of Huddleson and Hallman (1929), Weigmann (1931), and Zeller, Beller, and Stockmayer (1934) suggest that melitensis and American suis strains are more virulent than abortus strains. (For reproduction of disease in the larger animals see Chapter 56.)

No exotoxin is formed, but the intraperitoneal inoculation of mice with very large numbers of virulent *Br. abortus* brings about death within a few days. The toxicity of the organisms is said to be destroyed by heating to 56° for 20 minutes, and to be related to the virulence of the strain (Priestley and McEwen 1938). Redfearn (1960) submitted *Br. abortus* and *Br. melitensis* to the hot phenol water method developed by

Westphal, Luderitz, and Bister (1952) for extracting lipopolysaccharide, and found that the endotoxic fraction resided in the phenol phase, differing in this respect from the endotoxin of most other gram-negative bacteria which is present in the aqueous phase. Preliminary treatment of the cells with acetone did not affect the result (Jones and Berman 1976).

Pathogenicity in laboratory animals

Br. melitensis *Guinea-pigs* may die within a few days of intracerebral inoculation (Durham 1898, Eyre 1905), or intraperitoneal inoculation with large doses, but the disease set up by intramuscular or cutaneous inoculation is chronic, retrogressive, and rarely proves fatal. Small numbers of organisms cannot be relied on to cause infection. Most of the animals continue to gain weight slowly. If they are killed 6 weeks after inoculation, the following lesions may be found. Occasionally there is a local abscess containing creamy pus. The regional and the more distal lymphatic glands often show a certain amount of hyperplasia of the pale bloodless variety. The spleen may show a variable degree of enlargement, and may contain a number of circular, greyish-yellow necrotic foci, 0.1–0.5 mm in diameter, rarely projecting above the surface. Similar foci, usually few in number, may be present in the liver. Sometimes abscesses are found in connection with the joints or bones. The organisms can be cultivated most readily from the glands, spleen, and bone marrow. The blood usually contains agglutinins in fairly high titre. The lesions are very variable, and may not be detectable by naked-eye examination. The diagnosis must always be made by testing the blood serum for agglutinins and by cultivation of the causative organism from the tissues. A titre of 25 or over is strongly suggestive of infection. The intradermal test with a nucleoprotein extract may be used during life for diagnostic purposes, but reliance must never be placed on it alone. After 6 weeks the diagnosis becomes less easy, because the infection tends to retrogress. Guinea-pigs may also be infected by feeding, by conjunctival or nasal instillation, and by implantation on the scarified skin. Sometimes they contract the disease naturally from their fellows. (For a description of the histopathology of the disease, see Fabyan 1912.)

Rabbits appear to be rather less susceptible, but otherwise the infection runs much the same course as in guinea-pigs. The lymphatic system is less affected, and the organisms can rarely be demonstrated in the blood stream. Agglutinin formation is common, but the intradermal reaction is negative. There is little information about the effect of inoculation of *rats* or *mice* with *Br. melitensis*, but there is reason to believe that these animals are slightly susceptible to infection (see Singer-Brooks 1937). *Monkeys* may be infected either by feeding or by subcutaneous inoculation. Frequently an intermittent fever is set up, simulating in many respects undulant fever. If the animals are killed after a few weeks, there may be some enlargement of the lymph glands and spleen; occasionally necrotic lesions are found in the lungs or liver. Agglutinins are demonstrable in the serum, and the organisms can often be recovered from the tissues (see Bruce 1893, Hughes 1893, Horrocks and Kennedy 1906, Huddleson and Hallman 1929, Weigmann 1931, Zeller, Beller, and Stockmayer 1934).

Br. abortus In *guinea-pigs* a disease is set up closely resembling that caused by *Br. melitensis*, though the animals tend to suffer less constitutional disturbance (Braude 1951). After intramuscular inoculation a local suppurating lesion is rare, but abscess formation in the testis or epididymis is not uncommon. A mild infection can be produced in *rabbits*, *rats* and *mice*. The morbid anatomy and histopathology of the disease were described by Fabyan (1912). *Mice* develop a chronic non-fatal infection. They are more susceptible to inoculation by the intravenous or intracerebral than by the intramuscular or intraperitoneal route. After intramuscular injection the organisms can be demonstrated in the regional lymphatic glands and the spleen for a month or so, and agglutinins are present in the blood serum. They may also be infected by feeding with large doses. *Rats* may be infected by feeding as well as by intraperitoneal inoculation, and the organisms may be excreted for a time in the urine and faeces, but unless very large doses are used the infection retrogresses (Ber 1936, Sandholm 1938, Bosworth 1938). According to

Table 40.3 Characters of the species and biotypes of *Brucella**

Species and biotype		CO_2 requirement	H_2S production	Growth in presence of 1 in 50 000: thionin	fuchsin	Agglutination by monospecific serum†: A	M	R	Common host reservoirs	Remarks
melitensis	1	–	–	+	+	–	+	–	Sheep, goat	Typical melitensis
	2	–	–	+	+	+	–	–	,, ,,	Intermediate type
	3	–	–	+	+	+	+	–	,, ,,	
abortus	1	(+)	+	–	+	+	–	–	Cattle	Typical abortus
	2	(–)	+	–	–	+	–	–	,,	Wilson II type
	3	(+)	+	+‡	+	+	–	–	,,	Rhodesian type
	4	(+)	+	–	(+)	–	+	–	,,	
	5	–	–	+	+	–	+	–	,,	British melitensis
	6	–	(–)	+‡	+	+	–	–	,,	
	7	–	(+)	+	+	+	+	–	,,	
	9	–	+	+	+	–	+	–	,,	
suis	1	–	+	+	(–)	+	+	–	Pig	American suis
	2	–	–	+	+	+	–	–	Pig, hare	Danish suis
	3	–	–	+	+	+	–	–	Pig	USA
	4	–	–	+	(–)	+	+	–	Reindeer	
	Unclassified	–	–	+	–	–	+	–	Rodents	
neotomae		–	+	–	–	–	–	–	Desert wood rat	
canis		–	–	+	(–)	–	–	+	Dog	
ovis		+	–	+	(–)	–	–	+	Sheep	

(+) = Usually positive; (–) = usually negative.
* For oxidative metabolism, see Table 40.1; for phage susceptibility, see Table 40.2.
† A = abortus; M = melitensis; R = rough.
‡ *Br. abortus*; biotype 3 grows in the presence of 1 in 25 000 thionin; biotype 6 does not.

Emmel and Huddleson (1929, 1930), *fowls* can be infected by feeding or parenteral inoculation. The birds stop laying and suffer from severe diarrhoea. There is a gradually increasing pallor of the head, comb, and wattles, emaciation, and often paralysis and death. The course of the disease ranges from about 2 to 14 weeks. *Post mortem*, the main lesions consist of a necrotic enteritis and degenerative changes in the liver and kidneys. The majority of other workers, however, who have studied this question, have found that, on the whole, fowls are resistant to infection except with large doses administered parenterally (McNutt and Purwin 1930, 1932, van Roekel *et al.* 1932, Beller and Stockmayer 1933). With smaller doses the organisms can rarely be recovered from the tissues. In the absence of direct cultural experiments, a rise in the agglutinin titre cannot be interpreted as necessarily indicative of infection. The conclusions of Emmel and Huddleson require confirmation before being accepted. The *chick embryo* is said to be very susceptible to *Br. abortus*; even the attenuated S19 strain proves fatal in 3–9 days after injection into the yolk sac (Blitek, Parnas and Zuber 1957). *Monkeys* may be infected with *Br. abortus*, but they are less susceptible to it than to infection with *Br. melitensis*. (For further references to pathogenicity of *Br. abortus* for small animals see Schroeder and Cotton 1911, Morales-Otero 1930, Olin and Lindström 1934, Thomsen 1934, de Ropp 1945*b*, Cruickshank 1957.)

Br. suis Experimentally, this organism gives rise in *guinea-pigs* to a disease closely resembling that caused by *Br. melitensis*, though it tends to be slightly more pathogenic. Local abscess formation is rare, but in infections by the American type large suppurating lesions, few in number, are not uncommon in the spleen, liver, lymph glands, testicles, and joints. After subcutaneous or intraperitoneal injection the organisms are often found in the semen. Infection can be established by the respiratory route with aerosols (Druett *et al.* 1956). The Danish type appears to be less virulent than the American type. *Br. suis* appears to resemble *Br. abortus* in its infectivity for *mice*, *rabbits*, and *fowls*, but there is little exact information available. For *mice Br. suis* is said to be more virulent than *Br. abortus* (Singer-Brooks 1937). For *monkeys* it appears to be perhaps even more virulent than *Br. melitensis*. (References to pathogenicity of *Br. suis* for small animals: Smith 1926*a*, Hardy *et al.* 1930, Cotton 1932, Thomsen 1934, Huddleson 1934, Feldman and Olson 1935, Elberg and Henderson 1948, Hillaert *et al.* 1950.)

For reproduction of brucella infections in larger animals, see Chapter 56.

Classification and identification

According to Morgan and Corbel (1976) the classification of an organism as a member of the *Brucella* genus should be based on the following criteria. Gram-negative coccobacillary morphology; a G+C content for the DNA of 56–58 moles per cent; a minimum of 90 per cent homology with the DNA of reference strains in hybridization tests; disk electrophoretograms of acid-phenol soluble proteins of a pattern identical with that of reference strains; a cytochrome C absorption spectrum with absorption maxima between 522–530 nm and 552–560 nm; and extensive serological cross-reactions of intracellular antigens with those of reference strains. Subdivision into species depends upon principal natural host, oxidative metabolic pattern, and phage sensitivity. Classification into biotypes is based on CO_2 requirements, H_2S production, dye sensitivity, and agglutination by monospecific sera to M and A antigens. In practice, help in identification is afforded by cultural, biochemical, and pathogenic properties (see Robertson *et al.* 1980).

Subtypes are often associated with some special topographical distribution. *Br. melitensis* strains, for example, from Malta may differ from those met with in Palestine and the south of France. Rhodesian abortus strains were found to differ from those from Europe and America. American suis strains differ from Danish suis strains; and so on. Whether the members of the group should be classified into three main species and called *Br. abortus*, *Br. melitensis*, and *Br. suis*; or whether multiple species should be recognized; or whether one species only with multiple subspecies, were questions debated for a long time. The *Brucella* Subcommittee of the International Committee on Bacterial Nomenclature recommended the recognition of the three species just named, each with a number of biotypes (Report 1967*b*, 1971). It is now generally accepted that there are three biotypes of *Br. melitensis*, eight of *Br. abortus*, and at least four of *Br. suis* (see Table 40.3 and Corbel *et al.* 1978). *Br. suis* biotype 5 is no longer recognized (Corbel 1973); the species *suis* now comprises: biotypes 1 and 2, respectively the American and Danish strains just referred to; biotype 3, found mainly in the USA and highly pathogenic for man; biotype 4, isolated from reindeers and Eskimos, and differing from biotype 3 mainly in serological properties (Meyer 1964, 1966*a*); and a group of unclassified strains isolated from rodents. Some of the rodent strains, notably those isolated from members of the family Cricetidae, have the characters shown in Table 40.3 and an atypical pattern of oxidative metabolism, and may constitute a separate biotype; other rodent strains resemble biotype 3 more closely.

Three other species of *Brucella* are now recognized. Of these, *Br. neotomae* conforms closely to the general characters of the genus but the other two show considerably greater differences from these.

Br. canis differs from other members of the group mainly in its failure to use erythritol and in its rough antigenic structure. It has the metabolic properties of *Br. suis* biotypes 3 and 4, and Meyer (1969*a*) would recognize it as a *Brucella* and allocate it to biotype 5 of *Br. suis*. Most workers, however, prefer to regard it as a new species and to name it *Br. canis* (Diaz *et al.* 1968*a*, Jones *et al.* 1968*a*, Morisset and Spink 1969).

Br. ovis differs in rather more particulars. It uses neither glucose nor erythritol; it is said not to reduce nitrate to nitrite (Meyer 1969*b*); its oxidation of amino acid substances and carbohydrates is peculiar; and it is antigenically rough. On the other hand its water-soluble antigens show a close similarity to those of *Br. melitensis* (Diaz *et al.* 1967); and it has the same G + C ratio and much the same polynucleotide sequences as other members of the *Brucella* group (Hoyer and McCullough 1968).

(Useful reviews will be found in papers by Kristensen 1931, Zeller 1933, Habs 1933, Wilson 1933, Huddleson 1934, 1952, Thomsen 1934, Taylor *et al.* 1934, Olin and Lindström 1934, Lembke and Körnlein 1950, Renoux 1952*a*, Cruickshank 1954, Gerhardt 1958, Wundt 1961, 1967, Report 1967*a*, 1971, Meyer 1974, 1976, Regamey *et al.* 1976, and Stableforth and Jones 1977.) Practical methods for the identification of brucellae are given by Corbel, Gill and Thomas (1978), Corbel (1979), Corbel and Thomas (1980) and Robertson and his colleagues (1980). (For further information on the *Brucella* group, see Blobel and Schliesser 1982.)

The main diagnostic features of the six species of *Brucella* are summarized in Tables 40.1, 40.2 and 40.3. Other characters of *Br. melitensis*, *Br. abortus* and *Br. suis* are described in the text; additional notes on the three other species are given below.

Brucella neotomae

Isolated by Stoenner and Lackman (1957) from desert wood rats (*Neotoma lepida*) in Utah. Requires no extra CO_2 for growth. Forms H_2S for 3 or 4 days, and gives a strong urease reaction. Metabolically it resembles *Br. abortus* and *Br. melitensis* in its utilization of amino acids (Cameron and Meyer 1958). Antigenically it resembles *Br. abortus*. Is pathogenic for white mice, in which it gives rise to great enlargement of the spleen. Is less pathogenic for guinea-pigs. Neither mice nor guinea-pigs show any signs of illness, but the organisms persist in the visceral organs and lymph nodes for some months (Stoenner 1965, Thorpe *et al.* 1967). (See also Parnas *et al.* 1967.)

Brucella canis

Isolated by Carmichael and Bruner (1968) from dogs in the United States, where it gives rise to widespread abortion among beagles. Grows on media without enrichment and requires no extra CO_2. Forms rough mucoid colonies on agar and gives a ropy growth in liquid media. H_2S not produced for several days. Forms oxidase, catalase, and urease, but not indole. Reduces nitrate to nitrite. Growth completely inhibited by mitomycin C in a concentration of 1 μg/ml. Antigenically it lacks the lipopolysaccharide associated with the smooth O brucella agglutinogen, but is agglutinated by antisera prepared against rough brucellae including *Br. ovis*. In its growth characters resembles *Br. suis* biotype 3 and, apart from its failure to use erythritol, has the metabolic pattern of *Br. suis* biotypes 3 and 4. Pathogenic to guinea-pigs, mice and rats, giving rise after intraperitoneal injection to enlargement of the spleen, which contains discrete nodules glistening like tapioca grains, peritoneal and pleural effusion, enlargement of the lymph nodes, and sometimes a Straus reaction. Very susceptible to tetracycline.

Brucella ovis

Described by Buddle and Boyes (1953) and Buddle (1956) in New Zealand, who isolated it from sheep, in which it caused epididymo-orchitis in the rams, abortion in the ewes, and neonatal deaths in the lambs. This organism is fastidious. Requires extra CO_2 for growth and 10 per cent blood or serum for primary isolation (see Brown *et al.* 1971). Forms little or no H_2S and is urease negative. Said not to reduce nitrate to nitrite (Meyer 1969*b*). Differs metabolically from other species in its utilization of amino acid substrates and its oxidation of carbohydrates (Meyer 1969*b*). Does not use glucose or erythritol. Growth inhibited completely by mitomycin C in a concentration of 1 μg/ml (Myers and Dobosch 1969). Produces roughish colonies on solid media, is antigenically rough, is agglutinated by antiserum prepared against rough but not against smooth melitensis strains. By agar-gel precipitation and immunoelectrophoresis tests the water-soluble antigens show close similarity to those of *Br. melitensis* (Diaz *et al.* 1967). Pathogenicity for laboratory animals very low. Does not produce characteristic lesions of *Brucella* in guinea-pigs or stimulate the production of smooth brucella antibodies. Its undoubted affinity to rough melitensis strains and its possession of much the same polynucleotide sequences as *Brucella* (Hoyer and McCullough 1968) entitle it to be regarded as a member of the *Brucella* group.

(For further information on pathogenicity, see Isayama *et al.* 1977.)

References

Abdussalam, M. and Fein, D. A. (1976) See Regamey *et al.* p. 9.
Ahvonen, P., Jansson, E. and Aho, K. (1969) *Acta path. microbiol. scand.* **75,** 291.
Alessandrini, A. and Sabatucci, M. (1931) *Ann. Igiene (Sper.)* **41,** 29, 852.
Bang, B. (1897) *Z. Thiermed.* **1,** 241.
Bang, O. (1931) *2me Congr. int. Path. comp.* **i,** 95.
Beller, K. and Stockmayer, W. (1933) *Dtsch. tierärztl. Wschr.* **41,** 551.
Ber, A. (1936) *C. R. Soc. Biol.* **122,** 845.
Bevan, L. E. W. (1930) *Brit. med. J.* **ii,** 267.
Blitek, D., Parnas, J. and Zuber, S. (1957) *Ann. Inst. Pasteur* **92,** 146.
Blobel, H. and Schliesser, T. [Eds.] (1982) *Handbuch der Bakteriellen Infektionen bei Tieren.* Gustav Fischer, Jena.
Bobo, R. A. and Eagon, R. G. (1968) *Canad. J. Microbiol.* **14,** 503.
Börger, K. (1963) *Zbl. Bakt.* **190,** 47.
Bosworth, T. J. (1938) *J. comp. Path.* **50,** 345.
Boyd, D. M. and Casman, E. P. (1951) *Publ. Hlth Rep., Wash.* **66,** 44.
Braude, A. I. (1951) *J. infect. Dis.* **89,** 76.
Braun, W. (1950) See Report 1950, p. 26.
Brown, G. M., Ranger, C. R. and Kelley, D. J. (1971) *Cornell Vet.* **61,** 265.

Bruce, D. (1887) *Practitioner*, **39,** 161; (1893) *Ann. Inst. Pasteur* **7,** 289.
Buddingh, G. J. and Womack, F. C. (1941) *J. exp. Med.* **74,** 213.
Buddle, M. B. (1956) *J. Hyg., Camb.* **54,** 351.
Buddle, M. B. and Boyes, B. W. (1953) *Aust. vet. J.* **29,** 145.
Cameron, H. S. (1932) *Cornell Vet.* **22,** 212; (1933) *Rep. N.Y. St. vet. Coll.* 1931–32, No. 18.
Cameron, H. S. and Meyer, F. (1958) *J. Bact.* **76,** 546.
Carmichael, L. E. and Bruner, D. W. (1968) *Cornell Vet.* **58,** 579.
Christophorov, L. and Peschkov, J. (1969) *Zbl. Bakt.* **209,** 497.
Corbel, M. J. (1973) *J. Hyg., Camb.* **71,** 271; (1977) *Ann. Sclavo* **19,** 99; (1979) In: *Identification Methods for Microbiologists*, 2nd edn. Ed. by F. A. Skinner and D. W. Lovelock, Academic Press, London; (1982) WHO/BRUC/82. *372.*
Corbel, M. J. and Cullen, G. A. (1970) *J. Hyg., Camb.* **68,** 519.
Corbel, M. J., Gill, K. P. W. and Thomas, E. L. (1978) *Methods for the Identification of Brucella.* Ministry of Agriculture, Fisheries and Food, Pinner, Middlesex.
Corbel, M. J. and Thomas, E. L. (1976) See Regamey *et al.* p. 38; (1980) *The Brucella-phages, their Properties, Characterisation and Applications.* Ministry of Agriculture, Fisheries and Food, Pinner, Middlesex.
Cotton, W. E. (1932) *J. agric. Res.* **45,** 705.
Crouch, D. and Elberg, S. S. (1967) *J. Bact.* **94,** 1793.
Cruickshank, J. C. (1948) *J. Path. Bact.*, **60,** 328; (1954) *J. Hyg., Camb.* **52,** 105; (1955) *Ibid.* **53,** 305; (1957) *Ibid.* **55,** 140.
Diaz, R., Jones, L. M., Leong, D. and Wilson, J. B. (1968*b*) *J. Bact.* **96,** 893.
Diaz, R., Jones, L. M. and Wilson, J. B. (1967) *J. Bact.* **93,** 1262; (1968*a*) *Ibid.* **95,** 618.
Dranovskaya, E. A. and Vershilova, P. A. (1977) *Ann. Sclavo.* **19,** 109.
Druett, H. A., Henderson, D. W. and Peacock, S. (1956) *J. Hyg., Camb.* **54,** 49.
Dubray, G. (1972) *Ann. Inst. Pasteur* **123,** 171.
Duncan, J. T. (1928) *Trans. roy. Soc. trop. Med. Hyg.* **22,** 269.
Durham, H. E. (1898) *J. Path. Bact.* **5,** 377.
Elberg, S. S. and Henderson, D. W. (1948) *J. infect. Dis.* **82,** 302.
Ellwood, D. C., Keppie, J. and Smith, H. (1967) *Brit. J. exp. Path.* **48,** 28.
Emmel, M. W. and Huddleson, I. F. (1929) *J. Amer. vet. med. Ass.* **75,** 578; (1930) *Ibid.* **76,** 449.
Evans, A. C. (1918) *J. infect. Dis.* **22,** 580.
Eyre, J. W. H. (1905) *Rep. Comm. Medit. Fev., Lond.* Part II, p. 67.
Fabyan, M. (1912) *J. med. Res.* **26,** 441.
Feeley, J. C. (1969) *J. Bact.* **99,** 645.
Feldman, W. H. and Olson, C. (1935) *J. infect. Dis.* **57,** 212.
Forni, G. (1927) *G. Batt. Immun.* **2,** 823.
Francis, E. and Evans, A. C. (1926) *Publ. Hlth Rep., Wash.* **41,** No. 26, 1273.
Gallut, J. (1950) *Ann. Inst. Pasteur* **79,** 335; (1953) *Ibid.* **85,** 261.
Gerhardt, P. (1958) *Bact. Rev.* **22,** 81.
Gerhardt, P. and Gee, L. L. (1946) *J. Bact.* **52,** 261.
Gerhardt, P. and Wilson, J. B. (1948) *J. Bact.* **56,** 17.
Gladstone, G. P., Fildes, P. and Richardson, G. M. (1935) *Brit. J. exp. Path.* 16, 335.
Goodpasture, E. W. and Anderson, K. (1937) *Amer. J. Path.* **13,** 149.
Habs, H. (1930) *Zbl. Bakt.*, **116,** 89; (1933) *Zbl. ges. Hyg.* **28,** 481.
Habs, H. and Sievert, L. (1935) *Dtsch. med. Wschr.* **61,** 1398.
Hansen, K. and Köster, H. (1936) *Dtsch. tierärztl. Wschr.* **44,** 739.
Hardy, A. V., Jordan, C. F., Borts, I. H. and Hardy, G. C. (1930) *Nat. Inst. Hlth Bull.* No. 158.
Henry, B. S. (1933) *J. infect. Dis.* **52,** 374, 403.
Hillaert, E. L., Hutchings, L. M. and Andrews, F. N. (1950) *Amer. J. vet. Res.* **11,** 84.
Hines, W. D., Freeman, B. A. and Pearson, G. R. (1964) *J. Bact.* **87,** 438.
Holth, H. (1911) *Z. InfektKr. Haustiere* **10,** 207.
Horrocks, W. H. and Kennedy, J. C. (1906) *Rep. Comm. Medit. Fev., Lond.* Part IV, p. 37.
Hoyer, B. H. (1950) See Report 1950, p. 9.
Hoyer, B. H. and McCullough, N. B. (1968) *J. Bact.* **95,** 444.
Huddleson, I. F. (1921) *Cornell Vet.* **11,** 210; (1929) *Mich. State College, agric. Exp. Sta., Tec. Bull.* No. 100; (1931) *Amer. J. publ. Hlth* **21,** 491; (1934) *Brucella Infections in Animals and Man.* Commonwealth Fund, New York; (1952) *Mich. St. Coll., Agric. Exp. Sta.* Memo No. 6; (1956) *Proc. 56th gen. Mtg., Soc. Amer. Bact.* p. 53; (1964) *J. Bact.* **88,** 540.
Huddleson, I. F. and Abell, E. (1927) *J. Bact.* **13,** 13; (1928) *J. infect. Dis.* **43,** 81.
Huddleson, I. F. and Hallman, E. T. (1929) *J. infect. Dis.* **45,** 293.
Huddleson, I. F., Hasley, D. E. and Torrey, J. P. (1927) *J. infect. Dis.* **40,** 352.
Huddleson, I. F. and Stahl, W. H. (1943) *Tec. Bull. Mich. agric. Exp. Sta.* No. 182, p. 57.
Hughes, M. L. (1893) *Ann. Inst. Pasteur* **7,** 628.
Hurvell, B. and Lindberg, A. A. (1973) *Acta path. microbiol. scand.* **B83,** 113.
Hutchings, L. M., McCullough, N. B., Donham, C. R., Eisele, C. W. and Bunnell, D. E. (1951) *Publ. Hlth Rep., Wash.* **66,** 1402.
Isayama, Y., Azuma, R., Tanaka, S. and Suto, T. (1977) *Ann. Sclavo* **19,** 89.
Jacotot, H. and Vallée, A. (1956) *Ann. Inst. Pasteur* **90,** 121.
Jadassohn, W., Riedmüller, L. and Schaaf, F. (1934) *Klin. Wschr.* **13,** 897.
Jones, L. M. (1958) *Bull. World Hlth Org.*, **19,** 177; (1964) *J. Bact.* **88,** 1527.
Jones, L. M. and Berman, D. T. (1976) See Regamey *et al.* p. 62.
Jones, L. M., Merz, G. S. and Wilson, J. B. (1968*a*) *Appl. Microbiol.* **16,** 1179.
Jones, L. M., Montgomery, V. and Wilson, J. B. (1965) *J. infect. Dis.* **115,** 312.
Jones, L. M., Zanardi, M., Leong, D. and Wilson, J. B. (1968*b*) *J. Bact.* **95,** 625.
Joshi, D. V. and Prakash, O. (1971) *Indian J. med. Res.* **59,** 1556.
King, N. B. (1957) *J. Amer. vet. med. Ass.* **131,** 349.
King, R. O. C. (1934) *Aust. vet. J.* **10,** 93.
Kristensen, M. (1931) *Zbl. Bakt.* **120,** 179.
Lembke, A. and Körnlein, M. (1950) *Zbl. Bakt.* (I. Abt., Ref.) **147,** 449.

McAlpine, J. G. and Slanetz, C. A. (1928) *J. infect. Dis.* **43,** 232.
McCullough, N. B. and Beal, G. A. (1951) *J. infect. Dis.* **89,** 266.
McCullough, N. B., Eisele, C. W. and Beal, G. A. (1948) *J. infect. Dis.* **83,** 55.
McNutt, S. H. and Purwin, P. (1930) *J. Amer. vet. med. Ass.* **30,** 350; (1932) *Ibid.* **31,** 641.
Markenson, J., Sulitzeanu, D. and Olitzki, A. L. (1962) *Brit. J. exp. Path.* **43,** 67.
Meyer, K. F. and Zobell, C. E. (1932) *J. infect. Dis.* **51,** 72.
Meyer, M. E. (1961) *J. Bact.* **82,** 950; (1962) *Bull. World Hlth Org.* **26,** 829; (1964) *Amer. J. vet. Res.* **25,** 553; (1966*a*) *Ibid.* **27,** 353; (1966*b*) *Ibid.* **27,** 424; (1969*a*) *Ibid.* **30,** 1751; (1969*b*) *Ibid.* **30,** 1757; (1974) *Advanc. vet. Sci. comp. Med.* **18,** 231, Academic Press, New York and London; (1976) *Amer. J. vet. Res.* **37,** 203.
Meyer, M. E. and Cameron, H. S. (1958) *Amer. J. vet. Res.* **19,** 754; (1959) *J. Bact.* **78,** 130; (1961) *Ibid.* **82,** 387, 396.
Miles, A. A. (1933) *Brit. J. exp. Path.* **14,** 43; (1939) *Ibid.* **20,** 63.
Miles, A. A. and Pirie, N. W. (1939*a*) *Brit. J. exp. Path.* **20,** 83, 109, 278; (1939*b*) *Biochem. J.* **33,** 1709, 1716.
Moody, M. D., Biegeleisen, J. Z. and Taylor, G. C. (1961) *J. Bact.* **81,** 990.
Morales-Otero, P. (1929) *Porto Rico J. publ. Hlth. trop. Med.* **5,** 144; (1930) *Ibid.* **60,** 3; (1933) *J. infect. Dis.* **52,** 54.
Moreira-Jacob, M. (1963) *J. Bact.* **86,** 599; (1968) *Nature., Lond.* **219,** 752.
Morgan, W. J. B. and Corbel, M. J. (1976) See Regamey *et al.* p. 27.
Morisset, R. and Spink, W. W. (1969) *Lancet* **ii,** 1000.
Munger, M. and Huddleson, I. F. (1938) *J. Bact.* **35,** 255.
Myers, D. M. and Dobosch, D. (1969) *Appl. Microbiol.* **18,** 511.
Nagy, L. K. (1968) *Res. vet. Sci.* **9,** 197.
Neiva, C. (1934) *Brazil-Medico* **48,** 421.
Nowak, J. (1908) *Ann. Inst. Pasteur* **22,** 541.
Olin, G. (1935) *Studien über das Undulantfieber in Schweden.* Isaac Marcus Boktryckeri-Aktiebolag, Stockholm.
Olin, G. and Lindström, B. (1934) *Zbl. Bakt.* **131,** 457.
Olitzki, A. L. and Sulitzeanu, D. (1958) *Brit. J. exp. Path* **39,** 219.
Pacheco, G. and Mello, M. T. de. (1950) *J. Bact.* **59,** 689.
Painter, G. M., Deyoe, B. L. and Lambert, G. (1966) *Canad. J. comp. Med.* **30,** 218.
Pampana, E. J. (1931) *Ann. Igiene* (*Sper.*) **41,** 537.
Pandit, S. R. and Wilson, G. S. (1932) *J. Hyg., Camb.* **32,** 45.
Parnas, J. (1967) *Zbl. Bakt.* **204,** 383.
Parnas, J., Feltynowski, A. and Bulikowski, W. (1958) *Nature, Lond.* **182,** 1610.
Parnas, J., Zalichta, S. and Sidor-Wójtowicz, A. (1967) *Zbl. VetMed. B.* **14,** 634.
Pearce, J. H., Williams, A. E., Harris-Smith, P. W., Fitzgeorge, R. B. and Smith, H. (1962) *Brit. J. exp. Path.* **43,** 31.
Pennell, R. B. and Huddleson, I. F. (1937) *Tec. Bull. Mich. agric. exp. Sta.* No. 156, 1; (1938) *J. exp. Med.* **68,** 73, 83.
Petris, S. de, Karlsbad, G. and Kessel, R. W. I. (1964) *J. gen. Microbiol.* **35,** 373.
Pickett, M. J. and Nelson, E. L. (1955) *J. Bact.* **69,** 333.
Preisz, H. (1903) *Zbl. Bakt.* **33,** 190.
Priestley, F. W. (1933) *J. comp. Path.* **46,** 38; (1938) *Vet. Rec.* **50,** 137.
Priestley, F. W. and McEwen, A. D. (1938) *J. comp. Path.* **51,** 282.
Pullinger, E. J. (1935) *Lancet* **i,** 1342.
Püschel, J. (1936) *Klin. Wschr.* **15,** 375.
Redfearn, M. S. (1960) *An Immunochemical Study of Antigens of Brucella Extracted by the Westphal Technique.* Ph.D thesis, Univ. Wisconsin.
Regamey, R. H., Hulse, E. C. and Valette, L. (Eds.) (1976) *49th International Symposium on Brucellosis.* S. Karger, Basel.
Renoux, G. (1952*a*) *Ann. Inst. Pasteur* **82,** 289; (1952*b*) *Ibid.* **82,** 556.
Renoux, G. and Quatrefages, H. (1951) *Ann. Inst. Pasteur* **80,** 182.
Renoux, G., Renoux, M. and Guillaumin, J. M. (1976) See Regamey *et al.*, p. 100.
Report. (1905–7) *Comm. Mediterranean Fever*, Parts I, III, IV. Harrison and Sons, London; (1950) *Symposium on Brucellosis.* Amer. Ass. Advancement Science, Wash., D.C.; (1967*a*) *Monogr. Ser. World Hlth Org.* No. 55; (1967*b*) *Int. J. syst. Bact.* **17,** 371; (1971) *Int. J. syst. Bact.* **21,** 126.
Robertson, L., Farrell, I. D. and Hinchliffe, P. M. (1973) *J. med. Microbiol.* **6,** 549.
Robertson, L., Farrell, I. D., Hinchliffe, P. M. and Quaife, R. A. (1980) *Benchbook on Brucella. PHLS Monograph Series No. 14*, HMSO, Lond.
Rode, L. J., Oglesby, G. and Schuhardt, V. T. (1950) *J. Bact.* **60,** 661.
Roekel, H. van, Bullis, K. L., Flint, O. S. and Clarke, M. K. (1932) *J. Amer. vet. med. Ass.* **80,** 641.
Ropp, R. S. de. (1944) *J. comp. Path.* **54,** 53; (1945*a*) *Ibid.* **55,** 70; (1945*b*) *Ibid.* **55,** 85.
Sanders, E. and Huddleson, I. F. (1950) *Amer. J. vet. Res.* **11,** 70, 75.
Sanders, E. and Warner, J. (1951) *J. Bact.* **62,** 591.
Sandholm, A. (1938) *Z. InfektKr. Haust.* **53,** 201.
Schroeder, E. C. and Cotton, W. E. (1911) *Bur. Animal Ind., 28th Ann. Rep.* p. 139.
Schuhardt, V. T., Rode, L. J., Foster, J. W. and Oglesby, G. (1949) *J. Bact.* **57,** 1.
Serre, A., Lacave, C., Asselineau, J. and Bascoul, S. (1969) *Ann. Inst. Pasteur* **117,** 871.
Sievert, L. (1936) *Z. ImmunForsch.* **89,** 249.
Singer-Brooks, C. H. (1937) *J. infect. Dis.* **60,** 265.
Smith, H. (1968) *Bact. Rev.* **32,** 164.
Smith, H. and Fitzgeorge, R. B. (1964) *Brit. J. exp. Path.* **45,** 174.
Smith, H., Keppie, J., Pearce, J. H., Fuller, R. and Williams, A. E. (1961) *Brit. J. exp. Path.* **42,** 631.
Smith, H., Keppie, J., Pearce, J. H. and Witt, K. (1962) *Brit. J. exp. Path.* **43,** 538.
Smith, J. (1934) *J. Hyg., Camb.* **34,** 242.
Smith, T. (1924) *J. exp. Med.* **40,** 219; (1926*a*) *Ibid.* **43,** 207; (1926*b*) *Ibid.* **43,** 317.
Stableforth, A. W. and Jones, L. M. (1977) *Ann. Sclavo* **19,** 151.
Stafseth, H. J. (1920) *Mich. agric. exp. Sta., Tec. Bull.* No. 49.
Stoenner, H. G. (1965) *Amer. J. vet. Res.* **26,** 347.
Stoenner, H. G. and Lackman, D. B. (1957) *Amer. J. vet. Res.* **18,** 947.
Tamura, J. T. and Gibby, I. W. (1943) *J. Bact.* **45,** 361.
Taylor, R. M., Vidal, L. F. and Roman, G., (1934) *C. R. Soc. Biol.* **116,** 132.

Thomsen, A. (1931) *Rev. gén. Méd. vét.* **40,** 457; (1933) *Zbl. Bakt.* **130,** 257; (1934) *Brucella Infection in Swine. Acta path. microbiol. scand.*, Suppl. No. 21.

Thorpe, B. D., Sidwell, R. W. and Lundgren, D. L. (1967) *Amer. J. trop. Med. Hyg.* **16,** 665.

Traum, J. E. (1914) *Rep. Chief, Bur. Anim. Industry*, p. 30.

Veazie, L. and Meyer, K. F. (1936) *J. infect. Dis.* **58,** 280.

Verger, J. M. and Grayon, M. (1977) *Ann. Sclavo* **19,** 45.

Weigmann, F. (1931) *Zbl. Bakt.* **121,** 318.

Westphal, O., Lüderitz, O. and Bister, F. (1952) *Z. Naturforsch.* **7b,** 148.

White, P. G. and Wilson, J. B. (1951) *J. Bact.* **61,** 239.

Williams, A. E., Keppie, J. and Smith, H. (1964) *J. gen. Microbiol.* **37,** 285.

Wilson, G. S. (1930) *Brit. J. exp. Path.* **11,** 157; (1931*a*) *Ibid.* **12,** 88; (1931*b*) *Ibid.* **12,** 152; (1933) *J. Hyg., Camb.* **33,** 516; (1934) *Ibid.* **34,** 361.

Wilson, G. S. and Miles, A. A. (1932) *Brit. J. exp. Path.* **13,** 1.

Wundt, W. (1958*a*) *Z. Hyg. InfektKr.* **145,** 235; (1958*b*) *Zbl. Bakt.* **171,** 166; (1961) *Ergebn. Mikrobiol.*, **34,** 120; (1967) *Zbl. Bakt.* **205,** 234.

Zeller, H. (1933) *Münch. tierärztl. Wschr.* **84,** 337, 349, 361, 373, 389.

Zeller, H., Beller, K. and Stockmayer, W. (1934) *Münch. tierärztl. Wschr.* **85,** 143.

Zobell, C. E. and Meyer, K. F. (1932) *J. infect. Dis.* **51,** 91, 99, 109, 344, 361.

41

Bacillus: the aerobic spore-bearing bacilli

Graham Wilson

Introductory

Definition

Aerobic spore-bearing rods. Gram positive or gram variable. Often occur in long threads, and form rhizoid colonies. Form of rod changed at sporulation only in some species. Usually motile, by peritrichate flagella. Liquefy gelatin. Catalase positive. Mostly saprophytes. G+C content of the DNA: 32–50 moles per cent.

The type species is *Bacillus subtilis*.

The family Bacillaceae comprises rod-shaped or occasionally coccoid organisms that are usually gram positive and are all characterized by the formation of endospores. The anaerobic members of the family include *Clostridium* (Chapter 42) and the aerobic rod-shaped members form the genus *Bacillus*. Aerobic spore-bearing bacilli are widely distributed in air, water, soil, dust, wool, faeces and many other situations. Numerous species of *Bacillus* are devoid of pathogenicity, and only one—the anthrax bacillus—is a common cause of disease in man or other mammals. In consequence, comprehensive knowledge of the genus as a whole is of recent date. (For reviews, see Smith *et al.* 1952, Seidel 1962, Bonde 1975.)

Morphology and staining

Members of this group are rod-shaped organisms varying in size from about 3 μm × 0.4 μm to 9 μm × 2 μm. The sides are parallel, the axis straight or slightly curved, the ends either truncated, as in *B. anthracis*, or more usually convex. Their arrangement varies considerably; though single and diplobacillary forms predominate, they may be arranged in chains, often of considerable length, or in groups. Long, unjointed filaments are characteristic of some species, notably of the anthrax bacillus. Irregular forms, consisting mainly of poorly stained thin bacilli, or of club- or bottle-shaped bacilli, are not uncommon. With few exceptions, of which *B. anthracis* is the most important, all the members are motile by about 4–12 peritrichous flagella. Spores are present in all; and are formed only in the presence of oxygen; they vary in shape from spherical to ellipsoidal, and may appear at the equator, subterminally, or at the very end of the bacillus. In some members their diameter does not exceed that of the bacillus, but in others the rod is swollen to resemble a clostridium. One member, *B. anthracis*, has a prominent capsule, most obvious when it has grown in the animal body, and readily formed when it is grown in an excess of CO_2 on media containing 0.7 per cent bicarbonate (see, e.g. Burdon 1956) or better in 5 per cent CO_2 on serum or albumin agar (Meynell and Meynell 1964); capsules characterize some other strains. The organisms are usually gram positive, but considerable variation may be shown; some are strongly positive, others weakly positive, and a few frankly negative. When stained by Sudan black B, the cells of most species show the presence of fat globules (see Gibson and Topping 1938, Smith *et al.* 1952, Burdon 1956). In *B. megaterium* they are large and numerous and the lipid appears to be intimately mixed with the cytoplasm. *B. anthracis*, *B. cereus*, and *B. mycoides* are alike in having numerous discrete round droplets. Relatively few cells in cultures of *B. subtilis*, *B. pumilus*, and *B. licheniformis* contain fat, and this is mainly at the ends of the rods. *B. sphaericus* contains strips of lipid lying along the length of the cell. The fatty material is evident in cultures on ordinary media; Gibson and Topping recommend nutrient agar containing glucose or glycerol. None of the vegetative bacilli is acid-fast, though spores are moderately so.

Cultural reactions

Growth is free on all the ordinary media. Single colonies on agar are generally two to several millimetres in diameter. Some have a finely granular, mealy appearance, others are membranous and thrown into wrinkles. In broth a surface scum may be formed, with or without turbidity, or a heavy flocculent or membranous deposit. Gelatin is usually liquefied rapidly. In a stroke agar culture the growth is raised and confluent, and generally of membranous consistency, rendering emulsification difficult. Growth is sometimes improved by glucose, but not by blood or serum. Variant colonial types have been described for several members of the group. Some members form motile colonies (see *B. circulans*, p. 436).

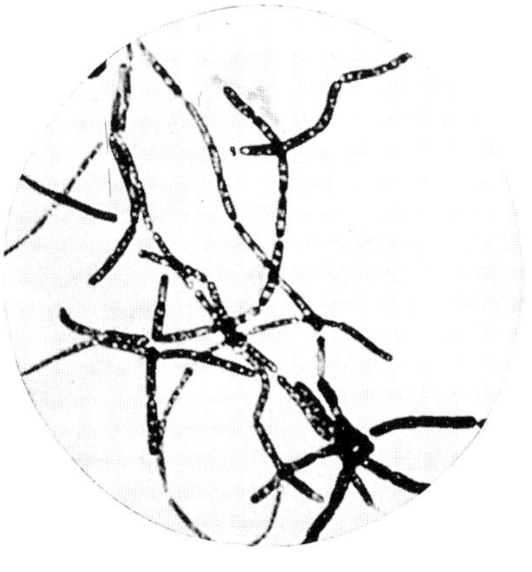

Fig. 41.1 *B. megaterium*. From an agar slope culture (× 1000).

Resistance

In the vegetative condition the bacilli of most species are killed by moist heat at a temperature of 55° in 1 hr. The spores vary greatly in resistance; some, like those of *B. anthracis*, are destroyed by boiling for about 10 min; others, like those of *B. subtilis*, may withstand boiling for hours. Most are killed by steam under pressure at 120° in 40 min, but spores embedded in garden earth are said to survive up to 6 hr (Kurzweil 1954). The spores of thermophilic organisms, which are abundant in granulated sugar, syrup, and spices, are mostly very resistant, withstanding boiling for 8 to 16 hr or so. Similarly with disinfectants, the resistance varies; $HgCl_2$ even in a 1 in 1000 solution may fail to kill anthrax spores in less than 70 hr (Poppe 1922). Potassium permanganate, on the other hand, in a 4 per cent solution kills them in 15 min, and a 3 per cent solution of hydrogen peroxide in 1 hr. Generally speaking, the spores are extremely resistant to chemical disinfection, except by oxidizing agents and by dyes. The anthrax bacillus is very susceptible to penicillin, though occasional resistant strains have been met with. The penicillin resistance of other aerobic spore bearers is variable; some form penicillinase (see Seidel 1962). In dry earth anthrax spores have been found active after 60 years (J. B. Wilson and Russell 1964); and in canned foods thermophilic spore bearers have survived for over 100 years (G. S. Wilson and Shipp 1939).

Metabolic and biochemical reactions

Some members form pigment, generally brownish-yellow, occasionally pink or black. The formation of red, brown and one type of black pigment (*B. subtilis* var. *aterrimus*) is enhanced by carbohydrate in the medium; another type of black pigment (*B. subtilis* var. *niger*) results from the action of tyrosinase (Clark and Smith 1939). On the whole, pigment formation is not a striking character; it tends to be variable and to appear late.

The optimum temperature for growth varies from 25°–37°, tending to be higher with the small-celled species such as *B. subtilis* than with the large-celled species such as *B. cereus*. The ability to grow at 55° varies not only with the species but with different members of the same species. Clearly, therefore, species cannot be defined by temperature requirements. All members grow aerobically, and some, such as *B. cereus*, *B. anthracis*, *B. licheniformis*, and *B. coagulans*, can grow under strictly anaerobic conditions (Knight and Proom 1950). The anthrax bacillus spores best in an earth extract medium (Seidel 1962). A high partial pressure of CO_2 tends to inhibit growth of *B. subtilis* (Levine 1936), and to promote capsulation but to restrict sporulation of *B. anthracis* (Sterne 1937*a*).

On carbohydrate media most members form acid only, but a few, such as *B. polymyxa* and *B. licheniformis*, produce gas; this is conveniently demonstrated by using the semi-anaerobic medium of Gibson and Abd-el-Malek (1945). Glucose, maltose, and sucrose are commonly fermented, mannitol and salicin less often. Lactose is rarely fermented. The end-products of glucose fermentation vary greatly. Some species produce lactic acid; others, such as *B. subtilis*, *B. licheniformis*, and *B. cereus*, form 2,3-butanediol and glycerol; *B. polymyxa* forms 2,3-butanediol, ethanol and hydrogen; and *B. macerans* forms chiefly ethanol, acetone, acetic and formic acids (Gibson and Gordon 1974). Burdon (1956) recommends tryptose agar slopes as the medium of choice. Fermentation reactions tend to be weak and irregular; Seidel (1962) regards them as of little value for differential purposes. A diastatic ferment capable of inverting starch is secreted by some. A proteolytic ferment for gelatin is produced by nearly all, and by a few for blood serum (see, e.g. Wieland *et al.* 1960). A true rennet-like clot is often formed in litmus milk, and is subsequently digested; the litmus is reduced. Some strains peptonize milk without actually clotting it. The reaction becomes alkaline. Both the catalase test and the oxidase reaction described by Gordon and McLeod (1928) are usually positive. Methylene blue is reduced in broth. Some members are able to produce H_2S, and some to reduce nitrate to nitrite. Indole is not produced. Some members give a positive Voges-Proskauer reaction when tested by Barritt's modification. *B. cereus*, *B. cereus* var. *mycoides*, and to a less extent *B. anthracis*, are distinguished by the synthesis of lecithinases which produce a thick milky turbidity in egg-yolk broth and an opaque zone around colonies on egg-yolk agar (McGaughey and Chu 1948, Costlow 1958). Like the lecithinase (α-toxin) of *Cl. perfringens* (see Chapter 42), these enzymes are activated by Ca and Mg ions, are moderately thermostable, give a Nagler reaction in human serum, lyse red blood corpuscles, are toxic to mice, produce dermonecrosis in guinea-pigs, and are neutralized by specific serum (Chu 1949). Many other species, such as *B. pumilus*, *B. polymyxa*, and *B. macerans*, produce a much fainter reaction on egg-yolk agar, in which the zone of opalescence is restricted to the medium directly under the colony and usually visible only when the growth is scraped off (Knight and Proom 1950). Collagenase is produced by some members (Evans and Wardlaw 1953).

Knight and Proom (1950), who studied the nutritional requirements of the mesophilic species, found that some organisms, such as *B. subtilis*, *B. megaterium*, and *B. licheniformis*, used ammonia as their sole nitrogen source and required no added growth factors; some, such as *B. cereus*, required no added growth factors but needed a mixture of amino acids; some, such as *B. pumilus* and *B. polymyxa*, grew with ammonia in the presence of added biotin; and some, such as *B. circulans*, required amino acids, biotin, and aneurin. Strains of *B. pasteuri* had even more complex requirements. A species isolated from soil, called *B. pantothenicus*, required aneurin, biotin, and pantothenic acid. Whereas *B. anthracis* and *B. cereus*, for example, each formed a nutritionally homogeneous group, *B. circulans* was so heterogeneous as to suggest inadequate classification (Proom and Knight 1955). According to Adams and Stokes (1968) the psychrophilic species generally require niacin, thiamine, and sometimes biotin for growth; the mesophilic species thiamine or biotin or both; and the thermophilic species thiamine, biotin, and niacin singly or in combination. (See also White 1972.)

A powerful filtrable haemolysin is formed by one member of the group—*B. megaterium*; like the α-haemolysin of *Staph. aureus* (see Chapter 30) it is inactivated by heat at 60° but can be reactivated by heating to 100° (Dreyer 1904). Many other species form extracellular haemolysins (Poppe 1922); that formed by *B. cereus* is oxygen-labile, resembling streptolysin O (Bernheimer and Grushoff 1967). Highly active bactericidal substances, including tyrocidin, gramicidin and licheniformin (see Chapter 5), are formed by several different members of the group.

Antigenic structure

Much of the serological work has been directed to separating *B. anthracis* from other *Bacillus* spp. Antisera to *B. anthracis* react to some extent with 'pseudo-anthrax' bacilli (see, e.g., Poppe 1922); a reliable serological differentiation within the group demands careful definition of the various antigens. Sievers and Zetterberg (1940) attempted, with some slight success, to classify other members of the aerobic spore-bearing group by precipitation and complement-fixation tests. Howie and Cruickshank (1940), working with *B. cereus*, *B. mesentericus*, and certain other members of the group, showed that the vegetative bacilli and the spores contained different antigens, thus confirming the previous general conclusions of Defalle (1902) and Mellon and Anderson (1919). Doak and Lamanna (1948), however, by treating the spores with alkali found that the specific spore antigens were concentrated at the surface, but that deeper down were antigens common to those of the vegetative cells. Lamanna (1940*b*) showed that *B. subtilis*, *B. vulgatus*, *B. mesentericus* and *B. agri*, belonging to the small-celled species, were distinguishable by their spore antigens. The separation of large-celled species by this means was less successful, but a broad distinction could be drawn between *B. cereus* and *B. megaterium*. The antigenic heterogeneity in the group is evident in the work of Tomcsik and his colleagues who used both agglutination reactions and 'specific cell-wall' reactions between antisera to polysaccharide cell-wall components and lysozyme-treated cells. In *B. megaterium*, they distinguished 38 cell-wall antigenic types in 54 strains (Baumann-Grace and Tomcsik 1957); and though they found only one cell-wall type in 12 strains of *B. anthracis*, there were in 23 strains of *B. cereus* 13 cell-wall types, four of which were related to *B. anthracis* (Tomcsik and Baumann-Grace 1959*b*; see also Ivánovics and Földes 1958, Baumann-Grace *et al.* 1959). By the analogous 'specific exosporium reaction' five spore serotypes were found in 36 strains of *B. megaterium* (Tomcsik and Baumann-Grace 1959*a*).

Study of the base composition of the deoxyribonucleates showed that the G+C content of *B. anthracis* and *B. cereus* was about 32–34 moles per cent, of *B. licheniformis* 43 per cent, and of *B. subtilis* 46 per cent; there was remarkable homogeneity within the species, all the strains showing identical base compositions (McDonald *et al.* 1963, Hill 1966). The figures given by Bonde and Jackson (1971), it may be noted, are higher than these, e.g. *B. cereus* 36.8 per cent, *B. licheniformis* 49.6 per cent, *B. subtilis* 47.5 per cent. (For further figures, see Holländer and Pohl 1980 and Table 41.3.)

Pathogenicity

The anthrax bacillus is highly pathogenic for most animals; most other members are non-pathogenic for all animals. This statement, however, must be qualified. Under natural conditions the anthrax bacillus gives rise to disease in man, cattle, sheep, and certain other of the domesticated animals; under experimental conditions it is pathogenic for the laboratory animals. Under natural conditions other species of *Bacillus* rarely give rise to disease, but an exception must be made for *B. subtilis*, which may cause severe eye lesions, notably iridocyclitis and panophthalmitis (Axenfeld 1908), and which may occasionally invade the blood stream of patients whose powers of resistance are lowered by the attack of some fatal disease (Sweany and Pinner 1925). Occasionally, too, infections such as meningitis and pneumonia may be due to pseudoanthrax bacilli (Senge 1913, Wilamowski 1912), as well as infections of the urinary and gastro-intestinal tracts (Melles *et al.* 1969). There is evidence that some members, when allowed to grow excessively in food, may produce toxic substances capable of giving rise on ingestion by man to gastro-enteritis (see Chapter 72). Experimentally, the pseudoanthrax bacilli are without effect on any laboratory animal except mice, and are pathogenic for these animals only when injected intraperitoneally in a large dose—1–3 loopfuls of an agar culture. *B. megaterium* is, however, definitely toxic, and is able to kill guinea-pigs in less than 24 hr when given intraperitoneally. This is due to the formation of a haemolysin. The occasional reports of pseudoanthrax bacilli that were pathogenic to man and animals may have been instances of aberrant strains of *B. anthracis* of low virulence. Nevertheless, authentic strains of the closely related *B. cereus* and *B. cereus* var. *mycoides* may be pathogenic to mice and guinea-pigs, and of *B. cereus* to rabbits (Nordberg 1953, Burdon and Wende 1960) and rats (Bonventre 1965). Apart from pathogenicity it may be mentioned that, as Seidel (1962) points out, many of the aerobic spore bearers cause immense economic loss by reason of the damage they do to foodstuffs, especially meat.

Classification

Numerous criteria have been proposed as a basis of classification, such as swelling of the bacillary cell by the spore (de Soriano 1935, Smith *et al.* 1952), the size of the bacterial cell and mode of germination of the spore (Lamanna 1940*a*), and the rhizoid nature of the growth and the Voges-Proskauer reaction (Gibson and Topping 1938). None of these criteria is sufficient of itself to draw a clear line of distinction between different species. Many organisms, particularly those of *B. cereus* and the *B. subtilis* group, are extremely variable, and the only sensible course at present is to recognize certain broad groups on the basis of several different characters.

The studies of Gibson (1944), of Gordon and Smith (1949), and of Smith, Gordon and Clark (1946, 1952), reveal a great degree of variation in individual species. Many organisms that had been regarded as valid species on the basis of cultural appearances, pigment formation, or biochemical reactions, seem to be no more than varieties of other species. Even *B. anthracis* is regarded by Smith and his colleagues as a virulent variant of *B. cereus*; the evidence for and against this is discussed by Seidel (1962).

In the 8th Edition of *Bergey's Manual*, Gibson and Gordon (1974) list 22 widely accepted species, for convenience distinguishing *B. anthracis* and the insect pathogen *B. thuringensis* from the closely related *B. cereus*, but regarding *B. mycoides* as a variety of *B. cereus*. The primary division is made on the shape and position of the spore, on the presence or absence of swelling of the parent form, the sporangium, containing the spore, and on the products of fermentation of glucose. The importance of the spore as a taxonomic character is emphasized by the observation of Bradley and Franklin (1958) that the electronmicrographic appearance of the spore is either a species characterstic, or at most common to two related species. It was noteworthy that Sneath's (1962) Adansonian classification, based on recorded characters, largely coincided with that given in the then current edition of *Bergey's Manual* (Breed *et al.* 1957; see also Seki *et al.* 1978). Bonde (1975) submitted 460 strains, mostly isolated by himself, to 67 different tests, and divided them by cluster analysis into 8 bigger and 2 smaller groups. As a result, he recommended that the description of the genus *Bacillus* should be amended.

We are concerned here with the pathogen *B. anthracis* and the saprophytic strains found in laboratory environments and in material from man and animals (see Burdon 1956), and with a few species of general biological interest. A simplified classification of this dozen or so species is as follows.

Group 1. *Sporangia not swollen. Gram positive.*

(*a*) Large-cell subgroup. Diameter of rod 0.9 μm or more.
B. megaterium
B. cereus
B. cereus var. *mycoides*
B. anthracis

(*b*) Small-cell subgroup. Diameter of rod less than 0.9 μm.
B. licheniformis
B. pumilus
B. subtilis
B. coagulans

Group 2. *Sporangia swollen by oval spores. Gram variable.*
B. circulans
B. polymyxa
B. stearothermophilus

Group 3. *Sporangia swollen by circular spores. Gram variable.*
B. pantothenicus
B. sphaericus
B. pasteuri

B. stearothermophilus and *B. coagulans* are obligatory thermophilic organisms, capable of growing at 55° or over. Gordon and Smith (1949) suggest that the term 'thermophilic' should be applied to *cultures* capable of growing at 55° or over, but that it should never be used to describe species. Allen (1953), whose review of the thermophilic aerobic spore-bearing bacilli should be consulted, points out that because thermophilic strains of almost any mesophilic *Bacillus* may occur, there is little justification for a separate classification of the thermophiles.

Knight and Proom (1950) draw attention to some useful differential criteria. For instance, growth under strictly anerobic conditions on ordinary nutrient agar is of value in distinguishing *B. cereus* from *B. megaterium*; these two organisms likewise differ in their ability to form lecithinase. The so-called restricted reaction on egg-yolk agar, in which the zone of opalescence is confined to the medium under the colony, serves to distinguish between *B. subtilis* and *B. pumilus*, and between *B. circulans* and other members of group 2. Knight and Proom also find that, whereas organisms belonging to groups 1 and 2 have their nutritional requirements satisfied by ammonia, amino acids, aneurin and biotin—and often by only one or two of these components—most members of group 3 have considerably more complex requirements.

To the medical and veterinary student the chief interest is in the separation of *B. anthracis* from the non-pathogenic species; this is discussed under the description of *B. anthracis*. (See also Table 41.3.)

Bacillus anthracis

Named Bacteridium by Davaine (1864) and *B. anthracis* by Cohn (1875*b*). This bacillus is non-motile, forms capsules in the animal body and sometimes on artificial media, and grows on agar in characteristic long, segmented, parallel or interwoven chains. The spores are ellipsoidal or oval, are formed equatorially, and germinate by polar rupture. The anthrax bacillus was the first micro-organism in which the presence of resistant spores was demonstrated. Spores are never found in the animal body during life, and in culture appear more slowly than those of other members of the group. They seem to be formed under conditions unfavourable to continued growth of the vegetative bacilli. Their appearance can be hastened by the addition of distilled water, 2 per cent sodium chloride, and other salts (Bongert 1903). According to Bordet and Renaux (1930), sporulation is inhibited by the presence of calcium chloride and promoted by its absence. Cultures grown on oxalated agar often come to consist mainly of spores, whereas those grown on agar to which $CaCl_2$ has been added may lose their spore-forming power completely. Capsulation is promoted by growth in air containing 10–25 per cent CO_2 (Ivánovics 1937). Under conditions of artificial culture, capsules are formed when the organisms are grown in inactivated fluid horse, ox, or sheep serum; they are at their best after about 6 hr at 37° and begin to disintigrate after 10 hr (Nordberg 1951).

The curled hair-lock appearance of single colonies on agar or gelatin is characteristic, but may be closely simulated by *B. cereus*. Microscopically this is seen to be due to the growth of the bacilli in long interwoven chains. Growth, particularly in broth, at a temperature of 42.5° for some days, causes the appearance of several different *variants*; some have tough, well defined capsules, give rise on agar to the typical curled colonies, and are highly virulent; some have soft, poorly defined capsules, form thin, shining colonies on agar, and are slightly virulent; others are non-capsulated, give rise to smooth, round, convex, glistening,

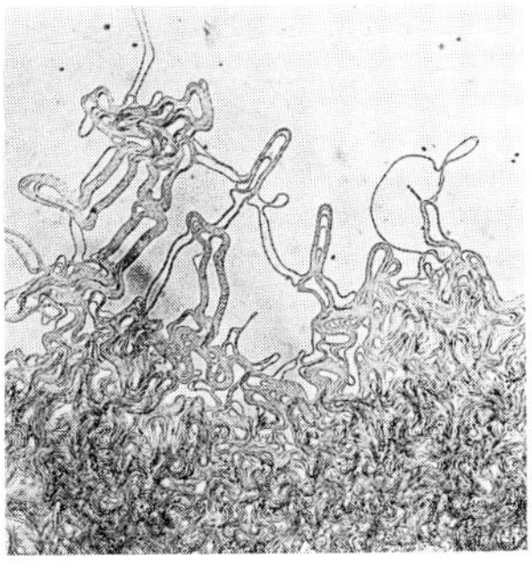

Fig. 41.2 *B. anthracis.* Edge of colony on agar to show curled hair-lock appearance (× 18).

mucoid colonies on agar, and are entirely avirulent (Preisz 1911, Bordet and Renaux 1930). Some variants are asporogenous; it was these which attracted Pasteur's attention, and which he considered to be avirulent. Preisz (1911), however, showed that, though there is a close association between capsule formation and virulence, there is none between spore formation and virulence (see also Nordberg 1951). Virulent strains may be either sporogenous or asporogenous; similarly with avirulent strains. Asporogenous varieties may appear spontaneously in cultures incubated at the usual temperature (Behring 1889), or in cultures containing weak antiseptics, such as 0.05 per cent potasium dichromate, or 0.1 per cent phenol (Roux 1890). Such varieties, when arising from a virulent strain, are themselves fully virulent, though prolonged contact with weak antiseptics may lower their virulence. The normal highly virulent bacillus forms a large, rough colony of frosted glass appearance with a curled edge; morphologically it consists of bacilli arranged in chains (Figs. 41.4, 41.6). The avirulent bacillus forms a smaller, smoother type of colony, with a slightly crenated edge; morphologically it consists of bacilli arranged singly, in pairs end-to-end, or in small bundles (Figs. 41.3, 41.5).

Though the anthrax bacillus as originally described does not exemplify the association, evident in the enterobacteria, of smooth colony with virulence, its true S forms are probably smooth. When the ordinary Medusa-head type of colony is transferred to 50 per

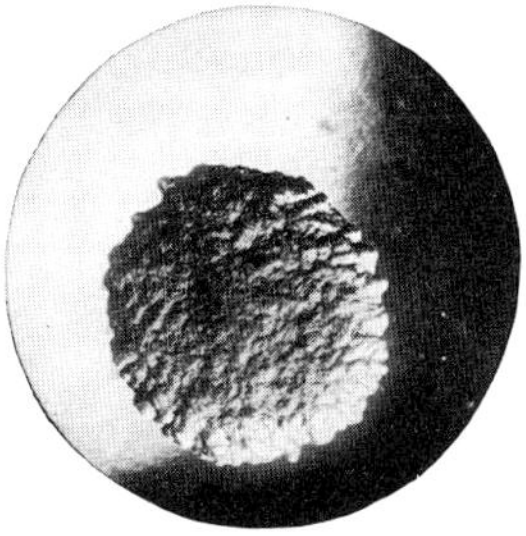

Fig. 41.3 *B. anthracis.* Smooth type of colony. Agar, 24 hr, 37° (× 8).

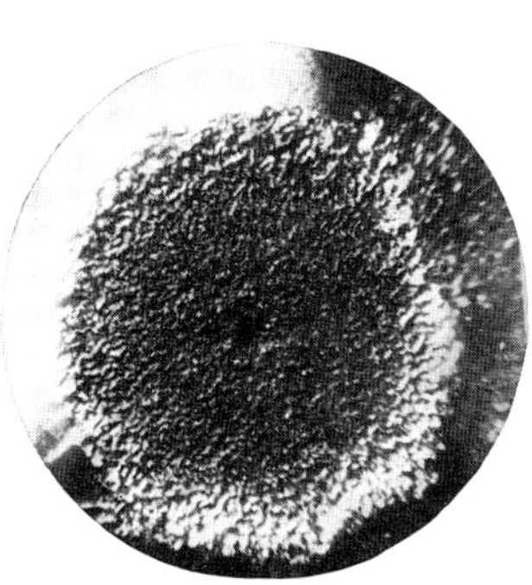

Fig. 41.4 *B. anthracis.* Rough type of colony. Agar, 24 hr, 37° (× 8).

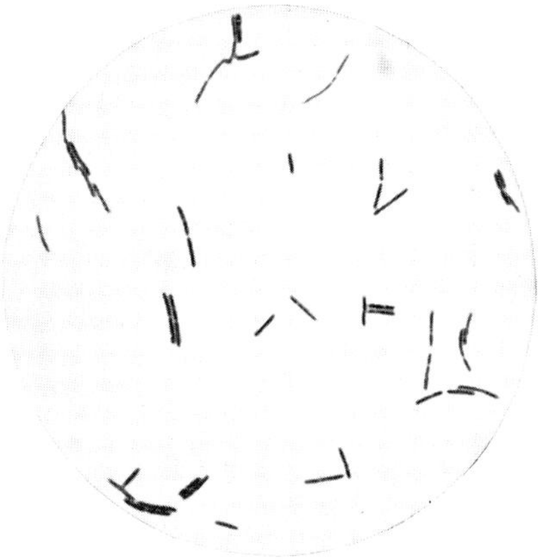

Fig. 41.5 *B. anthracis.* From smooth colony on agar, 24 hr, 37° (× 1000).

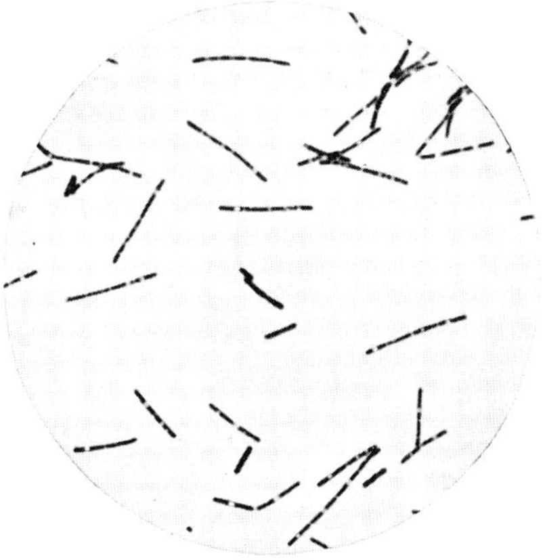

Fig. 41.6 *B. anthracis.* From rough colony on agar, 24 hr, 37° (× 1000).

cent serum agar and grown in air containing 65 per cent of CO_2, a smooth mucoid growth appears consisting of fully virulent capsulated organisms. On further incubation a rough outgrowth often occurs, subcultures from which are roughish and non-mucoid and consist of non-capsulated avirulent bacilli (Sterne 1937*a*, *b*). Sterne (1937*a*, *b*) and Chu (1952) regard the mucoid growth of virulent capsulated bacilli induced in this way as the normal S form, and the rough-colony form as the avirulent R form. (See also Sterne and Proom 1957.)

Resistance

In the dry state the spores may remain alive for 12 years or more (Pasteur 1881*a*). They are killed by dry heat at 150° in 60 min (Stein and Rogers 1945). In the frozen state at $-5°$ to $-10°$ they may survive for four years (Stein 1943). Moist heat sterilizes a saline suspension containing 1 million spores per ml in 15 to 45 min at 90°, in 10 to 25 min at 95°, and in 5 to 10 min at 100° (Murray 1931). In bundles of horsehair or bristles not more than $2\frac{1}{2}$ inches thick the spores are destroyed by autoclaving at 121° in 15 min (Schneiter and Kolb 1948). It is worth noting that anthrax spores may survive the usual process of heat-fixation and staining in ordinary film preparations (Soltys 1948). Of the various chemical agents tested, formaldehyde is one of the most active (see Chapter 4). In the unopened carcass of an animal dead from anthrax the bacilli are said to remain alive in the bone marrow for a week and in the skin for a fortnight (Minett 1950).

Antigenic structure

As early as 1921 Kramár obtained evidence that the capsule contained a glycoprotein probably belonging to the class of pseudomucins. A soluble specific carbohydrate, incapable of giving rise to antibodies on injection into animals, but reacting to a high titre with anti-anthrax serum, was extracted from anthrax bacilli by Combiesco, Soru and Stamatesco (1929) and by Schockaert (1929). It was, however, mainly Tomcsik (1930), Tomcsik and Szongott (1932, 1933), Tomcsik and Bodon (1934, 1935), Bodon and Tomcsik (1934), Sordelli and Deulofeu (1930, 1933) and Sordelli, Deulofeu and Ferrari (1932) who laid the foundation of our knowledge of the antigenic structure of this organism. There appeared to be two main serologically reactive substances, a polypeptide in the capsule, and a polysaccharide in the cell wall, but this distinction may have been an over-simplification. The microscopic changes in these structures when the bacilli are treated with appropriate antibody indicate that the capsule of *B. anthracis* is largely polypeptide; but in *B. megaterium*, for example, it consists usually of polypeptide interlaced with prolongations of cell-wall material perpendicular to the bacterial surface, intensified at the junction of two cells in a chain, and forming a cap at the free pole of the bacillus; strains occur in which the capsule is wholly polysaccharide (Baumann-Grace and Tomcsik 1958, Tomcsik and Baumann-Grace 1959*c*). It is therefore possible that, even with what appears to be a wholly polypeptide capsule, elements of cell-wall polysaccharide reach to the surface of the capsule. Ivánovics (1937, 1953). Ivánovics and Erdös (1937), and Ivánovics and Bruckner (1937*a*, *b*, 1938) found that the capsular substance contained a group-specific hapten consisting of a D-glutamic acid polypeptide, antisera to which reacted with polypeptides from other members of the *Bacillus* group. It reacted in high titre with a precipitating serum prepared by the injection into rabbits of heat-killed capsulated anthrax bacilli, but could not by itself stimulate the formation of antibodies (Tomcsik and Ivánovics 1938). Hanby and Rydon (1946) identified the purified capsular material as a long-chain molecule consisting of α-peptide chains of 50–100 D-glutamic acid residues joined together by γ-peptide chains. The molecular weights of the polymer, as isolated from bacilli and body fluids of the infected guinea-pig, were respectively 75 000 and 216 000 (Record and Wallis 1956*a*). The serological cross-reacting polypeptide of *B. subtilis* consists of a mixture of γ polymers of D- and L-glutamate (Watson *et al.* 1947*b*, Ivánovics 1958, Thorne and Leonard 1958); and that of *B. megaterium* a γ co-polymer of D- and L-glutamate (Utsumi *et al.* 1959). Leonard and Thorne (1961) question the specificity of the serum reactions of isolated polysaccharides, which in many cases appear to be due to basic serum proteins.

The main somatic polysaccharide consists largely of equal amounts of N-acetyl glucosamine and galactose (Ivánovics 1940, Smith *et al.* 1956*a*) with a molecular weight of 29 000 (Record and Wallis 1956*b*). A second polysaccharide is described as being also present in some strains, containing galactose, xylose, ribose and uronic acids (Combiesco *et al.* 1929, Barber *et al.* 1957). It is related antigenically to that of *B. cereus* (Baumann-Grace *et al.* 1958).

Toxins, aggressins and immunizing antigens

Several observers described 'toxic albumoses' in broth cultures of the anthrax bacillus (Hankin 1889, Marmier 1895, Standfusz and Schnauder 1925). Aoki and Yamamoto (1939) found an 'endotoxin' which increased the infectivity of spore suspensions. Disintegration of whole bacilli grown *in vitro* by shaking with glass beads yields no toxin (King and Stein 1950), but an aggressin that inhibits phagocytosis and increases the infectivity of anthrax bacilli *in vivo* (Keppie *et al.* 1953).

Culture filtrates contain a soluble antigen that elicits active immunity in rabbits, and antisera which confer passive protection (Gladstone 1946, 1948). It is heat-labile, being destroyed by heating to 57° for 30 min, by trypsin (Watson *et al.* 1947*a*), and by enzymes in the culture. In cultures of non-proteolytic variants of *B. anthracis*, it persists longer (Wright *et al.* 1951). The antigen is produced in chemically defined media (Wright *et al.* 1954, 1962, Strange and Belton 1954, Boor 1955, Strange and Thorne 1958).

The relation of these various substances to the pathogenic effects of the bacillus is complex. In the oedema fluid of infected rabbits Cromartie and his colleagues (1947) found a protective antigen and an 'inflammatory factor', which in other animals induced the characteristic oedema of natural infection.

Smith and Stoner (1967), in a study of bacilli and bacillary substances in the body fluids harvested from heavily infected guinea-pigs, found the oedema factor in fair abundance in the plasma, and separated it into three factors, I, II and III. None of these factors is toxic when injected alone. Factor I is a chelating agent containing phosphorus. It is non-immunogenic, but enhances the virulence of attenuated strains. Factor II, a protein, is immunogenic and anti-phagocytic. A combination of factors I and II is lethal on intravenous, and powerfully oedema-producing on intracutaneous, injection. The addition of the minor factor III reduces oedema production. The toxin is the major immunogen of the anthrax bacillus; and passively administered antitoxin is protective, contributing to the destruction of the infecting bacilli at the site of their primary lodgement. The non-toxic immunogen isolable from an in-vitro culture can replace factor II in reconstituting the lethal toxin. The capsular polyglutamate also enhances the virulence of attenuated strains; it is antiphagocytic, apparently because it inhibits opsonization of the bacilli; but it is neither lethal, oedema-producing nor immunogenic (Smith *et al.* 1955*b*, Smith *et al.* 1956*a*, *b*, Stanley and Smith 1961, Smith and Stanley 1962, Keppie *et al.* 1963). (Table 41.1.)

In the United States, Beall, Taylor and Thorn (1962) likewise separated the toxin into three fractions corresponding to those of the British workers. Factor I they called the oedema factor, factor II the protective factor, and factor III the lethal factor, because it was found to be rapidly fatal to Fischer rats when injected intravenously. As Smith and Stoner (1967) point out, this nomenclature is confusing; none of the three factors by itself is toxic, and a combination of factors II and III, in the absence of factor I, gives rise to oedema of the lung. Moreover, Mahlandt and his colleagues (1966) found that in rats, as opposed to guinea-pigs, the so-called lethal factor III was more protective against toxin than the so-called protection factor II.

The three factors have been purified by column chromatography. Exposure to a temperature of 30° for 120 hr destroyed their biological but not their serological activity; this was destroyed, however, by boiling for 10 min (Wilkie and Ward 1967, Fish *et al.* 1968).

Toxin can be extracted from cultures, but with difficulty owing to its instability. Evans and Shoesmith (1954) produced a toxic culture filtrate; and a toxin corresponding to that in Smith's in-vivo material was formed in cultures containing 90 per cent guinea-pig serum (Harris-Smith *et al.* 1958) and in serum-free media; though in these circumstances it was adsorbed to the sintered glass filters employed, from which it could be eluted with mild alkali (Thorne *et al.* 1960). Its toxic effect may be quantitatively assayed by intravenous injection in Fischer rats (Haines *et al.* 1965). Toxin is produced by both virulent and attenuated strains (Harris-Smith *et al.* 1958). Its toxicity for leucocytes is in general highest for cells of the species most susceptible to infection with *B. anthracis* (Kashiba *et al.* 1959). Its main action seems to be (*a*) to increase vascular permeability, as manifested by the production of oedema in the guinea-pig skin after intradermal injection and in the lungs of mice and Fischer rats after intravenous injection (Smith and Stoner 1967) and, (*b*) to cause capillary thrombosis (Dalldorf and Beall 1967). How far it is responsible for production of the disease is still uncertain. Ward and his colleagues (1965), for example, in experiments on guinea-pigs found that in immunized animals that had died after injection with a highly virulent strain of

Table 41.1 Effect of the three factors of anthrax toxin on guinea-pigs when injected intracutaneously and mice when injected intravenously, together with the immune response of guinea-pigs (Smith and Stoner 1967)

	Toxicity		
Factors	Oedema of skin	Death of mice	Immunogenicity for guinea-pig
I Chelating agent	–	–	–
II Protein	–	–	++
III Protein	–	–	–
I+II	++++	+	+++
I+III	–	–	+
II+III	–*	++	++
I+II+III	++	+++	++

* Produce pulmonary oedema in rats.

B. anthracis no toxin could be detected in the blood. Antitoxin, which was demonstrable in the blood, had not prevented death, nor was there any close relation between the amount of antibody in the blood before challenge and the survival time. According to Vick and his colleagues (1968) anthrax toxin in monkeys is responsible for death apparently by paralysis of the respiratory centre in the brain.

A haemolysin is formed for the red cells of sheep, goats and rabbits, but not for cattle or horses (Poppe 1922).

Bacteriophages

Lysogenic strains of *B. anthracis* occur, and (Cowles 1931) phages lysing the bacillus. McCloy (1951, 1958) isolated from a lysogenic strain of *B. cereus* a phage, W, highly specific for *B. anthracis*, though inactive against mucoid strains; and derived two phage mutants, α and β, of which β was the more specific. From phage W, Brown and Cherry (1955) derived a variant, γ, active against capsulated strains not attacked by W. Not all strains of the anthrax bacillus are susceptible to the γ phage. Buck and his colleagues (1963), who tested six different phages, found that 85 per cent of the anthrax strains tested were lysed by the γ phage and 95 per cent by one or more of the other phages. Lantos and her colleagues (1960) classified 8 soil phages into three serotypes, A_1, A_2 and A_3, type A_1 corresponding to McCloy's W phage. (See also Ciucă *et al.* 1959.)

Pathogenicity

The anthrax bacillus is naturally pathogenic mainly to the herbivora and to man, but occasionally it attacks other animals. Experimentally it proves fatal to the mouse, guinea-pig, and rabbit, less often to the rat. Hamsters are very susceptible (Stamatin and Cristescu 1958). The larger the dose, the shorter the time to death. Subcutaneous injection of a loopful of an 18-hr agar culture kills mice, guinea-pigs, and rabbits in 12 to 30 hr; a smaller dose, 1/100 of a loopful, kills them in 30 to 40 hr; a still smaller dose, 1/2 000 000 of a loopful, kills mice in 96 hr, guinea-pigs in 56 hr, and rabbits in 104 hr (Sobernheim 1897). The prolongation of survival time in mice is a good index of attenuation (Roth *et al.* 1956). With attenuated strains, doses of 0.2×10^6 to 0.9×10^6 spores kill only a proportion of infected mice (Burdon and Wende 1960). Members of other species of *Bacillus* never kill in such small doses.

Post mortem, there is a gelatinous haemorrhagic local oedema; the viscera are congested, the blood is dark red and coagulates less firmly than usual; the spleen is enlarged, dark red, and very friable. Microscopically, the bacilli are found in large numbers in the local lesion, in the blood, and in the thoracic and abdominal viscera; they are confined almost entirely to the interior of the capillaries, where their numbers may be so great as to cause obstruction to the blood flow; the tissues themselves are rarely penetrated. Though the disease terminates in a septicaemia, it is not till 4 or 5 hr before death, as a rule, that the bacilli gain access to the blood stream.

Infection may also be accomplished by cutaneous, intracutaneous, intramuscular, intraperitoneal or intravenous injection, or by feeding, as well as by exposure of the animals to a cloud of anthrax spores in a closed chamber (Young *et al.* 1946). The most certain route is the intramusuclar, the least certain the oral (Sobernheim and Murata 1924). In doses greater than 10^5 spores, *B. anthracis* produces a fatal infection after inhalation, terminating on the 2nd–6th day (see Henderson *et al.* 1956, Gochenour *et al.* 1962). In general it requires a large dose to produce a fatal infection by the mouth (Giovanardi 1931). *Post mortem* in cases of oral infection, in addition to the enlargement of the spleen and the occurrence of septicaemia, the intestinal mucosa is seen to be covered with small furuncular swellings. Of the three animals the mouse is the most susceptible and the rabbit the least, the guinea-pig occupying an intermediate position. This difference in susceptibility is scarcely noticeable, except with a strain of weakened virulence (see Chapter 54).

The later stages of infection in the guinea-pig are characterized by a state of secondary shock, with lowered blood pressure, decreased blood volume and haemoconcentration. There is a late fall in blood glucose, and rise in blood urea, and an accompanying nephrosis, evident in the lower nephron (Smith *et al.* 1955*a*; see also Slein and Logan 1960). When the number of bacilli in the blood is about 1/300th of that at the terminal stage, usually some 6–7 hr before death, the fatal outcome is determined and cannot be reversed by treatment with antibiotics (Keppie *et al.* 1955). In the rabbit the terminal stage of infection is marked by advanced haemolysis, extremely low oxygen tension in the blood, and death from asphyxia (Nordberg *et al.* 1961). That the low oxygen tension is not due to bacilli in the blood is shown by its occurrence in penicillin-treated animals in which bacillaemia is absent (Nordberg *et al.* 1964). The pathology of the disease in monkeys infected by the intradermal route is described by Berdjis, Gleiser and Hartman (1963).

Rats are more difficult to infect than other rodents, but are said to succumb easily when fatigued by continuous exercise on a revolving drum (Charrin and Roger 1890). A chronic disease may develop after subcutaneous injection, which does not prove fatal for 4 or 5 weeks. Intravenous injection in Wistar albino rats may prove fatal in a few hours (Eckert and Bonventre 1963). Fischer rats, however, are far more susceptible, and die within 1 to 3 hr of respiratory failure; *post mortem*, pulmonary oedema and hydrothorax are present (Beall *et al.* 1962, Beall and Dalldorf 1966). Both in Fischer rats and in rabbits and guinea-pigs the terminal asphyxia is attributed by Dalldorf and Beall (1967) to capillary thrombosis. Dogs may be infected by subcutaneous injection, though not uniformly. Birds, with the exception of sparrows and young doves, and cold-blooded animals, are comparatively resistant (see Davaine 1863*a*, *b*, 1864, Koch 1877, Chauveau 1880*a*, *b*, Sobernheim 1897, Poppe 1922, Sanarelli 1925; and, for the pathology of experimental anthrax in different animal hosts, see Gleiser 1967).

Pasteur (1881*b*, *c*) found that by growing the anthrax bacillus at 42.5° for about a month, he was able to lower its virulence to such an extent that it proved harmless to all animals except newborn guinea-pigs. By successive passage through these animals, the bacillus gradually regained its virulence till it was able to kill 2-, 3-, and 4-day-old and later adult guinea-pigs; eventually its virulence was entirely restored. Preisz (1911) showed that the process was one of selection of avirulent variants at the higher temperature of incubation. It appears that an attenuated strain is either a mixture of avirulent with a few virulent bacilli, or perhaps a population of avirulent bacilli in which mutation to virulence is occurring frequently enough for selection to take place after injection into a susceptible animal.

Differentiation of anthrax from pseudoanthrax bacilli

The main criteria of value for distinguishing *B. anthracis* from pseudoanthrax bacilli are given in Table 41.2.

Many of the differential tests have essentially a negative value; for example, an organism that ferments salicin rapidly is not *B. anthracis*, but the converse does not hold true (see Stein 1944). An inverted fir-tree growth in gelatin is given by some strains of pseudoanthrax bacilli, but the branches are thick and interlaced, quite different from the regular, delicate, lateral outgrowths of *B. anthracis*. Far the most important distinguishing character of the anthrax bacillus is its peculiar pathogenicity; no other member of the *Bacillus* group will kill a guinea-pig with 48 hr when injected subcutaneously in a dose of 0.5 ml of a 24-hr broth culture.

When freshly isolated from the animal body, the anthrax bacillus rarely causes difficulty in identification, but after prolonged subculture in the laboratory it may lose several of its important characteristics, such as capsule formation, the inverted fir-tree growth in stab gelatin, and its pathogenicity for laboratory animals, and may then be very difficult to classify. Nevertheless bacilli have been described which have given rise to an anthrax-like disease in man and other animals, yet which have not conformed to the usual criteria of *B. anthracis* (Schulz 1901, Wilamowski 1912, Senge 1913). Such bacilli have usually been classed as pseudoanthrax bacilli, but it is probable that some at least have been variants of the real *B. anthracis*, similar to those described by Preisz (1911).

Bacilli more or less closely resembling the anthrax bacillus have been isolated by numerous workers from such substances as soil, water, meat-, fish- and bone-meal, wool, dust, oil-cake, and less often from animals and man. These organisms have frequently been termed *B. pseudoanthracis* or *B. anthracoides*. From the available records it is evident that more than one species has been described. We do not propose to consider these organisms except to deal with their differential diagnosis from the anthrax bacillus. It is sufficient to point out that most of them are motile, non-capsulated, form spores abundantly within 24 hr on agar, produce an even turbidity or a surface pellicle in broth, give rise to colonies on agar which are less curled and have fewer and less regular outgrowths at the edges, are more resistant to heat, form alkali in litmus milk, and are generally non-pathogenic, though sometimes they may be fatal to mice and even

Table 41.2 Differences between *B. anthracis* and the pseudoanthrax bacilli

B. anthracis	Anthrax-like or pseudoanthrax bacilli
1. Non-motile.	Generally motile.
2. Capsulated.	Non-capsulated.
3. Grows in long chains.	Grow in short chains.
4. No turbidity or pellicle in broth.	Often turbidity and pellicle in broth.
5. No growth on penicillin agar (10 μg/ml).	Usually good growth on penicillin agar.
6. Inverted fir-tree growth in gelatin.	Fir-tree growth absent or atypical.
7. Methylene blue reduced weakly.	Methylene blue usually reduced strongly.
8. Haemolysis of sheep cells weak.	Haemolysis of sheep cells often strong.
9. Liquefaction of gelatin slow.	Liquefaction of gelatin usually rapid.
10. Lecithinase reaction weakly positive.	Lecithinase reaction strongly positive with *B. cereus* group.
11. Ferments salicin slowly.	Often ferment salicin rapidly.
12. Polysaccharide precipitin reaction strongly positive.	Polysaccharide precipitin reaction weakly positive.
13. Produces toxin, neutralized by *B. anthracis* antitoxin.	Any toxic substances produced not neutralized by *B. anthracis* antitoxin.
14. Pathogenic to laboratory animals.	Mostly non-pathogenic. If pathogenic, produce disease unlike anthrax.
15. Susceptible to γ phage.	Insusceptible to γ phage.
16. Culture filtrates non-toxic to tissue culture cells.	Culture filtrates (*B. cereus*) toxic to tissue culture cells (Bonventre 1965).

guinea-pigs on intraperitoneal inoculation in fairly large doses. The classification of these organisms is impossible; they seem to have ranged from avirulent variants of *B. anthracis* on the one hand, through pathogenic forms of *B. cereus* and *B. cereus* var. *mycoides*, to virulent variants of *B. subtilis* on the other. For some of the strains which have been described, see Schulz (1901), Wilamowski (1912), Pokschischewsky (1914) and Poppe (1922).

Burdon and Wende (1960), in their distinction of *B. anthracis* from *B. cereus*, note that young, rapidly transferred haemolytic strains of authentic *B. cereus* may cause fatal generalized infection, but the invading organisms do not become capsulated; and on subcutaneous injection *B. cereus* regularly produces a localized skin ulcer. They also add that in excess CO_2 in media containing 0.7 per cent bicarbonate, virulent strains of *B. anthracis*, unlike any strain of *B. cereus*, produce mucoid colonies of capsulated bacilli; that *B. anthracis* produces a reddish pigment on bicarbonate media (Chao and Williams 1959); and that on solid media containing 0.5 μg of penicillin per ml, young cultures consist of chains of rounded forms—the 'string of pearls' reaction of Jensen and Kleemayer (1953). Ivánovics and Földes (1958) add the hydrolysis of phenophthalein phosphate by *B. cereus*, but not by *B. anthracis*. (See Brown *et al.* 1955, 1958, Leise *et al.* 1959, Seidel 1962.)

For the rapid identification of *B. anthracis*, selective media containing lysozyme and haemin (Pearce and Powell 1951, see also McGaughey and St. George 1955), propamidine (Morris 1955), polymyxin (Gillissen and Scholz 1961), and lysozyme, polymyxin, and thallous acetate (Knisely 1966) have been devised. Three-hour-old colonies may be identified by microscopically observed lysis with γ phage (Chadwick 1959).

Cherry and Freeman (1959) recommend 'staining' the organism with fluorescent specific antisera; and to overcome the difficulty of preparing an antiserum monospecific for all *B. anthracis* strains, Dowdle and Hansen (1961) recommend treating the culture to be identified with species-specific γ phage, and identifying those bacilli to which it has been absorbed by treatment with fluorescent antibody to the phage. For a detailed account of the anthrax bacillus, as knowledge then existed, the reader is referred to the authoritative article by Soberheim (1913). For the rapid identification of the mesophilic aerobic spore bearers other than *B. anthracis* reference may be made to the scheme used by Franklin, Williams and Clegg (1956).

Bacillus anthracis

Synonyms Bactéridie du charbon, Milzbrandbacillus.

Habitat Parasitic in man, cattle, sheep and other animals.

Morphology Rods, 3–8 μm × 1–1.2 μm. Straight or slightly curved, ends truncate; on agar plates arranged characteristically in very long, segmented, parallel, or interwoven chains. Unjointed filiaments not infrequent in cultures. In blood of animals mostly in pairs or chains of 3 or 4. Spores equatorial, ellipsoidal, not bulging; germination by absorption of spore coat; not formed in animal body. Non-motile. Capsule formed in animal body, on serum media, and on bicarbonate media in an excess of CO_2; lost on agar; surrounds entire chain of bacilli. Gram positive. Non-acid-fast.

Agar plate Irregularly round, 2–3 mm in diameter, raised, dull, opaque, greyish-white, plumose colonies, with a tessellated or reticular structure, an uneven surface and a curled edge. Membranous consistency, emulsifiability difficult; colony consists of parallel interlacing chains of bacilli,and is characteristic. After about a week irregular round scales appear on the surface of the colony. On bicarbonate media in excess of CO_2, virulent strains form smooth mucoid colonies.

Agar slope Thick, raised, spreading, greyish-yellow growth, with an uneven surface and an undulate edge showing little curled projections; moist and slightly glistening. Looks as if there were innumerable tiny air bubbles beneath the surface. After about a week irregular round scales appear on the surface of the growth.

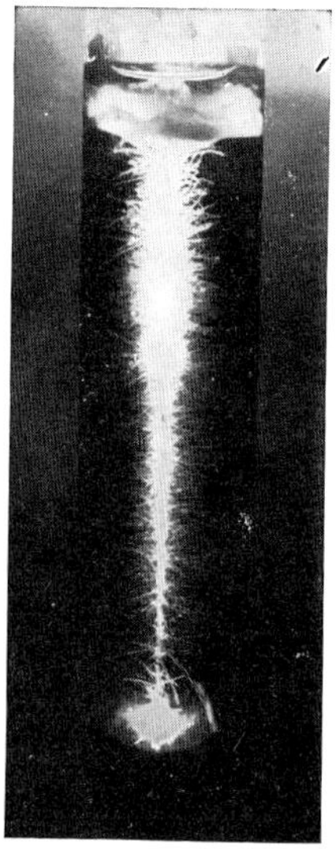

Fig. 41.7 *B. anthracis*. In gelatin stab culture, 3 days, 22° showing inverted fir-tree growth, with commencing liquefaction.

Gelatin stab Poor filiform growth followed by outgrowth of delicate lateral extensions, longest at the upper part of the culture, giving an inverted fir-tree or lamp-brush effect. Liquefaction crateriform; occurs very slowly.

Broth No turbidity, or very fine floccular turbidity; moderate floccular deposit, consisting of interwoven threads, and disintegrating partly on shaking. No surface growth.

Blood serum Abundant, creamy-yellow, confluent, curled growth with uneven surface. No liquefaction.

Resistance Spores killed by boiling in 10 min. In dry state remain alive for years.

Metabolic Aerobic; facultative anaerobe. Opt. temp. 37°. Limits 12° to 44°. Some strains are stated to haemolyse sheep's red cells. Requires amino acids and aneurin for growth; grows well on ordinary media; growth not improved by blood or serum and only slightly by glucose. Reddish pigment on bicarbonate media.

Biochemical Acid, no gas, in glucose, maltose, sucrose, and later in salacin; final pH 5.5–5.9. Indole −; MR ±; VP ±; nitrate reduced to nitrite. H_2S −; NH_3 + +; methylene blue reduced weakly; catalase +. Litmus milk coagulated and decolorized; later peptonized. Weak lecithinase reaction.

Antigenic structure Capsular polypeptide reacting with *B. anthracis* antisera. Somatic polysaccharide, apparently species-specific, though with some relation to that of certain strains of *B. cereus*. Immunogenic complex protein toxin, oedema-producing and lethal. G+C content of DNA 32–34 moles per cent.

Pathogenicity See p. 430.

Bacillus cereus

This organism, which was described by Frankland and Frankland (1887), is widely distributed and is said to be the commonest aerobic spore-bearer in soil. Like *B. subtilis*, it is more the centre of a group of closely related organisms than a narrowly defined species. The following description is based on that given by Smith, Gordon, and Clark (1952). Variant colonies with different bacillary morphology have been described by a number of workers. Soule (1928), who worked with the Michigan type of *B. subtilis*, now

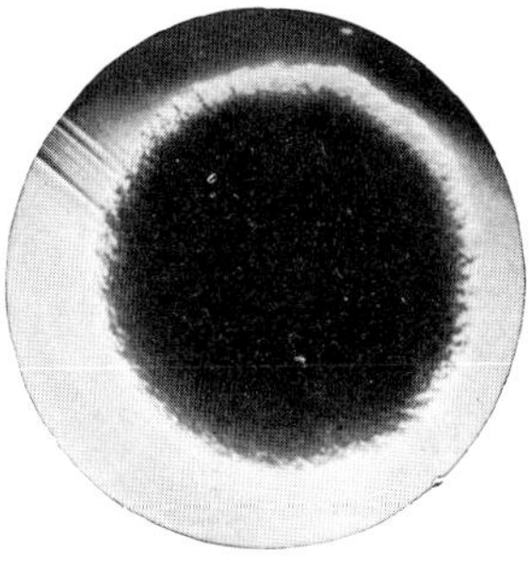

Fig. 41.8 *B. cereus*. Smooth type of colony. Agar, 24 hr, 37° (×8).

recognized as *B. cereus*, described a rough and a smooth type closely simulating the corresponding types of *B. anthracis*. Graham (1930) recognized four variants, two of which were motile and two usually non-motile; all four shared a common heat-stable somatic antigen, and the two motile variants had a common heat-labile flagellar antigen.

Tomcsik and Baumann-Grace (1959*b*) found 13 antigenically distinguishable cell-wall types among 23 strains. Has several synonyms, such as *B. mycoides*, *B. robur*, *B. albolactis*, *B. lactis*, *B. tropicus*. Resembles *B. anthracis* in most respects. Differs in the following ways: it is motile and non-capsulated. In gelatin stab abundant filiform growth with rapid liquefaction. Gives a positive Nagler test in serum broth, forms a powerful lecithinase, rapidly liquefies litmus milk, hydrolyses starch, and forms two haemolysins (Coolbaugh and Williams 1978) and a toxic phospholipase (Slein and Logan 1965). Antigenically, at least 18 serotypes based on flagellar antigens, some associated with the emetic form of food poisoning, others with the diarrhoeal form (Taylor and Gilbert 1975, Gilbert and Parry 1977; see also Chapter 72). A serotyping scheme based on the flagellar antigens was devised

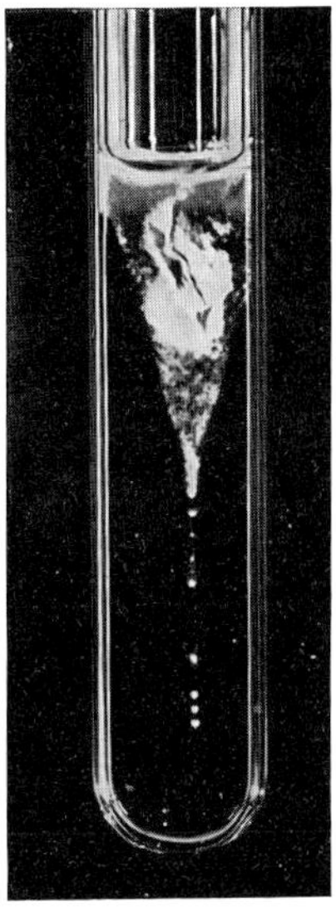

Fig. 41.9 *B. cereus*. In gelatin stab culture, 4 days, 20°, showing infundibuliform liquefaction.

by Taylor and Gilbert (1975) for strains from cases of food poisoning. Produces a toxin or toxins acting on the intestine (see Spira and Goeppert 1975, Melling and Capel 1978, Melling *et al.* 1978), and a skin-permeability factor in the presence of glucose (Spira and Silverman 1979). In man it causes two forms of food poisoning (see Chapter 72), and various systemic diseases (Turnbull *et al.* 1979). In cattle it gives rise to mastitis, occasionally fatal. Some strains cause localized ulceration in guinea-pigs and mice on intracutaneous or intraperitoneal injection (Burdon *et al.* 1967). Immunization against *B. anthracis* does not protect against infection with *B. cereus*. According to Seki and his colleagues (1978) the insect parasite *B. thuringiensis* must, on the grounds of DNA homology, be regarded as indentical with *B. cereus*. G+C content of DNA 36–39 moles per cent.

Bacillus cereus var. *mycoides*

First described by Flügge (1886). Is now regarded as a variant of *B. cereus*. Non-motile; gives a rhizoid growth on agar, sometimes lost on subculture. Maximum temperature for growth 35–40°, minimum 10–15°. Is highly proteolytic, and in soil plays an important part in denitrification. G+C content of DNA 35–37 moles per cent. (See Figs. 41.10–41.12.)

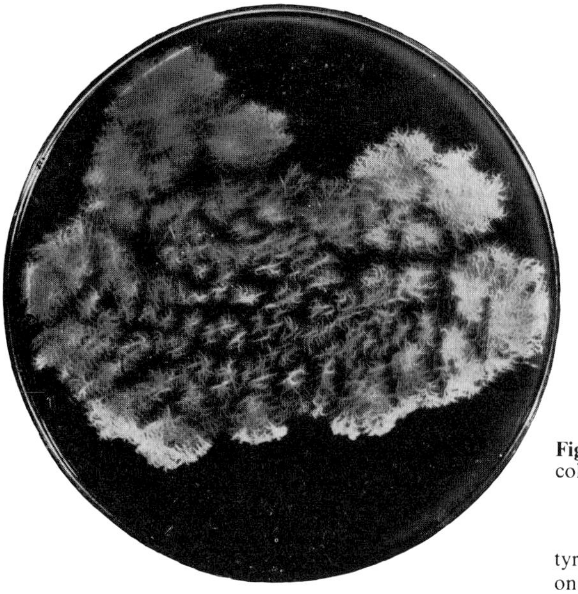

Fig. 41.10 *B. cereus* var. *mycoides*. Surface growth of rhizoid type on agar, 3 days, 30° (4/5ths natural size).

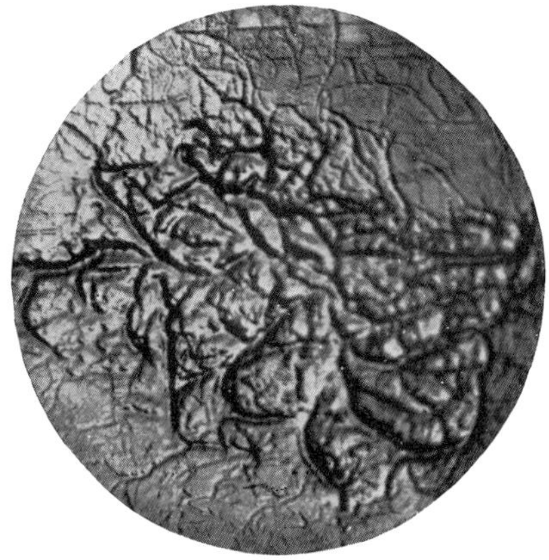

Fig. 41.12 *B. cereus* var. *mycoides*. Central part of a surface colony on agar, 3 days, 30°, showing rhizoid structure (×8).

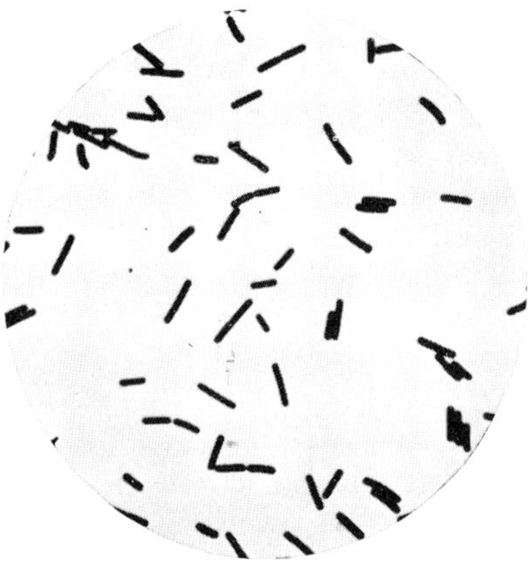

Fig. 41.11 *B. cereus* var. *mycoides*. From an agar slope culture, 2 days, 30° (×1000).

Bacillus megaterium

Synonyms: B. graveolens, B. petasites. Described by de Bary (1884). Found in dust, soil, water, milk. Is one of the largest members of the *Bacillus* group, 2–6 μm × 1.2–1.5 μm. Pleomorphism affected by environmental conditions, especially oxygen starvation (Rettger and Gillespie 1935). Motile by peritrichate flagella. Spores 1.5–2 μm × 1.0–1.2 μm, equatorial, oval, or ellipsoidal, not bulging; germination by absorption of spore coat. Usually non-capsulated. Gram positive. Morphological and colonial variants described by Knaysi (1933). Forms on agar round, low convex, greyish-white, opaque colonies with entire edge and glistening surface, butyrous or membranous consistency; irrecular scales appear on the surface of the colony after a week, and later the colonies become brownish. In gelatin stab, abundant filiform growth with infundibuliform or saccate liquefaction. On serum agar abundant creamy growth without liquefaction. Spores are killed at 20 lb pressure in 1 hr. Maximum temperature 35–45°. Aerobic. Sensitive to lysozyme; no action on egg-yolk. Powerful haemolysin apparently identical with the toxin, formed in broth at 37° (Todd 1901, 1902, Warden *et al.* 1921), reaching its maximum on the 6th or 7th day. Acid produced in glucose, maltose, mannitol, sucrose, salicin, arabinose, and usually xylose; 3 biochemical groups

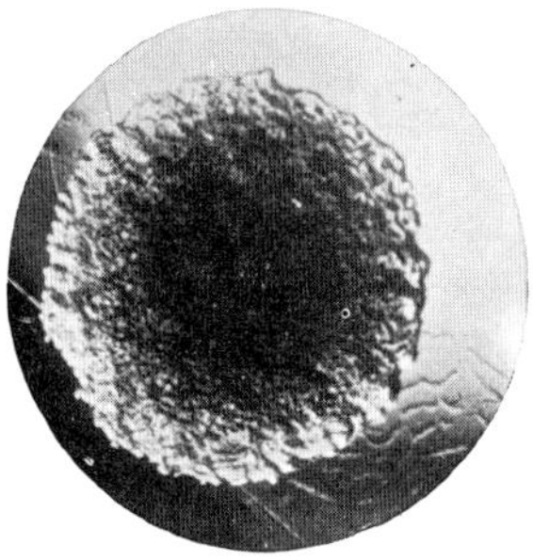

Fig. 41.13 *B. megaterium*. Surface colony on agar, 36 hr, 37° (×8).

found. Nitrate assimilated, but no free nitrite detectable. Hydrolyses starch. Catalase formed. Litmus milk peptonized and decolorized. Casein digested. VP negative. Antigenically, several cell-wall polysaccharide and 5 spore serotypes are known. The complex structure of the polyglutamate capsule and its relation to the cell-wall polysaccharide are discussed on p. 428. G+C content of DNA 37.3–43.0 moles per cent (Candeli *et al.* 1979). Non-pathogenic under natural conditions. Haemolysin fatal to mice and guinea-pigs when injected intraperitoneally; at necropsy, haemorrhagic exudate in peritoneum and pleura. (For review, see Gordon, Haynes and Pang 1973.)

Bacillus subtilis

Great confusion has prevailed over this organism. It was described by Ehrenberg in 1838, who found it in hay infusion, as *Vibro subtilis*, and by Cohn (1875*a*) as *Bacillus subtilis*. For a long time two different organisms passed under the same name—one forming small spores which germinate equatorially (Marburg type), the other forming larger spores showing polar germination (Michigan type) (see Conn 1930, Soule 1932). At the second International Microbiological Congress the Marburg type was officially accepted as the type strain (St. John-Brooks and Breed 1937). The The large type is now regarded as *B. cereus*. The following description is based mainly on that of Gibson (1944) and Smith, Gordon and Clark (1952). We have accepted Gibson's conclusion that the various organisms described under the names *B. mesentericus*, *B. mesentericus fuscus*, *B. mesentericus ruber*, and *B. mesentericus niger* are all strains of *B. subtilis*. (See also Lamanna 1942.)

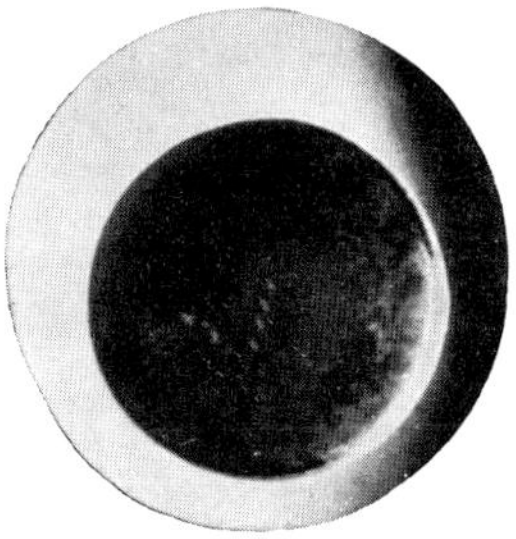

Fig. 41.14 *B. subtilis*. Smooth type of colony. Agar, 24 hr, 37° (×8).

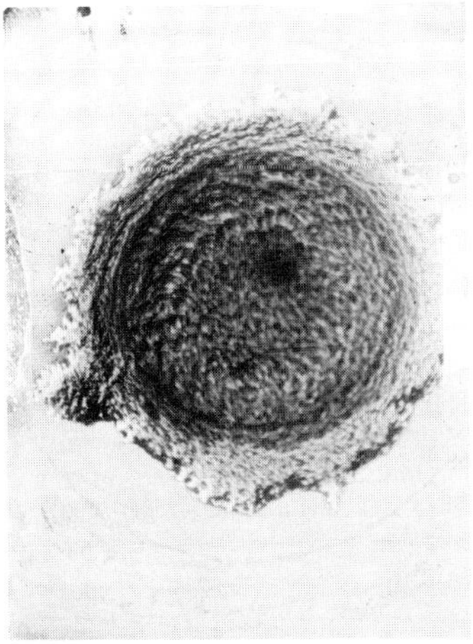

Fig. 41.15 *B. subtilis*. Same colony as in Fig. 41.14, after 48 hours' incubation, showing irregularity of surface and edge, and raised peripheral ring (×8).

Synonyms and Variants. *B. aterrimus*, *B. globigii*, *B. mesentericus*, *B. panis*, *B. vulgatus*.

Found in hay, dust, milk, soil, water. Smallish rods, 1.5–3 µm × 0.5–0.8 µm. Ovoid spores 1–1.5 µm × 0.6–0.9 µm, equatorial or subterminal, not distending the cell, germinating by lateral emergence of bacillary rod, with splitting of spore case. Gram positive. Motile by peritrichate flagella. Variants may be capsulated and non-motile. On agar colonies are round, irregular, spreading, 3.5 mm in diameter, slightly raised, with a darker centre and a lighter periphery; surface finely granular. In gelatin stab: slight growth down stab, with stratiform or saccate liquefaction, and often a tough wrinkled pellicle. Liquefies serum. Some strains give β-haemolysis on blood agar and form a soluble haemolysin (Büsing 1950). Acid formation in glucose, maltose, mannitol, sucrose, salicin, xylose, and arabinose. Litmus milk decolorized with rapid digestion of casein. Optimum temperature for growth 37°; limits 12–55°. Catalase and VP positive. Grows in 7 per cent NaCl, hydrolyses starch, uses citrate as sole source of carbon, reduces nitrate to nitrite, decomposes casein. Fails to grow anaerobically, to hydrolyse hippurate, to use propionate, to decompose tyrosine, or to give the egg-yolk reaction (Gordon *et al.* 1973). Antigenically, heat-labile flagellar and heat-stable somatic antigens, both different from spore antigen, which is species-specific. Capsular material is a mixture of D- and L-glutamic acid. G + C content of DNA 43–46 moles per cent. May give rise to conjunctivitis, iridochoroiditis and panophthalmitis in man, and occasionally invades the blood stream in cachectic diseases. Fatal to mice injected when intraperitoneally. Produces haemorrhagic necrotic lesions in the skin of rabbits (Evans and Wardlaw 1952).

The following two organisms, *B. licheniformis* and *B. pumilus*, together with *B. amyloliquefaciens*, which we do not describe, are separable from *B. subtilis* by pyrolysis gas-liquid chromatography and by DNA-DNA hybridization, and are considered by O'Donnell and his colleagues (1980) as different species, indistinguishable, however, by the usual biochemical tests.

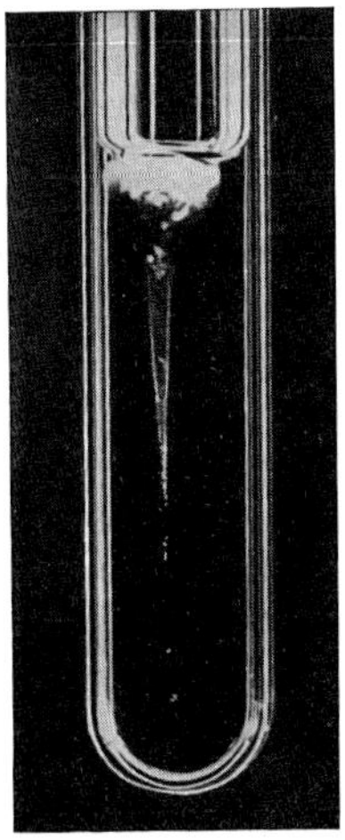

Fig. 41.16 *B. subtilis*. In stab gelatin, 4 days, 20°, showing napiform liquefaction.

Bacillus licheniformis

This organism, together with *B. pumilus*, forms part of the *B. subtilis* group. Its distinguishing characters are not altogether clear. Differences in colonial appearance noted by Gibson (1944) and the mode of germination by Burdon (1956) were found by Gordon, Haynes and Pang (1973) to be unreliable criteria. On the other hand, anaerobic production of gas from glucose and from nitrate, and possibly the use of propionate as a source of carbon, are regarded as more constant by Gordon and her colleagues (1973).

The general properties of *B. licheniformis* are as follows. Small rods 1.5–3.0 μm by 0.6–0.8 μm. Motile by peritrichate flagella. Spores ellipsoidal or cylindrical, not swelling the sporangium. Gram positive. Maximum temperature of growth 50–55°, minimum 15°. Grows anaerobically with production of gas from glucose and nitrate; catalase positive; VP positive; grows in 7 per cent NaCl; produces acid from glucose, arabinose, xylose, and mannitol; hydrolyses starch; uses propionate as a source of carbon; reduces nitrate to nitrite; forms alkali in litmus milk; and decomposes casein. On the negative side are the yolk reaction, resistance to lysozyme, hydrolysis of hippurate, and decomposition of tyrosine. G+C content of DNA 43–45 moles per cent. Appears to be responsible for some outbreaks of food poisoning (see Chapter 72), and is thought to be associated with erythrasma (Partridge and Jackson 1962).

Bacillus pumilus

Originally described by Gottheil (1901). Like *B. licheniformis*, this organism is a member of the *B. subtilis* group; its claim to specific rank is questionable. According to Gordon and her colleagues (1973), its properties are as follows. Rods 2–3 μm by 0.6–0.7 μm; motile and Gram positive. Spores ellipsoidal or cylindrical, central, not distending the sporangium; maximum temperature of growth 40–50°; minimum 5–15°; catalase and VP positive; grows at pH 5.2 and in 7 per cent NaCl; acidifies glucose, arabinose, xylose, and mannitol; hydrolyses hippurate; uses citrate and decomposes casein. Fails to grow anaerobically, to hydrolyse starch, to give the egg-yolk reaction, to reduce nitrate to nitrite, or to decompose tyrosine. G+C content of DNA 46.9 moles per cent. Is closely related to *B. coagulans*. Grilione and Carr (1960) describe a bacteriophage active against this organism.

Bacillus polymyxa

Synonyms: B. asterosporus, B. aerosporus.

Occurs in soil, water, faeces, sewage, and decaying vegetables. Rods 2–5 μm by 0.6–0.8 μm. Gram variable. Motile by peritrichate flagella. Spores ellipsoidal, equatorial to terminal, with heavily ribbed surface, swelling the sporangia. Thin, smooth, effuse, translucent, inconspicuous colonies, spreading over entire plate. Scanty to filiform growth in gelatin stab, with crateriform liquefaction. Fruity odour on potato. Maximum temperature of growth 35–45°, minimum 5–10°. Facultative anaerobe. Catalase and VP positive. Acid and usually gas from glucose, arabinose, xylose, mannitol, lactose, and sucrose. Nitrate reduced to nitrite. Forms a group of antibiotics—the polymyxins. Acid and usually gas in litmus milk with coagulation, decolorization, and partial peptonization. Starch hydrolysed. Indole negative. Antigenically heterogeneous. Spore antigen said to be common to all strains; somatic and flagellar antigens complex. Flagellar antigens have a type and group phase (Davies 1951). G+C content of DNA 44–48 moles per cent. Non-pathogenic to man and animals. (For differences from other species of *Bacillus*, see Table 41.3 and Gordon *et al.* 1973.)

Bacillus macerans

This organism, which was described by Schardinger (1905), is of industrial importance. It produces acetone and ethanol from carbohydrates, and is associated with the process of retting. It differs from *B. polymyxa* in growing very slowly on potato, in having a higher temperature range for growth—20° to 50°—in failing to liquefy gelatin, in not coagulating or peptonizing milk, in being VP negative, in its antigenic constitution, and in forming crystalline dextrins from starch (Porter *et al.* 1937). G+C content of DNA 54 moles per cent.

The *Bacillus circulans* group

Under this name we may include a group of organisms characterized by the formation of motile colonies on agar (see also *B. sphaericus*). *B. circulans* was described originally by Jordan in 1890 (see Ford 1916). A similar organism isolated by Cheshire and Cheyne (1885) from the hives of honey-bees affected with foulbrood was called *B. alvei*. Several other named organisms, such as *B. laterosporus* and *B. brevis*, appear to belong to the *B. circulans* species, which, like *B. cereus*, seems to be the parent of numerous variants. *B. circulans* is a rod of moderate size, 2–5 μm × 0.5–0.7 μm, slightly curved, with rounded or pointed ends, usually occurring singly. Motile. Fairly large oval spores, subterminal or terminal, causing bulging of the cell. Gram variable. Colonies on agar are effuse, transparent, spreading, and sometimes hardly visible; small rotating colonies may become detached from the edge. Gelatin is liquefied slowly or not at all. Milk is not peptonized. Acid is produced in glucose and many other sugars. VP negative. Egg-yolk reaction negative. Starch is hydrolysed. G+C content of DNA 36.7 moles per cent. *B. alvei* differs from *B. circulans* mainly in being VP positive, in giving a 'restricted' egg-yolk reaction, and in failing to attack the pentoses. The organism studied by Turner and Eales (1941) formed colonies that migrated at 37° at the rate of 2.5 mm per minute. An organism resembling *B. laterosporus*, producing a reddish-brown pigment, possessing proteolytic properties, and isolated from powdery scale disease of honey-bee larvae, was described by Katznelson (1950) under the name of *B. pulvifaciens*. G+C content of DNA 46 moles per cent.

B. pantothenicus, described by Proom and Knight (1950), resembles *B. circulans* in some ways, but does not form motile colonies, requires pantothenic acid as well as biotin and aneurin, does not ferment glucose in an inorganic medium, is proteolytic, grows better in the presence of 4 per cent NaCl, and gives a 'restricted' egg-yolk reaction. G+C content of DNA 36.7 moles per cent.

Bacillus sphaericus

This name was given by Neide (1904) to an aerobic bacillus having a round terminal spore. Morphologically, it is a rod of moderate size, 1–7 μm × 0.6–1 μm, with rounded or

Table 41.3 Some of the differential reactions of species of *Bacillus* (modified from Knight and Proom 1950)

Species	Anaerobic growth	Citrate utilization	Egg-yolk reaction	Gas (Gibson & Abd-el-Malek Test)	Glucose	Arabinose	Xylose	Starch	Casein	Gelatin	Nitrate reduction	VP	Urease production	G+C moles %	Remarks
B. anthracis	+	+	+	−	A	−	−	+	+	+	±	+	−	32–34	Pathogenic
B. cereus	+	±	+	−	A	−	−	+	+	+	±	+	∓	32–37[a]	Sometimes pathogenic
B. cereus var. *mycoides*	+	±	+	−	A	−	−	+	+	+	±	+	∓	39	
B. megaterium	−	+	−	−	A	±A	±A	+	+	+	∓	−	var.	37–38[b]	
B. subtilis	−	±	−	−	A	A	A	+	+	+	+	+	−	43[c]	
B. pumilus	−	+	sl. R	−	A	A	A	−	+	+	−	+	−	39–41[d]	
B. licheniformis	+	±	−	+	A	A	A	+	+	+	+	+	−	43–46[e]	Grows at 55°
B. polymyxa	+	−	var. R	+	AG	AG	AG	+	+	+	+	+	−	47–48	Ribbed spores
B. macerans	+	−	sl. R	±	AG	AG	AG	+	−	+	var.	−	−	50–54	Ribbed spores
B. circulans	+	∓	−	−	A	A	A	+	±	±	var.	−	∓	35	
B. alvei	+	−	+R	−	A	−	−	+	+	+	−	+	−	33	
B. laterosporus	+	−	+	−	A	A	A	−	+	+	+	−	−	40	
B. brevis	−	±	sl. R	−	±A	−	−	−	+	+	var.	−	−	43–45	
B. pantothenicus	+	?	+R	−	−	−	−	+	+	+	var.	−	−	−	Growth stimulated by 4% NaCl
B. sphaericus	−	−	−	−	−	−	−	−	−	±	−	−	−	37[f]	
B. pasteuri	−	−	−	−	−	−	−	−	−	±	+	−	+	−	
B. coagulans	+	−	−	−	−	−	−	+	−	−	−	+	−	−	Grows at 55°
B. stearothermophilus	±	−	?	−	±A	var.	var.	+	var.	+	±	+	?	44–46	Grows at 65°

Sl. = slight; var. = variable; R = restricted to medium immediately beneath growth.
a 36.8; *b* 45.0; *c* 47.5; *d* 44.6; *e* 49.6; *f* 39.4. (Figures given by Bonde and Jackson 1971, but see also those given in the text.)
Reactions in glucose, arabinose and xylose are those observed in an inorganic basal medium.

pointed ends. Motile. Spores are spherical, 0.7–1.3 μm in diameter, terminal or subterminal, causing bulging of the cell. Forms thin, smooth, translucent, spreading colonies on agar which, like those of *B. circulans*, show some degree of motility. Produces uniform turbidy in broth. Grows up to 40° or 45°. Slowly liquefies gelatin. Ferments no sugars. VP negative. Does not reduce nitrate to nitrite. Does not hydrolyse starch, peptonize milk or produce urease. Egg-yolk reaction negative. Requires amino acids, aneurin, and sometimes biotin for growth. G + C content of DNA 37 moles per cent.

Roberts (1935) described an organism closely related to *B. sphaericus* under the name of *B. rotans*. Its colonies display two sorts of co-ordinated motility—rotation and migration. Rotation may occur either clockwise or anti-clockwise, and is common in the early stages of colony formation. Later the whole colony migrates, pursuing an involved and sometimes spiral course, leaving behind it a few cells to mark its snail-like track. Unlike *Proteus*, this organism moves freely even on a dry surface. *B. rotans* grows best at 22° and does not grow above 35°. Smith, Gordon, and Clark (1952) regard it is a variant of *B. sphaericus*—*B. sphaericus* var. *rotans*. G + C content of DNA 38–40 moles per cent.

Another variant, *B. sphaericus* var. *fusiformis*, differs in producing urease.

Bacillus pasteuri

This organism does not grow on ordinary media; it requires urea or free ammonia, as well as aneurin and biotin, or nicotinic acid. Its other properties resemble those of *B. sphaericus*, though the colonies show little tendency to spread, and it reduces nitrate to nitrite (see Gibson 1935).

Bacillus coagulans

This organism was described by Hammer (1915), who isolated it from evaporated milk that had become coagulated. It has been studied particularly by Gordon and Smith (1949). Owing to the ability of most strains to grow at 55°, and of some strains to grow at 60°, it may be regarded as thermophile. Resembles *B. subtilis* morphologically, but differs in growing anaerobically, in not liquefying gelatin or digesting casein, in not growing in the presence of 5 per cent NaCl, in not utilizing citrate, and in failing to reduce nitrates, as well as in its ability to grow at higher temperatures than most strains of *B. subtilis*. G + C content of DNA 45–47 moles per cent.

Bacillus stearothermophilus

This organism tends to be slightly thicker than *B. coagulans*, measuring 2.5–3.5 μm long by 0.9–1.0 μm broad. Motile. Gram variable. Spores are oval, subterminal to terminal, and swell the sporangia. The organism grows at 65°; some strains grow at 75°. VP positive. Hydrolyses starch and gelatin. Usually reduces nitrate to nitrite. Ability to ferment sugars varies with different strains. Does not utilize citrate. Very resistant to heat. G + C content of DNA 44–52 moles per cent.

References

Adams, J. C. and Stokes, J. L. (1968) *J. Bact.* **95,** 239.
Allen, M. B. (1953) *Bact. Rev.* **17,** 125.
Aoki, K. and Yamamoto, K. (1939) *Z. ImmunForsch.* **95,** 374
Axenfeld, T. (1908) *The Bacteriology of the Eye*. London.
Balteano, L. (1922) *Ann. Inst. Pasteur* **36,** 805.
Barber, C., Zilisteane, E. and Nafta, I. (1957) *Arch. roum. Path. exp. Microbiol.* **16,** 573.
Bary, H. A. de (1884) *Vergleichende Morphologie und Biologie der Pilze Mycetozoen, und Bakterien.* Leipzig.
Basset, J. (1925) *C. R. Soc. Biol.* **93,** 1513, 1515, 1517.
Baumann-Grace, J. B., Kovács, H. and Tomcsik, J. (1958) *Schweiz. Z. allg. Path.* **22,** 158; (1959) *Ibid.* **22,** 158.
Baumann-Grace, J. B. and Tomcsik, J. (1957) *J. gen. Microbiol.* **17,** 227; (1958) *Schweiz. Z. allg. Path* **21,** 906.
Beall, F. A. and Dalldorf, F. G. (1966) *J. infect. Dis.* **116,** 377.
Beall, F. A., Taylor, M. J. and Thorne, C. B. (1962) *J. Bact.* **83,** 1274.
Behring. (1889) *Z. Hyg. InfektKr.* **7,** 171.
Berdjis, C. C. Gleiser, C. A. and Hartman, H. A. (1963) *Brit. J. exp. Path.* **44,** 101.
Bernheimer, A. W. and Grushoff, P. (1967) *J. Bact.* **93,** 1541.
Bodon, G. and Tomcsik, J. (1934) *Proc. Soc. exp. Biol., N.Y.* **32,** 122.
Bonde, G. J. (1975) *Dan. med. Bull.* **22,** 41.
Bonde, G. J. and Jackson, D. K. (1971) *J. gen. Microbiol.* **69,** Pt. 3, p. vii.
Bongert, J. (1903) *Zbl. Bakt.* **34,** 497, 623, 772.
Bonventre, P. F. (1965) *J. Bact.* **90,** 284.
Boor, A. K. (1955) *J. infect. Dis.* **97,** 194.
Bordet, J. and Renaux, E. (1930) *Ann. Inst. Pasteur* **45,** 1.
Bradley, D. E. and Franklin, J. G. (1958) *J. Bact.* **76,** 618.
Breed, R. S., Murray, E. D. G. and Smith, N. R. (1957) *Bergey's Manual of Determinative Bacteriology*, 7th edn. Baillière, Tindall and Cox, London.
Brown, E. R. *et al.* (1955) *J. Bact.* **69,** 590; (1958) *Ibid.* **75,** 499.
Brown, E. R. and Cherry, W. B. (1955) *J. infect. Dis.* **96,** 34.
Buck, C. A., Anacker, R. L., Newman, F. S. and Eisenstark, A. (1963) *J. Bact.* **85,** 1423.
Burdon, K. L. (1956) *J. Bact.* **71,** 25.
Burdon, K. L., Davis, J. S. and Wende, R. D. (1967) *J. infect. Dis.* **117,** 307.
Burdon, K. L. and Wende, R. D. (1960) *J. infect. Dis.* **107,** 224.
Büsing, K. H. (1950) *Arch. Hyg., Berl.* **133,** 63.
Candeli, A. *et al.* (1979) *Int. J. syst. Bact.* **29,** 25.
Chadwick, P. (1959) *J. gen. Microbiol.* **21,** 631.
Chao, K. C. and Williams, R. P. (1959) *Texas Rep. Biol. Med.* **17,** 197.
Charrin, A. and Roger, G. H. (1890) *Arch. Physiol. norm. path.* **22,** 273.
Chauveau, A. (1880*a*) *C. R. Acad. Sci* **91,** 33; (1880*b*) *Ibid.* **91,** 648.
Cherry, W. B. and Freeman, E. M. (1959) *Zbl. Bakt.* **175,** 582.
Cheshire, F. R. and Cheyne, W. W. (1885) *J. R. micr. Soc.*, N.S. **5,** 581.
Chu, H. P. (1949) *J. gen. Microbiol.* **3,** 255; (1952) *J. Hyg., Camb.* **50,** 433.

Ciucă, M., Zilisteanu, E. and Nafta, I. (1959) *Arch. roum. Path. exp. Microbiol.* **18,** 9.

Clark, F. E. and Smith, N. R. (1939) *J. Bact.* **37,** 277.

Cohn, F. (1875*a*) *Cohn's Beitr. Biol. Pflanz.* **1,** Heft 2, p. 175; (1875*b*) *Ibid.* **2,** Heft 3, p. 141.

Combiesco, D., Soru, E. and Stamatesco, S. (1929) *C. R. Soc. Biol.* **102,** 124.

Conn, H. J. (1930) *J. infect. Dis.* **46,** 341.

Coolbaugh, J. J. and Williams, R. P. (1978) *Canad. J. Microbiol.* **24,** 1289.

Costlow, R. D. (1958) *J. Bact.* **76,** 317.

Cowles, P. B. (1931) *J. Bact.* **21,** 161.

Cromartie, D. W. *et al.* (1947) *J. infect. Dis.* **80,** 1, 14.

Dalldorf, F. G. and Beall, F. A. (1967) *Arch. Path.* **83,** 154.

Davaine, C. (1863*a*) *C. R. Acad. Sci.* **57,** 220; (1863*b*) *Ibid.* **57,** 351; (1864) *Ibid.* **59,** 393.

Davies, S. N. (1951) *J. gen. Microbiol.* **5,** 807.

Defalle, W. (1902) *Ann. Inst. Pasteur* **16,** 756.

Doak, B. W. and Lamanna, C. (1948) *J. Bact.* **55,** 373.

Dowdle, W. R. and Hansen, P. A. (1961) *J. infect. Dis.* **108,** 125.

Dreyer, G. (1904) *Brit. med. J.* **ii,** 564.

Eckert, N. J. and Bonventre, P. F. (1963) *J. infect. Dis.* **112,** 226.

Ehrenberg. (1838) *Infusionsthierchen als vollkommene Organismen.* Leipzig.

Evans, D. G. and Shoesmith, J. G. (1954) *Lancet* **i,** 136.

Evans, D. G. and Wardlaw, A. C. (1952) *J. gen. Microbiol.* **7,** 397; (1953) *Ibid.* **8,** 481.

Fish, D. C., Mahlandt, B. G., Dobbs, J. P. and Lincoln, R. E. (1968) *J. Bact.* **95,** 907.

Flügge, C. G. F. W. (1886) *Die Mikroorganismen.* Leipzig.

Ford, W. W. (1916) *J. Bact.* **1,** 273.

Frank, G. and Lubarsch, O. (1892) *Z. Hyg. InfektKr.* **11,** 259.

Frankland, G. C. and Frankland, P. F. (1887) *Philos. Trans. B.* **178,** 257.

Franklin, J. G., Williams, D. J. and Clegg, L. F. L. (1956) *J. appl. Bact.* **19,** 46.

Gibson, T. (1935) *J. Bact.* **29,** 491; (1944) *J. Dairy Res.* **13,** 248.

Gibson, T. and Abd-el-Malek, Y. (1945) *J. Dairy Res.* **14,** 35.

Gibson, T. and Gordon, R. E. (1974) In: *Bergey's Manual of Determinative Bacteriology*, 8th edn, p. 529. Ed. by R. E. Buchanan and N. E. Gibbons. Williams and Wilkins, Baltimore.

Gibson, T. and Topping, L. E. (1938) *Proc. Soc. agric. Bact.* p. 43.

Gilbert, R. J. and Parry, J. M. (1977) *J. Hyg., Camb.* **78,** 69.

Gillissen G. and Scholz, H. G. (1961) *Zbl. Bakt.* **182,** 232.

Giovanardi, A. (1931) *Krankheitsforschung* **9,** 13.

Gladstone, G. P. (1946) *Brit. J. exp. Path.* **27,** 394; (1948) *Ibid.* **29,** 379.

Gleiser, C. A. (1967) *Fed. Proc.* **26,** 1518.

Gochenour, W. S., Gleiser, C. A. and Tigertt, W. D. (1962) *J. Hyg., Camb.* **60,** 29.

Gordon, J. and McLeod, J. W. (1928) *J. Path. Bact.* **31,** 185.

Gordon, R. E., Haynes, W. C. and Pang, C. H. N. (1973). *The genus Bacillus*, Agric. Handbk, No. 427. U.S. Dept Agric., Washington, D.C.

Gordon, R. E. and Smith, N. R. (1949) *J. Bact.* **58,** 327.

Gottheil, O. (1901) *Zbl. Bakt.*, IIte Abt. **7,** 430, 449, 529, 627, 680, 717.

Graham, N. C. (1930) *J. Path. Bact.* **33,** 665.

Grilione, P. L. and Carr, J. H. (1960) *J. Bact.* **80,** 47.

Haines, B. W., Klein, F. and Lincoln, R. E. (1965) *J. Bact.* **89,** 74.

Hammer, B. W. (1915) *Iowa agr. Exp. Sta., Res. Bull.* No. 19.

Hanby, W. E. and Rydon, H. N. (1946) *Biochem. J.* **40,** 297.

Hankin, E. H. (1889) *Brit. med. J.* **ii,** 810.

Harris-Smith, P. W., Smith, H. and Keppie, J. (1958) *J. gen. Microbiol.* **19,** 91.

Henderson, D. W., Peacock, S. and Belton, F. C. (1956) *J. Hyg., Camb.* **54,** 28.

Hill, L. R. (1966) *J. gen. Microbiol.* **44,** 419.

Holländer, R. and Pohl, S. (1980) *Zbl. Bakt., IAbt. Orig.* **A246,** 236.

Howie, J. W. and Cruickshank, J. (1940) *J. Path. Bact.* **50,** 235.

Ivánovics, G. (1937) *Zbl. Bakt.* **138,** 211, 449; (1940) *Z. ImmunForsch.* **97,** 402; (1953) *Zbl. Bakt.* **159,** 178; (1958) *Tetrahedron* **2,** 236.

Ivánovics, G. and Bruckner, V. (1937a) *Z. ImmunForsch.* **90,** 304; (1937*b*) *Ibid.* **91,** 175; (1938) *Ibid.* **93,** 119.

Ivánovics, G. and Erdös, L. (1937) *Z. ImmunForsch.* **90,** 5.

Ivánovics, G. and Földes, J. (1958) *Acta microbiol. Acad. Sci. hung.* **5,** 89.

Jensen, J. and Kleemayer, H. (1953) *Zbl. Bakt.* **159,** 494.

Kashiba, S. *et al.* (1959) *Biken's J.* **2,** 97.

Katznelson, H. (1950) *J. Bact.* **59,** 153.

Katzu, S. (1925) *Zbl. Bakt.* **94,** 165.

Keppie, J., Harris-Smith, P. W. and Smith, H. (1963) *Brit. J. exp. Path.* **44,** 446.

Keppie, J., Smith, H. and Harris-Smith, P. W. (1953) *Brit. J. exp. Path.* **34,** 486; (1955) *Ibid.* **36,** 315.

King, H. K. and Stein, J. H. (1950) *J. gen. Microbiol.* **4,** 48.

Knaysi, G. (1933) *J. Bact.* **26,** 623.

Knight, B. C. J. G. and Proom, H. (1950) *J. gen. Microbiol.* **4,** 508.

Knisely, R. F. (1966) *J. Bact.* **92,** 784.

Koch, R. (1877) *Cohn's Beitr. Biol. Pflanz.* **2,** 277.

Kramár, E. (1921) *Zbl. Bkt.* **87,** 401.

Kurzweil, H. (1954) *Z. Hyg. InfectKr.* **140,** 29.

Lamanna, C. (1940*a*) *J. Bact.* **39,** 593; (1940*b*) *Ibid.* **40,** 347; (1942) *Ibid.* **44,** 611.

Lantos, J., Varga, I. and Ivánovics, G. (1960) *Acta microbiol. Acad. Sci. hung.* **7,** 32.

Leise, J. M. *et al.* (1959) *J. Bact.* **77,** 655.

Leonard, C. G. and Thorne, C. B. (1961) *J. Immunol.* **87,** 175.

Levine, P. P. (1936) *J. Bact.* **31,** 151.

McCloy, E. W. (1951) *J. Hyg., Camb.* **49,** 114; (1958) *J. gen. Microbiol.* **18,** 198.

McDonald, W. C., Felkner, I. C., Turetsky, A. and Matney, T. S. (1963) *J. Bact.* **85,** 1071.

McGaughey, C. A. and Chu, H. P. (1948) *J. gen. Microbiol.* **2,** 334.

McGaughey, C. A. and St. George, C. (1955) *Vet. Rec.* **67,** 132.

Mahlandt, B. G., Klein, F., Lincoln, R. E., Haines, B. W., Jones, W. I. and Friedman, R. H. (1966) *J. Immunol.* **96,** 727.

Marmier, L. (1895) *Ann. Inst. Pasteur* **9,** 533.

Melles, Z., Nikodémusz, I. and Ábel, A. (1969) *Zbl. Bakt.* **212,** 174.

Melling, J. and Capel, B. J. (1978) *F.E.M.S. Microbiol. Lett.* **4,** 133.

Melling, J., Capel, B. J., Witham, M. D. and Gilbert, R. J. (1978) *J. appl. Bact.* **45,** xxv.

Mellon, R. R. and Anderson, L. M. (1919) *J. Immunol.* **4,** 203.

Meynell, E. and Meynell, G. G. (1964) *J. gen. Microbiol.* **34,** 153.

Minett, F. C. (1950) *J. comp. Path.* **60,** 161.

Morris, E. J. (1955) *J. gen. Microbiol.* **13,** 456.

Muller, L. (1925) *C. R. Soc. Biol.* **93,** 1243.

Murray, T. J. (1931) *J. infect. Dis.* **48,** 457.

Neide, E. (1904) *Zbl. Bakt.*, IIte Abt. **12,** 1, 161, 337, 539.

Nordberg, B. (1951) *Studies of Bacillus anthracis etc.* Gernandts Boktryckeri, Stockholm.

Nordberg, B. K. (1953) *Nord. vet. Med.* **5,** 915.

Nordberg, B. K., Schmitterlöw, C. G., Bergrahm, B. and Lundström, H. (1964) *Acta path. microbiol. scand.* **60,** 108.

Nordberg, B. K., Schmitterlöw, C. G. and Hansen, H. J. (1961) *Acta path. microbiol. scand.* **53,** 295.

O'Donnell, A. G. *et al.* (1980) *Int. J. syst. Bact.* **30,** 448.

Oppermann. (1906) *J. comp. Path.* **19,** 264.

Partridge, B. M. and Jackson, F. L. (1962) *Lancet* **i,** 591.

Pasteur, L. (1881*a*) *C. R. Acad. Sci.* **92,** 209; (1881*b*) *Ibid.* **92,** 429; (1881*c*) *Ibid.* **92,** 666.

Pearce, T. W. and Powell, E. O. (1951) *J. gen. Microbiol.* **5,** 387.

Pokschischewsky, N. (1914) *Arb. ReichsgesundhAmt.* **47,** 541.

Poppe, K. (1922) *Ergebn. Hyg. Bakt.* **5,** 597.

Porter, R., McCleskey, C. S. and Levine, M. (1937) *J. Bact.* **33,** 163.

Preisz, H. (1911) *Zbl. Bakt.* **58,** 510.

Proom, H. and Knight, B. C. J. G. (1950) *J. gen. Microbiol.* **4,** 539; (1955) *Ibid.* **13,** 474.

Record, B. R. and Wallis, R. G. (1956*a*) *Biochem. J.* **63,** 443; (1956*b*) *Ibid.* **63,** 453.

Rettger, L. F. and Gillespie, H. B. (1935) *J. Bact.* **30,** 213.

Roberts, J. L. (1935) *J. Bact.* **29,** 229.

Roth, N. A., Dearman, N. A. and Lively, O. H. (1956) *J. Bact.* **72,** 666.

Roux, E. (1890) *Ann. Inst. Pasteur* **4,** 25.

St. John-Brooks, R. and Breed, R. S. (1937) *J. Bact.* **33,** 445.

Sanarelli, G. (1925) *Ann. Inst. Pasteur* **39,** 209.

Schardinger, F. (1905) *Zbl. Bakt.* Abt. 2, **14,** 772.

Schneiter, R. and Kolb, R. W. (1948) *Publ. Hlth Rep., Wash.* Suppl. No. 207.

Schockaert, J. (1929) *Arch. int. Méd. exp.* **5,** 155.

Schulz, R. (1901) *Zbl. Bakt.* **30,** 582.

Seidel, G. (1962) *Beit. Hyg. Epidemiol.* **17,** 1–131.

Seki, T. *et al.* (1978) *Int. J. syst. Bact.* **28,** 182.

Senge, J. (1913) *Zbl. Bakt.* **70,** 353.

Sievers, O. and Zetterberg, B. (1940) *J. Bact.* **40,** 45.

Slein, M. W. and Logan, G. F. (1960) *J. Bact.* **80,** 77; (1965) *Ibid.* **90,** 69.

Smith, H., Keppie, J. and Stanley, J. L. (1955*b*) *Brit. J. exp. Path.* **36,** 460.

Smith, H. and Stanley, J. L. (1962) *J. gen. Microbiol.* **29,** 517.

Smith, H. and Stoner, H. B. (1967) *Fed. Proc.* **26,** 1554.

Smith, H., Strange, R. E. and Zwartouw, H. T. (1956*a*) *Nature, Lond.* **178,** 865.

Smith, H., Zwartouw, H. T. and Harris-Smith, P. W. (1956*b*) *Brit. J. exp. Path.* **37,** 361.

Smith, H. *et al.* (1955*a*) *Brit. J. exp. Path.* **36,** 323; (1956*a*) *Ibid.* **37,** 263.

Smith, N. R., Gordon, R. E. and Clarke, F. E. (1946) *Aerobic mesophilic sporeforming bacteria. Misc. Publ.*, No. 559, U.S. Dept. Agric., Wash. D.C.; (1952) *Aerobic Sporeforming Bacteria.* Agric. Monograph No. 16. U.S. Dept. Agriculture.

Sneath, P. H. A. (1962) *Symp. Soc. gen. Microbiol.* **12,** 289.

Sobernheim, G. (1897) *Z. Hyg. InfektKr.* **25,** 301; (1913) In: *Kolle und Wassermann's Handbuch der pathogenen Mikroorganismen*, 2te Aufl., Bd **3,** p. 583.

Sobernheim, G. and Murata, H. (1924) *Z. Hyg. InfektKr.* **103,** 691.

Soltys, M. A. (1948) *J. Path. Bact.* **60,** 253.

Sordelli, A. and Deulofeu, V. (1930) *C. R. Soc. Biol.* **105,** 721; (1933) *Folia biol.* No. 26–27, p. 121.

Sordelli, A., Deulofeu, V. and Ferrari, J. (1932) *Folia biol.* No. 11, p. 45; No. 20, pp. 93 and 94.

Soriano, A. M. de (1935) *Rev. Inst. bact., B. Aires* **6,** 507.

Soule, M. H. (1928) *J. infect. Dis.* **42,** 93; (1932) *Ibid.* **51,** 191.

Spira, W. M. and Goeppert, J. M. (1975) *Canad. J. Microbiol.* **21,** 1236.

Spira, W. M. and Silverman, G. J. (1979) *Appl. environm. Microbiol.* **37,** 109.

Stamatin, N. and Cristescu, P. (1958) *Ann. Inst. Pasteur* **94,** 243.

Standfusz, R. and Schnauder, F. (1925) *Zbl. Bakt.* **95,** 61.

Stanley, J. L. and Smith, H. (1961) *J. gen. Microbiol.* **26,** 49.

Stein, C. D. (1943) *Vet. Med.* **38,** 130; (1944) *Amer. J. vet. Res.* **5,** 38.

Stein, C. D. and Rogers, H. (1945) *Vet. Med.* **40,** 406.

Sterne, M. (1937*a*) *Onderstepoort J. vet. Sci.* **8,** 271; (1937*b*) *Ibid.* **9,** 49.

Sterne, M. and Proom, H. (1957) *J. Bact.* **74,** 541.

Strange, R. E. and Belton, F. C. (1954) *Brit. J. exp. Path.* **35,** 153.

Strange, R. E. and Thorne, C. B. (1958) *J. Bact.* **76,** 192.

Sweany, H. C. and Pinner, M. (1925) *J. infect. Dis.* **37,** 340.

Taylor, A. J. and Gilbert, R. J. (1975) *J. med. Microbiol.* **8,** 543.

Thorne, C. B. and Leonard, C. G. (1958) *J. biol. Chem.* **233,** 1109.

Thorne, C. B., Molnar, D. M. and Strange, R. E. (1960) *J. Bact.* **79,** 450.

Todd, C. (1901) *Lancet* **ii,** 1663; (1902) *Trans. path. Soc., Lond.* **53,** 196.

Tomcsik, J. (1930) *Z. Hyg. InfektKr.* **111,** 119.

Tomcsik, J. and Baumann-Grace, J. B. (1959*a*) *J. gen. Microbiol.* **21,** 666; (1959*b*) *Schweiz. Z. allg. Path.* **22,** 144; (1959*c*) *Proc. Soc. exp. Biol., N.Y.* **101,** 570.

Tomcsik, J. and Bodon, G. (1934) *Z. ImmunForsch.* **83,** 426; (1935) *Ibid.* **84,** 308.

Tomcsik, J. and Ivánovics, G. (1938) *Z. ImmunForsch.* **93,** 196.

Tomcsik, J. and Szongott, H. (1932) *Z. ImmunForsch.* **76,** 214; (1933) *Ibid.* **78,** 86.

Turnbull, P. C. B. *et al.* (1979) *J. clin. Path.* **32,** 289.

Turner, A. W. and Eales, C. E. (1941) *Aust. J. exp. Biol.* **19,** 161.

Utsumi, S. *et al.* (1959) *Biken's J.* **2,** 165.

Vick, J. A., Lincoln, R. E., Klein, F., Mahlandt, B. G., Walker, J. S. and Fish, D. C. (1968) *J. infect. Dis.* **118,** 85.

Ward, M. K., McGann, V. G., Hogge, A. L., Huff, M. L., Kanode, R. G. and Roberts, E. O. (1965) *J. infect. Dis.* **115,** 59.

Warden, C. C., Connell, J. T. and Holly, L. E. (1921) *J. Bact.* **6,** 103.
Watson, D. W., *et al.* (1947*a*) *J. infect. Dis.* **80,** 28; (1947*b*) *Ibid.* **80,** 121.
White, P. J. (1972) *J. gen. Microbiol.* **71,** 505.
Wieland, T. *et al.* (1960) *Arch. Mikrobiol.* **35,** 415.
Wilamowski, B. I. (1912) *Zbl. Bakt.* **66,** 39.
Wilkie, M. H. and Ward, M. K. (1967) *Fed. Proc.* **26,** 1527.
Wilson, G. S. and Shipp, H. L. (1939) *Historic Tinned Foods,* Publ. No. 85, 2nd ed., p. 49. Int. Tin Res. Develop. Coun., Engl.
Wilson, J. B. and Russell, K. E. (1964) *J. Bact.* **87,** 237.
Wright, G. C., Hedberg, M. A. and Feinberg, R. J. (1951) *J. exp. Med.* **93,** 523.
Wright, G. C., Hedberg, M. A. and Slein, J. B. (1954) *J. Immunol.* **72,** 263.
Wright, G. C., Puziss, M. and Neely, W. B. (1962) *J. Bact.* **83,** 515.
Young, G. A., Zelle, M. R. and Lincoln, R. E. (1946) *J. infect. Dis.* **79,** 233.

42

Clostridium: the spore-bearing anaerobes

A. Trevor Willis

Introductory

Definition

Anaerobic or aerotolerant rods, producing endospores, which are usually wider than the vegetative organisms in which they arise—so-called clostridium forms. Generally gram positive. Often vigorously decompose proteins and often ferment carbohydrates. Several species form exotoxins, and several are pathogenic. G+C content of DNA: 22–32 moles per cent.

Type species is *Clostridium butyricum*.

Before the war of 1914–18, the study of the spore-bearing anaerobes had been undertaken fitfully and by imperfect methods, with much attention to their pathogenicity, but little to their general biology. The same organism often received many names, and many organisms of the same name undoubtedly belonged to different species. The only indubitable organisms were the two that formed a highly potent toxin, recognizable by the effects they produced in animals—namely *Cl. tetani* and *Cl. botulinum*. It was not till the exigencies of war rendered an intensive study of the anaerobes necessary, and till the introduction of McIntosh and Fildes's jar (1916) made it relatively easy to obtain pure cultures, that the obscurity surrounding this group was dispersed.

Most of the earlier workers had failed to realize the difficulty of obtaining pure cultures of the anaerobic bacilli. Plate cultures by the new technique revealed at once the impurity of many of the classical strains, and facilitated single-colony subcultures. For the first time a distinctive account was provided of the main species; and fresh species were discovered. (For references on the production of anaerobiosis see the earlier editions of this book, Chapter 20 in the present edition, and Willis 1977, and to multiple articles on the isolation of anaerobes from various human, animal, and vegetable sources see Shapton and Board 1971.)

Habitat

The anaerobes are widely distributed in nature, but their main habitat is undoubtedly the soil. Some of them appear to be common inhabitants of the intestinal canal of man and animals. *Cl. perfringens*, for example, is uniformly present in human faeces and *Cl. difficile* is commonly present in the faeces of breast-fed infants. In domestic animals *Cl. tetani* is found in 10 to 40 per cent of faecal specimens, *Cl. sporogenes* often, and *Cl. histolyticum* occasionally. It has been held by some that the intestinal canal is the main habitat of certain of the anaerobes and that their presence in the soil is due to faecal contamination. It is more likely that the primary habitat of most anaerobes is the soil; that they are ingested frequently with vegetable foods; and that some of them have adapted themselves temporarily or permanently to a life in the intestinal canal. They frequently appear in dust, milk and sewage. Though they usually lead a saprophytic existence, several species are causally related to well recognized diseases in man and animals. It is with the pathogenic species, and those found in close association with man and animals, that we are mainly concerned in this chapter.

Morphology

Many of the clostridia are so pleomorphic that their identification on a morphological basis is difficult. The morphology of a strain varies widely, both within a culture, and from culture to culture.

Like the aerobic spore-bearing bacilli, they are large, rod-shaped organisms. In length they range from about 3 μm to 7 or 8 μm, but long filamentous forms are quite common. Their breadth ranges from about 0.4 to 1.2 μm. The vegetative bacilli are straight or curved, their sides parallel, and their ends rounded or somewhat truncated. Most are arranged singly, but some occur in pairs or in chains, others in bundles the members of which are arranged parallel to each other. Irregular forms include navicular, or boat-shaped, organisms; citron forms shaped like a lemon with a small knob at each end; large, swollen, non-sporing rods; snake-like filaments; deeply stained bulb-shaped types; and a great variety of so-called involution forms varying both in shape and in depth of staining. Autolysis frequently sets in when sporulation begins so that shadow forms are numerous particularly in certain species. L-forms occur in clostridia, including *Cl. tetani* (Scheibel and Assandri 1959, Gonzalez 1961) and *Cl. perfringens* (Kawatomari 1958, Mahony 1977).

Sporulation is common to all members, but there is considerable variation in the readiness with which it occurs. *Cl. sporogenes*, for example, spores readily on all media; *Cl. perfringens* only in particular media (see Bethge 1947, Labbe and Duncan 1975), and then inconstantly. All the pathogenic members are able to form spores in the animal body, though *Cl. perfringens* does so rarely.

The anaerobes are classifiable according to the shape of the spore and its position in the rod: those with (*a*) an equatorial or subterminal spore; (*b*) an oval terminal spore; and (*c*) a spherical terminal spore. The division is useful, but must not be used too rigidly. It is common, for instance, to find organisms that usually form subterminal spores giving rise to strictly terminal spores; and it may be hard to distinguish a spherical from an oval subterminal spore.

For encouraging sporulation in the laboratory, particularly of *Cl. perfringens*, the media of Ellner

(1956), of Duncan and Strong (1968) and of Sacks and Thompson (1978) are recommended. The germination of spores depends on a number of factors. For example, with *Cl. perfringens* Ahmed and Walker (1971) found that it was optimal at pH 6.0 at a temperature of 30°, and required preliminary heat activation at 75° for 20 min.

The spores of most members are wider than the vegetative bacilli; they therefore confer on the organism a distinctive appearance according to the position in which they arise. When they are formed at the equator the clostridium is spindle-shaped; when subterminally, club-shaped; with an oval terminal spore the organism may look like a tennis racket; with a spherical terminal spore like a drum-stick.

With the exception of *Cl. perfringens* and a few saprophytic species, all the members are motile, by peritrichate flagella. Motility, however, is often difficult to demonstrate, especially in artificial cultures and in strains that have been subcultured for some time. Young cultures in fluid medium, not more than 6 to 24 hr old, are the most suitable for examination. Examinations are best made in a closed capillary tube filled with a young broth culture and kept at 37° for about half an hour. Motility is rarely obvious, and is usually slow and stately.

Cl. butyricum and *Cl. perfringens* are the only members with a capsule; that of *Cl. perfringens* is noticeable in the animal body, and sometimes in cultures containing serum.

Staining reactions

All members stain readily with the usual dyes. Great irregularity is noticeable in the depth of staining, especially in cultures more than a day or two old. Sometimes metachromatic granules or points of more intense coloration are seen, especially when acetone is used for decolorization. In young cultures, the bacilli are all gram positive. Some species rapidly lose this property, and some are readily decolorized by the ethanol. In the early stages of spore formation, the position of the spore is often marked by an area of intense staining; as it matures, however, the spore presents a colourless centre surrounded by a peripherally stained ring.

Cultural appearances

On solid media growth is relatively slow, and sometimes takes the form of a thin, effuse, often spreading film, which may be difficult to distinguish from the underlying medium.

Film formation is promoted by moisture. On first isolation *Cl. septicum*, and particularly *Cl. tetani*, tend to spread rapidly over a moist surface. *Cl. tetani* inoculated into the condensation water of an agar slope, or laterally on a blood agar plate, in the course of a day spreads over the whole medium; the film is so thin that were it not for its dentate spreading edge it might easily escape detection. Advantage may be taken of this property in the isolation of the organism (Fildes 1925*a*). The spreading of clostridia is inhibited by certain chemicals; most of them, however, are to some extent bacteristatic; inhibition of spreading without bacteristasis may be effected by increasing the concentration of agar up to about 3 per cent (Miles and Hayward 1943), or by the addition of agglutinating antibody to the medium (Williams and Willis 1970). Certain clostridia also produce on agar motile daughter colonies, which rotate and wander over the surface of the medium (Turner and Eales 1941). Concentrated agar is less effective as an inhibitor of this type of spread.

Agar plates Single colonies are rounded, generally effuse, and have crenated, fimbriate, or rhizoid edges. *Cl. perfringens*, which is one of the less strict anaerobes, forms low convex colonies with an entire edge; *Cl. sporogenes* and *Cl. histolyticum* may form umbonate colonies with a raised centre and a flat periphery. The colonial appearances are often characteristic, but some species give rise to variants which not only are unlike the typical colony, but which strongly suggest the occurrence of contamination; and the colonies of some aerobic spore-bearing bacilli, growing anaerobically, simulate those of clostridia. Several different types of colony may be formed, for example, by *Cl. sporogenes*.

Glucose agar shake cultures Except near the surface, growth occurs throughout the medium; this is frequently disrupted and blown upwards by the development of gas. Single colonies are rounded or lenticular, and lenticular forms may later form irregular tufts at the edge or the poles of the lenses; sometimes they are differentiated into an opaque centre and a translucent periphery; their edge may be entire, but is more often woolly, erose, or has the curious reticular appearance of a cigarette thrown into water. There is a general correspondence between the form of surface and deep colonies. Thus round, entire-edged and raised surface colonies usually correspond to opaque and lenticular colonies in deep agar; irregular or coarsely rhizoidal to opaque and lumpy; delicately rhizoidal to fluffy; spreading colonies to deep colonies like a snowflake; and surface colonies with central papillae to deep colonies with a central opacity.

Deep cultures in soft agar Most clostridia grow in media with enough agar to prevent gross convection movements in the liquid. The medium devised by Brewer (1940) containing thioglycollic acid is one of this type. Separate diffuse 'colonies' may be found, but these readily become confluent, especially with motile forms. Mandia (1950) observed 'motile' colonies of the non-motile *Cl. perfringens* in these circumstances.

Blood agar plates On these, both colonial form and the degree and type of haemolysis may be characteristic. Haemolysis is well developed after 3 days' incubation at 37°; when the plates are stored in a dark cupboard at room temperature it often continues to increase. With a thick seeding the whole plate may be completely decolorized. The nature and extent of haemolysis produced by different organisms may vary according to the species of erythrocyte used (see Willis 1969*b*).

Many organisms give α-prime haemolysis after 3 days'

incubation (see Chapter 29); after a further 3 days this passes into the fully developed β variety. In some cases, it is possible to specify the haemolytic factors concerned. For instance, the relatively wide zone of haemolysis produced by toxigenic strains of *Cl. perfringens* on routine horse blood agar is usually due to the θ toxin (p. 453); when the action of θ toxin is suppressed by θ antitoxin, a narrower, ill defined zone of partial haemolysis is revealed, due to α toxin. On horse blood agar containing added calcium chloride (0.5%) enhancement of α toxin activity leads to the characteristic appearance of 'target' haemolysis (Evans 1945).

Human serum agar and egg-yolk agar These media are very useful for the detection of the lecithinases and lipolytic enzymes that characterize certain pathogenic clostridia. For the distinction of the different enzymes, part of the medium may be treated with antisera that specifically neutralize the enzymic effect (Nagler 1939, Macfarlane *et al.* 1941, Willis and Hobbs 1958, 1959, Willis 1960, 1977, Willis and Gowland 1962; see also Sebald and Prévot 1960).

Cooked meat medium Most of the members grow well in this medium. All render the fluid turbid and most produce gas. The proteolytic members turn the meat black and may obviously digest it; the saccharolytic members do not digest the meat, and frequently turn it pink. The proteolytic members form characteristic foul and pervasive odours, whereas with the saccharolytic members the odour is not foul, or is undetectable.

Coagulated serum and coagulated egg These media are used for testing the proteolytic powers. None of the saccharolytic organisms is able to liquefy them.

Gelatin At 23° most members grow poorly. At 37° growth is better and is generally accompanied by gelatinolysis.

(For formulae of media for clostridia see Spray 1936, and Willis 1977; and for an illustrated description of the morphological and colonial appearance of the pathogenic anaerobes see Batty and Walker 1966.)

Resistance

In the sporing stage all the members have a pronounced but variable resistance to heat, drying and disinfectants. In the vegetative state, clostridia are about as resistant to heat and disinfectants as non-sporing aerobes, but Savolainen (1948) records a slightly greater susceptibility of anaerobes to a large number of organic bactericides. The spores of *Cl. botulinum* may withstand boiling for 3 or 4 hr, and even at 105° may not be killed completely in less than 100 min. *Cl. novyi* is a little less resistant than *Cl. botulinium* (Hoyt *et al.* 1938). Spores of most strains of *Cl. perfringens* are destroyed by boiling in less than 5 min (Headlee 1931), but those of the food-poisoning type are usually more heat resistant (see Hobbs 1965). *Cl. sporogenes* can survive exposure for 8 days to a 5 per cent phenol solution. Among hospital disinfectants the greatest sporicidal activity is shown by alcoholic hypochlorite and glutaraldehyde; iodophors, formalin and phenolics are less effective (Kelsey *et al.* 1974). In dried earth or dust *Cl. tetani* may live for years. Stock cultures of most members in cooked meat medium remain viable for months; some, such as *Cl. fallax* and *Cl. cochlearium*, are more delicate and require transferring frequently.

In-vitro susceptibility to chemotherapeutic agents

The clostridia associated with gas gangrene are susceptible to the in-vitro action of the sulphonamides; sulphathiazole is one of the most effective drugs, and *Cl. novyi* one of the least susceptible of the organisms. These clostridia are susceptible to penicillin, especially *Cl. perfringens* and *Cl. septicum* (Abraham *et al.* 1941, McKee *et al.* 1943, Nagler 1945*b*). In general, though with some exceptions, penicillin, metronidazole, clindamycin and the tetracyclines are most effective, chloramphenicol rather less and the aminoglycosides much less so (Willich 1952, Andersen *et al.* 1953, Lennert-Petersen 1954, Lavergne *et al.* 1956, Bittner *et al.* 1961). Both metronidazole and the aminoglycosides are of special interest in relation to anaerobes. Metronidazole is bactericidal for all anaerobic microorganisms, but is inactive against aerobes and facultative anaerobes. The aminoglycosides, on the other hand, show broad-spectrum activity against aerobes and facultative anaerobes, but are inactive against anaerobes. Under highly reduced conditions the activity of aminoglycosides is greatly diminished or abolished (Gillespie and Guy 1956, Freeman *et al.* 1968, Kislak 1972, Martin *et al.* 1972, Savage 1974, Dornbusch *et al.* 1975). Selective media for the isolation of clostridia are based on the relative lack of inhibitory action of polymyxin B (Hirsch and Grinsted 1954), neomycin (Lowbury and Lilly 1955, Wetzler *et al.* 1956, Willis, 1957), crystal violet (Narayan 1966) and phenylethyl alcohol (Dowell *et al.* 1964).

Metabolism

It was originally believed that clostridia were unable to grow except when oxygen was rigidly excluded from the medium. Though free oxygen inhibits their growth, and may kill organisms in the non-sporing state, anaerobic bacteria can be grown in the presence of air provided a sufficiently low oxidation-reduction potential is established in the medium. This can be done by reducing substances, some of which act mainly by absorbing oxygen, others establishing a low Eh after the molecular oxygen has been nearly used up or removed by mechanical means. Sulphites, reduced iron compounds, unsaturated fatty acids, alkaline glucose, cysteine, glutathione, ascorbic acid, thioglycollic acid and metallic iron are examples of some of the substances commonly used (see, e.g. Linzenmeier 1959). Cooked meat medium affords excellent conditions for anaerobic growth even when incubated aerobically. Its virtue lies in its containing (1) unsaturated fatty acids, which take up oxygen, the reaction being catalysed by the haematin of the muscle, and (2) —SH compounds, which bring about a negative O-R potential corresponding to an Eh of about −0.2 volt (Lepper and Martin 1929, 1930). For the germination of tetanus spores an Eh in the medium approximating to +0.01 volt at pH 7.0 is required (Fildes 1929). It is probable that similar conditions determine the growth of most other anaerobes. There is also evidence that in the presence of oxygen, clostridia produce peroxides, which, because these organisms lack catalases and peroxidases, accumulate to lethal concentrations (McLeod and Gordon 1923, 1925, Gordon *et al.* 1953). Strong preparations of catalase promote growth in the presence of air (Holman 1955). *Cl. histolyticum*, *C. tertium* and *Cl. carnis*, however, are exceptional in growing to a limited extent aerobically, though they are said to form no spores under these conditions (Hall and Duffett 1935).

Substantial protection against oxygen toxicity is afforded to aerobes and facultative anaerobes by their possession of superoxide dismutase. Since this enzyme was not found in obligate anaerobes by McCord and his colleagues (1971), it appeared that the somewhat arbitrary distinction between obligate and facultative anaerobes could be more clearly defined in biochemical terms. It is now clear, however, that many organisms may possess substantial superoxide dismutase activity and yet be incapable of growth in air (Hewitt and Morris 1975, Carlsson *et al.* 1977). Tally and his associates (1977) related the degree of superoxide dismutase activity to oxygen tolerance; aerotolerant organisms such as *Cl. perfringens* contain superoxide dismutase, whereas highly oxygen-sensitive organisms such as *Cl. haemolyticum* have little if any of the enzyme. Thus, it is possible that variation in oxygen tolerence among anaerobes may be related to their content of superoxide dismutase, but further clarification is required. Even obligate anaerobes display a spectrum of oxygen intolerance ranging from those bacteria for which oxygen is apparently bactericidal at very low concentrations to those that can tolerate limited exposure to air which is reversibly bacteristatic rather than bactericidal (O'Brien and Morris 1971, Morris 1976, Rolfe *et al.* 1978).

Once growth has started, most anaerobic organisms appear to bring about a rapid fall in the O-R potential of the medium, probably owing to the production of a more active reducing system than that present in the medium itself. The Eh frequently falls to below −0.4 volt. Although there are upper limiting Eh values for growth (see, e.g. Hanke and Bailey 1945), within these limits the low Eh is probably not a condition for growth, but the result of an accumulation of reducing substances by the metabolizing bacteria, to the point where multiplication can take place. A reducing substance added to the medium probably protects the young cells from molecular oxygen until they have established their own mechanisms for dealing with it (Grunberg 1948). For surface growth an agar medium made with heart infusion broth, yeast extract, liver digest, with or without 5 per cent of blood, may be recommended (Khairat 1966).

We have discussed the nutritive requirements of certain clostridia and the problems of anaerobiosis in Chapter 3. The earlier work (Fildes 1935, Fildes and Knight 1933, Knight and Fildes 1933, Fildes and Richardson 1935, Pappenheimer 1935, Stickland 1934, 1935, Knight 1936) on essential nutrients of clostridia and their use has developed to the point where it is clear that most pathogenic clostridia are heterotrophs, requiring a battery of amino acids, carbohydrates, and vitamins for growth in artificial media. Essential nutrients have been defined, in some cases completely; as, for example, with strains of *Cl. tetani* (Mueller and Miller 1954, 1955, 1956, Kaufman and Humphries 1958), *Cl. perfringens* (Boyd *et al.* 1948, Fuchs and Bonde 1957, Roberts 1957), *Cl. botulinum* (Kindler *et al.* 1956*a*, Holdeman and Smith 1965), *Cl. butyricum* (Cummins and Johnson 1971), *Cl. septicum* (Bernheimer 1944) and *Cl. bifermentans* (Smith and Douglas 1950). The energy-producing mechanisms, especially of those clostridia that depend mainly on amino-acid breakdown for their energy, have been studied in some detail (see, for example, Gale 1940, Woods and Trim 1942, Clifton 1942, Guggenheim 1944). Some clostridia obtain all their energy by the fermentation of a single amino acid (Cardon and Barker 1946). A small concentration of CO_2 is essential for the growth of some anaerobic as it is for so many of the aerobic bacteria (Gladstone *et al.* 1935). In addition, the growth of some clostridia is greatly improved by a concentration of CO_2 of the order of 2–10 per cent (Aitken *et al.* 1936, Watt 1973, Reilly 1980). Some organisms, such as *Cl. perfringens*, are indifferent to CO_2.

Glucose promotes the growth of the saccharolytic species; blood or serum that of all. The optimum H-ion concentration for growth is about pH 7.0 to 7.4 (Reddish and Rettger 1924).

Most of the members with which we are dealing here grow best at about 37°, though many of them are capable of growing at temperatures of 20° and even lower. There is a group of thermophilic clostridia which have an optimum temperature about 50°–60°, and which sometimes do not grow at all below 30°.

Biochemical reactions

The action on sugars constitutes one basis of classification. The results are often irregular and must be repeated two or three times before they can be relied on. Some clostridia decolorize indicators irreversibly, so that the formation of acid in a fermentation tube should always be tested by the addition of fresh indicator to a sample of the culture.

Both indole formation from tryptophan, and the reduction of nitrate to nitrite, depend on the relative rates of breakdown of the original substrates, and of substances formed from them. Thus, most clostridia reduce both nitrate and nitrite, and only when the reduction of nitrate is the quicker of the two processes will a positive test for nitrite be obtained (Reed 1942,

Prévot and Enescu 1946). The addition of vanillin in ethanol, and then of acid, to peptone water cultures of *Cl. botulinum* types A and B and of *Cl. sporogenes* results in a characteristic deep violet colour, due perhaps to skatol-like compounds (Spray 1936).

The clostridia break down glucose by the fermentative method (Hugh and Leifson 1953). Many of them produce large amounts of gas, even in media free from fermentable carbohydrates (Wolf and Harris 1917). The CO_2: H_2 ratio is high with proteolytic and low with saccharolytic clostridia (Anderson 1924). The gases are formed from a wide variety of substrates. Volatile acids are produced from carbohydrates and proteins; the type of acid differs with the predominant fermentative power of the species, and in some cases, with the conditions of culture. Pappenheimer and Shaskan 1944) record in *Cl. perfringens* a change from an aceto-butyric to a lactic fermentation, induced by lowering the concentration of iron in the medium. Identification of different species by the patterns of their volatile fatty acids was first suggested by Prévot (1957) and Beerens *et al.* (1962). This new approach to bacterial taxonomy was facilitated by the development of gas-liquid chromatography, and the technique is now widely used in the study of anaerobes (Moore *et al.* 1966, Moore 1970, Moore and Holdeman, 1972).

Litmus milk was once used for differentiation (see Wolf and Harris 1917, Report 1919, Anderson 1924). Spray (1936) introduced iron-litmus milk and iron-gelatin as differential media, in which reactions with the iron, notably blackening of the medium, provide several useful distinctive characters among clostridia (see Table 42.3). Willis and Hobbs (1959) recommend the use of a glucose gelatin medium to detect glucose fermentation and gelatinase activity in a single tube, and a lactose egg-yolk milk agar to provide evidence of proteolysis in addition to the Nagler and fermentation reactions. The best method for detecting gelatinase activity is that of Frazier (1926; see Chapter 20).

Adamson (1919–20) states that none of the anaerobes forms catalase, an observation consistent with the demonstrable absence of any of the cytochrome enzymes in the clostridia (Chapter 3); but the statement may need modification, because, as we have ourselves noted, positive 'catalase' tests are obtainable with *Cl. sporogenes* and *Cl. histolyticum*. None of the clostridia forms oxidase.

For the most part the toxic and haemolytic substances synthesized by pathogenic clostridia are large-molecular and antigenic, and are discussed below. The pathogenic significance of small-molecular products has received little study; but *Cl. perfringens* type C, associated with necrotic enteritis in man, produces from histidine large amounts of histamine, which may play a part in the pathogenesis of the lesions in the natural disease (Koslowski *et al.* 1951). Fredette and Frappier (1946) and Chagnon and Fredette (1960) describe in *Cl. perfringens* a large-molecular substance with some of the properties of adrenaline that promotes the growth and the in-vivo invasive capacity of the bacterium.

Antigenic structure

Two groups of clostridial antigens have been studied in detail: the antigens of the flagella and the bacterial bodies, and those, toxic or other, found in filtrates of cultures. The second group, though originally examined for their pathogenic significance, are, like the first, useful in determining the relationships among clostridia.

Agglutination reactions are possible, but in some cases difficult because the bacterial suspensions are auto-agglutinable. Like many aerobic bacteria, the motile species of clostridia have thermolabile flagellar (H) antigens and thermostable somatic (O) antigens (Felix and Robertson 1928). Types and groups among strains and species of clostridia are variously identifiable, sometimes by the H, and sometimes by the O, antigens. Spore antigens are distinct from H and O antigens.

Henderson (1934) divided *Cl. septicum* into four O antigenic groups with subdivisions according to the H antigens. As regards *Cl. chauvoei*, there is an O antigen common to the ovine and bovine strains, but the H antigen is complex, differing according to the animal and country of origin of the strain (Roberts 1931, Henderson 1932). Antigenically, *Cl. septicum* and *Cl. chauvoei* appear to be related (Weinberg *et al.* 1929, Kreuzer 1939), and to possess many antigens in common (Katitch *et al.* 1967). Moussa (1959) divided 37 *Cl. septicum* strains into six groups, based on two O and five H antigens, and 35 *Cl. chauvoei* into two groups based on a single O antigen and two H antigens; two other *chauvoei* strains were distinct but related to *Cl. septicum*. All had a common spore antigen. *Cl. septicum* cells contain a haemagglutinin for ox, sheep and fowl RBC, which is inhibited by homologous antisera, but not by antisera to *Cl. chauvoei* (Dafaalla and Soltys 1951, Gadalla and Collee 1967).

Tulloch (1919) described 5 agglutinating types of *Cl. tetani*. Ten types, I–X, were later distinguished. All ten have a common O antigen (Batty and Walker 1964), and a second O antigen is present in toxigenic, but not in non-toxigenic types II, IV, V and IX. Except in type VI, which is a rather heterogeneous collection of non-motile strains, type-specificity depends on a flagellar antigen. There are cross-reactions between the heat-stable antigens of *Cl. tetani* and of several other clostridia (p. 466). Complement fixation reveals low-titre cross-reactions due to the common O antigen, strong cross-reactions among types II, IV, V and IX, which have the additional O antigen, and high-titre type-specific reactions due to the flagella of types I, III, VII and VIII (Gunnison 1947). Meisel and Rymkiewicz (1957) record spore antigens, distinct from those of vegetative forms, specific for each type.

Attempts to form serological groups of *Cl. perfringens* have met with varying success. Test antisera usually react fully with the homologous strain, and only with a few heterologous strains (Henriksen 1937, Duffett 1938, Orr and Reed 1940). Strains are classifiable into five main types on the basis of toxins and other biologically active antigens found in culture filtrates (see next section); and there is some evidence that the first four of these 'toxigenic' types (A, B, C and D) differ in their bacterial antigens (Kreuzer 1939), but each type is antigenically heterogeneous. Thus Henderson (1940) distinguished two kinds of somatic antigen, a heat-stable O, and a heat-labile L, antigen. He found strain-specific O antigens in type A, but no L antigen; 13 strains of type B fell into two O-antigenic groups, and into 7 L-antigenic groups; all his type C strains had a common O antigen but no L antigen; and there were various O and L antigenic groups in type D. Rodwell (1941) found a similar variety of O antigens in the four types (see also Bergmann and Rapp 1953). Type A strains associated with food poisoning (Chapter 72) are at present divisible into 75 serological types on the basis of somatic antigens that are capsular polysaccharides. This important subdivision was originally restricted to heat-resistant strains of *Cl. perfringens* (Hobbs *et al.* 1953). It subsequently became clear that heat-sensitive strains were also a frequent cause of food poisoning (Taylor and Coetzee 1966, Sutton and Hobbs 1968), so that the present 'set' of typing antisera is made up of 32 from heat-resistant strains and 50 from heat-sensitive ones (Stringer *et al.* 1980). Extracted polysaccharides are antigenically as heterogeneous as whole bacilli (Meisel 1938, Orr and Reed 1940, Svec and McCoy 1944), although among food-poisoning strains, heat-sensitive organisms are much more diverse in their surface antigens than heat-resistant ones. The reactive polysaccharides of the six types have uronic acid, glucose and galactose in common, and, though the content of other constituent sugars varies, it is not type-dependent (Meisel-Mikolajczyk 1959, 1960). The S → R variation in *Cl. perfringens* entails a loss of specific O antigen (Henderson 1940). Gürtürk (1952) records a number of antigenically distinguishable haemagglutinins for fowl RBC, distributed among strains of *Cl. perfringens*. The properties of *Cl. perfringens* haemagglutinin are reported upon by Collee (1965*a*, *b*).

The different toxigenic types of *Cl. botulinum* (p. 450) fall into three distinct serological groups on the basis of heat-stable somatic antigens (Walker and Batty 1964, Boothroyd and Georgala 1964). All type A strains and proteolytic strains of types B and F have agglutinating antigens in common; there is also some sharing of antigens with *Cl. tetani* and *Cl. histolyticum*. A second serological group comprises all type E strains and non-proteolytic strains of types B and F. Similarly, types C and D are related serologically and are distinguishable from type E. Some limited cross-reactions occur between strains of *Cl. botulinum* and *Cl. sporogenes* (McClung 1937, Batty and Walker 1967, Solomon *et al.* 1969, 1971, Lynt *et al.* 1971). The somatic serology of type G has not been studied.

As just stated, McClung and others found a somatic antigen common to certain strains of *Cl. botulinum* and *Cl. sporogenes*; and Smith (1937) and Hoogherheide (1937) an antigenic as well as a biochemical and cultural relation between R forms of *Cl. histolyticum* and *Cl. sporogenes*. Mandia explored proteolytic clostridia in terms of H antigens, and of somatic antigens, L and O, that are respectively labile and stable for 2 hr at 100° (Mandia and Bruner 1951, Mandia 1951, 1952) and proposed (1955) five serological groups, four of which cover *Cl. sporogenes*, *Cl. botulinum*, *Cl. histolyticum* and *Cl. tetani*. All have an O antigen (IV) in common, and each species has a distinctive O antigen. The groups are further divided analogously to those of the Kauffmann-White scheme for salmonellae (Chapter 37) on the basis of the 'labile' and the flagellar antigens, which are not shared between the four species, though in various combinations of two or three they distinguish a number of serotypes in each species, including the ten types of *Cl. tetani* noted above.

Cl. novyi strains have two O antigens in common, one of which is shared by *Cl. haemolyticum* (Turner and Eales 1943, Keppie 1944, Batty and Walker 1963, 1964, Batty *et al.* 1964); for typing purposes a haemolysin-neutralization test is said to be superior to the lecithovitellin test (Rutter and Collee 1969). *Cl. bifermentans* is closely related to *Cl. sordelli* (Clark and Hall 1937, Tardieux and Nisman 1952, Tataki and Huet 1953); and their spore antigens are so alike that Walker (1963) would regard the two organisms as belonging to one species. The spore antigens of *Cl. sporogenes* are the same in all strains (Walker 1963). Studying by the agar-gel method the neutralization of their soluble antigens by antisera, Ellner and Green (1963) were able to group and identify the ten most common pathogenic species of *Clostridium* (see also the review by McCoy and McClung 1938).

The *G + C content* of the DNA of clostridia ranges from about 22 to 28 moles per cent (Cummins and Johnson 1971), 26 to 32 per cent (Hill 1966). *Bacteriophages* have been described for many of the clostridia and their properties investigated (see Guélin *et al.* 1966, Prescott and Altenbern 1967, Sebald and Popovitch 1967, Mahony and Kalz 1968, Dolman and Chang 1972, Grant and Riemann 1976, Paquette and Fredette 1977). Betz and Anderson (1964), besides grouping the phages of *Cl. sporogenes*, found that many strains of this organism produced bacteriocine-like substances acting on other strains of the same species. Bacteriocines are also produced by a number of other clostridia, for example, *Cl. perfringens* (Mahony and Butler 1971, Mahony and Swantee 1978) and *Cl. botulinum* (Kautter *et al.* 1966). As might be expected, bacteria of unrelated species sometimes produce bacteriocines active upon the cells of others; Smith (1975*a*) isolated from soil strains of *Bacillus cereus*, *Cl. sporogenes* and *Cl. perfringens* that produced bacteriocines active against *Cl. botulinum*. Inoue and Iida (1971) noted a close relation between lysogenicity and toxigenicity among strains of *Cl. botulinum* types C and D, and succeeded in converting non-toxigenic into toxigenic strains by lysogenization. Not only can phages effect conversion of one type of *Cl. botulinum* to another, but they can also bring about interspecific conversion from *Cl. botulinum* type C to *Cl. novyi* type A (Eklund *et al.* 1974). A strain of *Cl. botulinum* type C was cured of its phage and thus converted into a non-toxigenic variant. On infection with one or two phages from *Cl. novyi* type A it produced *Cl. novyi* lecithinase (α toxin).

Toxin formation

The clostridia produce a variety of powerful exotoxins. Two of them—*Cl. botulinum* and *Cl. tetani*—give rise to neurotoxins more poisonous than any other substances with which we are acquainted. Purified toxins from *Cl. tetani* and *Cl. botulinum* type A contain per mg about 30 million LD50 for mice.

Little is known of the mode of formation of these protein exotoxins, or of their release into the culture medium. In cultures of *Cl. tetani* there is no fixed relation between toxin formation and either proteolytic activity of filtrates (Amoureux 1945) or protein content of the cells (Eisler 1952). Nevertheless, proteolytic activity plays a part in the extracellular activation of the prototoxins of *Cl. perfringens* types D and E and most of the botulinum neurotoxins, which appear in filtrates in this inert form. In most species, the active toxin is probably formed intracellularly and released into the medium. Substantial amounts of toxin can be extracted from the washed cells of *Cl. botulinum* type D, and of toxigenic *Cl. bifermentans*, but little from cells of *Cl. tetani*, *Cl. perfringens* type D and *Cl. chauvoei* (Raynaud 1951, Chamsy and Raynaud 1951, Saissac and Raynaud 1951, Boroff *et al.* 1952). However, some is extractable from *Cl. tetani* (Raynaud *et al.* 1955) and may be released from cells by lysozyme (Stone 1952, 1953). When the toxicity of washed cells is measured in mice treated with penicillin to prevent bacterial growth, it is clear that in-vitro intracellular toxin is formed early during growth of *Cl. tetani*, and reaches a maximum within a few days, just before autolysis of the cells; with autolysis, the toxin is released into the culture fluid (Mueller and Miller 1948, Miller *et al.* 1959, 1960); an inert prototoxin may be formed intracellularly in the early stage of growth (Seki *et al.* 1958). For the nutrients in casein digest essential for toxin production, see Mueller and Miller (1954, 1955, 1956), and Latham and co-workers 1962). The release of *Cl. botulinum* neurotoxins A, B and C is also associated with autolysis of the cells (Boroff 1955, Bonventre and Kempe 1960); A and B appear to be formed from a large-molecular precursor by the action of proteolytic enzymes. With type E toxin, trypsin increases the toxicity of filtrates 20- to 40-fold (Duff *et al.* 1956, Dolman 1957)—partly by fragmenting the molecule and exposing more toxic groups (Gerwing *et al.* 1965)—and liberates large amounts of toxin from whole cells (Sakaguchi and Sakaguchi 1959). Type A neurotoxin formed in the cells appears to be more firmly held than tetanus toxin by cells of *Cl. tetani*; whole cells are non-toxic *in vivo*, though the toxin can be released by sonic disruption (Kindler *et al.* 1956*b*). *Cl. perfringens* type A enterotoxin is synthesized during sporulation of cells and it is released on lysis of the sporangia; there is some evidence that the toxin might occur in cell extracts in a form similar to ε-prototoxin. It is not inactivated by trypsin. Enterotoxin has also been demonstrated in type C strains from cases of enteritis necroticans in New Guinea, and by single strains of type D (Skjelkvale and Duncan 1975, Uemura and Skjelkvale 1976). Enterotoxin produced by all strains appears to be identical. *Cl. difficile* toxin is released upon autolysis of the cells; it is inactivated by proteinases. (For the general properties of the toxins see Chapter 13 and reviews and monographs by Bonventre *et al.* 1967, Lamanna and Carr 1967, Raynaud 1967, van Heyningen 1970, Smith 1977, Arbuthnott 1978, and Bizzini 1979, the series of contributions edited by Kadis *et al.* 1971, and Reports 1975, 1978, 1982.)

Many of the clostridial toxins, like the diphtheria toxin, are detoxified by formaldehyde and other substances to form immunogenic toxoids (see Chapters 63, 64). The preparation of toxins and toxoids requires attention to many factors with which we have no space to deal. But the properties of the exotoxins as found in culture filtrates must be discussed along with other biologically active antigenic substances found in these filtrates, some of which are important as auxiliary factors in the pathogenesis of clostridial infections, and all of which are useful as characters for the classification of these bacilli.

Biologically active antigens in filtrates of cultures

The main toxins of *Cl. tetani* and *Cl. botulinum* are readily identified by their specific pharmacological effects. The toxins of the gas-gangrene organisms, and some of the other antigenic components in culture filtrates of clostridia, produce biological effects that may be common to several substances; and they can be separately identified only by using special test objects—like a given species of red blood cell for the haemolysins—or by their being neutralized by known monospecific antitoxins or other antisera. Care must be taken when testing for haemolysins to buffer the systems to the pH of optimum activity (Walbum 1938). Lethal effects are conveniently determined by intravenous, necrotizing effects by intracutaneous injection. Lecithinases, collagenases, gelatinases and so forth, are tested against appropriate substrates *in vitro*. Fibrinolysins are formed by some species (Carlen 1939, Reed *et al.* 1943); these may sometimes be kinases rather than proteinases.

It should be noted that non-toxigenic strains of toxigenic clostridia are not uncommon, especially among laboratory strains; and that toxin production may be profoundly affected by the culture medium. The inherent toxigenic capacity of a strain may be labile. Clones from individual cells in a culture may vary widely in toxigenesis; though the capacity of the culture as a whole may be maintained by animal passage (Wildführ 1949*a*,*b*, 1950). Observations by various workers indicate that with many clostridial species there is an inverse relation between toxigenesis and spore formation; the less toxigenic the strain, the greater is its potency for sporulation (see Nishida and Nakagawara 1965).

Cl. tetani Filtrates contain the classical neurotoxin, and a haemolysin. The haemolysin is relatively heat-labile, inactivated by oxygen, and is present in strains that do not form the neurotoxin (Kerrin 1930). It resembles other oxygen-labile haemolysins, such as the θ toxin of *Cl. perfringens* and the O streptolysin of *Str. pyogenes*, in being neutralized by antisera to any of these lysins. Injected intravenously it causes rapid death with pulmonary oedema in mice, intravascular haemolysis in rabbits and monkeys, and altered myocardial function in monkeys and mice (Hardegree *et al.* 1971). Tetanolysin produced by the different Tulloch types is similar (Lucain and Piffaretti 1977).

The neurotoxicity of filtrates is variable; a potent filtrate will kill a mouse in a dose of 10^{-8} to 10^{-9} ml. Mueller and Miller (1945, 1947, 1948) obtained high-titre filtrates from a highly toxigenic variant of *Cl. tetani*, in a medium containing a rich hydrolysate of protein, reduced iron, and glucose; and from such filtrates Pillemer isolated electrophoretically homogeneous, crystalline, carbohydrate-free protein, containing 6.6×10^7 mouse MLD per mg nitrogen (Pillemer 1946, Pillemer *et al.* 1946, 1948, Dunn *et al.* 1949). Further purification is possible by gel filtration (Latham *et al.* 1965); and when this is followed by chromatographic separation on Sephadex G-100, what is believed to be the pure monomer may be obtained containing 15×10^7 mouse MLD per mg nitrogen (Murphy and Miller 1967) and having a molecular weight of 150 000 (Mangalo *et al.* 1968). The neurotoxin, or tetanospasmin, is relatively heat-labile, being destroyed at 65° in 5 min; it is stable in the dry state. It is not usually absorbed from the alimentary canal, and is apparently destroyed by the digestive juices. It is neutralized by specific antitoxin and is fixed by gangliosides (van Heyningen and Mellanby 1968, van Heyningen 1976). The toxins produced by different strains of *Cl. tetani* all have identical pharmacological effects, and are all neutralized by a single antitoxin. The different preparations of toxin nevertheless change in their relative toxicity when tested in different species of laboratory animal; and the proportional potencies of different antitoxins change when tested against different toxin preparations (Ipsen 1940–41, Smith 1941–42, Petrie 1942–43). Some of the observed discrepancies may be due to heterogeneity of antitoxic antibodies of a serum (see Largier 1958); this may be because the toxin carries at least four antigenic determinants to each of which antitoxin may be formed (Nagel and Cohen 1973). The toxin may exist in several states of aggregation, ranging from a size corresponding to a sedimentation constant of 7S on extraction from the cells down to 2.3S after treatment (Raynaud *et al.* 1960). Murphy and Miller (1967) give a figure of 6.4S for the monomeric preparation. Tetanospasmin has been reviewed by Bizzini (1979). (For references to the discovery and early study of tetanus toxin see Behring and Kitasato 1890, Behring 1892, Madsen 1899, Rosenau and Anderson 1908.)

Cl. botulinum Seven main types, A–G, of this species are known, characterized by antigenically distinct, though pharmacologically similar toxins. Type C toxin contains at least two components, Cα and Cβ, and is serologically related to the D toxin. Haemolysins are present in filtrates (Schoenholz 1928). Guillaumie and Kréguer (1950) report an oxygen-labile haemolysin in type C, similar to that in *Cl. tetani* and to the θ toxin of *Cl. perfringens*, and Brygoo (1950) reports an oxygen-labile haemolysin in type D. On egg-yolk agar, colonies of all types produce an opacity in the surrounding medium and a 'pearly layer', suggesting a lipase (Willis 1960). A few strains of type C produce a lecithinase and are therefore culturally indistinguishable from *Cl. novyi* type A on egg-yolk agar. Type A produces a haemagglutinin for horse RBC, which forms a complex with the toxin; it is separable from the toxin in alkaline solutions of low ionic strength (Lamanna and Lowenthal 1951, Lowenthal and Lamanna 1953), and remains after toxin is destroyed by proteolytic enzymes (Meyer and Lamanna 1959).

Types B–E also produce haemagglutinins; that of type B is serologically similar to that of A; both differ from those of types C and D, which are serologically similar, and from that of type E (see Johnson *et al.* 1966).

The toxins of *Cl. botulinum* vary in heat resistance; in 2 min type A toxin is destroyed at 60°, types B and E at 70°, and types C and D at 90° (Prévot and Brygoo 1953*b*); type F toxin rapidly loses potency at 37° (Dolman and Murakami 1961). They are non-dialysable, highly potent, and like the ε and ι toxins of *Cl. perfringens* are absorbed from the alimentary canal. Most of the neurotoxins are elaborated as non-toxic prototoxins. Some from proteolytic strains of types A, B and F are fully activated by the organisms' own proteinases; some require additional *exogenous* proteolytic activation for full toxicity. Toxins from most type E strains and from non-proteolytic type B and F are not activated at all by *endogenous* enzymes.

Type A toxin, when purified by chromatography on Sephadex G-200, is said to have two components, both with a molecular weight of about 140 000 (Hauschild and Hilsheimer 1969); one of them is associated with the haemagglutinin, which has a higher molecular weight (Das Gupta and Boroff 1968). Before final purification the toxin with a molecular weight of 900 000 contains 2.2×10^8 mouse LD50 per mg nitrogen. There are no peculiarities in its amino-acid composition to account for its toxicity (see Lamanna and Carr 1967). The toxins of types B, C, D, and E appear to be of the same order of size as type A toxin. The toxin of type B has a molecular weight of 167 000, and is composed of two subunits with molecular weights of 59 000 and 104 000 (Beers and Reich 1969). The molecular weight of purified type E toxin is given as 18 600 by Gerwing and her colleagues (1964), but this awaits confirmation (Lamanna and Sakaguchi 1971). The mouse toxicities of types A, B, D, and E toxins are similar; type C toxin is about one-tenth as toxic (Duff *et al.* 1957, Cardella *et al.* 1957, 1960). Bioassay is carried out by determining the highest dilution of toxin that will kill 50 per cent of mice within 4 days of intraperitoneal injection. A much more rapid method is based on the intravenous injection of concentrated toxin with measurement of the survival time in minutes (Boroff and Fleck 1966); the results, however, are influenced by the molecular size of the toxin (Lamanna *et al.* 1970). All the seven toxins are antigenically distinct, except for minor antigens common to C and D toxins (see Prévot and Brygoo 1953*a*).

The purified toxins are readily detoxified by formalin, with the production of immunogenic toxoids. Non-toxic strains may be rendered toxic by phage conversion (Oguma *et al.* 1976). (For a series of papers given at the Moscow symposium of 1966, see Ingram and Roberts 1967; and for reviews see Lamanna and Sakaguchi 1971, and Smith 1977).

The main toxic components of the next three organisms we describe, namely *Cl. perfringens*, *Cl. septicum*, and *Cl. novyi* are similar in a moderate resistance to heat, being destroyed in 30 min at 70°, and to weak acid. In guinea-pigs and mice they all induce gelatinous oedema and necrosis, especially a myonecrosis, and greatly increase the permeability of small blood vessels (see Aikat and Dible 1956, 1960, Elder and Miles 1957, Craig and Miles 1961).

Cl. perfringens Wilsdon (1931, 1933) divided the species into four types, A, B, C, and D, depending on the ability of their antisera to neutralize the lethal effects of their toxins.

These types, and the diseases with which they are associated (see Chapter 63), are: the classical *Cl. perfringens* and human gas gangrene: '*Cl. agni*' and lamb dysentery; '*Cl. paludis*' and struck in sheep; '*Cl. ovitoxicum*' and enterotoxaemia of sheep and cattle. Bosworth (1943) described a new type, E, isolated from a calf dying of enterotoxaemia, that produced a lethal toxin not neutralizable by types A–D antisera. The designation type F was proposed for strains isolated by Zeissler and Rassfeld-Sternberg (1949) from cases of fatal necrotic enteritis in West Germany; these strains resembled those of type C, but differed in being heat resistant and in lacking the minor antigens, δ, θ, and κ. Later, Murrell and Roth (1963) encountered similar strains from cases of necrotizing jejunitis in Papua New Guinea, differing however in possessing the θ haemolysin and in not being unduly resistant to heat. Since neither of these groups formed lethal toxins other than those neutralizable by type C antisera, and since heat resistance and the presence or absence of minor antigens are insufficient to justify the establishment of a new type, Sterne and Warrack (1964) suggested the abolition of type F and the transference of the necrotic enteritis strains to type C.

Glenny and his colleagues (1933) identified five separate toxic components in culture filtrates of Wilsdon's types: α, β, γ, δ, and ε. The existence of these components was confirmed by a number of workers (Bosworth and Glover 1935, Borthwick 1935, Mason 1935, Weinberg and Guillaumie 1936, Dalling and Ross 1938, Duffett 1938, Stewart 1940, Taylor and Stewart 1941). *Cl. perfringens* type A was at first thought to contain only α. Prigge (1936, 1937), Ipsen (1939), Ipsen and Davoli (1939) and Ipsen *et al.* (1939) found two components α and ζ, and in one strain, a third component, which they designated η (see also Nagler 1940). Prigge's ζ toxin is equivalent to Glenny's α, and British workers (see Dalling and Stephenson 1942) adopted a convention whereby Prigge's α and ζ are designated θ and α respectively. The η toxin of Ipsen retains its original designation. Oakley and his colleagues added four more serologically identifiable substances, κ, λ, μ, ν (Oakley *et al.* 1948, Oakley and Warrack 1951, Warrack *et al.* 1951). There are thus 12 components by which *Cl. perfringens* may be typed (Table 42.1). They are identified by their biological effects, as with haemolysin or hyaluronidase, and by the characteristic lesions they produce on intradermal injection in the guinea-pig (Oakley 1967). When the same effect, such as haemolysis or skin necrosis, is produced by more than one component, neutralization tests with monospecific antitoxins may be necessary (Oakley and Warrack 1953). The purification of α toxin is described by Ito (1968), Casu *et al.* (1971) and Mollby and Wadstrom (1973), of θ toxin by Mitsui *et al.* (1973) and Smyth (1975), and of κ toxin by Kameyama and Akama (1971); and the separation of the four toxins of type D, namely, α, ε, θ, and κ, by Hauschild (1965).

Brooks and her colleagues (1957), in a survey of 307 strains, subdivided types A, B and C on the basis of these components, and added two more haemolysins, for ox and horse RBC, designated 'non-$\alpha\delta\theta$' because they are not neutralized by antibodies to α, δ or θ lysins. From this scheme, reproduced in Table 42.1, it will be seen that the α component predominates in type A; β in type B; β and sometimes δ in type C; ε in type D and ι in type E. The heat-resistant food-poisoning members of the sub-group in type A are characterized by absence of θ. In type B three strains from the intestinal contents of Iranian sheep were distinguished by producing κ, and no μ or λ. The Colorado strains of type C, from calves (Griner and Bracken 1953) and lambs (Griner and Johnson 1954) with enterotoxaemia, differ from the classical type C in producing no γ or δ (see also Buddle 1954); the piglet strains likewise produce no γ and have the non-$\alpha\delta\theta$ haemolysins. Four-day cultures of the heat-resistant food-poisoning strains of type A, and of Zeissler and Rassfeld-Sternberg's (1949) necrotic enteritis type C-strains, unlike those of the rest, resist 1–3 hours' heating at 100° (p. 464).

Table 42.1 indicates that some of the antigens are not consistently produced by a given type, and that both quantitative and qualitative variability are to be expected (see, e.g. Orlans and Jones 1958, Ellner and Bohan 1962). Loss variants are not uncommon; Mason (1935) records the loss of ability to produce ε toxin in type B strains, Borthwick (1937) a similar loss in type D strains, Taylor (1940) a loss of β-toxigenicity in a type B strain, and Nakamura *et al.* (1976) and Pinegar and Stringer (1977) a loss of α-toxigenicity in type A strains.

The α component is thermostable, lethal for mice, guinea-pigs, rabbits, pigeons and sheep, and, when given intradermally, produces a necrotic lesion. It is haemolytic for the red cells of most laboratory animals excepting the horse and the goat, and is a powerful lecithinase. The lecithinase activity of α toxin is the first known instance of an exotoxin dependent upon a demonstrable enzyme for its activity. Nagler (1939) found that toxic filtrates of types A–D produced an opalescence in human sera proportionally to their toxin content, and that the reaction was specifically inhibited by antisera to type A filtrates (see also Seiffert 1939). Macfarlane, Oakley and Anderson (1941) demonstrated a similar action of α toxin on extract of egg yolk, and suggested that both phenomena were due to an enzymic splitting of lipoprotein complexes in serum and egg yolk respectively. Macfarlane and Knight (1941) demonstrated a quantitative splitting of lecithin by α toxin into phosphorylcholine and a diglyceride, and the necessity for Ca or Mg ions in the reaction (Klein *et al.* 1975). The enzyme is therefore classified as a lecithinase C, more correctly called a phospholipase C. The binding, *in vitro* or *in vivo*, of divalent metal ions by a chelating agent renders the enzyme non-toxic (Moskowitz 1956). The lecithinase activity of a filtrate may be used as a measure of its α toxin content, and for comparisons of the neutralizing power of α antitoxins (for details see Nagler 1939, Oakley and Warrack 1941, van Heyningen 1941*a*, Crook 1942, Zamecnik *et al.* 1947). Lecithinase production in fluid and on agar cultures of *Cl. perfringens* may be detected by incorporating human serum or egg-yolk extract in media containing a sufficiency of free calcium (Hayward 1943, McClung and Toabe 1947, Willis and Hobbs 1958, 1959, Willis and Gowland 1962). (For factors affecting the production and activity of lecithinase see Nakamura *et al.* 1969; and for the effect of inorganic ions in the medium on the production of the various toxins of *Cl. perfringens* see Murata *et al.* 1968, 1969, Soda *et al.* 1969, Sato and Murata 1973, Smyth and Arbuthnott 1974). Other bacteria, mainly spore-bearing bacilli, produce lecithinase-like substances (Nagler 1939, Hayward 1941, Crook 1942, Oakley *et al.* 1947) but with the exceptions of *Cl. sordelli* and *Cl. bifermentans* (Hayward 1943, Miles and Miles 1950), none is neutralized by α antitoxin. The α toxin is haemolytic by reason of its attack on the phospholipins of the red cell surface (Macfarlane 1950*b*). It may be assayed by the intravenous injection into mice of two dilutions of the toxin sample and the reference toxin and determination of the survival time in minutes (Yamamoto *et al* 1972).

Table 42.1 Properties and distribution of biologically active antigens in culture filtrates of *Cl. perfringens* (after Brooks, Sterne and Warrack 1957, and Sterne and Warrack 1964)

		Major lethal antigens				Minor lethal and non-lethal antigens								Non-$\alpha\delta\theta$ haemolysis		
		α	β	ε	ι	γ	δ	η	θ	κ	λ	μ	ν			
Type	Occurrence	Lethal, necrotizing Ca^{++} dependent lecithinase	Lethal, necrotizing	Lethal, necrotizing	Lethal, necrotizing	Lethal	Lethal, haemolytic	Lethal	Haemolytic oxygen-labile (? lethal)	Collagenase	Proteinase	Hyaluronidase	Deoxyribo-nuclease	'Ox'	'Horse'	Heat resistance
A	Gas gangrene (man) Intestinal commensal (man and animals) Putrefactive processes Soil, etc.	+ + +	–	–	–	–	–	(+)	+ + –	+ + –	–	+ –	+ + –	(+)	(+)	–
	Patients and food in outbreaks of food poisoning	+ + +	–	–	–	–	–	–	(+)	+ + –	–	+ –	+ +	–	–	+ + +
B	Lamb dysentery Enterotoxaemia of foals	+ + +	+ + +	+ + –	–	+ +	(+)	–	+ +	–	+ + +	+ + +	+ +	–	+ –	–
	Haemorrhagic enteritis of goats and sheep, Iran	+ + +	+ + +	+ + +	–	?	–	–	+ + +	+ + +	–	–	+ –	–	–	–
C	'Struck' in sheep	+ + +	+ + +	–	–	+ +	+ + +	–	+ + +	+ + +	–	–	+ –	–	–	–
	Neonatal haemorrhage enteritis in calves and lambs	+ + +	+ + +	–	–	?	–	–	+ + +	+ + +	–	–	+ –	–	–	–
	Enteritis in piglets	+ + +	+ + +	–	–	?	–	–	+ + +	+ + +	–	(+)	+ +	+ + +	+ + +	–
	Necrotic enteritis of man in Germany (formerly type F)	+ + +	+ + +	–	–	+ + +	–	?	–	–	–	–	+ + +	+ +	+ +	+ + +
	Necrotizing jejunitis of man in Papua New Guinea	+ + +	+ + +	–	–	?	–	?	+ +	+ +	–	+ +	–	?	?	–
D	Enterotoxaemia in adult sheep, lambs, goats and cattle Intestines of man	+ + +	–	+ + +	–	–	–	–	+ +	+ +	+ +	+ + –	+ + –	(+)	+ –	–
E	Sheep, cattle (? pathogenic)	+ + +	–	–	+ + +	–	–	–	+ +	+ +	+ +	(+)	+ + –	+ –	+ –	–

+ + + = Produced by all strains; + + = produced by most strains; + + –, + – – = produced by some strains; (+) = produced by very few strains.

The θ component is prominent in type A and, though less so, in types B–E. It is strongly haemolytic and moderately lethal. As Todd (1941) showed, it is thermolabile and oxygen-labile, and the oxygen-inactivated material is reactivated by reducing agents containing sulphydryl groups. This property of reversible oxidation is shared by the O streptolysin of *Str. pyogenes* (Chapter 29), to which θ toxin is serologically related. This toxin may be removed from mixtures with α toxin by adsorption on to susceptible red cell stromata (van Heyningen 1941*b*; see also Gale and van Heyningen 1942, Roth and Pillemer 1953, 1955, Mitsui *et al.* 1973, Smyth 1975).

The β toxin is not haemolytic; it is lethal to mice, producing on intravenous injection a spasmodic twitching rapidly followed by death; given intradermally in guinea-pigs and rabbits it produces necrotic lesions. It is the β-toxin of human type C strains of *Cl. perfringens* that is the responsible factor for enteritis necroticans ('pig-bel') in the native population of Papua New Guinea (see Chapter 63). The γ toxin is neither necrotizing nor haemolytic. Its existence in filtrates can be proved only by its lethal activity in mice after the other toxins have been neutralized by appropriate antitoxins. The δ toxin is lethal and actively haemolytic (see Turner and Eales 1944); and, like γ toxin, is detected after neutralization of other components by antitoxins. The ε toxin, which predominates in type D filtrates, is not haemolytic, but is both lethal and necrotizing. Large doses in mice produce spasmodic twitching. Smaller doses produce paralysis after several days. The ε toxin is produced by type D strains as a thermostable, inert 'prototoxin', which is transformed by trypsin and other proteolytic enzymes into the thermolabile toxin (Gill 1933, Bosworth and Glover 1935). In disease produced by type D strains (see Chapter 63) the activating enzyme is apparently supplied by the infected animal, though according to Turner and Rodwell (1943) in favourable conditions extracellular proteinases of the bacillus itself will activate the toxin.

The η component is lethal, but neither haemolytic nor necrotizing. Like the ε toxin, the ι toxin is thermostable. It is lethal and necrotizing but not haemolytic; and it is elaborated as an inert prototoxin, activable by trypsin (Bosworth 1943, Ross *et al.* 1949). It must not be confused with an ill defined, lethal but not necrotizing antigen in types B and D, named ι toxin by Guillaumie, Kréguer and Fabre (1946*b*). The κ component is an enzyme attacking the collagen and reticulin of muscle, and gelatin. It is necrotizing and apparently lethal (Oakley *et al.* 1946, 1948, Bidwell and van Heyningen 1948, Kameyama and Akama 1971); it also digests decalcified tooth dentine (Evans and Prophet 1950). The λ component, first thought to be a collagenase because of its ability to digest powdered hide, proved to be a non-lethal, non-necrotizing proteinase and gelatinase (Oakley *et al*, 1948, Bidwell 1950). The designations μ and ν were proposed respectively for the hyaluronidase, first found in culture filtrates of type A (McClean 1936; see Chapter 63), and for deoxyribonuclease, which is detected by its power of digesting the nuclei of methanol-fixed leucocytes (Oakley and Warrack 1953).

The *enterotoxin* (see Chapter 72) differs from all the other toxins. Injected into the ligated intestinal loop of the rabbit it leads to distension with fluid (Duncan *et al.* 1968). Injected intravenously into mice and guinea-pigs it proves fatal within 20 min. Injected into the skin of the rabbit or guinea-pig it produces erythema without necrosis. It is a heat-labile, non-diffusible protein with a molecular weight of about 36 000, capable of stimulating the production of antitoxin. It is destroyed by Pronase but not by trypsin (Duncan and Strong 1971, Hauschild 1971, Craig 1972). Subsequent studies have shown that *Cl. perfringens* enterotoxin acts very rapidly, in minutes rather than hours. It causes tissue damage in the gut, inhibits metabolic processes in intestinal tissue and alters transport of fluid, ions and glucose. Moreover, it can affect basic function and structure by inhibiting protein, RNA and DNA synthesis, and causing membrane damage to individual cells (McDonel 1979).

Strains of A–D produce haemagglutinins for a variety of RBC, including those of man, sheep, cattle, pigs and fowl, the pattern of whose neutralization by type antisera has been made the basis of attempts to distinguish strains of *Cl. perfringens* (Gürtürk 1952, Dafaalla and Soltys 1953, Katitch 1954). According to Wickham (1956), the haemagglutinin is identical with the enzymes responsible for destroying the blood group substance and the virus receptors on group A human cells (but see Collee 1961); and, antigentically, is group- and not type-specific. *Cl. perfringens* neuraminidase has been further considered by Fraser and Smith (1975) and Fraser and Collee (1975). Type A filtrates contain an antigenic phagocytosis-inhibiting and adrenaline-potentiating factor (Ganley *et al.* 1955).

Cl. novyi Toxic filtrates of *Cl. novyi* are the most potent among those produced by the gas gangrene clostridia; the average MLD for a mouse is about 0.0002 ml; of *Cl. perfringens* toxin the MLD is about 0.25 ml; and of *Cl. septicum* toxin about 0.005 ml. Filtrates of *Cl. novyi* contain a potent toxin that produces intense gelatinous oedema in muscle. A lethal intravenous dose of toxin has few gross effects, but degenerative changes, particularly in the spleen and kidney, occur (Pasternack and Bengtson 1940). Three types of *Cl. novyi*. A, B and C, were distinguished by Scott, Turner and Vawter (1933) on the basis of glycerol fermentation and toxin production. This basis proved to be inadequate (Hayward and Gray 1946), but the three types were firmly established by Oakley and his colleagues (1947) on the basis of six antigens found in culture filtrates. Oakley and Warrack (1959), taking into account two more antigens, η and θ, respectively a tropomyosinase described by Macfarlane (1956) and another substance producing opalescence in egg-yolk media, proposed the classification, set out in Table 42.2, adding under type D the strains hitherto classified as *Cl. haemolyticum*, the organism associated with icterohaemoglobinuria in cattle. Of these antigens α is the only frank toxin and is lethal and necrotizing, but not haemolytic. The pathological significance of the other components is doubtful. The β and γ antigens are haemolytic lecithinases, present in small amounts, and are of the same enzyme type as *Cl. perfringens* α toxin, but all three differ antigenically and in the species of RBC they attack (Macfarlane 1948*b*, 1950*b*). The β component is also produced in large amounts by the related *Cl. haemolyticum* (Macfarlane 1950*a*). The δ component is an oxygen-labile haemolysin, of the same kind as tetanolysin and the *Cl. perfringens* θ toxin. The ε component is a lipase attacking simple fats, and is responsible for the formation of the characteristic pearly layer which develops on the surface of egg-yolk agar plate cultures of type A strains (Willis 1960, Rutter and Collee 1969). Bard and McClung (1948) describe a haemolytic lysolecithin in type B and in *Cl. haemolyticum*. Since the lecithovitellin reaction is not entirely reliable, Rutter and Collee (1969) advocate for typing purposes the simultaneous use of a haemolysis-neutralization test.

Cl. septicum Toxic filtrates of *Cl. septicum* produce

Table 42.2 Properties and distribution of biologically active antigens in culture filtrates of *Cl. novyi* (after Oakley and Warrack 1959)

Designation	Activity	Occurrence in filtrates of *Cl. novyi* types			
		A	B	C	D
α	Lethal, necrotizing	+	+	−	−
β	Haemolytic, necrotizing lecithinase	−	+	−	+
γ	Haemolytic lecithinase	+	−	−	−
δ	Haemolytic (oxygen-labile)	+	−	−	−
ε	Opalescence, pearly layer in egg yolk (lipase)	+	−	−	−
ζ	Haemolytic	−	+	−	−
η	Tropomyosinase	−	+	−	+
θ	Opalescence in egg yolk (lipase)	−	tr.	−	+

liquefactive necrosis of muscle; lethal doses, when given intravenously, produce intense capillary engorgement and interstitial haemorrhages in the heart, with hyaline degeneration of the muscle fibres, and a toxic nephrosis in the kidney of experimental animals (Pasternack and Bengtson 1936). Electrocardiographic observations on the mouse and the rabbit reveal the existence of two toxins acting on heart muscle, both of them lethal: one, not dialysable, is specifically neutralized by an antiserum; the other passes through the dialysis membrane and is not neutralized by antiserum (Bittner *et al.* 1970) Filtrates may also contain a haemolysin and a hyaluronidase. The haemolysin has usually been regarded as distinct from the lethal toxin; but in one strain Bernheimer (1944) found a strict parallelism between the toxic and the haemolytic activity. *Cl. septicum* also produces a hyaluronidase, and (Evans 1947) a collagenase. Warrack, Bidwell and Oakley (1951) give the name β toxin to a deoxyribonuclease in *Cl. septicum*, antigenically distinct from the ν component of *Cl. perfringens*, which may be active in infected tissues as a cytotoxin. The designations γ and δ are given respectively to a hyaluronidase and an oxygen-labile haemolysin related to that of *Cl. chauvoei* (Moussa 1958). Studies by Princewill and Oakley (1972*a*, *b*, 1976) show that the hyaluronidases of *Cl. septicum* and *Cl. chauvoei* are identical; and there is evidence of two types of deoxyribonuclease that are shared by the two organisms. There is also an antigenic relation between *Cl. septicum* α toxin and the α toxin of *Cl. histolyticum* (Guillaumie *et al.* 1946*a*, 1952, Guillaumie 1960, Sterne and Warrack 1962). *Cl. septicum* also produces a haemagglutinin and a neuraminidase (Gadalla and Collee 1967, 1968).

Cl. chauvoei Most of our modern knowledge of the soluble substances produced by *Cl. chauvoei* is due to the work of Moussa (1958). The α-toxin is an oxygen-stable haemolysin that is also necrotizing; it appears to be unrelated to *Cl. septicum* β-toxin. The deoxyribonuclease (β-toxin), the hyaluronidase (γ-toxin) and an oxygen-labile haemolysin (δ-antigen) are all antigenically related to the corresponding antigens produced by *Cl. septicum*.

Cl. histolyticum usually produces a weakly toxic, proteolytic filtrate. Oakley and Warrack (1950) describe three components: α, a lethal and necrotizing toxin; β, a collagenase (see also Jennison 1945, Tytell and Hewson 1950, MacLennan *et al.* 1953) which also attacks dentine (Evans and Prophet 1950); and γ, a cysteine-activated proteinase attacking azocoll (see also van Heyningen 1940, Kocholaty and Krejci 1948, Lepow *et al.* 1952, Mandl *et al.* 1953). Guillaumie (1944) reports a labile haemolysin in *Cl. histolyticum*. Other designated antigens are a proteolytic enzyme δ, not activated by cysteine (MacLennan *et al.* 1958, Oakley and Warrack 1958) which is an elastase (Oakley and Banerjee 1963), and ε, an oxygen-labile haemolysin neutralized by antisera to other oxygen-labile haemolysins (Howard 1953, Hobbs 1958). Gelatinases and peptidases are also produced (see Howes *et al.* 1960).

Cl. sordelli* and *Cl. bifermentans Culture filtrates of many strains of *Cl. sordelli* are moderately toxic. The toxin is non-haemolytic, necrotizing and lethal, and induces a gelatinous oedema. Both *Cl. sordelli* and the related *Cl. bifermentans* produce a phospholipase C resembling the α toxin of *Cl. perfringens*, and antigenically related to it (Miles and Miles 1947, 1950; Lewis and Macfarlane 1953; see also Meisel 1956).

Cl. difficile Strains of *Cl. difficile* produce a toxin which is the cause of antibiotic-associated pseudomembranous colitis in man and antibiotic-associated ileocaecitis in laboratory animals. It is neutralized by *Cl. sordelli* antitoxin, is heat-labile and inactivated by proteinases. It is lethal for the hamster by the orogastric and intraperitoneal routes; in tissue cultures it causes cytopathic effects in a variety of cell lines including Y1 adrenal cells, Chinese hamster ovary cells and HeLa cells (Rifkin *et al.* 1978, Rolfe and Finegold 1979, Donta and Shaffer 1980). (See Chapter 63.)

Pathogenicity

The pathogenicity of the anaerobes appears to depend almost entirely on their toxin production. *Cl. tetani*, for example, multiplies locally, and does not invade the body. *Cl. botulinum* is rarely a parasite; it is unable to grow easily in the tissues, and its pathogenic effects are usually determined by the formation of toxin in food substances before their ingestion; growth and toxin production in the lumen of the gut occurs in infant botulism, and wound botulism is an uncommon but well recognized syndrome. *Cl. novyi* remains almost confined to the site of inoculation. *Cl. perfringens* and *Cl. septicum* may become generalized in the final stages of an infection, but before this they multiply only locally. Tetanus, botulism and to a significant extent gas gangrene are intoxications. Many of the clostridia form neuraminidase, which may contribute to their pathogenicity (Müller 1971).

Pathogenicity for laboratory animals

Cl. botulinum

Monkeys Van Ermengem (1897) fed a *Macacus rhesus* with 5 ml of a preparation of macerated ham, which was known to be toxic. Symptoms developed in 12 hr, and consisted of restlessness, crying, coughing, and sneezing; later there was a secretion of viscid mucus in the nose and mouth, leading to transient suffocation; the pupils were dilated, reacting weakly to light. The animal became motionless, its head drooped, its eyes were fixed and half covered by the lids. Death occurred 24 hr after the time of feeding. At necropsy the stomach, the bases of the lungs, and the meninges were congested, and petechial haemorrhages were noticed on the arachnoid and throughout the brain and medulla.

Mice and guinea-pigs are highly susceptible. For the demonstration of botulinum toxin the mouse is usually chosen. A few hours after intraperitoneal inoculation, the earliest symptom is generally dyspnoea, the respiration being mainly costal. This is evidenced by an indrawing of the abdomen which gives a wasp-waist appearance, and by laboured 'bellows' breathing. Thereafter, various flaccid paralyses develop, or a generalized paralysis, so that the animal lies motionless with limbs outstretched.

Cats are also highly susceptible to intoxication, but dogs are much more resistant.

The effect of botulinum toxin given by mouth is dependent not only on the susceptibility of the animal species but also on the ability of the toxin to pass across the intestinal wall. The proportion of toxin that enters the circulation is indicated by the oral/intravenous ratio, which is the ratio between the minimal lethal amount of toxin given orally and that injected intravenously. This ratio varies, not only from one animal species to another, but also from type to type of botulinum toxin (see Smith 1977).

Cl. tetani

Tetanus can be reproduced by the inoculation of pure cultures, or of the toxin, into mice, rats, guinea-pigs, rabbits, goats, horses and monkeys. Cats and dogs are more resistant; birds and cold-blooded animals are highly resistant. The most susceptible animal, calculated on the amount of toxin per gram of body weight necessary to prove fatal on injection, is the horse. On this basis the horse is about 12 times, the guinea-pig 6 times, and the monkey 4 times as susceptible as the mouse (von Lingelsheim 1912, Sherrington 1917). On the other hand, the rabbit is twice, the dog 50 times, the cat 600, and the hen 30 000 times as resistant as the mouse (Kitasato 1891; see also Wright 1955).

Mice After the *subcutaneous* injection of a small quantity of toxin or culture near the root of the tail, symptoms develop in 12 to 24 hr. The spasms start near the site of the injection, and spread to the rest of the body, till the animal dies in a state of general tonic contraction. The first symptom is a stiffening of the tail, which becomes erect and is turned towards the side of inoculation; the hind leg of that side becomes stiff, and later the opposite leg. The contractions pass to the muscles of the trunk, and the mouse develops kyphosis or pleurothotonos. Next, the fore-legs are affected, and finally trismus and opisthotonos set in. The contractions are spasmodic, with remissions during which the animal can be readily excited by the slightest touch or a breath of air. Death follows in about 24 hr. *Post mortem*, there is little but slight congestion and oedema round the site of inoculation, and the spleen may be somewhat enlarged. Fluid, sometimes blood-stained, may be seen in the pleura or peritoneum. After injection of culture, the bacilli are cultivable from the site of inoculation, but are difficult to find under the microscope. The heart blood and viscera are sterile.

Guinea-pigs The experimental disease in guinea-pigs follows much the same course in about the same time as in mice.

Rabbits After subcutaneous or intramuscular injection the incubation period in rabbits is at least 24, and generally 36 hr; death does not occur for 3 or 4 days. The general tetanic spasms are more severe than in mice or guinea-pigs (Rosenbach 1886).

Cl. perfringens

Intramuscular injection of about 0.2 ml of an 18-hr glucose broth culture into the thigh of a guinea-pig usually results in gas gangrene with death in 12–48 hr. *Post mortem*, there is a large, brawny, crepitant swelling at the site of inoculation, covered with a dark-red, tense layer of skin. The muscle is pale and is undergoing liquefactive necrosis. In the subcutaneous tissue around the local lesion and spreading up to the abdomen, reaching sometimes to the sternum and over to the opposite thigh, is a collection of slightly blood-stained fluid, usually laden with globules of free fat, and gas smelling of hydrogen sulphide. The adrenal glands are often congested, so that the normally sharp differentiation of cortex from medulla becomes obscured. Microscopically, the organisms are present in large numbers in the local effusion and in much smaller numbers in the blood stream. Sporing forms are absent. An even more typical picture of gas gangrene is obtained in pigeons (Bull and Pritchett 1917*a*). Mice are less susceptible than guinea-pigs.

Cl. septicum

Intramuscular injection of about 0.1 ml of a 36-hr glucose broth culture into a guinea-pig causes death in 12 to 24 hr. *Post mortem*, there is a blood-stained gaseous oedema at the site of inoculation, spreading up over the abdominal wall, with collections of gas in the groins and axillae. The thigh and abdominal muscles are soft and deep red in colour. In the pericardial and peritoneal cavities there may be some fluid; the adrenals are congested, but not so noticeably as in animals infected with *Cl. perfringens*. Microscopically the exudate shows large numbers of motile rods and usually the characteristic navicular or citron forms. Most characteristic are the long curved filaments found on the peritoneal surface of the liver. In mice the experimental disease is similar.

Cl. novyi

Intramuscular injection of about 1 ml of a 24-hr glucose broth culture of types A and B into a guinea-pig or mouse produces death in 1 to 2 days. *Post mortem*, the muscles at the site of inoculation are very congested, purplish-red in colour, and infiltrated with small bubbles of gas. There is a massive spreading, gelatinous oedema, sometimes slightly blood-tinged, extending over the thigh. So extensive is the oedema in mice that the animals are flattened, and are twice their normal width across the pelvic region, assuming the shape of a squashed pear. The abdominal muscles are unaltered. Microscopically, bacilli are found in small numbers in the oedema fluid, and on the peritoneal surface of the liver; cultures from the heart blood may or may not be positive.

The action of these last three organisms differs. *Cl. perfringens* gives rise to a large amount of gas, *Cl. novyi* to very little. The oedema fluid of *Cl. novyi* infections is practically clear, of *Cl. perfringens* infections slightly blood-tinged, and of *Cl. septicum* infections strongly blood-tinged. With *Cl. perfringens* the muscles are pale pink, with *Cl. novyi* purplish-red, and with *Cl. septicum* intensely and deeply red. Human cases of gas gangrene differ too in certain respects; as a rule either oedema or, more rarely, gas production is dominant; occasionally both are apparent. The particular form in each patient is determined by the nature of the organisms present.

Classification

There is a large variety of anaerobic spore-bearing bacilli. Besides those of immediate interest to the medical and veterinary bacteriologist, there are many organisms of biological and economic importance; some with striking biochemical properties, like the thermophilic clostridia, or the clostridia attacking fats, cellulose, or pectin; some are important in industrial processes, like those associated with food spoilage and other forms of putrefaction. With the exception of *Cl. butyricum*, which is the type species of the genus, we have not discussed these anaerobes.

Although a number of clostridia have been fully investigated and deserve the rank of species accorded to them, a larger number are inadequately described. In these circumstances, classification that would include all named strains are clearly tentative.

A primary grouping suggested soon after the 1914–18 war (Report 1919; see also Weinberg and Séguin 1918, Heller 1921, and Hall 1922) was based on the position of the spore within the vegetative cell and possession of proteolytic and saccharolytic properties. In this context 'proteolytic' refers to the digestion of coagulated proteins and 'saccharolytic' to the frank fermentation of sugars. They do not necessarily reflect an organism's capacity for using proteins and carbohydrates. For example, *Cl. tetani* is usually classed as non-saccharolytic; but in the presence of iron salts a strain of *Cl. tetani* has been shown to produce CO_2 and ethanol from glucose (Lerner and Pickett 1945). The distinction is unsuitable for a primary classification because, although a frankly saccharolytic species like *Cl. butyricum* and a frankly proteolytic species like *Cl. sporogenes* are easily recognized, there are borderline cases where the significance of slight activity is in doubt. Moreover, within some species, proteolytic strains throw non-proteolytic variants, and vice versa.

There are no convenient and outstanding characters for a detailed subdivision of the genus, but some insight into the relation between the different species may be obtained by studying the composition of their cell walls, their nucleotide sequences in preparations of DNA, and their nutritional requirements (Cummins and Johnson 1971); and the nature of the amino acids used and their degradation products (Mead 1971, Elsden and Hilton 1979). Identification of volatile fatty acids in fluid cultures by gas–liquid chromatography is also of considerable importance in the taxonomic study of anaerobic bacteria (see Moore *et al.* 1966, Moore 1970, Moore and Holdeman 1972, Nakamura *et al.* 1983). Table 42.3 lists some important characters of the named species with which we are concerned.

There are many named species, some well established and some whose status in the *Clostridium* group is open to question (see Buchanan and Gibbons 1974). Prévot (1938, 1948) considers the genus *Clostridium* to be incapable of covering all the anaerobic spore-bearing bacilli. He (1966) proposes two orders—Clostridiales and Plectridiales. Within the Clostridiales are two families, the second of which contains three genera, the *Clostridium* genus having eight sub-genera and a miscellaneous group. Within the Plectridiales are two families, each with two genera. This classification in our opinion places too much weight on trivial and, in some cases, insufficiently established differences. The recognition of variation of the $S \rightarrow R$ type and detailed serological studies are helping to resolve some taxonomic difficulties, but it is clear that a large number of clostridia so far studied are antigenically heterogeneous, and that the variety of antigens, H, thermostable O, labile O, etc., is probably as complex as that displayed by the salmonellae.

Another difficulty in classification is the lack of any systematic study of variation among the clostridia. Rough and mucoid colony variants, with associated antigenic changes, occur in *Cl. perfringens*, and as we have noted, atoxic variants of toxigenic strains are not uncommon. Organisms in pure culture may occur in two or more phases from time to time, showing minor differences in morphological, cultural or biochemical characters. Pure cultures of anaerobes are not always easy to maintain, so that sudden or insidious changes and variations in the properties of a culture should always raise the suspicion of contamination.

Table 42.3 Reactions of some commonly encountered clostridia (after Willis 1977)

Organism	Acid from				Gelatinase activity	Indole production	Milk agar: digestion	Egg-yolk agar		Patho-genicity for laboratory animals
	glucose	maltose	lactose	sucrose				Lecithinase C activity	Lipase activity	
Cl. butyricum	+	+	+	+	−	−	−	−	−	−
Cl. perfringens (A–E)	+	+	+	+	+	−	−	+	−	v
Cl. tertium	+	+	+	+	−	−	−	−	−	−
Cl. fallax	+	+	+	+	−	−	−	−	−	v
Cl. carnis	+	+	+	+	−	−	−	−	−	+
Cl. chauvoei	+	+	+	+	+	−	−	−	−	+
Cl. septicum	+	+	+	−	+	−	−	−	−	+
Cl. paraputrificum	+	+	+	+	−	−	−	−	−	−
Cl. sphenoides	+	+	+	−	−	+	−	−	−	−
Cl. botulinum (A, B, F)	+	+	−	+	+	−	+	−	+	+
(C, D, E)	+	+	−	v	+	−	−	−	+	+
(G)	−	−	−	−	+	−	+	−	−	+
Cl. novyi (A)	+	+	−	−	+	−	−	+	+	+
(B)	+	+	−	−	+	−	−	+	−	+
(C)	+	−	−	−	+	−	−	−	−	−
(D)	+	−	−	−	v	+	−	+	−	+
Cl. bifermentans	+	+	−	−	+	+	+	+	−	−
Cl. sordelli	+	+	−	−	+	+	+	+	−	v
Cl. sporogenes	+	+	−	−	+	−	+	−	+	v
Cl. cadaveris	+	−	−	−	+	v	−	−	−	−
Cl. difficile	+	−	−	−	+	−	−	−	−	+
Cl. cochlearium	−	−	−	−	−	−	−	−	−	−
Cl. tetani	−	−	−	−	+	+	−	−	−	v
Cl. histolyticum	−	−	−	−	+	−	+	−	−	+
Cl. putrificum	+	−	−	−	+	−	+	−	−	−
Cl. hastiforme	−	−	−	−	+	−	+	v	−	−
Cl. ramosum	+	+	+	+	−	−	−	−	−	+

v = Some strains give a positive reaction; others do not.

(For keys to the identification of the pathogenic clostridia see Spray 1936, Reed and Orr 1941, Oakley 1956, Buchanan and Gibbons 1974, Cowan 1974, Smith 1975*b*, Willis 1977, Holdeman *et al.* 1977; for methods for their detection, isolation, and differentiation Batty 1967, Takacs 1967, Shapton and Board 1971, Walker and Batty 1980. and for more general reviews see von Hibler 1908, Wolf and Harris 1917, Weinberg and Séguin 1918, Report 1919, Weinberg and Ginsbourg 1927, Stickland 1934, Knight 1936, Sterne and Warrack 1964, Prévot 1966, Fredette 1967, Willis 1969*a,b*, Cummins and Johnson 1971, Mead 1971, Sterne and Batty 1975, Smith 1975*b*, 1977.)

It must not be thought that the following summarized descriptions cover all the clostridia that may be found in pathological material. Any investigation of anaerobic flora will yield clostridia which resemble neither well established nor ill described species, and their identification may be difficult or impossible. That most of them appear to have no pathogenic significance aggravates the difficulty, because their proper investigation is for that reason usually neglected, with a consequent perpetuation of the taxonomic confusion within this genus. *Cl. butyricum* is included, since it is the type species of the genus; though we may note that its identity with the organism originally described by Pasteur is dubious. The specific nomenclature adopted here is largely from the Approved Lists of Bacterial Names recommended by the Judicial Commission of the International Committee on Systematic Bacteriology (Skerman *et al.* 1980), which also closely parallels that used in Buchanan and Gibbons (1974). We have in this edition reduced *Cl. haemolyticum* to the status of type D *Cl. novyi* (Brooks *et al.* 1957).

Summarized descriptions of species

Clostridium butyricum

Apparently first described by Prazmowski (see Report 1919). Possibly identical with Pasteur's *Vibrion butyrique*, described fully by Winogradsky in 1902. Widely distributed in soil.

Morphologically, rods 3–4 μm × 0.7 μm; parallel sides, flattened ends, axis straight or slightly curved; arranged singly or in pairs end-to-end; considerable variation in length. Motile by peritrichate flagella. Spores oval, subterminal, measuring 1.6 μm × 1.3 μm; rod becomes spindle-shaped. Germination polar. Capsule formed on agar. Cells store glycogen; stain yellow with iodine. Gram positive.

Culturally, colonies are circular 0.5–1.0 mm in diameter, low convex, amorphous, faintly translucent or opaque, greyish-white, with smooth glistening surface and entire edge; butyrous consistency and easily emulsifiable. Non-haemolytic on horse blood agar, and egg-yolk negative. Grows poorly on any medium in the absence of a fermentable carbohydrate. Is otherwise not exacting, although it is an obligate anaerobe. Does not require amino acids or vitamins, other than biotin, and thus grows well in a mineral salts solution

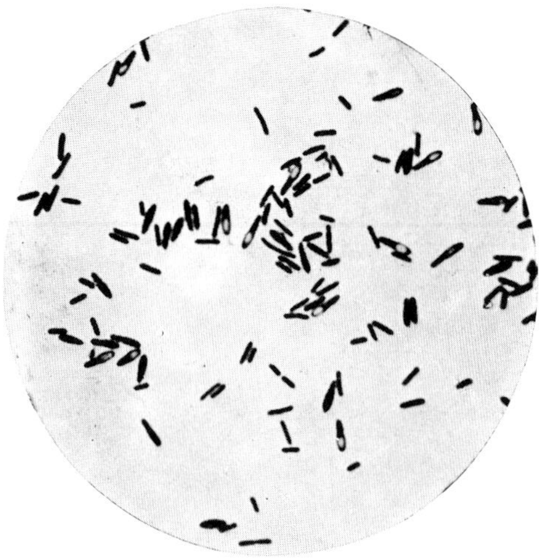

Fig. 42.1 *Clostridium butyricum*. From a surface agar culture, anaerobically, 5 days, 37° (× 1000).

containing biotin and glucose. Grows poorly, if at all, in cooked meat medium.

Glucose, maltose, lactose and sucrose are fermented, and starch is hydrolysed, but the organism is entirely non-proteolytic. Nitrate is not reduced, and neither H_2S nor indole is produced. Produces a typical stormy fermentation reaction in milk medium. Products of metabolism include acetic and butyric acids, and butanol.

Entirely non-pathogenic.

G + C content of DNA 28 moles per cent. Only sugar in cell wall is glucose (Cummins and Johnson 1971). Differs from *Cl. tyrobutyricum*, which is responsible for faulty fermentation in cheese (van Beynum and Pette 1935, Kutzner 1963).

Clostridium sporogenes

Described by Metchnikoff in 1908. Widely distributed in soil, and in faeces of man and animals.

Morphologically, a rod, 3–6 μm × 0.5 μm, with parallel sides, rounded ends, axis straight or slightly curved, arranged singly, in pairs, and small groups; long filaments occasionally formed. Spores freely; spores are oval, subterminal, and wider than bacillus; free spores numerous. Motile. No capsule. Strongly gram positive, except in old cultures. Spores are moderately to highly resistant, withstanding moist heat at 100° for 10–150 min, 105° for 45 min and 110° for 10 min.

Colonies are irregularly round, growing from a central focus; 2–6 mm in diameter, effuse or slightly umbonate, and rhizoid; surface covered by arborescent ridges, edge rhizoid; rather dull, greyish-yellow by reflected, bluish-grey by transmitted light; butyrous and easily emulsifiable; differentiated into compact brownish opaque centre and bluish translucent periphery. On horse blood agar small zones of pseudohaemolysis may be present under the colonies and extending up to 1 mm beyond them; there is considerable strain variation.

Cl. sporogenes is gelatinolytic and fibrinolytic, hydrolyses a wide variety of complex protein substrates including cooked meat, coagulated serum and egg, and casein. Native collagen not attacked. Proteinase activity accompanied by a putrid odour. In cooked meat medium there is heavy growth with dense turbidity; gas is produced and the meat digested

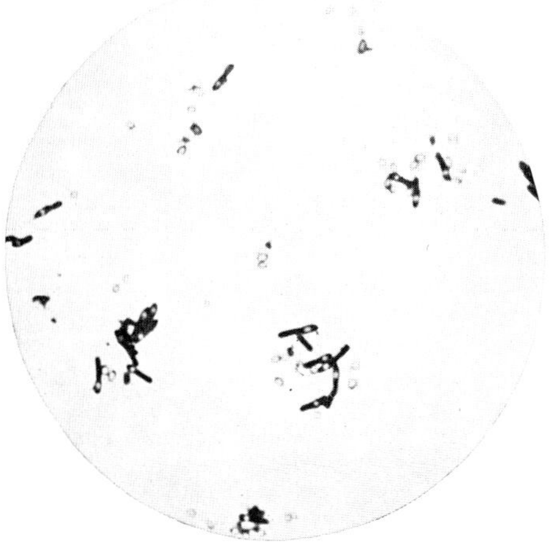

Fig. 42.2 *Clostridium sporogenes.* From a surface agar culture, anaerobically, 2 days, 37° (× 1000).

and blackened. Although the organism has some essential growth requirements, these are relatively simple and are provided in most ordinary media, upon which luxuriant growth occurs. Produces lipase, so that colonies on egg-yolk agar media show an underlying intense opacity coextensive with growth, and a superficial pearly layer; these are due to deposits of fatty acids derived from free simple fats in the medium. Does not produce lecithinase or cause an opacity in human serum agar media unless the serum is rich in fat.

Only glucose and maltose are fermented with the production of acid and gas. Indole is not formed and nitrate is not reduced, but H_2S is formed in abundance. As with some other proteolytic clostridia, volatile products of metabolism are multiple and complex, a wide range of both acids and alcohols being produced. Major acids include acetic, propionic, isobutyric, butyric, isovaleric and isocaproic.

Antigenically heterogeneous, but classifiable by heat-stable and heat-labile antigens. *Cl. botulinum*, *Cl. histolyticum*, and *Cl. tetani* share a minor heat-stable antigen. Spore antigens are distinct from H and O, which are not represented in the spore (Walker 1963). G+C content of DNA 26 moles per cent.

Does not produce any protein toxins and is not naturally pathogenic. Experimentally is non-pathogenic to laboratory animals, but enhances the pathogenicity of other anaerobes, such as *Cl. perfringens* in mixed infections and modifies the appearance of the local lesion. Filtrate of broth culture toxic to guinea-pigs in a dose of 1 ml, owing apparently to ammonia derived from deamination of amino acids, or possibly to toxic amines resulting from decarboxylation of amino acids.

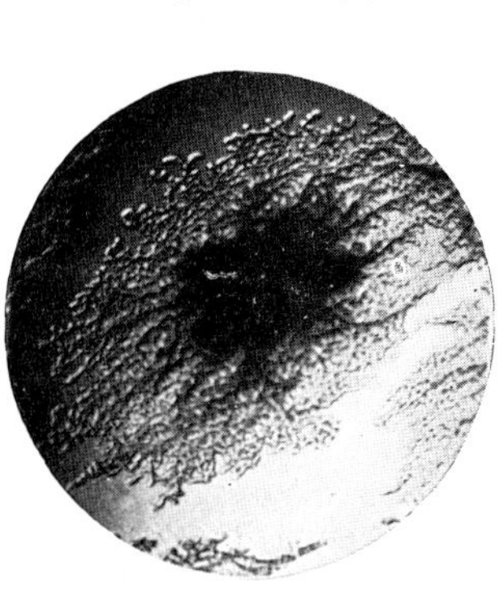

Fig. 42.3 *Clostridium sporogenes.* Surface colony on agar, anaerobically, 4 days, 37° (× 8).

Clostridium histolyticum

Described by Weinberg and Séguin in 1916 (1916, 1918). Widely but sparsely distributed in soil and probably also in the intestinal canal of man and animals.

Morphologically, a rod, 3–5 μm × 0.5–0.8 μm; parallel sides, rounded ends, axis generally straight. In cultures more than a day old and under aerobic conditions of culture irregular forms appear—long curved filaments, and irregularly stained forms. Spores are readily formed in all media but not under aerobic conditions. Spores are moderately heat resistant, being killed in 5 min at 105°; they are oval, subterminal, and wider than the bacillus; become free in old cultures. Motile by about 20 peritrichate flagella. Gram positive in young cultures. No capsule.

Culturally, although colonial appearances are variable, colonies are roughly circular in shape, about 1 mm in diameter, opaque and greyish-white with a shiny surface and an entire edge. No tendency for cultures to spread over the surface of the medium. Colonies of smooth strains are dewdrop-like, but those of rough variants may closely resemble *Cl. sporogenes*. Restricted growth occurs aerobically, a property *Cl. histolyticum* shares with *Cl. tertium* and *Cl. carnis*; aerobic colonies are small transparent domes, and the organisms may thus be regarded as an 'anaerobe and facultative aerobe'.

Though the nutritional requirements of *Cl. histolyticum* are moderately complex, good growth is obtained on most simple media. The organism is haemolytic on horse blood agar; shows neither lipase nor lecithinase activity on egg-yolk agar. Strongly proteolytic, however, and no other described organism rivals it in its ability to hydrolyse native proteins; some strains produce as many as nine different proteinases (Webster *et al.* 1962). It is gelatinolytic and fibrinolytic, and hydrolyses casein, haemoglobin, coagulated albumen and serum, and collagen both modified and raw. The organism ferments no sugars, does not produce indole or reduce nitrate, but produces abundant H_2S. In cooked meat medium there is abundant growth, the meat digested and slightly blackened. Gas is produced, and a deposit of white tyrosine crystals occurs, increasing with age. In milk medium casein is precipitated and digested. The only volatile product of metabolism produced in any amount is acetic acid.

Antigenically, the organism has a relatively simple structure, all strains sharing one common O antigen. On the basis of heat-labile antigens, *Cl. histolyticum* is divisible into two groups (Mandia 1955). There are minor antigenic affinities between *Cl. sporogenes*, *Cl. tetani* and *Cl. botulinum* types A, B, and F. Culture filtrate contains at least five antigens: a lethal and necrotizing toxin (α), a collagenase (β), a cysteine-activated proteinase (γ), an elastase (δ) and an oxygen-labile haemolysin (ε).

Exotoxin is said to be formed in very young cultures. *Natural pathogenicity* doubtful; appears often in gangrenous processes in man. *Experimentally*, strains vary in pathogenic-

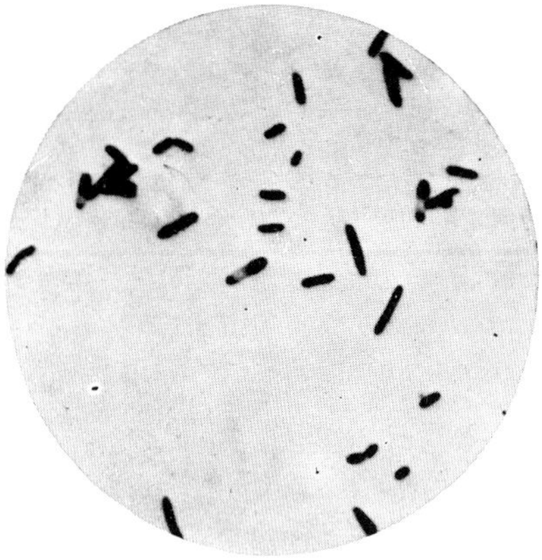

Fig. 42.4 *Clostridium histolyticum.* From a surface agar culture, anaerobically, 6 days, 37° (× 1000).

ity; susceptible animals are guinea-pig, rabbit and mouse. Bacillus is actively proteolytic and digests living tissue. 1 ml of a young broth culture injected intramuscularly into a guinea-pig causes digestion of skin and muscles, and a haemorrhagic liquefaction of the softer parts of the limb. This digestion may spread over the abdomen, and death occur during the next 12 to 24 hr; or recovery may follow with more or less complete necrosis of the limb, with auto-amputation at the hip joint following digestion of the joint ligaments and capsule. The fluid contains no gas and is not putrid. Cultures in gelatin contain large amounts of ammonia, sufficient to be acutely lethal on injection.

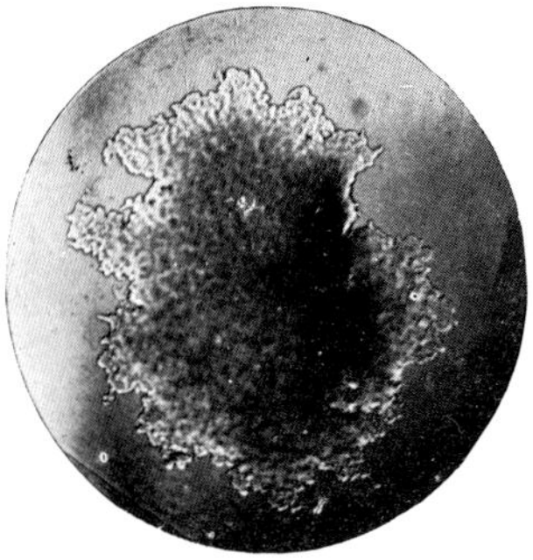

Fig. 42.5 *Clostridium histolyticum.* Surface colony on agar, anaerobically, 2 days, 37° (× 8).

Clostridium botulinum

Isolated by van Ermengem from ham in 1896 (1896, 1897). Several other organisms have since been isolated from botulism-like diseases in animals. Some of these were once referred to as *Cl. parabotulinum*, but are now given the letters A–G to refer to the toxigenic types. Types C_α was isolated by Bengtson (1922*a*, *b*, 1923) and by Graham and Boughton (1923*a*, *b*) in the United States from chickens and ducks, types C_β by Seddon (1922) in Australia from cattle; type D by Theiler and his colleagues (1926) in South Africa from cattle; type E from human botulism (Gunnison *et al.* 1936); type F (Dolman and Murakami 1961) in association with human botulism by Møller and Scheibel (1960); and type G (Gimenez and Ciccarelli 1970) from Argentinian soil (see also Ciccarelli *et al.* 1977). The following description refers chiefly to toxigenic strains of types A and B, to which types E and F are similar. The organism is widely distributed in soil, both virgin and cultivated, and irregularly in marine deposits. Not infrequently present in intestinal tract of animals and birds, in which types C and D are probably obligate parasites.

Morphologically, are large, stout rods, 4–6 μm × 0.9 μm; axis straight, parallel sides, and slightly rounded ends. Variations in depth of staining. Spores are oval, wider than the bacillus, thick-walled, and situated at or near the end. Free spores may be numerous, although highly toxic strains of all types tend to form few spores (Skulberg and Hausken 1965). Some strains spore readily, others hardly at all. Spores formed best in sugar-free media. Sluggishly motile by 4–8 peritrichate flagella. Spores are destroyed by dry heat at 180° in 5 min; moist heat at 100° destroys them in 5 hr, at 105° in 100 min and at 120° in 5 min (but see Chapter 72). Spores of types A, B and F are highly resistant, those of C and D intermediate, and of type E of low resistance. No capsule. Gram positive in young cultures.

Culturally, *Cl. botulinum* is conveniently divided into four groups (Smith 1977): *group I*: strains of type A and proteolytic strains of types B and F; *group II*: non-proteolytic strains of types B and F, and all type E strains; *group III*: non-proteolytic strains of types C and D; *group IV*: strains of type G.

Group I strains form colonies 3–8 mm in diameter, similar in appearance to those of *Cl. sporogenes*. Rounded, with an opaque compact centre and a somewhat rhizoidal margin; surface is matt or semi-glossy; growth often tends to spread over the surface of the medium.

Colonies of Group II organisms are smaller, 1–3 mm in diameter, irregularly circular, greyish and semitranslucent, and with a ground-glass surface.

Group III colonies are 1–3 mm in diameter, circular and entire, with a slightly undulate edge, raised, greyish-white and semitranslucent with a smooth matt surface.

Colonies of type G are small domes, about 0.5–1 mm in diameter, greyish and translucent, entire and with a shiny surface.

On horse blood agar, colonies of proteolytic strains may be surrounded by a narrow zone of haemolysis. On egg-yolk agar, all types except type G are lipolytic. A few group III strains produce lecithinase, and these are culturally indistinguishable from *Cl. novyi* type A on egg-yolk agar. Group I strains on egg-yolk media are similar in appearance to *Cl. sporogenes*.

Biochemically, group I organisms ferment glucose, do not attack mannose, and vary in their activity against maltose and sucrose. Group II organisms attack glucose, sucrose and mannose, but vary against maltose. Group III strains ferment glucose, maltose and mannose, but not sucrose. Group IV strains are entirely non-saccharolytic. None of the types ferments lactose. Production of abundant ammonia from peptone by group I strains may mask fermentation reactions.

Strains of all groups are gelatinolytic, but only those of groups I and IV attack casein and coagulated proteins, and produce H_2S. None of the organisms produces indole or reduces nitrate. All types grow well in cooked meat medium, but only strains of groups I and IV digest the meat particles.

Fermentation products vary from group to group. Strains of groups I and IV produce acetic, propionic, isobutyric, butyric and isovaleric acids; those of group II mainly acetic

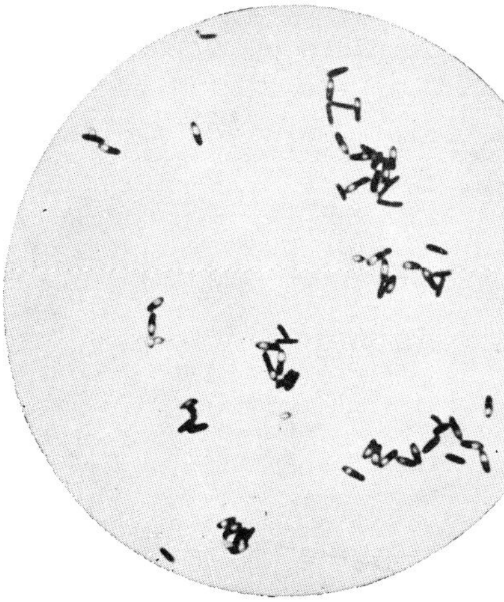

Fig. 42.6 *Clostridium botulinum*. From a surface agar culture, anaerobically, 2 days, 37° (× 1000).

and butyric acids, those of group III acetic, propionic and butyric acids.

Seven *antigenic types*, A–G, distinguished by their toxin production; toxins antigenically distinct, except for a minor antigen common to C and D. Filtrates also contain a haemagglutinin for human and other RBC. The toxigenic types A and B are divisible into several subtypes by agglutination reactions of bacillary suspensions. An antigenic division depending on heat-stable and heat-labile antigens does not distinguish A and B; but establishes a relation with *Cl. sporogenes*, *Cl. histolyticum* and *Cl. tetani*. In Type E the somatic antigens are homogeneous, the flagellar antigens strain-specific (Lynt *et al.* 1967).

Strains of all types produce serologically similar protease (Tjaberg and Fossum 1973). The validity of subdivision of the species into groups is emphasized by the high DNA and RNA homology values within each group, and differences in sugar content of cell walls (Smith 1977). G + C content of the DNA is 22–28 moles per cent.

Types A, B, E and F cause botulism in man. Type C_α causes limberneck in chickens and ducks; C_β causes one type of forage poisoning in horses in Australia and USA; C also causes botulism in horses in South Africa; D causes lamsiekte in cattle in South Africa. (For fuller description see Chapter 72.) The organism is a saprophyte but can multiply in the body to cause *wound botulism* and *infant botulism*. Injected subcutaneously, a broth culture of type A or B is fatal to guinea-pigs, mice, rabbits, cats, monkeys, and often chickens in 1 to 4 days; symptoms are muscular paralysis, dilatation of the pupils, shallow breathing, intense salivation, prostration and death. Toxicity varies widely with both the type of toxin and the test animal; in susceptible animals the toxins are the most potent known.

(For further references see Chapter 72.)

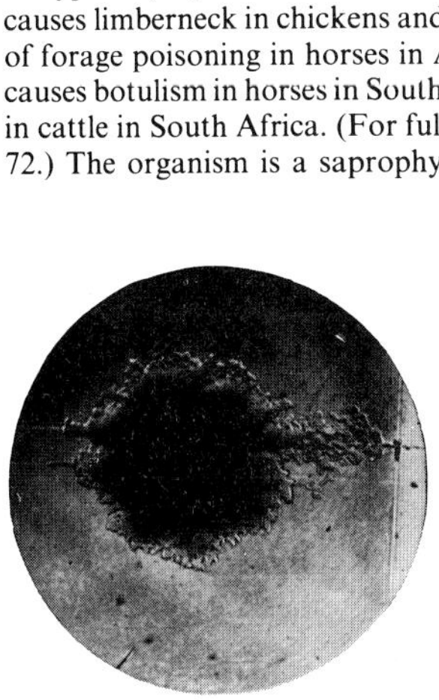

Fig. 42.7 *Clostridium botulinum*. Surface colony on agar.

Clostridium novyi (*oedematiens*)

The organism was described by Weinberg and Séguin in 1915, and is probably identical with Novy's *B. oedematis maligni* II, or *Cl. novyi* I (Novy 1894), Zeissler and Raszfeld's (1929) *B. gigas*, and Kraneveld's (1930) bacillus of osteomyelitis bacillosa bubalorum. Scott, Turner and Vawter (1933) proposed the terms type A, B and C for classical *Cl. novyi*, *B. gigas* and Kranefeld's bacillus respectively; and Oakley and Warrack (1959) proposed that the causal organism of bacillary haemoglobinuria of cattle in the United States (Vawter and Records 1926, 1931), given the species name *haemolyticum* (see Hall 1929), should be classed as *Cl. novyi* type D. Widely distributed in soil.

It is rod-shaped 3–10 μm × 0.8–1.0 μm, not unlike *Cl. perfringens*, but longer; type B and C strains may be as large as 4–20 μm × 1–2 μm; sides parallel; ends rounded; axis straight or curved; arranged singly, in pairs or chains; jointed filaments not uncommon. Spores formed freely in all media; they are large, oval and subterminal, generally free. They are destroyed by moist heat at 105° in 5 min. Motile by 20 or more peritrichate flagella; but motility is observed only under strictly anaerobic conditions. No capsule. Gram positive in young cultures.

Culturally, surface growth develops as irregularly round colonies, 2–3 mm in diameter, effuse, filamentous or curled, glistening, translucent with finely sponge-like surface and irregularly lobate edge with very fine dentations; greyish-yellow by reflected, greyish-blue by transmitted light; butyrous and easily emulsifiable. There is a tendency for growth of types A and B to spread over the surface of the medium, but colonies of type D are exceedingly small and compact (mere pin-points) in 24-hr cultures. On horse blood agar a zone of β-haemolysis coextensive with the colony is produced, except by type D strains which produce extensive haemolysis in 24 hr, and type C strains which are non-haemolytic. On egg-yolk agar media type A produces a diffuse lecithinase-C opacity and a restricted pearly layer due to lipolysis. Types B and D produce diffuse lecithinase reactions only; type C is egg-yolk negative.

Growth in cooked meat medium is of the saccharolytic type. All types produce gelatinase but do not attack more complex proteins. H_2S is produced, especially by type D, which also produces large amounts of indole; types A, B and C are indole negative. Fermentation reactions are different for different types; types A and B ferment glucose and maltose, but types C and D ferment glucose only (Rutter 1970). The major products of metabolism include propionic and butyric acids, with lesser amounts of acetic and valeric acids.

Cl. novyi is a strict anaerobe. Type A grows readily on ordinary enriched media, but types B, C and D are extremely demanding both in their nutritional requirements and in their intolerance of oxygen. They are conveniently cultured on the

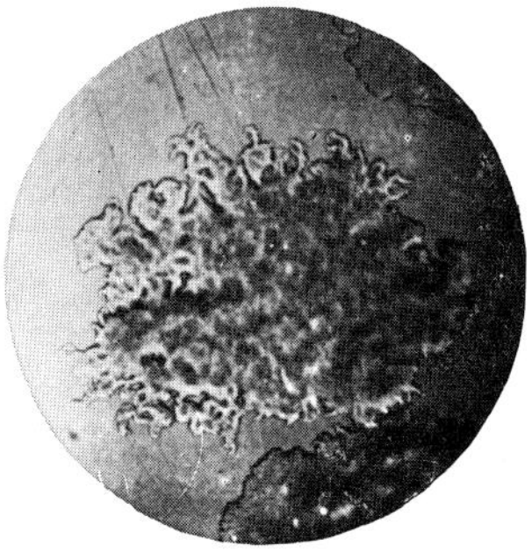

Fig. 42.8 *Clostridium novyi.* Surface colony on agar, anaerobically, 2 days, 37° (× 8).

glucose blood agar medium of Moore (1968), which contains cysteine and dithiothreitol. Type D is the most exacting and fastidious of all clostridia.

The somatic *antigenic structure* of *Cl. novyi* is comparatively homogeneous, two O antigens being shared by types A, B and C in varying proportions. Strains of type D share a common O antigen, but are antigenically distinct from other types. G + C content of the DNA 26–29 moles per cent.

Pathogenically, types A, B and D produce potent exotoxin. Culture filtrates may contain at least six antigens: α, the lethal toxin; β and γ, haemolytic lecithinases; δ, an oxygen-labile haemolysin; ε, a lipase; ζ, a haemolysin; η, a tropomyosinase, and θ, a substance producing opalescence in egg-yolk media that is probably a lipase. Species divisible into types A–D according to the presence of these antigens (Table 42.2). The organism is a cause of gas gangrene in man; it is responsible for one type of braxy in Europe, for black disease in Australia, for a non-fatal osteomyelitis of the humerus and femur of Dutch East Indian buffaloes and for haemoglobinuria of cattle. Types A, B and D are pathogenic for guinea-pigs, mice and rabbits; 0.25–1 ml of a 24-hr broth culture injected intramuscularly into a guinea-pig causes death in 24 to 48 hr. *Post mortem*, muscles are red and softened; little gas production, but a spreading gelatinous oedema. Bacilli found at site of inoculation, and occasionally on surface of liver. Blood cultures may or may not be positive.

(For further references see Chapter 63 and Sordelli *et al.* 1930, Djaenoedin and Kraneveld 1936, Haines and Scott 1940, Smith *et al.* 1956.)

Clostridium chauvoei

First distinctive description by Arloing, Cornevin and Thomas in 1879 (Arloing *et al.* 1887). The principal habitats of *Cl. chauvoei* are the intestinal tract of cattle and sheep, and soil.

Morphologically, rod-shaped, 3–8 μm × 0.6 μm, with parallel sides and rounded ends. Short filaments are not uncommon. Pleomorphism is the rule, navicular and swollen forms being common, as is variation in depth of staining, especially in the presence of serum or fresh tissue. Axis straight or slightly curved; arranged singly, in small groups, or in short chains. Spores are elongated, oval, subterminal, and wider than the bacillus. Clostridial forms are lemon- or pear-shaped. Spores may be destroyed by moist heat at 100° in 40–50 min. Motile. Gram positive, but weakly so after 4 days. No capsule. On surface of liver of infected animals it is found singly or in pairs, not in chains or filaments like *Cl. septicum*.

Culturally, colonies are irregularly round and entire, 2–4 mm in diameter after 48 hours' incubation, commonly umbonate, shiny greyish-white or semitranslucent. There is no tendency to swarm over the surface of the medium. On horse blood agar *Cl. chauvoei* is usually non-haemolytic but is β-haemolytic on sheep blood agar. It produces no reactions on egg-yolk agar. *Cl. chauvoei* is a strict and demanding anaerobe which grows poorly on ordinary media; growth is greatly improved by the addition of glucose and liver extract. Growth in cooked meat medium is of the saccharolytic type, with little or no turbidity.

Glucose, maltose, lactose and sucrose are fermented, but not salicin. H_2S is produced, but not indole. The organism produces a deoxyribonuclease and is gelatinolytic, but it does not attack more complex proteins. Major products of metabolism are acetic and butyric acids with lesser amounts of formic acid.

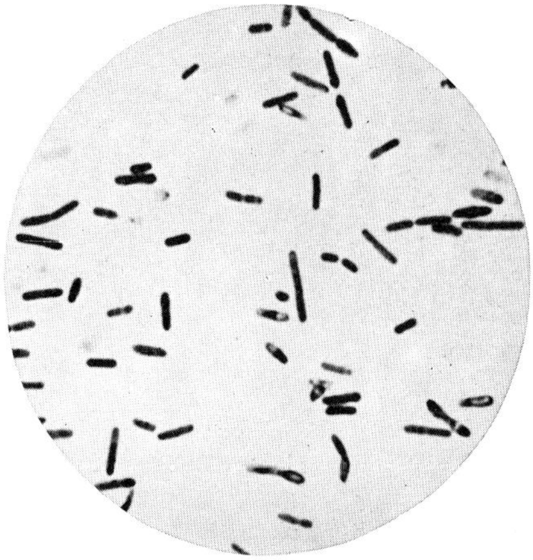

Fig. 42.9 *Clostridium novyi.* From a surface agar culture, anaerobically, 2 days, 37° (× 1000).

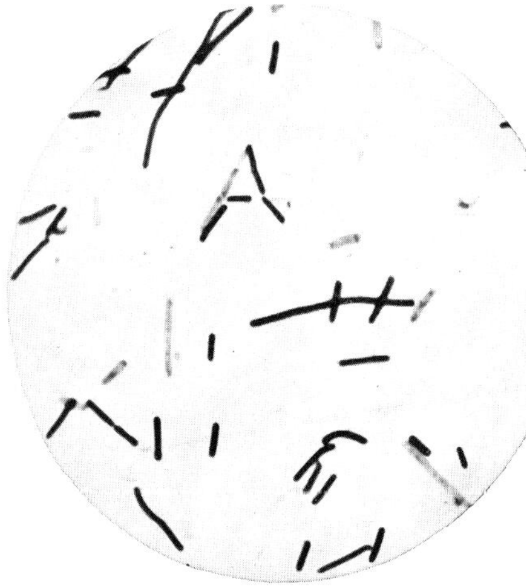

Fig. 42.10 *Clostridium chauvoei.* From a surface agar culture, anaerobically, 2 days, 37° (× 1000).

Antigenically, the organism is divisible into two groups on basis of one O and two H antigens; shares a spore antigen with *Cl. septicum*. Antitoxin is specific, protecting against *Cl. chauvoei* but not against *Cl. septicum*; *Cl. septicum* antitoxin provides not only homologous protection but also protection against intoxication by *Cl. chauvoei*, owing to the fact that it produces not only antigenically similar toxins to *Cl. chauvoei*, but also some additional ones. G+C content of the DNA 26–32 moles per cent.

Exotoxin produced by *Cl. chauvoei* which causes blackleg in cattle, and less often in sheep, against which protection is afforded by vaccination with formolized whole cultures. Non-pathogenic to man. Experimentally it is fatal to guinea-pigs and less often to mice; rabbits and pigeons are fairly resistant; 0.25 ml of a 24-hr culture intramuscularly kills a guinea-pig in 24 to 48 hr. *Post mortem*, slightly blood-stained serous exudate at site of injection; abdominal muscles are deep red and contain numerous small gas bubbles. *Cl. chauvoei* can be recovered from local lesion, peritoneal cavity, and heart blood.

(For further references see Chapter 63 and Kitt 1887, No card and Roux 1887, Markoff 1911, Landau 1917, Haslam and Lumb 1919, Heller 1920, Goss *et al.* 1921.)

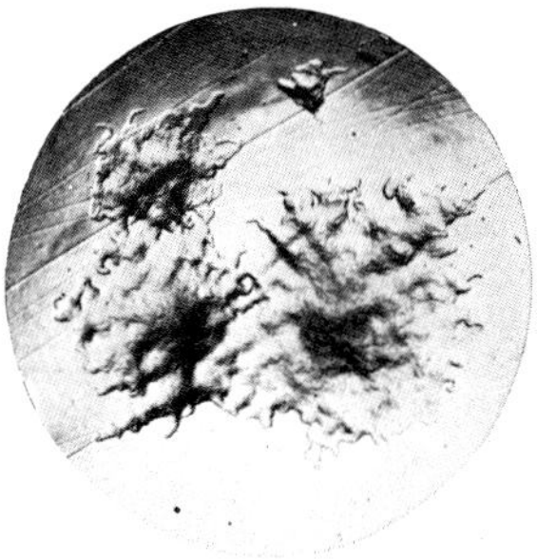

Fig. 42.11 *Clostridium chauvoei*. Surface colonies on agar, anaerobically, 6 days, 37° (×8).

Clostridium septicum

Described by Pasteur and Joubert in 1877, and synonymous with the *Vibrion septique* of Pasteur. Found widely distributed in soil but is also commonly present in the intestinal tract of animals and man.

Morphologically, the bacillus is of variable length and thickness; on agar cultures, 2–6 μm × 0.4–0.6 μm; sides parallel, ends rounded, axis straight or curved, arranged singly, in pairs, and in short chains. On peritoneal surface of dead guinea-pig it forms long jointed filaments. In tissue exudates and in fluid media containing fresh tissue there are navicular or citron forms with pale swollen bodies and deeper-staining pointed extremities. In agar cultures there is striking pleomorphism; organisms vary in size, shape, and depth of staining, large numbers of shadow forms are seen. Spores readily formed, and are oval, subterminal, and slightly wider than bacilli; often found free. Motile by 4–16 peritrichate flagella. No capsule. Gram positive in young cultures, but often frankly gram negative in older cultures.

Culturally, colonies are irregularly round, having a general cigarette-in-water appearance, 10 mm in diameter at 36 hr, effuse, filamentous, translucent colonies, with finely honeycombed surface due to crossing of numerous filaments,

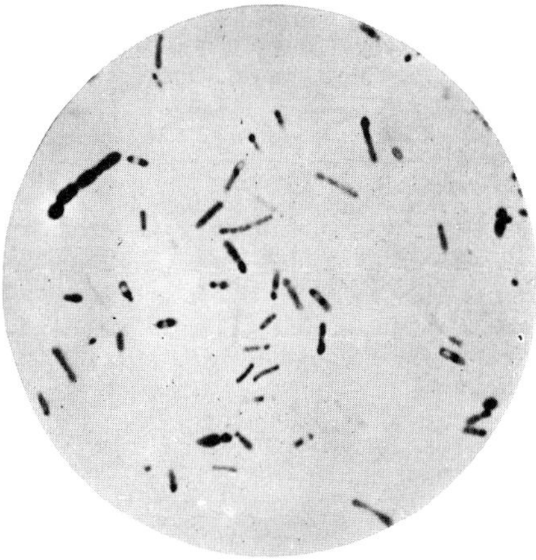

Fig. 42.12 *Clostridium septicum*. From a surface agar culture, anaerobically, 2 days, 37° (×1000).

and fimbriate edge; greyish by reflected light; butyrous and easily emulsifiable. The organism tends to swarm over the surface of the medium, similarly to *Cl. tetani*, although the film of growth is thicker and coarser. On horse blood agar the organism causes β-haemolysis, but on egg-yolk agar is non-lipolytic and lecithinase-negative. Growth in cooked meat medium is of the saccharolytic type.

Cl. septicum is a strict anaerobe, but less demanding than *Cl. chauvoei*, and grows moderately well on ordinary media. It ferments glucose, maltose, lactose and salicin, but not sucrose. Neither H_2S nor indole is produced. The organism produces a deoxyribonuclease, fibrinolysin and gelatinase, but does not attack more complex proteins. Major products of metabolism are acetic and butyric acids with lesser amounts of formic acid.

Antigenically, the organism is divisible into six groups on the basis of two O and five H antigens; it shares a spore antigen with *Cl. chauvoei*. The α-toxin is specific, except for a minor antigenic relation to the α-toxin of *Cl. histolyticum*. Culture filtrates contain at least four antigens: α, the major toxin, which is lethal and haemolytic; β, a deoxyribonuclease with leucocidal activity; λ, a hyaluronidase; and δ an

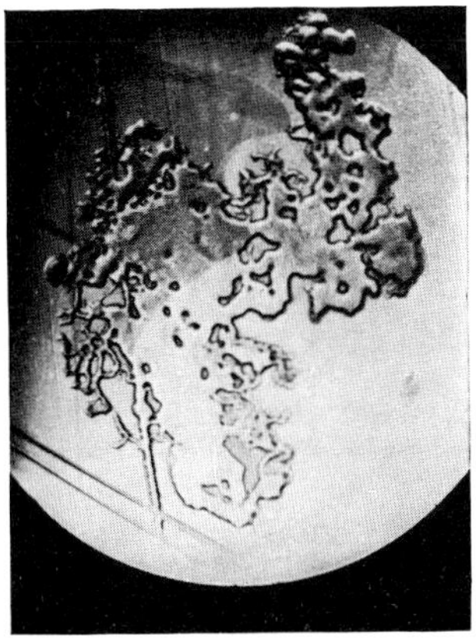

Fig. 42.13 *Clostridium septicum*. Surface colony on agar, anaerobically, 2 days, 37° (×8).

oxygen-labile haemolysin with necrotizing activity. The latter three antigens are related to those produced by *Cl. chauvoei*. G + C content of the DNA 24 moles per cent.

Pathogenic by virtue of the exotoxins formed; a cause of gas gangrene in man, malignant oedema in cattle and braxy in sheep, and sometimes blackleg in cattle. Experimentally, it is pathogenic to guinea-pigs, mice, rabbits, and pigeons. Pathogenicity is retained for years in subculture. 0.01–0.5 ml of a 24-hr glucose broth culture injected intramuscularly into guinea-pigs causes death in 12 to 24 hr. *Post mortem*, blood-stained oedema and gas production; muscles intense deep red in colour and softened; sometimes fluid in peritoneum and pericardium. Motile rods and navicular forms at site of injection, and long, jointed, snake-like filaments on peritoneal surface of liver.

Clostridium perfringens (*welchi*)

First complete description by Welch and Nuttall in 1892, who isolated it from a cadaver. Three organisms closely resembling *Cl. perfringens*, but differing from it in type of toxin production, have been isolated from diseased sheep, and called the lamb dysentery bacillus (Dalling 1926), *Cl. paludis* (McEwen 1930), and *Cl. ovitoxicum* (Bennetts 1932). To these, respectively designated types A, B, C and D by Wilsdon (1931), are now added type E from enterotoxaemia of calves (Bosworth 1943), and two type-C variants from enterotoxaemia in man (Zeissler and Raszfeld-Steinberg 1949, Oakley 1949, Lawrence and Walker 1976). *Cl. perfringens* is ubiquitous in nature and in the intestinal tract of man and animals.

Morphologically, it forms stout rods, varying considerably in length; 4–8 μm × 0.8–1.0 μm; sometimes shorter and more slender; filaments not uncommon. Parallel sides, ends truncated or slightly rounded; axis straight. Arranged singly, often side by side forming small bundles. Variation in depth of staining; involution forms—clubs, filaments, tadpoles, granular forms—frequent in old cultures. Spores large, oval, and subterminal but are rarely seen in artificial culture. Sporulation occurs more readily with some strains than with others, and is favoured by an alkaline reaction; does not occur below pH 6.6, and hence is unusual in media containing a fermentable carbohydrate. Spores are seldom seen in cultures. Non-motile. Capsules formed in animal body.

Culturally, it forms two main types of surface colony. One is round, 2–4 mm in diameter, low convex, amorphous, greyish-yellow, opaque, with smooth surface and entire edge; butyrous and easily emulsifiable. Other is umbonate, and is differentiated into an opaque brownish centre and a lighter, more translucent, radially striated periphery with a crenated edge (Fig. 42.15). Other variant forms occur, differing in morphology, colonial appearance, and sometimes toxicity. On horse blood agar, colonies are surrounded by a zone of β-haemolysis and commonly also by an outer zone of partial lysis ('target haemolysis') owing to the action of θ- and α-toxins respectively. Non-haemolytic strains on horse blood are not uncommon but most are highly haemolytic on sheep blood agar. On egg-yolk agar *Cl. perfringens* produces a diffuse lecithinase-C opacity, which is inhibited by *Cl. perfringens* antitoxin; no lipolysis is produced. In cooked meat medium early growth develops; gas is evolved and the meat particles are turned pink; no digestion; sour odour. In plain milk medium rapid fermentation causes the characteristic stormy fermentation reaction (Fig. 42.16). The organism is gelatinolytic but most strains do not attack more complex proteins. Grows fairly well on ordinary media, but growth is greatly improved by 1 per cent glucose. Green fluorescence on MacConkey plate.

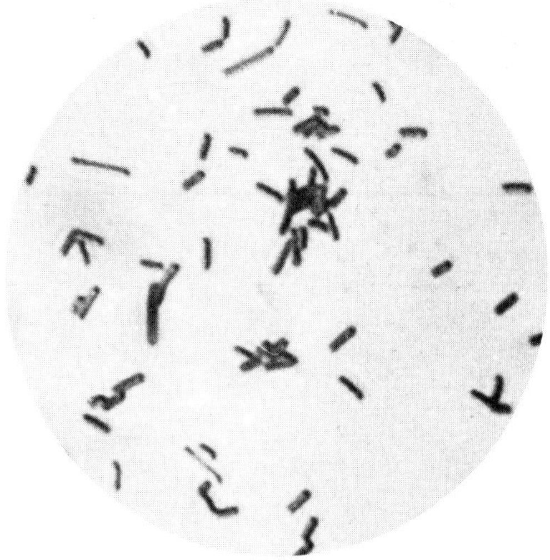

Fig. 42.14 *Clostridium perfringens*. From a surface agar culture, anaerobically. (Photomicrograph kindly supplied by Dr Betty Hobbs.)

Cultures in fermentable carbohydrate media die in a few days owing to the effect of the acid produced. In sugar-free protein media, in which spores have formed, the organisms may live for months. A suspension containing a million spores per ml is sterilized in 30 min at 90° and in 5 min or less at 100° (Headlee 1931), but some type A strains of the food-poisoning type and some type C strains may withstand boiling for 1 to 3 hr.

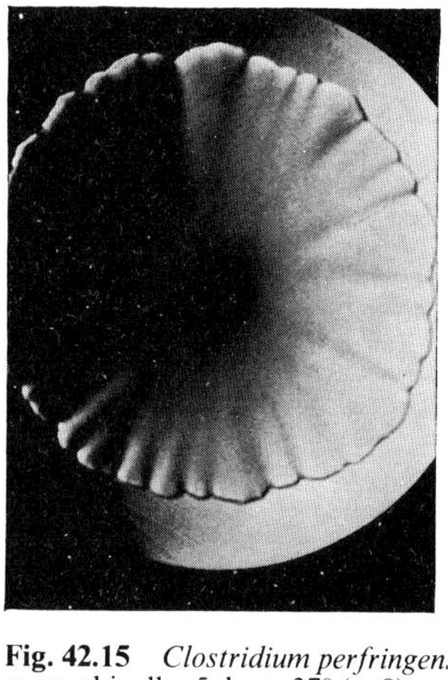

Fig. 42.15 *Clostridium perfringens*. Surface colony on agar, anaerobically, 5 days, 37° (× 8).

Fairly tolerant anaerobe. Opt. temp. 37°. Forms fibrinolysis, hyaluronidase, collagenase, gelatinase, haemolytic and lethal lecithinase, deoxyribonuclease and oxygen-stable and oxygen-labile haemolysins, haemagglutinin, and leucocidins.

Acid and gas in glucose, maltose, lactose, and sucrose.

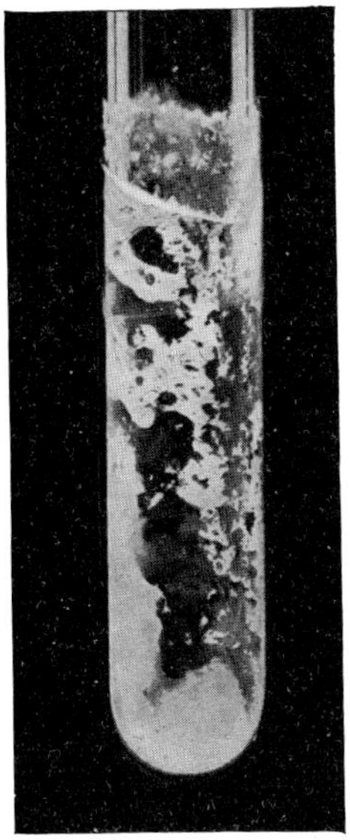

Fig. 42.16 *Clostridium perfringens.* Culture in litmus milk, anaerobically, 24 hr, 37°, showing stormy fermentation.

Indole −; nitrate slight reduction, greater with washed suspensions; NH_3 slight +; H_2S + +. Major products of metabolism are acetic and butyric acids; butanol is sometimes also produced.

Antigenically, by agglutination no clear cut grouping is apparent except among food-poisoning strains which are divisible into 75 serological types on the basis of surface polysaccharides. Considerable overlapping of antigens. No obvious relation between agglutinogens and types determined biochemically or by filtrate antigens, of which 12 are known: α, the lethal toxin which is a lecithinase; β, γ and η, lethal toxins; δ, lethal haemolysin; ε and ι lethal and necrotizing toxins, activated by trypsin; θ, oxygen-labile haemolysin; and κ, λ, μ, ν, respectively collagenase, gelatinase, hyaluronidase and deoxyribonuclease. The species is divisible into types A, B, C, D and E according to the presence of these antigens (Table 42.1). G+C content of DNA 24 moles per cent.

Phages of *Cl. perfringens* are described (Guélin 1950) and are isolable from lysogenic strains of types A, B and C, though so far not from types C, D and E (Smith 1959). They appear to be type- and even strain-specific (see also Mahony and Kalz 1968).

Pathogenically, chief agent of gas gangrene in man; also responsible for enteritis, food-poisoning, puerperal fever, septic abortion, and a variety of less common infections, such as meningitis, cholecystitis and panophthalmitis. Causes gas gangrene in animals, especially sheep. Experimentally, great variation in pathogenicity of different strains. Washed bacilli or spores are non-pathogenic. 0.1–1.0 ml broth culture injected intramuscularly into guinea-pig causes local tumefaction, spreading oedema, and death in 24 to 48 hr. Also pathogenic to mice, pigeons, and less so to rabbits. B type is responsible for lamb dysentery, C type for 'struck', an enteritis of sheep, and for human necrotizing jejunitis, D type for an enterotoxaemic disease and for pulpy kidney in sheep, and E type for an enterotoxaemia of calves (see Chapters 63 and 72). Extensive studies by Bullen and his colleagues (see Bullen *et al.* 1978) indicate the important role played by iron in gas gangrene infections, and the protective effect afforded by the iron-binding proteins lactoferrin and transferrin.

(For further references see Kamen 1904, Simonds 1951*a,b,c*, Robertson 1916, Humphreys 1924, Howard 1928, Torrey *et al.* 1930, Mason *et al.* 1931, McGaughey 1933, Brooks 1961.)

Clostridium tetani

Cl. tetani was described by Nicolaier (1884) and isolated by Kitasato (1889*b*). It is widely distributed in soil—especially cultivated soil—and in the intestine of man and animals.

Morphologically, it forms rods, 2.5 μm × 0.5 μm; considerable variation in length; long, curved, filamentous forms are not uncommon. Axis straight, sides parallel, ends rounded. Variation in depth of staining. Spores spherical, terminal, and wider than the bacillus, giving characteristic drum-stick appearance. Sluggishly motile; peritrichate flagella. No capsule. Strongly gram positive in young cultures. In early stages, spores stain solidly; later, only the thin wall stains;

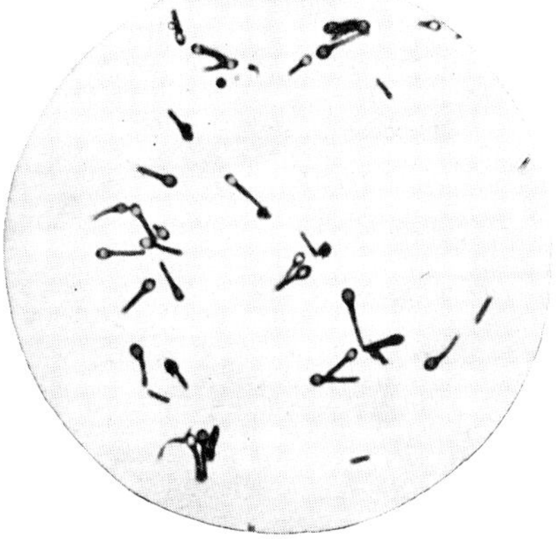

Fig. 42.17 *Clostridium tetani.* From a surface agar culture, anaerobically, 7 days, 37°, showing ring form of staining (× 1000).

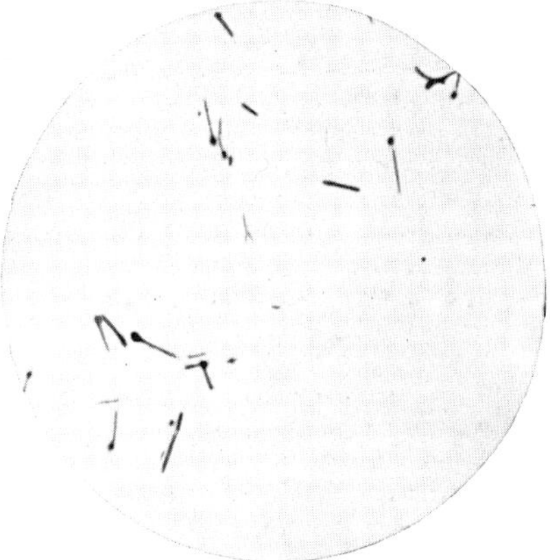

Fig. 42.18 *Clostridium tetani.* From a broth culture, anaerobically, 7 days, 37°, showing the spores stained solidly (× 1000).

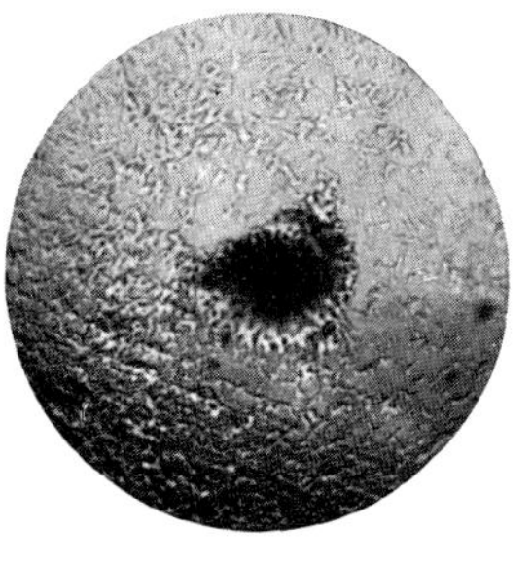

Fig. 42.19 *Clostridium tetani*. Surface colony on agar, anaerobically, 2 days, 37°, showing parent colony surrounded by an effuse growth (× 8).

spores rarely become free. The spores may resist heating for 15–90 min at 100°, but are destroyed at 105° in 3–30 min.

Colonies are irregularly round, 2–5 mm in diameter, effuse, glistening, translucent and greyish-yellow with irregularly granular surface, and ill defined edge, showing filamentous, curled projections; structure very finely granular or filamentous; butyrous consistency, emulsifying easily. Some strains form colonies differentiated into thicker, translucent, yellowish-brown centre, and thinner, transparent, almost colourless periphery. Whole colony has a fuzzy appearance. Isolated colonies of the normal motile type of *Cl. tetani* are extremely difficult to obtain, owing to the tendency of the organisms to spread as a delicately rhizoidal film over the surface of the medium; but non-motile variants give rise to separate discrete colonies.

The organism grows well on ordinary media, since its nutritional requirements are not exacting, although it is a strict anaerobe; growth is improved by the addition of blood or serum. On horse blood agar α-haemolysis is produced, later passing into β-haemolysis, owing to the production of tetanolysin, an oxygen-labile haemolysin. On egg-yolk agar there is no opalescence or pearly layer. Old cultures in cooked meat medium show slight blackening, but this is not due to proteolytic activity.

Cl. tetani does not attack sugars or complex proteins but forms a gelatinase. It causes a diffuse opacity in milk agar owing to precipitation of casein, presumably by a rennin-like enzyme (Willis and Williams 1970). Some rare aberrant strains ferment glucose, and about 50 per cent of isolates are fibrinolytic. Nitrate is not reduced, but indole is formed by most strains. H_2S is not formed. Metabolic products are acetic, propionic and butyric acids, ethanol and butanol.

Antigenically, *Cl. tetani* is differentiated by agglutination and complement fixation into at least 10 types, of which types I and III appear to be commonest in the British Isles. Toxins formed by all types are pharmacologically identical and are neutralized by antitoxin prepared against any one type. Antibacterial sera contain agglutinins and opsonins specific for each type. Heat-stable antigens related to those of *Cl. sporogenes*, *Cl. botulinum* and *Cl. histolyticum*. G + C content of the DNA 25–26 moles per cent.

Pathogenically, a potent exotoxin formed. Naturally pathogenic to man and horses in particular. Experimentally, mice, guinea-pigs, and rabbits are susceptible, dying of tetanus in 1 to 4 days after subcutaneous injection of a broth culture. Washed spores or bacilli are innocuous. Birds are resistant. (See Chapter 64.)

(For further references see Kitasato and Weyl 1890, Corbitt 1918, Tschertkow 1929, Mutermilch *et al.*, 1933.)

Clostridium difficile

Cl. difficile was described by Hall and O'Toole (1935) and Snyder (1937). It can be isolated from the faeces of 40 per cent of normal infants. It is rarely found in normal adults (George *et al.* 1978, Larson *et al.* 1978), which probably reflects a sparse presence rather than a restricted distribution.

Morphologically, the organism is a large gram-positive bacillus with terminal elongated spores slightly wider than the bacillary body. Cells measure 6–8 μm in length and 0.5 μm in width. It is uniformly gram positive in young cultures, but may become gram negative after 24–48 hr.

Culturally, *Cl. difficile* is a strict anaerobe. Colonies are 2–3 mm in diameter, irregularly circular, flat to slightly raised, semi-translucent and white with a glossy but rough and often pitted surface. The organism is non-haemolytic on horse blood agar, and produces neither lipase nor lecithinase on egg-yolk agar. It is gelatinolytic but does not attack more complex proteins. Glucose is fermented, but not maltose, lactose or sucrose. Indole and H_2S are not produced, and nitrate is not reduced. The organism is tolerant of *p*-cresol, which it produces during growth (Hafiz and Oakley 1976) from tyrosine; it has recently been shown that very large amounts of *p*-cresol are produced when its precursor, *p*-hydroxyphenyl acetic acid, is added to agar media at a concentration of 0.1 per cent (Phillips and Rogers, personal communication). Products of fermentation are multiple and complex, and include small amounts of acetic, isobutyric, isovaleric, valeric, butyric and isocaproic acids.

Pathogenically, some strains produce a thermolabile toxin which induces local oedema and convulsions in guinea-pigs. The toxin is lethal on injection into the cat, dog, rat, guinea-pig, rabbit and pigeon, but has no effect by mouth on the rat, guinea-pig and dog. The bacteria are antigenically heterogeneous; the toxin appears to be antigenically homogeneous. Whether or not this lethal toxin is identical with the toxin responsible for pseudomembranous colitis in man and hamsters remains to be established. The toxin responsible for pseudomembranous colitis is neutralized by *Cl. difficile* antitoxin and also by *Cl. sordelli* antitoxin, and its presence may be established and assayed by cytotoxicity testing in a variety of tissue culture cell lines, and by countercurrent immunoelectrophoresis (Welch *et al.* 1980).

Fig. 42.20 *Clostridium difficile*. From a surface blood agar culture, anaerobically, 2 days, 37° (× 1000).

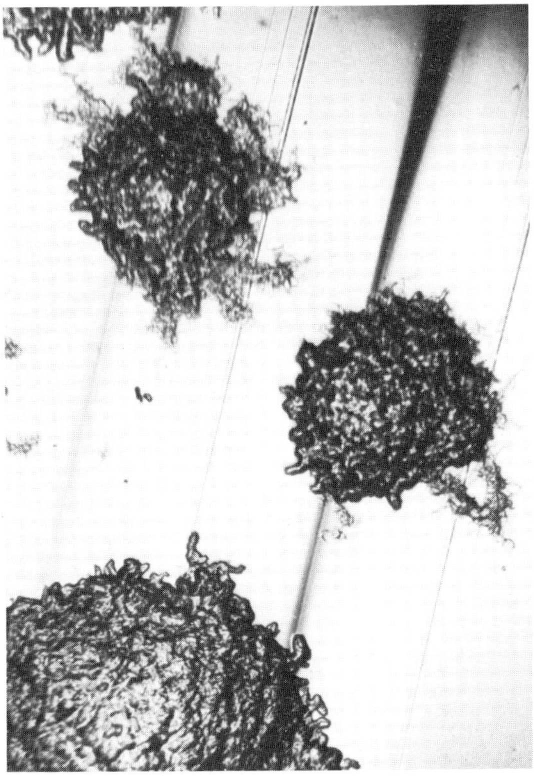

Fig. 42.21 *Clostridium difficile*. Surface colonies on blood agar, anaerobically, 2 days, 37° (×8).

Notes on less important species (see Table 42.3).

Cl. bifermentans Isolated in 1902 by Tissier and Martelly from putrefying meat, and named *B. bifermentans sporogenes*. So called from its being the first anaerobe shown to decompose both sugars and proteins. Probably identical with *Cl. centrosporogenes*, described by Hall (1922), McCoy and McClung (1936) and Clark and Hall (1937) (see also Weinberg, Davesne and Lefranc 1931, Hall and Scott 1927). Gram-positive rod, 3.6–6.0 μm × 1.2–1.5 μm, motile by peritrichate flagella, readily forming central or subterminal oval spores. Uniform turbidity in broth with filament production, and viscous deposit, which is easily disintegrated on shaking; nauseous odour. Surface colonies on horse blood agar are round, crenated, or irregular, and occasionally haemolytic. Gelatinolytic and actively proteolytic. Cooked meat medium digested, blackened, with a late deposit of small rounded masses of white crystals. Ferments glucose, maltose, mannose, sorbitol and salicin, but not lactose or sucrose. Urease –; indole +. Forms a fibrinolysin. Most strains form a lecithinase, partly neutralized by the α antitoxin of *Cl. perfringens* (Hayward 1943). Lipase is not formed. Major products of metabolism include large amounts of acetic, isobutyric and isovaleric acids, with smaller amounts of propionic and isocaproic acids. Vegetative forms antigenically heterogeneous. Spores antigenically distinct from those of *Cl. sordelli*. G+C content of DNA 24 moles per cent. (See Humphreys and Meleney 1927–8, Hall, Rymer and Jungherr 1929, Hall and Scott 1931, Levenson 1936.)

Cl. sordelli Isolated by Sordelli (1922, 1923) from a case of human gas gangrene in Buenos Aires; and regarded (see, e.g. Stewart 1938) as a toxigenic variant of *Cl. bifermentans*, but the two are distinguishable antigenically in both vegetative and spore forms and by urease production by *Cl. sordelli* (Tardieux and Nisman 1952, Tataki and Huet 1953, Meisel 1956). Morphologically like *Cl. bifermentans*. Colonies tend to spread in a continuous film; when discrete, are greyish-white and translucent, becoming more opaque as spores form, with crenated or rhizoid margins. May be haemolytic. Ferments glucose and maltose, but not mannose, sorbitol, salicin, lactose or sucrose. Urease +; indole +. Gelatin liquefied. Coagulated proteins digested. Cooked meat medium digested and blackened. Many strains form a lecithinase, partly neutralized by the α antitoxin of *Cl. perfringens*. Products of metabolism similar to those of *Cl. bifermentans*. May be toxigenic or non-toxigenic. Toxigenic strains lethal on intravenous or intramuscular injection in guinea-pigs; intracutaneously they produce a purple-brown necrosis, with underlying massive oedema. Though Japanese workers (Nishida *et al.* 1964, Tamai and Nishida 1964) were apparently able to convert some strains of *Cl. sordelli* into *Cl. bifermentans* by heat treatment, the assignment of the two organisms to different species is justified by differences in the composition of their cell walls and in their production of amines in culture. Thus the cell walls of urease-negative strains of *Cl. bifermentans* contain glucose, mannose, and rhamnose or glucose, galactose, and rhamnose, whereas those of urease-positive strains of *Cl. sordelli* contain only glucose (Novotny 1969); and in cooked meat medium, *Cl. bifermentans* produces within 6 hr β-phenylethylamine and tryptamine, whereas *Cl. sordelli* produces neither (Brooks *et al.* 1969). Confusion may be caused by the fact that some strains of *Cl. sordelli* possessing these characters nevertheless fail to form urease (Brooks *et al.* 1970). G+C content of the DNA 26 moles per cent.

Cl. cadaveris—formerly known as *Cl. capitovale*—was isolated by Snyder and Hall (1935) from various situations including the pleural fluid of a sheep that had died of gas gangrene, the heart blood and peritoneal fluid at necropsy of cases of septic infection in human beings, and the faeces of normal infants. It is a slender, motile, gram-positive rod with rounded ends, measuring 2.0–2.5 μm × 0.5–0.8 μm, and forming terminal oval spores. It gives rise on blood agar to small, circular or irregularly circular, transparent, non-haemolytic surface colonies. It is egg-yolk negative and ferments glucose, but not maltose, lactose or sucrose. It liquefies gelatin, is mildly proteolytic, clots milk irregularly and sometimes digests the clot. Products of metabolism include acetic and butyric acids. It is non-pathogenic to guinea-pigs and rabbits inoculated subcutaneously, and appears to be antigenically homogeneous.

Cl. carnis Described by Klein (1903). A similar bacillus, pathogenic for rabbits, 'von Hibler vi', was isolated from soil by von Hibler (1908). Hall and Duffett (1935) consider the two to be identical. (See also Sompolinsky 1950). Gram-positive rod, sluggishly motile by peritrichate flagella, 1.5 to 4.5 μm × 0.5–0.7 μm. Readily forms subterminal elongated spores, slightly wider than the bacillus, which later appear to be terminal. Aerotolerant. Surface colonies round and transparent, later flat and lobate. Uniform turbidity in broth, and later, granular, then mucoid deposit. Ferments glucose, maltose, lactose and sucrose. Litmus milk unchanged; egg-yolk negative. Coagulated serum and gelatin not liquefied. Indole−; H_2S-. Produces a soluble exotoxin. G+C content of the DNA 28 moles per cent. Pathogenic to mice, white rats and rabbits, with production of an oedematous and congested local lesion.

Cl. cochlearium Described by McIntosh (1917) as type III C, and named *B. cochlearis* by Douglas, Fleming and Colebrook (1920) on account of its likeness to a spoon. Its claim to the rank of a species confirmed by MacLennan (1939). Slender rod, weakly gram positive only in young cultures, 3–5 μm × 0.5–0.6 μm, actively motile by peritrichate flagella. Spores formed late and in small numbers; are oval, terminal and twice the width of the bacillus. Surface colonies are translucent, round, with delicately crenated edge. Non-haemolytic, non-saccharolytic, non-proteolytic, egg-yolk negative. NH_2-; indole−; H_2S+. Volatile products of metabolism include a major amount of butyric acid with smaller amounts of acetic and propionic acids. G+C content of the DNA 28 moles per cent. Antigenically homogeneous. Non-pathogenic to guinea-pigs.

Cl. fallax was found by Weinberg and Séguin (1918) in infected wounds and gas gangrene. It is a motile, gram-positive bacillus, 1.2–5 μm × 0.6 μm in diameter, with rounded ends, straight axis, arranged singly. Spores, which are rarely formed, are subterminal or central, oval and wider than the bacillus. Surface colonies round and transparent, later raised, irregular and more opaque. Uniform turbity in broth; later granular, then mucoid, deposit. Ferments glucose, maltose, lactose and sucrose. Litmus milk, coagulation in 7 days; non-proteolytic. Indole−; H_2S-. Major products of metabolism are lactic, acetic and butyric acids (see Duffett 1935). Produces a soluble toxin. When freshly isolated, is pathogenic for mice and guinea-pigs. G+C content of the DNA 22–26 moles per cent.

Cl. hastiforme was isolated by Cunningham (1930–31, 1931) and by MacLennan (1939), who accorded it the rank of a species. Probably identical with *Cl. subterminale*. A gram-positive rod, 2–6 μm × 0.3–0.6 μm, sluggishly motile by peritrichate flagella. Readily produces oval subterminal spores which, together with a minute terminal tip of bacillary protoplasm, constitute the spear-head shape from which the organism is named. Surface colonies are non-haemolytic, minute, transparent and round, later becoming irregular. Gelatin liquefied 7–10 days; coagulated serum not liquefied. Non-saccharolytic. H_2S+; indole−; NH_3-. Non-lipolytic, but some strains reported to produce lecithinase on egg-yolk agar. Antigenically homogeneous. Hayward (personal communication) has met with strains resembling *Cl. hastiforme* that are haemolytic on horse blood agar. Fermentation products multiple and complex and include major amounts of acetic, butyric and isovaleric acids. G+C content of the DNA 28 moles per cent.

Cl. putrificum Described by Bienstock (1884, 1899, 1901), who isolated it from faeces. Appears to have been a slender gram-positive rod with spherical or oval terminal spores, which digested proteins but had no action on carbohydrates. Its identity has been in doubt (for critical discussion see Hall and Snyder 1934, Hartsell and Rettger 1934, Morgan and Wright 1934). Many workers regarded it as identical with *Cl. cochlearium*, but this was disputed by Hartsell and Rettger and by MacLennan (1939). The name now applies to the isolate described by Reddish and Rettger (1922). *Cl. putrificum* is a slender bacillus, gram positive only in young cultures, 0.3–0.5 μm in diameter, oval, with terminal spores. Motile by peritrichous flagella. Colonies are irregularly round, transparent, haemolytic; egg-yolk negative; H_2S+; indole−; ferments glucose only with the production of multiple fatty acids, mainly acetic, isobutyric, butyric and isovaleric. Liquefies coagulated serum and gelatin in 7–20 days and blackens cooked meat medium slightly. G+C content of DNA 27 moles per cent.

Cl. paraputrificum According to Hall and his colleagues (Hall and Snyder 1934, Hall and Ridgway 1937) this organism, which was described by Bienstock (1906), is probably identical with Escherich's 'Köpfchenbakterien', von Hibler's ix bacillus, Rodella's iii bacillus, and Kleinschmidt's (1934) *B. innutritus*. It is found in the faeces, particularly of infants, both normal and ill nourished, and is a slender, motile, gram-variable bacillus with large terminal oval spores. It is non-proteolytic; it ferments glucose, maltose, lactose and sucrose. Volatile products of metabolism are major amounts of lactic, acetic and butyric acids. It is non-pathogenic for guinea-pigs and rabbits. G+C content of the DNA 26–27 moles per cent.

Cl. sphenoides Isolated by Douglas, Fleming and Colebrook (1920) from wounds. So called from the wedge shape of the sporing bacillus. Small, motile, weakly gram positive; vegetative bacilli are fusiform in shape and arranged in pairs end-to-end. Spores are large and round, appear subterminally, but soon become strictly terminal. Surface colonies are round with entire edges. Haemolytic, but egg-yolk negative. Non-proteolytic, but ferments glucose, maltose and lactose. Indole+; H_2S-. Major products of metabolism are acetic and formic acids. G+C content of DNA 41–42 moles per cent. Non-pathogenic.

Cl. tertium Described by Henry (1917). Thin, slightly curved bacillus, 3–5 μm long, sluggishly motile, gram positive, often showing granular staining. Spores freely, giving rise to large, oval, elongated terminal spores. Surface colonies are rounded, delicate, iridescent, and almost transparent, with entire or slightly crenated edge. Ferments glucose, maltose, lactose and sucrose. Entirely non-proteolytic. Indole−. Main products of metabolism include acetic and lactic acids with lesser amounts of butyric and formic acids. Non-pathogenic to guinea-pigs. (See von Hibler 1908, Hall and Matsuma 1924.) According to Hall and Duffett (1935), both *Cl. tertium* and *Cl. histolyticum* are aerotolerant rather than strictly anaerobic, but spores are formed only under anaerobic conditions. G+C content of the DNA 24 moles per cent.

Cl. ramosum This organism, said by Veillon and Zuber (1898) to be one of the commonest organisms in appendicitis, and reported in cases of gas gangrene and bacteraemia by Lemierre, Reilly and Bloch-Michel (1937), was for many years known as *Ramibacterium ramosum*. The findings, however, that it produces heat-resistant spores and has a G+C content of 27 moles per cent demands its transfer to the genus *Clostridium* (Holdeman *et al.* 1977). Gram positive, non-motile, arranged in pairs or short chains. V− and Y− forms are common, and pseudofilaments in culture. Terminal spores are formed scantily and resist heating at 80° for 10 min. Surface colonies are circular, convex, with an entire or scalloped edge. Uniform turbidity in broth and a muddy deposit. Cultures have a foetid odour, but very little gas is formed. Strict anaerobe. Ferments glucose, maltose, lactose and sucrose, but is entirely non-proteolytic. H_2S –; indole –; egg-yolk negative. Main products of metabolism are acetic and formic acids with smaller amounts of lactic acid. Inoculated intravenously into rabbits it causes death in some days by intoxication and cachexia; subcutaneously it causes abscess formation with death in 8 to 10 days.

References

Abraham, E. P., *et al.* (1941) *Lancet* **ii,** 177.

Adamson, R. S. (1919–20) *J. Path. Bact.* **23,** 241.

Ahmed, M. and Walker, H. W. (1971) *J. Milk Fd Technol.* **34,** 378.

Aikat, B. K. and Dible, J. H. (1956) *J. Path. Bact.* **71,** 461; (1960) *Ibid.* **79,** 227.

Aitken, R. S. Barling, B. and Miles, A. A. (1936) *Lancet* **ii,** 780.

Amoureux, G. (1945) *C. R. Soc. Biol.* **139,** 251, 252.

Anderson, A. A., Michener, H. D. and Olcott, H. S. (1953) *Antibiot. Chemother.* **3,** 561.

Anderson, B. G. (1924) *J. infect. Dis.* **35,** 213, 244.

Arbuthnott, J. P. (1978) *J. appl. Bact.* **44,** 329.

Arloing, S. *et al.* (1887) *Le Charbon Symptomatique du Boeuf*, 2nd edn. Paris.

Bard, R. C. and McClung, L. S. (1948) *J. Bact.* **56,** 665.

Batty, I. (1967) See Fredette, V., p. 85.

Batty, I., Buntain, D. and Walker, P. D. (1964) *Vet. Rec.* **76,** 1115.

Batty, I. and Walker, P. D. (1963) *Bull. Off. Int. Epizoot.* **59,** 1499; (1964) *J. Path. Bact.* **88,** 327; (1966) In: *Identification Methods for Microbiologists*, p. 81. Ed. by B. M. Gibbs and F. A. Skinner. Academic Press, London; (1967) In: *Internat. Symp. Immunol. Meth. Biol. Standardization, Royaumont 1965*, **4,** 73. Karger, Basel.

Beerens, H., Castel, M. M. and Fievez, L. (1962) *Proc. Int. Cong. Microbiol.* p. 120. Montreal.

Beers, W. H. and Reich, E. (1969) *J. biol. Chem.* **244,** 4473.

Behring. (1892) *Z. Hyg. InfektKr.* **12,** 45.

Behring and Kitsato. (1890) *Dtsch. med. Wschr.* **16,** 1113.

Bengtson, I. A. (1920) *Bull, U.S. hyg. Lab.* No. 122; (1921) *Publ. Hlth Rep., Wash.* **36,** 1665; (1922*a*) *Ibid.* **37,** 164; (1922*b*) *Ibid.* **37,** 2252; (1923) *Ibid.* **38,** 340; (1924*a*) *Bull. U.S. hyg. Lab.* No. 136; (1924*b*) *Ibid.* No. 139.

Bennetts, H. W. (1932) *Aust. Counc. sci. industr. Res., Bull.* No. 57.

Bergmann, J. and Rapp, G. (1953) *Zbl. Bakt.* **159,** 500.

Bernheimer, A. W. (1944) *J. exp. Med.* **89,** 309, 321, 333.

Bethge, J. (1947) *Z. Hyg. InfektKr.* **127,** 452.

Betz, J. V. and Anderson, K. E. (1964) *J. Bact.* **87,** 408.

Beynum, J. van and Pette, J. W. (1935) *Zbl. Bakt.*, II de Abt. **93,** 198.

Bidwell, E. (1950) *Biochem. J.* **46,** 589.

Bidwell, E. and Heyningen, W. E. van. (1948) *Biochem. J.* **42,** 140.

Bienstock, B. (1884) *Z. klin. Med.* **8,** 1; (1899) *Arch. Hyg.* **36,** 335; (1901) *Ibid.* **39,** 390; (1906) *Ann. Inst. Pasteur* **20,** 407.

Bittner, J. Portelli, C and Teodorescu, G. (1970) *Z. Immun-Forsch.* **140,** 317.

Bittner, J., Voinesco, V. and Antohi, S. (1961) *Arch roum. Path. exp. Microbiol.* **20,** 63.

Bizzini, B. (1979) *Microbiol. Rev.* **43,** 224.

Bonventre, P. F. and Kempe, L. L. (1960) *J. Bact.* **79,** 18, 24.

Bonventre, P. F., Lincoln, R. E. and Lamanna, C. (1967) *Bact. Rev.* **31,** 95.

Boothroyd, M. and Georgala, G. L. (1964) *Nature* **202,** 515.

Boroff, D. A. (1955) *J. Bact.* **70,** 363.

Boroff, D. A. and Fleck, U. (1966) *J. Bact.* **92,** 1580.

Boroff, D. A. Raynaud, M. and Prévot, A. R. (1952) *J. Immunol.* **68,** 503.

Borthwick, G. R. (1935) *Zbl. Bakt.* **134,** 289; (1937) *Brit. J. exp. Path.* **18,** 475.

Bosworth, T. J. (1943) *J. comp. Path.* **53,** 245.

Bosworth, T. J. and Glover, R. E. (1935) *Proc. R. Soc. Med.* **28,** 1004.

Boyd, M. J., Logan, M. A. and Tytell, A. A. (1948) *J. biol. Chem.* **174,** 1013.

Brewer, J. H. (1940) *J. Bact.* **39,** 10.

Brooks, J. B., Dowell, V. R., Farshy, D. C. and Armfield, A. Y. (1970) *Canad. J. Microbiol.* **11,** 1071.

Brooks, J. B., Moss, C. W. and Dowell, V. R. (1969) *J. Bact.* **100,** 528.

Brooks, M. E. (1961) *J. gen. Microbiol.* **26,** 231.

Brooks, M. E., Sterne, M. and Warrack, G. H. (1957) *J. Path. Bact.* **74,** 185.

Brygoo, E. R. (1950) *Ann. Inst. Pasteur* **78,** 795.

Buchanan, R. E. and Gibbons, N. E. (1974) *Bergey's Manual of Determinative Bacteriology*, 8th edn. Williams and Wilkins, Baltimore.

Buddle, M. B. (1954) *J. comp. Path.* **51,** 217.

Bull, C. G. and Pritchett, I. W. (1917*a*) *J. exp. Med.* **26,** 119; (1917*b*) *Ibid.* **26,** 867.

Bullen, J. J., Rogers, H. J. and Griffiths, E. (1978) *Curr. Topics Microbiol. Immunol.* **80,** 1.

Cardella, M. A. *et al.* (1957) *J. Bact.* **75,** 360; (1960) *Ibid.* **79,** 372.

Cardon, B. P. and Barker, H. A. (1946) *J. Bact.* **52,** 629.

Carlen, S. A. (1939) *Proc. Soc. exp. Biol., N.Y.* **40,** 39.

Carlsson, J., Wrethen, J. and Beckman, G. (1977) *J. clin. Microbiol.* **6,** 280.

Casu, A., Pala, V., Monacelli, R. and Nanni, G. (1971) *Ital. J. Biochem.* **20,** 166.

Chagnon, A. and Fredette, V. (1960) *Ann. Inst. Pasteur* **98,** 261.

Chamsy, M. and Raynaud, M. (1951) *Ann. Inst. Pasteur* **81,** 90.

Ciccarelli, A. S., Whaley, D. N., McCroskey, L. M., Gimenz, D. F., Dowell, V. R. and Hatheway, C. L. (1977) *Appl. environ. Microbiol.* **34,** 843.

Clark, F. E. and Hall, I. C. (1937) *J. Bact.* **33,** 23.

Clifton, C. E. (1942) *J. Bact.* **44,** 179.

Collee, J. G. (1961) *J. Path. Bact.* **81,** 297; (1965*a*) *J. Path. Bact.* **90,** 1; (1965*b*) *Ibid.* **90,** 13.

Corbitt, H. B. (1918) *Bull. U.S. hyg. Lab.* No. 112.

Cowan, S. T. (1974) *Cowan and Steel's Manual for the Identification of Medical Bacteria.* Cambridge University Press, London.

Craig, J. P. (1972) In: *Microbial Pathogenicity in Man and Animals*, p. 129. Cambridge University Press, London.

Craig, J. P. and Miles, A. A. (1961) *J. Path. Bact.* **81,** 481.

Crook, E. M. (1942) *Brit. J. exp. Path.* **23,** 37.

Cummins, C. S. and Johnson, J. L. (1971) *J. gen. Microbiol.* **67,** 33.

Cunningham, A. (1930–1) *Zbl. Bakt.* IIte Abt. **82,** 25, 481; (1931) *Ibid.* **83,** 1, 22, 219.

Dafaala, E. N. and Soltys, M. A. (1951) *Brit. J. exp. Path.* **32,** 510; (1953) *Nature, Lond.* **172,** 38.
Dalling, T. (1926) *J. comp. Path.* **39,** 148.
Dalling, T. and Ross, H. E. (1938) *J. comp. Path.* **51,** 235.
Dalling, T. and Stephenson, M. (1942) *Nature, Lond.* **149,** 56.
Das Gupta, B. R. and Boroff, D. A. (1968) *J. biol. Chem.* **243,** 1065.
Djaenoedin, R. and Kraneveld, F. C. (1936) *Ned-Ind. Bl. Diergeneesk* **48,** 290.
Dolman, C. E. (1957) *Canad. J. publ. Hlth* **48,** 187.
Dolman, C. E. and Chang, E. (1972) *Canad. J. Microbiol.* **18,** 67.
Dolman, C. E. and Murakami, L. (1961) *J. infect. Dis.* **109,** 107.
Donta, S. T. and Shaffer, S. J. (1980) *J. infect. Dis.* **141,** 218.
Dornbusch, K., Nord, C.-E. and Dahlback, A. (1975) *Scand. J. infect. Dis.* **7,** 127.
Douglas, S. R., Fleming, A. and Colebrook, L. (1920) *Spec. Rep. Ser. med. Res. Coun., Lond.* No. 57.
Dowell, V. R., Hill, E. O. and Altemeier, W. A. (1964) *J. Bact.* **88,** 1811.
Duff, J. T., Wright, G. G. and Yarinsky, A. (1956) *J. Bact.* **72,** 455.
Duff, J. T. *et al.* (1957) *J. Bact.* **73,** 42, 597.
Duffett, N. D. (1935) *J. Bact.* **29,** 573; (1938) *Univ. Color. Stud.* **26,** 46.
Duncan, C. L. and Strong, D. H. (1968) *Appl. Microbiol.* **16,** 82; (1971) *Infect Immun.* **3,** 167.
Duncan, C. L., Sugiyama, H. and Strong, D. H. (1968) *J. Bact.* **95,** 1560.
Dunn, M. S., Camien, M. N. and Pillemer, L. (1949) *Arch. Biochem.* **22,** 374.
Eisler, M. (1952) *Z. Hyg. InfektKr.* **135,** 577.
Eklund, M. W., Poysky, F. T., Meyers, J. A. and Pelroy, G. A. (1974) *Science, N.Y.* **186,** 456.
Elder, J. M. and Miles, A. A. (1957) *J. Path. Bact.* **74,** 133.
Ellner, P. D. (1956) *J. Bact.* **71,** 495.
Ellner, P. D. and Bohan, C. D. (1962) *J. Bact.* **83,** 284.
Ellner, P. D. and Green, S. S. (1963) *J. Bact.* **86,** 1084, 1098.
Elsden, S. R. and Hilton, M. G. (1979) *Arch. Microbiol.* **123,** 137.
Ermengem, E. van. (1896) *Zbl. Bakt.* **19,** 442; (1897) *Z. Hyg. InfektKr.* **26,** 1.
Evans, D. G. (1945) *J. Path. Bact.* **57,** 75; (1947) *J. gen. Microbiol.* **1,** 378.
Evans, D. G. and Prophet, A. S. (1950) *J. gen. Microbiol.* **4,** 360.
Felix, A. and Robertson, M. (1928) *Brit. J. exp. Path.* **9,** 6.
Fildes, P. (1925*a*) *Brit. J. exp. Path.* **6,** 62; (1925*b*) *Ibid.* **6,** 91; (1929) *Ibid.* **10,** 151; (1935) *Ibid.* **16,** 309.
Fildes, P. and Knight, B. C. J. G. (1933) *Brit. J. exp. Path.* **14,** 343.
Fildes, P. and Richardson, G. M. (1935) *Brit. J. exp. Path.* **16,** 326.
Fraser, A. G. and Collee, J. G. (1975) *J. med. Microbiol.* **8,** 251.
Fraser, A. G. and Smith, J. K. (1975) *J. med. Microbiol.* **8,** 235.
Frazier, W. C. (1926) *J. infect. Dis.* **39,** 302.
Fredette, V. (1967) *The Anaerobic Bacteria.* Montreal University.
Fredette, V. and Frappier, A. (1946) *Rev. canad. Biol.* **5,** 1.
Freeman, W. A., McFadzean, J. A. and Whelan, J. P. F. (1968) *J. appl. Bact.* **31,** 443.
Fuchs, A. R. and Bonde, G. J. (1957) *J. gen. Microbiol.* **16,** 317.
Gadalla, M. S. A. and Collee, J. G. (1967) *J. Path. Bact.* **93,** 255; (1968) *J. Path. Bact.* **96,** 169.
Gale, E. F. (1940) *Bact. Rev.* **4,** 135.
Gale, E. F. and van Heyningen, W. E. (1942) *Biochem. J.* **36,** 624.
Ganley, O. H., Merchant, D. J. and Bohr, D. F. (1955) *J. exp. Med.* **101,** 605.
George, W. L., Sutter, V. L. and Finegold, S. M. (1978) *Curr. Microbiol.* **1,** 55.
Gerwing, J., Dolman, C. E. and Ko, A. (1965) *J. Bact.* **89,** 1176.
Gerwing, J., Dolman, C. E., Reichmann, M. E. and Bains, H. S. (1964) *J. Bact.* **88,** 216.
Gill, D. A. (1933) *Vet. J.* **89,** 399.
Gillespie, W. A. and Guy, J. (1956) *Lancet* **i,** 1039
Giménez, D. F. and Ciccarelli, A. S. (1970) *Zbl. Bakt.* **215,** 212, 221.
Gladstone, G. P. Fildes, P. and Richardson, G. M. (1935) *Brit. J. exp. Path.* **16,** 335.
Glenny, A. T. *et al.* (1933) *J. Path. Bact.* **37,** 53.
Gonzalez, C. (1961) *Leeuwenhoek ned. Tijdschr.* **27,** 19.
Gordon, J., Holman, R. A. and McLeod, J. W. (1953) *J. Path. Bact.* **66,** 527.
Goss, L. W., Barbarin, R. E. and Haines, A. W. (1921) *J. infect. Dis.* **29,** 615.
Graham, R. and Boughton, T. B. (1923*a*) *Abstr. Bact.* **1,** 29; (1923*b*) *Ibid.* **7,** 30.
Grant, R. B. and Riemann, H. P. (1976) *Canad. J. Microbiol.* **22,** 603.
Griner, L. A. and Bracken, F. K. (1953) *J. Amer. vet. med. Ass.* **122,** 99.
Griner, L. A. and Johnson, H. W. (1954) *J. Amer. vet. med. Ass.* **125,** 125.
Grunberg, M. (1948) *Ann. Inst. Pasteur* **74,** 216.
Guélin, A. (1950) *Ann. Inst. Pasteur* **79,** 447.
Guélin, A., Beerens, H. and Petitprez, A. (1966) *Ann. Inst. Pasteur* **111,** 141.
Guggenheim, K. (1944) *J. Bact.* **47,** 313.
Guillaumie, M. (1944) *Ann. Inst. Pasteur,* **70,** 86; (1960) *Rev. Immunol., Paris* **24,** 581.
Guillaumie, M. and Kréguer, A. (1950) *Ann. Inst. Pasteur* **78,** 467.
Guillaumie, M., Kréguer, A. and Fabre, M. (1946*a*) *Ann. Inst. Pasteur* **72,** 814, 818; (1946*b*) *Ibid.* **72,** 908.
Guillaumie, M., Kréuguer, A., Geoffroy, M. and Réade, G. (1952) *Ann. Inst. Pasteur* **83,** 360.
Gunnison, J. B. (1947) *J. Immunol.* **57,** 67.
Gunnison, J. B., Cummings, J. R. and Meyer, K. F. (1936) *Proc. Soc. exp. Biol. N.Y.* **35,** 278.
Gürtürk, S. (1952) *Z. Hyg. InfektKr.* **133,** 573.
Hafiz, S. and Oakley, C. L. (1976) *J. med. Microbiol.* **9,** 129.
Haines, R. B. and Scott, W. J. (1940) *J. Hyg., Camb.* **40,** 154.
Hall, I. C. (1922) *J. infect. Dis.* **30,** 445; (1929) *Ibid.* **45,** 156.
Hall, I. C. and Duffett, N. D (1935) *J. Bact.* **29,** 269.
Hall, I. C. and Matsuma, K. (1924) *J. infect. Dis.* **35,** 502.
Hall, I. C. and Ridgway, D. (1937) *J. Bact.* **34,** 631.
Hall, I. C., Rymer, M. R. and Jungherr, E. (1929) *J. infect. Dis.* **45,** 42.

Hall, I. C. and Scott, A. L. (1927) *J. infect. Dis.* **41,** 329; (1931) *J. Bact.* **22,** 375.
Hall, I. C. and Snyder, M. L. (1934) *J. Bact.* **28,** 181.
Hall, I. C. and O'Toole, E. (1935) *Amer. J. Dis. Child.* **49,** 390.
Hanke, M. E. and Bailey, J. H. (1945) *Proc. Soc. exp. Biol., N.Y.* **59,** 163.
Hardegree, M. C., Palmer, A. E. and Duffin, N. (1971) *J. infect. Dis.* **123,** 51.
Hartsell, S. E. and Rettger, L. F. (1934) *J. Bact.* **27,** 497.
Hartwigk, H., Schick, H. D. and Plettner, B. (1970) *Zbl. Bakt.* **213,** 382.
Haslam, T. P. and Lumb, J. W. (1919) *J. infect. Dis.* **24,** 362.
Hauschild, A. H. W. (1965) *J. Immunol.* **94,** 551; (1971) *J. Milk Fd Technol.* **34,** 596.
Hauschild, A. H. W. and Hilsheimer, R. (1969) *Canad. J. Microbiol.* **15,** 1129.
Hayward, N. J. (1941) *Brit. med. J.* **i,** 811; (1943) *J. Path. Bact.* **55,** 285.
Hayward, N. J. and Gray, J. A. B. (1946) *J. Path. Bact.* **58,** 11.
Hazen, E. L. (1937) *J. infect. Dis.* **60,** 260; (1942) *Proc. Soc. exp. Biol., N.Y.* **50,** 112.
Headlee, M. T. (1931) *J. infect. Dis.* **48,** 468.
Heller, H. H. (1920) *J. infect. Dis.* **27,** 385; (1921) *J. Bact.* **6,** 521; (1922*a*) *J. infect. Dis.* **30,** 18; (1922*b*) *Ibid.* **30,** 33.
Henderson, D. W. (1932) *Brit. J. exp. Path.* **13,** 412; (1934) *Ibid.* **15,** 166; (1940) *J. Hyg., Camb.* **40,** 501.
Henriksen, S. D. (1937) *Acta. path. microbiol. scand.* **14,** 570.
Henry, H. (1917) *J. Path. Bact.* **21,** 344.
Hewitt, J. and Morris, J. G. (1975) *FEBS Lett.* **50,** 315.
Heyningen, W. E. van. (1940) *Biochem. J.* **34,** 1540; (1941*a*) *Ibid.,* **35,** 1246; (1941*b*) *Ibid.* **35,** 1257; (1970) *Zbl. Bakt. Abt. I. Orig.* **212,** 191; (1976) *FEBS Lett.* **68,** 5.
Heyningen, W. E. van and Mellanby, J. (1968) *J. gen. Microbiol.* **52,** 447.
Hibler, E. von. (1908) *Untersuchungen über die pathogenen Anäeroben.* Jena.
Hill, L. R. (1966) *J. gen. Microbiol.* **44,** 419.
Hirsch, A. and Grinsted, E. (1954) *J. Dairy Res.* **21,** 101.
Hobbs, B. C. (1965) *J. appl. Bact.* **28,** 74.
Hobbs, B. C. *et al.* (1953) *J. Hyg., Camb.* **51,** 75.
Hobbs, G. (1958) Cited by Willis, A. T. (1977).
Holdeman, L. V., Cato, E. P. and Moore, W. E. C. (1977) *Anaerobe Laboratory Manual*, 4th edn. Virginia Polytechnic Institute, Blacksburg, Va.
Holdeman, L. V. and Smith, L. D. (1965) *Canad. J. Microbiol.* **11,** 1009.
Holman, R. A. (1955) *J. Path. Bact.* **70,** 195.
Hoogerheide, J. C. (1937) *J. Bact.* **34,** 387.
Howard, A. (1928) *Ann. Inst. Pasteur* **42,** 1403.
Howard, J. G. (1953) *Brit. J. exp. Path.* **34,** 564.
Howes, E. L., Mandl, I. and Zaffuto, S. (1960) *J. Bact.* **79,** 191.
Hoyt, A., Chaney, A. L. and Cavell, K. (1938) *J. Bact.* **36,** 639.
Hugh, R. and Leifson, E. (1953) *J. Bact.* **66,** 24.
Humphreys, F. and Meleney, F. L. (1927–8) *Proc. Soc. exp. Biol., N.Y.* **25,** 611.
Humphreys, F. B. (1924) *J. infect. Dis.* **35,** 282.
Ingram, M. and Roberts, T. A. (1967) *Botulism 1966*. Chapman and Hall Ltd, London.
Inoue, K. and Iida, H. (1971) *Jap. J. med. Sci. Biol.* **24,** 53.
Ipsen, J. (1939) *Bull. Hlth Org., L.o.N.* **8,** 825; (1940–1) *Ibid.* **9,** 44, 452.
Ipsen, J. and Davoli, R. (1939) *Bull. Hlth Org., L.o.N.* **8,** 833.
Ipsen, J., Smith, M. L. and Sordelli, A. (1939) *Bull. Hlth. Org., L.o.N.* **8,** 797.
Ito, A. (1968) *Jap. J. med. Sci. Biol.* **21,** 379.
Jennison, M. W. (1945) *J. Bact.* **50,** 349.
Johnson, H. M., Brenner, K., Angelotti, R. and Hall, H. E. (1966) *J. Bact.* **91,** 967.
Kadis, S., Montie, T. C. and Ajl, S. J. (1971) *Microbial Toxins*, Vol. ii A. Academic Press, New York.
Kamen, L. (1904) *Zbl. Bakt.* **35,** 686.
Kameyama, S. and Akama, K. (1971) *Jap. J. med. Sci. Biol.* **24,** 9.
Katitch, R., Gadjanski, G., Dimitrievitch, M., Voukitchevitch, Z. and Militch, D. (1967) *Rev. Immunol.* **31,** 89.
Katitch, R. V. (1954) *Rev. Immunol., Paris* **18,** 199.
Kaufman, L. and Humphries, J. C. (1958) *Appl. Microbiol.* **6,** 311.
Kautter, D. A., Harmon, S. M., Lynt, R. K. and Lilly, T. (1966) *Appl. Microbiol.* **14,** 616.
Kawatomari, T. (1958) *J. Bact.* **76,** 227.
Kelsey, J. C., Mackinnon, I. H. and Maurer, I. M. (1974) *J. clin. Path.* **27,** 632.
Keppie, J. (1944) *A Study of the Antigens of Cl. oedematiens and Cl. gigas*. PhD Thesis, Cambridge.
Kerrin, J. C. (1930) *Brit. J. exp. Path.* **11,** 153; (1934) *J. Path. Bact.* **38,** 219.
Khairat, O. (1966) *Canad. J. Microbiol.* **12,** 323.
Kindler, S. H. *et al.* (1956*a*) *J. gen. Microbiol.* **15,** 386; (1956*b*) *Ibid.* **15,** 394.
Kislak, J. W. (1972) *J. infect Dis.* **125,** 295.
Kitasato, S. (1889*a*) *Z. Hyg. InfecktKr.* **6,** 105; (1889*b*) *Ibid.* **7,** 225; (1890) *Ibid.* **8,** 55; (1891) *Ibid.* **10,** 267.
Kitasato, S. and Weytl, T. (1890) *Z. Hyg. InfektKr.* **8,** 41, 404.
Kitt, T. (1887) *Zbl. Bakt.* **1,** 684, 716, 741.
Klein, E. (1903) *Zbl. Bakt.* **35,** 459.
Klein, R., Miller, N., Kemp, P. and Laser, H. (1975). *Chem. Phys. Lipids* **15,** 15.
Kleinschmidt, H. (1934) *Mschr. Kinderheilk.* **62,** 14.
Knight, B. C. J. G. (1936) *Spec. Rep. Ser. med. Res. Coun., Lond.* No. 210.
Knight, B. C. J. G. and Fildes, P. (1930) *Biochem. J.* **24,** 1496; (1933) *Brit. J. exp. Path.* **14,** 112.
Kocholaty, W. and Krejci, L. E. (1948) *Arch. Biochem.* **18,** 1.
Koslowski, L., Schneider, H. H. and Heise, C. (1951) *Klin. Wschr.* **29,** 29.
Kraneveld, F. C. (1930) *Ned-Ind. Bl. Diergeneesk.* **42,** 564.
Kreuzer, E. (1939) *Z. ImmunForsch.* **95,** 345.
Kutzner, H. J. (1963) *Zbl. Bakt.* **191,** 441.
Labbe, R. G. and Duncan, C. L. (1975) *Appl. Microbiol.* **29,** 345.
Lamanna, C. (1959) *Science* **130,** 763.
Lamanna, C. and Carr, C. J. (1967) *Clin. Pharmacol. Ther.* **8,** 286.
Lamanna, C. and Lowenthal, J. P. (1951) *J. Bact.* **61,** 751.
Lamanna, C. and Sakaguchi, G. (1971) *Bact. Rev.* **35,** 242.
Lamanna, C. Spero, L. and Schantz, E. J. (1970) *Infect. Immun.* **1,** 423.
Landau, H. (1917) *Zbl. Bakt.* **79,** 417.
Largier, J. F. (1958) *Arch. Biochem. Biophys.* **77,** 350.

Larson, H. E., Price, A. B., Honour, P. and Borriello, S. P. (1978) *Lancet* **i,** 1063.
Latham, C. W., Bent, D. F. and Levine, L. (1962) *Appl. Microbiol.* **10,** 146.
Latham, W. C. *et al.* (1965) *J. Immunol.* **95,** 487.
Lavergne, E. *et al.* (1956) *Ann. Inst. Pasteur* **91,** 631.
Lawrence, G. W. and Walker, P. D. (1976) *Lancet* **i,** 125.
Leclainche, E. and Vallée, H. (1900) *Ann. Inst. Pasteur* **14,** 202.
Lemierre, A., Reilly, J. and Bloch-Michel, H. (1937) *Bull. Acad. Méd., Paris* **117,** 322.
Lennert-Petersen, O. (1954) *Acta path. microbiol. scand.* **35,** 59.
Lepow, I. H., Katz, S., Pensky, J. and Pillemer, L. (1952) *J. Immunol.* **69,** 435.
Lepper, E. and Martin, C. J. (1929) *Brit. J. exp. Path.* **10,** 327; (1930) *Ibid.* **11,** 137, 140.
Lerner, E. M. and Pickett, M. J. (1945) *Arch. Biochem.* **8,** 183.
Levenson, S. (1936) *C. R. Soc. Biol.* **121,** 221.
Lewis, G. M. and Macfarlane, M. G. (1953) *Biochem., J.* **54,** 138.
Lingelsheim, von. (1912) See *Kolle and Wassermann's Hbd. path. Mikroorg.* IIte Aufl. (1912–13) 4, 737.
Linzenmeier, G. (1959) *Arch. Hyg., Berl.* **143,** 537, 561.
Lowbury, E. J. L. and Lilly, H. A. (1955) *J. Path. Bact.* **70,** 105.
Lowenthal, J. P. and Lamanna, C. (1953) *Amer. J. Hyg.* **57,** 46.
Lucain, C. and Piffaretti, J. C. (1977) *FEMS Microbiol. Lett.* **1,** 231.
Lynt, R. K., Solomon, H. M. and Kautter, D. A. (1971) *J. Food Sci.* **36,** 594.
Lynt, R. K., Solomon, H. M., Kautter, D. A. and Lilly, T. (1967) *J. Bact.* **93,** 27.
McClean, D. (1936) *J. Path. Bact.* **42,** 477.
McClung, L. S. (1937) *J. infect. Dis.* **60,** 122.
McClung, L. S. and Toabe, R. (1947) *J. Bact.* **53,** 139.
McCord, J. M., Keele, B. B. and Fridovich, I. (1971) *Proc. nat. Acad. Sci. Wash.* **68,** 1024.
McCoy, E. and McClung, L. S. (1936) *J. Bact.* **31,** 557; (1938) *Bact. Rev.* **2,** 47.
McDonel, J. L. (1979) *Amer. J. clin. Nutr.* **32,** 210.
McEwen, A. D. (1930) *J. comp. Path* **43,** 1.
Macfarlane, M. G. (1948*a*) *Biochem. J.* **42.** 587; (1948*b*) *Ibid.* **42,** 590; (1950*a*) *Ibid.* **47,** 267; (1950*b*) *Ibid.* **47,** 270; (1956) *Ibid.* **61,** 308.
Macfarlane, M. G. and Knight, B. C. J. G. (1941) *Biochem. J.* **35,** 884.
Macfarlane, R. G., Oakley, C. L. and Anderson, C. G. (1941) *J. Path. Bact.* **52,** 99.
McGaughey, C. A. (1933) *J. Path. Bact.* **36,** 263.
McIntosh, J. (1917) *Spec. Rep. Ser. med. Res. Coun., London,* No. 12.
McIntosh, J. and Fildes, P. (1916) *Lancet* **i,** 768.
McKee, C. M., Hamre, D. M. and Rake, G. (1943) *Proc. Soc. exp. Biol., N.Y.* **54,** 211.
MacLennan, J. D. (1939) *Brit. J. exp. Path.* **20,** 371.
MacLennan, J. D. *et al.* (1953) *J. clin. Invest.* **32,** 1317; (1958) *J. gen. Microbiol.* **18,** 1.
McLeod, J. W. and Gordon, J. (1923) *J. Path. Bact.* **26,** 332; (1925) *Ibid.* **21,** 147.
Madsen, T. (1899) *Z. Hyg. InfektKr.* **32,** 214.
Mahony, D. E. (1977) *Infect. Immun.* **15,** 19.
Mahony, D. E. and Butler, M. E. (1971) *Canad. J. Microbiol.* **17,** 1.
Mahony, D. E. and Kalz, G. G. (1968) *Canad. J. Microbiol* **14,** 1085.
Mahony, D. E. and Swantee, C. A. (1978) *J. clin. Microbiol.* **7,** 307.
Mandia, J. W. (1950) *J. Bact.* **60,** 275; (1951) *J. Immunol.* **67,** 49; (1952) *J. Infect. Dis.* **90,** 48; *J. Immunol.* **69,** 451; (1955) *J. infect. Dis.* **97,** 66.
Mandia, J. W. and Bruner, D. W. (1951) *J. Immunol.* **66,** 497.
Mandl, I., Maclennan, J. D. and Howes, E. L. (1953) *J. clin. Invest.* **32,** 1323.
Mangalo, R., Bizzini, B., Turpin, A. and Raynaud, M. (1968) *Biochim. biophys. acta* **168,** 583.
Markoff, W. N. (1911) *Zbl. Bakt.* **60,** 188.
Martin, W. J., Gardner, M. and Washington, J. A. (1972) *Antimicrob. Agents Chemother.* **1,** 148.
Mason, J. H. (1935) *Onderstepoort J. vet. Sci.* **5,** 363.
Mason, J. H., Ross, H. E. and Dalling, T. (1931) *J. comp. Path.* **44,** 258.
Mead, G. C. (1971) *J. gen. Microbiol.* **67,** 47.
Meisel, H. (1938) *Z. ImmunForsch.* **92,** 79; (1956) *Ann. Inst. Pasteur* **91,** 25.
Meisel, H. and Rymkiewicz, D. (1957) *Bull. Acad. polon. Sci.* **5,** 331.
Meisel-Mikolajczyk, F. (1959) *Bull. Acad. polon. Sci.* **7,** 387; (1960) *Ibid.* **8,** 97.
Metchnikoff, E. (1908) *Ann. Inst. Pasteur* **22,** 929.
Meyer, E. A. and Lamanna, C. (1959), C. (1959) *J. Bact.* **78,** 175.
Miles, A. A. and Hayward, N. J. (1943) *Lancet* **ii,** 116.
Miles, E. M. and Miles, A. A. (1947) *J. gen. Microbiol.* **1,** 385; (1950) *Ibid.* **4,** 22.
Miller, P. A., Eaton, M. D. and Gray, C. T. (1959) *J. Bact.* **77,** 733.
Miller, P. A., Gray, C. T. and Eaton, M. D. (1960) *J. Bact.* **79,** 95.
Mitsui, K., Mitsui, N. and Hase, J. (1973). *Japan. J. exp. Med.* **43,** 377.
Mollby, R. and Wadstrom, T. (1973) *Biochim. biophys. acta* **321,** 569.
Møller, V. and Scheibel, I. (1960) *Acta path. microbiol. scand.* **48,** 80.
Moore, W. B. (1968) *J. gen. Microbiol.* **53,** 415.
Moore, W. E. C. (1970) *Int. J. syst. Bact.* **20,** 535.
Moore, W. E. C., Cato, E. P. and Holdeman, L. V. (1966) *Int. J. syst. Bact.* **16,** 383.
Moore, W. E. C. and Holdeman, L. V. (1972) *Amer. J. clin. Nutr.* **25,** 1306.
Morgan, E. L. and Wright, H. D. (1934) *J. Path. Bact.* **39,** 457.
Morris, J. G. (1976) *J. appl. Bact.* **40,** 229.
Moskowitz, M. (1956) *Proc. Soc. exp. biol., N.Y.* **92,** 706.
Moussa, R. S. (1958) *J. Bact.* **76,** 538; (1959) *J. Path. Bact.* **77,** 341.
Mueller, J. H. and Miller, P. (1945) *J. Immunol.* **50,** 377; (1947) *Ibid.* **56,** 143; (1948) *J. Bact.* **55,** 421, **56,** 97, 219; (1954) *Ibid.* **67,** 271; (1955) *Ibid.* **69,** 634; (1956) *J. biol. Chem.* **223,** 185.
Müller, H. E. (1971) *Z. ImmunForsch.* **142,** 31.
Murata, R., Soda, S., Yamamoto, A. and Ito, A. (1968) *Jap. J. med. Sci. Biol.* **21,** 55; (1969) *Ibid.* **22,** 133.
Murphy, S. G. and Miller, K. D. (1967) *J. Bact.* **94,** 580.

Murrell, T. G. C. and Roth, L. (1963) *Med. J. Aust.* **i,** 61.
Mutermilch, S., Belin, M. and Salamon, E. (1933) *C. R. Soc. Biol.* **114,** 1005.
Nagel, J. and Cohen, H. (1973) *J. Immunol.* **110,** 1388.
Nagler, F. P. O. (1936) *Z. ImmunForsch.* **89,** 477; (1937) *Ibid.* **91,** 457; (1939) *Brit. J. exp. Path.* **20,** 473; (1940) *Z. ImmunForsch.* **97,** 273; (1945*a*) *Aust. J. exp. Biol. med. Sci.* **23,** 59; (1945*b*) *Brit. J. exp. Path.* **26,** 57.
Nakamura, S., Kimura, I., Yamakawa, K. and Nishida, S. (1983) *J. gen. Microbiol.*, **129,** 1473.
Nakamura, S., Sakurai, M., Nishida, S., Tatsuki, T., Yanagase, Y., Higashi, Y. and Amano, T. (1976) *Canad. J. Microbiol.* **22,** 1497.
Nakamura, M., Schulze, J. A. and Cross, W. R. (1969) *J. Hyg., Camb.* **67,** 153.
Narayan, K. G. (1966) *Zbl. Bakt* **201,** 57.
Nicolaier (1884) *Dtsch. med. Wschr.* **10,** 842.
Nishida, S. and Nakagawara, G. (1965) *J. Bact.* **89,** 993.
Nishida, S., Tamai, K. and Yamagishi, T. (1964) *J. Bact.* **88,** 1641.
Nocard and Roux. (1887) *Ann. Inst. Pasteur* **1,** 257.
Novotny, P. (1969) *J. med. Microbiol.* **2,** 81.
Novy, F. G. (1894) *Z. Hyg. InfektKr.* **17,** 209.
Oakley, C. L. (1949) *Brit. med. J.* **i** 269; (1956) *J. appl. Bact.* **19,** 112; (1967) See Fredette, V., p. 209.
Oakley, C. L. and Banerjee, N. G. (1963) *J. Path. Bact.* **85,** 489.
Oakley, C. L. and Warrack, G. H. (1941) *J. Path. Bact.* **53,** 335; (1950) *J. gen. Microbiol.* **4,** 365; (1951) *J. Path. Bact* **63,** 45; (1953) *J. Hyg., Camb.* **51,** 102; (1958) *J. gen. Microbiol* **18,** 9; (1959) *J. Path. Bact.* **78,** 543.
Oakley, C. L., Warrack, G. H. and Clarke, P. H. (1947) *J. gen. Microbiol.* **1,** 91.
Oakley, C. L., Warrack, G. H. and van Heyningen, W. E. (1946) *J. Path. Bact.* **58,** 229.
Oakley, C. L., Warrack, G. H. and Warren, M. E. (1948) *J. Path. Bact.* **60,** 495.
O'Brien, R. W. and Morris, J. G. (1971) *J. gen. Microbiol.* **68,** 307.
Oguma, K, IIda, H. Shiozaki, M. and Inoue, K. (1976) *Infect. Immun.* **13,** 855.
Orlans, E. S. and Jones, V. E. (1958) *Immunol.* **1,** 268.
Orr, J. H. and Reed. G. B. (1940) *J. Bact.* **40,** 441.
Pappenheimer, A. M. (1935) *Biochem. J.* **29,** 2057.
Pappenheimer, A. M. and Shaskan, E. (1944) *J. biol. Chem.* **155,** 265.
Paquette, G. and Fredette, V. (1977) *Rev. Canad. Biol.* **36,** 205.
Pasternack, J. G. and Bengtson, I. A. (1936) *Nat. Inst. Hlth Bull.* No. 168; (1940) *Publ. Hlth Rep., Wash.* **55,** 775.
Pasteur and Joubert (1877) *Bull. Acad. Méd.* **6,** 781.
Petrie, G. F. (1942–43) *Bull. Hlth Org., L.o.N.* **10,** 113.
Pillemer, L. (1946) *J. Immunol.* **53,** 237.
Pillemer, L. Grossberg, D. B. and Wittler, R. G. (1946) *J. Immunol* **54,** 213.
Pillemer, L. *et al.* (1948) *J. exp. Med.* **88,** 205.
Pinegar, J. A. and Stringer, M. F. (1977) *J. clin. Path.* **30,** 491.
Prescott, L. M. and Altenbern, R. A. (1967) *J. Bact.* **93,** 1220.
Prévot, A. R. (1938) *Ann. Inst. Pasteur* **61,** 71; (1948) *Manuel de Classification et de Détermination des Bacéries Anaérobies*, 2nd edn, Masson et Cie, Paris; (1957) *Ibid.*, 3rd edn; (1966) *Manual for the Classification and Determination of the Anaerobic Bacteria.* Henry Kimpton, London.
Prévot, A. R. and Brygoo, E. R. (1953*a*) *Ann. Inst. Pasteur* **84,** 1037; (1953*b*) *Ibid.* **85,** 544.
Prévot, A. R. and Enescu, M. (1946) *C. R. Soc. Biol.* **140,** 76.
Prigge, R. (1936) *Z. ImmunForsch.* **89,** 477; (1937) *Ibid.* **91,** 457
Princewill, T. J. T. and Oakley, C. L. (1972*a*) *Med. Lab. Tech.* **29,** 243; (1972*b*) *Ibid.* **29,** 255; (1976) *Med. Lab. Sci.* **33,** 105.
Raynaud, M. (1951) *Ann. Inst. Pasteur,* **80,** 356; (1967) See Fredette, V., p. 175.
Raynaud, M. *et al.* (1955) *Ann. Inst. Pasteur* **88,** 24.
Raynaud, M., Turpin, A. and Bizzini, B. (1960) *Ann. Inst. Pasteur* **99,** 167.
Reddish, G. F. and Rettger, L. F. (1922) *Abstracts Bact.*, **6,** 9; (1924) *J. Bact.* **9,** 13.
Reed, G. B. and Orr, J. H. (1941) *War Med.* **1,** 493.
Reed, G. B., Orr, J. H. and Brown, H. J. (1943) *J. Bact.* **46,** 475.
Reed, R. W. (1942) *J. Bact.* **44,** 425.
Reilly, S. (1980) *J. med. Microbiol* **13,** 573.
Report (1919) *Spec. Rep. Ser. med. Res. Coun. Lond.* No. 39; (1975) *Jap. J. med. Sci. Biol.* **28,** 55; (1978) *Ibid.* **31,** 1955; (1982) *Ibid.* **35,** 105.
Rifkin, G. D., Silva, J. and Fekety, R. (1978) *Gastroenterology* **74,** 52.
Roberts, J. E. (1957) *J. Bact.* **74,** 439.
Roberts, R. S. (1931) *J. comp. Path.* **44,** 246.
Robertson, M. (1916) *J. Path. Bact.* **20,** 327.
Rodwell, A. W. (1941) *Aust. vet. J.* **17,** 58.
Rolfe, R. D. and Finegold, S. M. (1979) *Infect. Immun.* **25,** 191.
Rolfe, R. D., Hentges, D. J., Campbell, B. J. and Barrett, J. T. (1978) *Appl. environ. Microbiol.* **36,** 306.
Rosenau, M. J. and Anderson, J. F. (1908) *Bull. U.S. hyg. Lab.* No. 43.
Rosenbach (1886) *Arch. klin. Chir.* **34,** 306.
Ross, H. E., Warren, M. E. and Barnes, J. M. (1949) *J. gen. Microbiol.* **3,** 149.
Roth, F. B. and Pillemer, L. (1953) *J. Immunol.* **70,** 533; (1955) *Ibid.* **75,** 50.
Rutter, J. M. (1970) *J. med. Microbiol.* **3,** 283.
Rutter, J. M. and Collee, J. G. (1969) *J. med. Microbiol.* **2,** 395.
Sacks, L. E. and Thompson, P. A. (1978) *Appl. environ. Microbiol.* **34,** 189.
Saissac, R. and Raynaud, M. (1951) *Ann. Inst. Pasteur.* **80,** 431, 434.
Sakaguchi, G. and Sakaguchi, S. (1959) *J. Bact.* **78,** 1.
Sato, H. and Murata, R. (1973) *Infect. Immun.* **8,** 360.
Savage, G. M. (1974) *Infection* **2,** 152.
Savolainen, T. (1948) *Ann. Med. exp. Biol. fenn.* Suppl.
Scheibel, I. and Assandri, J. (1959) *Acta path. microbiol. scand* **46,** 333.
Schoenholz, P. (1928) *J. infect. Dis.* **42,** 40.
Schoenholz, P. and Meyer, K. F. (1925) *J. Immunol.* **10,** 1.
Scott, J. P., Turner, A. W. and Vawter, L. R. (1933) *Proc. 12th int. vet. Congr. N.Y.*, p. 168.
Sebald, M. and Popovitch, M. (1967) *Ann. Inst. Pasteur* **113,** 781.
Sebald, M. and Prévot, A. R. (1960) *Ann. Inst. Pasteur* **99,** 386.
Seddon, H. R. (1922) *J. comp. Path.* **35,** 147.
Seiffert, G. (1939) *Z. ImmunForsch* **96,** 515.
Seki, T. *et al.* (1958) *Med. J. Osaka Univ.* **8,** 639.

Shapton, D. A. and Board, R. G. (1971) *Isolation of Anaerobes*. Academic Press, London.

Sherrington, C. S. (1917) *Lancet* **ii,** 964.

Simonds, J. P. (1915*a*) *J. infect. Dis.* **16,** 31; (1915*b*) *Ibid.* **16,** 35; (1915*c*) *Monogr. Rockefeller Inst.* No. 5.

Skerman, V. B. D., McGowan, V. and Sneath, P. H. A. (1980) *Int. J. syst. Bact.* **30,** 225.

Skjelvale, R. and Duncan, C. L. (1975) *Infect. Immun.* **11,** 563.

Skulberg, A. and Hausken, O. W. (1965) *J. appl. Bact.* **28,** 83.

Smidt, F. P. G. de. (1924) *J. Hyg., Camb.* **22,** 314.

Smith, H. W. (1959) *J. gen Microbiol.* **21,** 622.

Smith. L. (1937) *J. Bact.* **34,** 409.

Smith, L.DS, (1975*a*) *Appl. Microbiol.* **30,** 319; (1975*b*) *The Pathogenic Anaerobic Bacteria*, 2nd edn. Thomas, Springfield; (1977) *Botulism. The Organism. Its Toxin. The Disease*. Thomas, Springfield.

Smith, L. DS., Claus, K. D. and Matsuoka, T. (1956) *J. Bact.* **72,** 809.

Smith, L. DS. and Douglas, H. C. (1950) *J. Bact.* **60,** 9.

Smith, M. L. (1941–2) *Bull. Hlth Org., L.o.N.* **10,** 104.

Smyth, C. J. (1975) *J. gen. Microbiol.* **87,** 219.

Smyth, C. J. and Arbuthnott, J. P. (1974) *J. med. Microbiol.* **7,** 41.

Snyder, M. L. (1937) *J. infect. Dis.* **60,** 223.

Snyder, M. L. and Hall, I. C. (1935) *Zbl. Bakt.* **135,** 290.

Soda, S., Sato, H. and Murata, R. (1969) *Jap. J. med. Sci. Biol.* **22,** 175.

Solomon, H. M., Lynt, R. K., Kautter, D. A. and Lilly, T. (1969) *J. Bact.* **98,** 407; (1971) *Appl. Microbiol.* **21,** 295.

Sompolinsky, D. (1950) *Ann. Inst. Pasteur* **79,** 204.

Sordelli, A. (1922) *C. R. Soc. Biol.* **87,** 838; (1923) *Ibid.* **89,** 53.

Sordelli, A., Ferrari, J. and Prado, M. (1930) *Rev. Inst. bact., B. Aires* **5,** 797.

Spray, R. S. (1936) *J. Bact.* **32,** 135.

Sterne, M. and Batty, I. (1975) *Pathogenic Clostridia*. Butterworths, London.

Sterne, M. and Warrack, G. H. (1962) *J. Path. Bact.* **84,** 277; (1964) *Ibid.* **88,** 279.

Stewart, S. E. (1938) *J. Bact.* **35,** 13; (1940) *Publ. Hlth. Rep., Wash.* **55,** 753.

Stickland, L. H. (1934) *Biochem. J.* **28,** 1746; (1935) *Ibid.* **29,** 288, 889.

Stone, J. L. (1952) *J. Bact.* **64,** 299; (1953) *Yale J. Biol. Med.* **25,** 239.

Stringer, M. F., Turnbull, P. C. B. and Gilbert, R. J. (1980) *J. Hyg., Camb.* **84,** 443.

Sutton, R. G. A. and Hobbs, B. C. (1968) *J. Hyg., Camb.* **66,** 135.

Svec, M. H. and McCoy, E. (1944) *J. Bact.* **48,** 31.

Takacs, J. (1967) See Fredette, V., p. 101.

Tally, F. P., Goldin, B. R., Jacobus, N. V. and Gorbach, S. L. (1977) *Infect. Immun.* **16,** 20.

Tamai, K. and Nishida, S. (1964) *J. Bact.* **88,** 1647.

Tardieux, P. and Nisman, B. (1952) *Ann. Inst. Pasteur* **82,** 458.

Tataki, H. and Huet, M. (1953) *Ann. Inst. Pasteur* **84,** 890.

Taylor, A. W. (1940) *J. comp. Path.* **53,** 50.

Taylor, A. W. and Stewart, J. (1941) *J. Path. Bact.* **53,** 87.

Taylor, C. E. D. and Coetzee, E. F. C. (1966) *Mon. Bull. Minist. Hlth publ. Hlth Lab. Serv.* **25,** 142.

Theiler, A. *et al.* (1926) *11th and 12th Rep., Director vet. Educat. Res., S. Africa*, Part ii. p. 821.

Tissier, H. and Martelly. (1902) *Ann. Inst. Pasteur* **16,** 865.

Tjaberg, T. R. and Fossum, K. (1973) *Acta vet. scand.* **14,** 700.

Todd, E. W. (1941) *Brit. J. exp. Path* **22,** 172.

Torrey, J. C., Kahn, M. C. and Salinger, M. H. (1930) *J. Bact.* **20,** 85.

Tschertkow, L. (1929) *Z. ImmunForsch.* **63,** 262.

Tulloch, W. J. (1919) *J. Hyg., Camb.* **18,** 103.

Turner, A. W. and Eales, C. E. (1941) *Aust. J. exp. Biol. med. sci.* **19,** 167; (1943) *Ibid.* **21,** 79; (1944) *Ibid* **22,** 215.

Turner, A. W. and Rodwell, A. W. (1943) *Aust. J. exp. Biol. med. Sci.* **21,** 17, 27.

Tytell, A. A. and Hewson, K. (1950) *Proc. Soc. exp. Biol., N.Y.* **74,** 555.

Uemura, T. and Skjelvale, R. (1976) *Acta path. microbiol. scand.* **B84,** 414.

Vawter, L. R. and Records, E. (1926) *J. Amer. vet. med. Ass.* **68,** 494; (1931) *J. infect. Dis.* **48,** 51.

Veillon, A. and Zuber, A. (1898) *Arch. Méd. exp.* **10,** 517.

Walbum, L. E. (1938) *J. Path. Bact.* **46,** 85.

Walbum, L. E. and Reymann, C. G. (1933) *J. Path. Bact.* **36,** 469.

Walker, P. D. (1963) *J. Path. Bact.* **85,** 41.

Walker, P. D. and Batty, I. (1964) *J. appl. Bact.* **27,** 140; (1980) In: *Identification Methods for Microbiologists.* Ed. by F. A. Skinner and D. W. Lovelock, Academic Press, London.

Warrack, G. H., Bidwell, E. and Oakley, C. L. (1951) *J. Path. Bact.* **63,** 293.

Watt, B. (1973) *J. med. Microbiol.* **6,** 307.

Webster, M. E., Altieri, P. L., Conklin, D. A., Berman, S., Lowenthal, J. P. and Gochenour, R. B. (1962) *J. Bact.* **83,** 602.

Weinberg, M., Davesne, J. and Lefranc, M. (1931) *C. R. Soc. Biol.* **107,** 506.

Weinberg, M., Davesne, J., Mihailesco, M. and Sanchez, C. (1929) *C. R. Soc. Biol.* **101,** 907.

Weinberg, M. and Ginsbourg, B. (1927) *Données Récentes sur les Microbes Anaérobies et leur Rôle en Pathologie*. Masson et Cie, Paris.

Weinberg, M. and Guillaumie, M. (1936) *C. R. Soc. Biol.* **121,** 1275.

Weinberg and Séguin, P. (1915) *C. R. Soc. Biol.* **78,** 274; (1916) *C. R. Acad. Sci.* **163,** 449; (1918) *La Gangrène Gazeuse*. Paris.

Welch, D. F., Menge, S. K. and Matsen, J. M. (1980) *J. clin. Microbiol.* **11,** 470.

Welch, W. H. and Nuttall, G. H. F. (1892) *Johns Hopk. Hosp. Bull.* **3,** 81.

Wetzler, T. F., Marshall, J. D. and Cardella, M. A. (1956) *Amer. J. clin. Path* **26,** 418.

Wickham, N. (1956) *J. comp. Path.* **66,** 62, 71.

Wildführ, G. (1949*a*) *Z. ImmunForsch.* **106,** 369; (1949*b*) *Zbl. Bakt.* **154,** 211; (1950) *Z. Hyg. InfektKr.* **130,** 521, 529.

Williams, K. and Willis, A. T. (1970) *J. med. Microbiol.* **3,** 639.

Willich, G. (1952) *Z. Hyg. InfektKr.* **134,** 573.

Willis, A. T. (1957) *J. Path. Bact.* **74,** 113; (1960) *J. Path. Bact.* **80,** 379; (1969*a*) *Technique for the Study of Anaerobic Spore-forming Bacteria*. In: *Methods in Microbiology*, Vol. 3B, p. 79. Eds J. R. Norris and D. W. Ribbons. Academic Press, London; (1969*b*) *Clostridia of Wound Infection*. Butterworths, London; (1977) *Anaerobic Bacteriology. Clinical and Laboratory Practice*, 3rd edn. Butterworths, London.

Willis, A. T. and Gowland, G. (1962) *J. Path. Bact.* **83,** 219.
Willis, A. T. and Hobbs, G. (1958) *J. Path. Bact.* **75,** 299; (1959) *Ibid.* **77,** 511.
Willis, A. T. and Williams, K. (1970) *J. med. Microbiol.* **3,** 291.
Wilsdon, A. J. (1931) *2nd Rep. Director, Inst. Anim. Path., Camb.* p. 53; (1933) Ibid., *3rd Rep.*, p. 46.
Winogradsky, S. (1902) *Zbl. Bakt.*, IIte Abt. **9,** 43, 107.
Wolf, C. G. L. and Harris, J. E. (1917) *J. Path. Bact.* **21,** 386.
Woods, D. D. and Trim, A. R. (1942) *Biochem. J.* **36,** 501.
Wright, G. P. (1955) *Pharmacol. Rev.* **7,** 413.
Wynne, E. S. and Harrell, K. (1951) *Antibiot. Chemother.* **1,** 198.
Yamamoto, A. *et al.* (1972) *Jap. J. med. Sci. Biol.* **25,** 15.
Zamecnik, P. C., Brewster, L. E. and Lipmann, F. (1947) *J. exp. Med.* **85,** 381.
Zeissler, J. and Raszfeld, L. (1929) *Arch. wiss. prakt. Tierheilk.* **59,** 419.
Zeissler, J. and Raszfeld-Sternberg, L. (1949) *Brit. med. J.* **i,** 267.

43

Miscellaneous bacteria

Graham Wilson

We include in this chapter a number of organisms which, for one reason or another, cannot justifiably be allotted to any of the named groups that we have so far considered.

Bifidobacterium

Definition

Pleomorphic gram-positive rods showing true and false branching. Non-motile and non-sporing. Strictly anaerobic. Growth between 20 and 45° with optimum at 38°. Aciduric, but killed by heat at 60° in 5 min. Die rapidly in fluid culture media. Require sugar for growth. Ferment various carbohydrates with production of lactic and acetic acid but no gas. No proteolytic activity. Do not form catalase, oxidase, urease, indole or H_2S, or reduce nitrate. Have distinctive DNA composition, with G+C content of *ca* 60 moles per cent. Non-pathogenic to man and animals. Natural intestinal bacterium, predominating in breast-fed infants.

Type species: *Bifidobacterium bifidum*.

The first organism of this group was isolated from infants' stools by Tissier (1900), who called it *Bacillus bifidus*. Similar organisms, but varying in their properties, make up a high proportion of the bacterial flora of the faeces of breast-fed infants and, to a less extent, of artificially fed infants and of adults (see Dehnert 1957, 1961, Seeliger and Werner 1962, Reuter 1963).

As a detailed description follows of *Bifidobacterium bifidum*, there is no need here to do more than draw attention to certain properties of the organisms and to discuss their classification.

In the past much confusion has been caused by the failure to work with pure cultures. Bifidobacteria have been mixed with lactobacilli, with anaerobic corynebacteria, or with some other closely related organism (Gyllenberg 1958, Dehnert 1961). The establishment of pure cultures is often a long and difficult task requiring repeated testing till consistent results are obtained.

The pleomorphic morphology of the bifidobacteria is characteristic and distinctive, but changes with different media. Strains belonging to some groups are less characteristic than others and are not so easy to distinguish from strains of related genera.

The nutritional requirements are still imperfectly known. According to Gyllenberg and Carlberg (1958), some strains need only biotin, pantothenic acid and cysteine as essential growth factors. Others are dependent in addition on riboflavin, and others on purine and pyrimidine bases.

Biochemically one of the more striking features is the formation of acetic acid alone among the volatile acids produced during the fermentation of glucose. This property and the failure to form detectable gas separate the bifidobacteria from the heterofermentative group of lactobacilli (Beerens *et al.* 1957; see Chapter 29 for other distinctions between bifidobacteria and lactobacilli). The absence of propionic acid from the products of fermentation and the negative catalase reaction separate *Bifidobacterium* from *Propionibacterium*. Most of the bifidobacteria are able to hydrolyse cholic acid and conjugated bile acids such as sodium glycocholate and sodium taurocholate (Drasar and Hill 1974, Ferrari *et al.* 1980).

Antigenic structure is complex, and most workers have found a fairly high degree of strain specificity by agglutination. The antigen taking part in this reaction is heat-stable; the results are the same whether living or boiled organisms are used for the preparation of antisera. Unlike the lactobacilli, bifidobacteria do not seem to contain an extractable precipitinogen.

The amino-acid content of the cell wall varies with different strains. Cummins, Glendinning and Harris (1957), who studied eight strains, found that the chief amino acids were alanine, lysine and glutamic acid, but some strains had also serine, glycine, threonine or aspartic acid. The main sugars were rhamnose, galactose, glucose and often mannose. Glucosamine and muramic acid constituted the amino sugars; there was no galactosamine. The acid-resistant peptide of lysine and aspartic acid, noted by Cummins and Harris (1956) in lactobacilli, was absent, suggesting that the cell walls of the lactobacilli and of the bifidobacteria are distinct.

Study of the DNA composition reveals a G + C content of 57.2–64.2 moles per cent, with a mean of 60.1. These figures differ from the much lower ones of 33.0–52.5 for the lactobacilli, and the higher ones of 66.4–70.4 for the propionibacteria (Sebald *et al.* 1965, Werner *et al.* 1965, Werner 1967). (For DNA homology, see Scardovi *et al.* 1971.)

The classification of the bifidobacteria presents serious difficulties, partly because they have been incompletely studied. Most workers are of the opinion that they should be classed in a genus of their own, as suggested by Orla-Jensen (1924). Within the genus several biotypes or species have been described by Dehnert (1957, 1961), Gyllenberg and Carlberg (1958), Seeliger and Werner (1963), and Mitsuoka (1969). Reuter (1963, 1971) recognized and named eight separate species. By means of DNA hybridization, however, Scardovi and his colleagues (1971) found that some of these species were homologous, and they therefore reduced the number to five. They added themselves two other species that were not in Reuter's list, making a total of seven, namely: *Bif. bifidum*, *Bif. infantis* (*lactensis*), *Bif. parvulorum* (*breve*), *Bif. longum*, *Bif. adolescentis*, *Bif. ruminale* (*thermophilum*) and *Bif. pseudolongum* (*globosum*). Distinction between the species rested mainly on saccharolytic activity, antigenic structure, and G + C content of the DNA. Numerous other strains were studied, such as *Bif. suis* (Mateuzzi *et al.* 1971), but were not assigned to any of these species. Subsequently four new species were added, mainly on the evidence of DNA homology: *Bif. cuniculi* from rabbits, *Bif. choerinum* from pigs, *Bif. boum* from cattle, and *Bif. pseudocatenulatum* from man and calves (Scardovi *et al.* 1979).

Bifidobacterium bifidum

Habitat Common in the faeces of breast-fed, and much less common in those of bottle-fed infants. Sometimes present in the faeces of adults and of animals. In breast-fed infants during the first few weeks of life it may form 99 per cent of the faecal flora.

Morphology In faeces it is a delicate bacillus, about 4 μm long and 0.7 μm broad, with tapering pointed ends; arranged in pairs end-to-end, with the distal ends pointed and the proximal ends swollen; they generally lie parallel to one another, rarely intertwined. Two or three bacilli often radiate from a single point, forming a Y-shaped structure; clubbed forms and forms ending in knobs are not uncommon. Often arranged in palisades or Chinese letters. General appearance is not unlike a diphtheroid bacillus (Fig. 43.1). In culture its

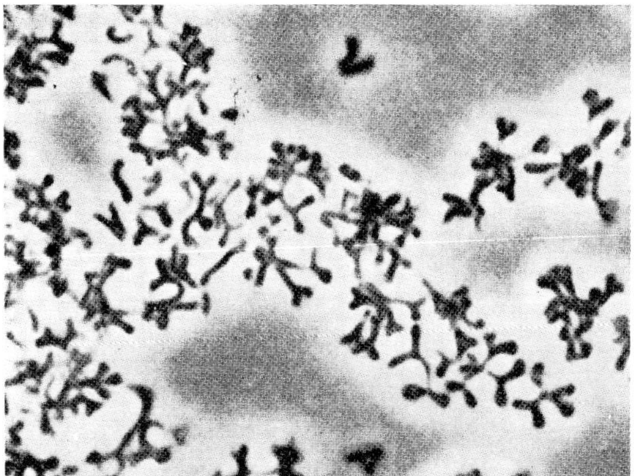

Fig. 43.1. *Bifidobacterium bifidum*. Showing the characteristic arrangement of branched club-ended bacilli. (From photomicrograph kindly supplied by Dr. Martin Kludas.)

morphology varies with the medium. Its characteristic appearance is that of a highly pleomorphic bacillus showing irregularly clubbed forms with branching often at both ends. Geniculate forms, forms ending in knobs, forms with lateral buds, bladder forms, and candle-flame forms may be seen. Absence of chain formation and of long filaments is noteworthy. Non-motile and non-sporing. Stain uniformly in

Table 43.1 Characters of the different groups of *Bifidobacterium* (after Dehnert 1961)

Group	Colonial form on milk agar 48 hr, 37°	Morphology on milk agar	Growth in Tomarelli's medium plus human milk	Per cent volatile acid produced	Acid formed in								No. of serological types
					arabinose	maltose	sucrose	raffinose	cellobiose	trehalose	xylose	salicin	
I/II	Low convex, rough, granular, with matt, uneven surface; up to 4 mm q	Clubbed and branched forms conspicuous 0.7–1.0 μm q	Poor: no turbidity	60–70	−	slight	slight	−	−	−	−	−	2
III	Raised; gritty, slightly wavy surface; 2–3 mm q	Forms with clubbed and swollen ends, with some branching 0.7 μm q	Moderate growth and turbidity	50	−	+	+	+	+	±	−	+	4
IV	Convex or pearl-like smooth, glistening surface; 1–2 mm q	Slender and unbranched. Ends tapering, rounded or swollen, granular staining. 0.4–0.6 μm q	Rich growth with dense turbidity	35–45	−	+	+	+	−	±	+	−	4
V	Low convex; granular or mucoid with finely waved glistening surface; 5–8 mm q	Clubbed and branched forms conspicuous 0.7–1.0 μm q	Fairly good growth with moderate turbidity	19	+	+	+	±	±	±	+	−	4

All strains form D-lactic acid, acidify milk with little or no coagulation, ferment glucose and lactose but not rhamnose, dulcitol, erythritol, adonitol or sorbose.

young cultures, but in older cultures irregular or granular staining not uncommon. Gram-positive in young cultures; later gram-negative forms appear. Non-acid-fast. (For description of morphology with illustrations see Orla-Jensen 1943, Kludas and Dobberstein 1959, Dehnert 1957, 1961.)

Growth characters Organic nitrogen and sugar required for growth. No growth occurs on nutrient agar. On glucose agar and human milk agar, colonies are usually low convex, greyish-brown, lenticular or ovoid, 0.5–2 mm in diameter, showing under the microscope a delicately granular structure, a brownish opaque centre, a thinner translucent periphery, and a finely crenated edge (Fig. 43.2). In glucose broth growth is moderate; in 3 or 4 days the organisms fall to the bottom of the tube forming an abundant loose flocculo-granular deposit, easily disintegrated on shaking. Very slight or no growth occurs at 20°; optimum temperature 38–39°; upper limit 45°. Optimum pH 6.5–7.0. CO_2 beneficial. Growth improved by liver, yeast extract, and faecal extract (Orla-Jensen 1943), and also by skim human milk, which contains a factor that is practically absent from cows' milk (György *et al.* 1954, Gauhe *et al.* 1954). For isolation from faeces Tanaka and Mutai (1980) recommend Petuely's (1956) synthetic medium supplemented by riboflavin, nucleic acid bases, pyruvic acid, and nalidixic acid.

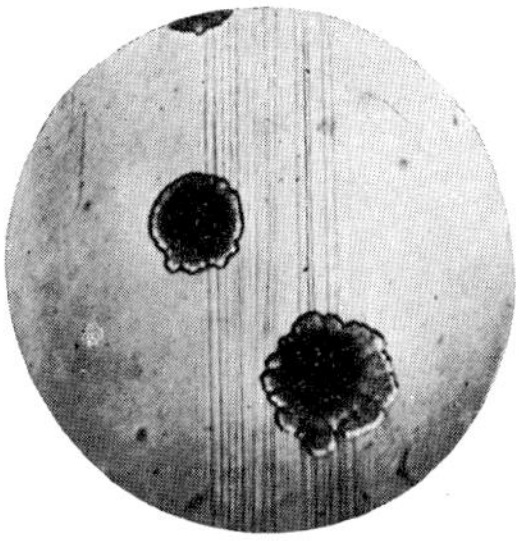

Fig. 43.2. *Bifidobacterium bifidum.* Surface colonies on glucose agar, 6 days, 37° (×8).

Biochemical activities Glucose and lactose are broken down to about equal amounts of D-lactic acid and acetic acid; no gas evolved. Maltose and sucrose only weakly fermented. No formation of oxidase, catalase, urease, indole, or H_2S, and no reduction of nitrate or hydrolysis of arginine.

Resistance In fluid cultures the organisms are often dead within 3 days. Killed by heat at 55° in half an hour and at 60° in 5 min. Resistant to acids. For preservation they should be subcultured daily, or suspended in skim human milk and freeze-dried.

Antigenic structure Incompletely studied. Agglutinated by sera prepared against live or heat-killed organisms. G+C content of DNA is about 60.1 mol per cent.

Pathogenicity Non-pathogenic to man or laboratory animals.

(For a study of bifidobacteria in the human intestine, see van der Wiel-Korstanje 1973; and for *Bifidobacterium* bacteriophages, see Matteuzi *et al.* 1971.)

Veillonella

Definition

Small, gram-negative, spherical, non-motile cocci occurring in pairs, masses and short chains. Grow poorly in ordinary media but well in special media containing trypticase. Anaerobic. Oxidase negative. Ferment no sugars. Indole negative. Nitrate reduced to nitrite or beyond. Common parasites present in the saliva of man and animals. G+C content of DNA 36 moles per cent.

Type species: *Veillonella parvula.*

The name *Veillonella* was given by Prévot after Veillon and Zuber (1898), who described the first species. The organisms are present abundantly as commensals in the mouth, and in the intestinal and respiratory tracts of man and animals; rarely they invade the bloodstream after operations on the mouth (McEntegart and Porterfield 1949, Khairat 1966; see also Koornhof and Robinson 1963). They differ from the neisseriae in growing anaerobically and in failing to form oxidase.

Morphologically they are spherical cocci, 0.3–0.5 μm in diameter, arranged in small clusters, short chains, and pairs (Sims 1960; Rogosa 1965). According to Langford, Faber and Pelczar (1950), they form circular colonies on trypticase soy agar, 1–1.5 mm in diameter after 48 hr at 35°, almost transparent with a light bluish or greenish tinge by transmitted light and a pearly appearance by reflected light. In shake tubes of cystine trypticase agar they form a band 1 cm below the surface. Cultures die rapidly in air. No sugars are fermented, but acid is produced in 1 per cent sodium lactate solution, and some gas may be liberated in peptone media. H_2S is formed, nitrate is reduced to nitrite, and nitrite to ammonia, depending on the presence of a suitable hydrogen donor (Yordy and Delwicke 1979). Gelatin is not liquefied. The indole, oxidase, and catalase reactions are negative. There is no haemolysis. Some growth occurs in the presence of 5 per cent bile. Selective media containing trypticase for isolation are described by Rogosa (1956) and Rogosa and others (1958; see also Rogosa and Bishop 1964). Growth occurs best at 30 to 37°; there is no growth at 18° or 24°. Growth is improved by CO_2 and inhibited by polymyxin B. The organisms are sensitive to most antibiotics, but are resistant to streptomycin. The composition of the cell walls is said to resemble that of the neisseriae but to contain rather more mucopeptide (Graham and May 1965). The presence of specific lipopolysaccharide endotoxins enables seven serological groups to be defined (Rogosa 1965; see also Hofstad and Kristoffersen 1970) and four chemotypes (Hofstad 1978). Rogosa recognizes two species, *V. parvula* and *V. alcalescens*, distinguishable on nutritional, biochemical and serological grounds.

Gardnerella vaginalis

This organism, which was described by Leopold (1953) and placed in the *Haemophilus* group by Gardner and Dukes (1955), is a gram-variable rod that requires neither X nor V factor. It is clearly not a species of *Haemophilus*, and was for some years regarded as a *Corynebacterium* sp. (Zinnemann and Turner 1963, Dunkelberg *et al.* 1970). Criswell and her colleagues (1971) dispute this. They consider it to be a gram-negative rod that cannot be assigned to the *Corynebacterium*, *Butyribacterium*, *Lactobacillus*, or any other group. They therefore propose to transfer it to a new genus, *Gardnerella*, containing gram-negative or gram-variable rod-shaped organisms having laminated cell walls. This proposal is strongly supported by Piot and his colleagues (1980), who point out that it differs from *Corynebacterium* spp. in its low G+C content, and in its cell-wall structure and composition. From *Propionibacterium* spp. it differs in that the end-product of fermentation is acetic acid rather than lactic and propionic acids. (See Greenwood and Pickett 1980.) The description of the type species of the genus, *Gardnerella vaginalis*, varies to some extent with different observers. In general, however, it is given as follows.

It is a gram-variable rod 1–2 μm $\times$ 0.3–0.6 μm in size, non-motile and non-capsulated. When stained by, for example, Albert's stain, intracellular granules are seen. During exponential growth on Loeffler's serum the rods are generally gram positive. At 37° it forms small round colonies on blood agar; β-haemolysis appears on plates of human or rabbit blood agar but not on sheep blood agar. Most strains grow well in humidified air plus 5 per cent CO_2. There is no growth on ordinary nutrient agar. Dunkelberg and co-workers (1970) recommend the use of a peptone starch glucose medium on which it forms characteristic white colonies with a slightly darker centre and shows evidence of starch hydrolysis. It is resistant to sulphonamides and nalidixic acid but sensitive to trimethoprim 5 mg per l; nearly all strains are sensitive to metronidazole 128 mg per l (Taylor and Phillips 1983). It produces acid from glucose, maltose, dextrin, starch and glycogen, and hydrolyses hippurate; and it gives negative reactions in the catalase, oxidase, urease and nitrate-reduction tests. Seven antigenic groups have been identified by means of precipitation tests. It requires five B vitamins as well as purine-pyrimidine compounds. The G+C content of the DNA is 42 moles per cent.

For the isolation of *G. vaginalis*, plates of human blood agar or peptone starch glucose agar containing nalidixic acid 30 mg per l, gentamicin 4 mg per l and amphotericin 2 mg per l, are recommended (Ison *et al.* 1982). The following reactions are useful for identifying the organism: characteristic cellular and colonial morphology, a negative catalase reaction, hydrolysis of starch and hippurate, haemolysis on human blood agar, sensitivity to trimethoprim and to a high concentration of metronidazole, failure to grow on nutrient agar or in the presence of 2 per cent NaCl or to form lactic acid from glucose (Dunkelberg *et al.* 1970, Mickelsen *et al.* 1977, Greenwood and Pickett 1979, Bailey *et al.* 1979, Wells and Goei 1981, Leighton 1982, Piot *et al.* 1982, Yong and Thompson 1982, and Taylor and Phillips 1983).

In women the organism is associated with vaginitis, cervicitis, and leucorrhoea, and in men with urethritis. Its aetiological significance is not fully established, but there is evidence to show that it is mildly pathogenic (Lewis *et al.* 1972). In vaginal swabs, squamous epithelial 'clue cells' may be seen covered with great numbers of the bacilli (see Brzin, 1969, Dunkelberg *et al.* 1970). (See also Chapter 57.)

Eikenella corrodens

The aerobic and facultatively anaerobic gram-negative bacilli that corrode the surface of agar media and are now known as *Eikenella corrodens* (Jackson and Goodman 1972) were first described by Henriksen (1948) in cultures from three patients with deep abscesses. They are distinct from the anaerobic corroding bacilli known as *Bacteroides ureolyticus* (Chapter 26) but the two species have been confused in the past (Jackson *et al.* 1971).

Eik. corrodens is a small gram-negative coccobacillus, 0.5 $\times$ 1–3 μm; non-motile; does not produce spores. It grows rather slowly on blood agar or heated (chocolate) blood agar at 35–37°C; growth is usually improved by 5–10 per cent CO_2 in the atmosphere and may be stimulated by haemin 5–20 mg per l. After 18–24 hr, colonies are only pin-point; with longer incubation they become dry and flat with a raised centre and an irregular, spreading edge. Most strains typically corrode the surface of the agar but non-corroding variants have been described. Slightly haemolytic; colonies may produce a small, variable zone of α-haemolysis. Oxidase positive and reduces nitrate to nitrite, but catalase negative; does not produce H_2S or indole, hydrolyse urea, digest gelatin or produce acid from carbohydrates. The G+C content of the DNA is 57–58 moles per cent (Jackson *et al.* 1971).

E. corrodens is part of the commensal flora of mucous membranes in man (Marsden and Hyde 1971); it is isolated most commonly from the upper respiratory tract but may be found in the gastro-intestinal and genito-urinary tracts. It has been isolated as the sole pathogen from infections such as endocarditis, central nervous system infections and a variety of deep abscesses, but most clinical isolates from human sources are from infections of the head and neck, with a smaller proportion from post-operative infections of the abdomen and perineum; these are invariably mixed infections with *Eik. corrodens* and other bacteria such as streptococci and enterobacteria (Marsden and Hyde 1971; Zinner *et al.* 1973; Dorff *et al.* 1974; Granato 1977).

'Bacterium granulosis'

This small gram-negative bacillus, which was isolated by Noguchi (1927) from American Indians suffering from trachoma and described in previous editions of this book, is, according to Professor P. H. A. Sneath (pers. comm.) no longer recognizable.

Legionella

The organism now known as *Legionella pneumophila* was first recognized as a cause of severe pneumonia that affected about 200 persons gathered for an American Legion convention in Philadelphia in 1976 (see Chapter 73). It was isolated from specimens of lung collected at necropsy by intraperitoneal injection into guinea-pigs; suspensions of guinea-pig spleen, liver or lung tissue were then inoculated into fertile eggs. The chick embryos died in 4–7 days, and smears of the yolk sacs stained by the Giménez method showed the presence of small gram-negative bacilli together with occasional filamentous forms (McDade *et al.* 1977). It was later found that this organism could also cause a short febrile illness without pneumonia (Pontiac fever; Chapter 73). Several other species of legionellae, each responsible for pneumonia in man, have since been described.

Morphology and cellular constituents

The organism is a short rod with tapering ends (Fig. 43.3*a*), 1–2 μm long and about 0.5 μm in greatest width. By electron microscopy, fimbriae arranged all round the organism and a single polar or subpolar flagellum can be seen (Rodgers *et al.* 1980; see Fig. 43.3*b*). The organism appears not to be capsulated. The cells have a structure typical of gram-negative bacteria: the cell envelope is composed of two triple-unit membranes with an intervening space and encloses a cytoplasm rich in ribosomes and in nuclear elements (Chandler *et al.* 1979*a*, Rodgers 1979, Rodgers and Davey 1982); some peptidoglycan can be demonstrated in the cell envelope. The cell contains branched-chain fatty acids of unusual type; determination of their composition by gas chromatography is a valuable means of identifying the organisms (Moss *et al.* 1977). The fatty acids are said to have weak endotoxic activity (Wong *et al.* 1979).

In tissue specimens stained by Gram's method the organisms are barely visible, but when only the first step of the gram procedure is used (the 'half-gram' method) they show up clearly (de Frietas *et al.* 1979). They can also be demonstrated in lung tissue by the Dieterle (1927) silver-impregnation method (Chandler *et al.* 1977). However, their identity needs to be confirmed antigenically by direct immunofluorescence

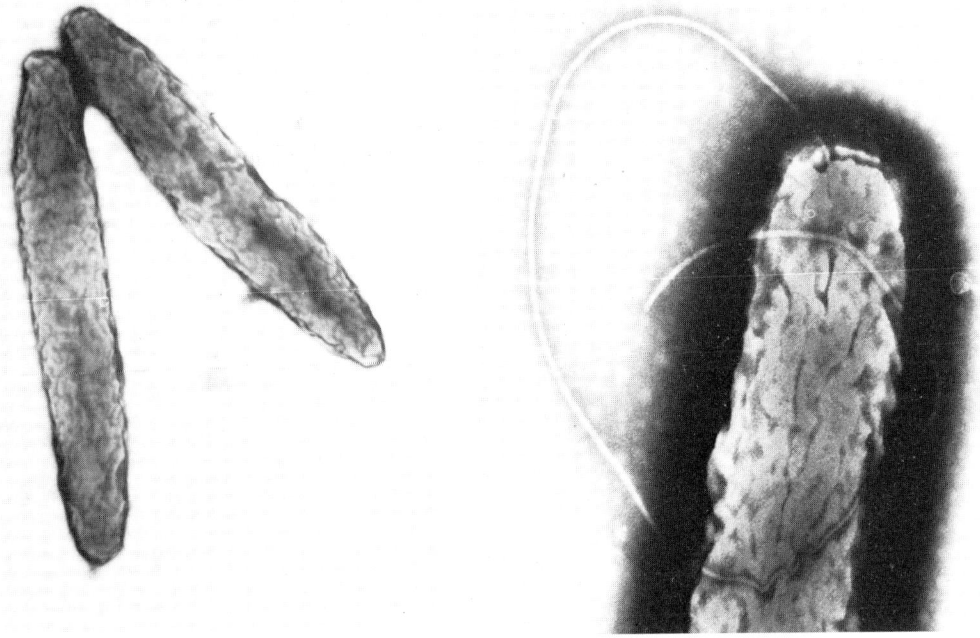

(a) (b)

Fig. 43.3. *Legionella pneumophila*, electronmicrographs of preparations stained with 1 per cent phosphotungstic acid. (a) Strain Bloomington 2 (serogroup 3), two organisms showing characteristic appearance, with tapered ends and 'ruffled' surface (× 25 000). (From *J. gen. Microbiol.*, 1982, **128,** 1547.) (b) Strain Cambridge 2 (serogroup 5), part of one organism, showing polar attachment of flagellum (× 66 000). (From *J. clin. Path.*, 1980, **33,** 1184.)
(By courtesy of Dr F. G. Rodgers.)

staining (Thomason *et al.* 1979) or immunoferritin electronmicroscopy (Rodgers 1982).

Requirements for growth

The organisms will not grow on blood agar; they were first cultivated on a nutrient agar containing haemoglobin and a commercial growth supplement. Essential components of this medium included iron and L-cysteine. Pine and co-workers (1979) described a chemically defined medium; they found that pyruvate and α-ketoglutarate stimulated growth, that cysteine and methionine were necessary, and that various amino acids served as sources of energy (see also Ristroph *et al.* 1981, and, for metallic requirements, Reeves *et al.* 1981). Two classes of medium are widely used: charcoal yeast agar (Feeley *et al.* 1979), usually with added α-ketoglutarate; and an enriched blood agar containing ferric pyrophosphate and L-cysteine (Greaves 1980). The organisms are very susceptible to changes of pH in culture, and special measures to control these have been recommended (Pasculle *et al.* 1980, Dennis *et al.* 1981).

Several semi-selective and enrichment media have been described for the isolation of the organisms from human tissue; these usually contain polymyxin and a fungistatic compound along with one or more other antibacterial agent, such as vancomycin, cefamandole or a sulphonamide (Edelstein and Finegold 1979, Greaves 1980, Edelstein 1981). Dilution of lung specimens and brief pretreatment of them with acid (HCl, pH 2, followed by neutralization after 1 hr) have also been advocated.

Legionellae grow best at 35° in air with the addition of 2.5–5.0 per cent of CO_2, and at pH 6.9. There is no growth anaerobically.

Colonial appearances

Visible growth usually appears after 2–3 days; colonies are at first few in number but more appear on further incubation. Most of them are pinpoint in size, grey in colour and glistening, but a proportion of larger ones, up to 2 mm in diameter, may develop in some strains. *Leg. pneumophila* and members of some other species form a brown diffusible pigment (Fisher-Hoch 1980). This is best seen on media that do not contain haemoglobin, especially when tyrosine is added. Several legionellae, but not *Leg. pneumophila* or *Leg. micdadei*, show a blue fluorescence with ultraviolet light. Haemolysis around colonies on enriched blood agar has been described and is said to be most pronounced when guinea-pig or rabbit blood is used (Baine *et al.* 1979), but some workers have experienced difficulty in demonstrating this property (see Orrison *et al.* 1981).

Resistance

Legionellae survive in water for at least a year (Skaliy and McEachern 1979) and on charcoal yeast agar for several months (Hébert 1980). They are susceptible to low concentrations of disinfectants (Wang *et al.* 1979) including a number of compounds used to control the multiplication of micro-organisms in stored water (Skaliy *et al.* 1980), but the performance of these agents in the field often does not correspond with that observed *in vitro* (England *et al.* 1982). They die in 30 min at 58° (Müller 1981).

The multiplication and long survival in water of such a delicate and nutritionally exacting organism is difficult to understand. According to Tison and his colleagues (1980), it can survive in certain heavily polluted waters only in association with blue-green algae. Rowbotham (1980) showed that free-living amoebae found in soil and water ingest *P. pneumophila*, which may then multiply within them. Some heavily infected amoebae die, but others survive, even though they contain many endoplasmic vacuoles packed with legionellae. Rowbotham suggests that the intracellular bacteria may be protected from drying, and that the infected amoeba, which may contain many hundreds of organisms, or the endoplasmic vacuoles released from it, may be the vehicle of infection for man.

Antibiotic susceptibility Legionellae are generally sensitive *in vitro* to therapeutically attainable concentrations of erythromycin, chloramphenicol, rifampicin, ticarcillin, cefoxitin and various aminoglycosides but are resistant to clindamycin and vancomycin. All species except *Leg. micdadei* form β-lactamase. The minimal inhibitory concentrations of penicillins and cephalosporins are to a considerable extent characteristic of individual species and cover a fairly wide range. (See Edelstein and Meyer 1980, Saravolatz *et al.* 1980, Pasculle *et al.* 1981.)

Biochemical activities

Legionellae do not use glucose, reduce nitrate or degrade urea. They are catalase positive and many of them give a weak oxidase reaction. With rare exceptions they form a protease that causes liquefaction of gelatin and attacks casein but not elastin (Thompson *et al.* 1981). Various esterases are formed; the ability to hydrolyse diacetylfluorescein is said to be a constant feature of legionellae but to be rare in other bacterial genera (Orrison *et al.* 1981). Lipase, deoxyribonuclease and ribonuclease have been reported (Thorpe and Miller 1981). Hippurate is hydrolysed by *Leg. pneumophila* but not by other species of *Legionella* (Hébert 1981).

Antigenic properties

Six serotypes of *Leg. pneumophila* have been distinguished by direct immunofluorescence staining (Taylor and Harrison 1979, Morris *et al.* 1980) and by bacterial agglutination and co-agglutination on slides (Wilkinson and Fikes 1981). Each serotype appears to have a specific antigen and cross-reacting antigenic components common to the species. Members of other species are all antigenically distinct.

Material of high molecular weight ($>4 \times 10^6$) removed from the cells of *Leg. pneumophila* by washing them with phosphate-buffered saline has type specificity (Johnson *et al.* 1979, Smith *et al.* 1981); immunization with it is said to induce type-specific immunity in guinea-pigs (Elliott *et al.* 1981). Type-specific soluble antigen can be detected in the urine of sufferers from Legionnaires' disease (Kohler *et al.* 1981, Mangiafico *et al.* 1981).

Antibody response in man McDade and his colleagues (1977) first used the indirect immunofluorescence method to demonstrate antibody in the serum of survivors from Legionnaires' disease. The value of this test is enhanced by the examination of serial samples and by the separate measurement of IgG and IgM antibody (Nagington *et al.* 1979). However, false-positive reactions, may occur in other infections (for references, see Lattimer and Cepil 1980); these can be very much reduced in frequency by the use of formolized yolk-sac antigen (Taylor *et al.* 1979).

Pathogenicity

The natural disease in man is described in Chapter 73.

Intraperitoneal injection of infective material into guinea-pigs gives rise to a febrile disease that proves fatal in about a week. Post-mortem examination shows a diffuse peritonitis, together with necrotic foci in the spleen, liver, lungs, lymph nodes and other areas. The bacteria are numerous in the peritoneal exudate but less abundant in the focal lesions. Phagocytosis by peritoneal macrophages is particularly noticeable (Chandler *et al.* 1979*b*) and the very large number of organisms present in each cell suggests that they have multiplied after being ingested. The inhalation of 10^4–10^5 cells of *Leg. pneumophila* as a fine-particle aerosol leads in guinea-pigs to a pyrexia and pneumonia which is uniformly fatal in 3 days (Baskerville *et al.* 1981); but intranasal instillation does not cause clinical illness. Rhesus monkeys infected by inhalation suffer a pneumonic infection, but this is less severe than that seen in guinea-pigs.

Leg. pneumophila tends to lose virulence after repeated subculture on artificial media; according to Wong and his colleagues (1981) this may be restored by passage in tissue culture. However, serial passage of freshly isolated strains in guinea-pigs occasionally fails, possibly because of inhibitory substances in tissue homogenates. *Leg. pneumophila* is taken up by human monocytes in tissue culture and subsequently multiplies intracellularly 10 000–100 000-fold (Horwitz and Silverstein 1980). It also multiplies intracellularly in human embryo-lung fibroblasts and causes severe cytopathic changes in them (Wong *et al.* 1980). Specific antibody and complement are not bactericidal; human granulocytes ingest sensitized organisms but do not kill them (Horwitz and Silverstein 1981).

Classification

Legionellae, as far as they have been studied, appear to differ from all other recognized bacterial species. Brenner and his colleagues (1979) classified the agent of Legionnaires' disease in a new genus *Legionella* as *Leg. pneumophila*. The G+C content of its DNA is about 39 moles per cent (McDade *et al.* 1979). Five other species have since been described, initially on the grounds of low percentages of DNA homology with *Leg. pneumophila* and with each other, and of differences in their cellular lipids. These species are: *Leg. micdadei* (which includes the Tatlock and HEBA strains and the Pittsburgh pneumonia agent) named after McDade (Hébert *et al.* 1980); *Leg. bozemani* and *Leg. dumoffi* (Brenner *et al.* 1980); *Leg. gormani* (Morris *et al.* 1980) and *Leg. longbeachi* (McKinney *et al.* 1981). Garrity and his colleagues (1980) consider that the differences between the species of legionellae are sufficient to warrant the recognition of three genera in a family Legionellaceae: *Legionella* (including *Leg. pneumophila*), *Tatlockia* (including *Tat. micdadei*) and *Fluoribacter* (including *Fluor. bozemani*).

The species of legionellae differ in several cultural characters that are of help in routine identification (see Orrison *et al.* 1981). These include: hydrolysis of hippurate only by *Leg. pneumophila*; absence of β-lactamase production and of formation of a brown pigment on tyrosine-containing medium by *Leg. micdadei*; negative oxidase reaction in *Leg. bozemani* and *Leg. dumoffi*; absence of fluorescence with ultraviolet light in *Leg. pneumophila* and *Leg. micdadei*.

(For general accounts of legionellae, see Symposium 1979, Broome and Fraser 1979, and Grimont and Bercovier 1980.)

Bartonella, Grahamella, Anaplasma, Haemobartonella and *Eperythrozoon*

Various workers have described bodies in close association with red blood corpuscles in man and animals suffering often, but not always, from infections in which anaemia is a prominent feature. The bodies stain poorly with aniline dyes and are gram negative, but they can be readily demonstrated by Giemsa's

Table 43.2. Differential characters of Bartonellaceae and Anaplasmataceae

Character	Bartonellaceae: *Bartonella*	Bartonellaceae: *Grahamella*	Anaplasmataceae: *Anaplasma* and related genera*	Anaplasmataceae: *Haemobartonella*	Anaplasmataceae: *Eperythrozoon*
Morphology	Mainly rods; lophotrichate flagella	Mainly rods	Mainly cocci	Mainly cocci	Mainly ring forms
Situation	In or on RBC; and in cytoplasm of endothelial cells	In RBC	In RBC; form inclusions containing a number of 'initial bodies'	Most on indents on surface of RBC; some intracellular	On surface of RBC and free in plasma
Motility	+	−	−	−	−
Cell wall	+	+	−	−	−
Growth *in vitro*	+	+	−	−	−
Sensitive to:					
penicillin and streptomycin	+	?	−	−	−
tetracycline	+	?	+	+	+
arsenicals	−	−	−†	+	+
Natural infection in:	*man only*: Oroya fever and verruga peruana	*mole, deer-mouse*: probably no ill effects‡	*various ruminants, and birds*: anaemia‡	*mice and rats*: parasitaemia after splenectomy, etc.; *dogs*: anaemia after splenectomy; *cats* anaemia in previously healthy animals‡	*mice*: parasitaemia after splenectomy; *sheep and pigs*: anaemia in previously healthy animals‡

RBC = red blood cells; + = character present; − = character absent or not yet reported.
**Paranaplasma* and *Aegyptianella* (see text).
†In *Anaplasma*.
‡More than one host-specific parasite.

method or with other Romanowsky-type stains. *Bartonella bacilliformis*, which infects man, and species of *Grahamella*, found in rodents and other small mammals, have distinct cell walls and have been cultivated in artificial media. These are almost certainly bacteria and have been placed in the family Bartonellaceae. Others, which have a cell membrane but no cell wall and have not yet been grown *in vitro*, are of more doubtful status; they have been grouped together in a separate family Anaplasmataceae.

Table 43.2 sets out some of the differential characters of these organisms; for reviews of them the reader may consult Weinman (1944, 1968, 1974), Peters and Wigand (1955), Nauck (1958), Wigand (1958), Tanaka *et al.* (1965), Tedeschi *et al.* (1967), Kreier and Ristic (1968, 1974), and Ristic and Kreier (1974).

Bartonella bacilliformis

This organism is a small bacillus, which destroys the red blood cells, and is responsible for Peruvian *Oroya fever* and for *verruga peruana* (see Chapter 73). It was called *Bartonella bacilliformis* by Strong and his colleagues (1915) in honour of Barton, who was one of the first to observe the bacillus in the red cells. Noguchi and Battistini in 1926 cultivated the organism, and reproduced a disease in monkeys bearing a close resemblance to the natural disease in human beings. The organism was recovered in pure culture from the blood of the injected monkeys.

Morphologically in young cultures it occurs mainly in the form of short rods arranged singly, in pairs, chains, and clumps. In older cultures coccoid forms predominate. There is a considerable degree of pleomorphism. Its size is about 0.3–1.5 μm long and 0.2–0.5 μm broad. It is motile, and a tuft of flagella, at least 10 in number, at one pole of the rod, can be demonstrated (Peters and Wigand 1952). Ultra-thin sections of red corpuscles examined by the electron microscope show that the organisms are situated inside the corpuscles as well as on the surface, and that they have a characteristic cell wall similar to that of bacteria. In man the organism is present not only in red blood cells but also in large numbers in the cytoplasm of endothelial cells. Cultivation can be effected on a variety of media, such as the semi-solid serum haemoglobin agar medium used for leptospirae, or a blood glucose cystine agar (Jiménez 1940), or a 2 per cent proteose agar to which 25 per cent of fresh defibrinated blood or serum from the rabbit or sheep is added, together with 0.2 per cent of an ascorbic acid glutathione solution (Geiman 1941). Though a high proportion of natural animal protein favours growth, it does not appear to be essential. Jiménez (1940) for example, states that good growth was obtained on 1 per cent glycerol infusion agar provided the X, though not the V, factor (see Chapter 39) was added. On solid media growth may occur in 4–5 days, either as minute, circular, clear, mucoid colonies, or as an opaque, finely granular, mucoid film that has a tendency to outgrow the original boundaries of the inoculum (Jiménez 1940). The organism is a strict aerobe, grows well at 25–37°, though best at about 25–28°, prefers a pH of 7.8, ferments no sugars, forms no haemolysin, and survives in semi-solid medium at −70° for years and at 25–28° for several weeks (Weinman 1968). It is sensitive to penicillin—which causes the formation of L forms (Sharp 1968)—and to streptomycin and tetracycline but not to arsenical compounds such as neosalvarsan, or to sulphonamides.

Pinkerton and Weinman (1937) found that in tissue cultures the organisms, unlike *Rickettsia*, grew extracellularly as well as intracellularly. Growth occurs also in the allantoic fluid of the developing chick embryo incubated at 25–28°, but this medium is unsuitable for serial cultivation (Jiménez and Buddingh 1940). Judged by the complement-fixation test, different strains appear to be antigenically homogeneous (Reese *et al.* 1950). Injected intravenously into young *rhesus* monkeys the organism may give rise to a peculiar, irregularly remittent type of fever, sometimes accompanied by severe anaemia; injected intradermally into the eyebrow, it gives rise to a nodule rich in cellular elements and capillary formation. Small laboratory animals are not susceptible to experimental infection. (For further description see Noguchi 1926, Kikuth 1931, Pittaluga 1938, Wigand 1958, Colichón and Bedón 1972.)

Grahamella

This parasite was first described by Graham-Smith (1905) at Cambridge, who observed it in the red blood cells of 10 per cent of moles that were being examined for *Piroplasma*. Since then it has been found in the blood of several other animals. The organisms are predominantly rods that resemble *Bartonella* but are rather less polymorphic and tend to be stained blue rather than a reddish tint by Giemsa's method. Motility has not been demonstrated, and they are found mainly within the erythrocytes and not in other tissues. Only a minority of red cells are infected (Graham-Smith 1905) but each may contain a number of organisms. They can be cultivated *in vitro* (Jettmar 1932, Tyzzer 1942); growth is favoured by haemoglobin. Like *Bartonella* they are unaffected by organic arsenicals. Two species have been described: *G. talpae* from moles and *G. peromysei* from deer-mice; whether the strains seen in other animals belong to different species has not been established (Kreier and Ristic 1968). The organisms appear to be non-pathogenic and to have no effect on the health of the host. In the rat Vassiliadis (1935) was able to transmit infection from one animal to another in series. In most animals splenectomy has little or no influence on infection, though the rat is said to constitute an exception (Vassiliadis 1935).

Anaplasma and related genera

These include parasites that form inclusions within erythrocytes and are found in no other part of the body. In Giemsa preparations, dense, round structures, reddish-violet in colour and up to 1 μm in diameter are seen. Electronmicroscopy shows that these are inclusions, surrounded by a limiting membrane and containing a number of subunits each 0.3–0.4 μm in diameter (Ristic and Watrach 1961). The subunits or 'initial bodies' appear to be the infecting unit and multiply within the inclusion by binary fission. In *Paranaplasma*, the inclusion bodies are indistinguishable from those of *Anaplasma* in Giemsa-stained preparations, but by special techniques can be shown to have tail-, loop- or ring-like appendages. Motility has not been demonstrated. None of these organisms has been grown in culture or in the chick embryo but infection can be transmitted to susceptible animals by parenteral routes. The organisms are resistant to penicillin, streptomycin and sulphonamides but sensitive to tetracycline. *Anaplasma* is said to be resistant to organic arsenicals.

Anaplasma contains several species that cause anaemia in cattle and other ruminants and are distinguished mainly by their host specificity and virulence; laboratory animals are refractory to experimental infection. *Paranaplasma* affects cattle only and *Aegyptianella* is a pathogen of birds. The natural diseases are tickborne. (For further information, see Ristic and Kreier 1974.)

Haemobartonella

These organisms usually occur on indentations of the surface of the red blood cell, but some are found in intracellular vacuoles; very few are free in the plasma. They are predominantly coccoid, and rod-shaped bodies are probably chains of cocci. They are non-motile, have not been cultivated *in vitro*, and are sensitive to organic arsenicals and tetracycline but not to penicillin or streptomycin. They cause anaemia in various animals, but often only after splenectomy. Several species have been described, each associated with one or a few related animal hosts.

Haemobartonella muris

This organism was first described by Mayer (1921), who found it in the blood cells of rats experimentally infected with trypanosomes. Morphologically it is a non-motile spherical coccus, 0.35–0.7 μm in diameter, having a single limiting membrane enclosing granules and some filaments, and without a cell wall (Fig. 43.4). According to Wigand (1958), the cytoplasm contains both DNA and RNA. The organisms are destroyed by exposure to a temperature of 57° for 30 min (Ford and Eliot 1928), but remain virulent in the frozen state for at least 11 weeks (Kessler 1942). Unlike *Bartonella* and *Rickettsia*, they can be dissolved by trypsin (Nauck 1958). The organism is fairly common in rats, causing an infection which normally remains latent, but which can be activated by splenectomy, by poisons such as toluylenediamine, phenylhydrazine, and by certain infections. When infected red cells are injected into haemobartonella-free splenectomized rats the organisms appear microscopically in the blood in 1–5 days and reach their maximum in 3–7 days. The same time

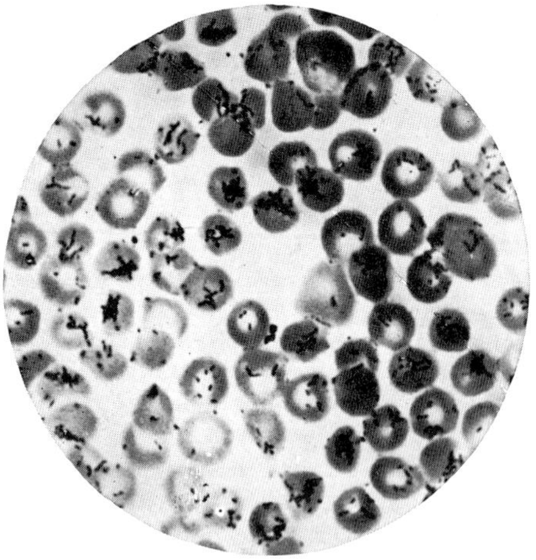

Fig. 43.4. *Haemobartonella muris.* Organisms in red blood cells of rat (× 1000). [From specimen kindly supplied by the late Professor J. G. Thomson.]

sequence is seen in naturally infected uninoculated rats after splenectomy. The organisms soon disappear from the blood. In animals, however, that survive relapses invariably occur. Experimentally rats, mice and hamsters can be infected. (For general descriptions see Lauda and Marcus 1928, Marmorston-Gottesman and Perla 1932, Kikuth 1931, 1934, Wigand and Peters 1950, Wigand 1958, Tanaka *et al.* 1965.)

Haemobartonella canis

This organism, which was described by Kikuth (1928), is responsible for infectious anaemia in splenectomized dogs, though healthy animals are seldom affected. (See Kikuth 1929, Pérard 1929, Lwoff and Provost 1929, Regendanz and Reichenow 1932, and Chapter 73.)

Haemobartonella felis

This organism resembles *H. canis* but, unlike it, causes parasitaemia and severe anaemia in apparently healthy cats. Experimentally it can be transmitted to cats by parenteral and oral routes, and intrauterine infection has been reported. Infections have occurred after the injection of pooled blood from apparently normal cats (Splitter *et al.* 1956), suggesting that latent infection may occur. Splenectomy has little effect on susceptibility. Seamer and Douglas (1959) found that, after intraperitoneal injection of infected blood into cats, the parasites appeared in the blood in 10–12 days: and lasted for two weeks or more. (See also Flint *et al.* 1958.)

Eperythrozoon

The eperythrozoa are parasites usually found on the surface of red blood cells and at a similar frequency free in the plasma. In Giemsa-stained films they are round and often exhibit a ring form. They are sensitive to organic arsenicals and tetracycline but not to penicillin and streptomycin.

Eperythrozoon coccoides

Ep. coccoides, which was discovered independently by Dinger (1928, 1929) and Schilling (1928), is a common parasite of mice, though the organisms are rarely found unless the spleen is removed. One to three days after splenectomy they appear in the blood in the form of rings or disks stuck on the red cells and staining violet with Giemsa. They reach their maximum numbers 3–7 days after splenectomy; and persist in the blood for a variable period—2 days to 7 weeks or more—reappearing from time to time during relapses. Electron-micrographs reveal a diameter of 0.35–0.6 μm and a limiting membrane but no cell wall. They have not yet been cultured *in vitro*, but according to Seamer (1959) they may be propagated in the chick embryo by injection intravenously or into the yolk sac. Seamer also states that they survive best when stored in 10 per cent glycerol at $-79°$. The disease produced in mice is very mild and is characterized by splenomegaly and general lympadenopathy (Stansky and Neilson 1966); haemolysis and anaemia are inconspicuous. The eperythrozoa may, however, activate other parasites, such as the mouse hepatitis virus, the lymphocytic choriomeningitis virus, and some gram-negative bacteria (Niven *et al.* 1952, Gledhill *et al.* 1965). Natural infection is spread by the louse, *Polyplax serrata*. Mice infected with *Ep. coccoides* are susceptible to infection with *H. muris*, indicating that the two organisms are distinct (see also McCluskie and Niven 1934, Schwetz 1934, Marmorston 1935, Wigand 1958). Rats and golden hamsters can both be infected with *Ep. coccoides*. Complement-fixing antisera against *Ep. coccoides* react to some extent with *H. muris*, but antisera against *H. muris* are said not to react with *Ep. coccoides* (Wigand 1958). Unlike rats infected with *H. muris*, mice may develop a degree of immunity that will protect them against a second infection some months after the first (Derrick *et al* 1954). Gledhill, Niven and Seamer (1965) were able to eliminate infection from a colony of mice by preventing louse infestation.

Numerous other animals are liable to infection with eperythrozoa (see Kreier and Ristic 1968). *Ep. ovis* causes mild disease in normal sheep, and *Ep. suis* an economically important and sometimes fatal disease in normal pigs (Splitter 1950). In both of these infections splenectomy intensifies the disease. Several other species of eperythrozoa have been described (see Kreier and Ristic 1974).

References

Bailey, R. K., Voss, J. L. and Smith, R. F. (1979) *J. clin. Microbiol.* **9,** 65.
Baine, W. B., Rasheed, J. K., Mackel, D. C., Bopp, C. A., Wells, J. G. and Kauffmann, A. F. (1979) *J. clin. Microbiol.* **9,** 453.
Baskerville, A., Fitzgeorge, R. B., Broster, M., Hambleton, P. and Dennis, P. J. (1981) *Lancet* **ii,** 1389.
Beerens, H., Gérard, A. and Guillaume J. (1957) *Ann. Inst. Pasteur, Lille* **9,** 77.
Brenner, D. J. *et al.* (1979) See *Symposium*, p. 656; (1980) *Curr. Microbiol.* **4,** 111.
Broome, C. V. and Fraser, D. W. (1979) *Epidem. Rev.* **1,** 1.
Brzin, B. (1969) *Zbl. Bakt.* **210,** 202.
Chandler, F. W., Hicklin, M. D. and Blackmon, J. A. (1977) *New Engl. J. Med.* **297,** 1218.
Chandler, F. W. *et al.* (1979*a*) *Amer. J. clin. Path.* **71,** 43; (1979*b*) See *Symposium*, p. 671.

Colichón, H. and Bedón, C. F. (1972) *Rev. lat.-amer. Microbiol.* **14,** 203.

Criswell, B. S., Marston, J. H., Stenbock, W. A., Black, S. H. and Gardner, H. L. (1971) *Canad. J. Microbiol.* **17,** 865.

Cummins, C. S., Glendinning, O. M. and Harris, H. (1957) *Nature, Lond.* **180,** 337.

Cummins, C. S. and Harris, H. (1956) *J. gen. Microbiol.* **14,** 583.

Dehnert, J. (1957) *Zbl. Bakt.* **169,** 66; (1961) *Über die Bedeutung der Intestinalbesiedlung beim Menschen. Bakteriologische Untersuchungen als Beitrag zum Bifidoproblem.* Habilistationsschrift, Heidelberg.

Dennis, P. J., Taylor, J. A. and Barrow, G. I. (1981) *Lancet* **ii,** 636.

Derrick, E. H., Pope, J. H., Chong, S. K., Carley, J. G. and Lee, P. E. (1954) *Aust. J. exp. Biol. med. Sci.* **32,** 577.

Dieterle, R. R. (1927) *Arch. Neurol. Psychiat.* **18,** 73.

Dinger, J. E. (1928) *Ned. Tijdschr. Geneesk.* No. 48, **72,** 5903; (1929) *Zbl. Bakt.* **113,** 503.

Dorff, G. J., Jackson, L. J. and Rytel, M. W. (1974) *Ann. intern. Med.* **80,** 305.

Drasar, B. S. and Hill, M. J. (1974) *Human Intestinal Flora.* Academic Press, London.

Dunkelberg, W. E., Skaggs, R. and Kellogg, D. S. (1970) *Amer. J. Path.* **53,** 370.

Edelstein, P. H. (1981) *J. clin. Microbiol.* **14,** 298.

Edelstein, P. H. and Finegold, S. M. (1979) *J. clin. Microbiol.* **10,** 141.

Edelstein, P. H. and Meyer, R. D. (1980) *Antimicrob. Agents Chemother.* **18,** 403.

Elliott, J. A., Johnson, W. and Helms, C. M. (1981) *Infect. Immun.* **31,** 822.

England, A. C. III, Fraser, D. W., Mallison, G. F., Mackel, D. C., Skaliy, P. and Gorman, G. W. (1982) *Appl. environm. Microbiol.* **43,** 240.

Feeley, J. C., Gorman, G. W., Weaver, R. E., Mackel, D. C. and Smith, H. W. (1978) *J. clin. Microbiol.* **8,** 320.

Feeley, J. C. *et al.* (1979) *J. clin. Microbiol.* **10,** 437.

Ferrari, A., Pacini, N. and Canzi, E. (1980) *J. appl. Bact.* **49,** 193.

Fisher-Hoch, S. (1980) *Lab. Lore* **9,** 671.

Flint, J. C., Roepke, M. H. and Jensen, R. (1958) *Amer. J. vet. Res.* **19,** 164.

Ford, W. W. and Eliot, C. P. (1928) *J. exp. Med.* **48,** 475.

Freitas, J. L. de, Borst, J. and Meenhorst, P. L. (1979) *Lancet* **i,** 270.

Gardner, H. L. and Dukes, C. D. (1955) *Amer. J. Obstet. Gynecol.* **69,** 962.

Garrity, G. M., Brown, A. and Vickers, R. M. (1980) *Int. J. syst. Bact.* **30,** 609.

Gauhe, A. *et al.* (1954) *Arch. Biochem.* **48,** 214.

Geiman, Q. M. (1941) *Proc. Soc. exp. Biol., N.Y.* **47,** 329.

Gledhill, A. W., Niven, J. S. F. and Seamer, J. (1965) *J. Hyg., Camb.* **63,** 73.

Graham, R. K. and May, J. W. (1965) *J. gen. Microbiol.* **41,** 243.

Graham-Smith, G. S. (1905) *J. Hyg., Camb.* **5,** 453.

Granato, P. A. (1977) In: *Clinical microbiology*, vol. 1, p. 207. (CRC Handbook Series in Clinical Laboratory Science, Section E.) Ed. by A. von Graevenitz. CRC Press, Cleveland, Ohio.

Greaves, P. W. (1980) *J. clin. Path.* **33,** 581.

Greenwood, J. R. and Pickett, M. J. (1979) *J. clin. Microbiol.* **9,** 200; (1980) *Int. J. syst. Bact.* **30,** 170.

Grimont, P. A. D. and Bercovier, H. (1980) *Bull. Inst. Pasteur* **78,** 267.

Gyllenberg, H. (1958) *Acta path. microbiol. scand.* **44,** 293.

Gyllenberg, H. and Carlberg, G. (1958) *Acta path. microbiol. scand.* **44,** 287.

György, P., Norris, R. F. and Rose, C. S. (1954) *Arch. Biochem. Biophys.* **48,** 193.

Hébert, G. A. (1980) *J. clin. Microbiol.* **12,** 807; (1981) *Ibid.* **13,** 240.

Hébert, G. A., Steigerwalt, A. G., and Brenner, D. J. (1980) *Curr. Microbiol.* **3,** 255.

Henriksen, S. D. (1948) *Acta path. microbiol. scand.* **25,** 368.

Hofstad, T. (1978) *Acta path. microbiol. scand.* **B86,** 47.

Hofstad, T. and Kristoffersen, T. (1970) *Acta path. microbiol. scand.* **B78,** 760.

Horwitz, M. A. and Silverstein, S. C. (1980) *J. clin. Invest.* **66,** 441; (1981) *J. exp. Med.* **153,** 386, 398.

Ison, C. A., Dawson, S. G., Hilton, J., Csonka, G. W. and Easmon, C. S. F. (1982) *J. clin. Path.* **35,** 550.

Jackson, F. L. and Goodman, Y. E. (1972) *Int. J. syst. Bact.* **22,** 73.

Jackson, F. L., Goodman, Y. E., Bel, F. R., Wong, P. C. and Whitehouse, R. L. S. (1971) *J. med. Microbiol.* **4,** 171.

Jettmar, H. M. (1932) *Z. Parasitenk.* **4,** 254.

Jiménez, J. F. (1940) *Proc. Soc. exp. Biol., N.Y.* **45,** 402.

Jiménez, J. F. and Buddingh, G. J. (1940) *Proc. Soc. exp. Biol., N.Y.* **45,** 546.

Johnson, W., Pesanti, E. and Elliott, J. (1979) *Infect. Immun.* **26,** 698.

Kessler, W. R. (1942) *Proc. Soc. exp. Biol., N.Y.* **49,** 238.

Khairat, O. (1966) *J. dent. Res.* **45,** 1191.

Kikuth, W. (1928) *Klin. Wschr.* **7,** 1729; (1929) *Zbl. Bakt.* **113,** 1; (1931) *Z. Immun Forsch.* **73,** 1; (1934) *Proc. R. Soc. Med.* **27,** 1241.

Kludas, M. and Dobberstein, H. (1959) *Zbl. Bakt.* **175,** 520.

Kohler, R. B. *et al.* (1981) *Ann. intern. Med.* **94,** 601.

Koornhof, H. J. and Robinson, R. G. (1963) *S. Afr. J. Lab. clin. Med.* **9,** 95.

Kreier, J. P. and Ristic, M. (1968) *Infectious Blood Diseases of Man and Animals,* Vol. ii, p. 387. Ed. by D. Weinman and M. Ristic. Academic Press, New York and London; (1974) In: *Bergey's Manual of Determinative Bacteriology*, 8th edn., p. 910. Ed. by R. E. Buchanan and N. E. Gibbons, Williams and Wilkins, Baltimore.

Langford, G. C., Faber, J. E. and Pelczar, M. J. (1950) *J. Bact.* **59,** 349.

Lattimer, G. L. and Cepil, B. A. (1980) *J. clin. Path.* **33,** 585.

Lauda, E. and Marcus, F. (1928) *Zbl. Bakt.* **107,** 104.

Leighton, P. M. (1982) *Canad. J. publ. Hlth.* **73,** 335.

Leopold, S. (1953) *U.S. Forces med. J.* **4,** 263.

Lewis, J. F., O'Brien, S. M., Ural, U. M. and Burke, T. (1972) *Amer. J. Obstet. Gynecol.* **112,** 87.

Lwoff, A. and Provost, A. (1929) *C. R. Soc. Biol.* **101,** 8.

McCluskie, J. A. W. and Niven, J. S. F. (1934) *J. Path. Bact.* **39,** 185.

McDade, J. E. *et al.* (1977) *New Engl. J. Med.* **297,** 1197; (1979) See *Symposium*, p. 659.

McEntegart, M. G. and Porterfield, J. S. (1949) *Lancet* **ii,** 596.

McKinney, R. M. *et al.* (1981) *Ann. intern. Med.* **94,** 739.

Mangiafico, J. A., Hedlund, K. W. and Knott, A. R. (1981) *J. clin. Microbiol.* **13,** 843.

Marmorston, J. (1935) *J. infect. Dis.* **56,** 142.
Marmorston-Gottesman, J. and Perla, D. (1932) *J. exp. Med.* **56,** 763.
Marsden, H. B. and Hyde, W. A. (1971) *J. clin. Path.* **24,** 171.
Matteuzzi, D., Crociani, F., Zani, G. and Trovatelli, L. D. (1971) *Z. allgem. Mikrobiol.* **11,** 387.
Mayer, M. (1921) *Arch. Schiffs- u. Tropenhyg.* **25,** 150.
Mickelsen, P. A., McCarthy, L. R., and Mangum, M. E. (1977) *J. clin. Microbiol.* **5,** 488.
Mitsuoka, T. (1969) *Zbl. Bakt.* **210,** 32, 52.
Morris, G. K. *et al.* (1980) *J. clin. Microbiol.* **12,** 718.
Moss, C. W., Weaver, R. E., Dees, S. B. and Cherry, W. B. (1977) *J. clin. Microbiol.* **6,** 140.
Müller, H. E. (1981) *Zbl. Bakt.* **B172,** 524.
Nagington, J., Wregitt, T. G., Tobin, J. O'H. and Macrae, A. D. (1979) *J. Hyg., Camb.* **83,** 377.
Nauck, E. G. (1958) *Forsch. u. Forschr.* **32,** 129.
Niven, J. S. F., Gledhill, A. W., Dick, G. W. A. and Andrewes, C. H. (1952) *Lancet* ii, 1061.
Noguchi, H. (1926) *J. exp. Med.* **44,** 533, 697, 715, 729; (1927) *J. Amer. med. Ass.* **89,** 739.
Noguchi, H. and Battistini, T. S. (1926) *J. exp. Med.* **43,** 851.
Orla-Jensen, S. (1924) *Le Lait* **4,** 469; (1943) *Die echten Milchsäurebakterien.* Ejnar Munksgaard, Copenhagen.
Orrison, L. H., Cherry, W. B., Fliermans, C. B., Dees, S. B., McDougal, L. K. and Dodd, D. J. (1981) *Appl. environm. Microbiol.* **42,** 109.
Pasculle, A. W. *et al.* (1980) *J. infect. Dis.*, **141,** 727; (1981) *Antimicrob. Agents Chemother.* **20,** 793.
Pérard, C. H. (1929) *C. R. Soc. Biol.* **100,** 1111.
Peters, D. and Wigand, R. (1952) *Z. Tropenmed. Parasit.* **3,** 313; (1955) *Bact. Rev.* **19,** 150.
Petuely, F. (1956) *Zbl. Bakt.* **166,** 95.
Pine, L., George, J. R., Reeves, M. W. and Harrell, W. K. (1979) *J. clin. Microbiol.* **9,** 615.
Pinkerton, H. and Weinman, D. (1937) *Proc. Soc. exp. Biol., N.Y.* **37,** 587.
Piot, P., van Dyck, E., Goodfellow, M. and Falkow, S. (1980) *J. gen. Microbiol.* **119,** 373.
Piot, P., van Dyck, E., Totten, P. A. and Holmes, K. K. (1982) *J. clin. Microbiol.* **15,** 19.
Pittaluga, G. (1938) *Bull. Inst. Pasteur* **36,** 961.
Reese, J. D., Morrison, M. E. and Fowler, E. M. (1950) *J. Immunol.* **65,** 355.
Reeves, M. W., Pine, L., Hutner, S. H., George, J. R. and Harrell, W. K. (1981) *J. clin. Microbiol.* **13,** 688.
Regendanz, P. and Reichenow, E. (1932) *Arch. Schiffs- u. Tropenhyg.* **36,** 305.
Reuter, G. (1963) *Zbl. Bakt.* **191,** 486; (1971) *Int. J. syst. Bact.* **21,** 273.
Ristic, M. and Kreier, J. P. (1974) In: *Bergey's Manual of Determinative Bacteriology*, 8th edn., p. 906. Ed. by R. E. Buchanan and N. E. Gibbons, Williams and Wilkins, Baltimore.
Ristic, M. and Watrach, A. (1961) *Amer. J. vet. Res.* **22,** 109.
Ristroph, J. D., Hedlund, K. W. and Goroda, S. (1981) *J. clin. Microbiol.* **13,** 115.
Rodgers, F. G. (1979) *J. clin. Path.* **32,** 1195; (1982) *J. med. Microbiol.* **15,** 181.
Rodgers, F. G. and Davey, M. R. (1982) *J. gen. Microbiol.* **128,** 1547.
Rodgers, F. G., Greaves, P. W., Macrae, A. D. and Lewis, M. J. (1980) *J. clin. Path.* **33,** 1184.
Rogosa, M. (1956) *J. Bact.* **72,** 533; (1965) *Ibid.* **90,** 704.
Rogosa, M. and Bishop, F. S. (1964) *J. Bact.* **87,** 574.
Rogosa, M., Fitzgerald, R. J., Mackintosh, M. E. and Beaman, A. J. (1958) *J. Bact.* **76,** 455.
Rowbotham, T. J. (1980) *J. clin. Path.* **33,** 1179.
Saravolatz, L. D., Pohlod, D. J. and Quinn, E. L. (1980) *Scand. J. infect. Dis.* **12,** 215.
Scardovi, V., Trovatelli, L. D., Biarati, B. and Zani, G. (1979) *Int. J. syst. Bact.* **29,** 291.
Scardovi, V., Trovatelli, L. D., Zani, G., Crociani, F. and Matteuzzi, D. (1971) *Int. J. syst. Bact.* **21,** 276.
Schilling, V. (1928) *Klin. Wschr.* **7,** 1853.
Schwetz, J. (1934) *Zbl. Bakt.* **132,** 211.
Seamer, J. (1959) *J. gen. Microbiol.* **21,** 344.
Seamer, J. and Douglas, S. W. (1959) *Vet. Rec.* **71,** 405.
Sebald, M., Gasser, F. and Werner, H. (1965) *Ann. Inst. Pasteur.* **109,** 251.
Seeliger, H. P. R. and Werner, H. (1962) *Z. Hyg. Infektkr.* **148,** 383; (1963) *Ann. Inst. Pasteur.* **105,** 911.
Sharp, J. T. (1968) *Proc. Soc. exp. Biol. Med.* **128,** 1072.
Sims, W. (1960) *Brit. dent. J.* **108,** 73.
Skaliy, P. and McEachern, H. V. (1979) *Ann. intern. Med.* **90,** 662.
Skaliy, P., Thompson, T. A., Gorman, G. W., Morris, G. K., McEachern, H. V. and Mackel, D. C. (1980) *Appl. environm. Microbiol.* **40,** 697.
Smith, R. A., DiGiorgio, S., Darner, J. and Wilhelm, A. (1981) *J. clin. Microbiol.* **13,** 637.
Splitter, E. J. (1950) *Amer. J. vet. Res.* **11,** 324.
Splitter, E. J., Castro, E. R. and Kanawyer, W. L. (1956) *Vet. Med.* **51,** 17.
Stansky, P. G. and Neilson, C. F. (1966) *Nature, Lond.* **211,** 1203.
Strong, R. P., Tyzzer, E. E., Brues, C. T., Sellards, A. W. and Gastiaburu, J. C. (1915) *Rep. 1st Expedition S. America, 1913.* Harvard University Press, Cambridge, Massachusetts.
Symposium (1979) *Ann. intern. Med.* **90,** 491–707.
Tanaka, H., Hall, W. T., Sheffield, J. B. and Moore, D. H. (1965) *J. Bact.* **90,** 1735.
Tanaka, R. and Mutai, M. (1980) *Appl. environm. Microbiol.* **40,** 866.
Taylor, A. G. and Harrison, T. G. (1979) *Lancet* **ii,** 47.
Taylor, A. G., Harrison, T. G., Dighero, M. W. and Bradstreet, C. M. P. (1979) *Ann. intern. Med.* **90,** 686.
Taylor, E. and Phillips, I. (1983) *J. med. Microbiol.* **16,** 83.
Tedeschi, G. G., Amici, D., Murri, O. and Paperelli, M. (1967) *Ann. Sclavo* **9,** 28.
Thomason, B. M., Harris, P. P., Lewallen, K. R. and McKinney, R. M. (1979) *Curr. Microbiol.* **2,** 357.
Thompson, M. R., Miller, R. D. and Inglewski, B. H. (1981) *Infect. Immun.* **34,** 299.
Thorpe, T. C. and Miller, R. D. (1981) *Infect. Immun.* **33,** 632.
Tison, D. L., Pope, D. H., Cherry, W. B. and Fliermans, C. B. (1980) *Appl. environm. Microbiol.* **39,** 456.
Tissier, H. (1900) *Recherches sur la Flore Intestinale des Nourrissons*, Paris.
Tyzzer, E. E. (1942) *Proc. Amer. philosoph. Soc.* **85,** 359.
Vassiliadis, P. (1935) *Ann. Soc. belge Méd. trop.* **15,** 279.
Veillon, A. and Zuber, A. (1898) *Arch. Méd. exp.* **10,** 517.
Wang, W. L. L. *et al.* (1979) See *Symposium*, p. 614.
Weinman, D. (1944) *Trans. Amer. philosoph. Soc.* **33,** Pt. III, 243; (1968) *Infectious Blood Diseases of Man and Animals*, **ii**, p. 3. Ed. by D. Weinman and M. Ristic. Academic Press,

New York and London; (1974) In: *Bergey's Manual of Determinative Bacteriology*, 8th edn., p. 903. Ed. by R. E. Buchanan and N. E. Gibbons, Williams and Wilkins, Baltimore.
Wells, J. I. and Goei, S. H. (1981) *J. clin. Path.* **34,** 917.
Werner, H. (1967) *Zbl. Bakt.* **205,** 210.
Werner, H., Gasser, F. and Sebald, M. (1965) *Zbl. Bakt.* **198,** 504.
Wiel-Korstanje, J. A. A. van der (1973) *Bifidobacteriën en Enterococcen in de Darmflora van de Mens.* Drukkerij Elinwijk, Utrecht.
Wigand, R. (1958) *Morphologische biologische und serologische Eigenschaften der Bartonellen.* Georg Thieme Verlag, Stuttgart.
Wigand, R. and Peters, D. (1950) *Z. Tropenmed. Parasit.* **2,** 206.
Wilkinson, H. W. and Fikes, B. J. (1981) *J. clin. Microbiol.* **14,** 322.
Wong, M. C., Ewing, E. P. Jr, Callaway, C. S. and Peacock, W. L. Jr (1980) *Infect. Immun.* **28,** 1014.
Wong, M. C., Peacock, W. L. Jr, McKinney, R. M. and Wong, K.-H. (1981) *Curr. Microbiol.* **5,** 31.
Wong, K. H. *et al.* (1979) In *Symposium*, p. 624, 634.
Yong, D. C. T. and Thompson, J. S. (1982) *J. clin. Microbiol.* **16,** 30.
Yordy, D. M. and Delwicke, E. A. (1979) *J. Bact.* **137,** 905.
Zinnemann, K. and Turner, G. C. (1963) *J. Path. Bact.* **85,** 313.
Zinner, S. H., Daly, A. K. and McCormack, W. H. (1973) *Appl. Microbiol.* **25,** 705.

44

The spirochaetes

Graham Wilson, A. E. Wilkinson and Joyce Coghlan

Introductory

The name Spirochaete was first given by Ehrenberg in 1838 to a large flexible motile organism occurring in water; it is now used as a general term for all elongated flexible organisms which lack cross striations, which are motile without having external flagella, and which appear to be spiral or helical in form. Though the spirochaetes vary greatly in size—especially in length—they possess certain features in common: thus they have no external flagella but they do have two or more axial fibrils or filaments (see below) which are now considered to be analogous to flagella; they have no antero-posterior polarity, i.e. they can move backwards and forwards; they contain no readily detected colouring matter and no cyanophycin granules; they contain no circumscribed nuclei; they divide by transverse fission; and they exhibit no indisputable sexual phenomena of reproduction. In the few spirochaetes which have been investigated, the RNA:DNA ratio is 2:1 (Siefert 1958). These characters have led to the classification of the spirochaetes with the bacteria. Indeed, as Dobell (1912) points out, there is only one feature that distinguishes them from most bacteria, namely motility without flagella.

Formerly the genera which were included in the Order Spirochaetales were subdivided into two families—Spirochaetaceae and Treponemataceae—on the

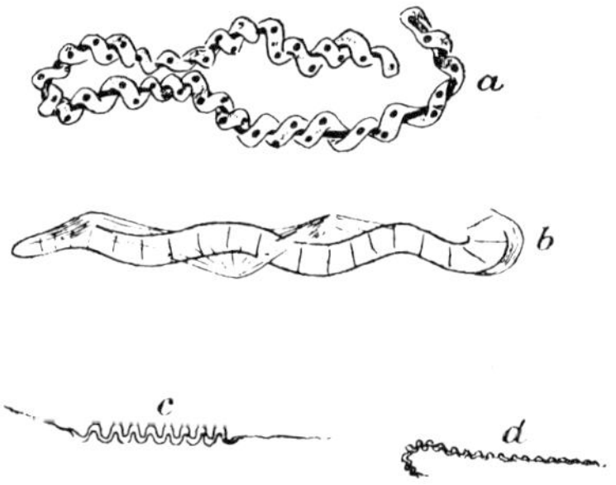

Fig. 44.1 Diagram of the spirochaetes. *a*, *Spirochaeta*; *b*, *Cristispira*; *c*, *Treponema*; *d*, *Leptospira*. (After Noguchi.)

basis of their length. However, Canale-Parola, Udris and Mandel (1968) have shown that the length of the organism depends on the stage of growth in culture and, perhaps, on environmental conditions in nature. They favour the inclusion of five genera—*Spirochaeta*, *Cristispira*, *Borrelia*, *Treponema*, and *Leptospira*—in one family, Spirochaetaceae. *Saprospira*, at one time included with these genera, has been transferred to Flexibacteriales (Lewin 1965): it is cross-striated and does not have axial filaments. The Subcommittee on the Taxonomy of Leptospira agreed on the formation of the family Leptospiraceae within the order Spirochaetales, which will include the genus *Leptospira* (Skerman *et al.* 1980).

Spirochaeta

The members of this group are free-living in fresh and marine waters and in sewage. They have an axial fibre round which the body is twisted helically, just as a spiral staircase is built round the newel. The organisms have a series of permanent primary spirals; during motion secondary waves may be superimposed on these. Volutin granules are distributed uniformly throughout the length of the organisms. Culturally, they are obligatory or facultative anaerobes. The type species, *Spirochaeta plicatilis* Ehrenberg is usually 200–500 μm in length and 0.5–0.7 μm in thickness (Wenyon 1926). The number of primary spirals is 100–250, the distance between successive turns, i.e. the wave-length, being about 2 μm. So far, no members of this group are known to cause disease. The G+C content of their DNA appears to be between 50 and 67 moles per cent (Canale-Parola *et al.* 1968, Canale-Parola 1978).

Cristispira

The chief organism in this group, *C. balbiani*, is found in the crystalline style of the oyster. It is a large spirochaete, 50–100 μm by 0.5–3.0 μm, having what seems to be a thin periplast membrane running spirally along the cell as a crest. Electronmicroscopic studies reveal the characteristic features of a spirochaete, and show that it differs from *Borrelia* and *Treponema* in its greater size, in the much larger number of fibrils constituting the locomotor system, and in the presence at the periphery of the cytoplasm of numerous vesicles limited by a double membrane (Ryter and Pillot 1965). The organism has never been cultivated *in vitro*, and dies out rapidly after removal from its natural habitat. Other members of the genus are found in the digestive-tract fluid of marine and freshwater molluscs.

Borrelia and Treponema

These two genera are distinguished by the ready staining of the members of *Borrelia* spp., whose refractive index is very like that of bacteria and whose spirals are usually coarse, shallow and irregular. *Treponema* spp. stain with Giemsa or by silver impregnation but not with the usual bacterial stains. They have finer spirals, which may be regular or irregular, and a low refractive index, which makes them difficult to see except by dark-ground or phase-contrast microscopy. Axial structures and the like are revealed by electronmicroscopy. One or both ends of borreliae and of pathogenic treponemes are pointed; and in both groups some members have a thin drawn-out filament at the ends representing the remains of the connecting bridge of cytoplasm seen during transverse fission.

Members of the two genera are widely distributed. Numerous species have been described in water, in the gut of certain insects such as white ants and cockroaches, and in the large gut of the toad; in human beings they are found in the mouth, sometimes in the alimentary tract and in the bronchi, around the urethral orifice, in certain ulcerating conditions of the skin,

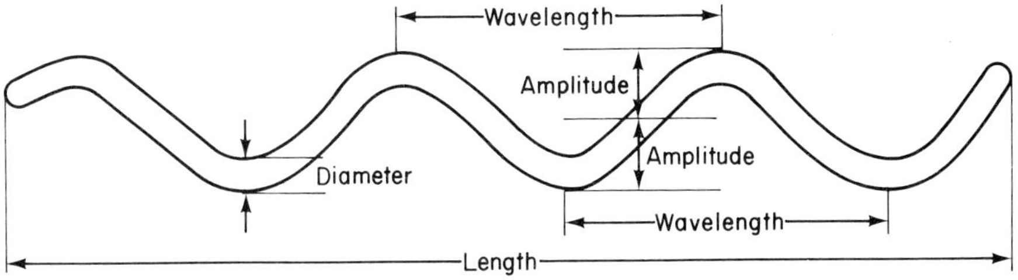

Fig. 44.2 Diagrammatic representation of a spirochaete, showing definition of length, diameter, wavelength and amplitude. (By courtesy of Dr Russell C. Johnson (1976) *The Biology of Parasitic Spirochetes*. Academic Press, New York.)

in condylomata, in the blood of patients with relapsing fever, and in the manifold lesions of syphilis. The range of size is considerable; Dobell (1912) describes a *Tr. termitidis* as large as 20–60 μm × 0.5 μm and a *Tr. parvum* as small as 3 μm × 0.2 μm. The usual dimensions of *Borrelia* are 3–20 μm by 0.2–0.5 μm. The members of both genera are anaerobic or microaerophilic. The G + C content of the DNA is given as 46.4 moles per cent for *Tr. vincenti* and 37.8 to 41.3 per cent for *Tr. pallidum* (Canale-Parola *et al.* 1968). The type species are respectively *Borr. anserina* and *Tr. pallidum*.

Morphology

Figure 44.2 represents diagrammatically the general shape of spirochaetes. Electron micrographs show that these organisms have a central protoplasmic cylinder covered by two membranes—an inner cytoplasmic membrane and an outer cell wall—separated by an interspace. This three-layered envelope is made up of polygonal sub-units (Swain 1955, Pillot and Ryter 1965, Jackson and Black 1971) (Fig. 44.3). An external slime layer or capsule, probably a mucopolysaccharide, staining with ruthenium red has been reported on *Tr. pallidum*, but not on the cultivable Reiter treponeme (Zeigler *et al.* 1976). A band of parallel fibres, previously referred to as axial filaments but now regarded as internal flagella, is attached to each end. In pathogenic, i.e. non-cultivable, treponemes, these number three, in borreliae 15–20 (Fig. 44.4). The flagella from each end wind round the central cylinder in the interspace between the cytoplasmic membrane and the cell wall, and overlap in the middle region of the cell (Hougen 1974). At the ends they are attached to the cytoplasm by knob- and hook-like bodies. Structurally, they resemble the flagella of gram-positive bacteria. In treponemes the shafts of the flagella are sheathed; in this respect they differ from those of borrelias in which no sheath can be discerned (Hougen 1976). In addition to flagella, treponemes contain bundles of cytoplasmic fibrils that wind around the core; their function is unknown (see Eipert and Black 1979). In cross-section ribosomes and nuclear regions are detectable by the electron microscope. (See also Figs. 44.5, 44.6, and 44.7.)

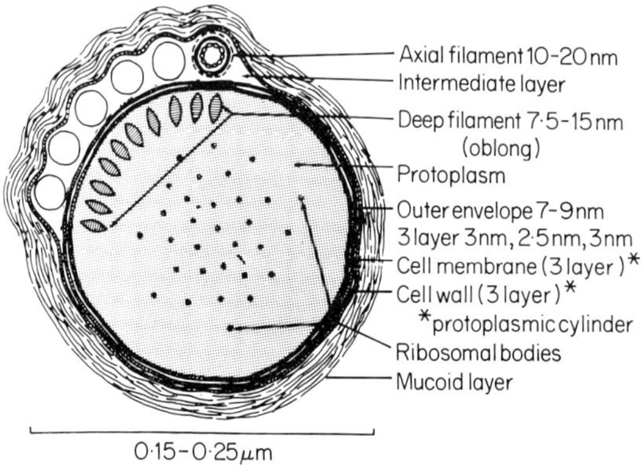

Fig. 44.3 Schematic drawing of a cross section of *Treponema pallidum* with various structures designated. (Reproduced from Wiegand *et al.* 1972 by kind permission of the authors.)

It should be noted that the helical shape is not universally accepted. Both DeLamater and his colleagues (1950) and Sequeira (1956) are of the opinion that at least some treponemes, including *Tr. pallidum*, have the form of a flat wave. This view is supported by Cox (1972), who observed 3–5 changes in the pitch of the flat wave in the Nichols strain of treponeme. A form of motion occurs in which a series of waves pass along the organism, each loop rotating through

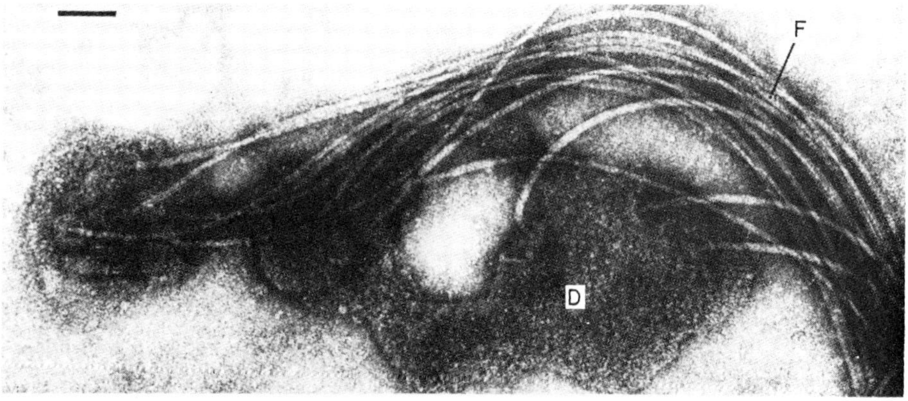

Fig. 44.4 Cell of *Borr. merionesi* treated with *Myxobacter* AL 1 protease I. Only membranous debris (D) and flagella (F) are left after the treatment. × 90 000. (By courtesy of Dr Russell C. Johnson (1976) *The Biology of Parasitic Spirochetes*. Academic Press, New York.)

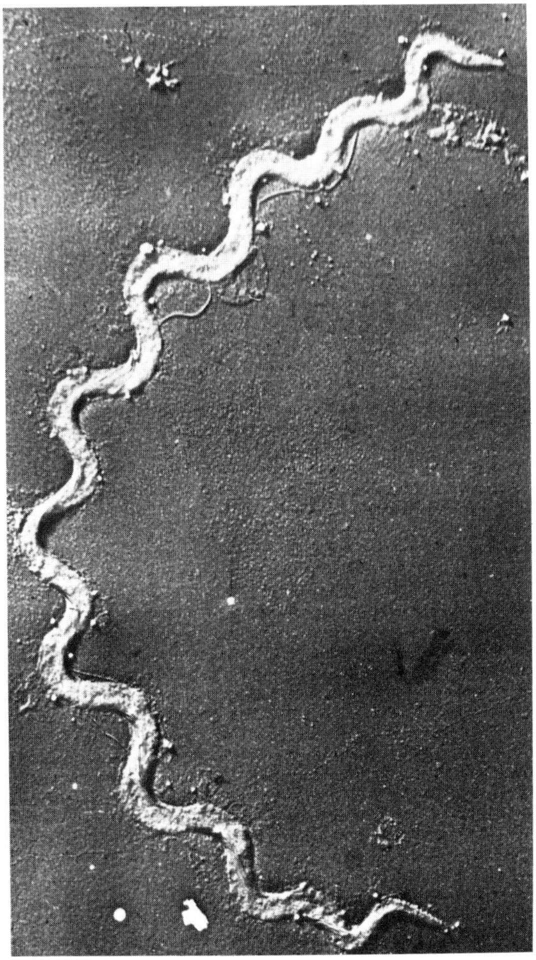

Fig. 44.5 Electronmicrograph of *Borr. recurrentis* (Swain 1955) × 10 300.

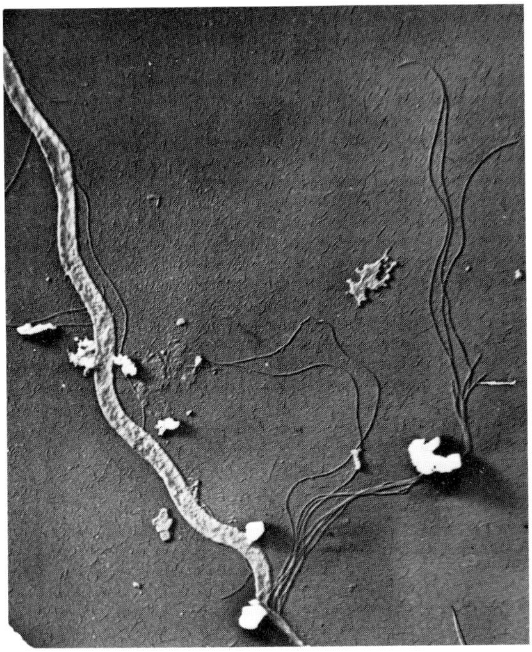

Fig. 44.7 Electronmicrograph of *Borr. recurrentis* after treatment with trypsin, showing the fibrils composing the axial filament detached from the spirochaete except at one end (Swain 1955). × 8 500.

about 180 degrees. The curves appear to lie in one plane; but during rapid rotation they give the impression of being helical.

Three kinds of movement are described. (1) Flexion: as a rule the natural shape of a spirochaete at rest is straight, but during movement all sorts of twists and turns may develop, one form following another in rapid succession, but each one tending to return to the normal straight form. During flexion the primary spirals remain unaltered. Sequences of flexion movements amounting to undulation have been observed (e.g. Noguchi 1917*b*, DeLamater *et al.* 1950). (2) Rotation round the long axis: this is difficult to see unless the ends of the organism are bent at an angle to the main axis. When the rotation is very rapid and when, as in *Leptospira*, the ends are hooked, the spirochaete may appear as a spiral thread with a button-hole at each end. (3) Translation: by this the organism moves from one place to another. Different spirochaetes vary in their activity. Some exhibit very active movements of flexion, lashing furiously in various directions but making little progress from their original position; others dart about rapidly, rendering their ocular pursuit almost impossible. The cause of movement in spirochaetes is doubtful—whether as a result of undulation, rotation, or something else. The observed rotation is not necessarily the cause, for a helical structure moved in the direction of its axis by other means would rotate.

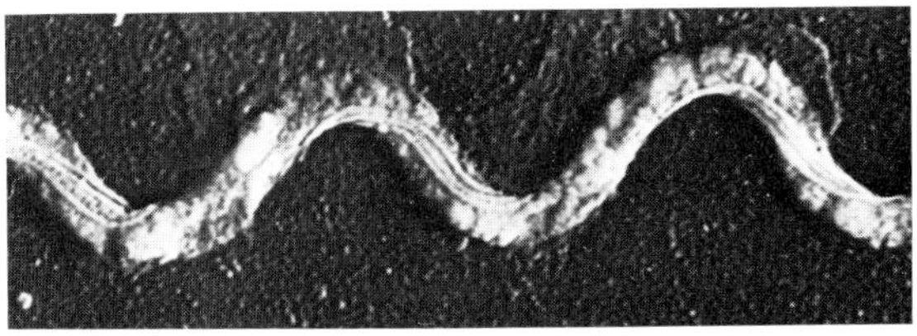

Fig. 44.6 Electronmicrograph of part of *Trep. pallidum* after treatment with trypsin, showing the three fibrils composing the axial filament (Swain 1955). × 39 000.

Unlike the borreliae, treponemes do not stain with the usual bacterial stains such as methylene blue. They can be stained by Giemsa's method but prolonged exposure is necessary. Treponemes can be demonstrated in secretions or tissue sections by silver staining; various techniques for this have been described (Collart *et al.* 1962, Walter *et al.* 1969). Immunofluorescence staining methods for their detection are reviewed by Kellogg and Mothershed (1969).

According to Schaudinn (see Nisbett 1976), the presence of *Tr. pallidum* in syphilitic lesions was overlooked by medical bacteriologists because they examined stained specimens in which the morphology of the organisms was obscured. Schaudinn, being a zoologist, examined living specimens and noted at once their characteristic corkscrew appearance and lively motility.

Multiplication is by transverse fission, which may sometimes be asymmetrical. The rate of division of *Tr. pallidum* has been estimated as 30–33 hr (Magnuson *et al.* 1948, Cumberland and Turner 1949). The cultivable treponemes multiply rapidly: the division time of the Reiter treponeme is 6.9–8.8 hr (Gelperin 1949, Rose and Morton 1952).

Observations of small granules attached to the sides of treponemes and of larger cystic bodies (DeLamater *et al.* 1951) have led to the suggestion that treponemes have a complicated life cycle, even including submicroscopic forms (Manouélian 1940). The large and often contradictory literature on the subject has been reviewed by Pillot, Dupouey and Ryter (1964) and by Willcox and Guthe (1966). The view now generally held is that multiplication occurs only by transverse fission, and that the structures put forward in support of a life cycle are degenerative forms produced by adverse environmental conditions. Cystic forms usually appear in ageing cultures, but Hardy and Nell (1961) showed that they could be produced in young cultures of Reiter treponemes by osmotic imbalance of the medium, and that magnesium ions had a protective effect; the spheres were thought to be degenerative forms and not to be viable.

Some spirochaetes, particularly the leptospires, are tenuous enough to pass through the usual bacterial filters. Measurements by Hindle and Elford (1933) made with graded collodion membranes showed that the width of *Tr. pallidum* was about 0.2 μm and of *Lepto. biflexa* 0.1 μm (see also Schmidt 1955). Use is often made of their filtrability to separate them from contaminating bacteria. (For the fine morphology of spirochaetes, see Babudieri 1972.)

Cultivation

Some of the pathogenic species of spirochaete, such as the leptospires, can be readily grown *in vitro* (see p. 504), but others, such as the borreliae, are more demanding. Virulent strains of *Tr. pallidum* have not been grown outside the animal body. The so-called cultivable strains of *Tr. pallidum* (Noguchi, Nichols, Kazan, Reiter) are thought not to be *Tr. pallidum* (Robinson and Wichelhausen 1946, Eagle and Germuth 1948, Gelperin 1951, but see also Reiter 1960). These strains differ morphologically from *Tr. pallidum*, grow well in artificial media, and are not pathogenic for animals.

Spirochaetes *in vitro* evidently have complex nutritional requirements. Nearly all the methods that have proved successful entail the use of a medium containing native animal protein, such as blood, rabbit serum, or ascitic fluid. Rabbit serum seems to act by virtue partly of its albumin content and partly of its long-chain fatty acids. Many of the treponemes use amino acids and do not attack sugars. Others, for example, *Tr. hyodysenteriae*, require glucose, which is broken down to acetic and succinic acids. Whatever basic medium is used, nutritional supplements, such as co-carboxylase, isobutyric acid, or glutamine must be added. Reiter's treponeme, a non-pathogenic spirochaete cultivated from a syphilitic lesion. grows in a defined medium containing a large number of amino acids, vitamins, purines, pyrimidines, glucose, acetate, ammonia, trace metals and crystalline serum albumin (Eagle and Steinman 1948, Steinman *et al.* 1952). The albumin appears to supply a lipid growth factor replaceable by oleate, and at the same time to detoxify the lipid (Oyama *et al.* 1953). In a defined medium, fourteen amino acids, either uracil or cytosine, and ammonia were essential as nutrients; and niacin, biotin, and pantothenate as growth factors (Steinman *et al.* 1954). Power and Pelczar (1959) record a threefold stimulation of growth of Reiter's spirochaete in a thioglycollate broth by a mixture of palmitic, stearic and oleic acids. Special precautions are needed for the growth of the anaerobic, or microaerophilic (see Fitzgerald *et al.* 1977), treponemes. Smibert (1976) uses a culture system of pre-reduced media containing oxygen-free nitrogen or carbon dioxide. The optimum Eh for growth is said to be -190 mv, and the pH 6.5–7.5. Cysteine or thioglycollic acid may be used as reducing agents.

So far, attempts to grow the pathogenic non-cultivable treponemes in tissue culture have failed, though some progress has undoubtedly been made (see Fitzgerald *et al.* 1977). The oxygen needed for growth of the tissue cells is, of course, inhibitory to the spirochaetes. This incompatibility is not as serious as it appears at first sight, because the toxic effect seems to be neutralized by the tissue cells in much the same way as it must be in the oxygen-rich tissues of the living body. The organisms attach themselves to the cells in the tissue culture, and actually gain entrance to the cells, but they lose their virulence within 24 or 48 hr. Whether this is because the oxygen is insufficiently neutralized or because some special nutrient factors are required is not clear.

Though the borreliae of relapsing fever can be grown in fertile eggs, their cultivation *in vitro* was not possible until Kelly (1971) devised a medium for *Borr. hermsi*. The complete medium was complex, but its chief constituents were proteose-peptone, tryptone, albumin, gelatin, glucose and *N*-acetyl-glucosamine in a mixture of mineral salts. Glucose was metabolized freely, and growth was limited by acid production unless the medium was heavily buffered by phosphate. The organisms proved to be microaerophilic and would not grow unless a low concentration of oxygen was present. Subcultures retained their ability to infect mice for at least eight months. Since then, other relapsing fever spirochaetes have been successfully cultivated from blood withdrawn from infected animals (Kelly 1976). With incubation at 35°, the organisms have a generation time of about 18 hr and reach a maximum density of 3–5 × 10^9 per ml in 6–7 days. Some species require modifications of the original medium. Fatty acids were found to be essential for growth, and were obtained from the albumin in the culture medium. Working with a modified Kelly medium, Stoenner (1974) found that *Borr. hermsi* underwent changes under continuous cultivation affecting the morphology and biology of the organisms.

Spirochaetes have been grown mainly in liquid or semisolid media. The early workers had difficulty in obtaining colonial growth on solid media; but Hardy, Lee and Nell (1963), who reviewed their earlier work, described the colonial appearances of 14 strains of cultivable treponemes, and Christiansen (1964*b*) those of the Reiter and Kazan strains.

The isolation of spirochaetes in pure culture from material contaminated by bacteria is difficult. Advantage may be taken of their ability to migrate from the line of inoculation in stab cultures. A more successful method is to place the inoculum on a Millipore membrane of pore diameter 0.2 μm lying on the surface of a plate of solid medium (Loesche and Socransky 1962). Spirochaetes grow through the filter and after about a week produce a zone of hazy growth in the underlying medium; bacteria are usually retained by the filter. Plugs from areas of spirochaetal growth are removed with a Pasteur pipette and transferred to semisolid medium, and, when growth is established, subcultured into liquid medium. Growth is slow at first, producing only a faint haze, but strains become adapted with repeated subculture. In young cultures the organisms are actively motile and show typical spiral morphology. In older cultures they lose their motility, become clumped and granular; cystic forms appear and may eventually come to predominate, but on subculture to fresh medium spiral forms grow out.

An alternative method of isolation is the use of animals. This has proved successful in the cultivation of *Tr. pallidum* in the testicle of the rabbit, and of *Borr. recurrentis* in the blood of rats and mice and in the developing chick embryo.

Stock cultures of established strains of treponemes can be kept in semisolid medium in tightly stoppered tubes and subcultured at intervals of one to two months. After extraction from infected animal tissues, virulent *Tr. pallidum* can be preserved for short periods under strict anaerobic conditions in the survival medium described by Nelson (1948); the organisms remain motile but do not multiply. Pyruvate, sodium chloride, and a sulphydryl compound were found to be absolute requirements for survival; and ox serum albumin, hydrolysed gelatin or serum ultra-filtrate, and inorganic phosphate for prolonged survival by Weber (1960). Optimal requirements of inorganic ions for survival are described by Doak, Freedman and Clark (1959*a*, *b*, 1961).

The pathogenic, non-cultivable treponemes are maintained *in vivo* in rabbits or mice, in which the infection becomes latent. They can be recovered from emulsions of the popliteal nodes of infected rabbits by transfer to the testis of fresh animals. Turner (1938) had little success in the preservation of treponemes by lyophilization, but maintained for long periods suspensions in 10 per cent inactivated normal rabbit serum in saline containing 15 per cent glycerol stored at −70° (Turner and Hollander 1957). Hanson and Cannefax (1964), on the other hand, recovered both treponemes and borrelias from lyophilized cultures after a year. Chorpenning and his colleagues (1952) successfully stored infected rabbit testes for use in the treponemal immobilization test at −78°; and Hart (1970) found that *Borr. anserina* remained alive and virulent for at least 150 days in fowl serum or citrated blood stored in liquid nitrogen.

Resistance and metabolism

The resistance of spirochaetes to inimical agencies is no greater and generally less than that of the vegetative bacteria. Dry heat, moist heat, and desiccation prove quickly fatal, as do comparatively low concentrations of the chemical disinfectants. Some of the highly parasitic members, such as *Tr. pallidum*, are unable to survive outside the animal body for more than an hour or two; though some authors report the preservation, for weeks and months, of motile forms in body fluids (McNabb *et al.* 1933, Lumsden 1947). Indeed this particular organism is extremely susceptible to heat, being destroyed in an hour at 41.5°. Advantage is sometimes taken of this property to sterilize the organisms in the tissues by exposure to fever-heat temperatures (see Chapter 75).

The Nichols (*Tr. refringens*) and Reiter strains, and *Tr. pallidum* are susceptible to penicillin, erythromycin, and the tetracyclines, but the faecal strains from swine (p. 501) tend to be more resistant. The antibiotics cause loss of motility, elongation of the cells without division, and death (see Beerman 1947, Morton and Ford 1953, Viaseleva 1957*a*, Smibert 1976). Against the borreliae the tetracyclines and chloramphenicol are more effective than penicillin (Sanford 1976). Populations of cultivable treponemes apparently cannot be trained to substantial antibiotic resistance (see e.g. Viaseleva 1957*b*, Gastinel *et al.* 1959, Mutermilch *et al.* 1959).

Treponemes are anaerobic, leptospires aerobic and borreliae microaerophilic. Leptospires do not have fermentative activity (Fulton and Spooner 1956); but treponemes and borreliae break down glucose without using oxygen (Scheff 1935). Although McKee and Geiman (1950) detected some oxygen uptake in these conditions, from the work of Fulton and Smith (1960) and Smith (1960), it is clear that *Borr. recurrentis* has no oxidative cycle; it breaks down glucose by a phosphorylative type of anaerobic glycolysis to lactate and pyruvate, with traces of butyric and valeric acids, and without any evolution of CO_2. According to Kelly (1976), cultures of *Borr. hermsi*, *Borr. parkeri* and *Borr. turicatae* all ferment glucose, maltose, starch and trehalose by the glycolytic pathway.

The Reiter treponeme can synthesize glycine, serine, alanine, aspartic acid, glutamic acid, proline, and possibly ornithine but not phenylalanine, arginine, lysine, leucine, iso-leucine, valine, threonine or histidine; glucose is a major source of energy (Allen *et al.* 1971). Treponemes are rich in lipid. Meyer and Meyer (1971) found that the Reiter treponeme could neither synthesize nor desaturate fatty acids; it depended for these on an outside source. *Tr. zuelzerae*, a free-living treponeme isolated from mud, was able to synthesize fatty acids. Pickett and Kelly (1974) found that relapsing fever strains of *Borrelia* possessed some degree of lipolytic activity as demonstrated by the presence of specific enzymes against lysolethicin. Tauber and his colleagues (1962) studied strains of cultivable spirochaetes and divided them into five groups on the basis of their enzyme systems. The peptidase activity of the Reiter treponeme was also studied by Szewczuk and Metzger (1970); and the end products of metabolism of oral spirochaetes were investigated by Socransky and his co-workers (1969).

Antigenic structure

Injection of spirochaetes results in the formation of antibodies producing agglutination, complement fixation, immobilization, lysis, and immune adherence of the organisms, and protective antibodies demonstrable by animal inoculation tests. Injection of pathogenic treponemes also produces antibody reacting with extracts of tissue lipids in complement-fixation and flocculation tests.

The antigenic structure of the treponemes has been investigated largely with the aim of providing specific antigens for diagnostic purposes to replace the non-specific tests in which tissue-extract antigens are used. Nelson and Mayer (1949) showed that suspensions of virulent *Tr. pallidum* when incubated under anaerobic conditions with syphilitic serum and complement were immobilized and killed, the organisms eventually being lysed. This reaction, the *treponemal immobilization test*, is specific for infections with the pathogenic treponemes but does not distinguish between them. *Tr. pallidum*, when sensitized with antibody, adheres to the surface of human or primate erythrocytes in the presence of complement (Nelson 1953)—an example of the immune adherence phenomenon described by Rieckenberg (1917). The agglutination of suspensions of *Tr. pallidum* by syphilitic sera was investigated by Hardy and Nell (1955, 1957). They found that Wassermann antibody could agglutinate *Tr. pallidum*; and that two other specific antibodies could be demonstrated, one of which reacted with treponemes that had been heated or subjected to tryptic digestion. The agglutination of erythrocytes sensitized with extracts of *Tr. pallidum* was also described by Tomizawa and Kasamatsu (1966) and Rathlev (1967). The union of antibody with *Tr. pallidum* can likewise be demonstrated by immunofluorescence techniques (Deacon *et al.* 1960, Hunter *et al.* 1964). (Diagnostic tests based on these methods are considered in Chapter 75.)

When first extracted from infected rabbit testis *Tr. pallidum* is not susceptible to the action of syphilitic serum and complement; immobilization begins only after a lag period of several hours (Nelson and Diesendruck 1951). Suspensions are inagglutinable at first, but become agglutinable on storage at 4° or heating (Hardy and Nell 1957). They rapidly become susceptible to immobilization or agglutination by syphilitic antibody if they are first treated with lysozyme. This suggests that, when present *in vivo*, *Tr. pallidum* has a surface component which protects it against the action of antibody. Cultivable treponemes are rapidly immobilized by antibody (Guest *et al.* 1967), indicating that they lack such a protective substance. Bharier and Rittenberg (1971*b*) found that, though antisera against whole *Tr. zuelzerae* would immobilize the organisms, antisera prepared against purified axial filaments could not do so.

Both species-specific and shared antigens have been demonstrated in treponemes by complement-fixation, gel-diffusion and immunofluorescence techniques (Deacon and Hunter 1962, Meyer and Hunter 1967, Dupouey 1963). The Reiter treponeme has been specially studied; Wallace and Harris (1967) reviewed the extensive literature on this and allied organisms. D'Alessandro and Dardanoni (1953) demonstrated four antigens in the Reiter treponeme: a heat-labile protein, a heat-stable polysaccharide and two lipids—one corresponding to the ubiquitous tissue lipid, and the other to an organ-specific brain lipid. Dardanoni and Censuales (1957) and Cannefax and Garson (1959) showed that the protein antigen was common to the Reiter treponeme and *Tr. pallidum*; De Bruijn (1961) found it in the free-living *Tr. zuelzerae*. Lipids, a lipoprotein complex, and a thermolabile protein have been isolated from disrupted cells of the Nichols strain of *Tr. pallidum*, the last two being antigenic in the rabbit (Siefert and Steiger 1959, Siefert 1960). Heat-labile and heat-stable antigens specific for *Tr. pallidum*, as well as antigens related to the lipopolysaccharide and protein components of the Reiter treponeme, have also been demonstrated in ultrasonically disintegrated *Tr. pallidum* (Miller *et al.* 1966). Antibody reacting with the common group protein antigen is produced during infection with the pathogenic treponemes; a complement-fixation test with protein from the Reiter treponeme has proved a useful means for detecting this type of antibody.

Polysaccharide components of the Reiter treponeme were studied by Christiansen (1962*a*, *b*, 1964*a*) and by Pillot and Dupouey (1964). Nell and Hardy (1966) isolated an immunologically homogeneous polysaccharide which fixed complement with antisera to the Reiter treponeme, three Kazan strains, two oral treponemes, and one strain of *Borrelia*, indicating the presence of similar, if not identical, antigens in these strains. It did not react with antisera to the Noguchi, cultivable Nichols, or pathogenic Nichols strain of *Tr. pallidum*. These strain relations are similar to those reported by Robinson and Wichelhausen (1946) and Eagle and Germuth (1948). De Bruijn (1961) showed that lipopolysaccharides obtained from the Reiter treponeme and *Tr. zuelzerae* were antigenically distinct. Specific differences between strains of treponemes may be due to the difference in their polysaccharide antigens.

Antigenically, the louse-borne *Borr. recurrentis* differs from the tick-borne species, such as *Borr. duttoni* and *Borr. turicatae*. This may well be associated with a difference in surface structure, for Baltazard (1947) noted that by dark-ground illumination *Borr. recurrentis* appeared uniformly refractile, whereas the tick-borne borrelias had a 'double contour'.

Pathogenicity

As already noted, members of the *Spirochaeta* and *Cristispira* genera are non-pathogenic. On the other hand, numerous members of the *Borrelia* and *Treponema* genera are pathogenic to man, animals, or birds. For the sake of convenience the host susceptibility of the various species will be dealt with in the summarized description that follows. Further information on the reproduction of disease in various animals is given in Chapters 74 and 75. The virulence of the parasitic members of these two groups is liable to change as the result of residence in the body of the host, or, with those members that can be grown *in vitro*, with subcultivation. Virulent strains may become avirulent, and virulence may be restored by animal passage. In relapsing fever the strains in the blood at the second or third relapse may differ antigenically from the strain responsible for the original attack, and by virtue of this change are able to multiply in the tissues of a host that has become immune to the original infecting strain.

An organism that is pathogenic to one species of animal may be commensal in another, and an organism that is highly pathogenic at one time to a particular host may at another give rise to no more than a latent infection. The mechanism underlying these changes still awaits elucidation.

Classification

The classification of the Spirochaetales given at the beginning of this chapter into the five genera—*Spirochaeta, Cristispira, Borrelia, Treponema* and *Leptospira*—needs some elaboration. Excluding *Leptospira*, which is considered separately (p. 502), and the non-pathogenic *Spirochaeta* and *Cristispira*, we are faced with the difficulty of applying the usual methods of classification to the *Borrelia* and *Treponema* genera, because too little is known of the characters of the organisms themselves to permit of their identification. Other methods have therefore to be resorted to. Borreliae, for example, are classified partly on the nature of the host infected, and partly on the arthropod vectors of the disease they cause. Treponemata are classified on the nature of the host infected and on the type of disease to which they give rise, i.e. acute infectious or chronic granulomatous diseases (Hardy 1976). Alternatively, they may be classified into the cultivable and non-cultivable, or into those that cause syphilis and related diseases such as yaws and pinta, and those that do not. Moreover, the cultivable ones may be subdivided into those capable and those incapable of fermenting carbohydrates (Smibert 1974).

We append a description of some of the members that are of most interest to the student of medical and veterinary bacteriology. Further information on the different groups will be found in Chapters 74 and 75.

Summarized description of some species of *Borrelia* and *Treponema*

Borrelia recurrentis

Observed by Obermeier (1873) in the blood of patients suffering from European relapsing fever. Actively motile spiral organisms usually 10–16 μm long, with one or both ends pointed (Babudieri and Bocciarelli 1948), 5–10 fairly regular primary spirals, each 2–3 μm long with an amplitude of about 1.5 μm (Fig. 44.8). Gram negative. Stain purplish-red with Giemsa. First cultivated by Noguchi (1912*c*) from infected rat or mouse blood in a narrow tube containing 15 ml of ascitic or hydrocele fluid and a small piece of sterile rabbit kidney. Strains of human origin can be grown in a complex medium devised by Kelly (1971, 1976) (see p. 494), and in the developing chick embryo. Microaerophilic. Resistance is similar to that of vegetative bacteria. Said to remain viable in clotted blood for 6 days at room temperature and for at least 100 days at 0° (Wynns and Beck 1935). Ferment glucose and several other carbohydrates with the production of lactic acid. Infection with *Borr. recurrentis* is borne by lice, but numerous other species are borne by ticks of the genus *Ornithodoros* and are named according to the species of the vector and the part of the world in which infection occurs, such as *Borr. hermsi, Borr. hispanica, Borr. duttoni, Borr. parkeri, Borr. turicatae, Borr. brasiliensis, Borr. persica*, and so on. All of these give rise to relapsing fever in man and in monkeys and many are pathogenic to rats and mice. Antigenically, relapse strains may differ from the parent strain. After intraperitoneal inoculation of mice with *Borr. recurren-*

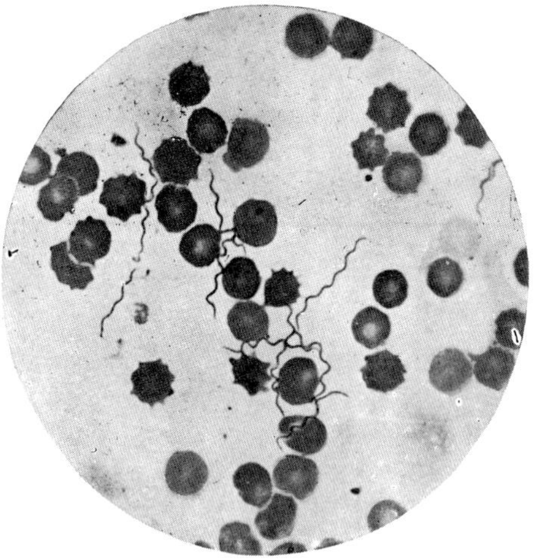

Fig. 44.8 *Borrelia recurrentis*. In film of blood. In one place the spirochaetes shows a tendency to agglutination in rosette form. Giemsa. (× 1000.) (From specimen kindly supplied by the late Prof. J. G. Thomson.)

tis the organisms appear in the blood within 24 hr and persist for 3 or 4 days. They then disappear for several days, after which a relapse usually occurs; three or four relapses may follow each other at intervals of about 7 days (Novy and Knapp 1906). As many as 10–50 organisms may be present per field of the microscope during the first infection, falling to 1 or 2 during the relapses. Rats may be infected, but never relapse (Novy and Knapp 1906; see also Eidmann *et al.* 1959).

Borrelia anserina

Described by Sakharoff (1891), who observed in it the blood of infected geese. Morphologically it closely resembles *Borr. recurrentis*. In blood its mean length is about 14 μm, and the mean number of coils about five (Knowles *et al.* 1932). The organism was cultivated by Noguchi (1912*e*), using his ascitic-fluid rabbit kidney medium. Growth maximum about the 5th day, after which degeneration sets in; all are usually dead in 3 weeks. Growth occurs best at 30°. Subcultures should be made every 4 days. In culture the organism is said to be 8–16 μm long, 0.3 μm wide, and to show rounded spirals, each of which is about 1.8 μm long and 1 μm in amplitude. According to Landauer (1931) and to Knowles and his colleagues (1932), one of the best media for its cultivation is that of Galloway (1925)—coagulated egg white to which dilute inactivated serum is added. In early cultures blood is advantageous. No growth anaerobically. *Borr. anserina* is pathogenic for birds, but rodents, lambs and lizards are resistant (McNeil *et al.* 1949). Intramuscular inoculations of fowls with 0.5 ml of infected blood gives rise to acute spirochaetosis in 24 hr. A high mortality results. Spirochaetes are numerous in the blood (Knowles *et al.* 1932). Cross-spirochaeticidal and cross-protection tests in chickens suggest that fowl spirochaetes may be subdivided into different antigenic groups (Kligler *et al.* 1938). Antigenically distinct from *Borrelia* species found in mammals. Transmitted by ticks of the *Argas* genus.

Borrelia cobayae

Found by Knowles and Basu (1935) in the blood of guinea-pigs. Blood parasite belonging to the relapsing fever group. Thin, delicate spirochaete, 13.5–23 μm in length, with finely tapering ends; average length of spirals 3.6 μm. Can be cultivated in Galloway's medium. Inoculation of guinea-pigs with infected blood is followed, after an incubation period of 2 to 6 days, by a febrile disease accompanied by the presence of spirochaetes in the blood. Fully virulent strains kill 30–60 per cent of inoculated animals. Relapses may occur in animals that recover from the first attack. White rats and rabbits are also susceptible to infection.

Blood spirochaetes have been described in other animals, such as the rabbit and the mouse (see Knowles and Basu 1935).

Treponema pallidum

Isolation Described by Schaudinn and Hoffmann (1905), who observed it in chancres and inguinal glands of syphilitic patients (see Schuberg and Schlossberger 1930).

Morphology Thin, delicate spirochaete with tapering ends. Its length varies from 4–14 μm and its breadth is about 0.2 μm. It contains a number of regular primary spirals, which appear rather sharp and angular, and each of which is a little over 1 μm in length. During motion secondary curves may appear and disappear in rapid succession, but the primary spirals remain undisturbed. The organism is actively motile; the movements were originally described by Schaudinn and Hoffmann (1905) as being of 3 types: (1) rotation round the longitudinal axis; (2) backward and forward movements; (3) flexion movements of the whole body, resulting in the production of secondary waves. Rotation, which is not very frequent, is responsible for the backward or forward movements. Whether these are caused by the primary spirals acting like the blades of a propeller, or by the three axial fibres or internal flagella attached to each end is not clear, though the probability is that both are concerned. Noguchi (1912*c*) described three types of *Tr. pallidum*—the thicker, the normal, and the thinner type. Whether these types are constant, or merely represent fluctuations round a mean, is not known. The organisms stain rose-red with Giemsa. They are held back by gradocol membranes having a pore size of 0.4 μm; their narrowest diameter is therefore 0.2 μm (Hindle and Elford 1933, Tilden 1937).

Cultivation Schereschewski (1909) and Noguchi (1912*b*) both claimed to have grown *Tr. pallidum in vitro* in tubes of serum containing fragments of tissue to promote anaerobiosis but their results have not been reproduced. Early studies on attempts at cultivation have been reviewed by Willcox and Guthe (1966). In the liquid media described by Nelson (1948) and Weber (1960) the organisms remain motile for several days but do not multiply. In such media the optimal pH for survival is 7.2 to 7.3 with an Eh of about −240 mV. To provide the latter, reducing agents such as cysteine, glutathione and thioglycollic acid are added to the media, although the viability of *Tr. pallidum* in superficial skin lesions suggests that it can tolerate low concentrations of oxygen, possibly because of the action of tissue enzymes such as superoxide dismutase. Recent work suggests that it is, in fact, microaerophilic (Graves and Billington 1979). Freshly isolated *Tr. pallidum* utilizes oxygen at a rate similar to that of aerobes (Cox and Barber 1974); and Lysko and Cox (1977) have shown the presence of a cytochrome system.

A promising approach has been the cultivation of *Tr. pallidum* in the presence of tissue cells under low concentrations of oxygen (1.5 to 3 per cent); this has been found to promote survival of the treponemes, as does the supernatant of tissue-culture medium in which cells have been grown (Fitzgerald, Miller and Sykes 1975; Sandok, Knight and Jenkin 1976; Sandok *et al.* 1976, 1978). Some studies in such systems have suggested a possible increase in the number of treponemes (Jones *et al.* 1976, Sandok *et al.* 1978, Fieldsteel, Stout and Becker 1979), although Foster and his colleagues (1977) were not able to reproduce the results recorded in the first of these reports. Fieldsteel, Cox and Moeckli (1981) cultured *Tr. pallidum* with monolayers of cotton-tail rabbit epithelium under 1.5 per cent oxygen. Although enumeration of treponemes is difficult because of clumping and adhesion to the cells, they found an average 49-fold increase in the number of treponemes, which reached a peak of about 2×10^8 organisms after 9 to 12 days' incubation; no further increase in numbers occurred. The amount of treponemal DNA increased in parallel with the number of organisms, and these were shown to have retained their virulence for rabbits. The interaction between treponemes and cells is not a passive one; the treponemes become firmly attached to the cell surface at their pointed ends and tend to form microcolonies on the cells with the accumulation of material thought to be an acidic mucopolysaccharide; whether this is a product of the cells or the treponemes is not certain. Dead *Tr. pallidum* or commensal treponemes do not attach to cells, and the attachment of living *Tr. pallidum* can be blocked by treatment with immune serum (Fitzgerald 1981*a*). For a review of recent attempts to grow *Tr. pallidum* in cell-culture systems see Fitzgerald (1981*b*).

Resistance Very susceptible to heat. According to Boak, Carpenter and Warren (1932), saline suspensions of infected rabbit testicle are sterilized by exposure to 39° for 5 hr, 40° for 3 hr, 41° for 2 hr, and 41.5° for 1 hr. Dies out rapidly in stored blood, unless frozen, so that the chances of transmitting syphilis with stored blood plasma or serum are very small (see Selbie 1943).

Antigenic structure Little is known. Serologically reactive protein, polysaccharide, and lipid components are present. Protein antigens are shared with other species and with *Borrelia*.

Pathogenicity of *Treponema pallidum* for Animals For general reviews see Gastinel and Pulvenis (1934), Turner and Hollander (1957), Wilcox and Guthe (1966) and Collart (1970).

Rabbits The rabbit is very susceptible to infection and is the most widely used experimental animal. Haensell (1881) produced a keratitis by inoculation of material into the anterior chamber of the eye; this was confirmed by Bertarelli (1907, 1908) who showed that the disease could be transferred from one animal to another. Intratesticular inoculation was first described by Parodi (1907); this route is widely used for the isolation of strains and for their propagation. After an incubation period depending on the size of the inoculum an acute orchitis develops and may spread to involve the cord and epididymis, and along the needle track to produce a chancre of the overlying skin. The lesions later regress and the testis atrophies. Intradermal injection is followed by the appearance of an indurated chancre. Infection by any route may be followed by the appearance of generalized lesions; Chesney and Schipper (1950) found these in 5 out of 8 rabbits inoculated intracutaneously, in 14 out of 17 inoculated into the testis, and in 28 out of 29 inoculated intravenously. There may be skin lesions varying from erythematous macules to papules; they appear chiefly on the extremities, ears, base of the tail and the nose, that is, in areas where the skin temperature is lowest. Periostitis of the bones of the foreleg, metatarsals, or nose may occur, and less frequently a keratitis. Orchitis of the opposite testis to that inoculated may be seen. Administration of cortisone during the early stage of active lesions produces an enormous proliferation of treponemes in the lesions, which contain large amounts of mucoid material but show little cellular response (Turner and Hollander 1954). The infection eventually becomes latent in the rabbit but treponemes persist in the lymph nodes (Brown and Pearce 1921) and can be recovered by the inoculation of suspensions of the popliteal and inguinal nodes into the testis of fresh rabbits. Rabbits may have a naturally acquired infection with *Tr. cuniculi*; this affords substantial protection against other pathogenic treponemes. Infection with this organism should be excluded by clinical and serological examination before rabbits are used for experimental work in this field.

Monkeys The experiments of Metchnikoff and Roux (1903, 1904*a*, *b*, 1905) amplified the earlier observations of Klebs in 1875–77 (see Klebs 1932), and showed that syphilis might be transmitted to the anthropoid apes, and with less certainty to monkeys. Of the apes the *chimpanzee* appeared to be the most susceptible. All of 22 chimpanzees (*Troglodytes niger* and *T. calvus*) inoculated with syphilitic material, either from man or experimental animals, were infected. Inoculation was performed by scarification of the genitals, the thigh, or the eyebrow. Primary and secondary lesions developed but most of the animals died of broncho-pneumonia after a short period of observation. Macacques and green monkeys have been successfully inoculated (Turner and Hollander 1957) but are not thought to be well suited for experimental research on treponemes. Owl monkeys (*Aotus trivirgatus*) can be infected and are easier to maintain and handle than the large sub-human primates (Clark and Yobs 1968). According to Uhlenhuth and Mulzer (1913) apes are more difficult to infect than rabbits. These workers were successful in conveying human syphilis to rabbits, from rabbits to monkeys, and from a monkey back to rabbits.

Mice Kolle and Schlossberger (1926, 1928) infected mice and rats with material from a rabbit chancre and showed that three months later mouse lymph nodes and brain were infectious for rabbits. Infection remains latent in the mouse, although Bessemans and Potter (1930) described the occurrence of lesions in the anal region. Most tissues have been shown to be infectious for rabbits. Treponemes, however, are scanty and difficult to find by staining techniques; they are said to be densest in the nasal mucosa, spleen and lymph nodes (Bessemans and Moore 1937, Rosahn *et al.* 1948). The infective dose for mice is thought to be considerably greater than that for rabbits; mouse lymph nodes have been found infective for rabbits 12 weeks after inoculation with 10^5 treponemes. *Tr. pallidum* may persist in mouse tissues for long periods; Rosahn (1952) recovered treponemes from mice as long as 872 days after infection. Treponemal immobilizing antibody may be found late in the infection, but cardiolipin-type antibodies do not appear to be formed. For a review of syphilis in the mouse see Gueft and Rosahn (1948).

Rats and *guinea-pigs* can be infected but the disease is latent. *Hamsters* can be infected by scarification of the skin of the groin (Bessemans *et al.* 1935). Skin lesions are not produced, but treponemes can be found in the enlarged inguinal lymph nodes. In contrast, *Tr. pertenue* produces ulcerative lesions of the skin when inoculated by this route (Vaisman *et al.* 1967, Miao and Fieldsteel 1980).

Treponema cuniculi

First observed by Bayon (1913). Responsible for a disease known as rabbit syphilis. Morphologically very similar to *Tr. pallidum* but tends to be slightly longer and thicker. According to Noguchi (1922) dimensions are: length 7–30 μm, average 13 μm; width 0.25 μm; length of spirals 1–1.2 μm; amplitude of spirals 0.6–1.0 μm. Like *Tr. pallidum* it stains rose-red with Giemsa. Not yet cultivated *in vitro*. Can be propagated by intratesticular inoculation of rabbits. Causes a latent infection in mice, guinea-pigs and hamsters. Inoculation of infected material on to the scarified skin of the genital region is followed, after an incubation period of 2 to 8 weeks, by characteristic lesions (see Chapter 75). (See review of literature by Smith and Pesetsky 1967.)

Notes on certain other treponemes found in the human or animal body

Treponema pertenue Described by Castellani (1905). The cause of yaws. Morphologically indistinguishable from *Tr. pallidum*. Exact relation to this organism not fully understood. Not yet cultivated *in vitro*. Virulent strains can be maintained by intratesticular inoculation of rabbits. Treponemes closely resembling *Tr. pertenue* have been found in lymph nodes of wild African baboons from areas in which yaws was present (Fribourg-Blanc *et al.* 1966). The organisms produce yaws-like lesions in susceptible monkeys (Sepetjian *et al.* 1969) but their relation to *Tr. pertenue* is not yet established.

Treponema carateum Described by Léon y Blanco (1940). The cause of pinta. Morphologically indistinguishable from *Tr. pallidum* (Angulo *et al.* 1951). Exact relation to this organism is not fully understood. Not yet cultivated *in vitro*. Infection has been transmitted to chimpanzees by inoculation of the skin (Kuhn *et al.* 1970).

Treponema refringens Described by Schaudinn and Hoffmann (1905) in their original report on the discovery of *Tr. pallidum*. Probably the same as the *Tr. calligyrum* of Noguchi (1913), and the likewise cultivable and non-pathogenic Nichols strain. Grows in an ascitic fluid agar stab medium, forming under anaerobic conditions white pinpoint colonies on the surface; also on a peptone yeast extract soft-agar serum medium. In culture the organisms are 6–24 μm long by 0.5–0.75 μm wide. Primary spirals are regular and deep, with an amplitude of 1 to 1.5 μm. The G+C content of the DNA is 39–43 moles per cent. Is a normal inhabitant of the male and female genitalia, and is non-pathogenic to man or animals.

Several spirochaetes have been described in ulcerative lesions round the genital region, such as *Tr. phagedenis*, *Tr. balanitidis*, *Tr. pseudo-pallidum*, and *Tr. gangrenosa nosocomialis* (see Noguchi 1912*d*).

Treponema vincenti

Described by Vincent (1896, 1899), who observed it in the throat of patients with Vincent's angina (Chapter 62). This organism is very delicate, about 5–10 μm long, and has 3 to 8 irregular spirals. In cultures filamentous forms are common. It stains poorly but uniformly, is gram negative, and is actively motile. It can be cultivated under anaerobic conditions in serum agar or in serum broth. Growth occurs most readily at 37°; there is no growth at room temperature. In serum agar, colonies appear in 3 days, and are very tiny and tenacious (Ellermann 1904). Under special conditions in hormone agar containing ascitic fluid and cysteine hydrochloride it forms a haze (Hampp 1945). Injected subcutaneously into guinea-pigs, the organisms are generally without effect (Tunnicliff 1906). It is not clear whether *Tr. vincenti* is responsible for the necrotic lesions in human beings in which it is found, or whether it is a mere secondary invader. Since the organism may sometimes be demonstrated in the depths of the infected tissues, it may possess actual invasive properties (Ellermann 1907). Very frequently found in association with a characteristic fusiform bacillus, likewise described by Vincent (1896, see Chapter 26). It was suggested (Tunnicliff 1906) that *Tr. vincenti* and the fusiform bacillus represented two phases of the same organism, but this view is no longer accepted.

Purulent material containing spirochaetes, from the mouth of patients with Vincent's angina, produces a mixed infection on subcutaneous injection into guinea-pigs. The infection can be maintained by serial passage. Fusiform and other bacilli, vibrios and streptococci, as well as spirochaetes, are isolable from the lesions, and although various combinations of the isolated strains are infective for the guinea-pig, none reproduces the typical fusobacterial-spirochaetal abscess (see Rosebury *et al.* 1950). Pure cultures, adapted to aerobic growth *in vitro*, are not pathogenic for laboratory animals, but will induce transient focal lesions on subcutaneous injection in man (Berger 1956, 1958).

Spirochaetes in the human mouth Spirochaetes of different types have been described in the mouth; they can generally be seen in scrapings from between the teeth. Sometimes organisms morphologically indistinguishable from *Tr. pallidum* are found. Noguchi (1912*a*) succeeded in cultivating what he regarded as two separate species. *Tr. microdentium* is a short spirochaete about 3–4 μm long by 0.25 μm wide, having shallow rectangular curves of constant size. The ends are drawn out and pointed. In culture it is said by Séguin and Vinzent (1938) to be an actively motile organism, 4–7 μm long, having 6–12 well defined regular spirals. In serum agar tissue medium it forms a haze near the bottom of the tube, gradually becoming denser and spreading upwards till it is within 2–3 cm of the surface. Growth is anaerobic. *Tr. macrodentium* is a larger organism, varying from 3–8 μm long by 0.7–1.0 μm broad in young cultures, and having 2–8 irregular shallow curves; the ends taper off abruptly. In older cultures the organisms are longer and thinner. In serum agar tissue medium growth occurs under anaerobic conditions in the form of a faint almost transparent haze. For methods of isolating and culturing the mouth spirochaetes, reference may be made to papers by Séguin and Vinzent (1938), Kast and Kolmer (1940), Wichelhausen and Wichelhausen (1942), and Hampp (1945).

Vinzent and Daufresne (1934), working mainly with pure cultures, provisionally classified the mouth spirochaetes into groups, which they labelled A to G. Group B corresponds to *Tr. microdentium* and group F to *Tr. macrodentium*.

For the serological analysis of mouth spirochaetes and of the cultivable strains of supposed *Tr. pallidum* by agglutination, complement-fixation and precipitin tests, the papers of Kolmer, Kast and Lynch (1941) and Robinson and Wichelhausen (1946) should be consulted.

Intestinal spirochaetes have been described coating the surface epithelium of the colon in man. They are $3 \times 0.2\,\mu m$, stain with haematoxylin and eosin, appear to be non-pathogenic (Lee *et al.* 1971). Among those infecting animals, one is *Treponema hyodysenteriae*, which is responsible, at any rate in part, for swine dysentery. Can be grown on agar or in semi-solid media (Taylor and Alexander 1971, Songer *et al.* 1976) under anaerobic conditions, and in a special liquid medium having a base of trypticase soya broth and rabbit serum (Lemcke *et al.* 1979). Grows better at 42° than at 37°. Produces a haemolysin for sheep, rabbit and pig RBC (Lemcke and Burrows 1982). A similar but non-enteropathogenic treponeme is found in the intestine of pigs and dogs. It has been called *Treponema innocens* by Kinyon and Harris (1979). It differs from *Tr. hyodysenteriae* in being only weakly haemolytic, in fermenting fructose, and in failing to produce indole. What relation *Tr. innocens* has to the treponeme described by Turek and Meyer (1978) is doubtful. Both *Tr. hyodysenteriae* and *Tr. innocens* have a G+C content of 25.7 to 25.9 moles per cent—very different from that of 37 to 52 mol per cent of other treponemes.

(For a review of the anatomy and chemistry of spirochaetes, see Holt 1978; and for a monograph on the biology of the parasitic spirochaetes, see Johnson 1976.)

Leptospira

The members of the genus *Leptospira* are flexible, helicoidal, motile organisms. They differ from those of other genera of spirochaetes in having closely set primary coils that are regular and permanent, and in their characteristic movements.

Some leptospires are saprophytic and non-pathogenic. These are widely distributed in fresh surface water and have been isolated from tap water, from sewage effluent and occasionally from sea water. Parasitic and potentially pathogenic strains occur in the tissues of a wide variety of wild and domestic animals, some of which remain unaffected and act as the reservoir or carrier hosts. In the early leptospiraemic stage of infection the organisms may be isolated from the blood. Later they tend to be localized in the convoluted tubules of the renal cortex from which they are excreted in the urine. By contaminating the environment they provide a source of infection to man and other animals and they, unlike the carrier hosts, may suffer from manifest disease.

History

The earliest investigations of leptospirosis were conducted on cases of spirochaetal jaundice known as Weil's disease, named after Adolph Weil (1886), who recognized it as a specific clinical syndrome long before the discovery of the causative organism. Stimson (1907) was apparently the first to see and describe leptospires. He observed them in the renal tubules of a man in New Orleans who had died from a febrile illness with jaundice, thought mistakenly at that time to be yellow fever (see Noguchi 1928, Sellards 1940). Stimson named the organism (*?Spirochaeta*) *interrogans*, the specific epithet indicating that its shape resembled a question mark. He prepared sections of kidney obtained *post mortem* and stained them by Levaditi's silver-impregnation method.

In 1914 in Japan, Inada and his colleagues (1916) saw spirochaetes in the liver of guinea-pigs that had been inoculated with the blood of patients suffering from Weil's disease. They noted that the organisms were present in the patients' urine for periods up to 25 days after onset. They succeeded in culturing the organisms and showed that they grew better at 22°-25° than at 15° or 37° and that they stimulated the formation of antibodies in the patients' blood. They named the organism *Spirochaeta icterohaemorrhagiae*. Ido and his colleagues (1917) found similar virulent organisms in the kidneys of 40 per cent of 86 rats. Independently in Germany and at a slightly later date, the causative organism of Weil's disease was transmitted from man to guinea-pigs by Hübener and Reiter (1915, 1916) and by Uhlenhuth and Fromme (1915, 1916*a*, *b*). These workers named it respectively *Spirochaeta nodosa* and *Spirochaeta icterogenes*. Their findings were confirmed by British workers (Stokes and Ryle 1916 and Stokes *et al.* 1917).

The earliest record of a non-pathogenic leptospire is that of Wolbach and Binger (1914). They isolated an organism, which they named *Spirochaeta biflexa*, from filtered pond water in Massachusetts, USA. They described its morphology and movements but were unable to maintain it in culture.

By comparing the various strains of spirochaetes that had been isolated from man and rats, Noguchi (1917*a*) concluded that they were sufficiently alike to justify their inclusion in a single genus—*Leptospira* ('fine coil'). The type species of Inada was therefore called *L. icterohaemorrhagiae*. Later, Noguchi (1919, 1920) isolated a leptospire which he called *L. icteroides* from cases of Weil's disease, mistakenly diagnosed on clinical grounds as yellow fever. This organism proved subsequently to be the same as *L. icterohaemorrhagiae*.

Classification

Two species of *Leptospira* are now officially recognized, namely *L. interrogans* which includes the parasitic and potentially pathogenic strains, and *L. biflexa*, the saprophytic, non-pathogenic strains. A number of biological tests can be used to determine to which species a strain belongs. For example, saprophytic strains can grow at 13° and in the presence of the purine analogue 8-azaguanine (200 μg per ml), whereas the pathogenic strains cannot do so (Johnson & Harris 1967, 1968; see also Faine and Stallman 1982). There is no genetic homology between strains of *L. biflexa* and *L. interrogans*. An organism with biological characters similar to but not identical with those of parasitic and saprophytic leptospires and said to represent a new species called *L. parva*, has been described (Hovind-Hougen *et al.* 1981).

Serological grouping and typing

Leptospires possess many antigens, some of which are common to all strains of both species, i.e. they are genus-specific, whereas others are restricted to certain types of strain. Strains with the same antigenic composition are allocated to a particular serological type (serotype or serovar) according to a scheme devised by Wolff & Broom (1954). This was based on the

Table 44.1 List of some serotypes (serovars) of *Leptospira interrogans* arranged alphabetically in serogroups (based on Report 1967, with subsequent amendments)

Serogroup	Serotype	Reference strain	Principal carriers
Australis	*australis*	Ballico	Hedgehog, field mice, opossum
	bratislava	Jez bratislava	Pigs
Autumnalis	*autumnalis*	Akiyami A	Voles, field mice
Ballum	*ballum* *arboraea* *castellonis*	MUS 127 Arborea Castellon 3	Mice (white mice), field mice, rats, pigs
Bataviae	*bataviae* *djatzi*	Van Tienen HS 26	Mice, rats
Canicola	*canicola* *schuffneri*	Hond Utrecht IV Vleermuis 90C	Dogs, pigs, jackals
Celledoni	*celledoni* *whitcombi*	Celledoni Whitcomb	Unknown
Cynopteri	*cynopteri*	3522 C	Bats
Djasiman	*djasiman*	Djasiman	Rats
Grippotyphosa	*grippotyphosa* *valbuzzi*	Moskva V Valbuzzi	Voles, field mice
Hebdomadis	*hebdomadis* *borincana*	Hebdomadis HS 622	Voles, field mice
Icterohaemorrhagiae	*icterohaemorrhagiae* *copenhageni*	RGA M20	Rats
Javanica	*javanica* *poi*	Veldrat Bataviae 46 Poi	Rats, mongoose, dogs, cats, shrews
Louisiana	*louisiana*	LSU 1945	Armadillo
Mini	*mini* *georgia*	Sari LT 117	Voles, field mice Voles, field mice
Panama	*panama*	CZ 214K	Opossum
Pomona	*pomona* *monjakov*	Pomona Monjakov	Pigs, cattle Field mice
Pyrogenes	*pyrogenes* *biggis*	Salinem Biggs	Rats, coypu
Sejroe	*sejroe* *hardjo*	M 84 Hardjoprajitno	Voles, field mice Cattle
Shermani	*shermani*	LT 821	Rats, opossum
Tarassovi	*tarassovi* *bakeri*	Perepelicin LT 79	Pigs

technique of agglutination and agglutinin-absorption developed by Schüffner and Mochtar (1927) in Amsterdam. The definition of serotype as restated by the World Health Organization (Report 1967) is as follows: 'two strains are considered to belong to different serotypes if, after cross-absorption with adequate amounts of heterologous antigen, 10 per cent or more of the homologous titre regularly remains in at least one of the two antisera in repeated tests'. Some serotypes have common antigens that cause them to cross-react in serological tests. These serotypes are assigned to the same group (serogroup). An officially recognized list of serotypes in which the pathogenic serotypes are arranged in 16 serogroups and the saprophytes in 2 serogroups is also given in the same WHO report. An extract with modifications is shown in Table 44.1.

Each serogroup is named after the earliest discovered serotype belonging to it. Strain RGA is the reference strain of serotype *icterohaemorrhagiae* of serogroup Icterohaemorrhagiae. It is the original strain (then called *Spirochaeta icterogenes*) that was isolated from a human case of spirochaetal jaundice by Uhlenhuth and Fromme in 1915. In fact, an earlier strain known as Ictero 1 was described by Inada and his colleagues (1916), but although maintained it was never distributed. Since the 1967 report many new pathogenic serotypes have been reported from different parts of the world. It has been suggested that some of the larger serogroups of *L. interrogans* should be further divided to provide additional serogroups (Dikken and Kmety, 1978). Similarly, many additional serotypes of *L. biflexa* have been discovered and many more serogroups of saprophytic serotypes are now recognized (Cinco and Petelin 1970). As already mentioned, a proposed new species, *L. parva*, with properties in between parasitic and saprophytic species has been described (Hovind-Hougen *et al.* 1981).

Antigenic-factor analysis. An alternative method of classifying the leptospires is based on an analysis of their antigenic factors (Kmety 1967). Reference antisera are absorbed with multiple antigenic suspensions until each reacts with only one or several related serotypes. This technique has shown that leptospires possess a mosaic of major and minor agglutinogens. Each serogroup is associated with a number of major antigens which give rise to the cross-agglutination on which the serogrouping is based; other major antigens are serotype-specific. Each new isolate can be typed qualitatively by using the appropriate factor sera of the serogroup to which it belongs. Minor antigens are present in various combinations not only in serotypes within a single serogroup, but also in serotypes of different serogroups; this explains the cross-reactions at low titres with strains of heterologous serogroups in agglutination tests.

Habitat

Pathogenic leptospires, after entering the body, are spread by the bloodstream and tend to localize in the convoluted tubules of the kidney. The renal carrier state may be transitory or it may persist for a long time, constituting the reservoir or maintenance state. Rats, particularly *Rattus norvegicus* are in this respect the chief hosts, but in certain localities other animals may be important. Saprophytic strains are found in pools, ditches, streams and in the slime that collects at the end of taps and on the roofs of mines. Strains have also been isolated from sea-water.

Morphology

Leptospires are very fine spiral bacteria the morphology of which can be studied only by dark-ground, phase-contrast or electron microscopy. They stain poorly with the usual dyes. They may be demonstrated by silver-deposition techniques but these obscure the primary coils. The organisms vary in length from 6 μm to 20 μm; division is by transverse fission (Fig. 44.9).

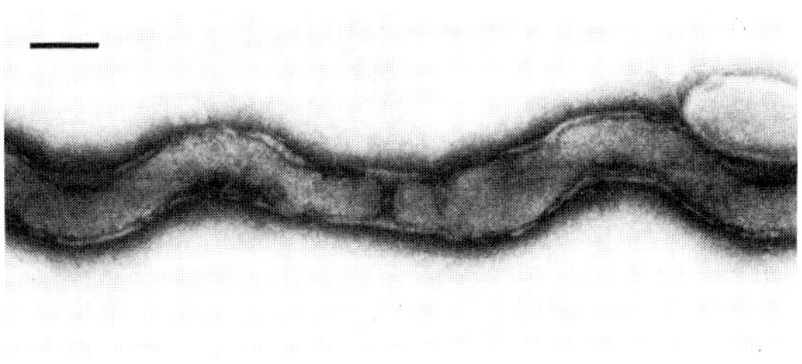

Fig. 44.9 *Leptospira interrogans*, serotype *icterohaemorrhagiae*, showing division of the cell. The bar on the print represents 100 μm. (By courtesy of Dr Kari Hovind-Hougen.)

The individual organisms appear straight and rigid with one or both ends characteristically hooked or bent. The diameter of the cell is 0.1 μm. The primary coils, which are fixed, regular and much closer set than in other spirochaetes, are of amplitude 0.1 to 0.15 μm and wavelength about 0.5 μm. The organisms are so thin that they can pass through a membrane filter of pore diameter 0.22 μm, and readily migrate through 1.0 per cent agar gel. Electronmicroscopy reveals a helicoidal protoplasmic cylinder delineated by a cytoplasmic membrane-peptidoglycan complex. An outer envelope membrane surrounds the whole organism. Two flagella (axial filaments) are located between the cytoplasmic membrane and the outer envelope, and are entwined around the cell body. Each is attached, subterminally at one or other end of the protoplasmic cylinder, to a basal body, similar to the basal bodies of the flagella of gram-negative bacteria (Fig. 44.10). Their free ends are directed towards the middle region of the organism, but do not usually

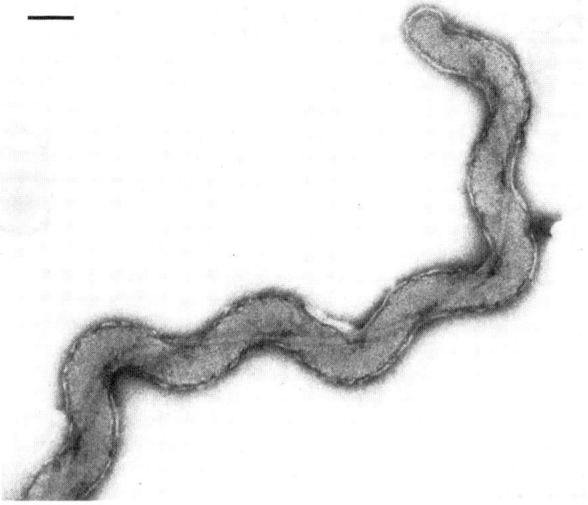

Fig. 44.10 *Leptospira interrogans* serotype *icterohaemorrhagiae*, showing one of the axial filaments attached to the basal body. The bar on the print represents 100 μm. (By courtesy of Dr Kari Hovind-Hougen.)

overlap. Nuclear material, ribosomes and mesosomes can be distinguished in the protoplasmic cylinder (Birch-Andersen *et al.* 1973). The G + C content of the DNA ranges from 35.3 to 41 moles per cent. The motility of leptospires is of various types depending on the milieu. In liquid medium they appear to rotate or oscillate about their long axis and the hooked ends then appear like buttonholes, while secondary coils appear and disappear (Fig. 44.11). The organisms

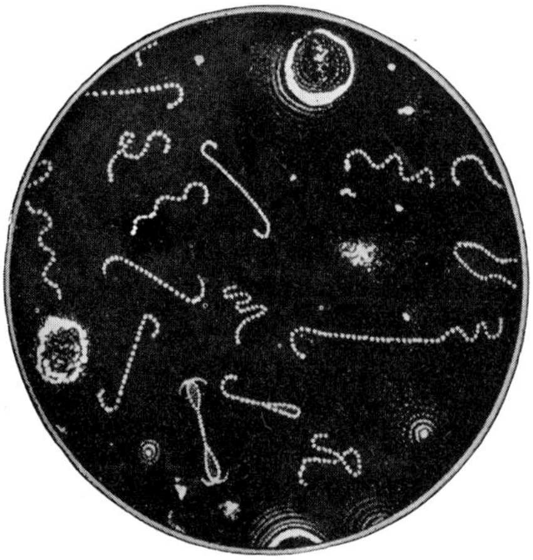

Fig. 44.11 Diagrammatic drawing of leptospires showing primary and secondary spirals, and hooked and button-hole ends. (After Wenyon.)

move backwards and forwards without polar differentiation, and dividing cells often flex sharply and vigorously at the point of impending divisions. In semi-solid medium, blood or tissue emulsions the

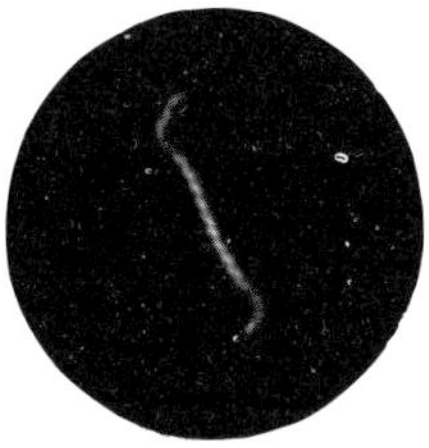

Fig. 44.12 *Lepto. biflexa.* Dark-ground illumination (× 1500 ca.).

organisms display boring or serpentine movements. (See also Fig. 44.12)

Cultivation

Unlike other spirochaetes, leptospires are easily grown in liquid medium of suitable composition. Various media, such as those of Fletcher (1928), Korthof (1932), and Stuart (1946) composed of salts, with or without peptone, buffered at pH 7.3–7.4 and enriched with 7–10 per cent (v/v) of sterile, heat-inactivated normal rabbit serum were and are still used. Most laboratories, however, now use a semi-synthetic medium in which the rabbit serum is replaced by bovine albumin, fraction V, and polysorbate (Tween 80), which provides the fatty-acid requirement. This medium known as EM medium was devised by Ellinghausen and McCullough (1965), and in a modified form (EMJH) described by Johnson and Harris (1967, 1968) is commercially available. A few strains of *Leptospira* do not thrive in such a medium but grow well if 2–3 per cent of rabbit serum is added to it. Satisfactory growth can be obtained with some but not all strains in chemically defined media, for example that of Shenberg (1967). Such media are used for research purposes, for metabolic studies, and for vaccine production when a protein-free medium is essential. All media may be used in liquid, semisolid and solid form. A semisolid medium, used for isolating leptospires, and for maintaining cultures for long periods, is prepared by adding agar in the proportion of 0.2–0.5 per cent (w/v). A solid medium (Cox 1966) contains 0.8–1.3 per cent of agar and is used mainly for separating leptospires from contaminating micro-organisms and for cloning strains. The optimal pH for growth is 7.2–7.6 (range 6.8–7.8) and the optimal temperature 28–30°. Growth occurs more slowly at lower temperatures. As already mentioned, saprophytic strains grow at 13°, but pathogenic strains do not. The generation time is 7–16 hr. The rate of multiplication is speeded up by shaking or stirring liquid cultures. In liquid medium growth is faintly turbid and when the tube is gently shaken has an appearance of 'shot silk'. In semisolid media the upper part becomes turbid as leptospires multiply; one or more characteristic flat disks develop 1–3 cm below the surface (Dinger 1932).

Metabolism

Leptospires are aerobic and require oxygen for growth. They are chemo-organotrophs. They require long-chain fatty acids as a major source of carbon and energy. Ammonium salts but not amino acids can be used as a source of nitrogen. Purines but not pyrimidines are utilized. Essential vitamins are thiamine (vitamin B_1) and cyanocobalamine (vitamin B_{12}). Leptospires form catalase and either peroxidase or esterase or both. Some strains produce a haemolysin. There is no evidence of exo- or endotoxins, but the histopathological effects suggest that the organisms exert some toxic action.

Resistance

Pathogenic leptospires can survive and even multiply outside the animal body provided the external conditions are favourable. They are sensitive to temperature, acidity and bacterial contamination. They are destroyed by human gastric juice, by bile and by human and cows' undiluted raw milk (Kirschner *et al.* 1957). Experiments carried out by Chang and his colleagues (1948) showed that the thermal death points in distilled water were 45° for 25–30 min, 50° for 5–10 min, 60° for 10 sec and 70° for less than 10 sec. In sterile tap water leptospires remained viable for over 4 weeks provided the pH was neutral but for only 2 days at pH 5. In water exposed to bacterial contamination, the survival time at atmospheric temperature was only half that in sterile water (18–20 days). An increase in temperature reduced the survival time by increasing bacterial multiplication. In undiluted sewage, leptospires survived for only 12–14 hr. However, when the sewage was aerated they survived for 2–3 days, suggesting that the adverse effect on survival was due to the anaerobic conditions and to the production of acid by bacteria. The same authors showed that leptospires were killed by 1 p.p.m. of residual chlorine in 3 min at neutral pH. Cationic detergents were highly lethal but anionic ones were not, except in high concentration. It may be concluded that leptospires are slightly more sensitive to disinfectants and heat than most non-sporing bacteria. Pathogenic leptospires survive for only 18–20 hr in sea-water. Some saprophytes, on the other hand, can survive in salt water and are indeed sodium dependent (Cinco del Fabbro and Sottocasa 1976).

Leptospires are sensitive *in vitro* to most antibiotics, including penicillin. They may be preserved and their virulence maintained by placing capillaries of the actively growing cultures in the storage unit (vapour phase) of a liquid nitrogen container at −148°.

Pathogenicity

Leptospiral infections in man and animals result in clinical manifestations that range from subclinical to fatal. The features that govern their pathogenicity are not clear but seem to depend partly on the serotype of the organism and partly on the resistance of the individual host. That the host resistance is of importance is suggested by the apparent variation in virulence of the same organism in different geographical areas. Thus, strains of serotype *bataviae*, which give rise to an acute febrile illness in Europe, may result in the much more serious Weil's syndrome in Far Eastern countries.

Serotypes within the Icterohaemorrhagiae serogroup are highly pathogenic for young guinea-pigs, golden hamsters, gerbils and 1–2 day old chicks. Rabbits, rats, mice and voles do not apparently suffer from the infection. Cats, sheep, hens, pigeons and monkeys are also said to be refractory (Uhlenhuth and Fromme 1916*a*, Martin and Pettit 1919). Dogs, especially kennel dogs, calves, and young pigs may become accidentally infected from natural sources and suffer from illness ranging in severity from a very mild to a rapidly fatal form of jaundice.

Young guinea-pigs can be used experimentally for isolating the organisms from human blood or other sources. For the diagnosis of human infection, intraperitoneal injection of 1–2 ml of human blood into guinea-pigs during the acute phase of the disease results in a severe clinical disease which terminates in the death of the animal after 5 to 12 days. Fever begins the day after inoculation and reaches its acme in a few days. The temperature falls to normal, and finally to subnormal just before death. Jaundice becomes visible with the fall in temperature, usually on the 4th or 5th day, and increases till death. It is often accompanied by choluria. Conjunctival congestion is frequent, and external haemorrhages from the rectum, nose, and genitals may occur. Blood counts reveal a lymphocytosis during the first few days of the disease, and an anaemia (Buchanan 1927). Spirochaetes appear in the blood between the 2nd and 6th days and can usually be found by dark-ground microscopy (Taylor and Goyle 1931). *Post mortem*, the animal shows generalized jaundice; there are haemorrhages into various parts of the body, particularly the lungs, intestinal walls, retroperitoneal tissues, and fatty tissues of the inguinal region. The haemorrhages in the lungs form irregular spots of varying size, sharply demarcated from the surrounding tissue, giving the lungs a resemblance to the mottled wings of a butterfly (Inada *et al.* 1916). The spleen is enlarged and congested; the kidneys are affected with an acute parenchymatous inflammation and capsular haemorrhages; the adrenals are often enlarged and haemorrhagic. Histologically, the chief lesions are cloudy swelling of the liver, sometimes accompanied by focal necroses, acute parenchymatous nephritis, endothelial-cell proliferation in the spleen and lymph glands, and haemorrhages in practically every structure of the body. Spirochaetes are most numerous in the liver and are best demonstrated by silver-impregnation methods. They occur in the spaces between the cells; when numerous they are arranged about the cells like a garland. Their appearance differs from that seen in culture under dark-ground illumination; they are short and thick; the primary spirals and tapering extremities are not evident, and numerous irregular undulations are seen. They are found in smaller numbers in the kidneys and adrenals. According to Wylie (1946) death of the animal results more from the renal than from the hepatic damage.

By serial passage in guinea-pigs the virulence of the organism can apparently be increased. Stokes and his colleagues (1917), found that the average time to death of animals inoculated intraperitoneally with human blood was 10 days, but that when passaged strains were used it was only 5 days (see also Noguchi 1917*a*).

Spirillum minus

Sometimes referred to as *Spirochaeta morsus muris*. Described by Futaki and his colleagues (1916, 1917) as the cause of rat-bite fever in man. According to Robertson (1924), it is a spirillum and not a spirochaete, and its correct name is *Spirillum minus*. Appears to be a natural parasite of rats, which act as healthy carriers of the organism. Morphologically the spirillum is short, rather thick, and has tapering ends, provided with one or more flagella. It is 2–5 μm long, motile, and has regular rigid spirals, each of which is about 1 μm in length (Fig. 44.13). The movements are rapid—like those of a vibrio. It is readily stained by ordinary aniline dyes, such as Loeffler's methylene blue, and by Giemsa. Cultures may be obtained in Shimamine's medium, but successive transfers have not been successful. The organism gives rise to one type of rat-bite fever in man. Intraperitoneal inoculation of infective human material into mice is followed by no clinical evidence of disease, but spirilla appear in the blood after 5 to 14 days. They are scarce at first, but later they increase, though they never become numerous; it is uncommon to find two organisms in the same field (Theiler 1926). They persist indefinitely, though only in small numbers. Rats behave like mice, but the number of spirilla in the blood is fewer. Intraperitoneal inoculation of guinea-pigs produces a febrile disease. After an incubation period of 6 to 15 days spirilla appear in small numbers in the blood, and pyrexia sets in accompanied by enlargement of the lymph glands. There may be considerable inflammation of the subcutaneous tissue in the ano-genital region, involving the scrotal sacs, perianal tissue, and prepuce in males and the labia and perianal tissue in females. Later, after 3 or 4 weeks, alopecia, ulceration of the skin, and chronic conjunctivitis and keratitis may occur. The disease is generally chronic, lasting from about 2 to 4 months, but sometimes death occurs in the first 5 weeks (Ishiwara *et al.* 1917). Spirilla can be demonstrated in the blood, lymph glands, spleen, kidney, adrenal, and subcutaneous tissue. In Robertson's (1924) experience spirilla were never demonstrable in the blood, even by mouse inoculation, nor did any of the guinea-pigs die. Rabbits may be infected, but are less suitable for diagnostic purposes than mice or guinea-pigs. Monkeys are also susceptible (Inada *et al.* 1916).

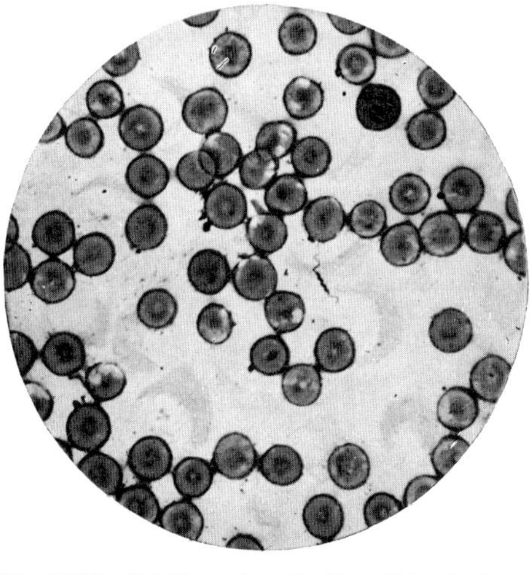

Fig. 44.13 *Spirillum minus*. In film of blood of experimentally infected mouse. Giemsa. (×1000.) (From specimen kindly supplied by the late Prof. J. G. Thomson.)

References

Allen, S. L., Johnson, R. C. and Peterson, D. (1971) *Infect. Immun.* **3,** 727.

Angulo, J. J., Watson, J. H. L., Wedderburn, C. C., Léon-Blanco, F. and Varela, G. (1951) *Amer. J. trop. Med.* **31,** 458.

Babudieri, B. (1972) *The Fine Morphology of Spirochaetes.* Leonardo Edizioni Scientifiche, Roma.

Babudieri, B. and Bocciarelli, D. (1948) *J. Hyg., Camb.* **46,** 438.

Baltazard, M. (1947) *Bull. Soc. Path. exot.* **40,** 77.

Bayon, H. (1913) *Brit. med. J.* ii, 1159.

Beerman, H. (1947) *Amer. J. med. Sci.* **214,** 442.

Berger, U. (1956) *Z. Hyg. InfektKr.* **143,** 23; (1958) *Beitr. Hyg. Epidem.* **12,** 139.

Bertarelli, E. (1907) *Zbl. Bakt.* **43,** 238, 448; (1908) *Ibid.* **46,** 51.

Bessemans, A. and Moore, A. de. (1937) *Derm. Ztschr.* **75,** 57.

Bessemans, A., Moore, A. de and Rigge, A. de. (1935) *C. R. Soc. Biol., Paris* **129,** 503.

Bessemans, A. and Potter, F. de. (1930) *C. R. Soc. Biol.* **104,** 818.

Bharier, M. A. and Rittenberg, S. C. (1971*a*) *J. Bact.* **105,** 422; (1971*b*) *Ibid.* **105,** 430.

Birch-Andersen, A., Hougen, K. H. and Borg-Petersen, C. (1973) *Acta path. microbiol. scand.* **B81,** 665.

Boak, R. A., Carpenter, C. M. and Warren, S. L. (1932) *J. exp. Med.* **56,** 725.

Brown, W. H. and Pearce, L. (1921) *J. exp. Med.* **34.** 185.
Bruijn, J. H. De. (1961) *Leeuwenhoek ned, Tijdschr.* **27,** 98.
Buchanan, G. (1927) *Spec. Rep. Ser. med. Res. Coun., Lond.* No. 113.
Canale-Parola, E. (1978) *Annu. Rev. Microbiol.* **32,** 69.
Canale-Parola, E., Udris, Z. and Mandel, M. (1968) *Arch. Mikrobiol.* **63,** 385.
Cannefax, G. R. and Garson, W. (1959) *J. Immunol.* **82,** 198.
Castellani, A. (1905) *Brit. med. J.* **ii,** 1280.
Chang, S. L., Buckingham, M. and Taylor, M. P. (1948) *J. infect. Dis.* **82,** 256.
Chesney, A. M. and Schipper, G. J. (1950) *Amer. J. Syph.* **34,** 18.
Chorpenning, F. W., Sanders, R. W. and Kent, J. F. (1952) *Amer. J. Syph.* **36,** 401.
Christiansen, A. H. (1962*a*) *Acta. path. microbiol. scand.* **56,** 166; (1962*b*) *Ibid.* **56,** 177; (1964*a*) *Ibid.* **60,** 123; (1964*b*) *Ibid.* **60,** 234.
Cinco, M. and Petelin, N. (1970) *Trop. geogr. Med.* **22,** 237.
Cinco del Fabbro, M. and Sottocasa, G. L. (1976) *Zbl. Bakt.* **A236,** 354.
Clark, J. W. and Yobs, A. R. (1968) *Brit. J. vener. Dis.* **44.** 208.
Collart, P. (1970) *Rev. Méd. Hyg. Outre-mer* **22,** 1265, 1279.
Collart, P., Borel, L.-J. and Durel, P. (1962) *Ann. Inst. Pasteur* **102,** 596.
Cox, C. and Barber, M. (1974) *Infect. Immun.* **10,** 123.
Cox, C. D. (1966) *Ann. Soc. belge Méd. trop.* **46,** 193; (1972) *J. Bact.* **109,** 943.
Cumberland, M. C. and Turner, T. B. (1949) *Amer. J. Syph.* **33,** 201.
D'Alessandro, G. and Dardanoni, L. (1953) *Amer. J. Syph.* **37,** 137.
Dardanoni, L. and Censuales, S. (1957) *Riv. Ist. sieroterap. ital.* **32,** 489.
Deacon, W. E., Freeman, E. M. and Harris, H. A. (1960) *Proc. Soc. exp. Biol., N.Y.* **103,** 827.
Deacon, W. E. and Hunter, E. F. (1962) *Proc. Soc. exp. Biol., N.Y.* **110,** 352.
DeLamater, E. D. Haanes, M., Wiggall, R. H. and Pillsbury, D. M. (1951) *J. invest. Dermatol.* **16,** 213.
DeLamater, E. D., Wiggall, R. H. and Haanes, M. (1950) *J. exp. Med.* **92,** 239.
Dikken, H. and Kmety, E. (1978) In: *Methods in Microbiology*. Vol. 11, p. 259. Ed. by T. Bergen and J. R. Norris, Academic Press, London.
Dinger, J. E. (1932) *Geneesk. Tijschr. Ned.-Ind.* **72,** 1511.
Doak, G. O., Freedman, L. D. and Clark, J. W. (1959*a*) *J. Bact.* **77,** 322; (1959*b*) *Ibid.* **78,** 703; (1961) *Ibid.* **82,** 909.
Dobell, C. (1912) *Arch. Protistenk.* **26,** 117.
Dupouey, P. (1963) *Ann. Inst. Pasteur* **105,** 725.
Eagle, H. and Germuth, F. G. (1948) *J. Immunol.* **60,** 223.
Eagle, H. and Steinman, H. G. (1948) *J. Bact.* **56,** 163.
Ehrenberg, C. G. (1838) *Die Infusionsthierchen als vollkommene Organismen.* L. Voss, Leipzig.
Eidmann, L., Lippelt, H. and Proespodihardjo, S. (1959) *Z. Tropenmed. Parasit.* **10,** 339.
Eipert, S. R. and Black, S. A. (1979) *Arch, Microbiol.* **120,** 205.
Ellermann, V. (1904) *Zbl. Bakt.* **37,** 729; (1907) *Z. Hyg. InfektKr.* **53,** 453.
Ellinghausen, H. C. and McCullough, W. G. (1965) *Amer. J. vet. Res.* **26,** 45.
Faine, S. and Stallman, N. D. (1982) *Int. J. syst. Bact.* **32,** 461.
Fieldsteel, A. H., Cox, D. L. and Moeckli, R. A. (1981) *Infect. Immun.* **32,** 908.
Fieldsteel, A. H., Stout, J. G. and Becker, F. A. (1979) *Infect. Immun.* **24,** 337.
Fitzgerald, T. J. (1981*a*) *Annu. Rev. Microbiol.* **35,** 29; (1981*b*) *Bull. World Hlth Org.* **59,** 787.
Fitzgerald, T. J., Johnson, R. C., Sykes, J. A. and Miller, J. N. (1977) *Infect. Immun.* **15,** 444.
Fitzgerald, T. J., Miller, J. N. and Sykes, J. A. (1975) *Infect. Immun.* **11,** 1133.
Fletcher, W. (1928) *Trans. R. Soc. trop. Med. Hyg.* **21,** 265.
Foster, J. W., Kellogg, D. S., Clark, J. W. and Balows, A. (1977) *Brit. J. vener. Dis.* **53,** 338.
Fribourg-Blanc, A., Niel, G. and Mollaret, H. H. (1966) *Bull. Soc., Path. exot.* **59,** 54.
Fulton, J. D. and Smith P. J. C. (1960) *Biochem. J.* **76,** 491.
Fulton, J. D. and Spooner, D. F. (1956) *Exp. Parasit.* **5,** 154.
Futaki, K., Takaki, I., Taniguchi, T. and Osumi, S. (1916) *J. exp. Med.* **23,** 249; (1917) *Ibid.* **25,** 33.
Galloway, I. A. (1925) *C. R. Soc. Biol.* **93,** 1074.
Gastinel, P., Collart, P. and Dunoyer, F. (1959) *Ann. Inst. Pasteur* **96,** 381.
Gastinel, P. and Pulvenis, R. (1934) *La syphilis expérimentale.* Masson et Cie, Paris.
Gelperin, A. (1949) *Amer. J. Syph.* **33,** 101; (1951) *Ibid.* **35,** 1.
Graves, S. R. and Billington, T. (1979) *Brit. J. vener. Dis.* **55,** 387.
Gueft, B. and Rosahn, P. D. (1948) *Amer. J. Syph.* **32,** 59.
Guest, W. J., Nevin, T. A., Thomas, M. L. and Adams, J. A. (1967) *J. Bact.* **93,** 1190.
Haensell, P. (1881) *v. Graefes Arch. Ophthal.* **27,** 93.
Hampp, E. G. (1945) *Amer. J. publ. Hlth.* **35,** 441; (1954) *J. dent. Res.* **33,** 660.
Hanson, A. W. and Cannefax, G. R. (1964) *J. Bact.* **88,** 811.
Hardy, P. H. (1976) In: *The Biology of Parasitic Spirochetes.* p. 107. Ed. by R. C. Johnson. Academic Press, New York.
Hardy, P. H., Lee, Y. C. and Nell, E. E. (1963) *J. Bact.* **86,** 616.
Hardy, P. H. and Nell, E. E. (1955) *J. exp. Med.* **101,** 367; (1957) *Amer. J. Hyg.* **66,** 160; (1961) *J. Bact.* **82,** 967.
Hart. L. (1970) *Aust. vet. J.* **46,** 455.
Hindle, E. and Elford, W. J. (1933) *J. Path. Bact.* **37,** 9.
Holt, S. C. (1978) *Microbiol. Rev.* **42,** 114.
Hovind-Hougen, K., Ellis, W. R. and Birch-Andersen, A. (1981) *Zbl. Bakt.*, I Abt. Orig. **250,** 343.
Hougen, K. H. (1974) *Acta path. microbiol. scand.* **B82,** 799; (1976) *Ibid.* Suppl. No. 255.
Hübener, E. A. and Reiter, H. (1915) *Dtsch. med. Wschr.* **41,** 1275; (1916) *Ibid.* 42, 131.
Hunter, E. F., Deacon, W. E. and Meyer, P. E. (1964) *Publ. Hlth Rep., Wash.* **79,** 410.
Ido, Y., Hoki, R., Ito, H. and Wani, H. (1917) *J. exp. Med.* **26,** 341.
Inada, R., Ido, Y., Hoki, R., Kaneko, R. and Ito, H. (1916) *J. exp. Med.* **23,** 377.
Ishiwara, K., Ohtawara, T. and Tamura, K. (1917) *J. exp. Med.* **25,** 45.
Jackson, S. and Black, S. A. (1971) *Arch. Mikrobiol.* **76,** 308, 325.
Johnson, R. C. (1976) [Ed.] *The Biology of Parasitic Spirochetes.* Academic Press, New York; (1977) *Annu. Rev. Microbiol.* **31,** 89.

Johnson, R. C. and Harris, V. G. (1967) *J. Bact.* **94,** 27; (1968) *Appl. Microbiol.* **16,** 1584.

Jones, R. H., Finn, M. A., Thomas, J. J. and Folger, C. (1976) *Brit. J. vener. Dis.* **52,** 18.

Kast, C. C. and Kolmer, J. A. (1940) *Amer. J. Syph.* **24,** 671.

Kellogg, D. S. and Mothershed, S. M. (1969) *J. Amer. med. Ass.* 207, 938.

Kelly, R. T. (1971) *Science* **173,** 443; (1976) In: *The Biology of Parasitic Spirochetes.* p. 87. Ed. by R. C. Johnson. Academic Press, New York.

Kinyon, J. M. and Harris, D. L. (1979) *Int. J. Syst. Bact.* **29,** 102.

Kirschner, L., Maguire, T. and Bertaud, W. S. (1957) *Brit. J. exp. Path.* **28,** 357.

Klebs, A. C. (1932) *Science* **75,** 101.

Kligler, I. J., Hermoni, D. and Perek, M. (1938) *J. comp. Path.* **51,** 206.

Kmety, E. (1967) *Faktorenalyse von Leptospiren der Ictero-haemorrhagiae und einiger Verwandter Serogruppen.* Thesis, Slovak Academy of Sciences, Bratislava.

Knowles, R. and Basu, B. C. (1935) *Indian J. med. Res.* **22,** 449.

Knowles, R., Gupta, B. M. D. and Basu, B. C. (1932) *Indian J. med. Res.* Memoir No. 22.

Kolle, W. and Schlossberger, H. (1926) *Dtsch. med. Wschr.* **52,** 1245; (1928) *Ibid.* **54,** 129.

Kolmer, J. A., Kast, C. C. and Lynch, E. R. (1941) *Amer. J. Syph.* **25,** 300, 412.

Korthof, G. (1932) *Zbl. Bakt.* **125,** 429.

Kuhn, U. S. G., Medina, R., Cohen, P. G. and Vegas, M. (1970) *Brit. J. vener. Dis.* **46,** 311.

Landauer, E. (1931) *Ann. Inst. Pasteur* **47,** 667.

Lee, F. D., Kraszewski, A., Gordon, J., Howie, J. G. R., McSeveney, D., and Harland, W. A. (1971) *Gut.* **12,** 126.

Lemcke, R. M., Bew, J., Burrows, M. R. and Lysons, R. J. (1979) *Res. vet. Sci.* **26,** 315.

Lemcke, R. M. and Burrows, M. R. (1982) *J. Med. Microbiol* **15,** 205.

León Y. Blanco, F. (1940) *Rev. Med. trop., Habana* **6,** 5, 13.

Lewin, R. A. (1965) *Canad. J. Microbiol.* **11,** 77, 135.

Loesche, W. J. and Socransky, S. (1962) *Science* **138,** 139.

Lumsden, C. E. (1947) *Lancet* **ii,** 827.

Lysko, P and Cox, C. (1977) *Infect. Immun.* **16,** 885.

McKee, R. W. and Geiman, Q. M. (1950) *Fed. Proc.* **9,** 201.

McNabb, A. L., Matthews, G. and McClure, A. D. (1933) *Canad. publ. Hlth J.,* **24,** 405.

McNeil, E., Hinshaw, W. R. and Kissling, R. E. (1949) *J. Bact.* **57,** 191.

Magnuson, H. J., Eagle, H. and Fleischman, R. (1948) *Amer. J. Syph.* **32,** 1.

Manouélian, Y. (1940) *Ann. Inst. Pasteur* **64,** 439.

Martin, L. and Pettit, A. (1919) *Spirochétose ictérohémorragique.* Paris.

Metchnikoff, E. and Roux, E. M. (1903) *Ann. Inst. Pasteur* **17,** 809; (1904*a*) *Ibid.* **18,** 1; (1904*b*) *Ibid.* **18,** 657; (1905) *Ibid.* **19,** 673.

Meyer, H. and Meyer, F. (1971) *Biochim. biophys. Acta* **231,** 93.

Meyer, P. E. and Hunter, E. F. (1967) *J. Bact.* **93,** 784.

Miao, R. M. and Fieldsteel, A. H. (1980) *J. Bact.* **141,** 427.

Miller, J. N., De Bruijn, J. H., Bekker, J. H. and Onvlee, P. C. (1966) *J. Immunol.* **96,** 450.

Morton, H. E. and Ford, W. T. (1953) *Amer. J. Syph.* **37,** 529.

Mutermilch, S., Gérard, S. and Delaville, M. (1959) *Ann. Inst. Pasteur* **96,** 402.

Nell, E. E. and Hardy, P. H. (1966) *Immunochemistry* **3,** 233.

Nelson, R. A. (1948) *Amer. J. Hyg.* **48,** 120; (1953) *Science* **118,** 733.

Nelson, R. A. and Diesendruck, J. (1951) *J. Immunol.* **66,** 667.

Nelson, R. A. and Mayer, M. M. (1949) *J. exp. Med.* **89,** 369.

Nisbett, A. (1976) *Konrad Lorenz.* J. M. Dent & Sons Ltd, London.

Noguchi, H. (1912*a*) *J. exp. Med.* **15,** 81; (1912*b*) *Ibid.* **15,** 90; (1912*c*) *Ibid.* **15,** 201; (1912*d*) *Ibid.* **16,** 261; (1912*e*) *Ibid.* **16,** 620; (1913) *Ibid.* **17,** 89; (1917*a*) *Ibid.* **25,** 755; (1917*b*) *Amer. J. Syph.* **1,** 261; (1919) *J. exp. Med.* **29,** 565; (1920) *Ibid.* **31,** 135; (1922) *Ibid.* **35,** 391; (1928) In: *The Newer Knowledge of Bacteriology and Immunology,* p. 452. Ed. by E. O. Jordan and I. S. Falk, University of Chicago Press, Chicago.

Novy, F. G. and Knapp, R. E. (1906) *J. infect. Dis.* **3,** 291.

Obermeier, O. (1873) *Berl. klin. Wschr.* **10,** 152, 378, 391, 455.

Oyama, V. I., Steinman, H. G. and Eagle, H. (1953) *J. Bact.* **65,** 609.

Parodi, U. (1907) *Zbl. Bakt.* **44,** 428.

Pickett, J. and Kelly, R. (1974) *Infect. Immun.* **9,** 279.

Pillot, J. and Dupouey, P. (1964) *Ann. Inst. Pasteur* **106,** 456.

Pillot, J., Dupouey, P. and Ryter, A. (1964) *Ann. Inst. Pasteur* **107,** 663.

Pillot, J. and Ryter, A. (1965) *Ann. Inst. Pasteur* **108,** 791.

Power, D. A. and Pelczar, M. J. (1959) *J. Bact.* **77,** 789.

Rathlev, T. (1967) *Brit. J. vener. Dis.* **43,** 181.

Reiter, H. (1960) *Brit. J. vener. Dis.* **36,** 18.

Report (1967) *Tech. Rep. Ser. World Hlth Org.* No. 380.

Rieckenberg, H. (1917) *Z. ImmunForsch.* **26,** 53.

Robertson, A. (1924) *Ann. trop. Med. Parasit.* **18,** 157.

Robinson, L. B. and Wichelhausen, R. H. (1946) *Johns Hopk. Hosp. Bull.* **79,** 436.

Rosahn, P. D. (1952) *Arch. Derm. Syph., Chic.* **66,** 547.

Rosahn, P. D., Gueft, B. and Rowe, C. L. (1948) *Amer. J. Syph.* **32,** 327.

Rose, N. R. and Morton, H. E. (1952) *Amer. J. Syph.* **36,** 17.

Rosebury, T., Clark, A. R., MacDonald, J. B. and O'Connell, D. C. (1950) *J. infect. Dis.* **87,** 234.

Ryter, A. and Pillot, J. (1965) *Ann. Inst. Pasteur* **109,** 552.

Sakharoff, N. (1891) *Ann. Inst. Pasteur* **5,** 564.

Sandok, P. L., Jenkin, H. M., Graves, S. R. and Knight, S. T. (1976) *J. clin. Microbiol.* **3,** 72.

Sandok, P. L., Jenkin, H. M., Matthews, H. M. and Roberts, M. S. (1978) *Infect. Immun.* **19,** 421.

Sandok, P. L., Knight, S. T. and Jenkin, H. M. (1976) *J. clin. Microbiol.* **4,** 360.

Sanford, J. P. (1976) In: *The Biology of Parasitic Spirochetes,* p. 307. Ed. by R. C. Johnson, Academic Press, New York.

Schaudinn, F. and Hoffmann, E. (1905) *Arb. Reichsgesundh Amt.* **22,** 527.

Scheff, G. (1935) *Zbl. Bakt.* **134,** 35.

Schereschewsky, J. (1909) *Dtsch. med. Wschr.* **35,** 835, 1260, 1652.

Schmidt, K. (1955) *Zbl. Bakt.* **162,** 280.

Schuberg, A. and Schlossberger, H. (1930) *Klin. Wschr.* **9,** 499.

Schüffner, W. and Mochtar, A. (1927) *Zbl. Bakt.* **101,** 405.

Séguin, P. and Vinzent, R. (1938) *Ann. Inst. Pasteur* **61,** 255.

Selbie, F. R. (1943) *Brit. J. exp. Path* **24,** 150.

Sellards, A. W. (1940) *Trans. [R.] Soc. trop. Med., Hyg.* **33,** 545.
Sepetjian, M., Tissot Guerraz, F., Salussola, D., Thivolet, J. and Monier, J. C. (1969) *Bull. Wld. Hlth Org.* **40,** 141.
Sequeira, P. J. L. (1956) *Lancet* **ii,** 749.
Shenberg, E. (1967) *J. Bact.* **93,** 1598.
Siefert, G. (1958) *Arch. Mikrobiol.* **29,** 406; (1960) *Z. ImmunForsch.* **119,** 120.
Siefert, G. and Steiger, I. (1959) *Z. ImmunForsch.* **117,** 147.
Skerman, V. B. D., Mc-Gowan, V. and Sneath, P. H. A. (1980) *Int. J. syst. Bact.* **30,** 225.
Smibert, R. M. (1974) In: *Bergey's Manual of Determinative Bacteriology*, 8th Edn, p. 175. Ed. by R. E. Buchanan and N. E. Gibbons, Williams and Wilkins, Baltimore; (1976) In: *The Biology of Parasitic Spirochetes*, p. 49. Ed. by R. C. Johnson. Academic Press, New York.
Smith, J. L. and Pesetsky, B. R. (1967) *Brit. J. vener. Dis.* **43,** 117.
Smith, P. J. C. (1960) *Biochem. J.* **76,** 500, 508, 514.
Socransky, S. S., Listgarten, M., Hubersack, C., Cotmore, J. and Clark, A. (1969) *J. Bact.* **98,** 878.
Songer, J. G., Kinyon, J. M. and Harris, D. L. (1976) *J. clin. Microbiol.* **4,** 57.
Steinman, H. G., Eagle, H. and Oyama V. I. (1952) *J. Bact.* **64,** 265.
Steinman, H. G., Oyama, V. I. and Schulze, H. O. (1954) *J. Bact.* **67,** 597.
Stimson, A. M. (1907) *Publ. Hlth Rep., Wash.* **22,** 541.
Stoenner, H. G. (1974) *Appl. Microbiol.* **28,** 540.
Stokes, A. and Ryle, J. A. (1916) *Brit. med. J.* **ii,** 413.
Stokes, A., Ryle, J. A. and Tytler, W. H. (1917) *Lancet* **i,** 142.
Stuart, R. D. (1946) *J. Path. Bact.* **58,** 343.
Swain, R. H. A. (1955) *J. Path. Bact.* **69,** 117.
Szewczuk, A. and Metzger, M. (1970) *Arch. Immunol. Ther. exp.* **18,** 643.
Tauber, H., Cannefax, G. R., Hanson, A. W. and Russell, H. (1962) *Exp. Med. Surg.* **20,** 324.
Taylor, D. J. and Alexander, T. J. (1971) *Brit. vet. J.* **127,** 58.
Taylor, J. and Goyle, A. N. (1931) *Indian med. Res. Memoirs* No. 20.
Theiler, M. (1926) *Amer. J. trop. Med.* **6,** 131.
Tilden, E. B. (1937) *J. Bact.* **23,** 307.
Tomizawa, T. and Kasamatsu, S. (1966) *Jap. J. med. Sci. Biol.* **19,** 305.
Tunnicliff, R. (1906) *J. infect. Dis.* **3,** 148.
Turek, J. J. and Meyer, R. C. (1978) *Infect. Immun.* **20,** 853.
Turner, T. B. (1938) *J. exp. Med.* **67,** 61.
Turner, T. B. and Hollander, D. H. (1954) *Amer. J. Syph.* **38,** 371; (1957) *Biology of the Treponematoses. World Health Organisation Monograph Series.* No. 35.
Uhlenhuth, P. and Fromme, W. (1915) *Med. Klinik* **44,** 1202; (1916*a*) *Berl. klin. Wschr.* **53,** 269; (1916*b*) *Z. ImmunForsch.* **25,** 317.
Uhlenhuth, P. and Mulzer, P. (1913) *Arb. ReichsgesundhAmt.* **44,** 307.
Vaisman, A., Paris-Hamelin, A. and Dunoyer, F. (1967) *Bull. Wld Hlth Org.* **36,** 339.
Viaseleva, S. M. (1957*a*) *J. Microbiol., Epidem., Immunobiol.* **28,** 57; (1957*b*) *Ibid.* **28,** 573.
Vincent, H. (1896) *Ann. Inst. Pasteur* **10,** 488; (1899) *Ibid.* **13,** 609.
Vinzent, R. and Daufresne, M. (1934) *C. R. Soc. Biol.* **116,** 490.
Wallace, A. L. and Harris, A. (1967) *Bull. Wld Hlth Org.* **36,** Supplement 2.
Walter, E. K., Smith, J. L., Israel, C. W. and Gager, W. E. (1969) *Brit. J. vener. Dis.* **45,** 6.
Weber, M. M. (1960) *Amer. J. Hyg.* **71,** 401.
Weil, A. (1886) *Dtsch. Arch. klin. Med.* **39,** 209.
Wenyon, C. M. (1926) *Protozoology* **ii,** Baillière, Tindall & Cox, London.
Wichelhausen, O. W. and Wichelhausen, R. H. (1942) *J. dent. Res.* **21,** 543.
Wiegand, S. E., Strobel, P. L. and Glassman, L. H. (1972) *J. invest. Derm.* **58,** 186.
Willcox, R. R. and Guthe, T. (1966) *Bull. Wld Hlth Org.* **35,** Supplement.
Wolbach, S. B. and Binger, C. A. L. (1914) *J. med. Res.* **80,** 23.
Wolff, J. W. and Broom, J. C. (1954) *Docum. Med. geograph. trop.* **6,** 78.
Wylie, J. A. H. (1946) *J. Path. Bact.* **58,** 351.
Wynns, H. L. and Beck, M. D. (1935) *Amer. J. publ. Hlth.* **25,** 270.
Zeigler, J. A., Jones, A. M., Jones, R. H. and Kubica, A. M. (1976) *Brit. J. vener. Dis.* **52,** 1.

45

Chlamydia

L. H. Collier

Introductory

Definition

Spherical or ovoid gram-negative bacteria that undergo a well defined life cycle in the cytoplasm of their host cells. Have not been grown in artificial media. The non-infective reticulate (or initial) bodies are 600–1000 nm in diameter and divide by fission to form infective elementary bodies 200–

300 nm in diameter. Stain well with Romanowsky stains. Are pathogenic for man and a wide range of animal and avian hosts. G+C content of DNA: 41–45 moles per cent.

Type species. *Chlam. trachomatis*

In 1907, Halberstaedter and von Prowazek, working in Java, described the transmission of trachoma from man to orang-utangs by inoculating their eyes with conjunctival scrapings. In Giemsa-stained epithelial cells from human subjects and apes they observed intracytoplasmic inclusions ('Einschlüssen') containing large numbers of minute particles, now known as elementary bodies. These workers proved to be correct in suggesting that the inclusions represented the causal agent of trachoma; they were, however, mistaken in classifying them between the bacteria and protozoa. Halberstaedter and von Prowazek called the newly discovered microorganisms Chlamydozoa (Gr. χλαμύς, a mantle) because of the blue-staining matrix in which the eosinophilic elementary bodies are embedded. Soon afterwards Lindner (1909*b*, 1910) described another form of inclusion, the 'initial body', which he rightly inferred was an early stage in the development of the mature inclusion.

In 1909, Stargardt and Schmeichler independently described inclusions morphologically identical with those of trachoma in the conjunctival cells of babies with non-gonococcal ophthalmia neonatorum; this infection could be transmitted to baboons, in which the inclusions were again demonstrable (Lindner 1909*a*, Fritsch *et al.* 1910). Similar inclusions were found in cervical scrapings from the mothers of some of these babies (Halberstaedter and von Prowazek 1909) and in the urethra of certain patients with non-gonococcal urethritis (Fritsch *et al.* 1910).

It thus became apparent that trachoma, inclusion conjunctivitis of the newborn and an infection of the adult genital tract were caused by similar or identical infective agents capable of passing filters that retained bacteria (Nicolle *et al.* 1912, Thygeson 1934). This property, coupled with an inability to grow in artificial media, led to the erroneous belief that these agents were viruses.

In 1929–30, widespread outbreaks of psittacosis in a number of countries evoked interest in the causal agent of this infection. In 1930, Levinthal described minute basophilic intracellular particles in tissues from infected parrots; in the same year Coles (1930) and Lillie (1930) reported similar appearances in reticuloendothelial cells from infected birds and human subjects. Very soon afterwards, Bedson and co-workers (1930) proved the aetiological relationship of these particles with psittacosis. Bedson and Bland (1932) then showed that the psittacosis agent had a well defined developmental cycle in the host-cell cytoplasm, during which the large basophilic 'initial bodies' underwent a series of fissions to form the smaller eosinophilic elementary bodies that are the infective form of the microorganism. With great perspicacity Bedson referred to the psittacosis agent as 'an obligate intracellular parasite with bacterial affinities', a concept that was not to be generally accepted for another 30 years. In due course it was grown in the chick-embryo chorioallantoic membrane (Burnet and Rountree 1935) and yolk sac (Yanamura and Meyer 1941).

At about this time the causal agent of lymphogranuloma venereum (LGV), a sexually transmitted infection of man, was passaged serially in the brain of monkeys (Hellerström and Wassén 1930) and mice (Levaditi *et al.* 1931) and then in the chick-embryo chorio-allantoic membrane (Miyagawa *et al.* 1935) and yolk sac (Rake *et al.* 1940).

As early as 1934, Thygeson drew attention to morphological resemblances between the agents of trachoma and inclusion conjunctivitis—as seen in infected conjunctival cells—and that of psittacosis; and during the next decade Thygeson, Rake and others developed the thesis that these agents and those of lymphogranuloma venereum and mouse pneumonitis formed a unique group of infective agents, often referred to as 'atypical viruses'. Observations by light microscopy suggested that all these agents underwent similar life cycles in the host cell cytoplasm, and the notion of their relationship was further strengthened by the finding of a common complement-fixing antigen (Rake *et al.* 1942).

The psittacosis and LGV agents were cultivated in chick embryos not long after their discovery: but those causing trachoma and inclusion conjunctivitis, although the first to be observed microscopically, were the last to be propagated in the laboratory. In 1957 however, T'ang and his colleagues isolated agents similar to those of psittacosis and LGV from trachoma patients in Peking by inoculating conjunctival material into the chick-embryo yolk sac. The Chinese workers' success, where many others had failed, was largely due to their realization that penicillin—employed to prevent bacterial contamination—also inhibited the growth of the trachoma agent; and that incubation of the chick embryos at 35° gave better results than the more usual temperature of 37°. These findings were soon confirmed and extended by Collier and Sowa (1958), who isolated similar agents from trachoma patients in The Gambia, demonstrated by electron microscopy their similarity to those causing psittacosis and LGV and showed them to contain the group complement-fixing antigen. Their aetiological relationship with trachoma was proved by inoculation of volunteers (Collier *et al.* 1958, 1960). In 1959, Jones and co-workers isolated apparently identical agents in London from a neonate with inclusion conjunctivitis and from the uterine cervix of the mother of such an infant. Their aetiological relationship with inclusion conjunctivitis was confirmed by inoculation into the eyes of volunteers and baboons.

Taxonomy and nomenclature

The classification and nomenclature of the chlamydiae were for long unsettled and subject to the introduction of designations that failed to conform with accepted rules. These organisms have been referred to, *inter alia*, as *Miyagawanella*, *Bedsonia* and colloquially as the psittacosis-lymphogranuloma-trachoma (PLT) group. Page (1966) proposed an amended and now generally accepted description of the genus *Chlamydia* (Jones *et al.* 1945). Buchanan and Gibbons (1974) classified *Chlamydia* with the 'Rickettsias'; they are however separated from the true rickettsiae (Order Rickettsiales) by several criteria, in particular, differences in metabolic activities. The Order Chlamydiales contains one family, Chlamydiaceae, and one genus, *Chlamydia*; there are two species, *Chlam. trachomatis* and *Chlam. psittaci*.

Chlam. trachomatis strains isolated from *tr*achoma, *i*nclusion *c*onjunctivitis or associated infections of the genital tract are sometimes referred to acronymically as TRIC agents and those from lymphogranuloma venereum as LGV agents. Individual strains are often given designations based on the system of Gear and his colleagues (1963); in full, this follows the formula a/b/c/d, where a = serotype, b = country or area where isolated, c = laboratory and strain number, and d = clinical syndrome. Thus A/WAG/MRC-17/OT is a serotype A strain isolated in West Africa, Gambia; the strain number is MRC (Medical Research Council laboratory) 17 and it was isolated from a case of ophthalmic trachoma. The suffixes Cx, U and ON indicate respectively strains isolated from the cervix uteri, urethra and ophthalmia neonatorum.

Relation of *Chlamydia* to other microorganisms

A major contribution to our understanding of *Chlamydia* was made by Moulder (1964), who marshalled compelling evidence for regarding these agents as bacteria and not as 'atypical viruses'. Chlamydiae contain both RNA and DNA, maintain their physical identity throughout their growth cycle, possess cell walls, undergo binary fission and are susceptible to a range of antimetabolites that do not affect viruses; they can also undertake some biosynthetic activities. In short, there is a fundamental discontinuity between the chlamydiae and even the largest viruses. Although the chlamydiae and rickettsiae differ in their morphology and metabolism, it seems likely that both evolved from free-living bacteria which became well adapted to an intracellular environment, losing in the process the ability to perform many of the metabolic functions undertaken by extracellular microbes. Moulder (1964) refuted the notion that such organisms represented a form of degenerative evolution in these terms:

> 'Intracellular parasites are often looked down on as metabolic weaklings lacking something required for growth outside of the cells, while few stop to think that their more rebust extracellular cousins lack something required for life inside the cell.'

The identity of the putative ancestral bacteria must remain a matter of speculation; but it is perhaps worth noting that members of the genus *Acinetobacter* (Chapter 32) possess antigens that fix complement with antibodies to chlamydiae (see Storz 1971, p. 30).

Relation between *Chlam. trachomatis* and *Chlam. psittaci*

Although members of the two species are closely similar in morphology and mode of replication and possess a common complement-fixation antigen, only 10 per cent or less of their DNA sequences are homologous (Kingsbury and Weiss 1968). The main differences between the species are listed in Table 45.1.

Table 45.1. Differences between *Chlam. trachomatis* and *Chlam. psittaci*

	Chlam. trachomatis	*Chlam. psittaci*
Inclusions		
Morphology	Compact, rigid	Vacuoles diffused throughout cytoplasm
Contain glycogen	Yes	No
Sensitive to sulphonamides	Yes	No
G+C content of DNA (average)*	45	41
Predominant hosts	Man	Birds and animals
Main syndromes	Infections of the eye and genital tract, mainly localized	Mainly generalized infections involving lungs, joints, CNS, gut, placenta

* Moles per cent.

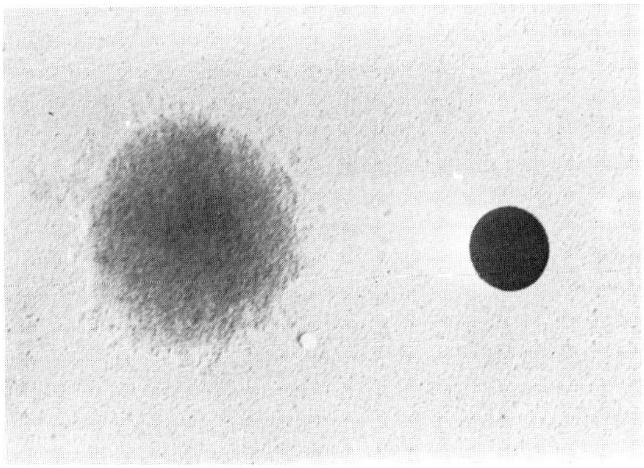

Fig. 45.1. Reticulate body of *Chlam. trachomatis*, palladium-shadowed. The latex sphere (right) is 260 nm in diameter. (Electron micrograph by R. Valentine.)

Fig. 45.3. Purified cell envelopes of *Chlam. psittaci* elementary bodies, electron micrograph, negative contrast stain (× 32 000). (Kindly provided by Dr G. P. Manire)

Fig. 45.2. Elementary bodies of *Chlam. trachomatis* palladium-shadowed (× 30 000). (Electron micrograph by R. Valentine)

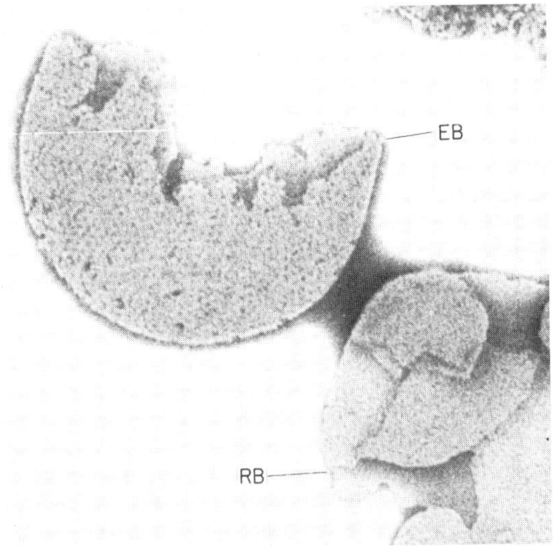

Fig. 45.4. Cell membranes of *Chlam. psittaci* reticulate body (RB) and elementary body (EB). The latter has a hexagonal array of subunits. Electron micrograph, negative contrast stain (× 70 000). (Kindly provided by Dr G. P. Manire)

There are also important serological differences; even so, the resemblances between *Chlam. trachomatis* and *Chlam. psittaci* outweigh their dissimilarities. For comprehensive reviews see Storz (1971) and Schachter and Dawson (1978). The former book provides much detailed information about chlamydiae of veterinary interest, i.e. *Chlam. psittaci*; for collected reports of research on *Chlam. trachomatis* see Symposium (1962, 1967, 1971, 1977).

Morphology and staining properties

Electron microscopy

The reticulate body (RB) This is the non-infective basophilic 'initial body' of Lindner (1909*b*). In shadowed preparations of free particles, the flattened RB measure 1000–1500 nm across and are comparatively featureless (Fig. 45.1). In thin sections of inclusions, they are 500–1000 nm in diameter with a reticulate internal structure surrounded by what appear to be a cell wall and a cytoplasmic membrane (Tamura *et al.* 1971).

The elementary bodies (EB) are derived from the RB by a series of fissions. In shadowed preparations they appear as round particles 200–300 nm in diameter containing an irregular electron-dense central area (the nucleoid) surrounded by a collapsed membrane (Fig. 45.2). By density-gradient centrifugation and treatment with enzymes and sodium dodecylsulphate (Manire 1966) these membranes can be isolated (Fig. 45.3). Their structure resembles that of a gram-negative bacterial cell wall; on the inside surface there is a hexagonally disposed array of subunits (Fig. 45.4). In thin sections of inclusions, a heterogeneous collection of forms intermediate between RB and EB can be seen, including some undergoing fission (Fig. 45.5). In such preparations, the mature EB is seen to possess a well defined cell wall and an internal electron-dense area containing particles with the appearance of ribosomes.

Light microscopy

The chlamydiae are gram negative. They stain blue by Castañeda's method and red by those of Macchiavello and Giménez; the latter two stains or Giemsa are most often used to demonstrate the organisms in impression preparations of the chick-embryo yolk sac, whereas Giemsa is the stain of choice for inclusions in cell cultures. In Giemsa-stained preparations viewed by dark-field microscopy, the EB stand out as brilliant yellow particles; this technique is useful both for detecting inclusions in cell cultures and for counting EB in suspensions (Reeve and Taverne 1962*a*). Giemsa staining also has the advantage of distinguishing between the small reddish-purple EB—which, measuring 300 nm in diameter or less, are just at the limit of resolution of the light microscope—and the larger, more basophilic RB.

The staining reactions with Giemsa suggest that the EB contain DNA and the RB a preponderance of RNA; this is confirmed by acridine orange staining with which these particles stain respectively yellow-green and orange (Pollard and Tanami 1968).

In intact mature *Chlam. trachomatis* inclusions, the elementary bodies are surrounded by a matrix containing glycogen, which stains red with PAS and coppery-brown with iodine. The iodine method is a simple and rapid way of demonstrating inclusions in

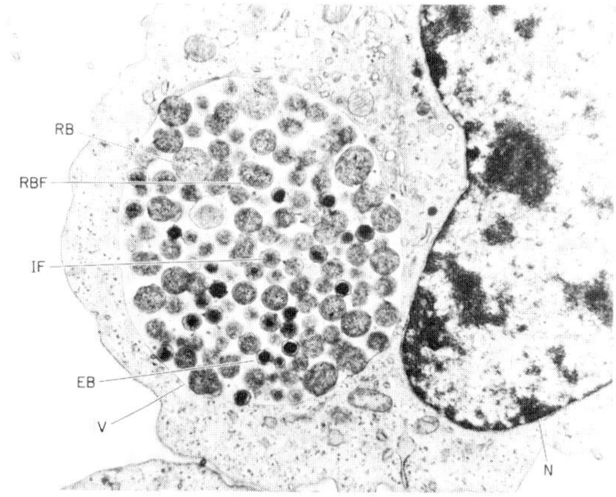

Fig. 45.5. Inclusion of *Chlam. psittaci* 32 hr after infection. Electron micrograph of thin section (× 4300). RB = reticulate body; RBF = RB undergoing fission; IF = intermediate form; EB = mature elementary body; V = inclusion vacuole; N = cell nucleus (From G. P. Manire, 1980: In *Non-gonococcal urethritis and related infections*, p. 172, by courtesy of the American Society of Microbiology)

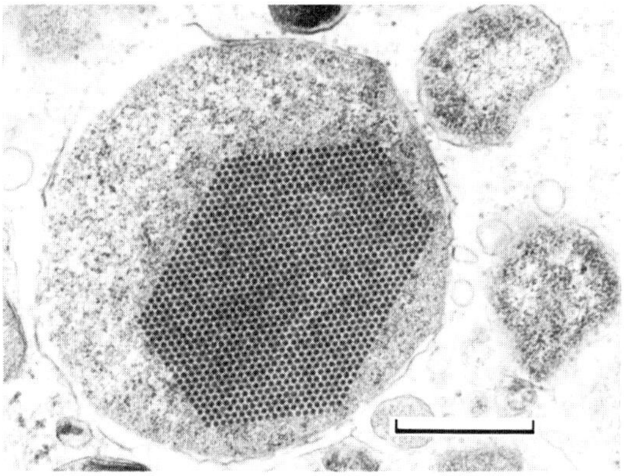

Fig. 45.6. Thin section of part of a chlamydial inclusion 40 hr after infection of McCoy cells with *Chlamydia psittaci* showing distended reticulate body with crystalline array of virions about 20 nm in diameter. Bar marker = 500 nm. (Electron micrograph by courtesy of Mrs P. Stirling)

conjunctival scrapings or in cell cultures; it is of no value for impression smears of yolk sac, in which the inclusions are disrupted.

Chlamydiophages Objects resembling bacteriophages have been seen in chlamydiae. By electron microscopy Harshbarger and co-workers (1977) found structures similar to chlamydiae in marine molluscs; some of the structures that resembled RB contained crystalline arrays of phage-like particles 50 nm in diameter. Similar but smaller (22 nm) particles were reported by Richmond and her colleagues (1982) in RB of two strains of *Chlam. psittaci* isolated in England from ducks (Fig. 45.6). Assessment of the implications of these findings, which may be important, awaits further investigation.

Interactions with host cells

Replication

The way in which chlamydiae replicate has been largely elucidated by a variety of techniques, including light and electron microscopy, infectivity assays, the use of antimetabolites, and chemical analysis of organisms purified at various stages (see Storz 1971). It should be borne in mind that some variations in the interactions between host cell and parasite may occur, depending on the type of cell and the particular chlamydia (see Schachter and Caldwell 1980).

The first event is the attachment of an infective elementary body (EB) to a host cell by means that are not fully understood (see Pearce *et al.* 1981). The cell receptor sites are destroyed by trypsin and appear to be protein in nature; those attaching some strains of *Chlam. trachomatis* are inactivated by neuraminidase. The EB is then phagocytosed; some cells—but not macrophages—phagocytose EB selectively, and this suggests that EB have a surface property that favours their uptake by the host cell. After ingestion, EB evade digestion by lysosomal enzymes by a mechanism similar to that of certain other non-viral intracellular parasites: the phagosomes containing the organisms do not for some reason become fused with lysosomes, at least until late in the reproductive cycle (Friis 1972, Lawn *et al.* 1973). Replication of the organism is confined to the cytoplasm of the host cell and takes place within the phagosome. During the 8 hr following its ingestion, the EB becomes reorganized into a larger non-infective body (RB). This is the vegetative form within which, during the next 12 hr, synthesis of RNA, DNA and structural components takes place. During this period, the RB undergo a series of fissions (without the formation of cross-walls) to form smaller particles with an increased content of DNA. By 20 hr, some small dense-centred particles are in evidence and eventually greatly outnumber the RB and intermediate forms within the inclusion. These are the new infective EB which are eventually released from the host cell, either over a period or by sudden rupture of the inclusion. The process is not synchronous, and for much of the growth cycle typical RB, EB and intermediate forms can be seen together (Fig. 45.5). (Figure 45.7 is an unusual picture showing four stages in the formation of *Chlam. trachomatis* inclusions within the cytoplasm of two conjunctival epithelial cells.) The replication cycle is depicted schematically in Fig. 45.8 (Storz 1971). (See pp. 516–17 for Figs 45.8 and 45.9.)

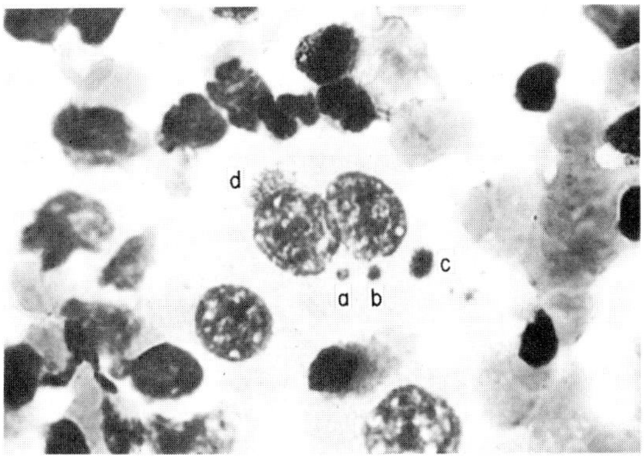

Fig. 45.7. Conjunctival inclusions stained Giemsa (× 1000). Four stages of development are seen, within two cells, from initial body (a) to mature inclusion (d).

Figure 45.9 (Furness *et al.* 1962) is a one-step growth curve in HeLa cells, and reflects the development of infectivity within the inclusions throughout the growth cycle. After adsorption, there is a rapid loss of infectivity followed by a period during which none is detectable; this behaviour superficially resembles the eclipse phase of viruses, but the analogy is invalid because chlamydiae retain their physical identity throughout the replication cycle. Between 16 and 20 hr, infectivity is again detectable and rises to a peak at about 38 hr, after which time infective EB start to be released into the supernatant fluid. In this experiment, the content of RNA and DNA at various points in the cycle was assessed by acridine-orange staining.

Effects on the host cell

These are reviewed by Storz (1971) and Schachter and Caldwell (1980). In brief, infection may interfere with the synthesis of host-cell nucleic acids and with progress from the G1 to the S phase of cell division; in logarithmically growing cells, protein synthesis is depressed. It seems clear, however, that at least *in vitro*, the magnitude of the effect on host cell functions varies directly with the multiplicity of infection; and very high doses of chlamydiae are liable to kill host cells before replication of the organism can take place.

Interferon

Live—but not heat-inactivated—chlamydiae induce interferon in cell cultures and in mice given the organ-

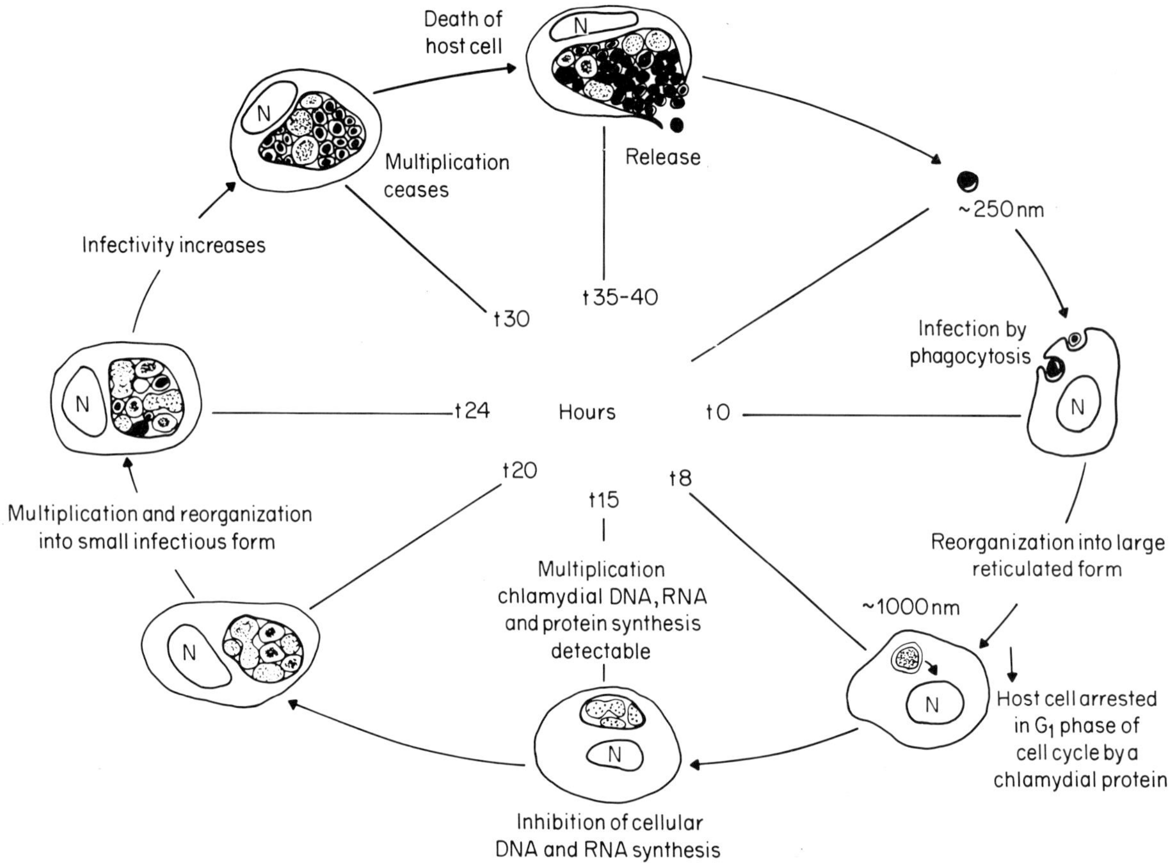

Fig. 45.8. Schematic representation of the developmental cycle of chlamydial agents. N = nucleus; t = time. (From J. Storz, 1971, *Chlamydia and chlamydia-induced diseases*, by courtesy of Charles C Thomas, Springfield, Ill.)

isms intravenously (Merigan and Hanna 1966; Jenkin and Lu 1967). The interferon is of higher molecular weight (50 000) than that induced by viruses, and the amount induced depends on the dose of chlamydiae employed. These findings are not surprising, because other non-viral microbes such as bacteria and rickettsiae also induce interferon. It is of greater interest that chlamydial replication can be inhibited by interferon induced either by chlamydiae or by vesicular stomatitis virus (Hanna *et al.* 1967). The mechanism is not clear and it is not known whether interferon plays any part in immunity to chlamydial infection.

Latency

Inapparent or subclinical infections are common in chlamydial disease (see Moulder 1964; Storz 1971) but the mechanism by which they are maintained *in vivo* is not clear. In the laboratory, the replication cycle may be extended, sometimes for long periods, by depriving cells of essential nutrients or by treatment with inhibitors such as penicillin, sulphonamides or D-cycloserine. If the missing nutrients are restored, or the influence of the antimetabolites is counteracted (by penicillinase, folic acid or D-alanine in these examples) replication resumes. By serial passage of material taken during the early phase of replication, Litwin (1959) was able to transfer pneumonitis agent in the chick-embryo chorio-allantoic membrane indefinitely without the appearance of normal infective EB; these however reappeared when incubation was extended.

Whether any of these laboratory manipulations are relevant to the prolonged maintenance of infectivity in the intact animal seems doubtful; it is more probable that immune processes keep the infection in check. It is not clear whether chlamydiae become truly latent *in vivo*, in that they survive in cells for long periods in a non-replicative form, or whether multiplication continues but at a very low rate.

Metabolic activities and chemical composition

The chlamydiae have many of the features of free-living bacteria and their metabolic processes and structure follow the same general patterns; the main difference is their dependence for certain functions

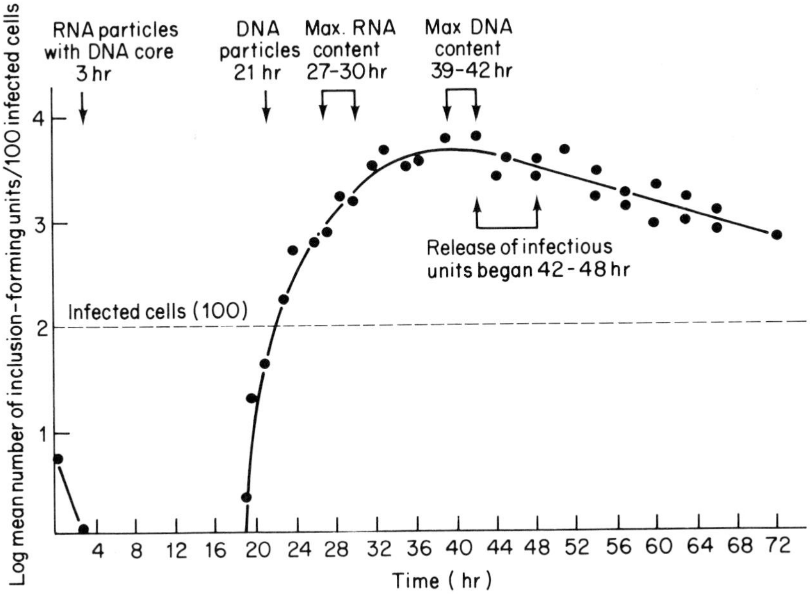

Fig. 45.9. The formation of RNA, DNA and infectious units in HeLa-cell monolayers infected with a strain of *Chlam. trachomatis*. (From Furness *et al.*, 1962, by courtesy of the Society of General Microbiology.)

upon the host cell. The elucidation of chlamydial metabolism thus presents problems similar to those posed by viruses; but its disentanglement from that of the host cell is if anything even more difficult because the metabolism of these prokaryotic microbes more nearly resembles that of their eukaryotic hosts. This topic is well reviewed by Moulder (1964), Storz (1971), Becker (1978), and Schachter and Dawson (1978).

EB are metabolically inactive and are somewhat analogous to bacterial spores; their tough cell membrane enables them to survive outside host cells, at least for limited periods. RB are the intracellular vegetative form within which most replicative functions take place; lacking a rigid cell wall they are much more fragile than EB (Tamura and Manire 1967).

Although chlamydiae possess some enzymes and synthesize a number of compounds that cannot be made by their host cell they have little or no capacity for generating the energy needed for these reactions: in view of this, Moulder (1964, p. 78) aptly referred to them as 'energy parasites'.

Glucose metabolism provides an example of this limitation. Chlamydiae contain enzymes that generate CO_2 from two of the carbon atoms of glucose, including carbon 1 but not carbon 6; there is, however, no evidence that this reaction yields useful energy (Weiss *et al.* 1964). Moulder and co-workers (1965) showed that the meningopneumonitis agent degrades glucose by the pentose pathway and concluded that it can synthesize the necessary enzymes. Nevertheless, Weiss (1965) found that glucose utilization depends on added ATP and Mg^{++} and is enhanced by adding NADP. Chlamydiae thus have the unusual ability to utilize the energy provided by these highly charged compounds, a property that implies permeability of the vegetative form to molecules that do not normally pass freely into intact microbial cells.

Folic acid One of the criteria for distinguishing *Chlam. trachomatis* from *Chlam. psittaci* is its inhibition by sulphonamides; this inhibition is reversed competitively by *p*-aminobenzoic acid and non-competitively by folic acid. By contrast, *Chlam. psittaci* is fully resistant to sulphonamides. This suggests that *Chlam. trachomatis* strains synthesize their folic acid from *p*-aminobenzoic acid, pteridine and glutamate, whereas sulphonamide-resistant chlamydiae convert host-cell folic acid to the specific forms needed for their own metabolism. It is likely that folic acid functions as a coenzyme in the synthesis of amino acids, purines, pyrimidines and other molecules essential for replication.

Nucleic acids and proteins EB contain RNA and DNA in approximately equal proportions (Tamura and Higashi 1963); the RNA has the characteristics of bacterial ribosomal RNA; 30S and 50S ribosomal subunits and 70S ribosomes can be isolated from RB (Tamura 1967) and EB (Higashi *et al.* 1975). The DNA is a circular molecule and is probably contained in the central nucleoid. Estimates of its molecular weight vary somewhat, but Higashi (1975) and Becker (1978), working respectively with *Chlam. psittaci* and *Chlam. trachomatis* both obtained data suggesting a value of 660×10^6; this is equivalent to 11.9×10^5 nucleotide base pairs. Such a molecule could code for more than 600 proteins, i.e. about 25 per cent of the information encoded in *Esch. coli* DNA (Sarov and Becker 1969).

The much greater degree of nucleic acid and protein synthesis in the host cell makes it difficult to analyse the events taking place in the RB during replication. It appears however that soon after the EB start to transform into RB, early mRNA is synthesized by an endogenous DNA-dependent RNA polymerase and the ribosomes contained in the EB start to make early proteins. DNA is replicated by a DNA-dependent DNA polymerase, some of the nucleotide precursors being provided by the host cell. Studies with *Chlam. psittaci* suggest that chlamydiae possess their own enzymes for a series of reactions resulting in the synthesis of thymidine from uridine provided by the host cell. Synthesis of DNA, various RNA species and proteins proceeds actively in the newly formed RB which in due course start to divide by fission. By 20 hr after infection, the RB contain 3–4 times more RNA than DNA, and ribosomal subunits and 70S ribosomes can be isolated from them. The rate of protein synthesis increases toward the end of the cycle, when progeny EB are being formed from the dividing RB.

Cell-wall components

Muramic acid The observation that penicillin interferes with the transformation of RB into EB suggested that, like free-living bacteria, chlamydiae might possess a cell wall containing peptidoglycan. Since the muramic acid component of peptidoglycan is peculiar to prokaryotes, the question whether it is present in *Chlamydia* is of some taxonomic significance. Perkins and Allison (1963) detected muramic acid in several chlamydiae, but Manire and Tamura (1967) failed to find it in intact EB or in purified cell walls of *Chlam. psittaci* (meningopneumonitis agent). The work of Garrett and his colleagues (1973) suggested the presence of traces of muramic acid in a strain of *Chlam. trachomatis*; but they concluded that its concentration in the cell wall could not exceed 0.04 per cent, a value much less than in bacteria. The balance of evidence suggests that chlamydiae do contain small quantities of muramic acid, but that peptidoglycan does not play an important part in conferring rigidity upon the chlamydial cell wall. Penicillin inhibits transpeptidation of bacterial peptidoglycans; Garrett and co-workers (1973) suggested that it might affect chlamydial replication by inhibiting the cross-linking of peptides that are not attached to a polysaccharide containing muramic acid.

D-alanine The formation of mature EB from RB is prevented by D-cycloserine, the effect on *Chlam. trachomatis* being greater than on *Chlam. psittaci*. D-cycloserine inhibits reactions in which D-alanine is the substrate, and its effect on chlamydiae is competitively reversed by the amino acid. Since D-alanine is concerned in the final transpeptidation reaction whereby the components of bacterial peptidoglycan are cross-linked, the effect of D-cycloserine infers that this amino acid is also involved in formation of the EB cell wall; indeed, the production of abnormally large RB in the presence of D-cycloserine and their failure to develop into EB is strikingly similar to the effect of penicillin.

Lipids Chlamydiae contain 30–50 per cent of lipid; the content of phospholipid is about 7.5 per cent, the major components being phosphatidyl cholamine and phosphatidyl ethanolamine. Branched-chain fatty acids not found in host cells are a characteristic feature (Jenkin *et al.* 1970). The lipid components are contained in the EB cell wall and possess haemagglutinating activity.

Inhibition by antibiotics and other antimetabolites

Drugs inhibiting chlamydiae fall into three categories, according to whether they interfere with folic acid synthesis, protein synthesis or cell-wall formation. The susceptibility of *Chlam. trachomatis* and resistance of *Chlam. psittaci* to sulphonamides has already been mentioned. With this important exception it is generally true that all chlamydiae possess the same order of sensitivity or resistance to a given antimicrobial drug. On the other hand, antibiotics with similar modes of action may differ significantly in their ability to inhibit chlamydial replication, a finding perhaps explicable in terms of permeability differences in the chlamydiae or host cells.

Ridgway and co-workers (1978) tested a range of antibiotics and other drugs against 11 strains of *Chlam. trachomatis* and divided them into groups with high, medium and low activity; the minimal inhibitory concentrations (MIC) were respectively <0.1, 0.25–4.0 and ≥ 64 mg/l. The activities of the drugs now to be considered are assigned on this basis.

Inhibitors of folic acid synthesis The sulphonamides inhibit *Chlam. trachomatis* by competitively preventing the utilization of *p*-aminobenzoic acid. The only *Chlam. psittaci* agent susceptible to sulphonamides is an atypical and long-established laboratory strain, 6BC. In cell cultures infected with a susceptible strain and treated with a sulphonamide any inclusion vacuoles formed are abnormally small, and contain few or no chlamydial particles; this appearance is quite different from that of the giant RB induced by penicillin. The sulphonamides possess medium inhibitory activity.

Inhibitors of protein synthesis The antibiotics in this category with the highest antichlamydial activity are rifampicin (MIC 0.007 mg/l) followed by the tetracyclines and erythromycin (MIC 0.03–0.06 mg/l). The tetracyclines are the antibiotics of choice for therapy, whether by topical application to the conjunctiva, as in the mass treatment of trachoma, or by mouth. Chloramphenicol possesses medium activity against chlamydiae.

Inhibitors of cell-wall synthesis Penicillin prevents

the maturation of EB by interfering with cell-wall formation. Its effects on a *Chlam. psittaci* agent (meningopneumonitis) were as follows (Tamura and Manire 1968, Matsumoto and Manire 1970): there was no apparent effect during the first 12 hr after infection, and EB were converted normally to RB; thereafter, RB failed to undergo fission and developed into large abnormal forms with an excess of membrane material. The chemical composition of the RB envelopes was not, however, altered by exposure to penicillin, and removal of the antibiotic 20 hr after infection permitted normal maturation of infective EB to resume. This may partly explain why penicillin is less effective *in vivo* than certain other antibiotics. The penicillins and cephaloridine possess only medium antichlamydial activity and are not used for therapy.

Drug-resistant mutants

Resistance to all three categories of antichlamydial drugs can be induced in the laboratory, but drug resistance has not so far presented problems in clinical practice. Strains of chlamydiae made resistant to sulphonamides probably acquire the ability to utilize host-cell folic acid; the mechanisms whereby resistance to inhibitors of proteins and cell wall synthesis is induced are not known.

Drugs not inhibiting chlamydiae

Chlamydiae are highly resistant to a number of antibiotics, notably the aminoglycosides, polymyxins, lincomycin, vancomycin and mycostatin. Some of these antibiotics, alone or in combination, are thus useful for suppressing contamination by bacteria, fungi and yeasts during the isolation of chlamydiae from clinical specimens or during routine passage in the laboratory.

Resistance to physical and chemical agents

RB lack a cell wall and hence are more readily disrupted by freezing or by ultrasonic treatment than are EB. Since only the latter are infective, studies of stability under various conditions have been confined almost entirely to EB. The results obtained by various workers depend on the state of purification of their test suspensions and it is here possible only to give a rather general account of representative experiments.

Heat and cold In general, suspensions of chlamydiae lose infectivity within hours at 35–37°. Survival of purified suspensions can be improved by using a buffer at pH 7.2 containing 0.4M sucrose and 0.2% bovine albumen. At 4°, some infectivity persists for periods measured in days, but purified organisms deteriorate more rapidly than crude suspensions. Infectivity is maintained for months or years at subzero temperatures, and many laboratories store chlamydiae at −70° or in liquid nitrogen. Freeze-drying gives variable results and is not widely used.

(For descriptions of the stability of *Chlam. trachomatis* at various temperatures and in different media, see T'ang *et al.* 1957; Jawetz and Hanna 1960; Weiss and Dressler 1967; Collier 1962*b*.)

Chemicals Nabli and Tarizzo (1967) reported the action of a range of chemicals—including a number of antiseptics in common use—on crude yolk-sac suspensions of a strain of *Chlam. trachomatis*. Ether, alcohols, iodine, potassium permanganate, sodium hypochlorite and silver nitrate at concentrations used for disinfection all destroyed or greatly reduced infectivity at room temperature within 1 min; the action of 1 per cent phenol was less rapid.

Survival in water and on fomites Nabli and Tarizzo (1967) found that infectivity could still be demonstrated after suspension in chlorinated swimming-pool water (amount of chlorine not stated) for 72 hr, in tap water for 24 hr and in sea water for 1 hr, all at room temperature. Cotton cloth lightly contaminated with *Chlam. trachomatis* remained infective after 45–90 min at room temperature (Sowa *et al.* 1965).

These experiments show that although chlamydiae do not generally retain their infectivity for long periods outside their host cells, recently infected water and fomites could serve as sources of infection. *Chlam. psittaci* strains may survive quite well at ambient temperatures in dust, feathers and faeces, a factor of obvious importance in the spread of infection.

Propagation in the laboratory

Chick embryos

Yolk sac All chlamydiae grow well in the yolk sac of 6–8-day-old chick embryos. They multiply in the endothelial cells and can be demonstrated in tissue sections or, more conveniently, in impression preparations stained by the Giemsa, Giménez or Macchiavello methods; Castañeda's stain is less satisfactory. After inoculation, the relative humidity should be maintained at 50–60 per cent. All *Chlam. trachomatis* strains grow at 35° and this temperature should be used for isolating them from clinical material (T'ang *et al.* 1957). Some laboratory strains can be propagated at 37°. *Chlam. psittaci* strains, on the other hand, may grow better at 39° (Page 1965). Jawetz (1962) and others described seasonal insusceptibility to chlamydiae of chick embryos inoculated by the yolk-sac route; the reason remains obscure.

After inoculation, chick embryos die from 3–4 days onward, the time depending on the strain and dose. The yolk sac is abnormally thin-walled and hyperaemic; the embryos are also hyperaemic in appearance and their feet may show developmental abnormalities. Deaths are probably caused by a toxin elaborated by the replicating organisms. After repeated passage in yolk sacs, some strains of *Chlam. trachomatis* acquire the ability to kill chick embryos more rapidly with a given dose (Reeve and Taverne 1963; 1967 *a*, *b*). This change, which is probably due to the emergence of a mutant, is accompanied by some or all of the following: a higher ratio of infective to total particles; increased ability to grow in cell culture; loss of pathogenicity for the primate conjunctiva; and antigenic changes. Such altered strains are designated *f* (for 'fast killing').

Yolk-sac inoculation has been much used to isolate chlamydiae from clinical material but is now largely superseded by cell culture methods. It is however valuable for preparing antigens used in serological tests.

Infectivity titrations can be performed by inoculating groups of chick embryos with serially diluted suspensions of chlamydiae; the time from inoculation to death may be used as a measure of the infective titre, but determinations of the 50 per cent infective or lethal end-points are more accurate.

Chorio-allantoic membrane Some chlamydiae will grow on this tissue with the production of small pock-like lesions; adaptation procedures may be necessary and growth is usually irregular.

Allantoic cavity Certain strains have been adapted to grow in the allantoic cavity; the method is useful for obtaining relatively pure suspensions for research purposes.

Cell cultures

Chlamydiae can be grown in a variety of cells from chick embryos and mammals; *Chlam. psittaci* has been propagated in fish and lizard cells (see Storz 1971, p. 82). Some strains induce a virus-like cytopathic effect, and *Chlam. psittaci* forms plaques; but the formation of cytoplasmic inclusions is by far the most commonly used criterion of infection and is widely employed both for isolating these agents and for studying their properties. Stained with Giemsa, mature inclusions appear as vacuoles containing many eosinophilic EB; some RB or intermediate forms may also be present. Under dark-field illumination, the EB show brilliant yellow autofluorescence. Mature *Chlam. trachomatis* inclusions contain glycogen and thus stain coppery-brown with iodine. In *Chlam. trachomatis* inclusions, the EB are contained in a single vacuole (Fig. 45.10); this is true even of cells infected with more than one EB, in which several early inclusions coalesce by 30 hr after infection (Blyth and Taverne 1972). The *Chlam. trachomatis* inclusions are rigid vacuoles that can displace the cell nucleus; by contrast, those of *Chlam. psittaci* rupture early during development and the organisms multiply throughout the cytoplasm (Fig. 45.11).

The readiness with which chlamydiae form inclusions depends greatly on the strain, type of cell and conditions of inoculation. For example, some laboratory strains form inclusions after simple inoculation of a cell monolayer whereas others, notably recent isolates, need special techniques for increasing the efficiency of infection; depending on the conditions, the ratio of infective to total particles may vary from 1: > 10 000 to a value approaching unity (Reeve and Taverne 1967*b*).

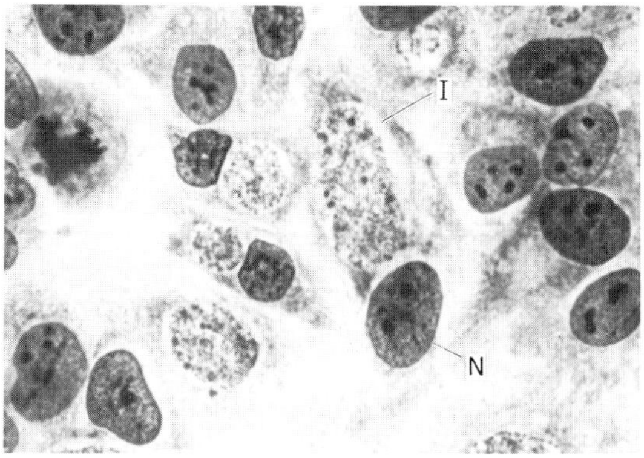

Fig. 45.10. Single *Chlam. trachomatis* inclusions in BHK-21 cells. Stained Giemsa (× 1100). I = inclusion; N = cell nucleus.

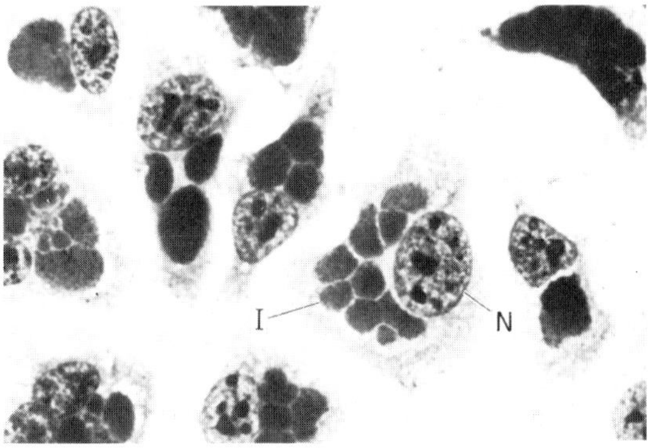

Fig. 45.11. Multiple, densely-stained *Chlam. psittaci* inclusions in irradiated McCoy cells. I = inclusion; N = cell nucleus. (Kindly provided by Dr J. H. Pearce.)

Centrifugation of chlamydial suspensions into monolayers of suitable cells at about 3000 g considerably increases the number of cells becoming infected; this appears to be due to a combination of increased pressure and directional force, with consequent changes in the cell surface (Pearce *et al.* 1981). *DEAE-dextran* treatment of the cell monolayer increases its susceptibility to chlamydial infection (Wang *et al.* 1975) probably by altering the surface charge. *Metabolic inhibitors*, such as X-irradiation (Gordon *et al.* 1969), 5-iodo-2-deoxyuridine (Wentworth and Alexander 1974), cycloheximide (Alexander 1968), or cytochalasin B (Sompolinsky and Richmond 1974), increase the susceptibility of cells to infection.

In practice, a combination of centrifugation with one of the cell treatments mentioned above is used to secure the maximum efficiency of infection; the chemical methods are more convenient than irradiation and have superseded it in many laboratories. The choice of cell is also important: the HeLa 229 line of human carcinoma and cells of murine origin are widely used.

Infectivity titrations The enumeration of cells containing inclusions is a rapid method for titrating the infectivity of chlamydiae; with careful control of the experimental conditions a standard error as low as ±10 per cent may be achieved (Furness *et al.* 1960).

Pathogenicity for laboratory animals

The susceptibility of intact mammalian and avian hosts to artificial infection with chlamydiae varies in degree and kind with the species and strain of organism (see Storz 1971). Some laboratory animals and birds are subject to natural infections with chlamydiae which are often inapparent and this possibility must be borne in mind during investigational work.

Mice

Natural infection The only *Chlam. trachomatis* agent that naturally infects a host other than man is one that causes pneumonia in mice (Nigg 1942). This agent has been discovered in many mouse colonies; serial passage of lung tissue by the intranasal route may be needed to detect it. Inapparent lung infections with *Chlam. psittaci* also occur (Ata *et al.* 1971).

Intranasal inoculation Many chlamydiae of either species cause pneumonitis when inoculated intranasally. Acute bronchiolitis with alveolar exudate may occur as early as 24 hr after inoculation; within a few days, greyish or red consolidation is visible, varying from discrete foci to involvement of whole lobes or lungs depending on the severity of infection. The mice are obviously ill and may die within a week or so.

Intraperitoneal inoculation All chlamydiae induce infection by this route, but its severity varies greatly with the strain of organism. Page (1959) characterized virulence in terms of the ratio between the 50 per cent infective and lethal doses. By this criterion, strains isolated from psittacine birds are more virulent than those from poultry. Chlamydiae isolated from mammals are generally of low virulence and pathogenicity.

Intracerebral inoculation The susceptibility of mice by this route is variable, and several 'blind' passages may be needed to establish overt infection, which is manifested as acute meningoencephalitis.

Intravenous inoculation Some strains of both species of *Chlamydia* contain a heat-labile toxin that kills mice 3–24 hr after intravenous inoculation; surviving animals may die several days later as a result of infection, so that a biphasic death curve results from inoculating a group of mice with a suitable dose of organisms. Early deaths due to toxin can be prevented by prior immunization with an antigenically similar strain (Bell *et al.* 1959).

Guinea-pigs

Natural infection A *Chlam. psittaci* agent causing mild conjunctivitis lasting about a month has been detected in a number of laboratory stocks; inclusions are readily demonstrable in conjunctival scrapings. The possibility of guinea-pig inclusion conjunctivitis (GP-IC) infection and its antibody must be considered when these animals are used for experimental work or as a source of complement.

Ophthalmic and urogenital inoculation Guinea-pigs artificially infected with GP-IC in the eye (Murray and Charbonnet 1971) or genital tract (Ozanne and Pearce 1980) have been used as models for the corresponding *Chlam. trachomatis* infections and immune responses in man.

Intraperitoneal inoculation has been used to isolate *Chlam. psittaci* from animals and birds, but this method is now superseded by cell-culture methods. The infection is manifested by peritonitis with lesions in the liver and spleen.

Primates

In view of their almost ubiquitous distribution among mammals, it is surprising that natural infections with neither species of *Chlamydia* seem to occur in apes and monkeys. Nevertheless, these animals are susceptible to experimental infection. They can be infected with LGV by intracerebral inoculation; but by far their most important role has been in connection with research on the trachoma-inclusion conjunctivitis (TRIC) agents. After the isolation of chlamydiae from trachoma (T'ang *et al.* 1957), the induction of follicular conjunctivitis with inclusions was widely used to confirm the identity of TRIC agents isolated from the human eye or genital tract. Baboons and monkeys have also played an important part in studies of the pathogenesis and immunology of infections caused by TRIC agents.

Ophthalmic inoculation Baboons (Collier 1962*a*), rhesus and cynomolgus monkeys (Dawson *et al.* 1962), macacus (Wang and Grayston 1962) and owl monkeys (Bell and Fraser 1969) are, *inter alios*, susceptible to experimental ophthalmic infections by TRIC agents isolated from the human eye or genital tract. The clinical signs vary somewhat with the species of primate and strain of organism; typically, however, acute conjunctivitis is noted 3–7 days after inoculation: it is characterized by oedema of the lid and purulent discharge; eversion of the lid reveals a hyperaemic and oedematous conjunctiva. These signs reach their maximum 10–14 days after inoculation and resolve within a month. Follicular hyperplasia is apparent a week or so after inoculation and may persist for some months. The cytology of conjunctival scrapings resembles that in man: the epithelial cells show degenerative changes; many neutrophil leukocytes and lymphocytes are present and, after the onset of follicular hyperplasia, primitive lymphoid cells appear. Inclusions are most numerous during the first 2 weeks but thereafter become increasingly scanty.

In baboons and monkeys, chlamydiae isolated from patients with inclusion conjunctivitis or infections of the genital tract tend to cause more severe infections than those isolated from ophthalmic trachoma. In previously uninoculated animals, the corneal lesions and conjunctival cicatrization characteristic of trachoma in man do not appear, and the infection runs a benign self-limiting course more characteristic of inclusion conjunctivitis than trachoma, whatever the source of the inoculated organisms. However, Taiwan monkeys (*Macaca cyclopis*) previously inoculated in the eye or given experimental trachoma vaccines may develop pannus after reinoculation of the eye (Wang and Grayston 1967); in two such monkeys, followed for 10 years, all the lesions of trachoma were seen, including pannus, keratitis, conjunctival scarring, trichiasis and entropion (Gale *et al.* 1971). The occurrence of corneal lesions in reinoculated animals was attributed to a hypersensitivity reaction.

Baboons and monkeys are subject to follicular hyperplasia of the conjunctiva, with or without signs of inflammation; the aetiology is unknown. Account must be taken of this factor in interpreting the response to experimental inoculation of chlamydiae.

Inoculation of the genital tract Alexander and Chiang (1967) inoculated the uterine cervices of pregnant *Macaca cyclopis* monkeys with TRIC agents and were able to isolate the micro-organisms within the next 2 weeks from about 25 per cent of them. There was no evidence of overt ophthalmic infection in the nine infants born subsequently and TRIC agent could not be demonstrated in their eyes. Nevertheless, these infants were all susceptible to conjunctival challenge 7 months after delivery and six of them developed pannus, a lesion that was not seen in four infants of uninfected mothers challenged similarly. The authors attributed this finding to sensitization following an inapparent infection acquired at birth.

Inoculation of the urethra of male baboons and macacus monkeys with TRIC agents isolated from the human urogenital tract resulted in follicular urethritis, from which chlamydiae could be isolated for 2–3 months after inoculation (Digiacomo *et al.* 1975).

Antigenic composition

This topic was reviewed by Schachter and Caldwell (1980) and by Lamont and Nichols (1980).

Genus-specific antigens

Chlam. trachomatis and *Chlam. psittaci* share a heat-stable complement-fixing genus ('group') antigen (Bedson 1936; Bedson *et al.* 1949). It is soluble in ether and partly destroyed by periodate; the fractions respectively sensitive and resistant to periodate are a lipopolysaccharide and a protein (Benedict and O'Brien 1956). The antigen is contained in the cell wall and is similar to the lipoprotein-polysaccharide complex in the cell walls of free-living gram-negative bacteria. The serologically active part of the complex is a water-soluble polysaccharide (M.W. $0.2–2.0 \times 10^6$) with an acidic component resembling the 2-keto-3-deoxyoctanoic acid in salmonella lipopolysaccharides (Dhir *et al.* 1972).

The genus-specific antigen is formed early in the replication cycle (Reeve and Taverne 1962*b*) and can be detected on purified RB by immunofluorescence (Yong *et al.* 1979). Being soluble, it diffuses out of the inclusion vacuole and into the host-cell cytoplasm, where it can also be demonstrated by fluorescence microscopy (Richmond 1980). It evokes antibodies detectable by complement fixation, but this test is less sensitive than immunofluorescence for obtaining evidence of infection (Richmond and Caul 1975).

Chlamydial haemagglutination reactions (Gogolak 1954) are mediated by antigen that is present in the cell wall of EB but not RB (Tamura and Manire 1974). It contains both lipid and polypeptide components and may well be identical with the group antigen.

Species-specific antigens

Chlam. trachomatis can be readily distinguished from *Chlam. psittaci* by various serological tests including agglutination of elementary bodies (Bernkopf 1962), immunodiffusion (Katzenelson and Bernkopf 1967), immunofluorescence (McComb and Bell 1967) and complement fixation. Ross and Jenkin (1962) found that some at least of these species-specific activities were associated with the cell wall: after extracting genus-specific antigen by treatment with deoxycholate and trypsin, species-specific antigen, resistant to periodate and heat, remained within the cell wall and was detectable by complement fixation and by reactions with antibodies that neutralized toxicity and infectivity. The toxin lethal for mice is an integral component of the EB cell wall; in experiments with meningopneumonitis agent (*Chlam. psittaci*), Christoffersen and Manire (1969) found it to be absent from RB.

Two-dimensional immunoelectrophoresis of *Chlam. trachomatis* (LGV) and *Chlam. psittaci* (meningopneumonitis) treated with Triton X100 revealed respectively 19 and 16 antigenic components of which only one—presumably the genus-specific antigen—gave reciprocal cross-reactions (Caldwell *et al.* 1975*a*). These authors concluded that some at least of these antigens were membrane-associated. They cited the high degree of antigenic heterogeneity as further evidence for the phylogenetic dissimilarity between the two species of *Chlamydia*.

The same workers used similar techniques to isolate an antigen that reacted specifically with sera from patients with a variety of *Chlam. trachomatis* infections of the eye and genital tract, including LGV. Conversely, a monospecific antiserum to the antigen reacted in indirect immunofluorescence tests with a similar range of *Chlam. trachomatis* agents grown in cell culture. *Chlam. psittaci* strains reacted in neither type of experiment; and it is particularly interesting that the antigen could not be demonstrated in a strain of mouse pneumonitis, the only *Chlam. trachomatis* agent known to occur in a species other than man (Caldwell *et al.* 1975*b*).

Type-specific antigens

This term refers to the antigenic determinants that permit the grouping of strains within a species into serotypes. Although overlapping occurs, serotyping is useful in studies of epidemiology and pathogenesis, particularly with *Chlam. trachomatis* infections, which have been much more extensively studied in this respect than those due to *Chlam. psittaci*.

Chlam. trachomatis

Bell and co-workers (1959) defined two serotypes of trachoma agent on the basis of the ability of formolized antigens to protect mice against toxic death caused by intravenous challenge with a high dose of live elementary bodies. With a similar test, Alexander and his colleagues (1967) examined 80 TRIC agents; 62 of 64 strains from ophthalmic trachoma belonged to one of three serotypes (A, B, C) whereas all 16 strains from infantile inclusion conjunctivitis or from the genital tract fell into three further serotypes (D, E, F). Nichols and McComb (1964) typed *Chlam. trachomatis* strains by direct immunofluorescence and later by the indirect method (McComb and Bell 1967). The good correspondence between the results of the mouse and immunofluorescence (IF) tests suggests that they define the same antigens. The indirect IF technique is however easier to perform and discriminates much more finely than the mouse test (Wang and Grayston 1971, Treharne *et al.* 1972). The micro-method devised by Wang (1971) made possible the serotyping of large numbers of isolates; with the aid of a template several sets, each of up to 16 antigens, are placed on a slide with ordinary pen nibs. Each set can then be stained simultaneously with an appropriate antiserum followed by an anti-species conjugate. The test may be used both for identifying known isolates and for titrating antibodies of various classes in sera or eye secretions. With its aid, 12 serotypes of TRIC agent (A, B, Ba, C to K) and three serotypes of LGV (L1, L2, L3) have been identified (Wang *et al.* 1977). Of the serotypes isolated from ophthalmic trachoma, nearly all type A strains come from the Middle East and North Africa, whereas types B and C have a worldwide distribution. In both trachoma-endemic and non-endemic areas, TRIC agents isolated from inclusion conjunctivitis or from the genital tract are mostly of the closely related types D and E; G and F are next in frequency.

The relations between the various TRIC serotypes are somewhat complex and cross-tests between the various antigens and their corresponding antisera are needed to obtain the maximum information. By analogy with the immunological relationships between influenza strains (Fazekas de St Groth 1969), Wang and Grayston (1971) postulated the existence of 'senior' and 'junior' strains of *Chlam. trachomatis*: 'senior' strains elicit antibodies that react to some extent with 'junior' strains, whereas the converse does not hold. 'Senior' strains are held to be those of more recent origin; on this hypothesis, the LGV agents are the oldest, followed by the 'oculogenital' serotypes such as D and E, and finally by the A, B and C serotypes that cause ophthalmic trachoma.

Chlam. psittaci

Strains from sheep and cattle can be clearly divided into two serotypes (1 and 2) by the inhibition of plaque formation in L 929 cells by sera from hyperimmunized roosters (Schachter *et al.* 1974, 1975). Type 1 strains cause abortion and enteric infections, whereas type 2 strains cause polyarthritis, encephalomyelitis and conjunctivitis. Four strains isolated from rabbits, mice and guinea-pigs were all type 1; it was suggested that these chlamydiae may have been environmental contaminants rather than naturally occurring pathogens. Neither type 1 nor 2 is serologically related to chlamydiae isolated from birds. Differences in the growth characteristics of types 1 and 2 in L 929 mouse cells suggest important differences in their metabolic requirements (Spears and Storz 1979).

Relation of serotype to pathogenicity

In general, the various syndromes caused by both species of *Chlamydia* tend to be associated with specific serotypes or groups of serotypes. One of the clearest examples is LGV, in which both the serotypes and the clinical syndrome are sharply demarcated from those associated with other *Chlam. trachomatis* infections; it has indeed been suggested on these grounds that LGV strains should be regarded as a separate species or sub-species within the genus. The two serotypes of *Chlamydia psittaci* isolated from sheep and cattle are also associated with different syndromes. These clinical observations are reinforced by laboratory findings. *Chlam. trachomatis* serotypes isolated from ophthalmic trachoma tend to be less pathogenic for the simian than for the human eye, whereas the converse is true for the serotypes from inclusion conjunctivitis and from the genital tract (Collier 1962*a*). Again, 'fast-killing' TRIC agents that have become altered during laboratory passage lose their capacity to infect the simian eye and acquire the ability to grow readily in cell cultures; such strains behave like LGV agents in immunofluorescence tests.

These considerations make it likely that antigenic determinants conferring type-specificity are the same as or closely associated with the factors that determine both virulence and pathogenicity for chick embryos, cell cultures and intact mammalian and avian hosts.

(For a review of the pathogenicity of chlamydiae to man and animals, see Wachendörfer and Lohrbach 1980; and for a description of the properties of the genital chlamydiae, see Richmond 1978.)

References

Alexander, E. R. and Chiang, W. T. (1967) *Amer J. Ophthal.* **63,** 1145.

Alexander, E. R., Wang, S. P. and Grayston, J. T. (1967) *Amer. J. Ophthal.* **63,** 1469.

Alexander, J. J. (1968) *J. Bact.* **95,** 327.

Ata, F. A., Stephenson, E. H. and Storz, J. (1971) *Infect. Immun.* **4,** 506.

Becker, Y. (1978) *Microbiol. Rev.* **42,** 274.

Bedson, S. P. (1936) *Brit. J. exp. Path.* **17,** 109.

Bedson, S. P., Barwell, C. F., King, E. J. and Bishop, L. W. J. (1949) *J. clin. Path.* **2,** 241.

Bedson, S. P. and Bland, J. O. W. (1932) *Brit. J. exp. Path.* **13,** 461.

Bedson, S. P., Western, G. T. and Simpson, S. L. (1930) *Lancet* **i,** 235, 345.

Bell, S. D. and Fraser C. E. O. (1969) *Amer. J. trop. Med.* **18,** 568.

Bell, S. D., Snyder, J. C. and Murray, E. S. (1959) *Science* **130,** 626.

Benedict, A. A. and O'Brien, E. (1956) *J. Immunol.* **76,** 293.

Bernkopf, H. (1962) *Ann. N.Y. Acad. Sci.* **98,** 345.

Blyth, W. A. and Taverne, J. (1972) *J. Hyg., Camb.* **70,** 33.

Buchanan, R. E. and Gibbons, N. E. (1974) *Bergey's Manual of Determinative Bacteriology*, 8th edn. Williams and Wilkins, Baltimore.

Burnet, F. M. and Rountree, P. M. (1935) *J. Path. Bact.* **40,** 471.

Caldwell, H. D., Kuo, C. C. and Kenny, G. E. (1975*a*) *J. Immunol.* **115,** 963; (1975*b*) *Ibid.* **115,** 969.

Christoffersen, G. and Manire, G. P. (1969) *J. Immunol.* **103,** 1085.

Coles, A. C. (1930) *Lancet* **i,** 1011.

Collier, L. H. (1962*a*) *Ann. N.Y. Acad. Sci.* **98,** 188; (1962*b*) *Ibid.* **98,** 259.

Collier, L. H., Duke-Elder, S. and Jones, B. R. (1958) *Brit. J. Ophthal.* **42,** 705; (1960) *Ibid.* **44,** 65.

Collier, L. H. and Sowa, J. (1958) *Lancet* **i,** 993.

Dawson, C. R., Mordhorst, C. H. and Thygeson, P. (1962) *Ann. N.Y., Acad. Sci.* **98,** 167.

Dhir, S. P., Hakomori, S., Kenny, G. E. and Grayston, J. T. (1972) *J. Immunol.* **109,** 116.

Digiacomo, R. F., Gale, J. L., Wang, S. P. and Kiviat, M. D. (1975) *Brit. J. vener. Dis.* **51,** 310.

Fazekas de St Groth, S. (1969) *Bull. World Hlth Org.* **41,** 651.

Friis, R. R. (1972) *J. Bact.* **110,** 706.

Fritsch, H., Hofstätter, A. and Lindner, K. (1910) *v Graefes Arch. Ophthal.* **76,** 547.

Furness, G., Graham, D. M. and Reeve, P. (1960) *J. gen. Microbiol.* **23,** 613.

Furness, G., Henderson, W. G., Csonka, G. W. and Fraser, E. F. (1962) *J. gen. Microbiol.* **28,** 571.

Gale, J. L., Wang, S. P. and Grayston, J. T. (1971) In: *Trachoma and Related Disorders Caused by Chlamydial Agents*, p. 489. Ed. by R. L. Nichols. Excerpta Medica, London.

Garrett, A. J., Harrison, M. J. and Manire, G. P. (1973) *J. gen. Microbiol.* **80,** 315.

Gear, J. H. S., Gordon, F. B., Jones, B. R. and Bell, S. D. (1963) *Nature, Lond.* **197,** 26.

Gogolak, F. M. (1954) *J. infect. Dis.* **95,** 220.

Gordon, F. B. *et al.* (1969) *J. infect. Dis.* **120,** 451.

Halberstaedter, L. and Prowazek, S. von (1907) *Arb. Gesundh-Amt., Berl.* **26,** 44; (1909) *Berl. klin. Wschr.* **46,** 1839.

Hanna, L., Merigan, T. C. and Jawetz, E. (1967) *Amer. J. Ophthal.* **63,** 1115.

Harshbarger, J. C., Chang, S. C. and Otto, S. V. (1977) *Science* **196,** 666.

Hellerström, S. and Wassén, E. (1930) *C.R. int. Derm. Syph., Paris* 1147.

Higashi, N. (1975) *Rep. Inst. Virus Res. Kyoto* **18,** 3.

Higashi, N., Nagatomo, Y. and Matsumoto, A. (1975) *Rep. Inst. Virus Res. Kyoto* **18,** 149.

Jawetz, E. (1962) *Ann. N.Y. Acad. Sci.* **98,** 31.

Jawetz, E. and Hanna, L. (1960) *Proc. Soc. exp. Biol., N.Y.* **105,** 207.

Jenkin, H. M. and Lu, Y. K. (1967) *Amer. J. Ophthal.* **63,** 1110.

Jenkin, H. M., Makino, S., Townsend, D., Riera, M. C. and Barron, A. L. (1970) *Infect. Immun.* **2,** 316.

Jones, B. R., Collier, L. H. and Smith, C. H. (1959) *Lancet* **i,** 902.

Jones, H., Rake, G. and Stearns, B. (1945) *J. infect. Dis.* **76,** 55.

Katzenelson, E. and Bernkopf, H. (1967) *Amer J. Ophthal.* **63,** 1483.

Kingsbury, D. T. and Weiss, E. (1968) *J. Bact.* **96,** 1421.

Lamont, H. C. and Nichols, R. L. (1980) In: *Immunology of Human Infection*, p. 441. Ed. by A. J. Nahmias and R. J. O'Reilly. Plenum, London.

Lawn, A. M., Blyth, W. A. and Taverne, J. (1973) *J. Hyg., Camb.* **71,** 515.

Levaditi, C., Ravant, P., Lépine, P. and Schoen, R. (1931) *C. R. Soc. Biol., Paris* **107,** 1525.

Levinthal W. (1930) *Klin. Wschr.* **9,** 654.

Lillie, R. D. (1930) *Publ. Hlth Rep., Wash.* **45,** 773.

Lindner, K. (1909*a*) *Wien. klin. Wschr.* **22,** 1555; (1909*b*) *Ibid.* **22,** 1697; (1910) *v Graefes Arch. Ophthal.* **76,** 559.

Litwin, J. (1959) *J. infect. Dis.* **105,** 129.

McComb, D. E. and Bell, S. D. (1967) *Amer. J. Ophthal.* **63,** 1429.

Manire, G. P. (1966) *J. Bact.* **91,** 409.

Manire, G. P. and Tamura, A. (1967) *J. Bact.* **94,** 1178.

Matsumoto, A. and Manire, G. P. (1970) *J. Bact.* **101,** 278.

Merigan, T. C. and Hanna, L. (1966) *Proc. Soc. exp. Biol., N.Y.* **122,** 421.

Miyagawa, Y. *et al.* (1935) *Jap. J. exp. Med.* **13,** 733.

Moulder, J. W. (1964) *The Psittacosis Group as Bacteria.* (Ciba Lectures in Microbial Biochemistry) John Wiley, New York.

Moulder, J. W., Grisso, D. L. and Brubaker, R. R. (1965) *J. Bact.* **89,** 810.

Murray, E. S. and Charbonnet, L. T. (1971) In: *Trachoma and Related Disorders Caused by Chlamydial Agents*, p. 369. Ed. by R. L. Nichols. Excerpta Medica, London.

Nabli, B. and Tarizzo, M. L. (1967) *Amer. J. Ophthal.* **63,** 1541.

Nichols, R. L. and McComb, D. E. (1964) *J. exp. Med.* **120,** 639.

Nicolle, C., Blaisot, L. and Cuénod, A. (1912) *C.R. Acad. Sci., Paris* **155,** 241.

Nigg, C. (1942) *Science* **95,** 49.

Ozanne, G. and Pearce, J. H. (1980) *J. gen. Microbiol.* **119,** 351.

Page, L. A. (1959) *Avian Dis.* **3,** 23; (1965) *Bull. Wildlife Dis. Ass.* **1,** 49; (1966) *Int. J. syst. Bact.* **16,** 223.

Pearce, J. H., Allan, I. and Ainsworth, S. (1981) In: *Adhesion and Microorganism Pathogenicity* (Ciba Founda-

tion Symposium), p. 234. Ed. by K. Elliott, M. O'Connor and J. Whelan. Pitman Medical, Tunbridge Wells.
Perkins, H. P. and Allison, A. C. (1963) *J. gen. Microbiol.* **30,** 469.
Pollard, M. and Tanami, Y. (1968) *Amer. J. Ophthal.* **63,** 50.
Rake, G., McKee, C. M. and Shaffer, M. F. (1940) *Proc. Soc. exp. Biol., N.Y.* **43,** 332.
Rake, G., Shaffer, M. F. and Thygeson, P. (1942) *Proc. Soc. exp. Biol., N.Y.* **49,** 545.
Reeve, P. and Taverne, J. (1962*a*) *Nature, Lond.* **195,** 923; (1962*b*) *J. gen. Microbiol.* **27,** 501; (1963) *J. Hyg., Camb.* **61,** 67; (1967*a*) *Amer. J. Ophthal.* **63,** 1162; (1967*b*) *Ibid.* **63,** 1167.
Richmond, S. J. (1978) *What's New, Publ. Hlth Lab. Serv.*, March No.; (1980) *FEMS Microbiol. Lett.* **8,** 47.
Richmond, S. J. and Caul, E. O. (1975) *J. clin. Microbiol.* **1,** 345.
Richmond, S. J., Stirling, P. and Ashley, C. R. (1982) FEMS *Microbiol. Lett.* **14,** 31.
Ridgway, G. L., Owen, J. M. and Oriel, J. D. (1978) *Brit. J. vener. Dis.* **54,** 103.
Ross, M. R. and Jenkin, H. M. (1962) *Ann. N.Y. Acad. Sci.* **98,** 329.
Sarov, I. and Becker, Y. (1969) *J. molec. Biol.* **42,** 581.
Schachter, J., Banks, J., Sugg, N., Sung, M., Storz, J. and Meyer, K. F. (1974) *Infect. Immun.* **9,** 92; (1975) *Ibid.* **11,** 904.
Schachter, J. and Caldwell, H. D. (1980) *Annu. Rev. Microbiol.* **34,** 285.
Schachter, J. and Dawson, C. R. (1978) *Human Chlamydial Infections*, P.S.G. Publishing Co., Littleton, Mass.
Schmeichler, L. (1909) *Berl. klin. Wschr.* **46,** 2057.
Sompolinsky, D. and Richmond, S. J. (1974) *Appl. Microbiol.* **28,** 912.
Sowa, S., Sowa, J., Collier, L. H. and Blyth, W. (1965) *Spec. Rep. Ser. med. Res. Coun.* no. 308. H.M. Stationery Office, London.
Spears, P. and Storz, J. (1979) *J. infect. Dis.* **140,** 959.
Stargardt, K. (1909) *v Graefes Arch. Ophthal.* **69,** 524.
Storz, J. (1971) *Chlamydia and Chlamydia-induced Diseases.* Charles C Thomas, Springfield, Illinois.
Symposium (1962) The biology of the trachoma agent. *Ann. N.Y. Acad. Sci.* **98** (Art. 1.), 1; (1967) Conference on trachoma and allied diseases. *Amer. J. Ophthal* **63,** 1027; (1971) *Trachoma and Related Disorders Caused by Chlamydial Agents.* Excerpta Medica, London; (1977) *Nongonococcal urethritis and related infections.* Ed. by D. Hobson and K. K. Holmes. *Amer. Soc. Microbiol.*, Washington DC.
Tamura, A. (1967) *J. Bact.* **93,** 2008.
Tamura, A. and Higashi, N. (1963) *Virology* **20,** 596.
Tamura, A. and Manire, G. P. (1967) *J. Bact.* **94,** 1184; (1968) *Ibid.* **96,** 875; (1974) *Ibid.* **118,** 144.
Tamura, A., Matsumoto, A., Manire, G. P. and Higashi, N. (1971) *J. Bact.* **105,** 355.
T'ang, F. F., Chang, H. L., Huang, Y. T. and Wang, K. C. (1957) *Chin. med. J.* **75,** 429.
Thygeson, P. (1934) *Amer. J. Ophthal.* **17,** 1019.
Treharne, T. D., Davey, S. J., Gray, S. J. and Jones, B. R. (1972) *Brit. J. vener. Dis.* **48,** 18.
Wachendörfer, G. and Lohrbach, W. (1980) *Berl. Münch. tierärtzl. Wschr.* **93,** 248.
Wang, S. P. (1971) In: *Trachoma and Related Disorders Caused by Chlamydial Agents*, p. 273. Ed. by R. L. Nichols. Excerpta Medica, London.
Wang, S. P. and Grayston, J. T. (1962) *Ann. N.Y. Acad. Sci.* **98,** 177; (1967) *Amer. J. Ophthal.* **63,** 1133; (1971) In: *Trachoma and Related Disorders Caused by Chlamydial Agents*, p. 303. Ed. by R. L. Nichols. Excerpta Medica, London.
Wang, S. P., Grayston, J. T., Kuo, C. C., Alexander, E. R. and Holmes, K. K. (1977) In: *Nongonococcal Urethritis and Related Infections*, p. 237. Ed. by D. Hobson and K. K. Holmes. American Society for Microbiology, Washington.
Wang, S. P., Kuo, C. C. and Grayston, J. T. (1975) In: *Genital Infections and their Complications*, p. 39. Ed. by D. Danielsson, L. Jublin and P-A. Mårdh. Almqvist and Wiksell International, Stockholm.
Weiss, E. (1965) *J. Bact.* **90,** 243.
Weiss, E. and Dressler, H. R. (1967) *Ann. N.Y. Acad. Sci.* **98,** 250.
Weiss, E., Myers, W. F., Dressler, H. R. and Chun-Hoon, H. (1964) *Virology* **22,** 551.
Wentworth, B. B. and Alexander, E. R. (1974) *Appl. Microbiol.* **27,** 912.
Yanamura, H. Y. and Meyer, K. F. (1941) *J. infect. Dis.* **68,** 1.
Yong, E. C., Chinn, J. S., Caldwell, H. D. and Kuo, C. C. (1979) *J. clin. Microbiol.* **10,** 351.

46

The rickettsiae

Barrie P. Marmion

Introductory

This group of organisms comprises rod-like or cocco-bacillary prokaryotes, similar to but substantially smaller than bacteria, that, with one exception, are obligate intracellular parasites. The cell-dependent members grow or are found in the cytoplasm or nuclei of living cells; all multiply by binary fission (Ormsbee 1969, Weiss, 1973, Kazar *et al.* 1978).

In this chapter, the common name, rickettsiae, is used as a convenient broad designation for a large group of biologically similar organisms that make up the family Rickettsiaceae. We shall devote most of our attention to three genera of rickettsiae, considered by some workers to form the tribe Rickettsieae, which contain organisms pathogenic for man: *Rickettsia*, *Rochalimaea* and *Coxiella*. Certain rather similar organisms, members of the genera *Ehrlichia*, *Cytoecetes*, *Cowdria* and *Neorickettsia*, are associated with disease in other mammals and will be described briefly at the end of the chapter.

The cell dependence of the rickettsiae, and the fact that some of them will pass through bacteria-retaining filters (Davis and Cox 1938) led to an initial confusion with viruses, from which, however, they are quite distinct. (See Moulder 1962, 1966 and Hanks 1966 for discussion of the biochemistry of intracellular parasitism as it applies to the differentiation of viruses, bacteria and 'intermediate' organisms.) Like the bacteria they have a single chromosome in the form of unbounded, fibrillary nuclear material—DNA—either dispersed through the cell or arranged centrally in coiled nucleoid filaments; there are also ribosomes, a trilaminar cytoplasmic membrane and a cell wall of bacterial type. The cell wall contains the amino acid, diaminopimelic acid, and the amino sugars, muramic acid and glucosamine, and it is sensitive to muramidases (lysozyme). Some of the organisms have a capsule or slime layer. Most species are gram negative, but *Coxiella burneti* is gram positive when alcoholic iodine is used as mordant (Giménez 1965).

The lipid, polysaccharide and amino-acid composition of the rickettsial cell wall resembles that of gram-negative bacteria. This similarity, and the fact that many of the organisms parasitize the intestinal cells of arthropods such as lice, ticks, fleas or mites, led to the suggestion that rickettsiae may have evolved from similar organisms that exist as gut commensals in arthropods. Although this is an attractive notion, little is known of the structure and physiology of these arthropod commensals; comparisons of rickettsiae with *Wolbachia persica*, one reasonably well studied commensal isolated from insects, reveals differences in morphology and physiology (Suitor and Weiss 1961, Weiss *et al.* 1962, Suitor 1964). Recent studies of '*Wolbachia*-like' organisms from five species of tick have shown that the morphology varies substantially from one species to another. They have a multilaminar, sinuous, membrane-type cell wall with a periplasmic space and a well differentiated cytoplasm containing ribosomes and a dense central body. Although differentiation of the organisms from rickettsiae may be difficult by light microscopy, differences in fine structure enable a distinction to be made (Hayes and Burgdorfer 1981).

Rickettsiae are distinguished from another group of cell-dependent prokaryotes, the chlamydiae, by their mode of reproduction, metabolic capabilities and requirements, and by their host and vector range (Chapter 45). *Bartonella bacilliformis*, of the family Bartonellaceae, has some biological features in common with rickettsiae, and grows in cell-free media containing blood (Chapter 43), as does *Ro. quintana*.

Rickettsia, *Rochalimaea* and *Coxiella*

The genus *Rickettsia* comprises ten or more species which fall into three distinct biological groups—the typhus, spotted-fever and scrub-typhus groups (Table 46.1). Most of these species are clearly associated with disease in man, but the evidence for this is still incomplete in respect of *Rick. canada* and *Rick. parkeri*,

Table 46.1. Species of *Rickettsia*, *Rochalimaea* and *Coxiella*; rickettsial diseases of man; vertebrate hosts and arthropod vectors.

					Natural infection cycle	
Biological Group	Infective agent	Disease in man*	Transmission to man	Geographical distribution	Arthropod vector	Vertebrate host
Typhus	*Rick. prowazeki*	Epidemic typhus	Infected louse faeces into broken skin or by inhalation	Worldwide	Body louse, squirrel louse, and flea	Man, flying squirrel (USA)
	Rick. prowazeki	Brill-Zinsser disease (recrudescent typhus)	Reactivation of latent infection	Worldwide	...	Man
	Rick. typhi (syn. *mooseri*)	Murine typhus	Infected flea faeces into broken skin	Worldwide	Rat flea	Rat
	Rick. canada	Equivocal; see text	Not defined	North America	Tick	Rabbits (?), birds
Spotted fever	*Rick. rickettsi*	Rocky Mountain spotted fever	Tick bite	Western Hemisphere	Tick	Small and medium-sized wild mammals, birds, dogs
	Rick. conori	Boutonneuse fever	Tick bite	Mediterranean countries, Africa, India	Tick	Small wild animals, dogs
	Rick. australis	Queensland tick typhus	Tick bite	Australia	Tick	Small wild rodents, marsupials
	Rick. sibirica	Siberian tick typhus (North Asian tick-borne rickettsiosis)	Tick bite	Siberia, Mongolia	Tick	Wild and domestic animals
	Rick. akari	Rickettsialpox (vesicular rickettsiosis)	Mite bite	North-eastern USA, USSR	Gamasid mite	House mouse
	Rick. parkeri and other isolates†	Not defined	Not defined	...	Ticks	Not defined
Scrub typhus	*Rick. tsutsugamushi*	Scrub typhus (tsutsugamushi disease, mite-borne typhus)	Mite bite	Asia, Australia, Pacific Islands	Trombiculid mite	Small wild rodents, birds
Trench fever	*Ro. quintana*	Trench fever (Wolhynian fever, five-day fever)	Infected louse faeces into broken skin	Europe, Middle East, North Africa, Mexico	Body louse	Man
Q fever	*Cox. burneti*	Q fever	Inhalation of infected aerosol or dust; also by ticks	Worldwide	Tick	Small wild mammals; and cattle, sheep, and goats without arthropod vector

* (And synonyms.)
† *Rick. parkeri*, *Rick. montana*, *Rick. rhipicephali*, Western Montana U strain, JC-880 and TT118 are members of spotted fever group but ecology and pathogenicity are uncertain.

which are included in the table for completeness. The scrub-typhus group comprises only one species, *Rick, tsutsugamushi*, but this has at least eight serotypes, a diversity as great as that among the species of the spotted-fever group. On the other hand, the genera *Rochalimaea* and *Coxiella* each contain one species only, respectively *Ro. quintana* which causes trench fever and *Cox. burneti* Q fever (but see vole agents, Table 46.3).

The initial grouping of the rickettsiae into genera and species was made according to their biological properties, supported by limited serological studies by means of the Weil-Felix reaction (p. 534). The biological characters employed were (1) the range of natural infection in vertebrate hosts, and the identity of the arthropod vector that transmitted the organism between natural hosts or to man, or the lack of dependence on an arthropod vector (Table 46.1), and (2) the pattern of disease in experimental animals (p. 536). These broad groupings were further substantiated by differences in antigenic constitution (p. 534) and by the ability of *Ro. quintana* to grow in cell-free media. More recently, classification has been refined and in some instances modified by comparative studies of the molecular biology of the organisms (p. 536).

(For a history of the discovery of the rickettsiae, and for accounts of the early biochemical and serological studies of them, the reader is referred to previous editions of this book.)

Morphology and chemical composition

Members of the genus *Rickettsia* measure 0.3 to 0.7 μm in width but vary considerably in length. Filamentous forms of *Rick. prowazeki* may be as long as 4.0 μm; other members of the typhus and spotted-fever groups average 1.2 to 1.8 μm in length. *Cox. burneti* on the other hand is generally shorter, not more than 1.0 μm long, more regular in size, and is frequently seen as short rods or coccobacilli.

In general electron micrographs of thin sections of rickettsiae (Fig. 46.1) show a cell wall 7–10 nm thick, with three layers. With higher resolution it is possible to detect a five-layered structure similar to that found in *Esch. coli* (Anderson *et al.* 1965, Ito and Vinson 1965, Anacker *et al.* 1967, Burton *et al.* 1975). The outer envelope of *Rick. tsutsugamushi* differs from the envelopes of *Rick. prowazeki* and *Rick. rickettsi* (Silverman and Wisseman 1978). Analysis of the cell wall of *Rick. typhi* shows a high protein (60 per cent) and a low reducing-sugar content (9.3 per cent). Fifteen amino acids are present and include the range of aliphatic, aromatic and heterocyclic amino acids; diaminopimelic acid, the characteristic cell-wall amino acid of bacteria other than the gram-positive cocci, is present in all genera of the Rickettsiaceae, including *Ro. quintana*. The amino sugars muramic acid and glucosamine are also present, but teichoic acid, a cell-

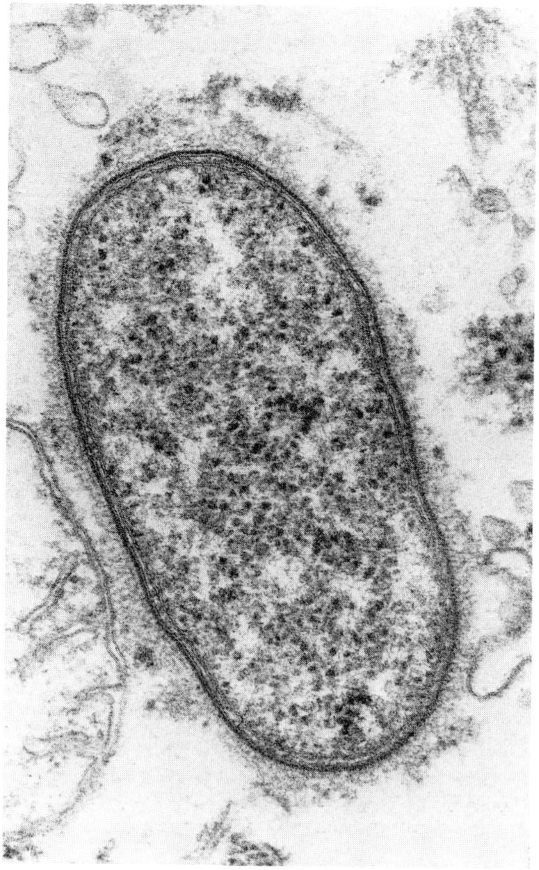

Fig. 46.1. Ultrathin section of *Rickettsia tsutsugamushi*, Gilliam strain (× 66 000), grown in a culture of chick-embryo fibroblasts and then homogenized with homologous antiserum, to show the slime-like layer on the surface. (By courtesy of Drs David J. Silverman and C. L. Wisseman Jr.)

wall component of gram-positive bacteria, is not found even though *Cox. burneti* is gram positive under certain staining conditions (Perkins and Allison 1963, Myers *et al.* 1967*a*, Wood and Wisseman 1967, Osterman *et al.* 1974). The rickettsial cell wall, like that of gram-negative bacteria, may have the properties of an endotoxin (Baca and Paretsky 1974, Chan *et al.* 1976).

Outside the cell wall there may be loosely attached capsular material, e.g. in the typhus and spotted-fever groups of organisms, that can be extracted with solvents such as ethyl ether—the so-called soluble antigens (Ito and Vinson 1965, Anacker *et al.* 1967, Palmer *et al.* 1974*a*,*b*, Silverman *et al.* 1974, 1978). In organisms such as *Cox. burneti* there is no obvious capsule but a 20-nm-thick layer of fibrillary material can be demonstrated on the surface by staining with ruthenium red and by examination of thin sections in the electron microscope (Burton *et al.* 1975, Čiampor *et al.* 1972).

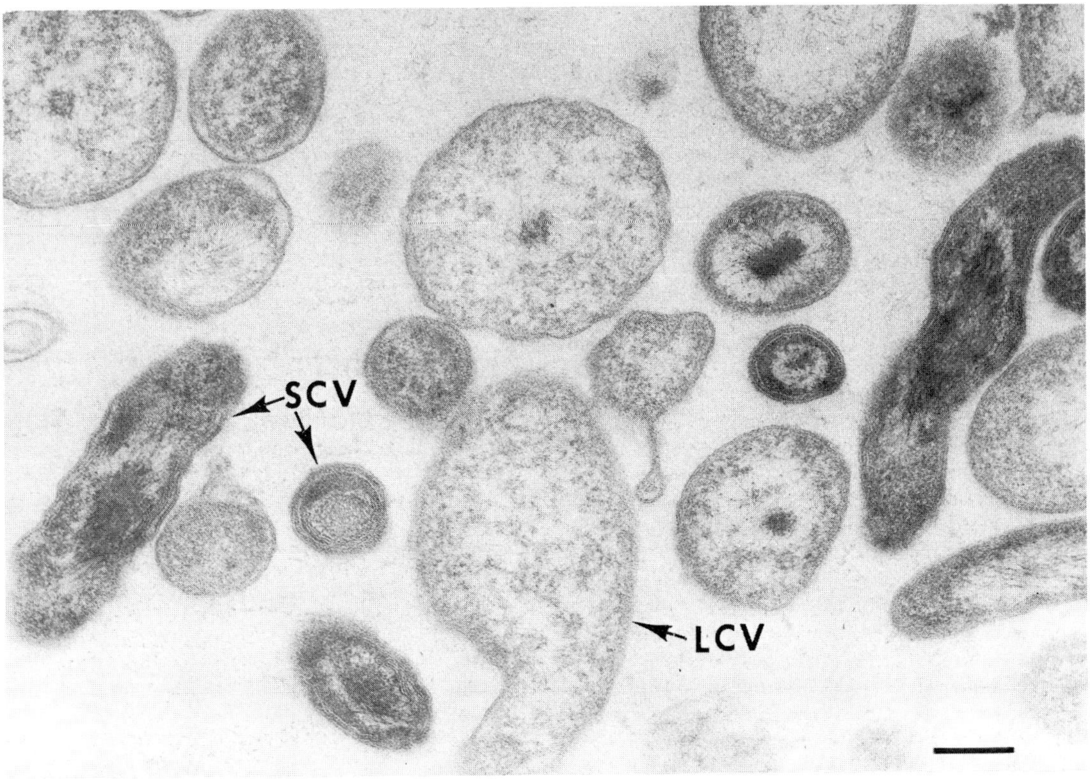

Fig. 46.2. Ultrathin section of *Coxiella burneti* to show large (LCV) and small (SCV) cell variants. Note the condensed structure and central nucleoid in SCV. Bar marker = 0.1 μm. (By courtesy of Dr Thomas F. McCaul.)

The underlying cytoplasmic membrane, 5–7 nm thick, has a typical trilaminar structure. Inside the cytoplasmic membrane there are electron-dense granules 7–20 nm diameter—ribosomes—and strands of DNA. The nuclear DNA fibrils are distributed through the cell in *Rickettsia* spp; the degree of dispersion and thickness of the fibrils depends to some extent on the fixatives used before electron microscopy. In *Cox. burneti*, regardless of the fixation method, they have a clear separation from the area of cytoplasm containing the ribosomes. The rickettsial ribosomes are 70S in size, similar to those of bacteria, and protein synthesis on them is inhibited by broad-spectrum antibiotics and the aminoglycosides (see p. 533). The cytoplasm may also contain electron-lucent spheres, 4–40 nm in diameter, and granules and intracytoplasmic membranous organelles—perhaps invaginations of cytoplasmic membrane.

McCaul and co-workers (1981) describe a large and a small cell variant of *Cox. burneti* (Fig. 46.2); the former is sensitive to osmotic shock but the latter, although metabolically active, is resistant and corresponds to an endospore. Large forms may contain the endospore; 'sporulation' may be related to the resistance of the organism to heat and other adverse influences, and to the filtrable forms described by Kordová (1960) (see also p. 531).

The nuclear DNA in rickettsiae is organized into a single chromosome of molecular weight $100–150 \times 10^6$; the mole percentage G+C ranges from 29 in the typhus group through 32.5 in the spotted-fever group, to 43 in *Cox. burneti* (for further information about the rickettsial genome, see p. 536). RNA is readily leached from resting or damaged rickettsiae, and early reports of high DNA: RNA ratios in rickettsial cells were misleading for this reason.

Staining properties

Although rickettsiae are gram negative, or weakly gram positive in the instance of *Cox. burneti* (Giménez 1965), the gram stain is not of practical value in the demonstration of organisms in impression smears from infected yolk sacs, cell cultures or the organs of infected experimental animals. Giemsa's method, particularly with prolonged contact ('overnight Giemsa') and with the stain buffered at a pH that will distinguish between cells and organisms, is commonly used. Under these circumstances the organisms appear as dark purple coccobacilli or rods. Yolk-sac granules, or mast-cell granules, may stain in a fashion that leads to confusion with rickettsiae and a differential stain is of value. Several methods have been described and depend on the acid-fast properties of most groups of rickettsiae when stained with basic fuchsin and decolorized with weak organic acids or differentiated with buffer solutions. The Macchiavello and Giménez methods are commonly used, but modifications of the Ziehl-Neelsen stain are used by veterinarians and some rickettsial workers (Zdrodovskii and Golinevich 1960). The Giménez stain (Giménez 1964) is probably the most satisfactory and employs an alcoholic phenolized basic fuchsin solution, diluted just before use in sodium phosphate buffer, pH 7.45, and counter-staining with malachite green. Scrub-typhus rickettsiae require a variant of the technique with fast green-ferric nitrate as counterstain. The method gives red rickettsiae against a green background, but some bacteria are also stained and the method is not completely specific for rickettsiae (for practical details see Elisberg and Bozeman 1979). Rickettsiae may also be detected in smears and tissue sections with specific antisera by immunofluorescence techniques (Woodwood *et al.* 1976.) Finally, use may also be made of fluorochromes such as acridine orange to detect the organisms by their nucleic-acid content (Silberman and Fiset 1968, Silverman *et al.* 1979).

Multiplication

Many attempts have been made to grow rickettsiae in cell-free media but, with the exception of *Ro. quintana* and the related Baker's vole agent (p. 535), these have been unsuccessful. The two organisms multiply slowly on a special blood agar medium; human strains of *Ro. quintana* also appear to multiply extracellularly in the louse gut. Weiss (1973) explores the possible reasons for the failure of other rickettsiae to multiply in cell-free media.

Early workers with rickettsiae cultivated them in arthropods such as lice, or in experimental animals such as guinea-pigs (typhus, spotted-fever, Q-fever strains) or in mice (scrub-typhus strains). Later, in a notable advance, Cox (1938, 1941) showed that rickettsiae grew prolifically in the chick-embryo yolk sac, an excellent source of organisms for diagnostic antigens and vaccines, and of material for biochemical and structural studies.

Growth in cell culture In natural infections of man and other mammals rickettsiae appear to have a sharply restricted host-cell range: vascular endothelial cells in the instance of *Rickettsia* and macrophages in *Coxiella*. Despite this, a remarkably wide range of cells can be infected in cell culture. These include epithelial or fibroblastic cells from mammals (human, non-human primate, rodent), fish, birds and insects. Various transformed cell lines, umbilical endothelial cells, macrophages and lymphocytes, also support rickettsial growth (Wisseman 1981, Elisberg and Bozeman 1979).

Cell cultures are of less value for the primary isolation of rickettsiae than is animal inoculation, but they have proved of considerable value in studies of the replication of the organisms and of interactions of parasite and host cell. A recent technical advance, rivalling in importance Cox's work with the chick-embryo yolk sac, is the development of plaque assays for rickettsiae (Kordová 1966; McDade *et al.* 1969, Ormsbee and Peacock 1976). A major consequence of this has been to make possible purification of rickettsial clones from mixtures of rickettsial species, or of strains of rickettsiae contaminated with bacteria, viruses or mycoplasmas picked up from the animal, chick embryo, arthropod or cell-culture systems in which they had been isolated or serially propagated. Stringent purification of rickettsial strains now includes three or more serial 'clonings' by subculture of single plaques in monolayer culture of chick-embryo cells from specific pathogen-free (SPF) eggs, followed by production of seed cultures in the yolk sac of SPF eggs (Wisseman 1981). Seed cultures for the production of rickettsial vaccines should be derived in this way and shown to be free of mycoplasmal and bacterial contaminants and avian leucosis virus. Strain purification is also essential in biochemical studies, and the 'plaquing' technique is central to work on antibiotic sensitivity and population genetics, for example, phase variation in *Cox. burneti*, and in numerous other ways.

Growth in animal tissues As with cells in culture, the range of insect and vertebrate hosts that can be infected experimentally, or that are infected in nature, is also very wide. Various species of monkeys, birds, dogs, cattle, sheep and goats, as well as small mammals such as white rats, cotton-rats, inbred and outbred mice, deer mice, voles, gerbils, ferrets, squirrels, rabbits and hamsters, may be infected. In nature *Cox. burneti* infects at least 39 species of ixodid and argasid ticks within 10 different genera. In many arthropods rickettsiae are transmitted vertically from one generation to the next via the egg, e.g. members of the spotted-fever group and *Rick. canada*. On the other hand, transovarial transmission of *Cox. burneti* has been successful with only four of ten species of ticks tested, and it does not occur with *Rick. prowazeki* in the louse or *Rick. typhi* in the flea. With *Rick. rickettsi* the burden of parasitism for the tick *Dermacentor andersoni* appears slight and detectable mainly by a decrease in the number of eggs hatched. On the other hand infection with *Rick. prowazeki* is invariably fatal in the louse, perhaps representing an imperfect parasitism of more recent origin.

Some small animals are said to be more sensitive to infection than the guinea-pig and to have advantages for the primary isolation of rickettsiae. Thus, meadow voles (*Microtus pennsylvanicus*) are more sensitive for the isolation of *Rick. prowazeki*, *Rick rickettsi* and other spotted-fever rickettsiae (see Elisberg and Bozeman 1979), and hamsters are more sensitive for the isolation of *Cox. burneti*.

Fertile hens' eggs, inoculated in the yolk sac during the 5th or 6th day of development, are highly susceptible to infection; with many rickettsiae there is a linear relationship between mean survival time of the chick embryo and dose of rickettsiae (Weiss 1973). The optimal temperature of incubation also varies: 35°C for most rickettsiae; 33°C for those of the spotted-fever group. Optimal yields of rickettsia per egg are 10^7–10^9 particles per g with most rickettsiae and may be as high as 10^{11} per g with *Cox. burneti*.

Dynamics of intracellular growth The ratio of total to infectious organisms varies with the assay system, the species of rickettsia and the physiological state of the micro-organism, loss of ATP and ions tending to lower infectivity. Ormsbee and co-workers (1978) describe the relation of counts of rickettsiae, determined by fluorescence microscopy, to plaque-forming units in chick-embryo fibroblasts and the ID50 in the chick-embryo yolk sac, mice and guinea-pigs. With a resistant organism such as *Cox. burneti* in phase 1 there was a close correspondence between rickettsial count and the ID50 for chick embryos, mice and guinea-pigs—few organisms were required to infect. With organisms such as *Rick. typhi*, *Rick. prowazeki* and members of the spotted-fever group there was a difference of one or two log dilutions between rickettsial count and ID50.

Observation of the growth of rickettsiae in cells can be made by combinations of phase-contrast microscopy, time-lapse cinematography and thin-section electron microscopy. A useful cell for this purpose is the chick-embryo entodermal cell prepared by explanting the avascular portion of the yolk sac of 4-day-old embryos. These yield monolayers of cells with a large faintly staining cytoplasm that allows of ready cytological observation after infection. Close contact between cells and rickettsiae aids penetration and is effected by centrifuging the inoculum on to the cells. In general, only viable rickettsiae penetrate into host cells; efficiency of penetration is increased by factors, such as protein in medium, that prevent inactivation of extracellular rickettsiae, or others (glutamic acid, α-ketoglutaric acid) that enhance metabolic activity and, in particular, the production of ATP (Weiss 1973).

There are striking differences in the sites of replication of rickettsiae in cells. All members of the genus *Rickettsia* multiply in the cytoplasm of cells outside of vacuoles, having passed through the cytoplasmic membrane of the cell or the limiting membrane of phagosomes—a process that may be assisted by phospholipase-A activity (Winkler and Miller 1981). *Rick. tsutsugamushi* accumulates in the perinuclear region; *Rick. rickettsi* and other spotted-fever rickettsiae in cytoplasmic processes and also in the nucleus. Intranuclear growth or migration of organisms into the nucleus is, however, present only in a minority of infected cells and appears to be commoner in mammalian than in avian or arthropod cells. *Rick. canada*, which shares antigens with *Rick. typhi* and has been placed in the typhus group, is found in the nucleus of cells and in this respect resembles the spotted-fever group of organisms. Studies of the DNA, however, make it clear that it is not closely related to either group (p. 536). *Cox. burneti*, on the other hand, is taken into the cell by phagocytosis and does not break out of the phagolysosome but replicates within it, as do some mycobacteria (Burton *et al.* 1978). The phagolysosome contains hydrolytic enzymes and is at a pH of 4.5–5.0 and, significantly, the enzymic activity of *Cox. burneti* is much more active at pH 4.0 than at neutrality (Hackstadt and Williams 1981). The growth of the organism in the phagolysosome continues until the cell is converted into one large vacuole with the nucleus stretched over its circumference. Finally, *Ro. quintana* and the Baker's vole agent grow not only in cell-free media but on the surface of cells in culture (pericellular growth), a location also favoured by the cell-dependent mycoplasmas. *Ro. quintana* is also found on the surface of the mid-gut cells of its louse vector (Ito and Vinson 1965, Merrell *et al.* 1978).

Observations on the multiplication of individual rickettsiae in cultured cells by time-lapse cinematography and phase-contrast microscopy confirm the belief that the organisms multiply by transverse binary fission—findings in line with the static observation of dividing forms in smears of infected yolk sacs, arthropod tissues and of electronmicrographs of infected tissues. Stages of replication similar to those in bacteria have been detected in *Rick. prowazeki*: a lag phase, an exponential phase during which the organisms increase in length and then divide, with a generation time around 9 hr, and then a stationary phase.

A developmental cycle, as with chlamydiae, has not been observed, although Kordová and co-workers (Kordová 1960, Kordová and Kováčova 1967) have suggested that *Cox. burneti* and *Rick. prowazeki* replicate through a suborganismal particle. The evidence for this depends on retention of the infectivity of *Cox. burneti* after filtration through collodion membranes that should hold back the whole organism, and the disappearance of recognizable organisms in cell culture shortly after infection, followed subsequently by the appearance of fine immunofluorescent-staining granules, then finally the appearance of typical rickettsiae. Other workers (Anacker *et al.* 1964), using electronmicroscopy and ferritin-labelled antibody have not been able to confirm these observations but the recent description of an 'endospore' in *Cox. burneti* suggests that the matter needs reinvestigation.

Rickettsiae appear to grow best in well nourished cells. However, the cells do not have to multiply and nuclear function is not essential; various species of rickettsiae have been observed to multiply in cells arrested in metaphase by colchicine, in enucleated cell cytoplasm, or in cells in which DNA synthesis is inhibited by irradiation or treatment with mitomycin C. Nor does host-cell protein synthesis appear to be essential, because rickettsiae will grow in cycloheximide-treated cells (Wisseman 1981).

The mode of liberation of rickettsiae from the host cell varies from species to species. *Cox. burneti* grows in the phagolysosome vesicle until this bursts from the mass of organisms and the ingress of fluid. Although *Rick. prowazeki* is also liberated by the breakdown of the host cell, *Rick. rickettsi* appears to escape via fine cell processes to infect contiguous cells, at least in the early stages of infection. Cell-to-cell spread of rickettsiae in a monolayer gives a 'plaque' similar to that produced by continuous spread of viruses in cell sheets. The differing kinetics of growth of the organisms in the host cells and modes of spread determine the size and other characteristics of the plaques. Primary chick-embryo cells have been used most extensively for plaque assays, but continuous cell lines such as HeLa, Vero and one from the mosquito *Aedes albopictus* have also been used. Incubation times for plaque production vary from 5 to 19 days with different species of rickettsiae.

Growth of *Ro. quintana* The mode of growth of this organism differs strikingly from that of other rickettsiae. As we have seen, it multiplies pericellularly. It was originally isolated in lice and produced a low-grade fever on inoculation into guinea-pigs, but it cannot be passaged serially in these animals, or in mice or chick-embryo yolk sacs. However, it grows well on a cell-free medium made up of agar, beef heart infusion, tryptose and NaCl (Difco blood agar base) supplemented with 6 per cent of heat-inactivated horse serum and 4 per cent of lysed horse erythrocytes. Blood from trench-fever patients inoculated on to this medium and incubated in air plus 5 per cent CO_2 yields minute colonies 65–200 μm in diameter after 12–14 days. Subculture is effected by emulsification of colonies in sucrose phosphate glutamate medium and by plating on to fresh growth medium. Crystalline haemoglobin or haemin may be used instead of the lysed horse erythrocytes and starch, and charcoal or bovine albumin may be used instead of the horse serum (Myers *et al.* 1969, 1972). Smears of the colonies may be stained by the Giemsa or Giménez techniques or by immunofluorescence. A complement-fixing antigen can be prepared in the corresponding liquid medium.

Metabolism and biochemical activities

The metabolic activities of rickettsiae have not been studied systematically, species by species; and it is difficult to make reliable generalizations from the fragmentary evidence that is available.

In general there appear to be two patterns. In the first—exhibited, for example, by *Rick. prowazeki*, *Rick. typhi* and *Rick. rickettsi*—glucose, glucose-6-phosphate, lactate or sucrose is not metabolized, but oxygen uptake is greatly stimulated by glutamate and to a lesser extent by pyruvate and the reaction products of glutamate breakdown—α-ketoglutarate, succinate, and other dicarboxylic acids of the citric-acid cycle (Weiss 1973, 1981). Tracer experiments with glutamate radiolabelled at various carbon atoms indicate the existence of a tricarboxylic acid cycle. Starved rickettsiae, held at 36° for 3 hr, form no ATP, whereas incubation with glutamate raises the level of ATP synthesized, particularly when adenylic acid is also added.

Rickettsiae, unlike chlamydiae, are not parasitic on the host cell for ATP ('energy parasites'). Zahorchak and Winkler (1981) find that *Rick. prowazeki* contains a membrane-bound, energy-transducing ATPase, as in bacteria, that couples respiration to ATP synthesis. The ATPase is inhibited by dicyclohexylcarbodiimide which has a similar action in bacteria, mitochondria and chloroplasts. The biological advantage of this enzyme is not at once clear because rickettsiae can transport cellular ATP from their intracellular environment, but it may play a part in extracellular activities or in supplementing ATP production when the cell is damaged by the infection. Spectrophotometric examinations of purified suspensions of *Rick. typhi* have indicated the presence of iron-containing respiratory enzymes, cytochromes a and b and a flavin-containing enzyme, presumably in the terminal respiratory pathway as in other bacteria.

The second pattern, seen in *Cox. burneti*, differs from that in *Rickettsia*. Early investigations (Ormsbee and Peacock 1964) with intact organisms revealed rates of respiration substantially lower than those of typhus rickettsiae and stimulated largely by pyruvate. On the other hand, Paretsky (1968) found that disrupted preparations of *Cox. burneti* contained a hexokinase, glucose-6-phosphate dehydrogenase and other enzymes of the Embden-Meyerhof pathway which suggested that, contrary to the limited reactions observed with intact organisms, there might be metabolism of glucose. Most of the enzymes of the citric-acid cycle were also detected.

These conflicting observations have recently been reconciled, at least partly, by the observations of Hackstadt and Williams (1981) that transport, catabolism and incorporation of glutamate and glucose by *Cox. burneti* are highly stimulated in the pH range (4–5) found in the macrophage phagolysosome, the site of multiplication of the organism; pH optima for glutamate metabolism by *Rick. typhi* and *Chlamydia psittaci*, organisms multiplying in the cytoplasm of cells rather than in phagocytic vacuoles, were around 6.5–7.0. Other experiments showed that radiolabelled nucleotides are incorporated by *Cox. burneti* into nucleic acid and that the process is inhibited by actinomycin D. Once again, the incorporation appeared more effective at acid than at neutral pH. These observations go some way towards explaining why *Cox. burneti* multiplies only in phagolysosomes, but do not explain its obligate intracellular existence. Gonzales and Paretsky (1981), in exploring this problem, point out that, while *Cox. burneti* has the autonomous enzyme capacity to make some precursors of cell-wall peptidoglycan, it seems to lack the capacity to synthesize glucosamine and *N*-acetylglucosamine; the host cell might provide these essential components.

Aside from glycolytic and citric-acid cycle reactions yielding high energy bonds, it has been shown that acetate is incorporated into rickettsial lipid and radiolabelled amino acids into protein. The latter observation tallies with the inhibition of rickettsial metabolism by chloramphenicol, tetracycline, erythromycin and aminoglycosides and the failure of cycloheximide, an inhibitor of protein synthesis on eukaryotic ribosomes, to affect rickettsial protein synthesis.

The metabolic activities of *Ro. quintana* appear to resemble more closely those of *Rickettsia* than of *Cox. burneti*. *Ro. quintana* does not use glucose but metabolizes pyruvate (Huang 1967, Weiss *et al* 1978). There is also evidence of ornithine decarboxylase activity and use of amino acids as substrates (Weiss 1981).

Resistance

Stability and reversible inactivation

The modest levels of endogenous ATP and macromolecular synthesis exhibited by rickettsiae *in vitro* may be partly a result of the damage they suffer when extracted from host cells and suspended in aqueous media. Except for hardy organisms such as *Cox. burneti*, infectivity is rapidly lost, and nucleotides and macromolecules such as RNA leach from the rickettsial cell. The stability of purified rickettsial suspensions is increased in media with a high K^+ and a low Na^+ content that mimic the intracellular environment in this respect, but this requirement may be one that is exhibited by rickettsiae with membrane damage rather than intact rickettsiae; *Ro. quintana*, which multiplies extracellularly, does not require a high K^+ concentration. Other substances or conditions that increase the stability of rickettsiae include a pH of 7.0, sucrose, bovine serum albumin, Mg^{++} and Mn^{++} ions, and glutamate; these requirements are met in the sucrose-phosphate-glutamate (SPG) fluid widely used as a transport medium for rickettsiae and chlamydiae. It should be noted, however, that medium with a high potassium content may be toxic when inoculated into animals and that brain heart infusion broth maintains the viability of typhus and spotted-fever rickettsiae as well if not better.

The loss of intracellular metabolites, enzymes and coenzymes such as NAD and Co-A, may be connected with reactivation of spotted-fever rickettsiae in ticks. Early observations showed that unfed adult *Dermacentor andersoni* contained rickettsiae that would immunize guinea-pigs but not produce illness. When ticks from the same batch were given a blood meal they became capable of causing disease in guinea-pigs.

The stability of *Cox. burneti* outside the animal cell is in marked contrast to that of other rickettsiae. Reference has already been made to the possibility that *Cox. burneti* forms an endospore. It will survive on wool for 7 to 9 months at 20°, in dried blood for at least 182 days at room temperature, and in tick faeces at room temperature for at least 586 days. It survives 63° for 30 min and is at the borderline of inactivation in the high-temperature short-time process for the pasteurization of milk. Its antigens in phase 1 are stable on autoclaving, and the cells are not broken up by chemical treatments that rupture *Rick. prowazeki* and gram-negative bacteria. In crude, poorly homogenized, yok-sac suspensions it will tolerate 1 per cent formalin for 24 hr. This hardiness underlies its survival in dust and in aerosols and its success as an airborne infective agent.

(The statements made in this section are supported by references cited by Stoker and Marmion 1955, Weiss 1973, and Elisberg and Bozeman 1979.)

Sensitivity to antibiotics and other inhibitors

Penicillin and streptomycin inhibit the growth of *Rickettsia*, *Coxiella* and *Rochalimaea* to a small extent, as might be expected given their possession respectively of a bacterial-type cell wall and of ribosomes. Although plaque reduction is demonstrable in cell culture, however, these antibiotics are ineffective in the treatment of rickettsial infections of man or animals. Penicillin (10^5–10^6 units per l) and streptomycin (100–500 mg per l) may be added to specimens inoculated into mice or the chick-embryo yolk sac but not when isolation is being attempted in cell culture or guinea-pigs. Gentamicin may be inhibitory at 50 mg per l. All three genera of rickettsiae are sensitive to the broad-spectrum antibiotics, chloramphenicol, tetracycline, doxycycline, erythromycin and minocycline, but *Cox. burneti* is significantly less sensitive to chloramphenicol and erythromycin than are members of the other two genera (Smadel *et al.* 1947, Ormsbee and Pickens 1951, Ormsbee *et al.* 1955, Wisseman *et al.* 1974).

Thiocymetin, a synthetic compound closely related to chloramphenicol, has no effect on *Cox. burneti* at doses just tolerated by the experimental hosts but is quite effective in suppressing growth of typhus and spotted-fever rickettsiae (Ormsbee 1969). A recent reassessment (Spicer *et al.* 1981) of antibiotic activity showed that rifampicin, trimethoprim, doxycycline and oxytetracycline were active against *Cox. burneti* in chicken embryos, whereas clindamycin, erythromycin, viomycin, cycloserine and cephalothin were ineffective; the action was rickettsistatic rather than rickettsicidal. Rifampicin, doxycycline, oxytetracycline and erythromycin were effective and rickettsicidal against *Rick. typhi* and *Rick. rickettsi* in the chick-embryo yolk sac; in contrast to the findings with *Cox. burneti*, however, trimethroprim was ineffective. Rifampicin has a low minimal inhibitory dose for the rickettsiae. This fact, and power to penetrate into cells, make it a potentially interesting agent for treatment of persistent or chronic rickettsial infection, e.g., Brill-Zinsser disease and Q-fever endocarditis. Drug resistance can be induced to

Table 46.2 Antigens of rickettsiae (modified from Elisberg and Bozeman 1979)

			Serological activity						
			Soluble antigens		Corpuscular antigens				Weil-Felix
Group	Species	Number of sero-types	CF	Haemagg-lutination (ESS)	CF	IF	Agglut-ination	'Toxin' neutral-ization	reaction: agglutination of *Proteus* strain
Typhus group	*Rick. prowazeki*	1	GS	GS	SS	SS	SS	SS	OX19
	Rick. typhi	1	GS	GS	SS	SS	SS	SS	OX19
	Rick. canada	1	GS	...	SS	SS	SS	SS	...
Spotted-fever group	*Rick. rickettsi*	>4*	GS	GS	SS	SS	SS	SS	OX19 or OX2
	Rick. sibirica	1	GS	...	SS	SS	...	SS	OX19 or OX2
	Rick. conori	1	GS	GS	SS	SS	SS	SS	OX19 or OX2
	Rick. akari	1	GS	GS	SS	SS	SS	None	None
	Rick. australis	1	GS	...	SS	SS	...	None	OX19 or OX2
Scrub-typhus group	*Rick. tsutsuga-mushi*	At least 8	Mainly TS	None	Mainly TS	Mainly TS	...	†	OXK
Trench fever	*Ro. quintana*	Probably 1	SS	None	SS	...	...	None	No reaction
Q fever	*Cox. burneti*	Two phases	None	None	SS	SS	SS	None	No reaction

CF = complement fixation; ESS = erythrocyte-sensitizing substance; IF = immunofluorescence.
GS = group specific; SS = species specific; TS = type specific; ... = no information.
* Cross-reacting serotypes (*Rick. montana*, 363D, etc.), see Philip *et al.* (1981).
† Results difficult to interpret.

chloramphenicol or erythromycin by serial passage of organisms in the presence of the drug, but antibiotic-resistant strains of rickettsiae have not been encountered with any frequency in nature.

The rickettsiae are resistant to sulphonamides; cotrimoxazole has been claimed to have an effect in Q-fever endocarditis (see Chapter 77). On the other hand, most rickettsiae, but not *Cox. burneti*, are inhibited by para-aminobenzoic acid. This unexplained inhibitory effect has no therapeutic significance.

Classification

The present classification of the rickettsiae that infect man, and of certain strains so far isolated only from animals that appear to be closely related to them, was set out in Table 46.1. It rests heavily upon the biological characters exhibited by the organisms in their natural environment. Evidence obtained from other sources, particularly from studies of their antigenic structure (Table 46.2) and of the rickettsial genome, generally confirms this picture and in some ways extends it.

Antigenic structure

The typhus and spotted-fever groups of *Rickettsia* each possess distinct 'soluble antigens' common to all members of the respective group. They also have 'corpuscular antigens' of narrower specificity that in most cases characterize named species; the exception is *Rick. rickettsi*, which has the group antigen of the spotted-fever group but includes serotypes defined mainly by corpuscular antigens and recognized by micro-immunofluorescence (Philip *et al.* 1981). In the scrub-typhus group, strains from various geographical areas are antigenically heterogeneous, but the results of cross-protection tests in mice indicate that they form a distinct group. Preparations of both soluble and corpuscular antigens give both group- and strain-specific reactions in in-vitro tests, and individual strains appear to possess group- and strain-specific antigenic material in varying amount. These reactions have been used to recognize a number of serotypes which are deemed to belong to a single species, *Rick. tsutsugamushi*. It is at present difficult to assign separate specific names to scrub-typhus rickettsiae differentiated on grounds of antigenic structure. There is thus a confusing lack of uniformity in the classification and nomenclature in the three goups of *Rickettsia*, but this is sanctioned by usage and must be accepted for the time being. *Rochalimaea* and *Coxiella* are antigenically distinct from each other and from *Rickettsia*.

Soluble antigens Extraction with ether (p. 528) yields a substance that gives a group-specific complement-fixation (CF) reaction in members of the typhus and spotted-fever groups and a species-specific reaction in *Ro. quintana*; reactions in *Rick. tsutsugamushi* are mainly type specific. Heating members of the typhus and spotted-fever groups under alkaline conditions releases an *erythrocyte-sensitizing substance* (ESS) that, when adsorbed to human group O or sheep red cells, gives a group-specific haemagglutination reaction; this substance does not react in the CF test.

Corpuscular antigens Repeatedly washed and purified suspensions of rickettsiae give reactions in CF, immunofluorescence and micro-agglutination tests that are species specific in the typhus and spotted-fever groups (except in *Rick. rickettsi*), and in *Ro. quintana* and *Cox. burneti*; reactions in *Rick. tsutsugamushi* are partly type and partly group specific.

The Weil-Felix reaction This is a bacterial-agglutination reaction that exhibits partial group specificity. Certain strains of *Proteus* are agglutinated by the sera of patients with rickettsial diseases, or of animals infected experimentally with rickettsiae; it is attributable to extrageneric cross reactions between the carbohydrate haptens of the O antigens of *Proteus* and an antigenic component of the rickettsiae (see Chapter 35). The rickettsial component is distinct from the ESS.

Neutralization of 'toxins' Giving concentrated suspensions of some rickettsial strains intravenously to white mice leads to increased capillary permeability, shock, and death in a few hours. This so-called 'toxic' effect is neutralized by antirickettsial sera, which give species-specific protection against it in the typhus and spotted-fever groups; however, normal monkey or human sera may sometimes neutralize the 'toxin' non-specifically (for references see Elisberg and Bozeman 1979). It differs from the action of rickettsial endotoxin in that it is caused only by viable organisms.

Cross-protection tests Challenge of animals convalescent from one rickettsial infection with a different strain of the organism generally reveals broad antigenic relationships. Thus, cross-protection can be demonstrated not only between members of the typhus group but also between them and members of the spotted-fever group. However, cross-protection cannot be demonstrated between *Rickettsia* and *Coxiella*.

Rickettsia

The typhus group *Rick. prowazeki* and *Rick. typhi* have a common soluble antigen and ESS but distinct corpuscular antigens. Sera of patients suffering from epidemic typhus and murine typhus contain agglutinins for *Proteus* strain OX19.

Strains of *Rick. prowazeki* isolated from flying squirrels (see Bozeman *et al.* 1975) are antigenically indistinguishable from classical strains of this species. The organism named *Rick. canada.* isolated from ticks, *Haemaphysalis leporispalustris*, in Ontario, Canada (McKiel *et al.* 1967), are included in this group despite certain aberrant features. Serological evidence of infection in man has been brought forward (Bozeman *et al.* 1970), but the organism has not been isolated from man, and the possibility of cross-reacting antibody from exposure to *Rick. prowazeki* cannot be excluded. *Rick. canada* forms the group-specific CF antigen but has a distinct corpuscular antigen, and 'toxin'-neutralization tests support its separate identity. However, growth in the cell nucleus, transmission through the ovary in ticks, and the characters of the genome (p. 536) make it unlikely that it should be included in the typhus group.

The spotted-fever group Relations in this group follow similar lines to those in the typhus group. There is a single group-specific CF antigen and, as far as has been ascertained, a common ESS; the corpuscular antigens are species or type specific. Weil-Felix reactions, when present, are with strains OX19, OX2, or both. Five species (*Rick. rickettsi*, *Rick. sibirica*, *Rick. conori*, *Rick. akari* and *Rick. australis*) are well established as pathogens in man.

A growing number of other isolates with distinct corpuscular antigens but of uncertain pathogenicity for man and ecology have been reported. These include, among others, *Rick. parkeri* (Parker *et al.* 1939), the Montana (Bell *et al.* 1963) and Western Montana U strain (Price 1953), the JC880 and TT118 strains (Robertson and Wisseman 1973), *Rick. rhipicephali*, the 364D-like organisms (Lane *et al.* 1981) and a recent isolate, WB-8-2 from *Amblyomma americanum* (Burgdorfer *et al.* 1981). In part these isolates are the products of new techniques for the immunofluorescent demonstration of rickettsiae in tick haemolymph and of the isolation of the organisms of low guinea-pig pathogenicity in cell culture. Philip and co-workers (1983) describe a new species—*Rick. bellii*—that is widely distributed in the USA and said to be distinct from the spotted-fever and typhus groups.

The scrub-typhus group The single species, *Rick. tsutsugamushi*, contains at least eight distinct serotypes.

Soluble antigens can be extracted with ether but their reactions are mainly strain specific; an ESS has not been demonstrated. The fragility of the organisms makes it difficult to prepare satisfactory washed suspensions; corpuscular antigens partly purified by means of cation-exchange resins give both species-specific and strain-specific reactions. The CF reactions of corpuscular antigens of strains isolated in Japan have shown three serotypes represented respectively by the Karp, Gilliam and Kato strains; elsewhere in Asia there are different serotypes. The pattern appears to be one of a range of different antigens, with some cross-reactions, represented in different proportions in isolates from different geographical areas (for references, see Elisberg and Bozeman 1979 and Shirai *et al.* 1979). Suspensions of *Rick. tsutsugamushi* are 'toxic' for mice. This effect is specifically neutralized by antiserum, but it is difficult to obtain heavy suspensions of the organism; neutralization tests are therefore not an effective means of classifying stains. Mice convalescent from experimental infections show solid immunity against subsequent challenge with heterologous scrub-typhus strains. Patients develop agglutinins against *Proteus* strain OXK.

Rochalimaea quintana

A CF antigen can be prepared from *Ro. quintana* by cultivation on the blood agar medium (p. 533) or in a fluid version of this (Mason 1970). Extraction with ether releases a soluble antigen (Vinson and Campbell 1968). ESS has not been detected. There appears to be only one serotype of *Ro. quintana* but small differences have been detected between other properties of strains. Somewhat similar organisms isolated from voles resemble *Ro. quintana* antigenically (Weiss 1981).

Coxiella burneti

This organism is antigenically distinct from each of the three groups of *Rickettsia* and does not stimulate agglutinins for *Proteus* strains. Extraction with ether does not liberate a soluble antigen and does not alter the activity of suspensions in the CF reaction although it may remove minor components reactive in immunofluorescence tests. Rickettsial suspensions are not toxic for mice.

Phase variation of the CF antigen (Stoker and Fiset 1956) is a phenomenon unique to *Cox. burneti*. On first isolation, strains from man, animals and anthropods are in phase 1—an antigenic state in which they react, by complement fixation, with 'late' post-infection sera from guinea-pigs. When phase 1 strains have been passed serially in the chick-embryo yolk sac, organisms with the phase 2 antigen are selected and become dominant in the population. These react in the CF test with 'early' post-infection sera from guinea-pigs. When strains propagated in the yolk sac are passaged in animals, the minority population of phase 1 organisms again becomes dominant.

Phase 1 and 2 antigens can be detected by other serological methods such as micro-agglutination and immunofluorescence (Fiset *et al.* 1969, Goldwasser and Halevy 1972). The organisms with the different antigens can be separated by density-gradient centrifugation. Treatment of phase 1 organisms with trichloracetic acid or sodium metaperiodate destroys the phase 1 antigen and reveals phase 2 reactivity (Fiset *et al.* 1969, Schramek *et al.* 1972). Phase 1 antigen can be liberated from the organism by ultrasonic treatment and can then become adsorbed to sheep erythrocytes—an analogue to the ESS. The phase 1 antigen is heat stable and is a carbohydrate or glycolipid and is responsible for provoking protective antibody; a phase 1 vaccine is some 300 times more effective at stimulating protective immunity than a phase 2 vaccine of similar numbers of organisms (Ormsbee *et al.* 1964). *Cox. burneti* in phase 1 has other interesting properties resembling those of *Corynebacterium parvum*; it is a powerful activator of macrophages (Kelly 1977), stimulates tumour immunity (Kelly *et al.* 1976) and immunity to *Babesia* spp. and *Plasmodium* spp. in mice (Clark 1979).

Characters of the genome

Table 46.3 summarizes information about genetic similarities and dissimilarities between rickettsiae derived from a number of studies (Smith and Stoker 1951, Shramek 1968, 1972, Wisseman 1973, Tyeryar *et al.* 1973, Myers *et al.* 1979, 1980, Myers and Wisseman 1980). Myers and Wisseman (1981) discuss the significance of these findings for classification.

Whereas the degree of sequence homology between various strains of *Rick. prowazeki*, including those recently isolated from flying squirrels, approaches 100 per cent, that between *Rick. prowazeki* and *Rick. typhi* is around 70 per cent. *Rick canada* has been grouped with the typhus rickettsiae, but its genome is larger than those of members of the typhus and spotted-fever groups, and the degree of sequence homology both with members of the typhus group and with *Rick. rickettsi* is low (40–50 per cent). This lends further support to the view that *Rick. canada* does not belong to any of the present biological groups of *Rickettsia*. Sequence homology between *Ro. quintana* and the vole agent also suggests that there are substantial differences between them (Myers *et al.* 1979).

Gel electrophoresis of proteins Analysis of the patterns of cell protein by one- or two-dimensional polyacrylamide-gel electrophoresis is a very discriminating method of classification that might be expected to reflect the characters of the genome. The results obtained support the distinctions made between *Rick. canada* and the members of the typhus group, and between *Ro. quintana* and the vole agent (Weiss 1981). It also confirms the heterogeneity of *Rick. tsutsugamushi* (Hanson and Wisseman 1981). In addition, two-dimensional polyacrylamide-gel electrophoresis reveals fine differences between the virulent Breinl strain of *Rick. prowazeki* and the Madrid E strain that is used as a vaccine (Oaks *et al.* 1981).

Animal pathogenicity

The arthropod and animal reservoirs of the rickettsiae are described in Chapter 77. The present description is concerned with experimental infections in laboratory animals. A wide range of experimental animals—non-human primates, guinea-pigs, mice, rabbits, hamsters, cotton-rats and voles—are susceptible (see earlier editions of this book) but in practice the adult male guinea-pig and the adult white mouse are the animals of choice for primary isolation of rickettsiae. Meadow voles, hamsters and cotton-rats have been considered by some workers to have a greater susceptibility for certain purposes, e.g. the isolation of rickettsiae from cases of Brill's disease or Q fever.

Adult male guinea-pigs are given intraperitoneal injections of 2–4 ml of blood from a febrile patient (or a tissue suspension from another animal). The main response of the guinea-pig to rickettsial infection is fever (rectal temperature of 40° or higher). *Rick. typhi*

Table 46.3 Genetic relatedness among rickettsias: genome size, G + C content, and nucleic acid hybridization (compiled from various authors)

Group	Species	Strain designation	Genome size (mol. wt × 10^6 ± SD)	G + C content (moles per cent)	Degree of sequence homology between designated pairs of organisms: percentage DNA/DNA hybridization
Typhus group	*Rick. prowazeki*	Breinl	106 ± 6.6	29.0	*Rick. prowazeki*/*Rick. typhi* 70–77
	Rick. prowazeki	Flying squirrel	114 ± 5.8	29.5	...
	Rick. typhi	Wilmington	109 ± 7.9	29.0	*Rick. canada*/*Rick. typhi* 47–52 *Rick. canada*/*Rick. prowazeki* 44–45
	Rick. canada	2678	149 ± 4	29.2	*Rick. canada*/*Rick. rickettsi* 39–43
Spotted fever group	*Rick. rickettsi*	Sheila Smith	130 ± 10	32.6	*Rick. rickettsi*/*Rick. prowazeki* 47–53
	Rick. conori	MHLISH	...	33.3	...
	Rick. akari	MK	...	32.4	...
Q fever	*Cox. burneti*	Nine mile, phase 1	104 ± 4	42.7	...
Trench fever and related organisms	*Ro. quintana*	Fuller	95 ± 4	38.5	*Ro. quintana*/*Ro. quintana* 95
	Vole agent	Vole 1 & 2	133 ± 5	39.0	Vole agent/Vole agent 100 Vole agent/*Ro. quintana* 31–42

Table 46.4 Experimental infection in guinea-pigs and mice

Rickettsia	Experimental animal	Incubation period (days)	Clinical response	Mortality
Rick. prowazeki	Male guinea pig	5–12	Fever	None
Rick. typhi	Male guinea pig	3–10	Fever, scrotal reaction	None
Rick. rickettsi	Male guinea pig	3–10	Fever, scrotal reaction, haemorrhages	Common
Rick. akari	White mouse	6–9	Inactivity, rough fur	Usual
Rick. tsutsugamushi	White mouse	3–7	Inactivity, rough fur, ascites	Common
Cox. burneti	Guinea-pig	3–12	Fever	Rare

and members of the spotted-fever group produce an intense inflammation of the testes and scrotum; this is not present in infections with *Rick. prowazeki*, *Cox. burneti* or *Ro. quintana*.

Strains vary in virulence, and the response in guinea-pigs may range from fever and seroconversion to, for example in infections with *Rick. rickettsi*, haemorrhages in scrotum, testes, foot-pads and ears, and the death of a proportion of animals. Heparinized blood taken at the onset of fever, suspensions of the tunica vaginalis or of liver and spleen may be used to pass the rickettsia to fresh animals.

Although a few strains of *Rick. tsutsugamushi* and *Rick akari* will produce some signs of illness in guinea-pigs, white mice are more uniformly susceptible. After an incubation period of 3–10 days the mice become sluggish, with ruffled fur, with ascites and rapid breathing and eventually die. At necropsy subcutaneous oedema and lymphadenitis may be present; also a serofibrinous haemorrhagic exudate in the peritoneal cavity. The spleen may be enlarged.

Rickettsiae may be demonstrated, by Giemsa's or other stains or by immunofluorescence, in impression smears from tunica, spleen or liver of infected animals. The responses of guinea-pigs and mice to the various species of rickettsiae are summarized in Table 46.4. Note that *Ro. quintana* may produce a slight fever in guinea-pigs but cannot be passaged to fresh animals and does not produce a continuing infection.

For further information on rickettsiae and rickettsial diseases, see the Proceedings of the Rickettsiology Conference held in Maryland, USA, in March 1979 (see Philip *et al.* 1980).

Ehrlichia, Cytoecetes, Cowdria and *Neorickettsia*

According to some systematists these genera form a second tribe, the Ehrlicheae, composed of intracytoplasmic rickettsiae that are pathogenic for certain mammals. With one possible exception noted below they appear not to cause disease in man.

Ehrlichia These organisms are associated with monocytes and lymphocytes in dogs, cattle and sheep. *Eh. canis* causes canine rickettsiosis (canine typhus), an acute febrile disease that is widely distributed around the Mediterranean and in Africa, India and the USA. The severity of the disease is said to be enhanced by concurrent infection with *Babesia*, *Leishmania* and *Bartonella*-like organisms (Ewing 1969). Bovine and ovine infections are milder and of less clinical significance than canine infections. *Eh. canis* is transmitted to dogs by the tick, *Rhipicephalus sanguineus*. It varies in size from 0.2 μm to 1.5 μm long and is found in clusters in the cytoplasm of circulating monocytes by the Giemsa method but it does not retain the fuchsin of the Macchiavello stain. It is sensitive to tetracycline. Infection is not transmissible to small laboratory animals or to the chick embryo. In naturally infected animals, agglutinins against *Proteus* strains OX19 or OXK may appear; specific antibody may also be detected by immunofluorescence on cultured monocytes infected with *Eh. canis* (Nyindo *et al.* 1971, Ristic *et al.* 1972).

Edlinger and co-workers (1980) have suggested that *Eh. canis* or a related organism may be responsible for Kawazaki's disease in children (Chap. 73), an acute illness of unknown aetiology characterized by conjunctivitis and a scarlatiniform rash.

Cytoecetes Members of this genus cause tick-borne fever in sheep and cattle. *Cyt. phagocytophila* (Foggie 1951) and *Cyt. bovis* are responsible for this in parts of the United Kingdom and northern Europe that are infested with the tick *Ixodes ricinus*. Bovine petechial fever (Ondiri disease) is probably also tick-borne and appears to be caused by a similar organism (Krauss *et al.* 1972), but the disease is more severe (Haig 1955, Danskin and Burdin 1963). *Cytoecetes* somewhat resembles *Ehrlichia*, with which some workers would classify it, but it is found in neutrophils, eosinophils and basophils as well as in monocytes. It does not retain the fuchsin of the Macchiavello stain but takes up the counterstain. Various forms of the organism, thought to be developmental forms, are seen in leucocytes; they include small spherical granules (0.3–0.5 μm) and large oval bodies (2.5–3.5 μm in greatest diameter) in which clusters of more darkly staining particles appear (morula forms) and which fragment into particles of irregular size—a pattern somewhat reminiscent of chlamydiae. In bovine petechial fever the organism also multiplies in vascular epithelium, producing thrombosis and haemorrhage. *Cytoecetes* is resistant to penicillin and streptomycin; it is sensitive to tetracycline, which may modify the disease, at least in transmission experiments.

Cowdria The single species in this genus, *Cow. ruminantium*, causes heartwater in cattle, sheep and goats in southern Africa and is transmitted by the tick *Amblyomma hebraeum*. The more acute form of the disease is characterized by fever and increased vascular permeability; in less acute cases there may be neurological signs. The organism is 0.2–0.8 μm in size and pleomorphic, forming spherical, oval or even short bacillary forms, and is found mainly in vascular epithelial cells. It is stained by aniline dyes as well as by Giemsa's method, and is gram negative. Like other rickettsiae it is sensitive to tetracycline. The organism can be demonstrated in blood films at the onset of fever and in smears of the intima of large blood vessels at necropsy; it can be cultivated in ovine choroid-plexus cells.

Neorickettsia *Neo. helminthoeca* causes an acute and often fatal disease (salmon poisoning) in dogs that have ingested the metacercariae of the fluke *Nanophyetus salmonicola* found in salmon and trout on the west coast of the USA. It is of interest as a rickettsial disease transmitted by a trematode rather than an arthropod (Philip *et al.* 1954, Kitao *et al.* 1973, Buxton and Fraser 1977). The organisms are found predominantly in the fixed reticuloendothelial cells of dogs. In preparations stained by Giemsa's method they appear as single bodies that are morula-like clusters of coccoid, oval or short bacillary forms each 0.3–0.5 μm in size. It should be noted that the term 'neo rickettsia' has been used by French workers to describe ill defined groups of rickettsiae and chlamydiae observed in Africa.

(For further information about members of these four genera, see Buxton and Fraser 1977.)

References

Anacker, R. L., Fukushi, K., Pickens, E. G. and Lackman, D. B. (1964) *J. Bact.* **88,** 1130.
Anacker, R. L., Pickens, E. G. and Lackman, D. B. (1967) *J. Bact.* **94,** 260.
Anderson, D. R. Hopps, H. E., Barile, M. F. and Bernheim, B. L. (1965) *J. Bact.* **90,** 1387.
Baca, O. G. and Paretsky, D. (1974) *Infect. Immun.* **9,** 959.
Bell, E. J., Kohls, G. M., Stoenner, H. G. and Lackman, D. B. (1963). *J. Immunol.* **90,** 770.
Bozeman, F. M., Elisberg, B. L., Humphries, J. W., Runcik, K. and Palmer, D. B. Jr. (1970) *J. infect. Dis.* **121,** 367.
Bozeman, F. M., Masiello, S. A., Williams, M. S. and Elisberg, B. L. (1975) *Nature, Lond.* **255,** 545.
Burgdorfer, W., Hayes, S. F., Thomas, L. A. and Lancaster, J. L. (1981) In: *Rickettsiae and Rickettsial Diseases*, p. 595. Ed. W. Burgdorfer, and R. L. Anacker. Academic Press, New York.
Burton, P. R., Stueckemann, J. and Paretsky, D. (1975) *J. Bact.* **122,** 316.
Burton, P. R., Stueckemann, J., Welsh, R. M. and Paretsky, D. (1978) *Infect. Immun.* **21,** 556.
Buxton, A. & Frazer, G. (1977) In: *Animal Microbiology*, p. 359. Blackwell Scientific Publications, Oxford.
Chan., M. L., McChesney, J. and Paretsky, D. (1976) *Infect. Immun.* **13,** 1721.
Čiampor, F., Schramek, S. and Brezina, R. (1972) *Acta virol.* **16,** 503.
Clark, I. A. (1979) *Infect. Immun.* **24,** 319.
Cox, H. R. (1938) *Publ. Hlth Rep., Wash.* **53,** 2241; (1941) *Science, N.Y.* **94,** 399.
Danskin, D. and Burdin, M. L. (1963) *Vet. Rec.* **75,** 391.
Davis, G. E. and Cox, H. R. (1938) *Publ. Hlth Rep., Wash.* **53,** 2259.
Edlinger, E. A., Benichon, J. J. and Labrune, B. (1980) *Lancet* **i,** 1146.
Elisberg, B. L. and Bozeman, F. M. (1979) In: *Diagnostic Procedures for Viral, Rickettsial and Chlamydial Infections*, p. 1061. Ed. by E. H. Lennette and N. J. Schmidt. Amer. Publ. Hlth. Ass., Washington, D.C.
Ewing, S. A. (1969) *Adv. vet. Sci. comp. Med.* **13,** 331.
Fiset, P., Ormsbee, R. A., Silberman, R., Peacock, M. and Spielman, S. H. (1969) *Acta virol.* **13,** 60.
Foggie, A. (1951) *J. Path. Bact.* **63,** 1.
Giménez, D. F. (1964) *Stain Tech.* **39,** 135; (1965) *J. Bact.* **90,** 834.
Goldwasser, R. A and Halevy, M. (1972) *Israel J. med. Sci.* **8,** 583.
Gonzales, F. R. and Paretsky, D. (1981). In: *Rickettsiae and Rickettsial Diseases*, p. 493. Eds. W. Burgdorfer and R. L. Anacker. Academic Press, London.
Hackstadt, T. and Williams, J. C. (1981) In: *Rickettsiae and Rickettsial Diseases*, p. 431. Ed. by W. Burdorfer and R. L. Anacker, Academic Press, London.
Haig, D. A. (1955) *Adv. vet. Sci.* **2,** 307.
Hanks, J. H. (1966) *Bact. Rev.* **30,** 114.
Hanson, B. and Wisseman, C. L. Jr. (1981). In: *Rickettsiae and Rickettsial Diseases*, p. 503. Ed. by W. Burgdorfer and R. L. Anacker. Academic Press, London.
Hayes, S. F. and Burgdorfer, W. (1981) In: *Rickettsiae and Rickettsial Diseases*, p. 281. Ed. by W. Burgdorfer and R. L. Anacker. Academic Press, London.
Huang, J. (1967) *J. Bact.* **93,** 853.
Ito, S. and Vinson, J. W. (1965) *J. Bact.*. **89,** 481.
Kazar, J., Ormsbee, R. A. and Tarasevich, I. M. (Eds.) (1978) *Rickettsiae and Rickettsial Disease*. Veda, Bratislava.
Kelly, M. T. (1977) *Cell. Immunol.* **28,** 189.
Kelly, M. T. *et al.* (1976) *Cancer Immunol. Immunother.* **1,** 189.
Kitao, T., Farreli, R. K. and Fukuda, T. (1973) *Amer. J. vet, Res.* **34,** 927.
Kordová, N. (1960) *Acta virol.* **4,** 56; (1966) *Ibid.* **10,** 278.
Kordová, N. and Kováčova, E. (1967) *Acta virol.* **11,** 252.
Krauss, H., Davies, F. G., Odegaard, O. A. and Cooper, J. E. (1972) *J. comp. Path.* **82,** 241.
Lane, R. S., Philip, R. N. and Casper, E. A. (1981) In: *Rickettsiae and Rickettsial Diseases*, p. 575. Ed. by W. Burgdorfer and R. L. Anacker. Academic Press, London.
McCaul, T. F., Hackstadt, T. and Williams, J. C. (1981). In: *Rickettsiae and Rickettsial Diseases*, p. 267. Ed. by W. Burgdorfer and R. L. Anacker. Academic Press, London.
McDade, J. E., Stakebake, J. R. and Gerone, P. J. (1969) *J. Bact.* **99,** 910.
McKiel, J. A., Bell, E. J. and Lackman, D. B. (1967) *Canad. J. Microbiol.* **13,** 503.
Mason, R. A. (1970) *J. Bact.* **103,** 184.
Merrell, B. R., Weiss, E. and Dasch, G. A. (1978) *J. Bact.* **135,** 633.
Moulder, J. W. (1962) *The Biochemistry of Intercellular Parasitism*, University of Chicago Press, Chicago; (1966) *Annu. Rev. Microbiol.* **20,** 107.
Myers, W. F., Baca, O. G. and Wisseman, C. L. Jr. (1980) *J. Bact.* **144,** 460.
Myers, W. F., Cutler, L. D. and Wisseman, C. L. Jr. (1969) *J. Bact.*. **97,** 633.
Myers, W. F., Ormsbee, R. A., Osterman, J. V. and Wisseman, C. L. Jr. (1967*a*) *Proc. Soc. exp. Biol. Med.* **125,** 459.
Myers, W. F., Osterman, J. V. and Wisseman, C. L. Jr. (1972) *J. Bact.* **109,** 89.
Myers, W. F., Provost, P. J. and Wisseman, C. L. Jr. (1967*b*) *J. Bact.* **93,** 950.

Myers, W. F. and Wisseman, C. L. Jr. (1980) *Int. J. syst. Bact.* **30,** 143; (1981) In: *Rickettsiae and Rickettsial Diseases* p. 313. Ed. by W. Burgdorfer and R. L. Anacker. Academic Press, New York.

Myers, W. F., Wisseman, C. L. Jr., Fiset, P., Oaks, E. V. and Smith, J. F. (1979) *Infect. Immun.* **26,** 976.

Nyindo, M. B. A., Ristic, M., Huxsoll, D. L. and Smith, A. R. (1971) *Amer. J. vet. Res.* **32,** 1651.

Oaks, E. V., Wisseman, C. L. and Smith, J. F. (1981). In: *Rickettsiae and Rickettsial Diseases*, p. 461. Ed. by W. Burgdorfer and R. L. Anacker. Academic Press, New York.

Ormsbee, R. A. (1969). *Annu. Rev. Microbiol.* **23,** 275.

Ormsbee, R. A., Bell, E. J., Lackman, D. B. and Tallent, G. (1964) *J. Immunol.* **92,** 404.

Ormsbee, R. A., Parker, H. and Pickens, E. (1955) *J. infect. Dis.* **96,** 162.

Ormsbee, R. A. and Peacock, M. G. (1964) *J. Bact.* **88,** 1205; (1976). In: *Tissue Culture Association Manual* **2,** 475.

Ormsbee, R., Peacock, M., Gerloff, R., Tallent, G. and Wike, D. (1978) *Infect. Immun.* **19,** 239.

Ormsbee, R. A. and Pickens, E. G. (1951) *J. Immunol.* **67,** 437.

Osterman, J. V., Myers, W. F. and Wisseman, C. L. Jr. (1974) *Acta virol.* **18,** 151.

Palmer, E. L., Mallavia, L. P., Tzianobos, T. and Obijeski, J. F. (1974*a*) *J. Bact.* **118,** 1158.

Palmer, E. L., Martin, M. L. and Mallavia, L. (1974*b*) *Appl. Microbiol.* **28,** 713.

Paretsky, D. (1968) *Zbl. Bakt.* I Abt. Orig. **206,** 283.

Parker, R. R., Kohls, G. M., Cox, G. W. and Davis, G. E. (1939) *Publ. Hlth Rep., Wash.* **54,** 1482.

Perkins, H. R. and Allison, A. C. (1963) *J. gen. Microbiol.* **30,** 469.

Philip, C. B., Hadlow, W. J. and Hughes, L. E. (1954) *Exper. Parasitol.* **3,** 336.

Philip, R. N., Lane, R. S. and Casper, E. A. (1981) *Amer. J. trop. Med.* **30,** 722.

Philip, R. N., Paretsky, D., Weiss, E. and Wisseman, C. L. (1980) *J. infect. Dis.* **141,** 112.

Philip, R. N. *et al.* (1983) *Int. J. syst. Bact.* **33,** 94.

Price, W. H. (1953) *Amer. J. Hyg.* **58,** 248.

Ristic, M., Huxsoll, D. L., Weisiger, R. M., Hildebrandt, P. K. and Nyindo, M. B. A. (1972) *Infect. Immun.* **6,** 226.

Robertson, R. G. and Wisseman, C. L. Jr. (1973) *Amer. J. Epidemiol.* **97,** 55.

Schramek, S. (1968) *Acta virol.* **12,** 18; (1972) *Ibid.* **16,** 447.

Schramek, S., Brezina, S. R. and Urvolgyi, J. (1972) *Acta virol.* **16,** 487.

Shirai, A., Robinson, D. M., Brown, D. W., Gan, E. and Huxsoll, D. L. (1979) *Jap. J. med. Sci. Biol.* **32,** 337.

Silberman, R. and Fiset, P. (1968) *J. Bact.* **95,** 259.

Silverman, D. J., Boese, J. L. and Wisseman, C. L. Jr. (1974) *Infect. Immun.* **10,** 257.

Silverman, D. J., Fiset, P., and Wisseman, C. L. Jr. (1979) *J. clin. Microbiol.* 9, 437.

Silverman, D. J. and Wisseman, C. L. Jr. (1978) *Infect. Immun.* **21,** 1020.

Silverman, D. J., Wisseman, C. L. Jr., Waddell, A. D. and Jones, M. (1978) *Infect. Immun.* **22,** 233.

Smadel, J. E., Jackson, E. B. and Gould, R. L. (1947) *J. Immunol.* **57,** 273.

Smith, J. D. and Stoker, M. G. P. (1951) *Brit. J. exp. Path.* **32,** 433.

Spicer, A. J., Peacock, M. G. and Williams, J. C. (1981) In: *Rickettsiae and Rickettsial Diseases*, p. 375. Ed. by W. Burgdorfer and R. L. Anacker. Academic Press, London.

Stoker, M. G. P. and Fiset, P. (1956) *Canad. J. Microbiol.* **2,** 310.

Stoker, M. G. P. and Marmion, B. P. (1955) *Bull. World Hlth Org.* **13,** 781.

Suitor, E. C. Jr. (1964) *J. infect. Dis.* **114,** 125.

Suitor, E. C. Jr., and Weiss, E. (1961) *J. infect. Dis.* **108,** 95.

Tyeryar, F. J., Jr., Weiss, E., Millar, D. B., Bozeman, F. M. and Ormsbee, R. A. (1973) *Science, N.Y.* **180,** 415.

Vinson, J. W. and Campbell, E. S. (1968) *Acta virol.* **12,** 54.

Weiss, E. (1973) *Bact. Rev.* **37,** 259; (1981) In: *Rickettsiae and Rickettsial Diseases*, p. 387. Ed. by W. Burgdorfer and R. L. Anacker. Academic Press, London.

Weiss, E., Dasch, G. A., Woodman, D. R. and Williams, J. C. (1978) *Infect. Immun.* **19,** 1013.

Weiss, E., Myers, W. F., Suitor, E. C. Jr., and Neptune, E. M. (1962) *J. infect. Dis.* **110,** 155.

Winkler, H. H. and Miller, E. T. (1981) In: *Rickettsiae and Rickettsial Diseases*, p. 327. Ed. by W. Burgdorfer and R. L. Anacker. Academic Press, London.

Wisseman, C. L. Jr. (1973) *Acta virol.* **17,** 443; (1981) In: *Rickettsiae and Rickettsial Diseases*, p. 293. Ed. by W. Burgdorfer and R. L. Anacker. Academic Press, London.

Wisseman, C. L. Jr., Waddell, A. D. and Walsh, W. T. (1974) *J. infect Dis.* **130,** 564.

Wood, W. H. Jr. and Wisseman, C. L. Jr. (1967) *J. Bact.* **93,** 1113.

Woodward, T. E., Pedersen, C. E. Jr., Oster, C. N., Bagley, L. R., Romberger, J. and Snyder, M. T. (1976) *J. infect. Dis.* **134,** 297.

Zdrodovskii, P. F. and Golinevich, H. M. (1960) *The Rickettsial Diseases*. Pergamon Press, Oxford.

Zahorchak, R. J. and Winkler, H. H. (1981). In: *Rickettsiae and Rickettsial Diseases*, p. 401. Ed. by W. Burgdorfer and R. L. Anacker. Academic Press, London.

47

The Mycoplasmatales: *Mycoplasma, Ureaplasma* and *Acholeplasma*

Geoffrey Smith

Definition

Microscopically visible, extremely pleomorphic organisms showing granules, rings, coccoid forms, filaments and other bizarre forms. Possess a limiting membrane but no rigid cell wall. The smallest visible forms can pass membrane filters of APDs between 220 and 450 nm. Non-motile in the usual sense. Stain poorly with ordinary bacterial stains but well with Giemsa. Gram negative. Grow in nutrient media in absence of living tissue cells. Facultative anaerobes. Growth often best in an atmosphere containing added CO_2. Most species on suitable solid media form characteristic minute colonies showing central downgrowth into the medium. Most species require animal protein; and *Mycoplasma* (but not *Acholeplasma*) species require sterol in the medium. Usually ferment carbohydrates or hydrolyse arginine or urea. Readily destroyed by heat. Antigenic specificity is usual and growth is inhibited by specific antiserum. Do not give rise to inclusion bodies in tissues. Considerable degree of host specificity. Some species pathogenic. Immunity following disease does not appear to be specially lasting. G+C content of DNA mostly 23–41 moles per cent.

Type species: *Mycoplasma mycoides* subsp. *mycoides* and *Acholeplasma laidlawi*.

Introductory

This chapter deals with the group of organisms previously known as the *pleuropneumonia group* because of their resemblance to the agent now called *Mycoplasma mycoides* subsp. *mycoides* (Borrel *et al.* 1910) responsible for contagious bovine pleuropneumonia (see Chapter 78). This agent was recognized and cultivated by Nocard and Roux as long ago as 1898, but it is only within more recent years that a number of closely related organisms—pathogens, commensals, and possibly saprophytes—have been described, and that the importance of this large group of organisms possessing unusual and distinctive properties has been realized. The complex morphology of the pleuropneumonia organism was described by Bordet (1910) and by Borrel and his colleagues (1910). The fact that Berkefeld filtrates often proved infective afforded ground for the belief that it was a filtrable virus. The observations, however, of Barnard (1926), Smiles (1926), Ørskov (1927), Nowak (1929), Wroblewski (1931), Ledingham (1933), Klieneberger (1934), Tang and his colleagues (1935, 1936), Turner (1935), and Merling-Eisenberg (1935) indicate that only the small forms, and some of the plastic filamentous forms, are capable of passing through coarse filters.

The second organism of this group was described by Bridré and Donatien (1923, 1925), who isolated it from sheep infected with contagious agalactia (see Chapter 78). From the mid-1930s onwards, the work of Klieneberger (later Klieneberger-Nobel) did much to illuminate the nature of mycoplasmas, and their pathogenicity, particularly for rats and mice (see Tully 1980). The first mycoplasma from man was reported by Dienes and Edsall in 1937; it came from a Bartholin's abscess. Later the Eaton Agent, originally found in cold-agglutinin positive cases of primary atypical pneumonia (Eaton *et al.* 1944), was shown to be a mycoplasma, *M. pneumoniae* (Clyde, 1961, Marmion and Goodburn 1961, Chanock *et al.* 1962). Mycoplasmas have now been found in a wide range of animal and bird species including most of the domestic and farmyard animals and certain wild species. Some of these organisms are primarily responsible for the disease with which they are associated. Others are looked upon as opportunist pathogens or secondary invaders, or as one element in the Faktorkrankheiten of German veterinary workers (Eichwald *et al.* 1971). Some appear to be simple commensals. They frequently occur as contaminants in animal tissue cultures, particularly in continuous cell lines. The genus *Spiroplasma* contains mycoplasmas of insects and plants (see Razin 1978, Whitcomb and Tully 1979). The existence of organisms thought to be saprophytic was demonstrated in sewage by Laidlaw and Elford (1936) and confirmed by Seiffert (1937*a*, *b*).

Classification

It would serve no useful purpose to discuss the various classifications suggested by the earlier workers (see 5th edition, p. 1149). In reference to the properties already mentioned, mycoplasmas differ from bacteria in their lack of a rigid cell wall and—with the exception of the acholeplasmas—in their requirement for cholesterol or other source of sterols; the G+C content of their DNA is low, and many of them (see below) possess a genome of a size smaller than 10^9 daltons—the smallest genome size of wall-covered bacteria. It is now generally agreed that these organisms should be grouped in the class Mollicutes, order Mycoplasmatales, to which the binary system of nomenclature should be applied (Edward *et al.* 1966, Report 1967, Freundt and Edward 1979). Members of the family Mycoplasmataceae have a genome size of 4.5×10^8 daltons and are classified in the genus *Mycoplasma*, which contains more than 60 named species, and the genus *Ureaplasma*. The ureaplasmas comprise the T-strains (T for tiny colony formation) of Shepard (1954); they form a distinctive collection of sterol-requiring organisms characterized by growth at low pH, ability to hydrolyse urea, and complex serological reactions (Report 1971, Shepard and Masover 1979). Human ureaplasmas belong to the species *U. urealyticum*, in which there are at least 14 serotypes (Lin and Kass 1980). Howard and his colleagues (1978) proposed 11 strains as representing the serological diversity of bovine ureaplasmas; these strains were antigenically distinct from *U. urealyticum* and should probably be regarded as belonging to a separate species. Members of the family Acholeplasmataceae (see Tully 1979) have a genome size of 10^9 daltons, are independent of sterol, and belong to the genus *Acholeplasma*, in which there are at least seven species. The genus *Spiroplasma* of the family Spiroplasmataceae contains organisms that consist of helical motile filaments with a genome size of 10^9 daltons. Most spiroplasmas cause disease in plants, but their primary hosts are probably insects (Maramorosch 1974, Razin 1978, Whitcomb and Tully 1979). The *suckling-mouse cataract agent* is a spiroplasma originally isolated from ticks; it is pathogenic for infant mice when inoculated intracerebrally (Tully *et al.* 1977). Unless specifically noted to the contrary, the term mycoplasma will be used in this chapter to apply to all members of the order Mycoplasmatales.

L-forms, which were described in Chapter 2, and which were at one time thought to belong to the *Mycoplasma* group (Klieneberger 1935, 1942), were shown by Dienes (1939, 1942) to be variant forms of bacteria and to be unrelated to the mycoplasmas. The evidence at present available is strongly against the identity of these two groups (see reviews by Hijmans *et al.* 1969, Somerson and Weissman 1969). Although their morphology and colonial appearance are similar, there are no confirmed reports of reversion of a mycoplasma to a bacterial phase; nor when relationships between them were suspected have these been confirmed by DNA base ratio or nucleic acid homology studies (McGee *et al.* 1967, Somerson and Weissman 1969, Neimark 1971). These and other differences between L-phase bacteria and mycoplasmas were reviewed by Edward and Freundt (1969*a*, *b*; see also Hijmans *et al.* 1969) who, supported by the Subcommittee on Taxonomy (Edward *et al.* 1966), considered there was no longer any reason to confuse the two groups. (For further information on classification and taxonomy see Freundt and Edward 1979, Report 1979.)

Cellular morphology

Individual cells of *Mycoplasma* and *Acholeplasma* species show extreme pleomorphism ranging from coccoid, coccobacillary, ring or signet ring, dumb-bell, and asteroid forms to long branching beaded or segmented filaments. Examples of some of the forms that may be seen are shown in Figures 47.1 to 47.5. Though

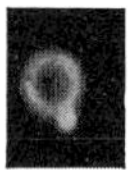

Fig. 47.1. *M. mycoides* subsp. *mycoides* signet ring form. Dark-ground illumination (× 3600). (After Turner.)

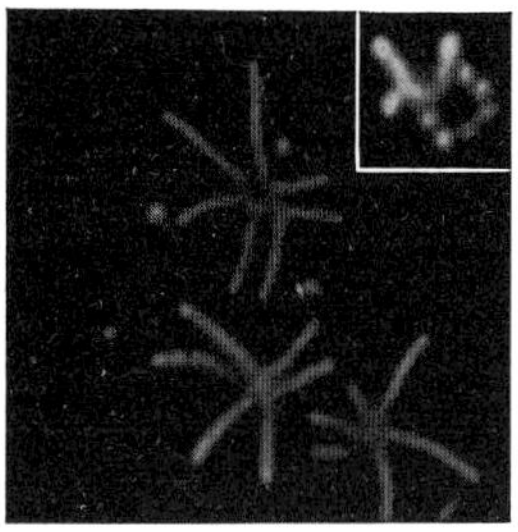

Fig. 47.2 (inset). *M. mycoides* subsp. *mycoides* ring form with developing filaments. Dark-ground illumination (× 3600).

Fig. 47.3. *M. mycoides* subsp. *mycoides*. Asteroid form. Dark-ground illumination (× 3600). (After Turner.)

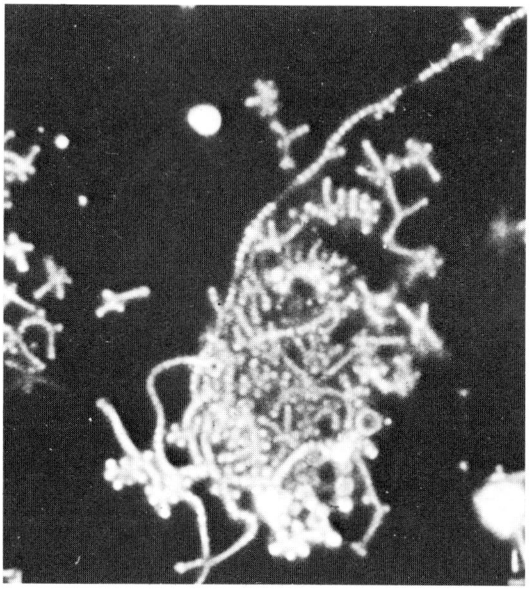

Fig. 47.4. *M. mycoides* subsp. *mycoides*; branching beaded filaments. Dark-ground illumination (× 2520). (By courtesy of A. W. Turner.)

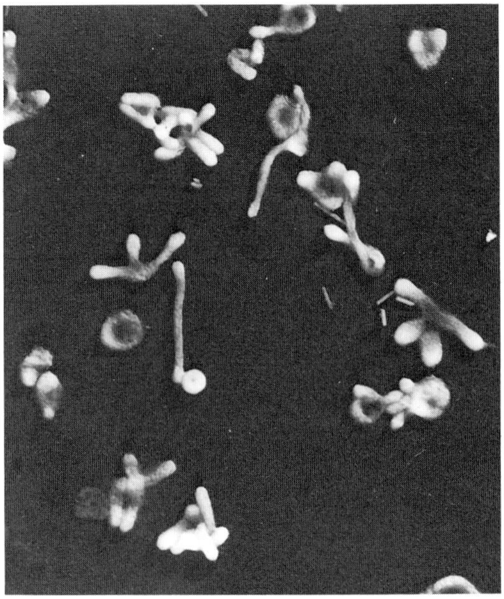

Fig. 47.5. Goat mycoplasma. From culture in liquid medium showing ring forms and short filaments, some branching. Electronmicrograph shadowed (× 10 000). (Kindly supplied by Dr Klieneberger-Nobel.)

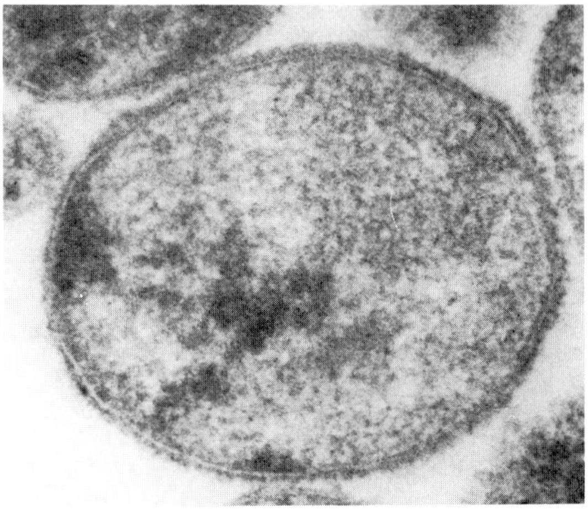

Fig. 47.6. Thin section electronmicrograph of *M. bovoculi* showing triple-layered cell membrane (× 120 000). (Kindly supplied by Mr F. Rodgers.)

many of these types of cell can be seen in cultures of most species, there is a tendency with some species for particular forms to predominate; experienced workers may sometimes find this helpful in preliminary identification. Numerous observers from Turner (1935) onwards are agreed that the large cells such as those seen at the periphery of the *M. hominis* colony illustrated in Fig. 47.10 often contain minute areas of concentrated material usually referred to as 'granules' (the significance of these will be considered below). *M. pneumoniae*, *M. pulmonis*, and to a lesser degree *M. gallisepticum*, show gliding motility on liquid-covered surfaces, and spiroplasmas show rotatory and flexional movement resembling that of spirochaetes (Bredt *et al.* 1970, Razin 1978, Bredt 1979).

Electronmicrographs show that the organisms are bounded by a triple-layered membrane, 7.5 to 10 nm wide, the middle layer being less electron-dense than the inner and outer layers (Fig. 47.6). (For a review of mycoplasma membranes see Razin 1975.) The cytoplasm appears to be mainly amorphous or finely granular (Domermuth *et al.* 1964, Maniloff *et al.* 1965). No cell wall of the type seen in bacteria can be distinguished. The absence of such a structure clearly accounts for the plasticity of the organisms. Ribosomes were identified in cells of *M. gallisepticum* by Morowitz and others (1962), and strands resembling DNA were seen by Anderson and Barile (1965) in thin sections of *M. hominis. M. gallisepticum* often shows blebs 80–100 nm in diameter at the end of elongated cells (Maniloff *et al.* 1965); and *M. pneumoniae* filaments may show rod-like thickenings at one end of the cell (Biberfeld and Biberfeld 1970). The function of these structures is not understood. Chu and Horne (1967) described surface projections in electronmicrographs of negatively stained *M. gallisepticum*. These were similar to but somewhat coarser than those seen in myxoviruses. The authors drew attention to the similarity between these organisms in their haemagglutination behaviour. Capsular material has been demonstrated in a number of mycoplasmas (Razin 1978). Cells of *M. mycoides* subsp. *mycoides* are covered with galactan, a substance probably associated with pathogenicity.

Mode of reproduction

The mode of reproduction has for long been a matter of controversy, and even until quite recently such workers as Freundt (1969) and Bredt (1968*a*) were not in complete agreement on the part played by binary fission, budding, and the formation of coccoid bodies within the filaments, or on the occurrence or not of a specific growth cycle.

The early workers (Turner 1935, Tang *et al.* 1935) postulated complex reproductive cycles for *M. mycoides* subsp. *mycoides* in an attempt to explain the peculiar morphology of these organisms, but the observations of Klieneberger and Smiles (1942), Klieneberger (1942), W. E. Smith and coworkers (1948), and Cuckow and Klieneberger-Nobel (1955) (see also reveiws by Freundt 1958, Klieneberger-Nobel 1962) led to simpler concepts based on the realization that, owing to their extreme plasticity, some of the bizarre morphological forms might be artefacts produced during the preparation of specimens. At that time it seemed probable that reproduction resulted from fragmentation or budding of spherical organisms and fragmentation of beaded filaments. In addition there was evidence suggesting that the granules of 100–150 nm seen in very large cells were released to grow into mature organisms. According to Morowitz and his colleagues (1967; see also Pirie 1969) the theoretical minimum size for a cell capable of reproduction, i.e. containing only the genome, a ribosome and a cell membrane, is about 130 nm. It therefore seems unlikely that the granules, being so close to this size, are viable reproductive units. Indeed there is now general agreement that the mycoplasmas have a minimal viable unit about 300 nm in diameter. Lemcke (1971), for instance, found that cells of *A. laidlawi* passed through a 220 nm APD filter only when high pressure was applied. Likewise the studies of Rakovskaja and co-workers (1973) by use of density-gradient centrifugation showed that most viable cells in broth culture of *A. laidlawi* were between 400 and 600 nm in diameter; less than 0.1 per cent had a diameter smaller than 250 nm. Keller and Morton (1954), studying growth curves, and Furness and his colleagues (1968) studying synchronously dividing cultures, obtained evidence of binary fission, but not of cell bursts liberating multiple elementary bodies. It seems reasonable, therefore, to abandon the concept of liberation of granules from large spherical cells as playing any part in the reproductive process. Similarly fragmentation of organisms seems to be excluded, but budding of spherical cells and the fragmentation of beaded filaments remain. Both of these processes can be regarded as modifications of binary fission, if it is assumed that cytoplasmic division may get out of step with genome replication, as suggested by Razin (1973). Bredt (1968*a*, *b*, 1970, 1972) and Bredt *et al.* (1970, 1971) studied the growth of several mycoplasma species by phase-contrast microscopy and time-lapse cinematography and found by each method evidence only of budding and fragmentation as the mode of reproduction. During the growth of *M. hominis* repeated changes in form were observed with frequent reversion of a given form to one of the previous forms (see Fig. 47.7).

Growth requirements

Mycoplasmas need a rich medium containing natural animal protein and, with the exception of the acholeplasmas, a sterol component. Moisture is essential, a soft agar is preferable, and added CO_2 and a low oxygen pressure are often advantageous. Of the natural animal proteins the most commonly used is blood serum in a concentration of 10 to 20 per cent. Serum supplies not only cholesterol but also saturated and unsaturated fatty acids, which the organisms are unable to synthesize (P. F. Smith 1964). A good basal medium is that of Hayflick (1965), consisting of the so-called PPLO medium of Morton and his colleagues (1951) to which are added 20 per cent of horse serum and 10 per cent of a 25 per cent extract of fresh bakers' yeast. The final concentration of agar is 1 per cent. A pH of 7.0–7.8 is advisable, except for T-strains, which prefer one of 6.0 (Shepard 1967). Incubation is best at 33–37°. Fluid media can be made up with the same constituents as those of solid media but without agar.

The addition of crude DNA (Edward 1954) or boiled blood extract (Klieneberger-Nobel 1962, Hederscheê 1963) to the medium improves the growth of some species. Others need modified media. For example, a medium was devised for *M. hyopneumoniae* that contained pig serum and lactalbumin hydrolysate in Hanks's solution in addition to broth and yeast extract (Goodwin and Whittlestone 1964, Goodwin *et al.* 1965, 1967), which proved suitable also for *M. dispar* and *M. flocculare*. *M. synoviae* requires nicotinamide adenine dinucleotide (NAD) in the medium (Chalquest 1962). Certain mycoplasmas—sometimes described as *non-cultivable*—grow with ease in tissue culture but with difficulty

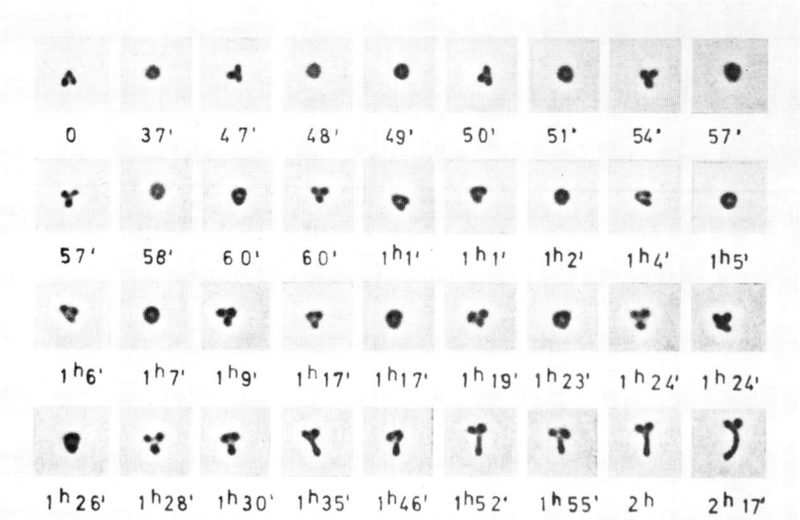

Fig. 47.7. Development of a single cell of *M. hominis*. Selected frames from phase-contrast time-lapse ciné film of 2 hr 17 min duration (magnification ×2500). (Kindly supplied by Professor W. Bredt and reproduced by permission of Springer Verlag, Heidelberg, from *Z. med. Microbiol. Immunol.*, 1970, **155,** 263.)

or not at all in cell-free media (Hopps *et al.* 1973, Freundt and Edward 1979).

Penicillin (100–1000 IU per ml) and thallous acetate (1 in 2000–1 in 4000) are usually incorporated in mycoplasma media to inhibit overgrowth by contaminant organisms. Thallous acetate, however, should be omitted or reduced for ureaplasmas (Shepard and Lunceford 1967). Andrews and his colleagues (1973) found that media containing ampicillin were better than those containing benzylpenicillin for the isolation of *M. dispar*. The sterol-independent *Acholeplasma* species grow readily on the standard broth yeast serum medium but the serum, which often promotes growth, is not essential. (See review by Lemcke 1965, Fallon and Whittlestone 1969.)

Colonial morphology

When grown on appropriate agar media most strains of *Mycoplasma* form minute colonies, 0.1–1 mm in diameter, only just visible to the naked eye. The size, however, varies greatly. *Ureaplasma* colonies, for example, are smaller and may be less than 0.01 mm in diameter (Shepard 1967), whereas colonies of bovine group 7 may exceed 2 mm (Leach 1973). Under low-power magnification (×25 to ×100) characteristic colonies are umbonate by reflected light. By transmitted light most species manifest the so-called fried-egg appearance with differentiation into a central dark area caused by downgrowth into the agar and a peripheral lighter area of surface growth (Razin and Oliver 1961, Domermuth *et al.* 1964, Knudson and Macleod 1970, Le Normand *et al.* 1971, Nakamura and Kawaguchi 1972). The relative proportions of the central and peripheral areas of the colonies of a given species vary on different media. Some species, such as *M. pneumoniae*, show practically no surface growth; others, such as *M. hyopneumoniae* and *M. pulmonis*, show little downgrowth. The surface growth of some species, *M. pulmonis* and *M. hominis* for example, usually has a vacuolated or lacy appearance owing to the presence of very large single cells in the periphery of the colony. Figures 47.8, 47.9 and 47.10 show respectively an umbonate colony, Giemsa-stained colonies of a goat mycoplasma having a finely granular structure, and *M. hominis* with a vacuolated appearance.

A warning must be given about the occurrence of pseudo-colonies in agar media containing a high proportion of serum. They consist of calcium and magnesium soaps, and give rise to fresh centres of crystallization when transferred to uninoculated serum agar plates, causing further confusion

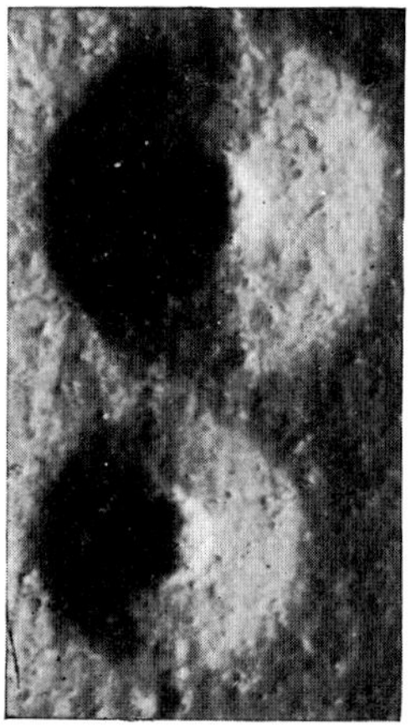

Fig. 47.8. *M. mycoides* subsp. *mycoides*. Umbonate surface colonies by reflected light (×140). (After Tang *et al.*)

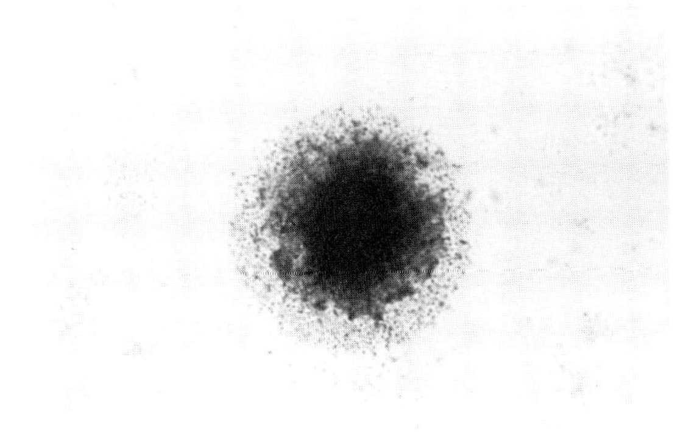

Fig. 47.9. Goat mycoplasma. Colony of fine granular type. Giemsa (× 500).

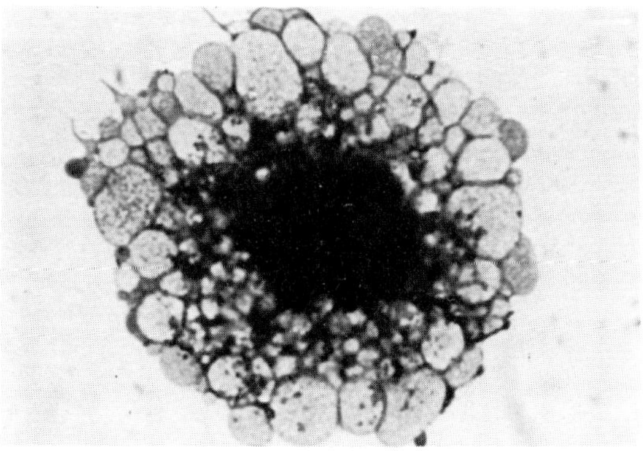

Fig. 47.10. *M. hominis* colony of 'vacuolated' type. Giemsa (× 750).

by the impression they give of being colonies of reproducible living bodies.

The techniques for the cultivation of mycoplasmas are in general similar to those for bacteria. However, since colonies on agar are often adherent to the medium, subculture is usually best done by cutting out a block of agar containing one or more colonies; this is inverted on a new plate and smeared over the surface, or alternatively is dropped into a small quantity of broth. A heavy inoculum is often essential in the early stages of adaptation to in-vitro growth, and at all stages plate cultures need to be protected from drying by incubation in sealed containers. A peculiarity of mycoplasmas is that, once maximum growth in solid or liquid media is obtained, the culture often loses viability very rapidly. An extreme example of this is provided by some ureaplasmas which may reach maximum titre in broth culture and then die out completely within 24 hours.

Chemical composition and metabolism

Little is known about the chemical composition of the mycoplasmas. The cell membrane consists almost entirely of lipoprotein (Kahane and Razin 1970, Razin 1973) and, unlike that of bacteria, contains cholesterol but lacks α, ε-diaminopimelic acid. The G + C content of the DNA ranges from 23 to 41 moles per cent, with the exceptionally high value of 40 per cent for *M. pneumoniae*. (For a list of DNA values see Neimark 1971, Holländer and Pohl 1980.) Cell-protein 'fingerprints', obtained by one- or two-dimensional gel electrophoresis (Rottem and Razin 1967, Morowitz and Terry 1969, Rodwell and Rodwell 1978) are of value in distinguishing and comparing species or subspecies, each of which gives a clearly recognizable and reproducible pattern. Neimark (1977) extracted an actin-like protein from *M. pneumoniae*; this was the first report of contractile protein in prokaryotes. In regard to enzymic activities Rodwell (1969), who designed a completely synthetic medium for the growth of a goat strain, concluded that wide variations existed in the energy-yielding metabolism and the metabolic patterns of different species.

For detailed discussion on the physiology of these organisms reference should be made to Hayflick (1969), Razin (1969, 1973, 1978), P. F. Smith (1971), and Barile and Razin (1979).

Resistance

According to Tang and his colleagues (1935) serum broth cultures of *M. mycoides* subsp. *mycoides* may remain viable for 45 days at 37° and for 98 days at 0–5°. At −20° they survive for 6 to 12 months (Klieneberger-Nobel 1962). Most species of *Mycoplasma*, however, die out much more rapidly than this at temperatures above freezing. They may be preserved by freeze-drying, and by storage at −70° or in liquid nitrogen. At −70° practically no loss of viability may occur after 42 months (Addey *et al.* 1970); and in liquid nitrogen, after some loss of viability during freezing, they remain stable for 3 to 9 years (Norman *et al.* 1970). As already noted, ureaplasmas may survive for only a few hours after the end of maximum growth (Ford 1962). Most if not all *Mycoplasma* species are killed if held at 56° for 30 min; for some species a much shorter time is sufficient.

The organisms are soluble in bile, but are said to be very resistant to ultraviolet rays and the photodynamic action of methylene blue (Tang *et al.* 1936), though on this point there is considerable doubt. They are resistant to the sulphonamides, and to thallium acetate in a concentration of about 1 in 1000 (Edward 1947). Most, but not quite all, strains resist penicillin. The reason for the occasional exceptions is obscure; since penicillin acts on the cell wall and the mycoplasmas are devoid of this structure, it must be assumed that some other factor is responsible (see Wright 1967). Nearly all strains are inhibited by tetracycline in low concentration, and usually also by kanamycin (see Newnham and Chu 1965). Their reaction to erythromycin and cycloserine is variable (Niitu *et al.* 1970). The growth of the sterol-requiring species is said to be inhibited by digitonin and sodium polyanethol sulphonate; this property is of aid in distinguishing strains of *Mycoplasma* from those of *Acholeplasma* (Freundt *et al.* 1973).

Biochemical reactions

Mycoplasmas differ widely in their biochemical activity and the metabolic pathways they use (Rodwell 1969). Some strains ferment glucose, others do not. Those that do often produce acid also from mannose, maltose, starch and glycogen, and usually fail to deaminate arginine. Non-fermenters, on the other hand, are mostly able to deaminate arginine. Ureaplasmas hydrolyse urea but not arginine, and fail to ferment glucose. Fermentative strains catabolize carbohydrates as a rule through the glycolysis pathway. For purposes of classification, sterol dependence inferred from failure to grow in serum-free media, inhibition of growth by digitonin, fermentation of sugars, and hydrolysis of arginine and urea are of some value. To these tests may be added the production of carotenoids and the hydrolysis of aesculin, which are of differential value in the acholeplasmas (Razin and Cleverdon 1965, Tully and Razin 1968), all of which ferment glucose (see also Williams and Wittler 1971). Tests for phosphatase and proteolytic activity, for tetrazolium reduction (Aluotto *et al.* 1970), for peroxide formation (Lind 1968, Sobeslavsky and Chanock 1968), and for the development of 'film and spots' in egg yolk medium (Fabricant and Freundt 1967) may also be of help. *M. pneumoniae* is distinctive among human strains in its ability to cause complete lysis of mammalian red corpuscles, though most other strains, human and animal, including T-strains, produce some degree of haemolysis in blood agar made up with red cells from the guinea-pig (Shepard 1969). Catalase is not formed. (For further information see Report 1972.)

Antigenic structure and serological behaviour

As pointed out by Purcell and his colleagues (1969) mycoplasmas resemble viruses (1) in being separable into a finite number of serotypes; (2) in the possession by certain species of haemagglutinating, haemadsorbing, and cell-attaching properties; and (3) in being inhibited or neutralized by specific antiserum.

The picture obtained of their antigenic structure is determined by the particular method used for their examination. Among the common methods available are the following. (*a*) Growth inhibition, meaning a diminution in the number of colonies formed on a solid medium or in the opacity produced in a liquid medium under the influence of incorporated antiserum. (b) Metabolic inhibition, which means determination of the inhibitory action of antiserum indirectly by observation on the interference it produces in a metabolic reaction, such as fermentation of a sugar, reduction of a dye, the liberation of ammonia from arginine, or some similar enzymic reaction. (c) Fluorescent-antibody test. (d) Gel-diffusion test. (e) Indirect-haemagglutination test. (f) Complement-fixation test. (g) Agglutination and latex-agglutination tests. These tests differ in both their sensitivity and their specificity. The most specific, namely the growth-inhibition, the metabolic-inhibition and the immunofluorescent tests pick out the specific antigens and show little cross-reaction between the species. Such tests, however, failed to distinguish *M. mycoides* subsp. *mycoides* from certain caprine and ovine mycoplasmas, whereas a cross-immunization test in mice succeeded (G. R. Smith *et al.* 1980). The indirect-haemagglutination and gel-diffusion tests are slightly less specific than growth inhibition, metabolic inhibition and immunofluorescence, and show some relation, for example, between some arginine-utilizing species, such as that between *M. hominis*, *M. salivarium* and *M. orale*, or between *M. fermentans* and *M. buccale*. The complement-fixation test is even less specific, bringing out a good deal of cross-reaction between the species. The complement-fixation and gel-diffusion tests are probably influenced by internal antigens that do not affect other tests. Ureaplasmas are, as already stated, antigenically heterogeneous; though related to one another, they each have their own specific identity (Purcell *et al.* 1969, Howard and Gourlay 1972). Some species of *Mycoplasma* are themselves heterogeneous, such as *M. hominis* (Hollingdale and Lemcke 1969, 1970) and *M. pulmonis* (Forshaw and Fallon 1972).

For the identification of strains the most useful tests are the immunofluorescence, the growth-inhibition and the metabolic-inhibition tests. Colony immunofluorescence (Rosendal and Black 1972) is now widely used. For serological diagnosis in man the indirect-haemagglutination and complement-fixation tests are recommended (Purcell *et al.* 1969); in cattle the complement-fixation, gel-diffusion and direct-agglutination tests are favoured. (For further information see Kenny 1979.)

Virus infection of mycoplasmas

Like bacteria, some mycoplasmas are susceptible to attack by bacteriophages, or what are perhaps more correctly referred to as viruses. The first report was by Gourlay in 1970 who isolated a virus from a strain of *A. laidlawi*. Since then, viruses have been isolated from other strains of the same organism. Liss and Maniloff (1971) isolated 9 viruses from 14 strains and found that 3 of them, though morphologically alike, differed in other properties from each other. In size these viruses are 86–90 nm long and 13–14 nm broad (Milne *et al.* 1972, Gourlay and Wyld 1972). Within the organism they complete their cycle of development in 30–60 min and are then released suddenly by bursting of the cell. On solid media they form plaques that are often indistinct. Sometimes the colonies present a 'cratered' appearance (Milne *et al.* 1972). Both clear and turbid plaques may be present, the latter suggesting that some cells of *A. laidlawi* are resistant. Three groups of mycoplasmal viruses have been described—naked bullet- or rod-shaped particles, spherical enveloped particles, and polyhedral particles with tails; all have been propagated on *A. laidlawi.* There is abundant morphological evidence that spiroplasmas are infected with viruses of several different types; Cole and his colleagues (1977) produced plaque formation on *S. citri* with one such virus. (For reviews of mycoplasmal viruses see Razin 1978, Cole 1979 and Maniloff *et al.* 1979.)

Pathogenicity

Our knowledge of the pathogenic role of the mycoplasmas is very incomplete. About the ability of the two subspecies of *M. mycoides* to give rise to pleuropneumonia in cattle and goats, of *M. agalactiae* to cause contagious agalactia in sheep and goats, of *M. hyopneumoniae* to cause enzootic pneumonia in pigs, of *M. gallisepticum* to cause chronic respiratory disease in chickens, and of *M. pneumoniae* to cause so-called atypical pneumonia in man there is little doubt. *M. pulmonis* produces chronic respiratory disease in rats and mice. Polyarthritis in rats is commonly due to *M. arthritidis*, and *M. neurolyticum* causes *rolling disease* in mice. *M. synoviae* is responsible for infectious synovitis and sometimes chronic respiratory disease in chickens and turkeys; and *M. meleagridis* is a common pathogen of young turkeys.

The pathogenicity of many other species is much less easy to determine. Part of the difficulty lies in their frequent presence in the respiratory and genito-urinary tracts, which may cast doubt on the significance of finding them in diseased tissues. In the human field *M. hominis* is regarded as a potential pathogen, being held responsible for pelvic disease in women and for puerperal septicaemia; and growing evidence suggests that ureaplasmas cause some cases of non-gonococcal urethritis in men. (For further information see Chapter 78.)

Characteristic disease may sometimes be reproduced experimentally in the larger domestic animals. Unnatural laboratory-animal hosts are usually refractory to mycoplasmal infection; there are some exceptions such as the ability of *M. pneumoniae* to produce pulmonary lesions in cotton-rats and hamsters (Eaton *et al.* 1944), and of *M. mycoides* subsp. *mycoides* to produce symptomless infections accompanied by mycoplasmaemia in mice (G. R. Smith 1971).

Motility, attachment mechanisms and toxicity may all be concerned in pathogenicity (see Archer 1979). Toxic factors include the galactan of *M. mycoides* subsp. *mycoides*, the exotoxin of *M. neurolyticum*, the polysaccharide inflammatory toxin of *M. bovis* (Geary *et al.* 1981), and metabolic products such as hydrogen peroxide and ammonia. Seid and his colleagues (1980) showed that lipopolysaccharides from acholeplasmas—organisms without a clearly defined pathogenic role—possessed toxic activity resembling that of bacterial endotoxin.

It should be added that mycoplasmas—especially *M. hyorhinis*, *M. orale*, *M. arginini* and *A. laidlawi*—are very prone to invade *tissue cultures*. Sometimes they interfere with the growth of cultured viruses; at others they themselves cause cytopathic damage (see Witzleb 1970, Witzleb *et al.* 1970). Their removal with antibiotics is often unsuccessful. More hopeful is treatment of the tissue cultures with a powerful antiserum against the invading mycoplasma (Schweizer *et al.* 1970); with 5-bromouracil, a fluorochrome, and light (Marcus *et al.* 1980); or by brief co-cultivation with mouse macrophages in the presence of antibiotic (Schimmelpfeng *et al.* 1980).

Experimental infection of animals

***M. mycoides* subsp. mycoides** The subcutaneous inoculation of cattle with 0.5–1.0 ml of infected lymph from an animal suffering from pleuropneumonia, or a virulent culture of subsp. *mycoides*, produces in 8 to 25 days a tense, hot, painful inflammatory swelling accompanied by high fever and often followed by death. Incision of the skin over the affected part permits the exudation of a clear straw-coloured fluid, sometimes amounting to several litres. *Post mortem*, the connective tissue meshes are distended with an immense quantity of clear yellow fluid, which is here and there coagulated into gelatinous trembling masses. Microscopical examination of the freshly collected fluid reveals the presence of forms similar to those seen in culture. A little serous exudate may be present in the pleural cavity, and the thoracic and inguinal glands may be affected. According to Daubney (1935) the typical disease (see Chapter 78) can be reproduced by inoculation into the jugular vein of lymph or culture mixed with a few millilitres of 10 per cent agar. The emboli are held up in the lungs and form the starting point of the disease. Cattle infected by endobronchial intubation will transmit the disease to others kept in close contact (Hudson and Turner 1963). Goats and sheep are susceptible to subcutaneous infection, but small laboratory animals are generally resistant (Turner 1959). Not all cultures of subsp. *mycoides* are suitable; only freshly isolated strains showing the filamentous phase of development are fully virulent, even for cattle (see review by Hudson 1971). A protracted mycoplasmaemia, without apparent illness, develops in mice infected intraperitoneally with virulent, but not avirulent, strains; this mycoplasmaemia is suppressed by pretreatment of the mice with immune antiserum (G. R. Smith 1968, Dyson and Smith 1975). *M. mycoides* subsp. *capri* is pathogenic for goats and sheep by intratracheal or subcutaneous inoculation (Watson *et al.* 1968).

M. agalactiae Contagious agalactia can be reproduced

Table 47.1. Properties of some species of human mycoplasmas

Species	Glucose	Arginine	Phosphatase	Film and spots (in egg yolk medium)	Tetrazolium reduction: Aerobic	Tetrazolium reduction: Anaerobic	Other	Disease association
*M. pneumoniae**	+	–	–	Usually –	+	+	Characteristic colonies	Primary atypical pneumonia.
M. fermentans	+	+	+	+	–	+	...	...
M. salivarium	–	+	–	+	–	Weak	...	...
M. hominis	–	+	–	–	–	–	Characteristic colonies Erythromycin resistant	Presumptive pathogen in genito-urinary tract (particularly female).
M. orale	–	+	–	–	–	–	Erythromycin sensitive	...
M. buccale	–	+	+	–	–	+	Erythromycin sensitive	...

* Gives clear β-Haemolysis.

Table 47.2. Properties of some species of bovine mycoplasmas

Species	Inhibition by digitonin	Glucose	Arginine	Phosphatase	Aerobic tetrazolium reduction	Film and spots (in egg yolk medium)	Other	Disease association*
M. mycoides subsp. *mycoides*	+	+	–	–	+	–	Digests serum	Contagious bovine pleuropneumonia
M. bovirhinis	+	+	–	–	V	+	β-Haemolysis	Pneumonia in calves
M. bovoculi	+	+	–	–	+	+	α-Haemolysis	Conjunctivitis
M. dispar	+	+	–	–	+	–	Special medium required	Pneumonia
M. bovigenitalium	+	–	–	+	V	+	Haemagglutination +	Mastitis and vesiculitis
M. bovis	+	–	–	+	+	+	Haemagglutination –	Mastitis, sometimes associated with arthritis; respiratory disease
M. arginini	+	–	+	–	–	–	...	Pneumonia
M. alkalescens	+	–	+	+	–	–	...	Not known
A. laidlawi	–	+	–	–	+	–	Aesculin + Carotenoid produced	Salpingitis
A. modicum	–	+	–	–	+	–	Aesculin – Carotenoid not produced	Not known

* Pathogenic significance not always clear.
V = variable.

experimentally in goats by inoculation with pure cultures of *M. agalactiae*. Sheep are less susceptible than goats. Subcutaneous inoculation of 0.5–1.0 ml is followed in 4 to 7 days by the appearance of a small local swelling which disappears during the next week. After a further incubation period of 1 to 4 weeks localizing lesions occur in the joints, cornea and, in lactating females, in the udder. The amount of milk secreted diminishes, and a yellowish purulent fluid takes its place. Laboratory animals are insusceptible.

M. neurolyticum This organism which, as Sabin (1938) showed, forms a powerful neurolytic toxin, gives rise on experimental inoculation of mice to the so-called *rolling disease*. Intracerebral injection is followed, usually after 2–3 days, by a disease characterized by rolling movements on the long axis of the body. Some mice die, and *post mortem* extensive necrosis and lysis of the posterior pole of the cerebellum are found.

M. gallisepticum This organism, which gives rise to chronic respiratory disease in chickens, can be cultivated in the yolk sac of 5–7 day chick embryos. Injection of infected yolk into 2-day-old chickens or 7-week-old turkeys by various routes was found by Yamamaoto and Adler (1958) to give rise to lesions in the joints, air sac and nervous system. Strains varied considerably in virulence. Fowl mycoplasmas were formerly known as coccobacilliform bodies (Nelson 1950*a*, *b*).

Appended are two tables, one showing some of the properties of human, the other of cattle mycoplasmas. (For further information see Tully and Whitcomb 1979.)

References

Addey, J. P., Taylor-Robinson, D. and Dimic, D. (1970) *J. med. Microbiol.* **3,** 139.
Aluotto, B. B., Wittler, R. G., Williams, C. O. and Faber, J. E. (1970) *Int. J. syst. Bact.* **20,** 35.
Anderson, D. R. and Barile, M. F. (1965) *J. Bact.* **90,** 180.
Andrews, B. E., Leach, R. H., Gourlay, R. N. and Howard, C. J. (1973) *Vet. Rec.* **93,** 603.
Archer, D. B. (1979) *Nature, Lond.* **277,** 268.
Barile, M. F. and Razin, S. (Eds.) (1979) *The Mycoplasmas*, vol. 1. Academic Press, New York and London.
Barnard, J. E. (1926) *J. roy. micr. Soc.* p. 253.
Biberfeld, G. and Biberfeld, P. (1970) *J. Bact.* **102,** 855.
Bordet, J. (1910) *Ann. Inst. Pasteur* **24,** 161.
Borrel, Dujardin-Beaumetz, Jeantet and Jouan. (1910) *Ann. Inst. Pasteur* **24,** 168.
Bredt, W. (1968*a*) *Zbl. Bakt.* **208,** 549; (1968*b*) *Path. et Microbiol., Basel* **32,** 321; (1970) *Z. med. Mikrobiol.*, **155,** 248; (1972) *Ibid.* **157,** 169; (1979) See Barile and Razin (1979), p. 141.
Bredt, W., Höfling, K. H. and Heunert, H. H. (1970) Film E 1633 (16 mm silent) (available from BMA Film Library); (1971) Film E 1813 (16 mm silent). *Encyclopaedia Cinematographica, Göttingen.*
Bridré, J. and Donatien, A. (1923) *C. R. Acad. Sci., Paris* **177,** 841; (1925) *Ann. Inst. Pasteur* **39,** 925.
Chalquest, R. R. (1962) *Avian Dis.* **6,** 36.
Chanock, R. M., Hayflick, L. and Barile, M. F. (1962) *Proc. nat. Acad. Sci., Wash.* **48,** 41.
Chu, H. P. and Horne, R. W. (1967) *Ann. N.Y. Acad. Sci.* **143,** 190.
Clyde, W. A. Jr. (1961) *Proc. Soc. exp. Biol., N.Y.* **107,** 715.
Cole, R. M. (1979) See Barile and Razin (1979), p. 385.
Cole, R. M., Mitchell, W. O. and Garon, C. F. (1977) *Science* **198,** 1262.
Cuckow, F. W. and Klieneberger-Nobel, E. (1955) *J. gen. Microbiol.* **13,** 149.
Daubney, R. (1935) *J. comp. Path.* **48,** 83.
Dienes, L. (1939) *J. infect. Dis.*, **65,** 24; (1942) *J. Bact.* **44,** 37.
Dienes, L. and Edsall, G. (1937) *Proc. Soc. exp. Biol., N.Y.* **36,** 740.
Domermuth, C. H., Nielsen, M. H., Freundt, E. A. and Birch-Andersen, A. (1964) *J. Bact.* **88,** 727, 1428.
Dyson, D. A. and Smith, G. R. (1975) *Res. vet. Sci.* **19,** 8.
Eaton, M. D., Meiklejohn, G. and Van Herick, W. (1944) *J. exp. Med.* **79,** 649.
Edward, D. G. ff. (1947) *J. gen. Microbiol.* **1,** 238; (1954) *Ibid.* **10,** 27.
Edward, D. G. ff. and Freundt, E. A. (1969*a*) See Hayflick (1969), p. 147; (1969*b*) *J. gen. Microbiol.* **57,** 391.
Edward, D. G. ff. *et al.* (1966) *Science* **155,** 1694.
Eichwald, C., Illner, F. and Trolldenier, H. (1971) *Die Mykoplasmose beim Tier*, p. 15, 60, 249. Jena, Gustav Fischer Verlag.
Fabricant, J. and Freundt, E. A. (1967) *Ann. N.Y. Acad. Sci.* **143,** 50.
Fallon, R. J. and Whittlestone, P. (1969) *Methods in Microbiology*, Vol. 3B. Ed. by J. R. Norris and D. W. Ribbons. Academic Press, London and New York.
Ford, D. K. (1962) *J. Bact.* **84,** 1028.
Forshaw, K. A. and Fallon, R. J. (1972) *J. gen. Microbiol.* **72,** 501.
Freundt, E. A. (1958) *The Mycoplasmataceae*, Munksgaard, Copenhagen; (1969). See Hayflick (1969), p. 281.
Freundt, E. A., Andrews, B. E., Ernø, H., Kunze, M. and Black, F. T. (1973) *Zbl. Bakt.* **225,** 104.
Freundt, E. A. and Edward, D. G. ff. (1979) See Barile and Razin (1979), p. 1.
Furness, G., Pipes, F. J. and McMurtrey, M. J. (1968) *J. infect. Dis.* **118,** 7.
Geary, S. J., Tourtellotte, M. E. and Cameron, J. A. (1981) *Science* **212,** 1032.
Goodwin, R. F. W., Pomeroy, A. P. and Whittlestone, P. (1965) *Vet. Rec.* **77,** 1247; (1967) *J. Hyg., Camb.* **65,** 85.
Goodwin, R. F. W. and Whittlestone, P. (1964) *Vet. Rec.* **76,** 611.
Gourlay, R. N. (1970) *Nature, Lond.* **225,** 1165.
Gourlay, R. N. and Wylde, S. G. (1972) *J. gen. Virol.* **14,** 15.
Hayflick, L. (1965) *Texas Rep. Biol. Med.* **23** (Suppl. 1), 285; (1969) *The Mycoplasmatales and the L-phase of Bacteria.* Amsterdam, North Holland Publishing Company, also New York, Appleton-Century Crofts.
Hederscheê, D. (1963) *Antonie van Leeuwenhoek* **29,** 154.
Hijmans, W., Boven, C. P. A. van and Clasener, H. A. L. (1969) In Hayflick (1969), p. 67.
Holländer, R. and Pohl, S. (1980) *Zbl. Bakt.* 1 Abt. Orig., **A246,** 236.
Hollingdale, M. R. and Lemcke, R. M. (1969) *J. Hyg., Camb.* **67,** 585; (1970) *Ibid.* **68,** 469.
Hopps, H. E., Meyer, B. C., Barile, M. F. and DelGuidice, R. A. (1973) *Ann. N.Y. Acad. Sci.* **225,** 265.
Howard, C. J. and Gourlay, R. N. (1972) *Brit. vet. J.* **128,** xxxvii.
Howard, C. J., Gourlay, R. N. and Collins, J. (1978) *Int. J. syst. Bact.* **28,** 473.
Hudson, J. R. (1971) *Review of Contagious Bovine Pleuropneumonia* (*FAO Agric. Studies No. 86*). FAO, Rome.
Hudson, J. R. and Turner, A. W. (1963) *Aust. vet. J.* **39,** 373.
Kahane, I. and Razin, S. (1970) *FEBS Lett.* **10,** 261.
Keller, R. and Morton, H. E. (1954) *J. Bact.* **67,** 129.
Kenny, G. E. (1979) See Barile and Razin (1979), p. 351.
Klieneberger, E. (1934) *J. Path. Bact.*, **39,** 409; (1935) *Ibid.* **40,** 93; (1942) *J. Hyg., Camb.* **42,** 485.
Klieneberger, E. and Smiles, J. (1942) *J. Hyg., Camb.* **42,** 110.
Klieneberger-Nobel, E. (1962) *Pleuropneumonia-like Organisms (PPLO) Mycoplasmataceae.* Academic Press, London and New York.
Knudson, D. L. and Macleod, R. (1970) *J. Bact.* **101,** 609.
Laidlaw, P. P. and Elford, W. J. (1936) *Proc. roy. Soc. B*, **120,** 292.
Leach, R. H. (1973) *J. gen. Microbiol.* **75,** 135.
Ledingham, J. C. G. (1933) *J. Path. Bact.* **37,** 393.
Lemcke, R. M. (1965) *Lab. Pract.* **14,** 712; (1971) *Nature, Lond.*, **229,** 492.
Le Normand, M., Gourrett, J. P. and Maillet, P.-L. (1971) *C. R. Acad. Sci. Paris*, **273,** 2016.
Lin, J.-S. L. and Kass, E. H. (1980) *Infection* **8,** 152.
Lind, K. (1968) *Acta. path, microbiol. scand.* **73,** 459.
Liss, A. and Maniloff, J. (1971) *Science*, **173,** 725.
McGee, Z. A., Rogul, M. and Whittler, R. G. (1967) *Ann. N.Y. Acad. Sci.* **143,** 21.
Maniloff, J., Das, J., Putzrath, R. M. and Nowak, J. A. (1979) See Barile and Razin (1979), p. 411.
Maniloff, J., Morowitz, H. J. and Barrnett, R. J. (1965) *J. Bact.* **90,** 193.
Maramorosch, K. (1974) *Annu. Rev. Microbiol.* **28,** 301.

Marcus, M., Lavi, U., Nattenberg, A., Rottem, S. and Markovitz, O. (1980) *Nature, Lond.* **285,** 659.

Marmion, B. P. and Goodburn, G. M. (1961) *Nature, Lond.* **189,** 247.

Merling-Eisenberg, K. B. (1935) *Brit. J. exp. Path.* **16,** 411.

Milne, R. G., Thompson, G. W. and Taylor-Robinson, D. (1972) *Arch. ges. Virusforsch.* **37,** 378.

Morowitz, H. J., Bode, H. R. and Kirk, R. G. (1967) *Ann. N.Y. Acad. Sci.* **143,** 110.

Morowitz, H. J. and Terry, T. M. (1969) *Biochim. biophys. Acta.* **183,** 276.

Morowitz, J. M., Tourtellotte, M. E., Guild, M. E., Castro, W. R. and Woese, C. (1962) *J. molec. Biol.* **4,** 93.

Morton, H. E., Smith, P. F. and Leberman, P. R. (1951) *Amer. J. Syph. Gonor. vener. Dis.* **35,** 361.

Nakamura, M. and Kawaguchi, M. (1972) *J. gen. Microbiol.* **70,** 305.

Neimark, H. C. (1971) *J. gen. Microbiol.* **63,** 249; (1977) *Proc. nat. Acad. Sci., Wash.* **74,** 4041.

Nelson, J. B. (1950*a*) *J. exp. Med.* **91,** 309; (1950*b*) *Ibid.* **92,** 431.

Newnham, A. G. and Chu, H. P. (1965) *J. Hyg., Camb.* **63,** 1.

Niitu, Y. *et al.* (1970) *J. Pediat.* **76,** 438.

Nocard and Roux (1898) *Ann. Inst. Pasteur* **12,** 240.

Norman, M. C., Franck, E. B. and Choate, R. V. (1970) *Appl. Microbiol.* **20,** 69.

Nowak, J. (1929) *Ann. Inst. Pasteur* **43,** 1330.

Ørskov, J. (1927) *Ann. Inst. Pasteur* **41,** 473.

Pirie, N. W. (1969). See Hayflick (1969), p. 3.

Purcell, R., Chanock, R. and Taylor-Robinson, D. (1969). See Hayflick (1969), pp. 221, 254.

Rakovskaja, I. V., Melnikov, V. A. and Andreev, B. M. (1973) *In Vitro*, ČSSR, 1/2, p. 12*b*.

Razin, S. (1969) *Annu. Rev. Microbiol.* **23,** 317; (1973) *Advanc. microb. Physiol.* **10,** 1; (1975) *Progr. surf. membr. Sci.* **9,** 257; (1978) *Microbiol. Rev.* **42,** 414.

Razin, S. and Cleverdon, R. C. (1965) *J. gen. Microbiol.* **41,** 409.

Razin, S. and Oliver, O. (1961) *J. gen. Microbiol.* **24,** 225.

Report. (1967) *Int. J. syst. Bact.* **17,** 105; (1971) *Ibid.* **21,** 151; (1972) *Ibid.* **22,** 184; (1979) *Ibid.* **29,** 172.

Rodwell, A. (1969). See Hayflick (1969), p. 413.

Rodwell, A. W. and Rodwell, E. S. (1978) *J. gen. Microbiol.* **109,** 259.

Rosendal, S. and Black, F. T. (1972) *Acta path. microbiol. scand.* **B80,** 615.

Rottem, S. and Razin, S. (1967) *J. Bact.* **94,** 359.

Sabin, A. B. (1938) *Science* **88,** 189, 575.

Schimmelpfeng, L., Langenberg, U. and Hinrich Peters, J. (1980) *Nature, Lond.* **285,** 661.

Schweizer, H., Witzleb, W. and Blumöhr, T. (1970) *Arch. ges. Virusforsch*, **30,** 130.

Seid, R. C., Smith, P. F., Guevarra, G., Hochstein, H. D. and Barile, M. F. (1980) *Infect. Immun.* **29,** 990.

Seiffert, G. (1937*a*) *Zbl. Bakt.* **139,** 337; (1937*b*) *Ibid.* **140,** Beiheft, p. 168.

Shepard, M. C. (1954) *Amer. J. Syph. Gonor. vener. Dis.* **38,** 113; (1967) *Ann. N.Y. Acad. Sci.* **143,** 505; (1969). See Hayflick (1969), p. 49.

Shepard, M. C. and Lunceford, C. D. (1967) *J. Bact.* **93,** 1513.

Shepard, M. C. and Masover, G. K. (1979) See Barile and Razin (1979), p. 452.

Smiles, J. (1926) *J. roy. micr. Soc.*, p. 257.

Smith, G. R. (1968) *J. comp. Path.* **78,** 267; (1971) *Proc. R. Soc. Med.* **64,** 6.

Smith, G. R., Hooker, J. M. and Milligan, R. A. (1980) *J. Hyg., Camb.* **85,** 247.

Smith, P. F. (1964) *Bact. Rev.* **28,** 97; (1971) *The Biology of Mycoplasmas*, Academic Press, New York.

Smith, W. E., Hillier, J. and Mudd, S. (1948) *J. Bact.* **56,** 589.

Sobeslavsky, O, and Chanock, R. M. (1968) *Proc. Soc. exp. Biol. Med.* **122,** 786.

Somerson, N. L. and Weissman, S. M. (1969) See Hayflick (1969), p. 201.

Tang, F. F., Wei, H. and Edgar, J. (1936) *J. Path. Bact.* **42,** 45.

Tang, F. F., Wei, H., McWhirter, D. L. and Edgar, J. (1935) *J. Path. Bact.* **40,** 391.

Tully, J. G. (1979) See Barile and Razin (1979), p. 431; (1980) Foreword, in *Memoirs*, by E. Klieneberger-Nobel. Academic Press, London and New York.

Tully, J. G. and Razin, S. (1968) *J. Bact.* **95,** 1504.

Tully, J. G. and Whitcomb, R. F. (Eds.) (1979) *The Mycoplasmas*, vol. 2. Academic Press, New York and London.

Tully, J. G., Whitcomb, R. F., Clark, H. F. and Williamson, D. L. (1977) *Science* **195,** 892.

Turner, A. W. (1935) *J. Path. Bact.* **41,** 1; (1959) In: *Infectious Diseases of Animals: Diseases due to Bacteria*, vol. 2, p. 437. Ed. by A. W. Stableforth and I. A. Galloway. Butterworth, London.

Watson, W. A., Cottew, G. S., Erdağ, O. and Arisoy, F. (1968) *J. comp. Path.* **78,** 283.

Whitcomb, R. F. and Tully, J. G. (Eds.) (1979) *The Mycoplasmas*, vol. 3. Academic Press, New York, San Francisco and London.

Williams, C. O. and Wittler, R. G. (1971) *Int. J. syst. Bact.* **21,** 73.

Witzleb, W. (1970) *Arch. ges. Virusforsch.* **30,** 113.

Witzleb, W., Blumöhr, T., Dziambor, H. and Schweizer, H. (1970) *Arch. ges. Virusforsch.* **30,** 121.

Wright, D. N. (1967) *J. Bact.* **93,** 185.

Wroblewski, W. (1931) *Ann. Inst. Pasteur* **47,** 94.

Yamamoto, R. and Adler, H. E. (1958) *J. infect. Dis.* **102,** 148.

Index

Most subjects that would normally be entered in the index are included in the contents list at the beginning of each chapter. The present index serves to indicate where subjects not obviously related to any special chapter will be found. To assist the reader who is looking for information on a particular organism we preface the index by a list of genera and species mentioned in this volume. The main genera figure in the contents lists, but some of the lesser known ones appear only in the text. These, however, are usually referred to in the general index.

Genera of bacteria

Species of bacteria

General Index